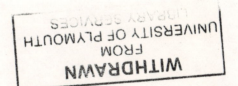

Asteroids

The cover is from a painting by W. K. Hartmann; it is a view from the surface of an Apollo asteroid. A bright comet is passing through the inner solar system, spewing out a linear Type I gas tail and a Type II tail of meteoric dust, against which Earth and Moon are silhouetted. The sun has just set behind the crater rim, revealing the corona and inner zodiacal light, and to the lower right there is a minor satellite.

The backcover is from a plate taken by C. T. Kowal with the Palomar 122-cm Schmidt; it is of Apollo asteroid 2063 Bacchus. The plate was taken on 25 April 1977, during 75 min on IIIa-J emulsion with Wratten 2C filter.

Asteroids

Edited by

TOM GEHRELS

With the assistance of
MILDRED SHAPLEY MATTHEWS

With 69 collaborating authors

THE UNIVERSITY OF ARIZONA PRESS
TUCSON, ARIZONA

SPACE SCIENCE SERIES

Tom Gehrels, Space Sciences Consultant

PLANETS, STARS AND NEBULAE, STUDIED WITH
PHOTOPOLARIMETRY, T. Gehrels Ed., 1974, 1133 pp.

JUPITER, T. Gehrels Ed., 1976, 1254 pp.

PLANETARY SATELLITES, J. A. Burns Ed., 1977, 598 pp.

PROTOSTARS AND PLANETS, T. Gehrels Ed., 1978, 756 pp.

ASTEROIDS, T. Gehrels Ed., 1979, 1181 pp.

COMETS, L. L. Wilkening, 1982, 766 pp.

THE SATELLITES OF JUPITER, D. Morrison Ed., 1982, 974 pp.

Second printing 1982

THE UNIVERSITY OF ARIZONA PRESS

This book was set in 10/12 IBM Press Roman
Manufactured in the U.S.A.

Library of Congress Cataloging in Publication Data

Asteroids.

Includes index.
1. Planets, Minor.   I. Gehrels, Tom, 1925–
II. Matthews, Mildred Shapley.
QB651.A85   523.4'4   79-19686

ISBN 0-8165-0695-7

CONTENTS

Part III — INTERRELATION

Part IV — CONFIGURATION

Part V — COMPOSITION

Part VI — EVOLUTION

Part VII — TABULATION

GLOSSARY, ACKNOWLEDGMENTS, AND INDEX

COLLABORATING AUTHORS

D. F. Bender, *1014*
R. P. Binzel, *443*
M. Blander, *809*
D. D. Bogard, *558*
E. Bowell, *132, 1108*
T. E. Bunch, *745*
J. A. Burns, *494*
A. G. W. Cameron, *992*
A. Carusi, *391*
C. R. Chapman, *25, 359, 528, 601, 655, 1064*
M. J. Cintala, *579*
D. R. Davis, *528*
J. Degewij, *417*
J. R. Dickel, *212*
A. Dollfus, *170*
M. J. Drake, *765*
J. L. Elliot, *98*
E. Everhart, *283*
M. J. Gaffey, *655, 688, 1064*
T. Gehrels, *3, 1108*
J. C. Gradie, *359*
R. Greenberg, *310, 528, 601*
A. W. Harris, *528*
W. K. Hartmann, *466*
J. W. Head, *579*
E. F. Helin, *253*
S. Herrick, *222*
K. R. Housen, *601*
R. F. Jurgens, *206*
J. F. Kerridge, *745*
C. T. Kowal, *436*
Y. Kozai, *334*
Ľ. Kresák, *289*
H. P. Larson, *724*
L. A. Lebofsky, *184*

K. Lumme, *132*
B. G. Marsden, *77*
D. L. Matson, *84*
T. B. McCord, *688*
R. L. Millis, *98*
D. Morrison, *184, 227, 1090*
J. Niehoff, *227*
G. H. Pettengill, *206*
F. Pilcher, *1130*
R. T. Reynolds, *822*
V. S. Safronov, *975*
H. Scholl, *310*
J. Schubart, *84*
E. R. D. Scott, *892*
E. M. Shoemaker, *253*
C. P. Sonett, *822*
R. C. Taylor, *480*
E. F. Tedesco, *443, 494, 1098*
P. Thomas, *628*
G. B. Valsecchi, *391*
T. C. Van Flandern, *443*
C. J. van Houten, *417*
G. J. Veeder, *724*
J. Veverka, *628*
J. T. Wasson, *926*
S. J. Weidenschilling, *528*
G. W. Wetherill, *926*
L. L. Wilkening, *61, 601*
J. G. Williams, *253, 359, 1040*
L. Wilson, *579*
R. F. Wolfe, *253*
J. A. Wood, *849*
S. P. Worden, *119*
B. Zellner, *170, 783, 1011, 1090, 1108*

PREFACE

"We may never do a better book." I know I have used this sentence before (for "Jupiter," University of Arizona Press, 1976), but we are learning to make better books. The goal is to make a new type of source book, a textbook aimed at graduate students, where the development of a new discipline needs such treatment. This one is better than previous books because the procedure has been improved and because it contains great surprises. Readers to whom the asteroids are new, will find a field of study that invites their participation in new exploration. Readers who believe they know the asteroids will find major contributions that are not published elsewhere.

The general procedure for making these books is described in the Introduction of "Protostars and Planets" (University of Arizona Press, 1978). For this book the procedure entailed the following steps: an international meeting of the authors and their colleagues (the participation in the book is not, however, dependent upon attendance of the meeting); open attendance and organization of meeting and book, with a large Organizing Committee consisting of volunteers rather than privately selected members; the principle of volunteering also applies to authorship, while the Organizing Committee may in addition invite essential reviewers; the Organizing Committee may even suggest "shotgun weddings" of authors who have different backgrounds but work on the same topic; the refereeing and editing of the chapters are done in a thorough manner; we shun duplication of publication, urging the authors to submit new material and new reviews without publication elsewhere; we invite summary chapters of introduction, make an extensive Glossary and Index, and add cross referencing among the chapters; finally, with the cooperation of referees, authors and Press, the book is prepared for fastest possible publication.

For this book, the meeting was held in Tucson, 6-10 March 1979, exactly eight years after the first asteroid conference in Tucson. The Executive Vice-President of the University of Arizona, A. B. Weaver, opened the meeting. There were 144 people in attendance, of whom 29 came from abroad. Twenty-eight contributed papers have been submitted to *Icarus* for a special asteroid issue (December 1979) with B. Zellner as Guest Editor.

I cannot begin to thank individuals because there are so many that worked so hard to make this come about; it is their book. Instead, there is an extensive list of acknowledgments toward the end of the book. We are grateful for Hartmann's painting reproduced on the cover and for Kowal's photograph on the back. It is a pleasure to acknowledge essential support from the University of Arizona Press, the University of Arizona Foundation, the National Science Foundation, and the National Aeronautics and Space Administration.

Tom Gehrels

PART I

Introduction

THE ASTEROIDS: HISTORY, SURVEYS, TECHNIQUES, AND FUTURE WORK

T. GEHRELS
University of Arizona

A chronology and historical sketch of asteroid studies are made for the period 1766-1978. A survey or overview is made of the various types of small bodies, all called asteroids, that occur between the orbits of Mercury and Pluto. The basic concepts are introduced for readers new to the field. Observational techniques, future studies, and topics that seem to be missing in this book are overviewed from an astronomical, observational perspective.

This chapter is followed by two other introductions, namely by Chapman for a discussion of origin and evolution and by Wilkening for the connections between asteroids and meteorites. We each concentrate on aspects closest to our own interests, but with a group of three reviews we hope to achieve a fair balance of coverage. My own emphasis is on the astronomical aspects; there will, in fact, be a few reports on my surveying with the Palomar 122-cm Schmidt telescope.

Readers new to the field might wish to peruse the extensive Glossary as well as to read the three introductions before settling down with Parts II-VI of this book.

In this chapter I will first mention some previous literature and historical development. Then, in Sec. III, the various kinds of asteroids will be described, how they are found and what kind of orbits they have. Section IV will overview the techniques of observation and the basic data which we have for the asteroids. In the final section I will mention future observations and topics that were not covered in this book.

I. LITERATURE AND CHRONOLOGY

A previous sourcebook on asteroids was assembled in 1971 (Gehrels 1971); on its page xxvi the older literature is summarized. Asteroid books were edited by Samoylova-Yakhontova (1973), Cristescu *et al.* (1974), Delsemme (1977) and by Morrison and Wells (1978). A review of physical and orbital studies was made by Chapman *et al.* (1978).

The following is a chronology, compiled with the help of some of our authors, for the principal events, or milestones, in physical studies of minor planets.

1766 Titius von Wittenburg formulated what has become known as the Titius-Bode law of planetary distances. The discovery of Uranus, by Herschel in 1781, further kindled the belief that there was a missing planet near 2.8 AU.

1794 Chladni inferred an extraterrestrial origin of meteorites. By 1803 Olbers proposed that the meteorites come from an asteroidal exploded planet.

1801 Piazzi at Palermo discovered asteroid 1 Ceres.

1867 Kirkwood pointed out that there are resonance gaps in the asteroid belt.

1891 Max Wolf introduced photography in the asteroid observations.

1898 433 Eros was discovered by Witt in Berlin, and von Oppolzer soon observed light variation which he explained by rotation of an object of irregular shape.

~1900 Centers for data on minor planets developed in Berlin and Kiel. After World War II this was continued in Heidelberg; also in Cincinnati by Herget and then transferred to Cambridge, Massachusetts, by Marsden in 1978. Since World War II, the *Ephemerides of Minor Planets* are published by the Institute of Theoretical Astronomy in Leningrad.

1906 The first Trojan, 588 Achilles, was discovered by Wolf.

1918 Hirayama published his first paper on asteroid families; the work was also continued by Brouwer, Arnold, Kozai and J.G. Williams.

1920 944 Hidalgo was discovered by Baade at Bergedorf.

1923 Prior published the first *Catalog of Meteorites*.

1932 Reinmuth discovered 1862 Apollo; it was lost, but recovered in 1973.

1944 O.J. Schmidt postulated the start of formation of a planet near 2.8 AU, by accumulation of particles and planetesimals, which was, however, interrupted by Jupiter's perturbations so that the asteroids resulted instead. The work is continued by Safronov and others.

1949 Kuiper initiated the Yerkes-McDonald Survey of asteroids and a series of photometric studies.

1950 Whipple proposed a "dirty snowball" model for the nuclei of comets.

1951 Öpik evaluated the effects of close approaches to planets on orbits of asteroids and calculated asteroid lifetimes.

1951 Gerling and Povlova measured the first K-Ar formation ages of meteorites.

1953 Piotrowski published a basic study of collisions among asteroids. The work was continued by Anders, Wetherill, Dohnanyi and others.

1957 Begemann and co-workers measured the first cosmic ray exposure age of a meteorite.

1964 Wänke attributed large amounts of helium in certain meteorites to solar wind implantation.

1964 Anders formulated a comprehensive theory of the origin of meteorites in asteroids.

1964 J.A. Wood calculated cooling rates of iron meteorites and found that they must have originated in objects of asteroidal size, 100-500 km in diameter.

1966 H.G. Hertz determined the mass of 4 Vesta. The work is continued by Schubart.

1967 Wetherill initiated a study of acceleration mechanisms to deliver meteorites from the asteroidal belt.

1968 1566 Icarus was observed by radar at both Goldstone and Haystack.

1969 McCord initiated a program of spectrophotometry of asteroids and other objects in the solar system. This method was applied to many asteroids by Chapman.

1969 The study of lunar samples clarified the processes which also form brecciated meteorites.

1970 The reflectivity of 1566 Icarus was determined with the polarimetric method. The method was developed in detail by Veverka and applied to many asteroids by Zellner.

1970 The Palomar-Leiden Survey of faint minor planets was published.

1970 D.A. Allen published the first infrared diameter of an asteroid, namely of 4 Vesta. Infrared radiometry was further developed by Matson, who discovered the wide range of albedos; the method was applied to many asteroids by Hansen and Morrison.

1971 "Physical Studies of Minor Planets" was published, based on an international conference held that year in Tucson, Arizona.

1971 Herrick proposed exploitation of asteroids; various aspects are promoted by O'Neil and O'Leary.

1972 A coordinated campaign of observations of 1685 Toro was made, and in 1975 a similar campaign on 433 Eros.

1972 The first assessment and advisory report on asteroid space missions was published by Alfvén and colleagues.

1975 Bowell initiated a large survey of UBV photometry of asteroids.

1975   Chapman, Morrison and Zellner combined the results of polarimetry, radiometry, and spectrophotometry and defined the broadband system of classification in terms of C types, S types, etc.

1975   Larson and Fink obtained the first Fourier transform spectroscopy, on 4 Vesta.

1976   Helin discovered the first Aten asteroid in a systematic survey of Apollo and Amor asteroids organized by Shoemaker.

1977   Kowal found 2060 Chiron.

1978   New developments were made regarding regoliths, satellites of asteroids, family members, taxomy and origins, as mentioned in the following section.

II. HISTORICAL SKETCH

"Inter Jovem et Martem interposui planetam" (Kepler 1596) marks the start of our asteroid studies. This was only a passing remark, however, because it was fundamental to Kepler's entire scheme that there should be only six planets (Gingerich, personal communication 1979). In any case, for asteroid studies there was a slow start because not until 170 yr later was the next step taken, namely by Titius von Wittenburg (1766). The law of planetary distances is more commonly known as the Titius-Bode law, or sometimes simply as Bode's law because it was J.E. Bode who stressed that relationship as being most important to infer that a planet was missing between Mars and Jupiter. (Titius may not have been the first to express that law anyway.) There are various expressions for it, such as $r = 0.4 + 0.3 \times 2^n$, the relationship is used even today in the understanding of planet formation (see Kuiper 1951, and Wetherill 1978).

Towards the end of the 18th century, there was much consideration of these topics especially after Herschel discovered Uranus at a distance from the sun that closely fitted Bode's law. Then the hunt for the missing planet was really on; during an astronomical conference in 1796 held in Gotha near Leipzig it was urged to make a systematic search. Piazzi was not partaking in that search but he was checking a star catalog, by Wollaston, when he observed on 1 January 1801 an object that he then noted as a star; on following nights, however, the object had moved and Piazzi believed he had discovered a comet. The object was observed over a period of 41 days. Soon the excitement in the scientific world was rising high because Bode and others believed that Piazzi had found the missing planet! It became an urgent problem in mathematics to develop a method by which predictions could be made for the recovery of that planet; Gauss solved the orbit and ephemeris problem so that Ceres was found again, on 7 December 1801.

The magnitude of the new planet was much fainter than that of its neighbors Mars and Jupiter. It was therefore perhaps not too surprising that other minor planets were found, three within the next six years (Pilcher, in

Part VII of this book, gives a listing of the discoveries). There was some discussion, by Olbers and others, of an exploded planet (see the Safronov chapter in this book), particularly because the first few of the discovered asteroids seemed to have intersecting orbits (this is not true; see Herget 1971).

The early days of asteroid science were exciting to the whole scientific community. However, another way to look at the period from 1807 to 1845 is that in 38 years no other asteroid was found. Better star catalogs and maps had to be made in order to find asteroids. The interest in these minor bodies may also have been lagging, perhaps because the hunt for a major missing planet was over. In any case, by 1845 a new era with new, rather specialized, people began, resulting in the finding of $\sim$4 minor planets per year; this increased to $\sim$7 by 1872, and $\sim$20 per year by 1891 after the new technique of photography was applied.

By the turn of the century, however, the astronomical community as a whole surely had lost interest in the minor planets. Spectroscopy and astrophysics were on the rise and these subdisciplines were applicable to hot gases and the understanding of stellar structure. It was too early for the study of star formation and the origin of the solar system, let alone the importance of planetesimals in such processes. The small amount of mass, $\sim$5 x 10^{-4} $M_\oplus$, between Mars and Jupiter was not to be understood in the grand scheme of solar system evolution until the present time. By the 1950's the malaise in asteroid studies had come to the point where it was improper at the major observatories to work on these "minor" bodies that were called "the vermin of the sky." Even the old-timers wondered how many more useless asteroids should be discovered (see, for instance, the Transactions IAU of 1952).

The minor-planet work had been left to a devoted few, who worked hard, for there was much to be done. It is easy to take plates and find new asteroids, but it requires time and attention to follow up with measuring precise positions, computing orbits and ephemerides and doing this over successive apparitions in order to make the orbits so precise that the object will not be lost again. In 1931, for example, 398 new asteroids were discovered (Putilin 1953), but only 159 were not lost again (Bowell, personal communication 1979). Some of the lost ones might be "discovered" later, but keeping track of all the discoveries, orbits and identifications was a large task. In Heidelberg, Leningrad, Berlin (later Cincinnati) and at a few other observatories such cataloging was done. In 1952 it was agreed that a yearly catalog of orbital elements and ephemerides should be published by the Institute of Theoretical Astronomy in Leningrad in close cooperation with the Minor Planet Center at Cincinnati Observatory (Cambridge, Massachusetts, since 1978).

By 1950, Kuiper and his associates made a beginning with a systematic study of the statistics of the asteroids and with a study of some physical parameters in photometry. It was not until 1970, however, that physical studies of minor planets became a discipline with the participation of a number of groups of people using a variety of techniques. This came about largely because of the space program, which by 1970 had reached the stage of planning missions beyond Mars. The flights of Pioneers 10 and 11 were at first announced, in 1968, as the Jupiter/Asteroid Missions.

Alfvén deserves credit for urging people and agencies to take an interest in the small and relatively undisturbed bodies for the study of the origin and evolution of the solar system. Alfvén urged asteroid missions (Stuhlinger *et al.* 1972) that might have resulted in a fast, but first, fly-by reconnaissance of ∼4 asteroids already in 1974, with existing spacecraft that were spin stabilized (see Russell and Tomasko 1976; Brahic *et al.* 1979). Alfvén also stimulated a conference and book on physical studies of minor planets (Gehrels 1971). The time was ripe because of the advances in meteoritics and new techniques, in the laboratory and at the telescope, such as the spectrophotometry of McCord and associates, the infrared radiometry and the photopolarimetry. It is rare in the history of science that the basic idea of one man changes a field as much as was the case with McCord in spectrophotometry and the understanding of asteroid surfaces.

By 1975 there began a clear development of taxonomy (i.e. assessment or classification) and mineralogical interpretation of the composition of individual asteroids. (The taxonomy is explained in the chapter by Zellner and the mineralogy by Gaffey and McCord.) We are learning to distinguish individual asteroids and their evolution, as they are impacted bodies, or fragmented, or originally differentiated, etc. Also and even more important, by 1978 the role of asteroidal planetesimals in the formation of stellar and solar systems became much more precisely modeled and described.

In 1978 the Minor Planet Center was moved to Cambridge, Massachusetts. There is new life in the theoretical recovery, with ∼100 newly numbered asteroids per year (see Marsden's chapter). The work of the Crimean Astrophysical Observatory should be mentioned in this connection, which regularly observes ∼700 minor planets per year. Several other observers are active — the field is truly alive again!

1978 was a remarkable year in which our image of an asteroid changed drastically. The dusty rock, like a meteorite in a museum, but covered with its own dust (we knew that from polarimetry), became a loosely agglomerated megaregolith that has been plowed over and under ("gardened") many times by mutual collisions, while one or more of its fragments just might be orbiting overhead as a minor satellite. The discovery of a probable satellite of Pluto, the reported discovery of satellites from asteroidal occultations and the interpretation of the pictures of Phobos made a deep impression. How lucky we were to have the Stickney impact just at the narrow range of energies between mere cratering and total fragmentation! Several chapters in this book will describe the new model (see Wilkening's introduction) and pictures are given in the chapter by Veverka and Thomas as well as on the cover of this book.

We are now nearing the culmination of physical studies of asteroids, with many refinements and many asteroids being better understood. Another discontinuity and great expansion is to be expected at the time when landing and sampling missions take place.

III. A SURVEY OF ASTEROIDS

We discussed in the previous section how near 2.8 AU from the sun a planet was believed to be missing, then it was discovered, but many more small planets were found. The name "minor planet" was used, and it still is used particularly in astrometry and celestial mechanics. (I have seen the name "planetoid" a few times in popular literature, but not in the manuscripts for this book.) The name "asteroid" has a connection with the appearance and it may for this reason be more popular in physical studies. The largest diameter seen for the asteroids is ~0.7 arcsec and, since only in exceptional seeing can 0.7 arcsec be resolved, the appearance at the telescope is therefore starlike (Greek: *Asteroeides*). This is usually the case; some of us have seen disks of Ceres, Pallas and Vesta.

The name "asteroid" is, however, used for a variety of objects with different orbital and physical characteristics. In this section, I will overview the asteroids in order of increasing distance from the sun, and I mention the surveys that have been made for these objects in order to study the degree of completion to which the various populations are known. As we begin with the possibility of Trojan asteroids of Mercury, it is necessary first to introduce a few concepts of orbits and Trojans.

The stability of orbits is discussed in the chapters by Everhart and by Greenberg and Scholl; the distinction is made between orbits that are stable on a time scale of $\sim 10^9$ yr and those that are not. The latter can be chaotic orbits, or they may be irregular retrograde orbits. It is noted in this context that the asteroids all move in the prograde direction, as do the planets, and that we have never yet observed a retrograde asteroid. In fact, the inclination, i, of the asteroid orbits is generally not larger than $\sim 20°$, a basic datum in the origin and evolution of the asteroids as is the fact that the inclinations are not zero. But note that the major surveys, described below, have been confined to low ecliptic latitude and that the knowledge of the high-inclination population is therefore seriously limited. The largest inclination for the numbered asteroids is 64°, for asteroid 2102.

Lagrange made a study of the stability of small bodies (see the chapter by Degewij and van Houten); two of his regions are especially known, namely in the planetary orbit *preceding* (L_4) or *following* (L_5) the planet by 60°. I would like to define L_4 and L_5 this way, even though in the literature there is some confusion; for instance, van Houten et al. (1970a) refer to the preceding point as L_5. Another type of orbit wanders between L_4 and L_5, away from the planet, and thereby describes a secular modification of the orbit that looks like a horseshoe; such objects are referred to as horseshoe objects.

Trojans or horseshoe objects of Mercury may be stable enough to exist (R. Greenberg, personal communication 1977), but a search from Earth is difficult because of the proximity to the sun and the light scattered by the earth's atmosphere. Trumpler (1923) reported surveys made in part during solar eclipse, and he concluded that there are no such objects $\gtrsim 60$ km in diameter. Chapman (next chapter) mentions infrared searching, even for objects within Mercury's orbit.

I have searched for Trojans of the earth with the Palomar 122-cm Schmidt to a limiting magnitude of B~19, over a region of 6° x 20° during two observing runs, carefully guiding for the predicted motion of such objects, but nothing was found. In the Earth-Moon system there also are two semi-stable regions L_4 and L_5 and there sometimes have been reports of clouds of particles seen near these regions, but nothing has been confirmed.

At the time of this writing (June 1979), four Aten objects are known, three of which discovered by Helin with the Palomar 46-cm Schmidt. The survey for objects near the earth is referred to in the chapter by Shoemaker *et al.* (see Helin and Shoemaker 1979) and their definitions of Aten, Apollo, and Amor asteroids are reproduced in the Glossary. In principle, following older and coarser definitions, the Atens remain near the earth, at least for some time, and do not cross the Mars orbit at present, the Apollos cross the earth as well as the Mars orbit, while the Amors cross only the Mars orbit.

The origin of these objects and their connection with the remainder of the asteroids is discussed in several of our chapters. The density of material at 2-4 AU may have been originally the same as that outside that zone, but at an early stage of the formation of the major planets great perturbations and collisions may have reduced the density down to what is presently observed. What is left in the asteroid belt has such a high degree of stability that, in fact, one of our major problems is to obtain meteorites from the belt. The present amount of material in the belt is only ~0.0005 $M_\oplus$ of which 0.0002 $M_\oplus$ is the mass of 1 Ceres (see the chapter by Schubart and Matson).

Ceres, Vesta, and Pallas are exceptional objects. Ceres is as large as about one-third the diameter of Pluto, Europa, or the moon. This is not to say that the size of Ceres is outside of the size distribution of the asteroids; Kresák (1978, personal communication) has concluded that the size of Ceres does not deviate from the asteroidal size statistics. It is noted, however, that Ceres, Pallas and Vesta have unusual compositions, while Pallas has an unusual orbit. Vesta is unique in composition, as described in Drake's chapter (see Gaffey and McCord's and Dickel's chapters, also for Ceres). Vesta seem differentiated, while Ceres does not: it is believed that Vesta at one stage in its evolution was heated and melted sufficiently that the heavier elements settled towards the center. Pallas' inclination is exceptionally large, and it may be a leftover sample of large objects that helped to partially clear out the asteroid belt at the time of formation of the major planets.

A good introduction to the asteroid belt can be obtained from studying Fig. 1 of Kresák's and Fig. 3 of the Wasson and Wetherill chapter. The main asteroid belt generally lies between 2.2 and 3.3 AU, but the Hungarias (1.9-2.0 AU) and Hildas near 4 AU are included. Kirkwood gaps and secular ν resonances are shown as basic features in these figures.

Kozai, in his chapter, reviews the origin of Hirayama families which are groupings of asteroids that have the same orbital elements a, e' and i' (or sin i'); the primes for e' and i' indicate that they are proper elements namely that they have been corrected for the effects of perturbations, ideally all perturbations by all the planets (see the chapter by Safronov). Family memberships are listed by Kozai in his chapter, as well as by Williams (in Part VII of this book), while Williams also lists the proper elements of the asteroids. The families are named after the lowest-numbered member; for instance, Kozai's family No. 41 is called the Koronis family after 158 Koronis. The distinction between groups of asteroids and families of asteroids is made at the beginning of the chapter by Gradie et $al.$; the name family should be used for fragments of a parent body. For a discussion and critique of jetstreams see Chapman's and Safronov's chapters.

A basic topic also is the difference between asteroids and comets; the latter probably formed at greater distances from the sun so that a mixture of ices and solids aggregated the cometary core. Comets are not discussed for their own sake in this book, but there is considerable interest in extinct cores because these may be hidden among the asteroids without our knowing the difference. The prevailing model of the comets is that they are icy conglomerates that show activity when their ices evaporate; that activity, in fact, is seen in the telescope as a fuzzy nebula around the nucleus; a coma and a tail or two make the distinction of the name. In other words, when a coma is seen the object is called a comet, but when no activity is ever observed the object is called an asteroid. The expectation is to find some activity, some slight coma, on Chiron, Hidalgo, and possibly Icarus and other asteroids, but as long as this has not been observed, these objects will be referred to as asteroids and be numbered and named as such (see Sec. IV).

After evaporation of the volatiles the extinct nucleus will remain somewhere in space, probably in a comet-like orbit, that is with appreciable eccentricity and inclination, but in due time the orbit may be modified by close planetary encounters so that it is not distinguishable among the asteroids. Jet action of escaping volatiles from the nucleus can also effect the orbit and one speaks then of nongravitational forces.

The completeness of the populations in the asteroid belt and Jupiter-Trojan regions is known from three sources, namely the Ephemerides catalog (issued yearly in Leningrad) which seems to be complete to about the 15th apparent opposition magnitude; The Yerkes-McDonald Survey which is complete, after some corrections, to about the 16th magnitude; and the

Palomar-Leiden Survey which seems to have fairly good completeness corrections, for low inclinations, to about the 19th magnitude.

The Yerkes-McDonald Survey (MDS; Kuiper et al. 1958) was made over a width of ±20° in ecliptic latitude and the completion for high inclination objects, that appear beyond that width, drops off sharply with inclination; this remark applies particularly to the Hungarias, Phocaeas, Apollos and Amor objects. The coverage in longitude is sufficient because the ecliptic belt was photographed nearly twice around in 1950-1952. In the MDS a 25-cm triplet-lens was used for taking 2400 plates, each 20 x 25 cm, 6°.5 x 8°.1 in size. The blinking (see Sec. IV), measuring of positions and magnitudes, reductions and identifications took ~14 man-years in addition to the observing (2 man-years), by well qualified and dedicated people.

The Palomar-Leiden Survey (PLS: van Houten *et al.* 1970*b*) was a different type of program because it covered only a small part of the sky, 12° in ecliptic latitude and 18° in longitude, and therefore the completion corrections, by extrapolation to the whole sky, become less certain. The topic is discussed by van Houten *et al.* (1970*b*), and also by Kiang, Kresák and others in Gehrels (1971). The principal advantage of the PLS over MDS is the improvement in limiting magnitude, to ~20, which was attained with the 122-cm Schmidt telescope on Palomar Mountain. While the observing took only a month, the careful reductions by the van Houtens at the Leiden Observatory still required 3 "man"-years, while in addition at the Cincinnati Observatory Herget made the computer reductions of the orbits. Much information was obtained on phase functions, groupings and on the overall statistics especially concerning the relationship that the number of asteroids increases by a factor of 2.5 for each fainter magnitude. A generally linear rise of the logarithm of the number as a function of magnitude was found, but when various sizes and compositional classes are considered in detail, and observational incompleteness ("bias") is removed, the distributions are far from linear (see Figs. 3-8 in the Zellner chapter, and the discussion in the next chapter by Chapman).

A very special group of asteroids occurs in the Lagrangian points of Jupiter. A rather unexpected estimate of 700 Trojans preceding Jupiter and only 200 following Jupiter down to a limit of ~15 km is announced and intrepreted, as are their physical characteristics, in the chapter by Degewij and van Houten.

The outer satellites of Jupiter are discussed in the chapters by Carusi and Valsecchi and by Degewij and van Houten. I made a special search with the Palomar 122-cm Schmidt in order to establish the statistics of the outer satellites. If they were simply captured asteroids, one would expect the asteroidal magnitude-frequency relation to be valid. The limiting magnitude of the old surveys, resulting in a total of 7 outer satellites, was ~18-19 mag, while now it is ~21. Kowal discovered a new satellite, J13 in 1974 and there are a few observations of a fourteenth, but a large number will never be found. I expect that even with the 4-meter telescopes at Kitt Peak or Cerro Tololo only a few additional Jovian satellites, if any, could be discovered. The size distribution of the outer Jovian satellites is most peculiar; with respect to the distribution of the asteroids in the main belt both the faint and the bright ones are missing. The interpretation seems straightfoward. Fragments larger than the present outer satellites could not have remained in capture by the nebula around proto-Jupiter because they were too massive and shot right through. The smallest fragments were, on the other hand, after the capture and breakup of the parent body, stopped "dead in their tracks" and fell into Jupiter. Even the fine detail of the size distributions of the two groups of outer satellites seems to confirm this reasoning: the outer group has generally smaller objects than the inner one, and the range of sizes is even more confined; this is understood with the outer nebula of proto-Jupiter being thinner at greater distance.

Phobos and Deimos also may be asteroids captured at the time when there still was the resisting medium outside proto-Mars. These two objects are discussed in the chapter by Veverka and Thomas, particularly in the vein of "what asteroids may look like."

Kowal maintains a regular Solar System Survey with the resulting discoveries of 2060 Chiron and several interesting asteroids and comets in addition to J13; this survey is described in his chapter. Chiron has a perihelion slightly inside the Saturn orbit and aphelion as far as the Uranus orbit. Chiron is discussed in detail in the chapter by Kowal, while various other chapters examine the nature of this unique object; there is a consensus that Chiron is a large comet, as is asteroid 944 Hidalgo which moves from a perihelion within the asteroid belt and an aphelion as far as the Saturn orbit.

IV. OBSERVATIONAL TECHNIQUES

For the discovery of moving objects on photographic plates the technique of blinking is used. A "blink" is a sturdy instrument to hold two plates side by side firmly while they can be adjusted, sideways and in rotation, to be inspected consecutively through microscope optics with fine superposition of the star images. One blinks by flipping a mirror so that one can look back and forth at the two plates. Any object in the solar system will be seen as moving, due to a relative motion of the earth and the object. When observed at opposition (that is, when opposite the sun as seen from the earth), the motion is closely related to the geocentric distance.

While comets are named after the discoverer (see Glossary) and meteorites after the place where they were found, the asteroids first get a preliminary designation which, after 1925, is made as follows: The year is followed by two letters, the first indicating the half month of observation (excepting I and Z), the second the order of discovery within that half month. When all two-letter combinations have been used, numbers are added.

Thus 1979AA, AB ... AZ, AA1, etc. A preliminary orbit is computed from at least three observations within that first apparition. If the minor planet is found again and confirmed, generally in two more apparitions, a permanent number is assigned. In the Ephemerides of 1979 there are 2042 permanently numbered asteroids. In addition to the above preliminary identification, there is a running number for asteroids found in the Palomar-Leiden Survey, for instance asteroid 2042 was 4633 P-L. After the assignment of a permanent number, the discoverer may name the asteroid; originally the custom was to put all asteroid names into feminine form and find the name from mythology. The objects in L_4 of the Sun-Jupiter system are named after Greek warriors, with 624 Hektor as the Trojan spy, and in L_5 after their Trojan counterparts, with 617 Patroclus as a Greek spy; we refer to both of these groups as the Trojans. Generally, however, the naming of asteroids has become messy because names suggested in the past by discoverers to recognize pets, friends and political heroes have been accepted.

In astrometry and celestial mechanics, minor planets are used to determine planetary masses, constants related to the motion of the earth, the solar parallax and astronomical unit and constants related to fundamental coordinate systems of star catalogs (see Rabe 1971 for an overview). Ceres, Pallas, Juno and Vesta have precisely determined orbits and their ephemerides are listed for every day of the year in *The American Ephemeris and Nautical Almanac* and similar ephemerides in other countries. The yearly catalog of all asteroids is issued by the Institute for Theoretical Astronomy in Leningrad; it is referred to as the *Ephemerides* or EMP, for Ephemerides of Minor Planets.

The remainder of this section deals with the techniques of physical studies by which we mean the investigation of physical parameters such as size, shape, composition, surface texture, spin rate, orientation of the spin axis in space, and including the description of any satellites. Magnitude-frequency relations for various compositions are included and so are dynamical studies, collisional probabilities, distributions of orbital parameters, the understanding of resonances, interrelations with comets and meteorites, and the origin and evolution of the asteroids and of their orbits.

I wrote an overview of the techniques of photometry of asteroids (Gehrels 1970) that may still be useful as an introduction to magnitude determinations, photometric systems, the Yerkes-McDonald Survey, the Palomar-Leiden Survey, lightcurve observations, phase and aspect variations, and perhaps even to the determinations of reflectivity. The developments since 1970 are described in the chapters by Bowell and Lumme and by Dollfus and Zellner; I will here introduce only a few terms of nomenclature and refer to various parts of the book.

Phase relations for various numbered asteroids are shown in Figs. 3, 5 and 8 of the chapter by Bowell and Lumme (also see the chapter by Veverka and Thomas for a discussion of phase effects). Between phase angles $7° \lesssim \alpha \lesssim 20°$ the magnitude-phase relation is nearly linear (see, however, the Bowell and Lumme chapter for a detailed discussion), and we speak of a phase factor or phase coefficient, β, in mag/deg. The extrapolation of the nearly-linear part to zero phase angle is the absolute magnitude $B(1,0)$. The nonlinear part of the phase relation, at $\alpha < 7°$, is named the opposition effect. Physical relations connected with these phase effects are discussed by Bowell and Lumme, who propose a new theory and symbolism; there is some criticism of their theory in a paper by Whitaker (1979). The mean opposition magnitude is not used much in this book, but it is listed in the Ephemerides and it is defined by $B(a,0) = B(1,0) + 5 \log[a(a-1)]$, where a is the semimajor axis of the asteroid's orbit.

When the asteroid is observed for its brightness variation during an extended period of time, one speaks of a lightcurve. The asteroids generally turn about their axis with a median rotation period on the order of 8 hr; the shortest period found is $\sim$2.5 hr and the longest exceptional one $\sim$85 hr; longer periods will probably be found as time goes on. The lightcurves, spin rates and body shapes are discussed in the chapter by Burns and Tedesco. When a set of lightcurves of the same asteroid is available over a wide range of ecliptic longitude (and/or latitude) and the timing of the lightcurve features is precisely intercompared one speaks of photometric astrometry from which it is possible to obtain the orientation of the spin axis in space; this topic is discussed by Taylor in his chapter.

There are two other techniques to study light variations. Bowell and Lumme describe how they observe the brightness of an asteroid on various nights and derive from the differences at least a statistical indication of the light variation and thereby of the body shape. Harris (his results are discussed in the chapter by Burns and Tedesco) observes several asteroids per night, returning to them repeatedly and thereby obtains for each a lightcurve although with few points.

Spectrophotometry began with the paper by Bobrovnikoff (1929) which was so far ahead of its time that it was overlooked. He obtained good spectra over the 0.4-0.5 μm range of wavelengths and the difference between Vesta and S-type asteroids such as 6 Hebe and 7 Iris is clearly seen. The chapters by Chapman and by Bowell and Lumme further refer to Bobrovnikoff's work. The spectrophotometry (with more than UBV filters) of asteroids was not done again until 1969, by McCord and his colleagues (see the chapter by Gaffey and McCord). The chapter by Chapman and Gaffey describes the observations and reductions in detail. The wavelength range of 0.3-1.1 μm is now being observed with photometers that have filters of various widths, ranging from the 25 filters of McCord and his associates, through the 8 filters mentioned in Zellner's chapter, to the wide bands of the standard UBVRI system.

The highest resolution spectroscopy is described in the chapter by Larson and Veeder; they explain their Fourier transform spectroscopy (FTS) technique. It is applicable to bright objects and it will be hard to observe the darker C types, nevertheless the infrared range of their instrument is a diagnostic one. The comparison of results obtained by the FTS with those at shorter wavelengths and in the far infrared is a field of rapid development.

Polarimetry is used in two different ways for the studies of asteroids, namely to determine the reflectivity and to study the surface textures; both these applications and the techniques are discussed in the chapter by Dollfus and Zellner.

Size determinations of the asteroids are of fundamental importance for the understanding of the reflectivities and the surface composition. At present we have 2 or 3 *indirect* methods for size determination, namely by polarimetry, by radiometry and also perhaps by straight photometry (see Table I of Bowell and Lumme), and 4 more *direct* methods namely by micrometer, disk meter, interferometry and occultation.

In polarimetry the basic principle is an inverse relationship between polarization and reflectivity, the so-called Umov effect, discussed in the chapter by Dollfus and Zellner. In radiometry at wavelengths near 10 μm, one compares the heat radiation, which depends on the size of the body, with the incident radiation by sunlight; the topic is reviewed in the chapters by Morrison and Lebofsky, who describe the technique, and by Schubart and Matson. The latter adopt a 987 ± 150 km diameter for Ceres, 538 ± 50 for Pallas and 544 ± 80 for Vesta.

The direct methods with the micrometer and the disk meter on the asteroids have been failures. Because the images are near the seeing resolution, one would set the wires of a double-wire micrometer too much towards the center of the light because the outer parts are made less visible by atmospheric turbulence. The resulting diameters therefore were too small.

The interferometry techniques (see Worden's chapter) have been available for a long time but the application to asteroids still is sparse (see Sec. V, below).

Occultations have not often been observed for the asteroids as yet, but results are important for size calibration and they are intriguing because of the reports of possible satellites. The occultation techniques and results are discussed in the Millis and Elliot chapter while the possibilities of satellites are treated in a more optimistic manner by Van Flandern *et al.*; the two chapters in fact debate the reality of satellites of asteroids.

One of the most desired parameters for the asteroids is the density for which one must have the mass in addition to the size; the topics are treated in the chapter by Schubart and Matson. Three techniques are possible for mass determinations namely from repeated close approaches of another asteroid, presence of a satellite, or flyby of a spacecraft. Schubart's discussion is on the first possibility, while the second is in the chapter by Van Flandern *et al.* and the third in the chapter by Morrison and Niehoff.

What is TRIAD? The Tucson Revised Index of Asteroid Data was initiated by Zellner as a collection of most if not all orbital and physical parameters on the asteroids. The complete TRIAD file is published here for the first time, in Part VII. It has a brief introduction by Zellner while in his chapter in Part V he uses the data to describe classifications under the general name of taxonomy. Other authors also use the TRIAD file although mostly from an earlier unpublished version since the present one was finally compiled only in June of 1979; files like these will, of course, be continually updated and improved as new observations are made.

The classifications themselves are in a state of flux. Of the wide-filter variety we have at the date of this writing the C, S, M, E, R, D classes and a designation U meaning unclassifiable. The D class, called RD in the Degewij and van Houten chapter, was added at the time of making this book! The classifications C, S, M, etc. are based entirely on observational parameters and do not imply any mineralogical or meteoritic identification. Classification, mineralogical interpretation, and meteoritic identification are three distinct steps. A closer correspondence between classification and meteoritic types is attempted in the taxonomic system of Gaffey and McCord (see their Table I). The advantage of the broad classification is that it can be applied relatively easily to many asteroids and used in statistical studies. For instance, 75% of all the asteroids are of the dark C type and there is an appreciable trend in the asteroid belt with the S becoming predominant in the inner belt.

I cannot even begin to describe the many observational laboratory techniques in meteoritics. I quote from a report from The Meteoritical Society ("Studies of Extraterrestrial Samples: Progress and Prospects," November 1978):

> "The return of the lunar samples created a need to examine and analyze the surfaces and interiors of minute grains and to measure differences in elemental or isotopic compositions, traces of remanent magnetization and other properties more precisely than ever before. New generations of mass spectrometers, electron and ion microprobes, and numerous other instruments were developed for the study of lunar samples and meteorites."

V. FUTURE WORK

In this section I will make some personal observations on future work and topics not covered in this book. These remarks are in addition to those on the future made in almost every one of the chapters, while the chapter by Morrison and Niehoff especially addresses the future. The major problems before us, of the origin and evolution of the asteroids and the identification of the meteorite parent bodies, are overviewed by Chapman and Wilkening. The role of the asteroids in the formation of the solar system is further studied in "Protostars and Planets" (Gehrels 1978).

The surveying of small bodies in the solar system that I referred to in Sec. III seems an important area for expansion. As many as possible near-Earth objects should be studied for future space missions as well as for their connection with meteorites and comets. There are three applications:

1. After landings on the moon and before those on Mars, manned exploration of nearby asteroids is the goal that Shoemaker *et al.* imply in their chapter. The target is 1943 Anteros, or preferably a lower-ΔV candidate (see the chapter by Morrison and Niehoff).
2. Mining of asteroids is not discussed in this book but extensively elsewhere (see e.g. O'Leary 1977, 1978). My own first reaction, years ago, was to consider it sacrilegious to damage an asteroid, but I now see that this possibility of pollution-free mining of resources must be investigated with high priority. How purely differentiated, how smelted, are the asteroids?
3. The controlled collision of an asteroid with Earth is discussed in Herrick's chapter. In Sudbury, Ontario in Canada, there is a major mining operation of nickel on a most peculiar circular area, presumably caused by the impact of an asteroid. The Tunguska event of 1908 is mentioned in Kresák's chapter. Should we not know, at least, which objects may impact the earth next?

These considerations for the future appear to need a dedicated search telescope to cover the sky to such a limited magnitude that suitable candidates would be known within the next decade. The University of Arizona Observatories have made a dome and site on Kitt Peak available for such a dedicated telescope.

A repetition of the Palomar-Leiden Survey of faint asteroids in the main belt is being considered in connection with the infrared astronomical satellite (IRAS). The purpose is to provide orbital information and optical magnitudes and thus radiometric albedos and diameters. The limit for IRAS is 7 or 8 in terms of a magnitude at 10 μm, which translates to 17 or 18 for a magnitude near 0.5 μm. That next survey would then be made to somewhat lesser limiting magnitude than the PLS but over a correspondingly larger area of the sky which will have the advantage of reducing the corrections for incompleteness (sec. III).

Further searches for Trojans of Jupiter, of the earth and even Mercury, possibly in infrared light, might be considered. Objects in Lagrangian points L_2 and L_3 — close to Jupiter on the Jupiter-Sun line — probably do not occur for they would have been found in the searches for outer Jovian satellites by Kowal and myself. Their distance from Jupiter is much greater than that of outer satellites, but they would appear close to Jupiter by foreshortening. Objects in the Lagrangian point L_1, on the Jupiter-Sun line but opposite Jupiter, have not been found in the Yerkes-McDonald Survey down to the 16th magnitude; a fainter search might be worthwhile.

Comets of large perihelion distance offer a special challenge, scientifically and observationally. If the parent molecule for a large part of the volatiles is water, then the comets are visible to distances from the sun of

~4 AU, but more volatile parent molecules can make the comets visible at greater distances. In that case, however, the apparent motion of the comets is small, the comets themselves appear to be small, and they cannot always be readily recognized by the usual tail or coma. The only way to find inactive distant comets is by taking plate pairs with an interval of several hours and to blink the pairs. In this manner I have found 5 comets with the Palomar 122-cm Schmidt and Kowal regularly discovers comets as a part of his survey (see his chapter), although so far the harvest of *distant* comets is still meager. The transition from active comets to extinct nuclei is an important topic of study for the understanding of the asteroids. In addition to the search for distant comets, various physical studies will have to be made in order to try to distinguish between the active and extinct cometary nuclei; this problem has not been solved as yet; we have not found a technique to determine the difference. This topic reminds us further of the origin and nature of 2060 Chiron which is discussed in the chapter by Kowal.

I believe that the Chiron population is sparse, because more would have been found on the many survey plates taken by Kowal and myself to date, but further searching may pay off especially if it can be done to fainter magnitude. Similarly for Trojans of Saturn and Neptune (see p. 654 of Gehrels 1971), fainter surveys are lacking. The density distribution in the present solar system drops off drastically beyond Neptune so that one wonders if more objects were not formed there in addition to Pluto. Observationally also, such searches seem attractive because the only complete survey for objects like Pluto, at the Lowell Observatory, was to the 16th, or perhaps only the 15th magnitude, while with a Big Schmidt the limiting magnitude can be ~21. It seems most likely that Kowal in his Survey will find other Plutos.

If a large number of new objects is discovered in the near future, the follow-up astrometry becomes urgent or we will be again in a situation (Sec. II) as in the early part of this century, when a large number of objects was discovered but only few of them could become permanently numbered asteroids. Hopefully the new astrometric observatory in Venezuela will undertake some of this work. At Indiana University also there is an interest in continuing an asteroid program with the 25-cm Cooke triplet of the Goethe Link Observatory.

Students in institutions that do not have large telescopes may especially note in various chapters the great progress that is being made in dynamical studies. Clearing of the Kirkwood gaps, delivery of material from the asteroid belt, the origin of the Apollos and Amors, these are topics for further study (see the next chapter by Chapman). Collisions may yield fragments that almost escaped, but that became satellites instead. How far from the primary can the satellite be? The crucial estimates are given in the chapter by Van Flandern *et al.*, but this topic may need further study. Van Flandern derived an approximate factor of 100 times the diameter of the primary for the maximum distance of the secondary. Some wide pairs (up to ~20 arcsec) might be seen on our photographic plates. We have not seen those, while some of us must have looked at thousands of asteroid images on photographic plates, but we did not consciously look for satellites and it might be good to keep this in mind in the future.

Direct evidence of satellites, if any, will come from occultations (see the chapter by Millis and Elliot) and from the interferometry described in the chapter by Worden. In speckle interferometry there is a theoretical promise of a resultion of 0.02 arcsec and it seems possible that this might be achieved, albeit with large telescopes, to the 10th magnitude. Regarding the asteroid occultations, we hope that a network of small telescopes with photoelectric photometers, as is beginning to be available in the United States, might also be established in India and Sri Lanka. The greatest difficulty lies in precise prediction of the events, and astrometry of the highest quality is needed a day or two in advance of the occultation. This requires close coordination with observatories that can do such astrometry.

Lightcurves are in demand again, for the study of satellites of asteroids (see the chapters by Hartmann, Van Flandern *et al.* and Taylor) in addition to the study of the rotation rate and the approximate shape of the asteroid (Burns and Tedesco). Some asteroids need extended coverage over a long rotation period and combination of observations at various observatories. Lightcurve observers seem to be in good contact with each other in order to plan such coordinated campaigns; for instance, asteroid 44 Nysa is a most puzzling case for which a coordinated campaign will be executed in 1979. Once the problems with inverted lightcurves are resolved, the technique of photometric astrometry (chapter by Taylor) can be used to determine the orientation of the rotation axis in space even for binary asteroids. A set of lightcurves obtained within the same opposition, but at different values of the phase angle, may give an immediate resolution to the question of the presence of satellites.

The variegation, i.e. the variation of characteristics over the surface, has only been mentioned in this book (chapter by Dollfus and Zellner), but not discussed in detail. There is a paper on the observations (Degewij *et al.* 1979), and there is the striking case of Vesta as the only asteroid that has a lightcurve with one maximum and one minimum just as Moon would have if it were set free. More of these studies seem possible, especially with double-beam photometers and polarimeters that can yield precisions an order of magnitude better than with a single beam. The interpretations need to take into account the effect of integration over the surface; when variegation is detected there must be *large* surface areas that are different from the mean.

Little has been mentioned in this book regarding the theoretical study of the internal constitution of the asteroids and the prediction of possible precessional effects on the spin axis. Theoretical work is also intriguing where the rotation rates are concerned. We do not even understand primordial rotation. Why do the asteroids rotate in ~8 hours? Burns and Tedesco give a further introduction to these basic questions.

The sense of rotation can be determined from infrared radiometry, as is explained in the chapter by Morrison and Lebofsky (also see Taylor's chapter), and this is a much needed job to be done for, say, 100 asteroids.

There is a continuing need for magnitudes and phase relations as is clear from the chapter by Bowell and Lumme. The phase relations are especially needed for faint asteroids. Nearly all asteroids redden with increasing phase. The only exception noted so far is 16 Psyche for which there is a bluing with increasing phase, possibly connected with a metallic composition. This is only a hunch, however, and it should be established with photometric measurements in the laboratory. There is little discussion in this book of the fundamental scattering laws, such as the Fresnel law and the applicability of the Mie theory. In the case of the lunar surface, I once tried to bring in the Mie scattering and met with frustration if not failure (Gehrels et al. 1964). There is little in this book about the micron-scale texture of the surfaces. It still is curious to me that the inversion angle — that is, the angle where negative polarization rises with increasing phase and becomes positive — ranges for asteroids between $15°$ and $23°$, whereas for lunar regions the angle is confined between $22°.6$ and $23°.6$. Bowell and Lumme, in their chapter, make a fresh approach, with a quantitative parametrization of asteroid surface texture: The inversion angle α_0 is primarily controlled by the refractive index n, increasing as n increases. Thus for Europa, $n = 1.33$, they would predict $\alpha_0 = 6°.6$ and for Mars, $n = 2.0$ of limonite, $\alpha_0 = 32°$. The greater range in α_0, and hence in n, for asteroids by comparison with the moon betokens greater compositional heterogeneity in asteroid surfaces (Bowell, personal communication 1979). A theory that can make these inferences and analyses deserves close study.

The surface texture of the asteroids can be studied from observations of color and polarization as a function of phase. The wavelength dependence of polarization has not yet been studied for asteroids; a fundamental study still seems needed. It has been noted that the phase coefficient, β, has a slight dependence upon wavelength (Gehrels and Tedesco 1979). The reddening with phase may affect the spectrophotometry and this effect seems to be neglected. Moreover, Fig. 9 of Gehrels et al. (1970) shows, near 0.8 μm, a 0.2 mag decline in brightness as the phase changes from $40°$ to $90°$. That paper contained spectrophotometry, albedo measurements and pole determination; such combined study campaigns should be executed for other nearby asteroids, for which the phase changes so drastically. Most scattering phenomena are strongly dependent on phase angle.

Most likely the proposal will be made to the International Astronomical Union in 1982 to change the definition of absolute magnitude of the asteroids. Whereas presently it is defined as the intercept of the extension of the linear part with $0°$ phase, the new definition will take the opposition effect into account. The asteroids will brighten by 0.3 mag overnight! There is a debate over the parameters that then should be used. Will it be the opposition effect and phase coefficient β, or the new parameters of Bowell and Lumme, or both? The polarimetric and radiometric size determinations need to be clearly (re)defined or the asteroids may also *grow* overnight.

Photoelectric checks of the present asteroid magnitude system have been made down to the 16th magnitude and it would be wise to extend that to the 20th. Over the 16-20 mag range, the magnitudes and calibrations have been determined photographically, based on photoelectric sequences in Selected Areas that have not even been properly published. It would not be surprising to find a systematic error in the present system of ~0.5 mag near the 20th magnitude.

I can see the need for another Ph.D. dissertation or two as that of Gradie on physical studies of Hirayama families. Various suggestions for further work on Hirayama families are made at the end of the chapter by Gradie *et al.* The photometric observations on family members are already being made in the 8-color programs, but the radiometry would have to be obtained in addition.

IRAS will be flying in 1982 and many asteroids will be observed, willy-nilly. Here is a great source of radiometry for Hirayama family members and for asteroids in general. The involvement of people interested in asteroids seems needed because of the large amount of IRAS data.

The International Ultraviolet Explorer (IUE) has not been discussed in this book. Some observations on asteroids are made by Zellner; Hapke makes laboratory studies of light reflected by various samples at wavelengths shorter than 0.3 μm. The combination could result in great progress because some asteroids have strong absorptions in the ultraviolet.

The radio and radar observations are coming into their own. Whereas Pettengill and Jurgens discuss (in their chapter) ~5 asteroids at this time, they believe that ~50 can be observed in the next decade. The observations at 3 mm-4 cm are described in the chapter by Dickel who plans to use the Very Large Array facility in New Mexico on asteroids. Worden describes new techniques in interferometry for the determination of diameters and the imaging of surface features and possible satellites. Because of the introduction of new detectors a great improvement in limiting magnitude is about to be realized and results for many asteroids are therefore possible. Students of asteroids should become involved in these new techniques.

The zodiacal light and the interplanetary particles have not been discussed in this book. The connection with the asteroids was considered in the book I edited before (Gehrels 1971). New information on the collection and analysis was reviewed by Brownlee (1978); the data seem to point more to a cometary origin than an asteroidal one. The results from the Pioneer missions through the asteroid belt were similarly summarized by Gehrels (1976).

Regarding space missions, there are only a few immediate possibilities (mentioned in the chapter by Morrison and Niehoff). The Space Telescope may be used for various observations outside of the earth's atmosphere, in Earth orbit, for high resolution imaging. Brahic *et al.* (1979) indicate the possibility of a French, or European, spacecraft flying by several asteroids. NASA, however, has no asteroid mission planned for the 1980's. Rendezvous and sample return missions do not seem likely until the third Millennium. Until then, physical studies of asteroids are made by remote sensing, the subject of this book.

Acknowledgment. I thank E.G. Bowell, J. Degewij, C.R. Chapman, B.G. Marsden, E. Roemer, E.F. Tedesco, L.L. Wilkening and B.H. Zellner for additions and deletions to the chapter and chronology, and Wilkening in addition for her help with the Glossary.

REFERENCES

Bobrovnikoff, N. T. 1929. The spectra of minor planets. *Lick obs. Bull.* 14: 18-27.

Brahic, A.; Breton, J.; Caubel, J.; Cazenave, A.; Cruvellier, P.; Dupuis, V.; Lago, B.; Minster, J. F.; Perret, A.; and Scribot, A. 1979. Feasibility study of a multiple fly-by mission of main belt asteroids. *Icarus* (special Asteroid issue).

Brownlee, D. E. 1978. Interplanetary dust: Possible implications for comets and pre-solar interstellar grains. In *Protostars and Planets,* ed. T. Gehrels (Tucson: University of Arizona Press), pp. 134-150.

Chapman, C. R.; Williams, J. G.; and Hartmann, W. K. 1978. The asteroids. *Ann. Rev. Astron. Astrophys.* 16: 33-75.

Cristescu, C.; Klepczynski, W. J.; and Milet, B. eds. 1974. *Asteroids, Comets, Meteoritic Matter* (Bucuresti: Editura Academiei Republicii Socialiste Romania).

Degewij, J.; Tedesco, E. F.; and Zellner, B. 1979. Albedo and color contrasts on asteroid surfaces. *Icarus* (special Asteroid issue).

Delsemme, A. H., ed. 1977. *Comets, Asteroids, Meteorites* (Toledo, Ohio: University of Toledo Press).

Gehrels, T. 1970. Photometry of asteroids. In *Surfaces and Interiors of Planets and Satellites,* ed. A. Dollfus (New York: Academic Press), pp. 317-375.

Gehrels, T., ed. 1971. *Physical Studies of Minor Planets* (NASA SP-267, Washington, D.C.: U.S. Government Printing Office).

Gehrels, T., ed. 1976. *Jupiter* (Tucson: University of Arizona Press), pp. 558-559.

Gehrels, T., ed. 1978. *Protostars and Planets* (Tucson: University of Arizona Press).

Gehrels, T.; Coffeen, T.; and Owings, D. 1964. Wavelength dependence of polarization. III. The lunar surface. *Astron. J.* 69: 826-852.

Gehrels, T.; Roemer, E.; Taylor, R. C.; and Zellner, B. H. 1970. Minor planets and related objects. IV. Asteroid (1556) Icarus. *Astron. J.* 75: 186-195.

Gehrels, T.; and Tedesco, E. F. 1979. Minor planets and related objects. XXVIII. Asteroid magnitudes and phase relations. *Astron. J.* 84. In press.

Helin, E. F., and Shoemaker, E. M. 1979. Palomar planet-crossing asteroid survey. *Icarus* (special Asteroid issue).

Herget, P. 1971. In Physical Studies of Minor Planets, ed. T. Gehrels (NASA SP-267 Washington, D.C.: U.S. Government Printing Office), p. xiv.

Kepler, J. 1596. Mysterium Cosmographicum. In *Prodromus Dissertationum Cosmographicum.*

Kuiper, G. P. 1951. On the origin of the solar system. In *Astrophysics,* ed. J. A. Hynek (New York: McGraw-Hill).

Kuiper, G. P.; Fujita, Y.; Gehrels, T.; Groeneveld, I.; Kent, J.; Van Biesbroeck, G.; and van Houten, C. J. 1958. Survey of asteroids. *Astrophys. J. Suppl.* 3: 289-428.

Morrison, D., and Wells, W. C., eds. 1978. *Asteroids: An Exploration Assessment,* NASA Conf. Publ. 2053.

O'Leary, B. 1977. Mining the Apollo and Amor asteroids. *Science* 197: 363-366.

O'Leary, B. 1978. Asteroid mining. *Astronomy* 6: 6-15.

Putilin, I. I. 1953. Minor planets (Moscow: State Publication of Technical-Theoretical Literature). In Russian.

Rabe, E. 1971. The use of asteroids for determination of masses and other fundamental constants. In *Physical Studies of Minor Planets,* ed. T. Gehrels (NASA SP-267, Washington, D.C.: U.S. Government Printing Office), pp. 13-23.

Russell, E. E., and Tomasko, M. G. 1976. Spin-scan imaging – application to planetary missions. In *Chemical Evolution of the Giant Planets,* ed. C. Ponnamperuma (New York: Academic Press), pp. 147-164.

Samoylova-Yakhontova, N. S. 1973. Minor Planets (Moscow: Academia Nauka). In Russian.

Stuhlinger, E.; Alfvén, H.; Arrhenius, G.; Bourke, R.; Doe, B.; Dwornik, S.; Friedlander, A.; Gehrels, T.; Guttman, C.; Strangway, D.; and Whipple, F. 1972. *Comets and Asteroids: A Strategy for Exploration,* NASA TM X-64677.

Titius von Wittenburg, J. D. 1766. *Betrachtung über die Natur,* vom Herrn Karl Bonnet, p. 7. Leipzig. (Translation into German from French.)

Trumpler, R. 1923. Search for small planets at the triangle points of Mercury and the sun. *Publ. Astron. Soc. Pacific* 35: 313-318.

van Houten, C. J.; van Houten-Groeneveld, I.; and Gehrels, T. 1970a. Minor planets and related objects. V. The density of Trojans near the preceding Lagrangian point. *Astron. J.* 75: 659-662.

van Houten, C. J.; van Houten-Groeneveld, I.; Herget, P.; and Gehrels, T. 1970b. The Palomar-Leiden Survey of faint minor planets. *Astron. Astrophys. Suppl.* 2: 339-448.

Wetherill, G. W. 1978. Accumulation of the terrestrial planets. In *Protostars and Planets,* ed. T. Gehrels (Tucson: The University of Arizona Press), pp. 565-598.

Whitaker, E. A. 1979. Implications for asteroidal regolith properties from comparisons with lunar brightness versus phase curves and theoretical considerations. *Icarus* (special Asteroid issue).

THE ASTEROIDS:
NATURE, INTERRELATIONS, ORIGIN, AND EVOLUTION

CLARK R. CHAPMAN
Planetary Science Institute

This chapter concerns the physical and chemical nature of asteroids, their physical and orbital distributions and interrelationships, and hypotheses about how asteroids formed and are evolving. A thematic synopsis is presented of these topics, all of which are covered in greater detail in other chapters of this book. Accepted interpretations are summarized, and also the new data and new alternatives that continue to make asteroid research an exciting endeavor.

"What are the asteroids like and how did they get to be that way?" has been a topic for speculation since the start of the 19th century when the first asteroid was discovered. But until the past decade of remarkable achievement in observing asteroids, ushered in by the first Tucson asteroid conference (Gehrels 1971), the data base was too limited to provide a firm foundation for hypotheses. Now a vast amount of data is available in the Tucson Revised Index of Asteroid Data (TRIAD), published in Part VII of this book.

During the mid-1970's a small group of asteroid researchers began synthesizing the data then available and developed preliminary but influential interpretations. Since then many more researchers have begun to study asteroids and the canonical interpretations have been subjected to sharp scrutiny. The second Tucson asteroid conference and the present book have served both to consolidate these views and to foster the development of serious critiques and alternatives. In this chapter I endeavor to summarize what we think we know about asteroids by presenting canonical views followed by critiques. This slightly artificial dialectic helps to separate my two tasks of reviewing the field as represented in the literature and of summarizing current research and controversy. Since this is a book of review chapters, I usually cross-reference other chapters wherein further references may be found, rather than refer the reader to the vast original literature.

I begin by discussing the physical and chemical properties of individual asteroids, building on the description of observing techniques in the previous chapter. Then I summarize interrelationships among the asteroids involving both orbital and physical properties. Finally I review the still immature considerations of origin and evolution — why asteroids are as they are, why an asteroidal planet does not exist, and what clues asteroids contain about early epochs of solar system history. I have felt too intimidated by the immense variety of data and often divergent interpretations to synthesize a model for the evolution of asteroids. In the following synopsis chapter, however, the reader will find a fine attempt to outline such a synthesis.

I. PHYSICAL PROPERTIES OF INDIVIDUAL ASTEROIDS

Our knowledge of the physical properties of asteroids is limited by our groundbased perspective. Despite sophisticated techniques for measuring the photometric behavior and spectral content of reflected and emitted radiation, asteroids remain mere star-like points of light in the largest telescopes. The gross shape of an asteroid can be measured only indirectly and we can only speculate about asteroidal "geology." As if we were observing a picturesque scene with tunnel vision, we can see or learn about some things very well, but we may be missing other major features.

The orthodox view of an asteroid is that of a single rocky body, perhaps roughly spherical, perhaps rather irregular in shape, spinning several times a day about an axis randomly oriented in space. Artists' conceptions of such bodies graced the frontispiece of the first asteroid conference book and have sailed through science fiction movies. Actual photographs by Mariner 9 and Viking of presumed asteroid-like bodies, the Martian satellites Phobos and Deimos (see the chapter by Veverka and Thomas), changed this image of an asteroid only in details. In 1978 and 1979, an alternative idea emerged that some or many asteroids may be double, multiple, or otherwise oddly shaped bodies, perhaps with swarms of small satellites orbiting about them. These models are caricatured in Fig. 1. As we shall see, the observational evidence for such asteroid models is suggestive at best and certainly not definitive, even for one asteroid, let alone for the majority. Nevertheless, current discussions of the possibility that asteroids may have such weird configurations plainly reveal that earlier data on which the adopted single-body model was based

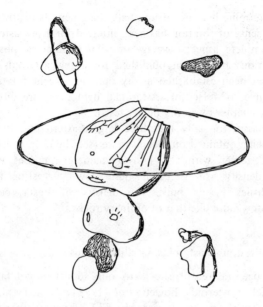

Fig. 1. Caricature of a multiple-body model for an asteroid.

were insufficient to specify asteroidal shapes uniquely.

Canonical model: size, shape, spin, mass

Diameter determinations from radiometry (see the chapter by Morrison and Lebofsky) and polarimetry (Dollfus and Zellner's chapter) have yielded sizes and geometric albedos for many asteroids; these indirect methods agree with the few direct determinations by stellar occultations (chapter by Millis and Elliot). Therefore, asteroidal diameters (or equivalent cross-sectional areas) and albedos are now well known. This is a major advance since the last decade when albedos were almost completely unknown and crude diameters were inferred from the measured magnitudes by assuming an albedo. The largest asteroid is Ceres with a diameter of ~ 1000 km. Pallas and Vesta have diameters a little more than half that of Ceres, and nearly 30 more bodies exceed 200 km diameter.

Lightcurve amplitudes suggest some asteroids are nearly spherical while others are elongated. A few asteroids have been observed over a sufficient range of aspect geometries that triaxial ellipsoidal fits to their shapes may be determined. There are even occasional hints of higher-order irregularities. But usually we know only the degree to which a particular asteroid is spherical or elongated and in individual instances we may even be fooled by the geometry. Lightcurve periods yield spin rates ranging from ~ 0.5 to 6 rotations per day, most commonly ~ 2.5 or 3. In a few cases, there may be a factor of two ambiguity in period and in other cases periods are difficult or impossible to determine because of unusually long periods or low lightcurve amplitudes. The sense of rotation has been inferred for a few asteroids from exceptionally complete aspect coverage or from radiometric phase effects. Polar axis directions have been published for a few asteroids, but their reliability has not been established in my opinion, except for one or two bodies, for example, 624 Hektor. Asteroidal lightcurves are discussed and interpreted in the chapter by Burns and Tedesco.

We know masses for only the three largest asteroids, although some information is being obtained on several more (see the chapter by Schubart and Matson). Together with the known diameters, masses yield mean densities. The density for Ceres is quite low, suggesting appreciable low-density volatiles in its bulk. The density of Vesta seems to be significantly greater, more like that of ordinary rocks.

Alternative models: double asteroids, satellite swarms

There has been a recent shock to our adoption-by-default of the single-body model of asteroids. Bowell *et al.* (1978) recorded a photoelectric record of an occultation of a star by 532 Herculina, confirming an independent visual observation, thus providing the first serious reason for considering that an asteroid might have a satellite. (Several earlier visual observations interpreted in terms of asteroid duplicity were generally discounted due to the known quirks of human vision and the lack of confirmation.) During the next year, several more photoelectric observations seemed to imply duplicity of, or satellites about, other asteroids. Binzel and Van Flandern (1979) amassed visual reports of many secondary occultations attributed to asteroidal satellites. Taken at face value, their statistics would imply that *most* asteroids are surrounded by *swarms* of satellites, including some bodies that are reasonably large in comparison with their primaries. See further discussion in the chapter by Van Flandern *et al.* in this book.

Perhaps some, or even most, asteroids have satellites. But do the reported data *require* that even *one* asteroid is double? There have been so many observations suggesting duplicity that, even if no single one is unambiguous, proponents believe some asteroids must really be double ("where there's smoke, there's fire"). Based on my own studies of the reliability of the human eye as an astronomical detector, however, I believe the visual reports of secondary events have no probative value enmasse. Some visual observations can be accurate and reliable, such as timings of known phenomena. But visual observers are notoriously poor at reporting unexpected or improbable phenomena, which characterize asteroidal occultations. Predictions have been so poor that any individual observer is lucky to see the main event; a wish to be the "lucky one" can be a powerful subconscious incentive to "see" a brief event in a twinkling star image.

Proponents of double asteroids also have argued that at least a few photoelectric observations are secure, within reasonable limits. Some experienced photoelectric observers (e.g. Reitsema 1979) believe, however, that confirmation of signal drop-outs by at least two photoelectric stations, both operating under nominal observing conditions, is a minimum requirement for documenting the existence of an asteroidal satellite. Even the Bowell observation of 532 Herculina, the best case yet of a "confirmed" observation, was made under extreme conditions and according to Bowell should be considered suggestive at best. I conclude, therefore, that we do not yet *know* that a satellite exists about any individual asteroid nor do we yet have any basis for concluding that the existence of asteroidal satellites is probable (see the chapter by Millis and Elliot).

Nevertheless, the observations cannot be totally discounted. They are sufficiently suggestive that satellites may exist that efforts to document a single unambiguous event should continue. The next step would be to determine the frequency of asteroidal satellites. For now, the most important effect of this subject has been the broadening of our horizons about the possibility of unusual asteroidal configurations. Already large-amplitude lightcurves of several asteroids have been interpreted as being consistent with a binary model (see Wijesinghe and Tedesco 1979).

It is becoming clear that few observational data, if any, rule out a double or multiple body as a possible configuration for any asteroid. As Bobrovnikoff (1929) wrote long ago, "If an occasional asteroid were not a single body but consisted of several pieces . . . we could never tell the difference." While hypotheses for the origin of double or multiple asteroids have yet to be worked out in detail (see Hartmann's chapter), there are apparently no theoretical reasons for doubting that some asteroids could be double. For now, we must remain open-minded about the possibility of double asteroids and be alert to the advantages that this alternative model might provide for understanding data that in the past would have been automatically interpreted in a single-body context.

II. SURFACE COMPOSITIONS OF INDIVIDUAL ASTEROIDS

What are asteroids made of? Apart from bulk densities of three asteroids, the only knowledge we have of asteroidal compositions is inferred from the hemispherically averaged radiation reflected and emitted from their *surfaces*. Evidence for regional variations has been sought by monitoring colors and polarizations of asteroids as they rotate. Degewij *et al.* (1979) summarize the results: most asteroids are uniform at this coarse scale of resolution but a few, including 4 Vesta, show small variations in color and/or albedo. Regional heterogeneities could even reflect underlying vertical variations in composition within an eroding or fragmented body, in the same way vertical stratigraphy is exposed in the Grand Canyon; but conclusions are precluded by the poor spatial resolution. Indeed, apparent lateral homogeneities could be due to the masking of any underlying variations by global regolith deposition. On the assumption that Hirayama family members are pieces of a single precursor body, statistics summarized by Gradie *et al.* in their chapter suggest that some bodies were volumetrically homogeneous and others not.

In order to decipher composition from the radiation reflected and emitted from the uppermost surface of an asteroid, it is important to know the physical state or texture of the surface. The presence of negative branches in polarization versus phase curves suggests all asteroids are coated with at least a partial monolayer of fine dust (see the chapter by Dollfus and Zellner). Radiometric data imply the presence of low-thermal-inertia particulate regoliths, of at least centimeter-scale depth, on most asteroids; but some bodies smaller than 10 km diameter (perhaps even the 20-km object Eros) may have largely rocky surfaces (Morrison and Lebofsky's chapter). In a few cases, radar data suggest even deeper regoliths (Pettengill and Jurgen's chapter) but most of what we know of asteroidal regoliths comes from theoretical considerations and evidence from meteorites hypothesized to have originated on asteroidal surfaces (see the chapter by Housen *et al.*).

Sunlight scattered through and reflected from the particulate surfaces of asteroids carries spectral information on composition. Especially diagnostic are absorption bands imposed on light transmitted through small, translucent surficial grains. Useful spectral features exist in an asteroid's *reflectance spectrum* throughout the measurable wavelength range 0.3 μm to 3.5 μm. Spectra are scaled to geometric albedos determined from radiometry and polarimetry; in the absence of albedo data, they are scaled to unity at 0.56 μm for purposes of comparison.

The most fundamental assessment of an asteroid's surface composition would be in terms of elemental abundances, which however cannot be inferred directly. Spectra are useful for interpreting asteroidal mineralogy, that is, the presence of particular crystalline and amorphous compounds, including rocks, metals, and ices. In principle, a precise assay of minerals and mineral compositions specifies also the chemical composition. But the same chemicals may form different minerals, depending on the temperatures and pressures under which they formed and on subsequent processes of alteration. So an assessment of asteroidal mineralogy, which we attempt to do from reflectance spectra, can determine the major chemical abundances *and* set constraints on processes of origin and evolution. As we shall see, however, Earth-based capabilities seem very limited, thereby restricting our understanding.

Asteroidal spectra have been measured by several observational techniques. Three-filter colorimetry, readily applicable to many objects, is not itself diagnostic of composition and has been used chiefly to group similar asteroids. The infrared part of the spectrum (from 0.9 to 3.5 μm) is most diagnostic of composition, due to numerous electronic and molecular absorptions. Beyond 0.8 μm, Fourier transform spectroscopy, supplemented by some mid-infrared filter photometry, can be applied to the brighter asteroids (see the chapter by Larson and Veeder). The largest spectral data set with sufficient spectral resolution for mineralogical interpretation is that of 277 spectra of Chapman and Gaffey (see their chapter) from 0.33 to 1.07 μm, a spectral range in which many minerals have electronic and charge-transfer absorptions. Albedo data are also useful in compositional interpretation, especially differences exceeding the factor of 2 that can result from petrological and grain-size variations that are unrelated to composition. Our best understanding of asteroidal mineralogies comes at present from the complementary application of all these approaches.

Simple colorimetry was sufficient to establish that asteroids are different from one another. In the early 1970's it was recognized from UBV photometry and albedo data that most observed asteroids fall into two broad groups (cf. Zellner's chapter): (1) a very low-albedo group (geometric albedo <0.065, typically 0.035 to 0.04) with flattish spectra (neutral colors), termed "C type"; and (2) a moderate-albedo group (geometric albedo 0.065 to 0.23, typically 0.14) with curving, reddish-sloping spectra, termed "S type". A much smaller third group, termed "M type", has moderate albedos and straight, slightly reddish spectra. Some asteroids (~10%) show unusual spectra, different from the C, S, and M types; a few fall into two rarer types (E and R) defined recently on the basis of color- and albedo-sensitive data, but most are designated "U" (for "unclassifiable" among classes so far defined). As in any taxonomy, a few individuals inevitably fall at boundaries between groups. Such is the case with Ceres; the errors on its well-determined spectral and albedo traits straddle the edge of the C field, so in the past it has been typed as both C and U; in this book (Part VII) it is again a C, albeit an unusual one. Despite the mnemonic association of the letters C, S, M, and E with specific compositions, the groups are defined by observational parameters only. Minerals have been inferred from more complete spectral data; extension to less well observed asteroids is made by hypothesis only.

A canonical view of asteroidal compositions was developed in some detail, during 1974-1978. It has formed the framework for most thinking about asteroidal evolution and relationships to meteorites. Since then, however, important new data and new interpretations of earlier data have begun to challenge many of the canonical interpretations. While some preferred alternatives are preliminary and will require elaboration before they could replace accepted views, others already have merit perhaps exceeding that of the widely accepted models. We have entered an exciting period in which we may hope that several experimental and theoretical programs will reduce the unacceptably wide range of interpretations of reflectance spectra currently being debated.

Canonical model: carbonaceous asteroids, stony-irons, and a few achondrite parent bodies

The methodology of inferring asteroidal mineralogy from reflectance spectra is outlined in the chapter by Gaffey and McCord. There are three chief approaches:

1. Spectra have been matched against standard libraries of laboratory spectra of powdered terrestrial rocks and meteorites.
2. From crystal physics the theory has been developed to explain how spectral features are imposed on light transmitted through crystals of different compositions and optical properties; assisted by computer simulation, the theory is being extended to the practical case of light being reflected from and scattered through a size distribution of particulates of mixed composition and random orientation.
3. As an extension of the first approach, systematic laboratory spectral measurements are being made of controlled mixtures of well-defined materials in specific physical states in order to parameterize and understand reflectance spectra.

Major mineral phases that have been tentatively or securely identified from asteroidal reflectance spectra include: ortho- and clino-pyroxenes of various compositions; olivine; nickel-iron alloys of various Ni:Fe ratios; non-metallic, low-albedo opaques (magnetite and/or carbonaceous compounds); and poorly characterized, often hydrous, phyllosilicates and clay minerals. Plagioclase and iron-free enstatite have been inferred to be present on a few asteroids. Most asteroids are each thought to be composed dominantly of mixtures of a few of the above minerals. Vesta provided a famous test case on the credibility of asteroid spectral interpretations. McCord et al. (1970) identified Vesta's 0.9 μm band as due to pyroxene and predicted that a 2 μm band should exist; the longer wavelength band was subsequently discovered by Larson and Fink (1975).

Vesta's surface is dominantly composed of plagioclase (in addition to the pyroxene) and is interpreted to be basaltic. Thus Vesta has received wide attention as a parent body for achondritic meteorites. (The best infrared spectrum of Vesta is shown by Larson and Veeder in their chapter; the literature on Vesta is reviewed in the chapter by Gaffey and McCord.) Another unusual asteroid, E-type 44 Nysa, has been interpreted (Zellner 1975) as being composed of iron-free achondritic enstatite, as in aubrite meteorites. There are a few other asteroids possibly of achondritic compositions, but in the canonical view they are rare.

The most common asteroids are the ultra-black C types. Some have geometric albedos of only 3%, requiring that a finely divided opaque material be distributed throughout the surface material in the amount of at least several, possibly 10, percent. Another mineral, presumed to be a layer-lattice silicate, is evidenced on most C types by the edge of an ultraviolet absorption feature. The widely accepted interpretation is that C types are analogous to carbonaceous chondritic meteorites. This inference was strengthened by Lebofsky's (1978) announcement of a hydrated phase on the C-type body Ceres, inferred from a 3 μm absorption similar to that measured in CM carbonaceous chondrites. Later Lebofsky (1979) reported the probable presence of the 3 μm feature for C types 19 Fortuna, 31 Euphrosyne, and 70 Panopaea, as well as for C-like U's 2 Pallas, 51 Nemausa, and 72 Feronia. Thus, in comparison with meteorite spectra, many C-type asteroids seem most similar to several types of carbonaceous chondrites. This canonical view of C-type asteroids is elaborated on by Gaffey and McCord in their chapter. We will see later, however, that other C-type asteroids differ from carbonaceous meteorites.

According to the canonical view, most S-type spectra indicate the presence of pyroxene and/or olivine mixed in various proportions with nickel-iron. S types are a diverse collection of bodies; over two dozen separate S-like spectral groups have been noted by Chapman and Gaffey (see their chapter). Even 12 Victoria and 433 Eros, noted in the chapter by Larson and Veeder as being so similar in the infrared, differ substantially in the visible. Most S types are interpreted to have metal-to-silicate ratios of order unity, ranging from less than 1/3 to at least 3. In terms of meteoritic analogs, most S-type asteroids would be classed as stony-irons (perhaps mesosiderites for pyroxene-rich cases, pallasites for olivine-rich cases). The extreme metal-poor end-members of the S class have been interpreted to approach the metal:pyroxene:olivine proportions of some ordinary chondrites. (However, the best analogs for the L and LL type ordinary chondrites tend to be classed as R's, not S's: e.g. 496 Gryphia.)

The uncommon M-type asteroids have spectra interpreted to be dominated by the reddish-sloping but otherwise featureless signature of metal (especially nickel-iron). Meteoritic analogs include nickel-irons and enstatite chondrites. The latter consist of nickel-iron imbedded in a relatively transparent, iron-free enstatite. One might think of M's as the metal-rich extension of the S class, except that there is a hiatus between the types in the U-V color index.

Critique of the canonical model

With what confidence may we accept these mineralogical interpretations? There is no general answer. Some minerals (e.g. iron-bearing orthopyroxene) have unique spectra with diagnostic absorptions. But the presence on asteroids of minerals such as enstatite, which lacks spectral features, must be regarded as speculative. Some mixtures of minerals are difficult to diagnose. For example, a mineral such as plagioclase may be present in considerable amounts without impressing its distinctive features on a spectrum swamped by another mineral. Relatively tiny quantities of a low-albedo, opaque substance (carbon black) can swamp even some strong spectral features of pyroxene.

There is continued theoretical and experimental research on how to interpret visible and infrared reflection spectra uniquely. In the meantime, some intellectual shortcuts have been adopted. Most interpretations of asteroidal spectra have been restricted explicitly or implicitly to cosmically abundant elements and mineral phases, chiefly as represented among known meteorites. From such a perspective, one might conclude that the mineralogically significant spectral features of an asteroid *are consistent with* one or more meteorite types, but that is a far cry from a unique interpretation. Conversely, a significant spectral difference between an asteroid and a meteorite rules out an exact equivalence, but the difference might not be important from a cosmochemical perspective. In the absence of double-blind tests of the uniqueness of the canonical spectral interpretations, there is ample room for alternatives. Below I try to evaluate the relative merits of alternative interpretations for the *C, S,* and *M* types.

Are C types carbonaceous chondrites? The widespread association of *C*-type asteroids with carbonaceous chondrites must be regarded as based on circumstantial evidence. Evidently a substance is required that is even blacker than the matrix of most primitive *C*-type meteorites! While plausibility arguments still lead one in the general direction of primitive *C*-like meteorites, the identification of the low-albedo substance, or the other minerals with which it is mixed on asteroidal surfaces, remains unresolved. Lebofsky (1979) reports that the 3 μm water absorption is *absent* from the spectra of several classical *C*-type asteroids (65 Cybele, 88 Thisbe, 324 Bamberga, and 554 Peraga). Larson and Veeder (in their chapter) report preliminary speculation that Bamberga, at least, might bear some relationship to the ureilites, a carbon-bearing kind of achondrite. The association cannot be taken literally since ureilites have albedos of 8% or higher, relatively deep 1 μm absorption features, and ultraviolet absorption edges at a longer wavelength than observed on Bamberga (Gaffey and McCord's chapter). But even the hint of similarity between a well-known *C* type and a differentiated assemblage is sufficient to highlight the challenge to the canonical interpretation of *C* types.

The interpretation of the water absorption feature on *C*-type asteroids as implying carbonaceous chondritic assemblages is also not secure. Larson and Veeder (see their chapter, and references therein) regard Ceres and Pallas as closely resembling CM chondrites, especially because of similarities in the infrared. But Ceres has an ultraviolet reflectance unlike any measured CM meteorite and *U*-type Pallas is even more extreme; Gaffey and McCord in their chapter characterize them as being unlike any known meteorite in composition.

I still cling to the idea, however, that there must be some association between the carbonaceous meteorites and many of the C-type asteroids. After all, C-type asteroids are ubiquitous, especially in the main belt but also in Earth-crossing orbits. Thus we probably have some fragments from them on Earth, which would certainly be among the carbonaceous types of meteorites. The detailed research necessary to understand the variety of C-type asteroids is still in its infancy.

Are S and M types really metal-rich? The presence of silicates on S types has not been disputed. But the canonical inference that these asteroids are metal-rich (hence differentiated) has long worried those who have hoped to find a plentiful supply of (undifferentiated) ordinary chondrites in the main asteroid belt. It has been suggested that some kind of weathering or maturation process might alter chondritic spectra in the manner attributed to abundant metal, namely reddening and linearizing the spectrum and weakening silicate absorption features. Physical maturation of lunar soils under micrometeorite bombardment reddens and linearizes the lunar spectrum, but it also darkens lunar soils. Matson *et al.* (1977) show that there is no correlation between albedo and the strength of the "metal signature" in asteroid spectra. Perhaps laboratory experimentation on meteoritic assemblages would reveal some as-yet-unknown weathering process that spectrally mimics abundant iron.

Anders (1978) has suggested that differential comminution of silicate and metallic grains and preferential mechanical sorting by regolithic processes might enhance the *apparent* metal content in a spectrum of an ordinary chondritic assemblage. Meteorite samples measured in the laboratory are normally powdered by mortar and pestle which does not comminute ductile metal grains. At temperatures in the asteroid belt, however, the metal is brittle (Remo and Johnson 1974) and might be finely comminuted by hypervelocity impacts. If temperature controlled the comminution of metal, one might expect S-like asteroids closer to the sun to look more like ordinary chondrites than those farther out. Figure 2 shows such a tendency, though it might instead reflect completely different physical or chemical processes (e.g. variation of nebular condensate composition with heliocentric distance). Such differential comminution would explain why a few asteroids show little linear reddening, which is hard to explain with the maturation analogy. For example, unreddened 4 Vesta is interpreted as having an achondritic composition and thus has *no* metal to be comminuted. Some other asteroids interpreted on independent grounds as having metal-free or low-metal composition show little linear reddening, but a clear relationship has not yet been established. In short, I believe that in our present state of ignorance, the metal-silicate ratio of comminuting mixtures of the two phases cannot now be determined with the precision necessary to separate chondritic values from stony-iron values.

One interpretation of M-type spectra is in terms of nickel-iron alloy, as in

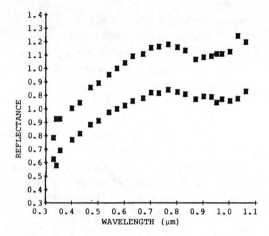

Fig. 2. Mean spectra for *S*-like asteroids as a function of semimajor axis. Means are for high-albedo asteroids ($p_V > 0.065$) *not* typed in TRIAD with types beginning with the letters *"C"* or *"M"*. TRIAD type file diameters of 50 km or more are included. The mean spectrum of 17 asteroids with $a < 2.4$ AU is at the top; below it is the mean spectrum of 51 asteroids with $a > 2.6$ AU.

S's but without silicates. Polarization data of some *M* types have been interpreted as supporting the presence of metal (see the chapter by Dollfus and Zellner). If either *M*- or *S*-type asteroids contained the canonical proportions of metal, they should be electrically conducting and would have a high cross-section to radar. While one or two such asteroids have been sought with radar, none has yet been detected (cf. the chapter by Pettengill and Jurgens). The reason for nondetection is not clear and more observations are planned. Another observational technique sensitive to high metal content is comparison of thermal fluxes at 10 and 20 μm (Morrison and Lebofsky's chapter). Although more extensive observations are required, preliminary data do not seem to deviate much from the adopted dielectric (nonmetallic) model. Radar or radiometric observations may soon provide a clear answer on the question of metal-rich asteroids.

A special case: 349 Dembowska

The methodology of interpreting asteroid spectra, and its uncertainties, may be exemplified by 349 Dembowska. Once the sole candidate for a main-belt asteroid of ordinary chondritic composition, Dembowska has more recently received attention as a possible melted achondritic parent body (discussed later in Sec. IV). The interpretation history of its spectrum illustrates the difficulties in understanding even a relatively simple three-component mixture. The three components involved in this story (pyroxene, olivine, and metal) include two that have prominent spectral features (see Fig. 3). The three components are also of great cosmochemical

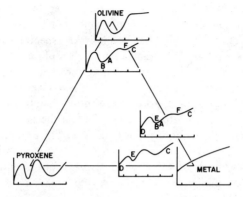

Fig. 3. Schematic ternary diagram of the pyroxene/olivine/metal system (volume percent). Reflectance spectra for the pure components are from Gaffey (1976), spanning the range 0.3 to 2.5 μm (tick marks at intervals of 0.5). Model spectra for three mixtures of the components illustrate features discussed in the text. Two components in each mixture contribute roughly equal spectral traits. Also the third component is present in each mixture in volumetric amounts of 5 to 10%. Note the spectral dominance of pyroxene; its features are prominent even though it is a minority constituent of all three mixtures. Modifications of the 0.9 μm pyroxene band due to olivine are: A, depression near 1.2 μm; B, shift of band center to longer wavelength. Modifications of the pyroxene spectrum by metal are: C, high 2 μm reflectance (also could be due to olivine); D, higher ultraviolet reflectance; E, shallower absorption band depth. A mixture of olivine and pyroxene makes the 2 μm band, F, shallower than the combined band near 1 μm.

interest since they separate several classes of undifferentiated chondrites from each other and from important metal-rich and metal-poor differentiated assemblages.

Dembowska's spectrum was first measured out to 1.06 μm by McCord and Chapman (1975), who noted its low infrared reflectance and its strong absorption band centered at a longer wavelength than is typical of orthopyroxene (see A and B in Fig. 3). They inferred an olivine/pyroxene ratio of ~ 2 and suggested an LL chondrite composition (L chondrite, with olivine/pyroxene <2, could not be ruled out). Later Veeder et al. (1978) measured a high 2 μm reflectance (C in Fig. 3) which they interpreted as due to reddening by a major metal component. A re-examination of the visible/near-infrared data by Gaffey and McCord (1978) gave evidence for some reddening (e.g., D in Fig. 3) but chiefly led to the conclusion that pyroxene must be in low abundance or absent.

Feierberg et al. (1978, 1979) obtained a Fourier transform spectrum of Dembowska from 0.9 μm to 2.5 μm (the spectrum is shown by Larson and Veeder). An absorption feature near 1.9 μm is uniquely attributed to orthopyroxene, ruling out a simple olivine or olivine/metal composition for Dembowska. Feierberg et al. (1979) propose that Dembowska is composed of a metal-poor assemblage with an olivine/pyroxene ratio of at least 2 and perhaps approaching 10, from which they infer that it is an achondritic assemblage probably not yet represented among meteorites.

Feierberg et al. conclude that Dembowska's spectral contrast is too great in the ultraviolet (D in Fig. 3) and that its absorption bands are too deep to permit more than 10% metal. Furthermore they believe that the weakness of the 1.9 μm feature is consistent with as little as 10% pyroxene. By comparing the relative strengths of the two absorption bands with those of pure pyroxene, Feierberg et al. further conclude that the short wavelength band is dominated by an olivine absorption (near A in Fig. 3), implying 70% to 90% olivine.

In my view none of the above interpretations is entirely satisfactory. Consider the olivine/pyroxene ratio: A good measure is from the center wavelength for the combined 1 μm absorption band, as calibrated by Gaffey (1976) and Chapman and Salisbury (1973). The ratio turns out to be 1:1 to 2:1, apparently incompatible with the Gaffey/McCord and Feierberg et al. interpretations. Also application of Feierberg et al.'s band-depth comparison approach to published spectra of LL–type ordinary chondrites (composed mainly of olivine and pyroxene in the ratio of ~2) would result in an inference that they are much more olivine-rich than they actually are.

The inference of metal content is also uncertain. While Dembowska's band is sufficiently deep to be consistent with some LL chondrites, which contain little metal, some other LL chondrites have deeper bands that could be made equivalent to Dembowska's by addition of up to 40% metal. It is appropriate to question whether chondritic assemblages of pyroxene, olivine, and metal are entirely satisfactory for modeling potentially achondritic assemblages of the same phases with different petrological characteristics (Feierberg, personal communication). There are even more substantial problems involved in understanding comminution and sorting of metal in asteroidal regoliths, as discussed above in the context of S-type asteroids. The high infrared reflectance of Dembowska (C in Fig. 3) could be due to high olivine content, to high-metal content, or to finely comminuted metal masquerading as high metal content. The apparent failure of the original LL chondrite interpretation of Dembowska to explain the infrared reflectance may be another manifestation of the S-type problem.

I am not convinced we have sufficient laboratory data to decide uniquely among any of the previously proposed assemblages: LL chondritic; high-metal olivine-rich; and low-metal olivine-rich. If one accepts the latter interpretation (that of Feierberg et al.) it is still not clear whether Dembowska is achondritic. LL chondrites are marginally within the limits established by Feierberg et al. (olivine/pyroxene ratios as high as 2.4 are known). Even more olivine-rich compositions might be compatible with a hypothetical chondritic composition proposed by Gaffey and McCord (1978): a carbon-free C3.

In view of the uncertainties exemplified by Dembowska, spectral interpreters should redouble their theoretical and laboratory efforts so that the full implications of the large set of spectral data on asteroids may be understood.

Compositional relationships between asteroids and meteorites

It has been advocated, from spectral evidence only, that there may be asteroidal counterparts for the following meteorite types: irons; mesosiderites; pallasites; diogenites; howardites; eucrites; aubrites; ureilites; enstatite chondrites; ordinary chondrites types H, L, and LL; and carbonaceous chondrites types 1 (C1), 2(CM), 3(CV and CO), and 4. Apparently not yet recognized in the belt are asteroids composed wholly of assemblages like chassignites, nakhlites, or angrites; of course, such highly differentiated assemblages might plausibly exist in only small localities on a large achondrite parent body. Some of the most common meteorites are extremely rare although present in the main belt (e.g. the ordinary chondrites, unless they can be identified with the S types).

In summary, I think it is a plausible but not proven working hypothesis that many C-type asteroids are like carbonaceous chondrites. There are two equally reasonable alternatives for most S types: (a) they are metal-rich silicate assemblages akin to stony-iron metorites, or (b) they are less metal-rich silicate assemblages akin to ordinary chondrites. In either case, there are major differences from asteroid to asteroid in the olivine/pyroxene ratio and the metal and silicates might be mixed on almost any spatial scale. Many M-type asteroids seem dominated by the spectral signature of metal; despite the mnemonic "M" it is quite plausible that they are chiefly made of colorless silicates, as are the enstatite chondrites. The basaltic composition of Vesta is well established, but achondritic interpretations of Nysa (and its family members) and of Dembowska are still speculative.

III. POPULATION CHARACTERISTICS, INTERRELATIONSHIPS, AND CORRELATIONS

The asteroids are intriguing not only as individual bodies but as a *group*. There are many asteroids — a few of appreciable size but myriads of tiny ones. They orbit the sun in specific locations, some specially related to the orbits of the major planets. They exhibit dynamical and physical relationships among themselves as well as between themselves and other small members of the solar system. The interrelationships may help us understand why an asteroidal planet failed to form and how the asteroids evolved over solar system history.

Orbits

The first one and two-thirds centuries of asteroidal research emphasized determination of orbits; orbital distributions are lavishly summarized by Brown *et al.* (1967). Most numbered asteroids have semimajor axes between

2.2 and 3.2 AU, in the *main belt*, but there are four *Atens* with semimajor axes near Earth's and other special asteroids ranging out to Hidalgo and distant Chiron (the latter is discussed by its discoverer, Kowal, in this book). Studies of the dynamical evolution of some Earth-approaching asteroids show their orbits to be short-lived compared with the age of the solar system, so they must have originated somewhere else. But, even if they were originally comets, they are now termed asteroids. Comet/asteroid relationships are discussed in the chapter by Kresák.

Of particular interest are the groups of asteroids beyond the main belt having particular orbital commensurabilities with Jupiter (see Fig. 4): the Hildas at 3:2, Thule at 4:3, and the Trojans at 1:1 (see the chapter by Greenberg and Scholl). Just beyond the inner and outer boundaries of the main belt are some stragglers which may fundamentally be main-belt objects that are isolated by gaps in the distribution. Around 3.4 AU, beyond the 2:1 gap, large asteroids are as common as in the inner third of the main belt (see Fig. 4). But the Palomar-Leiden Survey (van Houten *et al.* 1970; Gehrels' chapter) found almost no small asteroids in this region.

The *Kirkwood gaps* in the semimajor axis distribution of main-belt asteroids have long been known; many are centered on low-order commensurabilities with Jupiter (see Greenberg and Scholl's chapter). More recently, gaps have been noted in distributions of proper eccentricity and inclination (see Fig. 3 in the chapter by Gradie *et al.*); they also have been identified with resonant phenomena (cf. Greenberg and Scholl's chapter).

In addition to the boundaries and gaps in *a-e-i* distributions just described, there is a finer-scale clumpiness in the distribution of asteroidal proper elements, first recognized by Hirayama half a century ago. Roughly half the asteroids are now thought to be members of about 100 *families* (see the chapters by Kozai and by Gradie *et al.*), although fewer than half-a-dozen especially populous families are conspicuous on two-parameter plots of orbital elements. Clusterings in all five orbital elements ("jet streams") have also been reported, but observational bias appears to be responsible for such groupings (Kresák 1971).

During the 1970's, the major research efforts concerning asteroidal orbits have dealt with the general problem of stability. If some orbital configurations were known to be stable against planetary perturbations or other forces on time scales comparable to the age of the solar system, then objects found in such orbits could have formed there, or at least they might retain a kind of dynamical "memory" of their origin. Other orbital configurations are short-lived in the sense that bodies in such orbits would likely impact major planets, be ejected from the solar system, or be perturbed into radically different orbits on time scales short compared with the age of the solar system. The absence of asteroids in such "chaotic" orbits (as they are termed by Everhart in his chapter) could then be explained by the aforementioned depletion mechanisms and any asteroids found in such orbits

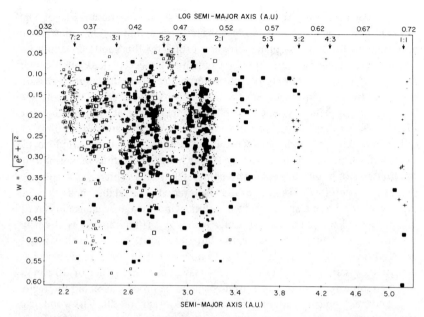

Fig. 4. Overview of the asteroid belt. Nearly all numbered asteroids are plotted as a function of (log) semimajor axis and a quantity related to departure from an in-plane circular orbit. High-velocity asteroids are toward the bottom, with exceptional ones, such as 2 Pallas, below the bounds of the diagram. All asteroids in the TRIAD type file are shown; three symbols indicate type: solid boxes, C, the actual or probable type; open boxes, S; and crosses, neither C nor S (mainly M and U). The symbols are plotted in two different sizes. The larger ones represent a nearly unbiased sample of asteroids larger than 80 km diameter; smaller symbols are for smaller asteroids, which are subject to bias against low-albedo types, especially in the outer half of the diagram. Vertical (Kirkwood) gaps in the main-belt population and groups of asteroids beyond the main belt are correlated with the Jupiter commensurabilities indicated at the top (e.g. Hildas, 3:2; Thule, 4:3; Trojans, 1:1). Prominent clusters of identical symbols are the more populous, homogeneous Hirayama families: Themis ($a = 3.14$, $w = 0.15$), Eos ($a = 3.01$, $w = 0.19$), Koronis ($a = 2.88$, $w = 0.06$), and Flora group ($a = 2.21$, $w = 0.16$). The Phocaea group (near $a = 2.4$, $w = 0.48$) is bounded above, below, and to the left by three secular resonances and to the right by the 3:1 Kirkwood gap. The C types increasingly predominate in the outer belt, but note the unusual distribution of types near 2.8 - 3.0 AU. The generally faint numbered asteroids for which no physical data are available (shown as small dots) are observationally biased toward the inner belt. The distribution of larger symbols shows that the larger asteroids are roughly as common near 3.4 AU (beyond the 2:1 gap) as in the inner third of the main belt (inside the 3:1 gap).

would have to be regarded as being in transition between other more long-lived configurations.

The most direct approach to the problem of stability of different types of orbits is to numerically integrate the orbits, including all pertinent perturbing bodies. Computer costs limit such studies to about 10^5 yr. Especially if bolstered by a theoretical understanding of pertinent resonance phenomena, such results might be extrapolatable to 10^7 yr, but not to 10^9

yr. Statistical approaches, based on the conceptual framework of Öpik (1951), have been used to study longer term stability. One may obtain approximate lifetimes of planet-crossing orbits by this method, taking into account (for example) the known excursions in eccentricity of planetary orbits. But potential effects of subtle resonance phenomena are necessarily excluded by the approach. Subtle mechanisms might serve to protect an asteroid from planetary encounters, thus increasing its expected orbital lifetime, as is marginally true for 1685 Toro (Williams and Wetherill 1973). Other mechanisms might conceivably introduce very long-period secular changes in orbits not recognizable in 10^5 yr integrations, thus rendering finite the lifetimes of bodies in apparently completely stable orbits.

Most asteroids exist in orbits believed to be stable and they are infrequent or absent in short-lived orbits. One area of uncertainty is the region beyond the main belt from 3.4 to 3.7 AU (Franklin 1979). Asteroids are noticeably depleted in this region but such orbits have been shown to be at least reasonably stable. It is not yet known whether such bodies have been depleted by currently operating forces (say, on a time scale of 10^8 yr) or whether some very early, short-lived cosmogonic process depleted the region.

Another topic of intense interest is the origin of the Kirkwood gaps. Orbits very near Jupiter commensurabilities are radically altered by Jupiter perturbations, but such librating objects are not necessarily removed. Asteroids nearer the edges of the gaps should not even be much affected by Jupiter. Perhaps the Kirkwood gaps might be explained by resonant motion modified by dissipative forces combined with destructive collisions, but an adequate model has yet to be described in detail (cf. the chapter by Greenberg and Scholl).

Study of the orbital stability and evolution of Earth-approaching asteroids is doubly important; not only are such bodies intimately connected with the delivery of meteorites to Earth, but they are also dominantly responsible for current cratering of the terrestrial planets. Since these objects are in short-lived orbits, the present population must represent at least a quasi-equilibrium population between a source (or sources) and the better understood sinks (generally planetary impact or ejection). In particular, Apollo asteroids larger than 1 km diameter must be supplied at a rate of about 15 per million years. Currently known methods of supplying such objects from the cometary reservoir (by the physical and orbital decay of short-period comets) and from main-belt asteroids (by a combination of collisional and orbital evolution) both approach the known rate to within roughly an order of magnitude. Dead comets may dominate the population, but estimates are still very uncertain (Wetherill 1979a, b; see also the chapters by Shoemaker et al. and by Wasson and Wetherill).

Distribution of compositional types

Fischer (1941) may have been the first to suggest that asteroids at the

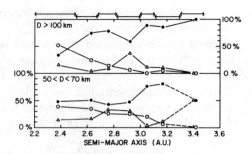

Fig. 5. The distribution of (probable) compositional types with semimajor axis. Percentages of bias-corrected asteroid populations in 7 zones (illustrated by the arrows) are plotted for large asteroids > 100 km diameter (top) and for asteroids between 50 and 70 km diameter (bottom). Solid symbols are C types, open symbols, S types, and triangles, neither C nor S (mainly M and U). The samples from which the percentages are computed are typically 20 to 40; the sample is only 4 for the outer zone of small asteroids, connected with dashed lines. Compare with Fig. 9 of Zellner's chapter. Zellner's bias correction factors were used in this analysis.

inner edge of the belt are redder than those in the outer belt. The trend was firmly established by the bias-corrected analysis of Chapman *et al.* (1975). The latest study is that of Zellner (see his chapter) who reports that, if certain large Hirayama families and groups are excluded, the fraction of asteroids of S type diminishes from about 60% near 2.2 AU to about 10% near 3.2 AU. The percentage of C types increases from a minority at the inner edge of the belt to over 95% at 3.4 AU. Zellner is impressed by the smoothness of the monotonic trend through the main belt and infers that a primordial process was responsible.

I do not think the compositional variation is so regular. There is a tendency, even in Zellner's Table IX, for an inversion in the trend in the middle of the belt near the 5:2 and 7:3 Kirkwood gaps (2.8 to 3.0 AU). Bias-corrected data for asteroids larger than 40 km diameter show that M's are especially enhanced in this region and, to a lesser extent, so are U's, relative to adjacent zones. While relatively few objects are involved, the anomaly is augmented if one includes the unusual Eos and, particularly, Koronis families in the statistics. Other peculiarities of this part of the belt include a relative lack of high eccentricities and a possibly higher percentage of inferred olivine-rich assemblages. Figure 5 highlights the anomaly in the middle of the belt (see also the discussion in Tedesco 1979). The figure also shows that the predominance of C types and the strong gradient in the distribution of S types are less obvious, though still present, among smaller asteroids. The unusual size distributions of some groups of asteroids are discussed further below.

Virtually all measured asteroids beyond about 3.4 AU (including the Trojans) have very low albedos. In addition, many of them tend toward the unusual traits epitomized by Degewij's RD types (Degewij and van Houten's chapter). These traits include: low visual albedo, relatively high ultraviolet reflectance, and a strongly reddish trend at wavelengths longer than 0.7 μm. Such spectra are extremely rare or absent in the main belt.

There is no prominent general correlation of taxonomic type with either eccentricity or inclination. When the *a-e-i* volume of the asteroid belt is finely dissected into separate Hirayama families, there are some families whose members clearly deviate in compositional mix from other asteroids with similar orbits (the chapter by Gradie *et al.*). Most asteroid families, however, have compositional mixtures not clearly different from that of the general population.

I have made preliminary attempts to correlate more subtle asteroidal spectral parameters with orbital elements. But aside from the relationship illustrated in Fig. 2, none of the borderline relationships is prominent.

Size Distributions

Do asteroidal sizes reflect processes of accretion or the effects of disintegration of a much larger body? Do they represent equilibrium amid continuing processes of mutual fragmentation (or accretion)? To the degree that different processes might yield different size distributions, it is important to establish the present distribution and investigate any variations with respect to orbital elements or physical properties. From the first comprehensive analysis of the McDonald Survey (Kuiper *et al.* 1958) through the first Tucson asteroid conference (cf. Dohnanyi 1971), sizes were approximated by astronomical magnitudes since albedos were not known. A matter of lively dispute in the early 1970's was whether the asteroids obeyed a single power-law size distribution (as argued by Dohnanyi) or exhibited kinks and variations in different parts of the belt. An underlying theoretical theme was that a simple power law would result from equilibrium fragmentation processes and that departures would indicate preservation of some part of an accretionary population. Theoretical expectations are no longer so simple (see below), nor are the data.

Above diameters of $\sim$ 30 km (less in the case of high-albedo or nearby asteroids), the size distribution may be determined from radiometry and polarimetry or from albedos inferred from taxonomic class, corrected somewhat for observational incompleteness (Zellner's chapter). Zellner plots log number versus log diameter for asteroids of different taxonomic classes and in different parts of the asteroid belt; a power law would be linear on such plots. It is clear that a power law does *not* represent the asteroids as a whole, nor most of the population subsets studied by Zellner. The dominant effect is that, relative to a power law, there is an excess of asteroids of diameter 70 to 150 km. The three largest asteroids (diameters greater than 500 km) have long been known to constitute another excess, which may have real physical importance but is of doubtful significance as a statistical attribute of the population.

Zellner's study also shows conclusively that the size distributions differ radically for different parts of the asteroidal population. For example, C-type asteroids in the outer third of the main belt show a much more nearly linear distribution than do C types elsewhere in the belt; especially in the middle third of the belt, there is a dearth of 50 km C types. In the inner third of the belt, S types of about 20 km diameter are only just as plentiful as S types of 100 km diameter, rather than being more numerous by factors of 10 or 20 as would be expected for hypothesized power-law distributions. It is not clear if size distributions for C and S asteroids are significantly different from each other; they seem to differ in at least the outer third of the main belt.

Diameters have been measured for too few members of individual Hirayama families to determine well their size distributions. The largest families do not depart drastically from the general population (see Zellner's chapter and that of Gradie et al.). Several authors (most recently Ip 1979) have studied the largest members of families; some have one large body and a swarm of tinier ones while others have a number of equal-sized large bodies.

For diameters between a few km and 30 km, the only information on main-belt size distributions comes from the Palomar-Leiden Survey (van Houten et al. 1970). The relatively steep slope of the magnitude distribution implies, in the absence of knowledge of the albedos of asteroids represented, that at least some types of asteroids rapidly increase in number toward smaller sizes. (There is no knowledge of the distribution of meteoroids and asteroids in the main belt in the centimeter to kilometer size range.) Taken together, all these data suggest a complicated picture of asteroid sizes.

Distributions of shapes and spins

Distributions of asteroidal orbits and sizes discussed above presumably reflect the gravitational and collisional interactions of asteroids with themselves and with planets. There are further clues about collisional interactions in lightcurve data on shapes and spin rates. Two approaches are summarized in this book. Burns and Tedesco address the inhomogeneous sample of lightcurve data for over 300 asteroids in the TRIAD file. Bowell and Lumme report a preliminary statistical analysis of a more homogeneous, but far less comprehensive and specific, photometric data set.

It is difficult to interpret correlations of lightcurve data with orbit, size, and taxonomic type due to inadequate statistics for both the largest and smallest asteroids and due to confusing correlations of the latter variables with each other. In the observed population, the (larger) C types spin more slowly than the (smaller) S types; it is unclear whether this effect is due to type or size. There is also a possibly real contrary tendency for the largest C types (over 175 km diameter) to spin faster than smaller ones. A small sample of M types may be the fastest spinners of all main-belt asteroids (Harris, personal communication). Small, Earth-approaching asteroids often have rapid spins.

An apparent tendency for the smaller asteroids in TRIAD to have larger lightcurve amplitudes, hence more elongated or irregular shapes, has not been verified by several observing programs directed towards this problem. These and further relationships are analyzed in detail in the chapter by Burns and Tedesco.

IV. ORIGIN AND EVOLUTION OF THE ASTEROIDS

The origin of the asteroids is intimately connected with the origin of the solar system. The main evidence about the age of the solar system comes from the ages of meteorites, at least some of which must be asteroidal fragments. Some of the asteroids appear to be stuck in resonances, or between resonances, in orbits which are reasonably expected to have been stable since the formation of the major planets. Even the most extreme hypothesis of asteroid formation — that they are fragments of a recently exploded planet — is motivated by Ovenden's (1972) belief that a planet must have formed, at the same time as the others, in the famous gap between Mars and Jupiter in the Titius-Bode "law" of planetary distances.

Several themes have emerged in the other chapters in this book. They include a blurring of some distinctions between comets and asteroids, an augmented role for Jupiter and other planets in forming the asteroidal population as we now observe it, and an increased perception of variety among asteroids and the novelty of processes affecting them. Before addressing the fundamental but most speculative issues about origin and early evolution, I summarize the processes that we think are affecting the asteroids today. Our understanding of these processes is still in an early stage of development so that dichotomies between well-formulated alternatives have not yet emerged.

Collisions

Asteroids collide with a size spectrum of bodies ranging from elementary nuclear particles to the major planets. Such collisions, especially with bodies of asteroidal size and larger, certainly have the dominant effect on the global character of asteroids. Other available sources of energy (e.g. sunlight or decay of long-lived radionuclides) are insufficient to drive geological processes. The collisions asteroids have with cosmic ray particles, solar wind particles, and micrometeorites are literally superficial, leaving remnants of their impact effects in tracks, implanted gases, and microcraters, all recognizable in meteorites (cf. Lorin and Pellas 1979; Goswami and Lal 1979). Such evidence of irradiation may have been volumetrically distributed within asteroids as their surfaces were buried, either during accretion (Wasson and Wetherill's chapter) or by development of a megaregolith (Housen *et al.*'s chapter).

The globally important impacts are the numerous cratering events, the larger of which may spall off outer layers (see the chapter by Cintala *et al.*) or produce fractures like those on Phobos (Thomas and Veverka 1979; see their chapter), and the still larger impacts that fragment and may disperse an asteroid altogether (Davis *et al.* discuss *catastrophic collisions* in their chapter). Energetic collisions approaching or exceeding the catastrophic limit may be predominantly responsible for establishing asteroidal shapes and configurations (cf. Hartmann's chapter) and, in principle, spins as well (chapter by Davis *et al.*). Collisional equilibrium of spin states is more likely to have been established for the smallest and largest asteroids. Although theoretical modeling may not yet be sufficiently sophisticated to explain the observed spins, Davis *et al.* consider it unlikely that asteroidal spins are now in equilibrium with collisional processes, contrary to Harris (1979).

The high random velocities of main-belt asteroids (~ 5 km sec^{-1}) and their space density assure that all of them have been subjected to major collisions during solar system history. The uncertainties in modeling collisional phenomena concern the physical outcomes of such collisions. Laboratory experiments on rock fragmentation fall many, many orders of magnitude short of representing the scale of such colossal asteroidal collisions. What work has been done (Davis *et al.* chapter) assumes that rocky asteroidal materials fail at roughly the same limiting energy density as do laboratory-scale materials and that the resulting size- and velocity-distributions of fragments are also similar. For most asteroids larger than ~50 km diameter, the material strength is less significant in holding them together than is self-gravity. The chief uncertainty in determining a gravity-dominated asteroid's collisional lifetime is in estimating the *efficiency* with which the kinetic energy of the impacting projectile may be converted into kinetic energy of escaping asteroidal fragments.

Until asteroid diameters (hence collisional cross-sections) were revised upwards in the early 1970's, it was thought that many original asteroids might still have escaped large collisions. Now that we know that collisions are common, the issue is whether their effects on asteroids of order 100 km diameter are primarily,

(1)   to break them up into smaller pieces, forming a steady-state size distribution of fragments (Chapman and Davis 1975), or
(2)   to severely alter their physical properties (e.g. by megaregolith formation) without changing their physical identity (the current preferred scenario of Davis *et al.* in their chapter).

In Case (1), the present asteroids could be the remnant of a vastly larger population of asteroids. As the lucky few that escaped earlier catastrophic destruction, such bodies might be expected to have preserved much of their original structure intact. But in the currently preferred Case (2), more of the energy is expended in crushing and redistributing asteroidal materials, while the asteroids are less readily destroyed. The present population might then represent most of the material extant when the present high velocities were established and the size distributions might retain "memory" of that early epoch. Concomitantly, the textures and structures of the original parent bodies would have become drastically altered by the cumulative impact history. Although some special massive early belt populations can evolve to the present population using the physics of Case (2), the number of large asteroids (exceeding a few hundred km diameter) could never have been much greater than observed today. Ceres, for example, is simply too large to be destroyed by any other asteroid, except possibly by Pallas.

Theoretical modeling of collisional processes now demonstrates, contrary to earlier models, that quasi-equilibrium size distributions need not be power laws. This is primarily due to the strong gravitational binding for asteroids of moderate size or larger, which extends the lifetimes of larger bodies longer than given by the $\sim\sqrt{D}$ dependence of the classical theory. Also the internal strength of such asteroids is modified by repeated subcatastrophic impacts, thus affecting the size distribution of fragments from such a body when it is ultimately disrupted. Davis *et al.* (their chapter), by modeling two interacting asteroid populations of different material strengths (e.g. *C*'s and *S*'s), are able to reproduce some features of the observed non-power-law distributions of asteroids given by Zellner (his chapter). But it is premature to invert the observed size distributions to derive unique starting conditions.

Orbital change

While collisions are practically the only events that change the physical character of asteroids today, the inexorable force of gravity always governs an asteroid's changing position in the solar system. (In principle, asteroids could also be moved by collisions, but ordinary rocky materials are far too weak to withstand big impulses; so an asteroid is more likely to be shattered than moved very far by large collisions.) Orbital nodes and apsides are gradually reoriented and, over longer durations, gravitational perturbations may change eccentricities and inclinations. Semimajor axes, dependent on orbital energy, are most invariant.

The orbital changes occurring now mainly involve the less stable, but still populated orbits. The least stable orbits have already been depopulated (see later discussion of origin of asteroids). The most stable, most populated orbits evolve only rarely, or slowly. In special cases collisions can alter an otherwise stable orbit that happens to be adjacent to a less stable state; such orbits are thus effectively unstable.

Evolving orbits are chiefly those whose eccentricities have been sufficiently enhanced, by proximity to a planet or planetary resonance, that they cross the orbits of one or more terrestrial planets. Combinations of resonances, close approaches to Jupiter, and collisions can move bodies from selected parts of the main belt into Earth-crossing orbits (see the chapter by Wasson and Wetherill). Since celestial mechanics also operates in reverse, the solely gravitational transport mechanisms can inject near-Earth bodies into certain parts of the asteroid belt. Some present Apollo-Amors may be injected into main-belt Mars-grazing orbits and implantation of bodies originally formed near the Earth could have been a much more important process in early times (Wetherill 1977).

On a longer time scale, all asteroidal orbits must surely be diffusing in *a-e-i* space due to mutual perturbations, subtle planetary perturbations, collisions, and possibly other nongravitational forces. But the nature and rates of any such diffusion are not known. To the degree that sharp boundaries in asteroidal distributions (e.g. Kirkwood gap edges) are thought to date from the epoch of planetary formation, they set stringent limits on diffusion rates. On the other hand, if the scale length of variation in proportion of S to C types is to be explained by orbital diffusion, then asteroids have been mixed over 0.5 AU or more.

Hirayama families

Families hold important clues for helping us understand asteroidal evolution. About half the asteroids are members of families (see the chapters by Gradie *et al.* and by Kozai). Indeed, 10% of the numbered asteroids and a similar percentage of the smaller Palomar-Leiden asteroids are members of the three most populous families. That some families, especially the populous ones, exhibit compositional homogeneity strongly suggests a genetic association among members. Interestingly, other families are heterogeneous. If families originate from the catastrophic collisional disruption of larger precursor bodies, then the distribution of orbits, sizes, and spin states of the fragments should be informative about the physics of disruption events. Distributions of spectral types and inferred mineralogies should shed light on interior compositions of the precursor bodies. Families could provide us the unprecedented opportunity to observe directly the interiors of planets (admittedly small planets, but nevertheless differentiated ones, in some cases).

There have been attempts to "put a family back together." The Eunomia family was discussed by Chapman (1976), the Nysa/Hertha families by Zellner *et al.* (1977), and other families by Gradie (1978) and Tedesco (1979) (see the chapter by Gradie *et al.* in this book). The problem of reconstructing geochemically plausible precursors makes the Humpty Dumpty case seem like a cinch. Not only must one account for the volumetric constraints imposed by the asteroidal shapes, but there are uncertainties in whether subsequent collisional evolution can account for the "missing mass." Finally there are confusing background interlopers. I am not satisfied with the geochemical and collisional plausibility of the reconstructions published so far.

Eos and Koronis families. As food for thought, I now sketch the elements of an unsolved problem involving the origin of two of the largest families (see Fig. 4). For another perspective on this problem, see Tedesco (1979). Orbits of Koronis family members are among the most circular, in-plane orbits of all. They are situated midway in the belt, as defined by the semimajor axis distribution of larger asteroids, and are bounded on either side by the 5:2 and 7:3 Jupiter commensurabilities. At intermediate velocities ($w = 0.2$), where $w = \sqrt{e^2 + i^2}$, are two clusters of asteroids; one cluster, composed mainly of Eos family members, is just exterior to the 7:3 gap while another cluster, not dominated by any single family, is just interior to the 5:2 gap. Relative to the C-dominated populations in this region of the belt, compositions of all three groups of asteroids are anomalous. The Koronis members themselves are unusually neutral-colored S types, while most of the asteroids in the symmetrically placed clusters are S or U types or unusually reddish C types (Eos family members could be physical mixtures of C- and S-type material: see the chapter by Chapman and Gaffey).

I have already noted the generally unusual compositional types in this region of the asteroid belt (see Fig. 5) as well as the dearth of high-velocity objects between the 5:2 and 7:3 resonances. The peculiar Budrosa family, discussed below, is located at the same semimajor axis as the Koronis family, between the two Kirkwood gaps, at still higher velocities ($w \sim 0.3$) than the two anomalous clusters of asteroids. Are all these relationships accidental? Are the several families and clusters somehow the result of linked collisional events? Perhaps they provide clues about the processes that cleared the two Kirkwood gaps. No one who has noted these peculiar relationships has offered a clear explanation; further research is certainly called for.

Achondrite parent body(ies). Perhaps the most important problem, on which progress is being made, that links asteroid astronomers with cosmochemists concerns the identification of the parent body(ies) for the achondrite meteorites. These meteorites constrain the early evolution of not only asteroids, but of the planets as well. They result from early geochemical fractionation, probably due to melting and gravitational segregation of materials within a relatively small body (see Drake's chapter; Stolper *et al.* 1979). Since asteroid-sized bodies radiate heat much more efficiently than larger planets, whatever process was responsible for melting differentiated meteorite parent bodies might well have had even more important effects on the moon and larger planets (see the chapter by Sonett and Reynolds for a discussion of primordial heating).

Vesta seems to be an *intact* differentiated body; it is apparently covered rather uniformly by basaltic materials (chapters by Drake and by Gaffey and McCord). But there are problems with deriving the achondritic meteorites from Vesta (Wasson and Wetherill's chapter). Not only is Vesta far from any resonance known to be capable of transporting ejecta into Earth-crossing orbits, but it is so large that ejecta perhaps cannot be accelerated to escape velocity without damaging the pristine basaltic textures evident in such meteorites as Ibitira. The problem of yield is even worse for other types of achondrites, which would be expected to exist at depth in the parent body mantle; such materials are clearly not widely exposed on the surface of Vesta, if at all.

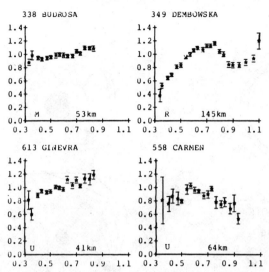

Fig. 6. Relative reflectances as a function of wavelength in μm for the four measured members of the Budrosa family (see the chapters by Chapman and Gaffey and by Gradie *et al.* in this book). Also listed are TRIAD taxonomic types and associated diameters.

An alternate source of achondrites may be the Budrosa family (Williams' No. 124), which is immediately adjacent to the 5:2 Kirkwood gap, a region shown by Scholl and Froeschlé (1977) to be a possible source region for meteorites. Spectra, diameters, and types of the largest 4 of the family's 6 members are shown in Fig. 6. The largest member is the unique object 349 Dembowska. Feierberg *et al.* (1978) propose that Dembowska is of an as-yet-unknown achondritic composition, poor in metal and rich in olivine, similar to the mantle material predicted for the basaltic achondrite parent body (Drake's chapter). However, I have noted earlier that alternative assemblages involving olivine, pyroxene, and metal cannot yet be ruled out. A metal-*rich* and olivine-rich assemblage (as in pallasites) would still be compatible with the interior of a basaltic achondrite parent body; Dembowska would then be identified with the lower mantle/core interface rather than with the upper mantle of the parent body. But if the second alternative — an LL chondritic assemblage — were correct, then Dembowska might be a primitive, undifferentiated body. In that case, it would be difficult to construct a geochemically plausible precursor body for the Budrosa family, just as it has been difficult to do for many other small Hirayama families.

A speculative scenario is offered in Fig. 7 for the evolution of the Budrosa family and the delivery of achondritic (and even iron) meteorites from 558 Carmen. Carmen is an unusual but spectrally poorly measured member of the Budrosa family, located very close to the 5:2 commensurability. I assume that part of Carmen's surface is basaltic, a possibility that is probably consistent with, but in no way demonstrated by, the spectral data in Fig. 6. Carmen's albedo seems discordant with basalts, so

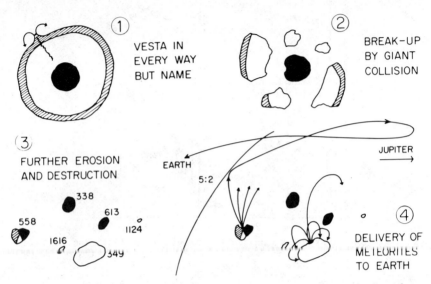

Fig. 7. A speculative scenario for the origin and evolution of the Budrosa family. A large body of primitive composition melts and differentiates. The resulting Vesta-like body then is fragmented and disrupted by a major collision. After further erosion and destruction, the six family members known today are left. Meteorites are derived from 558 Carmen via the proximate 5:2 Kirkwood gap. In this schematic drawing, the basaltic crust is indicated by shading, the mantle by white, and the metallic core-material by black.

I have drawn another material for its other side. For a variety of reasons, this *ad hoc* scenario is not very satisfying. Dembowska's greater distance from the 5:2 resonance is probably insufficient to prevent an appreciable yield of meteorites having the composition of the inferred mantle material. Yet no known meteorite has such a composition, although the Antarctic achondrite Alan Hills 77005 has a possibly related composition. (A small fraction of meteorites do, however, have the alternative pallasitic composition.) Since there are so few other asteroidal candidates for the achondritic meteorites, further research on the Budrosa family is essential.

Origin of the asteroids

It is natural to imagine that the asteroids are fragments of an exploded planet. Olbers (1805) may have been the first to suggest that the first few asteroids discovered might be fragments of the "missing planet" between Mars and Jupiter. Since 1950, however, the trend of cosmochemical and cosmogonical thought has been away from the exploded-planet hypothesis. Until the mid-1960's, it was thought that the properties of many meteorites required parent bodies of at least lunar dimensions. Since then, inferred pressures and cooling rates for meteorites have been interpreted as requiring bodies no bigger than the larger asteroids. Indeed Wood (see his chapter) and others now believe parent bodies might have been the size of smaller asteroids (see the review by Wilkening, next chapter). Earlier embarrassment about a lack of heat adequate to metamorphose and melt small parent bodies has been reduced with the discovery of magnesium anomalies, indicating an early presence of short-lived radionuclides in some meteoritic materials, and due to the discovery of additional heating processes (cf. Sonett and Reynolds' chapter). Meanwhile the planetesimal hypothesis for the origin of planets has gained considerable acceptance (Safronov's chapter), so that it now seems natural to find remnant populations of asteroid- and comet-sized bodies that failed to become incorporated into larger planets. Explosive fragmentation is not required. So most research on asteroidal origins has been directed toward obtaining the observed properties and distributions of asteroids as the natural outcome of planetary accretion. I return below to the modern version of the exploded-planet hypothesis, but consider now the cosmogonical approach.

The origin of the solar system is a lively field of research (Gehrels 1978), but the diverse observations are too few to give me much confidence that any of the theoretical constructs now popular must be correct. Yet some pieces of the problem are now understood. It is widely expected that a distended solar nebula cooled and various condensates, perhaps including never-vaporized materials from outside the solar system, agglomerated and rained down toward the midplane of the nebula. Cameron would have the major protoplanets already formed at this stage whereas Safronov would not (see their chapters in this book). But the asteroidal scenarios are similar in the two models. Local gravitational instabilities rapidly form planetesimals of roughly kilometer dimensions. These are building blocks of asteroids and comets. In Safronov's case, planetesimals are also the building blocks of the planets whereas Cameron would have them only augment the crustal layers of the already-extant, stripped protoplanetary cores.

Planetesimals should have formed throughout the midplane of the solar system, except (a) far inside the orbit of Mercury where it was too hot for condensation of solids prior to dissipation of the nebula, and (b) in the extremities where the nebular density presumably was low. Where are the planetesimals now? Most of them were either accreted into (or onto) the planets or ejected from the solar system by close encounters with the larger planets. There are only a limited variety of orbits that are believed to be stable on the time scale of the age of the solar system. In nearly all such zones in which small bodies would be observable, small bodies are now found. Objects in planetocentric orbits, deep within the spheres of gravitational influence of the larger planets, are constrained to stay there for rather long periods, although tidal processes cause some orbits to evolve. Indeed, most planets have satellites. The only heliocentric orbits inside the orbit of Mars that may be stable are those appreciably inside Mercury's orbit (Weidenschilling 1978). No such "vulcanoids" are known, but they would be difficult to see due to proximity to the sun. An infrared search now under way at the Planetary Science Institute may uncover such objects, if they condensed in the first place.

As previously discussed, a variety of main-belt asteroidal orbits are believed to be stable, as well as some resonant configurations closer to Jupiter. Still further out in the solar system, any asteroid-sized bodies become increasingly faint. The discovery of 2060 Chiron in the vicinity of Saturn and Uranus suggests the existence of a large population of bodies in the outer solar system. Although Chiron itself is in a chaotic orbit (cf. Everhart's chapter), several types of outer solar system orbits are hypothesized to be stable. Comets are an extreme example of the latter type, barely moving in the far outer reaches of the solar system. Since it is doubtful that nebular densities could have been sufficiently great at such distances to accrete comets, it is thought that comets are a population of planetesimals ejected by moderately close encounters with Uranus and Neptune, and perhaps other planets. Thus, they are a remnant group of outer solar system planetesimals in originally *unstable* orbits; they would be permanently hidden from view in the Oort cloud were they not so numerous that some are occasionally perturbed into the inner solar system and, because of their volatile-rich compositions, make a flashy display.

Thus it appears that observable small bodies exist in orbits known to be stable and do not exist or are rare in orbits known to be unstable. In cases where stability is uncertain, it takes only a small leap of faith to imagine that the presence or absence of bodies establishes whether particular orbits are stable or not. We must necessarily include in our concept of stability the effects of collisional destruction. Collisions are more likely and more destructive for bodies in high-velocity orbits compared with in-plane, circular orbits; so the former should be deemed relatively unstable. The absence of high-velocity asteroids probably does not represent current orbital instability, however, since the original accretion of asteroids required relatively low velocities — much lower than current values, which clearly result in fragmentation rather than accretion. The present distribution of small bodies in the solar system, then, is consistent with an original ubiquitous distribution of planetesimals throughout the median plane of the solar nebula. We now know how some of the planetesimals were rapidly removed; research continues on mechanisms for clearing other zones, such as the Kirkwood gaps.

Much attention has been given during the 1970's to the variation with heliocentric distance of the mean chemical compositions of solar system bodies, ranging from iron-rich Mercury, through the silicate-rich terrestrial planets, to the volatile-rich outer planets (cf. Lewis 1974). The asteroids are in between the volatile-poor and volatile-rich planets and they even exhibit their own compositional gradation from silicate-rich assemblages near 2 AU to hydrated-carbonaceous assemblages beyond 3 AU. The comets, hypothesized to have originated beyond Saturn's orbit, may be even more volatile-rich than are the C-type asteroids. I hasten to point out, however, that the surface layers of comets (obviously) devolatilize at asteroid-belt distances from the sun, so the failure of asteroids to exhibit comae need not imply that asteroidal *interiors* are volatile-poor. I can readily imagine that a catastrophic collision involving a C-type asteroid might produce an unparalleled display of asteroidal "comets" until near-surface layers had devolatilized.

The variation of S to C types with semimajor axis is considered to be smooth and thus necessarily a primordial feature by several chapter authors (e.g. Safronov, Zellner). I think there are too many irregularities in the distributions of size versus type versus heliocentric distance (cf. Fig. 5) to consider the S/C variation smooth (or necessarily primordial). Non-nebular processes might be responsible for the variation. It might reflect the efficacy of early heating processes (e.g. S types = differentiated assemblages; C types = primitive, unmodified assemblages). To the degree that accretion might have yielded semimajor axis variations in the sizes of proto-asteroids, the manifestation of size-dependent thermal processes could be an S/C gradation with semimajor axis.

The distinctions between comets and asteroids are becoming increasingly blurred, except for the extreme difference in where they have been *stored* during solar system history. (Wilkening, in her review chapter, has even gone so far as to suggest that extensive thermal modification might have occurred within cometary cores as well as within asteroids.) It has long been thought that unlike the comets, which retain no memory of their dynamical origin, main-belt asteroids have been quite stable for 4.5 Gyr, providing *in situ* evidence of primordial processes that occurred in that region. Wasson and Wetherill (in their chapter) describe attempts to exploit the stable spatial structure of the asteroid belt for inferring early nebular conditions. But Wetherill (1977) has advocated that a significant part of the population of the inner regions of the asteroid belt might have been implanted from the vicinity of Earth and Venus. In at least a limited sense, then, the asteroid belt has been a dumping ground for stray material from various parts of the solar system. The role of interzone mixing of materials condensed in various parts of the solar system has been increasingly recognized (Hartmann 1976; Cameron's chapter). But there are limits; we still do not know how to efficiently implant a comet into a main-belt asteroidal orbit.

Once kilometer-scale planetesimals had formed in low-eccentricity, low-inclination orbits, they began to accrete by mutual low-velocity collisions into larger objects. At some point in the evolution, the proto-asteroids in stable orbits were modified in two important ways: (a) they were drastically depleted, and (b) their velocities were enhanced. It is reasonable, but not necessary, to suppose that the two effects were caused by one process. Most scenarios blame Jupiter, at least indirectly. In the planetesimal hypothesis, it is expected that large bodies accreted more rapidly in Jupiter's zone than in the asteroidal zone due to the enhanced proportion of condensed volatiles at Jupiter's distance. Once Jupiter had grown to a significant fraction of its present size, its gravitational interactions with smaller objects in its vicinity would have scattered them into elliptical orbits, reaching into the asteroidal zone (see Safronov's chapter). The effects on the more slowly growing asteroids, which had not yet reached 1000 km diameter, might have been catastrophic.

Davis et al. (in their chapter) suggest that an Earth-mass object scattered by Jupiter into the asteroidal zone would pass sufficiently close to all asteroids to stir them up to the order of their present-day 5 km sec^{-1} velocities over a duration similar to the expected lifetime of the body prior to ejection from the solar system by Jupiter. Cameron (see his chapter) also invokes Earth-sized bodies passing through the asteroid region to stir up the velocities of asteroids; the major difference in his scenario from that of Safronov and Davis et al. is that the bodies, rather than being from Jupiter's zone, are excess protoplanetary cores formed in the inner solar system and stripped of their outer envelopes. They are also eventually ejected from the solar system by Jupiter.

Once the velocities of asteroids were stirred up to a fraction of their current values, accretion stopped and fragmental destruction began.[a] Perhaps the depletion of asteroids was accomplished by their mutual high-velocity collisions. Present evidence on size distributions and rotations is inconclusive about whether or not the present asteroid belt is a remnant of a much larger population, decaying by mutual collisions (see Davis et al.'s chapter). Alternatively, as preferred by Safronov in his chapter, the Jupiter-scattered planetesimals might have collided with, accreted, and swept out the proto-asteroids. Still another process of depletion would be collisional fragmentation of the proto-asteroids by a large population of smaller Jupiter-scattered planetesimals which might have accompanied the large ones that are hypothesized to be responsible for stirring up the velocities. The efficiency of these processes depends, of course, on the mass distribution of the Jupiter-scattered bodies and on the time interval they spent sweeping through the asteroid belt. Estimates of these parameters are model-dependent.

Safronov also partly relies on direct collisions to augment the relative velocities. I prefer to think that the eccentricities and inclinations were pumped up by close approaches of particularly large Jupiter-scattered bodies rather than by direct collisions by smaller bodies. Asteroids might well disintegrate long before their orbits were much changed, although

---

[a]Ideas of Alfvén and his associates (see several chapters in Gehrels 1971) that some asteroids might still be undergoing net accretion has no observational support and is at variance with much experimental data on the behavior of materials impacting at velocities of hundreds of meters to kilometers per second.

uncertainties in collisional physics discussed earlier make this an area for further research.

A particularly appealing aspect of this general approach to enhancing asteroidal velocities by gravitational interactions with an Earth-sized body is that it can explain readily the "strange case of Pallas," as Whipple *et al.* (1972) have called it. It is difficult to imagine other processes that could move Pallas intact into its highly inclined orbit ($i = 35°$). Nonetheless, these ideas are all quite new and research continues on other potential methods of enhancing the asteroids' velocities during primordial epochs. Whatever the true explanation is, it was the fundamental reason in the cosmogonical hypothesis for why an asteroidal planet failed to form.

The various cosmogonical models imply specific rates for formation of the several suites of planetesimals of different sizes. The rates, in turn, depend on highly uncertain estimates of conditions under which the solar system formed, not the least of which is the surface density of material in the nebula. Somehow a complete picture for the development of the asteroids and the meteorite parent bodies must factor in other pertinent time scales, including the injection time scale and decay half-life of ^{26}Al, time scales of possible T Tauri-like solar behavior, time scales for cooling and dissipation of the gaseous nebula, Jupiter's formation interval, time scales for melting and cooling asteroid-sized bodies, and so on. The meteorites provide some observational constraints on several of these time scales (cf. the review by Wilkening, next in this book). I am aware of no major irreconcilable differences in observed and theoretical time scales, but there are still too few constraints to allow the fashioning of a unique, self-consistent history for the formation of the asteroids.

Even though not pinned down, the cosmogonical approach to the origin of the asteroids appears to provide asteroids, comets, and meteorites naturally, with no insurmountable problems. With such an acceptable (though not necessarily correct) paradigm, there is little motivation to search for radically different explanations. Ovenden, however, uncovered what he regarded as an insurmountable problem in his research on Bode's law and solar system stability. He felt obliged (Ovenden 1972) to hypothesize the rather recent existence of a 90 Earth-mass planet in the asteroid belt. His work suggested that the asteroids might have resulted from the destruction of this planet some 16 million years ago. More recently Van Flandern (1978) uncovered some anomalies in the orbits of long-period comets (although his data are disputed) which he chose to interpret as confirming Ovenden's hypothesis. Still more recently, Ovenden (1976) has backed away from the 16 million year time scale, which was an important ingredient in Van Flandern's model.

The hypothesis has not been received with general favor and many criticisms have been leveled at it. The criticisms, which are often to the point, but sometimes not, keep the model's only two defenders busy responding.

Experts in celestial dynamics find Ovenden's "principle of least interaction action" intriguing, but are hardly persuaded that an asteroidal planet was required to keep the solar system stable over its age, as Ovenden believes. Even if Van Flandern's comet observations are valid, they seem easily explained by disintegration of a massive (say asteroid-sized) comet and do not require the much more difficult feat of disintegrating a 90 Earth-mass planet.

In my own view, we cannot be assured that the cosmogonical approach is necessarily correct nor can we be assured that there is no merit in some aspects of the Ovenden/Van Flandern hypothesis. But I do not think there are such serious problems with the former that it is worthwhile to divert much effort into the new idea; therefore the canonical approach to understanding asteroidal origins remains the cosmogonical hypothesis. But, as we have seen in earlier parts of this chapter, canonical views are easily challenged in asteroid science.

Acknowledgments. I thank many of my colleagues for assistance in preparing this chapter. Especially helpful reviews were given by E. Tedesco, J. Wasson, G. Wetherill, L. Wilkening, and my colleagues in the Tucson office of Planetary Science Institute. This work was supported by the National Aeronautics and Space Administration.

REFERENCES

Anders, E. 1978. Most stony meteorites come from the asteroid belt. In *Asteroids: An Exploration Assessment,* eds. D. Morrison and W. C. Wells, *NASA Conf. Publ.* 2053, pp. 57-75.

Binzel, R. P., and Van Flandern, T. C. 1979. Minor planets: The discovery of minor satellites. *Science* 203: 903-905.

Bobrovnikoff, N. T. 1929. The spectra of minor planets. *Lick Obs. Bulletin* 14, No. 407: 18-27.

Bowell, E.; McMahon, J.; Horne, K.; A'Hearn, M. F.; Dunham, D. W.; Penhallow, W.; Taylor, G. E.; Wasserman, L. H.; and White, N. M. 1978. A possible satellite of Herculina (abstract). *Bull. Amer. Astron. Soc.* 10: 594.

Brown, H.; Goddard, I.; and Kane, J. 1967. Qualitative aspects of asteroid studies. *Astrophys. J. Suppl.* 14: 57-123.

Chapman, C. R. 1976. Asteroids as meteorite parent-bodies: The astronomical perspective. *Geochim. Cosmochim. Acta* 40: 701-719.

Chapman, C. R., and Davis, D. R. 1975. Asteroid collisional evolution: Evidence for a much larger early population. *Science* 190: 553-556.

Chapman, C. R.; Morrison, D.; and Zellner, B. 1975. Surface properties of asteroids: A synthesis of polarimetry, radiometry, and spectrophotometry. *Icarus* 25: 104-130.

Chapman, C. R., and Salisbury, J. W. 1973. Comparisons of meteorite and asteroid spectral reflectivities. *Icarus* 19: 507-522.

Degewij, J.; Tedesco, E. F.: and Zellner, B. 1979. Albedo and color contrasts on asteroid surfaces. *Icarus* (special Asteroid issue).

Dohnanyi, J. S. 1971. Fragmentation and distribution of asteroids. In *Physical Studies of Minor Planets,* ed. T. Gehrels (NASA SP-267, Washington, D.C.: U.S. Government Printing Office), pp. 263-295.

Feierberg, M. A.; Larson, H. P.; Fink, U.; and Smith, H. A. 1979. Spectroscopic evidence for two achondrite parent bodies: Asteroids 349 Dembowska and 4 Vesta. Submitted to *Geochim. Cosmochim. Acta.*

Feierberg, M.; Larson, H.; Smith, H.; and Fink, U. 1978. Spectroscopic evidence for at least two achondrite parent bodies (abstract). *Bull. Amer. Astron. Soc.* 10: 595.

Fischer, H. 1941. Farbmessungen an kleinen Planeten. *Astron. Nachr.* 272: 127-147.

Franklin, F. A. 1979. Some long-term implications from the motions of outer belt asteroids. *Icarus* (special Asteroid issue).

Gaffey, M. J. 1976. Spectral reflectance characteristics of the meteorite classes. *J. Geophys. Res.* 81: 905-920.

Gaffey, M. J., and McCord, T. B. 1978. Asteroid surface materials: Mineralogical characterizations from reflectance spectra. *Space Sci. Reviews* 21: 555-628.

Gehrels, T. ed., 1971. *Physical Studies of Minor Planets*, (NASA SP-267, Washington, D.C.: U.S. Government Printing Office).

Gehrels, T. ed. 1978. *Protostars and Planets*, (Tucson: University of Arizona Press).

Goswami, A., and Lal, D. 1979. Formation of the parent bodies of the carbonaceous chondrites. *Icarus* (special Asteroid issue).

Gradie, J. C. 1978. An astrophysical study of the minor planets in the Eos and Koronis asteroid families. Ph.D. dissertation, University of Arizona.

Harris, A. W. 1979. Comments on asteroid rotation. II. A theory for the collisional evolution of rotation rates. *Icarus* (in press).

Hartmann, W. K. 1976. Planet formation: Compositional mixing and lunar compositional anomalies. *Icarus* 27: 553-559.

Ip., W.-H. 1979. On three types of fragmentation processes observed in the asteroid belt. *Icarus* (special Asteroid issue).

Kresák, L. 1971. Orbital selection effects in the Palomar-Leiden asteroid survey. In *Physical Studies of Minor Planets*, ed. T. Gehrels (NASA SP-267, Washington, D.C.: U.S. Government Printing Office), pp. 197-210.

Kuiper, G. P.; Fugita, Y.; Gehrels, T.; Groeneveld, I.; Kent. J.; van Biesbroeck, G.; and van Houten, C. J. 1958. Survey of asteroids. *Astrophys. J. Suppl.* 3: 289-428.

Larson, H. P., and Fink, U. 1975. Infrared spectral observations of asteroid (4) Vesta. *Icarus* 26: 420-427.

Lebofsky, L. A. 1978. Asteroid 1 Ceres: Evidence for water of hydration. *Mon. Not. Roy. Astron. Soc.* 182: 17-21.

Lebofsky, L. A. 1979. Infrared reflectance spectra of asteroids: A search for water of hydration. In preparation.

Lewis, J. S. 1974. The temperature gradient in the solar nebula. *Science* 186: 440-443.

Lorin, J. C., and Pellas, P. 1979. Pre-irradiation history of Djermaia H chondritic breccia. *Icarus* (special Asteroid issue).

Matson, D. L.; Johnson, T. V.; and Veeder, G. J. 1977. Soil maturity and planetary regoliths: The moon, Mercury, and the asteroids. *Proc. Lunar Sci. Conf. VIII* (Oxford: Pergamon Press), pp. 1001-1011.

McCord, T. B.; Adams, J. B.; and Johnson, T. V. 1970. Asteroid Vesta: Spectral reflectivity and compositional implications. *Science* 168: 1445-1447.

McCord, T. B., and Chapman, C. R. 1975. Asteroids: Spectral reflectance and color characteristics. II. *Astrophys. J.* 197: 781-790.

Olbers, W. 1805. Entdeckung eines beweglichen Sterne, den man gleichfalls für einem zwischen Mars und Jupiter sich aufhaltenden planetarischen Körper halten kann. *Berlin. Astr. Jahrbuch*, p. 108.

Öpik, E. J. 1951. Collision probabilities with the planets and the distribution of interplanetary matter. *Proc. Roy. Irish Acad.* 54A, pp. 164-194.

Ovenden, M. W. 1972. Bode's law and the missing planet. *Nature* 239: 508-509.

Ovenden, M. W. 1976. Principle of least interaction action. In *Long-Time Predictions in Dynamics*, eds. V. Szebehely and B. D. Tapley (Dordrecht: D. Reidel), pp. 295-305.

Reitsema, H. J. 1979. Reliability of minor planet satellite observations. Submitted to *Science*.

Remo, J. L., and Johnson, A. A. 1974. The ductile-brittle transition in meteoritic irons. *Meteoritics* 9: 209-213.

Scholl, H., and Froeschlé, C. 1977. The Kirkwood gaps as an asteroidal source of meteorites. In *Comets, Asteroids, Meteorites*, ed. A. H. Delsemme (Toledo, Ohio: University of Toledo Press), pp. 293-295.

Stolper, E. M.; McSween, H. Y.; and Hays, J. F. 1979. A petrogenetic model of the relationships among achondritic meteorites. *Geochim. Cosmochim. Acta* (in press).

Tedesco, E. F. 1979. A photometric investigation of the colors, shapes and spin rates of Hirayama family asteriods. Ph.D. dissertation, New Mexico State University.

van Flandern, T. C. 1978. A former asteroidal planet as the origin of comets. *Icarus* 36: 51-74.

van Houten, C. J.; Van Houten-Groeneveld, I.; Herget, P.; and Gehrels, T. 1970. The Palomar-Leiden survey of faint minor planets. *Astron. Astrophys. Suppl.* 2: 339-448.

Veeder, G. J.; Matson, D. L.; and Smith, J. C. 1978. Visual and infrared photometry of asteroids. *Astron. J.* 83: 651-663.

Weidenschilling, S. J. 1978. Iron/silicate fractionation and the origin of Mercury. *Icarus* 35: 99-111.

Wetherill, G. W. 1977. Evolution of the earth's planetesimal swarm subsequent to the formation of the earth and moon. *Proc. Lunar Sci. Conf. VIII* (Oxford: Pergamon Press), pp. 1-16.

Wetherill, G. W. 1979a. Steady state populations of Apollo-Amor objects. *Icarus* 37: 96-112.

Wetherill, G. W. 1979b. Apollo objects. *Scientific Amer.* 240 (3): 54-65.

Whipple, F. L.; Lecar, M.; and Franklin, F. A. 1972. The strange case of Pallas. In *L'Origine du Système Solaire,* ed. H. Reeves (Paris: Centre National de la Recherche Scientifique), pp. 312-319.

Wijesinghe, M., and Tedesco, E. F. 1979. A test of the plausibility of eclipsing binary asteroids. *Icarus* (special Asteroid issue).

Williams. J. G., and Wetherill, G. W. 1973. Minor planets and related objects. XIII. Long-term orbital evolution of (1685) Toro. *Astron. J.* 78: 510-515.

Zellner, B. 1975. 44 Nysa: An iron-depleted asteroid. *Astrophys. J.* 198: L45-L47.

Zellner, B.; Leake, M.; Morrison, D.; and Williams, J. G. 1977. The E asteroids and the origin of the enstatite achondrites. *Geochim. Cosmochim. Acta* 41: 1759-1767.

THE ASTEROIDS:
ACCRETION, DIFFERENTIATION, FRAGMENTATION, AND IRRADIATION

LAUREL L. WILKENING
University of Arizona

Various types of meteorites experienced processes of condensation, accretion, metamorphism, differentiation, brecciation, irradiation and fragmentation. A typical view of meteorite formation has been that the processes following accretion took place in a few asteroidal-sized (~ 100 km) objects. Discovery of decay products of now extinct ^{26}Al and ^{107}Pd in meteorites, discovery of isotopic heterogeneity among meteorite types, re-analysis of meteorite cooling rates, and continuing study of meteoritic compositions have led some meteoriticists (see especially chapters by Scott and Wood) to conclude that meteorites obtained their chemical, isotopic and some textural characteristics in objects initially <10 km in diameter. Such a scenario, which is described in this chapter, raises the possibility that some of these small planetesimals may have been "condensation nuclei" for the formation of comets as well as the precursors of asteroids.

When scientists from different disciplines assemble to discuss a scientific problem of mutual interest, there is always the hope that a ray of light will be shed on some particularly vexing aspect of a problem in one's own field by those who work on the problem from a different disciplinary perspective. Unfortunately this rarely happens in planetary sciences because of the complex nature of the problems and the inadequacy of the data base. It is like an intricate jigsaw puzzle in which the picture does not really emerge

[61]

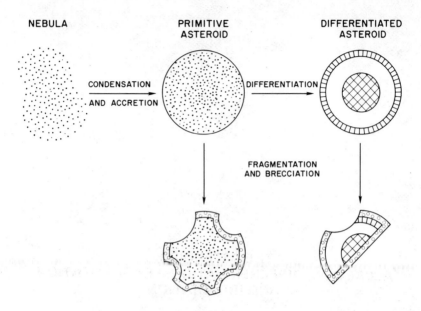

Fig. 1. An abbreviated conventional view of the formation of meteorite parent bodies or asteroids. Primitive objects which formed the population in the asteroid belt are on the order of 100 km in diameter. Differentiation is believed to have taken place in objects of this size.

until nearly all of the pieces are in place. In the case of the small bodies in the solar system, even when the data from all relevant fields are brought together, we seem to be far from having all of the pieces, much less having them in their proper places. Nevertheless, this has never kept us from trying to imagine what the picture must look like. It is in this spirit that this chapter is written.

Drawing almost exclusively from the data on meteorites, Anders (1964) constructed a scenario for the formation of meteorites in small bodies, i.e., those of asteroidal size. This was a marked departure from planet-sized or moon-sized objects postulated to be meteorite parent bodies by various of his predecessors and contemporaries. Anders clearly identified meteorite parent bodies with asteroids, in particular with the Mars-crossing asteroids. His ideas have guided the thinking of most meteoriticists over the past 15 years. Almost all subsequently obtained information on meteorites has been interpreted in the context of asteroids as meteorite parent bodies. From the study of meteorites, with insights gained from the study of the lunar samples and morphology of the lunar surface, the scenario for meteorite parent body formation grew and evolved. It is shown, considerably oversimplified, in Fig. 1. Of course, various objections to this scheme have been raised. All of these objections cannot be discussed here, but one recurring problem has been that an inadequate, steady-state supply of material is funneled from the asteroid belt into Apollo, Amor, Aten orbits, and thence to Earth. Because of this deficiency of mass, a cometary origin of at least some meteorites has been a recurring suggestion (Wetherill 1974, 1976).

This chapter is divided into three parts to briefly discuss the status of our knowledge of each step shown in Fig. 1: condensation and accretion, differentiation, and brecciation and fragmentation. At the end of the chapter I describe a revised picture of meteorite parent body formation which seems as consistent with the observations as the picture in Fig. 1.

I. CONDENSATION AND ACCRETION

Condensation and accretion are processes which are conceptually quite different, but about which we know too little to say whether or not the two processes were widely separated in space and time or whether they were contemporaneous and co-located in the solar system. As an example, D. D. Clayton (1978) has argued that dust grains which condensed in circumstellar shells and in other locations in the universe were at some later time introduced into the solar nebula as solids unrelated to nebular gas compositions. In such a view at least part of the condensation was widely separated in time and space from accretion. A more widely held view is that most, say 95% of the solid matter in the solar system condensed and accreted from our solar nebular over a time span of 10^4-10^5 yr. The extreme of this view is called heterogeneous accretion, in which condensation and accretion occurred concomitantly.

An outcome of an understanding of condensation processes which bears on our interests in asteroids and meteorites would be a prediction of the locations in which various meteorites formed. If we know the beginning location and the ending one with certainty, perhaps, it would be possible to deduce the path between them. At the very least the problem would be bounded. The condensation process was studied theoretically. By using solar system abundances, experimentally determined thermodynamic data and the principles of thermodynamics, a sequence of condensed solids arising from an initially gaseous nebula was calculated. The early successes of condensation theory (Larimer and Anders 1967; Lewis 1972; Grossman 1972) raised hopes that the sought after answers were within reach. But the discovery that the nebula was neither compositionally homogeneous nor totally gaseous (Clayton et al. 1973, 1977) raised difficulties with the homogeneous condensation theory. Although the initial discovery of refractory Ca-Al-rich inclusions in carbonaceous chondrites (Marvin et al. 1970) had generally confirmed calculations of the high-temperature portion of the condensation sequence, observations of low-temperature phases admixed with, included in, or layered over the high-temperature phases comprising Ca-Al-rich inclusions (hereafter called CAI's) indicated that even at high temperatures the condensation process was not so simple as first believed (Wark and Lovering

1977; El Goresy *et al.* 1978). Multiple condensation and/or reaction episodes may have taken place. Some (e.g., see Blander's chapter) believe that the assumption of equilibrium underlying the work described above is incorrect; others have attributed discrepancies between predictions and observations to subsequent processing of the original condensates. For these reasons and others, the anticipated fruits of condensation theory have not been realized. Specifically there are no clear predictions of the original spatial locations of meteorite parent bodies. The chapter by Wasson and Wetherill documents the case for meteorites having originated over a wide range of heliocentric distances. In their view remnants of the original population of parent bodies were perturbed into the asteroid belt and have survived there. Sears (1979) presented a similar view restricted to iron meteorites in which he argued that only one group of iron meteorites formed in the asteroid belt; the remaining groups formed in widely dispersed regions of the nebula. Blander (his chapter) was led to a similar conclusion for chondrites even after examining the influence of certain nonequilibrium effects on condensation theory. Larimer (1979) did not address specifically the question of the original solar system location of the formation of the meteorite parent bodies; however, his analysis led him to conclude that mixing of materials of different compositions is needed to explain the major element compositions of the chondrites, the earth, moon and the eucrite parent body. Wetherill (1975) and Hartmann (1976) have pointed out that the terrestrial planets probably have accreted materials which condensed in different locations. Hence there is a consensus among researchers studying condensation from quite different perspectives that mixing of material formed in different locations in the nebula, and perhaps even outside it, is necessary to explain the compositions of meteorites and the planets. It seems reasonable to conclude that materials condensed and accreted in various locations may now form subsets of the objects in the asteroid belt.

There is little evidence in meteorites bearing directly on the early process of accretion (Herndon and Wilkening 1978). Only two meteoritic components demonstrate well-defined sequential, textural relationships: sequences of layers or rims of differing compositions of CAI's (Wark and Lovering 1977; Lee *et al.* 1979) and rims on chondrules (Allen *et al.* 1979). To varying degrees the rims could reflect secondary condensation and/or reaction with the gas phase or accretion of fine-grained dust. The existence of CAI's and droplet chondrules as recognizable entities that could not have formed *in situ* in meteorites means that at least one stage of accretion involved objects in the mm-cm size range. There appears to be no recognizable record in meteorites of *early* accretion involving larger-sized objects.

II. DIFFERENTIATION

Co nposition and Interrelationships among Differentiated Meteorites

Most achondrites, irons and pallasites were formed by melting of parent silicate and iron materials, followed by segregation of the metallic and silicate phases. Some meteorites were formed by further geochemical differentiation of the silicate portions of the initially differentiated material (Consolmagno and Drake 1977; Stolper et al. 1979; Mittlefehldt 1979). An important goal of studying differentiation is to identify genetic relationships among meteorite groups. Two types of relationships are of interest. First, what differentiated meteorite groups have formed in the same parent body? Following disruption the fragments might be expected to be present as members of a presently extant Hirayama asteroid family. Secondly, what was the composition of the parent material that yielded the differentiated meteorites? Answers to the latter question could yield connections between primitive and differentiated objects and, perhaps, some insight regarding the heating mechanism responsible for the differentiation process (see below).

Considerable progress in understanding the evolution of achondritic meteorites, especially the eucrites has been made (see the chapter by Drake). Geochemical modeling of the formation of the eucrites has shown that a reasonable parent material would have been composed of the major rock-forming elements: Mg, Si, Al, Ca and Fe in cosmic proportions (Consolmagno and Drake 1977), such as would be found in various types of chondrites. However, the proportion of metallic iron or the amounts of volatile elements, either of which would permit a considerable narrowing of possibilities among possible parent types, are uncertain (Morgan et al. 1978). An additional problem concerns the rarity among meteorites of the olivine-rich material which should comprise the mantle of the eucrite parent body.

Spectroscopic data obtained for the asteroids have permitted the identification of the mineral assemblage on the surface of 4 Vesta with that in eucrites, howardites or shergottites. However 4 Vesta is not located in an orbit from which known mechanisms can efficiently place material in Earth-crossing orbits (chapter by Wasson and Wetherill). Hence there is still a challenge which must be met by either the dynamicists, by finding new transport mechanisms, or by observers finding new objects.

There are a number of other types of differentiated stony meteorites which should be amenable to the types of analysis carried out on the eucrites. Indeed, some studies of the diogenites (Consolmagno 1979) and other achondrites (Stolper et al. 1979) are underway. Stolper et al. (1979) have shown that basaltic achondrites, shergottites, nakhlites and chassignites could come from chemically related source regions, differing mainly in the volatile elements present. Scott (1977) studied the pallasites and has concluded that they are related to a particular group of iron meteorites, the IIIAB irons.

Certainly the studies of genetic relationships among differentiated meteorites will continue to be a fruitful area of research.

Heating and cooling

Heat sources, cooling rates, and formation ages are inseparable from any discussion of differentiation. Since most differentiated meteorites, with the exception of the nakhlites, are known to be very old, 4.5–4.6 Gyr, and some also contain products of the decay of now extinct radionuclides: ^{244}Pu ($t_{1/2}$ = 82 Myr), ^{129}I (16 Myr) and ^{107}Pd (6.5 Myr), the time scale for differentiation is tightly constrained to be relatively short.

Since the evidence points to a very early differentiation of the parent bodies of most achondrites and iron meteorites, it is natural to inquire about the heat sources responsible for the differentiation processes. The present candidate sources of heat are short-lived ^{26}Al and electromagnetic induction heating. These and other heat sources are reviewed in the chapter by Sonett and Reynolds and by Mittlefehldt (1979). The possibility that live ^{26}Al and other short-lived radioactivities were present when meteorites formed (Lee *et al.* 1976) has opened the possibilities of early heating of small objects (Lee *et al.* 1977). This is further discussed in Wood's chapter.

Cooling rates of differentiated meteorites also constrain the asteroid-meteorites parent body relationship. Cooling rates are calculated by assuming a simple cooling history with no reheating events. Cooling rate and related age data are reviewed in Wood's chapter. In order to reconcile old ages and measured cooling rates he concludes that cooling rates were initially rapid, then slowed to the measured meteoritic values of 1–10 deg/Myr in the temperature range of 900–650 K. Wood postulates that melting began in small objects which subsequently grew in size through continued accretion of dust, resulting in slower cooling rates. A somewhat similar conclusion was reached by Scott (see his chapter) in his review of iron meteorites. To explain the elemental compositions and cooling rates of irons, Scott concluded that all 12 groups and 50 grouplets of iron meteorites were formed by melting of chondrite-like material in separate locations, and all but two groups were produced in molten cores in km-sized bodies which subsequently accreted into larger bodies. Both Wood and Scott agree that the present day asteroids are not the objects in which the differentiation events which produced iron meteorite took place. This is certainly a new perspective, which contrasts with our starting cartoon (Fig. 1) in which differentiation took place in bodies on the order of 100 km in size.

There is evidence of less intense heating in chondrites, which may or may not have been contemporaneous with differentiation. Thermal metamorphism in chondrites may have been an early process; certainly it preceded brecciation of chondrites and their exposure to solar wind gases at the surface of meteorite parent bodies (chapter by Wasson and Wetherill). The

TABLE I

**Relative Ages of Meteorites and Meteoritic Components
from Ancient to More Recent**

Meteoritic Component or Type	Dating System	References
Ca-Al-rich inclusions	$^{26}Al - {}^{26}Mg$	Lee *et al.* 1976.
Iron meteorites, Pallasites(?)	$^{107}Pd - {}^{107}Ag$	Kelly and Wasserburg 1978.
Magnetite, Chondrules, Enstatite and Ordinary chondrites, Enstatite Achondrites	$\left\{ {}^{129}I - {}^{129}Xe \right\}$	Reynolds 1967. Drozd and Podosek 1976.
Basaltic achondrites	$\left\{ \begin{array}{l} {}^{129}I - {}^{129}Xe \\ {}^{244}Pu - Xe \end{array} \right\}$	Rowe 1967.
Nakhlites	K-Ar, Rb-Sr, Sm-Nd	Papanastassiou and Wasserburg 1974; Chapter by Bogard.

carbonaceous chondrites experienced aqueous alteration, which is thought to have been contemporaneous with brecciation (chapter by Kerridge and Bunch; Richardson 1978), but radiochronological dates of this aqueous activity have not yet become available. Compaction ages of 4.2 Gyr obtained from some CM chondrites support the possibility that these meteorites are not pristine samples formed 4.6 Gyr ago (Macdougall and Kothari 1976), and Rb-Sr data suggest CI and CM chondrites have been disturbed more recently than 3.6 Gyr ago (Mittlefehldt and Wetherill 1979). Note that impact cratering could be a heat source for localized metamorphism and production of hydrothermal fluids.

Time scales

The time scale over which condensation, accretion and differentiation occurred was short in cosmic terms. With only a few exceptions all types of meteorites have formation ages of 4.5 − 4.6 Gyr measured by several age-dating techniques. Extinct radioactivities permit some differences in formation ages to be resolved. A few generalizations from the studies of extinct radioactivities in meteorites are listed in Table I. The problem with such a table is that not every isotopic system has been, or can be, measured in all possible meteorite types. Hence certain positions in the sequence could be switched when additional measurements are made. A few of the generalizations are firm. The presence of ^{26}Mg formed by decay of ^{26}Al in CAI's (Lee *et al.* 1976) sets CAI's at the head of this list as the oldest material. Xenon has been measured in many meteorites, so the conclusion that most chondrites contain ^{129}Xe is well established. However, relative ages within the ^{129}Xe containing group (third time step) are not understood. Although listed as the second step, formation of group IVA iron meteorites dated by the presence of ^{107}Ag from extinct ^{107}Pd ($t_{1/2}$ = 6.5 Myr) could be contemporaneous with chondrite formation (Kelly and Wasserburg 1978). On the basis of radiogenic ^{129}Xe and fission xenon data it appears that basaltic achondrites post-date chondrites by about 100 Myr (Rowe 1967). However some of the other chronometers applied to achondrites and chondrites do not yield a concordant time scale (e.g., see discussion by Gray *et al.* 1973). It is well established that the nakhlites are much younger (age 1.3 Gyr) than any other meteorite class. Some meteorites or parts of meteorites have young ages due to impact (shock) melting or heating.

III. BRECCIATION AND FRAGMENTATION

Once formed either as differentiated objects or primitive agglomerates, meteorite parent bodies did not remain undisturbed. At some point in the evolution of the small bodies in the solar system, relative velocities increased to the point where mutual encounters resulted in net mass loss rather than net accretion (see Chapman's chapter). All objects in the solar system have been subjected to impacting debris. This last phase of solar system evolution has left its imprint in most classes of meteorites. A brecciated, clastic structure due to the shock accompanying impact events is found in virtually all mesosiderites, howardites, LL-chondrites, and CI chondrites, most eucrites and many diogenites, ureilites, aubrites, H-group, L-group and E-group chondrites. Comminution, crystal fracturing and distortion, veins, shock melting and metamorphism are the major effects of shock that can be observed in stony meteorites (Wahl 1952). Many iron meteorites also show effects of shock (see Scott's chapter).

During the period of regolith processing on the surfaces of stony meteorite parent bodies, solar wind gases and solar flare particles were implanted in the exposed surface layers of the regolith, agglutinates and microcraters were formed by micrometeoroids, and foreign debris was mixed into the surface. (For more details see chapters by Housen *et al.* and Wasson and Wetherill, as well as Rajan [1974] and Wilkening [1977].) Stony meteorites containing solar wind gases comprise 2–100% of the meteorites in the major stony meteorite classes; typical values for stony meteorites of various types are on the order of 10% (Anders 1975). Because of the widespread nature of these effects, understanding them is essential to disentangling impact features from those produced in earlier stages of evolution.

The effects of cratering on small bodies are reviewed in the chapter by Housen *et al.* Their conclusion is that regolith surfaces occur on all but the smallest, strongest bodies. A similar conclusion was reached in the chapters by Veverka and Thomas and by Cintala *et al.* as a result of their studies of

Phobos and Deimos.

Meteorites are clearly the products of a fragmentation process, at least the fragmentation event that released them from bodies greater than the 1–10 m sizes which are typical of meteorites. We know from cosmic ray exposure ages that most stony meteorites existed as meter-sized objects for 1–100 Myr (Bogard's chapter) and most iron meteorites existed as meter-sized objects for 100–1000 Myr (Scott's chapter). Some shocked L-group meteorites seem to have experienced an intense shock event 300–700 Myr ago. Several other examples of fragmentation and a fuller discussion can be found in Bogard's and Scott's chapters. One could summarize the data presented in those chapters as indicating that the fragmentation events which have yielded meteorites on earth over the past 10^5 yr took place over the past 10^9 yr. This is in approximate agreement with calculated orbital dynamics for possible source regions of meteorites (Wetherill 1974). Prior to ejection from their parent bodies, meteorites were subjected to a variety of geological processes. Some planet-wide igneous activity may have produced the nakhlite achondrites as recently as 1.3 Gyr ago. However, the possibility that nakhlites formed by igneous differentiation of a very large, impact-generated melt 1.3 Gyr ago remains open.

Because of the effects which igneous differentiation, metamorphism, brecciation and fragmentation have caused in meteorites, very few textural characteristics relate to earlier processes of condensation and accretion. Elemental and isotopic compositions are the best records, albeit imperfect ones, of the earliest processes of formation.

IV. REVISIONIST HISTORY: A NEW SCENARIO

On the basis of the material in this book, the preceding discussion, and other meteoritic evidence, a revised scenario for the formation of meteorite parent bodies may be constructed. In chronological order its basic steps are given below. A fuller discussion of some of the points follows this list.

(1) Condensation of "high-temperature" materials, Ca-Al-rich inclusions (CAI'S), Mg-silicates and Ni-Fe metal, occured.

(2) Some fraction of the high-temperature materials formed in (1) reacted with the nebular gas to produce feldspars, ferromagnesian silicates and FeS.

(3) Chondrules were formed from the materials of (1) and (2) by transient heating processes.

(4) Condensation (reaction at lower temperatures) proceeded to produce lower-temperature materials: FeNiS, polymeric hydrocarbons, and H_2O-bearing silicates. This general type of material is called carbonaceous chondrite matrix materials (CCMM). It probably was not a single, homogeneous material.

(5) CAI's and chondrules coated with CCMM agglomerated. Pure CCMM

agglomerated in regions deficient in chondrules and CAI's. The proportions of the three components which agglomerated varied. This yielded CCMM, CV, H3, L3 and LL3 planetesimals on the order of 5 km in radius (Goldreich and Ward 1973).

(6) Planetesimals probably suffered different degrees of heating as a function of the proportion of short-lived radionuclides which they contained, although other heating mechanisms such as induction heating are also possible. Whatever the heating process was, it was probably dependent on location in the solar system. Some planetesimals totally melted and differentiated, forming metal cores and silicate mantles. These differentiated, small bodies were the source of pallasites and irons (chapters by Scott and Wood). Fragments of their silicate mantles did not survive to be collected as meteorites due to their short lifetimes (as compared to metal) against collisional destruction. Some planetesimals were heated only to metamorphic temperatures, yielding CO, H4–6, L4–6, LL4–6 chondritic material. Some were not heated at all.

(7) Planetesimals accreted into larger, $\gtrsim$ 100 km diameter, objects, i.e. asteroidal objects. Differentiated planetesimals having formed primarily in one region of space preferentially accreted forming S-type asteroids; unheated or weakly heated planetesimals accreted to form C-type and ordinary chondrite-type asteroids. Cooling rates of differentiated and metamorphosed objects slowed significantly.

Or (7′) Condensation or accretion of ices occurred on planetesimals in Jupiter's vicinity and/or beyond to form comets. Comets could conceivably have primitive, metamorphosed or differentiated cores.

(8) A few first-formed, larger objects in the asteroid belt (like Vesta) underwent planet-wide differentiation producing basaltic achondrites and other types of achondrites.

(9) Relative velocities increased; collisions began to result in net erosion rather than net accretion. Shock heating produced transient liquid water in planetesimals consisting of CCMM or of chondrules and CAI's plus a high proportion of CCMM, yielding CI's and CM's respectively (Richardson 1978; McSween 1979; chapter by Kerridge and Bunch). Brecciation and irradiation produced gas-rich meteorites in the surfaces of all moderately sized parent bodies.

This scenario assumes some isotopic and chemical inhomogeneity in a nebula of approximately solar composition. The first accretionary processes which produced planetesimals occurred everywhere in the early solar system. Other processes were location-dependent, including (3), (5), (6), (7), (7′) and (8). A very simplified cartoon illustrating portions of this scenario is given in Fig. 2.

This scenario produces comets, S-asteroids, C-asteroids, extensively differentiated asteroids, and ordinary chondrite asteroids. Most or even all

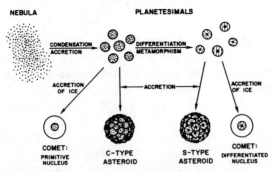

Fig. 2. An abbreviated, very schematic view of the formation of meteorite parent bodies, asteroids and comets as given in the text. The initial population is comprised of km-sized objects which grow through continued accretion. Formation of ordinary, metamorphosed chondrite parent bodies is not shown, but it would be similar to the formation of differentiated objects. Brecciation and fragmentation episodes which follow the illustrated steps are not depicted.

iron meteorites and pallasites are produced in differentiation events different from those which yielded the basaltic achondrites. The silicate mantles associated with most irons have not been (and will not be) sampled; the metallic cores and silicate mantles associated with basaltic achondrites may not have been sampled since they are buried deep in larger objects. In this model a variety of possibilities exist for cometary nuclei. Still not explained is the apparent paucity of ordinary chondrites in the asteroid belt. One *ad hoc* solution is to hypothesize that ordinary chondrite planetesimals formed in a place (near a planet?) from which they were readily perturbed into colder regions where ice accreted on them, making them preferentially into comets. Another solution is to imagine that S-type asteroid spectra are really due to ordinary chondrites (see Chapman's chapter).

This picture (Fig. 2) differs from the previous picture (Fig. 1) largely because evidence for the existence of extinct radionuclides in meteorites yields a heat source that can melt small bodies (Lee *et al.* 1977). However, it is important to note that this scenario calls for three separate heating events: (1) chondrule formation which requires heating of small ⩽1 mm particles in the nebula; (2) heating of planetesimals ~1 km (maybe up to 10 km) probably by extinct radionuclides; and (3) continued or subsequent heating of a few larger asteroids. The chondrule formation event is not unique to this picture; it is required by the existence of droplet chondrules and seems to call for a nebular heating mechanism. Several have been suggested (e.g., collisions in a turbulent nebula, lightning, magnetic reconnection); none is entirely satisfactory. The second heating event could depend entirely on short-lived radionuclides. The third step could also be due to heating by extinct radionuclides if the first-formed, larger objects could have accreted fast enough to retain sufficient heat from the diminished radioactivities. Alternatively, the electromagnetic heating mechanism (chapter by Sonett and Reynolds) could have contributed at this point. I would point out that some of the metamorphic features, especially textural features of ordinary chondrites, may be due to the effects of shock heating in step (9). Interpreting the scenario in terms of relative ages yields the following progression from oldest to youngest:

(a) CAI's,
(b) chondrules,
(c) CCMM, chondrule rims,
(d) H3–L3–LL3, proto-CV chondrites,
(e) irons, pallasites, CO, H4–6, L4–6, LL4–6 (extended cooling history, however),
(f) basaltic achondrites,
(g) CI and CM (low temperature aqueous altered phases); brecciated and shocked meteorites.

This scenario was, of course, constructed to agree with existing chronological data. Nevertheless an interesting consequence of the sequence is that by having basaltic achondrites forming the last, readily datable material, the entire scenario (except for 9) must be accomplished in less than ~ 200 Myr according to Pu-fission-xenon clock. Impact driven processes are the only geological processes that have affected most asteroidal-sized objects in the last 4.5 Gyr.

Acknowledgments. C. R. Chapman's unending challenges to the assumptions made by meteoriticists stimulated my thinking on this topic. In addition to Chapman I wish to thank W. V. Boynton, M. J. Drake, D. W. Mittlefehldt, D. Morrison, G. T. Sill, J. T. Wasson and J. A. Wood for helpful comments. This work was supported by a grant from the National Aeronautics and Space Administration.

REFERENCES

Allen, J. S.; Nozette, S.; and Wilkening, L. L. 1979. Chondrule rims: Composition and texture. *Lunar Science X.* The Lunar and Planet. Inst., pp. 27-29.

Anders, E. 1964. Origin, age and composition of meteorites. *Space Sci. Rev.* 3: 583-714.

Anders, E. 1975. Do stony meteorites come from comets? *Icarus* 24: 363-371.

Clayton, D. D. 1978. Precondensed matter: Key to the early solar system. *Moon and Planets* 19: 109-137.

Clayton, R. N.; Grossman, L.; and Mayeda, T. K. 1973. A component of primitive nuclear composition in carbonaceous meteorites. *Science* 182: 485-488.

Clayton, R. N.; Onuma, N.; Grossman, L.; and Mayeda, T. 1977. Distribution of the pre-solar components in Allende and other carbonaceous chondrites. *Earth Planet. Sci. Lett.* 34: 209-224.

Consolmagno, G. J. 1979. REE patterns versus the origin of the basaltic achondrites. *Icarus* (special Asteroid issue).

Consolmagno, G. J., and Drake, M. J. 1977. Composition and evolution of the eucrite parent body: Evidence from rare earth elements. *Geochim. Cosmochim. Acta* 41: 1271-1282.

Drozd, R. J., and Podosek, F. A. 1976. Primordial [129]Xe in meteorites. *Earth Planet. Sci. Lett.* 31: 15-30.

El Goresy, A.; Nagel, K.; and Ramdohr, P. 1978. The Allende meteorite: Frendlinge and their noble relatives. *Lunar Science IX.* The Lunar and Planet. Inst., pp. 282-284.

Goldreich, P., and Ward, W. R. 1973. The formation of planetesimals. *Astrophys. J.* 183: 1051-1061.

Gray, C. M.; Papanastassiou, D. A.; and Wasserburg, G. J. 1973. The identification of early condensates from the solar nebula. *Icarus* 20: 213-239.

Grossman, L. 1972. Condensation in the primitive solar nebula. *Geochim. Cosmochim. Acta* 36: 597-619.

Hartmann, W. K. 1976. Planet formation: Compositional mixing and lunar compositional anomalies. *Icarus* 27: 553-559.

Herndon, J. M., and Wilkening, L. L. 1978. Conclusions derived from the evidence on accretion in meteorites. In *Protostars and Planets,* ed. T. Gehrels (Tucson: University of Arizona Press), pp. 502-515.

Kelly, W. R., and Wasserburg, G. J. 1978. Evidence for the existence of [107]Pd in the early solar system. *Geophys. Res. Lett.* 5: 1079-1082.

Larimer, J. W. 1979. The condensation and fractionation of refractory lithophile elements, *Icarus* (special Asteroid issue).

Larimer, J. W., and Anders, E. 1967. Chemical fractionations in meteorites. II. Abundance patterns and their interpretation. *Geochim. Cosmochim. Acta* 31: 1239-1270.

Lee, T.; Papanastassiou, D. A.; and Wasserburg, G. J. 1976. Demonstration of [26]Mg in Allende and evidence for [26]Al. *Geophys. Res. Lett.* 3: 109-112.

Lee, T.; Papanastassiou, D. A.; and Wasserburg, G. J. 1977. Aluminum-26 in the early solar system: Fossil or fuel? *Astrophys. J.* 211: L107-L110.

Lee, T.; Russell, W. A.; and Wasserburg, G. J. 1979. A new member of the "FUN" family. *Lunar Science X.* The Lunar and Planet. Inst., pp. 713-716.

Lewis, J. S. 1972. Metal/silicate fractionation in the solar system. *Earth Planet. Sci. Lett.* 15: 266-290.

Macdougall, J. D., and Kothari, B. K. 1976. Formation chronology for C2 meteorites. *Earth Planet. Sci. Lett.* 33: 36-44.

Marvin, U. B.; Wood, J. A.; and Dickey, J. S., Jr. 1970. Ca-Al rich phases in the Allende meteorite. *Earth Planet. Sci. Lett* 7: 346-350.

McSween, H. Y., Jr. 1979. Are carbonaceous chondrites primitive or processed? – A Review. *Rev. Geophys. Space Phys.* (submitted).

Mittlefehldt, D. W. 1979. The nature of asteroidal differentiation processes: Implication for the primordial heat sources. *Proc. Lunar Sci. Conf. X* (Oxford: Pergamon Press). In press.

Mittlefehldt, D. W., and Wetherill, G. W. 1979. Rb-Sr studies of CI and CM chondrites. *Geochim. Cosmochim. Acta* 43: 201-206.

Morgan, J. W.; Higuchi, H.; Takahashi, H.; and Hertogen, J. 1978. A "chondritic" eucrite parent body: Inference from trace elements. *Geochim. Cosmochim. Acta* 42: 27-38.

Papanastassiou, D. A., and Wasserburg, G. J. 1974. Evidence for late formation and young metamorphism in the achondrite Nakhla. *Geophys. Res. Lett.* 1: 23-26.

Rajan, R. S. 1974. On the irradiation history and origin of gas-rich meteorites. *Geochim. Cosmochim. Acta* 38: 777-788.

Reynolds, J. H. 1967. Isotopic abundance anomalies in the solar system. *Ann. Rev. Nucl. Sci.* 17: 253-316.

Richardson, S. M. 1978. Vein formation in the C1 carbonaceous chondrites. *Meteoritics* 13: 141-159.

Rowe, M. W. 1967. Xenomalies. *Center for Meteorite Studies* (Monograph), Ariz. State Univ.

Scott, E. R. D. 1977. Geochemical relationships between some pallasites and iron meteorites. *Mineral Mag.* 41: 265-272.

Sears, D. W. 1979. Did iron meteorites form in the asteroid belt? Evidence from thermodynamic models. *Icarus* (submitted).

Stolper, E.; McSween, H. Y., Jr.; and Hays, J. F. 1979. A petrogenetic model of the relationships among achondritic meteorites. *Geochim. Cosmochim. Acta* 43: 589-602.

Wahl, W. 1952. The brecciated stony meteorites and meteorites containing foreign fragments. *Geochim. Cosmochim. Acta* 2: 91-117.

Wark, D. A., and Lovering, J. F. 1977. Marker events in the early evolution of the solar system: evidence from rims on calcium-aluminum-rich inclusions in carbonaceous chondrites. *Proc. Lunar Sci. Conf. VIII* (Oxford: Pergamon Press), pp. 95-112.

Wetherill, G. W. 1974. Solar system sources of meteorites and large meteoroids. *Annual Rev. Earth Planet. Sci.* 2: 303-331.

Wetherill, G. W. 1975. Late heavy bombardment of the moon and terrestrial planets. *Proc. Lunar Sci. Conf. VI* (Oxford: Pergamon Press), pp. 1539-1561.

Wetherill, G. W. 1976. Where do the meteorites come from? A re-evaluation of the Earth-crossing Apollo objects as sources of chondritic meteorites. *Geochim. Cosmochim. Acta* 40: 1297-1318.

Wilkening, L. L. 1977. Meteorites in meteorites: Evidence for mixing among the asteroids. In *Comets, Asteroids, Meteorites,* ed., A. H. Delsemme, (Toledo, Ohio: University of Toledo Press), pp. 389-396.

PART II

Exploration

THE WORK OF THE MINOR PLANET CENTER

BRIAN G. MARSDEN
Harvard-Smithsonian Center for Astrophysics

The work of the Minor Planet Center and the collaboration of the Institute for Theoretical Astronomy in Leningrad are described. The Center is responsible for collecting and disseminating astrometric observations and current orbital and ephemeris data on minor planets. It attends to the cataloguing of unidentified minor planets as they are discovered, and when satisfactory orbits can be determined from observations at several oppositions, it gives new permanent numbers to the minor planets. The information is largely distributed through the Minor Planet Circulars but is also collected generally in machine-readable form. The work of the Minor Planet Center complements that undertaken at the Institute for Theoretical Astronomy in Leningrad, which is responsible for the publication of Efemeridy Malykh Planet, the annual volume of ephemerides of numbered minor planets.

The Minor Planet Center, established by the International Astronomical Union (IAU) at the Cincinnati Observatory in 1947, moved to the Smithsonian Astrophysical Observatory upon the retirement of the director, P. Herget, in June 1978. The basic responsibilities of the Center have not been changed by the move, and the division of labor with the Institute for Theoretical Astronomy in Leningrad continues as before. This collaboration dates from the reconstruction of minor planet astronomy following World War II. For a review of earlier work the reader is referred to reports by Herget (1971) and Marsden (1979a). Herget was a pioneer in the introduction of automatic computing machines in astronomy, and the reader is also referred to his lively account (Herget 1966) of some of his early experiences in this area.

The most visible consequences of the collaboration are the annual volumes *Efemeridy Malykh Planet* (EMP), published by the Institute for Theoretical Astronomy, and the *Minor Planet Circulars* (MPC), published by the Minor Planet Center. The EMP volumes contain ephemerides with 1 arcmin precision for eight standard 10-day dates around opposition for each of the permanently numbered minor planets. The mean anomaly, apparent photographic magnitude (ignoring the phase effect), heliocentric and geocentric distances, and variations in position corresponding to a change of +1 degree in mean anomaly are provided for the fourth date in each ephemeris. More detailed quadrature-to-quadrature ephemerides are supplied for the brighter minor planets and, beginning with the 1979 EMP volume, these are given to 0.1 arcmin precision and for all objects brighter than magnitude 12.5 at mean opposition. Special ephemerides are also supplied for 26 unusual objects including 944 Hidalgo and the minor planets − except for the lost 719 Albert − of perihelion distances less than 1.3 AU.

The MPC contain notification of newly numbered minor planets and include ephemerides for these until they can be incorporated in the EMP volumes. Orbital elements and ephemerides are provided for selected unnumbered minor planets and the MPC also contain positional observations of minor planets, the assignment of provisional designations for unnumbered minor planets, the announcement of names, and other miscellaneous information. They are produced by photo-offset printing from computer output and since the Minor Planet Center moved to Cambridge they have been issued in regular monthly batches. *Minor Planet Circular 4000* was published in June 1976, and the March 1979 batch consisted of numbers 4649-4688.

More urgent minor planet information, generally that pertaining to discoveries of new Apollo-type objects, is distributed through the *IAU Circulars* (IAUC) and telegrams. Since the IAU Central Bureau for Astronomical Telegrams and the Minor Planet Center now operate together, in the interests of efficiency some of the less urgent cometary data that have in the past appeared in the IAUC are now published in the MPC (which can also be interpreted as "Minor Planets and Comets").

The Minor Planet Center is also responsible for collecting positional observations. By the time of the transfer from Cincinnati about 190,000 observations had been collected and filed on magnetic tape. The file is intended to be complete back to the beginning of 1939 and it also includes many earlier observations, notably the important and extensive series of photographic observations at Heidelberg (Reinmuth 1953, 1960), which extend back to the 1890's. Since the transfer, some 10,000 more observations have been collected, although because of format changes and conversion to a new computer, these have not yet been incorporated in the Cincinnati file. The computer program that generates the MPC recognizes observations and sorts them into a separate file, each observation being automatically referenced with the number of the MPC in which it will be published. Comet observations are handled in the same way, and it is planned to combine these with a magnetic-tape file of some 25,000 older comet observations that have been collected by the Central Bureau for Astronomical Telegrams. After fairly extensive checking of the data it is hoped that copies of both files can be made available to interested users.

Computations of orbits and ephemerides are made both at the Minor Planet Center and at the Institute for Theoretical Astronomy. Some duplication of effort is inevitable, but strict adherence to producing the computations necessary for the respective publications can minimize this. Under the supervision of Yu. V. Batrakov, astronomers at the Institute for Theoretical Astronomy make orbit improvements for typically 100 numbered minor planets each year, and a further few dozen orbits are improved at the Latvian State University in Riga. The 1979 EMP volume refers to the 2042 minor planets numbered up to June 1977, and some 58% of the orbits have been calculated during the past decade. Twenty-two numbered minor planets are considered lost − almost a tenfold reduction during the past 40 years. There are perhaps half a dozen minor planets with ephemerides now off by more than 10 arcmin and some three dozen with errors in the 5-10 arcmin range, which represents about a tenfold reduction in eight years (Marsden 1971). New minor planets are being numbered at an increasing rate − 346 in the past ten years; early in 1979 minor planet numbers extended to 2125. Most of the orbits of the new objects are computed by C. M. Bardwell, who has moved from Cincinnati to Cambridge and is now Assistant Director of the Minor Planet Center. With consideration also of the unnumbered minor planets and a few of the older numbered ones, orbit improvements are being made at the Minor Planet Center at about the same rate as at the Institute for Theoretical Astronomy. Herget in his retirement in Cincinnati is also continuing to recompute orbits of the older numbered minor planets.

Preliminary orbits are computed whenever possible for newly discovered minor planets and these are now routinely published in the MPC. Several hundred unpublished orbits have accumulated over the past few decades and we are now giving them, one line per orbit, in the MPC. The format includes the arc length and number of observations utilized, thereby enabling the user to judge the reliability of these orbits. The listings in the MPC are more or less complete back to 1970. Largely because the principal program of observations of minor planets is conducted by the Institute for Theoretical Astronomy with the 40-cm double astrograph at the Crimean Astrophysical Observatory (Chernykh and Chernykh 1974), many of these orbits are calculated both in Leningrad and Cambridge, and similar results are also produced by several amateur astronomers in Japan. The Crimean observing program has become the most sustained and intense ever conducted and minor planets are routinely being recorded down to magnitude 17.5 and fainter. Since 1975 provisional designations have been supplied at a rate of more than 1000 per year; although such an annual total is only half of what the one-time Palomar-Leiden Survey achieved in little over a month in 1960, it is more than twice of that achieved in any year prior to 1950.

Deliberate attempts are frequently made to use the preliminary orbits of minor planets, observed for a month or two, to recover the objects at their following oppositions since telescope time is often at a premium. The principal use of the preliminary orbits is, however, to try to establish identifications with minor planets already observed at previous oppositions but for which orbits are usually not available. The detective work involved in this can be quite engrossing, and the satisfaction one gets when a tentative link to observations at two oppositions makes single observations at half a dozen other oppositions to fit into place far surpasses that of completing a more mundane jigsaw puzzle. If the identifications refer to a lost numbered minor planet, the sense of achievement is even greater, as when we recently recognized (Marsden 1979b) that the preliminary orbit of 1975 EW1 showed that object undoubtedly to be 1465 Autonoma, not definitely observed since its discovery opposition in 1936.

Bardwell has been particularly successful in this work, and important contributions have also been made by O. Kippes at Wurzburg, by the amateurs H. Oishi and T. Urata in Japan, and, increasingly, by J. G. Williams and E. Bowell. It was Williams who pointed out to us the identification of the previously discovered Hungaria-type object 1978 WA with the minor planet that had just been numbered (2083). He in fact supplied us with a 1978 ephemeris. This embarrassing circumstance arose because the Cambridge Minor Planet Center had not yet made an up-to-date compilation of the orbits of the numbered minor planets in a suitable order (such as by the longitude of the ascending node); this has now been rectified.

The above example illustrates that a large amount of bookkeeping must be done to keep the Minor Planet Center working effectively and to respond to the many requests and queries which are received. The bookkeeping becomes a particular problem when observations of minor planets are incorrectly identified. In the announcement of the numbering of minor planet 1860, for example (Bardwell 1974), it is stated that "the 1971 Feb. 20 observation erroneously assigned to (294) belongs to (1860). The observation incorrectly designated 1971 BM1 on that same date does not belong." The January 27 discovery observation of 1971 BM1 does belong to (2060). An error of this type, when we have only two observations a month apart, is a frequent and understandable occurrence, and therefore we must be careful how the correction is documented. The erroneous 1971 BM1 observation was disposed of easily enough — but after a lapse of another four years — by redesignating it as 1971 DC2 and reprinting it in this way in a special list of identification changes in the MPC. The erroneous observation of (294) was not handled so satisfactorily, however, and if anyone wonders what has happened to it he simply has to know that it now belongs to (1860).

A slightly different example of error is given by a pair of August 1949 observations initially published in the MPC as being of (358) which in fact refer to (755). This was mentioned in 1978 as an erratum. The problem is that there are *two* pairs of 1949 observations, and the erratum refers to the wrong pair. So we shall presumably need an erratum to the erratum. There are dozens of examples of this type, and tracking them down is not very straightforward. Although it is a little cumbersome, the best procedure seems to be first to give a new provisional designation to the observations that have been incorrectly attributed to a particular numbered minor planet and then to identify that provisional designation with the correct numbered minor planet. Since provisional designations can be well documented, reconstruction of the situation is then quite simple. A couple of previously unpublished misidentifications are handled in this way on the latest batch of MPC's.

A detailed documentation of provisional designations generally up to 1961, with references, was published by Strobel (1963). We are preparing a detailed edition of this, including discoverers, as in the earlier edition by Stracke (1938), and also the discovery positions and magnitudes. Discovery information on numbered minor planets was included in the tabulation by Pilcher and Meeus (1973; see also Pilcher's chapter) but our investigations have revealed a significant number of errors and in many cases the situation was simply ambiguous.

The principal task of the Minor Planet Center is to increase the number of objects for which reliable orbital data can be provided. Williams (Chapman *et al.* 1978) has pointed out the need for these data since there are several Hirayama families that as of January 1979 have only a few known members. At the present and anticipated future rate of growth, it is not unreasonable to suppose that there will be 4000 numbered minor planets by the end of the century. Whether that will be the time to terminate this cataloguing is unclear. New types of discoveries are being made all the time: 132 Aethra, the first Mars crosser, in 1873 (but lost until 1922); 153 Hilda, the first 3:2 librator, in 1875; 279 Thule, most distant at the time and 4:3 librator, in 1888; earth-approacher 433 Eros and 434 Hungaria, in 1898; 588 Achilles, the first Trojan, in 1906; 887 Alinda, the first 3:1 librator, in 1918; 944 Hidalgo, the first Saturn crosser, in 1920; 1862 Apollo, the first earth crosser, in 1932 (lost until 1973); 1362 Griqua, the first 2:1 librator, and 1373 Cincinnati, the first argument-of-perihelion librator, in 1935; 1566 Icarus, Mercury crosser, in 1949; 1580 Betulia, with its record orbital inclination, in 1950; 2102 Tantalus, with an even higher inclination, in 1975; 2062 Aten, with the first mean distance of less than 1 AU, in 1976; and 2060 Chiron (see Kowal's chapter), ranging from the orbits of Saturn to Uranus, in 1977. Perhaps we should also include the unnumbered Pluto, the first Neptune crosser, discovered in 1930, and 1979 BA, found to have the highest orbital inclination of any non-earth-approaching object. Clearly we could concentrate on discovering just the unusual objects, but the statistics would become biased if the more ordinary, or seemingly ordinary, minor planets were ignored.

All those involved in observational work on minor planets have a need for reliable orbital and ephemeris data. The responsibility for providing these is divided between the Minor Planet Center and the Institute for Theoretical Astronomy. It is mainly the latter organization, however, that produces the data for the brighter and larger minor planets, which are of course the ones generally of interest for physical observations, radar detections, occultation predictions, and the planning of possible space missions. These new requirements differ from the former need for only approximate opposition ephemerides so that further astrometric observations could be made. Nowadays many users want detailed information on the distances of minor planets, phase angles, and such, and they often require it for times when the objects are far from opposition. Rather than expand the ephemerides in the EMP to incorporate all this additional information, a prodigious undertaking, it seems preferable to supply the data in a different form. The most appropriate solution appears to be for the EMP to supply for all the numbered minor planets high-precision osculating elements for a current standard epoch each year. This was the course recommended by IAU Commission 20 at the Grenoble meetings in 1976. A start has been made in the 1979 EMP, which gives all the newly improved orbits for the osculation date 1979 Nov. 23.0 ET. The orbits (except those of one-apparition objects) appearing in the MPC are also currently being given for this epoch.

When the recommendation has been fully implemented, presumably in EMP 1980 for the epoch 1980 Dec. 27.0 ET, users will generally need only a two-body computer program to produce ephemeris data that meet their own requirements. Somewhat greater sophistication is expected of those concerned with the predictions of occultations by minor planets, and it will probably always be necessary to do some last-minute astrometry when the planet and the star to be occulted are in the same field. Those involved in such work are already prepared to do this, however, as is attested to by the phenomenal successes in this area, particularly during 1978.

REFERENCES

Bardwell, C. M. 1974. Improved elements of minor planets. *Minor Planet Circ. 3657-3671.*

Chapman, C. R.; Williams, J. G.; and Hartmann, W. K. 1978. The asteroids. *Ann. Rev. Astron. Astrophys.* 16: 33-75.

Chernykh, N. S., and Chernykh, L. I. 1974. The program of minor planet observations at the Crimea Observatory. In *Asteroids, Comets, Meteoritic Matter,* eds. C. Cristescu,

W. J. Klepczynski, and B. Milet (Bucuresti: Editura Academiei Republicii Socialiste Romania), pp. 25-27.

Herget, P. 1966. The Minor Planet Center at the Cincinnati Observatory. *Bull. Cincinnati Hist. Soc.* 24: 175-187.

Herget, P. 1971. The work at the Minor Planet Center. In *Physical Studies of Minor Planets,* ed. T. Gehrels (NASA SP-267, Washington, D.C.: U.S. Government Printing Office), pp. 9-12.

Marsden, B. G. 1971. Precision of ephemerides for space missions. In *Physical Studies of Minor Planets,* ed. T. Gehrels (NASA SP-267, Washington, D.C.: U.S. Government Printing Office), pp. 639-642. *Wiss. Phys.-math. Kl.* 4.

Marsden, B. G. 1979a. The Minor Planet Center. *Celes. Mech.* (in press).

Marsden, B. G. 1979b. Orbital elements. *Minor Planet Circ. 4642.*

Pilcher, F., and Meeus, J. 1973. *Tables of Minor Planets,* (Jacksonville, Illinois: private publication).

Reinmuth, K. 1953. Katalog von 6500 genauen photographischen Positionen Kleiner Planeten. *Veröff. Sternw. Heidelberg-Königstuhl* 16.

Reinmuth, K. 1960. Katalog von 6000 genauen photographischen Positionen Kleiner Planeten. *Veröff. Sternw. Heidelberg-Königstuhl* 17.

Stracke, G. 1938. Identifizierungsnachweis der Kleinen Planeten. *Abh. Preuss. Akad. Wiss. Phys.-math. Kl.* 4.

Strobel, W. 1963. Identifizierungsnachweis der Kleinen Planeten. *Veröff Astron. Rechen -Inst.* 9.

MASSES AND DENSITIES OF ASTEROIDS

J. SCHUBART
Astronomische Rechen-Institut, Heidelberg

and

D. L. MATSON
Jet Propulsion Laboratory

Ceres, Pallas and Vesta allow a determination of mass from gravitational effects in the motion of another asteroid. An extended and partly corrected set of observations of 197 Arete leads to an increase of 15 % in the resulting mass of Vesta that was first determined by H. G. Hertz. Other possibilities of mass determination and estimates of the total mass of asteroids are mentioned. We review the available diameter determinations and adopt preferred values. The densities for Ceres, Pallas and Vesta are respectively 2.3 ± 1.1 g cm^{-3}, 2.6 ± 0.9 g cm^{-3} and 3.3 ± 1.5 g cm^{-3}.

The mass and size are fundamental properties of all astronomical bodies. The first direct determination of the mass of an asteroid, by H. G. Hertz, appeared in an IAU Circular in December 1966. Mass determinations are now available for Ceres, Pallas and Vesta. Earlier estimates of mass have been reviewed by Schubart (1971*a*).

The exact sizes of Ceres, Pallas and Vesta were poorly known from the time of their discovery in the first decade of the nineteenth century until the 1970's. By that time a variety of new methods had been developed and were being used for determinations of asteroid diameter. As a result we now possess a reasonable knowledge of their sizes. However, it is desirable to increase further the accuracy of size determinations. The resulting densities are needed to improve our knowledge of the bulk compositions, which in turn place constraints on theories of solar system formation.

I. MASSES

In a search for close encounters between two asteroids Fayet (1949) listed 13 pairs that gave evidence of fairly close approaches at some date not far from a common opposition. From this study it appeared that 197 Arete was close to 4 Vesta on 1939 Oct. 18. Later Hertz (1968) discovered that Arete approaches Vesta within 0.04 AU once every 18 yr, and he succeeded in determining the mass of Vesta from its gravitational effects in an orbital theory of Arete. The Vesta-Arete pair appears again in a list by Davis and Bender (1977) made in a search for encounter opportunities for mass determination in the interval 1970-90. This pair offers a unique opportunity due to its massive member Vesta that causes measurable effects, due to the early discovery of Arete, and due to the frequent repetition of long lasting encounters with comparable minimum distances, Δ_m, as is seen in the following list of Vesta-Arete encounters:

Date	Δ_m (AU)	Date	Δ_m (AU)
1885 May	0.018	1939 Oct.	0.032
1903 July	0.027	1957 Dec.	0.035
1921 Aug.	0.029	1976 Jan.	0.035

Evidently, the strongest gravitational action by Vesta occurred during the first encounter after the well-observed discovery opposition of Arete in 1879, and the more recent approaches are less important.

The two other massive asteroids of the main belt, Ceres and Pallas, show an observable gravitational interaction (Schubart 1971b), due to the approximate 1/1 ratio of their mean motions. An accumulation of small effects of attraction of Ceres occurs in the motion of Pallas, and vice versa. There is also a smaller interaction between Ceres and Vesta. Schubart (1974, 1975) has determined the masses of Ceres and Pallas from a well-prepared collection of observations (Schubart 1976). However, use was made of Hertz's (1968) mass determination of Vesta in computing its action on the motion of Ceres. Therefore it has appeared essential to confirm and to extend Hertz's work on Vesta and Arete; the results by Schubart are given here.

Until his death in 1976 Hertz continued to collect observations of 197 Arete after his 1968 solution (Schubart 1977), and the whole reliable collection of observations was kindly made available by his brother, R. H. Hertz. Special efforts were made to observe Arete in 1976/77. When Schubart repeated the Hertz determination of 1968 with nearly the same set of observations of Arete from 1879 to 1966, there was no change in the resulting mass of Vesta, although Ceres was added to the bodies attracting Arete. Schubart then added the more recent observations, but again this caused no change. However, an increase of 15 % resulted for the mass of Vesta, when observations for the period 1907-1962 were added which Hertz had collected subsequent to his determination, and one allows for systematic corrections to the FK4 system in nine of the 1879 observations that have suitable reference stars. According to Schubart (1975), this increase caused a small decrease in his result for Pallas. If these changes are considered, the interacting pairs Ceres-Pallas and Vesta-Arete lead to the following values of mass, in 10^{-10} solar units.

$$
\begin{array}{llll}
1 & \text{Ceres} & : & 5.9 \ \pm 0.3 \\
2 & \text{Pallas} & : & 1.08 \pm 0.22 \\
4 & \text{Vesta} & : & 1.38 \pm 0.12
\end{array}
$$

The given mean errors of the three values are larger than the formal mean errors of the respective solutions, to allow for unknown systematic errors in the observational basis of the solutions (c.f. Schubart 1974, 1975). The mean error for Ceres has been increased by about 50 % of its amount, and the formal mean errors in the results for the masses of Pallas and Vesta have been doubled. Especially in the case of the comparatively faint object Arete, the technique of observation has changed considerably during the period of interest, and the same may be true for possible systematic errors. It cannot be excluded that unknown encounters with comparatively large asteroids have caused small effects in the motion of the objects studied for mass determination. However, the approach to $\Delta_m = 0.02$ AU of 324 Bamberga to 1 Ceres in April 1944 did not cause observable effects in the motion of Ceres. Schubart has used the same methods of numerical integration and differential correction and also the same set of masses and orbits of the major planets (Schubart 1974) in the derivation of the three values of mass listed above. Geocentric positions depend upon Herget's (1953) geocentric coordinates of the sun.

The question arises whether there are further encounter opportunities or other possibilities of mass determination. Davis and Bender (1977), in conjunction with E. Bowell, are continuing a search program for close, low relative velocity encounters between one of the ten largest asteroids and another numbered asteroid. Several encounters have been found that may result in observable perturbations, such as 1 Ceres − 534 Nassovia, which came within 0.022 AU of each other at a relative speed of 2.8 km sec^{-1} in 1975. For comparison, the Vesta-Arete encounter in 1976 had a relative speed of 2.1 km sec^{-1}. An unusually low-velocity encounter of 0.8 km sec^{-1} occurred between 65 Cybele and 609 Fulvia in 1970. A low-velocity encounter favors the accumulation of gravitational effects, but none of the events found more recently has the frequent re-encounter period of Vesta-Arete. Schubart (1972) looked first for approaches to Ceres, Pallas and Vesta, but later other comparatively large bodies like 10 Hygiea were included in the searches. However, there was no real success in finding significant new opportunities for determining independent values of mass. In an encounter most asteroids pass each other too quickly for the generation of observable effects.

It is expected that tracking of suitable spacecraft will offer new opportunities for mass determination of large asteroids (Schubart 1977). Hellings and Standish (1979, personal communication) are exploring the possibilities of computing the masses of Ceres, Pallas and Vesta from range data between tracking stations and Mars orbiters and landers. A result at least for Ceres is expected. Asteroid missions, or even asteroid orbiters and landers will offer more possibilities in the future.

Morrison (1977b,c) and Kresák (1977) published independent estimates of the mass contribution of smaller bodies to the total mass of asteroids. Morrison's estimate of 3.1×10^{24}g for the bodies with diameters between 700 and 20 km is based on a mean density of 3.0 g cm^{-3} which may be too high. Kresák used different estimates of density for C-type and S-type asteroids, and his estimate of 3.0×10^{24}g refers to the total mass of all asteroids including the Trojan group.

II. SIZES

Many values have been published for the diameters of Ceres, Pallas and Vesta, especially in the years 1971 through 1979. It is the purpose of this section to sift through these values and to provide the best available data for the calculation of densities. A collection on early efforts to determine the diameters was published by H. Sadler and copied by Barnard (1895); this collection is now only of historic interest. Barnard's (1895, 1900a) measurements by micrometer of 1894-99, and the corresponding adopted values of diameter (Barnard 1900b, 1902), together with interferometric measurements on Vesta (Hamy 1899a,b), gave the basic information on size for a long period. However, the difficulties of micrometry in case of small disks close to the limit of atmospheric seeing were well known (Dollfus 1970, 1971).

Widorn (1967) proposed a polarimetric method that relies on an empirical relationship between albedo and the rate at which the polarization of the reflected light increases between certain phase angles (c.f. the chapter by Dollfus and Zellner in this book). Using a geometric albedo derived from the empirical relationship, the size of an asteroid can be computed from the absolute magnitude.

Allen (1970, 1971) and Matson (1971) introduced the infrared method, another indirect method for size determination of asteroids. In this method the insolation is balanced against the total of the reflected and thermally emitted radiation. In practice there is only limited infcrmation from measurements of the two types of radiation that leave the body and depend on its size and albedo. Especially the range of phase angles is restricted to values near the one of full phase in the case of the asteroids under consideration. Nevertheless it is possible to derive size and albedo, if a suitable model is adopted for the processes of radiation (c.f. the chapter by Morrison and Lebofsky in this book; Matson *et al.* 1978).

A collection of diameter determinations is shown in Table I. The values are tabulated on a uniform basis. As the various methods of diameter determination have been improved, it is only natural for an author to regard some of his earlier values as obsolete. Where this is known to us, we have indicated this by a footnote. Table I demonstrates the great amount of work done with respect to the diameters of the three large asteroids in the period from ~ 1971 to early 1979. Some of the measurements by double image micrometer (Dollfus 1971) were made during less favorable conditions, with respect to geocentric distance. The method of speckle interferometry (c.f. the chapter by Worden) was first applied to Vesta, an object of comparatively high albedo and then to Pallas. This gives rise to the hope that this method can be further developed and applied to all three large asteroids. The chapter by Millis and Elliot in this book describes the method that is based on the observation of a stellar occultation by a network of observers. An occultation by Pallas was successfully observed in this way in 1978 (Wasserman *et al.* 1979). According to the observed chords across the disk and to earlier lightcurve data, the body of Pallas appears to be nonspherical. A mean diameter has resulted from a simple model for the shape of the body.

Some of the methods mentioned above may achieve high apparent precision, but they are subject to various sources of systematic error, the magnitudes of which are not fully understood. We feel that the method of reducing occultation data has fewer sources of systematic error than do other methods. Since there are accurate observations of an occultation by Pallas from several stations, we feel that the resulting mean diameter is the best one available. Thus we adopt a value of 538 km for the diameter of Pallas.

For Ceres and Vesta we have to rely on other methods. Table II lists Barnard's (1900*a*) adopted diameters, and the following lines show values that were adopted by various authors in 1963-78. There is a great unexplained discrepancy between the results based on micrometry and on indirect methods in case of the largest body, Ceres. The scatter of Barnard's (1895, 1900*a*) measurements is too small to give an explanation. It is likely that subjective and systematic errors have affected the direct telescopic measurements of a disk of one arc-second or less in apparent size.

The measurements of Vesta by double image micrometer may be less affected by such errors, but the angular size of Vesta is smaller. We are unable to estimate the amount of the assumed errors that affect results from micrometry, and we suspect that the effects of these errors could be large.

TABLE I

Asteroid Diameter Determinations

No.	Diameters (km)			Method	References and Remarks
	1 Ceres	2 Pallas	4 Vesta		
1	781 ± 87[a]	490 ± 118[a]	391 ± 46[a]	Filar Micrometer	Barnard 1895, 1900a. Obs. of 1894-5.
2	——	——	390	Interferometer	Hamy 1899a,b.
3	706 ± 84[a]	——	347 ± 70[a]	Filar Micrometer	Barnard 1900b. Observations of 1897-8.
4	850	500	390	Polarimetric	Widorn 1967.
5	——	——	450	Double Image Micrometer	Dollfus 1970. Observers: Muller, Focas, Dollfus, 1967.
6	——	921 ± 256[c]	435 ± 73[c]	Double Image Micrometer	Dollfus 1971. Dollfus, Lecacheux and Camichel, 1969.
7	1160 ± 80	——	570 ± 10	Radiometric	Allen 1971.
8	1000 ± 100	$530 \,{}^{+100}_{-175}$	600 ± 60	Radiometric	Matson 1971.
9	——	——	$515 \,{}^{+95}_{-60}$	Polarimetric	Veverka 1971.
10	——	——	$512 \,{}^{+38}_{-22}$	Mass of Vesta and mean density of achondrites	Veverka 1971.
11	$1220 \,{}^{+120}_{-240}$	660 ± 110	$580 \,{}^{+70}_{-90}$	Polarimetric	Veverka 1973.
12	1080 ± 80	550 ± 50	540 ± 40	Radiometric	Cruikshank and Morrison 1973.
13	1020 ± 100	——	580 ± 60	Radiometric	Morrison 1973.
14	1060 ± 130	600 ± 75	550 ± 70	Polarimetric	Bowell and Zellner 1974. Also cited in Bowell et al. 1973.
15	1041	569	526	Radiometric	Morrison 1974. Using the observations of Cruikshank and Morrison 1973.

TABLE I (Continued)

Asteroid Diameter Determinations

No.	Diameters (km) 1 Ceres	2 Pallas	4 Vesta	Method	References and Remarks
16	—	—	553	Radiometric	Morrison 1974. Using new observations.
17	$1050^c/1000^c$	$560^c/530^c$	$515^c/530^c$	Polarimetric/Radiometric	Chapman and Morrison 1974.
18	1050	570	490	Polarimetric	Zellner et al. 1974.
19	$1173 \pm 104^{b,d}$	$754 \pm 34^{b,d}$	$602 \pm 51^{b,d}$	Radiometric	Hansen 1976.
20	914	573	496	Polarimetric	Zellner and Gradie 1976.
21	1019	597	538	Radiometric	Hansen 1977.
22	1018 ± 43^b	602 ± 57^b	531 ± 15^b	Radiometric	Morrison 1977a. Wt. av. of all determinations.
23	1017	585	531	Radiometric	Morrison 1977b. Wt. av. of selected determinations.
24	957	—	558	Polarimetric	Zellner et al. 1977. Implicit data.
25	—	—	513 ± 51	Speckle Interferometry	Worden et al. 1977.
26	—	673 ± 55	550 ± 23	Speckle Interferometry	Worden and Stein 1979.
27	—	538 ± 12	—	Stellar Occulation and Lightcurve	Wasserman et al. 1978, 1979; Elliot et al. 1978.

[a]Original source does not give an error bar. The standard deviation given here was calculated using data in the work cited.

[b]This diameter (mean and standard deviation) was calculated from a listing of data in the work cited. For weighted averages, the author's weights were used.

[c]Labeled as "preliminary" in reference cited.

[d]Author or senior author of work cited has designated this value as having been superseded by more recent work.

TABLE II

Asteroid Diameters Adopted by Various Authors

No.	Diameters in Kilometers			Method	References and Remarks
	1 Ceres	2 Pallas	4 Vesta		
1	769	490	385	Average weighted by the number of observations.	Barnard 1900b, 1902.
2	700	460	380	—	Allen 1963. Astrophys. Quant., 2nd ed.
3	770 ± 40	490 ± 50	420 ± 35	Barnard's values for Ceres and Pallas. Weighted average for Vesta (single wt. 3 and 4 of Table I; double wt. for No. 6).	Dollfus 1970.
4	770 ± 40	—	410	Barnard's value for Ceres. Average of 1, 2 and 6 of Table I for Vesta.	Dollfus 1971.
5	760	480	480 (sic)	Veverka's value (No. 12) for Vesta.	Allen 1973. Astrophys. Quant., 3rd ed.
6	1040	570	528	New radiometric model.	Jones and Morrison 1974.
7	1022[a]	558	503	Weighted average of polarimetric and adjusted radiometric diameters. See p. 114 in reference for discussion.	Chapman et al. 1975. Tabulated in their Table I.
8	955	—	—	(Same as for Line 7)	Chapman et al. 1975. Footnote to their Table I.
9	1003	608	538	Weighted average of selected radiometric and adjusted polarimetric diameters. See reference p. 203 for discussion.	Morrison 1977b. (Also adopted by Morrison 1977a,c,d, Zellner and Bowell 1977, and McCord 1978.)
10	1000	616	547	Average of recalibrated polarimetric diameter and the diameters in line No. 9 of this table.	Zellner et al. 1977.

TABLE II (Continued)

Asteroid Diameters Adopted by Various Authors

| No. | Diameters in Kilometers | | | Method | References and Remarks |
	1 Ceres	2 Pallas	4 Vesta		
11	1018	629	548	TRIAD.	Morrison 1978, p. 90.
12	1020	538	549	TRIAD — Value from occultation, line 13 this table.	Chapman and Zellner 1978. (Also adopted by Degewij 1978.)
13	— —	538 ± 12	— —	Single method. Stellar occultation and lightcurve analysis of published data.	Wasserman et al. 1978, 1979 Elliot et al. 1978.
14	987 ± 150	538 ± 50	544 ± 80	Occultation-lightcurve model for Pallas. Average of radiometric (Morrison 1977b) and polarimetric (Zellner et al. 1977) diameters.	This chapter. The assigned error bars are derived in the text.

[a]Senior author of work cited has designated this value as having been superseded by a more recent work.

Therefore we have decided not to use the results from micrometry. We have also omitted the diameters determined by speckle interferometry, because there is an unexplained difference between a result by this method and our adopted value for Pallas (see Table I). We adopt an unweighted average of the radiometric (Morrison 1977b) and the polarimetric (Zellner et al. 1977) diameters as the best estimate for Ceres and Vesta. Zellner et al. do not give polarimetric diameters explicitly, but we used their formula (their p. 1108) and the appropriate data in their Table 6 to calculate polarimetric diameters (see Table I). Our adopted values of diameter are listed in the last line of Table II, together with error bars to be explained below.

III. DENSITIES

Asteroidal volumes were calculated on the assumption that Ceres and Vesta were spherical. Our adopted diameter of Pallas is the diameter of a sphere that is equal in volume to the ellipsoidal model used for the shape of Pallas by Wasserman et al. (1979). The computation of the three volumes is therefore simple; but what are the accuracies of these volumes?

The formal error for Pallas' mean radius implies a volume error of 7 %. We feel that the real uncertainty of the volume is not described by this number. We know the limb profile of one projection of the body on the surface of the earth, and we have some additional information from photometry. However, we do not know details about the surface of Pallas; the shape of the body could be very irregular. Concerning the irregular body of Phobos (see the chapter by Veverka and Thomas) the surface coverage by imaging is essentially complete, but the relative uncertainty in the volume is still about 10 %, according to Veverka (1978, pp. 218-219), who mentions large uncertainty of the volume of Deimos, a body with a surface that is only known in part. In the case of Pallas, much depends upon the accuracy of the preliminary photometric determination of the pole of rotation (Schroll et al. 1976) and upon the assumed negligibility of albedo variations and topography. All considered we estimate the accuracy of Pallas' volume determination to be within about ± 30 %. Thus we attach a 10 % error bar to our adopted diameter.

Both Ceres and Vesta have lightcurves of small amplitude. Vesta's lightcurve has been found to be predominantly due to albedo variations (Degewij and Zellner 1978). These facts are compatible with a spherical shape for both asteroids. However, present data are not sufficient to rule out oblate spheroidal figures for either Ceres or Vesta with ratios of axial diameters as large as 1:1.2. This in itself contributes an additional 20 % uncertainty in the volume beyond the effect of errors inherent in the radiometric and polarimetric methods. Thus we adopt a 45 % uncertainty in the volumes of Ceres and Vesta. This corresponds to a 15 % error bar for our adopted diameter, as shown in Table II.

TABLE III

Asteroid Densities[a]

No.	Density (g cm^{-3})			Reference
	1 Ceres	2 Pallas	4 Vesta	
1	– –	– –	8 *1a*	Hertz 1968.
2	5 ± 1 *2a*	– –	5 *1b*	Schubart 1971a.
3	1.6 ± 0.7 *3c*	– –	2.5 ± 0.7 *1c*	D. Allen 1971.
4	2.1 ± 0.3 *4d*	2.8 ± 0.9 *4d*	3.1 ± 0.5 *1d*	Morrison 1974.
5	2.1 ± 1.0 *4e*	– –	3.0 ± 1.5 *1e*	Matson *et al.* 1976.
6	2.2 *4g*	1.9 *5g*	2.9 *1g*	Chapman *et al.*1978.
7	– –	2.8 ± 0.5 *5f*	– –	Wasserman *et al.* 1979.
8	2.3 ± 1.1 *6h*	2.6 ± 0.9 *6h*	3.3 ± 1.5 *6h*	This chapter.

[a]References for the data used in density determinations − Mass: (*1*) Hertz 1968, (*2*) Schubart 1971a, (*3*) Schubart 1970, (*4*) Schubart 1974, (*5*) Schubart 1975, (*6*) Sec. I of this chapter. Volume: (*a*) Barnard 1900*b*, (*b*) Dollfus 1970, (*c*) D. Allen 1971, (*d*) 1025 ± 50 km, 560 ± 30 km and 525 ± 25 km respectively for asteroids 1, 2 and 4, (*e*) average of Hansen 1976 and the polarimetric diameters in Chapman *et al.* 1975, (*f*) Wasserman *et al.* 1979, (*g*) Morrison 1977*b*, (*h*) Table II of this chapter.

The densities derived from our adopted asteroid masses and volumes are given in Table III. For comparison we also tabulate previously published values. The new density values reflect the continuing evolution and refinement of the methods for mass and size determination, and we believe that the values given are the best currently available.

If ratios of asteroidal diameters are formed, then the multiplicative calibration errors common to both diameter determinations will vanish. As a result, a density ratio can be computed which is more accurate than a direct ratio of the values in Table III. Morrison (1976) used this method in the computation of his Vesta/Ceres density ratio of 1.33 ± 0.17. We have considered the use of the ratio technique to obtain additional useful information. Unfortunately, the method does not mitigate the propagation of the principal errors that are in the figures (e.g. oblate spheroid) of Ceres and Vesta, in the absolute magnitudes, and in the masses. Of these the uncertainties in the figures are the most severe.

Future work is expected to improve greatly the densities of the asteroids. For example, the figures of Ceres and Vesta can be found from occultation observations and from direct imaging from space.

Acknowledgments. Following requests by H. G. Hertz and J. Schubart, many observers have made special efforts to obtain new observations of Arete or to re-reduce existing ones, P. K. Seidelmann, U. S. Naval Observatory, and R. Hertz, Roslyn Heights, N. Y., discovered the collection of observations of

Arete in the material left by H. G. Hertz, and a copy was kindly made available to Schubart. The IBM 360-44 and 370-168 computers of the University of Heidelberg's Rechenzentrum were used in the analysis. R. Lungwitz changed the programs developed for Schubart's earlier mass determinations to FORTRAN 4. We thank G. Veeder for many helpful discussions on asteroid size determinations. In part, this chapter presents the results of one phase of research carried out at the Jet Propulsion Laboratory, California Institute of Technology, sponsored by the National Aeronautics and Space Administration.

REFERENCES

Allen, D. A. 1970. Infrared diameter of Vesta. *Nature* 227: 158.
Allen, D. A. 1971. The method of determining infrared diameters. In *Physical Studies of Minor Planets*, ed. T. Gehrels (NASA SP-267, Washington, D. C.: U. S. Government Printing Office), pp. 41-44.
Barnard, E. E. 1895. Micrometrical determinations of the diameters of the minor planets Ceres (1), Pallas (2), Juno (3), and Vesta (4), made with filar micrometer of the 36-inch equatorial of the Lick Observatory, and on the albedos of those planets. *Mon. Not. Roy. Astron. Soc.* 56: 55-63.
Barnard, E. E. 1900*a*. On the diameter of Ceres and Vesta. *Mon. Not. Roy. Astron. Soc.* 60: 261-262.
Barnard, E. E. 1900*b*. The diameter of the asteroid Juno (3), determined with the micrometer of the 40-inch refractor of the Yerkes Observatory, with remarks on some of the other asteroids. *Mon. Not. Roy. Astron. Soc.* 61: 68-69.
Barnard, E. F. 1902. On the dimensions of the planets and satellites. *Astron. Nachr.* 157: 261-268.
Bowell, E.; Dollfus, A.; Zellner, B.; and Geake, J. E. 1973. Polarimetric properties of the lunar surface and its interpretation. Part 6. Albedo determinations from polarimetric characteristics. *Proc. Lunar Sci. Conf. IV.*, ed. W. A. Gose (New York: Pergamon Press), pp. 3167-3174.
Bowell, E., and Zellner, B. 1974. Polarizations of asteroids and satellites. In *Planets, Stars, and Nebulae Studied with Photopolarimetry*, ed. T. Gehrels (Tucson: University of Arizona Press), pp. 381-404.
Chapman, C. R., and Morrison, D. 1974. The minor planets, Sizes and mineralogy. *Sky and Tel.* 47: 92-95.
Chapman, C. R.; Morrison, D.; and Zellner, B. 1975. Surface properties of asteroids: A synthesis of polarimetry, radiometry, and spectrophotometry. *Icarus* 25: 104-130.
Chapman, C. R.; Williams, J. G.; and Hartmann, W. K. 1978. The asteroids. In *Ann. Rev. Astron. Astrophys.*, eds. G. Burbidge, D. Layzer and J. Phillips (Palo Alto, Calif.: Annual Reviews, Inc.) Vol. 16, pp. 33-75.
Chapman, C. R., and Zellner, B. 1978. Asteroid data. *Proc. Lunar Sci. Conf. IX.* (Oxford: Pergamon Press), "end paper."
Cruikshank, D. P., and Morrison, D. 1973. Radii and albedos of asteroids 1, 2, 3, 4, 6, 15, 51, 433, and 511. *Icarus* 20: 477-481.
Davis, D. R., and Bender, D. F. 1977. Asteroid mass determinations: A search for further encounter opportunities (abstract). *Bull. Amer. Astron. Soc.* 9: 502-503.
Degewij, J. 1978. *Photometry of faint asteroids and satellites.* Ph.D. dissertation, Leiden University.
Degewij, J., and Zellner, B. 1978. Asteroid surface variegation. (abstract). *Lunar Science IX.* The Lunar and Planet. Inst., pp. 235-237.
Dollfus, A. 1970. Diamètres des planètes et satellites. In *Surfaces and Interiors of Planets and Satellites,* ed. A. Dollfus (New York: Academic Press), pp. 45-139.
Dollfus, A. 1971. Diameter measurements of asteroids. In *Physical Studies of Minor Planets,* ed. T. Gehrels (NASA SP-267, Washington, D. C.: U. S. Government Printing Office), pp. 25-31.

Elliot, J. L.; Wasserman, L. H.; Millis, R. L.; and Bowell, E. 1978. The shape, albedo, and density of Pallas (abstract). *Bull. Amer. Astron. Soc.* 10: 596.
Fayet, G. 1949. Contribution à l'étude des proximités d'orbites dans le systeme solaire. *Ann. Bureau des Longitudes Paris* 12: B.1-B.156.
Hamy, M. 1899*a*. Mesure interférentielle des diamètres des satellites de Jupiter et de Vesta, effectuée au grand équatorial coudé de l'Observatoire de Paris. *Comptes Rendus* 128: 583-586.
Hamy, M. 1899*b*. Sur la mesure interférentielle des petits diamètres. *Bull. Astron.* 16: 257-274.
Hansen, O. L. 1976. Radii and albedos of 84 asteroids from visual and infrared photometry. *Astron. J.* 81: 74-84.
Hansen, O. L. 1977. An explication of the radiometric method for size and albedo determination. *Icarus* 31: 456-482.
Herget, P. 1953. Solar coordinates 1800-2000. *Astron. Papers Amer. Ephemeris, Washington* 14: 1-735.
Hertz, H. G. 1968. Mass of Vesta. *Science* 160: 299-300.
Jones, T. J., and Morrison, D. 1974. Recalibration of the photometric/radiometric method of determining asteroid sizes. *Astron. J.* 79: 892-895.
Kresák, L. 1977. Mass content and mass distribution of the asteroid system. *Bull. Astron. Inst. Czech.* 28: 65-82.
Matson, D. L. 1971. 1. Astronomical photometry at wavelengths of 8.5, 10.5, and 11.6 μm. 2. Infrared emilssion from asteroids at wavelengths of 8.5, 10.5 and 11.6 μm. Ph.D. Dissertation, California Inst. of Technology.
Matson, D. L.; Fanale, F. P.; Johnson, T. V.; and Veeder, G. J. 1976. Asteroids and comparative planetology. *Proc. Lunar Sci. Conf. VII* (Oxford: Pergamon Press), pp. 3603-3627.
Matson, D. L.; Veeder, G. J.; and Lebofsky, L. A. 1978. Infrared observations of asteroids from Earth and space. In *Asteroids: An Exploration Assessment,* eds. D. Morrison and W. C. Wells, NASA Conf. Publ. 2053, pp. 127-144.
McCord, T. B. 1978. Asteroid surface mineralogy: Evidence from Earth-based telescope observations. In *Asteroids: An Exploration Assessment,* eds. D. Morrison and W. C. Wells, NASA Conf. Publ. 2053, pp. 109-125.
Morrison, D. 1973. Determination of radii of satellites and asteroids from radiometry and photometry. *Icarus* 19: 1-14.
Morrison, D. 1974. Radiometric diameters and albedos of 40 asteroids. *Astrophys. J.* 194: 203-212.
Morrison, D. 1976. The densities and bulk compositions of Ceres and Vesta. *Geophys. Res. Lett.* 3: 701-704.
Morrison, D. 1977*a*. Radiometric diameters of 84 asteroids from observations in 1974-1976. *Astrophys. J.* 214: 667-677.
Morrison, D. 1977*b*. Asteroid sizes and albedos. *Icarus* 31: 185-220.
Morrison, D. 1977*c*. Sizes and albedos of the larger asteroids. In *Comets, Asteroids, Meteorites,* ed. A. H. Delsemme (Toledo, Ohio: University of Toledo Press), p.p. 177-184.
Morrison, D. 1977*d*. Diameters of minor planets. *Sky and Tel.* 53: 181-183.
Morrison, D. 1978. Physical observations and taxonomy of asteroids. In *Asteroids: An Exploration Assessment,* eds. D. Morrison and W. C. Wells, NASA Conf. Publ. 2053, pp. 81-97.
Schroll, A.; Haupt, H. F.; and Maitzen, H. M. 1976. Rotation and photometric characteristics of Pallas. *Icarus* 27: 147-156.
Schubart, J. 1970. The mass of Ceres, *IAU Circ.* 2268.
Schubart, J. 1971*a*. Asteroid masses and densities. In *Physical Studies of Minor Planets,* ed. T. Gehrels (NASA SP-267, Washington, D. C.: U. S. Government Printing Office), pp. 33-39.
Schubart, J. 1971*b*. The planetary masses and the orbits of the first four minor planets. *Celes. Mech.* 4: 246-249.
Schubart, J. 1972. Massenbestimmung von Planetoiden. *Mitt. Astron. Ges.* 31: 182-183.
Schubart, J. 1974. The masses of the first two asteroids. *Astron. Astrophys.* 30:

289-292.

Schubart, J. 1975. The mass of Pallas. *Astron. Astrophys.* 39: 147-148.

Schubart, J. 1976. New reduction and collection of meridian observations of Ceres and Pallas. *Astron. Astrophys. Suppl.* 26: 405-413.

Schubart, J. 1977. Present status of mass determination of minor planets (abstract). *Trans. IAU* 16B: 166.

Veverka, J. 1971. The polarization curve and the absolute diameter of Vesta. *Icarus* 15: 11-17.

Veverka, J. 1973. Polarimetric observations of 9 Metis, 15 Eunomia, 89 Julia, and other asteroids. *Icarus* 19: 114-117.

Veverka, J. 1978. Imaging asteroids: Some lessons learned from the Viking investigation of Phobos and Deimos. In *Asteroids: An Exploration Assessment,* eds. D. Morrison and W. C. Wells, NASA Conf. Publ. 2053, pp. 207-223.

Wasserman, L. H.; Millis, R. L.; Franz, O. G.; Bowell, E.; White, N. M.; Giclas, H. L.; Elliot, J. L.; Dunham, E.; Mink, D.; Baron, R.; Honeycutt, R. K.; Henden, A. A.; Kephart, J. E.; A'Hearn, M. F.; Reitsema, H.; Radick, R.; and Taylor, G. E. 1978. A Reliable Diameter for Pallas (abstract). *Bull. Amer. Astron. Soc.* 10: 595.

Wasserman, L. H.; Millis, R. L.; Franz, O. G.; Bowell, E.; White, N. M.; Giclas, H. L.; Martin, L. J.; Elliot, J. L.; Dunham, E.; Mink, D.; Baron, R.; Honeycutt, R. K.; Henden, A. A.; Kephart, J. E.; A'Hearn, M. F.; Reitsema, H. J.; Radick, R.; and Taylor, G. E. 1979. The diameter of Pallas from its occultation of SAO 85009. *Astron. J.* 84: 259-268.

Widorn, T. 1967. Zur photometrischen Bestimmung der Durchmesser der Kleinen Planeten. *Annal. Universitäts-Sternwarte Wien* 27: 3: 111-119.

Worden, S. P., and Stein, M. K. 1979. Angular diameter of the asteroids Vesta and Pallas determined from speckle observations. *Astron. J.* 84: 140-142.

Worden, S. P.; Stein, M. K.; Schmidt, G. D.; and Angel, J. R. P. 1977. The angular diameter of Vesta from speckle interferometry. *Icarus* 32: 450-457.

Zellner, B., and Bowell, E. 1977. Asteroid compositional types and their Distributions. In *Comets, Asteroids, Meteorites,* ed. A. H. Delsemme (Toledo, Ohio: University of Toledo Press), pp. 185-197.

Zellner, B.; Gehrels, T.; and Gradie, J. 1974. Minor planets and related objects. XVI. Polarimetric diameters. *Astron. J.* 79: 1100-1110.

Zellner, B., and Gradie, J. 1976. Minor planets and related objects. XX. Polarimetric evidence for the albedos and compositions of 94 asteroids. *Astron. J.* 81: 262-280.

Zellner, B.; Leake, M.; Lebertre, T.; Duseaux, M.; and Dollfus, A. 1977. The asteroid albedo scale. I. Laboratory polarimetry of meteorites. *Proc. Lunar Sci. Conf. VIII* (Oxford: Pergamon Press), pp. 1091-1110.

DIRECT DETERMINATION OF ASTEROID DIAMETERS
FROM OCCULTATION OBSERVATIONS

R. L. MILLIS
Lowell Observatory

and

J. L. ELLIOT
Massachusetts Institute of Technology

Of the available Earth-based techniques for determining asteroid diameters, observation of stellar occultations involving asteroids is clearly the most direct. The high degree of accuracy achievable by this method has already been demonstrated in the case of Pallas, whose mean diameter has been measured with a standard error of ±2 %. In this chapter the problems, results and prospects of the stellar occultation technique are reviewed. It is shown that, with the use of a network of small, portable telescopes, the method is currently applicable to a large number of asteroids. The best results can be expected for asteroids of large angular diameter and regular shape. The potential of lunar occultation observations for asteroid diameter measurements is also briefly discussed.

Size is a fundamental property of any planetary body which must be known accurately to compute albedo and bulk density. Because the apparent angular size of even the larger asteroids is less than the typical seeing disk, classical visual micrometric techniques of diameter determination have been applied to only a handful of asteroids (Dollfus 1971). It is difficult to assess the uncertainties associated with the visual measurements, but consideration of the Rayleigh limits of the telescopes employed suggests that the resulting diameters are uncertain by at least 10% to 50%, depending on the object.

In the early 1970's two indirect techniques for determining the size of small, airless bodies were developed (see chapters by Morrison and Lebofsky, and by Dollfus and Zellner). These methods — one based on infrared radiometry, the other on polarimetry — have been applied to nearly 200 asteroids (Morrison 1977). Although diameters derived by the radiometric and polarimetric methods are now, as a rule, mutually consistent, it must be remembered that significant systematic errors could be present in both (Millis *et al.* 1978). Therefore, a primary objective of direct asteroid diameter determinations is an accurate calibration of the radiometric and polarimetric techniques.

Classical visual techniques aside, there are at present three methods whereby asteroid diameters can be "directly" measured from Earth using optical telescopes. One of these, based on interferometry, is described in the chapter by Worden. A second involves photometric observation of occultations of asteroids by the moon. Strictly speaking, neither the interferometric technique nor the lunar occultation technique is truly a direct method, since both require the asteroid's shape, limb darkening and albedo distribution to be assumed in order to derive the diameter. The third technique, based on photometry of stellar occultations involving asteroids, is more direct and under favorable conditions can yield both size and shape without the need of any simplifying assumptions (Elliot 1979). The mean diameter of Pallas, for example, has been measured in this way to within a standard error of ±2 % (Wasserman *et al.* 1979).

In this review we will discuss the stellar occultation method, its results and prospects as applied to the problem of accurately determining diameters of asteroids. Application of the lunar occultation technique to this problem will also be addressed.

I. STELLAR OCCULTATIONS

A. Frequency

Each asteroid casts a shadow in the light of every star in the sky. Viewed in this way, it is not surprising that such shadows frequently intersect the earth, resulting in potentially observable occultations. The number of occultations to be expected per year can be readily estimated for any particular asteroid on the basis of its angular diameter, horizontal parallax, mean motion, and the density of stars in the relevant region of the sky (O'Leary 1972). In Table I we have listed the estimated annual number of stellar occultations for thirteen of the larger asteroids. The limiting stellar magnitude was chosen to give at least a 5 % change in the combined

TABLE I

Expected Mean Frequency of Asteroid Occultations

Name	No. of occultations per year	Search limiting magnitude
1 Ceres	2.0	10.2
2 Pallas	4.4	11.0
4 Vesta	0.6	9.1
10 Hygiea	13.8	12.2
31 Euphrosyne	16.6	12.4
704 Interamnia	18.2	12.5
511 Davida	20.5	12.6
65 Cybele	29.6	13.0
52 Europa	22.4	12.7
451 Patientia	9.6	11.8
15 Eunomia	7.1	11.5
16 Psyche	13.8	12.2
3 Juno	7.8	11.6

TOTAL   166.4

brightness of the star and asteroid at immersion and emersion. It is evident from the large number of occultations involving the limited sample considered in Table I that observable occultations of stars by asteroids are a daily occurrence.

The cross-section of the asteroid's shadow has the same size and shape as the apparent profile of the asteroid. The shadow sweeps across the surface of the earth, mapping out an occultation track similar to the one shown in Fig. 1. Assuming clear skies, any observer within the track will see an occultation, the duration of which is determined by the location of the observer relative to the centerline of the track, the apparent velocity of the asteroid relative to the star, and the size and shape of the asteroid. In any case, each observed duration yields the length of one chord across the asteroid. If the asteroid's profile is circular, two properly spaced chords suffice to determine the diameter. If the profile is elliptical, at least three observing sites are needed. An occultation involving a highly irregular asteroid must be observed at many sites scattered across the track, if size and shape are to be accurately derived.

B. Predictions

Widespread application of the stellar occultation technique is at present constrained by the limited accuracy and scope of available predictions (see Dunham *et al.* 1979*a*). Potential occultations are most easily identified by

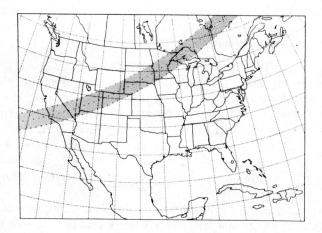

Fig. 1. Predicted nominal track of an occultation of AGK3+0°1022 by Juno on 11 December 1979 (Taylor 1978). This track is based solely on the catalog position of the star and the ephemeris of Juno. Photographic astrometry near the time of the occultation is required to determine the precise location of the track.

comparison of asteroid ephemerides with star catalogs. Taylor of the Royal Greenwich Observatory has provided predictions based on catalog searches for several years (e.g., Taylor 1978). It should be stated, in fact, that most stellar occultations involving asteroids which have been observed were first predicted by Taylor. There is no doubt that this approach to the prediction problem will continue to be very fruitful. At the same time, one must admit that catalog searches suffer from a basic limitation: namely, that observable occultations involving stars fainter than the catalog limits will be missed. It is apparent in Table I that many occultations in this category are to be expected. Bowell and Wasserman, in an effort to identify some of these occultations, have begun measuring the positions of candidate stars along the paths of selected asteroids on photographic plates. Their measurements yielded, for example, three potentially observable occultations during 1979/80 involving Ceres (Bowell and Wasserman 1979). Two of the three occultations involved stars not in the SAO or AGK3 catalogs.

Regardless of the method used to initially predict an occultation, careful refinement of the prediction a short time prior to the occultation is almost invariably required. An uncertainty of only ±0.2 arcsec in the predicted separation of the star and asteroid at closest approach will, for an asteroid at a typical distance from Earth of 2 AU, result in a cross-track uncertainty in the position of the occultation track of at least 300 km. Many occultation tracks are narrower than this figure. Random and zonal errors in star catalogs and uncertainties in asteroid ephemerides combine to make accurate long-range prediction of stellar occultations extremely difficult. Usually one must wait until the two objects are close enough in the sky to be photographed on the same plate using a high-quality astrometric telescope.

C. Observational Requirements

Because of the narrowness of asteroid occultation tracks and the necessity of responding to weather conditions and last-minute changes in predictions, effective coverage of these events in most instances requires that permanent observatories be augmented with portable telescopes. This fact is illustrated by Fig. 1, which shows the nominal track of an occultation of AGK3+0°1022 by Juno on 11 December 1979, as predicted by Taylor (1978). Although the track is relatively wide and goes through the telescope-rich southwestern United States, it crosses only four or five large-to-medium-sized observatories.

The instrumental requirements for observing stellar occultations involving asteroids are minimal. Basically one wishes to determine the times of immersion and emersion to an accuracy of a few tenths of a second, a task which is easily accomplished with a simple photoelectric photometer equipped with a DC amplifier and a strip-chart recorder. More elaborate (and more capable) systems have been described by several authors (c.f., Elliot et al. 1975; Wasserman et al. 1977; Hubbard et al. 1977).

Visual Observations. Owing to the scarcity of photoelectrically equipped portable telescopes, many asteroid occultation observations have been made visually (O'Leary et al. 1976; Taylor and Dunham 1978). Because of the subjective nature of visual observations, it is difficult, especially for one who has never attempted such measurements, to assess their reliability. With this fact in mind, we have performed a simple experiment aimed at exploring some of the parameters affecting visual occultation observations.

In our experiment, an illuminated pinhole was focused on the entrance aperture of a conventional photoelectric photometer. Ten percent of the light from this artificial star was reflected into a viewing eyepiece by means of a beam splitter while the remainder fell on the photomultiplier. During the experiment, an observer would view the artificial star through the eyepiece while an operator in another room caused the brightness of the light source to vary in a step-wise fashion: first dimming, then returning several seconds later to full brightness. The "unocculted" brightness of the artificial star was comparable to that of a 5th-magnitude star viewed directly through a 25-cm telescope. When the observer perceived a change in brightness, he pressed a button. Both the amplified signal from the photomultiplier and the observer's responses were simultaneously recorded on a strip chart as seen in Fig. 2.

Each "test" lasted five to ten minutes, during which time several simulated occultations of varying lengths and depths were produced. Ten observational astronomers and four non-observers participated. The following general conclusions emerged: (1) Experienced observers performed better than non-observers. (2) Most observers signaled more than one false alarm,

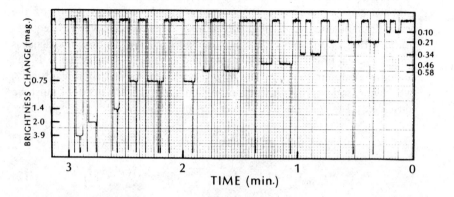

Fig. 2. Sample output from an experiment aimed at determining the minimum brightness change required for occultations to be reliably detected visually. The sharp negative spikes superimposed on the lightcurve are signals by the observer that he detected a brightness change. Note that the observer consistently detected changes at a level of 0.75 mag. Also notice the erroneous "detections" when the signal had not in fact changed.

i.e., they perceived a brightness change when none had occurred. (3) Performance varied among the experienced observers, but on average a brightness change in excess of 0.75 mag was required to be detected without fail. (4) Brightness changes as small as 0.2 mag were almost never detected, while variations smaller than 0.5 mag were frequently missed. (5) When the eyepiece was in a slightly awkward position, as often would be the case in an actual observing situation, performance was degraded.

Although we do not pretend that the results of this experiment are entirely conclusive, we believe that performance at the telescope in general is likely to be worse because of the effects of seeing, scintillation, wind, cold, fatigue, etc. and because the star will usually be fainter than the one simulated in our experiment. Consequently, visual occultation observations are unlikely to be consistently reliable unless the star is at least as bright as the asteroid, and then only if the observer is experienced.

Photoelectric Observations. While visual observations, in our judgment, can provide a useful supplement to photoelectric measurements, the latter are clearly to be preferred. Not only will properly designed photoelectric equipment give more accurate timing, but the resulting precise lightcurve will usually allow one to say without question whether an occultation did or did not occur. Additionally, many occultations for which the brightness change is too small to be reliably detected visually are easily recorded photoelectrically. The 29 May 1978 occultation of SAO 85009 by Pallas (Wasserman *et al.* 1979) is a case in point.

In evaluating the observability of any particular occultation, an intuitive estimate can be made simply by computing the expected depth of the occultation lightcurve. A quantitative judgment, however, requires consideration of the expected signal-to-noise ratio. In this case, the signal is the change in brightness at immersion and emersion, and the noise is the random variation in the combined brightness of the star and asteroid. Usually, two major sources of noise, photon statistics and scintillation, must be considered. The contribution of each can be estimated from two equations:

$$(S/N)_{photon} = \tfrac{1}{2}\sqrt{\pi q \Delta \lambda N_*} \; D \left(1+10^{0.4\,(m_* - m_a)}\right)^{-\frac{1}{2}} \tag{1}$$

$$(S/N)_{scint} = 43 \, D^{\frac{2}{3}} M^{-\frac{3}{2}} e^{\frac{h}{8000}} \left(1+10^{0.4\,(m_* - m_a)}\right)^{-1} \tag{2}$$

where (S/N) is the signal-to-noise ratio for a one-second integration, N_* is the number of photons per second per cm^2 per angstrom from the star outside the earth's atmosphere (see Tüg et al. 1977), q is the fraction of photons reaching the top of the atmosphere which actually result in output pulses from the detector, $\Delta\lambda$ is the passband in angstroms, D is the diameter of the telescope in centimeters, M is the air mass through which the observation is made, h is the observer's altitude in meters, and m_* and m_a are the apparent magnitudes in the passband of the observations of the star and asteroid, respectively. The derivation of Eqs. 1 and 2 follows Young (1974) and Elliot (1977).

Equations 1 and 2 do not take into consideration contributions to the noise from dark current or sky brightness. For well-designed equipment the former should be negligible, while the latter depends critically on the position and phase of the moon, the photometer's entrance aperture size, and the passband. For a dark sky, the skylight entering an aperture one arcmin in diameter is roughly equivalent to the intensity of a star with $V = +14$ mag.

Using signal-to-noise ratios obtained from Eq. 1 or 2, it is possible to estimate the accuracy with which the duration of an occultation can be measured. For low signal-to-noise ratios, the Fresnel fringes formed by the asteroid's limb will not be detectable and the occultation lightcurve at immersion and emersion can be represented by a discontinuous change in mean signal. Because of the discontinuity in the signal, the usual nonlinear least-squares method is not strictly applicable to finding the expected error in the time of immersion (or emersion). We have adopted the following approach. The data are averaged into bins of duration Δt such that the signal-to-noise ratio for the averaged data is 3. With a signal-to-noise ratio of 3, the immersion or emersion time, t_o, should be reliably determined within a bin width Δt. Hence, we can make the approximation that the uncertainty, δt_o, in t_o equals Δt. For data whose signal-to-noise ratio for a one-second integration, S/N, is given by Eq. 1 or 2, one can write

$$\delta t_{\mathrm{o}} = 9(S/N)^{-2} \tag{3}$$

Presumably the error in the times of immersion and emersion will be uncorrelated so that the error in the duration of the occultation will be $\delta t_{\mathrm{o}} \sqrt{2}$.

To obtain accurate asteroid diameters we shall require that the duration of the occultation, T, be determined with an uncertainty of 1 % or less. This requirement can be expressed as

$$\frac{\delta t_{\mathrm{o}} \sqrt{2}}{T} \leqslant 0.01. \tag{4}$$

Substituting Eq. 3 into Eq. 4 and solving for S/N, we find that the signal-to-noise ratio for a one-second integration must satisfy the condition

$$S/N \geqslant \frac{36}{\sqrt{T}} \cdot \tag{5}$$

This criterion can be related to the appearance of the data by reference to Fig. 3. This figure shows four simulated occultation lightcurves, each having a duration, T, of 30 seconds. Individual data points represent 0.1-sec integrations. Different amounts of random noise have been added to each lightcurve. Applying the condition expressed by Eq. 5 to these data, one finds that the signal-to-noise ratio for a one-second integration must be 6.6 or larger. Hence, the lightcurves with $S/N = 10$ or 20 should yield durations of the required accuracy.

The stellar magnitude limits at which S/N will be 100, 20, and 4, assuming a 35-cm telescope and a 1000-Å passband centered at 5500 Å, are plotted in Fig. 4 as a function of the V magnitude of the occulting asteroid. Note that occultation observations with a telescope of this size are scintillation noise limited for asteroids brighter than about 8 mag. Typically, occultations of stars by asteroids have durations of a few tens of seconds which, according to Eq. 5, means that useful results can be derived when the signal-to-noise ratio is ~ 4 or greater. One can therefore conclude from an examination of Fig. 4 that only when the star is much fainter than the asteroid or when both are very faint is a relatively small portable telescope incapable of giving high-quality results.

The validity of the above analysis of the expected signal-to-noise ratio and timing accuracy can be tested by comparison with actual observations of the 29 May 1978 occultation of SAO 85009 by Pallas. Observations of this occultation by Reitsema with a 35-cm telescope at Behlen Observatory gave the occultation duration with a quoted uncertainty of 0.1 % (Wasserman *et al.* 1979). Substituting this degree of uncertainty on the right-hand side of Eq. 4 and proceeding as described in the derivation of Eq. 5, one finds that a signal-to-noise ratio of 20 was required, in good agreement with that expected for this event from Fig. 4.

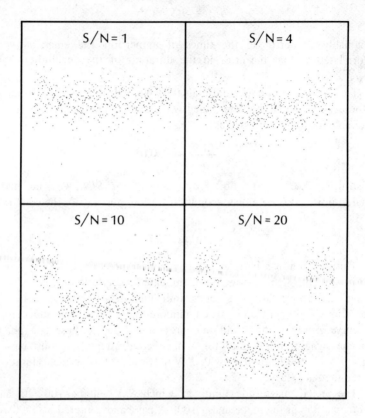

Fig. 3. Artificially generated occultation lightcurves. Data points represent 0.1-second integrations, while the indicated signal-to-noise ratios are for 1-second integrations.

D. Method of Analysis

In analyzing stellar occultation observations it is convenient to define a rectangular coordinate system whose origin is at the center of the earth and whose z-axis is instantaneously parallel to a line connecting the occulting body with the star. The x,y plane (see Fig. 5) is called the "fundamental plane," and the "shadow" cast on this plane has at all times the same size and shape as the apparent profile of the occulting body seen in the sky. It is easy to show (Smart 1960) that the coordinates of the center of the shadow on the fundamental plane are given by

$$x = \Delta \cos \delta_a \sin (\alpha_a - \alpha_*) \tag{6}$$

$$y = \Delta [\sin \delta_a \cos \delta_* - \cos \delta_a \sin \delta_* \cos (\alpha_a - \alpha_*)], \tag{7}$$

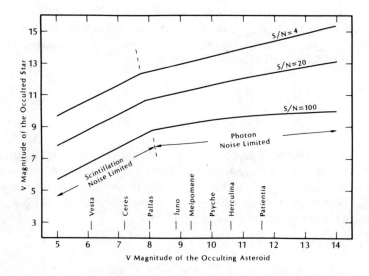

Fig. 4. V magnitude of the occulted star at which 35-cm telescopes yield signal-to-noise ratios for a 1-second integration of 4, 20, and 100 plotted as a function of the V magnitude of the occulting asteroid.

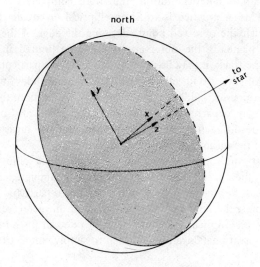

Fig. 5. Illustration of the fundamental or Besselian plane and the coordinate system used in occultation analyses. The positive z-axis points continuously toward the star to be occulted. The shaded plane is orthogonal to the z-axis. The positive x-axis points eastward, while the y-axis is the projection of the earth's axis of rotation onto the fundamental plane.

where Δ is the asteroid's geocentric distance in units of the earth's equatorial radius, and α_a, δ_a and α_*, δ_* are the right ascension and declination of the asteroid and star, respectively. Similarly, the projected coordinates on the fundamental plane (ξ,η) of an observer are given by

$$\xi = \rho \cos \phi' \sin H \qquad (8)$$

$$\eta = \rho \left[\cos \delta_* \sin \phi' - \sin \delta_* \cos \phi' \cos H\right] \qquad (9)$$

where ρ is the observer's geocentric distance in units of the earth's equatorial radius, ϕ' is the observer's geocentric latitude, and H is the hour angle of the star. If we then transform to an x,y coordinate system in the fundamental plane whose origin is fixed with respect to the center of the shadow, the projected coordinates of an observer (ξ',η') are

$$\xi' = \xi\text{-}x \qquad (10)$$

$$\eta' = \eta\text{-}y. \qquad (11)$$

The coordinates (ξ',η') of an observer at the time of immersion or emersion define a point on the edge of the shadow or equivalently on the limb of the asteroid. Obviously, the more observers there are distributed across the path of an occultation, the more completely will the limb be mapped. The usual practice is to fit an elliptical or circular profile by least squares through the observed points defining the edge of the shadow in the fundamental plane. If the observations are very limited in number or in coverage of the track, it may be necessary to assume a particular value for the eccentricity of the fitted ellipse. While the amplitude of the asteroid's lightcurve can provide guidance in making this assumption, the true shape of the asteroid may depart substantially from that assumed. Consequently the formal error resulting from the fit is not an accurate measure of the true uncertainty in the mean diameter of the asteroid.

If the asteroid's profile is in fact approximately elliptical and if the coverage of the occultation track is nearly complete, both size and eccentricity can be derived (see Fig. 6). On the other hand, especially for smaller asteroids, the limb may be very irregular. In such cases it makes little sense to force an elliptical solution. The best one can do is to strive for high density of observational coverage and then to simply connect the points.

E. Summary of Results

Existing published observations of occultations of stars by asteroids are summarized in Table II. The first two events were observed at just a single site and therefore resulted only in lower limits on the diameters of the two

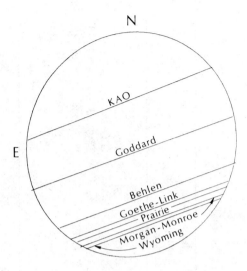

Fig. 6. Seven chords across Pallas derived from the 29 May 1978 occultation of SAO
85009 and the best-fitting elliptical limb profile. The labels indicate the observatories
from which observations were made (see Wasserman *et al.* 1979). The KAO is the
Kuiper Airborne Observatory, a 91.5-cm telescope on a C 141 aircraft.

asteroids involved (Taylor 1962). In view of the extreme narrowness of the
occultation track, it at first glance seems remarkable that the 24 January
1975 Eros event was observed at all. However, as discussed in Sec. I. G below,
it is the angular diameter of the asteroid, not its linear diameter, which
determines the precision with which an occultation track can be located. Eros
was at the time only 0.15 AU from Earth and subtended nearly 0.2 arcsec.
Eight visual observers did record the occultation; but, because of the
relatively short duration of the occultation (~3 sec) and the
less-than-optimum distribution of observers, an accurate size for Eros could
not be derived (O'Leary *et al.* 1976). The 5 March 1977 occultation of γ Ceti
A by Hebe was again observed exclusively by visual means. Observations were
made at two sites, with two observers at one site and three at the other. The
resulting diameter has a small formal uncertainty, but depends strongly on
the assumed shape (Taylor and Dunham 1978).

   Without a doubt the best-observed stellar occultation involving an
asteroid was that of SAO 85009 by Pallas on 29 May 1978 (Wasserman *et al.*
1979). Seven photoelectric observations distributed across the occultation
track are well fitted by an elliptical profile having semimajor and semiminor
axes of 279.5 ± 2.9 km and 262.7 ± 4.5 km, respectively (see Fig. 6). At the
time of the occultation Pallas was seen nearly pole on so that, to a good
approximation, the two axes of the apparent ellipse obtained from the
occultation data equal the longer and shorter equatorial axes of the asteroid.
Through an analysis of the published lightcurve maxima and minima when

TABLE II

Summary of Observed Stellar Occultations Involving Asteroids

Asteroid	Occultation	Occulted Star	No. of Chords[a]	Mean Diameter (km)	Reference
3 Juno	19 Feb 58	BD+6°808	1v	>110	Taylor 1962
2 Pallas	2 Oct 61	BD-5°5863	1p	>430	Taylor 1962
433 Eros	24 Jan 75	κ Gem A	8v	12-23	O'Leary et al. 1976
6 Hebe	5 Mar 77	γ Ceti A	2v	186[b]	Taylor & Dunham 1978
2 Pallas	29 May 78	SAO 85009	7p	538±12	Wasserman et al. 1979
532 Herculina	7 Jun 78	SAO 120774	1p/2v	217[b]	Bowell et al. 1978
3 Juno	19 Jul 78	SAO 144070	1v	>256	Dunham & Sheffer 1979
18 Melpomene	11 Dec 78	SAO 114159	6p/2v	135[b]	Dunham et al. 1979

[a]p = photoelectric; v = visual

[b]The actual uncertainty in these values is difficult to evaluate because of limited observational coverage and/or irregular shape. The formal uncertainties quoted by the authors are based on assumed shapes.

Pallas has appeared nearly equator on, the length of the polar radius was found to be 266 ± 15 km. This error is somewhat larger than those for the radii obtained directly from the occultation data because it is not certain how much of the brightness variation is due to the shape of Pallas and how much is due to nonuniform albedo of its surface. This question could be resolved by observation of an occultation by Pallas at a time when its polar axis is nearly perpendicular to the Earth-Pallas line.

Accounting for the uncertainty in the polar axis, the mean diameter of Pallas was found to be 538 ± 12 km by Wasserman *et al.* (1979). We believe that this error is a realistic estimate of the true uncertainty in the mean diameter of Pallas because of the good coverage of the occultation track, the internal consistency of the occultation data, and the procedure used for obtaining the polar axis. In our view, the ±50-km error assigned to the occultation diameter of Pallas by Schubart and Matson (see their chapter in this book) has been unnecessarily inflated. For the revised mass of Pallas given by Shubart and Matson, this diameter leads to a density estimate for Pallas of 2.6 ± 0.5 gm cm^{-3}.

The 7 June 1978 occultation of SAO 120774 by Herculina was observed from three sites, all located well south of the center of the track (Bowell *et al.* 1978). Because of the limited coverage, the diameter quoted in Table II is sensitive to the assumed elliptical shape. Only a single visual chord was obtained during the 19 July 1978 occultation of SAO 144070 by Juno (Dunham and Sheffer 1979), but the resulting lower limit to Juno's diameter is sufficiently close to the radiometric/polarimetric value (see Table III) to be useful. Better coverage was obtained of the 11 December 1978 occultation of SAO 114159 by Melpomene (Dunham *et al.* 1979b). The observations indicate an irregularly shaped body, making a precise diameter determination difficult. An interesting aspect of this event, however, is that the occulted star is a binary, thereby giving two chords from each observing site.

F. Satellites of Asteroids

An unexpected by-product of asteroid occultation observations has been the discovery of evidence that some asteroids may have satellites. The first photoelectric evidence surfaced during the 7 June 1978 occultation of SAO 120774 by Herculina. A few minutes prior to the main occultation, the flat-bottomed dip shown in Fig. 7a occurred on a strip-chart record of the occultation obtained with Lowell Observatory's 1.06-m reflector (Bowell *et al.* 1978). This dip "lines up" with one of six secondary occultations reported by McMahon, a visual observer at Boron, California. Six months later, during the 11 December 1978 Melpomene event, Williamon of Fernbank Science Center, who was well outside the main occultation track, recorded an apparent secondary occultation, shown in Fig. 7b (Williamon 1978).

TABLE III

Number of Observing Sites Required

Asteroid	Diameter (km)	Mean opposition distance from Earth (Δ, AU)	Q (for $\sigma = 0''.05$)	No. of observing sites required (N_T) for prediction uncertainty (σ)		
				$\sigma = 0''.05$	$\sigma = 0''.2$	$\sigma = 0''.5$
1 Ceres	1025	1.77	0.13	4	5	7
4 Vesta	555	1.36	0.18	4	6	9
2 Pallas	538[a]	1.77	0.24	4	6	11
10 Hygiea	443	2.14	0.35	5	8	14
704 Interamnia	338	2.06	0.44	5	9	17
511 Davida	335	2.18	0.47	5	9	18
65 Cybele	311	2.43	0.57	5	10	20
52 Europa	291	2.10	0.52	5	10	19
451 Patientia	281	2.07	0.53	5	10	19
31 Euphrosyne	270	2.15	0.58	5	10	21
15 Eunomia	261	1.64	0.46	5	9	17
324 Bamberga	256	1.69	0.48	5	9	18
107 Camilla	252	2.49	0.72	6	12	25
87 Sylvia	251	2.48	0.72	6	12	25
45 Eugenia	250	1.72	0.50	5	9	18
3 Juno	249	1.67	0.49	5	9	18
16 Psyche	249	1.92	0.56	5	10	20
24 Themis	249	2.13	0.62	5	11	22
532 Herculina	217[a]	1.77	0.59	5	11	21
6 Hebe	186[a]	1.42	0.55	5	10	20
18 Melpomene	135[a]	1.30	0.70	6	12	24
433 Eros	23[a]	0.46	1.45	8	21	47

[a]From Table II. Other diameters are from the TRIAD file (Part VII in this book).

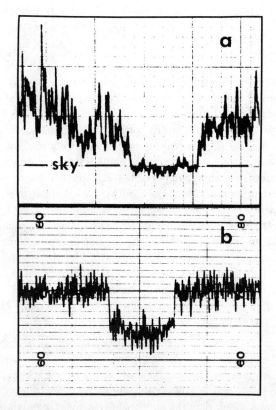

Fig. 7. Two lightcurves of a possible secondary occultation. Fig. 7a, lightcurve observed by Bowell *et al.* (1978) prior to the occultation of SAO 120774 by 532 Herculina. Fig. 7b, lightcurve observed by Williamon (1978) during the 11 December 1978 appulse of 18 Melpomene and SAO 114159.

Although some have contended that these observations constitute essentially definite proof of the existence of satellites around Herculina and Melpomene (Dunham *et al.* 1979b; Binzel and Van Flandern 1979; see the chapter by Van Flandern *et al.*), it should be remembered that both observations were made under difficult circumstances. In the case of the Lowell observations, photometric conditions were excellent, but Herculina was only 3° above the horizon. Williamon's measurements were made on a night of variable cloudiness. The photometric record indicates that the transparency was stable at the time of the apparent occultation, but conditions deteriorated shortly thereafter, preventing an accurate determination of the sky brightness level. For this reason there is no way of determining whether the occultation lightcurve has the expected depth. In our opinion, the question of the existence of minor planet satellites cannot be conclusively resolved on the basis of existing observations. The question is nevertheless intriguing and will be answered eventually through occultation observations, interferometry, or direct imaging from spacecraft.

G. Prospects

The viability of determining asteroid diameters by observation of stellar occultations has been clearly established by the four successful observations of such events during 1978 (see Table II). Furthermore, it is interesting to note that one individual, A'Hearn, observed three of these occultations. Looking to the future, we can expect more frequent, highly accurate diameter determinations as occultation prediction methods improve and as more photoelectrically equipped portable telescopes become available.

Clearly, accurate predictions and a network of portable telescopes are the keys to efficient observations of stellar occultations involving asteroids. The number of telescopes required to insure adequate coverage of an occultation is proportional to the uncertainty in the predicted angular separation of the asteroid and star at closest approach, σ, divided by the angular diameter of the asteroid. This ratio, which we will call Q, is readily computed from Eq 12, where Δ is the Earth-asteroid separation (in km) at the time of occultation, d is the estimated diameter of the asteroid (in km), and σ is expressed in radians:

$$Q = 2\sigma\Delta/d. \tag{12}$$

Assuming that three telescopes spaced across the track at intervals of one-third the track width provide the minimum acceptable coverage for a diameter determination, then the number of telescopes required is given by

$$N_T = 3(Q+1). \tag{13}$$

Table III lists N_T as a function of σ for a number of asteroids. We have set Δ equal to the mean opposition distance and have taken the diameters from Table II or from the TRIAD file (see Part VII in this book). The numbers given in Table III are in a sense typical values. Circumstances of individual occultations may differ significantly, particularly if the asteroid is in an eccentric orbit or if it is far from opposition. It is apparent in Table III that the error associated with occultation predictions must in general be well under $\pm0''.5$ if one is to have a reasonable chance of obtaining adequate observational coverage. Fortunately, experience indicates that an uncertainty of about $\pm0''.2$ is routinely achievable, while an uncertainty as small as $\pm0''.05$ is possible. Therefore, it appears that stellar occultations involving almost all the asteroids listed in Table III, as well as many others, are realistically observable with a modest number of portable telescopes.

In searching for future occultations, particular attention should be given to Ceres and Vesta. Not only is Q very small for these objects, but their masses are known (Hertz 1968; Schubart 1974; see the chapter by Schubart and Matson), and they are large enough that substantial departures from a regular shape are unlikely. Likewise, 10 Hygiea is an excellent candidate for occultation studies, and observable occultations involving this asteroid should occur frequently (Table I).

As one goes to smaller and smaller asteroids, the observational difficulties increase, depending on the distance of the asteroid as described above. In addition, the difficulty of interpreting the observations may also increase due to the higher probability of the asteroid having an irregular shape (see the chapter by Burns and Tedesco). Although it is possible, in principle, to determine the size and shape of the apparent profile of an irregular asteroid, a large number of telescopes is required. The value of such a measurement is doubtful anyway, since it would pertain only to one aspect of the asteroid and could not be used to compute mean density, nor could it be used to calibrate the radiometric and polarimetric methods unless simultaneous observations were obtained by those techniques. The amplitude and shape of an asteroid's lightcurve provide evidence of any sizeable departures from a regular shape. It will probably be most productive in occultation work to concentrate on asteroids with low-amplitude lightcurves. Table IV lists the more promising candidates for diameter determination using the stellar occultation technique.

II. LUNAR OCCULTATIONS OF ASTEROIDS

Accurate diameters of small solar system bodies can also be obtained from photometric observations of lunar occultations. This technique has already been applied to several satellites of Saturn (Elliot et al. 1975), the Galilean satellites (Vilas et al. 1977; Buarque 1978), and Ceres (Dunham et al. 1974). Unfortunately, the Ceres observations were done visually and are therefore of relatively low accuracy.

Two factors limit the accuracy of asteroid diameters determined by lunar occultations. The first is that both the shape and the degree of limb darkening of the occulted body must usually be known to derive a diameter. Secondly, the occulted object should be as bright and subtend as large an angle as possible. For asteroids these parameters peak rather sharply around opposition at which time the occulting moon will be near full phase rendering any observation difficult. In view of these problems it appears that the lunar occultation technique is primarily applicable to only a few of the brighter asteroids.

Acknowledgments. The authors wish to thank D. Dunham for his very helpful comments and suggestions regarding this chapter. We have also benefited from discussions with several colleagues including G. E. Taylor, L. Wasserman and E. Bowell. R. Williamon and E. Bowell kindly provided their observations of possible minor planet satellites in advance of publication.

TABLE IV

Selected Candidates for Occultation Diameter Measurements[a]

Asteroid	Taxonomic type	Diameter (km)	V at mean opposition	No. of telescopes needed for 3 chords $\sigma = 0.2$	Maximum light curve amplitude (mag)	Pole position available?
1 Ceres	C	1025	6.1	5	0.04	—
4 Vesta	U	555	4.8	6	0.14	c
2 Pallas	U	538±12	6.8	6	0.15	b
10 Hygiea	C	443	9.3	8	0.21	—
704 Interamnia	U	338	9.9	9	0.11	—
511 Davida	C	335	10.2	9	0.25	b,c
65 Cybele	C	311	11.3	10	0.06	—
52 Europa	C	291	10.2	10	0.09	—
451 Patientia	C	281	10.4	10	0.1	—
31 Euphrosyne	C	270	11	10	—	—
15 Eunomia	S	261	8.0	9	0.53	c
324 Bamberga	C	256	9.7	9	0.07	—
107 Camilla	C	252	11.6	12	0.53	—
87 Sylvia	CMEU	251	11.6	12	0.42	—
45 Eugenia	U	250	10.2	9	0.33	—
3 Juno	S	249	8.3	9	0.15	c
16 Psyche	M	249	9.5	10	0.32	—
24 Themis	C	249	10.8	11	0.14	—

[a]Data from TRIAD file (Part VII in this book).

[b]Pole position available in TRIAD file.

[c]Pole position given by Vesely (1971); for an update on pole positions see the chapter by Taylor in this book.

This work was supported by the National Aeronautics and Space Administration.

REFERENCES

Buarque, J. A. 1978. Occultations of Jupiter satellites. *Publ. Astron. Soc. Pacific* 90: 117-118.

Binzel, R. P., and Van Flandern, T. C. 1979. Minor planets: The discovery of minor satellites. *Science* 203: 903-905.

Bowell, E.; McMahon, J.; Horne, K.; A'Hearn, M. F.; Dunham, D. W.; Penhallow, W.; Taylor, G. E.; Wasserman, L. H.; and White, N. M. 1978. A possible satellite of Herculina (abstract). *Bull. Amer. Astron. Soc.* 10: 594.

Bowell, E., and Wasserman, L. H. 1979. Occultations of stars by solar system objects. I. Predictions for Ceres, 1979/80. *Astron. J.* (in press).

Dollfus, A. 1971. Diameter measurements of asteroids. In *Physical Studies of Minor Planets*, ed. T. Gehrels (Washington, D. C.: U.S. Government Printing Office), pp. 25-31.

Dunham, D. W.; Killen, S. W.; and Boone, T. L. 1974. The diameter of Ceres from a lunar occultation (abstract). *Bull. Amer. Astron. Soc.* 6: 432-433.

Dunham, D. W., and Sheffer, Y. 1979. Juno larger than expected: Probable graze by distant satellite. *Occultation Newsletter* 2: 12.

Dunham, D. W.; Taylor, G. E.; Bowell, E.; Wasserman, L. H.; and Franz, O. G. 1979a. Prediction and astrometry of occultations involving asteroids. *Icarus* (special Asteroid issue).

Dunham, D. W.; Van Flandern, T. C.; Schmidt, R.; A'Hearn, M. F.; Skillman, D.; Williamon, R.; Poss, H.; and Erickson, G. 1979b. More evidence for satellites from an occultation by 18 Melpomene. In preparation.

Elliot, J. L. 1977. Signal-to-noise ratios for stellar occultations by the rings of Uranus, 1977-1980. *Astron J.* 82: 1036-1038.

Elliot, J. L. 1979. Stellar occultation studies of the solar system. In *Annual Reviews of Astronomy and Astrophysics*, Vol. 17, eds. G. Burbidge, D. Layzer and J. Phillips (Palo Alto, Calif.: Annual Reviews, Inc., in press).

Elliot, J. L.; Veverka, J.; and Goguen, J. 1975. Lunar occultation of Saturn. I. The diameters of Tethys, Dione, Rhea, Titan, and Iapetus. *Icarus* 26: 387-407.

Hertz, H. G. 1968. The mass of Vesta. *Science* 160: 299-300.

Hubbard, W. B.; Coyne, G. V.; Gehrels, T.; Smith, B. A.; and Zellner, B. H. 1977. Observations of Uranus occultation events. *Nature* 268: 33-34.

Millis, R. L.; Bowell, E.; Wasserman, L. H.; and Elliot, J. L. 1978. A critique of asteroid diameter measurements (abstract). *Bull. Amer. Astron. Soc.* 10: 596-597.

Morrison, D. 1977. Asteroid sizes and albedos. *Icarus* 31: 185-200.

O'Leary, B. 1972. Frequencies of occultations of stars by planets, satellites, and asteroids. *Science* 175: 1108-1111.

O'Leary, B.; Marsden, B. G.; Dragon, R.; Hauser, E.; McGrath, M.; Backus, P.; and Robkoff, H. 1976. The occultation of κ Geminorum by Eros. *Icarus* 28: 133-146.

Schubart, J. 1974. The masses of the first two asteroids. *Astron. Astrophys.* 30: 289-292.

Smart, W. M. 1960. *Textbook on Spherical Astronomy*, 4th ed. (London: Cambridge University Press), pp. 368-378.

Taylor, G. E. 1962. The diameters of minor planets. *J. Brit. Astron. Assoc.* 72: 212-214.

Taylor, G. E. 1978. Possible occultations of stars by asteroids in 1979. *IAU Comm. 20 Working Group on Predictions of Occultations by Satellites and Minor Planets. Bull. 12.*

Taylor, G. E., and Dunham, D. W. 1978. The size of minor planet 6 Hebe. *Icarus* 34: 89-92.

Tüg, H.; White, N. M.; and Lockwood, G. W. 1977. Absolute energy distributions of Alpha Lyrae and 109 Virginis from 3295 Å to 9040 Å. *Astron. Astrophys.* 61: 679-684.

Vesely, C. D. 1971. Summary on orientations of rotation axes. In *Physical Studies of Minor Planets*, ed. T. Gehrels (Washington, D. C.: U. S. Government Printing Office), pp. 133-140.

Vilas, F.; Millis, R. L.; and Wasserman, L. H. 1977. Lunar occultations of Io and Ganymede (abstract). *Bull. Amer. Astron. Soc.* 9: 464.

Wasserman, L. H.; Millis, R. L.; Franz, O. G.; Bowell, E.; White, N. M.; Giclas, H. L.; Martin, L. J.; Elliot, J. L.; Dunham, E.; Mink, D.; Baron, R.; Honeycutt, R. K.; Henden, A. A.; Kephart, J. E.; A'Hearn, M. F.; Reitsema, H.; Radick, R. R.; and Taylor, G. E. 1979. The diameter of Pallas from its occultation of SAO 85009. *Astron. J.* 84: 259-268.

Wasserman, L. H.; Millis, R. L.; and Williamon, R. T. 1977. Analysis of the occultation of ε Geminorum by Mars. *Astron. J.* 82: 506-510.

Williamon, R. M. 1978. *IAU Circular No. 3315.*

Young, A. T. 1974. Other components in photometric systems. In *Methods of Experimental Physics, Vol. 12-Part A. Astrophysics,* ed. N. Carleton (New York: Academic Press), pp. 95-122.

INTERFEROMETRIC DETERMINATIONS OF ASTEROID DIAMETERS

S. P. WORDEN
Sacramento Peak Observatory

A promising Earth-based technique for directly determining diameters of asteroids is based on new developments in interferometry. Until 1978 application of interferometric techniques to asteroids was limited to the very brightest objects by the low sensitivity of available detectors. Results have been published only for Pallas and Vesta. However, modern photoelectric detectors are now being used in these observations and diameter measurements for a number of minor planets will be forthcoming.

The resolution of optical telescopes is classically given by the Rayleigh criterion and is inversely proportional to the telescope diameter. However, as is well known, turbulence in the earth's atmosphere or "seeing" degrades virtually all telescope images to about one arcsec. Since the largest asteroid angular diameters are about 0.5 arc sec, conventional telescope images are limited for measuring asteroid sizes and shapes. In principle, telescopes of 4–5 meter diamter not limited by seeing could obtain angular resolution of 0.02 arcsec, translating into about 30-km resolution in the asteroid belt.

Recent techniques, speckle and amplitude interferometry (or single aperture interferometry) offer great promise for removing the degrading effects of the atmosphere. Interferometry was first employed to measure the diameter of the asteroid Vesta by Hamy in about 1899 (see Dollfus 1971). In 1977 Vesta and Pallas were observed by Worden *et al.* (1977) and later by

Worden and Stein (1979) using speckle interferometry, but extension of this work to fainter asteroids has been hindered by the low sensitivity of available detectors. With suitable application of these methods it now appears possible to directly measure angular sizes and even produce images for several hundreds of the larger asteroids. In this chapter I discuss the limited work already done on asteroids with speckle interferometry and more extensive work in progress to apply single aperture interferometry to asteroid studies.

I. SPECKLE INTERFEROMETRY

Small-scale temperature inhomogeneities in the earth's atmosphere produce changes in the index of refraction. These refractive index changes cause phase delays along an incoming plane wave, which may be light from a stellar point source. This is represented schematically in Fig. 1. Without phase errors, optical systems produce the image shown in Fig. 1A, which is said to be "diffraction limited," where a point-source image is the classical Airy disk for a circular telescope aperture. The size of this image is inversely proportional to the telescope diameter. With any phase errors telescope resolution is degraded to that appropriate for an optical system only as large as the scale over which there is some phase coherence (i.e., the phase is the same). Since the atmosphere breaks an incoming plane wave into about 10-cm fragments, all telescopes produce images with resolution no better than that of a 10-cm telescope, namely one arcsec. This process is shown in Fig. 1B.

Labeyrie (1970) proposed a method to recover some information down to large telescope diffraction limits. He pointed out that short-exposure ($\Delta t \approx$ 0.01 sec) photos "freeze" the turbulence in the atmosphere. Although the phase coherence size in this "frozen" system is still only 10 cm, there will be some 10-cm patches scattered over the entire aperture which are at the same phase. These portions act in concert as a form of "multiple" aperture interferometer which provides some information down to the diffraction limit of the *entire* telescope aperture. As shown in Figure 1C, the image of a point source seen through a multiple-aperture interferometer is a series of nearly diffraction-limited images modulated by a one arcsec seeing disk. This process is known as "speckle interferometry" since the short-exposure photos, as shown in Fig. 2, look like laser speckle photos.

II. AMPLITUDE INTERFEROMETRY

An alternate approach to stellar interferometry, suggested by Currie (1967) and Currie et al. (1974), is similar to Michelson interferometry. Known as "amplitude interferometry," this technique uses a device like that shown in Fig. 3. The individual collection apertures are smaller than the 10-cm coherence length in order to reduce the correction due to atmospheric degradation to a negligible level. As the atmosphere modulates the relative

121

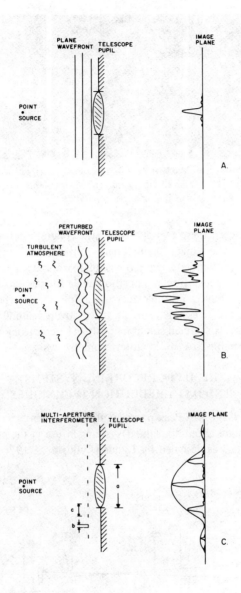

Fig. 1. Schematic diagram of image formation through a turbulent atmosphere: (A) diffraction-limited optics, no atmospheric turbulence; (B) image formation through a turbulent atmosphere; and (C) multiple-aperture interferometer.

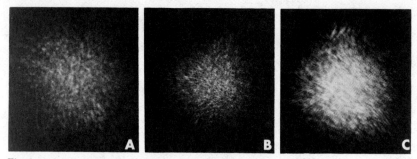

Fig. 2. Speckle photos from the Kitt Peak 4-m telescope. Each image is about 1 arcsec in extent. (A) α Orionis, a resolved supergiant star, (B) γ Orionis, a point-source star, and (C) α Aurigae, a binary star with 0.05-arcsec separation.

phase shifts between these two apertures, the coherence properties (and thus angular size) of the object as it appears outside the atmosphere can be learned. To obtain complete two-dimensional size and shape information the observer varies the separation and position angle for the two apertures. Currie has proposed and built a multiple-aperture amplitude interferometer system, that allows the full telescope aperture to be covered simultaneously and all Fourier components sampled simultaneously. The efficiency of such a system should be comparable to a speckle interferometry system.

III. DATA RECORDING SYSTEMS
AND DATA REDUCTION TECHNIQUES

A diagram of the Kitt Peak photographic speckle interferometer is shown in Fig. 4. There are about six similar systems in use at the present time. The Kitt Peak camera was designed by Lynds (Lynds *et al.* 1976; Breckinridge *et*

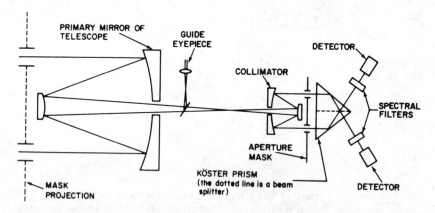

Fig. 3. Diagram of Currie's amplitude interferometer.

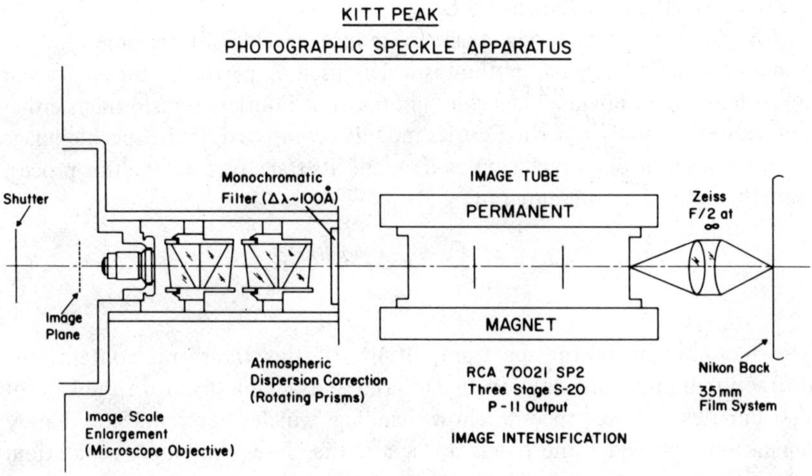

KITT PEAK

PHOTOGRAPHIC SPECKLE APPARATUS

Fig. 4. Diagram of Lynds' speckle interferometer.

al. 1978). As shown in the figure, light from the telescope passes through a shutter and focuses at the telescope image plane. The shutter is necessary to insure exposures shorter than the atmospheric change time, typically 20 msec. The telescope image is relayed and magnified by a microscope objective. The magnification is set to provide a pixel resolution oversampling the telescope diffraction spot size by at least a factor of four. For the Kitt Peak 4-m telescope, this provides a final image scale of 0.2 arcsec/mm. Atmospheric dispersion blurs speckle image patterns in the sense that the "red" portion of the image focuses at a slightly different position than the "blue" portion. Since this may be significant for even 200 Å bandpass photos, a set of rotating atmospheric-compensating prisms is included to counteract the dispersion. Since there are about 20 orders of optical interference across a speckle photo, a narrow band ($\Delta\lambda \approx 200$ Å) interference filter is used to preserve coherence across the entire speckle photo. If this were not included, the "speckles" near the edge of the photos would be elongated. A three-stage image tube intensifies the image enough to allow photographic data recording. A transfer lens relays the intensified image to a data recording system, in this case a 35-mm film camera.

The speckle photos in Fig. 2 were taken with the Kitt Peak system. The different character of these photos is readily apparent. This is understandable from the analogy to a multiple-aperture interferometer. Each speckle should be a diffraction-limited image of the object. Indeed, the binary star (α Aur) speckles are double, the point-source speckles roughly diffraction spots, and the resolved star (α Ori) speckles somewhat larger. This aspect led Lynds *et al.* to a direct speckle image reconstruction scheme whereby individual speckles were identified and co-added to produce a nearly diffraction-limited image for the special case of stars like α Ori.

A number of methods exist to reduce speckle interferometry data. Labeyrie's (1970) original method is widely used, in particular for binary star measurements. Individual speckle photos are Fourier transformed either optically or digitally and the Fourier modulus computed. If the speckle image is represented in one dimension as $i(x)$, and its transform as $I(s)$, this process is mathematically represented by

$$I(s) = \int_{-\infty}^{\infty} i(x)e^{-2\pi i x s}dx. \tag{1}$$

The modulus or power spectrum, $|I(s)|^2$, of this transform contains the diffraction-limited information in an easily extractable form. In the case of the binaries, power spectra show banding which represents the binary separation; the wider the bands are apart, the closer the binary separation. The orientation of these bands represents the position angle of the binary system.

The residual effects of the seeing must be removed to yield the maximum accuracy. Even though the bands (fringes) are readily visible in raw speckle power spectra, their spacing is affected by the residual seeing effects. Labeyrie's method uses observation of point-source stars to determine these seeing effects and remove them. If $P_i(x)$ are point-source speckle photos with a mean power spectrum $<|P(s)|^2>$, and $<|I(s)|^2>$ the mean power spectrum of the object photos $i_i(x)$, then the diffraction-limited power spectrum of the object is given by

$$|O(s)|^2 = \frac{<|I(s)|^2>}{<|P(s)|^2>}. \tag{2}$$

Point-source data are usually derived from speckle observations of point-source stars situated near the program objects on the sky. Since these point-source objects are not in general observed within the same isoplanatic angle and not at the same time, their power spectrum can only represent the residual seeing effects in a statistical sense.

Worden et al. (1977) have developed a method to calibrate for residual seeing effects using the same set of speckle photos as used to study the program objects. We illustrate this method in Fig. 5. The method proceeds as follows: the mean autocorrelation function of a series of speckle, $i_i(x)$ photos is computed:

$$<AC(\Delta x)> = <\int_{-\infty}^{\infty} i_i(x) \cdot i_i(x - \Delta x)dx>$$
$$= <i_i(x) \star i_i(x)>. \tag{3}$$

(The autocorrelation is the Fourier transform of the power spectrum: see Bracewell 1965 for details.) As we see in Fig. 5, the mean autocorrelations are

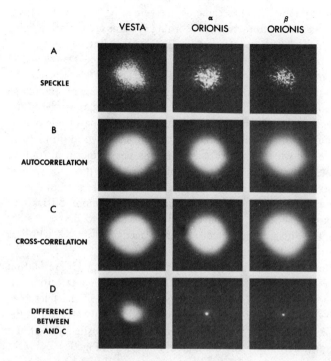

Fig. 5. Illustration of the Worden *et al.* (1977) method for reducing speckle interferometry data.

dominated by the seeing background. This background may be accurately removed by computing and subtracting the mean cross-correlation *between* consecutive speckle photos of the *same* set of data used to compute the autocorrelation. The cross-correlation (XC) between the ith and $i^{th} + 1$ speckle photo is given by

$$\langle XC(\Delta x)\rangle = \langle \int_{-\infty}^{+\infty} i_i(x) \cdot i_{i+1}(x\text{-}\Delta x)dx \rangle$$

$$= \langle i_i(x) \star i_{i+1}(x)\rangle. \tag{4}$$

Welter and Worden (1978) showed that the resulting subtraction is the object autocorrelation as it would appear with virtually all seeing effects removed. This "diffraction-limited" autocorrelation contains information on such quantities as angular diameter in easily extractable form. For example, the angular diameter of an asteroid is simply the distance between the autocorrelation maximum and the point where the autocorrelation falls to zero.

Current photographic speckle cameras are generally limited to objects brighter than +7 mag. The photographic recording systems are therefore being replaced with high quantum efficiency digital recording systems. The University of Arizona speckle camera uses a Charge Injected Device (CID) television system to record photon arrivals. This system simply replaces the photographic emulsion, and it can record data for objects faint enough so that only a few photons arrive in a 20-msec exposure. In Fig. 6 we show data from this system for Saturn's moon Rhea, which is a 10th-magnitude object. The limiting requirement for this method is that at least two photons lie in the same speckle. If we can record two photons per frame in a 20-msec exposure in some of the frames these two photons will lie in the same speckle and contribute to our diffraction limited signal. This translates to about a +16 stellar magnitude limit. Although angular diameters are more difficult to derive than binary separations, we have used this system to derive angular diameters for a 10.5 magnitude object (Iapetus) accurate to ± 5% with less than five minutes total observing time.

The amplitude interferometer obtains the high angular resolution information in a somewhat different fashion than the speckle interferometer. In this case, the light is sampled at the entrance aperture of the telescope, where the effect of the atmosphere has been to introduce only an error in the phase delay. The light from two separate apertures on opposite sides of the telescope is then interferometrically combined, as shown in Fig. 3.

In order to permit the observation of fainter objects, we wish to simultaneously use all the light entering the telescope aperture, i.e., the data from many thousands of pairs of apertures (a Multiple Aperture Amplitude Interferometer or MAAI). This may be done by replacing each of the two photomultipliers with a "television camera" in which each resolution element acts as a separate channel interferometer.

Results on solar system objects have been limited, largely due to the bright limiting magnitude of existing photographic speckle systems and amplitude interferometers. With the advent of efficient photoelectric and television systems this situation is changing. Table I lists data published on the asteroids Vesta and Pallas (Worden et al. 1977; Worden and Stein 1978). In these diameter determinations, circular objects, uniform albedo, and no limb darkening have been assumed. With a photographic limiting magnitude of +7, Vesta and Pallas are the only asteroids which have been observable so far. Amplitude interferometry has had an even more stringent limit, about +3 mag. I note that the speckle results for Vesta match other values for that asteroid's diameter very well, although Pallas' diameter is about 25% larger than the accurate occultation results for this object (Wasserman et al. 1979; see the chapter by Millis and Elliott). However, as alluded to earlier, photoelectric systems now make it possible to observe objects as faint as stellar magnitude +16.

We have used a prototype television system (Strittmatter and Woolf

TABLE I

Diameters of Solar System Objects from Speckle Interferometry

Object	Date (UT)	Angular Diameter (arcsec)	Diameter (km)
Vesta	1 Dec 76	0.400 ± 0.040	513 ± 51
Vesta	3 Feb 77	0.470 ± 0.020	550 ± 23
Pallas	3 Feb 77	0.730 ± 0.060	673 ± 55
Rhea	17 Apr 78	0.234 ± 0.005	1487 ± 40
Iapetus	17 Apr 78	0.189 ± 0.021	1200 ± 132

1978) on Steward Observatory's 2.3-m telescope to observe Saturn's satellites Rhea (~9.5 mag) and Iapetus (~10.5 mag). Figure 6 shows a speckle photograph from this system for Rhea: individual photon events are readily apparent. The angular diameters of these objects are listed in Table I. These values match lunar occultation values quite well (Elliot *et al.* 1975). A system developed by Boksenberg at the University College London has been used to

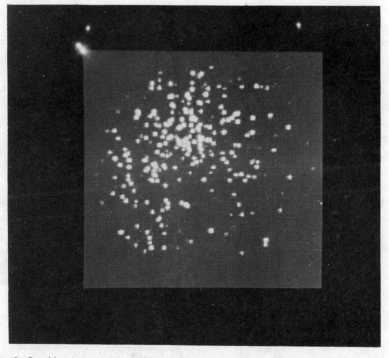

Fig. 6. Speckle data obtained with a digital camera for Rhea showing individual photons.

observe the asteroids 511 Davida and 40 Harmonia at about stellar magnitudes +11.5 and +12, respectively. Although these results have not been accurately calibrated yet, the ratio of the two objects' diameters of about 3, reported by other investigators, is confirmed.

The accuracies reported in Table I exceed the telescope diffraction limits by a significant factor. The quoted errors are based on diameter determinations from independent subsets of the data. The diameters themselves are derived by convolving various diameter asteroid profiles with point source speckle data obtained at the same time until a best match with the observed asteroid speckle profile is reached. Stellar diameters accurate to ± 3% internal error have been derived in this manner for objects close to the diffraction limit in size (Worden 1975) and diameters for objects considerably smaller than the diffraction limit are accurate to ± 30% (Worden 1976). These high precisions follow directly from the fact that the half width of Gaussian or other similar profile may be determined very accurately. As discussed by Worden (1976), this accuracy is considerably greater than the telescope diffraction limit. This problem is somewhat analogous to the ability of stellar astrometrists to measure star positions far more accurately than the half width of a stellar seeing disk.

There are, however, sources of systematic error in interferometric asteroid diameters, the assumption of uniform albedo, no limb darkening and spherical shape being among the most serious. Based on our stellar speckle results (Welter and Worden 1979), up to a 20% increase in angular diameter results from a fully limb darkened disk as compared to a uniform disk. However, based on space probe planetary and satellite images obtained to date, large limb darkening is highly unlikely. Indeed, as shown by McDonnell and Bates (1976) limb darkening may be a free parameter to be fit in the diameter fitting procedure described above, given accurate enough speckle data. In a similar manner other parameters such as elongated shape may be fit. Occultation results for Pallas show some elongation, so it will be necessary in future work to consider the effects of parameters other than uniform disk diameter. Until we have perfected these methods, interferometric diameters represent only the diameter of a uniform sphere which would produce the same light as the asteroid being observed. The accuracies reported therefore refer to our ability to determine this useful, but not complete parameter. To fully assess errors caused by this assumption we ultimately require actual high resolution images of each asteroid being studied.

The exciting new possibility of actual reconstructed images for asteroids appears to be within grasp. The methods for reducing data discussed above produce only a power spectrum or Fourier amplitude of the true image. Although size and shape may be derived from the Fourier amplitude, the Fourier phase is needed to reconstruct actual images. Several schemes have been developed to estimate the phase, (Bates 1978; Baldwin and Warner 1978) the most promising having recently been proposed by Fienup (1978).

The method involves an iterative scheme to guess phase values and determine whether the resulting image is consistent (i.e., all values positive and the object has a diameter matching the known value). We have tried this method on our power spectrum results from photographic Vesta data. With a 4-m telescope, a diffraction-limited image of Vesta would have over 100 resolution elements. The resulting image shown in Fig. 7, has the right diameter. This is encouraging since the diameter is a free parameter in Fienup's method as it is applied here. The mean noise in this image is ±13%, based on several sets of data taken within five minutes. We see no surface structure larger than this, although this is not a stringent limit. We do see a slight elongation with the long axis along position angle $16° \pm 4°$ relative to east-west. The diameter ratio from the longest to shortest seems to be 1.19 ± 0.02. The reality of this elongation is open to question since polarimetric results (Gradie *et al.* 1978) indicate Vesta is spherical. Moreover, while the derived elongation is 20%, other speckle results from Kitt Peak often show up to a 10% elongation. We must therefore regard this image as a very preliminary attempt at asteroid imaging. Nonetheless, it is an encouraging development which clearly warrants further study.

Another promising aspect of interferometric methods is the possibility of

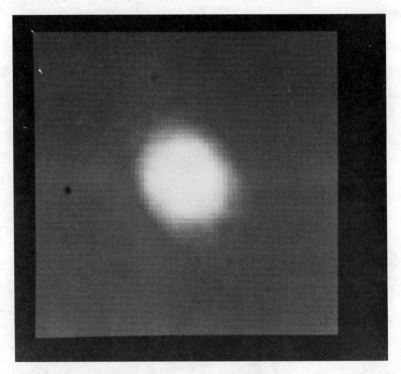

Fig. 7. Reconstructed image for Vesta from 3 February 1977 4-m Kitt Peak speckle data. The asteroid disk is 0.470 arcsec in diameter.

detecting close asteroid pairs. Binary star speckle has already been demonstrated for objects as faint as +12 mag by our group at the University of Arizona. It may therefore be possible to detect asteroid satellites using interferometric methods and work is now underway to accomplish this.

There are several limitations to interferometric methods. The +16-mag limit is set by the requirement of at least two photons per frame in a 0.01-sec exposure. To make interferometric methods work at all, light from all points of the object must pass through the same column of turbulent atmosphere. This is called the "isoplanatic" requirement, and it has been measured to be about 3–5 arcsec. Consequently, asteroid studies should not be affected by isoplanatic problems. More critically, the Worden *et al.* (1977) correlation method begins to break down when objects approach the seeing limit, 1 arcsec. This may help explain the discrepancy between speckle diameters for Pallas and other values, since Pallas was about 0.7 arcsec in diameter when observed.

REFERENCES

Baldwin, J. E., and Warner, P. J. 1978. Phaseless aperture synthesis. *Mon. Not. Roy. Astron. Soc.* 182: 411-422.

Bates, R. H. T. 1978. On phase problems. II. *Optik.* 51: 223-234.

Bracewell, R. 1965. *The Fourier Transform and Its Application* (New York: McGraw Hill).

Breckinridge, J. B.; McAlister, H. A.; and Robinson, W. G. 1978. The Kitt Peak Speckle Camera. *Appl. Opt.* (in press).

Currie, D. G. 1967. Amplitude interferometry. In *Woods Hole Summer Study on Synthesic Aperture Optics,* Vol. 2 (Washington, D. C.: NAS/NRC).

Currie, D. G.; Knapp, S. L.; and Liewer, K. M. 1974. Four stellar-diameter measurements by a new technique:

Dollfus, A. 1971. Diameter measurement of asteroids. In *Physical Studies of Minor Planets,* ed. T. Gehrels (NASA SP-267, Washington, D.C.: U.S. Government Printing Office), pp. 25-31.

Elliot, J. L.; Veverka, J.; and Goguen, J. 1975. Lunar occultation of Saturn. I. The diameters of Tethys, Dione, Rhea, Titan and Iapetus. *Icarus* 26: 387-407.

Fienup, J. 1978. Reconstruction of an object from the modulus of its Fourier transform. *Opt. Lett.* 3:27-31.

Gradie, J.; Tedesco, E.; and Zellner, B. 1978. Rotational variants in the optical polarization and reflection spectrum of Vesta (abstract). *Bull. Amer. Astron. Soc.* 10: 595.

Labeyrie, A. 1970. Attainment of diffraction limited resolution in large telescopes by Fourier analyzing speckle patterns in star images. *Astron. Astrophys.* 6: 85-87.

Lynds, C. R.; Worden, S. P.; and Harvey, J. W. 1976. Digital image reconstruction applied to Alpha Orionis. *Astrophys. J.* 207: 174-180.

McDonnell, M. J. and Bates, R. H. T. 1976. Digital restoration of an image of Betelgeuse. *Astrophys. J.* 208: 443-452.

Strittmatter, P. A. and Woolf, N. J. 1978. Image reconstruction using large astronomical telescope. *Air Force Geophysics Lab. Rept.* AFGL-TR-78-0167 (Hanscom Air Force Base).

Wasserman, L. H.; Millis, R. L.; Franz, O. G.; Bowell, E.; White, N. M.; Giclas, H. L.; Martin, L. J.; Elliot, J. L.; Dunham, E.; Mink, D.; Baron, R.; Honeycutt, R. K.; Henden, A. A.; Kephart, J. E.; A'Hearn, M. F.; Reitsema, H.; Radick, R. R.; and

Taylor, G. E. 1979. The diameter of Pallas from its occultation of SAO 85009. *Astron. J.* (in press).

Welter, G. L., and Worden, S. P. 1978. A method for processing stellar speckle interferometry data. *J. Opt. Soc. Amer.* 68: 1271-1275.

Welter, G. L., and Worden, S. P. 1979. The angular diameters of super-giant stars from speckle interferometry. *Astrophys. J.* (in press).

Worden, S. P. 1975. The angular diameter of Alpha Herculis A. *Astrophys. J.* 201: L69-L70.

Worden, S. P. 1976. Digital analysis of speckle photographs: The angular diameter of Arcturus. *Pub. Astron. Soc. Pacific* 88: 69-72.

Worden, S. P., and Stein, M. K. 1979. Angular diameter of the asteroids Vesta and Pallas determined from speckle observations. *Astron J.* 84: 140-142.

Worden, S. P.; Stein, M. K.; Schmidt, G. D.; and Angel, J. R. P. 1977. The angular diameter of Vesta from speckle interferometry. *Icarus* 32: 450-457.

COLORIMETRY AND MAGNITUDES OF ASTEROIDS

EDWARD BOWELL
Lowell Observatory

and

KARI LUMME
University of Helsinki

A new, rather general theory of multiple scattering in complex surfaces has been used to interpret 1500 UBV observations of asteroids. Phase curves are shown to consist of a surface-texture-controlled component due to singly scattered light and a component due to multiple scattering. The shapes of phase curves can be characterized by a single parameter, Q, the proportion of multiply scattered light; as Q increases the relative importance of the opposition effect is diminished. Asteroid surfaces are particulate and strikingly similar in texture, being moderately porous and moderately rough on a scale greater than the wavelength of light. In consequence, Q (and also the phase coefficient) correlate well with geometric albedo, and there exists a purely photometric means of determining albedos and diameters. Also, mean phase curves for C, S, M and other taxonomic types differ from one another. We estimate the mean Q, $\bar{Q}$, appropriate for the numbered main-belt population and show that an incorrect choice of $\bar{Q}$ results in systematic errors in diameters and albedos determined from radiometry and polarimetry. For C and S asteroids mean color index and geometric albedo do not vary with diameter and semimajor axis. Q is also independent of semimajor axis but does depend weakly on diameter. Thus C and S asteroids form two populations that are, photometrically, extremely homogeneous on a belt-wide scale. C asteroids are probably,

[132]

on the average, slightly less spherical than S asteroids. We see no brightness-related change in B−V with rotation at a threshold of ±0.005 mag, so we cannot photometrically detect the presence of impacting S material on C asteroids and vice versa.

In the early 1970's, when the first summaries of photoelectric photometry of asteroids were given by Gehrels (1970) and Taylor (1971), UBV color indices were available for barely 50 objects. In the intervening years there has been a rapid acquisition of data, so that we now have data for 735 asteroids in the TRIAD file given in Part VII of this book. There are good reasons for this apparently profligate observational activity, which has greatly advanced our knowledge of the physical constitution of the asteroid belt. B−V and U−B indices have helped classify asteroids into various taxonomic types; and magnitudes, when allowance is made for the effect of solar phase angle, have been used in the radiometric and polarimetric methods of diameter determination. Technical reasons for the vigorous pursuit of asteroid UBV photometry are that elaborate equipment is not required and faint objects may be observed readily and quickly.

Applications of asteroid UBV photometry can be categorized into two overlapping and interdependent areas. In the first, which can be labeled *taxonomy,* color indices and magnitudes have to be interpreted with the aid of physically diagnostic data from other techniques. In contrast, there are applications which do not, *a priori,* require observations by other techniques. In this area, a purely photometric description of the asteroid belt is sought. The observational data yield rotation periods, body shapes, spin axis orientations, absolute magnitudes, phase curves, information on surface texture, and albedos and diameters. As with taxonomy, the photometric inferences may be applied either to individual asteroids or in a statistical way to large numbers of asteroids.

In this chapter we confine our attention to broadband photometry in the UBV system. The small number of data in the R and I bands and at other wavelengths is excluded. We concentrate largely on the interpretation of asteroid magnitudes and color indices using a newly developed theory of multiple scattering; and we summarize work which is planned to be published in a series of papers having the general title "Photometry and Polarimetry of Atmosphereless Bodies."

We have introduced a new definition of the absolute magnitude, and we have parametrized the shapes of asteroid phase curves using a single quantity, which we call the multiple-scattering factor, rather than the two quantities previously used (the linear phase coefficient and the opposition effect). In an attempt to avoid confusion, we use the new and old terminology side by side.

Our data base consists in part of published magnitudes and color indices for particular asteroids (where good phase angle coverage is available) but mainly of the large corpus of UBV observations made by Bowell (1979) during the course of a four-year photometric survey of bright asteroids. A general aim of the survey has been to observe a substantial fraction of the numbered asteroids down to opposition magnitude B = 14.0; to date, about 600 asteroids have been observed; a total of 1500 individual observations, mostly in three-color, has been amassed. A deliberate attempt has been made to observe each asteroid more than once and at different phase angles, and observations have generally been spaced at least several days apart so as to randomize the effects of rotational brightness changes. A substantial number of asteroids have been observed at more than one apparition. The survey data constitute a rather homogeneous sampling of the entire main belt down to a diameter of about 50 km. We have also included photometry of 45 bright asteroids reported by Zellner et al. (1975).

Color indices and absolute magnitudes of many other asteroids are given in the listing of the TRIAD file in Part VII of this book. Most of these asteroids have been observed during the course of special-purpose surveys, and the data are therefore not always suitable for our purposes in this chapter, but for completeness we list the principal references here. Degewij (1978), Degewij et al. (1978), and Zellner et al. (1977a) have observed faint asteroids; Gradie (1978) and Tedesco (1979b) have studied Hirayama family members; Hansen (1976) has observed a number of asteroids in the B, V, R, and I bands in connection with a program of radiometry; and Veeder et al. (1978) give magnitudes of 30 asteroids observed through broadband filters centered on 0.56, 1.6, and 2.2 μm.

I. TAXONOMY

The distribution of points in a color-color plot (e.g. B–V versus U–B) for asteroids exhibits distinct clumpings, and these have long been attributed to compositional differences (see, e.g., Kitamura 1959; Hapke 1971). It is, in principle, possible to devise a taxonomic system based entirely on these clusterings, but such a system would offer little or no physical insight into the nature of asteroid surfaces. Thus the first real attempt at classifying asteroids (Chapman et al. 1971) relied in part on relating UBV photometry to narrowband spectrophotometry; this work set the pattern for the subsequent development of taxonomic systems. More recently, further parameters have been added as survey-type observational programs of asteroids have been undertaken. For example, asteroid albedos have been estimated from infrared radiometric and optical polarimetric observations.

Even though a B–V, U–B diagram contains little intrinsic physical information about asteroids, it is clear, when account is taken of the albedo data, that useful inferences may be drawn about most asteroids' albedos from color indices alone. Similarly, when mineralogical interpretations from spectrophotometric data are incorporated, B–V and U–B color indices suffice to give good separation of S and C asteroids. Color indices themselves are weak mineralogical discriminants, however, because the most compositionally diagnostic region of the spectrum lies to the red of the V band.

The important role played by UBV photometry in the taxonomic classification of asteroids can be appreciated by considering the number of asteroids classified on the basis of color indices alone. In the work of Chapman et al. (1975; see Zellner's chapter), where the so-called CSM taxonomic system was first defined, the color indices were not separately considered to be taxonomically diagnostic; but in the expanded sample used by Zellner and Bowell (1977), about 160 of the 336 asteroids were classified using color indices alone. In the work of Bowell et al. (1978), the number had risen to about 300 of 521 asteroids; and in the TRIAD file dated March 1979, 382 of 752 asteroids are so classified.

Figures 1 and 2 are B–V, U–B plots from the TRIAD color file. It can be seen from Fig. 1 that color indices discriminate well between C and S asteroids, but poorly among C, M, and E asteroids. Comparison of Fig. 2 with Fig. 1 gives the impression that the separation between C and S asteroids has somewhat broken down, but this coalescence is very likely due to larger observational errors in the data used for Fig. 2 which, on average, pertain to fainter asteroids.

Taxonomic applications of UBV photometry have been important in the following ways: (a) they have determined the observationally unbiased distributions of the various asteroid types according to diameter and semimajor axis (Zellner and Bowell 1977; Zellner's chapter); (b) have provided good evidence that the larger Hirayama families result from the collisional breakup of rather homogeneous parent bodies (see the chapters by Gradie et al.; Tedesco 1979b); (c) have led to the realization that close Earth-approaching asteroids form a compositionally heterogeneous group presumably having a variety of origins and histories (see the chapter by Shoemaker et al.); (d) have helped provide some first insights into the structure of the Hilda and Trojan regions (see the chapter by Degewij and van Houten); and (e) have been a useful means of discovering unusual or unique asteroids.

II. PHASE CURVES

In the most general terms, the observed brightness of an asteroid depends on its distance from the sun and earth, its size and shape, and the scattering properties of its surface material. Of these components, the first is purely geometric and can be allowed for easily, but to understand the contributions of the others, even qualitatively, one must make a number of assumptions. Our purpose in this section is to examine what these assumptions might be in terms of a newly developed theory of multiple scattering and thereby determine just what information about the physics of asteroid surfaces is contained in observations of their brightness.

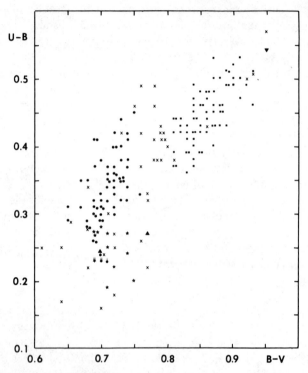

Fig. 1. Color-color plot for 189 asteroids with measured geometric albedos. Symbols correspond to different taxonomic types: ●C, ■S, ★M, ▲E, ▼R, and X U. Asteroids 446 and 863, both R types, are off scale at B–V = 1.03, U–B = 0.61 and 1.06, 0.56, respectively.

A basic tool is the phase curve. This is a plot of the *reduced magnitude,* that is, the brightness of an object at a hypothetical location 1 AU from the sun and earth, versus the solar *phase angle,* the solar elongation of the earth seen from the asteroid. For main-belt asteroids, phase angles are confined to the range $0° \leqslant \alpha \lesssim 30°$, although Earth-approachers may attain much larger values. It has been known since the last century that away from zero phase the magnitudes of asteroids drop off very nearly linearly with phase angle (see e.g., Müller 1897), and the rate of dimming has traditionally been called the *phase coefficient* (or phase factor) and expressed in units of mag/deg. Near zero phase, the phase curve exhibits a nonlinear surge in brightness which has become known as the *opposition effect.*

In what follows we assume that the phase curves to be analyzed are free from the effects of rotational brightness variations either because the lightcurve is explicitly known and allowed for or because a large number of observations made at random rotational phases is used. We also assume that, if a phase curve is delineated during one apparition, the effect of changing

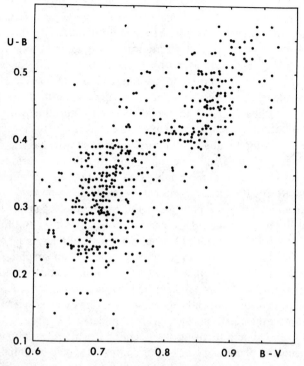

Fig. 2. Color-color plot for 503 asteroids for which no albedo data are available. Asteroid 1658 is off scale at B—V = 0.96, U—B = 0.61.

aspect angle (the angle between the rotation axis and the line of sight) is small. This is tantamount to assuming that the average cross-sectional area of an asteroid is constant over the observed phase curve, or that the phase curve represents the brightness change due to surface structure only.

A. The multiple-scattering theory

The theoretical model for multiple scattering in the surfaces of atmosphereless solar system bodies is described by Lumme and Bowell (1979a), and the confrontation of the model with selected astronomical data is treated in detail by Lumme and Bowell (1979b). An attempt has been made to make the model as rigorous and general as possible, while at the same time seeking useful approximations that allow ready comparison between theory and observations. A guiding principle has been to identify and allow only the most important effects of light scattering in rough surfaces. Thus, it has been clearly demonstrated that the contribution due to multiple scattering is significant for all but the darkest asteroids, whereas an effect such as diffraction can safely be neglected since diffraction is only important in the far-field approximation of radiative transfer theory (where scattering particles are far apart and the volume density is very low, as in Saturn's rings).

It is not our intention to describe the multiple-scattering theory in detail here, but a phenomenological description is appropriate. In what follows, we make a number of assertions without supporting proof; for a rigorous treatment the reader is referred to Lumme and Bowell (1979a,b). At least four independent parameters must be used to describe light scattering in a surface. Those chosen are D the volume density (together with $1-D$, the porosity), ρ the roughness, g the Henyey-Greenstein asymmetry factor, and $\widetilde{\omega}_0$ the single-scattering albedo of a particle. It is shown that the surfaces of all the bodies concerned must be made up of particles of various sizes and shapes, with interconnecting voids. (It is clear, for example, that a non-particulate rough surface alone cannot explain the observed opposition effect in asteroids.) For the purpose of this review the roughness is taken to refer to all scales larger than the wavelength of light. In the full formulation of the theory we do distinguish between the effects of microroughness (a few particle diameters) and macroroughness (large topographic features). Roughness at all scales is parametrized as the mean height to radius ratio of surface asperities. The Henyey-Greenstein asymmetry factor describes the relative amounts of forward and backward scattering by a particle.

Light-scattering equations have been formulated for a plane surface and integrated over a sphere to give disk-integrated brightness as a function of ρ, D, g, $\widetilde{\omega}_0$, and the phase angle α. In this context it should be noted that limb darkening is allowed for implicitly in the calculation of disk-integrated brightness by means of the four fundamental parameters and does not have to be treated as a separate phenomenon. A full treatment of limb darkening is planned by Lumme (1979b).

We have used a probabilistic method to describe scattering in a dark surface. It has been assumed in most earlier works that all scatterers in a surface are spheres, perhaps of equal size; but in the new theory, surfaces are described in terms of statistical slopes whose surface elements need not be made up of particles of regular shape. In a very dark surface – one for which the geometric albedo p tends to zero and in which there is no multiple scattering – $\widetilde{\omega}_0$ tends to zero and g is undefined. A dark surface is one, therefore, for which the amount of emergent light depends only on the porosity and roughness. It follows that the phase curve of a low-albedo asteroid ought to contain directly accessible information on these quantities. We have been able to show that the effects of porosity and roughness are well separable; porosity controls the shape and extent of the opposition effect, whereas roughness controls the slope of the phase curve away from zero phase. (We point out below that, where multiple scattering is active, albedo effects are dominant in controling the slope of the phase curve.)

In September 1977, the M asteroid 69 Hesperia passed very close to the anti-solar point at opposition, attaining the unusually small phase angle of

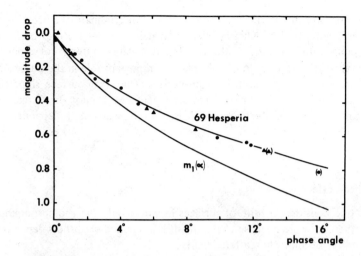

Fig. 3. V-filter observations of Hesperia. The upper of the two curves is a theoretical fit to the observations, using Eq. (5), with $V(0°) = 7.17$ mag and $Q = 0.16$. The rms residual is 0.02 mag. Symbols denote observations made before (●) and after (▲) opposition. Uncertain observations, in parentheses, were not used in the curve fitting. The lower curve represents the phase function for a surface with no multiple scattering.

$0°.03$. The possibility of measuring the brightness of an asteroid at smaller phase angles than hitherto provided an excellent opportunity to make a quantitative determination of the porosity and roughness of an asteroid surface, and thereby to estimate the phase curve of a surface in which there is no multiple scattering.

In practice, deriving the phase curve of a very dark asteroid from that of a moderate-albedo object like 69 Hesperia is not straightforward because multiple scattering must be allowed for. Fortunately, the amount of multiply scattered light does not vary rapidly with phase angle, and so the shape of Hesperia's phase curve has enabled us to deduce the shape of the "zero-albedo" phase curve, even though we may have misestimated the amount of multiple scattering appropriate to Hesperia's surface. (Actually, we also used Gehrels' (1956) observations of 20 Massalia, which has a phase curve identical to Hesperia's.) Figure 3 (from Bowell et al. 1979) shows V-band observations of Hesperia together with the calculated phase curve m_1, resulting from single scattering only and normalized at $\alpha = 0°$. Numerical values of the mean porosity D and mean large-scale roughness ρ can be determined from m_1. Preliminary values are $D \approx 0.4$ and $\rho \approx 1$, and these imply that Hesperia's surface is moderately porous (uniformly packed smooth spheres would have $D = 0.74$) and moderately rough on a scale larger than the wavelength of light. Our assertion that the optically active surface layer is made up of separate particles is consistent with an inference that has been repeatedly drawn from optical polarimetry (e.g. Dollfus et al. 1977).

Photometry by itself cannot be used to discriminate the sizes of the particles involved, although it is clear that neither relief features on a scale comparable with the size of the asteroid (e.g. large craters) nor asperities smaller than the wavelength of light play an important role in the phase-angle dependence of the light-scattering properties. The mean surface slope (and therefore ρ) associated with large-scale features cannot be large, and significant numbers of Mie scatterers cause a strong wavelength dependence in the surface brightness.

B. Albedo effects

If one considers a suite of surfaces in which only the geometric albedo varies, then, at a given phase angle, the disk-integrated brightness depends on the intensity of multiply scattered light:

$$L_{obs}(\alpha) \sim L_1(\alpha) + L_M(\alpha) \tag{1}$$

where L_1 is the contribution from single scattering and L_M that from multiple scattering. This may be normalized to zero phase and rewritten as

$$\Phi_{obs}(\alpha) = \frac{L_1(\alpha) + L_M(\alpha)}{L_1(0°) + L_M(0°)} = Q + (1-Q)\,\Phi_1(\alpha) + r(\alpha) - 1 \tag{2}$$

where

$$\Phi_{obs}(\alpha) = \frac{L_{obs}(\alpha)}{L_{obs}(0°)}, \quad \Phi_1(\alpha) = \frac{L_1(\alpha)}{L_1(0°)},$$

$$Q = \frac{L_M(0°)}{L_1(0°) + L_M(0°)}, \quad r(\alpha) = \frac{L_1(0°) + L_M(\alpha)}{L_1(0°) + L_M(0°)}. \tag{3}$$

Of these quantities, only Q has a real physical significance, and we have named it the *multiple-scattering factor;* it is the ratio of multiply scattered light to the total scattered light at zero phase angle. Now, $L_M(\alpha)$ is insensitive to phase angle in the range of interest ($0° \leqslant \alpha \lesssim 30°$) because multiply scattered light emerges in more-or-less random directions, so that

$$r(\alpha) \approx 1$$

and

$$\Phi_{obs}(\alpha) \approx Q + (1-Q)\,\Phi_1(\alpha). \tag{4}$$

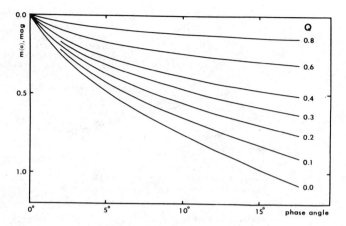

Fig. 4. Theoretical phase curves for surfaces with differing multiple-scattering factor Q, but with fixed $\rho = 0.4$, $D = 1.0$, and $g = 0$.

Converting intensities into magnitudes:

$$10^{-0.4 m(\alpha)} = a_0 + a_1 \, 10^{-0.4 m_1(\alpha)}$$

where

$$a_0 = Q \, 10^{-0.4 m(0°)} \,, \quad a_1 = (1-Q) \, 10^{-0.4 m(0°)} \,. \quad (5)$$

Here, $m(\alpha)$ is the theoretical representation of the phase curve of an object with multiple-scattering factor Q; a_0 and a_1 are constants for the object; and $m_1(\alpha)$ is the phase curve for a very dark body having the same surface texture. A good numerical approximation for m_1 is:

$$m_1(\alpha) = 0.067 \, \alpha^{0.785} + \alpha/(1.36\alpha + 14.73) \,; \quad 0° \leqslant \alpha \leqslant 30°. \quad (6)$$

As already stated, the kernel of the scattering theory is the rigorous derivation of this function in terms of D and ρ; the numerical coefficients have been derived from values of these parameters that originate from observations of Hesperia and Massalia.

Equations (5) and (6) represent a family of phase curves that describe the variation in brightness of a hypothetical suite of surfaces in which only the geometric albedo (and therefore the multiple-scattering factor) varies. The curves are plotted in Fig. 4. As the multiple-scattering factor increases, the slope of the phase curve decreases, most noticeably away from zero phase. The maximum slope occurs as Q tends to zero.

We have compared the observed phase curves of more than 60 asteroids, the Galilean satellites, some satellites of Saturn, the moon, and Mercury, with

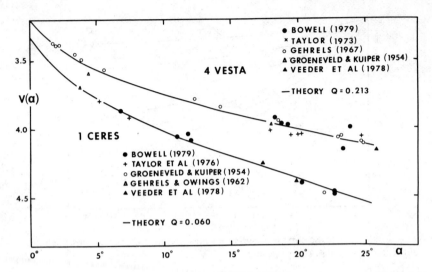

Fig. 5. The v-band observations of Ceres and Vesta, and fitted phase curves. No distinction is made between magnitudes averaged over rotation and individual observations.

Eq. (5); we find *without exception* that theory fits data within the observational error. Figure 5 shows observational data and theoretical phase curves for the dark *C* asteroid Ceres and the moderately high-albedo *U* asteroid Vesta, and Fig. 6 similarly treats two-color observations for the ice-covered Galilean satellite Europa.

C. Photometric albedos and diameters

Using observed phase curves, and diameters derived from radiometry, polarimetry, and other methods, we next consider the relationship between multiple-scattering factor and geometric albedo. Figure 7 is a plot of these quantities for Mercury, various satellites, and asteroids. The geometric albedos pertain explicitly to zero phase and are numerically larger than those conventionally used by a factor of 1.34. The notation $p(0°)$ is used for the zero-phase albedo in order to distinguish it from the albedo currently adopted (see below and Sec. II-F).

The correlation between Q and p $(0°)$ is quite good for almost all the objects considered.[a] This implies that, of our four model parameters, $D, \rho, g,$ and $\tilde{\omega}_o$, only $\tilde{\omega}_o$ differs greatly from object to object. There appears to be a small difference in the Henyey-Greenstein asymmetry factor between

[a]Two intransigent exceptions are Io, whose surface scattering properties behave peculiarly with wavelength, and 64 Angelina, discussed below. In contrast, we note the unique satellite S8 (Iapetus). The extreme photometric difference between its leading and trailing hemispheres can be attributed entirely to large-scale albedo variegation, in complete confirmation of Zellner's (1972) finding from polarimetry.

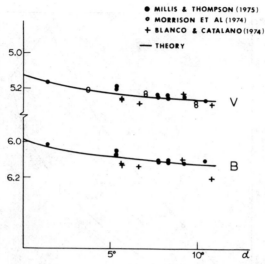

Fig. 6. V— and B—band observations of Europa at orbital phases $65° \leqslant \theta \leqslant 115°$, and fitted phase curves.

asteroids $(g \sim -0.1)$ and the larger solar system objects considered $(g \sim +0.1)$, so that albedo loci for the two classes of object follow slightly different branches in the $Q,p(0°)$ plane.[b] However, this difference in g is small when one realizes that the theoretical bounds are ± 1.0 (a Lambert sphere has $g = -0.3$). We therefore state that, to a first approximation, the surfaces of almost all asteroids have similar scattering properties because their surface structures, characterized by porosity and roughness, are similar.

In many ways this uniformity among asteroids is disappointing; but as a recompense, the good correlation between Q and $p(0°)$ does afford a purely photometric method of estimating geometric albedos and diameters. Using radiometric and polarimetric diameters for asteroids with well-determined phase curves and absolute magnitudes, we derive:

$$p(0°) = 1.252 \, Q + 0.039 \pm 0.04; \quad 0.05 < Q < 0.35. \tag{7}$$

The "uncertainty" term reflects the fact that, for asteroids of all albedos, there seems to be some real dispersion about a linear relationship. (In other words, there do appear to be *small* differences in scattering properties among asteroids.) The nonzero ordinate could result from a systematic variation of g with $p(0°)$ or from a slight miscalibration of the scale of Q (see below). For the other solar system objects represented in Fig. 7, an equation slightly different from (7) should be used.

[b]A possible explanation could be that asteroid surfaces comprise slightly more opaque particles than those of the other objects involved. As particle transmission increases, so too do both $\widetilde{\omega}_0$ and g.

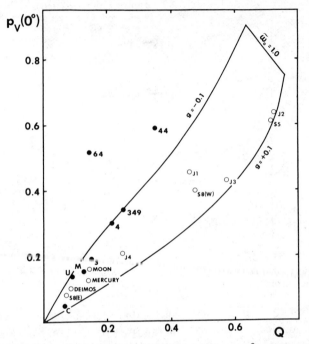

Fig. 7. Multiple-scattering factor Q versus geometric albedo $p(0°)$ for planetary satellites and Mercury (○), and asteroids (●) in the V band. Mean parameters are shown for large samples of C, U, M and S asteroids. Loci for constant values of the Henyey-Greenstein asymmetry factor g, and the upper bound for the single-scattering albedo of a particle $\tilde{\omega}_0$ are also shown.

To derive photometric albedos and diameters, one proceeds as follows: From Eq. (5), Q and $m(0°)$ are derived by least squares or graphically (using Fig. 4, for example). Then Eq. (7) is used to estimate the geometric albedo, and a relationship of the type

$$\log d = k - 0.5 \log p(0°) - 0.2m(0°) \qquad (8)$$

serves to determine the photometric diameter d. k is a wavelength-dependent constant; if d is in km, then $k = 3.122$ for the V band and 3.248 for the B band (from Bowell and Zellner 1974). Obviously, from Eq. (7), the albedos and diameters of C asteroids $[p_V(0°) \approx 0.05]$ cannot be determined by this method.

Table I lists photometric diameters for a number of asteroids. Only high-quality photometric data are quoted, and low-albedo asteroids are excluded. The mean 1σ uncertainties in Q_V and $V(0°)$ are ± 0.015 and ± 0.032 mag, respectively. For comparison, diameters derived from radiometry, polarimetry, or the stellar occultation method are given, and in general the

TABLE I

Photometric Diameters of Asteroids

Asteroid	Type	Q_V	$V(0°)^a$	d_{phot}	d_{other}^b
2 Pallas	U	0.05	4.09	618 ± 139	538
4 Vesta	U	0.21	3.20	564 ± 61	555
6 Hebe	S	0.15	5.71	201 ± 26	206
12 Victoria	S	0.12	7.13	114 ± 19	135
15 Eunomia	S	0.10	5.21	296 ± 55	261
20 Massalia	S	0.15	6.50	141 ± 21	140
22 Kalliope	M	0.14	6.52	142 ± 21	175
29 Amphitrite	S	0.14	5.87	195 ± 30	199
63 Ausonia	S	0.16	7.72	83 ± 16	94
64 Angelina	E	0.14	7.44	92 ± 15	60
69 Hesperia	M	0.16	7.17	99 ± 10	108
79 Eurynome	S	0.14	7.86	76 ± 9	80
110 Lydia	U	0.12	7.80	83 ± 12	102
349 Dembowska	R	0.25	5.99	141 ± 12	144

[a]True zero-phase magnitude, calculated from Eq. (5) and at mean brightness (averaged over rotation).

[b]Polarimetric and/or radiometric diameters from the TRIAD file, except for 2 Pallas, where the diameter is from Wasserman et al. (1979; see the chapter by Millis and Elliot).

two values agree within the supposed errors of the various methods. For S asteroids, the photometric method appears to yield diameters within 20%. We do not advise that the photometric method of diameter determination — at least in the form presented here — be used to estimate asteroid diameters in the way that the radiometric and polarimetric methods have been. Indeed, photometry alone does not lead to diameters that are any more reliable than those calculated from albedos assumed on the basis of taxonomic type.

For a small sample of asteroids, there exist radiometric, polarimetric, and good photometric albedos. We have examined linear regressions among the three albedo scales, and find them all to be highly correlated. Somewhat surprisingly, the highest correlation exists between the photometry and polarimetry — even higher than that between radiometry and polarimetry — but this difference may not be significant. However, it is possible that, since optical wavelengths are involved in both cases, similar physical parameters of asteroid surfaces control the polarimetry and photometry, yet do not have the same effect at radiometric wavelengths. Lumme (1979a) is exploring the interrelationship between polarimetry and photometry.

One unexpected discrepancy between photometric and other diameters concerns 64 Angelina, an E asteroid (Table I). Radiometry and polarimetry both lead to a zero-phase albedo near 0.45, but a preliminary value of the

TABLE II

**Relationships Among the Multiple-Scattering Factor Q,
the Linear Phase Coefficient β, and the Zero-Phase
Geometric Albedo $p(0^\circ)$ for Asteroids**

Q	β (mag/deg)	$p(0^\circ)$
0.00	0.042	0.00
0.05	0.037	0.07
0.10	0.033	0.14
0.15	0.029	0.20
0.20	0.026	0.27
0.30	0.020	0.41
0.40	0.016	0.55

multiple-scattering factor suggests an albedo of only 0.20, similar to that of S asteroids (Fig. 7). We note that Q and $p(0^\circ)$ for the other E asteroid, 44 Nysa, also seem to be "discordant", but poorly known values ($Q_V = 0.35 \pm 0.12$ and $p_V(0^\circ) = 0.60 \pm 0.10$) could be the cause.

A corollary of the good correlation between Q and $p(0^\circ)$ is that the slope of the phase curve away from zero phase is albedo controlled. Thus there exists a close relationship between the phase coefficient β and albedo. Table II lists values of β and $p(0^\circ)$ corresponding to various Q. Q and $p(0^\circ)$ are connected by equation (7), and β is defined by

$$\beta = [m(20^\circ) - m(10^\circ)]/10 \tag{9}$$

where $m(20^\circ)$ and $m(10^\circ)$ are calculated from Eqs. (5) and (6). Good numerical approximations to the relationship between Q and β (in mag/deg) are

$$Q = 0.821 - 30.76\,\beta + 267.6\,\beta^2\,;\ \beta > 0.015 \text{ mag/deg,}$$
$$\beta = 0.0417 - 0.0936\,Q + 0.0739\,Q^2\,;\ Q < 0.5. \tag{10}$$

There has been lively discussion in the literature for many years as to whether phase coefficients are related to albedo. An empirical relationship developed by Stumpff (1948) has been used by many workers. Veverka's (1971) pessimistic view, that the phase coefficient of an asteroid is roughness -controlled and therefore not diagnostic of albedo, is the most carefully reasoned recent discussion. If he is correct, then a plot of data in the Q, $p(0^\circ)$ plane (Fig. 7) could resemble a scatter diagram. Essentially, we agree with Veverka, since we predict that differences from object to object in the large-scale roughness or the Henyey-Greenstein asymmetry factor would destroy

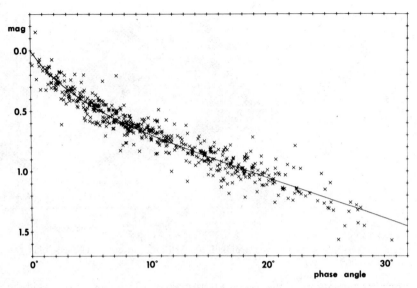

Fig. 8. Mean phase curve for C asteroids. From 512 observations of 166 asteroids, $Q = 0.064 \pm 0.008$.

the correlation between Q and $p(0°)$ but these differences appear to be small.

As already mentioned (Sec. II-A), we may have misestimated the amount of light multiply scattered from Hesperia, and therefore all asteroids. Indeed, there is evidence from other phase curves that this is so; we note from Eq. (6) and Table II that the maximum possible phase coefficient is 0.042 mag/deg, a value that is exceeded in the case of a few low-albedo asteroids. The values of Q given in this chapter should therefore be taken as provisional. In a final analysis Q will be slightly rescaled using a linear transformation. None of the conclusions given here will be qualitatively altered by this adjustment.

D. Statistics of phase curves

A first attempt at describing the phase curves of a large number of asteroids in a statistical way was made by Bowell (1977). Using a much larger data base, we elaborate on the earlier findings and, additionally, use the multiple-scattering theory as an interpretative tool. First, we establish that there are characteristic values of Q_V for each of the principal taxonomic types, and then we use subsets of the data to investigate the dependences on diameter and semimajor axis.

Figures 8 and 9 show theoretical phase curves fitted to the complete data sets of C and S asteroids, respectively. Equations (5) and (6) describe the fitted functions illustrated. We have verified that at phase angles larger than $10°$ the multiple-scattering theory fits the data significantly better than the linear approximation hitherto used. Scaltriti and Zappalà (1979) also refer to

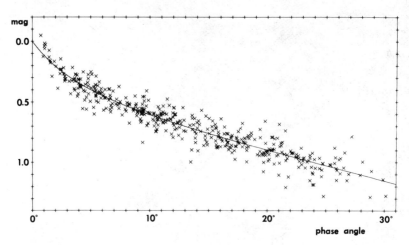

Fig. 9. Mean phase curve for S asteroids. From 416 observations of 132 asteroids, $Q = 0.150 \pm 0.009$.

a possible curvature, on the order of 0.01 mag, of asteroid phase curves in the range $10° < \alpha < 20°$. The higher-albedo S asteroids obviously conform to a less steep phase curve than C asteroids, in accordance with our findings in Secs. II-B and C.

The method of determining the optimum value of Q is as follows: A trial value of Q is selected and individual observations $V(\alpha)$ are combined to obtain $V(0°)$ for each asteroid. Calculated $V(\alpha)$ are then compared with the observations, and rms magnitude residuals characterize the closeness of fit. The results for the entire sample of asteroids are combined. (Of course asteroids for which there are only single observations are excluded.) In this way each observation is assigned equal weight, but obviously observations at widely separated phase angles have greater "leverage". The process is repeated for various Q until a best-fit $\bar{Q}$ is established. The uncertainty in this value has been determined by estimating the variance in $\bar{Q}$ for a number of subsets of the total sample, and by relating the variance to the number of observations in each subset.

Table III lists mean values of Q_V, β_V and $p_V(0°)$ for the total samples of C, U, M, S, R and E asteroids. These values have been derived from a total of 356 asteroids, and more than 1100 individual observations have been considered. As expected, Q and β are strongly correlated with $p(0°)$. Table IV gives numerical values for the theoretical mean phase curves of C, M, and S asteroids in the V passband.

None of the data presented above give any indication whether there is significant intrinsic dispersion in Q for individual asteroids of a given taxonomic type. We estimate, from individual well-determined phase curves, that the maximum departure from $\bar{Q}$ for C, M, and S asteroids is about 0.05. It can be seen from Table III that this implies a continuum of values for Q from

TABLE III

Mean Multiple-Scattering Factor $\bar{Q}_V$, Linear Phase Coefficient $\bar{\beta}_V$, and Geometric Albedo $p_V(0°)$ for Different Taxonomic Types

Type	No. of Asteroids	No. of Obs. per Sample	Q_V	$\beta_V{}^a$ (mag/deg)	$p_V(0°)$
C	166	512	0.064 ± 0.008	0.0359 ± 0.0007	0.049
U	40	128	0.090 ± 0.022	0.0337 ± 0.0027	0.136
M	16	56	0.126 ± 0.042	0.0310 ± 0.0032	0.161
S	132	416	0.150 ± 0.009	0.0293 ± 0.0006	0.191
R^c	1	10	0.25 ± 0.02	0.023 ± 0.002	0.34
E^d	1	—	0.35 ± 0.12	0.018 ± 0.006	0.59

[a] $\bar{\beta}_V$ is defined by Eq. (9) and derived from Eqs. (5) and (6).
[b] The $\bar{p}_V(0°)$ are mean zero-phase albedos for asteroids observed by means of radiometry and/or polarimetry.
[c] For R asteroids, data for 349 Dembowska by Zappalà et al. (1979) are used.
[d] For E asteroids, data for 44 Nysa by Gehrels and Owings (1962), Groeneveld and Kuiper (1954), and Veeder et al. (1978) have been analyzed.

TABLE IV

Theoretical Mean Phase Curves (V Passband)
for *C, M* and *S* Asteroids[a]

Phase Angle (deg)	Taxonomic Type		
	C	*M*	*S*
0	0.000	0.000	0.000
0.5	0.068	0.063	0.062
1	0.128	0.119	0.116
2	0.230	0.213	0.206
3	0.314	0.290	0.281
4	0.385	0.355	0.344
5	0.448	0.412	0.398
6	0.504	0.463	0.447
8	0.603	0.551	0.531
10	0.689	0.627	0.604
15	0.877	0.791	0.760
20	1.048	0.937	0.897
25	1.215	1.076	1.027
30	1.385	1.214	1.154

[a]Entries give the brightness drop in magnitudes measured from zero phase.

0.0 to at least 0.2. A decision about the best estimate for the average phase curve of main-belt asteroids is made below.

We divide asteroid semimajor axes into fifteen zones, following Zellner (see his chapter), and we treat *C* and *S* asteroids separately. Most of our data pertain to zones 6 (main-belt I), 7 (main-belt II), 11 (main-belt III), and 12 (main-belt IV); we lack sufficient observations to say anything about the outermost zones (13, Hilda and 14, Trojans), although some findings have been given in the chapter by Degewij and van Houten. For our total samples (i.e. all diameters) of both *C* and *S* asteroids, we see no significant trend in $\bar{Q}$ with semimajor axis. Thus the scattering properties of the majority of asteroids in the belt do not seem to vary significantly with heliocentric distance. There may be rare exceptions; for example, Themis family members (zone 10) appear to have larger *Q* (smaller phase coefficient) than other low-albedo asteroids; but the difference, although pronounced, is not statistically significant. Jonathan Gradie (personal communication) points out that 171 Ophelia has $p_V(0°) = 0.133$. Perhaps the Themis family does not consist entirely of normal *C* asteroids.

In order to investigate the dependence on diameter, it must be assumed that the asteroid light-scattering properties are independent, not only of semimajor axis, but also of other orbital parameters. We have not yet verified that this is so. We have subdivided the data for *C* and *S* asteroids into three diameter ranges, each containing almost equal numbers of observations.

TABLE V

Variation of Mean Multiple-Scattering Factor $\bar{Q}$
with Median Diameter

Type	Diameter Range (km)	Median Diameter (km)	$\bar{Q}$
C	1025–150	194	0.087 ± 0.018
	149–104	126	0.057
	103–27	76	0.033
S	261–96	138	0.170 ± 0.021
	95–57	72	0.132
	54–10	38	0.140

Table V indicates significant correlations between multiple-scattering factor and median diameter for both C and S asteroids. In both cases $\bar{Q}$ is larger for the larger asteroids, and the rate of change of $\bar{Q}$ with diameter appears to be the same for both C and S asteroids. Interpretation of this result in terms of the multiple-scattering theory is not straightforward, however. We consider two plausible hypotheses and examine their ramifications. For simplicity, only C asteroids are discussed.

First, we postulate that small C asteroids are, on the average, darker than large ones, in such a way that Eq. (7) is satisfied. The average V-band albedo of 100-km-diameter C asteroids would then be 0.051, and that of 200-km-diameter asteroids would be 0.061. This difference is small enough that its reality cannot clearly be decided on the basis of existing observations. In particular, a plot of radiometric albedo versus diameter does not exclude such an effect, but data are almost completely lacking for asteroids smaller than 100 km in diameter. There is a further complication; radiometric albedos of C asteroids depend on absolute magnitudes derived from observations fitted by a common phase curve, and this simplification conflicts with our finding here. If we repeat our diameter-dependence analysis holding Q fixed, we find that there is no longer a correlation between geometric albedo and diameter. We therefore suggest that the postulated variation of albedo with diameter could be an artificial result of the way in which the data have been reduced (see also Sec. II-F). In any case, it is difficult to think of any underlying physical cause for such a finding.

We are then led to suppose that the albedos of C asteroids are independent of diameter. This leads to a horizontal locus in the $Q,p(0°)$ plane, in "violation" of the correlation between Q and $p(0°)$ found when asteroids of all diameters are considered (Eq. [7]). Clearly, this result requires that there are surface textural differences that depend on diameter. In terms of the multiple-scattering theory, the finding would be consistent with an increase in the roughness with decreasing diameter. This has also been suggested by Tedesco (1978), but the effect we are considering here is several times smaller. Perhaps the micro-roughness is gravity controlled; on the larger asteroids, gravitationally induced regolith compaction could be perceived photometrically as reduced surface roughness. However, there are other possible explanations; for example, variation in the distribution of particle sizes with asteroid diameter or the presence of different but *small* quantities of sub-micron particles (Mie scatterers) could be responsible. The latter possibility is outside the bounds of the multiple-scattering theory in its present state of development.

Finally, we ask the question: What is the "average" phase curve for the main belt? We seek to determine some mean Q for asteroids having semimajor axes in the range $2.2 < a < 3.4$, as a function of limiting mean opposition magnitude $V_{lim}(0°)$. Evidently, the factors to be taken into account are the relative proportions of S, M, and C asteroids (the other types are rare and can be neglected), the variations in these abundances with semimajor axis and diameter, and the diameter itself. Bias-corrected distributions of asteroids over diameter and semimajor axis according to Zellner and Bowell (1977) are used, and three assumptions are necessary: (1) that the distributions over diameter for S and C asteroids are independent of semimajor axis;[a] (2) that the trends of Q with diameter (Table V) are linear; and (3) that, in view of their similar albedos and phase curves, S and M asteroids can be grouped together.

Smoothed values of $\bar{Q}$ resulting from our analysis are shown in Fig. 10. Also indicated are average diameters sampled throughout the belt for C and $S+M$ asteroids. The curves have slopes that are controlled by the variation of $\bar{Q}$ with diameter (Table V); and, since they diverge slightly with increasing $V_{lim}(0°)$, the variance in $\bar{Q}$ for the total observable sample also increases at fainter limiting magnitudes. We have good sampling statistics down to $V_{lim}(0°) \approx 15$. The horizontal line at $\bar{Q} = 0.036$ corresponds to the effective multiple-scattering factor for the unbiased population. Thus, if the true ratio of $S+M$ to C asteroids is independent of diameter, this line should asymptotically meet the curve labeled "all", appropriate to the observable population, as $V_{lim}(0°) \rightarrow \infty$.

Interestingly, the line $\bar{Q} = 0.036$, for which $\beta_V = 0.038$ mag/deg, corresponds well with van Houten *et al.*'s (1970) mean phase coefficient ($\beta = 0.039 \pm 0.002$ mag/deg) for small, faint Palomar-Leiden survey asteroids. Also, from the numbers of asteroids analyzed in Table III, one might infer that Bowell's (1979) UBV survey of asteroids is "complete" to mean opposition magnitude $V_{lim}(0°) = 12.75$; fainter asteroids have been observed, but their number is

---

[a]This has recently been found by Zellner (see his chapter), whose results were not available to us at the time our chapter was assembled. Incorporation of Zellner's bias-corrected type frequencies does not significantly change what follows.

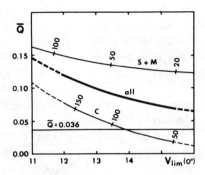

Fig. 10. Mean multiple-scattering factor $\bar{Q}$ as a function of limiting magnitude $V_{lim}(0°)$ for C, $S+M$, and all observable numbered main-belt asteroids. Numbers on curves indicate mean diameters in km. Dashed portions of curves pertain to diameter ranges for which statistics are poor. $Q = 0.036$ corresponds to the effective multiple-scattering factor for an unbiased asteroid population.

offset by unobserved brighter asteroids. Calculated $\bar{Q}$ are 0.060 for C and 0.147 for $S+M$ asteroids, in almost perfect agreement with values from Table III (0.064 and 0.147, respectively).

For the 1941 asteroids comprising the numbered main-belt population (as of March 1979), the mean opposition magnitude is $V(0°) = 14.0$ (well within the range of good sampling statistics), and so one might assign $\bar{Q}_V = 0.085$, corresponding to $\bar{\beta}_V = 0.034$ mag/deg. This phase coefficient is somewhat smaller than that chosen by Gehrels and Tedesco (1979) for the calculation of absolute magnitudes; we discuss their value in Sec. II-G. We estimate the rms dispersions in $\bar{Q}_V$ and $\bar{\beta}_V$ to be ±0.05 and ±0.004 mag/deg, respectively; these arise from the difference between the C and $S+M$ taxonomic components, from integrating over diameter, and from intrinsic variety in surface scattering properties.

E. The opposition effect

As remarked in Sec. II-A, it is unusual that asteroids can be observed close enough to zero phase for the opposition effect to be properly delineated. Gehrels' (1956) fine observations of 20 Massalia were the first proof of such an effect for an asteroid, and subsequent observations of other asteroids established that a surge in brightness near zero phase is a general phenomenon. The phase curve of 69 Hesperia (Fig. 3, from Bowell *et al.* 1979) is the best-defined example of the opposition effect.

Gehrels and Tedesco (1979) and Scaltriti and Zappalà (1979), following earlier work by Gehrels and Taylor (1977), have shown that the opposition effect is very similar for all asteroids for which good data exist. However, there is no accepted definition of the opposition effect. Gehrels and Tedesco define it as the nonlinear part of the phase curve at phase angles less than 7°

whereas Scaltriti and Zappalà find that the effect pertains to phase angles out to 8°.5. Several empirical equations have been derived to match asteroid phase curves at small phase angles, all of them being power laws of one sort or another. An excellent two-parameter representation of asteroid phase curves is $m(\alpha) = u \sin^v \alpha$. Here $m(\alpha)$ is the drop from zero phase and u and v are constants.

The view that we wish to put forward in this chapter is that the opposition effect for asteroids does vary from asteroid to asteroid, although only by small amounts, and that its size and extent vary with albedo. We have analyzed observations of seven well-observed asteroids, made at phase angles less than 8°, to calculate values of Q. Although the data are noisy, the correlation between Q and $p(0°)$ is highly significant and quite similar to that expressed by Eq. (7).

It should be pointed out that, if phase curves are matched by eye, rather than analytically, differences will be hard to perceive because of the restricted range of phase angles. As an example, it will be seen from Table IV that the drop in light predicted between $\alpha = 3°$ and $\alpha = 8°$ is 0.289 mag for C and 0.250 mag for S asteroids. Matching the curves by aligning them at their midpoints, one will see a difference of only $0.5(0.289-0.250) = 0.020$ mag, whereas if the true zero phase magnitude were known in each case, the light drops would be 0.603 mag for C and 0.531 mag for S asteroids – the difference then being 0.072 mag.

It is intriguing to note, as Scaltriti and Zappalà (1979) have done, that although the predicted differences in magnitude - drop from zero phase are discernibly different for different taxonomic types, the excess brightness near zero phase over a linear extrapolation from larger phase angles is very similar for all asteroids. Thus, if the magnitude is extrapolated by a line passing through V(10°) and V(20°), the excess at $\alpha = 0°$ is predicted to be 0.33 mag for C and 0.31 mag for S asteroids (see also Sec. II-F). This similarly is a direct consequence of phase curves being made up of a surface-texture-controlled component (the zero-albedo phase curve m_1) and a component due to multiple scattering.

F. Absolute magnitudes

In this section we examine the difference between our definition of the absolute zero-phase magnitude V(0°) and the currently used V(1,0). We also determine the effects of the choice of phase coefficient (or multiple-scattering factor) on the absolute magnitudes of asteroids, and therefore on diameters and albedos derived from radiometry and polarimetry.

Customarily, the absolute magnitude at zero phase has been calculated by an extrapolation of the psuedo-linear part of the phase curve ($\alpha \gtrsim 8°$). For asteroids with unknown phase coefficients, an average value has been chosen. Thus, in the compilations of Gehrels (1970) and Gehrels and Gehrels (1978),

$\beta_B = 0.023$ mag/deg was used, based on photoelectric and photographic photometry of bright asteroids. In the most recent compilation (Gehrels and Tedesco 1979), a larger phase coefficient (0.039 mag/deg) was chosen to represent the phase-dependent variation in brightness of a population of asteroids that is not biased by observational selection; where known, other phase coefficients were used for individual asteroids.

Although the opposition effect is explicitly excluded from definition of B(1,0), it is allowed for, in an average way, when individual magnitudes are incorporated. Thus, for each observation (or magnitude averaged over a rotation, where known), Gehrels and Tedesco express the magnitude drop from zero phase to phase angle α (in degrees) as

$$B(1,\alpha) - B(1,0) = -0.538 + 0.134 \, \alpha^{0.714} + 7\beta_B; \quad 0 \leqslant \alpha < 7°. \qquad (11)$$

For observations made at larger phase angles

$$B(1,\alpha) - B(1,0) = \alpha\beta_B; \quad \alpha \geqslant 7°. \qquad (12)$$

Gehrels and Tedesco's "standard" phase coefficient $\beta_B = 0.039$ mag/deg corresponds to a multiple-scattering coefficient $Q = 0.029$, and so we may compare their phase relationship with that given by Eqs. (5) and (6) using these values. The agreement is very good; between $0°$ and $20°$ phase angle the two phase functions have almost identical shapes, but differ in absolute value by 0.32 ± 0.03 (s.d.) mag. This offset reflects the difference in definition of zero-phase magnitude. Moreover, if comparisons between the two functions are made using various Q in the range $0 < Q < 0.15$ and corresponding β_B, it is found that the offset is constant to within 0.01 mag. The difference between the phase functions is, not unexpectedly, confined to small phase angles and is a consequence of the weak albedo dependence of the opposition effect predicted by Eqs. (5) and (6) (and discussed in Sec. II-E). This dependence is not included in Gehrels and Tedesco's treatment.

As a good approximation, we have for all asteroids:

$$V(0°) = V(1,0) - 0.32 \text{ mag}. \qquad (13)$$

This equation also applies in other passbands. It follows that, for asteroids of given diameter, the geometric albedo calculated from $V(0°)$ is 1.34 times that calculated from $V(1,0)$. The mean albedos for different taxonomic types given in Table III are, in this respect, entirely consistent with albedos chosen by Bowell et al. (1978) in their taxonomic study.

In Sec. II-D we asserted that a phase coefficient smaller than 0.039 mag/deg (and therefore a multiple-scattering factor larger than 0.029) is appropriate to the numbered main-belt population. This is because the numbered population is biased by observational selection. In particular, C asteroids are under-represented, and objects larger than 100 km in diameter await discovery in the outer parts of the main belt. In fact, we estimate that the $C:C+S+M$ number ratio is 0.50, whereas in an unbiased sample this ratio is about 0.78. Thus only about half the numbered asteroids are likely to be C types, and $S+M$ asteroids are over-represented by a factor of two. As Fig. 10 suggests, only for very faint limiting magnitudes should the multiple-scattering factor for observable asteroids approach that of the total, unbiased population.

With these considerations, we ask: What are the effects of changing $\bar{Q}$(or $\bar{\beta}$) for the numbered asteroids? The amount of change $\Delta V(0°)$ in $V(0°)$ resulting from a change $\Delta\bar{Q}$ in $\bar{Q}$ is approximately proportional to the mean phase angle $\bar{\alpha}$ at which observations are made. Examining Eqs. (5) and (6), we find, for small $\bar{Q}$:

$$dV(0°) = -0.1\ \bar{\alpha}d\bar{Q} \tag{14}$$

where $\bar{\alpha}$ is in degrees. For illustration, we choose $\bar{\alpha} = 10°$, a not untypical value, and suppose that $\bar{Q}$ is changed from 0.029 (corresponding to $\beta = 0.039$ mag/deg, as in Gehrels and Tedesco 1979) to 0.085 ($\beta = 0.034$ mag/deg, Sec. II-D). Then $dV(0°) \approx -0.06$ mag.

A change in $V(0°)$ affects radiometric and polarimetric determinations of diameters and albedos. The photometric/radiometric method (Morrison 1977) is mainly sensitive to diameter, whereas polarimetry (Zellner *et al.* 1977*b,c*) discriminates albedo. Now, from Eqs. (8) and (14), the changes dD in diameter and $dp\ (0°)$ in geometric albedo are:

$$\frac{dD}{D} = -0.5\ \frac{dp(0°)}{p(0°)} \approx 0.05\ \bar{\alpha}dQ. \tag{15}$$

Continuing our numerical example, we see that diameters are increased by ~3% and albedos are decreased by 6%.

Clearly, since individual asteroids have multiple-scattering factors that depart from the mean, there will be errors in albedo and diameter caused by misestimating $V(0°)$. If an average multiple-scattering factor is used for the entire numbered main-belt population, then the errors will be both type-dependent and albedo-dependent. Figure 10 indicates that multiple-scattering coefficients appropriate to all numbered C and $S+M$ asteroids are 0.037 and 0.133, respectively. Thus, if a mean $\bar{Q} = 0.085$ is chosen for all asteroids, C asteroids could on the average have calculated diameters 3% *greater* than the best possible estimate, and $S+M$ asteroids could have calculated diameters 3% *less* than the best estimate. Thus, an error of 6% could be made in the diameters of C relative to $S+M$ asteroids. If, on the other hand, different mean values of the multiple-scattering factor are used for C and $S+M$

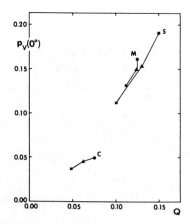

Fig. 11. Observed phase reddening of *C*, *M* and *S* asteroids plotted in the *Q,p* plane. Symbols indicate passband: ●V, ▲B, ■U.

asteroids, the type-dependent error should be largely removed. The remaining diameter-dependent error is likely to be about ±2% for all asteroids and have effect in the sense that the diameters of the largest asteroids would be overestimated and those of the smallest asteroids underestimated.

A comparison of diameters listed in the TRIAD file in Part VII of this book with those given by Bowell *et al.* (1978) indicates that the effects described above have been active; the TRIAD file diameters are, on the average, larger. Since all these ills are dependent on the mean phase angle of observation $\bar{\alpha}$, one should be especially cautious about diameters and albedos of asteroids for which $\bar{\alpha}$ is large and for which β is not known. An obvious remedy is to secure observations at small phase angles. Ben Zellner (personal communication) has pointed out that, in addition to the relative diameter errors discussed above, an absolute error in the radiometric and polarimetric diameter scales could have been introduced from a number of sources, including an error in the accepted absolute magnitude of the sun.

G. Phase reddening

According to Gehrels (1970), the discovery that asteroids become slightly redder as phase angle increases is due to Giclas (1951) and Haupt (1958). In recent years, evidence from an increasing number of phase-curve observations has led to the belief that the effect is general.

We have investigated phase reddening separately for *C,M,* and *S* asteroids, by determining best-fitting phase curves for observations in V, B, and U. The results are plotted in the $Q,p(0°)$ plane in Fig. 11. Clearly, loci of increasing wavelength are directed away from the origin, and this suggests that phase reddening is largely a differential albedo effect. Thus, at increasing wavelengths, as the geometric albedo increases, there is an accompanying increase in *Q*. (Loci of phase reddening in the B–V, U–B plane would be directed

TABLE VI

Observed and Calculated Phase Reddening
in B–V and U–B

Type	Q_V-Q_B Obs.	Q_V-Q_B Calc.	Q_B-Q_U Obs.	Q_B-Q_U Calc.
C	0.013 ±0.007	0.005	0.015 ±0.009	0.012
M	0.001 ±0.012	0.009	0.010 ±0.015	0.015
S	0.019 ±0.009	0.029	0.029 ±0.011	0.032

away from the point having solar colors.) Furthermore, the amount of phase reddening, characterized by the lengths of the lines in Fig. 11, depends on the color indices of asteroids. In U–V, for example, the phase reddening for S asteroids is two or three times that for C and M asteroids.

Table VI gives phase reddening in B–V and U–B as determined from the observations and as calculated on the assumption that differential albedo effects are responsible. The agreement is tolerably good. For the observationally biased numbered main-belt population we calculate average reddening with phase to be $Q_V-Q_B = 0.015$ and $Q_B-Q_U = 0.022$. The first of these corresponds to $\beta_B-\beta_V = 0.0012$ mag/deg, in good agreement with the estimate for the whole belt of 0.0015 mag/deg made by Gehrels and Tedesco (1979). However, it is clear from individual determinations of phase reddening that there are wide variations among asteroids.

H. Mean color indices

Table VII gives mean color indices of C, M and S asteroids from 902 observations of 301 asteroids. Also given are mean phase angles of observation and zero-phase color indices, obtained by correcting the observed color indices in accordance with the calculated phase reddening given in Table VI.

By dividing the samples of C and S asteroids into three diameter and three semimajor axis groups, we have determined that, within the expected statistical fluctuation (±0.003 mag), mean B–V and U–B color indices are independent of both diameter and semimajor axis.

III. EFFECTS OF ROTATION

Brightness variations of an asteroid on short time scales are mainly attributable to changes in apparent cross-sectional area due to rotation.

TABLE VII

Mean Color Indices of Asteroids

Type	$\bar{\alpha}$ (deg)	$\alpha = \bar{\alpha}$		$\alpha = 0°$	
		B–V (mag)	U–B (mag)	B–V (mag)	U–B (mag)
C	11.2	0.710	0.350	0.70	0.34
M	11.4	0.708	0.242	0.70	0.23
S	13.1	0.864	0.451	0.84	0.42

Another possible cause, large-scale albedo variegation, is evidently of only minor importance for the vast majority of asteroids. The two effects might be separable, either by Fourier analysis of the lightcurve or by synoptic observation using composition-or albedo-sensitive techniques. In the case of Fourier analysis, even-order terms are due to shape, and odd-order terms arise from albedo variations (Lacis and Fix 1971). Polarimetry, spectrophotometry, and broadband photometry are techniques that have been used to detect large-scale albedo and color patches (Degewij *et al.* [1979] summarize the observational results).

Lightcurve observations are used to determine rotation periods and shapes of asteroids (Harris and Burns 1979; see the chapter by Burns and Tedesco) and orientations of spin axes (see Taylor's chapter). The shapes of some lightcurves appear to be evidence for the existence of binary asteroids (Tedesco 1979a). The TRIAD file dated March 1979 lists lightcurve information on 310 asteroids, and so a broad range of diameters is well represented. In many respects, however, the data in this file are inhomogeneous and may not be suitable for statistical applications. Thus, as an example, the detection threshold of light variation for asteroids observed photographically is about 0.2 mag, which is far too coarse an increment for such observations to be useful for studying the distribution of lightcurve amplitudes.

The individual magnitudes in the data base used in this chapter contain lightcurve information because, for any given asteroid, they have generally been made at least a few days apart so as to randomize the effects of rotational brightness changes. They also constitute a very homogeneous data set since observations have always been made without prior knowledge of an asteroid's lightcurve and usually in ignorance of the taxonomic type. In this section, we examine two problems to which the data can be well applied: the frequency distributions of lightcurve amplitudes, and color variations with rotation.

A. Distribution of lightcurve amplitudes

When the effects of heliocentric and geocentric distances and phase angle variation have been removed from observations of an asteroid, the remaining differences can be attributed to rotational light variations. Departures from mean brightness ought to be some function of the lightcurve amplitude, the lightcurve shape, and the number of observations. This function can be determined, if certain assumptions are made, and one can therefore estimate the lightcurve amplitude sufficiently well for statistical purposes.

In a preliminary analysis, we have taken lightcurves to be sinusoidal, and we have assumed that all observations are equivalent to sampling sine curves at random rotational phases. In this case it can be shown that probable values of the lightcurve amplitude are 2.7 times the brightness difference between two observations, 1.6 times the greatest difference among three observations, 1.3 times for four observations, and so on. One's confidence in the estimate of the amplitude increases rapidly with the number of observations. We have removed the effects of phase angle using values of the multiple-scattering factors for C and S asteroids given in Table III, and we have excluded observations which span more than $5°$ in phase angle in order to reduce the effects of an incorrect choice of Q to less than about 0.02 mag, an amount comparable to the intrinsic accuracy of the photoelectric photometry.

We have also allowed for a small effect of multiple scattering in nonspherical objects. When a nonspherical object is viewed at maximum light (and cross-sectional area), the distribution of slopes differs from that at minimum light. We have modeled this effect for an isotropically scattering prolate spheroid having major axis a, minor axis b, and rotating about b. The minimum/maximum geometric area ratio is b/a, but the minimum/maximum brightness ratio differs from this by an amount that depends on b/a and the geometric albedo. The departure is always less than 10% for asteroids, but is nevertheless significant. If the polar flattening $f = 1-b/a$, we find, for small and moderate $p(0°)$:

$$f_{time} = f_{obs} \left[1 - 0.28\, p(0°)^{0.8} \, (1 - f_{obs}) \right] . \qquad (16)$$

Thus, the true flattening is smaller than that implied from lightcurve observations by an amount which increases with albedo and flattening. One may think of the effect in terms of limb darkening by considering center-to-limb surface brightness curves along the equator of a prolate spheroid; at maximum area the surface brightness is everywhere greater than at minimum area, except at the center of the disk and the limb, where it is equal. A treatment of this kind has been given by French (1979).

We assume that the quantity f_{obs} can usefully be equated to the observed lightcurve amplitude, for the purposes of this small correction, even if the true body shape is not spheroidal. Thus, if A_{obs} is the lightcurve amplitude in magnitudes, the "observed" flattening is given by

$$\log (1 - f_{obs}) = -0.4\, A_{obs} . \qquad (17)$$

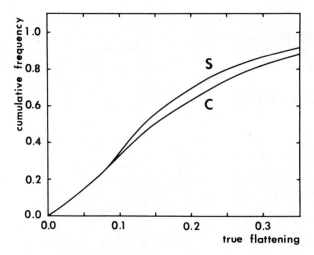

Fig. 12. Cumulative frequency distributions of body shape, expressed in terms of true flattening (see text), for C and S asteroids.

Results for 150 C and 112 S asteroids are shown in the smoothed cumulative frequency histograms of Fig. 12. Cumulative frequencies are thought to be reliable within ±0.03. The distributions of flattening for C and S asteroids are obviously rather similar, but C asteroids are probably, on the average, slightly less spherical than S asteroids. (Without the inclusion of the multiple-scattering effect the difference is smaller.) Median lightcurve amplitudes are 0.18 mag for C asteroids and 0.16 mag for S asteroids. Replotted in an incremental frequency diagram, the data are very noisy, but it is clear that the tail of the distribution extends significantly to large values of the flattening (in other words, very nonspherical asteroids are abundant). The shape of the incremental frequency distribution could be Maxwellian.

It is probably not prudent to draw detailed conclusions about body shapes from these statistical lightcurve data, but it is certain that true body shapes depart more from sphericity than is predicted by Eqs. (16) and (17) for two reasons: (1) the shapes of most asteroid lightcurves are not sinusoidal; there tend to be flat maxima in lightcurves, and these would lead one to underestimate the true lightcurve amplitude; and, (2) perhaps more importantly, most observations have been made at single apparitions during which a given asteroid is viewed at an almost fixed aspect angle.

Asteroids that appear to have exceptionally large or small lightcurve amplitudes can be readily identified from the UBV survey data. Since it is not appropriate to calculate lightcurve amplitudes for individual asteroids, we list the greatest range in observed brightness in Tables VIII and IX. The true lightcurve amplitude should be greater than the range in most cases. We exclude from Tables VIII and IX asteroids known to have larger lightcurve

TABLE VIII

Asteroids With Lightcurve Amplitudes >0.4 Mag

Asteroid	Type	No. of Obs.	Range (mag)
17	S	4	0.43
19	C	6[a]	0.46
48	U	4	0.43
52	C	4[a]	0.42
84	C	3[a]	0.51
95	C	5[a]	0.41
121	C	3	0.56
182	S	4[a]	0.48
382	CMEU	2	0.46
423	C	3[a]	0.50
498	U	3[a]	0.44
535	C	2	0.43
572	C	2	0.47
647	CMEU	2	0.61
804	C	4[a]	0.55
925	S	3[a]	0.52
2000	S	4	0.47

[a]Observed at more than one apparition.

TABLE IX

Asteroids Thought to Have Lightcurve Amplitudes ≤0.02 Mag

Asteroid	Type	No. of Obs.	Range (mag)
96	U	2	0.02
123	S	2	0.01
180	S	2	0.02
207	C	2	0.02
213	CMEU	2	0.01
221	U	2	0.00
245	S	2	0.02
255	CMEU	2	0.01
293	U	2	0.00
331	C	2	0.01
386	C	4	0.02
445	C	2	0.00
487	S	2	0.01
489	C	2	0.00
530	CMEU	2	0.00
674	S	3	0.02
697	C	2	0.02
717	CMEU	2	0.00
731	C	2	0.00
764	C	2	0.00
772	C	2	0.00
834	U	2	0.01
1140	S	2	0.02
1306	S	2	0.02
1329	S	2	0.02
1390	S	2	0.02

amplitudes from the TRIAD lightcurve file. It is notable that a majority of asteroids thought to have large lightcurve amplitudes (Table VIII) have been observed at more than one apparition, whereas those thought to be nearly spherical (Table IX) have poorer observational coverage. Almost certainly, some of the entries in Table IX are erroneous because observations were fortuitously made at similar rotational phases.

B. Color patches

Bobrovnikoff (1929) seems to have been the first to detect color changes on an asteroid as it rotates. He accomplished this by comparing sequences of spectra of Vesta and, incidentally, was able to deduce a rotation period close to that just recently accepted (5.34 hr according to Gradie et al. [1978]). However, he was fortunate to choose Vesta for observation since, as Degewij et al. (1979) point out, only a few asteroids appear to show any evidence for rotational color changes. For the most part, those that do exhibit such changes are redder at maximum brightness than at minimum.

We noticed that, used statistically, the data base considered in this chapter should be suitable for searching out such systematic rotational color differences to a smaller threshold than that applicable to observations of individual lightcurves. Mean B–V color indices were computed for 160 C and 127 S asteroids, both when the asteroids were brighter and fainter than average in V. For both taxonomic types there appeared to be clear evidence that asteroids are indeed redder when brighter than average by several thousandths of a magnitude. However, it was pointed out by J. G. Williams (personal communication) that such a result could be a reflection of measurement errors: thus if V is underestimated, B–V is overestimated. On further checking we found that the color-brightness relationship was largely, but not completely, destroyed when the average B magnitudes of asteroids were considered.

We conclude, conservatively, that there is no statistical evidence for color changes of a systematic type with rotation, either for C or S asteroids, at the 0.005 mag level in B–V. (However, more careful analysis of the observations is necessary to exclude the possibility that significant color changes are present at smaller thresholds in B–V.) Of course, this finding neither rules out small-scale mixing of C and S materials, nor does it exclude large-scale mosaicking if, on the average, the color index is uncorrelated with the rotational phase. Gradie (personal communication) states that impact mechanics mitigate against finding S material on C asteroids and vice versa. An impact at velocities greater than 5 km sec^{-1} would more-or-less melt and vaporize the projectile. Slower impacts would leave projectile material, but the blanketing effect of the impact ejecta (on small bodies) might cover up or substantially dilute the projectile material.

IV. DISCUSSION

We have analyzed a large body of UBV observations by means of a new multiple-scattering theory. The theory gives a number of clear insights into the optical nature of asteroid surfaces that have only been hazily, if at all, recognized in the past. For example, we now have clear ideas about what factors control the shapes of asteroid phase curves. Two components are active. The first, due to singly scattered light, depends on the surface texture. Specifically, the bulk density (or porosity) affects the size and width of the opposition effect, and the roughness determines the slope of the phase curve away from opposition. The second component is characterized by the proportion of multiply scattered light, and its effect is almost independent of phase angle. There is good evidence that the textures of almost all asteroid surfaces are closely similar; they are moderately porous and moderately rough on a scale greater than the wavelength of light and they appear to be particulate. This similarity implies that only the multiple-scattering component of phase curves varies from asteroid to asteroid. In other words, differences among asteroid phase curves are due solely to variations in the amount of multiply scattered light. As the proportion of multiply scattered light increases, the slope of the phase curve decreases, and the opposition effect is diluted. It also follows that the geometric albedo is strongly correlated with phase curve shape, and this affords a purely photometric means of estimating asteroid albedos and diameters.

One inescapable consequence of using the multiple-scattering theory has been either the redefinition or abandonment of some long-used parameters. Thus, we have absorbed the meaning of the phase coefficient into a quantity called the multiple-scattering factor and adopted a new definition for absolute magnitude. These are important conceptual changes from previous practice but we believe them to be both useful and unifying. Use of $V(0°)$ rather than $V(1,0)$ presents no problem. To a good approximation, the two quantities differ by a constant amount for all asteroids (Eq. 13), and the resulting effect on geometric albedo is to increase presently accepted albedos by a constant factor of 1.34.

The advantages we perceive in using Q and $V(0°)$ to describe phase curves rather than β and $V(1,0)$ are as follows:

1. A single parameter, the multiple-scattering factor Q, suffices to describe the shapes of phase curves for a wide variety of atmosphereless objects, from the darkest C asteroid to ice-covered satellites. The phase curves of these very dissimilar objects do not have to be treated in different ways.
2. Q has a physical meaning that is well defined.
3. Q applies at all phase angles. Thus, if observations are available only at small or large phase angles, the whole phase curve is still defined.
4. The absolute magnitude $V(0°)$ and the geometric albedo $p(0°)$ are explicitly specified at zero phase; there is no uncertainty in extrapolation to zero phase. The new definition of $p(0°)$, being surface-texture independent (which the currently accepted definition is not), is both physically meaningful and more in the spirit of the original definition.
5. Since there are only two parameters, Q and $V(0°)$, which describe phase curves, only two observations at different phase angles (and free from the effects of rotational brightness variations) are required to delineate a phase curve completely. Moreover, the uncertainties in Q and $V(0°)$ can be formally stated when more than two observations are available.

To calculate asteroid magnitudes $V(\alpha)$ using Q and $V(0°)$, the following recipe can be used: (i) starting with Eq. (13), derive $V(0°)$ from $V(1,0)$; (ii) use Q appropriate to β (Table II), or to type (Table III), or $Q = 0.085$ if the type is unknown; (iii) calculate $m(\alpha) = V(\alpha)$ from Eqs. (5) and (6) $[m(0°) = V(0°)$, of course].

In Sec. II-F we discussed in detail the consequences of choosing various values of Q to describe the brightness variation of large numbers of asteroids with phase angle, and showed in particular that an increase in Q leads to a decrease (brightening) in $V(0°)$, which in turn results in an increase in diameters derived from radiometry and polarimetry. At present, absolute magnitudes of asteroids having unknown phase coefficients are computed using a mean value. We have pointed out some of the disadvantages of this oversimplification; for example, the relative diameters of C and S asteroids, and of large and small asteroids, can be misestimated by several percent. Also, since the numbered main-belt population is biased by observational selection (small, low-albedo asteroids are under-represented), we have good reasons to believe that the mean phase coefficient of this population is different from that currently adopted. In consequence, there are systematic errors in diameters and albedos, sometimes amounting to several percent.

It is not our intention at this time to propose remedies for these problems. Indeed, it is clear that over the years there has been a steady approach to the best possible characterization of the phase dependence of asteroid magnitudes. However, we do identify four possible options for consideration in the future. In order of increasing complexity, these are:

1. Preserve $Q_V = 0.029$ ($\beta_V = 0.039$ mag/deg), currently adopted for all asteroids with unknown phase curves. Disadvantage: systematic errors of several percent exist in $V(0°)$, d, and $p(0°)$.

2. Use $Q_V = 0.085$ ($\beta_V = 0.034$ mag/deg) for the numbered population. Advantage: removes systematic errors in $V(0°)$, etc., for the numbered asteroids. Disadvantage: fails to allow for known differences among asteroids and perpetuates relative, type- and diameter-dependent, systematic errors.

3. Use $Q_V = 0.037$ ($\beta_V = 0.038$ mag/deg) for C asteroids, $Q_V = 0.133$ ($\beta_V = 0.030$ mag/deg) for S asteroids, and $Q_V = 0.085$ for other or unknown types. Advantage: gives an excellent overall representation for the numbered asteroids. Disadvantage: fails to allow for diameter-dependent effects.

4. As option 3, but additionally, allow Q_V to depend on diameter (when known) or $V(0°)$ (when the diameter is unknown). Advantage: gives the best representation possible with available data. Disadvantage: too complicated?

In Sec. II-G an analysis of color-index data indicated that reddening with phase can be ascribed largely to differential albedo effects. Agreement of observation with predictions by the multiple-scattering theory, although quite good, by no means proves this. And furthermore, there are insufficient data to decide whether phase reddening has any dependence on diameter. We are therefore hesitant, at this time, to suggest that color-index observations of asteroids be corrected to zero phase, either individually or statistically. The errors incurred by not making such corrections appear to be, on average, quite small; color indices adopted in the TRIAD file are probably about 0.01 and 0.02 too red in B–V and U–B, respectively.

For C and S asteroids, which make up the majority of the main belt, it is clear that none of the photometric parameters discussed in this chapter varies with semimajor axis. Thus the mean phase curves (as characterized by $\bar{Q}$), the mean color indices, and the mean geometric albedos for these taxonomic types seem constant. Adding to this extraordinary uniformity of C and S asteroids, one might also cite the apparently general invariability of the B–V color index with rotation. It therefore appears, on photometric grounds, that the materials constituting C and S asteroids are very similar over a zone at least 1 AU wide. Combining this inference with Zellner and Bowell's (1977) finding regarding the smoothly changing proportion of S to C asteroids with semimajor axis, one is led to speculate that the present mix of C and S asteroids in the main belt has come about by a diffusion process acting between two extremely homogeneous populations.

The invariability of the mean color indices for C and S asteroids with semimajor axis and diameter, and the apparent constancy of B–V with rotation further suggest that we cannot photometrically detect the presence of impacting material on asteroid surfaces. This conforms with current thinking about the mantling of asteroids by material from impacting projectiles. The diameter-dependent effects that are perceived photometrically are minor and are probably related either to gravitational influences on roughness or to differences in particle size distribution.

There is much work to be done in the future. We have reliable photometric statistics for C asteroids down to 50 km in diameter and S asteroids down to 25 km in diameter, but we do not know the (photoelectric) photometric properties of smaller asteroids. One observational datum urgently needed is the determination of a phase curve for a kilometer-size asteroid. If, for example, the surface structure on such a small body is not particulate, we should readily be able to detect higher volume density by modeling the shape of the opposition effect. Conversely, we should also be able to detect a gravitationally induced increase in the roughness on smaller bodies. We do not understand the scattering properties of the E asteroid 64 Angelina (and perhaps also those of 44 Nysa), and we have not explored the small diameter-dependent differences in the scattering properties of asteroids. Whether there exist characteristic light-scattering properties for individual Hirayama families is unknown; as noted in Sec. II-D, the multiple-scattering factor for Themis family members may possibly be unrepresentative of C asteroids. It is possible (see the chapter by Gradie *et al.*) that the large degree of homogeneity among the larger Hirayama families may be anomalous. UBV observations on smaller families could resolve this. In this chapter we have concentrated on correlating photometric properties of asteroids with taxonomic type, albedo, semimajor axis and diameter; there are other parameters, including orbital inclination and eccentricity, that might also be considered. We do not yet know whether phase reddening depends on diameter, nor can we say what causes the apparently large dispersion in phase reddening for objects of similar albedo and diameter. Our treatment of the statistics of lightcurve amplitudes is very rudimentary; the data presumably contain fairly detailed information on the distribution of body shapes, on how that distribution varies with diameter and, perhaps, on the distribution of spin axis directions.

There are a number of theoretical developments, that the multiple-scattering theory is amenable to, that might help resolve some of the problems mentioned above. Immediately foreseeable are: extension of the theory to larger phase angles; allowance for the presence of submicron particles (Mie scatterers); investigation of the relationship between lightcurve amplitude and phase angle in a statistical way; and prediction of limb darkening for bodies of different albedos. The multiple-scattering theory is being incorporated into a theory of the polarization of light scattered in rough surfaces (Lumme 1979a). There is also the possibility that it could be extended to other wavelength ranges (infrared, radar).

Acknowledgments. We thank J. Gradie and B. Zellner for perceptive review comments. This research has been largely funded by grants from the National Aeronautics and Space Administration.

REFERENCES

Blanco, C., and Catalano, S. 1974. On the photometric variations of the Saturn and Jupiter satellites. *Astron. Astrophys.* 33: 105-111.

Bobrovnikoff, N. T. 1929. The spectra of minor planets. *Lick Obs. Bull.* 14: 18-27.

Bowell, E. 1977. UBV photometric survey of asteroids (abstract). *Bull. Amer. Astron. Soc.* 9: 459.

Bowell, E. 1979. UBV survey of asteroids. *Icarus* (special Asteroid issue).

Bowell, E.; Chapman, C. R.; Gradie, J. C.; Morrison, D.; and Zellner, B. 1978. Taxonomy of asteroids. *Icarus* 35: 313-335.

Bowell, E.; Martin, L. J.; Poutanen, M.; and Thompson, D. T. 1979. Photometry and polarization of atmosphereless bodies. I. Photometry of 69 Hesperia. To be submitted to *Astron. J.*

Bowell, E., and Zellner, B. 1974. Polarizations of asteroids and satellites. In *Planets, Stars, and Nebulae Studied with Photopolarimetry,* ed. T. Gehrels (Tucson: University of Arizona Press), pp. 381-404.

Chapman, C. R.; Johnson, T. V.; and McCord, T. B. 1971. A review of spectrophotometric studies of asteroids. In *Physical Studies of Minor Planets,* ed. T. Gehrels (NASA SP-267, Washington, D.C.: U.S. Government Printing Office), pp. 51-65.

Chapman, C. R.; Morrison, D.; and Zellner, B. 1975. Surface properties of asteroids: A synthesis of polarimetry, radiometry, and spectrophotometry. *Icarus* 25: 104-130.

Degewij, J. 1978. Photometry of faint asteroids and satellites. Ph.D. dissertation, University of Leiden.

Degewij, J.; Gradie, J.; and Zellner, B. 1978. Minor planets and related objects. XXV. UBV photometry of 145 faint asteroids. *Astron. J.* 83: 643-650.

Degewij, J.; Tedesco, E. F.; and Zellner, B. 1979. Albedo and color contrasts on asteroid surfaces. *Icarus* (special Asteroid issue).

Dollfus, A.; Geake, J. E.; Mandeville, J. C.; and Zellner, B. 1977. The nature of asteroid surfaces, from optical polarimetry. In *Comets, Asteroids, Meteorites*, ed. A. H. Delsemme (Toledo, Ohio: University of Toledo Press), pp. 243-251.

French, L. 1979. Photometric studies of carbonaceous chondrites. Ph.D. dissertation (in preparation), Cornell University.

Gehrels, T. 1956. Photometric studies of asteroids. V. The lightcurve and phase functions of 20 Massalia. *Astrophys. J.* 123: 331-338.

Gehrels, T. 1967. Minor planets. I. The rotation of Vesta. *Astron. J.* 72: 929-938.

Gehrels, T. 1970. Photometry of asteroids. In *Surfaces and Interiors of Planets and Satellites*, ed. A. Dollfus (New York: Academic Press), pp. 319-375.

Gehrels, T., and Gehrels, N. 1978. Minor planets and related objects, XXVI. Asteroid magnitudes. *Astron. J.* 33: 1660-1674.

Gehrels, T., and Owings, D. 1962. Photometric studies of asteroids. IX. Additional light-curves. *Astrophys. J.* 135: 906-924.

Gehrels, T., and Taylor, R. C. 1977. Minor planets and related objects. XXII. Phase functions for (6) Hebe. *Astron. J.* 82: 229-237.

Gehrels, T., and Tedesco, E. F. 1979. Minor planets and related objects. XXVIII. Asteroid magnitudes and phase relations. *Astron. J.* (in press).

Giclas, H. L. 1951. The project for the study of planetary atmospheres, *Lowell Observatory Report No. 9*, pp. 46-56.

Gradie, J. C. 1978. An astrophysical study of the minor planets in the Eos and Koronis families. Ph D dissertation, University of Arizona.

Gradie, J.; Tedesco, E.; and Zellner, B. 1978. Rotational variations in the optical polarization and reflection spectrum of Vesta (abstract). *Bull. Amer. Astron. Soc* 10: 595.

Groeneveld, I., and Kuiper, G. P. 1954. Photometric studies of asteroids. I. *Astrophys. J.* 120: 200-220.

Hansen, O. L. 1976. Radii and albedos of 84 asteroids from visual and infrared photometry. *Astron. J.* 81: 74-84.

Hapke, B. 1971. Inferences from optical properties concerning the surface texture and composition of asteroids. In *Physical Studies of Minor Planets*, ed. T. Gehrels (NASA SP-267, Washington, D.C.: U.S. Government Printing Office), pp. 67-77.

Harris, A. W., and Burns, J. A. 1979. Asteroid rotation. I. Tabulation and analysis of data. *Icarus* (in press).

Haupt, H. 1958. Photoelektrish-photometrische Studien an Vesta. *Mitt. Sonnobs, Kanzelhöhe* 14: 172-173.

Kitamura, M. 1959. Photoelectric study of colors of asteroids and meteorites. *Publ. Astron. Soc. Japan* 11: 79-89.

Lacis, A. A., and Fix, J. D. 1971. Lightcurve inversion and surface reflectivity. In *Physical Studies of Minor Planets*, ed. T. Gehrels (NASA SP-267, Washington, D.C.: U.S. Government Printing Office), pp. 141-146.

Lumme, K. 1979a. Photometry and polarization of atmosphereless bodies. IV. Theory for polarization. To be submitted to *Astron. J.*

Lumme, K. 1979b. Photometry and polarization of atmosphereless bodies. V. Limb darkening. To be submitted to *Astron. J.*

Lumme, K., and Bowell, E. 1979a. Photometry and polarization of atmosphereless bodies. II. Theory for the photometric function. To be submitted to *Astron. J.*

Lumme, K., and Bowell, E. 1979b. Photometry and polarization of atmosphereless bodies. III. Interpretation of phase curves. To be submitted to *Astron. J.*

Millis, R. L., and Thompson, D. T. 1975. UBV photometry of the Galilean satellites. *Icarus* 26: 408-419.

Morrison, D. 1977. Asteroid sizes and albedos. *Icarus* 31: 185-220.

Morrison, D.; Morrison, N. D.; and Lazarewicz, A. R. 1974. Four-color photometry of the Galilean satellites. *Icarus* 23: 399-416.

Müller, G. 1897. *Die Photometrie der Gestirne* (Leipzig: Engelmann), pp. 375-381.

Scaltriti, F., and Zappalà, V. 1979. The similarity of the opposition effect among asteroids. Submitted to *Astron. Astrophys.*

Stumpff, K. 1948. Über die Albedo der Planeten und die photometrische Bestimmung von Planetoidendurchmessern. *Astron. Nachr.* 276: 118-126.

Taylor, R. C. 1971. Photometric observations and reductions of lightcurves of asteroids. In *Physical Studies of Minor Planets,* ed. T. Gehrels (NASA SP-267, Washington, D.C.: U.S. Government Printing Office), pp. 117-131.

Taylor, R. C. 1973. Minor planets and related objects. XIV. Asteroid (4) Vesta. *Astron. J.* 78: 1131-1139.

Taylor, R. C.; Gehrels, T.; and Capen, R. C. 1976. Minor planets and related objects. XXI. Photometry of eight asteroids. *Astron. J.* 81: 778-786.

Tedesco, E. F. 1978. Asteroid phase coefficient size dependence (abstract). *Bull. Amer. Astron. Soc.* 10: 597.

Tedesco, E. F. 1979a. Binary asteroids: Evidence for their existence from lightcurves. *Science* 203: 905-907.

Tedesco, E. F. 1979b. Photometric investigation of the colors, shapes and spin rates of Hirayama family asteroids. Ph.D. dissertation, New Mexico State University.

van Houten, C. J.; van Houten-Groeneveld, I.; Herget, P.; and Gehrels, T. 1970. The Palomar-Leiden survey of faint minor planets. *Astron. Astrophys. Suppl.* 2: 339-448.

Veeder, G. J.; Matson, D. L.; and Smith, J. C. 1978. Visual and infrared photometry of asteroids. *Astron. J.* 83: 651-663.

Veverka, J. 1971. The physical meaning of phase coefficients. In *Physical Studies of Minor Planets,* ed. T. Gehrels (NASA SP-267, Washington, D.C.: U.S. Government Printing Office), pp. 79-90.

Wasserman, L. H.; Millis, R. L.; Franz, O. G.; Bowell, E.; White, N. M.; Giclas, H. L.; Martin, L. J.; Elliot, J. L.; Dunham, E.; Mink, D.; Baron, R.; Honeycutt, R. K.; Henden, A. A.; Kephart, J. E.; A'Hearn, M. F.; Reitsema, H.; Radick, R. R.; and Taylor, G. E. 1979. The diameter of Pallas from its occultation of SAO 85009. *Astron. J.* 84: 259-268.

Zappala, V.; van Houten-Groeneveld, I.; and van Houten, C. J. 1979. Rotation period and phase curves of asteroids 349 Dembowska and 354 Eleonora. *Astron. Astrophys. Suppl.* 35: 213-221.

Zellner, B. 1972. On the nature of Iapetus. *Astrophys. J.* 174: L107-L109.

Zellner, B.; Andersson, L.; and Gradie, J. 1977a. UBV photometry of small and distant asteroids. *Icarus* 31: 447-455.

Zellner, B., and Bowell, E. 1977. Asteroid compositional types and their distributions. In *Comets, Asteroids, Meteorites,* ed. A. H. Delsemme (Toledo, Ohio: University of Toledo Press), pp. 185-197.

Zellner, B.; Leake, M.; Lebertre, T.; Duseaux, M.; and Dollfus, A. 1977b. Laboratory polarimetry of meteorites and the asteroid albedo scale. *Proc. Lunar Sci. Conf. VIII,* (Oxford: Pergamon Press), pp. 1091-1110.

Zellner, B.; Lebertre, T.; and Day, K. 1977c. Laboratory polarimetry of dark carbon-bearing silicates and the asteroid albedo scale. *Proc. Lunar Sci. Conf. VIII* (Oxford: Pergamon Press), pp. 1111-1117.

Zellner, B.; Wisniewski, W.; Andersson, L.; and Bowell, E. 1975. Minor planets and related objects. XVIII. UBV photometry and surface composition. *Astron. J.* 80: 968-995.

OPTICAL POLARIMETRY
OF ASTEROIDS AND LABORATORY SAMPLES

A. DOLLFUS
Observatoire de Paris

and

B. ZELLNER
University of Arizona

Interpretations of astronomical polarization data for more than 100 minor planets are summarized with reference to laboratory data for lunar, meteoritic, and terrestrial samples. All observed asteroids, including objects only a few kilometers in diameter, have microscopically intricate surfaces. Detailed comparisons between laboratory measurements of basaltic achondrites and telescopic results for Vesta show that its surface is particulate with a broad range of particle sizes, including a component of fine (1 – 5 μm) dust. The surface soils have not, however, undergone marked optical alterations as is the case for the lunar fines. For albedos greater than about 0.06, the albedo and hence the diameter can be determined with some reliability from the slope of the ascending branch of the polarization-phase curve. Many minor planets have surfaces that are remarkably uniform in albedo and texture on a hemispheric scale. A notable exception is 4 Vesta, which shows about ten percent albedo variegation between hemispheres.

The linear polarization of light reflected from a solid surface is a function of the scattering geometry, the surface refractive index (indirectly, the albedo and composition), and the surface texture. Astronomically it can be measured with high precision, given only an adequate photon count, and it may carry information to which other remote-sensing techniques are essentially blind. Telescopic results for more than 100 minor planets are listed in the TRIAD tables (Part VII of this book). As noted by Zellner (see his chapter in this book), the polarimetry provides parameters which can separate out types of asteroids that are spectrally almost indistinguishable. Its most powerful application for the asteroids, however, is concerned with questions of the surface texture as described in Sec. II below. It also provides one of the few available techniques for remote determination of asteroid albedos and diameters, as we describe in Sec. I, and is the most sensitive technique available for detection of subtle albedo variations on the surface of a spinning asteroid (Sec. III). In Sec. IV we discuss some profitable avenues for future work.

Any study of the polarimetric properties of atmosphereless solar system bodies must begin with the work of Lyot (1929). Subsequent work predating the large-scale observations of minor planets has been reviewed by Dollfus (1955, 1961, 1971), Veverka (1970), and Bowell and Zellner (1974). A long series of papers from the Paris Observatory (e.g., Dollfus and Geake 1975) dealt with laboratory studies of lunar materials and implications of the results for the surfaces of the moon, the asteroids and other objects. Most of the telescopic observations of minor planets were made at the University of Arizona (Zellner and Gradie 1976) and the interpretations were summarized in some detail by Zellner et al. (1977a,b) and by Dollfus et al. (1977).

Figure 1 illustrates elementary angles for the study of any reflection process in the laboratory, and Fig. 2 illustrates the same angles on the surface of a sphere in space. The *phase angle* α is measured in the plane of scattering, which contains the source, sample and detector. For disk-integrated observations of a spherical planet circular polarization cannot arise and the dominant electric vector is constrained by symmetry to lie either in the plane of scattering or else perpendicular to that plane. The latter case prevails for simple Fresnel reflection at any phase, and is spoken of as *positive* polarization. In terms of intensity components $I_\perp$ and $I_\parallel$, respectively measured perpendicular and parallel to the plane of scattering, the degree of polarization is defined by

$$P = \frac{I_\perp - I_\parallel}{I_\perp + I_\parallel} \tag{1}$$

a quantity which may be positive, negative, or zero.

For the moon and for the relatively dark, particulate surfaces indicated for most of the asteroids, limb effects are minor. That is, the polarization (and the photometric brightness, at small phase) is almost constant over the disk and depends only on α. Thus it usually suffices to make laboratory measurements

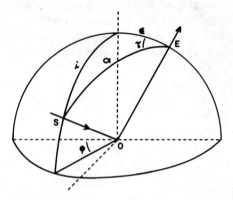

Fig. 1. Angles of incidence i, emergence ϵ, azimuth φ, and phase α for a horizontal sample at point O. The light source is in direction S, and the detector at direction E. The significance of the angle τ is best understood with reference to Fig. 2. (From Zellner 1977.)

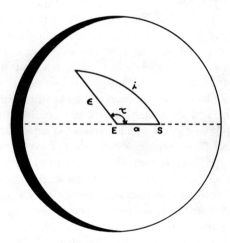

Fig. 2. Angles as in Fig. 1, projected on the apparent disk of a spherical body in space. S is the subsolar point, and E the subearth point. (From Zellner 1977.)

of horizontal powder surfaces at $\tau = 0$. For critical work, however, it is possible to mimic the entire surface of a hemisphere at any phase by summing $I_\perp$ and $I_\parallel$ over appropriate combinations of i, ϵ, and τ.

The polarization of light from a diffusely reflecting surface is intimately connected with its geometric albedo p, defined as the ratio of apparent surface brightness to that of an ideal Lambert screen, located at the same position and oriented perpendicular to the incident light. The Lambert screen is a mathematical abstraction but it can be approximated quite closely in the

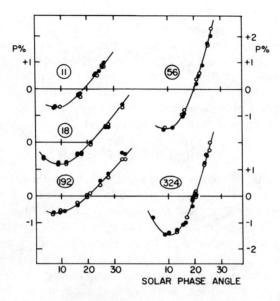

Fig. 3. Polarization-phase curves for the *S*-type asteroids 11 Parthenope, 18 Melpomene, and 192 Nausikaa (left), and the *C* objects 56 Melete and 324 Bamberga. (From Zellner *et al.* 1977*a*.)

laboratory at optical wavelengths by a metal plate thickly coated with MgO or $BaSO_4$ smoke. Traditionally geometric albedos are measured at about 5° phase, which approximates the linear extrapolation of the photometric magnitude-phase relation to zero phase (see the chapter by Bowell and Lumme in this book).

No really adequate theory of optical polarization by rough or particulate surfaces has been made available. Formulations of the problem by Wolff (1975) and by Lumme (1979, in preparation) each appear promising. Unfortunately they also appear to be mutually contradictory in several important respects, and neither has been developed to the point of providing useful remote-sensing information. For interpretations of the polarization-phase curves of minor planets we are still largely dependent upon laboratory simulations. That such simulations can be quite successful was demonstrated in the case of the moon, for which the surface texture was deduced with remarkable fidelity more than a decade before the first lunar landings (e.g., Dollfus 1971).

Figure 3 illustrates polarization-phase curves for several asteroids. Like the moon, Mercury, and the atmosphereless natural satellites, the asteroids invariably show *negative* polarizations at small phases. The phenomenon is a signature of rough, porous, or particulate surfaces. The curves are described by the parameters P_{min}, which is the maximum depth of the negative branch, usually found at about $10°$ phase; the inversion angle α_0 where the polarization changes sign, near $20°$ for the asteroids; and the slope h of the ascending branch, measured at the inversion angle. The phase coverage available from the earth ranges from $0°$ to about $30°$ for the inner mainbelt objects but only to about $12°$ for the Trojans. The peak polarization P_{max} at large phases is highly diagnostic of surface texture, but is in principle observable only for objects in earth-crossing orbits, or from space probes, and in fact has not yet been observed for any asteroid.

I. THE SLOPE-ALBEDO LAW

The existence of a reciprocal relationship between the geometric albedo of a particulate surface and the degree of linear polarization of the reflected light is known as the Umov law. As shown by Dollfus and Titulaer (1971) and by Bowell and Zellner (1974), it is a consequence of the approximation that the intensity of *polarized* light (being essentially due to simple Fresnel reflection) is constant from one sample to another, while variations in albedo are due to variations in the *unpolarized* component (generated by diffuse processes such as multiple scattering and edge diffraction). A Umov relationship is readily demonstrable for P_{max} and P_{min} for structurally homogeneous materials, but both parameters are also critically dependent on surface texture (e.g., Bowell *et al.* 1972).

Following suggestions by Widorn (1967), KenKnight *et al.* (1967), Veverka (1970) and others, Bowell and Zellner (1974) demonstrated from laboratory and telescopic data that the polarimetric slope h is, to a good first approximation, independent of composition, texture, or wavelength *per se*, and is correlated with geometric albedo only. The relationship can be written

$$\log p = -C_1 \log h + C_2 \tag{2}$$

where h is measured in percent polarization per degree of phase angle. The constants can be determined empirically from telescopic observations of the moon, Mercury, etc. or by laboratory measurements of terrestrial, lunar, and meteoritic materials. Using only high-quality data for meteorites crushed to approximately the correct surface texture for the asteroids as described below, Zellner *et al.* (1977a) adopted $C_1 = 0.93$ and $C_2 = -1.78$.

Thus we can obtain the geometric albedo of an asteroid surface from its polarization-phase curve. Ignoring possible complications due to limb effects, we may then compute asteroid diameters according to

$$2 \log d = 6.244 - 0.4 \, V(1,0) - \log p_V \tag{3}$$

where d is the diameter in kilometers. The slope-albedo law is not perfectly obeyed for any reasonably diverse suite of laboratory materials, and it may

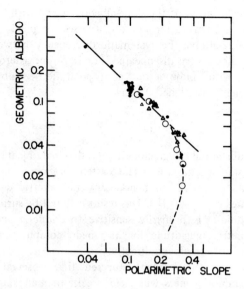

Fig. 4. Geometric albedo versus slope of the ascending branch of the polarization-phase curve, both in green light, for artificial carbon-bearing silicates (O), carbonaceous chondrites (▲), and asteroids (●). Results for pure carbon black are off scale at the bottom. The solid line is the relationship $\log p = -0.93 \log h -1.78$ adopted by Zellner *et al.* (1977*a*) from laboratory polarimetry of meteorites. Radiometric albedos of asteroids are adapted from visual-wavelength data by Morrison. (From Zellner *et al.* 1977*b*.)

have no fundamental physical significance, but it does provide a useful tool.

Polarimetric albedos were reported most extensively by Zellner and Gradie (1976), who at that time used a calibration inferior to the one described above. There was a worrisome discrepancy with respect to results from the independent thermal-radiometric technique (see the chapter by Morrison and Lebofsky in this book); the polarimetric albedos were generally the higher and in particular did not reproduce the very low radiometric values for the C-type asteroids. The data for the polarimetric calibration were restricted to albedos > 0.05, and indeed lower albedos are found for very few natural materials (and no meteorites) in powdered form.

Figure 4 is reproduced from work by Zellner *et al.* (1977*b*) on the optical properties of some very dark synthetic carbon-bearing silicates. It was shown that the slope-albedo law becomes saturated, and hence the polarimetric albedos become unreliable, for albedos less than ~ 0.06. Also it is the case that much of the asteroid polarimetry was done in blue light, and it is now realized that reliable *visual* albedos cannot be derived from such data. Accordingly the TRIAD tables (see Part VII in this book) list polarimetric slopes for 52 objects, but albedos and diameters are computed for only 27.

With the recalibration of the slope-albedo law and with the general slight increase of tabulated asteroid brightnesses implied by the new magnitude-phase relations, the systematic discrepancy between radiometric and polarimetric albedos has disappeared. For 18 S-type asteroids listed in the TRIAD tables with radiometric and polarimetric albedos both of observational weight 2 or higher, we have

$$\langle p_{pol} / p_{rad} \rangle = 1.00 \pm 0.18. \tag{4}$$

There are still substantial disagreements for individual objects, as reflected by the rather large standard deviation. The S asteroids show a substantially larger albedo range $(0.09 - 0.22)$ in high-quality radiometric data than in the polarimetric values $(0.12 - 0.18)$. This result may not be surprising, since the radiometric technique is primarily sensitive to *diameter* and is dependent upon a photometric magnitude for its albedo computation, whereas the polarimetry is sensitive to albedo directly.

For some time it has been expected that observations of stellar occultations by minor planets will yield precise diameters and hence either confirm the polarimetric and radiometric albedo scales or else indicate the need for recalibration (see the chapter by Millis and Elliot in this book). The one really satisfactory occultation diameter now available is that for 2 Pallas. The albedo turns out to be higher than either the polarimetric or the radiometric predictions, but much closer to the latter. Pallas is such an unusual object, however, that we are not rushing to apply a recalibration based on this result to the more common types of minor planets. Occultation results for 532 Herculina, though rather model-dependent, provide evidence that both scales are essentially correct (Gradie *et al.* 1978*a*).

The radiometric method, unlike the polarimetric technique, is capable of reliable albedo determinations for dark and distant objects. Given first-class equipment, it is also the faster technique in terms of telescope time, and hence is the method of choice for extended asteroid surveys. Polarimetric albedo measurements should be continued in special cases and particularly for the small cis-Martian objects, for which the thinness of the regolith may cause uncertainties in the radiometric models (Lebofsky and Rieke 1979).

II. SURFACE TEXTURE

The shape of the negative polarization branch (described by its depth P_{min} and its width α_o) together with the albedo p are related to the asteroid surface texture (Dollfus 1955, 1961, 1971). The three parameters P_{min}, α_o and p were measured on a large number of terrestrial, lunar and meteoritic samples, in natural conditions, or chipped, crushed, sieved or artificially prepared. Figure 5 shows two 3-dimensional models relating these 3 parameters.

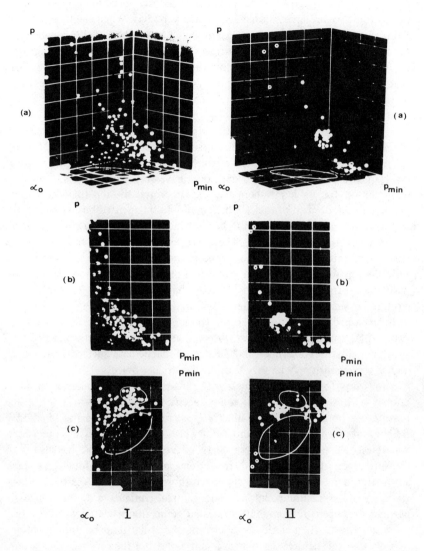

Fig. 5. Model constructed with the three parameters: inversion angle α_0, polarization minimum P_{min} and albedo p for phase angle $5°$. The Model I on the left (pictures a, b and c) is made with laboratory measurements on terrestrial, lunar and meteoritic samples in natural or artificial textures. The Model II on the right is made with telescopic observations on atmosphereless solar system objects. Pictures a: p is vertical from 0 % (at bottom) to 75 %; P_{min} is toward the right in units of 10^{-3} from 0 (at corner), to 20; α_0 is the third coordinate from $0°$ (at left) to $28°$ (at corner); Pictures b: model seen from the left, with p vertical and P_{min} horizontal; Pictures c: model seen from above, with P_{min} horizontal and p vertical.

The model on the left is constructed from laboratory measurements on samples. Lunar regolith fines are the natural product of meteoroid bombardments on large silicaceous planetary bodies. When the model is seen from above, picture I(c), lunar soils are the white beads circled in a delineated domain at top. Solid silicaceous surfaces (natural, chipped or with grain-sizes not smaller than 1 mm) are the gray beads confined in another larger domain near the center of picture I(c). The powders made of pulverized silicates lie in between these two areas; they are strongly depleted in micron-size grains when compared to the lunar soil.

The model of Fig. 5 on the right is constructed from telescopic observations of atmosphereless solar system objects. The moon and Mercury lie in the domain of the lunar regolith as seen in II(c). The three inner Galilean satellites (beads with a dark dot) have very high albedos. Callisto is documented for its hemisphere leading its orbital motion (half bead) and belongs to the domain of rocks, which could be fragmented but without dust.

The S asteroids are the white beads which are compactly clustered in all the three pictures of Model II. The C asteroids are grouped in another area clearly separated at right of the S-type cluster. Two E-types and Vesta have high albedos which separate them at left. For all asteroids the results indicate particulate surfaces, distinguishable from bare rock but also distinguishable from lunar surface fines with their glassy, radiation-darkened agglutinates and well-developed fairy-castle structure. The largest S-type asteroids in the main belt (~250 km) and the smallest observed among the Mars-crossers (~5 km) show equally the polarimetric evidence of particulate surfaces, though of course the regolith may be very thin for the smaller objects.

In cases for which the meteoritic analog for a particular asteroid is known with some confidence, more specific statements about the surface texture can be made. Figure 6 illustrates a sample of the basaltic achondrite Bereba, prepared by LeBertre and Zellner (1978) exactly to match the albedo and telescopic polarization-phase curve of Vesta. The sample has also been measured for its near-infrared reflection spectrum, and found to show absorption band strengths very similar to those observed for Vesta (Feierberg 1979, personal communication). The sample contains a broad range of particle sizes up to 200 μm, with the larger grains thinly coated with very fine ($\lesssim 5$ μm) dust. The polarimetry places no particular limit on the maximum particle size, but neither the coarser grains alone, nor the very fine dust alone, is correct for Vesta.

We are convinced that, if a sample of Vesta's surface could be examined in a 1-cm sample tray, it would look much like Fig. 5. Let us emphasize that all the known optical properties of Vesta can be duplicated by crushing a stony meteorite, with no further alterations. Lunar soils, by contrast, cannot be produced by simple crushing of lunar rocks. The size distribution of particles on Vesta is superficially like that on the moon (as is the gross composition), but it is not necessary to invoke the various complex processes

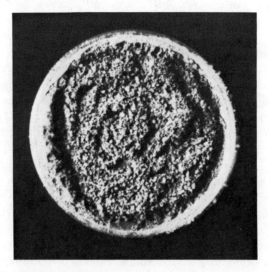

Fig. 6. Photomicrograph of a sample of the eucrite Bereba, crushed to a surface texture yielding the best match with the telescopic polarization-phase curve of Vesta (Le Bertre and Zellner 1978). The sample tray is 1 cm in diameter.

that have gone into producing the lunar soils.

The asteroid surface soils seem to be rather coarser than the lunar fines, which have 20 % of their mass in grains smaller than 10 μm and only 40 % in grains larger than 100 μm. By contrast the powders best simulating the asteroid surfaces have most of their mass in grains of diameter ~100 μm and less than 10 % in the form of 10 μm and finer dust particles. The differences are doubtless related in part to the lower average impact velocities in the main belt (~5 km sec^{-1}), but they are more specifically attributable to the small size and low surface gravity of the asteroids. As discussed by Housen *et al.* in their chapter in this book, asteroid regolith particles tend to lie undisturbed for a period of time and then be excavated and lost in subsequent impacts; the very extensive gardening processes that are characteristic of the lunar surface do not occur.

As illustrated in Fig. 5(b), the depth P_{min} of the negative branch can be used as an indicator of albedo, though such usage is hazardous in view of its sensitivity to surface texture. P_{min} ranges from about 0.3 % for the high-albedo E asteroids to values slightly in excess of 2 % for some of the dark C objects. Both the theory of Wolff and that of Lumme predict an upper limit of ~ 2.5 % for P_{min}; the largest value reliably measured in the laboratory is about 2.2 %. Like the polarimetric slope, the negative branch saturates for albedos in the range 0.02 − 0.05 and is weaker for very dark surfaces. Thus for dark, distant objects like the Trojans, not observable at phases larger than 12°, the polarimetry can be used only to place upper limits on the albedo. For the main-belt asteroids P_{min} is a good indicator of taxonomic type, with which a mean albedo may be associated for statistical purposes (see Zellner's chapter in this book). We do not, however, list asteroid albedos derived directly from P_{min} alone.

Dollfus et al. (1979) have succeeded in generating polarization-phase curves like those observed for the M asteroids by using iron filings of diameter ~30 μm. Presumably even metallic surfaces will be shattered and pulverized, producing particulate regoliths, under the hypervelocity impacts. Because of the low temperatures encountered in the asteroid belt, the ferro-nickels are no longer ductile but brittle. Such processes are very difficult to simulate in the laboratory. The enstatite chondrites are spectrally indistinguishable from pure metal, and could be an alternative explanation for the correct mineralogy for the M asteroids; but E-type chondrite samples matching the polarization of M asteroids have not been identified.

Another puzzle is provided by the C3 chondrites, which measured in the laboratory have invariably shown polarimetric inversion angles in the range $23^\circ - 28^\circ$, unlike anything observed in the asteroid belt. Further work on the C3 samples is in progress. Finally, there is the large, dark, unclassifiable object 704 Interamnia, with inversion angle $16°$, far smaller than is observed for all the rest of the asteroids. Is Interamnia alone dust-free? It seems most unlikely, and the explanation may be associated with its peculiar and presently unknown composition.

III. SURFACE VARIEGATION

As minor planets rotate, any differences in mean albedo or surface texture from one hemisphere to another should be revealed by changes in the optical polarization. According to results summarized by Degewij and Zellner (1978), most asteroids are in fact remarkably uniform in disk-integrated measurements. The Amor object 433 Eros shows undetectable polarization variations at the level of one part in 40, and for 1 Ceres the variations are less than one part in 200. For both objects the photometric lightcurves are due entirely to shape and shadowing and not to albedo spots. Similar results, though of lower precision, are found for several other objects, yet each asteroid can be recognized as nearly unique if well-enough observed. Apparently, under impact bombardment, each asteroid blankets itself in a well-mixed layer of its own debris.

Vesta provides a notable exception, as demonstrated in Fig. 7 from work reported by Gradie et al. (1978b). Vesta shows polarization variations of about one part in ten, in exact synchronization with and very close mirror-image to its photometric lightcurve. The planet can be regarded as a spheroid with the brightness changes being fully attributable to albedo variegation. Whether the differences are functions of surface texture, composition, or contamination by impacting foreign material can be elucidated in principle from detailed studies of the reflection spectrum and of

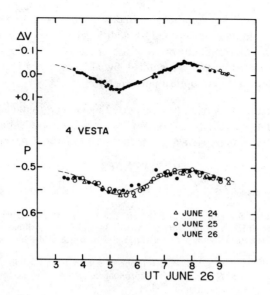

Fig. 7. Variations of asteroid 4 Vesta in photometric magnitude (top) and in polarization (bottom) on three nights in June 1978.

the polarization near the inversion angle. In UBV colors Vesta is known to be slightly redder near maximum light (Blanco and Catalano 1979), but the supplementary polarimetry has not been done and attempts to detect spectral changes at longer wavelengths have been inconclusive. Vesta with its surface markings remains a challenge for the observer.

IV. FUTURE WORK

We can now discuss with some confidence the albedos and surface textures of minor planets as indicated by optical polarimetry. Some pieces are still missing, however. The laboratory work is incomplete with respect to C3 and enstatite chondrites and with respect to metal-rich meteorites. Polarimetrically we really ought to be able to distinguish between stony-irons and H chondrites for the S asteroids, and between enstatite chondrites and nickel-irons for the M objects, but the problems of sample preparation are formidable. It is possible that some of the answers will come from theoretical work. Attempts to model the polarization of light reflected from particulate surfaces have broken the spirits of more than one theoretician, but good people are still trying.

Observationally, we do not expect that the number of minor planets with well-defined polarization-phase curves will substantially increase in the near future. Large telescope time for asteroid work is too limited, and the need for an extended spectrophotometric survey of the minor planet population is too compelling. Rather, we see the polarimetry playing a supporting role, for examination of unusual objects and for additional detailed studies of asteroid spottedness. Finally, any objects that are chosen for spacecraft flyby and rendezvous missions should be studied by all available groundbased techniques, including extensive optical polarimetry.

Acknowledgments. This work was supported by Observatoire de Paris, Centre National de la Recherche Scientifique (CNRS) and Centre National d'Etudes Spatiales (CNES), and by NASA grants to the University of Arizona.

REFERENCES

Blanco, C., and Catalano, S. 1979. UBV Photometry of Vesta. *Icarus* (special Asteroid issue).

Bowell, E.; Dollfus, A.; and Geake, J. E. 1972. Polarimetric properties of the lunar surface and its interpretation. Part 5: Apollo 14 and Luna 16 lunar samples. *Proc. Lunar Sci. Conf. III* (Cambridge: Massachusetts Institute of Technology Press), pp. 3103-3126.

Bowell, E., and Zellner, B. 1974. Polarizations of asteroids and satellites. In *Planets, Stars, and Nebulae Studied with Photopolarimetry,* ed. T. Gehrels (Tucson: University of Arizona Press), pp. 381-404.

Degewij, J., and Zellner, B. 1978. Asteroid surface variegation, *Lunar Sci. IX,* Lunar and Planetary Institute, Houston. pp. 235-237.

Dollfus, A. 1955. Etude des planètes par la polarisation de leur lumière. Thesis, University of Paris. NASA Tech. Transl. F-188, 1964.

Dollfus, A. 1961. Polarization studies of planets. In *Planets and Satellites,* eds. G. P. Kuiper and B. M. Middlehurst (Chicago: Univ. of Chicago Press), pp. 343-399.

Dollfus, A. 1971. Physical studies of asteroids by polarization of the light. In *Physical Studies of Minor Planets,* ed. T. Gehrels (NASA SP-267, Washington, D. C.: U. S. Government Printing Office), pp. 95-116.

Dollfus, A., and Geake, J. E. 1975. Polarimetric properties of the lunar surface and its interpretation: Part 7. Other solar system objects. *Proc. Lunar Sci. Conf. VI* (Oxford: Pergamon Press), pp. 2749-2768.

Dollfus, A.; Geake, J. E.; Mandeville, J. C.; and Zellner, B. 1977. The nature of asteroid surfaces, from optical polarimetry. In *Comets, Asteroids, Meteorites,* ed. A. H. Delsemme (Toledo, Ohio: University of Toledo Press), pp. 243-251.

Dollfus, A.; Mandeville, J.C.; and Duseaux, M. 1979. The nature of the M-type asteroids from optical polarimetry. *Icarus* 37: 124-132.

Dollfus, A., and Titulaer, C. 1971. Polarimetric properties of the lunar surface and its interpretation. Part 3. Volcanic samples in several wavelengths. *Astron. and Astrophys.* 12: 199-209.

Gradie, J.; Lebofsky, L.; and Zellner, B. 1978a. Radiometric and polarimetric diameter and albedo of 532 Herculina (abstract). *Bull. Amer. Astron. Soc.* 10: 596.

Gradie, J.; Tedesco, E.; and Zellner, B. 1978b. Rotational variations in the optical polarization and reflection spectrum of Vesta (abstract). *Bull. Amer. Astron. Soc.* 10: 595.

KenKnight, C. E.; Rosenberg, D. L.; and Wehner, G. K. 1967. Parameters of the optical properties of the lunar surface powder in relation to solar wind bombardment. *J. Geophys. Res.* 72: 3105-3129.

LeBertre, T., and Zellner, B. 1978. The surface texture of Vesta. *Lunar Sci. IX,* The Lunar and Planetary Institute, pp. 642-644.

Lebofsky, L. A., and Rieke, G. H. 1979. Thermal properties of 433 Eros. *Icarus* (in press).

Lyot, B. 1929. Recherches sur la polarisation de la lumière des planètes et de quelques substances terrestres. Doctoral Thesis, University of Paris. NASA Tech. Transl. F-187, 1964.

Veverka, J. 1970. Photometric and polarimetric studies of minor planets and satellites. Ph.D. dissertation, Harvard University.

Widorn, T. 1967. Zur photometrischen Bestimmung der Durchmesser der Kleinen Planeten. *Ann. Univ. Sternw. Wien* 27: 112-119.

Wolff, M. 1975. Polarization of light reflected from rough planetary surface. *Appl. Optics* 14: 1395-1405.

Zellner, B. 1977. Optical polarimetry of particulate surfaces. In *Optical Polarimetry, Instrumentation and Applications*, eds. R. M. A. Azzam and D. L. Coffeen, Proc. Soc. Photo-Optical Instrum. Eng., vol. 112, pp. 168-175.

Zellner, B., and Gradie, J. 1976. Minor planets and related objects. XX. Polarimetric evidence for the albedos and compositions of 94 asteroids. *Astron. J.* 81: 262-280.

Zellner, B.; Leake, M.; LeBertre, T.; Duseaux, M.; and Dollfus, A. 1977a. The asteroid albedo scale. I. Laboratory polarimetry of meteorites. *Proc. Lunar Sci. Conf. VIII* (Oxford: Pergamon Press), pp. 1091-1110.

Zellner, B.; LeBertre, T.; and Day, K. 1977b. The asteroid albedo scale. II. Laboratory polarimetry of dark carbon-bearing silicates. *Proc. Lunar Sci. Conf. VIII*, (Oxford: Pergamon Press), pp. 1111-1117.

RADIOMETRY OF ASTEROIDS

DAVID MORRISON
University of Hawaii

and

LARRY LEBOFSKY
University of Arizona

Measurements of the thermal emission of asteroids can be used to derive albedos and diameters and to infer information on thermophysical properties of the upper layer of asteroid surfaces. Radiometric diameters and albedos, derived on the basis of standard thermal models for low-conductivity regoliths, have been determined for approximately 200 asteroids, with results in agreement with determinations by polarimetry and stellar occultations. For objects to which these standard models apply, diameters can be determined with an accuracy of ±10%. In several cases of small asteroids, however, there is evidence that the standard models do not apply, presumably due to the absence of an insulating regolith. Thus the interpretation of radiometric diameters is in question and the utility for objects with diameters less than ~ 30 km is diminished. In these cases the thermal observations potentially can provide information on the ability of low-gravity objects to retain regoliths.

[184]

In many areas of planetary astronomy, measurements of thermal radiation are used to deduce surface temperatures. However, if the effective size of the emitting area is not known, a thermal flux measurement is inadequate to define the temperature. In principle, a thermal spectrum could be used to determine temperature, but in practice variations of emissivity with wavelength of naturally occurring surfaces, as well as the variation of temperature with location on the surface, make it difficult to achieve the desired precision. Rather, the thermal flux can be used, in combination with a photometric measurement of reflected sunlight, to derive the size and the albedo. In the study of asteroids, diameters and albedos are fundamental, but not easily measured, parameters. Radiometry has therefore made an important contribution as the simplest and fastest way to measure diameter and albedo with precision adequate for all but the most demanding tasks, such as derivation of bulk density.

The first application of thermal radiometry to derive an asteroid diameter was made by Allen (1970), who used 10 μm observations to find what he called an "infrared diameter" for 4 Vesta. Matson (1971 a,b) independently carried out a program to measure diameters for about 20 asteroids from observations in three infrared bandpasses in the 8 to 14 μm window. Both of these authors concluded that Vesta was larger than had been thought from visual observations, and Matson made the additional discovery that at least one asteroid, 324 Bamberga, was extremely dark, with geometric albedo < 0.04. The publication of both Allen's (1971) and Matson's (1971a) preliminary results stimulated considerable interest and controversy.

In the eight years since 1971, radiometry has earned an important place in the study of physical properties of asteroids. When Chapman et al. (1975) introduced the *CSM* taxonomy (see Zellner's chapter in this book), and summarized the then known properties of asteroids, 47 radiometric diameters had been published. Two years later, when Morrison (1977b) reviewed the field, a total of 187 objects had been measured. In the TRIAD file (Bender et al. 1978) presented in Part VII of this book, radiometric diameters are given for 195 asteroids. At this writing, the general survey work needed for reconnaisance and classification of bright main-belt asteroids has been largely completed; radiometry will, however, continue to play an important role in the characterization of special groups, or of objects found from color data to be of particular interest. In addition, there is the prospect of a comprehensive survey of most of the numbered asteroids and faint asteroids in general from space in a very few years, with the anticipated launch of the Infrared Astronomical Satellite (IRAS; see the chapter by Morrison and Niehoff).

In this review we will discuss the standard models for the interpretation of asteroid radiometry (Sec. I), other nonstandard thermal models that appear to be required in some cases (Sec. II), and finally the data base and assumptions that are used to generate the TRIAD file of standard asteroid diameters and albedos.

I. STANDARD THERMAL MODELS

Basic Principles

The principle of the radiometric determination of size and albedo is simple. The visible brightness of an asteroid is proportional to the product of geometric albedo and cross-section. The total thermal emission is proportional to the product of the absorbed insolation ($1 - A$, where A is Bond albedo) and cross-section. For a surface in equilibrium with the insolation, the reflected and emitted radiation together must equal the total solar radiation intercepted. Thus measurements of both the reflected and emitted components, together with appropriate assumptions concerning the photometric properties of the surface and the relationship between geometric and Bond albedos, is sufficient to determine both the size and the albedo of the asteroid.

The reflected component of energy is easily measured by a broadband visible photometric observation, using typically the B and V of the standard UBV system. The absorbed and reradiated component is best measured in the infrared, near the blackbody peak; for asteroids, broadband photometry near 10 and 20 μm is most convenient. Since the visible and infrared components respond in a complementary way to differences in albedo, even a relatively crude measurement of the instantaneous difference in visible and infrared magnitude (e.g., a precision of 10–20%) is adequate to determine the asteroid diameter to within ~ 10%.

In principle it is easy to determine the relationship of geometric albedo to bolometric Bond albedo. The geometric albedo, p_λ or p (λ), at a given wavelength, is the ratio of the back-scattered light from an asteroid or planet compared with a flat Lambertian (diffuse reflecting) disk of unit albedo, while the bolometric Bond albedo is the ratio of total incident to total reflected radiation integrated over the entire surface of the object and integrated over all wavelengths. Since the brightness of an asteroid or planet will vary with viewing angle and illumination, the geometric albedo will also vary. Therefore, to relate the geometric and bolometric Bond albedos we must also introduce the phase integral q which was first defined by Russell (1916) as

$$q\ (\lambda) = 2 \int_0^\pi \phi(\lambda,\alpha) \sin \alpha \ d\alpha \qquad (1)$$

where $\phi(\lambda,\alpha)$ is the disk integrated brightness of the asteroid at phase angle α, relative to its brightness at $\alpha = 0°$. We can thus define Bond albedo at any given wavelength as

$$A_\lambda = q\,(\lambda)\,p\,(\lambda) \qquad (2)$$

and finally A_B (or simply A) as the bolometric Bond albedo

$$A_B = \overline{A_\lambda}\,. \qquad (3)$$

However, since we can observe asteroids over only a limited range of phase angles ($< 25°$) we cannot determine q explicitly and thus it must be assumed or modeled.

One of the major unknowns which must be assumed or modeled for the radiometric method is the relationship between geometric and Bond albedo. However, for the typical dark asteroid it turns out that derived diameters and geometric albedos are remarkably insensitive to this relationship. If nearly all the incident sunlight is absorbed, the surface is close to blackbody temperature, and even a large change in Bond albedo (e.g. a factor of 2) makes little difference to the surface temperature or the emitted infrared flux. The main parameter influencing infrared brightness is simply diameter, and an infrared magnitude by itself comes very close to determining the size. The visible brightness, on the other hand, is directly proportional to geometric albedo, whether the albedo is large or small. Thus the diameter, together with the visual magnitude, yields the geometric albedo. In actual practice, the variables are not as totally separable as this discussion has implied, but for dark objects it well illustrates that radiometry yields a diameter that is only weakly dependent on either V-magnitude or assumptions of photometric properties, whereas the geometric albedo is directly related to the visual magnitude, and errors in V transform directly into errors in albedo.

The techniques that have been used to derive radiometric diameters and albedos all require assumptions concerning the photometric properties and the thermal properties of asteroid surfaces. Such assumptions must be made, since the measurements of reflected and emitted energy necessarily refer to only a limited geometry, usually at small phase angles, whereas the balance between absorbed and emitted energy is a global property of the asteroid. In the remainder of this section we describe the standard models that have been used during the past decade.

Standard Models

A number of authors have described models, all of which are roughly equivalent, that can be characterized as standard or lunar-like. Early work included Allen (1970), Matson (1971a,b), Morrison (1973), Jones and Morrison (1974), and Hansen (1976). More recent descriptions, which include review and criticism of earlier work, are given by Morrison (1977b), Hansen (1977a), and Matson et al. (1978). Other useful but nonstandard approaches will be discussed in Sec. II.

Treatment of the visible photometric properties of asteroids is generally rather simple, since the computed diameters and geometric albedos are only weakly dependent on the values adopted. Adopted linear phase coefficients of 0.02 to 0.04 mag/deg are used (Gehrels and Tedesco 1979; see Gehrels' chapter), and the opposition surge is neglected in computing geometric albedo. The V-band brightness is taken as representative of the integrated, or bolometric magnitude, although when color information has been available it has sometimes also been used (e.g. Hansen 1976). The most important parameter that must be assumed is the ratio (q, the phase integral) of Bond to geometric albedo. Morrison and his collaborators have generally adopted for all asteroids a value $q = 0.6$, the same as the measured phase integral of the moon and Mercury. Matson and his collaborators have used a semi-empirical value of q computed from the albedo based on the correlation of phase integral to albedo derived for the moon. This method appears to be valid, based on the recent theoretical work of Bowell and Lumme (see their chapter in this book). In practice the value of q in the range 0.5–0.8 has negligible influence on the computed diameters unless the albedo is high; e.g. > 0.20 (cf. Jones and Morrison 1974; Morrison 1977b).

The most important assumptions in the standard models concern the relationship between the infrared magnitude and the total thermal emission of an asteroid. The first problem is the absolute calibration of the infrared photometric scale. Uncertainties in the adopted magnitudes of standard stars and in the definition of zero magnitude in terms of absolute radiance levels may be as high as 10 to 15%. In addition, small but significant relative differences in the magnitudes reported by different observers are attributable to the slightly different photometric systems used by each. In part these differences are illusory, since differences in reported magnitudes do not necessarily imply equal differences in the derived monochromatic radiance levels, which are the numbers used in modeling asteroid diameters.

We summarize in Table I the standard stars used by recent asteroid surveys: Morrison (U. of Hawaii/Kitt Peak), Hansen (Caltech/Cerro Tololo), and Rieke/Lebofsky (U. of Arizona). While there are no large systematic differences, the uncertainty of absolute calibration remains a primary limit on the accuracy of asteroid radiometric measurements.

A more basic problem in the interpretation of the radiometric measurements lies in the unknown thermal properties of the asteroids themselves. As a result of differing assumptions concerning the emissivity and thermal inertia of asteroid surfaces, different workers have occasionally derived albedos differing by as much as a factor of 2 from the same or similar data. The "standard model" in the title of this sub-section refers essentially to a particular set of these assumptions that has proved highly useful in deriving diameters and albedos of main-belt asteroids.

Following Matson et al. (1978), we can write for a spherical object in equilibrium with the insolation:

TABLE I

Comparison of Absolute Calibration Systems

Group	Star	$m_{\lambda 10}$[a]	$m_{\lambda 20}$[b]	$F_{\lambda 10}$[c]	$F_{\lambda 20}$[c]	$F_{10.0}$[d]	$F_{20.0}$[d]
				(W-cm^{-2}-μm^{-1})		(W-cm^{-2}-μm^{-1})	
Caltech/Cerro	α Lyr	0.00	0.00	1.14×10^{-16}	8.07×10^{-18}	1.16×10^{-16}	7.49×10^{-18}
Tololo	α Ori[e]	-5.17	-5.73	1.33×10^{-14}	1.58×10^{-15}	1.35×10^{-14}	1.47×10^{-15}
(Hansen)	γ Cru	-3.29	-3.43	2.36×10^{-15}	1.90×10^{-16}	2.40×10^{-15}	1.76×10^{-16}
	α Sco[e]	-4.65	-5.01	8.26×10^{-15}	8.14×10^{-16}	8.39×10^{-15}	7.55×10^{-16}
	β Gru	-3.46	-3.36	2.76×10^{-15}	1.78×10^{-16}	2.80×10^{-15}	1.65×10^{-16}
Hawaii–Kitt Peak	α Boo	-3.2	-3.3	2.2×10^{-15}	1.5×10^{-16}	2.2×10^{-16}	1.5×10^{-16}
(Morrison)	α Her[e]	-4.0	-4.3	4.6×10^{-15}	3.6×10^{-16}	4.6×10^{-16}	3.6×10^{-16}
	β Peg[e]	-2.5	-2.7	1.3×10^{-15}	9.0×10^{-17}	1.3×10^{-16}	9.0×10^{-17}
	α Tau	-3.1	-3.2	2.0×10^{-15}	1.4×10^{-16}	2.0×10^{-16}	1.4×10^{-16}
	α Ori[e]	-5.2	-5.7	1.4×10^{-14}	1.4×10^{-15}	1.4×10^{-14}	1.4×10^{-15}
	β Gem	-1.30	---	3.87×10^{-16}	---	3.87×10^{-16}	---
Arizona	β And	-2.06	-2.23	6.47×10^{-16}	5.07×10^{-17}	8.13×10^{-16}	6.16×10^{-17}
(Rieke–Lebofsky)	α Ari	-0.75	-0.85	1.94×10^{-16}	1.42×10^{-17}	2.43×10^{-16}	1.73×10^{-17}
	α Tau	-2.99	-3.12	1.52×10^{-15}	1.15×10^{-16}	1.92×10^{-15}	1.40×10^{-16}
	α Aur	-1.90	-1.93	5.58×10^{-16}	3.85×10^{-17}	7.02×10^{-16}	4.67×10^{-17}
	α CMi	-0.66	-0.57	1.78×10^{-16}	1.10×10^{-17}	2.24×10^{-16}	1.34×10^{-17}
	α Hya	-1.38	---	3.46×10^{-16}	---	4.35×10^{-16}	---
	α Boo	-3.12	-3.30	1.72×10^{-15}	1.36×10^{-16}	2.16×10^{-15}	1.65×10^{-16}
	γ Dra	-1.45	---	3.69×10^{-16}	---	4.64×10^{-16}	---

TABLE I (Continued)

Comparison of Absolute Calibration Systems

Group	Star	$m_{\lambda 10}$[a]	$m_{\lambda 20}$[b]	$F_{\lambda 10}$[c]	$F_{\lambda 20}$[c]	$F_{10.0}$[d]	$F_{20.0}$[d]
Arizona	γ Aql	-0.72	---	1.88×10^{-16}	---	2.37×10^{-16}	---
	β Gem	-1.20	-1.28	2.93×10^{-16}	2.11×10^{-17}	3.68×10^{-16}	2.57×10^{-17}

[a] Standard star magnitude, 10 μm bandpass. Caltech: $\lambda_{10} = 10.04$ μm; Hawaii: $\lambda_{10} = 10.0$ μm; Arizona: $\lambda_{10} = 10.6$ μm.
[b] Standard star magnitude 20 μm bandpass. Caltech: $\lambda_{20} = 19.63$ μm; Hawaii: $\lambda_{20} = 20.0$ μm; Arizona: $\lambda_{20} = 21.0$ μm.
[c] Standard star fluxes 10 and 20 μm bandpasses. Zero magnitude fluxes 10 μm (20 μm); Caltech: 1.14×10^{-16} (8.07×10^{-18}); Hawaii: 1.17×10^{-16} (7.3×10^{-18}); Arizona: 9.7×10^{-17} (6.5×10^{-18}).
[d] Standard star fluxes standardized to 10.0 μm and 20.0 μm. Zero magnitude fluxes 10.0 μm (20.0 μm): Caltech: 1.16×10^{-16} (7.49×10^{-18}); Hawaii: 1.17×10^{-16} (7.3×10^{-18}) Arizona: 1.22×10^{-16} (7.9×10^{-18}).
[e] Stars listed in variable star catalog as having visual variabilities greater than 0.5 mag.

$$\pi r^2 (1-A) S_o = \beta \epsilon \, \sigma \, r^2 \int\limits_{0}^{2\pi} \int\limits_{-\pi/2}^{\pi/2} T^4(\theta,\phi) \cos\phi \, d\phi \, d\theta \qquad (4)$$

where r is the radius, A is the bolometric Bond albedo, S_o is the solar radiant flux, β is a normalization constant (of order unity) related to the angular distribution of thermal emission, ϵ is the bolometric emissivity, σ the Boltzmann constant, and $T(\theta,\phi)$ is the effective temperature of a point on the surface at longitude θ and latitude ϕ. The earth and sun both are assumed to lie along the normal to the surface at $\theta = 0$, $\phi = 0$; i.e., the asteroid is observed at zero phase. The standard model, implicitly or explicitly, needs to consider three things: the value of ϵ and β, and the distribution of surface temperature.

The infrared emissivities of natural dielectric materials are generally ~ 0.9, the value usually associated with the standard model. Since infrared measurements are typically made at 10 or 20 μm, near the peak of the asteroid thermal emission, the solutions are only weakly dependent on choice of ϵ; the main effect of changing ϵ is to change the temperature and hence the wavelength dependence of brightness, but not to alter significantly the emitted flux in the passbands observed.

The distribution of temperature is primarily a function of the thermal inertia of the surface and the rotation period of the asteroid. If the inertia is large (high thermal conductivity) and the rotation rapid, the surface layers continue to radiate substantial energy as they rotate into the unobserved night hemisphere, and the infrared brightness seen toward small phase angles is decreased. The standard models are often called "lunar-like" models because they assume low thermal inertia, corresponding to a loose, particulate regolith. With such lunar thermal properties, an asteroid with a typical rotation period of several hours radiates only a few percent of the absorbed insolation from its dark hemisphere. The standard model simply assumes:

$$T = T_{max} \cos^{\frac{1}{4}} \theta \cos^{\frac{1}{4}} \phi \qquad (5)$$

for the illuminated hemisphere and $T = 0$ for the dark hemisphere.

The normalization constant β allows for the possibility of enhanced infrared emission at small phase angles, such as has been seen for the moon (Saari and Shorthill 1972); in this case, $\beta < 1$. It can also be used to express, in a convenient way, departures from the temperature distribution adopted for the standard model. If the conductivity is so high that little cooling takes place, β approaches a value of π.

The standard models of Jones and Morrison (1974), Hansen (1976, 1977a), Matson et al. (1978) and Lebofsky et al. (1978) all provide differing rationales for their choice of degree of enhancement of emission toward zero phase and of the temperature distribution over the surface, but for purposes of computing asteroid diameters and albedos these differences can be expressed simply as differences in choice of the normalization constant β. Jones and Morrison attempted to calibrate their models from observations of the Galilean satellites and the moon, leading to a value for β of ~ 0.9. Hansen (1976) did not incorporate peaking and also assumed significant dark-side emission, yielding $\beta \sim 1.3$. In a later detailed model in which the emission properties of a rough, cratered surface were considered, Hansen (1977a) derived $\beta \sim 1.0$, thus obtaining diameters and albedos similar to those of Jones and Morrison. Matson et al. and Lebofsky et al. have explicitly considered models yielding a wide range in β, as will be discussed in Sec. II.

In re-examining the controversy that surrounded radiometric diameters in the years 1974-1977, we have computed new thermal models to study Hansen's (1976) assumptions. Hansen assumed that the temperature at the terminator was 60% of that at the subsolar point, and thus that about 40% of the asteroid's thermal energy is radiated from the unlit side. His model was therefore not really lunar-like; for a surface of lunar soil, only 3 to 5% is radiated from the dark hemisphere. In addition, we have found an error introduced by his treatment of the emission spectrum as that of a single blackbody rather than a composite of radiation from many temperatures. Because he made this assumption in combining his 10 and 20 μm data, Hansen's derived emissivities are systematically too low (37 out of 66 of his computed emissivities were less than 0.8, the lowest being 0.35). When these problems are corrected in Hansen's 1976 model, much of the discrepancy between his values and those of Morrison are eliminated, even without his later crater model (Hansen 1977a).

Morrison has made both 10 and 20 μm measurements of asteroids. When the diameters derived from these are compared, the 20 μm diameters are in general slightly larger ($\sim 4\%$) than the 10 μm diameters. Since Morrison does assume a surface temperature distribution and does not use a single blackbody (see, e.g., Jones and Morrison 1974), the discrepancies could be due to uncertainties in the relative 10 and 20 μm flux calibration (20 μm flux calibration high relative to 10 μm flux calibration).

The standard model, then, assumes the asteroid to be exactly spherical, with a surface composed of a rough dielectric of low thermal conductivity, to a depth of at least a few centimeters. The photometric properties are those of a dark, dusty surface, as is suggested by the observed phase functions and polarimetric behavior of asteroids. In the infrared, the emissivity is near unity, there is a peaking or beaming of thermal emission at small phase angles, and no more than a few percent of the absorbed insolation is radiated from the night hemisphere.

Calibration of the Standard Model

The most important application of radiometry to date has been for the derivation of diameters and albedos. The values actually derived depend on a

TABLE II

Comparison of the Occultation and Radiometric Diameters of Asteroids

Asteroid	D (occultation)[a]	D (radiometric)	Rad/Occ
2 Pallas	538±11	583±58	1.1±0.1
6 Hebe	186± 9	206±21	1.1±0.1
532 Herculina	217±15	219±11	1.0±0.1

[a]Estimated errors are 1σ residuals of fit of geometric model to observations.

number of assumed model parameters, described above. We now consider the final choice of the normalization factor β or its equivalent to yield the best values for asteroid diameters.

Historically, different infrared observers made somewhat arbitrary choices of model parameters, derived diameters and albedos, and then compared their results. Much of the discrepancy centered on the reality of the normalization for zero-phase peaking introduced by Jones and Morrison (1974) and Morrison (1974) and subsequently enlarged still more by Morrison and Chapman (1976). Their scale yielded higher albedos than those calculated by Matson (1971b) and Hansen (1976), although still not high enough to agree with the polarimetric albedos of Zellner and colleagues at that time. Subsequently, Hansen (1977a), using the model of Winter and Krupp (1971) showed how a cratered surface could generate the beaming effect. Since 1977 a consensus has emerged in favor of a standard model, for at least the larger asteroids, similar to that of Morrison and colleagues.

Two independent sources of asteroid data can be invoked to test the radiometric diameter scale. First are the polarimetric albedos, obtained from laboratory calibration of an empirical polarization slope-albedo law (e.g. Bowell and Zellner 1974, and the chapter by Dollfus and Zellner in this book). A comprehensive recent re-analysis of these albedos by Zellner et al. (1977a,b) yields a scale for asteroid albedos in agreement with the radiometric values (Morrison 1977b) for $p_v > 0.06$, and explains discrepancies at lower albedos in terms of a breakdown of the polarization slope-albedo relationship for low-albedo materials. Based on 25 asteroids observed well by both techniques, the systematic difference between the two scales is well under 5% in albedo, and half as great in diameter.

The recent determination of several asteroid diameters from stellar occultations (Bowell et al. 1978b; Taylor and Dunham 1978; Wasserman et al. 1979; see the chapter by Millis et al. in this book) provides the potential for an even more secure calibration for the larger asteroids. Table II compares the occultation diameters for Pallas, Hebe, and Herculina with the radiometric diameters calculated by Morrison from TRIAD. The two sets of diameters agree within their stated uncertainties, although there is a slight suggestion

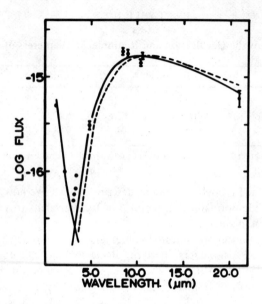

Fig. 1. The observed flux of 324 Bamberga from 1.25–21 μm. A solar spectrum has been fit through the K-band measurement and a least squares fit has been made to the thermal infrared measurements (5–21 μm). Solid line: dusty model. Dashed line: rocky model.

that the radiometric diameters may be a few percent too large. Pallas should provide the best test, but unfortunately the radiometry for this object is not all of high quality, with a wide range among separate observations. Presumably one or more of the observations is bad, but new observations are required to resolve the matter. Until that time, any use of the occultation diameters to adjust the radiometric scale is premature.

Another approach to testing the standard model is to obtain radiometric observations over a broad spectral range. The predictions of the standard model concerning the thermal emission spectrum can then be tested. Figure 1 illustrates observations of this type for 324 Bamberga, made in 1978 by Lebofsky. The standard thermal model fits this spectrum very well.

In summary, the radiometric calibration adopted in 1974 by Morrison and his colleagues, with a value of $\beta \sim 0.9$, appears to yield results for the larger asteroids in agreement with both polarimetric albedos and occultation diameters. For those asteroids with thermal and electrical surface properties consistent with the assumptions of the standard model, radiometric diameters should be accurate to ± 10%, and albedos to ± 20%.

II. NONSTANDARD THERMAL MODELS

The agreement between the diameters and albedos obtained from radiometry, polarimetry, and occultations suggests that the standard model, with its lunar-like surface layer, applies to most of the asteroids studied radiometrically. Typically, these are main-belt objects with diameters of 50 km or more. However, based on our understanding of meteorites and their relationship to asteroids, it is probable that some asteroids do not have the assumed lunar thermal properties, or that others might have metallic rather than rocky surfaces. Thus we need to consider what the radiometric behavior might be for asteroids that do not conform to the assumptions of the standard model.

The need for nonstandard modeling of asteroids has been recently demonstrated in the observations by Lebofsky *et al.* (1978) of 1580 Betula and by Lebofsky *et al.* (1979) of 2100 Ra-Shalom, 1978RA. In both cases, differences between radiometric diameters obtained with the standard model and diameters inferred from other techniques suggest that these small asteroids lack a low-conductivity regolith (i.e., have rock surfaces). In addition, the extensive observations made of 433 Eros in 1975 (Lebofsky and Rieke 1979) suggest that it has a partially rocky surface.

Departures from Standard Models

There are three main ways in which the assumptions of the standard model may break down for real asteroids. The first and perhaps simplest case is that of a nonspherical asteroid. If the object is greatly elongated, as is the case for 433 Eros and 624 Hektor, the temperature distribution must be calculated for a more realistic geometry, such as two spheres in contact or a cylinder with hemispherical ends. In reality, of course, the asteroid will have a more complex shape, but such simple models are adequate to specify the only parameter of interest for calculating diameters, which is the distribution of orientation of surface elements with respect to insolation.

Of greater potential interest are failures in the assumptions concerning thermal and electrical properties of the surface. If there is no regolith, the thermal inertia may be high and as much as 50% of the absorbed insolation may be radiated from the dark hemisphere. In that case, radiometry from small phase angles will underestimate the total absorbed energy, and a diameter calculated with the standard model will be too small. In the extreme case of an isothermal asteroid, the error could be a factor of 2 in albedo, or of $\sqrt{2}$ in diameter.

If an asteroid does not have a dielectric surface, as might be the case for the parent body of a metal meteorite, both the thermal and electrical properties may be anomalous. Metal has low infrared emissivity (0.1), as well as high thermal conductivity. The low emissivity would raise the surface temperature, shifting the bulk of the thermal radiation to shorter wavelengths. Depending on the infrared wavelength bands observed, the failure of the standard model could be substantial.

In Table III, adopted from Matson *et al.* (1978), we compare two nonstandard models with the standard, lunar-like model (Model I). Model II

TABLE III

Model Parameters

Parameter	Model I (Standard Model)	Model II	Model III
Analogy	Lunar surface	Rock	Iron meteorite
Thermal response	Low thermal inertia	High thermal inertia	High thermal inertia
Rotation	Slow (nonrotating)	Rapid	Rapid
β	0.9	π	π
Emissivity, ϵ	0.9	0.9	0.1
$T(\theta,\phi)$, $\|\theta\| \leqslant 90°$	$T_{\max} \cos^{1/4}\theta \cos^{1/4}\phi$	$T_{\max} \cos^{1/4}\phi$	$T_{\max} \cos^{1/4}\phi$
$\|\theta\| > 90°$	0	$T_{\max} \cos^{1/4}\phi$	$T_{\max} \cos^{1/4}\phi$

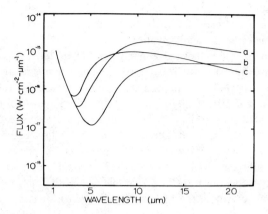

Fig. 2. Three model surfaces as originally discussed by Lebofsky *et al.* (1978) and Matson *et al.* (1978). All surfaces have the same albedo. (a) Model I: $\beta = 0.9$, $\epsilon = 0.9$, nonrotating, dusty model. (b) Model II: $\beta = \pi$, $\epsilon = 0.9$, rapid rotating, rocky model (isothermal). (c) Model III: $\beta = \pi$, $\epsilon = 0.1$, rapid rotating, metal model.

corresponds to an extreme "rocky" surface in which the thermal conductivity is so high, and the rotation so rapid, that there is no temperature variation with phase. (Note, however, that the surface is not strictly isothermal; temperature still varies with latitude.) A true rocky surface with no dust, while it would display some diurnal temperature variation, might approach the conditions of this model for a rapidly rotating asteroid. Model III corresponds to a metallic object. Again, there is no variation of temperature with phase, but in addition the emissivity is extremely low. A true asteroid might approach these properties even if the metal were mixed with ordinary silicates, so long as the connected metal masses were large compared with a thermal wavelength (tens of centimeters).

Figure 2 illustrates the spectrum calculated with each model for an asteroid with $p_v = 0.055$ at a distance from the sun of 2.77 AU (1 Ceres). The absolute difference between the curves for Model I and Model II, resulting from the fact that half the energy in Model II is emitted unseen from the dark side, is obvious in this plot. However, if the difference had to be inferred from the *shape* of the spectrum, very careful measurements would be required at several wavelengths. In contrast, the shape of the spectrum for Model III, the metal asteroid, could be distinguished from the others on the basis of 10 and 20 μm observations alone.

It is clear that there are three main ways of detecting departures from the standard models:

1. Inconsistency of the radiometric diameter with values derived by other means, such as polarimetry or radar. In principle such discrepancies could be seen with even a single infrared measurement (as in the cases of

Betulia and Ra-Shalom); however, it would be difficult to demonstrate that the problem really lay with the thermal asteroid model and was not simply some error in the observation.

2. Inconsistency of the thermal emission spectrum. Most departures from the standard model result in an apparent color temperature that differs from that expected. In the simplest case, evidence for such a failure could be seen if 10 and 20 μm radiometry, interpreted with the standard model, yielded substantially different diameters. Broad spectral coverage now exists for two asteroids (Eros and Bamberga) discussed below.

3. Unusual dependence of the thermal emission on phase angle. If an asteroid could be observed over a wide enough range of phase, such effects might be obvious. For example, an isothermal, metal asteroid would show no drop in thermal emission at increasing phase angles.

Observational Evidence for Nonstandard Behavior

The primary evidence for departures of real asteroids from the lunar-like properties assumed in the standard models comes from recent observations of two small, Earth-approaching objects. In 1977, Lebofsky et al. (1978) measured 1580 Betulia at 10 μm and derived a radiometric diameter that was inconsistent (smaller diameter) with the results of polarimetry (Tedesco et al. 1978) and radar (Pettengill et al. 1979) as well as with the spectral type suggested by its color. Because of this discrepancy, they had to invoke a model in which Betulia had a bare rock surface. To fit the polarimetric and radar data, Betulia required a combination of Models I and II with about 40% bare rock. More recently (Lebofsky et al. 1979), a similar discrepancy was found for 2100 Ra-Shalom, but the polarimetry was too imprecise to determine an accurate model. It may be that smaller asteroids with low gravity do not have well enough developed regoliths for the standard model to apply, but the available data do not as yet support such a sweeping generalization.

It is possible that we are seeing the crossover point between the "large" asteroids and the "small" asteroids in the Eos and Koronis families, for which Gradie (1978; see the chapter by Gradie et al.) has obtained extensive UBV and radiometric observations. Although these objects are much larger than Apollo, Amor and Aten asteroids, such as Betulia and Ra-Shalom, they are significantly smaller on the average than the main-belt asteroids studied previously. The Koronis family was homogenous and of S type; however, an anomalously large percentage of the Eos family has peculiarities leading to a designation of "unclassifiable." These peculiarities include radiometric albedos (0.06 to 0.09) that tend to fall between the majority of C and S objects. Thus, while the standard model seems to give results consistent with other measurements for the Koronis family, this may not be the case for the Eos family, implying surfaces that may not be lunar-like.

Other attempts to look for departures from the standard model have yielded negative results. Morrison (1974, 1977*a,b*) looked at the 10 and 20 μm color temperatures for about a hundred asteroids and found no major anomalies of the sort suggested by Model III in Fig. 2. As discussed above, the broad spectral coverage of 324 Bamberga also confirms the validity of the standard model for slow rotating, large main-belt asteroids. Several discrepancies between radiometric and polarimetric diameters have sprung up over the years, such as that for 532 Herculina noted by Chapman *et al.* (1975) and Morrison (1977*b*) but these have all been resolved when new data appeared (in the case of Herculina the absolute magnitude B(1,0) was in error by more than one magnitude). Finally, in the few cases where a large asteroid has been observed at a variety of phase angles (Matson 1971*b*; Morrison 1977*b*), the observed phase dependence of infrared emission is consistent with a lunar thermal inertia. Small differences between the brightness at the same phase angle seen before and after opposition are attributable to the fact that in one case the morning side is seen, and in the other the warmer afternoon side faces Earth. These differences are too small to permit a solution for the value of the surface thermal inertia, but their existence has been used to infer the sense of rotation for half a dozen asteroids by Morrison (1977*b*) and Hansen (1977*b*); see also the chapter by Taylor in this book.

The Special Cases of 324 Bamberga and 433 Eros

Broad spectral coverage (0.3–21 μm) now exists for two asteroids. These two asteroids represent extremes: (1) 324 Bamberga, a large main-belt asteroid with a very long rotation period (29.4 hr; Zappalà, personal communication), and (2) 433 Eros, a small Earth-approaching asteroid with a typical rotation period (5.2 hr). These observations have given us our first chance to test the validity of the standard thermal model and also to investigate the nonstandard modeling techniques. As noted before (Fig. 1), the standard model gives the best fit to the observations of Bamberga, suggesting that the standard model is valid for the large main-belt asteroids, at least when their rotation rate is slow.

During the close approach to Earth in 1975, many studies were made of 433 Eros (cf. Zellner 1976), including extensive radiometry by both Morrison at Hawaii and Rieke at Arizona. Special modeling efforts have permitted these data to be used to test some of the assumptions of the standard thermal model. Morrison (1976) interpreted his observations in two ways. First, he used the radiometry to determine diameter and albedo. His geometric model was that of a cylinder with hemispherical ends. At minimum, one of the hemispheres nearly faces Earth, and the standard model was used without modification; at maximum, corrections were required to account for the large fraction of the total area normal to the sunlight.

Lebofsky and Rieke (1979) obtained measurements of high precision over a wide range of wavelengths, and their analysis makes use of the

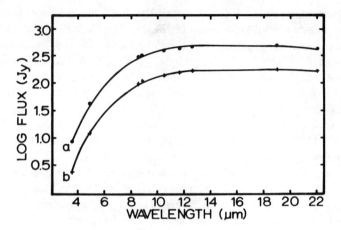

Fig. 3. Best calculated fit to the observations for Eros using the dust plus rock model (solid lines). (a) Maximum light (·). (b) Minimum light (+). (From Lebofsky and Rieke 1979.)

increased leverage inherent in such data. They calculated models that explicitly included the geometry of Eros (assumed a cylinder with hemispheric ends) and allowed for the possibility of surface components of differing thermal inertia. Their model was able to explain (even with a lunar-like surface material) the amplitude difference between the visual and infrared lightcurves and the infrared phase lag. Figure 3 illustrates the best fits obtained to both shape and absolute level of the spectra at maximum (curve *a*) and minimum (curve *b*). The curves correspond to models in which there is a combination of materials: either a mixture of about equal parts of "dust" and "sand", or primarily "dust" with a few percent bare rock. A pure lunar-type model is somewhat inconsistent with the observed spectra.

With the nonstandard thermal model indicated by the emission spectrum, Lebofsky and Rieke derived an albedo that is lower and a size that is larger than that of Morrison. Table IV compares the albedos and sizes for Eros as obtained by polarimetry, radar and these two radiometric models. As indicated, the standard radiometry appears to agree with the polarimetry (although the polarimetric albedo may be high due to the elongated shape of Eros [B. Zellner, personal communication]), but the radar diameter is higher, and the diameter obtained by Lebofsky and Rieke higher still.

It is difficult to generalize from Eros to the more general problem of regolith development on asteroids. Eros is a small, irregularly shaped asteroid in an Earth-approaching orbit, unlike the typical bright main-belt asteroids that have been studied by the standard thermal modeling techniques. However, we also have a much more typical asteroid, 324 Bamberga, which does appear to have a well-developed regolith. Only with more detailed observations of asteroids of various sizes, from Bamberga down to Eros and

TABLE IV

Albedo and Mean Diameter of 433 Eros

	P_v	D_o(km)
Morrison (1976)	0.18±0.03	22
Zellner *et al.* (1977 *a,b*)	0.17±0.02	23
Jurgens and Goldstein (1976)	0.14±0.02	25
Lebofsky and Rieke (1979)	0.125±0.025	27

smaller, can we begin to answer general questions about regolith development on asteroids.

III. THE TRIAD FILE OF RADIOMETRIC DIAMETERS AND ALBEDOS

In this section we describe the asteroid diameters and albedos assembled for the TRIAD file (see Part VII) and used by several authors in this book for statistical studies. This discussion is an update of the review by Morrison (1977*b*).

The Data Base

The radiometric observations used for the TRIAD file consist of broadband radiometry at 10 μm (the N band) or 20 μm (Q band). All observations published since 1972 have been used, (except for a few exceptions noted specifically below). The observations previous to 1978 are those compiled by Morrison (1977*b*). The computed diameters have been updated by the use of the new values of V(1,0) available from TRIAD.

All of these observations have been interpreted with the standard model described in Sec. I, using the computer code of Jones and Morrison (1974). The calibration corresponds in their notation to T_o = 408 K, and β = 0.9. In each case, the infrared observations themselves were placed on a uniform magnitude scale and reduced to zero phase with an assumed radiometric phase coefficient of 0.01 mag/deg, as described in more detail by Morrison (1977*a*). The primary modifications introduced for this chapter are (1) the addition of new radiometric observations by Degewij (1978), Gradie (1978) and Hartmann and Cruikshank (1978), and (2) the adoption of revised absolute magnitudes V(1,0).

The new radiometric measurements are primarily of small objects in the Eos and Koronis families. Gradie (1978) has reduced these observations (including those by Degewij, 1978), made with the University of Arizona telescopes, to the same photometric system used by Morrison, and he also observed several bright asteroids to confirm that his techniques yielded results consistent with the rest of the data base.

In order to derive geometric albedos from the radiometry, accurate visual magnitudes are required. Except for a very few special cases discussed by

Morrison (1977*b*), the observations were reduced with the standard values of V(1,0). Even when simultaneous UBV photometry was available, the final values for diameters and albedos apply to the standard absolute magnitude. These albedos correspond to a linear phase coefficient, neglecting the opposition surge, in conformity with standard usage.

The new B(1,0) magnitudes were supplied by Gehrels (TRIAD file in Part VII of this book). Larger phase coefficients were used than in previous compilations, resulting in systematically brighter B(1,0) values and consequently a small increase, on a statistical basis, in the computed asteroid albedos. Since the albedos quoted are in V rather than B, the value of B(1,0) had to be converted to V(1,0). Nearly all of these asteroids have recently been observed in UBV by Bowell, who supplied the B-V colors (TRIAD file); in the few cases where no direct color observations were available, the colors used were the averages for the appropriate *CSM* class (see Zellner's chapter).

The only new radiometry not included in the TRIAD file is that of 1580 Betulia and 2100 Ra-Shalom (Lebofsky *et al.* 1978, 1979) discussed in Section II. Since there are discrepancies between the standard model results and those obtained by other techniques that are used to infer something about the asteroid regolith, there is no true independent diameter that can be calculated from this radiometry.

Results

The TRIAD file contains radiometric albedos and diameters for 195 asteroids (Table V). Each asteroid is assigned a quality code, according to the quality and consistency of the data, following the scheme described by Morrison (1977*a*). Code 1 is assigned to single observations or to data with large internal uncertainties; these values may not be individually reliable and should be used with caution. Code 2 is assigned in a broad range of situations, including at the minimum two independent measurements on a good photometric night. This code indicates basically secure, believable results. Code 3 is assigned to excellent, consistent, multiply-verified data, almost always at both 10 and 20 μm. Usually observations from several nights are required to justify Code 3. While it is not possible to associate a code directly with uncertainties, the three codes correspond roughly to internal uncertainties in the diameters of $\pm 20\%$, $\pm 10\%$, and $\pm 5\%$, respectively. At present the possibility exists of additional errors of up to 10% in the overall diameter calibration, due to the absolute flux calibration.

Figure 4 is a histogram of the measured albedos; filled bars refer to results with Code 2 or 3, and open bars include all observations. The measured albedos range from 0.48 for 44 Nysa down to 0.021 for 596 Scheila, or possibly to 0.019 for 95 Arethusa, if we wish to count a Code 1 result. As in past figures of this type, the broad albedo division between *C* types and all others is clear. Since the uncertainties in the Code 2 and 3

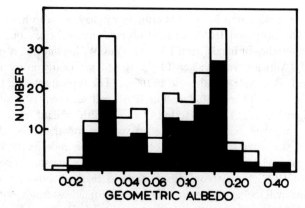

Fig. 4. Histogram of measured asteroid albedos. Filled bars refer to results with Code 2 or 3 (see text), and open bars include all observations.

results are no greater than the widths of the histogram bars, the spread in each broad peak represents a real diversity of asteroid albedos, even within a single *CSM* class. The median value of p_v for the C asteroids is about 0.035, and for the S asteroids it is perhaps 0.14.

In general, the targets for radiometry have been selected according to brightness, and many large, dark asteroids in the outer belt have not been measured. Corrections for these sampling biases have been discussed by a number of authors, including Chapman *et al.* (1975), Morrison (1977*a*), and Zellner and Bowell (1977), and they are treated in detail in the chapter by Zellner in this book.

The Future

The period of 1974 to 1977 saw a great boom in asteroid radiometry. Once 150 or so diameters and albedos were measured, however, it became possible to associate albedos reliably with other optical properties, and Bowell *et al.* (1978*a*) and others have found that most asteroids can be classified, and approximate albedos assigned, on the basis of UBV colors alone (having established a correlation between UBV colors and albedo). Radiometry remains the best technique for measuring asteroid albedos and diameters, but recently it has been applied selectively to special problems, as with the work on families by Gradie and on the Trojans by Cruikshank (1977).

An additional recent thrust in asteroid radiometry deals with observations of small, Earth-approaching objects. Here the observations by Lebofsky and his colleagues suggest departures from the standard models. Radiometry of small asteroids may be used as a tool to study regolith development, rather than to derive diameters and albedos. Also with broad spectral coverage of high accuracy, as in the case of 324 Bamberga, it may be possible to refine the thermal modeling techniques as well as the infrared calibration system.

The next major step in asteroid radiometry may come from observations in space. The proposed IRAS infrared observatory will carry out an all-sky survey in a number of bands from 10 to 100 μm. With a limiting N magnitude of 6 or 7 (Aumann and Walker 1977), it will see nearly every numbered asteroid and thousands of others in addition. (The typical asteroid has a V–N color index of 10–11 magnitudes; thus IRAS will see essentially all asteroids brighter than V $\sim$17). The extraction of this wealth of data from the survey remains a problem, since from the perspective of the original goal of obtaining a sky survey, the asteroid observations are undesirable; but this is clearly a case where one person's signal is another person's noise. If the potential of IRAS for asteroid observations is realized, however, diameters and albedos will be available in a few years for all asteroids with known orbits, and additional statistical information may also be obtained on the distribution of faint asteroids.

Acknowledgment. This work is supported by the National Aeronautics and Space Administration.

REFERENCES

Allen, D. A. 1970. The infrared diameter of Vesta. *Nature* 227: 158-159.

Allen, D. A. 1971. The method of determining infrared diameters. In *Physical Studies of Minor Planets,* ed., T. Gehrels (NASA SP-267, Washington, D.C.: U.S. Printing Office), pp. 41-44.

Aumann, H. H., and Walker, R. G. 1977. Infrared astronomical satellite. *Opt. Eng.* 16: 537-543.

Bender, E.; Bowell, E.; Chapman, C.; Gaffey, M.; Gehrels, T.; Zellner, B.; Morrison, D.; and Tedesco, E. 1978. The Tucson Revised Index of Asteroid Data. *Icarus* 33: 630-631.

Bowell, E.; Chapman, C. R.; Gradie, J. C.; Morrison, D.; and Zellner, B. 1978a. Taxonomy of asteroids. *Icarus* 35: 313-335.

Bowell, E.; McMahon, J.; Horne, K.; A'Hearn, M. F.; Dunham, D. W.; Penhallow, W.; Taylor, G. E.; Wasserman, L. H.; and White, N. M. 1978b. A possible satellite of Herculina (abstract). *Bull. Amer. Astron. Soc.* 10: 594.

Bowell, E., and Zellner, B. 1974. Polarizations of asteroids and satellites. In *Planets, Stars* and *Nebulae Studied* by *Photopolarimetry,* ed., T. Gehrels (Tucson: University of Arizona Press), pp. 381-404.

Chapman, C. R.; Morrison, D.; and Zellner, B. 1975. Surface properties of asteroids: A synthesis of polarimetry, radiometry, and spectrophotometry. *Icarus* 25: 104-130.

Cruikshank, D. P. 1977. Radii and albedos of Trojan asteroids and Jovian satellites 6 and 7. *Icarus* 30: 224-230.

Degewij, J. 1978. Photometry of faint asteroids, satellites and cometary nuclei. Ph.D. dissertation, Leiden University.

Gehrels, T. and Tedesco, E. F. 1979. Minor planets and related objects, XXVIII. Asteroid magnitudes and phase relations. *Astron. J.* (in Press).

Gradie, J. C. 1978. An astrophysical study of the minor planets in the Eos and Koronis asteroid families. Ph.D. dissertation, University of Arizona.

Hansen, O. L. 1976. Radii and albedos of 84 asteroids from visual and infrared photometry. *Astron. J.* 81: 74-84.

Hansen, O. L. 1977a. An explication of the radiometric method for size and albedo determination. *Icarus* 31: 456-482.

Hansen, O. L. 1977b. On the prograde rotation of asteroids. *Icarus* 32: 458-460.

Hartmann, W. K., and Cruikshank, D. P. 1978. The nature of trojan asteroid 624 Hektor. *Icarus* 36: 353-366.

Jones, T. J., and Morrison, D. 1974. A recalibration of the radiometric/photometric method of determining asteroid sizes. *Astron. J.* 79: 892-895.

Jurgen, R. F., and Goldstein, R. M. 1976. Radar observations at 3.5 and 12.6 cm wavelength of asteroid 433 Eros. *Icarus* 28: 1-16.

Lebofsky, L. A.; Lebofsky, M. J.; and Rieke, G. H. 1979. Thermal properties of Apollo, Amor, and Aten objects. *Astron. J.* (in press).

Lebofsky, L. A.; Lebofsky, M. J.; and Rieke, G. H. 1979. Radiometry and surface properties of Apollo, Amor, and Aten objects. *Astron. J.* 84: 885-888.

Lebofsky, L. A.; Veeder, G. J.; Lebofsky, M. J.; and Matson, D. L. 1978. Visual and radiometric photometry of 1580 Betulia. *Icarus* 35: 336-343.

Matson, D. L. 1971a. Infrared observations of asteroids. In *Physical Studies of Minor Planets* ed., T. Gehrels (NASA SP-267, Washington, D.C.: U.S. Government Printing Office), pp. 45-50.

Matson, D. L. 1971b. I. Astronomical photometry at wavelengths of 8.5, 10.5 and 11.6 μm. II. Infrared emission from asteroids at wavelengths of 8.5, 10.5 and 11.6 μm. Ph.D. dissertation, California Institute of Technology.

Matson, D. L.; Veeder, G. J.; and Lebofsky, L. A. 1978. Infrared observations of asteroids from earth and space. In *Asteroids: An Exploration Assessment*, eds. D. Morrison and W.C. Wells, NASA Conf. Publ. 2053. pp. 127-144.

Morrison, D. 1973. Determination of radii of satellites and asteroids from radiometry and photometry. *Icarus* 19: 1-14.

Morrison, D. 1974. Radiometric diameters and albedos of 40 asteroids. *Astrophys. J.* 194: 203-212.

Morrison, D. 1976. The diameter and thermal inertia of 433 Eros. *Icarus* 28: 125-132.

Morrison, D. 1977a. Radiometric diameters of 84 asteroids from observations in 1974-76. *Astrophys. J.* 214: 667-677.

Morrison, D. 1977b. Asteroid sizes and albedos. *Icarus* 31: 185-220.

Pettengill, G. H.; Ostro, S. J.; Shapiro, I. I.; Marsden, B. G.; and Campbell, D. B. 1979. Radar observations of 1580 Betulia. *Icarus* (Special Asteroid issue).

Russel, H. N. 1916. On the albedo of the planets and their satellites. *Astrophys. J.* 43: 173-195.

Saari, J. M., and Shorthill, R. W. 1972. The sunlit lunar surface. I. Albedo studies and full Moon temperature distribution. *The Moon* 5: 161-178.

Taylor, G. E., and Dunham, D. W. 1978. The size of minor planet 6 Hebe. *Icarus* 34: 89-92.

Tedesco, E. F.; Drummond III, J. D.; Candy, B.; Birch, P.; Nikoloff, I.; and Zellner, B. 1978. The Amor asteroid 1580 Betulia: An unusual asteroid with an extraordinary lightcurve. *Icarus* 35: 344-359.

Wasserman, L. H.; Millis, R. L.; Franz, O. G.; Bowell, E.; White, N. M.; Giclas, H. L.; Martin, L. J.; Elliot, J. L.; Dunham, E.; Mink, D.; Baron, R.; Honeycutt, R. K.; Henden, A. A.; Kephart, J. E.; A'Hearn, M. F.; Reitsema, H.; Radick, R. R.; and Taylor, G. E. 1979. The diameter of Pallas from its occultation of SAO 85009. *Astron. J.* 84: 259-268.

Winter, D. F., and Krupp, J. A. 1971. Directional characteristics of infrared emission from the Moon. *Moon* 2: 279-292.

Zellner, B. 1976. Physical properties of 433 Eros. *Icarus* 28: 149-153.

Zellner, B., and Bowell, E. 1977. Asteroid composition types and their distributions. In *Comets, Asteroids, Meteorites,*. ed., A. H. Delsemme (Toledo, Ohio: University of Toledo Press), pp. 185-197.

Zellner, B.; Leake, M.; Lebertre, T.; Duseaux, M.; and Dollfus, A. 1977b. The asteroid albedo scale. I. Laboratory polarimetry of meteorites. *Proc. Lunar Sci. Conf. VIII* (Oxford: Pergamon Press), pp. 1091-1110.

Zellner, B.; Lebertre, T.; and Day, K. 1977a. The asteroid albedo scale. II. Laboratory polarimetry of dark carbon-bearing silicates. *Proc. Lunar Sci. Conf. VIII* (Oxford: Pergamon Press), pp. 1111-1117.

RADAR OBSERVATIONS OF ASTEROIDS

G. H. PETTENGILL
Massachusetts Institute of Technology

AND

R. F. JURGENS
Jet Propulsion Laboratory

*Radar reveals information concerning asteroids which is comple-
mentary to that available from optical and infrared observations.
Specifically, the distance and radial velocity, deduced from radar,
nicely complement the angular position of the object as determined
from telescopic measurement. Since radar coherently illuminates the
target, the surface scattering properties at radio wavelengths as a
function of angle and polarization are directly determined. Five
asteroids have been observed by radar prior to 1979; more than 50 may
be observed over the next decade using radar equipment currently
available.*

Radar has been used to study the asteroids in much the same way it has been
used to study the larger planets and their satellites (Pettengill 1978).
Specifically, one seeks information on the surface scattering characteristics
(radar albedo, angular scattering law, polarizing properties and the variation
of these with rotational phase), on the orbital distance and velocity, and the
size and shape of these small bodies. Five asteroids have been detected so far
by radar: 1566 Icarus, 1685 Toro, 433 Eros, 1580 Betulia, and I Ceres. A
survey by Jurgens and Bender (1977) indicates that the Arecibo *S*-band radar

(λ = 12.6 cm) should be capable of detecting 60 asteroids, and the Goldstone X-band (3.54 cm) 18 asteroids, during the next ten years.

These asteroids are primarily of two types: (1) the largest main-belt asteroids, and (2) those having Mars-crossing orbits that bring them closer than a few tenths of one AU from Earth. Of the asteroids detected so far, only 1 Ceres falls in the first group. Asteroids in the second group are often more easily detectable than those of the first, because the radar detectability depends inversely on the fourth power of distance but only on the 3/2 power, directly, of the diameter. The radar scattering properties and the radar albedo are also important in establishing a detection. Table I gives the peak radar cross sections for each asteroid observed prior to 1979 as a function of radar wavelength. Only in the case of 433 Eros (Jurgens and Goldstein 1976) was the variation in radar cross-section observed as a function of rotational phase for a sufficiently long time that the rotational period could be determined. Observations of 1 Ceres obtained in March and April of 1977 by Ostro *et al.* (1979) were distributed over a wide range of rotational phase angles, but these show relatively little variation in cross-section, except for one night. Cross-polarized observations were made only for 433 Eros at 12.6- and 3.5-cm wavelengths (Jurgens and Goldstein 1976); the circularly polarized echo component having the same rotational sense as that transmitted was found to be approximately four times weaker at both wavelengths than its orthogonal partner.

The majority of radar observations of asteroids have used simple CW (continuous-wave) waveforms, with transmission lasting for the duration of the round-trip delay. The echo is received, spectrum-analyzed and integrated for a similar period.

The way in which the spectral information may be related to target scattering properties is shown in Fig. 1, where actual data from observing Mercury are given. The lower spectrum results from the analysis of echoes received in the polarization sense corresponding to coherent reflection; the upper spectrum corresponds to the polarization sense orthogonal to the lower, and is not sensitive to the coherently scattered power. Thus, having data from both senses of received polarization allows the relative importance of coherent (i.e. quasi-specular) and incoherent (i.e. diffuse) scattering from the target to be assessed.

As may be seen in Fig. 1, there is a *minimum* angle of incidence at which power observed at a given frequency is scattered. Note that quasi-specular scattering tends to concentrate the echo in a direction satisfying the constraints for classical coherent reflection. In the present case, with illumination and observation from the same direction, the peak of the received coherent component occurs at right angles to the surface, and thus at the frequency containing the center of the visible disk.

Unlike the situation for the inner planets and the moon, however, the radar spectra observed for asteroids do not exhibit a pronounced quasi-

TABLE I

Radar Cross Sections, Scattering Law and Radius for Radar-Observed Asteroids

Target and Date	Wavelength (cm)	Peak radar cross-section (km^2)	n	Radius (km)	Reference
1566 Icarus (6/68)	12.6	0.1±0.03	_[a]	≥0.5	Goldstein 1969
	3.8	0.08±0.04	_[a]	≥0.25	Pettengill et al. 1969
1685 Toro (8/72)	12.6	1.3±0.2	_[a]	≥1.7	Goldstein et al. 1973
433 Eros (1/75)	70	39±15	_[a]	-/16±4[b]	Campbell et al. 1976
	12.6	38±6	1.2±0.3	_[a]	Jurgens & Goldstein 1976
	3.5	30±7	1.2±0.3	8/18±1[b]	Jurgens & Goldstein 1976
1580 Betulia (5/76)	12.6	3.4±0.9	1.0±0.5	≥2.9	Pettengill et al. 1979
1 Ceres (3/77)	12.6	(0.04±0.01)[c]	7±3	_[a]	Ostro et al. 1979

[a] No radar data available.

[b] Eros has been modeled as a tri-axial ellipsoid; the two entries given correspond to the intermediate and largest semiaxes, respectively.

[c] Units of physical cross-section: $\pi a^2 \simeq 8 \times 10^5$ km^2.

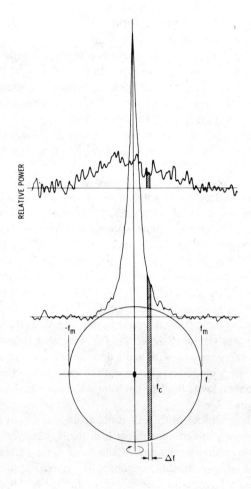

Fig. 1. Diagram showing the relationship between Doppler frequency and surface location for components of a radar echo spectrum scattered from a rotating spherical target. The lower, sharply peaked spectrum was obtained in the sense of polarization which maximizes sensitivity to coherently reflected power. In the example (drawn from radar observations of Mercury) shown here, considerable coherent, or quasi-specularly reflected, power may be seen. The upper spectrum is obtained in the polarization sense orthogonal to that below and effectively excludes coherently scattered power. The upper data provide a direct measure of the amount of incoherently scattered power and thus an estimate of the amount of wavelength-sized surface structure (i.e. roughness).

specular scattering component, but appear much like the spectra seen in the upper part of Fig. 1. We infer from this that the surfaces of asteroids are very rough as compared to those of the inner planets.

The width of the spectrum (from edge to edge) is dependent upon the radius, the rotational period, and the direction of the target's pole. If the latter two quantities are known from optical observations, the radius can be determined from a determination of the width between band edges. Only in the case of 1580 Betulia were the signals strong enough to determine distinctly the location of the edges of the spectrum. In all cases listed in Table I the rotational period has been determined from optical lightcurves, but the rotational pole position is known only for 433 Eros and 1 Ceres. For the latter object, the radar echoes were too weak to be useful in determining the radius. The determination of the radius of 433 Eros is further complicated by the known irregularity in the asteroid's shape. The major problem is that long integration times are needed to obtain a good signal-to-noise ratio; these tend to blur the instantaneous shape of the spectrum, however, much as long time exposures reduce the amplitude of lightcurves. To accommodate this problem, models of radar spectra based on figures other than simple spherical models are required. Jurgens and Goldstein (1976) used a rotating triaxial ellipsoidial model to estimate two axes of 433 Eros. Estimates of the third axis were determined from lightcurve and occultation data. Both Jurgens and Goldstein (1976) and Campbell *et al.* (1976) were able to show that their measurements were consistent with radii determined from timing of stellar occultations. If the pole direction is not known, the observed spectral broadening sets only a lower limit to the asteroid's radius as shown in the size estimates quoted in Table I.

Because of uncertainties in the values of radius and, except for 433 Eros, of the cross-polarized echo component, it is difficult to convert the observed cross-sections into geometric albedos. If the lower limit for radius is used, however, and the cross-polarized component is ignored, an upper limit for albedo is obtained which in no case exceeds ~0.08. For 1 Ceres, where a relatively well-established value of radius (510 km; see Bowell *et al* 1978, and Part VII of this book) exists from optical and infrared measurements, the corresponding (single-polarization) albedo is only about 0.01, a low value by any standard. We thus conclude that the surfaces of all asteroids seen so far contain little or no ice and do not have substantial amounts of exposed metallic material. The low albedo found for Ceres is hard to explain, but may reflect the presence of a relatively thick layer of loosely compacted low-density regolith material.

Estimates of the surface-scattering law have also been made in a few cases where circumstances appear to warrant. A scattering law proportional to $\cos^n \theta$ was assumed, where θ is the angle of incidence to the local mean surface. Generally, a spherical model is adequate except as pointed out for 433 Eros. The value of n is strongly correlated with the measurement of spectral width, since it controls the shape of the spectrum near the band edges. Therefore, unless the signal-to-noise ratio is sufficiently large that the edges can be seen distinctly, one tends to estimate an effective bandwidth given by $2f_m n^{-\frac{1}{2}}$, where f_m is the true spectral bandwidth from center to edge. Thus the separation is generally possible only if other information yielding values of the radius, rotation rate, and pole direction is available. In the case of 433 Eros, estimates of n were also constrained by failure to observe an apparent displacement of the spectral center frequency as a function of rotational phase. Table I contains estimates of the exponent n where available. Low values of n (i.e. very diffuse scattering laws) suggest very rough surfaces.

No radar observation of an asteroid has yet achieved resolution in delay, either because of weak signals (Ceres) or because of the relatively small size of those (Mars-crossing) objects from which echoes have been strong. The delay coordinate, which has provided so much insight into the surface properties of the terrestrial planets, will no doubt soon become available from new studies of asteroids using improved instrumentation. In view of the extremely diffuse angular scattering laws seen in the radar observations of asteroids, it should prove relatively easy using delay-Doppler measurements to locate features fixed to the surface of an asteroid, and to track the apparent motion of these features as the target rotates. In this way, improved estimates of radius and rotational pole position may be obtained, quantities of considerable value in the interpretation of optical and infrared measurements as well as of substantial interest in themselves.

REFERENCES

Bowell, E.; Chapman, C. R.; Gradie, I. C.; Morrison, D.; and Zellner, B. 1978. Taxonomy of asteroids. *Icarus* 35: 313-335.

Campbell, D. B.; Pettengill, G. H.; and Shapiro, I. I. 1976. 70-cm radar observations of 433 Eros. *Icarus* 28: 17-20.

Goldstein, R. M. 1969. Radar observations of Icarus. *Icarus* 10: 430-431.

Goldstein, R. M.; Holdridge, D. B.; and Lieske, J. H. 1973. Minor planets and related objects. XII. Radar observations of 1685 Toro. *Astron. J.* 78: 508-509.

Jurgens, R. F., and D. F. Bender 1977. Radar detectability of asteroids: A survey of opportunities for 1977 through 1987. *Icarus* 31: 483-497.

Jurgens, R. F., and Goldstein, R. M. 1976. Radar observations at 3.5 and 12.6-cm wavelengths of asteroid 433 Eros. *Icarus* 28: 1-15.

O'Leary, B.; Marsden, B. G.; Dragon, R.; Hauser, E.; McGrath, M.; Bacus, P.; and Robkoff, H. 1976. The occulation of κ Geminorum by Eros. *Icarus* 28: 133-146.

Ostro, S. J.; Pettengill, G. H.; Shapiro, I. I.; Campbell, D. B.; and Green, R. R. 1979. Radar observations of asteroid 1 Ceres. *Icarus* (in press).

Pettengill, G. H. 1978. Physical properties of the planets and satellites from radar observations. *Ann. Rev. Astron. Astrophys.* eds. G. Burbidge, D. Layzer and J. Phillips (Palo Alto, Calif.: Annual Reviews, Inc.), 16: 265-292.

Pettengill, G. H.; Ostro, S. J.; Shapiro, I. I.; and Campbell, D. B. 1979. Radar observations of asteroid 1580 Betulia. *Icarus* (in press).

Pettengill, G. H.; Shapiro, I. I.; Ash, M. E.; Ingalls, R. P.; Rainville, L. P.; Smith, W. B.; and Stone, M. L. 1969. Radar observations of Icarus. *Icarus* 10: 432-435.

RADIO OBSERVATIONS OF ASTEROIDS:
RESULTS AND PROSPECTS

JOHN R. DICKEL
University of Illinois

Radio observations of the asteroids can provide information on the thermal and dielectric properties of the surface materials and, because the radio emission arises somewhat below the surface, the data give some indication of layering. Observational difficulty has limited the investigations to only 6 asteroids. 1 Ceres and 324 Bamberga appear to have a layer of dust covering a more compacted material; the data on 4 Vesta cannot be matched by any current models for the surface; and the results for 18 Melpomene, 31 Euphrosyne and 433 Eros are too incomplete for firm conclusions. Future possibilities include more accurate radiometry of a few selected asteroids of different taxonomic classes and actual resolution of some of the larger objects by aperture synthesis techniques.

Radio observations of small planetary bodies provide unique information on physical parameters of the material in their subsurface layers. The radio emission is of thermal origin and arises on the order of several wavelengths below the surface. Thus the observed brightness depends upon the inward conduction of the heat from the sun and the outward transfer of the radiation. These processes depend upon the properties of the material, particularly its compaction and so a comparison of the data with the brightness of model asteroids can give a measure of the properties.

Because radio data are difficult to obtain and give only a single integrated flux density, they are clearly complementary to those obtained at optical and infrared wavelengths. The observational information necessary for a full interpretation of the radio results is discussed in Sec. I, the development of models for comparison with the data is described in Sec. II and the results to date are given in Sec. III. Finally, some future prospects using new techniques are presented in Sec. IV.

I. REQUIRED INFORMATION

Radio Data

Because the asteroids are small, none have yet been resolved with a radio telescope and we can measure only their integrated flux densities. In order to study the heat transfer, we need the temperatures. The flux density and temperature are related through the Planck law and the solid angle of the object:

$$S_\nu = \frac{2h\,\nu^3}{c^2} \frac{1}{e^{h\nu/kT}-1} \; \Omega \tag{1}$$

where S_ν is the flux density at a given frequency ν. The flux density is usually given in units of Jansky (Jy) where 1 Jansky $= 10^{-26}$ Wm^{-2} Hz^{-1}. h is Planck's constant, c is the speed of light, T the temperature, and Ω the solid angle subtended by the object. Thus we need the diameter of the asteroid to actually obtain a temperature. A knowledge of possible multiplicity is also important in evaluating the size.

Because the asteroids are typically black bodies at temperatures near 200 K, their spectra are peaked in the infrared, and the radio intensities are very low. This fact, coupled with the small diameters, makes their detection difficult. The observations require many hours of integration with the world's largest radio telescopes and careful subtraction of the sky background.

The new data presented in Table I were obtained in December 1978 with the 100-meter telescope of the Max Planck Institute for Radioastronomy in Bonn. I observed at a wavelength of 2 cm using a cooled radiometer which was continually switched between two beams separated by 3 arcmin in the sky. The procedure was to scan across the source in the direction of the beam separation so that the difference signal between the two beams produced first a negative response and then a positive response, with the characteristic beam pattern. A total of about 30 observations of 50 scans each were made per asteroid. This required about 20 hours of telescope time each, including some period for calibration. The primary calibration standard was 3C48 for which a flux density of 2.0 Jy at the 2-cm wavelength was adopted. Atmospheric extinction was monitored by regular observations of secondary calibration sources very near the asteroids in the sky.

A final requirement for the radio observations is accurate ephemerides. The telescopes typically have half-power beamwidths of about 1-arcmin and

TABLE I

Radio Observations of Asteroids

Object	T_B	λ	Phase Angle	Dist. from Sun (AU)	Type	Assumed Diam. (km)	Remarks	Ref.
1 Ceres	<150>	3 mm-3.7 cm	8°	2.6	C	980	dust on rock	1,2,3
4 Vesta	210±27	3 mm	<~5°	2.4	U	500	no good fitting models	3
18 Melpomene	<300	2 cm	~20°	2.1	S	152	–	5
31 Euphrosyne	151±30	2 cm	~18°	2.4	CM	333?	need infrared data	5
324 Bamberga	188±50	2 cm	-23°	2.0	C	251	probably thin compacted dust	5
433 Eros	<460	2 cm	<~5°	1.1	S	12x24	–	4

References:

1. Andrew (1974).
2. Briggs (1973).
3. Conklin *et al.* (1977).
4. Pauliny-Toth *et al.* (1976).
5. This chapter.

so we must track the asteroids to within a few arcsec. Because we cannot see them directly but must integrate blindly for several hours, the predicted positions must be precise.

Other Data on the Asteroids

The models for the study of heat transfer depend upon the insolation which varies with time because of asteroid rotation. We thus need the rotation period and orientation of the pole, based generally upon optical photometry.

The thermal budget also depends on the amount of energy available, or the albedo of the object. This requires good optical polarimetry and/or optical plus infrared photometry. Finally infrared and radio flux densities — preferably as a function of phase angle or at least at the same phase angle — are important to determine the emissivity of the material. Much of the support information can be obtained from the TRIAD file (see Part VII of this book).

Properties of Materials

In the following model analysis we shall relate the heat transfer to a parameter called the thermal inertia which is given by $(k\rho s)^{1/2}$ where k is the thermal conductivity in cal cm^{-1} sec^{-1} K^{-1}, ρ is the density in g cm^{-3}, and s is the specific heat of the material in cal g^{-1} K^{-1}. The values of these parameters are quite uncertain for many materials. We generally use values within the ranges given by Fountain and West (1970), Robie and Hemingway (1971), Cremers (1972), Cremers and Hsia (1973) and Hemingway et al. (1973) for typical terrestrial and lunar basalts in solid and loose states of compaction. To investigate the transfer of the emergent radiation, the dielectric constant ϵ and the electrical loss tangent tan Δ must be set to appropriate values for each material (Campbell and Ulrichs 1969; Bassett and Shackleford 1972). If ice is important, the properties of this material can be taken from Evans (1965).

II. THE MODELS

Structure

The general approach in modeling the radio emission from an asteroid is to adopt a two layer surface for the object: a base region of rock or other dense substance, with an overlying layer of less compacted material. The thickness of the top layer can be varied as well as the thermal and electrical properties of each region. Sample properties for a model of Ceres to match observations by Conklin et al. (1977) are listed in Table II.

TABLE II

Parameters for Ceres

OBSERVED

Radius	490 km
Albedo	0.04
Rotation Period	9 hr
Heliocentric Distance	2.72 AU (Dec 1975)
Geocentric Distance	1.77 AU (Dec 1975)
Phase Angle	$8°2$ (Dec 1975)
Observing Wavelength	3.33 mm
Observed Flux Density	0.374×10^{-23} erg sec^{-1} cm^{-2} Hz^{-1}

MODEL	**UPPER LAYER**	**LOWER LAYER**
Composition	dust	basalt
Thickness	0.5 cm	— —
Absorption Length	2.1 cm	0.36 cm
Dielectric Constant	2.9	7.2
Loss Tangent	0.015	0.054
Density	1.0 g cm^{-3}	2.6 g cm^{-3}
Specific Heat	0.09 cal $\text{g}^{-1} \text{K}^{-1}$	0.10 cal $\text{g}^{-1} \text{K}^{-1}$
Thermal Conductivity	2×10^{-6} cal $\text{cm}^{-1} \text{sec}^{-1} \text{K}^{-1}$	4×10^{-3} cal $\text{cm}^{-1} \text{sec}^{-1} \text{K}^{-1}$
Infrared Emissivity	0.99	
Scale Depth	2.9 cm	

PREDICTED BRIGHTNESS TEMP.　　142 K

Temperature

The procedure for analysis is to integrate the equation of conductive transport of the heat from the incoming solar radiation downward into the planet. For the numerical integration, the planet can be typically divided into zones 30° in latitude by 30° in longitude over which the temperature and insolation are averaged. If the time step in the integrations is set equal to 1/400 rotation and 8 full rotations of the planet are completed before the final temperatures are read, the averaging procedures are found to be accurate to within 2 %. The step size in depth should be a small fraction of the thermal wavelength given by $L_t = (Pk/\rho s \pi)^{1/2}$ where P is the rotation period and the other symbols are as defined above. Figure 1 shows sample profiles of the temperature distribution with depth at various phase angles for the parameters given in Table II for a model of Ceres. At a given spot, the input of heat necessary to raise the temperature a certain amount is given by $(\rho s/k)^{1/2}$ but the conductivity also enters to carry the heat away so that the final temperature is governed by the thermal inertia $(k\rho s)^{1/2}$. This quantity thus measures the effective resistance of the medium to heating. Note that the low thermal inertia in the upper layer causes large variations in the surface

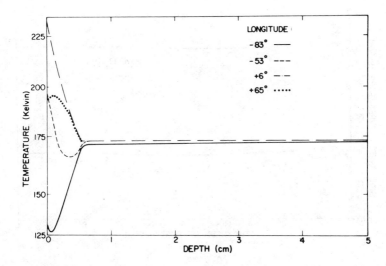

Fig. 1. Profiles of the temperature distribution with depth into Ceres for several phase angles using the model parameters given in Table II.

temperature with phase angle and a steep gradient with depth but then the thermal wave literally hits a stone wall at the interface between the layers. In the rock, the greatly increased thermal inertia allows a much deeper thermal wave but of much lower amplitude. The rapid spin of the asteroids never allows the thermal wave to penetrate very deeply into the body. Further, these bodies have sufficiently low mean temperatures so that possible radiative transfer of energy (e.g., Linsky 1966) will be negligible and only conduction need be considered.

Radiation

Next the equation of radiative transfer must be integrated outward through the medium taking into account reflections at the interfaces of the different media. The dielectric constant ϵ affects the reflection and emissivity at each interface. In addition, within each zone there is a phase change in the wave as it propagates, which makes self-interference or an effective absorption. This is dependent upon the loss tangent, $\tan \Delta = 2\sigma\lambda/c\epsilon$ where σ is the electrical conductivity, λ the wavelength, and c is the speed of light. With these values of the absorptivity and emissivity for each depth at its given temperature, plus the reflectivities we can determine the emergent intensity at each point on the planet. Finally, the intensity is integrated over the visible disk to give the expected brightness temperature at the phase angle of interest.

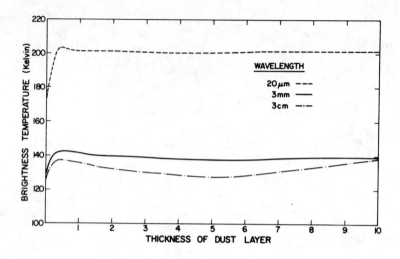

Fig. 2. The brightness temperature of Ceres at a phase angle of $0°$ as a function of the thickness of the surface dust layer at 3 different wavelengths.

Discussion

To see the effect of some of the parameters, let us look at Fig. 2 which is a plot of the apparent brightness temperatures at different wavelengths for a phase angle of $0°$ as a function of the depth of the top dust layer. We first note that the infrared temperature is much higher than the radio ones, and this is expected since the infrared emission arises much closer to the surface where the thermal wave has its greatest amplitude. The infrared temperature would, of course, be much lower than the radio ones when the sun is not illuminating the surface near phase angles of $180°$. A large difference in temperature is found between models with pure rock and those with a dust cover, even at the long wavelengths which arise in deep layers where little thermal variation is experienced. This is because a pure rock surface with its higher heat capacity and also greater conductivity never has a chance to heat up as much as does one with even a thin dust cover, so that a planet without dust has a lower mean temperature throughout.

In a dusty zone, the dielectric constant is low, producing lower reflectivity and higher emissivity and thus a higher apparent brightness temperature. Although $\tan \Delta$ is proportional to $1/\epsilon$, the dust has a much lower electrical conductivity and thus a lower loss tangent. The net result is that the penetration depth of the radio wave, $L_R = (2\pi \epsilon^{1/2} \tan \Delta)^{-1}$ is large. The emergent wave will arise deep down but the low thermal inertia limits the thermal wave to a very shallow depth. In rock, the parameters go the opposite way, however, and with a dust layer of about 0.5-1 cm thickness on top of

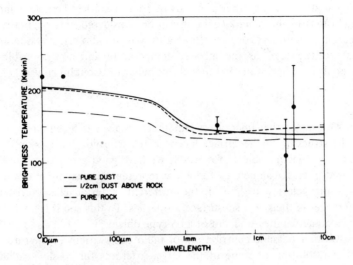

Fig. 3. Brightness temperature spectrum of 1 Ceres. The observed values represented by the points are from: 10 μm and 20 μm – Cruikshank and Morrison (1973); 3 mm – Conklin *et al.* (1977); 2.8 cm – Andrew (1974); 3.4 cm – Briggs (1973). The models represented by the lines are for various thicknesses of dust overlying rock as described in the text.

rock, we find that the net thermal and radio depths in the mixed medium are about equal ($L_R/L_T\lambda \sim 1$). This configuration produces a maximum value in the observed brightness temperature. With more dust, the thermal wave is damped before reaching the depth where radio emission occurs – with less dust, the thermal wave is so deep that it has very small amplitude.

III. RESULTS

Because of the observational problems discussed above, the results of radio observations to date are not many. They are all presented in Table I. Ceres is the only object for which more than one radio measurement exists and its other parameters are quite reliably determined. The comparison of the observational data with various models of this asteroid is illustrated in Fig. 3 (Conklin *et al.* 1977). Although variations of the thermal and dielectric properties of a given layer by as much as a factor of two generally affect the results by less than the uncertainties in the observations, we can draw some conclusions. Clearly pure rock cannot reproduce the observed values, but it is not possible to establish the thickness of the dust layer. As can be seen from Fig. 3, only a very accurate measurement at a wavelength of 10 cm or longer can provide some discrimination of the depth of the dust. Furthermore, the difference between rock and very compacted regolith cannot be distinguished.

The data for the other asteroids in Table I are less complete and so the conclusions in the remarks column can be considered only qualitative. Vesta is certainly unusual, however. Its relatively high radio brightness temperature is about the same as the infrared temperature and no reasonable surface materials have been found to model this behavior (Conklin *et al.* 1977).

IV. PROSPECTS

We can see that microwave radiometry is a viable technique to give clues on the structure of the surface layers of the asteroids, but it is only practical for a few specific objects for which we have good geometrical, optical and infrared data. It can be used to look at representative examples of the various taxonomic classes to see if the different exterior characteristics show differences in their near-subsurface properties. To this end it is important that we get an actual measurement of an *S*-type object.

Another possible contribution of radio observations in the future would be the resolution of some of the larger asteroids for possible satellites and binary pairs. The Very Large Array of radio telescopes has an angular resolution of about 0.1 arcsec at a wavelength of 1.3 cm and should be able to detect any asteroid with a diameter greater than about 0.2 arcsec. Thus any pair of objects satisfying the following conditions should be measurable:

1) resolution: $\left[\dfrac{\text{separation (in km)}}{\text{distance from earth (in AU)}}\right] > 200$

2) sensitivity: $\left[\dfrac{\text{diameter (in km)}}{\text{distance from earth (in AU)}}\right]^2 > 25{,}000$

3) preferably an orbital period of the satellite of several days.

Acknowledgments. I gratefully acknowledge the help of W. Altenhof and R. Marcum in making the 2-cm observations a success. B. Marsden provided ephemerides and P. Stumpf helped with both ephemerides and general problems with the computer of the Max Planck Institute. D. T. Ther participated actively in the model analysis. Financial assistance was provided by a grant from the National Aeronautics and Space Administration and by the Netherlands-America Commission for Eductional Exchange through the Fulbright-Hays program during my sabbatical leave at the Leiden Observatory.

REFERENCES

Andrew, B. H. 1974. Radio observations of some minor and major planets. *Icarus* 22: 454-458.
Bassett, H. L., and Shackleford, R. G. 1972. Dielectric properties of Apollo 14 lunar

samples at microwave and millimeter wavelengths. *Proc. Third Lunar Sci. Conf.* Vol. 3, ed. D. R. Criswell (Cambridge: Massachusetts Inst. Technology Press), pp. 3157-3160.

Briggs, F. H. 1973. Radio emission from Ceres. *Astrophys. J.* 184: 637-639.

Campbell, M. J., and Ulrichs, J. 1969. Electrical properties of rocks and their significances for lunar radar observations. *J. Geophys. Res.* 74: 5867-5881.

Conklin, E. K.; Ulich, B. L.; Dickel, J. R.; and Ther, D. T. 1977. Microwave brightnesses of 1 Ceres and 4 Vesta. In *Comets, Asteroids, and Meteorites,* ed. A. Delsemme (Toledo, Ohio: University of Toledo Press), pp. 257-260.

Cremers, C. J. 1972. Thermal conductivity of Apollo 14 fines. *Proc. Third Lunar Sci. Conf.,* Vol. 3, ed. D. R. Criswell (Cambridge: Massachusetts Inst. Technology Press), pp. 2611-2617.

Cremers, C. J., and Hsia, H. S. 1973. Thermal conductivity and diffusivity of Apollo 15 fines at low density. *Proc. Fourth Lunar Sci. Conf.,* Vol. 3, ed. W. A. Gose (New York: Pergamon Press), pp. 2459-2464.

Cruikshank, D. P., and Morrison, D. D. 1973. Radii and albedos of asteroids 1, 2, 3, 4, 6, 15, 51, 433, and 511. *Icarus* 20: 477-481.

Evans, S. 1965. Dielectric properties of ice and snow. A review. *J. Glaciology* 5: 773-792.

Fountain, J. A., and West, E. A. 1970. Thermal conductivity of particulate basalt as a function of density in simulated lunar and Martian environments. *J. Geophys. Res.* 75: 4063-4069.

Hemingway, B. S.; Robie, R. A.; and Wilson, W. H. 1973. Specific heats of lunar soils, basalt, and breccia from the Apollo 14, 15 and 16 landing sites, between 90 and $350°$K. *Proc. Fourth Lunar Sci. Conf.,* Vol. 3, ed. W. A. Gose (New York: Pergamon Press), pp. 2481-2487.

Linsky, J. L. 1966. Models of the lunar surface including temperature-dependent thermal properties. *Icarus* 5: 606-634.

Pauliny-Toth, I. I. K.; Witzel, A.; and Dickel, J. R. 1976. Upper limit to the 2-cm brightness temperature of asteroid 433 Eros. *Icarus* 28: 147.

Robie, R. A., and Hemingway, B. S. 1971. Specific heats of the lunar breccia (10021) and olivine dolerite (12018) between 90° and 350° Kelvin. *Proc. Second Lunar Sci. Conf.,* Vol. 3, ed. A. A. Levinson (Cambridge; Massachusetts Inst. Technology Press), pp. 2361-2365.

EXPLORATION AND 1994 EXPLOITATION
OF GEOGRAPHOS[a]

SAMUEL HERRICK

University of California at Los Angeles

1. AVANT-PROPOS

A mid-day brilliance streaking the midnight skies of Quito, Bogotá, Medellín; a massive tremor of ground and air radiating from the jungle wasteland of northwestern Colombia; for the first time the conscious efforts of man will have tapped the potential energy of the planetary system and augmented the Earth's waning and decreasingly accessible resources of basic materials. The year: 1994; the day, August 25. The products: the excavation of a new canal from sea to sea; and the stowage of a treasure — a measured part of the minor planet Geographos — estimated to be worth some 900 billion dollars in nickel and the heavier elements that are mostly locked in the earth's core: rhenium, osmium, iridium, platinum, gold, etc. The problems: enormous . . . challenging . . . constructive.

2. INTRODUCTION

The close-approach minor planets Geographos, Icarus, Betulia, Toro, and others yet to be discovered or recovered, are not merely scientific curiosities,

---

[a]Editorial Note: This was written as a Preliminary Draft in March 1971 for "Physical Studies of Minor Planets," NASA SP-267. It was judged, by referees and Editor alike, "outrageously innovative" and "premature" and it was therefore not published. I thank Betulia Toro Herrick for safekeeping this manuscript and other memories of her husband, a scholar who was ahead of his time. — T. Gehrels.

nor does their interest stop even with the wealth of physical and historical information that we shall be extracting from them. They have a very special interest to all of us, scientist and layman alike, because inexorably, sooner or later, they will collide with the earth. The effect of such a collision upon the surface of the earth may be described briefly as follows: if on land, it might produce a crater 100 kilometers in diameter, violent seismic effects beyond that of the greatest recorded earthquake, perhaps vulcanism, a shock wave in the atmosphere capable of destructive effects for perhaps 1000 kilometers; and if by sea, it would add great sea waves (tsunami or "tidal wave") perhaps 1000 meters high.

When? If we employ the statistical (Monte Carlo) approach by the methods of Öpik, Arnold, Wetherill, we gain a feeling for the inexorability of collision, but over periods measured in millions of years. Evidently we must also seek evidence through alternative methods that are more dynamical than statistical. A first crude calculation indicates that for Geographos, because of the ominous present orientation of its orbit, the "sooner or later" may well be in the Third Millenium. Evidently we must gird ourselves to protect the whole Earth rather than just our lives, our environment, and our ecology, by devoting a part of our space program to detecting, reducing, controlling, and utilizing the predetermined fate of these little planets.

The present document constitutes a report of work performed and work planned with the aid of a 1969 grant from the National Geographic Society climaxing a continuous effort of the previous 18 years. The report naturally divides itself into: (a) a brief report on the "navigational" work on Geographos, its relationship to current scientific purposes, and its extension to the undertaking proposed; (b) a scientific and engineering program for the undertaking; (c) a human and environmental program for the undertaking; and (d) the phasing of the programs.

3. THE ORBIT

Procedures and techniques have been carefully refined and programmed for the prediction of ephemeral positions, velocities, and observable data for Geographos, and for the "updating" observational correction of the orbit upon which these predictions are based. The orbit of Geographos, in preparation for its close approach of 1969, was brought to the same state of excellence as that of Icarus in 1968, of which Goldstein of Goldstone reported with pleased surprise that it enabled him immediately to locate and identify the returning radar signal, within a matter of seconds. The orbit was improved on the basis of 1969 observations, and the accurate prediction of the 1994 close approach resulted therefrom. We have this orbit scheduled for further definitive correction in order to verify and continuingly refine the effects of the 1969 close approach.

Procedures and programs are highly advanced for the physical correction of the orbit, explosive cleaving and launching of a reduced and measured portion of Geographos into a rendezvous orbit, and for its corrective thrust-guidance thereafter. The first of these operations is envisaged as taking place about 5 years before rendezvous. The second involves a series of vernier corrections. Both operations are to be readied by the most refined theory and sophisticated programming to take full advantage of the natural perturbing forces acting upon the object, and to establish the engineering requirements for targeting the object with an acceptably small "footprint."

Additional programs are being readied to answer design questions, including those concerned with the date of natural impact of both the Toro group of Earth-orbit-crossers[a] and the Betulia group of potential Earth-orbit-crossers.[b]

4. THE SCIENTIFIC AND ENGINEERING PROGRAM

The foregoing navigational portions of the total scientific and engineering program are concerned with the bringing of reduced portions of Geographos to predetermined points on the surface of the Earth where the impact effects will be constructive and tolerable. It is evident that this goal, before it can be effectuated, requires answers to a host of sub-questions that require the cooperative effort of scientists and engineers in a great many fields. But also it gives rise to other important questions that demand answers almost in reverse order, starting with:

How can we delimit and utilize the physical consequences of impact? At the present time, thanks to the researches of Shoemaker, Chao, French and others, it seems likely that we can calculate better than other effects the size of the crater that would be produced by a given fragment of Geographos. We are less certain of the elongation of the crater that would result from the very shallow angle of descent; rough calculations indicate that the axis would be SSW to NNE. Both the size of the crater desired and the value of the cargo urge that the selected portion of Geographos be not too small. The upper limit on its size will probably be set by the air temblor rather than the seismic one. Unfortunately the effects of the blast of air radiating from the great Siberian meteorite of 1908 are better known than is its mass. The challenge of the calculation of such effects, however, will without doubt produce very reliable figures from competent experts now ready in this field. At the present time important indications favor the location of the interocean Crater-Canal on the Atrato River site in northwestern Colombia — which was specified in 1540 in the first proposal for a Central American canal!

What are the compositions and value of the close-approach minor

---

[a]Including also Icarus and Geographos, each of which is a unique sub-group of itself, and the Apollo sub-group of Apollo, Adonis, Hermes, and others.

[b]Not including Eros, according to J. G. Williams, but possibly including 1968 AA as well as Betulia.

planets, and of Geographos in particular, in terms of the rare heavy elements? If they are like the iron meteorites, they will be of great value. If they are like the stony meteorites, the immediate value of the undertaking may rest rather in the giant engineering project, a forerunner of others modifying desolate areas to support life systems. If they are the nuclei of dead comets, what may they, or may they not, contain? Evidently we have need for a program that unites the labors of those studying the physical properties of these objects from the observatory with the information to be gleaned from flybys and landings, unmanned and manned. Thus the economic and scientific motivations for landings on the minor planets coincide and become clear and pressing.

Where are the other Earth-impacting minor planets — those that have not been discovered and those that need to be recovered? It is no longer possible to regard close-approach minor planets as mere curiosities, to be reported on if one has time, interest, and equipment to measure a likely streak on a photographic plate. Astrophysical observatories, whose minds and hearts are understandably elsewhere, must be alerted to the needs associated with the ever-present possibility of the discovery of a fast-moving minor planet, and must be programmed to respond. The recent discovery of such an object at Kitt Peak illustrates our need to establish such a program. Amazingly, if I have the story correctly, the discovery was *visual* and no photographic equipment was accessible; but at that very moment E. Roemer was making photographic observations at Catalina Station, only ~50 miles away. Obviously our program should include the notification of all possible observers, by telephone. Obviously also it should include notification of persons like myself who can produce orbit and ephemeris at a moment's notice, out of one measure of position and length of trail, even though approximate, or out of two positions, before the classic three exact observations are available. Asteroid impacts, unlike earthquakes at the present time, are predictable and controllable, *if we develop an adequate program for discovery, observation, and orbit determination.*

5. THE HUMAN AND ENVIRONMENTAL PROGRAM

The Space Program is suffering at the present time from the fact that the voter and the politician do not understand its announced goals. Perhaps we scientists are at fault in expecting always-enthusiastic support for the advancement of human knowledge, and the contribution of tax dollars for areas that are not understood as having even an eventual return. For one who hardly knows what he owes to the compass and the earth's magnetic field it is difficult to appreciate the "relevance" of the moon's magnetic field, and it is easy to regard a moon-landing as a stunt that does not have to be repeated.

Perhaps we have forgotten that the greatest achievements in science have come from long-term, silent work, occasionally with popular approval but more often not. And perhaps we have forgotten that many of them were incidental to and supported by more understandable goals. The proposed operation upon Geographos involves three such goals:

1. The construction of a new Central American Canal, as a token unlocking of the riches of the earth for the benefit of all mankind.
2. The replenishment of the earth's resources, whose exhaustibility, even in the face of the most strenuous efforts at strict conservation, will cause panic and dislocation when they become well understood.
3. The removal of a menace that, however remote it may be for a given asteroid, might come upon us at any time and find us without skills or preparedness.

The layman will be keenly interested in seeing how well we tackle the problem, especially in its human and ecological aspects. He will judge us by how we safeguard human life and plan to preserve unique species. He will even be interested in the accuracy of aim, and the means developed to prevent mixing of Pacific and Caribbean waters in the event it is found necessary to do so. It is evident that it is necessary to organize the project in such a way that the reasoning processes involved are simple and open to review. A broad participation of agencies, disciplines, and competitive approaches will not only insure success to the project, but also the simplicity and understandability that will prevent confusion, misunderstanding, and sensational misinterpretation.

6. PHASING

It is evident that a Geographos program of the character proposed involves an early stage of quiet working collaboration and interchange of advice and recommendations between individual researchers, aided by a steering committee, before the total group of interested advisors can determine that it should be pursued as an active project with a thoroughgoing phasing of an engineering character.

FUTURE EXPLORATION OF THE ASTEROIDS

DAVID MORRISON
University of Hawaii

and

JOHN NIEHOFF
Science Application Inc.

Astronomical studies of asteroids have greatly advanced our understanding during the past few years; today, in 1979, they are at their peak, but within a few years current techniques will have reached a point of diminishing returns. New opportunities will be provided from Earth orbit by the Infrared Astronomical Satellite (IRAS), the Space Telescope (ST) and Spacelab Infrared Telescope Facility (SIRTF), but the next truly major step will require direct study by space probes. In order to sample the diversity of the asteroid population, a multi-target capability is required for a first mission. We discuss various mission modes and propulsion systems. A multi-asteroid ballistic flyby would provide a useful reconnaissance, but for detailed study of individual asteroids as global entities, including chemical analysis of their surfaces, a rendezvous or orbiter mode is required. Ion-drive propulsion systems now under development will provide the capability to carry out a variety of exciting multi-target rendezvous missions during the 1980's. We discuss target selection, science objectives, and possible science payloads for missions of this class.

[227]

The asteroids are a unique component of the solar system. Large in numbers but small in mass, they are generally thought to be a collisionally evolved remnant population of the planetesimals that once filled the solar system, accreted to form the planets and satellites, and were the agents of the heavy bombardment that abundantly cratered the surfaces of the larger bodies four billion years ago. Like the comets, many asteroids apparently are made of chemically primitive material, preserving a record of conditions in the solar nebula at the time the solar system was formed. In addition, however, there are among the asteroids objects that have been thermally modified in ways analogous, if not identical, to those that have so greatly altered the larger bodies in the solar system. Finally, the accessability of the meteorites for detailed laboratory study has stimulated great scientific interest in the parent bodies of these fragments, at least some of which must be found among today's asteroids.

Before about 1970, the asteroids received relatively scant attention from planetary scientists. Although the discovery of new minor planets and determination of their orbits had been a significant and often visible branch of astronomy for a century and a half, few physical studies had been carried out and even the presumptive relationship between meteorites and asteroids were tenuous. The first Tucson asteroid conference (Gehrels 1971), however, came at a time when the physical investigation of asteroids was on the brink of a decade of rapid progress. The present book reaps much of the fruit of this growth: asteroid science is a mature field, with many connections to other branches of planetary science, and with a broadly-based group of active researchers. From our present peak, we need to assess the future as well as the past, and to ask where asteroid science is going, and should go, from here.

The purpose of this chapter is to suggest some future directions for the study of asteroids. We briefly examine the role of traditional astronomical techniques, but these are extensively treated elsewhere in this book and do not need lengthy discussion here. Our main focus is on the prospects that exist to extend our vision beyond the earth, utilizing the techniques of space missions to undertake direct exploration of the asteroids. We recognize that the challenge of supporting an ambitious program of asteroid exploration is a major one, and that no missions will actually be launched until many more years of planning and preparation have passed. But we believe that we will never develop a satisfactory understanding of the asteroids until we have actually visited several of them, and that no first-order reconnaissance of the solar system can be considered complete without direct exploration of these objects.

This review draws heavily upon several studies of the potential of asteroid exploration in general, and of space missions in particular. Among these are particularly the papers in Morrison and Wells (1978), by Chapman and Zellner, Fanale, Niehoff, and Shoemaker and Helin; by Wright *et al.* (1979); and by Friedlander *et al.* (1979).

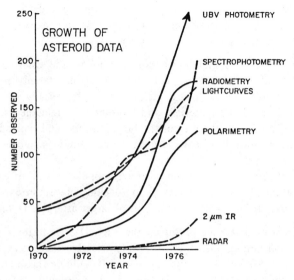

Fig. 1. Schematic representation of the growth of asteroid data from astronomical observations. From Chapman and Zellner (1978), their Fig. 1.

I. GROUND-BASED RESEARCH

During the past decade, a variety of astronomical techniques has been applied to permit a first-order characterization of the asteroid population in terms of physically significant parameters, i.e., parameters that are related to the chemistry and peculiar history of individual objects, in addition to their orbits. The most fruitful techniques have been UBV photometry (colors, hence classification into broad compositional types), visible-infrared spectrophotometry (more precise compositional classification related to mineralogy), radiometry (diameters and, in conjunction with UBV photometry, albedos), and polarimetry (surface texture; albedos and, in conjunction with UBV photometry, diameters). In addition, a strong program of photometric studies continues to yield lightcurves (hence rotation periods and an index of shape) and phase coefficients (related to surface optical properties).

Figure 1, taken from Chapman and Zellner (1978), schematically illustrates the explosive growth in numbers of objects studied by astronomical techniques. The TRIAD file (Bender *et al.* 1978; Part VII of this book), summarizes the data available in 1979. The total number of asteroids for which some physical classification is possible is 752 (see the chapter by Zellner) – more than a third of the numbered asteroids. The observed sample extends to all parts of the asteroid belt and includes significant numbers from several of the major families. Fewer data exist for the Apollo, Amor and Aten objects, but even here the information is not negligible. Although the compositional inferences that can be drawn from these data are limited (cf. Bowell *et al.* 1978; Gaffey and McCord 1978), the existing data base is sufficient to show that:

1. Many different mineralogies are present in the asteroid belt, with members of different classes reasonably dispersed spatially.
2. The inferred mineralogies include both chemically primitive and thermally modified materials, and in general the mineral assemblages are similiar to those of the meteorites, both primitive and evolved.
3. Genetic relationships exist among some family members, suggesting origin in a fragmentation event.
4. There are significant differences among populations in different parts of the belt, with the known Earth-approaching objects in particular being nonrepresentative of the population of larger asteroids in the main belt.

The surveys that have produced the TRIAD data during the past eight years are continuing, but their future is clearly limited; at the present observing rate, 90% of the numbered asteroids could be classified in the *CSM* system within a very few years. The primary new development in this survey work has been the extension of the UBV photometry to near-infrared wavelengths through the 8-color system introduced by Zellner, Tedesco and others. Their observations now allow the routine investigation for faint objects of the compositionally diagnostic reflectivities near 1 μm. With the application of this system to a thousand or more asteroids during the next three years, we may expect the survey mode to be at the point of diminishing returns. We will then have physical classification of essentially all numbered asteroids larger than 25 kilometers in diameter and will have identified those few that deserve more extensive study on the basis of their anomalous colors.

Other ground based techniques are available for the detailed study of special objects or of type-members of the broad taxonomic groups that result from surveys. These include extensions of reflectance measurements to wavelengths longer than 1.1 μm, determination of high-resolution infrared spectra by Fourier spectroscopy, and measurements of thermal emission at a variety of wavelengths (including millimeter- and centimeter-wave radio emission) and phase angles. A particularly noteworthy groundbased technique that has not yet been widely applied is radar. Differences in radar reflectivity may prove highly diagnostic of the amount of free metal in the surface of an asteroid, a property not easily determined from spectral reflectance measurements. Radar ranging also has a long-term potential to contribute to mass determinations, as does increased attention to asteroid astrometry (see the chapter by Schubart and Matson).

While current observing programs are rapidly surveying the main-belt asteroids, the reconnaissance of those with large and small semimajor axes has been less complete. There may be a major population of objects beyond Jupiter; perhaps these are more closely related to the comets than to the main-belt asteroids, but in any case our current knowledge is quite meager. At the inner edge of the distribution, much progress in discovery of Apollo and Amor objects has been made, but the recent identification of an entirely new orbital group, the Aten asteroids, demonstrates the incompleteness of our knowledge. Similarly, a great deal needs to be done in physical observations of Earth-approaching objects; basic questions of the genetic relationships among these asteroids, main-belt asteroids, comets and meteorites are amenable to investigation by continuing groundbased observation. Acquiring the necessary data requires access to large telescopes on very short notice, but progress is being made as planetary astronomers generally seem to be more successful in competing for time on large instruments.

Today may mark the golden age of asteroid astronomy. For the first time we have a true perspective on the asteroid population, and the TRIAD data constitute a resource that is sure to generate new ideas and interpretations. Additional observations are coming in at a rapid rate. But the success of this effort bears the seeds of its own demise. Within less than a decade, current observational approaches will have run their course. The next major step in understanding the asteroids may well depend on new techniques, including both observations from above the earth's atmosphere and direct visits to a few asteroids by spacecraft.

II. OBSERVATIONS FROM EARTH ORBIT

Both improved spatial resolution and extended spectral capability are available to telescopes in space. Three Earth-orbiting facilities planned by NASA for the decade of the 1980's may contribute significantly to asteroid studies. These are, in chronological order, IRAS, Space Telescope, and SIRTF.

The Infrared Astronomical Satellite (IRAS) is an Explorer-class satellite scheduled for launch about 1982. Equipped with a cryogenically cooled telescope and detectors for broadband photometry from 10 μm to 100 μm, this survey instrument is expected to reach an N (10 μm) magnitude of $\sim$ 7 (Aumann and Walker 1978). Typically, asteroids have V$-$N color indexes of 10 magnitudes; thus IRAS should be capable of detecting a thermal signal from virtually every asteroid with V $<$17, corresponding to a diameter of only $\sim$10 km for a C asteroid at a distance of 2 AU. Indeed, IRAS not only *can* see such objects, it *will* see them, since it operates in a sky-survey mode. The original software design for the survey would reject all asteroid signals, along with other sources of noise, but it appears feasible to modify the ground-processing system to preserve the asteroid data. If this can be done, we can expect to have multi-spectral thermal emission data, and hence diameters and albedos, for nearly all of the numbered asteroids within the next five years.

The Space Telescope (ST) is a 2.4 meter astronomical telescope scheduled for Shuttle launch in late 1983, with a projected long lifetime; the initial instrumentation will consist of two spectrographs, a high-speed photometer, and two imaging systems. There will be no infrared instrument, although this capability is likely to be added later in the decade. Ultraviolet spectra of asteroids could be obtained with ST, if such data are deemed necessary to determine surface composition (cf. Wagner *et al.* 1979), but the prime use of ST for asteroid research will probably be in direct imaging with the Planetary Camera, as analyzed by Smith and Reitsema (personal communication 1979).

Operating at f/30 with four 800 x 800 charge couple device (CCD) detectors, the Planetary Camera will have a resolution of 0.1 arcsec per line pair, or 80 km at a range of 1 AU. The largest asteroids will fill up to about a hundred pixels, permitting some study of albedo markings on Ceres, Pallas, Vesta and a few others. However, this resolution is insufficient to distinguish geologic structure, or even to contribute any increase in the precision of diameter measurements over that expected from other techniques in the mid 1980's. In one respect, however, ST imaging can have a major impact: the identification of close companions. The suggested satellite of Herculina (see the chapter by Van Flandern *et al.*), for instance, would be separated from the asteroid by more than 10 pixels, and the 3.6 mag difference in brightness is well within the range of the CCD detector. If asteroid satellites or companions are real, ST will be able to see them, opening a fertile new field of asteroid research.

Later in the 1980's NASA plans to launch a Spacelab Infrared Telescope Facility (SIRTF), consisting of a 1.5 meter cryogenically cooled telescope. Such an instrument could extend the IRAS observations, and it might provide an extremely sensitive way to search for very faint asteroids, for instance in the Jovian Trojan clouds.

It should be emphasized that the competition for telescope time on ST and SIRTF will be extremely keen; in practice, it is unrealistic to expect that many asteroid observing programs will be approved. These facilities will be useful tools, but they seem unlikely to generate by themselves a major revolution in asteroid science.

III. RATIONALE FOR SPACE MISSIONS

The history of the past two decades has repeatedly demonstrated the revolutionary advances in understanding that have accompanied the exploration of the planets by spacecraft. Each time a mission has visited a planet, perspectives have changed radically; indeed, the major fields of planetary geology and geochemistry could hardly exist today if there were no space program. There is every reason to expect a similar revolution from the first asteroid missions.

The asteroids differ from other space targets in their number and diversity. The step from a mission that studies *an asteroid* to an understanding of *the asteroids* is a major one. Careful attention must be given to the selection of mission targets and to integration of space data with the results of groundbased and Earth-orbital observations.

When serious discussion of asteroid missions first began about a decade ago (e.g., Stuhlinger *et al.* 1972), very little was known about the physical nature of asteroids, and there was little basis for selecting one object over another as a target. In fact, Anders (1971) effectively put the case against an early mission as follows:

"Ground-based research on asteroids and meteorites is nowhere near exhaustion; on the contrary, it is moving at an impressive pace. If we maintain this pace for another decade or two, we will not only have answered most of the questions posed for an *early* mission, but will be able to come up with a more worthwhile, more informative mission . . . Some crucial questions will undoubtedly remain when all ground-based studies have been pushed to their limit, and at that stage, perhaps ten years from now, further progress will require space missions. We do not know what sort of a target will have highest scientific interest at that time . . . Any choice we make now is likely to seem trivial or uninformative a decade hence."

Nearly a decade has now passed since Anders' analysis, and the progress in groundbased studies that he predicted has come to pass. It will probably be another decade before the first spacecraft arrives at an asteroid, but now is the right time to begin serious planning for a mission.

Without actually visiting an asteroid, we will never understand these objects as global entities – as true planets, deserving of study in their own right. Even the increased spatial resolution of the Space Telescope will not be sufficient to tell us what an asteroid really looks like in a geologically useful way (compare with the Pioneer 10 picture of Ganymede; Gehrels 1976). Most questions of cratering and fragmentation history, regolith depth and texture, local heterogeneity in surface materials, or of any other geological characterization, will never be successfully answered from the earth.

A second fundamental area of investigation that requires space missions is the characterization of the bulk properties of an asteroid. Both masses and volumes are extremely difficult to measure from Earth, and there is no prospect of every obtaining these data for any but the very largest asteroids. An understanding of asteroids as planets, and of their relationships to the processes acting at the time of formation of the solar system, requires a measurement of the bulk density of a sample of objects to a precision of a few percent. Also within the capability of an asteroid orbiter would be determination of gravity harmonics and search for intrinsic magnetization.

The third field of study that requires an asteroid mission is geochemistry. Remote sensing from Earth has yielded important inferences on surface mineralogy (hemispherically averaged), but no way exists to measure elemental abundances without a close approach to the object under study. Today, the presumptive relationships between meteorites and asteroids tempt one to conclude that the study of a particular meteorite type may yield information on an individual, identifiable asteroidal parent, but with the possible exception of Vesta (e.g., Consolmagno and Drake 1977, and Drake's chapter in this book) such a firm identification seems remote. Direct measurements of asteroid chemistry are required, both to advance our understanding of the chemical processes in the solar nebula and to increase the utility of meteorite research by providing tighter links between meteorite and asteroid compositions (Wilkening 1979).

More generally, the acquisition of direct, reliable information on asteroid surface chemistry, geology, and bulk properties would broadly impact our understanding both of the processes of formation of the solar system and of the subsequent thermal and chemical evolution of the planets. Fanale (1978) has well summarized the broad questions that can be answered, at least in part, through investigations carried out on asteroid missions as follows:

1. "What were physical and chemical conditions in the solar system during planetary accretion like?

   a. What were the *physical* interactions among solid bodies of all sizes like during accretion of our planetary system? This includes processes of accretion, fragmentation, and dynamic rearrangement. What do these physical condition imply for the formation and initial state of very large objects like the Earth and their subsequent bombardment history?

   b. What *chemical* fractionation processes operated during condensation/accretion to produce differences in *bulk* composition *among* asteroids and could these same processes account for apparent differences in bulk compositions among the terrestrial planets? Did these condensation/accretion processes produce "ready-made" or *initial* zonal layering *within* asteroidal or planetary bodies in the solar system?

2. What magmatic processes operated within accreted bodies to produce internal differentiation? When did these processes operate and what were the energy sources (short-lived nuclides, solar electromagnetic interaction, etc.)? Why did they seemingly affect some asteroids and not others? Did they affect the Earth and the other planets as well?

3. "What are the genetic relationships among small bodies in the solar system? Are there parental relationships among (a) various orbital families of asteroids, (b) various spectral classes of asteroids, (c) comets, (d) meteorites, (e) planetary satellites, and (f) interplanetary or interstellar dust? In what context does this place the vast library of isotopic, geochemical, textural, and other information we have already accumulated on meteorites and what, in turn, does this tell us about planetesimal/planetary genesis?

4. "What is the potential of the asteroids as sources of raw materials? What variety of raw materials are available? Is mining from asteroids

of any of these materials for any application preferable to mining, processing, and launching from Earth, or mining from non-asteroidal extraterrestrial sources such as the Moon?"

In Sec. V of this chapter, we summarize possible instruments that could be used to acquire the data needed to attack these problems.

In all mission planning, an overriding concern is the diversity of the asteroid population: large and small, primitive and evolved, close to or far from the sun, etc. It has seemed clear to recent groups studying mission possibilities that no one asteroid, no matter how carefully selected or intensively studied, will yield information sufficient to characterize the population generally. Therefore, *a first mission must include visits to several asteroids if it is to sample adequately the diversity inherent in these objects.* In the next section we will examine a variety of possible mission modes, but each of these, to be acceptable for a first mission, must include a multi-asteroid capability.

IV. MISSION MODES AND PROPULSION REQUIREMENTS

The basic asteroid mission modes are quite similar to their planetary counterparts: flybys, orbiters, landers, and sample return. As a matter of terminology, the name usually given to the comet or asteroid mission mode in which the spacecraft remains a long time in the close vicinity of the target is rendezvous, but in practice for all but the smallest asteroids (diameter < 10 km) rendezvous spacecraft are, in fact, also orbiters.

Flyby

The flyby of an asteroid on a ballistic trajectory typically involves velocities of about 5 km sec^{-1}, similar to the relative velocities of the main-belt asteroids themselves. Consider the capability of such a flyby, taking imaging as a first example. The Voyager exploration of the Galilean satellites suggests that resolution of about 10 km is required for a minimum characterization of surface geology on a first mission; for a camera with 1500 mm focal length, this resolution is reached at a range of 500,000 km; the spacecraft is within this range for 2×10^5 sec, or about two days. For typical rotation periods, complete longitudinal coverage is thus provided. The resolution reaches 1 km, of course, for only about 4 hours, and it is unlikely to achieve much better due to smearing from the large relative velocities. The level of imaging is thus quite similar to that obtained by Voyager for the Galilean satellites, and it could provide extremely rewarding reconnaissance for the largest asteroids. The smaller the target is, however, the more limited the rewards. Figure 2, the best Voyager image of Amalthea, illustrates the results to be expected for a typical, moderately large (265 km long) asteroid seen at a range of 500,000 km. Global coverage of this quality would be available from a flyby; at closest approach, about 40% of the surface could be

Fig. 2. Photograph of Amalthea from Voyager 1. Taken 4 March 1979 at a range of 425,000 km. The resolution in this image is 8 km per line pair. Amalthea, with dimensions of 265 X 140 km and a low albedo, is representative of a moderately large asteroid. (JPL P-21223 B/W.)

imaged at ten times this resolution.

For measuring bulk properties, also, the flyby mode works best for large asteroids. For an object several hundred kilometers in diameter, a close flyby could yield a good mass value, but for small objects the interaction is too weak. Finally, however, the capability to perform chemical analysis from a flyby is extremely limited. Gamma ray spectroscopy, for instance, can be carried out effectively only within about one diameter of the target. In our example, the time available would be three minutes for Vesta, and ten seconds for Eros – clearly inadequate for any meaningful analysis. This inability to carry out chemical investigation is the most severe drawback of flyby missions.

A main-belt multiple flyby mission is within the capability of a number of current launch vehicles of both the United States and other nations (Brooks and Hampshire 1971; Niehoff 1978). Under active study in Europe is a mission of this type that could be launched with the Ariane rocket (Brahic et al. 1979a,b). Trajectory analyses show that, once the spacecraft is in an asteroid-type orbit, a flyby of one of the numbered asteroids can be achieved about every five months, or four per orbit. Compared with the multi-rendezvous missions to be discussed below, the flyby mode carries out its mission very quickly. Indeed, the rapid progress from one asteroid to the next is a principal advantage of the flyby approach; with the use of two spacecraft, one to include Ceres and the other Vesta, in their tours, flybys of these two large asteroids plus about eight smaller ones could be completed 3 to 4 yr after launch.

The capability of a flyby is increased if the encounter velocity is reduced. If, instead of 5 km sec^{-1}, the relative velocity is 0.5 km sec^{-1}, the spacecraft can be within 50,000 km (corresponding to 1 km imaging resolution) for a full rotation period; if the velocity is further reduced to 0.05 km sec^{-1}, some gamma ray or X ray chemical experiments begin to be feasible. However, the propulsion capability needed to slow the spacecraft to such speeds and then accelerate it again for transfer to the text target is nearly the same as that required for rendezvous and orbit.

Rendezvous (Orbiter)

Several recent panels and studies examining multi-asteroid mission modes, particularly a NASA-sponsored workshop at the University of Chicago in January 1978 (see Morrison and Wells 1978) and a Summer Study conducted by the U.S. National Academy of Sciences in Aspen, Colorado in August 1978, have recommended rendezvous as the mission mode of choice, and studies conducted for NASA during the past year by Jet Propulsion Laboratory (Wright *et al.* 1979) and Science Applications, Inc. (Friedlander *et al.* 1979) have been directed toward this approach.

In a rendezvous, the spacecraft matches velocity with the target and then either maintains position with it ("station keeping") or goes into orbit. Once rendezvous is achieved, the duration of studies can be extended indefinitely, and the capability of remote sensing experiments is not limited by either range or time. For instance, from a 1000 km orbit around Vesta the entire surface could be mapped at an imaging resolution of a few meters, comparable to the best Viking pictures of Phobos and Deimos (see the chapter by Veverka and Thomas). With long integration times, also, the full potential of chemical remote sensing instruments could be realized. The penalty imposed by rendezvous is one of energy; while existing chemical propulsion systems allow (marginally) an orbit of a single chosen asteroid, multi-asteroid rendezvous missions are beyond the capability of any present or planned chemical propulsion systems.

Multi-target rendezvous missions are made possible by the development of low-thrust ion-drive propulsion systems, now under active study for a comet rendezvous mission. A minimum ion-drive system of 25 kW power with flat solar-cell arrays, employed with a 600 kg mission module (including 100 kg of science instruments), will yield 3 rendezvous with numbered asteroids for virtually any launch date, and up to 5 with properly selected opportunities (Wright *et al.* 1979). The addition of 2:1 concentrators to improve solar cell efficiency beyond 2 AU will yield numerous missions to 6 asteroids, and possible missions to as many as 7 or 8 (Wright *et al.* 1979). Further improvements are possible with higher power solar arrays, but a situation of diminishing returns is quickly reached due to the long times required. Indeed one of the primary problems with multi-target rendezvous missions is their duration, with typical intervals between encounters of about 20 months. A six-asteroid mission could easily extend to more than a decade.

In Sec. VI of this chapter we will discuss specific multi-target missions and describe the science returns to be anticipated from rendezvous. First, however, we consider two additional mission modes: landers and sample return.

Lander (Docking)

Because of the small gravitational fields of asteroids, we will consider as a single exploration mode any scenario that involves physical contact with the surface, whether by high-speed penetrator, hard lander, or soft lander. All of these systems introduce the potential for *in situ* chemical analysis, direct measurements of the physical properties of the surface, extremely high resolution imaging (including microscopy), and possible investigation of bulk properties by techniques such as seismic sounding. Soft landers (Viking and Surveyor) and hard landers (Ranger) have been used to date in the U.S. planetary program. Hard landers have been included especially in the USSR Luna and Venera missions. Penetrators have been extensively developed by the U.S. military for terrestrial application.

Penetrator and hard lander systems appear to be small enough to be included in an initial multi-asteroid mission, either flyby or rendezvous. (Difficulties with communications probably would make their use impractical for a flyby, however; the parent spacecraft moves out of range too quickly to serve as a relay, and the lander is unlikely to be able to return data directly to Earth.) Sophisticated soft landers, on the other hand, represent a major increase in mission mass and complexity, approaching that required for a sample return.

In the recent studies of multi-asteroid mission capabilities by Friedlander *et al.* (1979) and Wright *et al.* (1979), up to four penetrators of about 100 kg each were considered as part of the standard payload. Planned ion-drive systems can accommodate the additional mass; approximately one asteroid in a tour is given up for each two penetrators carried. We discuss possible landed science instruments in the next section.

Sample Return

The return to Earth of a sample of one or more asteroids is a generally agreed upon long-term goal for a program of exploration of the asteroids. The capabilities of terrestrial laboratories for chemical and mineralogical characterization, including particularly isotopic analysis and age dating, far exceed that of any likely robot laboratory. However, the cost of sample return is much higher than that of flyby or rendezvous missions, and with currently planned propulsion systems, only a single asteroid could be sampled in a given mission. Thus, sample return for a potential first asteroid mission suffers from many of the problems cited by Anders (1971) quoted above: we do not know how to select wisely a single, optimum target; and we run the risk that our selection will turn out to be trivial, particularly if the sample returned is identical to a meteorite already in our collection. The Chicago Asteroid Workshop (Morrison and Wells 1978) concluded: "We recommend that the exact role and timing of sample return be judged after the results of prerequisite rendezvous missions are available."

The probable capabilities of sample return have been estimated in several recent studies carried out at Science Applications, Inc. (Friedlander *et al.* 1979), based in part on earlier work on Mars sample return. The sample return mission module presupposes a 600 kg orbiter spacecraft of the type discussed for a rendezvous mission. The Lander/Ascent/Rendezvous module has a mass of 300 to 500 kg, including its chemical propulsion system. A 1.0 kg sample is collected in a 30 kg canister; the canister is returned from the surface to the orbiter, where the ion drive system can provide the propulsion power for return to Earth orbit, and is retrieved by the Shuttle. A 33 kW ion drive with 2:1 concentration could achieve such a sample return from a typical low-inclination main-belt asteroid with a round trip time of 3−4 yr.

Sample return from Apollo/Amor/Aten objects is easier in terms of trajectory energy requirements than from main-belt asteroids. In the case of 1943 Anteros, which is the most accessible object discovered to date, a 1.0 kg sample could be returned with a chemical propulsion system; for a 1992 launch, the launch energy C_3 is only 35 km^2 sec^{-2}, the post-launch velocity change is only 2.71 km sec^{-1}, and the round trip time is 3.0 yr (Friedlander *et al.* 1979). It is reasonable to expect that the discovery of additional Earth-approaching asteroids will reveal even easier targets for sample return; however, it is worth noting that favorable launch opportunities for the most accessible objects, which have orbits similar to that of the earth, are spaced many years apart.

V. SCIENCE PAYLOADS FOR RENDEZVOUS MISSIONS

In this section we discuss a basic, 100 kg payload of instruments for the rendezvous spacecraft, and an additional instrument package for a small penetrator or hard lander. The instruments themselves are similar to those recommended at the University of Chicago Workshop (Morrison and Wells 1978), and further amplified by Friedlander *et al.* (1979).

Table I lists the proposed instruments, together with their masses and heritage from previous missions. The capabilities of each are summarized below.

Imaging. Imaging provides the eyes of the mission; its role is essential for acquisition of the target and navigation to orbit as well as for science. Imaging

TABLE I

Typical Orbiter Science Payload

Instrument	Mass (kg)	Heritage
Imaging camera	30	Galileo, Voyager, Comet
Reflectance spectrometer	15	LPO,[a] Galileo, Comet
Gamma ray spectrometer	12	LPO
X ray spectrometer	10	LPO
Thermal radiometer	6	Mariner 10, Viking, LPO
Radar altimeter	7	LPO
Magnetometer	5	Voyager, Galileo
Plasma particle detector	6	LPO
Sub total:	91	
Contingency (10%)	9	
TOTAL	100 kg	

[a]LPO=Lunar Polar Orbiter.

is the central instrument for geological studies; it also provides bulk measurements of dimensions, shape, rotation axis, and volume (hence density). The candidate system contains two cameras: for Voyager, the focal lengths are 1500 mm and 200 mm, yielding fields of view of 7 and 50 mrad, respectively. The sensor is an 800 $\times$ 800 CCD, similar to those being developed for the Space Telescope and the "Galileo" mission to the Galilean satellites; spectral range is from 0.3 to 1.1 μm.

The wide angle camera can achieve a resolution of 50 m at an orbital altitude of 400 km; complete 6-color coverage of a large ($D > 300$ km) asteroid requires about 5000 frames, or 2.5×10^{10} bits, distributed over perhaps 100 orbits. At the same time, the 1500 mm camera might be used to map 10–20% of the surface at a resolution of about 10 m. Additional frames would be needed to photograph the same areas at different illumination angles, and to obtain stereo coverage of limited regions. In general, a full imaging program for a large asteroid might require 30 to 60 days in orbit, and the acquistion and transmission of ten thousand frames, about half of the Voyager picture budget for a single Jupiter encounter.

Reflectance Spectrometer. This instrument measures the reflectance as a function of wavelength from the visible through the infrared (to perhaps 5.0 μm). A number of minerals, such as pyroxene, plagioclase, and olivine have strong absorption bands in this spectral region, as do several ices and water of hydration. Reflectance spectroscopy is, of course, one of the main tools of groundbased asteroid research; on a rendezvous spacecraft, such an instrument would operate with high spatial resolution and could determine the mineralogy of small units. Of particular interest are "windows" to interior layers that might be revealed by large impacts, or possibly stratigraphy exposed by fragmentation early in the asteroid's history.

A spectrometer was proposed for the Lunar Polar Orbiter covering from 0.35 to 2.5 μm in 256 channels; a mapping spectrometer being built for Galileo covers 0.7 to 5.0 μm, in 200 channels. Either instrument, or designs now under consideration for a comet rendezvous, could serve as the basis for an asteroid instrument. The field of view would probably be about 10 mrad, with typical integration times of a second or so; the corresponding spatial resolution is on the order of 1 km. About 10^5 individual spectra would be needed for complete coverage of a moderate-sized asteroid, providing data on mineralogy necessary to understand the chemical and fragmentation history of the body.

Gamma-Ray Spectrometer. This instrument measures the elemental abundance of a number of species, including the important radioactive elements, over the uppermost half-meter of the surface. For a one-hour integration, the sensitivity to concentration of various elements is (Haines *et al.* 1976; Arnold 1978); U (0.1 ppm); Th (0.5 ppm); K (0.03%); Cl (0.3%); H (0.8%); Ti (0.9%); Ni (1%); Fe (2%); Mn (2%); Si (3%); Mg (3%); C (5%); Al (6%); O (7%), S (7%).

Gamma ray spectrometers were flown on Apollo and were proposed for the Lunar Polar Orbiter. The large germanium detector is uncollimated, yielding an effective field of view about equal to the spacecraft altitude; thus a low orbit is required to realize any substantial spatial resolution. Global abundances could be obtained in a few hours from a station-keeping position; from a later orbit at an altitude of 25% of the asteroid radius, the surface could be mapped into about 60 resolved areas in a few days' observing. With longer times, of course, higher precision can be achieved. Determination of the abundance of these elements will permit a reliable identification to be made with particular meteorite classes, if the meteorite analog exists in our collections; if not, the chemical characterization obtained will be crucial for determining the relationship of the asteroid to chemical formation processes in the solar nebula, or to the subsequent differentiation history of the object if it was thermally cooled.

X Ray Spectrometer. An X ray fluorescence spectrometer determines elemental abundances in the top 0.1 mm of the asteroid surface material by detecting X rays stimulated by the solar X ray flux. Since X rays can be focused, it is possible to obtain angular resolutions of as high as 100 mrad. The primary elements to be measured are Mg, Al, and Si; also probably measurable are C, N, O, Na, Ca, S, and Fe. X ray spectrometers were flown on Apollos 15 and 16 and were proposed for the Lunar Polar Orbiter. Proportional counters are used as detectors. The expected sensitivity should allow measurements of Mg, Al, and Si to ±1% of total concentration, with integrations of several minutes. Complete surface coverage could be achieved in a few days with a close orbit. These data provide an important extension of the chemical analysis obtained by the gamma ray spectrometer.

Thermal Radiometer. Surface temperature measurements can be made with either an infrared or a microwave radiometer; obtained at high spatial resolution, these data will yield the thermal inertia of differing geologic units in the surface, which are basic indicators of regolith depth and, indirectly, of cratering history. In combination, infrared and microwave systems might also be used to measure internal heat flow, as was proposed for the Lunar Polar Orbiter.

Radar Altimeter. The distance between the spacecraft and the surface is needed to derive the figure of the asteroid and to measure its volume, as required for a good determination of density. The physical nature of the surface also influences the reflectance for microwaves. Either a microwave radar or a laser system could be used.

Magnetometer. An asteroid could have associated with it a magnetic field from any of three sources: an internal dynamo, an induced field, or a remanent field. Thus a magnetometer offers the potential for discovery of a past or present field as a bulk property of the asteroid, as well as for study of the interaction of the body with the solar wind. The required measurements need to be made from a low orbit; the hardware is, however, conventional and has flown on many previous missions.

Plasma Particle Detector. This instrument complements the magnetometer. It can be used to measure the bow shock where the solar wind flow is retarded upstream from the asteroid, and it can carry out searches from low orbit for electrons reflected from surface magnetic anomalies. Similar measurements have been carried out for the moon, and an instrument of the type proposed for the Lunar Polar Orbiter would be suitable for an asteroid mission.

Instrumentation for a Hard Lander. Three types of measurements are compatible with a simple penetrator or hard lander, such as might be developed as part of an initial, multi-target rendezvous mission, with a total science payload of less than 10 kg.

The first and probably most important lander instrument is an alpha-particle and X ray spectrometer (Economou and Turkevich 1976; Turkevich and Economou 1978). An alpha ray system was used on the lunar surface by Surveyor, and an X ray system on Mars by Viking. Used together, these instruments can provide a much better chemical analysis, in the spot where they land, than can be achieved from orbital data alone. The abundances can be measured to 1 wt% for the principal elements C, O, Na, Mg, Al, Si, K, Ca, Ti and Fe. Substantially higher precisions can be achieved for trace elements, such as Rb, Sr, Y, Zr, Ba, La, Ce, Nd, and Sm. With such data, it should be possible to make an unambiguous identification between an asteroid and a meteorite, whereas orbital data alone probably will only establish relationships between an asteroid and a meteorite class. Additional hard lander instruments might include an accelerometer/seismometer, both to sense the surface bearing strength from the force of impact and to search for possible internal seismic activity. Another small instrument that might yield exciting data would be a facsimile camera to image the environment of the lander with spatial resolutions as great as 1 mm. While providing only a "worms's eye" perspective, such an image might be extremely informative, both for its own sake and as a means of evaluating the sample being analyzed chemically by the alpha-particle/X ray instrument. Indeed, pictures on this scale could provide a direct link between the asteroid surface and the texture and crystal sizes frequently used to characterize terrestrial meteorites.

VI. TARGET SELECTION AND MISSION EXAMPLES

Criteria for Target Selection

The primary requirement for target selection in a multi-asteroid mission is to investigate a sufficient variety of objects to carry out the broad comparative goals of asteroid exploration. Thus, any group of targets should include both large and small objects and should include examples of the major taxonomic types. In addition, it would be highly desirable to visit several dynamically related objects that appear to be products of the breakup of a single parent body. Because they are fragments of originally larger bodies, such family members provide a unique opportunity to study the interiors of planets directly. It is probably not an appropriate criterion to select asteroids that are thought to be specific meteorite parent bodies; indeed, it could be argued that we will learn most by visiting at least some objects that are not represented in our meteorite collections.

Some examples of asteroids or classes of asteroids recommended as suitable targets at the Chicago Workshop (Morrison and Wells 1978) are:

1. Ceres (largest; presumably unfractured; relatively primitive, but probably experienced some thermal evolution; may have bound H_2O on the surface; may resemble original planetesimals).
2. Vesta (third largest; presumably unfractured; differentiated and thermally evolved; may be typical of original parent bodies of differentiated meteorites; may hold important clues to lunar evolution).
3. Two or more members of a Hirayama family (to examine fragments of fractured parent body for data on internal structure and accretion history).
4. A small very dark C type (to examine primitive material, investigate accretion history).
5. Typical members of compositional classes, e.g., an S, a metallic surface,

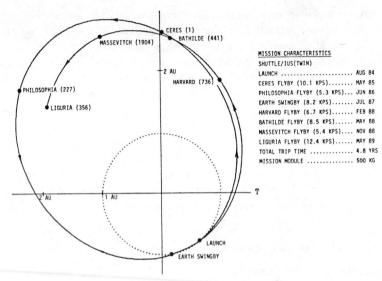

Fig. 3. Trajectory for a multi-asteroid flyby mission that includes Ceres, Bathilde, and four additional objects. From Niehoff (1978).

an enstatite chondrite surface (to trace varied differentiation and thermal evolution, study geologic processes on a variety of compositions).

6. An Apollo or Amor (key link between main belt and meteorites; possible dead comet nucleus).

7. A Trojan (unique types; possibly different initial conditions and evolution from main belt).

These are only examples to illustrate the diversity of possible targets; undoubtedly, continuing observations and interpretations will modify and refine this list before any actual mission commitment is made. In addition, of course, availability of targets from the celestial mechanics viewpoint further constrains the choice; Pallas, for instance, is a very difficult target because of its high inclination. Within the main belt, however, there are a great many possible tours, and it should prove relatively simple to find trajectories that yield a number of satisfactory encounters for essentially any launch year.

A number of searches for candidate asteroid tours have been carried out. We will now discuss several specific cases as examples of what can be accomplished with the space propulsion systems of the 1980's.

Main-Belt Flyby Tours

We begin with multi-asteroid flyby missions. Niehoff (1978) gives one example, a 5 yr mission launched in August 1984 that includes flybys of Ceres and six additional objects, including one of class M (441 Bathilde). Figure 3 illustrates the trajectory and lists the time and relative velocity for

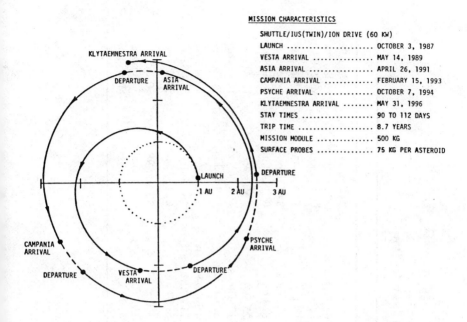

MISSION CHARACTERISTICS

SHUTTLE/IUS(TWIN)/ION DRIVE (60 KW)
LAUNCH .... OCTOBER 3, 1987
VESTA ARRIVAL .... MAY 14, 1989
ASIA ARRIVAL .... APRIL 26, 1991
CAMPANIA ARRIVAL .... FEBRUARY 15, 1993
PSYCHE ARRIVAL .... OCTOBER 7, 1994
KLYTAEMNESTRA ARRIVAL .... MAY 31, 1996
STAY TIMES .... 90 TO 112 DAYS
TRIP TIME .... 8.7 YEARS
MISSION MODULE .... 500 KG
SURFACE PROBES .... 75 KG PER ASTEROID

Fig. 4. Trajectory for a multi-asteroid rendezvous mission requiring a 60 kW ion drive. This tour includes Vesta, Psyche and three additional objects. From Bender (1977), as quoted in Niehoff (1978).

each flyby. There are two fly-throughs of the main belt, separated by a re-encounter of the earth three years after launch. The total post-launch propulsion impulse required is only 1.7 km sec^{-1}

Main-Belt Rendezvous Tours

Figure 4 (Niehoff 1978) illustrates a multi-target rendezvous mission generated by Bender (1977). This particular tour includes encounters with Vesta, Psyche and three other asteroids, with a duration of 8.7 yr. However, it demands a 60 kW ion-drive system, larger than is likely to be available. More recent studies have searched for appropriate tours that can be accomplished with smaller propulsion systems.

Table II summarizes several five-asteroid tours that are within the capability of the 33 kW ion drive with 2:1 concentrators and 600 kg mission module (Friedlander *et al.* 1979). A 30 day stay time is allowed at each asteroid. It is clear that even within the limited time and parameters of this search, a wide variety of tours is possible. One of the most extensive is illustrated in Fig. 5. With a launch on 1 January 1988, rendezvous is possible with eight asteroids over a 12 yr period.

The specific tours listed here are purely examples. At any point in a given tour, there are several accessible targets, any one of which can be chosen.

TABLE II

Examples of Rendezvous Tours with a 33 kW Ion Drive

Launch	1	2	3	4	5
30 Mar 87	19 Nov 88 8 Flora S, 151 km	30 Sep 90 149 Medusa U, 16 km	21 Jun 92 64 Angelina E, 56 km	21 Jun 94 90 Antiope C, 124 km	12 Mar 96 16 Psyche M, 250 km
1 Jan 88	22 Aug 89 21 Lutetia M, 115 km	14 May 91 308 Polyxo U, 138 km	17 Jul 92 533 Sara S, 37 km	12 Jun 93 34 Circe C, 114 km	4 Mar 95 4 Vesta U, 538 km
19 Jan 89	31 Oct 90 19 Fortuna C, 215 km	1 Jan 92 11 Parthenope S, 150 km	25 Sep 93 17 Thetis S, 109 km	9 Mar 95 308 Polyxo U, 138 km	17 Apr 97 163 Erigone C, 76 km
31 Jan 89	11 Jan 91 4 Vesta U, 538 km	22 Oct 92 799 Gudula U, 32 km	4 Jul 94 119 Althaea S, 57 km	5 Jan 96 46 Hestia C, 133 km	8 Jul 97 533 Sara S, 37 km
17 Nov 91	17 Oct 93 11 Parthenope S, 150 km	17 Oct 95 45 Eugenia C, 226 km	5 Dec 97 103 Hera S, 90 km	30 Mar 99 206 Hersilia C, 101 km	30 Oct 00 77 Frigga M, 65 km

[a]Table from Friedlander *et al.* 1979.

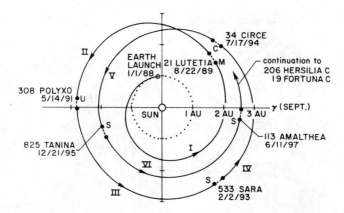

Fig. 5. Example of a multi-target rendezvous tour that visits 8 asteroids. From Friedlander *et al.* (1978).

Thus, even after launch, the tour can be varied, proceeding down any of a number of branches. As a practical matter, example tours have been generated according to rather arbitrary algorithms, e.g., by specifying the first asteroid to be visited, and then requiring that each successive target be of a different taxonomic group from the previous. With more than 3000 asteroids to select from, finding candidate tours becomes a game. Only when more specific and realistic criteria are specified, can the large number of possible missions be reduced to a manageable size.

While the vast number of possible tours may be a frustration to the mission planner, it illustrates the richness of the asteroid population and the wealth of opportunity for exciting missions. Current NASA projections for planetary exploration suggest that, by about 1986, only Neptune, Pluto, and the asteroids will *not* have been visited by spacecraft. It seems clear that the time for direct exploration of the asteroids is rapidly approaching.

Main-Belt Sample Return

Although probably not suitable as an initial mission, sample return has frequently been considered as the logical second-generation mission in a long-term program of asteroid exploration. Figure 6 illustrates a specific example: a Vesta sample return launched in 1993. As analyzed by Friedlander *et al.* (1979), this mission could return a 1.0 kg sample to Earth orbit 3.4 yr after launch, requiring only the 33-kW ion-drive already developed for a asteroid rendezvous tour.

VII. ASTEROIDS AS POTENTIAL RESOURCES

The asteroids have been suggested as a potential source of raw materials for space industrialization in the 21st century in a variety of books and

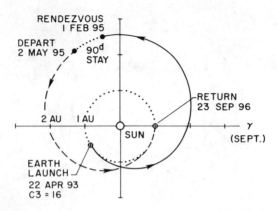

Fig. 6. Ion Drive sample-return mission to Vesta. This example trajectory needs a 25 kW solar array and 2:1 concentrators. From Friedlander *et al.* (1979).

articles by O'Neil and others, and some recent technical studies have been made to examine this possibility (e.g. O'Leary 1977; Arnold and Duke 1978; O'Leary *et al.* 1979). Asteroid retrieval which would bring an entire asteroid into Earth orbit might be called the ultimate form of sample return. Considering the size of the object which must be brought back, it is necessary to consider some new options for both trajectories and propulsion systems. For reducing the total impulse required for the return trajectory, O'Leary (1979) suggests double lunar gravity assists, planetary gravity assist and grazing reentry through the earth's atmosphere into a capture orbit. One propulsion option uses chemical fuels derived from the asteroid itself. Another, called a mass driver, uses solar energy to drive a linear electromagnetic accelerator that expels asteroid material carried in buckets (the buckets are recycled). Clearly, before asteroid retrieval missions can be attempted, sample return missions must be performed and considerable effort must go into trajectory and propulsion system design and into economic studies.

Return of a small sample from an Earth-approaching asteroid is a necessary first step in assessing the potential of these objects for industrial use, as well as being of considerable scientific interest (Shoemaker and Helin 1978). The propulsion requirements for sample return from these objects depends critically on the orbit, with much greater variation than is the case for mainbelt missions. Table III gives trajectory and performance data for these relatively easy missions. Shoemaker and Helin have also noted that the most accessible of these objects are the logical first targets for manned space missions beyond the immediate Earth-Moon system. Thus, quite apart from questions of scientific strategy, the Earth-approaching asteroids may play a central role in an expanded space program in the 21st century.

Acknowledgments. W. Wells provided valuable assistance in preparing this

TABLE III

Apollo/Amor Sample Return: Ballistic Flight Mode[a]

	Anteros	Eros	1976AA
Trajectory Data			
Launch date	26 May 1991	4 Apr 1993	16 Jul 1993
Launch C_3, km^2 sec^{-2}	35	51	71
Spacecraft ΔV, km sec^{-1}	2.71	3.61	7.75
Stay time, day	177	319	30
Round trip, yr	3.0	3.0	1.0
Mass Performance Data[b]			
Number of shuttle launches	1	2	4
Spacecraft propulsion type[c]	E/S	E/S	S/S
Propulsion system mass, kg	1870	3180	7000
Rendezvous payload margin, kg	180	70	400

[a]From Friedlander et al. (1979).
[b]Nominal payload = 600 (orbiter) + 300 (experiments) + 50 (sample canister) = 950 kg
[c]E/S is Earth-storable; S/S is space-storable

chapter; useful comments were also received from C. Pilcher, L. Wilkening and B. Zellner. B. Smith and H. Reitsema kindly made available information on the capability of the Planetary Camera on the Space Telescope for asteroid studies. Financial support was provided by NASA in grants to the University of Hawaii and contracts to Science Applications, Inc.

REFERENCES

Anders, E. 1971. Interrelations of meteorites, asteroids, and comets. In *Physical Studies of Minor Planets,* ed., T. Gehrels, (NASA SP 267, Washington, D.C.: U.S. Government Printing Office), pp. 429-446.

Arnold, J. R. 1978. Remote geochemical sensing of asteroids. In *Asteroids: An Exploration Assessment,* eds., D. Morrison and W. C. Wells, NASA Conf. Publ. 2053, pp. 275-278.

Arnold, J. R., and Duke, M. B., eds. 1978. *Summer Workshop in Near-Earth Resources,* NASA Conf. Publ. 2031.

Aumann, H. H., and Walker, R. G. 1977. Infrared astronomical satellite. *Opt. Eng.* 16: 537-543.

Bender, D. F. 1977. Ion-drive multi-asteroid rendezvous trajectories. Unpublished.

Bender, D.; Bowell, E.; Chapman, C.; Gaffey, M.; Gehrels, T.; Zellner, B.; Morrison, D.; and Tedesco, E. 1978. The Tucson Revised Index of Asteroid Data. *Icarus* 33: 630-631.

Bowell, E.; Chapman, C. R.; Gradie, J. C.; Morrison, D.; and Zellner, B. 1978. Taxonomy of asteroids. *Icarus* 35: 313-335.

Brahic, A. et al. 1979a. Asteroids: A mission proposal to ESA. Unpublished.

Brahic, A.; Breton, J.; Caubel, J.; Cazenave, A.; Cruvellier, P.; Dupuis, Y.; Lago, B.; Minster, J.F.; Perret, A.; and Scribot, A. 1979b. Feasibility of a multiple fly-by mission of main-belt asteroids. *Icarus* (special Asteroid issue).

Brooks, D. R., and Hampshire, W. F., II. 1971. Multiple asteroid flyby missions. In
    Physical Studies of Minor Planets, ed. T. Gehrels (NASA SP-267, Washington, D. C.:
    U.S. Government Printing Office), pp. 527-537.
Chapman, C. R., and Zellner, B. H. 1978. The role of earth-based observations of
    asteroids during the next decade. In *Asteroids: An Exploration Assessment*, eds., D.
    Morrison and W. C. Wells, NASA Conf. Publ. 2053, pp. 183-192.
Consolmagno, G. J., and Drake, M. J. 1977. Composition and evolution of the Eucrite
    parent body: Evidence from rare earth elements. *Geochim. Cosmochim. Acta* 41:
    1271-1282.
Economou, T. E., and Turkevich, A. L. 1976. An alpha-particle instrument with alpha,
    proton, and x-ray modes for planetary chemical analysis. *Nuc. Instr. Meth.* 134:
    391-400.
Fanale, F. P. 1978. Science rationale for an initial asteroid-dedicated mission. In
    Asteroids: An Exploration Assessment, eds., D. Morrison and W. C. Wells, NASA
    Conf. Publ. 2053, pp. 193-206.
Friedlander, A. L.; Wells, W. C.; Davis, D. R.; Housen, K.; and Wilkening, L. L. 1979.
    Asteroid Mission Study. *Science Applications, Inc. Report* No. 1-120-839-M11.
Gaffey, M. J., and McCord, T. B. 1978. Asteroid surface materials: Mineralogical
    interpretations from reflectance spectra. *Space Sci. Rev.* 21: 555-628.
Gehrels, T. ed. 1971. *Physical Studies of Minor Planets* (NASA SP-267, Washington,
    D.C.: U.S. Government Printing Office).
Gehrels, T. 1976. Picture of Ganymede. In *Planetary Satellites*, ed., J. A. Burns (Tucson:
    University of Arizona Press), pp. 406-411.
Haines, E. L.; Arnold, J. R.; and Metzger, A. E. 1976. Chemical mapping of planetary
    surfaces. *IEEE Trans. on Geosci Electron.* GE-14: 141-153.
Morrison, D., and Wells, W. C., eds. 1978. *Asteroids: An Exploration Assessment*, NASA
    Conf. Publ. 2053.
Niehoff, J. C. 1978. Asteroid mission alternatives. In *Asteroids: An Exploration Assess-
    ment*, eds., D. Morrison and W. C. Wells, NASA Conf. Publ. 2053, pp. 225-244.
O'Leary, B. T. 1977. Moving the Apollo and Amor asteroids. *Science* 197: 363-366.
O'Leary, B. T. 1979. Asteroid prospecting and retrieval. *Amer. Inst. Aeronautics
    Astronautics* Paper No. 79-1432.
O'Leary, B.; Gaffey, M. J.; Ross, D. J.; and Salkeld, R. 1979. Retrieval of asteroidal
    materials. In *Space Resources and Space Settlements*, eds. J. Dillingham, W. Gilbreath
    and B. O'Leary (NASA SP-428, Washington, D. C.: U.S. Government Printing
    Office), pp. 173-189, in press.
Shoemaker, E. M., and Helin, E. F. 1978. Earth-approaching asteroids as targets for
    exploration. In *Asteroids: An Exploration Assessment*, eds., D. Morrison and W. C.
    Wells, NASA Conf. Publ. 2053, pp. 245-256.
Stuhlinger, E.; Alfvén, H.; Arrhenius, G.; Bourke, R.; Doe, B.; Dwornik, S.; Friedlander,
    A.; Gehrels, T., Guttman, C.; Strangway, D.; and Whipple, F. 1972. *Comets and
    Asteroids: A Strategy for Exploration*, NASA TM-64677.
Turkevich, A., and Economou, T. 1978. Experiments on asteroids using hard landers. In
    Asteroids: An Exploration Assessment, eds., D. Morrison and W. C. Wells, NASA
    Conf. Publ. 2053, pp. 285-294.
Wagner, J.; Cohen, A.; and Hapke, B. 1979. Vacuum ultraviolet reflectance spectra of
    gray H chondrites. *Proc. Lunar Sci. Conf. X* (Oxford: Pergamon Press), in press.
Wilkening, L L. 1979. Asteroid missions: Goals for elemental abundance analysis. *Icarus*
    (special Asteroid issue).
Wright, J. L.; Bender, D. B.; Staehle, R. L.; and Spradlin, F. A. 1979. Asteroid multiple
    rendezvous mission concept. *Jet Propulsion Lab. Report* No. 710-26.

PART III

Interrelation

EARTH-CROSSING ASTEROIDS:

ORBITAL CLASSES, COLLISION RATES WITH EARTH, AND ORIGIN

E. M. SHOEMAKER, J. G. WILLIAMS, E. F. HELIN, AND R. F. WOLFE
California Institute of Technology

The term Earth-crossing asteroid is here taken to mean a minor planet on an orbit which, as a consequence of secular perturbations, can intersect the orbit of the earth. Known classes of Earth-crossing asteroids include Aten asteroids ($a < 1.0$ AU, $Q > 0.983$ AU), Apollo asteroids ($a \geqslant 1.0$ AU, $q \leqslant 1.017$ AU), and Amor asteroids ($a > 1.0$ AU, 1.017 AU $< q \leqslant 1.3$ AU). All three known Atens, all but one of the 23 known Apollos, and half of the 20 known Amors are Earth crossers. The total population of Earth-crossing asteroids to $V(1,0) = 18$ is estimated at $\sim 1.3 \times 10^3$, of which $\sim 8\%$ are Atens, $\sim 50\%$ are Apollos, and $\sim 40\%$ are Amors. A wide variety of physical types is represented among the Earth crossers, including four objects with UBV colors in the C field, several S-type objects, and several objects of distinctive colors that fall outside the C and S fields. The Earth crossers probably are of diverse origin; some probably have been derived from a residual population of old Mars crossers, some from widely separated regions of the main asteroid belt, and some from short-period comets. The principal sources appear to be extinct comet nuclei and collision fragments from regions in the main asteroid belt bordering the $\dot{\nu}_5$ and ν_6 secular resonances and the 3:1 and 5:2 commensurabilities with Jupiter. The present collision rate of Earth-crossing asteroids with the earth is estimated at ~ 3.5 objects, to absolute magnitude 18, per million years. Comparison of this rate with the record of impact cratering on the earth and the moon suggests that the present population of Earth crossers may be larger than the average population over the past 3.3 billion years.

[253]

The term Earth-crossing is here applied to include all asteroids on orbits which, as a consequence solely of secular perturbations, can intersect the orbit of the Earth. This definition is, at once, both broader and more exclusive than common past usage, where "Earth-crossing asteroid" has been taken to mean an object on an orbit that currently overlaps the orbit of Earth. Rigorous application of the concept of Earth- or planet-crossing orbits has been dependent, however, on the development of appropriate theoretical techniques (Williams 1969) or on practicable numerical integration by high-speed computer to solve the secular variation of asteroid orbits. Only recently have sufficient calculations been completed, primarily by Williams (1979 and unpublished), to confidently identify the known Earth-crossing asteroids.

It has long been recognized (Öpik 1951) that an asteroid orbit which overlaps the orbit of a planet may, as a result of secular advance of the apsides, intersect the orbit of the planet. The topological relationship between overlapping orbits has two possible states: linked and unlinked. If overlap of a planetary orbit of low eccentricity by a highly eccentric asteroid orbit is temporally continuous and deep, the two orbits must be linked (looped through one another like two links of a chain) when the argument of perihelion of the asteroid orbit ω is 0 and π. At $\omega = \pi/2$ and $3\pi/2$, on the other hand, the two orbits are unlinked. In a complete cycle of advance of ω, therefore, the transition between linked and unlinked states occurs four times; at each of these transitions the two orbits must intersect. Intersections of crossings, as given by solution of the polar equation of the ellipse, occur at

$$\omega = \cos^{-1} \pm \frac{1}{e} \left[\frac{a(1-e^2)}{\rho} - 1 \right] \qquad (1)$$

where a is the semimajor axis of asteroid orbit, e is the eccentricity of asteroid orbit at time of each intersection, and ρ is the radius to the planet's orbit along the line of nodes at the time of intersection. Because, in general, the planet's orbit is not circular and its eccentricity also varies as a consequence of secular perturbation, ρ will have four different values in one cycle of ω, which must be found by simultaneous solution of Eq. (1) and the polar equation for the elliptical orbit of the planet; e_0 and ω_0, the eccentricity and argument of perihelion of the planet's orbit, must be known at the time of each crossing. For a perfectly circular orbit of Earth ($e_0 = 0$) all values of ρ are 1.0 AU; at the other extreme of e_0, 0.933 AU $\leqslant \rho \leqslant$ 1.067 AU. In the case of continuous deep overlap of orbits, the four crossings given by Eq. (1) can fail to occur only if ω librates within a restricted range. If the overlap of the orbit of an asteroid with that of a planet is sufficiently shallow, on the other hand, less than four crossings may occur in a complete cycle of advance at ω. Crossings can be missed when ω and ω_0, the argument of perihelion of the planet, have the same phase. Secular variation of either e or e_0, moreover, can lead to loss of overlap during parts of the cycle of ω. At and very close to the times of any crossing there is a finite and calculable probability of collision of the asteroid with the planet unless encounter is prevented by exact average commensurability between the mean motion of the asteroid and mean motion of the planet.

Secular perturbations cause periodic oscillations in e and γ, the inclination of an asteroid orbit to the invariable plane, sometimes in a, and also in e_0 and γ_0, the inclination of the orbit of the planet, in addition to changes in ω. As a consequence of these perturbations, the orbits of less than half of the known Earth-crossing asteroids overlap the orbit of Earth continuously. Orbits of the other known Earth crossers overlap the orbit of the earth part of the time and part of the time lie entirely outside the orbit of Earth, primarily as a result of secular variation of e. Similarly, orbits which lie entirely inside Earth's orbit part of the time are possible, though no objects have yet been found on such orbits. Intersections of asteroid orbits with the orbit of Earth can occur which are associated with transitions from the condition of overlap to the condition of nonoverlap and also the reverse transitions. The intersections occur when r_a, the radius to the node of the asteroid orbit, equals ρ.

Secular variation of e, as a general rule, is correlated in phase with secular variation of ω. Most commonly, e is at a maximum at $\omega = \pi/2, 3\pi/2$. If the amplitude of oscillation of e is relatively small, periodic oscillation of r_a, for ascending and descending nodes, occurs mainly as a result of secular advance of ω. This type of oscillation of r_a is illustrated for the asteroid 2062 Aten in Fig. 1. If the asteroid is a relatively deep crosser, as is the case for Aten, four crossings of r_a at ρ will occur in each cycle of ω. This type of Earth-crossing asteroid is referred to here as a *quadruple crosser*. If the amplitude of oscillation of e is sufficiently high, however, the orbit of the asteroid can change from a condition of nonoverlap, at $\omega = 0,\pi$, to relatively deep overlap at $\omega = \pi/2, 3\pi/2$. In this case, two crossings can occur with each $\pi/2$ advance of ω, one controlled mainly by the large change of e and one mainly by the advance of ω. This type of oscillation of r_a is illustrated for the asteroid 1974 MA in Fig. 2; a total of eight crossings occur in each cycle of ω. We refer to this type of asteroid as an *octuple crosser*. 1580 Betulia was the first asteroid recognized to have this type of crossing behavior (Wetherill and Williams 1968).

In some cases, where γ is sufficiently high, secular perturbations do not lead to continuous advance in ω. Instead, ω librates around $\pi/2$ or $3\pi/2$; large periodic oscillation of e occurs during the libration cycle, with e reaching both a maximum and a minimum at the central value of ω. For deep crossers, the combined effect of oscillation of both e and ω produces four crossings of r_a at ρ in each libration cycle of ω. We refer to asteroids exhibiting this type of crossing behavior, such as 1973 NA (Fig. 3), as *quadruple crossing ω*

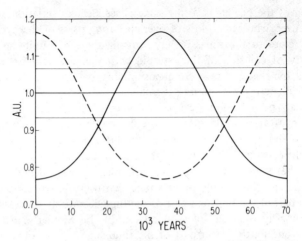

Fig. 1. Secular variation of radius to the node of the orbit of 2062 Aten for the zero-order state (forced oscillations in e are neglected). One cycle of advance of ω is represented, starting at $\omega = 0$. Ascending node shown with solid line and descending node with dashed line. Extreme values of aphelion and perihelion distances of the earth are shown by horizontal lines lying equal distances above and below 1 AU. Aten is an example of a quadruple crosser.

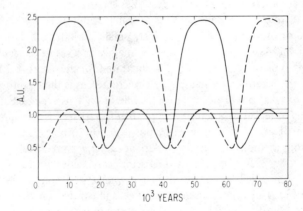

Fig. 2. Secular variation of radius to the node of the orbit of 1974 MA, based on forward integration of the motion of the asteroid. About 1.5 cycles of advance of ω are represented. Ascending node shown with solid line and descending node with dashed line. Extreme values of aphelion and perihelion distances of the earth are shown by horizontal lines lying equal distances above and below 1 AU. 1974 MA is an example of an octuple crosser.

librators.

A fourth type of crossing behavior is exhibited by Earth-crossing asteroids that librate about the 3:1 commensurability with Jupiter. The 3:1 resonant perturbation causes relatively high-frequency oscillation in both a

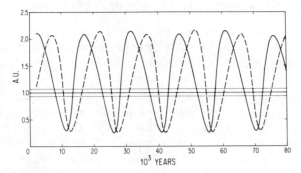

Fig. 3. Secular variation of radius to the node of the orbit of 1973 NA, based on forward integration of the motion of the asteroid. About five cycles of libration of ω are represented. Ascending node shown with solid line and descending node with dashed line. Extreme values of aphelion and perihelion distances of the earth are shown by horizontal lines lying equal distances above and below 1 AU. 1973 NA is an example of a quadruple crossing ω librator.

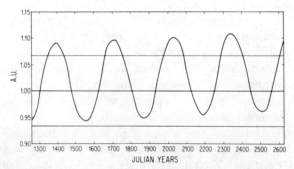

Fig. 4. Secular variation of radius to the ascending node of the orbit of 1915 Quetzalcoatl, based on forward and backward integrations of the motion of the asteroid by Marsden. About four cycles of libration of the mean motion of the asteroid about the 3:1 commensurability are represented. Extreme values of aphelion and perihelion distances of the earth are shown by horizontal lines lying equal distances above and below 1 AU. 1915 Quetzalcoatl is an example of a supercrosser.

and e, which can lead, in turn, to relatively high-frequency oscillation between conditions of overlap and nonoverlap of the orbit of the asteroid with the orbit of Earth. Slightly more than four cycles of oscillation of r_a in 1400 years and nine crossings of the orbit of Earth by the ascending node are illustrated for the asteroid 1915 Quetzalcoatl in Fig. 4. Asteroids exhibiting this type of crossing behavior are referred to here as *supercrossers*. Marsden (1970) has carried out 1400 year integrations of the motions of the supercrossers 1915 Quetzalcoatl and 887 Alinda, but much more numerical study must be done before the crossing behavior of these asteroids is fully understood.

I. CLASSES OF EARTH-CROSSING ASTEROIDS

The Earth-crossing asteroids are divided, for purposes of further discussion in this chapter, into three groups on the basis of their present osculating orbital elements: (1) Aten asteroids, (2) Apollo asteroids, and (3) Earth-crossing Amor asteroids. This classification has been used primarily because of its simplicity. However, the stable characteristics of the orbits of Apollo asteroids that overlap Earth's orbit part of the time and of the orbits of Earth-crossing Amors are not basically different. These two orbital classes are distinguished chiefly by the present phase of the cycle of variation of their perihelion distance. Most Earth-crossing Amors are only shallow Earth crossers, on the other hand, whereas the majority of Apollos are deep crossers. The orbits of most Earth-crossing Amors, moreover, overlap the orbit of the earth only a small fraction of the time. Therefore it seems useful to retain the traditional distinction between Apollo and Amor asteroids.

Aten Asteroids

Aten asteroids (Helin *et al.* 1978) have orbits with semimajor axes less than 1 AU which overlap the orbit of Earth near their aphelia. This class includes all asteroids with $a < 1.0$ AU and the aphelion distance of the asteroid, $Q \geqslant 0.983$ AU, where 0.983 AU is the present perihelion distance of Earth. Since Aten, the first member of this class, was discovered in 1976 (Helin and Shoemaker 1977) two more have been found (Table I). All three objects are relatively deep quadruple crossers whose orbital overlap with Earth's orbit is continuous or nearly continuous.

The total number of Aten asteroids to $V(1,0) = 18$, as will be shown, probably is on the order of 100. Many of these Atens may be expected to exhibit only part-time orbital overlap with Earth, just as many Apollos exhibit only part-time overlap at perihelion. Hence it may be expected that there are Earth crossers with orbits that are, at present, entirely inside the orbit of the Earth, just as there are Earth-crossing Amors with current osculating orbits entirely outside the orbit of the earth. As a rough guess, the population of this undiscovered class of Earth crossers with current aphelia less than 0.983 AU may be a few tens of objects to $V(1,0) = 18$. Wetherill (1979) has estimated the total population of asteroids with orbits inside Earth's at one to a few percent of the population of Apollos.

Apollo Asteroids

Apollo asteroids are defined by $a \geqslant 1.0$ AU, and perihelion distance, $q \leqslant 1.017$ AU, where 1.017 AU is the present aphelion distance of the earth. Apollo, the first such asteroid discovered, was found by K. Reinmuth in 1932. The orbits of Apollos overlap the orbit of Earth in the region of perihelion. Out of 22 known Apollos with reasonably well-determined orbits, 21 are Earth-crossing (Table I). 1866 Sisyphus is a doubtful crosser; on a

TABLE I

Earth-crossing Asteroids: Crossing Characteristics and Collision Parameters

	Orbital overlap with Earth	F_ℓ[a]	Linkage state[b]	Earth-crossing class[c]	a (AU)	e	γ (Deg)	$\|dr_a/dt\|$ $\rho = 1.0$ AU[e] (10^{-4} AU yr^{-1})	T_c[f] (10^4 yr)	P_s[g] (10^{-9} yr^{-1})	P_θ[h] (10^{-9} yr^{-1})
		Linkage with Earth's orbit			Orbital elements at $\rho = 1.0$ AU[d]						
ATEN ASTEROIDS											
1976 UA	Continuous	0.55	L	Quad.	0.844	0.424	6.27	0.28	9.54	14	14
2100 Ra-Shalom	Continuous	0.54	L	Quad.	0.832	0.465	13.1	0.32	9.91	6.3	6.7
2062 Aten	Continuous	0.73	L	Quad.	0.966	0.237	17.9	0.20	7.04	6.9	6.4
APOLLO ASTEROIDS											
1978 SB	Continuous	0.89	L	Quad.	2.164	0.888	10.5	32	0.83	0.41	1.2
1566 Icarus	Continuous	0.70	L	Quad.	1.078	0.828	20.2	2.2	5.30	1.6	2.0
1974 MA	Part time	0.61	L	Oct.	1.757	{0.446 / 0.772}	{53.4 / 32.8}	{0.66 / 3.6}	4.23	0.93	2.5
2101 Adonis	Continuous	0.95	L	Quad.	1.873	0.732	1.96	7.3	1.58	2.9	6.3
1864 Daedalus	Continuous	0.92	L	Quad.	1.461	0.640	15.9	2.1	3.69	1.0	1.6
1865 Cerberus	Continuous	0.80	L	Quad.	1.080	0.513	14.9	0.68	6.81	2.5	3.1
Hermes	Continuous	0.99	L	Quad.	1.639	0.622	5.62	2.6	2.57	2.2	3.7
1981 Midas	Part time	0.29	U	Q.O.L.	1.776	{0.586 / 0.652}	{45.5 / 41.3}	{0.42 / 0.53}	2.62	3.8	0.7
1862 Apollo	Continuous	0.91	L	Quad.	1.470	0.505	6.14	1.3	3.00	3.5	4.8
2063 Bacchus	Continuous	0.94	L	Quad.	1.077	0.320	8.78	0.49	4.41	7.4	7.7
1959 LM	?	(0.8)	L	Quad.?	(1.34)[j]	(0.379)[j]	(3.3)	—	—	—	(13)[j]
1685 Toro	Continuous	(0.8)	L	Quad.?	1.368	0.438	9.15	0.78	(3)[j]	(4)[j]	(4.2)[j]
2135 1977 HA	Part time	0.76	U	Q.&O.	1.600	(0.490)	(23.3)	(0.43)	(3.97)	(2.0)[k]	(1.5)[k]
6743 P-L	Continuous	0.73	U	Quad.	1.620	0.476	6.81	0.82	2.87	4.1	4.8
1620 Geographos	Continuous	0.84	U	Quad.	1.245	0.351	14.2	0.39	5.21	3.8	3.9
1976 WA	Part time	0.74	L	Oct.	2.407	{0.589 / 0.700}	{32.3 / 17.0}	{0.8 / 5.1}	1.19	(1.8)[k]	(2.1)[k]
1950 DA	Continuous	0.83	L	Quad.	1.683	0.535	10.7	1.3	2.66	1.8	2.4

TABLE I (Continued)

Earth-crossing Asteroids: Crossing Characteristics
and Collision Parameters

	Orbital overlap with Earth	Linkage with Earth's orbit F_ℓ^a	Linkage state[b]	Earth-crossing class[c]	Orbital elements at $\rho = 1.0$ AU[d] a (AU)	e	γ (Deg)	$\|dr_a/dt\|$ $\rho = 1.0$ AU[e] (10^{-4} AU yr^{-1})	T^f (10^4 yr)	P^g (10^{-9} yr^{-1})	P^h (10^{-9} yr^{-1})
1866 Sisyphus	Part time	—	U	D.C.	—	—	—	—	—	0?	0?
1973 NA	Part time	0.36	U	Q.O.L.	2.447	{ 0.672 / 0.891	67.9 / 51.8	{ 2.9 / 6.4	1.47	0.54	0.36
1978 CA	Continuous	0.90	U	Quad.	1.125	0.278	25.3	0.16	8.31	4.1	3.3
1863 Antinous	Part time	—	U	S.Oct.	—	—	—	—	—	—	—
2102 Tantalus	Part time	0.49	U	Q.O.L.	1.290	{ 0.312 / 0.744	62.5 / 49.1	{ 0.39 / 1.4	4.87	2.5	1.5
6344 P-L	Part time	—	U	S.Oct.	—	—	—	—	—	—	—
AMOR ASTEROIDS 1978 DA	Part time	—	U	S.Super.	—	—	—	—	—	—	—
2061 Anza	Part time	—	U	S.Quad.	—	—	—	—	—	—	—
1915 Quetzalcoatl	Part time	(0.4)	U	Super.	(2.49)	(0.60)	(19.7)	(15.4)	(0.031)	(3.5)ℓ	(1.5)ℓ
1917 Cuyo	Part time	—	U	S.Quad.	—	—	—	—	—	—	—
1943 Anteros	Part time	—	U	S.Quad.	—	—	—	—	—	—	—
1221 Amor	Part time	—	U	S.Quad.	—	—	—	—	—	—	—
1580 Betulia	Part time	0.51	U	S.Quad.	2.196	{ 0.564 / 0.792	48.5 / 26.4	{ 1.8 / 7.0	2.37	0.50	1.7
1972 RA	Part time	—	U	S.Quad.	—	—	—	—	—	—	—
1627 Ivar	Part time	—	U	S.Quad.	—	—	—	—	—	—	—
887 Alinda	Part time	—	U	S.Super.	—	—	—	—	—	—	—

[a] F_ℓ is the linkage fraction, which is defined as the average fraction of time that the orbit of the asteroid and the orbit of the Earth are linked. Figures shown in parentheses are derived from Eq. (2) or from incomplete numerical study of secular variation of orbit.

[b] Linkage state is the current condition of linkage of the orbit of the asteroid with the orbit of the earth: L = linked, and U = unlinked.

[c] Quad. = quadruple crosser; S. Quad. = shallow quadruple crosser; Oct. = octuple crosser; S. Oct. = shallow octuple crosser; Q. O. L. = quadruple crossing ω librator; Q. & O. = quadruple and sometimes octuple crosser; D. C. = doubtful crosser; Super. = supercrosser; S. Super. = shallow supercrosser.

d Except as otherwise noted, all values of e and γ are for the zero order state (i.e. only free oscillations of e and γ are considered) derived by interpolation from computer runs based on a first order theory of secular perturbation (Williams 1969). Node and γ are referred to the invariable plane. Values e and γ bar ω librators, for 1685 Toro, and for 1580 are average values obtained from numerical integration of the motion of the asteroid by Williams; for Quetzalcoatl e and γ were derived by us from a numerical integration by Marsden (personal communication, 1979).

e $|dr_a/dt|$ at $\rho = 1.0$ AU is the derivative with respect to time of the radius to the node when the radius is equal to 1.0 AU. This derivative is obtained by interpolation from computer runs indicated in footnote d.

f For quadruple and octuple crossers, T_c is the mean period of a complete cycle of advance of ω; for ω librators, T_c is the mean period of libration of ω; for supercrossers, T_c is the period of oscillation of r. The values of T_c are derived from computer runs indicated in footnote d.

g P_s is probability of collision with Earth based on Eqs. (6) and (7).

h P_o is probability of collision with Earth based on the equations of Öpik (1951), using the orbital elements listed in this table. For octuple crossers, the probabilities obtained from the two sets of e and γ were summed, and, for ω librators, the two solutions for the probability were averaged.

i Orbital elements based on short arc and not accurately known. Values listed for e and γ are the current osculating values and are used to obtain an approximate solution for P_o.

j Toro currently has a commensurable mean motion with the Earth. T_c is estimated from a numerical integration of the motion of Toro that spans only a fraction of the period of oscillation of r_a. P_s and P_o are calculated by assuming that the present state of commensurable motion is short lived.

k P_s and P_o listed for 1977 HA and 1976 WA should be considered lower limits, because forced oscillations of e will occasionally lead to very low values of $|dr_a/dt|$ at $\rho = 1.0$ AU.

l Collision parameters and collision probabilities for Quetzalcoatl are based on numerical integration of motion spanning only a small fraction of the secular variation of ω, hence the true mean collision parameters are not known.

278,000 year integration of the motion of this object by Williams, r_a did not become smaller than 1.07 AU. One other Apollo whose orbit has been determined only from a short arc, 1959 LM, very probably is an Earth crosser. Among the 21 established Earth crossers, 13 exhibit continuous or nearly continuous orbital overlap with the Earth and eight have part-time overlap. All 13 with continuous orbital overlap are quadruple crossers or a probable quadruple crosser. Among the Apollos that exhibit part-time overlap with the earth, four are octuple crossers, one is a quadruple crosser part of the time and an octuple crosser part of the time, and three are quadruple crossing ω librators.

Amor Asteroids

Amor asteroids have been defined simply on the basis of perihelion distance, $1.017 \, \text{AU} < q \leqslant 1.3 \, \text{AU}$ (Shoemaker and Helin 1978). These are objects that make relatively close approaches to the earth but do not, at present, overlap the earth's orbit. Traditionally, the name Amor, an asteroid discovered in 1932 by E. Delporte, has been given to this class, even though Eros, discovered in 1898, is a much better known member. The upper bound of q at 1.3 AU is arbitrary; it was chosen near a minimum in the radial frequency distribution of q for discovered objects. Somewhat less obviously, the distinction between Amor and Apollo asteroids is also rather arbitrary; some Apollos become Amors, and vice versa, as a consequence of secular perturbations. Had the Amor asteroid 1915 Quetzalcoatl been found in 1942 it would have been classed as an Apollo at the time of discovery. Of 20 known Amor asteroids, half are Earth-crossing, including 1221 Amor. Six Amors are quadruple crossers, three are supercrossers, and one is an octuple crosser (Table I). Only two of these objects, 1580 Betulia and 1915 Quetzalcoatl, are deep or even moderately deep Earth crossers, however.

II. LINKAGE STATES

For an asteroid orbit with deep overlap with Earth's orbit, constant e, and uniform $\dot{\omega}$, the fraction of time that the orbits are linked, F_ℓ, is given approximately by

$$F_\ell \approx \frac{2}{\pi} \cos^{-1} \left| \frac{1}{e} \left[\frac{a(1-e^2)}{\bar{\rho}} - 1 \right] \right| \qquad (2)$$

where $\bar{\rho} = 1$. The linkage fraction, F_ℓ, is close to 1 when the semilatus rectum, $a(1-e^2)$, is close to $\bar{\rho}$; F_ℓ will be much less than 1 when the semilatus rectum is either much greater or much less than $\bar{\rho}$.

If Apollos and Atens, the asteroids with orbits which currently overlap Earth's orbit, are combined, the mean predicted linkage fraction, taking into account secular variation of e and nonuniform $\dot{\omega}$, is 0.74. The observed degree of linkage is 16/23 or 0.70 (Table I), in satisfactory agreement with prediction. This close agreement suggests that there is no strong selection effect in the discovery of Earth-crossing asteroids with regard to the present radii to the nodes of the asteroids.

III. COLLISION RATES WITH EARTH

All known Earth-crossing asteroids, as defined in this chapter, have a finite probability of collision with Earth. Only in the case of 1685 Toro has a commensurable mean motion been found that might preclude collision at times of orbit intersection (Danielsson and Ip 1972; Janiczek *et al.* 1972; Ip and Mehra 1973). As shown by Williams and Wetherill (1973), however, even if Toro is able to avoid Earth at the present time, close encounters with Mars will tend to displace Toro from deep resonance in about 3×10^6 yr. Its probability of collision with Earth can be decreased only slightly by its present commensurable motion.

Close encounters with the planets cause large changes in the orbits of the planet-crossing asteroids. Many such changes usually occur in the orbit of an Earth crosser before it meets its ultimate fate – collision or ejection from the solar system (Arnold 1964, 1965; Wetherill and Williams 1968). Because of these changes, the probability of collision with Earth at any one time in the orbital evolution of a given asteroid does not ordinarily represent its long-term average probability of collision. The sum of the present probabilities of collision with Earth of all Earth crossers, on the other hand, does approximate the present collision rate with Earth, as the total number of Earth crossers is large enough for this sum to be statistically stable.

We will assume here that the orbits of the discovered Aten, Apollo and Earth-crossing Amor asteroids represent a relatively unbiased sample of the orbits of the entire population in each class of Earth crossers. The mean present probabilities of collision for each class, estimated from the available sample of orbits, multiplied by the estimated population will provide an estimate of the present collision rate for objects in that class. An estimate of the present collision rate on Earth, then, will be obtained from the sum of the collision rates for all classes of Earth-crossing objects.

The collision rate on Earth may be expected to change only slowly with time. Although the orbits of individual Earth crossers are scrambled by close planetary encounters, the statistical distribution of orbital characteristics of the entire population of Earth crossers probably is nearly in steady state. The dynamical relaxation time of the Earth-crossing asteroid population may be estimated by the harmonic mean lifetime of Earth crossers against collision or ejection. As found from Monte Carlo studies of close encounter and collision, typical lifetimes for Earth crossers are on the order of a few tens of millions of years (Wetherill and Williams 1968; Wetherill 1976). The present collision rate on Earth may be representative of the rate for the past 10^7 yr and possibly much longer, if the population of Earth crossers is also approximately in equilibrium. We will test this possibility by comparison of the present collision rate with the geological record of impact on the earth.

An approximate theory for calculating the probability of collision of planet-crossing bodies was first developed by Öpik (1951). By breaking down the dynamics of encounter into a series of two-body approximations, Öpik greatly simplified the calculation of impact probability. One of the colliding bodies generally is taken to be of major planetary dimensions and one much smaller, and the motion of the two bodies near encounter is taken as linear, which leads to still further simplification. The conditions under which the errors introduced by these approximations remain small has been discussed in detail by Öpik (1976). In the development of Öpik's theory, the orbit of the major planet was initially taken as circular and a correction applied to the final equations to account for a nonzero mean e_o. Moreover, e and γ for the minor body were assumed constant and ω, over a sufficiently long period of time, was assumed uniformly distributed. This last assumption is equivalent to taking $\dot{\omega}$ as constant or independent of ω. In order to deal with collisions between asteroids, Wetherill (1967) refined Öpik's theory by introducing the finite eccentricity of both bodies at the outset of the development of the problem. The result is a more precise but significantly more complex set of equations.

As will be seen, Öpik's assumption of constant e and uniform distribution of ω commonly leads to significant error, especially in the case of ω librators. Now that the necessary calculations have been carried out for the secular variation of the orbital elements of the Earth crossers, it is appropriate to introduce further revisions in Öpik's theory to take account of these variations. Our development of the theory closely parallels that of Wetherill's (1967). In place of an explicit formula for $d\Delta/d\omega$ at the time of crossing introduced by Öpik (1951), where $\Delta = r_a - r_p$ and r_p is the radius to the nearest node of the orbit of the planet, and the assumption of Öpik of constant $d\omega/dt$, however, we let $d\Delta/dt$ remain a variable to be determined from secular perturbation theory. The resulting expression for probability of collision of an asteroid with a planet per unit time P_s is

$$P_s = \frac{1}{4\pi T_c} \; \Sigma_i \; \left| \frac{d\Delta}{dt} \right|^{-1} \; \frac{\tau_i^2 \, U_i \, \rho_i}{a_o^2 \, (1-e_{oi}^2)^{\frac{1}{2}} \; a_i^2 \, (1-e_i^2)^{\frac{1}{2}} \sin i_i} \tag{3}$$

where T_c is the period of oscillation of Δ, U is the encounter velocity of the asteroid at the sphere of influence of the planet, τ is the capture radius of the planet, i is the inclination of the asteroid orbit referred to the plane of the planet's orbit, and all elements subscripted i as well as $|d\Delta/dt|$ are the values at the times of crossing:

$$U_i^2 = \frac{GM}{\rho_i^2} \left\{ a_i \frac{(2A_i - 1)}{A_i^2} + a_0 \frac{(2A_{0i} - 1)}{A_{0i}^2} \right.$$

$$\left. -2 \left[a_i a_0 (1 - e_i^2)(1 - e_{0i}^2) \right]^{\frac{1}{2}} (\cos i_i + \cot \alpha_i \cot \alpha_{0i}) \right\}$$

(4)

where G is the gravitational constant, M is the mass of the sun, A_i is the a_i/ρ_i, A_{0i} is the a_0/ρ_i, and

$$\cot^2 \alpha_i = \frac{A_i^2 e_i^2 - (A_i - 1)^2}{A_i^2 (1 - e_i^2)} , \cot^2 \alpha_{0i} = \frac{A_{0i}^2 e_{0i}^2 - (A_{0i} - 1)^2}{A_{0i}^2 (1 - e^2_{0i})}$$

(5)

$$\tau_i^2 = R^2 \left(1 + \frac{8 \pi^2 m}{R U_i^2 M} \right)$$

where R and m are respectively the radius and mass of the planet. For quadruple crossers and octuple crossers, T_c is the period of one cycle of advanced of ω; for quadruple crossing ω librators, T_c is the libration period; for supercrossers, T_c is provisionally taken as the libration period of a. The summation in Eq. (3) is to be carried out for the number of crossings in one period T_c, nominally four in the case of quadruple crossers, eight in the case of octuple crossers, and two in the case of supercrossers. Under conditions of shallow overlap of orbits, the number of crossings can be less than nominal. A rigorous evaluation of Eq. (3) would consist of a time-averaged value of P_s for all simultaneous combinations of a_i, e_i, i_i, e_{0i}, ρ_i and $|d\Delta/dt|_i$ that can occur at the times of crossing.

We are not yet prepared to offer a precise solution to Eq. (3) for any Earth crosser, but an approximate solution for deep crossers can be obtained by adopting $\rho = 1$ AU, which is very close to the average ρ for deep crossers. Mean values of a, e, i, and $|dr_a/dt|$ at $\rho = 1$ AU can then be estimated fairly readily from secular perturbation theory. Mean $|dr_a/dt|$ at $\rho = 1$ AU, moreover, is close to mean $|d\Delta/dt|$ at $\rho = 1$ AU. Octuple crossers and quadruple crossing ω librators exhibit two distinctly different types of crossing, and two sets of mean e, i, and $|dr_a/dt|$ have been estimated, one set for each type of crossing. Equations (3) and (4) then reduce to

$$P_s = \frac{\rho^2}{\pi T_c a_0^2 (1 - e_0^2)^{\frac{1}{2}}} \left[\left| \frac{dr_a}{dt} \right|_1^{-1} \frac{\tau_1^2 U_1}{a_1^2 (1 - e_1^2)^{\frac{1}{2}} \sin i_i'} \right.$$

(6)

$$\left. + \left| \frac{dr_a}{dt} \right|_2^{-1} \frac{\tau_2^2 U_2}{a_2^2 (1 - e_2^2)^{\frac{1}{2}} \sin i_2'} \right]$$

where e_o^2 is the mean squared eccentricity of Earth's orbit, $\sin i'$ is $(\sin^2 \lambda + \sin^2 \lambda_o)^{1/2}$, and $\sin^2 \lambda_o$ is the mean squared sine of inclination of Earth's orbit to invariable plane, and subscripts 1 and 2 refer to the two distinct types of crossing;

$$U^2 = \frac{GM}{\rho} \left\{ 4 - \frac{1}{A} - \frac{1}{A_o} - 2 \left[A A_o (1-e^2)(1-e_o^2) \right]^{\frac{1}{2}} \cos i' \right\} . \quad (7)$$

The solution for P_s given by Eq. (6) must be divided by 2 to obtain the correct probability for collision of quadruple crossing ω librators and supercrossers.

First estimates of the collision parameters for each asteroid, which are T_c and mean a, e, γ, and $|dr_a/dt|$ at $\rho = 1.0$ AU, can be obtained for most quadruple and octuple crossers from secular perturbation theory by considering only the free oscillations of the orbital elements. These estimates are shown in Table I. The free oscillations, which remain when the eccentricities and inclinations of the perturbing planets are reduced to zero, are referred to here as the zero-order state. Forced oscillations, due to the finite and varying eccentricities and inclinations of the planets, produce dispersion of e, γ, and r_a, about the values found for the zero-order state. Crossings of very shallow crossers occur only as a consequence of the forced oscillations of e; in these cases the zero-order state does not yield values for the collision parameters. The collision parameters found for deep crossers from the zero-order state are close to the time-averaged values. But, for shallow crossers, the values obtained from the zero-order state will be displaced significantly from the time-averaged means. In particular, $|d\Delta/dt|$ at the time of crossing occasionally will approach zero, in the case of shallow crossers, as a result of $\dot{r}_a$ and $\dot{r}_p$ being in phase or as a consequence of the forced oscillations of e or of secular variation of e_o. Under this circumstance, Eqs. (6) and (7) no longer provide a satisfactory solution for P_s. Hence no solutions of P_s are shown for shallow crossers in Table I.

Cases where $|d\Delta/dt|$ becomes very small occur, in general, for asteroid orbits with only part-time overlap of Earth's orbit, and the fraction of crossings with small $|d\Delta/dt|$ tends to be roughly inversely proportional to the fraction of time that there is overlap. Thus, even though collision probabilities of shallow crossers tend to be high at times of crossing, the frequency of such crossings tends to be low, and the net collision probability generally is of the same order of magnitude as that of deep crossers. A detailed treatment of this problem will be given in a separate paper in preparation.

Collision parameters for ω librators, the supercrosser Quetzalcoatl, and also for the octuple crosser Betulia, which is in a resonant orbit, have been estimated from numerical integrations (Table I). Although based on a limited number of cycles of r_a, these estimates are believed to be nearly as close to the time averaged means as the collision parameters estimated from the

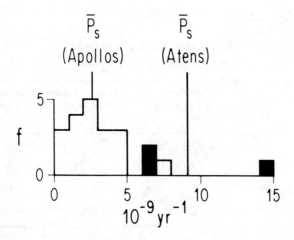

Fig. 5. Frequency distributions of collision probability with Earth, P_s, for Apollo (open bars) and Aten (solid bars) asteroids. Mean impact probabilities, $\bar{P}_s$, for Apollos and Atens are shown with vertical lines.

zero-order state.

It should be noted that the estimates of mean $|dr_a/dt|$ at $\rho = 1.0$ AU obtained either from the theoretical free oscillations of the orbital elements or from numerical integration tend to be maximum estimates. Low values of $|dr_a/dt|$ that occur at times of relatively shallow crossing (under conditions where the two-body linear approximation of Öpik is still good) are not represented in these estimated means. Thus all values of P_s calculated from Eqs. (6) and (7) (Table I) should be regarded as minimum estimates of the true collision probabilities.

Also shown in Table I are collision probabilities based on the equations of Öpik (1951), using the estimated mean orbital elements at $\rho = 1.0$ AU. It may be seen that P_o, the probability of collision with Earth based on Öpik's equations, is in good agreement with P_s for Aten asteroids. In some cases of Apollos and Earth-crossing Amors, however, p_o differs from P_s by factors 2 to 5. Most of the large differences are found among the octuple crossers and ω librators, but a difference factor of more than 2.5 was found for the quadruple crossers Adonis and 1978 SB.

The frequency distributions of computed collision probabilities, P_s, for Apollos and Atens are shown in Fig. 5. Amors are not illustrated in this figure because solutions for Eqs. (6) and (7) can be obtained for only two Amors. The mean probability for collision with Earth is 9.1×10^{-9} yr^{-1} for Atens and 2.6×10^{-9} yr^{-1} for Apollos (where $P_s = 0$ has been adopted for the Apollo asteroid 1866 Sisyphus). Only a rough estimate can by made for mean P_s for the Amors. The orbits of some Amors overlap the orbit of Earth a very small fraction of the time, but they tend to have fairly high probability of

TABLE II

Collision rate with Earth of known
classes of Earth-crossing asteroids

	Population to V(1,0)=18	Mean Collision Probability 10^{-9} yr^{-1}	Collision Rate to V(1,0)=18 10^{-6} yr^{-1}
Atens	~ 100	9.1	~ 0.9
Apollos	700 ± 300	2.6	1.8 ± 0.8
Earth-crossing Amors	~ 500	~ 1	~ 0.5
Total	~ 1300		~ 3.5

collision during these periods of overlap. The mean probability of collision with Earth for the known Earth-crossing Amors is not less than 0.4×10^{-9} yr^{-1} and probably is close to 1×10^{-9} yr^{-1}.

The population of Atens and Apollos combined is estimated by Helin and Shoemaker (1979) as 800 ± 300 to absolute visual magnitude V(1,0) = 18. As Atens constitute 3/26 or 12% of the combined set of known Atens and Apollos, the population of Atens is roughly estimated at ~ 100 to V(1,0) = 18, and the population of Apollos, by subtraction, is 700 ± 300. The population of Earth-crossing Amors can be estimated very roughly from the ratio of discovered Amors to discovered Apollos. A bias exists against discovery of Amors with large q and small a, however, both in systematic surveys and by accidental discovery (Helin and Shoemaker 1979). Therefore the ratio of discovered Amors to discovered Apollos, about 1:1, probably is lower than the ratio of the populations of these objects. Monte Carlo studies of evolution of orbits of Amor and Apollo asteroids suggest that the true ratio of the Amor population to the Apollo population is in the range of 1.5 to 3 (Wetherill 1979). Hence the Amor population to V(1,0) = 18 may be in the range of 1000 to 2000. Although half of the known Amors are Earth crossers, the Earth-crossing Amors tend to have relatively small q and the bias against their discovery is less than for the other Amors. We tentatively estimate the population of Earth-crossing Amors to absolute magnitude 18 at about 500.

When the estimated mean collision probabilities for the three known classes of Earth-crossing asteroids are multiplied by the estimates of the populations for each class (Table II) the total collision rate with Earth is found to be ~ 3.5 asteroids to absolute magnitude 18 per million years. On the average, one of these asteroids will be an Aten, two will be Apollos, and one half will be an Amor. To this list of colliding objects can be added the nuclei of active comets and the undiscovered class of Earth-crossing asteroids

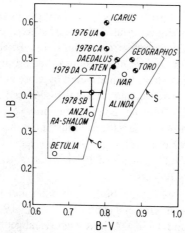

Fig. 6. UBV colors of Earth-crossing asteroids. Outer limits of color fields for C- and S-type asteroids are from Bowell *et al.* (1978). Error bars shown for 1978 SB (Bowell 1978) are probably fairly representative of errors of UBV observations of most other Earth-crossing asteroids. UBV data are from TRIAD data file (Part VII of this book). Solid dots are Atens, target symbols are Apollos, and open circles are Amors.

with orbits currently inside that of Earth. From a consideration of the observed flux of comets in the vicinity of Earth and unpublished measurements by E. Roemer of the magnitudes of comet nuclei when they are at large solar distances and relatively inactive, we estimate that the collision rate of still active comet nuclei is not more than about 10% of the collision rate of Earth-crossing asteroids. In our estimate of an upper bound for the collision rate of comets, a very large allowance is made for incompleteness of discovery of Earth-crossing comets. The contribution to the total collision rate of the undiscovered class of asteroids with very small orbits probably does not exceed ~ 5%.

IV. PHYSICAL CHARACTERISTICS OF EARTH-CROSSING ASTEROIDS

High-quality observations on physical characteristics have been obtained for about 40% of the known Earth-crossing asteroids. In six cases these observations have been made during the discovery apparition, all since January 1976. For some objects, only UBV observations have been made, but in addition, polarimetry, infrared radiometry, narrowband spectrophotometry and observations by radar have been obtained for a number of the Earth crossers. Diversity of UBV colors suggests that a variety of mineralogical compositions are represented among the Earth-crossing asteroids (Fig. 6). Four Earth crossers are *S*-type asteroids; two of these are Apollos and two are Amors. Four Earth crossers, including one Aten, one Apollo, and two Amors, have UBV colors in or on the boundary of the *C* field, as defined by Bowell *et al.* (1978). The colors of other Earth crossers lie outside the limits of the C and S fields, and some Earth crossers have UBV colors unlike those of any other asteroids. Two of these, the Aten asteroid 1976 UA and the Apollo asteroid 1566 Icarus, have closely similar UBV colors that are characterized by extreme values of U–B; Shoemaker and Helin (1978) suggested that the color and possibly other properties of the surfaces of these two asteroids may have been affected by close approaches to the sun.

Observations of Earth-crossing asteroids discovered in 1978 suggest that the concept of Earth crossers as predominantly S-type or high-albedo asteroids, based on earlier observations, must now be re-examined. When corrections are made for observational selection effects due to differences in albedo, it appears that the majority of Earth crossers, to any given size limit, may turn out to be dark objects.

Albedos have been estimated both from polarimetry and from infrared radiometry for six Earth crossers (Lebofsky *et al.* 1979). Three of these objects, Alinda, Aten and 1978 DA, have high visual geometric albedos (0.16 to 0.20), as derived from polarimetry as well as from radiometry, using a standard model of thermal inertia (Jones and Morrison 1974; Morrison 1977) for the surfaces of the asteroids. The other three objects, Betulia, Ra-Shalom and 1978 CA, have low polarimetric albedos; 1978 CA has a low radiometric albedo, as determined from the standard thermal inertia model, but radiometric albedos consistent with polarimetric observations are found for Betulia and Ra-Shalom only if the surfaces of these asteroids are assumed to be rocky (Lebofsky *et al.* 1978, 1979).

Perhaps the most remarkable solar system object, found in 1978, was 1978 SB. This Apollo asteroid, discovered by L. I. Chernykh, has an orbit somewhat like that of comet Encke and is probably the largest known Earth crosser. The UBV color of 1978 SB, measured by Bowell (1978), falls within the C color domain (Fig. 6); from measurements of 1978 SB reported by Bowell we estimated V(1,0) = 14.01, assuming a phase coefficient of 0.035 mag/deg. If a visual geometric albedo p_v near the mean for C-type asteroids is adopted for 1978 SB, its calculated diameter is greater than 10 km (Table III). As shown in Table III, if objects with UBV colors in the C field are assumed to have low albedo, following Bowell *et al.* (1978), then the bulk of the volume of photometrically observed Earth crossers is contained in these dark asteroids.

Clearly it is premature to draw firm conclusions about the relative abundance of difference physical types among the Earth-crossing asteroids from the small sample of physical observations now available. Many more observations are needed, especially polarimetric and radiometric observations of 1978 SB and 2061 Anza, the two largest Earth crossers. Discoveries and observations made in the past year do suggest, however, that the proportion of high-albedo asteroids among the Earth crossers is not as great as initially surmised. As will be seen, this leads to significant revision in the estimated present rate of impact crater production on the earth.

Estimated albedos and diameters of Earth-crossing asteroids

	p_v	Diameter (km)	Volume (km³)	Reference
Asteroids with UBV colors in the C field				
1580 Betulia	0.04^a	6.4	140	Lebofsky et al. (1978)
2061 Anza	$(0.037)^b$	(9.5)	(450)	Bowell et al. (1978)
2100 Ra-Shalom	0.04^a	3.4	21	Lebofsky et al. (1979)
1978 SB	$(0.04)^a$	(10.4)	(590)	this chapter
			1200	
S-type asteroids				
887 Alinda	0.16_5^a	4.1	37	Zellner et al. (1974)
1620 Geographos	0.21^c	2.0	4.2	Zellner et al. (1974)
1627 Ivar	$(0.14)^b$	(7.0)	(180)	Bowell et al. (1978)
1685 Toro	0.14_7^c	6.0	110	Zellner et al. (1974)
			330	
Asteroids with UBV colors outside the C and S fields				
1566 Icarus	0.17_9^c	1.4	1.4	Zellner et al. (1974)
1865 Daedalus	$(0.14)^b$	3.3	19	Bowell et al. (1978)
2062 Aten	0.20^a	0.9	0.4	Morrison et al. (1976)
1976 UA	$(0.14)^b$	(0.2)	(0.004)	Bowell et al. (1979)
1978 CA	0.06^d	1.9	3.6	Lebofsky et al. (1979)
1978 DA	0.16^d	0.9	0.4	Lebofsky et al. (1979)
			25	

[a] p_v based on polarimetry and radiometry
[b] p_v assigned on the basis of UBV colors
[c] p_v based on polarimetry
[d] p_v based on radiometry

V. SOURCES OF EARTH-CROSSING ASTEROIDS

The origin of Earth-crossing asteroids has been a subject of long-standing debate. As first noted by Öpik (1951), the Apollo asteroids are not likely to have survived as Earth-crossing objects from the time of formation of the planets; their average lifetime against collision with the planets or ejection from the solar system is very much less than the age of the system. The initial population would necessarily have to be enormous and there would have been a steady, roughly exponential decline in the population throughout geologic time. The impact record of the earth and moon shows that such a decline did not occur in the last ~3 × 10^9 yr. This same argument also applies to the Aten asteroids, which have still shorter lifetimes than Apollos. A few Earth-crossing Amors, with dynamic lifetimes on the order of 10^9 yr or longer could conceivably be surviving Earth planetesimals that have remained in the earth's neighborhood for 4.5 aeons. The vast majority of all Earth crossers, however, must have been injected from other regions of space into Earth-crossing orbits at relatively late times in solar system history.

Earth-crossing asteroids almost surely are of diverse origin; some probably have been derived from a residual population of old Mars crossers, some from widely separated regions of the main asteroid belt, and some from short-period comets. Öpik (1963) and Wetherill and Williams (1968) found that the population of Mars crossers would have to be several hundred times greater than the population of Apollos in order to maintain the Apollo population in quasi-steady state by deflection of Mars crossers. The actual ratio of Mars crossers to Apollos plus Atens, however, appears to be in the range of 10 to 60 (Shoemaker and Helin 1977; Helin and Shoemaker 1979). Part of the present population of Mars crossers, moreover, must consist of objects that have been injected into Mars-crossing orbits late in geologic time from the same source regions that yield the majority of Earth-crossing asteroids. Thus only a few percent of Apollos and Atens and a somewhat larger fraction of Earth-crossing Amors can be derived from the steadily dwindling supply of old Mars crossers (surviving Mars planetesimals).

Regions of the main asteroid belt bordering low-order commensurabilities and secular resonances are likely source regions for Earth-crossing asteroids. Large amplitude oscillations of certain orbital elements occur when asteroids are placed deep in these resonances; most such asteroids become Mars crossers and some would be occasional Earth crossers. The Kirkwood gaps at 2:1, 5:2, and 3:1 commensurabilities with Jupiter and zones centered on $\dot{\nu}_5$, $\dot{\nu}_6$ and $\dot{\nu}_{16}$ secular resonances of Williams (1969) are all greatly depleted in asteroids, a circumstance which suggests that asteroids have, indeed, been removed as a consequence of planetary encounters (Williams 1971). Zimmerman and Wetherill (1973), Williams (1973 *a, b*) and Scholl and Froeschlé (1977) have shown how meteorites could be delivered to Earth as collision fragments injected into the resonances from asteroids on stable orbits bordering either the 2:1 or 5:2 Kirkwood gaps or the $\dot{\nu}_5$ or $\dot{\nu}_6$ secular resonances. More recently Wetherill (1977, 1979) and Wetherill and Williams (1979) have examined the interplay between oscillation of orbital elements caused by the $\dot{\nu}_6$ resonance and the changes in orbit caused by close encounter with Mars. They found that collision fragments from asteroids bordering the $\dot{\nu}_6$ resonance can be delivered fairly efficiently into Earth-crossing orbits as a result of this interplay. Many objects initially injected into a shallow part of the resonance are later placed deep in the resonance as a result of Mars encounters (see the chapter by Wasson and Wetherill).

Once an asteroid becomes Earth-crossing, further orbital changes due to encounters with Earth and Venus tend to obscure its origin. It is of interest, nonetheless, to examine the present relationship of the known Earth crossers to resonances, in order to see whether these relationships provide any clues as to origin. Three Earth-crossing Amors currently librate around the 3:1 commensurability with Jupiter, but no known Earth crossers are near the 2:1 or 5:2 commensurabilities. This suggests that the 3:1 commensurability may play a substantial role in transferring asteroids into Earth-crossing orbits, although the precise mechanisms have not been studied. Probably there is a "synergistic" interaction between the resonant perturbations of the 3:1 commensurability and changes in orbit due to Mars encounter, as found by Wetherill (1977, 1979) for the $\dot{\nu}_6$ secular resonance. Regions of the main asteroid belt bordering both the 3:1 and 5:2 commensurabilities may be significant sources of Earth crossers.

With regard to the secular resonances, five Earth crossers are deep in the $\dot{\nu}_5$ resonance, one is in the $\dot{\nu}_{16}$ resonance, but none are known in the $\dot{\nu}_6$ resonance. Unlike the $\dot{\nu}_6$ resonance, $\dot{\nu}_5$ extends far into the region of the terrestrial planets. The occurrence of an Earth crosser in $\dot{\nu}_5$ does not, therefore, necessarily indicate its place of origin, especially for objects of small semimajor axis such as Icarus, Ra-Shalom and 1978 CA. These asteroids probably were placed in the $\dot{\nu}_5$ resonance by close planetary encounters at a late stage in their orbital evolution. On the other hand, Betulia and 1974 MA, which are also in $\dot{\nu}_5$, have semimajor axes of 2.20 and 1.76 AU, within the inner part of the main asteroid belt. Hence it is possible that they were derived from regions in the main belt close to $\dot{\nu}_5$. Daedalus is located in the $\dot{\nu}_{16}$ resonance, 1977 HA is close to $\dot{\nu}_{16}$ and 1974 MA is not only in $\dot{\nu}_5$ but also close to $\dot{\nu}_{16}$. It is not yet clear how $\dot{\nu}_{16}$ might have assisted the transfer of these objects into Earth-crossing orbits, but the asteroid belt is known to be depleted in the vicinity of $\dot{\nu}_{16}$. Probably the $\dot{\nu}_5$, $\dot{\nu}_6$ and $\dot{\nu}_{16}$ secular resonances all play significant roles in the transfer of collision fragments from main-belt asteroids into Earth-crossing orbits. On the basis of the present relationship of known Earth crossers to the resonances, $\dot{\nu}_5$ appears to be at least as important as $\dot{\nu}_6$ in the orbital evolution of Earth crossers, even though the density of potential parent belt asteroids is greater along the margin of $\dot{\nu}_6$.

From a quantitative assessment of the production of collision fragments, as well as the efficiency of their transfer to Earth-crossing orbits, Wetherill (1979) concluded that the best estimate of the yield of objects from the margin of the $\dot{\nu}_6$ resonance is an order of magnitude less than that required to maintain the Earth-crossing asteroid population. If all plausible sources in the main belt are considered, perhaps several tens of percent of the Earth crossers can be accounted for as collision fragments of belt asteroids. Most of the remaining Earth-crossing asteroids probably are derived from extinct short-period comets.

Most short-period comets have extremely short dynamical lifetimes, because they are Jupiter-crossing. Therefore they are unlikely to be captured by encounters with the terrestrial planets into orbits like those of the Earth-crossing asteroids. A few comets have aphelia inside the orbit of Jupiter, however, and one, P/Encke, has an aphelion as small as 4.1 AU, so that it does not pass within the sphere of influence of Jupiter. Evidently P/Encke has arrived in this orbit safe from Jupiter encounter as a consequence of nongravitational forces arising from the evaporation of volatile constituents during perihelion passages (Sekanina 1971). Historical decay of the nongravitational acceleration of P/Encke suggests that it might become extinct in a period as short as 60–70 yr (Sekanina 1972) leaving a kilometer-sized inactive nucleus that will be observed in the future as an Earth-crossing asteroid. A few other comets, in less stable orbits, also appear to be nearly extinct, including P/Arend-Rigaux and P/Neujmin 1 (see Kresák's chapter). Although the Jupiter-crossing object Hidalgo has been asteroidal in appearance at all times that it has been observed, its unusual orbit indicates that it is very probably an extinct comet. Hence there is little doubt that some comets can evolve into asteroidal objects and that a few can be placed in orbits like those of the Earth-crossing asteroids. As found by Wetherill (1979) from Monte Carlo studies, when the orbit of P/Encke is chosen as a model starting orbit, further orbital evolution resulting from close encounters with the terrestrial planets produces an equilibrium distribution of orbits like that observed among the Earth-crossing asteroids.

The supply of comets which become extinct in orbits safe from Jupiter encounter appears to be adequate to maintain the population of Earth crossers. That one such comet is observed in the process of decaying into an Earth-crossing asteroid is evidently a matter of luck. On the average, only one comet like P/Encke is required every few tens of thousands of years to sustain the Earth-crossing asteroid population, whereas the average period of activity of short-period comets may be no greater than a few thousand years.

At least two Earth-crossing asteroids, 1978 SB and 1973 NA, are on "comet-like" orbits and may represent recent additions to the Earth-crossing asteroid population. The maximum aphelion of 1978 SB is 4.09 AU, like that of P/Encke, just inside the limit where it is safe from Jupiter encounter. As noted by Kresák (1977) the orbit of 1973 NA is comparable, in certain respects, to that of many periodic comets. There is a large periodic oscillation of eccentricity of 1973 NA; at times its aphelion exceeds 4.6 AU. Because of its high inclination and restricted range of ω, however, it is also safe from Jupiter encounter.

VI. EVIDENCE FOR FLUCTUATION OF THE EARTH-CROSSING ASTEROID POPULATION IN LATE GEOLOGIC TIME

The history of the Earth-crossing asteroid population is reflected in the geologic record of impact craters on the earth and on the moon. The population and associated impact rate must be translated into an equivalent cratering rate in order to interpret that record. To do this it is necessary to determine sizes and volumes for asteroids of a given magnitude and to make estimates of asteroid bulk densities. Kinetic energies of impact for asteroids of a given mass can then be solved, with the aid of Eq. (7), by taking proper account of the acceleration of the asteroids in the gravity fields of the earth and the moon. Finally, the kinetic energies are related to impact crater diameters through an appropriate scaling relationship.

Diameters of spherical bodies corresponding in brightness to asteroids of a given absolute magnitude are given by

$$\log d = 3.122 - 0.2V(1,0) - 0.5 \log p_v \tag{8}$$

where d is the diameter in km. The constant in this equation is based on $V = -26.77$ for the sun, as adopted by Gehrels $et\ al.$ (1964). For asteroids of $V(1,0) = 18$, Eq. (8) reduces to

$$\log d_{18} = -0.5 \log p_v - 0.472. \tag{9}$$

We will evaluate the cratering rate for three different assumptions about the distributions of p_v in the Earth-crossing asteroid population:

1. All the Earth crossers are bright (mean $p_v = 0.14$, equal to the mean for S-type asteroids.
2. All the Earth crossers are dark (mean $p_v = 0.037$, equal to the mean for C-type asteroids).
3. Half of the Earth crossers are bright ($p_v = 0.14$) and half are dark ($p_v = 0.037$).

This set of assumptions more than spans the plausible range of mean albedos for the Earth crossers. At $p_v = 0.14$, $d = 0.89$ km and at $p_v = 0.037$, $d = 1.73$ km, for asteroids of absolute visual magnitude 18.

Bulk densities may be estimated on the basis of analogies drawn between various types of asteroids and meteorites. The material of S-type asteroids is here assigned a density of 3.5 g cm^{-3}, comparable to the density of ordinary chondrites, and that of C-type asteroids 2.5 g cm^{-3}, comparable to the density of carbonaceous chondrites. A correction to the density is then applied to account both for void space due to brecciation and for nonspherical shapes of the asteroids.

Many or most asteroids in the size range of Earth crossers may be expected to contain significant void space due to brecciation resulting from collisions. Generally about 25% new void space, the so-called bulking factor, is created by fragmentation of rock in mining and quarrying operations. Observed bulking factors for ejecta from cratering experiments in dense rock range from 4 to 70%, with most values in the range of 20 to 30% (Nugent and Banks 1966; Frandsen 1969). A volume expansion of 28% was adopted as a best value for the ejecta from solid rock by Ramspott (1970). Comparable or somewhat smaller amounts of excess void space occur in uncemented breccias beneath impact craters, as indicated by gravity observations on both Earth and Moon (Innes 1961; Dvorak 1979).

Observations of Phobos provide the only available check on void space, presumably due to brecciation, in a body close to the size of Earth-crossing asteroids. Photometric observations of Phobos show that it has a low albedo and a spectral reflectance like that of the C-type asteroids and certain types of carbonaceous meteorites (Pang *et al.* 1978; Pollack *et al.* 1978). Hence it is reasonable to suppose that Phobos is composed of material with a density of $\sim$2.5 g cm^{-3}, like that of the largest C-type asteroid, Ceres (Morrison 1976; see the chapter by Schubart and Matson), or the meteorites with comparable optical characteristics. From data presented by Mason (1963), the mean density of CI meteorites is found to be 2.33 g cm^{-3}, that of CII meteorites 2.73 g cm^{-3}, and that of CIII meteorites 3.48 g cm^{-3}. Only the CI and CII meteorites are similar optically to Phobos, Ceres, or other C-type asteroids. Therefore we adopted 2.5 g cm^{-3}, a density intermediate between that of CI and CII meteorites, as the best estimate for the density of the material of all C-type objects. But the estimated density of Phobos is 1.9 $\pm$ 0.6 g cm^{-3} (Tolson *et al.* 1978). While this estimate just overlaps our adopted density of 2.5 g cm^{-3}, at one standard deviation, it is interesting that the central value is 24% below the adopted density, about what would be expected for a small brecciated body. The observed distribution of craters on Phobos (Thomas *et al.* 1979; see the chapter by Veverka and Thomas) suggests to us that it is thoroughly fragmented. A void space of 24% of the total volume is here taken as the best estimate of the bulking factor due to impact brecciation of small asteroids.

An average correction for the nonspherical shapes of small asteroids, taking into account the types of irregularities observed on Phobos and the evidence from lightcurves for marked elongation of some Earth-crossing objects, is roughly estimated at 8%. This correction may be thought of as "external void space." The total correction is equivalent to a reduction of the bulk density by 32%. Final calculated masses are 0.87 $\times$ 10^{15} g for S-type asteroids and 4.5 $\times$ 10^{15} g for C-type asteroids at V(1,0) = 18.

TABLE IV

Estimates of Present Cratering Rate on Earth

Calculated production of craters $\geqslant 10$ km diameter 10^{-14} km^{-2} yr^{-1}		Last 1/2 Gyr record on North America 10^{-14} km^{-2} yr^{-1}	
All objects bright ($p_V = 0.14$)	~ 1.5		
Half bright, half dark	~ 2.3	$1.4_4 \pm 0.4$	(Grieve and Dence 1979)
All objects dark ($p_V = 0.037$)	~ 3.5	2.2 ± 1.1	(Shoemaker 1977)
Equivalent cratering rate on Earth from last 3.3 Gyr record on Moon		0.3	(derived from Neukum *et al.* 1975)
		1.1 ± 0.5	(Shoemaker 1977)

Crater diameters are obtained from the scaling relationship

$$D_e = 74 \, W^{\frac{1}{3.4}} \tag{10}$$

where D_e is the rim diameter of the crater on Earth in meters, W is the kinetic energy of the asteroid in kilotons TNT equivalent (1 kt TNT equivalent = 4.185×10^{19} ergs), which is based on nuclear cratering experiments (Shoemaker *et al.* 1963; Shoemaker 1977). For craters on Earth larger than about 3 km, D_e given by Eq. (10) is multiplied by 1.3 to account for crater collapse. This is a conservative correction for crater collapse; a best value may be closer to 1.35 (Shoemaker 1962; 1977). Use of Eq. (10), including the correction for collapse, yields crater diameters within 5% of those obtained from the scaling relationship of Dence *et al.* (1977) for terrestrial craters larger than 3 km. For craters on the moon, the diameters given by Eq. (10) are scaled for the difference in gravitational acceleration (Gault and Wedekind 1977) by

$$D_m / D_e = (g_e / g_m)^{\frac{1}{6}} \tag{11}$$

where D_m is the rim diameter of a crater on Moon in meters, g_e is the surface gravity on Earth and g_m is the surface gravity on the moon. Correction for crater collapse is not required for most lunar craters smaller than 15 km diameter.

Cratering rates on Earth for different assumptions about the distribution of p_V are given in Table IV. The calculated cratering rates are based on the collision rates of various classes of Earth-crossing asteroids listed in Table II and an rms impact velocity, weighted by collision probability, of 20.1 km sec^{-1} for all classes of Earth crossers. The cumulative frequency of craters produced by asteroid impact was assumed to be proportional to $D^{-1.7}$, consistent with the observed distribution of post-mare lunar craters larger than 3 km diameter (Shoemaker *et al.* 1963; Baldwin 1971).

Assuming that half of the Earth-crossing asteroids are similar in albedo to *S*-type asteroids and half are similar in albedo to *C*-type, the estimated present production of craters $\geqslant$ 10 km diameter on Earth is $\sim$ 2.3 $\times$ 10^{-14}km^{-2}yr^{-1}. This cratering rate calculated from observations of Earth-crossing asteroids is essentially indistinguishable from the average production of impact craters in the last half billion years on North America estimated by Shoemaker (1977) from impact structures in the United States (Table IV). A somewhat lower rate, equivalent to 1.4$_4$ $\pm$0.4 $\times$ 10^{-14}km^{-2}yr^{-1} for craters $\geqslant$ 10 km diameter, was found by Grieve and Dence (1979) for the Phanerozoic crater record of the structurally stable part of North America.

The cratering rates estimated from asteroid data and from the Phanerozoic geologic record in North America have comparable uncertainties. Both types of estimates are likely to be minimum values, as they depend on completeness of survey of regions sampled. The cratering rate estimate assuming half of the Earth-crossing asteroids are bright and half are dark and the estimate from the 0.5 Gyr North American cratering record by Shoemaker are twice as high as the equivalent average cratering rate on Earth derived from the density of craters on 3.3 Gyr old surfaces on the moon adopted by Shoemaker (1977). Corrections are made in deriving the terrestrial cratering rate from the lunar record for differences in the capture cross-sections of the earth and moon, for differences in crater scaling in the gravity fields of the two bodies, and for collapse of craters >3 km diameter on the earth. The estimate of long-term average cratering rate derived from the lunar record is a maximum, owing to the fact that craters older than 3.3 Gyr may be erroneously counted as formed on 3.3 Gyr lava surfaces (Neukum *et al.* 1975). Neukum *et al.* (1975) estimate a density for craters $\geqslant$ 10 km diameter more than three times lower than the density adopted by Shoemaker (1977) for 3.3 Gyr lava surfaces. Hence the cratering rates estimated from the Earth-crossing asteroid population and the North American cratering record may be more than twice as high as the equivalent terrestrial cratering rate implied by the 3.3 Gyr lunar cratering record.

Because the estimates of the modern cratering rate obtained for the Earth from asteroid observations (which are consistent with the North America geologic record for the Phanerozoic) are conservative and are, therefore, minimum estimates, whereas the estimate by Shoemaker (1977) of the 3.3 Gyr average cratering rate obtained from the moon is a maximum and probably errs on the high side, the difference between the two probably should be regarded as significant. An increase in cratering rate in the last half billion years implies a corresponding increase in the population of Earth-crossing asteroids. If Earth-crossing asteroids are derived primarily from extinct comets, an increase in Earth-crossing asteroids suggests that there has been an increase in the flux of long-period comets crossing the orbit of Jupiter. A sudden increase in the Earth-crossing asteroid population might arise from a catastrophic collision in the asteroid belt that injected large numbers of fragments into one of the secular resonances or commensurabilities. Such a perturbation in the population would be expected to decay in times on the order of a few tens of millions of years, however, whereas the consistency between the present Earth-crossing asteroid population and the North American cratering record suggests the average level of the population has been near the present level for times of the order of several 100 million years. Hence a change in the comet flux seems a more likely explanation for the apparent change in Earth-crossing asteroid population.

Fluctuations in the flux of comets in the inner solar system probably reflect changes in the number or mass of stars passing near the Sun that perturb the Oort cometary cloud. On the basis of the difference between the late terrestrial cratering record and the 3.3 Gyr lunar record, we tentatively postulate that the average flux of stars in the solar neighborhood has been higher during the last several hundred million years than during the preceding $\sim$ 3 billion years.

Acknowledgments. We thank B. G. Marsden for providing us with unpublished results from his numerical integration of the motion of 1915 Quetzalcoatl and for many other courtesies that he has extended to us in the course of this work. E. Roemer kindly provided unpublished photographic magnitudes of comets observed by her over a period of nearly ten years while she was at the Flagstaff station of the U.S. Naval Observatory. G. W. Wetherill and S. J. Weidenschilling made many helpful suggestions in reviewing this manuscript. Part of this research has been carried out at California Institute of Technology and part is the result of one phase of research carried out at the Jet Propulsion Laboratory, both under NASA contracts.

Note added in proof. The errors in calculating the rate of advance of the apsides and the node rate of earth-crossing asteroids are sufficiently large that the identification of asteroids occurring in the secular resonances should be treated with caution. At the present time it is not known with certainty whether any of the earth-crossing asteroids occur in secular resonances.

REFERENCES

Arnold, J. R. 1964. *The Origin of Meteorites as Small Bodies, Isotopic and Cosmic Chemistry*, N. Holland Publishing Co., (Amsterdam: North Holland Publ. Co.), pp. 347-364.

Arnold, J. R. 1965. The origin of meteorites as small bodies. 2. The model. *Astrophys. J.* 141: 1536-1547.

Baldwin, R. B. 1971. On the history of lunar impact cratering: The absolute time scale and the origin of planetesimals. *Icarus* 14: 36-52.

Bowell, E. 1979. 1978 SB. *IAU Circ.* 3284.

Bowell, E.; Chapman, C. R.; Gradie, J. C.; Morrison, D.; and Zellner, B. 1978. Taxonomy of Asteroids. *Icarus* 35: 313-335.

Bowell, E.; Chernykh, N. S.; Helin, E. F.; Kowal, C. T.; Marsden, B. G.; Niehoff, J. C.; Sebok, W. L.; Shoemaker, E. M.; Wetherill, G. W.; Williams, J. G.; and Zellner, B. 1979. Discovery and observations of asteroid 1976 UA. *Icarus* (to be submitted).

Danielsson, L., and Ip, W. H. 1972. Capture resonance of asteroid 1685 Toro by the earth. *Science* 176: 906-907.

Dence, M. R.; Grieve, R. A. F.; and Robertson, R. B. 1977. Terrestrial impact structures: Principal characteristics and energy considerations. In *Impact and Explosion Cratering: Planetary and Terrestrial Implications,* eds. D. J. Roddy, R. O. Pepin and R. B. Merrill (New York: Pergamon Press), pp. 247-275.

Dvorak, J. J. 1979. Analysis of small scale lunar gravity anomalies: Implications for crater formation and crustal history: Ph. D. dissertation, California Institute of Technology.

Frandsen, A. D. 1969. Engineering properties investigations of the Cabriolet Crater. *U.S. Army Engineers Nuclear Cratering Group Report* PNE-957.

Gault, P. E., and Wedekind, J. A. 1977. Experimental hypervelocity impact into quartz sand II. Effects of gravitational acceleration. In *Impact and Explosion Cratering,* eds. D. J. Roddy, R. O. Pepin, and R. B. Merrill (New York: Pergamon Press), pp. 1231-1244.

Gehrels, T.; Coffeen, T.; and Owings, D. 1964. Wavelength dependence of polarization. III. The lunar surface. *Astron. J.* 69: 826-852.

Grieve, R. A. F., and Dence, M. R. 1979. The terrestrial cratering record. II. The crater production rate. *Icarus* 38: 230-242.

Helin, E. F., and Shoemaker, E. M. 1977. Discovery of asteroid 1976 AA. *Icarus* 31: 415-419.

Helin, E. F., and Shoemaker, E. M. 1979. Palomar Planet-crossing Asteroid Survey, 1973-1978. *Icarus,* in press.

Helin, E. F.; Shoemaker, E. M.; and Wolfe, R. F. 1978. Ra-Shalom: Third member of the Aten class of Earth-crossing asteroids (abstract). *Bull Amer. Astron. Soc.* 10: 732.

Innes, M. J. S. 1961. The use of gravity methods to study the underground structure and impact energy of meteorite craters. *J. Geophys. Res.* 66: 2225-2239.

Ip, W. H., and Mehra, R. 1973. Resonances and librations of some Apollo and Amor asteroids with the earth. *Astron. J.* 78: 142-147.

Janiezek, P. M.; Seidelmann, P. D.; and Duncombe, R. L. 1972. Resonances and encounters in the inner solar system. *Astron. J.* 77: 764-773.

Jones, T. J., and Morrison, D. 1974. A recalibration of the radiometric/photometric method of determining asteroid sizes. *Astron. J.* 79: 892-895.

Kresák, L., 1977. Asteroid versus comet discrimination from orbital data. In *Comets, Asteroids, Meteorites,* ed. A. H. Delsemme (Toledo, Ohio: University of Toledo Press), pp. 313-321.

Lebofsky, L. A.; Lebofsky, M. J.; and Rieke, G. H. 1979. Radiometry and surface properties of Apollo, Amor, and Aten asteroids. *Astron. J.* 84: 885-888.

Lebofsky, L. A.; Veeder, G. J.; Lebofsky, M. J.; and Matson, D. L. 1978. Visual and radiometric photometry of 1580 Betulia. *Icarus* 35: 336-343.

Marsden, B. G. 1970. On the relationship between comets and minor planets. *Astron. J.* 75: 206-217.

Mason, B. 1963. The carbonaceous chondrites. *Space Sci. Rev.* 1: 621-646.

Morrison, D. 1976. The densities and bulk composition of Ceres and Vesta. *Geophys. Res. Lett.* 3: 701-704.

Morrison, D. 1977. Asteroid sizes and albedos. *Icarus* 31: 185-220.

Morrison, D.; Gradie, J. C.; and Reike, G. H. 1976. Radiometric diameter and albedo of the remarkable asteroid 1976 AA. *Nature* 260: 691.

Neukum, G.; König, B.; Fechtig, H.; and Storzer, D. 1975. Cratering in the Earth-moon system: Consequences for age determination by crater counting. *Proc. Lunar Sci. Conf. VI* (Oxford: Pergamon Press), pp. 2597-2620.

Nugent, R. C., and Banks, D. C. 1966. Engineering-geologic investigations, Project Danny Boy. *U.S. Army Engineers Nuclear Cratering Group Report PNE-5005.*

Öpik, E. J. 1951. Collision probabilities with the planets and distribution of interplanetary matter. *Proc. Roy. Irish Acad.* 54A: 165-199.

Öpik, E. J. 1963. The stray bodies in the solar system. Part 1. Survival of cometary nuclei and the asteroids: *Advan. Astron. Astrophys.* 2: 219-262.

Öpik, E. J. 1976. *Interplanetary Encounters: Close-Range Gravitational Interactions.* (New York: Elsevier).

Pang, K. D.; Pollack, J. B.; Veverka, J.; Lane, A. L.; and Ajello, J. M. 1978. The composition of Phobos: Evidence for carbonaceous chondrite surface from spectral analysis. *Science* 199: 64-66.

Pollack, J. B.; Veverka, J.; Pang, K. D.; Colburn, D.; Lane, A. L.; and Ajello, J. M. 1978. Multicolor observations of Phobos with the Viking lander cameras: Evidence for a carbonaceous chondritic composition. *Science* 199: 66-69.

Ramspott, L. C. 1970. Empirical analysis of the probability of formation of a collapsed crater by an underground nuclear explosion. *Univ. of Calif. Radiation Lab. Rept. UCRL-50883* (classified).

Scholl, H., and Froeschlé, C. 1977. The Kirkwood gaps as an asteroidal source of meteorites. In *Comets, Asteroids, Meteorites,* ed. A. H. Delsemme (Toledo, Ohio: University of Toledo Press), pp. 293-295.

Sekanina, Z. 1971. A core-mantle model for cometary nuclei and asteroids of possible cometary origin. In *Physical Studies of Minor Planets,* ed. T. Gehrels (NASA SP-267, Washington, D.C.: U.S. Government Printing Office), pp. 423-426.

Sekanina, Z. 1972. A model for the nucleus of Encke's Comet. In *The Motion, Evolution of Orbits, and Origin of Comets,* eds. G. S. Chebotarev and Kazimirchak-Polonskaya (Dordrecht: D. Reidel), pp. 301-307.

Shoemaker, E. M. 1962. Interpretation of Lunar Craters. In *Physics and Astronomy of the Moon,* ed. Z. Kopal (New York: Academic Press), pp. 283-359.

Shoemaker, E. M. 1977. Astronomically observable crater-forming projectiles. In *Impact and Explosion Cratering: Planetary and Terrestrial Implications,* eds. D. J. Roddy, R. O. Pepin, and R. B. Merrill (New York: Pergamon Press), pp. 617-628.

Shoemaker, E. M.; Hackman, R. J.; and Eggleton, R. E. 1963. Interplanetary correlation of geologic time. *Adv. Astronaut. Sci.* 8: 70-89.

Shoemaker, E. M., and Helin, E. F. 1977. Populations of planet-crossing asteroids and the relation of Apollo objects to main-belt asteroids and comets. In *Comets, Asteroids, Meteorites,* ed. A. H. Delsemme (Toledo, Ohio: University of Toledo Press), pp. 297-300.

Shoemaker, E. M., and Helin, E. F. 1978. Earth-approaching asteroids: Populations, origin and compositional types. NASA Conf. Publc. 2053, pp. 161-175.

Thomas, P.; Veverka, J.; and Chapman, C. R. 1979. Crater densities on the satellites of Mars. Submitted to *J. Geophys. Res.*

Tolson, R. H.; Duxbury, T. C.; Born, G. H.; Christensen, E. J.; Diehl, R. E.; Farless, D.; Hildebrand, C. E.; Mitchell, R. T.; Molko, P. M.; Morabito, L. A.; Palluconi, F. O.; Reichert, R. J.; Taraji, H.; Veverka, J.; Neugebauer, G.; and Findlay, J. T. 1978. Viking first encounter of Phobos: Preliminary results. *Science* 199: 61-64.

Wetherill, G. W. 1967. Collisions in the asteroid belt. *J. Geophys. Res.* 72: 2429-2444.

Wetherill, G. W. 1976. Where do the meteorites come from: A reevaluation of the Earth-crossing Apollo objects as sources of stone meteorites. *Geochim. Cosmochim. Acta.* 40: 1297-1317.

Wetherill, G. W. 1977. Fragmentation of asteroids and delivery of fragments to Earth. In *Comets, Asteroids, Meteorites,* ed. A. H. Delsemme (Toledo, Ohio: University of Toledo Press), pp. 283-291.

Wetherill, G. W. 1979. Steady state populations of Apollo-Amor objects. *Icarus* 37: 96-112.

Wetherill, G. W., and Williams, J. G. 1968. Evaluation of the Apollo asteroids as sources of stone meteorites. *J. Geophys. Res.* 73: 635-648.

Wetherill, G. W., and Williams, J. G. 1979. Origin of differentiated meteorites. In *Proc. 2nd Internat. Conf. on Origin and Abundance of the Elements,* ed. H. de la Roche (New York: Pergamon Press). In press.

Williams, J. G. 1969. Secular perturbations in the solar system: Ph.D. dissertation, University of California at Los Angeles.

Williams, J. G. 1971. Proper elements, families, and belt boundaries. In *Physical Studies of Minor Planets,* ed., T. Gehrels (NASA SP-267, Washington, D.C.: U.S. Government Printing Office), pp. 177-181.

Williams, J. G. 1973*a*. Meteorites from the asteroid belt? (abstract) *Eos: Trans. Amer. Geophys. Union* 54: 233.

Williams, J. G. 1973*b*. Secular resonances (abstract). *Bull. Amer. Astron. Soc.* 5: 363.

Williams, J. G. 1979. Classification of planet-crossing asteroids (abstract). *Lunar Sci. X.* Lunar and Planet. Inst. p. 1349.

Williams, J. G., and Wetherill, G. W. 1973. Physical studies of the minor planets. XIII. Long-term orbital evolution of 1685 Toro. *Astron. J.* 78: 510-515.

Zellner, B.; Gehrels, T.; and Gradie, J. 1974. Minor planets and related objects. XVI. Polarimetric parameters. *Astron. J.* 79: 1100-1110.

Zimmerman, P. D., and Wetherill, G. W. 1973. Asteroidal source of meteorites. *Science* 182: 51-53.

CHAOTIC ORBITS IN THE SOLAR SYSTEM

EDGAR EVERHART
University of Denver

Chaotic orbits are a class of unstable orbits of moderate inclination which change from one form to another. A typical chaotic orbit is described in detail. At times this orbit is like those of the short-period comets, at times it is like the orbit of Chiron, and in between it has a wide variety of different forms. Every object on a chaotic orbit is ultimately expelled to infinity. In the absence of dissipation or other nongravitational effects, there is no channel connecting chaotic orbits with the stable orbits of satellites.

These results on chaotic orbits are based on experience in numerical integration of many hundreds of randomly-started orbits, each followed for many revolutions — 3000 revolutions, or 10,000 or even 50,000 revolutions in some cases. In all, over a million orbit revolutions have been studied in a series of papers (Everhart 1973*a,b;* Oikawa and Everhart 1979). These are experimental results, not derived from theory.

There seem to be at least three general classes of orbits in the solar system, namely stable orbits, chaotic orbits and irregular retrograde orbits.

I. THE STABLE ORBITS

Stable orbits include the orbits of the planets, the orbits of most of the known satellites, the Jupiter Trojans, and (probably) the orbits of most asteroids in the main asteroid belt. There is no numerical integration, to our knowledge, demonstrating that any of these orbits are unstable. Of course, there is no entrance into and no exit from a stable orbit, otherwise it would not be stable. There may be other stable orbit forms.

II. THE CHAOTIC ORBITS

Chaotic orbits show an astonishing variety of forms, both quasi-stable and irregular, but they belong to the same class because there are evolutionary channels connecting all the chaotic orbit forms, with certain statistical probability. These include practically all unstable orbit forms where the inclination is small to moderate, such as:

(a)   Saturn Trojans

(b)   Jupiter and Saturn horeshoes

(c)   Generalized Trojans and generalized horseshoe orbits associated with Jupiter or Saturn

(d)   Long-lived circular orbits in two belts between the orbits of Jupiter and Saturn

(e)   A short-period orbit of oscillating inclination that lasts for hundreds of revolutions

(f)   Temporary satellite capture orbits around Jupiter and Saturn, which may be retrograde about the planet

(g)   Orbits similar to Chiron's present orbit

(h)   Librating orbits at submultiples and ratios to the period of a major planet

(i)   Long elliptical orbits

(j)   Near parabolic orbits or hyperbolas extending to or from infinity

(k)   Orbits making hyperbolic encounters with a planet

(l)   Other orbits too irregular to classify

We should note that classes (i), (j), (k) and (l) are not chaotic if they are of high inclination or retrograde, as discussed in Sec. III below. Some of these forms are well known, others are less so. All are described in Everhart (1973a). Lecar and Franklin (1973, see their Sec. III and Figs. 17-19) also have described the long-lived circular orbits.

A typical chaotic orbit is shown in Fig. 1, which is taken from Everhart (1973b). This is for a hypothetical object started in a circular orbit at 6.2 AU radius, inclination $7°.2$. Jupiter and Saturn were placed in fixed elliptical orbits at randomly chosen mean anomalies. The several orbital elements are plotted versus revolution number.

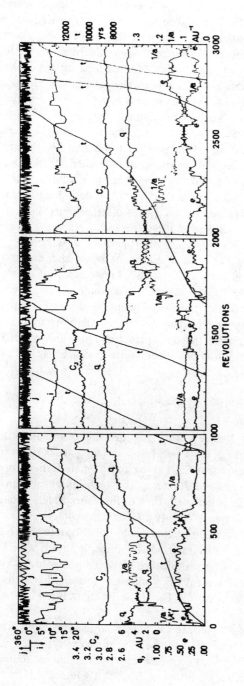

Fig. 1. A chaotic orbit interacting with Jupiter and Saturn.

The elements eccentricity e, perihelion distance q, Jacobi quantity C_j, inclination i labeled on the left, and reciprocal axis $1/a$ labeled on the right are self-explanatory. The time-line t traces 15,000 yr per sweep. Perhaps the only unfamiliar quantity is the Holetschek angle j plotted across the top edge. This is the planet-sun-object angle evaluated at each perihelion of the object.

After five close encounters with Jupiter, the perihelion q dropped to less than 2 AU by revolution 114, and the orbit became like that of an observable short-period comet. The inclination i oscillated in the range of 5° to 16°, as in form (e) listed above. A close encounter with Jupiter on the 478th revolution destroyed this pattern and raised the q-value to 4.2 AU. For the next 1000 revolutions the q-value continued to rise until it reached 10 AU. During this time the orbit was somewhat like the present orbit of Chiron. The detailed analysis of Fig. 1 continues in Everhart (1973b), describing a generalized Saturn horseshoe, a generalized Jupiter Trojan orbit, a Jupiter horseshoe, etc. Twice more during the 3000 revolutions the orbit reached a low enough perihelion placing the object in a typical short-period comet orbit.

Our recent study of the orbit of Chiron (Oikawa and Everhart 1979) shows this object to be in a chaotic orbit also. In one aspect of the study this orbit was integrated in 15th order taking into account all five outer planets. Figure 2 shows the variation of the Jacobi quantities during that integration.

The Jacobi quantity is invariant in the circularly-restricted 3-body problem. Thus if Jupiter were the only planet and its orbit about the sun were a circle, then a small third body would move in such a way as to keep C_j constant, where

$$C_j = a_j/a + 2[(q/a_j)(2-q/a)]^{\frac{1}{2}} \cos i. \tag{1}$$

Here a_j is the radius of Jupiter's orbit, a is the semimajor axis of the small body's heliocentric orbit, q is its perihelion distance and i its inclination. There is an exact expression, but Eq. 1 is the approximate form known as the Tisserand criterion. If one allows Jupiter's orbit to be elliptical then C_j is no longer constant. If Saturn is included, then C_j fluctuates widely. Nonetheless C_j is often used as a parameter to classify asteroids and comets.

Thus, C_j shows the variation of the Jacobi quantity referred to Jupiter, and C_s shows the same quantity referred to Saturn. If Saturn were the only planet and if it were in a circular orbit then C_s would be rigorously constant. We see that in the present many-body elliptical case neither C_s nor C_j is constant, but C_s varies less than C_j. This indicates that Chiron is now mostly under the control of Saturn.

There are two possible sources for objects in chaotic orbits where the storage time may be comparable to the age of the solar system. The first is Oort's cloud, the presumed source of new comets. The second source might be those few orbits of asteroids in the asteroid belt which are Mars crossers. Exceedingly rare close encounters with Mars might free some of these objects.

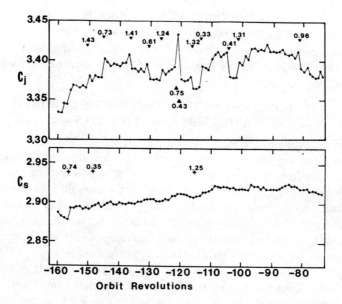

Fig. 2. A plot of the Jacobi quantities C_j referred to Jupiter and C_s referred to Saturn for an accurate integration. The object at the present epoch has the orbit of Chiron.

We would prefer to define chaotic orbits as those of objects *after* they have left these possible sources. The chance that an object in a chaotic orbit would re-enter one of these sources is not zero, but is exceedingly small.

Some generalizations can be made about chaotic orbits:

1. They are not stable.
2. In the absence of a nongravitational event, such as direct collision with another object or complete vaporization, they must ultimately exit to infinity on a hyperbola.
3. Chaotic orbits change rapidly and the exit to infinity occurs in a time very small compared to the age of the solar system.
4. The various forms of chaotic orbits are statistically connected by evolutionary channels of certain finite probabilities. For example, the chance that Chiron will some day be in an orbit similar to that of typical short-period comets is good. However, there is also a finite probability that the perturbations by Saturn or Jupiter will sooner eject Chiron on a hyperbolic orbit to infinity.
5. All chaotic orbits are heliocentric direct. We have seldom seen an inclination greater than $35°$, and usually it is less than $20°$. Never has a direct orbit evolved into a retrograde orbit.
6. One cannot use the Jacobi integral to classify orbits permanently. The values of C_s or C_j are of some use in describing an orbit for a short time.

III. THE IRREGULAR RETROGRADE ORBITS

There are also irregular and unstable retrograde orbits, such as those for the retrograde long-period comets. These are not evolutionarily connected with the chaotic orbits we have described and so must be members of this third class.

IV. CONCLUSIONS RELATING CHAOTIC ORBITS TO ASTEROIDS AND SATELLITES

If the orbits in the main asteroid belt are stable, then there would be no gravitational way that meteors or comets or other objects in chaotic orbits could come from the asteroid belt. Conversely, comets starting from chaotic orbits could not evolve to storage places in the asteroid belt. If one would maintain that asteroids are dead comets or that meteorites come from the asteroid belt, then one must show that the orbits in the main asteroid belt are not stable, or describe nongravitational mechanisms, such as collisions, which can remove pieces which then may move freely in chaotic orbits.

The satellites of any planet cannot be captured asteroids or comets by any gravitational process. There has been a considerable amount of work on this topic by several authors, which we will not review here, but the agreed result is that this capture is not possible in a purely gravitational solution (see the chapter by Carusi and Valsecchi). There must be considerable frictional drag before such a capture can take place. The only time in our opinion when such frictional drag could have been present was at the time of original formation and condensation from the solar nebula. The satellites of Mars, Phobos and Deimos (see the chapter by Veverka and Thomas), are not captured asteroids. They have been in orbit around Mars ever since the time of the solar nebula when the planets and sun were formed.

What is an asteroid? Is it an object with certain physical characteristics or one with certain orbital characteristics? Why cannot Phobos and Deimos have the same condensational history, the same cratering history, the same cosmic ray bombardment, the same kinds of rocks, etc. without being in the same orbit pattern as an asteroid? The orbit pattern has very little to do with the physical history (except for collisions) at the same distance from the sun. If we classify objects by their orbits then these satellites of Mars were never asteroids, at least not since the solar nebula condensed. But they may well be physically typical of other minor bodies, such as asteroids.

REFERENCES

Lecar, M. and Franklin, F. A. 1973. On the original distribution of the asteroids. I. *Icarus* 20: 422-436.

Everhart, E. 1973a. Horseshoe and Trojan orbits associated with Jupiter and Saturn. *Astron. J.* 78: 319-328.

Everhart, E. 1973b. Examination of several ideas of comet origins. *Astron. J.* 78: 329-337.

Oikawa, S., and Everhart, E. 1979. The past and future orbit of UB 1977, object Chiron. *Astron. J.* 84: 134-139.

DYNAMICAL INTERRELATIONS AMONG COMETS AND ASTEROIDS

LUBOR KRESÁK

Astronomical Institute of the
Slovak Academy of Sciences

Differences in the nature and origin of comets and asteroids are reviewed. The presence of an outer (with respect to the planetary region) source of comets and their limited active lifetimes, in contrast to the asteroid system stabilized on a much longer time scale, substantiates dynamical criteria for discriminating between these two kinds of objects even when the cometary nuclei become extinct. In general, the dividing line is set by the possibility or impossibility of closer planetary encounters. The minimum attainable distance from Jupiter, the aphelion distance, and the Tisserand invariant relative to Jupiter, yield useful quantitative criteria; the latter two, however, only in conjunction with the resonance and libration effects. It is attempted to specify those comets which can evolve, after final deactivation, into objects resembling genuine asteroids and those asteroids which appear more like extinct cometary nuclei. Such objects are rare but the range of their orbital characteristics is considerable, covering the whole boundary between the dynamical evolutionary paths of comets and the asteroid system. Possible evolutionary links between the marginal objects are discussed.

I. DIFFERENCES IN ORIGIN AND EVOLUTION

The basic difference between the two kinds of larger interplanetary objects, the asteroids and the comets, consists in the different place of their origin. Present asteroids and their predecessors formed in the inner part of the solar nebula, in the transition zone between the terrestrial and giant planets (see chapters by Safronov and by Cameron in this book). Even if similar objects did also accrete beyond the outskirts of the present asteroid belt and the Trojan clouds, they would have been swept off rather early by perturbations (Lecar and Franklin 1974; Froeschlé and Scholl 1979). Mutual collisions and encounters with planets would have fragmented and depleted the original population considerably before the present, rather stabilized state was reached (Chapman 1977, 1978; see also chapter by Davis *et al.*).

The comets would have formed farther away from the sun, in a region screened by the matter interposed between the inner and outer planets. The exact place of their origin is not yet specified, with alternatives ranging from the pre-planetary ring of Jupiter through the Uranus-Neptune region to the outskirts of the solar nebula and associated interstellar clouds (Öpik 1973; Whipple 1972; Safronov 1972; Cameron 1973*a,b;* Donn 1976). The main difficulty with the inner source is that there is too much wastage in overshooting during the transport of comets into the Oort cloud, which would require an excessive original mass — a drawback which can be mitigated by assuming two or more major mass-loss events in the sun's early evolution (Dermott and Gold 1978). The main argument against the outer source is the very low environmental density for the growth of cometary nuclei on a reasonable time scale. Conjectures of a recent or perpetual formation of comets within the planetary system (Mendis and Alfvén 1976; Vsekhsvyatskij 1977; Van Flandern 1977) receive little acceptance. For criticism of these contradictory alternatives see arguments by, e.g., Öpik (1973), Delsemme (1977) or Safronov (1977).

In any case, at present the comets form a large, rather homogeneous spherical cloud surrounding the planetary system, with a volume at least 10^{13} times larger than the asteroid ring. The depletion of the original population must have been even more drastic than in the case of asteroids, and is still continuing both by decay of individual objects and by their outward diffusion. The early periods of saturation (Whipple 1975) have been superseded by a quasi-steady state in the inner solar system. In contrast to the asteroids, the steadiness is not controlled by long survival times of a limited population but by the presence of an abundant external source (the Oort cloud) maintaining equilibrium with the dying-out individual objects in the planetary region.

The main imprint of the different place of origin and subsequent evolution is the higher density, the higher crushing strength, and the absence of volatiles on the asteroids. It is the volatile content of cometary nuclei which provides the most apparent distinction. Due to seasonal heating on the perihelion arc of the orbit, strong melting and outgassing takes place, with solid meteor particles ejected into interplanetary orbits and fine dust removed from the solar system by differential radiation pressure. The luminosity and diffuseness of the surrounding coma make a comet more easily detectable than an asteroid of similar size, and readily distinguishable from it.

At the same time, this response to solar radiation prevents the cometary nucleus from being directly observed because it is obscured by the coma. In spite of repeated attempts to distinguish the light reflected from the nuclear surface, we still do not have any dependable evidence regarding its optical properties. When a comet eventually becomes deactivated by having lost all volatiles or having developed a thick insulating crust, its motion remains the only means of discrimination from genuine asteroids. However, this situation may never occur, as a cometary nucleus can also disintegrate completely at the end of its active lifetime. Under these conditions, the interrelation of comets and asteroids is essentially a problem of the ultimate evolutionary phase of comets, of the size of their remnants, and of the orbits in which they move.

II. DYNAMICAL DIFFERENCES BETWEEN ASTEROIDS AND COMETS

A straightforward consequence of the presence of an outer source of comets (with respect to the planetary region where they are observed), and of the absence of an analogous source for the asteroids, is that the former move in unstable orbits and the latter in stable ones (see chapter by Everhart). The comets enter the planetary region in erratic orbits and their retention there is controlled by perturbational decelerations experienced during encounters with the planets. A comet may approach the planets repeatedly and be ejected again. While the nongravitational effects of progressive mass loss may assist some comets to settle in relatively stable orbits, the time scale of this process would be much too long in most cases compared with the active lifetime of the comet. The situation is the reverse for the asteroids. They can be injected into unstable orbits by collisions — but such events are apparently rare — partially destructive, and easily followed by ejection from the asteroid belt.

Orbital stability of both the comets and the asteroids is dominated by Jupiter. As documented by an analysis of the motion of 400 long-period comets (Everhart and Raghavan 1970), Jupiter was the main contributor to the change of the total energy, $1/a$, in 92% of cases, and Saturn in the remaining 8%. Moreover, even these 8% were due to Jupiter's perturbations being extraordinarily weak or cancelled out along the trajectory. Only in one case out of 400 was the effect of Saturn slightly stronger than the average effect of Jupiter! The outer planets may dominate if the perihelion distance is on the order of 10 AU; but when the comet is captured into a short-period orbit, the predominance of Jupiter becomes still more pronounced. As Jupiter is by far the most effective perturber of the asteroid system, significant results can be obtained by the restricted three-body approximation Sun-Jupiter-asteroid. The eccentricity of Jupiter's orbit would generally introduce more serious deviations from this simplified model than the presence of the other planets.

The simplest way to ensure stability is to avoid close approaches to Jupiter by locating the aphelion far enough *inside* Jupiter's orbit. Indeed, no known comet, but 96% of the numbered asteroids, have aphelion distances 1 AU less than that of Jupiter. An alignment of the lines of apsides, which impresses on the asteroid ring the eccentricity and orientation of the orbit of Jupiter, and the effect of inclination tend to increase the actual minimum separation distance appreciably.

Yet an approach to, or an intersection with, the *orbit* of Jupiter does not mean that encounters are really possible. An approximate resonance with the revolution period of Jupiter produces a semi-regular pattern of the jovicentric trajectories, with the recurrence of certain configurations and the absence of others. These oscillations are either stable or unstable according to the extremes of the libration argument, which in turn determine the minimum distance from Jupiter. A number of comets are librating around the resonance of 2:1 (Franklin *et al.* 1975), but these librations are invariably unstable, induced and destroyed again within a few centuries as the comets make close approaches to Jupiter. Even while the libration persists, approaches to within 0.5 AU of Jupiter would often occur near the extremes of the libration argument. Rather surprisingly, a better protection against encounters with Jupiter was established for two comets with resonance ratios 7:4 (P/Arend-Rigaux: closest approach distance $\rho > 0.9$ AU for 1000 years) and 2:3 (P/Neujmin 1: $\rho > 1.1$ AU for 4500 years). Marsden (1970) to whom most of the result on the long-term behavior of short-period comets are due, suggested that this type of motion may be characteristic of the transition phase between comets and asteroids.

The situation is entirely different for the asteroids. From among the 92 numbered objects with $\Delta = q_J - Q < 1.0$ (i.e., $Q > 3.95$), where Q is the aphelion distance of the asteroid and q_J the perihelion distance of Jupiter, 54 perform librations around the zero- or first-order resonances: 1:1 (22 Trojan asteroids), 4:3 (279 Thule), 3:2 (28 Hilda asteroids) and 2:1 (1362 Griqua, 1921 Pala, and 1922 Zulu). Only for two of these objects (334 Chicago and 1256 Normannia, both belonging to the Hilda group) the librations may be unstable (Schubart 1968). Librations other than about a period-to-period resonance can also become effective in increasing the minimum attainable distance from Jupiter, by keeping the aphelion point far from Jupiter's orbital plane at the critical moments (see chapter by Greenberg and Scholl). Such cases have been pointed out by Kozai (1962, 1979) and Marsden (1970). The argument of perihelion of 1373 Cincinnati librates with a small amplitude around a mean value of 90°: hence the aphelion, which can recede to the distance of Jupiter due to long-period perturbations, always remains far from the orbital plane of the perturbing planet. For other large-Q asteroids the circulation of the argument of perihelion can bring the aphelion into the node, but long-period perturbations let this only happen when the eccentricity and aphelion distance is small. Again, when the aphelion distance is large, the effect of inclination keeps the asteroid at a safe distance from Jupiter.

When all of these objects are subtracted, the remaining 34 numbered asteroids of $\Delta < 1$ exhibit a sharp cutoff just below $\Delta = 1$. In fact, one half of them have $0.93 < \Delta < 1.00$, and all but two have $0.78 < \Delta < 1.00$. The remaining two objects, 944 Hidalgo and 2060 Chiron, are quite peculiar; the latter cannot approach Jupiter either, as its perihelion distance is 3 AU larger than the aphelion distance of Jupiter. The change of the distribution with changing from the aphelion difference Δ to the minimum approach distance ρ is demonstrated by the upper two sections of Table I. The statistics include all asteroids as numbered by January 1, 1979, and all comets with well-determined orbits (Marsden 1975) extended to the same date.

In view of the dominant role of Jupiter in separating comets from asteroids it is natural to apply, as another criterion, the value of the Jacobian integral of the restricted three-body problem with Sun and Jupiter as primaries. If the semimajor axis of the orbit of Jupiter is adopted as a unit of length, the quantity which remains constant, the Tisserand invariant, reads

$$T = A^{-1} + 2 A^{\frac{1}{2}} \cos \phi \cos I \qquad (1)$$

where $A = a/a_J$, ϕ is the eccentricity angle ($e = \sin \phi$), and I is the inclination to the orbital plane of Jupiter. This expression obviously presumes the absence of perturbations by other planets and the absence of nongravitational forces. Furthermore, two secondary terms are neglected. One comes from the eccentricity of Jupiter's orbit (Vrcelj and Kiewiet de Jonge 1978), and becomes effective when significant perturbations of the elliptic heliocentric orbit occur at different true anomalies of Jupiter. The other is the part of the Jacobian integral with the jovicentric distance in the denominator, neglected in Tisserand's treatment. This becomes effective during close encounters but cancels out again afterwards, so that the heliocentric motion is only temporarily described by a reduced and variable value of T. In spite of these simplifications, the relative changes of T for observable short-period comets are normally 20 times smaller than those of their total energies, A^{-1}, and still 10 times smaller when orbital inclination is neglected by putting $\cos I = 1$ (Kresák 1972a).

Since the encounter velocity with respect to a circular motion of Jupiter, U, is given by

$$U = (3 - T)^{\frac{1}{2}} \qquad (2)$$

approaches to Jupiter are only possible, or can be induced by perturbations by the same planet, if $T < 3$ and if stable librations do not prevent this. As

TABLE I

Statistics for Δ, the Difference Between Perihelion Distance of Jupiter and Aphelion Distance of Comets or Asteroids; for ρ, the Closest Approach Distance to Jupiter; for T, the Tisserand invariant

	Comets		Asteroids		
	P > 20 yr	P < 20 yr	resonance 1:1	resonance 4:3, 3:2, 2:1	other
Δ < 0.00	561	83	22	1	2
0.00 < Δ < 0.25	0	9	0	12	0
0.25 < Δ < 0.50	0	4	0	11	1
0.50 < Δ < 0.75	0	0	0	7	2
0.75 < Δ < 1.00	0	1	0	1	33
1.00 < Δ	0	0	0	0	2026
ρ < 0.80		94	0	0	1
0.80 < ρ < 1.00		2	0	0	0
1.00 < ρ < 1.20		1	0	2	1
1.20 < ρ		0	22	30	2062
T < 2.90	561	74	10	1	2
2.90 < T < 2.95	0	12	3	2	0
2.95 < T < 3.00	0	5	9	7	2
3.00 < T < 3.05	0	6	0	19	15
3.05 < T	0	0	0	3	2045

demonstrated by the lower part of Table I, the definition of a *cometary* orbit as one of $T < 3$ without resonance, and of an *asteroidal* orbit as one which either has $T > 3$ or librates around a simple resonance ratio, sets a very good dividing line between the two populations. Passages across this dividing line can only take place by interference of another planet (e.g., by deceleration of a comet or acceleration of an asteroid at rare close encounters with inner planets near perihelion) or by onset of, or escape from, stable librations (e.g., due to collisions or nongravitational effects of mass loss). The evolutionary nongravitational effects in comets seem to make transitions into asteroidal orbits easier than vice versa. At the same time, the irregular character of these effects makes it impossible to trace the motion of an active comet over a longer time span with the required accuracy, to be sure about the degree of stability.

Figure 1 shows all numbered asteroids and all known Amor objects, Apollo objects, and short-period comets plotted in a diagram of semimajor axis vs. eccentricity. The regions of high concentration of the asteroids are only indicated by their boundaries. The critical value of $T_0 = 3$ (T approximated by putting cos $I = 1$) is represented by the dashed line separating the cometary region (B + C) from the planetary regions (A, D + E). A further subdivision delineates the following areas :

A : Transjovian region ($T_0 > 3$, $q > 5.2$).
B : Jupiter's domain of weak cometary activity ($T_0 < 3$, $q > 2.5$).
C : Jupiter's domain of strong cometary activity ($T_0 < 3$, $q < 2.5$).
D : Minor planet region ($T_0 > 3$, $1 < q < 5.2$).
E : Apollo region ($T_0 > 3$, $q < 1$).

The gaps separating individual populations are clearly seen, as well as the tendency to bridging of the comet/asteroid boundary in horizontal strips of low-order resonances, identified by the left-hand scale. In a three-dimensional representation, with orbital inclination as the third orthogonal co-ordinate, the cometary region B + C would be delineated by curved surfaces, slowly opening upwards. It is noteworthy, however, that the projection into the plane of $I = 0$ (i.e., the use of T_0 instead of T) makes the groups of related objects much more compact. For example, the range of T for the numbered Trojans is 2.67 to 2.99, while that in T_0 is 2.98 to 3.00 only.

III. DYING COMETS AND THEIR REMNANTS

There is no observational evidence that deactivated comets do leave remnants with longer survival times, large enough to be discovered. In fact, no object of asteroidal appearance has ever been found to move in an orbit typical for long-period comets. Estimates of the number of comets situated at any time within the distance of Jupiter indicate that less than 10% of them are of long period (Kresák and Pittich 1978; Kresák, 1979a). Hence, the

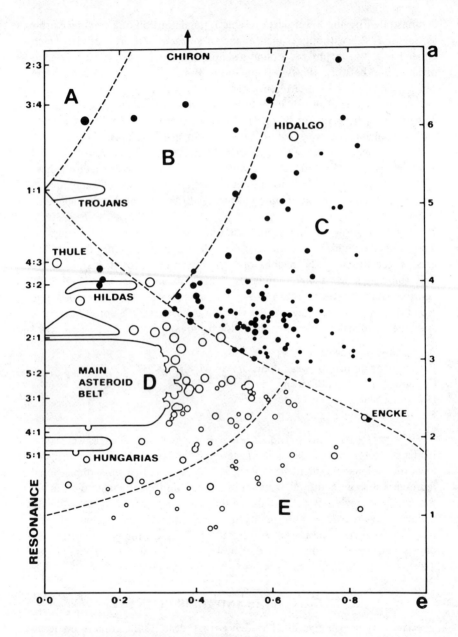

Fig. 1. Short-period comets (solid circles) and asteroids (open circles) plotted in a diagram of semimajor axis versus eccentricity. Increasing circle sizes distinguish the objects as follows : diameter less than 1 km or a lost object; diameter 1 to 3 km; diameter 3 to 10 km; diameter 10 to 30 km; diameter > 30 km. A indicates the transjovian Region, B Jupiter's domain of weak cometary activity, C Jupiter's domain of strong cometary activity, D the minor planet region, and E the Apollo region.

absence of asteroidal objects in nearly parabolic orbits appears less surprising than the lack of such objects moving in orbits similar to those of short-period comets. 944 Hidalgo is the only exception among 2118 numbered asteroids. The numbered asteroids are mostly objects with diameters exceeding 30 km. Therefore, in searching for extinct cometary nuclei we have to inquire what can be the upper limit of their sizes.

Our data on this issue are very meager because the history of reliable comet observations covers only 1/10,000 of the revolution period of a typical "new" comet coming from the Oort cloud. Moreover, all but a few of these comets escape detection, passing perihelia at great solar distances. A formal extrapolation of the absolute brightness distribution of long-period comets (Vsekhsvyatskij 1958) would predict the largest comets to be greater than Jupiter, if Oort's (1963) estimate of their total number is adopted, and still much greater than the earth for the lower estimate of Öpik (1973). Evidently, such an extrapolation is quite unrealistic, and an upper bound set by the original accretion process must exist. For the particular process of comet formation in pre-planetary rings according to Öpik (1973), the upper limit of nuclear diameter allowed by angular momentum is 40 km, and that allowed by dynamical acceleration versus damping is 60 km. This corresponds to an absolute magnitude H = 1 to 2. Vsekhsvyatskij's catalogue (1958) lists ten comets of $H < 1$ recorded during the last 600 years. It should be pointed out that two of them, P/Holmes in 1892 and P/Schwassmann-Wachmann 1, were short-period comets which increased their brightness enormously during repeated outbursts lasting two to three weeks. According to the observations in quiet periods (after having been missed in seven returns!) P/Holmes is a small object of less than 2 km in diameter, while P/Schwassmann-Wachmann 1 is the largest short-period comet on record, possibly about 20 to 25 km in diameter.

In most of the remaining eight cases of $H < 1$, similar surges of activity, albeit of longer duration, appear more likely than an excessive size of the nucleus. For comets 1402, 1500, 1744, and 1811 I, major irregular changes may be inferred from the comparison of individual magnitude estimates, and comets 1577 and 1747 were under observation for less than 3 and 4 months, respectively. Comet 1882 II, the brightest member of the Kreutz group, was undoubtedly very large; and Comet 1729, visible with unaided eye at a distance of 4 AU from the sun, was probably the largest comet ever observed. An analysis of more recent photometric data by Whipple (1979), made separately for the pre-perihelion and post-perihelion periods, includes four comets of $H < 1$. It is instructive to note that three of them, 1943 I, 1948 V and 1953 I, had a much lower post-perihelion absolute brightness (between H = 4 and H = 11 !) and that post-perihelion data are missing for the fourth comet, 1966 V. Thus 1729 and 1882 II are the only comets which appear to conflict with Öpik's estimates of the limiting nuclear diameter. The evidence is too poor to substantiate a definite revision, but it must be borne in mind that the sample of known comets is but a minute fraction of their total number, less than one millionth.

It is well established that the absolute brightness of "new" comets decreases steeply after the first passage near the sun (Whipple 1977). A pronounced brightness decrease also tends to follow each capture by Jupiter associated with a major reduction of perihelion distance (Kresák 1974). The long process of stepwise capture into short-period orbits (Everhart 1972, 1977) suggests that the faint short-period comets we observe are remnants of objects which were originally much larger, or much more resistant to break-up, than typical long-period comets. Accordingly, their chances of surviving as extinct asteroid-like nuclei appear much better. Parallel with this, the Tisserand invariants of short-period comets (median 2.79) are much closer to the asteroid limit than are those of the long-period comets (median 0).

Applying the dynamical criteria discussed in Sec. II, the following comets can be specified as potential candidates for evolving into extinct objects moving in asteroid-like orbits :

1. **Quasi-Hilda type** : P/Oterma (T = 3.04, Q = 4.53), P/Gehrels 3 (T = 3.03, Q = 4.65), P/Smirnova-Chernykh (T = 3.01, Q = 4.78). P/Oterma has already left this orbit, after an acceleration by Jupiter in 1963; P/Gunn (T = 3.00, Q = 4.73) and P/Longmore (T = 2.86, Q = 4.90) moved in similar orbits prior to the deceleration by Jupiter in 1965 and 1963, respectively. This type of motion arises when a comet is captured from a low-eccentricity orbit between Jupiter and Saturn (Region A in Fig. 1) and, restoring a high value of the Tisserand invariant after skipping over Region B, assumes an orbit with the aphelion well inside the orbit of Jupiter (Region D). The zone in which such transitions can occur is relatively narrow, running vertically along e = 0.15. Lower eccentricities are improbable due to the narrow range of encounter parameters resulting in nearly circular orbits; for higher eccentricities, unusually close encounters with Jupiter are necessary to produce the required change of energy, $1/a$. This is why these comets occupy, in the a/e phase-space, the region of the Hilda asteroids, and can become involved, like P/Oterma in 1937-1963, in a highly unstable 3:2 resonance. The similarity of the orbits is evident from Fig. 2. A substantial difference between the Hilda asteroids and these comets is that the former cannot approach Jupiter within less than ρ = 1.4 to 2.0 (Chebotarev and Shor 1978; Schubart 1979), whereas the three comets passed close to Jupiter recently : P/Oterma at a distance of 0.16 in 1936 and 0.10 in 1963 (Kazimirchak-Polonskaya 1972), P/Gehrels 3 at 0.0014 in 1970 and 0.04 in 1973, and P/Smirnova-Chernykh at 0.20 in 1955 and 0.47 in 1963 (Rickman 1979). Most of such comets will be ejected soon again, like P/Oterma, or perturbed into smaller orbits, like P/Gunn. However, a temporary capture into this type of orbit seems to be frequent indeed, betraying the existence of a ring of comets between the orbits of Jupiter and Saturn (Kresák 1972b). One can speculate whether or not a transition into a stable Hilda-type orbit

HILDA ASTEROIDS   **COMETS**

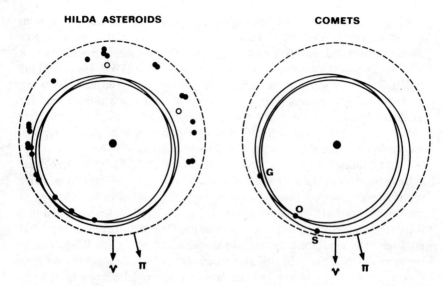

Fig. 2. A comparison of the orbits of Hilda asteroids and some comets. Dashed line −
the orbit of Jupiter, with the directions of vernal equinox and perihelion indicated by
arrows; dots − aphelia of stable librators; circles − aphelia of 334 Chicago and 1256
Normannia. The orbits of 1212 Francette, 1345 Potomac and 1439 Vogtia, as shown
on the left, reveal a striking resemblance to those of comets P/Oterma (O), P/Gehrels 3
(G) and P/Smirnova-Chernykh (S), shown on the right, except for the stability of
motion.

could sometimes occur. This would require that the comet avoids encounters
with Jupiter for a number of revolutions, during which nongravitational
effects assist the comet to reach a stable resonance.

2. **Intermediate type** : P/Gunn (T = 3.00, Q = 4.73),
P/Schwassmann-Wachmann 2 (T = 3.00, Q = 4.83), P/Clark (T = 2.99, Q =
4.69), P/Tempel 1 (T = 2.67, Q = 4.73), P/Tempel 2 (T = 2.97, Q = 4.68). Most
of these comets have probably experienced the same capture history as those
of the preceding group. A smaller size and a smaller perihelion distance can
make the nongravitational forces more effective, and a retrograde rotation of
the nucleus can make the apehlion recede from the orbit of Jupiter, and thus
stabilize the motion. It seems, however, that most of these comets will
become deactivated long before they reach a safe minimum distance from
Jupiter. In this case, and in the case of direct rotation, the evolution will be
very probably terminated by an ejection.

3. **P/Encke** : (T = 3.02, Q = 4.09). This is an extraordinary comet, and
gives us a unique demonstration that a change from a "cometary" into an
"asteroidal" orbit can really occur. The comet must have been originally very
large to survive its long dynamical history as an active object. It has
contributed an appreciable part of the interplanetary dust complex (Whipple

1967), including the annual showers of the nighttime and daytime Taurids and many bright fireballs. A study of secular perturbations of photographic meteors associated with this comet suggested that the direct parent body of some of them was a companion of P/Encke, now extinct (Whipple and Hamid 1952). The body which produced the well-known 1908 Tunguska event was probably another extinct fragment of P/Encke, not much smaller than the smallest known Apollo objects (Kresák 1978). After its complete deactivation, this comet will probably change into one or more asteroidal objects indistinguishable from the Apollo asteroids like 1978 SB.

4. **Librating comets** : P/Arend-Rigaux (T = 2.72, Q = 5.76) and P/Neujmin 1 (T = 2.16, Q = 12.16). On account of their relatively stable orbits avoiding approaches to Jupiter, very weak activity, and absence of measurable nongravitational effects, Marsden (1968, 1970) has suggested that these two comets may represent a transition phase between comets and asteroids. His arguments appear sound, especially as regards P/Neujmin 1. However, even if the orbits of these comets would remain stable over long time spans, they could not become typically asteroidal because of the low values of the Tisserand invariant. There is some orbital similarity between P/Neujmin 1 and the unique asteroid 944 Hidalgo.

IV. POSSIBLE EX-COMETS AMONG THE ASTEROIDS

1. **2060 Chiron** : (T = 3.36, Q = 18.88, q = 8.51). In this unique case it is unsuitable to speak about an ex-comet: if this object is of cometary nature, it has probably never been active. The asteroidal appearance gives no information on the presence of volatiles, in view of the large distance from the sun. The orbit is not unusual for the model evolutions of comets computed by Everhart (1977; see his chapter); while T is rather large (and incidentally equal to that of a typical main-belt asteroid) a stepwise capture by the outer planets would change this significantly. For example, if a capture by Saturn were to place the aphelion at the orbit of Saturn and the perihelion at the orbit of Jupiter, T would be reduced to a cometary value of about 2.9. Computations by Marsden (Kowal *et al.* 1979; see Kowal's chapter) and Scholl (1979) have demonstrated that close approaches to Saturn are inevitable on a time scale of 10^4 yr. Oikawa and Everhart (1979) have shown that there are 7 chances in 8 that Chiron's orbit will evolve inward so that it can interact strongly with Jupiter, and that there is an evolutionary path connecting its present orbit with the orbits of observable short-period comets. More of a problem is the great size of Chiron, for it is probably exceeded only by the three largest asteroids. Its diameter is apparently an order of magnitude greater than the upper limit accepted for cometary muclei by Öpik (1973). In view of the dissimilarity to any other asteroid or comet, a recognition of the nature of Chiron would not contribute too much to the problem of comet-asteroid relationships in the inner solar

system.

2. 944 Hidalgo : (T = 2.07, Q = 9.71, q = 2.01). This object moves in an orbit which is definitely more similar to those of some comets (P/Peters, P/Väisälä 1, P/Wild 1, P/Chernykh, P/Sanguin) than to that of any other asteroid. The motion is relatively stable but encounters with Jupiter are possible. As shown by Kozai (1979), the libration of Hidalgo's argument of perihelion is not dissimilar from the behaviour of a number of short-period comets. Attempts to detect any signs of cometary activity (Soderblom and Harlan 1976; see chapter by Degewij and van Houten) or of nongravitational dynamical effects associated with it (Marsden 1970) were unsuccessful. The surface properties of Hidalgo, with a resemblance to the D-type of asteroids and strong color variations (Degewij 1978; see chapter by Degewij and van Houten) appear peculiar, but not in the sense that is expected mostly for an extinct cometary nucleus. The diameter estimate of 39 km places the object within a reasonable size limit but well above any known short-period comet, inclusive of P/Schwassmann-Wachmann 1. This evidence is consistent with the absence of other similar objects, indicating that Hidalgo may be a remnant of an unusually large comet. A collisional ejection of an ordinary asteroid into this type of orbit cannot be ruled out entirely: yet Hidalgo is more likely to be an ex-comet than any other object we know.

3. The Trojans : (22 numbered objects of T = 2.67 to 2.99, Q = 5.31 to 5.96, q = 4.41 to 5.15; T for 1976 UQ is as low as 2.57). A stable resonance of 1:1 with libration cycles of about 150 years does not allow the Trojans to approach Jupiter to within ρ = 2.5 (Chebotarev *et al.* 1974). There is no reason to suppose that they are extinct comets. On the other hand, Rabe (1971, 1974) suggested that comets of the Jupiter family accreted, together with the Trojans, from primordial clouds associated with the triangular libration centers. Rabe advanced the following arguments for this opinion :

(a) The distributions of the Jacobi constants of the Trojans and short-period comets are almost identical.

(b) The 180° post-perihelion arc of Jupiter, which includes the denser Trojan cloud around the preceding libration center L5 (Gehrels 1977; see chapter by Degewij and van Houten) also contains more perihelia of short-period comets than the 180° pre-perihelion arc.

(c) The Jacobi constants of the comets with perihelia situated on Jupiter's post-perihelion arc are significantly lower.

A re-appraisal of the statistical arguments, using all data now available, shows that conclusions (b) and (c) are invalid. The greater abundance of cometary perihelia in the direction of Jupiter's post-perihelion arc can be fully accounted for by observational selection, and the distributions of T are essentially the same, with medians at 2.77 and 2.80, respectively (Kresák 1979*b*). There is no evidence of any known Trojan being able to escape from libration.

4. 279 Thule : (T = 3.03, Q = 4.40, q = 4.12). This is the only known librator in a resonance of 4:3, a relatively large object. It cannot approach Jupiter to within $\rho = 1.1$ (Marsden 1970). The possibility of a dynamical relation to comets is about the same as for the Hilda asteroids.

5. **The Hilda group** : (28 numbered objects of $T = 2.94$ to 3.06, $Q = 4.11$ to 5.10). A resonance of 3:2, with libration cycles of 250 to 300 years, does not allow the stable Hilda asteroids to pass within $\rho = 1.4$ of Jupiter (Chebotarev and Shor 1978). The interest in these objects is raised by the frequent intermingling of newly captured comets (see Figs. 1 and 2). Also, the population of this resonance is much more abundant than that of the neighbouring first-order resonances of 4:3 and 2:1. Libration preventing closer encounters with Jupiter is generally stable, the only doubtful cases being 334 Chicago and 1256 Normannia (Schubart 1968, 1979). Chicago can pass Jupiter within $\rho = 1.1$ (Marsden 1970). However, its low eccentricity and small aphelion distance makes the capture of a comet into a similar orbit very improbable.

6. **The Griqua group** : 1362 Griqua with $T = 2.96$, $Q = 4.40$; 1921 Pala with $T = 2.96$, $Q = 4.56$; and 1922 Zulu with $T = 2.72$, $Q = 4.81$; resonance 2:1, libration periods over 300 years). The position in the a/e plane, close to the main concentration of short-period comets (Fig. 1), and a pronounced cometary value of the Tisserand invariant, would classify Zulu as a first-rank candidate for a cometary origin. However, the resonance with Jupiter, combined with a high inclination, makes this object very successful in avoiding close encounters, the minimum attainable distance from Jupiter being as large as $\rho = 2.2$ according to Franklin *et al.* (1975), and $\rho = 2.6$ according to Schubart (1979). The other two asteroids cannot pass closer than $\rho = 2.0$ either. A search among unnumbered faint asteroids (Franklin *et al.* 1975) revealed three cases where much closer approaches are possible : 1968 HP ($\rho = 0.3$), PLS 9548 ($\rho = 0.6$) and PLS 2691 ($\rho = 0.8$). Unfortunately, the short observing time spans of $5 - 10$ days, with only 3 or 4 positions available, make the orbits of these objects too uncertain for any dependable conclusion.

7. **Other asteroids of T $<$ 3.0 or Q $>$ 4.2** : (1373 Cincinnati with $T = 2.70$, $Q = 4.51$; 692 Hippodamia with $T = 2.98$, $Q = 4.00$; 225 Henrietta with $T = 2.99$, $Q = 4.33$; 1006 Lagrangea with $T = 3.07$, Q 4.26). In each case where Q exceeds 4.2, an approach to Jupiter is prevented by some librational mechanism. Either the aphelion cannot approach the nodes at all, or the eccentricity drops to the minimum when this occurs (Kozai 1962; Marsden 1970). One can take the view that it is just this peculiar pattern of long-period perturbations which permits a transient and imperfect simulation of a cometary orbit. The frequency of these and similar types of orbits is greater among large asteroids (e.g., Nos. 31, 76, 319, 1036, 1144) which also makes an association with extinct comets improbable.

8. **Apollo and Amor objects** : ($T > 3.02$, $Q < 4.10$; $q < 1.00$ and $q <$ 1.30, respectively). Although the criteria for possible ex-comets mentioned in

Sec. II are rather unfavorable for these objects, there are several reasons for considering this possibility seriously :

(a) The dynamical criteria referred to Jupiter as the only significant source of perturbations can break down when encounters with other planets take place. Approaches to the terrestrial planets can decouple the aphelion from the orbit of Jupiter and increase the Tisserand invariant with respect to that planet.

(b) Earlier studies (Öpik 1963) suggested that the supply of new Apollo objects by Mars' perturbations acting on the inner boundary of the asteroid ring is much too small to account for the observed abundance of these objects, that have limited lifetimes. These conjectures have been somewhat weakened by considering a more remote source in the Kirkwood gaps. Nonetheless, a significant excess of the Apollo objects (especially relative to the Amor objects) is still disturbing (see chapter by Shoemaker *et al.*). This was originally the main reason for demanding a cometary origin for *most* of the Apollo objects (Öpik 1963), and even nowadays it is the main argument for defending this point of view (Wetherill 1976; see the chapter by Wasson and Wetherill).

(c) The density structure of the Apollo system suggests the presence of two distinct populations, one of which, the outer halo, might be of cometary origin (Kresák 1979*a*). A broad range of shapes (Marsden 1971) and surface textures (Gehrels 1971) is also indicative of an inhomogeneity.

(d) While most of the known main-belt asteroids appear much too large for extinct cometary nuclei, the sizes of the known Apollo objects are nearly what can be expected. Sekanina (1971) estimates the original diameters at the *beginning* of cometary activity, compatible with his core-mantle model, as 50 to 100 km on the average.

(e) Some meteor streams consisting, in all probability, of recent cometary ejecta move in orbits of Apollo type, and a compact meteor stream like the Geminids cannot have been perturbed into this type of orbit *after* the separation from its parent body (Kresák 1973). A few minor streams have been tentatively associated with individual Apollo asteroids (Marsden 1971; Sekanina 1973).

The most striking similarity between a cometary and an asteroidal orbit, that of P/Encke and 1978 SB, is shown in Fig. 3. Their numerical parameters compare as follows : T = 3.02 and 3.10, Q = 4.09 and 3.97, q = 0.34 and 0.36, i = 12.0° and 11.9°, difference of the perihelion longitude from that of Jupiter +146° and -138°. 1978 SB is probably somewhat larger than the nucleus of P/Encke. Another conspicuous unnumbered object is 1973 NA which, owing to an inclination of 68°, has a very low value of T = 2.54. The aphelion distance is 4.01. The only planetary encounters permitted by the present orientation of the orbit are those with the earth, at ρ = 0.1. The orbit determined from a 26-day arc does not seem to be accurate enough for assessing the long-term evolution.

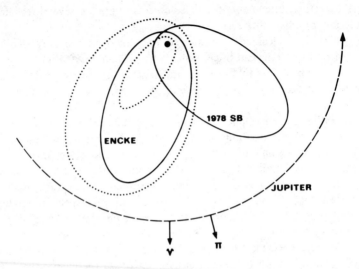

Fig. 3. A comparison of the orbits of the Apollo asteroid 1978 SB and Comet Encke. Dashed line − the orbit of Jupiter, with the directions of vernal equinox and perihelion indicated by arrows; dotted lines − the approximate boundaries of the Taurid meteor stream associated with P/Encke. (Adapted from Porubčan 1978.)

Among the numbered Apollo asteroids, the best candidates for a cometary origin seem to be 2101 Adonis, 1580 Betulia, 1866 Sisyphus, and 1981 Midas (Kresák 1977). Adonis is that asteroid for which a meteor stream association appears to be best established (Sekanina 1973), yet a chance coincidence cannot be ruled out. Extensive physical observations of Betulia during its 1976 apparition (Lebofsky et al. 1978; Tedesco et al. 1978) revealed a surprising number of unusual properties, but none of these can be definitely attributed to a cometary origin. Sisyphus and Midas are two of the four numbered asteroids for which Kozai (1979) finds the argument of perihelion to librate, a feature which is quite common among the comets.

The overlapping cometary and asteroidal characteristics of individual marginal objects are summarized in Table II. C means that the criterion yields a cometary value of the respective parameter, A an asteroidal value. CA and AC mean either a transitional value, or a group of objects extending over the adopted limit, with the dominating value given first. Objects having all characteristics the same, C or A, are omitted. The objects listed deserve special attention in physical observations.

When dealing with the interrelations among different types of interplanetary objects, the irregular satellites of the outer planets may not be omitted. As shown by Rickman (1979) and by Carusi, and Valsecchi this book), P/Gehrels 3 was moving in a satellite jovicentric orbit for more than 7 years during the transition into the Jupiter family of comets. In numerical

TABLE II

Statistics of Parameters that Appear to be Asteroidal (A) or Comet-like (C)

	Number of objects	Physical appearance	Aphelion distance	Planetary encounters	Tisserand invariant
Comets:					
Quasi-Hilda type	3	C	C	C	AC
Intermediate type	5:	C	C	C	CA
P/Encke	1	C	A	CA	AC
Librating comets	2	CA	C	CA	C
Asteroids:					
2060 Chiron	1	.	C	C	A
944 Hidalgo	1	A	C	C	C
Trojans	22	A	C	A	C
279 Thule	1	A	AC	A	AC
Hildas	28	A	C	A	AC
Griquas	3	A	CA	A	C
225, 692, 1006, 1373	4	A	AC	A	CA
Amors	15	A	A	CA	A
Apollos	14	A	A	CA	A
1973 NA	1	A	A	CA	C
1978 SB	1	A	A	CA	AC

experiments with simulated cometary orbits, a number of such temporary captures have been identified; among them, a satellite orbit around Jupiter persisting for 100 years (Carusi, and Pozzi 1979) and a satellite orbit around Saturn persisting for 80 years (Everhart 1973). Kazimirchak-Polonskaya (1974) has demonstrated the possibility of capture of a comet as a permanent satellite of Neptune. If this is possible, the reverse process can occur as well. Kowal *et al.* (1979; see Kowal's chapter) speculate about a possible genetic relation between 2060 Chiron and Saturn's satellite Phoebe.

However, the residence exchange with the satellite systems does not seem to be of major importance, at least at the present stage of the solar system's evolution. The number of smaller objects moving in planetocentric orbits is apparently a minute fraction of the population of the interplanetary systems of comets and asteroids. According to the computations of Everhart (1973; see Everhart's chapter in this book), captures into Trojan or horseshoe orbits − both by Jupiter and Saturn − are of much longer persistence, often on the order of $10^4 - 10^5$ yr. Consequently, librations avoiding planetary encounters may provide more effective means of temporary storage than planetocentric motions. Even when trapping into libration is a rare event, it may induce a long period of stability and prolong the survival time considerably. This makes a substantial difference between the occurrence rate of various evolutionary histories, and the proportion of objects found at any time at a stage characteristic for a given history.

REFERENCES

Cameron, A. G. W. 1973*a*. Accumulation processes in the primitive solar nebula. *Icarus* 18: 407-450.

Cameron, A. G. W. 1973*b*. The early evolution of the solar system. In *Evolutionary and Physical Properties of Meteoroids,* eds. C. L. Hemenway, P. M. Millman, and A. F. Cook (NASA SP-319, Washington, D.C.: U.S. Government Printing Office), pp. 347-353.

Carusi, A., and Pozzi, F. 1979. Planetary close encounters between Jupiter and about 3000 fictitious minor bodies. *Astrophys. Space. Sci.* (in press).

Chapman, C. R. 1977. The evolution of asteroids as meteorite parent bodies. In *Comets, Asteroids, Meteorites,* ed. A. H. Delsemme (Toledo, Ohio: Univ. of Toledo Press), pp. 265-275.

Chapman, C. R. 1978. Asteroid collisions, craters, regoliths, and lifetimes. In *Asteroids: An Exploration Assessment,* ed. D. Morrison, and W. C. Wells, NASA Conference Publ. 2053, pp. 145-160.

Chebotarev, G. A.; Belyaev, N. A.; and Eremenko, R. P. 1974. Orbits of Trojan asteroids. In *The Stability of the Solar System and of Small Stellar Systems,* ed. Y. Kozai (Dordrecht: D. Reidel), pp. 63-69.

Chebotarev, G. A., and Shor, V. A. 1978. The structure of the asteroid belt. *Fundamentals of Cosmic Physics* 3: 87-138.

Degewij, J. 1978. Photometry of faint asteroids and satellites, Ph. D. dissertation, Leiden Univ., pp. 71-74.

Delsemme, A. H. 1977. The origin of comets. In *Comets, Asteroids, Meteorites,* ed. A. H. Delsemme (Toledo, Ohio: Univ. of Toledo Press), pp. 453-467.

Dermott, S. F., and Gold, T. 1978. On the origin of the Oort cloud. *Astron. J.* 83: 449-450.

Donn, B. 1976. Comets, interstellar clouds and star clusters. In *The Study of Comets,* eds. B. Donn, M. Mumma, W. Jackson, M. A'Hearn, and R. Harrington (NASA SP-393, Washington, D.C.: U.S. Government Printing Office), pp. 663-672.

Everhart, E. 1972. The origin of short-period comets. *Astrophys. Lett.* 10: 131-135.

Everhart, E. 1973. Horseshoe and Trojan orbits associated with Jupiter and Saturn. *Astron. J.* 78: 316-328.

Everhart, E. 1977. The evolution of comet orbits as perturbed by Uranus and Neptune. In *Comets, Asteroids, Meteorites,* ed. A. H. Delsemme (Toledo, Ohio: Univ. of Toledo Press), pp. 99-104.

Everhart, E., and Raghavan, N. 1970. Changes in total energy for 392 long-period comets, 1800-1970. *Astron. J.* 75: 258-272.

Franklin, F. A.; Marsden, B. G.; Williams, J. G.; and Bardwell, C. M. 1975. Minor planets and comets in libration about the 2:1 resonance with Jupiter. *Astron. J.* 80: 729-746.

Froeschlé, C., and Scholl, H. 1979. New numerical experiments to deplete the outer part of the asteroid belt. *Astron. Astrophys.* 72: 246-255.

Gehrels, T. 1971. Future work. In *Physical Studies of Minor Planets,* ed. T. Gehrels, (NASA SP-267, Washington, D.C.: U.S. Government Printing Office), pp. 653-659.

Gehrels, T. 1977. Some interrelations of asteroids, Trojans and satellites. In *Comets, Asteroids, Meteorites,* ed. A. H. Delsemme (Toledo, Ohio: Univ. of Toledo Press), pp. 323-325.

Kazimirchak-Polonskaya, E. I. 1972. The major planets as powerful transformers of cometary orbits. In *The Motion, Evolution of Orbits, and Origin of Comets,* ed. G. A. Chebotarev, E. I. Kazimirchak-Polonskaya, and B. G. Marsden (Dordrecht: D. Reidel), pp. 373-397.

Kazimirchak-Polonskaya, E. I. 1974. Capture d'une comète par Neptune et son passage à l'orbite d'un satellite de la planète. In *Asteroids, Comets, Meteoric Matter,* eds. C. Cristescu, W. J. Klepczynski, and B. Milet (Bucuresti: Editura Academiei Republicii Socialiste Romania), pp. 205-221.

Kowal, C. T.; Liller, W.; and Marsden, B. G. 1979. The discovery and orbit of 2060 Chiron. In *Dynamics of the Solar System,* ed. R. L. Duncombe (Dordrecht: D. Reidel, in press).

Kozai, Y. 1962. Secular perturbations of asteroids with high inclinations and eccentricities. *Astron. J.* 67: 591-598.

Kozai, Y. 1979. Secular perturbations of asteroids and comets. In *Dynamics of the Solar System,* ed. R. L. Duncombe (Dordrecht: D. Reidel, in press).

Kresák, L. 1972a. Jacobian integral as a classificational and evolutionary parameter of interplanetary bodies. *Bull. Astron. Inst. Czechosl.* 23: 1-34.

Kresák, L. 1972b. On the dividing line between cometary and asteroidal orbits. In *The Motion, Evolution of Orbits, and Origin of Comets,* ed. G. A. Chebotarev, E. I. Kazimirchak-Polonskaya, and B. G. Marsden (Dordrecht: D. Reidel), pp. 503-514.

Kresák, L. 1973. The cometary and asteroidal origins of meteors. In *Evolutionary and Physical Properties of Meteoroids,* ed. C. L. Hemenway, P. M. Millman, and A. F. Cook (NASA SP-319, Washington, D.C.: U.S. Government Printing Office), pp. 331-341.

Kresák, L. 1974. The aging and the brightness decrease of comets. *Bull. Astron. Inst. Czechosl.* 25: 87-112.

Kresák, L. 1977. Asteroid versus comet discrimination from orbital data. In *Comets, Asteroids, Meteorites,* ed. A. H. Delsemme (Toledo, Ohio: Univ. of Toledo Press), pp. 313-321.

Kresák, L. 1978. The Tunguska object: a fragment of Comet Encke? *Bull. Astron. Inst. Czechosl.* 29: 129-134.

Kresák, L. 1979a. Three-dimensional distributions of minor planets and comets. In *Dynamics of the Solar System,* ed. R. L. Duncombe (Dordrecht: D. Reidel, in press).

Kresák, L. 1979b. On the distribution of aphelia of short-period comets. *Bull. Astron. Inst. Czechosl.* (in press).

Kresák, L., and Pittich, E. M. 1978. The intrinsic number density of active long-period comets in the inner solar system. *Bull. Astron. Inst. Czechosl.* 29: 299-309.

Lebofsky, L. A.; Veeder, G. J.; Lebofsky, M. J.; and Matson, D. L. 1978. Visual and radiometric photometry of 1580 Betulia. *Icarus* 35: 336-343.

Lecar, M., and Franklin, F. A. 1974. On the original distribution of the asteroids. In *The Stability of the Solar System and of Small Stellar Systems*, ed. Y. Kozai (Dordrecht: D. Reidel), pp. 37-56.

Marsden, B. G. 1968. Comets and nongravitational forces. *Astron. J.* 73: 367-379.

Marsden, B. G. 1970. On the relationship between comets and minor planets. *Astron. J.* 75: 206-217.

Marsden, B. G. 1971. Evolution of comets into asteroids? In *Physical Studies of Minor Planets*, ed. T. Gehrels, (NASA SP-267, Washington, D.C.: U.S. Government Printing Office), pp. 413-421.

Marsden, B. G. 1975. *Catalogue of Cometary Orbits* (Cambridge: Smithsonian Astrophys. Obs.).

Mendis, A., and Alfvén, H. 1976. On the origin of comets. In *The Study of Comets*, eds. B. Donn, M. Mumma, W. Jackson, M. A'Hearn, and R. Harrington (NASA SP-393, Washington, D.C.: U.S. Government Printing Office), pp. 638-659.

Oikawa, S., and Everhart, E. 1979. The past and future of 1977 UB, object Chiron. *Astron. J.* 84: 134-139.

Oort, J. H. 1963. Empirical data on the origin of comets. In *The Moon, Meteorites, and Comets*, eds. B. M. Middlehurst and G. P. Kuiper (Chicago: Univ. of Chicago Press), pp. 665-673.

Öpik, E. J. 1963. The stray bodies in the solar system. Part I. Survival of cometary nuclei and the asteroids. *Adv. Astron. Astrophys.* ed. Z. Kopal (New York: Academic Press), 2: 219-262.

Öpik, E. J. 1973. Comets and the formation of planets. *Astrophys. Space Sci.* 21: 307-398.

Porubčan, V. 1978. Dispersion of orbital elements within the Geminid and Taurid meteor streams. *Bull. Astron. Inst. Czechosl.* 29: 218-224.

Rabe, E. 1971. Trojans and comets of the Jupiter group. In *Physical Studies of Minor Planets*, ed. T. Gehrels, (NASA SP-267, Washington, D.C.: U.S. Government Printing Office), pp. 407-412.

Rabe, E. 1974. Perihelion longitudes and Jacobi constants of Jupiter group of comets as indicators of their possible origin from the two equilateral Trojan clouds. In *Asteroids, Comets, Meteoric Matter*, eds. C. Cristescu, W. J. Klepczynski, and B. Milet (Bucuresti: Editura Academiei Republicii Socialiste Romania), pp. 165-169.

Rickman, H. 1979. Recent dynamical history of the six short-period comets discovered in 1975. In *Dynamics of the Solar System*, ed. R. L. Duncombe (Dordrecht: D. Reidel, in press).

Safronov, V. S. 1972. Ejection of bodies from the solar system in the course of the accumulation of the giant planets and the formation of the cometary cloud. In *The Motion, Evolution of Orbits, and Origin of Comets*, ed. G. A. Chebotarev, E. I. Kazimirchak-Polonskaya, and B. G. Marsden (Dordrecht: D. Reidel), pp. 329-334.

Safronov, V. S. 1977. Oort's cometary cloud in the light of modern cosmogony. In *Comets, Asteroids, Meteorites*, ed. A. H. Delsemme (Toledo, Ohio: Univ. of Toledo Press), pp. 483-484.

Schubart, J. 1968. Long-period effects in the motion of Hilda-type planets. *Astron. J.* 73: 99-103.

Schubart, J. 1979. Asteroidal motion at commensurabilities treated in three dimensions. In *Dynamics of the Solar System*, ed. R. L. Duncombe (Dordrecht: D. Reidel, in press).

Sekanina, Z. 1971. A core-mantle model for cometary nuclei and asteroids of possible cometary origin. In *Physical Studies of Minor Planets*, ed. T. Gehrels, (NASA SP-267, Washington, D.C.: U.S. Government Printing Office), pp. 423-428.

Sekanina, Z. 1973. Statistical model of meteor streams. III. Stream search among 19303 radio meteors. *Icarus* 18: 253-284.

Soderblom, D. R., and Harlan, E. A. 1976. 944 Hidalgo. *IAU Circ. 3007*.

Tedesco, E.; Drummond, J.; Candy, M.; Birch, P.; Nikoloff, I.; and Zellner, B. 1978. 1580 Betulia : An unusual asteroid with an extraordinary lightcurve. *Icarus* 35: 344-359.

Van Flandern, T. C. 1977. A former major planet of the solar system. In *Comets, Asteroids, Meteorites,* ed. A. H. Delsemme (Toledo, Ohio: Univ. of Toledo Press), pp. 475-481.

Vrcelj, Z., and Kiewiet de Jonge, J. H. 1978. An invariant relation in the elliptic restricted problem of three bodies, I. *Astron. J.* 83: 514-521.

Vsekhsvyatskij, S. K. 1958. *Fizicheskie Kharakteristiki Komet* (Moskva: Gos. izd. fiz. mat. literatury).

Vsekhsvyatskij, S. K. 1977. Comets and the cosmogony of the solar system. In *Comets, Asteroids, Meteorites,* ed. A. H. Delsemme (Toledo, Ohio: Univ. of Toledo Press), pp. 469-474.

Wetherill, G. W. 1976. Where do the meteorites come from? A re-evaluation of the earth-crossing Apollo objects as sources of chondritic meteorites. *Geochim. Cosmochim. Acta* 40: 1297-1317.

Whipple, F. L. 1967. On maintaining the meteoritic complex. In *The Zodiacal Light and the Interplanetary Medium,* ed. J. L. Weinberg (NASA SP-150, Washington, D.C.: U.S. Government Printing Office), pp. 409-426.

Whipple, F. L. 1972. The origin of comets. In *The Motion, Evolution of Orbits, and Origin of Comets,* ed. G. A. Chebotarev, E. I. Kazimirchak-Polonskaya, and B. G. Marsden (Dordrecht: D. Reidel), pp. 401-408.

Whipple, F. L. 1975. A speculation about comets and the earth. In *Astrophysique et spectroscopie,* Mém. Soc. Roy. Sci. Liège, Coll. 8°, 6e Sér. 9: 101-111.

Whipple, F. L. 1977. The constitution of cometary nuclei. In *Comets, Asteroids, Meteorites,* ed. A. H. Delsemme (Toledo, Ohio: Univ. of Toledo Press), pp. 25-35.

Whipple, F. L. 1979. Cometary brightness variation and nucleus structure. *Astrophys. Space Sci.* (in press).

Whipple, F. L., and Hamid, S. E. 1952. On the origin of the Taurid meteor complex. *Helwan Obs. Bull. No. 41.*

RESONANCES IN THE ASTEROID BELT

RICHARD GREENBERG
Planetary Science Institute

and

HANS SCHOLL
Astronomischen Rechen-Institut, Heidelberg

The asteroids span a belt in the solar system rich in orbital resonances with Jupiter. We speculate that these resonances may have played a role in the prevention of growth of a full-sized planet and induced the subsequent collisional comminution. More certainly, the resonances have helped govern the distribution of material within the belt. Resonances generally tend to enhance eccentricities and inclinations so that their values oscillate about a "forced" value which is governed by the resonance, with an amplitude called the "free" or "proper" value which is a constant governed by initial conditions. Resonances also tend to be stable, i.e. to maintain themselves against outside disturbances. The last property makes it difficult to understand why so many resonant positions in the belt are depleted in observable asteroids. The answer has been elusive because it probably depends on collisional behavior coupled with the gravitational interactions of classical celestial mechanics. Purely gravitational theories so far fail to indicate any clearing mechanism, at least over the period of time of their validity $\sim 10^5$ yr. We do know that in a dissipative medium, semimajor axes tend to be driven out of resonance zones. Another hypothesis has been that destructive collisions may be more common at some resonances where the enhanced eccentricities and inclinations increase relative motion. However, detailed study indicates that this *hypothesis is not compelling. The cosmogonic hypothesis that planetesimals were never able to form in the gaps, has suffered from inadequacies in the understanding of how planetesimals form in dynamically more normal regions, although this remains a promising avenue of research.*

The asteroids span a belt in the solar system rich in orbital resonances with Jupiter. An asteroid's orbit is resonant if it implies some periodicity in the geometrical relationship with Jupiter, which results in an enhancement of Jupiter's perturbation. For example, if Jupiter's and the asteroid's orbital periods have a ratio of small whole numbers (a "commensurability"), conjunction of the two bodies must occur repeatedly at certain longitudes. The eccentricity forced by Jupiter will be much greater than in the nonresonant case. Possibly, by enhancing orbital eccentricities, such resonances played a role in the prevention of growth of a full-sized planet at the asteroid belt and helped induce the subsequent collisional comminution; the difficulties with that very speculative hypothesis will be discussed in Sec. IV of this chapter.

We can be more certain that resonances did help govern the distribution of material within the belt. Some resonances correlate with gaps in the distribution of orbital elements, such as the Kirkwood (1867) gaps in semimajor axis, while others correlate with concentrations. Figure 1, updated from Brouwer (1963), displays the relationship between radial distribution and resonance positions. Along the bottom scale is a histogram of numbers as a function of mean motion n, which is directly related to semimajor axis a by Kepler's third law. Along the top is a very crude representation of the strengths of the resonances that lie in this range of mean motions. The strength is roughly indicated here by the order of the commensurability q, defined as the difference between the numerator and the denominator in the whole number ratio. We shall show in the next section on resonance mechanisms why small q tends to imply strong resonance. This diagram clearly shows the Kirkwood gaps at the 3/1, 5/2, 7/3, and 2/1 resonances and the concentrations at 3/2 (Hilda group), 4/3 (Thule) and 1/1 (Trojans). After we have discussed the basic principles of resonance mechanisms, we shall address the status of theories on how resonances govern the distribution of asteroids. A critical objective of any theory is to explain why some resonances produce gaps and others concentrations, but an adequate explanation has yet to be found.

At this introductory stage, it may be useful to consider a phenomenological rather than physical approach to the problem. Suppose we construct an envelope around the lines that indicate resonance strength and position. Such an envelope, which has been drawn by eye rather than by any rigorous technique, is shown in Fig. 1. The curve has been smoothed over mean motion on a scale of $\sim n/30$. Such a smoothing seems appropriate at least near resonances with $q = 1$, which produce forced eccentricities

$$\frac{d\widetilde{\omega}}{dt} = (m'/M_\odot)\,(n\alpha/e)F_3 \cos \sigma \tag{3}$$

$$\frac{dn}{dt} = -3(m'/M_\odot)n^2\,\alpha\,e\,F_3 \sin \sigma \tag{4}$$

$$\frac{de}{dt} = -2(m'/M_\odot)n\,\alpha^2\,dF_0/d\alpha \tag{5}$$

where ϵ is the mean longitude at epoch. The variation of ϵ represents a small correction to the sidereal mean motion. We also have from the definition of σ:

$$\frac{d\sigma}{dt} = 2n'-n-\frac{d\epsilon}{dt}-\frac{d\widetilde{\omega}}{dt}. \tag{6}$$

If e is very small, n is nearly constant according to Eq. (4). In that case Eq. (6) takes the form

$$\frac{d\sigma}{dt} = A-(m'/M_\odot)\,(n\alpha/e)F_3 \cos \sigma \tag{7}$$

where $A \equiv 2n'-n-d\epsilon/dt$ is constant. If we take n to be the "corrected" sidereal mean motion, then $A \equiv 2n'-n$. Solving Eqs. (2) and (7), we find

$$e \sin \sigma = C \sin (At+\delta) \tag{8a}$$

$$e \cos \sigma = C \cos (At+\delta) + (m'/M_\odot)\,n\,\alpha\,F_3/A \tag{8b}$$

where amplitude C and phase δ are the constants of integration. On an e, σ polar plot (Fig. 2), this solution is the sum of two vectors:

i. One of magnitude $(m'/M_\odot)n\alpha F_3/(2n'-n)$ and direction $\sigma = 0$. The magnitude increases as the exact resonance is approached, although the analysis breaks down if the eccentricity gets large. This vector represents the "forced" component of the eccentricity. Note that if $2n'-n$ is less than 0, i.e., if the asteroid is closer to the sun than the exact resonance position, the amplitude changes sign, so the vector points toward $\sigma = 180°$.

ii. A "free" eccentricity has arbitrary magnitude and its direction circulates at rate $(2n'-n)$. Note that if the free e is greater than the forced e, σ circulates through $360°$; otherwise σ librates about that value 0 or $180°$ directed by the forced e.

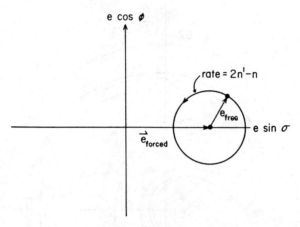

Fig. 2. Trajectory in $e\sigma$ space corresponding to the solution in Eq. (8).

Physically the analysis in the previous page describes a perturber, Jupiter, exerting a radial force outward at conjunction with an asteroid on a nearly circular orbit. The radial force affects both e and $\tilde{\omega}$ in varying degrees depending on the position of conjunction relative to perihelion. Depending upon the exact ratio of mean motions (i.e., upon the rate of precession of the longitude of conjunction), there will be some stable equilibrium combination of e and σ. If the initial conditions are close enough to equilibrium, e and σ oscillate about the equilibrium values. Otherwise σ may circulate. The behavior of σ is closely analogous to the motion of a swinging pendulum. The details of the physical interpretation in terms of force effects on e and $\tilde{\omega}$ are discussed by Greenberg (1977). An important physical question is whether restoration towards equilibrium is governed predominantly by exchange of orbital energy ($\propto -1/a$) or angular momentum ($\propto \sqrt{a(1-e^2)}$). The mechanism just described depends on variation of e with negligible variation of a, hence it might appear that the angular momentum exchange dominates. However, the slight variation of a given by Eq. (4) cancels out variation of angular momentum through second order in e. Thus, both energy and angular momentum are exchanged only slightly. The motion of the asteroid is really changed very little even as the e and $\tilde{\omega}$ of its nearly circular orbit vary.

The above analysis depended on the assumption that e was sufficiently small that n was nearly constant. In fact, as long as the free e is near zero, $e \sin \sigma$ and, hence, dn/dt are also nearly zero. Suppose the free e is too large for this approximation to be valid. If $e \sin \sigma$ starts near zero, it will begin to increase or decrease as the system follows the circular e,σ locus. As $e \sin \sigma$ deviates from 0, n will vary. This introduces a shift in the position of the center of the circular locus. The motion in e, σ space continues at any instant to follow a circle, but the center keeps moving. The trajectory is distorted

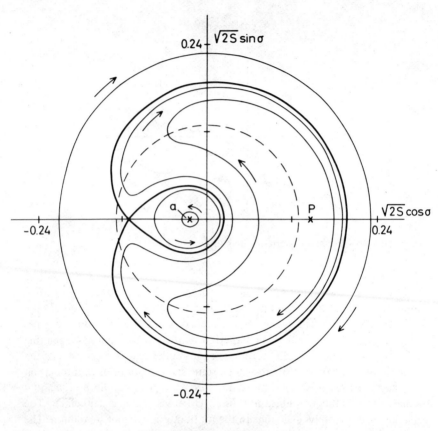

Fig. 3. Trajectories in σ versus $\sqrt{2S}$ space from Schubart's averaged circular model at the 2/1 commensurability for $K = 0.802$. The arrows indicate directions of motion in this space. The darker lines correspond to critical bifurcation trajectories. Paths immediately around point a are apocentric librators; those about p are pericentric librators. The dashed circle corresponds to the exact center of the resonance.

ultimately into a banana shape, such as those shown in Fig. 3. (The topology of such trajectories will be discussed in more detail in Sec. II).

In such a case, the libration amplitude of σ and variation of n may be quite large, while the variation of e is relatively small. Physically, in this larger e case, the force of Jupiter near conjunction can make substantial changes in the asteroid's orbital energy (and hence in its a and n). Such orbital energy exchange adjusts the mean motion such that conjunction tends to be driven towards the symmetrical equilibrium at perihelion ($\sigma = 0$). There is also an equilibrium at $\sigma = 180°$, but in this case of larger e, only $\sigma = 0$ is stable (see Greenberg 1977 for a physical explanation). It should be emphasized that angular momentum as well as energy is exchanged between Jupiter and asteroid during libration.

Inclination-type resonances are also possible. These tend to be weaker than the eccentricity-type described above because the out-of-plane components of the perturbing force which govern inclination and nodes are so much smaller than the in-plane components which govern e and $\widetilde{\omega}$.

Even if the asteroid is so far from a commensurability that only secular terms need be retained in R, resonances are possible. In this case, the important variation equations are

$$\frac{de}{dt} = +(m'/M_\odot)n\alpha e' \, F_2 \sin \phi \tag{9}$$

$$\frac{d\widetilde{\omega}}{dt} = (m'/M_\odot)n \, \alpha \left[2F_1 + (e'/e)F_2 \cos \phi \right] \tag{10}$$

where $\phi = \widetilde{\omega} - \widetilde{\omega}'$. Note that

$$\frac{d\phi}{dt} = \frac{d\widetilde{\omega}}{dt} - N' = N - N' + (m'/M_\odot) \, (n\alpha e'/e)F_2 \cos \phi \tag{11}$$

where N' is Jupiter's apsidal precession, due to other planets, and N is the value from Eq. (10) that the asteroid's precession would have if Jupiter's orbit were circular. In other words, N is the precession computed assuming Jupiter's mass were approximated by a circular ring around the sun. Eqs. (9) and (11) have the same form as Eqs. (2) and (7) so their solution is similar to Eq. (8):

$$e \sin \phi = C \sin \left[(N-N')t + \delta \right] \tag{12a}$$

$$e \cos \phi = C \cos \left[(N-N')t + \delta \right] - (m'/M_\odot)n \, \alpha \, e' \, F_2/(N-N') \, . \tag{12b}$$

Again, there is a forced eccentricity with magnitude proportional to m'. It is also proportional to e'. If $N \approx N'$, there is a resonant enhancement of the forced eccentricity. This is an example of a "secular" resonance. The free component of the eccentricity with magnitude C is called the "proper" eccentricity. It is a measure of the eccentric motion intrinsic to the asteroid, not that forced by Jupiter. If C is smaller than the forced eccentricity, $\phi = 0$ or $180°$; the asteroid's line of apsides librates about the orientation of Jupiter's. Physically, the reason for the similarity between the results for small-e commensurability resonance and secular resonances is that the variations in both cases are due to radial forces due to Jupiter at certain preferred longitudes: at conjunction in the former case and at Jupiter's perihelion in the latter. Similar relations hold for the inclinations and nodal lines as for eccentricities and apsidal lines.

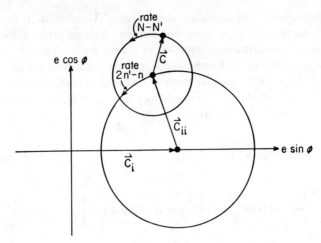

Fig. 4. Trajectory in e,ϕ space corresponding to the solution of Eq. (13). C_i is a fixed forced component of the eccentricity, C_{ii} is a rotating forced component; $\vec{C}$ is the free eccentricity.

In the above discussions of secular and commensurability resonances we considered each type of perturbation independently. In fact, near a commensurability the secular terms are present and, in certain circumstances, can have qualitatively important effects. If we include both secular and critical terms we obtain the more general solution:

$$e \sin \phi = C \sin (N-N')t + \left[(m'/M_\odot) n\alpha F_3/(A-N+N')\right] \sin At \quad (13a)$$

$$e \cos \phi = C \cos (N-N')t + \left[(m'/M_\odot)n\alpha F_3/(A-N+N')\right] \cos At \quad (13b)$$

$$-(m'/M_\odot) n\alpha F_2 e'/(N-N')$$

$$\sigma = (A-N')t - \phi \quad (13c)$$

where $t \equiv 0$ when $\sigma = -\phi$. In this case, the solution trajectory in e,ϕ polar space is traced by the sum of three vectors (Fig. 4):

(i)   a fixed vector;

(ii)   one of fixed magnitude, but direction varying at slow rate A; and

(iii) one of arbitrary constant magnitude C and direction varying at faster rate N-N'.

If the magnitudes of (i) and (ii) are nearly equal, and C is smaller than their sum, ϕ will appear to alternately librate and circulate. Moreover, Eq. (13)

implies that σ circulates when ϕ librates and vice versa. This behavior was termed "coupled" libration by Greenberg and Franklin (1975).

The above discussion applies to commensurabilities of form $(p + q)/p$ where $q = 1$ only, based on the specific example of the 2/1 case. The analysis would be quite similar for any other $q = 1$ case (e.g. 3/2, 4/3, etc.) except that σ would take the general form $(p + 1)\lambda' - p\lambda - \tilde{\omega}$. However, as a general rule, the coefficient of the critical term in the disturbing function Eq. (1) contains eccentricities and inclinations to a total power at least as great as q. Hence, higher q commensurabilities are harder to treat analytically and are generally treated numerically in specific asteroid cases. The higher orders of e and i in cases of larger q express the weakness of the resonance (Fig. 1). The physical explanation for this weakness is that conjunctions occur at q different longitudes. For example, in a 3/1 commensurability with small eccentricities, the asteroid would make 1½ revolutions to Jupiter's ½ revolution between conjunctions. Thus, any effects at one conjunction would be substantially neutralized at the next conjunction.

In the case of the 1/1 resonance, the Fourier expansion of the disturbing function breaks down; it is not valid for bodies of equal distance from the sun. The general way to handle this resonance is to consider an asteroid's motion in a reference frame rotating with Jupiter's orbital motion. In this frame, there are potential maxima at the points 60° behind and ahead of Jupiter. These positions are stable, however, due to the effects of Coriolis forces. Asteroids librating about those longitudes (the Trojans) maintain a 1/1 commensurability with Jupiter.

In the introduction, we described the most striking asteroid resonance phenomenon, the gaps and concentrations. We can describe other specific observed resonance phenomena (see Chapman *et al.* (1978), for a review). The asteroid at the 4/3 resonance (279 Thule) and most of those at the 3/2 resonance, the Hilda types, have a libration about a value near 0 (Takenouchi 1962; Sinclair 1969; Schubart 1968). Having conjunction with Jupiter thus constrained away from the asteroid's aphelion prevents close approach to Jupiter. Perhaps this enforced separation helps explain the concentrations at these resonances, due to resistance to scattering by Jupiter. For two members of the Hilda group, 334 Chicago and 1256 Normannia, σ circulates and therefore the dangerous configuration $\sigma = 180°$ occurs periodically. However, the orbital eccentricity reaches a minimum at such times (cf. Eq. 8), so that they are safe from close approach to Jupiter. Chapman *et al.* leave these asteroids out of the Hilda group because the σ's do not librate; this may be unnecessarily restrictive, since only a small difference in physical motion may sometimes make the difference between libration and circulation.

At the relatively unpopulated 2/1 resonance, there are several librators: 1362 Griqua, 1921 Pala, 1922 Zulu, and others with $\sigma = 0$ (Schubart 1966; Schweizer 1969; Sinclair 1969; Franklin *et al.* 1975). Franklin *et al.* also discovered that eight other asteroids near the 2:1 commensurability librate with $\sigma = 180°$. These "apocentric" librators have very small values of e, consistent with the discussion in the previous section. One, 9594 P-L, exhibits, in the numerical integration of its orbit, the alternating circulation and libration, coupled with ϕ behavior discussed above. At the 3/1 commensurability, 887 Alinda and 1915 Quetzalcoatl are known librators; both are Mars crossers as well. Numerical integration of the motion of 1685 Toro (Danielsson and Ip 1972; Janiczek *et al.* 1972; Williams and Wetherill 1973; Danielsson 1978) indicates that it alternates between librations in a 5/8 commensurability with Earth and a 5/13 commensurability with Venus. The librations tend to prevent close approaches to Earth or Venus, but the lifetime before a catastrophic approach to Mars is probably only 10^6 yr. There are no known individual examples of secular resonances; rather, pronounced gaps appear near those surfaces in a,e,i that correspond to secular resonances.

II. NUMERICAL STUDIES OF RESONANT MOTION

The heuristic description of the resonance mechanism presented in Sec. I applies in a strict sense only to special cases of small eccentricity. In order to gain insight into resonance mechanisms in more realistic cases, as well as to study the motion of specific resonant asteroids, numerical studies have been performed by various researchers. Behavior near commensurabilities has been investigated numerically with two different sets of differential equations:

A. Newton's equations of motion for the Sun-Jupiter-Saturn-Asteroid problem in three dimensions. The program which solves this set of equations will be called N-Body Program. The equations were numerically integrated by an Adams method modified by Schubart and Stumpff (1966).

B. Poincaré-Schubart equations for the planar elliptic Sun-Jupiter-Asteroid problem averaged over the corresponding commensurability period.

Set B was introduced by Poincaré (1902) and later modified twice by Schubart (1964, 1968) for the investigation of commensurable motion at a resonance. The averaging method has wider application for higher e orbits than the method of Sec. I, in which short period terms are simply neglected. Away from a resonance and in cases with approach closer than 1 AU to Jupiter, Set B cannot be applied, however. Generally, Set B is used to study resonant motion, with Set A used as a check.

Using his numerical methods, Schubart (1964) generated figures equivalent to the e, σ plot discussed in Sec. I, but applicable to higher values of eccentricity. In Fig. 3, trajectories are plotted in polar coordinates $\sqrt{2S}$ and σ, where

$$S \equiv \sqrt{a/a'}\,(1 - \sqrt{1 - e^2}). \tag{14}$$

For small e, $\sqrt{2S} \propto e$. The trajectories shown are for the 2/1 commensurability, with $e' = 0$. Schubart noted that the quantity

$$K = \sqrt{a/a'} \left(\frac{p+q}{q} - \sqrt{1-e^2} \right) \tag{15}$$

is a constant of integration. The curves shown in Fig. 3 are all for the same value of K, but differ in the value of the remaining integral of the motion $\bar{H}$, Schubart's time-averaged Hamiltonian. These are the trajectories which lie on a single constant K plane in $K, \sqrt{2S}, \sigma$ space. The analytic solutions in Eq. (8) describe the special case where $e \sin \sigma$ is small, i.e., the trajectories in immediate vicinities of points a and p. Trajectories about a are for "apocentric librators" (σ about $180°$); those about p are for "pericentric librators" (σ about $0°$). There is a critical bifurcated trajectory (shown as a darker line) which divides each constant K plane into three regions. When the inner portion of the critical curve encompasses the origin, as in the case of Fig. 3, the apocentric librators share one region with a class of circulating trajectories (inner circulators). Next is the region of generally banana-shaped pericentric librators. And finally, farthest from the origin is the region of outer circulators.

Schubart (1968) also investigated the 3/2 resonance where the Hilda group of asteroids is located. Scholl and Froeschlé (1974, 1975; see also Scholl and Giffen 1974) used the same methods for their systematic investigation of the 3/1, 5/2, 7/3, and 2/1 resonance, where the principal Kirkwood gaps are located. Typical periods covered by the numerical integrations were on the order of 10^4 yr. A few orbits were computed over 10^5 yr (see also Schubart 1970, 1978, 1979).

In general, the results of these numerical experiments are consistent with the qualitative considerations discussed earlier. The semimajor axes are found to be semiperiodic functions of time. In no case was secular behavior found, in accordance with Poisson's theorem. Therefore asteroids situated in any of the principal Kirkwood gaps do not leave their gaps. This statement is only valid for the applied model and for periods of 10^5 yr. Froeschlé and Scholl speculate that this statement might also be true for periods of 10^7 yr. However, for periods comparable to the age of the solar system of some 10^9 yr, no definitive statement can be made because over such a long time span, perturbations by other planets might be important. The effect of Saturn was investigated in an N-Body calculation over 10^5 yr. Good agreement was found with the corresponding orbits calculated with the averaging method. We therefore conclude that even in a realistic model no asteroid would escape out of a Kirkwood gap within 10^7 yr under the influence of purely gravitational forces.

The eccentricities of fictitious objects near resonance can be increased significantly. Given an initially very small eccentricity, the closer an asteroid

TABLE I

Eccentricity Variations (Initial $e \sim e_{min}$)

Resonance	e_{min}	e_{max}	Characteristic period in e (yr)
2/1	0.00	0.03	$\sim10^2$
2/1	0.02	0.04	$\sim10^2$
2/1	0.04	0.15	$\sim10^2$
2/1	0.08	0.20	~3000
2/1	0.14	0.25	~3000
3/1	0.00	0.03	$\sim10^2$
3/1	0.04	0.10	$\sim10^2$
3/1	0.08	0.15	~8000
3/1	0.14	0.25	~8000
5/2	0.00	0.03	$\sim10^2$
5/2	0.04	0.20	$\sim10^2$
5/2	0.08	0.35	$\sim50,000$
5/2	0.14	0.40	$\sim50,000$
7/3	0.00	0.02	$\sim10^2$
7/3	0.08	0.11	$\sim10^2$
7/3	0.14	0.18	$\sim10^2$

starts to the exact commensurability of mean motions, the greater are the subsequent variations in eccentricity. Our analysis in Sec. I is completely consistent with this more general numerical result. Moreover, numerical results show that the maximum variation in e does not occur at the exact resonance but slightly closer to the sun, also consistent with the analysis in Sec. I.

To give some rule of thumb for the typical range of e, Table I shows variations for randomly selected a values in Kirkwood gaps.

In the elliptical model ($e' \neq 0$), K is not an integral of motion, but a slowly changing variable. An orbit therefore cannot be represented by a curve in the $\sqrt{2S},\sigma$ plane. An orbit which starts on a curve in Fig. 3 does not necessarily remain in that plane; as it migrates into a new plane, it follows the curve for the same value of $\bar{H}$ in the new plane. We, therefore, have to superimpose sheets of figures with varying values of K in order to describe an orbit by the coordinates $\sqrt{2S},\sigma$. This approach is meaningful since in the elliptical problem, K varies slowly. Therefore, an orbit crosses the different sheets slowly. In other words, the elliptical model can be approached as a perturbed circular model with Jupiter's eccentricity as the perturbing parameter (cf. Froeschlé and Scholl 1977).

In the elliptical case, a trajectory can alternate between circulation and libration in two ways as it moves through the various constant-K sheets:

1. In the first way the trajectory remains in the same region of Schubart's topology, e.g. the transition from apocentric libration to inner circulation in Fig. 3. There is no profound variation in the physical motion of the asteroid through such a transition. The alternation between ϕ and σ libration discussed in Sec. I is an example of another way to look at such a transition.

2. Another transition involves change across the critical curve, as from pericentric libration to outer circulation. Here the trajectory suffers a significant bifurcation which shows up in numerical integration as a sudden change in amplitude of oscillation of a.

For the elliptical model, Giffen (1973) investigated commensurable motion from a different point of view. He applied the technique of Hénon and Heiles (1964) of "surface of section" for the 2/1 and 3/2 resonance in order to find another integral besides the averages Hamiltonian $\bar{H}$ that restricts motion to a limited sub-space of phase space (in the $e' = 0$ case, K served this purpose).

In Schubart's averaged model, motion is described in a four-dimensional phase space $(a,e,\sigma,\tilde{\omega})$. Since $\bar{H} = h(a,e,\sigma,\tilde{\omega})$ is an integral, motion is restricted to a three-dimensional surface. If another isolating integral $\bar{G}$ exists in the form $\bar{G} = g\ (a,e,\sigma,\tilde{\omega})$, motion is further restricted to a two-dimensional surface. In order to recognize an orbit as restricted on a two-dimensional surface in the four-dimensional phase space $(a,e,\sigma,\tilde{\omega})$, we cut the phase space by an arbitrarily chosen surface, such as a = constant.

The intersection of the constant-a surface with a two-dimensional surface defined by constant $\bar{H}$ and $\bar{G}$ is a one-dimensional curve in phase space. We can calculate an orbit numerically and, whenever the semimajor axis of the asteroidal orbit is equal to the selected value of a, record the point in the $e,\sigma,\tilde{\omega}$ space. We then plot the recorded values $e,\sigma,\tilde{\omega}$ in the $e-\sigma$, $\sigma-\tilde{\omega}$, and $e-\tilde{\omega}$ planes. If the points lie on a curve, we presume that the motion is in fact confined to a two-dimensional surface in the four dimensional phase space and hence that a second integral of motion exists. Otherwise, if the points are scattered on the planes, no second integral seems to exist; we call it a "wild" or nonintegrable orbit. This is the surface-of-section method.

Giffen found only integrable orbits for the 3/2 resonance, but both integrable and wild orbits for the 2/1 resonance. The wild orbits occurred for small starting eccentricities $e < 0.15$ and integrable orbits for large eccentricities $e > 0.3$. Scholl and Froeschlé (1974, 1975) made a systematic survey of integrable and nonintegrable orbits at the resonances 3/1, 5/2, 7/3 and 2/1. According to their results only a few nonintegrable orbits occur. For $e > 0.2$, all the orbits are integrable. For $e < 0.2$ both types, integrable and nonintegrable orbits occur. Since the large majority of orbits is integrable, Giffen's discovery of some nonintegrable orbits was exceptional. Giffen speculated that the nonintegrable orbits may play an important role in the question of the Kirkwood gaps. An asteroid situated on a wild orbit is not limited to a curve in the $e-\sigma$ plane as described above. The asteroid may increase its eccentricity so much that it will collide with a major planet and then be removed from its gap, according to this speculation.

For the integrable orbits, we note that they all stay in their corresponding Kirkwood gaps, since it is possible to determine the boundaries of motion by the surface-of-section method. The invariant curves which were found numerically for periods of some 10^4 yr determine these boundaries. However, we cannot exclude the possibility that these invariant curves might dissolve, over much larger periods, $\sim 10^8$ or 10^9 yr; asteroids might drift out of a gap on such extended time scales. That is still an unsolved problem.

Froeschlé and Scholl (1976) investigated the nonintegrable orbits in order to find out if these orbits can increase their eccentricities to 1 or if their motion is restricted in the phase space, $(a,e,\sigma,\tilde{\omega})$. As was stated above, an orbit is surely limited to a three-dimensional surface because $\bar{H}$ is an integral of motion. This integral $\bar{H}$ plays the same role as the well-known Jacobi integral in the nonaveraged circular restricted three-body problem. The Jacobi integral determines a zero velocity curve which divides the space into a forbidden and an allowed region for the orbit. For the averaged elliptic problem, $\bar{H}$ also determines a curve which divides the space into a forbidden and an allowed region. Here, the curve has to be obtained numerically and it cannot be interpreted as a zero velocity curve. Figure 5 describes the situation. The allowed region is open to the right. Therefore, the eccentricity does not appear confined to low values. However, invariant curves of different orbits with the same value for $\bar{H}$ prevent the eccentricities from exceeding 0.2. The orbit is trapped by the limiting "zero velocity" curve and by invariant curves of other orbits which all have the same zero velocity curve. Therefore, for periods up to 10^7 yr, wild orbits do not escape out of a Kirkwood gap.

As this statement is only true for the model described above, Froeschlé and Scholl (1979) have computed wild orbits in the Sun-Jupiter-Saturn model over 100,000 yr. Again, good qualitative agreement was found.

III. THE DISTRIBUTION OF ASTEROID ORBITS

Apparently, resonance effects determine to a large extent the distribution of asteroidal semimajor axes a. Since the orbital energy of an asteroid is given by a, we can also say that resonance effects due to Jupiter seem to govern the energy distribution of asteroidal orbits. In this section, we will investigate how and to what extent Jupiter controls this energy distribution. In particular, we will investigate the following question: Does Jupiter control completely this energy distribution in the sense that gaps and concentrations correspond to dynamically allowed and forbidden regions?

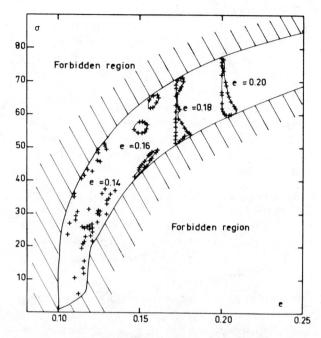

Fig. 5. Two-dimensional surface of section $a = 3.27$ AU for different starting values of eccentricity but with the same value for the common integral of motion $\overline{H}$.

The Dynamical Structure of the Outer Belt

The shape of the distribution function between 3.3 AU and 5.2 AU was first investigated by Lecar and Franklin (1973) over $\sim 10^4$ yr and later by Froeschlé and Scholl (1979) over $\sim 10^5$ yr. According to these numerical experiments, the emptiness of the region between the Hilda group at 3.9 AU and the Trojans at 5.2 AU can be explained by Jovian perturbations. The mean lifetime of an object with a typical asteroidal eccentricity and inclination is some 10^3 yr. Only at the 4/3 commensurability, where we observe one single object, 279 Thule, or in the case of the Trojans and the Hildas, can objects survive much longer. Thus, the stabilizing effect of these resonances, by preventing close approach to Jupiter as discussed earlier, may explain the survival of these objects within a wide zone which has been otherwise depleted. Hence, concentrations are to be expected at these resonances.

For the orbits of the Hildas and Thule, we can describe the stabilizing mechanism geometrically by the behavior of the critical argument σ. For the Trojans the properties of periodic solutions about the stable Lagrange points L_4 and L_5 are used. It is however uncertain over what periods these orbits remain stable in a more realistic model which includes all the planets.

Numerical integrations indicate stability for periods of $10^4 - 10^5$ yr, a result which might be extrapolated to 10^7 yr. For periods comparable to the age of the solar system of some 10^9 yr we are not able to make a definitive statement.

The Kirkwood Gaps

Next we consider the distribution function at $a < 3.3$ AU. Most attention has been given to the Kirkwood gaps. Since their discovery it was felt that they could be explained by resonance effects. The hypotheses which try to explain the Kirkwood gaps can be divided in four types, which will be discussed in detail below:

a. The Kirkwood gaps are a statistical phenomenon.
b. Asteroids that originally formed in the Kirkwood gaps were removed later by purely gravitational forces due to Jupiter.
c. The removal involved collisional as well as gravitational effects.
d. Asteroids could never have formed in a Kirkwood gap (cosmogonic hypothesis).

a. The Statistical Hypothesis. The statistical hypothesis assumes that asteroids librate around gaps, spending most of their time outside of the gaps, just as a pendulum spends most of its time farthest from its equilibrium point. Therefore, the probability of finding an asteroid in a gap at a given epoch is very low. In order to test this assumption, Schweizer (1969) calculated the orbits of 185 observed asteroids around the Hecuba gap at the 2/1 commensurability, 20 observed orbits around the 5/2 commensurability and 13 observed orbits around the Hestia gap at the 3/1 commensurability as well as the orbits of four asteroids situated inside of gaps. The numerical calculations covered a few hundred years in order to find the long period in mean motion n. Then Schweizer plotted the distribution function for the mean motions $\overline{n}$ averaged over their corresponding long periods. The resulting averaged distribution function still shows Kirkwood gaps, but they are somewhat narrower than in the distribution of nonaveraged mean motion.

Schweizer was well aware that the period covered by his calculations was too short to give a definitive disproof of the statistical hypothesis. Indeed, from numerical experiments by Scholl and Froeschlé (1974) we know that orbits can be found that stay on one side of a gap for 10,000 yr, then librate for a while through the gap and continue on the other side of the gap. These orbits, however, are very peculiar and exceptional. They involve transitions across the critical curve in S, σ space. We, therefore, suppose that even a continuation of Schweizer's calculation to periods of 10^5 yr would confirm his original result.

Wiesel (1976) investigated the statistical hypothesis analytically by the method of phase mixing. Starting with an initial normal distribution of the orbital elements a and e, he tried to determine the distribution of these elements in the a,e phase space for $t \to \infty$ on the average. The starting distribution is given by the probability density function

$$f_0 = N \cdot \exp \left[-\left(\frac{e - e_0}{e_f}\right)^2 \right] \qquad (16)$$

where N is a normalizing factor, and f_0 is assumed constant over a. The variance e_f and the mean value e_0 are regarded as parameters. Wiesel used three sets e_f, e_0 at the commensurabilities 3/1, 5/2, and 2/1 in order to calculate the corresponding time-averaged probability density function

$$\lim_{t \to \infty} \frac{1}{t} \int_0^t f_0 \, dt. \qquad (17)$$

The trick of the phase mixing method is to replace the averaging over time by an averaging over space. This procedure is justified by Birkhoff's ergodic theorem (Khinchin 1948). We give a simple example to see how this procedure works; we assume that the orbit of a particle fills a certain region R in the a,e phase space. In order to find the average position of the particle in R for $t \to \infty$, two methods can be applied which are supposed to yield the same result:

(1) We add up the positions $P_i(a,e)$ of the particle at certain instants of time t_i, $i = 1 \ldots n$, and take the average $\frac{1}{n} \sum_{i=1}^{n} P(a,e,t_i)$. Further refinement of t_i and further extension of t yields the time-averaged position.

(2) We plot the orbit on an a-e diagram and put a grid on this diagram. At each point (a_j, e_j) of the grid, we count how often the orbit has crossed it and multiply that number with the position. The sum total of these positions divided by the area yields a space-averaged position.

Replacement of the sums in (1) and (2) by integrals yields the final averaged positions. Wiesel applies this principle to the probability density function P_0 given above. For the calculation of "positions", he used Poincaré's method (1902).

The resulting time-averaged probability density functions, which were obtained by a space averaging, do not yield Kirkwood gaps. Wiesel found minima in the distribution of a which were slightly shifted to the side of the resonance closer to Jupiter. However, these minima were not deep enough to explain Kirkwood gaps. Wiesel's result is in good agreement with results obtained by Message (1966) who also used phase mixing, but who calculated the "positions" numerically. Therefore, the results of Message, Schweizer and Wiesel militate against the statistical hypothesis.

b. The Gravitational Hypothesis. For this hypothesis, we need a way to remove a large number of asteroids from the gaps at the commensurabilities 2/1, 3/1, 5/2, and 7/3, while at the same time leaving a sufficient number at 3/2 and 1/1. At the commensurability 4/3, we observe only one single object, a number which is comparable to what we observe in the Kirkwood gaps.

As we showed above, the analytic treatment of resonances involves small divisors which introduce singularities at exact commensurabilities. The forced *e* goes to infinity. Therefore it is still a wide-spread opinion that since there is a singularity, motion is not possible at commensurabilities. For a while, the Kolmogorov-Arnold-Moser (KAM) theory seemed to be suited to solve the problem. If certain conditions like "sufficiently small perturbations" are fulfilled, the KAM theorem guarantees the existence of quasi-periodic solutions. As Jefferys and Szebehely (1978) have pointed out, the KAM theory requires a very low boundary for permissible perturbations which makes the application of this theory very difficult. In addition, the KAM theory requires "poor approximation" of the ratio n/n_J by rationals. It is not clear how far away from the mathematically exact resonance (for instance 2/1) the KAM theory is applicable. Since the application of the KAM theorem is very doubtful for resonant motion, it does not seem relevant to the questions of gaps or concentrations in the asteroid belt.

Kiang (1978) has introduced a new principle to determine stable motion at resonances. For the 2/1 and 3/2 commensurability he used Schubart's circular model to calculate stability criteria using a Hill's equation. However, Aoki (1978) showed that Kiang's derivation of Hill's equation is wrong, since Kiang omitted necessary terms. Therefore, the equation which Kiang uses to explain the difference between Hilda and Hecuba type motion is not complete. Subsequently, "stable" and "unstable" motion in Kiang's sense does not mean that asteroids stay in the Hilda group and leave the Hecuba gap.

c. The Collisional Hypothesis. According to the collisional hypothesis, asteroids are removed from the gaps by collisions with neighboring asteroids. These collisions might result in a destruction of the gap-asteroids or in a change of mean motion which removes them from the gaps. The collision hypothesis is in part motivated by study of the distribution with *a* (Fig. 1). Around the Hilda group, and around the Trojans, the density of asteroids is low while the density around the Kirkwood gaps is much greater. Therefore the Hildas as well as the Trojans have a low probability of colliding with neighboring asteroids. Brouwer (1963) and Jefferys (1967) showed that if collisions among asteroids produce a smooth distribution of Brouwer's integral of motion Γ, V-shaped gaps are obtained in the distribution function for *a* at commensurabilities. Brouwer's integral of motion has the form

$$\Gamma = \frac{1}{2} L^{-2} + \left(\frac{p+q}{q} \right) \cdot L + R^* \qquad (18)$$

where $L = a^{-1/2}$, p and q are prime integers which determine the commensurability. R^* is the disturbing function after elimination of short-periodic terms. (Brouwer's integral cannot be directly identified with either of Schubart's because it represents a solution to a somewhat differently averaged problem.) Whether collisions really yield a smooth distribution for Γ remains problematical.

Heppenheimer (1975a) investigated this problem by a Monte Carlo simulation of collisions at the 2/1 commensurability. Orbits were computed by solving Lagrange's planetary equations with Brouwer's disturbing function R^*. The collision frequency was chosen at random in the interval $0 - 100$ asteroidal revolutions. After a collision, the orbit continued with slightly different velocities chosen at random. After 20 collisions, the random walk of an asteroid stopped. The evolution of 50 fictitious asteroids with a uniform initial distribution of a in the gap and of e in the range $0.15 - 0.35$ did not yield a gap. The asteroids remained in the Kirkwood gap. No smooth distribution for Brouwer's integral Γ was found and, therefore, the Brouwer-Jefferys collisional hypothesis was not supported. It might be worthwhile to repeat Heppenheimer's Monte Carlo calculation with a three-dimensional model. The collisional frequency should be related to the velocity or to the time an asteroid spends outside of the main belt where the density of asteroids is so low.

A second type of collisional hypothesis presumes that asteroids situated in a gap have a much larger collision probability compared with the surrounding asteroids. This assumption is based on the fact that gap-asteroids increase their eccentricities and therefore their velocities much more than nonresonant asteroids, and on the fact that collision probability depends on velocity. Heppenheimer (1975b) investigated this second hypothesis on the basis of Dohnanyi's (1969) and Wetherill's (1967) collision models for the asteroidal belt. Heppenheimer concluded that the fragmentation of gap-asteroids by collision is only slightly faster than for nonresonant asteroids outside of the gaps. Resonant motion does not yield a sufficient increase in velocity to pulverize asteroids in a comparatively short period.

Ip (1977) investigated the probability of collision for main belt asteroids. According to his results the probability of collision is not only a function of orbits intersected, but it is also a function of the time an asteroid remains outside the main belt. Therefore an asteroid with a small eccentricity which is situated at the 2/1 gap has a higher collision probability than an asteroid with a large eccentricity, since the latter asteroid may spend most of its time outside the main belt where there are only a few asteroids. This result works against the collision hypothesis.

Meteoritics suggests that material from the asteroid belt does reach the earth and other terrestrial planets. Most scenarios for transporting material begin with collisional debris being injected into a resonance zone, either a commensurability (Zimmerman and Wetherill 1973), or secular resonance

(Williams 1973). There, enhancement of e leads to Jupiter or Mars approaches and eventual transfer into orbits crossing those of other terrestrial planets. Such scenarios do seem to provide a viable way to deliver meteorites. Whether the same mechanism has been responsible for clearing the gaps remains problematical, because resonances can enhance eccentricities to the required extent only in a very narrow range of mean motions. According to Eq. (8), the forced eccentricity is only large enough when the mean motion differs from the exact commensurability value by a fraction less than $\sim m'/M_\odot$, but the observed gaps are ~ 30 times as wide. Perhaps collisions in this wider region have fed material into the narrow zone of large forced eccentricities.

   d. The Cosmogonic Hypotheses. The last type of hypothesis about the origin of the Kirkwood gaps presumes the formation of the gaps when the planetary system formed. Since our present ideas about the formation of the planetary system depend on theoretical models with minimal observational constraints, construction of a definitive cosmogonic theory for the Kirkwood gaps is difficult.

   Heppenheimer (1978) showed that Goldreich-Ward (1973) planetesimals could not have formed in gaps. These planetesimals break up for velocities larger than 100 m sec^{-1} according to experimental data collected by Greenberg et al. (1977). Heppenheimer's strategy, therefore, was to show that planetesimals situated in a gap exceed this critical velocity while planetesimals outside of gaps remain far below 100 m sec^{-1}. The model takes into account Jovian perturbations and the gravitational and drag effect produced by a solar nebula. Heppenheimer developed a secular perturbation theory which includes all these effects. Jupiter increases the eccentricities of growing planetesimals and can bring them, therefore, to the critical value of 100 m sec^{-1} for their velocities. The nebula, however, reduces the eccentricities of planetesimals outside of the gaps. For planetesimals inside of a gap the nebula is not able to prevent the eccentricities from increasing. The planetesimals exceed 100 m sec^{-1} and start to comminute. Therefore, according to this model, asteroids never formed in a gap. Heppenheimer's model requires an appropriate density, mass and temperature for the nebula. In addition, the model implies that all the asteroids are formed in almost circular orbits in a thin disk. Objects which we observe to be in a resonant motion like the Hildas must have originated at a different place, since they could not have formed at a resonance. In addition, we should observe asteroids in the Kirkwood gaps with low eccentricities, since low eccentric orbits at a resonance behave like nonresonant orbits according to the numerical experiments by Scholl and Froeschlé. Further investigations of orbital elements for planetesimals at an early stage of the solar system are needed to overcome these problems.

   Another possible mechanism for removal of material from resonance zones early in the history of the solar system was suggested by Greenberg (1978). If a resonant asteroid is subjected to collisions in an eccentricity-damping medium, such as smaller bodies in its neighborhood, a secular variation in semimajor axis is introduced. We can demonstrate this by adding an extra negative term to de/dt in Eq. (2). This changes the solution in such a way that the equilibrium value of sin σ is no longer zero, as it was in the solution in Eq. (8). This phase shift in the solution introduces a secular decrease in a through Eq. (4). For a plausible early mass distribution, most of the mass near the 2/1 commensurability is removed; the computed width of the cleared zone agrees well with the width of the 2/1 Kirkwood gap. At the 3/2 and 4/3 resonances, the density of material would be much lower due to direct removal of material by Jupiter as discussed earlier. Hence e-damping and subsequent removal of material would be ineffective.

IV. OTHER REMAINING PROBLEMS

While the Kirkwood gaps represent minima in the asteroid distribution, from a broader perspective, the asteroid belt itself is a gap in the distribution of planetary material. Is it possible that resonances played a role in increasing relative velocities, enhancing comminution, removing material and preventing growth of a full-sized planet? The difficulty with this hypothesis is that resonances are important only in the narrow zones near low-order commensurabilities ($\Delta a/a \lesssim m'/M_\odot$ according to Eq. 8). No quantitative mechanism for stirring the whole belt by resonances has yet been derived. (Other stirring mechanisms are discussed by Davis *et al.* in their chapter.) However, several ideas deserve consideration. First, if some a-changing mechanism, such as gas drag or radiation effects, acted to sweep material across resonances, subsequent enhancement of eccentricities and collisions might have stirred a significant portion of the material in the present asteroid belt. Second, the material removed from resonances by e-damping would be concentrated at the inner edge of resonance zones. There, gravitational instabilities might have created several close, large planetesimals whose mutual interactions may have had stirring consequences. Finally, as suggested by Chapman *et al.* (1978) and Heppenheimer (1979), positions of secular resonances in the solar system would have migrated as the early nebula cleared and planets grew, suggesting the possibility that such resonances swept across the asteroid belt, raising relative velocities.

Another major remaining problem is the question of the removal of material from the region 3.3 AU to 3.9 AU. Exterior to that, Lecar and Franklin (1973) showed removal by Jupiter with survival of stable orbits near resonances; interior to that, the distribution has been cut by the gaps at resonances according to various models. In this intermediate zone for now, we can only appeal to the phenomenological observation mentioned in the introduction. Perhaps the dense population of commensurabilities shown at the top of Fig. 1 between 3.3 and 3.9 AU was responsible for clearing this giant "gap," just as isolated resonances cleared the Kirkwood gaps. We have shown in this chapter that the latter mechanism is still poorly understood.

Whether the same mechanism operated over the broad intermediate zone remains an open question.

REFERENCES

Aoki, S. 1978. Stability problem of Kirkwood gaps. *Nature* 257: 568.

Brouwer, D. 1963. The problem of the Kirkwood gaps in the asteroid belt. *Astron J.* 68: 152-159.

Chapman, C.R.; Williams, J.G.; and Hartmann, W.K. 1978. The asteroids. *Ann. Rev. Astron. Astrophys.*, eds. G. Burbidge, D. Layzer and J. Phillips (Palo Alto, Calif.: Annual Reviews, Inc., Vol. 16, pp. 33-75.

Danby, J.M.A. 1962. *Fundamentals of Celestial Mechanics* (New York: MacMillan) p. 251.

Danielsson, L. 1978. The orbital resonances between the asteroid Toro and the Earth and Venus. *Moon and Planets* 18: 265-272.

Danielsson, L., and Ip, W.-H. 1972. Capture resonance of the asteroid 1685 Toro by the earth. *Science* 176: 906-907.

Dohnanyi, J.S. 1969. Collisional model of asteroids and their debris. *J. Geophys. Res.* 74: 2531-2554.

Franklin, F.A.; Marsden, B.G.; Williams, J.G.; and Bardwell, C.M. 1975. Minor planets and comets in libration about the 2:1 resonance with Jupiter. *Astron. J.* 80: 729-746.

Froeschlé, C., and Scholl, H. 1976. On the dynamical topology of the Kirkwood gaps. *Astron. Astrophys.* 48: 389-393.

Froeschlé, C., and Scholl, H. 1977. A qualitative comparison between the circular and elliptic Sun-Jupiter-Asteroid problem at commensurabilities. *Astron. Astrophys.* 57: 33-39.

Froeschlé, C., and Scholl, H. 1979. New numerical experiments to deplete the outer part of the asteroidal belt. *Astron. Astrophys.* 72: 246-255.

Giffen, R. 1973. A study of commensurable motion in the asteroidal belt. *Astron. Astrophys.* 23: 387-403.

Goldreich, P., and Ward, W.R. 1973. The formation of planetesimals. *Astrophys. J.* 183: 1051-1061.

Greenberg, R. 1977. Orbit-orbit resonances in the solar system: Varieties and similarities. *Vistas in Astronomy* 21: 209-239.

Greenberg, R. 1978. Orbital resonance in a dissipative medium. *Icarus* 33: 62-73.

Greenberg, R.; Davis, D.R.; Hartmann, W.K.; and Chapman, C.R. 1977. Size distribution of particles in planetary rings. *Icarus* 30: 769-779.

Greenberg, R, and Franklin, F.A. 1975. Coupled librations in the motions of asteroids near the 2:1 resonance. *Mon. Not. Roy. Astron. Soc.* 173: 1-8.

Hénon, M., and Heiles, C. 1964. The applicability of the third integral of motion: Some numerical experiments. *Astron. J.* 69: 73-79.

Heppenheimer, T.A. 1975a. Adiabatic invariants and phase equilibria for first-order orbital resonances. *Astron. J.* 80: 465-472.

Heppenheimer, T.A. 1975b. On the alleged collisional origin of the Kirkwood gaps. *Icarus* 26: 367-376.

Heppenheimer, T.A. 1978. On the origin of the Kirkwood gaps and of satellite-satellite resonances. *Astron. Astrophys.* 70: 457-465.

Heppenheimer, T.A. 1979. Secular resonances and the early history of the solar system. Submitted to *Icarus.*

Ip, W.H. 1977. On the orbital dependence of the asteroidal collision process. *Icarus* 32: 378-381.

Janiczek, P.M.; Seidelmann, P.K.; and Duncombe, R.L. 1972. Resonances and encounters in the inner solar system. *Astron. J.* 77: 764-773.

Jefferys, W.H. 1967. Nongravitational forces and resonances in the solar system. *Astron. J.* 72: 872-875.

Jefferys, W.H., and Szebehely, V.G. 1978. Dynamics and stability of the solar system. *Comments on Astrophys.* 8: 9-17.

Khinchin, A.I. 1948. *Mathematical Foundations of Statistical Mechanics* (New York: Dover).

Kiang, T. 1978. Kirkwood gaps and stability of conservative periodic systems. *Nature* 273: 734-736.

Kirkwood, D. 1867. *Meteoric Astronomy: A Treatise on Shooting-Stars, Fireballs, and Aerolites,* Ch. 13 (Philadelphia: J.B. Lippincott).

Lecar, M., and Franklin, F.A. 1973. On the original distribution of the asteroids. *Icarus* 20: 422-436.

Message, P.J. 1966. On nearly-commensurable periods in the restricted problem of three bodies, with calculations of the long-period variations in the interior 2:1 case. In *The Theory of the Orbits in the Solar System and in Stellar Systems,* ed. G. Contopoulos (London: Academic Press), pp. 197-222.

Poincaré, M.H. 1902. Sur les planètes du type d'Hécube. *Bull. Astron.* 19: 298-310.

Scholl, H., and Froeschlé, C. 1974. Asteroidal motion at the 3/1 commensurability. *Astron. Astrophys.* 33: 455-458.

Scholl, H., and Froeschlé, C. 1975. Asteroidal motion at the 5/2, 7/3, and 2/1 resonances. *Astron. Astrophys.* 42: 457-463.

Scholl, H., and Giffen, R. 1974. Stability of asteroidal motion in the Hecuba gap. In *The Stability of the Solar System and of Small Stellar Systems,* ed. Y. Kozai (Dordrecht: D. Reidel) pp. 77-79.

Schubart, J. 1964. Long-period effects in nearly commensurable cases of the restricted three-body problem. *Smithsonian Astrophys. Obs. Special Report* No. 149.

Schubart, J. 1966. Special cases of the restricted problem of three bodies. In *The Theory of Orbits in the Solar System and in Stellar Systems,* ed. G. Contopoulos (London: Academic Press), pp. 187-193.

Schubart, J. 1968. Long-period effects in the motion of Hilda-type planets. *Astron. J.* 73: 99-103.

Schubart, J. 1970. Minor planets on commensurable orbits with approaches to Jupiter. In *Periodic Orbits, Stability and Resonances,* ed. G.E.O. Giacaglia (Dordrecht: D. Reidel) pp. 45-52.

Schubart, J. 1978. New results on the commensurability cases of the problem Sun-Jupiter-Asteroid. In *Dynamics of Planets and Satellites and Theories of their Motion,* ed. V. Szebehely (Dordrecht: D. Reidel), pp. 137-143.

Schubart, J. 1979. Asteroidal motion at commensurabilities treated in three dimensions. In *Dynamics of the Solar System,* ed. R.L. Duncombe (Dordrecht: D. Reidel), in press.

Schubart, J., and Stumpff, P. 1966. On an N-body Program of high accuracy for the computation of ephemerides of minor planets and comets. *Veröff. Astron. Rechen-Inst.* No. *18.*

Schweizer, F. 1969. Resonant asteroids in the Kirkwood gaps and statistical explanation of the gaps. *Astron. J.* 74: 779-788.

Sinclair, A.T. 1969. The motions of minor planets close to commensurabilities with Jupiter. *Mon. Not. Roy. Astron. Soc.* 142: 289-292.

Takenouchi, T. 1962. On the characteristic motion and the critical argument of asteroid 279 Thule. *Ann. Tokyo Obs.* 7: 191-208.

Wetherill, G.W. 1967. Collisions in the asteroid belt. *J. Geophys. Res.* 72: 2429-2444.

Wiesel, W.E. 1976. A statistical theory of resonance motion in the Sun-Jupiter system. *Celes. Mech.* 13: 3-37.

Williams, J.G. 1973. Meteorites from the asteroid belt? *Eos: Trans. Amer. Geophys. Union* 54: 233.

Williams, J.G., and Wetherill, G. 1973. Long-term orbital evolution of 1625 Toro. *Astron. J.* 78: 510-515.

Zimmerman, P.D., and Wetherill, G.W. 1973. Asteroidal source of meteorites. *Science* 182: 51-53.

THE DYNAMICAL EVOLUTION OF THE HIRAYAMA FAMILY

YOSHIHIDE KOZAI
Tokyo Astronomical Observatory

First a review is made of the assumption in classical linear theory for the secular perturbations on which the Hirayama families are based, and of the behavior of the secular perturbations when higher degree and higher order terms are included. Particularly it is shown numerically how the secular motions of the perihelion and the ascending node will change; it is demonstrated that the sum of the proper longitude of the perihelion and that of the ascending node is definitely not a stable quantity as the classical theory would predict. In addition 72 families of the asteroids are tabulated with the numbers of member asteroids (according to the method of Kozai, 1979). The distribution of the numbered asteroids with respect to the semimajor axis is compared with that for asteroids that do not belong to any of the families.

Families of asteroids were discovered by Hirayama (1918, 1919, 1920, 1923, 1928) by grouping the asteroids according to their semimajor axis, the proper eccentricity and the proper inclination, and they were later re-examined by Brouwer (1951) and others (Arnold 1969; Lindblad and Southworth 1971; Williams 1971; see the chapter by Gradie *et al.* in this book).

The semimajor axis, which corresponds to the total energy in the two-body problem, is a stable quantity. Namely, the perturbations in it are only short-periodic with amplitude on the order of Jupiter's mass, except for asteroids whose mean motions are nearly commensurable with that of Jupiter. For such commensurable asteroids long-periodic perturbations also appear, with a period of about 200 yr and an amplitude on the order of the square root of Jupiter's mass. Both short- and long-periodic terms include the mean longitudes, of the asteroid in question and/or Jupiter, in the arguments of the trigonometric terms.

Besides the periodic perturbations, secular terms may appear in some of the orbital elements. In the expressions of the secular perturbations none of the mean longitudes of the asteroid and Jupiter appears. As the secular terms are of very long period or expressed by polynomials of time, the effects are accumulated as the time increases. As the secular perturbations do not appear in the expression of the semimajor axis, it is considered that the semimajor axis is stable.

The differential equations for the secular perturbations are reduced to those for a dynamical system with two degrees of freedom so that the semimajor axis can be regarded as constant; i.e. the mean longitude does not appear in the equations. When the eccentricity and the inclination of the asteroid are very small (as small as the square root of Jupiter's mass) and when all the terms on the order of the square root of Jupiter's mass are neglected, the system of the equations is reduced to two sets of linear differential equations of the second order, the two sets being independent of each other. The variables are $e \cos \bar{\omega}$ and $e \sin \bar{\omega}$ for one set and $\sin i \cos \Omega$ and $\sin i \sin \Omega$ for the other, where e, i, $\bar{\omega}$ and Ω are respectively the eccentricity, the inclination, the longitude of the perihelion and that of the ascending node.

For major planets the secular perturbations can be treated in a similar way, and the differential equations for this case are reduced to two sets of linear differential equations of order 18 for 9 planets; each of the variables can be expressed by sums of nine trigonometric terms whose frequencies can be determined by solving two sets of proper equations. Therefore, the variables for the major planets can be regarded as known functions of time when they appear in the equations for the secular perturbations of asteroids.

The variables for the asteroids are expressed by sums of free and forced oscillations which are caused by secular perturbations in the orbital elements of the major planets. The amplitude of the free oscillation for $e \cos \bar{\omega}$ and $e \sin \bar{\omega}$ is called the proper eccentricity, as it reduces to the eccentricity if there were no forced oscillations, and for the other two variables it is the proper inclination. Of course, the proper eccentricity and the proper inclination are constant in the theory and are stable. Therefore, when some asteroids with similar values of the semimajor axes, the proper eccentricities and the proper inclinations are found, it can be assumed that they have a common origin. In this manner Hirayama discovered the families.

I would like to make clear what Hirayama did as there have been several possible misquotings of his work in subsequent papers, except Brouwer's. To do this I shall quote the following sentences from three papers.

Arnold (1969) wrote, "Hirayama (1928, 1933) first showed the existence of families; that is, groups of asteroids whose semimajor axis a, eccentricity e, and inclination i (or sin i) are closely clustered around certain special values. Brouwer (1951) restudied the problem in terms of the proper elements corrected for periodic perturbations produced by secular variations of the major planets."

The description of Lindblad and Southworth (1971) is similar to that of Arnold as they wrote, "Hirayama (1918, 1928) has shown the existence of families, i.e., groups of asteroids with almost equal values of the orbital elements a, e and i. Brouwer (1951), who restudied this problem using the proper elements, has added several new families."

Williams (1971) wrote, "Except for the work of Hirayama, (1918) which used osculating elements, all the above studies (Hirayama 1923, 1928; Brouwer 1951; Arnold 1969) looked for clusterings of the semimajor axis a, the proper eccentricity e', and proper inclination i'."

The description by Williams, different from those by the others, is correct although I do not completely agree with him. It is true that Hirayama started his discussion by making a statistical study of the osculating elements of a, e and i in his first paper (1918) and found several clusterings. However, he then computed values of $p = i$ sin Ω and $q = i$ cos Ω, and $u = e$ sin $\bar{\omega}$ and $v = e$ cos $\bar{\omega}$ for asteroids belonging to the three clusterings which he later called Koronis, Eos and Themis families; he plotted points whose coordinates are q and p and v and u for each asteroid in q,p and v,u planes. Thus it was found that the points are distributed along a circumference whose center is nearly coincident with the pole of Jupiter's orbital plane in the q,p plane for each clustering and that in the v,u plane also the points are distributed in a similar way. He, of course, knew the physical meanings of their distributions. Although he did not use the technical terms of the proper eccentricity and the proper inclination in his first three papers (1918, 1919, 1920) he depended upon the idea of the secular perturbations of asteroids to derive his conclusion that the asteroids in the same clustering had a common origin and had originated from one parent asteroid. It was not before plotting the points that he decided to give the name of "family" to the clusterings. In this sense I have mentioned that Hirayama found the families by using the proper elements.

However, in his first three papers he computed the secular perturbations of asteroids by Jupiter's action, under the assumption that Jupiter's orbit was not perturbed by any other planet. In this way the proper inclination is equal to the inclination with respect to the orbital plane of Jupiter; this is the reason why the centers of the circumferences in the q,p planes could not coincide exactly with Jupiter's pole and those in the v,u planes did not coincide with his theoretical points.

There is no doubt that in his two main papers Hirayama (1923, 1928) computed the secular perturbations by including all planets which are perturbed by the other planets and derived values of the proper elements of the asteroids. Therefore, it is clear that Hirayama used the proper elements to group the asteroids.

The frequencies of the free oscillations are expressed by linear combinations of the masses of the disturbing major planets with coefficients depending on the semimajor axes only. It can be easily proved that the two frequencies are the same in absolute value and are opposite in sense in the classical linear theory. If the argument of the free oscillation for $e \cos \bar{\omega}$ and $e \sin \bar{\omega}$ is called the proper longitude of the perihelion, and that for $\sin i \cos \Omega$ and $\sin i \sin \Omega$ the proper longitude of the ascending node, it can be said that the proper longitude of the perihelion moves with the same speed as that of the ascending node but in opposite directions. Therefore, when Brouwer (1951) re-examined families of asteroids, he argued that the ages of the families are related to scatterings of the sums of the two proper longitudes, while Arnold (1969), and Lindblad and Southworth (1971) tried to find "jet streams" of asteroids, for which the sums have nearly equal values in addition to the clusterings of the other three elements.

However, when the secular perturbations are treated more carefully, namely by including higher-degree and higher-order terms for the eccentricities, the inclinations and the disturbing masses, the behavior of the solutions becomes quite different from that of the linear classical theory, particularly for asteroids with high eccentricity and/or high inclination. In fact, for four or more of the numbered asteroids the arguments of perihelion, i.e. the longitudes of the perihelia minus those of the ascending nodes, are found to librate (Kozai 1962, 1979). This means that for these asteroids the secular motions of the longitudes of the perihelia are the same in directions and in absolute values as those of the ascending nodes on the average. This result is quite different from that in classical linear theory. Even though there is no libration in the argument of perihelion for most of the asteroids, the secular motions, the eccentricity and the inclination vary rather strongly as functions of the argument of perihelion for high eccentricity and inclination asteroids. This also is not expected from the classical theory. Therefore, I have proposed an alternate method for grouping the asteroids into families (Kozai 1979).

Williams (1969) and Yuasa (1973) developed secular perturbation theories of asteroids by including higher-order terms of Jupiter's mass and higher-degree terms of the eccentricities and the inclinations of not only the concerned asteroid but also the disturbing planets. Their solutions clearly show that the secular motion of the proper longitude of the perihelion deviates considerably from that of the node in their absolute values.

In Table I numerical data for the secular motions of the proper longitudes of the perihelion and of the ascending node are given as functions of the semimajor axis, a, in astronomical units. The computations are based on Yuasa's theory and include effects of the earth, Mars, Jupiter and Saturn; however, the second-order part is computed for Jupiter only. Under the heading b the secular rate for the perihelion derived by the classical linear

TABLE I

Data for Secular Motions of Perihelion and Node.

a	b	A	B	C	A'	B'	C'	D	D'
3.60	158″01	188″87	497″21	-1833″00	-163″30	-1833″00	929″50	17″29	-4″18
3.55	147.59	184.72	418.89	-1595.14	-153.80	-1595.14	809.13	20.17	-4.42
3.50	138.08	180.71	353.56	-1392.94	-145.30	-1392.94	706.75	22.71	-4.74
3.45	129.36	182.78	298.85	-1220.32	-137.99	-1220.32	619.30	27.92	-5.28
3.40	121.36	202.60	252.89	-1072.34	-132.41	-1072.34	544.30	41.67	-6.36
3.35	113.99	298.43	214.16	-945.01	-130.77	-945.01	479.72	93.14	-9.10
3.30	107.20	1646.17	181.42	-835.04	-154.35	-835.04	423.92	770.30	-24.19
3.25	100.91	1182.30	153.68	-739.75	-63.73	-739.75	375.55	541.40	18.06
3.20	95.09	219.09	130.13	-656.90	-83.59	-656.90	333.47	62.62	5.29
3.15	89.69	134.75	110.08	-584.66	-83.50	-584.66	296.76	23.08	2.70
3.10	84.67	108.32	92.98	-521.48	-80.72	-521.48	264.64	12.31	1.63
3.05	79.99	94.91	78.38	-466.07	-77.25	-466.07	236.46	7.89	1.07
3.00	75.62	86.13	65.89	-417.35	-73.63	-417.35	211.68	5.64	0.74
2.95	71.54	79.50	55.20	-374.41	-70.05	-374.41	189.82	4.32	0.53
2.90	67.73	74.06	46.04	-336.47	-66.57	-336.47	170.51	3.47	0.39
2.85	64.15	69.41	38.19	-302.87	-63.24	-302.87	153.40	2.90	0.29
2.80	60.79	65.31	31.45	-273.06	-60.07	-273.06	138.22	2.50	0.22
2.75	57.64	61.67	25.67	-246.56	-57.06	-246.56	124.72	2.23	0.17
2.70	54.67	58.43	20.71	-222.96	-54.20	-222.96	112.69	2.07	0.14
2.65	51.87	55.60	16.46	-201.90	-51.48	-201.90	101.96	2.04	0.11
2.60	49.24	53.38	12.83	-183.08	-48.91	-183.08	92.37	2.23	0.09
2.55	46.75	52.83	9.72	-166.24	-46.45	-166.24	83.79	3.18	0.09

TABLE I (Continued)

Data for Secular Motions of Perihelion and Node.

a	b	A	B	C	A'	B'	C'	D	D'
2.50	44".40	-381".14	7".07	-151".16	-50".42	-151".16	76".11	-213".64	-3".05
2.45	42.18	39.21	4.83	-137.64	-42.09	-137.64	69.22	-1.37	0.01
2.40	40.08	39.08	2.93	-125.52	-40.00	-125.52	63.05	-0.39	0.02
2.35	38.10	37.68	1.35	-114.66	-38.03	-114.66	57.53	-0.11	0.02
2.30	36.22	36.04	0.04	-104.95	-36.18	-104.95	52.59	0.00	0.01
2.25	34.44	34.37	-1.01	-96.29	-34.43	-96.29	48.19	0.06	0.01
2.20	32.76	32.74	-1.82	-88.64	-32.77	-88.64	44.30	0.08	0.01
2.15	31.17	31.16	-2.40	-81.97	-31.22	-81.97	40.91	0.09	0.01
2.10	29.67	29.64	-2.71	-76.31	-29.79	-76.31	38.03	0.09	0.01
2.05	28.26	28.13	-2.71	-71.79	-28.53	-71.79	35.74	0.09	0.01
2.00	26.94	26.28	-2.27	-68.70	-27.84	-68.70	34.17	0.08	0.00
1.95	25.72	-6454.46	-1.13	-67.67	-6491.23	-67.67	33.65	0.07	0.00
1.90	24.61	23.92	1.28	-70.07	-25.40	-70.07	34.85	0.07	0.00
1.85	23.63	22.94	6.31	-79.11	-25.68	-79.11	39.39	0.06	0.00
1.80	22.84	21.23	17.59	-103.11	-30.37	103.11	51.44	0.06	0.00

theory is given in the unit of arc seconds per year. Here the secular rate for the perihelion is expressed by $A + Be^2 + C \sin^2 i$ and that for the node has primed coefficients. Under D and D' the second-order parts are given. However D and D' as well as the effects of the eccentricities and the inclinations of the disturbing planets and of the forced eccentricities and the forced inclinations of the asteroid are included in A and A'. For any asteroid with nearly commensurable mean motion with that of Jupiter the second-order terms are larger than or almost equal to the first-order terms, and therefore even if the eccentricity and the inclination of the asteroid are very small, the sum of the two proper longitudes is not a stable quantity at all and any conclusion on the age of the family cannot be derived by their scatterings. Large differences between b and A or A' near $a = 1.95$ are due to the secular commensurabilities between b and the secular motions of the perihelion($23''086$) or of the node ($-25''734$) of Mars which produce large forced oscillations and their effects appear in A and A'.

Kozai and Yuasa (1974) tried to determine the ages of some of the families by finding correlations between the sum of the proper longitudes and the sum of the two secular motions for asteroids belonging to the same family. However, the ages thus determined are not realistic, namely too short, on the order of 10^6 yr.

Kozai (1979) has tried to group the numbered asteroids into families according to clusterings of the semimajor axis, the minimum value of the inclination, i_m, and the value of $\Theta = (1-e^2)^{1/2} \cos i$, where the inclination is referred to the orbital plane of Jupiter. It is assumed that all the disturbing planets are moving on the same plane with circular orbits, Θ is constant, and the eccentricity and the inclination change as functions of the argument of perihelion. When the argument of perihelion is $0°$ and $180°$, the eccentricity is minimum and the inclination is maximum, and when the argument of perihelion is $90°$ and $270°$ the eccentricity is maximum and the inclination is minimum. Therefore, i_m is the value of the inclination when the argument of perihelion is $90°$ and $270°$. When Θ and i_m are known, the secular variations of the eccentricity and the inclination can be derived as functions of the argument of perihelion (Kozai 1979).

Of 2125 numbered asteroids I have grouped about three quarters into 72 families. In Table II the 72 families with necessary data are shown and in Table III the numbers of the member asteroids are given for each family. (Tables II and III can be found at the end of this chapter.) In Table II the first column shows the serial number in increasing order of the mean semimajor axis, the second column the name of the top member asteroid, that is, the least number in the family, the third column has the number of the member asteroids, the fourth column the volume of the space occupied by the family in the three-dimensional space of a, Θ and i_m, and the fifth column shows the number density. Then for the semimajor axis, Θ and i_m their boundary limits, the mean values and their standard deviations are given. The unit volume in

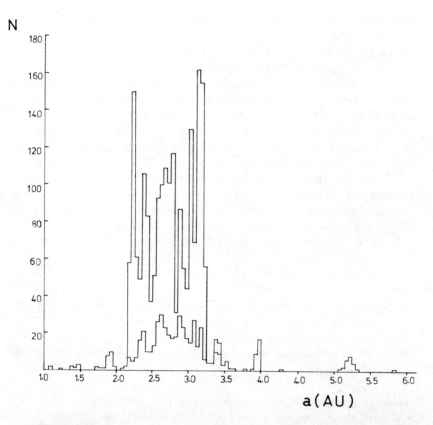

Fig. 1. Distribution of the numbered asteroids that are numbered in the asteroid *Ephemerides,* and of those not belonging to any of the 72 families (lower lines between 2.0 and 3.5 AU), with respect to the semimajor axis.

the space is defined by the cube whose sides are 0.1 AU in a, 0.01 in Θ and 2° in i_m. The average number density is 5 to 6 in denser parts, i.e., areas with the values of Θ nearly equal to 1 and small values of i_m and in the main asteroid belt; however, it is reduced as the value of Θ is decreased, the value of i_m is increased and as the value of the semimajor axis is deviated from that in the main belt.

Of the 72 families seven are those originally found by Hirayama. These are nos. 1, 6, 17, 31, 41, 49 and 56 which are respectively, Flora, Phocaea, Maria, Pallas, Koronis, Eos and Themis families. Of these Koronis, Eos and Themis families were discovered by him in his first paper (1918) and are among the densest families with the highest population, i.e. they are clustered densely in small volumes in phase space. The density is low, however, for the Pallas family, although it is a remarkable one (Kozai 1979) with low value of Θ and high value of i_m, and the Pallas as well as Phocaea and Maria families are clearly isolated from other asteroids in phase space. Flora family has the highest population among the 72 although it occupies a very wide region without any definite boundary. However, it could not be divided into sub-families as Brouwer (1951) did. As has been shown, the original seven families each have individual characteristics.

In the process of the grouping, clusterings with a small number of members, 3 or 4, are usually disregarded. However, there are still families in Table II with 3 or 4 members, because their densities are very high. Also it should be mentioned that the grouping is made in a rather arbitrary way and there is no definite criterion. It is therefore difficult to compare the new families here with those found by other authors.

Of the 72 families some are very compact whereas others are not so compact but rather loose without clear boundary. However, even such loose extended families are bounded in part by the surface where commensurable relations for either mean motions or secular motions are nearly satisfied (Williams 1969). This suggests that the origins of such families are related to some dynamical effects.

In Fig. 1 the distribution of the numbered asteroids with respect to the semimajor axis together with that of those not belonging to any of the families are given. In the latter distribution which is shown by lower lines between 2.0 and 3.5 AU, it seems to me that the shapes of the Kirkwood gaps are a little different from those in the whole distribution, and I would therefore like to examine these diagrams more carefully to understand the origins of the families.

Acknowledgment. Computations in this chapter were made by use of the FACOM 230-58 Computer at the Tokyo Astronomical Observatory.

TABLE II

Families of Asteroids and their Data

No		N	A	D	Semimajor axis		$(1-e^2)^{1/2}$	cos i	minimum of i	
1	8	259	15	17	2.164-2.301	2.229±30	0.966-0.999	0.986± 6	1.°26- 7.°91	4.°62±1.44
2	80	8	0.5	15	2.259-2.312	2.289±16	0.972-0.983	0.978± 4	8.16- 9.99	8.87±0.65
3	428	7	0.2	34	2.307-2.326	2.316± 6	0.974-0.988	0.981± 5	5.55- 7.10	6.17±0.42
4	12	8	0.4	23	2.321-2.350	2.334± 9	0.961-0.968	0.964± 2	5.59- 9.06	7.80±1.30
5	261	13	0.3	53	2.328-2.395	2.363±23	0.984-0.991	0.987± 2	2.48- 3.53	3.10±0.33
6	25	34	4	8	2.305-2.425	2.367±34	0.873-0.909	0.889± 9	20.92-22.91	21.80±0.57
7	51	5	0.14	35	2.354-2.384	2.367±13	0.974-0.985	0.981± 4	9.84-10.70	10.16±0.29
8	84	16	18	14	2.352-2.399	2.369±14	0.957-0.970	0.964± 4	8.08-11.94	10.06±1.19
9	4	40	3	16	2.345-2.402	2.372±14	0.970-0.991	0.984± 6	4.32- 8.61	6.20±1.14
10	30	4	0.08	49	2.365-2.394	2.379±11	0.994-0.998	0.995± 2	1.92- 3.34	2.61±0.54
11	6	8	1.5	5	2.377-2.431	2.408±19	0.941-0.959	0.952± 6	12.06-15.05	13.45±0.98
12	343	4	0.15	27	2.411-2.424	2.417± 6	0.965-0.975	0.970± 4	2.26- 4.53	3.41±0.94
13	470	6	0.2	30	2.404-2.464	2.426±19	0.983-0.988	0.985± 2	7.38- 8.72	7.83±0.46
14	42	16	2.4	7	2.402-2.469	2.435±20	0.961-0.979	0.972± 6	5.79- 9.82	7.56±1.14
15	11	68	4	19	2.402-2.490	2.445±23	0.981-0.999	0.989± 5	1.77- 6.21	3.89±1.20
16	421	12	0.3	39	2.532-2.550	2.542± 6	0.956-0.979	0.969± 7	6.35- 7.83	7.02±0.53
17	170	25	0.3	72	2.521-2.597	2.547±17	0.957-0.969	0.961± 3	14.35-15.11	14.80±0.25
18	29	21	1.7	12	2.514-2.594	2.552±22	0.983-0.996	0.991± 4	2.91- 6.17	4.87±1.07
19	5	7	0.2	36	2.530-2.578	2.554±16	0.972-0.976	0.975± 2	4.83- 6.80	5.73±0.65
20	101	13	1.7	8	2.548-2.599	2.575±15	0.958-0.978	0.968± 6	10.03-13.42	11.66±1.15
21	14	16	0.8	21	2.551-2.601	2.577±15	0.962-0.979	0.971± 5	7.50- 9.32	8.53±0.54
22	686	3	.015	196	2.565-2.589	2.580±10	0.9268-0.9282	0.927± 1	15.83-16.74	16.33±0.38

TABLE II (Continued)

Families of Asteroids and their Data

No	N	A	D	Semimajor axis		$(1-e^2)^{1/2}$	$\cos i$	minimum of i	
23	119	1.5	13	2.551-2.636	2.593±26	0.981-0.993	0.987± 3	6°.27- 9°.12	7°.90±0°.87
24	258	0.3	18	2.574-2.636	2.604±21	0.946-0.952	0.949± 2	12.07-13.88	13.18±0.67
25	15	4	15	2.605-2.689	2.542±25	0.957-0.980	0.966± 5	9.93-14.05	12.07±0.98
26	78	.06	18	2.616-2.687	2.648±27	0.961-0.976	0.969± 5	8.08- 9.24	8.44±0.51
27	3	2.1	9	2.653-2.698	2.670±13	0.935-0.958	0.949± 7	11.35-15.36	13.47±1.31
28	123	.08	88	2.675-2.699	2.689± 7	0.983-0.993	0.988± 3	6.89- 7.55	7.26±0.25
29	869	0.3	36	2.675-2.701	2.691± 8	0.963-0.977	0.970± 5	6.18- 7.85	6.94±0.60
30	26	13	8	2.611-2.802	2.718±52	0.972-1.000	0.989± 7	1.08- 5.95	4.14±1.27
31	2	13	0.6	2.630-2.785	2.723±57	0.779-0.822	0.804±13	26.93-30.74	28.92±1.11
32	146	.01	466	2.705-2.744	2.725±16	0.973-0.978	0.976± 2	11.93-12.04	12.00±0.04
33	156	.05	120	2.719-2.741	2.730± 6	0.949-0.955	0.952± 2	8.56- 9.32	8.91±0.27
34	54	4	4	2.708-2.796	2.753±29	0.944-0.960	0.952± 6	10.14-15.63	13.15±1.60
35	387	0.4	19	2.722-2.785	2.753±23	0.925-0.935	0.931± 3	14.43-15.57	15.13±0.35
36	1	6	12	2.703-2.810	2.763±30	0.969-0.992	0.980± 6	6.32-11.26	8.21±1.17
37	82	0.3	19	2.751-2.773	2.765± 8	0.961-0.970	0.966± 3	2.89- 5.54	3.98±1.32
38	143	0.6	13	2.761-2.803	2.778±18	0.960-0.975	0.968± 5	11.92-13.81	12.91±0.60
39	139	0.3	29	2.771-2.799	2.780± 9	0.961-0.970	0.966± 3	7.45- 9.93	8.57±0.84
40	174	.24	41	2.849-2.869	2.860± 7	0.962-0.977	0.971± 5	10.95-12.57	11.63±0.57
41	158	.15	364	2.840-2.905	2.873±18	0.996-1.000	0.998± 1	1.68- 2.80	2.24±0.26
42	293	.18	34	2.850-2.901	2.878±17	0.957-0.964	0.960± 2	13.75-14.75	14.25±0.35
43	22	0.5	24	2.893-2.939	2.911±14	0.968-0.979	0.975± 3	10.66-12.67	11.59±0.63
44	280	0.6	23	2.900-2.946	2.918±14	0.985-0.995	0.990± 2	5.44- 7.92	6.92±0.87

TABLE II (Continued)

Families of Asteroids and their Data

No	N	A	D	Semimajor axis		$(1-e^2)^{1/2} \cos i$		minimum of i	
45	16	.13	45	2.907-2.940	2.924±10	0.992-1.000	0.996± 2	2°.13- 3.13	2°.61±0.36
46	179	0.3	29	2.968-3.016	2.992±18	0.984-0.990	0.987± 2	7.41- 9.32	8.54±0.65
47	117	0.4	20	2.984-3.014	2.997± 8	0.958-0.970	0.964± 3	12.94-15.40	14.04±0.82
48	150	1.4	14	2.977-3.062	3.010±25	0.990-0.998	0.994± 2	2.52- 6.74	4.90±1.45
49	221	.17	554	2.993-3.030	3.015± 7	0.978-0.985	0.982± 1	9.47-10.78	10.08±0.30
50	69	3	6	2.979-3.080	3.028±28	0.961-0.978	0.972± 5	7.91-11.90	10.03±1.18
51	133	.13	37	3.042-3.073	3.057±11	0.978-0.984	0.981± 2	7.75- 9.20	8.37±0.52
52	399	.16	37	3.038-3.070	3.059±10	0.964-0.970	0.968± 2	12.39-14.07	13.25±0.77
53	1383	.07	55	3.056-3.076	3.066± 8	0.983-0.997	0.991± 6	1.39- 1.91	1.59±0.20
54	137	1.3	19	3.090-3.146	3.121±17	0.946-0.960	0.954± 4	13.04-16.32	14.47±0.97
55	181	0.8	14	3.112-3.138	3.122± 7	0.928-0.946	0.937± 5	16.19-19.63	17.64±1.19
56	24	0.7	113	3.093-3.178	3.139±23	0.980-0.995	0.988± 3	1.08- 2.18	1.71±0.30
57	250	1.6	14	3.109-3.179	3.139±21	0.961-0.976	0.967± 4	12.07-15.14	13.45±0.95
58	130	0.8	9	3.109-3.196	3.148±29	0.898-0.905	0.902± 2	18.35-20.99	19.64±0.92
59	886	1.0	6	3.114-3.176	3.148±22	0.921-0.939	0.931± 6	14.21-16.00	15.47±0.64
60	10	6	9	3.087-3.227	3.149±42	0.978-0.999	0.988± 5	2.20- 6.48	4.29±1.25
61	152	4	12	3.090-3.219	3.152±30	0.956-0.983	0.973± 8	9.66-11.90	10.70±0.60
62	227	.08	38	3.133-3.172	3.158±17	0.954-0.959	0.957± 2	8.77- 9.57	9.39±0.49
63	48	4	13	3.090-3.220	3.160±39	0.971-0.993	0.983± 6	6.60- 9.62	8.28±0.81
64	154	1.4	8	3.156-3.210	3.187±17	0.926-0.945	0.937± 7	16.95-19.75	18.46±1.00
65	199	0.8	8	3.162-3.217	3.189±18	0.946-0.960	0.953± 4	12.67-14.61	13.48±0.65
66	445	0.9	5	3.167-3.223	3.192±19	0.902-0.920	0.910± 7	21.03-22.84	21.90±0.70

TABLE II (Continued)

Families of Asteroids and their Data

No	N	A	D	Semimajor axis		$(1-e^2)^{1/2}$	$\cos i$	minimum of i	
67	15	1.5	10	3.157-3.228	3.197±21	0.943-0.960	0.954± 5	15°09-17.53	16°30±0.73
68	7	0.2	35	3.185-3.210	3.198±12	0.980-1.000	0.992± 7	1.15- 1.95	1.59±0.35
69	6	0.2	33	3.197-3.227	3.211±10	0.961-0.968	0.965± 2	12.87-14.59	13.56±0.70
70	4	.16	24	3.207-3.245	3.226±14	0.927-0.937	0.931± 3	20.54-21.40	21.11±0.35
71	5	0.4	13	3.377-3.422	3.405±18	0.989-0.996	0.992± 2	5.20- 7.61	6.53±0.93
72	8	0.6	13	3.386-3.445	3.409±23	0.988-0.998	0.994± 3	2.09- 4.11	3.14±0.70

TABLE III

Member Asteroids of the Families

No. 1														
8	40	43	72	149	207	244	254	270	281	291	296	298	315	317
336	341	352	364	367	376	422	440	443	453	496	525	540	548	553
574	641	685	700	703	707	711	722	728	730	736	749	763	770	782
800	802	809	810	813	819	822	823	825	827	831	836	837	841	843
851	857	864	871	883	901	905	913	915	929	935	937	939	951	956
960	963	967	985	1016	1026	1034	1037	1047	1052	1055	1056	1058	1060	1078
1088	1089	1110	1117	1120	1123	1126	1130	1131	1133	1141	1147	1150	1153	1156
1185	1188	1195	1216	1218	1219	1225	1249	1270	1274	1293	1307	1314	1324	1335
1338	1344	1365	1370	1374	1376	1377	1382	1387	1396	1399	1401	1405	1412	1415
1418	1419	1422	1446	1449	1451	1455	1472	1476	1480	1492	1494	1496	1500	1513
1514	1518	1523	1527	1530	1536	1549	1562	1563	1577	1590	1601	1602	1608	1610
1619	1621	1622	1631	1634	1636	1648	1651	1652	1661	1663	1666	1667	1675	1676
1682	1696	1699	1703	1704	1705	1707	1713	1717	1720	1729	1733	1736	1738	1739
1744	1752	1763	1785	1789	1790	1793	1797	1798	1806	1807	1810	1814	1818	1820
1822	1823	1829	1830	1831	1837	1842	1850	1855	1856	1857	1879	1885	1897	1899
1900	1905	1938	1944	1946	1950	1967	1978	1990	1991	1997	2004	2013	2017	2018
2019	2030	2031	2034	2036	2037	2056	2070	2071	2076	2080	2087	2088	2093	2094
2110	2112	2119	2121											
No. 2														
80	136	853	896	1182	1224	1594	2028							
No. 3														
428	646	689	1526	1959	2021	2024								

TABLE III (Continued)

Member Asteroids of the Families

Family															
No. 4	12	220	753	783	870	1507	1664	1710							
No. 5	261	554	1183	1378	1511	1559	1757	1846	1924	1932	2007	2012	2066		
No. 6	25	290	323	326	502	852	950	1164	1170	1192	1310	1318	1322	1367	1565
	1568	1573	1575	1591	1657	1660	1803	1816	1817	1883	1884	1934	1942	1963	1987
	2000	2014	2050	2055											
No. 7	51	172	230	287	1490										
No. 8	84	115	219	249	284	313	432	486	584	855	916	1504	1538	1694	1718
	1937														
No. 9	4	7	9	60	63	113	161	163	169	306	337	442	757	854	917
	933	1077	1217	1273	1279	1290	1432	1448	1454	1475	1646	1697	1709	1781	1800
	1906	1929	1933	1949	1979	1989	2011	2029	2045	2086					
No. 10	30	1544	1551	1976											

TABLE III (Continued)

Member Asteroids of the Families

Family															
No. 11	6	234	304	463	865	930	1236	1466							
No. 12	343	571	1080	1909											
No. 13	470	620	932	1478	1998	2002									
No. 14	42	67	118	192	198	474	585	622	647	889	1151	1155	1528	1665	1689
	1743														
No. 15	11	17	19	20	21	44	79	83	112	126	131	135	138	142	182
	189	248	299	302	335	435	477	495	556	557	650	684	750	752	877
	902	969	1012	1066	1076	1137	1145	1152	1190	1257	1267	1296	1311	1358	1375
	1381	1393	1394	1493	1586	1589	1643	1650	1740	1768	1770	1773	1804	1914	1923
	1928	1964	1966	1972	2026	2072	2081	2113							
No. 16	421	518	1018	1205	1325	1391	1501	1587	1644	1888	1931	2118			
No. 17	170	292	472	546	575	616	652	660	695	714	727	751	787	875	879
	897	994	1158	1160	1215	1368	1379	1677	1996	2089					

TABLE III (Continued)

Member Asteroids of the Families

No.															
No. 18	29	32	46	111	151	355	535	797	799	974	1044	1121	1292	1343	1430
	1447	1722	1772	1854	1907	1945									
No. 19	5	232	262	678	1460	1756	1948								
No. 20	101	134	157	402	409	550	672	712	817	1094	1106	1230	1347	2079	
No. 21	14	56	544	603	606	666	1096	1136	1226	1281	1426	1515	1639	1658	1714
	1860														
No. 22	686	726	1609												
No. 23	119	342	362	389	407	418	438	524	829	1017	1053	1119	1233	1237	1260
	1728	1925	1983	2116											
No. 24	258	397	404	737	923	1473									
No. 25	15	70	85	141	145	166	347	369	390	429	476	484	500	510	597
	625	630	657	682	789	801	812	815	839	920	971	995	997	999	1050

TABLE III (Continued)

Member Asteroids of the Families

1059	1098	1184	1187	1193	1231	1238	1275	1284	1329	1333	1346	1384	1392	1425
1431	1458	1495	1499	1503	1510	1531	1554	1775	1783	1886	1926	1927	1936	1994
2005														

No. 26

78	204	459	498	505	792	990	1179	1516	1843	1935

No. 27

3	97	98	99	218	246	455	591	593	771	779	989	1013	1313	1326
1402	1505	1719												

No. 28

123	377	614	1176	1201	1409	1813

No. 29

869	922	1178	1277	1294	1457	1479	1525	1560	2065	2109

No. 30

26	34	37	53	58	64	66	73	74	77	88	103	110	114	116
124	125	128	140	144	160	177	180	201	203	206	210	213	214	215
224	240	267	272	275	295	301	308	309	310	332	340	363	380	384
394	395	396	454	460	503	569	578	615	632	708	743	808	847	868
873	941	1007	1020	1039	1071	1084	1128	1135	1181	1228	1251	1327	1348	1352
1420	1450	1483	1502	1517	1541	1595	1638	1662	1680	1681	1692	1726	1766	1777
1791	1808	1825	1827	1858	1880	1968	2022	2042	2073	2085	2095	2125		

TABLE III (Continued)

Member Asteroids of the Families

No. 31														
2	531	594	945	1252	1301	1474	1508							
No. 32														
146	1341	1724	1891	1904										
No. 33														
156	187	410	521	1403	1534									
No. 34														
54	155	173	216	226	416	456	485	504	532	598	706	793	888	934
1048	1548													
No. 35														
387	547	599	687	1021	1484	1659								
No. 36														
1	28	38	39	45	55	59	68	93	127	197	200	205	236	237
255	264	278	312	327	346	351	359	371	374	378	417	424	441	444
446	464	479	481	519	527	539	559	560	638	670	716	741	824	858
872	919	953	984	1002	1214	1242	1248	1272	1299	1300	1385	1414	1427	1433
1603	1642	1716	1751	1836	1839	1893	1970	1975	1977	2053	2106			
No. 37														
82	288	821	947	1545										

TABLE III (Continued)

Member Asteroids of the Families

No. 38
143 188 266 365 412 715 860 1140

No. 39
139 322 356 668 675 1655 1730 1734 1795

No. 40
174 242 607 698 891 918 968 1067 1540 1597

No. 41
158 167 208 243 263 277 311 321 452 462 534 658 720 761 811
832 962 993 1029 1079 1100 1223 1245 1289 1336 1350 1363 1389 1423 1442
1482 1497 1570 1618 1635 1725 1741 1742 1745 1762 1774 1802 1824 1848 1878
1894 1912 1913 1955 2051 2092 2117 2123

No. 42
293 392 697 1521 1596 1930

No. 43
22 191 238 542 838 845 906 1149 1189 1194 1211 1334

No. 44
280 307 338 349 467 558 613 627 1068 1092 1124 1308 1616

No. 45
16 1010 1443 1553 1809 1840

TABLE III (Continued)

Member Asteroids of the Families

No. 46														
179	348	701	1046	1234	1411	1417	1532							

No. 47														
117	256	478	482	611	816	850	904	1614						

No. 48														
150	241	271	331	388	427	447	494	514	533	586	655	723	738	1062
1111	1285	1332	1481	1984										

No. 49														
221	320	339	450	513	520	529	562	573	579	590	592	608	633	639
651	653	661	669	742	766	775	798	807	833	876	890	1075	1087	1095
1105	1112	1129	1148	1174	1186	1199	1207	1210	1220	1265	1286	1287	1291	1297
1339	1353	1364	1388	1410	1413	1416	1434	1464	1485	1533	1552	1557	1588	1604
1605	1641	1649	1654	1711	1723	1732	1737	1753	1758	1767	1780	1786	1787	1801
1812	1826	1834	1844	1852	1861	1882	1887	1903	1957	1971	1992	2020	2027	2052
2091	2111	2115	2124											

No. 50														
69	360	368	537	576	634	676	691	731	740	948	952	972	992	1033
1349	1465	1488	1799	2090										

No. 51														
133	202	283	543	964										

TABLE III (Continued)

Member Asteroids of the Families

No. 52														
399	451	1041	1637	1690	1828									
No. 53														
1383	1645	2003	2032											
No. 54														
137	328	373	375	457	493	601	788	791	928	986				
1306	1369	1452	1498	1520	1571	1735	1794	1875	2040					
No. 55														
181	439	683	746	780	912	998	1035	1282	1679	1985				
No. 56														
24	62	90	171	184	222	223	268	316	379	383	431	461	468	492
515	526	555	561	621	637	656	710	767	846	848	936	938	946	954
981	988	991	996	1003	1027	1061	1073	1074	1082	1142	1247	1259	1302	1340
1440	1445	1462	1487	1539	1576	1581	1615	1623	1624	1633	1669	1674	1686	1687
1691	1698	1764	1778	1782	1788	1805	1815	1851	1895	1898	1939	1953	1986	2009
2016	2039	2046	2058											
No. 57														
250	357	489	508	640	681	690	762	818	894	977	1069	1113	1138	1213
1330	1398	1436	1461	1469	1524	1952								

TABLE III (Continued)

Member Asteroids of the Families

No. 58

130

No. 59

886

No. 60

10	49	86	100	104	106	108	122	147	159	175	196	212	245	257
305	333	517	538	577	580	604	609	734	755	758	820	830	835	885
959	1081	1085	1107	1109	1171	1209	1271	1298	1331	1438	1470	1489	1491	1542
1561	1599	1611	1673	1684	1731	1961	1969	2010	2041	2043				

No. 61

152	165	251	252	314	318	366	381	400	408	448	469	488	507	545
589	635	764	840	874	931	943	979	1001	1023	1114	1157	1232	1309	1359
1380	1456	1519	1564	1569	1572	1582	1617	1678	1755	1771	1881	1947	1958	1973
1974	1999													

No. 62

227	648	664

No. 63

48	52	92	94	120	209	259	297	303	325	382	490	530	552	567
583	629	642	645	671	744	769	803	844	861	866	942	976	1008	1015
1032	1043	1086	1161	1163	1254	1255	1258	1283	1315	1351	1356	1395	1408	1421
1424	1463	1558	1760	1776	1811	1859	1890	1918	2025	2069				

TABLE III (Continued)

Member Asteroids of the Families

No. 64	154	436	491	1005	1042	1303	1323	1444	1712	1765	2008	2104			
No. 65	199	511	696	786	805	842									
No. 66	445	1000	1051	1241	1276										
No. 67	57	286	595	618	921	973	983	1070	1175	1304	1371	1546	1567	1701	1832
No. 68	300	828	1229	1253	1956	1962	2114								
No. 69	777	806	859	927	1118	1357									
No. 70	581	892	1101	1838											
No. 71	168	420	940	1167	1280										
No. 72	65	229	566	570	1004	1154	1295	1841							

REFERENCES

Arnold, J. R. 1969. Asteroid families and "jet streams". *Astron. J.* 74: 1235-1242.

Brouwer, D. 1951. Secular variations of the orbital elements of minor planets. *Astron. J.* 56: 9-32.

Hirayama, K. 1918. Groups of asteroids probably of common origin. *Astron. J.* 31: 185-188. *Proc. Phys.-Math. Soc. Japan,* 2nd ser. 9: 354.

Hirayama, K. 1919. Further note on the families of asteroids. *Proc. Phys.-Math. Soc. Japan,* 3rd ser. 1: 52-59.

Hirayama, K. 1920. New asteroids belonging to the families. *Proc. Phys.-Math. Soc. Japan,* 3rd ser. 2: 236-240.

Hirayama, K. 1923. Families of asteroids. *Japan J. Astron. Geophys.* 1: 55-93.

Hirayama, K. 1928. Families of asteroids. Second paper. *Japan J. Astron. Geophys.* 5: 137-162.

Hirayama, K. 1933. Present state of the families of asteroids. *Proc. Japan Academy* 9: 482-485.

Kozai, Y. 1962. Secular perturbations of asteroids with high inclination and eccentricity. *Astron. J.* 67: 591-598.

Kozai, Y. 1979. Secular perturbations of asteroids and comets. In *Dynamics of the Solar System,* ed. R. L. Duncombe (Dordrecht: D. Reidel, in press).

Kozai, Y. and Yuasa, M. 1974. Secular perturbations for asteroids belonging to families. In *Stability of the Solar System and of Small Stellar Systems,* ed. Y. Kozai (Dordrecht: D. Reidel), pp. 81-82.

Lindblad, B. A., and Southworth, R.B. 1971. A study of asteroid families and streams by computer techniques. In *Physical Studies of Minor Planets,* ed. T. Gehrels (NASA SP-267, Washington, D.C.: U.S. Government Printing Office), pp. 337-352.

Williams, J. G. 1969. Secular perturbations in the solar system. Ph.D. dissertation, University of California at Los Angeles.

Williams, J. G. 1971. Proper elements, families, and belt boundaries. In *Physical Studies of Minor Planets,* ed. T. Gehrels, (NASA SP-267, Washington, D.C.: U.S. Government Printing Office), pp. 177-181.

Yuasa, M. 1973. Theory of secular perturbations of asteroids including terms of higher order and higher degree. *Publ. Astron. Soc. Japan* 25: 399-445.

FAMILIES OF MINOR PLANETS

JONATHAN C. GRADIE
Cornell University

CLARK R. CHAPMAN
Planetary Science Institute

JAMES G. WILLIAMS
Jet Propulsion Laboratory

Statistical studies of the distribution of orbits of the asteroids have revealed the existence of groups of minor planets with similar dynamical characteristics. These groups are divided into commensurability groups, regions of the asteroid belt isolated by resonances, and families.

Physical studies of individual family members show that at least the Themis, Eos, Koronis, Nysa/Hertha and Budrosa families are the result of the breakup of discrete parent bodies. Physical studies of asteroids in the Phocaea region indicate that the physical features of these objects are more characteristic of a random selection of main-belt asteroids than of fragments from a single parent body. The complexity of the Flora region may have resulted from considerable collisional evolution of the family members. The diameter frequency distribution of the Themis, Eos and Koronis families are not significantly different from that found for the surrounding field asteroids. The Budrosa and the Nysa/Hertha families are the best examples of families formed from highly differentiated parent bodies.

Physical studies of the smaller families that contain one large and many small objects are presently lacking and the breakup hypothesis can be only marginally confirmed. The origin of most families seems to be the breakup of discrete parent bodies but, in many respects, the large families of unusual but homogeneous composition appear to be anomalous.

The nonrandom distribution of the orbital elements among the minor planets was first noticed and explained by Kirkwood (1867). He identified the gaps in the frequency distribution of semimajor axis, *a,* of the asteroids as corresponding to commensurabilities with Jupiter. In the time since, the structure of the asteroid belt has been found to include distinctive groupings of minor planets (Fig. 1) with either unusual dynamic properties or nearly identical orbital elements. Among these groups are found the Trojan asteroids at two of the Lagrangian points of Jupiter, the Hilda-type asteroids near the 2/3 commensurability with Jupiter and in the main belt the Hirayama families composed of minor planets with nearly identical orbital elements (c.f., Arnold 1969).

Asteroids can be assigned to groups according to various schemes. According to van Houten (1971), the word "group," when used in the context of orbital elements of minor planets, can be used to discern two basic asteroidal divisions: those that have a dynamical cause and those that probably have little or no apparent dynamical cause. In the former group fall the Trojans, the commensurability groups (including the single object Thule at

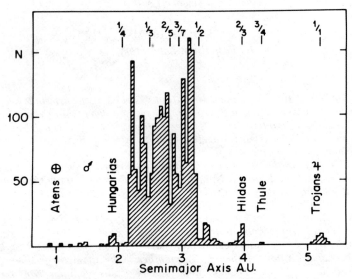

Fig. 1. The helocentric distribution of minor planets in increments of 0.05 AU. The positions of some of the important asteroid groups are indicated. Fractions indicate ratio of orbital periods for the principal dynamical resonances with Jupiter. Families are found in the main belt. Data are from the TRIAD files (Bender *et al.* 1978; see Part VII of this book). (Adapted from Zellner *et al.* 1977.)

the 3/4 commensurability); the Hilda group; and regions of the belt isolated by resonances, i.e., the Hungaria region and the Phocaea region (Williams 1971). In the latter group van Houten included Hirayama families under the assumption that they were not the result of dynamical processes.

The term "family" was coined by Hirayama (1923; see Kozai's chapter) to express his belief that family members in each family were created by the spontaneous fragmentation of a parent asteroid. However, the dynamical families are defined as concentrations of orbital elements in proper orbital element space and are not assumed to be composed of members of a common "lineage." It has been suggested that "group," "association," "cluster," "clumping," or "concentration" be used instead and that the term "family" be reserved only for those associations of asteroids that have common parentage. To resolve this problem and to retain van Houten's (1971) suggestions, it is proposed that a group be identified as an association of asteroids if it can be shown that it is not composed of fragments of a parent body or bodies.

I. IDENTIFICATION OF FAMILIES

The identification of an asteroid family and the determination of family membership requires accurate *proper orbital elements,* i.e., those elements which have been corrected for the perturbations of the major planets and are usually denoted as the primed quantities e', i', ω' and Ω'. The proper orbital elements should be corrected for all long-term and short-term periodic disturbances, a correction difficult to make in practice. Early corrections for the secular variations (e.g., Brouwer and van Woerkom 1950; Brouwer and Clemence 1961) involved only low-order expansions in the eccentricities and inclinations. Williams (1969), Yuaza (1973) and Kozai (1979; also see Kozai's chapter) have improved the precision of the proper elements by developing the theory to accurately handle high inclinations and large eccentricities.

The first discussion of asteroid groupings is found in a series of papers by Hirayama (1918, 1919, 1923, 1928, 1933) in which he presented the results of his search for minor planets in similar orbits. A total of nine obvious clusterings called "families" were found in a, e', and i'. Three of these families, the Themis (H1), the Eos (H2) and the Koronis (H3), are identified in Fig. 2.

Brouwer (1951) confirmed Hirayama's result using a much larger sample of asteroids and improved proper elements. Arnold (1969) extended these earlier studies by testing for statistically significant deviations from a random distribution of asteroid orbits using digital computer techniques.

The Palomar Leiden Survey (PLS) (van Houten et al. 1970) added a new dimension to the study of asteroid families when families were shown to be defined even among the km-sized objects. Five new families consisting mainly of small asteroids were reported. The limits of the family boundaries used in

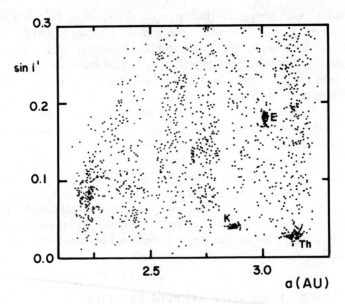

Fig. 2. The heliocentric distribution of the proper inclination for the main-belt asteroids. Three Hirayama families are indicated: Themis (Th), Eos (E), and Koronis (K). The Kirkwood gaps are visible. (Adapted from Brown, Goddard and Kane 1967.)

the PLS are generally wider than those used in most other studies. Family membership for some objects in that study is dubious; however, the overall effect of misidentifications is minor. Nearly half of the asteroids discovered by the PLS belong to families. Also reported in the PLS was the tentative identification of twin families: families that occur in nearly the same interval a and e' while differing somewhat in the intervals of i'. Subsequent studies have not called attention to their existence and their importance remains to be explored.

Detailed searches for families among the orbits of both the numbered asteroids and the high quality PLS objects were performed by Lindblad and Southworth (1971) and Williams (1971, 1975). Lindblad and Southworth searched for orbital similarity such as in meteor streams, and noted that the major difficulty encountered was the selection of the appropriate rejection criteria. Williams (1971) found that the Hungaria and Phocaea regions are not large families but regions of the asteroid belt isolated by resonances (Fig. 3). Williams (1975) reported finding 104 families among 2800 orbits of the numbered asteroids and PLS objects, and noted that nearly half of the asteroids are found in families. The proper element and family membership of the numbered asteroids are tabulated by Williams in Part VII of this book. However, some of these families are composed of few members so their uniqueness is questionable.

The identification of family members and boundaries has been

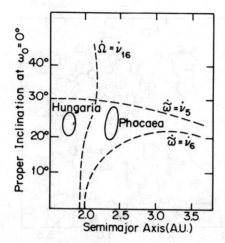

Fig. 3. The isolation of the Hungaria and Phocaea regions by resonances. The terms $\dot{\nu}_5$, $\dot{\nu}_6$ and $\dot{\nu}_{16}$ are the two eccentricity and one inclination frequency perturbation terms caused by Jupiter and Saturn (Brouwer and van Woerkom 1950). For an orbit in the asteroid belt, a resonance occurs where $\dot{\Omega}$ of the orbit equals $\dot{\nu}_{16}$ or where $\tilde{\omega}$ equals $\dot{\nu}_5$ or $\dot{\nu}_6$ (Figure adapted from Williams 1969.)

approached in a variety of ways. Generally, trial families are first identified as obvious groupings of orbital elements and boundaries are drawn somewhat subjectively after initial statistical tests. Final, rigorous tests are then applied to determine the statistical significance of a particular family with respect to a random distribution of asteroid elements. Arnold (1969) assumed the simplest form for the distribution function, i.e., a function that is equivalent to a uniform distribution of objects from 2.15 − 3.35 AU and from 0 − 0.3 in e' and sin i'. Williams (1975) modified this assumption and used a distribution function dependent upon a and sin i', but independent of e'. He considered a family significant only if the Poisson probability of finding the trial family in the random distribution is less than 1/3000.

A different approach to the statistical significance problem has been developed by Carusi and Massaro (1978). They used a cluster analysis that allowed the multivariate distribution of all orbital elements to be reduced to a univariate quasi-Gaussian distribution by means of successive transformations. Their results for the elements a and i' are shown in Fig. 4 and can be compared in a qualitative way with Fig. 2.

This break from the traditional scheme for computing the criteria of orbital similarity has produced some surprising results; only ten families were recognized by Carusi and Massaro. The many smaller families of the previous investigations did not have statistical relevance in their test. This apparent discrepancy between the two statistical methods is disconcerting since it is of vital importance that the dynamic relationship among the members of a family be defined before any physical studies can be given a plausible

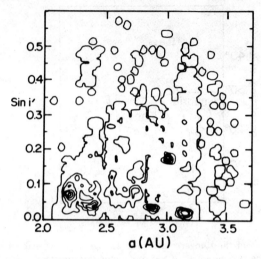

Fig. 4. Isodensity curves in semimajor axis and proper inclination for the cumulate population of PLS and numbered asteroids. The Flora region is located at about 2.2 AU. The Themis, Eos and Koronis families are also visible. (Adapted from Carussi and Massaro 1978.)

interpretation. Does the discrepancy reflect shortcomings in one of the search methods or does it reflect only differences in selection critera?

The Maria family was not found by Carusi and Massaro, although it has been identified in all previous studies. The Maria family was excluded from their list of families since the relative number of faint Maria members found by the PLS to numbered Maria members was less than for the other well-defined families. This argument is not wholly valid because the Maria family is located at an inclination of $15°$ where the coverage of the PLS is much poorer than for lower inclination orbits. This case parallels that of the Phocaea region ($i' \sim 22°$) where, as noted by van Houten *et al.* (1970), orbital selection effects play an important role. Also, the suggestion that the Maria family is scattered too much in e' to show up as a family (Carusi and Massaro 1978) is not consistent with the large scatter in e' seen for some other families which have a comparable number of members.

It is possible that the large intervals in semimajor axis (0.50 AU compared to 0.05 AU for other studies), inclination and eccentricity used by Carusi and Massaro increased the "noise level" from the background asteroids so as to mask some families. Distinguishing families from the background must be done cautiously since careful attention must be given to the parameters involved and to the specific situation. Also, boundaries can be somewhat difficult to define. For example, the Flora region is so complex that family boundaries become meaningless and arbitrary boundaries serve only to confuse any attempt at an interpretation of the physical properties of family members in terms of a parent body.

Aside from the major families, which are defined mainly in a, e', and i', are the elusive *jet streams*. Jet streams or asteroidal streams are defined as an assembly of orbits, showing similarity in all five orbital elements a, e', i', $\tilde{\omega}'$, and Ω', and are somewhat analogous to meteor streams. Alfvén (1969) reported three separate streams in the Flora family, denoted A, B, and C. Additional streams have been reported by Arnold (1969), Danielson (1969, 1971), and Lindblad and Southworth (1971). Very little agreement has been found between the various investigations. Several reasons have been suggested to account for this discrepancy, e.g., magnitude completeness limitations, discovery selection effects, and orbit selection effects. The differences between the studies of Lindblad and Southworth, and Arnold are probably the result of differences in selection criteria since the stream search program designed by Arnold did not emphasize the alignment of the orbital major axis.

Considerable controversy has evolved over the reality of jet streams. Various arguments have been raised that question the validity of stream identification due to selection effects and completeness corrections. The lack of agreement among the various investigations adds little credence for their existence.

Dynamical issues argue that the time scale necessary to randomize the nodes and perihelion directions of a family by planetary perturbations is on the order of 10^6 yr or about the time scale estimated by Alfvén and Arrhenius (1970) for the formation and contraction of asteroidal jet streams. Also, Monte Carlo approximations and computer simulations by Ip (1977) and Brahic (1975) indicate that stream formation by collisional processes produce flattened rings instead of compact contracting streams. These studies are not yet conclusive, although it may suffice to say that the role of jet streams in the present collision evolution of the asteroids is highly questionable.

The dynamical evolution of families is discussed by Ip (1979) and Kozai (see his chapter in this book). The evolution is apparently determined by interactions among family members themselves, the interaction of two families, and the interaction of family members with field objects as well as the perturbing effects of the planets, primarily Jupiter. Kozai suggests that the proximity of some families to the Jovian commensurabilities implies that some "trigger" mechanism connected with commensurabilities is responsible for their formation. This suggestion is intriguing, since it would appear that eccentricities, hence, relative velocities would be increased in these regions (see chapter by Greenberg and Scholl), permitting smaller bodies to produce more energetic collisions which have a greater chance of producing a family. However, the magnitude of the increased relative velocities is not large, and it is difficult to understand why it should cause major observable effects.

II. PHYSICAL STUDIES OF FAMILY MEMBERS AND
THE ORIGINS OF ASTEROID FAMILIES

The origin of the asteroid families is of paramount importance for the investigation and the understanding of asteroidal interiors. If families do indeed represent the debris of catastrophic collisions between asteroids, then the family members provide us with an unprecedented view of the interiors of the precursor asteroids. Those families, composed of several relatively equal-sized fragments, should be representative of the interior structure of the parent body. Ultimately, the geochemistry of that parent body could be reconstructed, whether or not the fragments display physical heterogeneity. Other families with one large member and several small members are probably the result of large cratering events and presumably can provide only limited information about the interior, since it is doubtful that many fragments would come from the deep interior.

It is still unclear whether all families result from the destruction of single bodies. Some families have been thought to be conglomerations of random asteroids collected by some dynamical means such as collisional accretion (Alfvén 1969). The identification of the true origin of a particular family must be based upon: (1) the interpretation of the physical data of the individual family members in the context of the differences and similarities with the surrounding field population; (2) the degree of similarity among orbits; and (3) whether the reconstructed body makes "cosmochemical sense."

The interpretation of the physical data, i.e., photometry, spectrophotometry, radiometry, etc., must be carried out with some caution. For example, relationship tests among family members using the *CSMERU* taxonomic scheme of Bowell *et al.* (1978; see the taxonomy chapter of Zellner) have limited applicability since these groupings are broadly defined and may mask the subtle differences that can have profound importance concerning the origin.

Initial studies of asteroid families met with only limited success. Chapman (1976) pointed out the spectrophotometry of the objects 15 Eunomia, 85 Io, and 141 Lumen contained in Williams family #140 required an interpretation of either the loss of a tremendous amount of mass, if all bodies were to have originated in the same parent body, or a collision between two large asteroids. The latter suggestion requires a very atypical event according to collision theory. Zellner *et al.* (1977) arrived at a similar conclusion concerning the Nysa family (Brouwer's family #24). If 44 Nysa, an *E* object, were to be related to the *M* object, 135 Hertha, on geochemical grounds, an inordinate amount of material had to be lost subsequent to family formation. Hansen (1977) pointed out that the *C* to *S* ratio in the larger families appeared to be different than the ratio found in the background or nonfamily population. He argued that these ratios were not inconsistent with the "breakup" hypothesis.

Gradie and Zellner (1977) and Gradie (1978) apparently have resolved the issue at least for the Eos and Koronis families. Both families appear to be the result of the collisional disruption of individual homogeneous parent bodies as shown from their studies of individual family members using UBV photometry and thermal radiometry. Tedesco (1979) has examined the nine major families found by Hirayama (1933) in the light of available UBV photometry, radiometry and visual photometry as well as his observations of lightcurves. He finds a probable origin by collisional fragmentation for the Themis, Maria, and Flora families, although not for the asteroids in the Phocaea region.

We will review the arguments presented for the collisional fragmentation or "breakup" origin for several families and identify some families where such an origin is not apparent at least with the present data base. The TRIAD files (Bender *et al.* 1978; see Part VII of this book) which contain UBV photometric, spectrophotometric, albedo and polarimetric data from various sources on nearly a third of the minor planets have been used in this analysis as well as the studies by Gradie and Zellner (1977), Gradie (1978), and Tedesco (1979).

Of the more than 100 families identified (see tabulation by Williams in this book) physical parameters exist in the TRIAD data files for two or more members of 48 families. By far the majority of data on family members is found in UBV photometric observations. Although UBV photometry is not diagnostic of a specific mineralogy, it can be used to rule out certain formation scenarios as well as provide evidence for homogeneity or heterogeneity among family members. The analysis of families used by Gradie and Zellner, Gradie, and Tedesco has been based on the philosophy that a color distribution among family members which is significantly dissimilar to the color distribution found in the surrounding nonfamily field population is indicative of an origin by the breakup of a larger precursor body. Likewise, a color distribution similar to the field population is indicative either of a family formed from a heterogeneous parent body whose internal composition matches that of the field population or of a family that was formed by some other process involving dynamics or multiple parent bodies.

Visible and near-infrared spectra that allow for specific mineral identifications have been measured for two or more members of 25 families (see the chapter by Chapman and Gaffey). This fact is no accident since the spectrophotometric-observing program has targeted a number of Williams' families for special attention. This chapter is in part a preliminary report of the Chapman/Williams collaborative effort. First, we discuss individual families in detail. Family numbers are those assigned by Williams (see tabulation by Williams in Part VII of this book).

Themis family (1)

The Themis family (1) located in the outer part of the belt ($a \sim 3.13$ AU) is one of the most populous and best-defined families. The family appears to consist of a "core" of large objects (except 268 Adorea) which is surrounded by a "cloud" of smaller objects. Objects 104 Klymene and 184 Dejopeja are not considered to be members of this family although they were included by Arnold (1969).

The observed UBV colors of the Themis asteroids are shown in Fig. 5 in which are delineated the location of the major taxonomic classes *C* and *S* as defined by Bowell *et al.* (1978). The clustering of objects in the *C*-region is suggestive of an overall commonality linked to an origin in a single parent body. However, the Themis family is located in a region of the belt dominated by low albedo, spectrally neutral *C*-type objects (Zellner and Bowell 1977). Tedesco (1979) argues that in spite of the problems with the background, the dispersion of colors among Themis family members is significantly smaller than the field, and that the family is more homogeneous than are the surrounding asteroids. Tedesco concludes that the Themis family resulted from the breakup of a single body calculated to be at least ~300 km in diameter. Such a body was comparable in size to the present-day *C*-object, 65 Cybele. However, the albedo of the one Themis object measured (171 Ophelia) is 0.10 which is definitely *not* in the *C*-class but is more typical of an *M*-object.

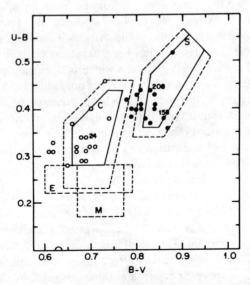

Fig. 5. The UBV colors of the Themis family (open circles) and of the Koronis family (filled circles). Data are from Gradie (1978) and Bender *et al.* (1978). Boundaries of *C*, *S*, *M* and *E* groups are from Bowell *et al.* (1978). Solar colors are marked at B-V=0.63 and U-B=0.10.

Spectra have been measured for six asteroids in the core of the Themis family and for one asteroid in the extended Themis family (Nos. 24, 62, 90, 468, 526, 846, 268). While the spectra are similar, there are substantial real variations in both the ultraviolet reflectances and in the slope of the spectra through the visible. The TRIAD-type file lists types for 20 Themis family members (see Table I) which, in general tend to be neurtrally colored C-types. Tedesco (1979) argues that the small differences in color between the larger and smaller members could be the result of differential "weathering" or else a slightly inhomogeneous parent body. It is not clear whether any inhomogeneity implies inhomogeneous accretion, internal thermal metamorphism, or some form of shocked induced metamorphism.

Eos family (2)

This large, well-defined family is found in the C-dominated outer region ($a \sim 3.02$ AU) of the belt (see chapter by Zellner). The UBV colors of the Eos family members are shown in Fig. 6. A lack of the more neutrally colored C-objects is evident as well as is a distinct clustering in colors near B-V = 0.82 and U-B = 0.40. A comparison between the radiometrically determined

TABLE I

Asteroid Families in the Outer Belt

Family No.	Distribution of Types	Overall Similarity	Comments
1	8 C, 2 CU, 2 CEU, 1 EU, 1 MEU, 6 U	Similar neutral-colored C	See text
2	12 S, 4 SU, 4 SM, 13 U, 6 C	Similar U (nearly S)	See text
101	1 U, 1 S	Similar; unusual	Spectra of 108 and 758 unusual but not identical
106	3 C, 1 U	Mainly C; 1 U	Spectra of 3 C's slightly different; see text
108	2 S, 1 $CMEU$, 1 M	Dissimilar	
111	1 C, 1 CEU	Could be similar C-like	
112	2 C	Similar C	
119	1 CU, 1 U, 1 $CMEU$	Could be similar C-like U's	
120	3 S	Similar S	

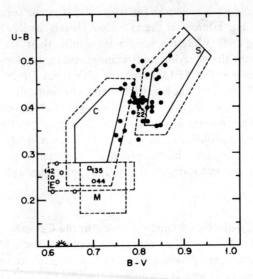

Fig. 6. The UBV colors of the Eos family (filled circles) and the Nysa family (open circles). The object 135 Hertha is indicated by the square. Data are from Gradie (1978) and Zellner *et al.* (1977). Boundaries and solar colors are the same as for Fig. 5.

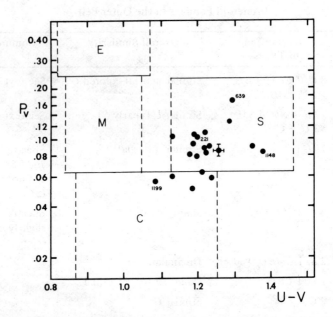

Fig. 7. The visual geometric albedo (p_v and U-V color distribution of the observed Eos family members (Gradie 1978). Boundaries of the *C, S, M,* and *E* taxonomic classes are from Bowell *et al.* (1978).

geometric albedo, p_v, and U-V color illustrated in Fig. 7 again demonstrates the abundance of S-like objects in this family in this C-dominated region of the belt.

Spectra have been measured for asteroids 221, 339, 513, 562, 579, 639, 1075, 1199, and 1364. In most respects the spectra are all quite similar. The greatest difference is in the BEND parameter (a measure of the curvature of the visible part of the reflection spectra) – the largest asteroids in the family (221 and 579) have much larger values of BEND than do other members. Although 639 and 1199 have very different UBV colors, the spectra show only modest differences for these two bodies. The tabulation of different types for the 39 Eos family members in TRIAD suggests great heterogeneity (Table I). But, in fact, nearly all Eos asteroids have surface properties roughly intermediate between S- and C-types, leaning toward the S-type (especially in albedo).

The color and albedo distribution of the Eos family is difficult to interpret. The largest objects 221 Eos and 579 Sidonia (for which an albedo is unavailable) appear to be identical but the smaller objects display continuous albeit limited ranges in albedos and colors. Gradie (1978) assumed this spread to be the effect of compositional differences among family members, hence compositional differences in the parent body. However, whether these differences reflect compositional zones in a differentiated body or compositional changes as a result of processes during or subsequent to the collision that formed the family remains to be demonstrated. It seems doubtful that the family-forming collision event was energetic enough to metamorphise a substantial fraction of the proto-Eos body.

Gradie (1978) concluded that the Eos parent body was at least 180 km in diameter but his calculation was done without regard to the observed compositional variations. Figure 8 illustrates the proto-Eos body schematically but with the correct relation between the sizes of the larger bodies. If we are to assume that most of the originated body still exists as fragments, then the reconstruction of the geochemistry presents a perplexing problem. There appears to be no obvious stratigraphic pattern. Members 221 Eos and 579 Sidonia should be examined closely for possible surface compositional variations.

Koronis family (3)

The Koronis family ($a \sim 2.87$ AU) is another well-defined family in the outer belt. The UBV colors of the Koronis family are shown in Fig. 5. As with the previous family, the Koronis family lacks the neutrally colored C-objects. Not only is there great similarity in UBV colors but in albedo as well (Fig. 9).

Spectra exist for 158, 167, 208, 243, 462, and 811. Except for 811, the spectra could be identical. The 18 Koronis members in the TRIAD-type file confirm that the members are quite similar, having spectral properties of

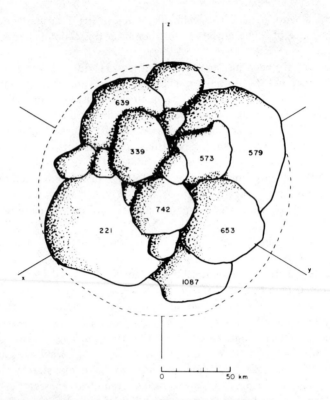

Fig. 8. A schematic reconstruction of the Eos parent body. Objects are positioned from geometrical consideration only and without regard to possible compositional differences. (From Gradie 1978.)

relatively neutral-colored *S*-types.

The Koronis undoubtedly represents the fragments of a highly homogeneous parent body presumably of *S*-type. Gradie (1978) estimates the minimum size of the parent to have been about 90 km in diameter. This parent body is drawn schematically in Fig. 10.

It appears that some *S*-objects smaller than ~100 km diameter may be uniform throughout. Do these bodies represent the cores of bodies whose mantles have been lost through collisional evolution? Or, are *S*-objects original accretions?

Structure of the Themis, Eos and Koronis Families

The Themis, Eos and Koronis families are the best-defined, most populous families found in the belt. In proper-element space the structure of the Themis family consists of a tight "core" of the largest objects surrounded by a cloud of small objects. The distributions in i' and e' for the Eos and

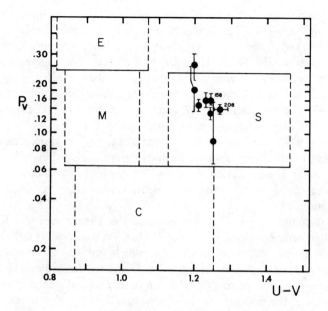

Fig. 9. The visual geometric albedo (p_V) and U-V color distribution of the observed Koronis family members (Gradie 1978). Boundaries of the *C, S, M,* and *E* taxonomic classes are from Bowell *et al.* (1978).

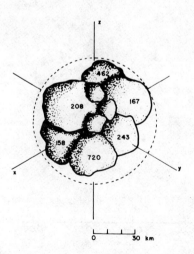

Fig. 10. A schematic reconstruction of the Koronis parent body. Objects are positioned from geometrical considerations only and without regard to possible compositional differences. (From Gradie 1978.)

Koronis families are strikingly different. The Eos family, composed of two ~95-km objects, a few ~50-km objects and a distribution of smaller objects, is spread in i' and e' as shown in Fig. 11. Several objects seemingly apart from the main body of the family are small PLS objects. There appears to be a correlation between e' and i' in the sense that objects with larger eccentricities have large inclinations. The Koronis family, composed of several objects ~35 km in diameter, is concentrated towards a single proper inclination but is spread in proper eccentricity (Fig. 12).

Wiesel (1978) examined the distribution of the proper orbital elements in the context of the distribution of fragments from exploded earth satellites. The similarities in distribution in semimajor axis lead him to conclude that these three families are consistent with a formation by catastrophic fragmentation.

The diameter frequency distribution of the Themis, Eos and Koronis families has been examined by Gradie (1978), Tedesco (1979) and Degewij (1978). For the diameter ranges for which substantial completeness corrections are not applicable, no significant variation from the field objects in the outer region of the belt is apparent. The diameter frequency distributions of the families and field objects are shown in Fig. 13. Diameters

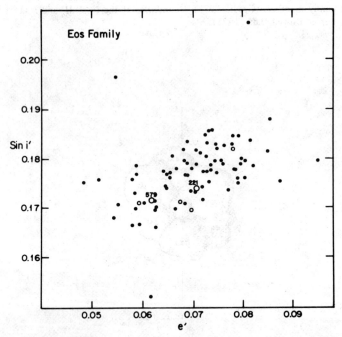

Fig. 11. The distribution of Eos family members in the proper elements e' and sin i'. Open circles represent the larger members; 221 Eos and 579 Sidonia are the two largest members (Gradie 1978).

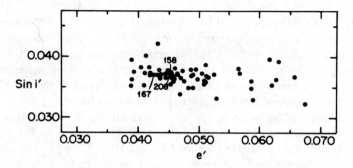

Fig. 12. The distribution of Koronis family members in the proper elements e' and sin i'.

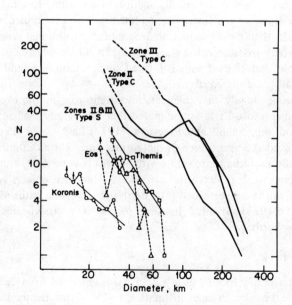

Fig. 13. The size distribution of the Themis, Eos and Koronis families (except for their largest members) compared to the size distribution for field objects in the same region of the belt. Data for families are from Gradie (1978) and Tedesco (1979). Arrows indicate limits where corrections for completeness are large. Data for field asteroids are from TRIAD (see chapter by Zellner). Zone II is $2.50 < a \leqslant 2.82$ AU and Zone III is $2.82 < a \leqslant 3.27$ AU. For both zones $e \leqslant 0.35$ and $i \leqslant 30°$.

of family members are calculated using assumed albedos: Themis, $p_v = 0.037$, Eos, $p_v = 0.09$, Koronis, $p_v = 0.13$. The arrows on the small diameter ends indicate points where completeness corrections become large and should be considered unreliable. Tedesco (1979) points out the possibility that the marginally smaller slope found in the size distribution of the Koronis family compared with the field may be real and may be related to the age of the family. This suggestion, however, remains to be demonstrated.

Any comparison between the size distributions of families and the surrounding field population must be done with some caution. In Fig. 13 the size distributions for zones II ($2.50 < a \leqslant 2.87$) and III ($2.82 < a \leqslant 3.27$) for types C and S are shown (see chapter by Zellner) for comparison. However, as Zellner points out, the distributions are valid only for diameters >50 km for Type C of Zone III and for diameters $\gtrsim 40$ km for Type S. A critical evaluation of the size frequencies distribution remains to be done.

It is interesting to note that the size distributions of the Themis and Eos families are so similar despite the differences in composition. Tedesco (1979) suggests that, to first order, the distribution of comminution energy is independent of material strength at least for bodies of composition similar to the parent bodies of the Themis and Eos families. The smaller slope found for the Koronis family is not statistically significant according to Gradie (1978) and the true slope of the distribution remains to be determined. Chapman (1978) suggests that in the production of families by collisional processes, the parent body may pass through a stage where it is thoroughly fractured into a "pile of rubble" before the family is formed. Such fracturing would mask any compositionally determined strength factors.

The Koronis family may play an important role in the evolution of asteroid fragments into Earth-bound meteorites. This family is adjacent to the 2:5 Kirkwood gap. Scholl and Froeschlé (1977) have noted that this gap should be considered as a source of meteorites since objects entering the gap may be rapidly perturbed into orbits with aphelia beyond 4 AU and perihelia which are Mars-crossing. Meteorites can evolve from such objects as collision ejecta which might be perturbed into Earth-orbit crossing paths as outlined by Wetherill (1978). It may be that pieces of proto-Koronis exist as meteorites on Earth.

Maria family (4)

The Maria family located at $a = 2.25$ AU is separated from this crowded region of the belt by its high inclination ($\sim 15°$). This family is not very populous and seems to be lacking in the smaller, km-sized objects. UBV colors are available for only four members but all objects fall in the S region of Fig. 14. Tedesco (1979) asserts that this sample is significant since the family is so small and he concludes that the "breakup" theory for the formation of the Maria family is plausible.

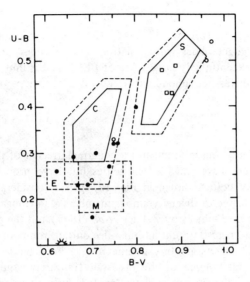

Fig. 14. The UBV colors of the Maria family (squares), the Hungaria region (open circles) and the Hilda group (closed circles). Data are from Bender *et al.* (1978). Boundaries and solar colors are the same as in Fig. 5.

Spectra for 170, 472, 660, 695, and 714 can all be described as *S*-like, but there are modest variations in R/B and BEND, and substantial differences in the infrared portions of the spectra. 170 Maria itself is formally called "U" because of its extremely linear trend into the infrared. One other family member listed in TRIAD is an *S*-type. The hypothetical parent body of the Maria family was somewhat inhomogeneous.

Flora Region

The Flora region is so complex that family boundaries become meaningless. Brouwer's (1956) family divisions were somewhat arbitrary. The Hirayama families (H6) through (H9) were considered together in Tedesco's (1979) analysis of the region. Because 98% of the Flora members measured have nearly identical, reddish UBV colors as compared to only 48% for the field asteroids, Tedesco concludes that this family is not a random collection of field asteroids but a family formed in some complex manner. It would seem that Flora, containing 75% of the volume of the hypothetical parent body, would represent the largest remaining fragment of the parent body that had undergone a near catastrophic collision. However, Tedesco points out that the collective diameter of the two largest members requires that any parent body reconstructed from the existing members must have been grossly irregular.

Flora Family

All seven spectra (for 8, 281, 341, 453, 496, 1055 and 1058) have reddish-*S*-like characteristics with relatively large values of BEND. 496 Gryphia has an exceptionally large value of BEND and is thus called *R*-type. The major real difference among the spectra is in the shape of the infrared absorption band.

Hungaria Region

The Hungaria region is found in the innermost region of the belt ($\sim$1.9 AU). Objects here have small eccentricities but large proper inclinations ($\sim$22°). The UBV colors, shown in Fig. 14, span a considerable range and indicate a wide degree of diversity among members. 434 Hungaria, itself, was considered to be an *E*-object by Zellner *et al.* (1977), but the presence of an absorption feature near 0.95 μm all but rules out an iron-poor composition.

All Hungaria objects are tiny (diameter $\lesssim$15 km) compared with most asteroids. The total volume of the observed Hungarias would fit into an equivalent sphere only 35 km in diameter. It is difficult to envision such a small body being highly differentiated. If the Hungarias did indeed come from a single body, then the present asteroids in this region must represent the last remaining fragments of a larger, differentiated precursor.

Hilda Region

The Hilda region bounds the outer side of the main belt. The colors of these objects are presented in Fig. 14. This region is discussed in detail in the chapter by Degewij and van Houten.

Phocaea Region

Colors of Phocaea-type asteroids are shown in Fig. 15. Since the Phocaea region is a region of the belt isolated by resonances and not a true family (Williams 1971), it is not surprising that a wide range in colors is found, and that the distribution of color types is like that found for the main belt. The largest object, 105 Artemis, is a low-albedo *C*-object. Included in this region is a group of four asteroids (323, 852, 1568, and 1575) that appear to form a family. The largest and brightest is the *S*-object, 323 Brucia. None of the other three members of this extremely small family have been studied.

Nysa (24) and Hertha (160) Families

The Nysa and Hertha families are close neighbors separated from each other by a narrow but distinct gap in proper inclination. Both families are composed of single large asteroids (44 Nysa and 135 Hertha) about 70 km in diameter and many small objects about 20 km or less in diameter. Many of

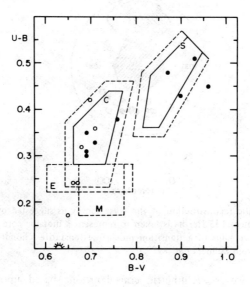

Fig. 15. The UBV colors of the Phocaea region members (filled circles) and Concordia family members (open circles). Data are from Bender *et al.* (1978). The boundaries and solar colors are described in Fig. 5.

these smaller asteroids are tiny PLS objects.

The UBV colors of Nysa family objects and 135 Hertha are plotted in Fig. 6. All objects fall in or close to the *E* and *M* domains. Spectra for asteroids 44, 750, and 1493 are all compatible with the *E*-type assigned to Nysa itself, although 1493 seems to have low reflectances in the extreme ultraviolet. The spectrum for asteroid 969 appears different, but it is of very poor quality. Including the three other TRIAD objects for which UBV colors are available, it is safe to suggest that all seven observed Nysa members could be consistent with Type *E*. But the UV color disparity for 1493 appears real as does the redder overall color for Nysa itself shown in both its spectrum and UBV colors. Recent radiometric measurements of 142 Polana of the Nysa family indicate that its albedo ($p_v = 0.04$) is more consistent with low albedo, neutrally colored *C*-objects although not classified as such in TRIAD. Polana is near the edge of the Nysa family and may be an interloper.

Zellner *et al.* (1977) have presented the first analysis of these families. From geochemical considerations they suggested that both families should be combined since the *E*-type object 44 Nysa, thought to be composed of some an original accretion not too different from an enstatite-chondritic mantle that once covered the metallic iron asteroid 135 Hertha as illustrated in Fig. 16. For such a case, the parent body was the differentiated remains of an original accretion not too different from an enstatite-achondritic

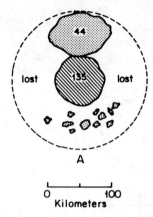

Fig. 16. A schematic reconstruction of the Nysa family as suggested by Zellner *et al.* (1977). The *M*-object 135 Hertha is taken to represent a metallic core and 44 Nysa is taken to be the remains of a transition metal-free (enstatite-achondritic?) material. (Figure adapted from Zellner *et al.* 1977.)

composition. However, volumetric considerations based upon chondritic Fe/Mg ratios and the size of the present-day Nysa and Hertha seem to rule out such a starting material. Asteroid 135 Hertha is not large enough to account for a mantle as thick as Nysa.

The volumetric considerations can be resolved in one of several ways: 1) the starting body was not purely enstatite-chondritic in composition; 2) the diameter estimates for high albedo, limb-darkened materials are inacccurate; or 3) most of the mass of the Nysa/Hertha system is now missing as a result of continuing collisional evolution.

The average diameter of Nysa now given in TRIAD (~66 km) is almost 25% smaller than the value used by Zellner *et al.* (1977). The calculated size of the original parent body of the Nysa/Hertha family is now thought to be considerably smaller. If the minimum diamater of Nysa (~50 km) is considered to be the maximum thickness of the mantle, then the present-day Hertha represents about 20% of the hypothetical core and Nysa less than 3% of the hypothetical mantle. The large ratio of remaining strong, metallic core to weaker mantle is consistent with collision theory which suggests that the smaller and the weaker fragments would be broken into still smaller fragments more rapidly than the larger and stronger fragments.

The 1:3 commensurability (Fig. 1) at 2.50 AU is adjacent to the Nysa/Hertha family (*a* ~ 2.42 AU). Some family members are so close to the commensurability that an ejection velocity of less than 100 m/sec is required for fragments to be thrown into the gap (Zellner *et al.* 1977). Although it is not clear exactly how material can be transported from this gap to the earth, it is possible that this family is an important source of some meteorites.

The Nysa/Hertha families need to be studied more in detail. In particular, the albedo of 142 Polana must be confirmed. The collisional evolution of the hypothetical, original family should be modeled to determine if meteorite production is possible.

Leto family (126)

This small, well-defined family is found in a C-dominated region of the belt (2.8 AU). The UBV colors of four of the largest members have been obtained and are all S-like. The spectra of asteroids 68, 236 and 858 could all be described as moderate S-types, but there are differences in the infrared. For example, the spectrum of 68 Leto itself shows a prominent olivine absorption while that of 236 Honoria has no band at all. There are no C types observed in the family.

The Leto family undoubtably was derived from a larger parent body or bodies. A single parent body would have been at least 200 km in diameter and displayed some interior inhomogeneities. The details of the inhomogeneity among family members remain to be studied.

Concordia family (132)

The Concordia family is a small but well-defined family in a crowded region of the belt. The UBV colors of six members are plotted in Fig. 15. Spectrophotometry shows that there are only modest differences among the three C-like spectra (for 58, 128, and 210) while 340 Eduarda has a typical S-type spectrum.

Most of the mass of the family is contained in the 190-km diameter, C-object 128 Nemesis. If this family is to be considered the result of the breakup of a single parent body (C-like, $\gtrsim$250-km diameter), then Eduarda might be a field object. Since Eduarda is on the outer edge of the family in terms of proper elements this possibility is likely.

Alexandra family (138)

The Alexandra family neighbors the Concordia family but has twice the eccentricity ($e' \sim 0.164$) and a higher inclination ($i' \sim 12°$). Five members have been observed with UBV photometry or spectrophotometry. The spectra of asteroids 54 and 145 are identical (relatively reddish C-types), but that of 166 is significantly different, and that of 1284 even more so. Another member of the family, 70 Panopaea, is a C-type. Overall, this family is rather heterogeneous and unusual.

Eunomia family (140)

The Eunomia family was first described as unusual by Chapman (1976) who recognized that the family could not be easily understood on the basis of the then available physical data. According to photometric, spectrophotometric, and radiometric studies, the largest object, 15 Eunomia, is of the type *S* (see Table II), whereas the next two largest objects, 85 Io and 141 Lumen belong to Type *C*. In a hypothetical parent body the *S*-object could be the core and the *C*-objects the remains of a mantle at least 150 km thick; the diameter of the whole body would be about 550 km. Two arguments have been used to rule out such a body. First, nearly 60% of the mass of the parent body is currently unaccounted for and was presumably lost. Second, the parent body of the Themis family was at least 300 km in diameter and physical studies of the present Themis members do not support this kind of interior construction.

It is possible that the evolution of the Eunomia family has been similar to that suggested for the Nysa/Hertha family and that a substantial amount of material has been lost since formation. It also may be possible that a 550-km diameter body evolves substantially differently from a 300-km body and that the Themis family cannot be used as a comparison.

It cannot be argued that the Eunomia family is not a family at all but is only an association of random asteroids, since further data in the TRIAD file for five other family members (Table II) reveals that this family is composed of equal mixtures of *S*- and *C*-types in a region where the *C*'s are four times more abundant than the *S*'s (see chapter by Zellner). This family can only be described as unusual.

TABLE II

Asteroid Families in the Middle Belt

Family No.	Distribution of Types	Overall Similarity	Comments
3	8 *S*, 2 *SU*, 8 *U*	Similar neutral-colored *S*	5 of 6 spectra appear identical; see text
4	5 *S*, 1 *U*	Similar *S*	Real differences among 5 spectra; see text
43	2 *U*	Could be similar *S*-like *U*'s	
67	4 *S*, 1 *R*, 1 *M*, 1 *C*	Dissimilar	Even the spectra of the 4 *S*-types differ significantly; see text
124	2 *U*, 1 *R*, 1 *M*	Dissimilar	See text
125	2 *C*, 1 *U*	Could be similar *C*	Spectrum exists for 293
126	4 *S*, 1 *U*	Mainly *S*; 1 *U*	Real spectral differences among *S*-types; see text

TABLE II (Continued)

Asteroid Families in the Middle Belt

Family No.	Distribution of Types	Overall Similarity	Comments
130	2 *U*, 1 *C*	Dissimilar	The 2 *U*'s are different also; see text
132	3 *C*, 1 *CEU*, 1 *CMEU*, 1 *U*, 1 *S*	Mainly *C*	Modest differences among 3 similar spectra; see text
133	1 *C*, 2 *CMEU*, 1 *U*, 1 *S*	Mainly *C*-like; 1 *S*	Spectra of 45 and 741 very similar, but significantly different; spectrum exists for *S*-type 197
134	1 *C*, 1 *S*	Dissimilar	
137	1 *S*, 1 *U*	Could be similar *S*-like	
138	3 *C*, 1 *U*, 1 *S*	Mainly reddish *C*'s	See text
140	3 *S*, 1 *SU*, 1 *U*, 1 *CU* 2 *C*	Dissimilar	See text
141	1 *S*, 1 *M*, 1 *C*, 1 *U*	Dissimilar	Spectra exist for 124 and 1595
142	2 *C*, 1 *S*	Dissimilar	Spectra exist for 37 and 66
148	1 *S*, 1 *C*	Possibly similar; unusual	Spectra for 402 and 510 actually similar; differ only in UV
149	2 *C*	Similar *C*	
150	1 *S*, 1 *RU*	Dissimilar	

Budrosa family (124)

This small family is composed of only six objects. It is located in a "rarefied" region of the belt, so in a sense is well defined.

The largest member of this family, 349 Dembowska, is an exceptional object suspected by Feierberg *et al.* (1978) of being achondritic in composition. 338 Budrosa is a rather typical *M*. Object 613 Ginevra does not appear too different from 338, but its UBV colors place it in the unclassified or *U* region. 558 Carmen appears quite unusual, despite the relatively poor quality of its spectrum. The spectrum *could be* consistent with the spectrum of 4 Vesta, although Carmen's albedo ($p_v \sim 0.09$) seems far too low to be

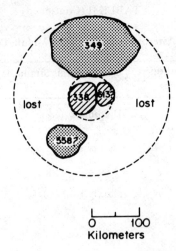

O   100
Kilometers

Fig. 17. A schematic reconstruction of the Budrosa family. The *M*-object 338 Budrosa and the possible *M*-object 613 Ginevra are taken to represent the remains of the iron core and 349 Dembowska the largest remaining achondritic-like fragment of the mantle of a differentiated achondritic asteroid.

identified as an achondritic object. Considering the suggestion (see chapter by Drake) that the mantle of the eucrite parent body should have an olivine-rich composition like that of Dembowska, the similarity in spectrum between Carmen and Vesta (the favored parent body for the eucrites) takes on special significance. A metal core (Budrosa?) is also consistent with the eucrite parent body model. A possible configuration of this parent body is illustrated schematically in Fig. 17.

The proximity of this family to the 2:5 Kirkwood gap, a possible escape hatch for meteorites according to Scholl and Froeschlé (1977; also see chapter by Greenberg and Scholl) makes this family an especially important one for further study.

Ceres family (67)

The largest member of this small family is the largest asteroid, 1 Ceres. Seven members of this family (1, 39, 264, 374, 407, 441, and 446) have been observed spectrally. These spectra are highly dissimilar, even if one were to assume that Ceres itself were an interloper.

Undina family (106)

Another small family is found in the outer parts of the belt (~3.2 AU). Spectra for four members of this family (92, 94, 490, and 1015) are identical in the visible, but there are real differences in the ultraviolet. In addition, 92 Undina has an *M*-like P_{min} while Aurora has a distinctly *C*-like albedo.

Lydia family (130)

The largest object in this family is the 139-km diameter U object 308 Polyxo. Spectra for three members of this family (110, 308, 363) reveal this to be an unusual family. Not only is the one C (363 Padua) different from the two U's (110 and 308), but the two U's are dissimilar from each other.

III. DISCUSSION

Summaries of the known types for 48 families are tabulated in Tables I, II and III. The middle column summarizes the family type if all or most members of a family have similar types. Otherwise the family is characterized as having types that are dissimilar. The final column lists comments pertinent to available spectra of two or more family members. In almost every case, asteroids of identical or similar type display significantly different spectra.

Several caveats are necessary for interpreting the tables. First, although Williams has applied fairly stiff statistical requirements for defining these families, they may not all be real. In some cases, the families certainly include one or more interlopers from the field asteroid population. Therefore, our sampling of the physical properties of just a few members of a family may occasionally give misleading results. But the statistics of Hirayama families taken as a whole should certainly be meaningful. In particular, there is no reason to believe that any of these families are significantly less real than the populous Eos, Koronis and Themis families; thus, their distinctive statistical traits, described below, suggest important physical differences between the smaller and larger families.

Of the sampled families, 26 of them (more than half) appear to have members of similar spectral character. Nine others are dominantly composed of one type but have one member (possibly an interloper) of markedly different type. Thirteen families are most accurately characterized as dissimilar, having two or more distinctive types and exhibiting no predominance for one type over another. The same qualitative results still hold if we restrict the sample to the 14 families for which we have studied at least five individual members. Thus, while there is a tendency for Hirayama families to be composed of similar members, consistent with origin by breakup of an approximately homogeneous body, a significant fraction of the families have a heterogeneous composition. When spectral properties are studied in detail (as with the 25-filter spectrophotometric technique), even the relatively homogeneous families nearly always reveal significant compositional differences among members.

Some of the larger families tend to have rather unusual spectra, often considerably different from the spectra typical of field asteroids in their part of the asteroid belt. Examples are the rather high-albedo Eos family in the outer belt, the S-type Koronis family in the middle belt, and the E-type Nysa

TABLE III

Asteroid Families in the Inner Belt

Family No.	Distribution of Types	Overall Similarity	Comments
24	1 *E*, 4 *EU*, 2 *U*	Could all be *E*	Real spectral differences between 44, 750, 1493; see text
157	1 *S*, 1 *CMEU*	Dissimilar	
158	1 *C*, 1 *M*, 1 *S*, 1 *U*	Dissimilar	Spectra exist for 19, 21, and 435
161	1 *C*, 1 *CMEU*, 1 *SM*	Dissimilar	
162	2 *S*	Similar *S*	Properties *not* identical
163	2 *S*, 1 *C*, 1 *SU*	Mainly *S*; 1 *C*	Spectra of 115 and 584 could be identical
165	2 *S*		
170	1 *S*, 1 *U*	Dissimilar	Both albedos and IR spectra differ for 9 and 113
171	3 *C*, 1 *CEU*, 1 *S*	Mainly *C*; 1 *S*	Spectra exist for 12, 84, 220, and 783
174	1 *C*, 1 *U*	Could be similar *C*-like objects	
175	2 *S*	Similar *S*	
180	1 *S*, 1 *SMRU*	Similar *S*	
183	3 *S*, 1 *RU*	Similar reddish *S*-types	
184	1 *U*, 1 *S*, 1 *SU* 1 *CMEU*	Mainly *S*-like *U*'s	
186	2 *S*	Similar *S*	
188	6 *S*	Similar *S*	
189	6 *S*, 1 *U*, 1 *R*	All *S*-like	Real differences in IR; 496 is *R*-type; see text
190	1 *U*, 1 *CMEU*	Similar?	Spectra of 434 and 1103 are both unusual
191	1 *R*, 1 *RU*, 1 *U*, 1 *C*	Mainly *R*; 1 *C*	Spectra for 1019 and 1656 appear similar but are not identical

TABLE IV

Minimum Diameters of the Parents Bodies of Some Families

Family	Minimum diameter of parent body	Parent body types[a]
1 Themis	300 km	$\sim C$
2 Eos	189 km	$S?$
3 Koronis	90 km	S
4 Maria	90 km	$\sim S$
24 Nysa/Hertha	200 km	$E + M$ differentiated
124 Budrosa	380 km	Vesta-like + M differentiated
126 Leto	200 km	$\sim S$
132 Concordia	250 km	$\sim C$
138 Alexandra	270 km	$C?$
189 Flora	165 km	$\sim S$

[a]See Tables I, II, III for discussion of component members.

family. These characteristics of larger families are not noticed in the smaller families. Most of the homogeneous families in the inner belt are composed of rather typical S-type asteroids. In the middle part of the belt, C- and S-type homogeneous families are equally common, and a larger percentage of families are heterogeneous. In the outer belt, C-type homogeneous families predominate. In spite of the observational bias toward brighter, higher-albedo asteroids in our observations, these statistics reflect well the overall increasing predominance of C-type asteroids with semimajor axis. In fact, except for the largest families, the distribution of types in the sampled families does not appear significantly different from random. One might conclude either (a) the smaller families have no real physical significance, or (b) the processes that produce the smaller asteroid families are similar to those that produce the general asteroid population. Since the families are believed to be real entities and the asteroids are believed to be collisionally evolved, it is likely that conclusion (b) is correct. Thus, the anomalies are the several large families of unusual but homogeneous composition.

The minimum sizes of the parent bodies for some families are given in Table IV. Such estimates are based upon the observed size distribution, the largest members in the family or in the case of the differentiated parent bodies an estimate of the mantle thickness and core diameter.

The study of families of minor planets is not complete by any means. Several future courses of study are crucial for our understanding of families and their parent bodies: 1) detailed physical studies of family members; 2) a precise mineralogical interpretation of the observational data to identify the geochemistry of the parent bodies; and 3) a study of the collisional evolution of family members.

Only a few families have been studied in detail. And even in those few cases, the full force of observational techniques have not been applied. Currently there are no polarimetric observation of members of the populous Themis, Eos and Koronis families. Unfortunately, most family members are small and faint, and observations will be difficult and time consuming for anything but broadband photometry. Broadband infrared photometry and high resolution infrared spectroscopy are two techniques that may prove useful.

Variegated surfaces are possible for members of some families. Simple broadband color lightcurves can be effective for demonstrating such effects. For example, in the Leto family the spectrum of 68 Leto has a prominent olivine absorption while 236 Honoria, only half the diameter of Leto, has none. Either object could be monitored in the olivine band and outside the band simultaneously to search for surface variations.

The identification of the diagnostic relationships between the mineralogy of a material and its optical and spectral properties is one of the most important yet more difficult aspects of asteroid studies. The ultimate goal of physical studies, either spectra or photometry, is either to identify the specific mineralogy of a particular minor planet or to definitely assign meteorite specimens to particular source bodies. Our understanding of the geochemistry and geochemical evolution of asteroids, as deduced from family members, is totally dependent on these relationships.

The collisional evolution of family members is of interest, also. It has been suggested (Tedesco 1979) that the size distribution in a family may be a function of the family age. Collision theory suggests that asteroid lifetimes are roughly proportional to the square root of the diameter, so the smaller fragments might tend to disappear rapidly. A relative lack of small family members may indicate an old family. The actual size distribution of most families may be difficult to obtain without special, and tedious, time-consuming observational programs.

Acknowledgements. The authors wish to thank E. Tedesco for much essential data in advance of publication and E. Bowell for a careful and helpful review. This chapter presents in part the results of one phase of research carried out at Jet Propulsion Laboratory, California Institute of Technology sponsored by the National Aeronautics and Space Administration.

REFERENCES

Alfvén, H. 1969. Asteroidal jet streams. *Astrophys. Space Sci.* 4: 84-102.

Alfvén, H., and Arrhenius, G. 1970. Origin and evolution of the solar system. II. *Astrophys. Space Sci.* 9: 3-33.

Arnold, J. 1969. Asteroid families and "jet streams." *Astron. J.* 74: 1235-1242.

Bender, E.; Bowell, E.; Chapman, C.; Gaffey, M.; Gehrels, T.; Zellner, B.; Morrison, D.; and Tedesco, E. 1978. The Tucson revised index of asteroid data. *Icarus* 33: 630-631.

Bowell, E.; Chapman, C. R.; Gradie, J. C.; Morrison, D.; and Zellner, B. 1978. Taxonomy of asteroids. *Icarus* 35: 313-335.

Brahic, A. 1975. A numerical study of a gravitating system of colliding particles: Applications to the dynamics of Saturn's rings and to the formation of the solar system. *Icarus* 25: 452-458.

Brouwer, D. 1951. Secular variations of the orbital elements of minor planets. *Astron. J.* 56: 9-32.

Brouwer, D., and Clemence, G. M. 1961. *Methods of Celestial Mechanics* (New York: Academic Press), pp. 598.

Brouwer, D., and van Woerkom, A. J. J. 1950. The secular variations of the orbital elements of the principal planets. *Astron. Papers Amer. Ephem.* 13: 81-107.

Carussi, A., and Massaro, E. 1978. Statistics and mapping of asteroid concentrations in the proper elements' space. *Astron. Astrophys. Suppl.* 34: 81-90.

Chapman, C. R. 1976. Asteroids as meteorite parent bodies: The astronomical perspective. *Geochim. Cosmochin. Acta* 40: 701-719.

Chapman, C. R. 1978. Asteroid collisions, craters, regoliths and lifetimes. In *Asteroids: An Exploration Assessment,* eds. D. Morrison and W. C. Wells, NASA CP-2053, pp. 145-160.

Danielson, L. 1969. Statistical arguments for asteroidal jet streams. *Astrophys. Space Sci.* 5: 53-58.

Danielson, L. 1971. The profile of a jet stream. In *Physical Studies of Minor Planets,* ed. T. Gehrels (NASA SP-267, Washington, D.C.: U.S. Government Printing Office), pp. 353-362.

Degewij, J. 1978. Photometry of faint asteroids, satellites and cometary nuclei. Ph.D. dissertation, Leiden Univ.

Feierberg, M.; Larson, H.; Smith, H.; and Fink, U. 1978. Spectroscopic evidence for at least two achondritic parent bodies (abstract). *Bull. Amer. Astron. Soc.* 10: 595.

Gradie, J. C. 1978. An astrophysical study of the minor planets in the Eos and Koronis asteroid families. Ph.D. dissertation, University of Arizona.

Gradie, J., and Zellner, B. 1977. Asteroid families: Observational evidence for common origins. *Science* 197: 254-255.

Hansen, O. 1977. Search for correlations between asteroid families and classes. *Icarus* 32: 229-232.

Hirayama, K. 1918. Groups of asteroids probably of common origin. *Proc. Phys.-Math. Soc. Japan* II: 354-361.

Hirayama, K. 1919. Further notes on the families of asteroids. *Proc. Phys.-Math. Soc. Japan* Ser. 3: 1-52.

Hirayama, K. 1923. Families of asteroids. *Jap. J. Astron. Geophys. Transactions* 1: 55-93.

Hirayama, K. 1928. Families of asteroids: Second paper. *Jap. J. Astron. Geophys. Transactions* 5: 137-162.

Hirayama, K. 1933. Present state of the families of asteroids. *Proc. Imp. Acad. Japan* 9: 482-485.

Ip, W.-H. 1977. Monte Carlo simulation of the jet stream process. *Proc. Lunar Sci. Conf. VII* (Oxford: Pergamon Press), pp. 67-77.

Ip, W.-H. 1979. On three types of fragmentation processes observed in the asteroid belt. *Icarus* (special Asteroid issue).

Kirkwood, D. 1867. *Meteoric Astronomy: A Treatise on Shootingstars, Fireballs, and Aerolites.* (Philadelphia: J. B. Lippincott and Co.), pp. 129.

Kozai, Y. 1979. Secular perturbations of asteroids and comets. In *Dynamics of the Solar System,* ed. R. L. Duncombe (Dordrecht: D. Reidel, in press).

Lindblad, B. A., and Southworth, R. B. 1971. A study of the asteroid families and streams by computer techniques. In *Physical Studies of Minor Planets,* ed. T. Gehrels (NASA SP-267, Washington, D.C.: U.S. Government Printing Office), pp. 337-352.

Scholl, H., and Froeschlé, C. 1977. The Kirkwood gaps as an asteroidal source of meteorites. In *Comets, Asteroids, Meteorites,* ed. A. H. Delsemme (Toledo, Ohio: University of Toledo Press), pp. 293-295.

Tedesco, E. F. 1979. Photometric investigation of the colors, shapes and spin rates of Hirayama family asteroids. Ph.D. dissertation, New Mexico State University).

van Houten, C. J. 1971. Descriptive survey of families, Trojans, and jet streams. In *Physical Studies of Minor Planets,* ed. T. Gehrels (NASA SP-267, Washington, D.C.: U.S. Government Printing Office), pp. 173-175.

van Houten, C. J.; van Houten-Groeneveld, I.; Herget, P.; and Gehrels, T. 1970. The Palomar-Leiden survey of faint minor planets. *Astron. Astrophys. Suppl.* 2: 339-448.

Wetherill, G. W. 1978. Dynamical evidence regarding the relationship between asteroids and meteorites. *Asteroids: An Exploration Assessment,* eds. D. Morrison and W. C. Wells, NASA CP-2053, pp. 17-35.

Wiesel, W. 1978. Fragmentation of asteroids and artificial satellites in orbit. *Icarus* 34: 99-116.

Williams, J. G. 1969. Secular perturbations in the solar system. Ph.D. dissertation, Univ. of California at Los Angeles.

Williams, J. G. 1971. Proper elements, families, and belt boundaries. In *Physical Studies of Minor Planets,* ed. T. Gehrels (NASA SP-267, Washington, D.C.: U.S. Government Printing Office), pp. 177-181.

Williams, J. G. 1975. Asteroid families (abstract). *Bull. Amer. Astron. Soc.* 5: 363.

Yuasa, M. 1973. Theory of secular perturbation of asteroids, including terms of higher orders and higher degrees. *Publ. Astron. Soc. Japan* 25: 399-445.

Zellner, B.; Leake, M.; Morrison, D.; and Williams, J. G. 1977. The E asteroids and the origin of the enstatite achondrites. *Geochim. Cosmochim. Acta* 41: 1759-1767.

NUMERICAL SIMULATIONS OF CLOSE ENCOUNTERS BETWEEN JUPITER AND MINOR BODIES

A. CARUSI

and

G. B. VALSECCHI
Laboratorio di Astrofisica Spaziale

Our numerical work on close encounters between Jupiter and fictitious small bodies is reviewed along with two related problems, i.e. the changes in populations of planet-crossers due to an encounter with Jupiter and the occurrence of temporary satellite captures. Temporary satellite captures are found to occur when the initial orbit of the minor body is nearly tangent to that of the planet. An improvement of the method used earlier for close-encounter computations, allowing for the variation of the number of interacting bodies, is presented. With this new method, the very close encounters of some fictitious bodies of previous studies have been recomputed taking into account the presence of the Galilean satellites and of the outer planets. While the effects of the latter on satellite captures, and in general on the orbital evolution, appear to be unimportant, the effects of the Galilean satellites are in some cases quite relevant. For comparison, the possible evolution forwards and backwards for a total of ~ 400 yr of four slightly different sets of orbital elements of comet P/Gehrels 3 is presented. In this computation the presence of the four outer planets and, during the 1970 encounter, also of the Galilean satellites was taken into account.

The availability of big computers has made possible the study of some of the most complex situations in celestial mechanics related to the general n-body problem. Among these, one of the most intriguing is close gravitational interaction which requires great accuracy in the integration of the motion equations.

The problem of close encounters, that has been widely discussed in the literature from a numerical point of view (see, e.g., Belyaev 1967; Everhart 1972, 1973, 1976; Kazimirchak-Polonskaya 1967, 1972, 1976; Rickmann 1979; Sitarski 1968), could not be treated in a purely Newtonian framework since celestial bodies do not behave exactly as mass points. However when the duration of these interactions is not too long, the effects of forces as tidal torques or those arising from the bodies' oblatenesses, can be disregarded without affecting the computational precision very much. Similarly, at least in special cases, the effects of nongravitational forces due to the presence of magnetic fields, or to the geochemical structure of the bodies, can be neglected.

We are presently involved in an extensive numerical study of the effects of close encounters between Jupiter and fictitious minor bodies. The research, that regards one of the most important problems of the past and future dynamical evolution of minor bodies in the solar system, was made possible by the use of a new method of integration of the motion equations based on Greenspan's discrete mechanics (Greenspan 1973; LaBudde and Greenspan 1976a,b). The details of the method, which is especially devoted to the treatment of close interactions, are reported in Carusi and Pozzi (1978a) and Carusi (1979).

The first part of the research has been the most general one, concerning the effects of a single close encounter with Jupiter on the members of three populations of about 1000 objects each (Carusi and Pozzi 1978b). The first population (ASTRID) was formed from a dynamical point of view by "asteroids" ($e\leqslant0.5$; $i\leqslant35°$); the second one (COMET) by "comets" ($0.5<e\leqslant1.$; $i\leqslant90°$); and the third one (RANDUN) was formed by intermediate objects ($e\leqslant0.9$; $i\leqslant35°$). The semimajor axes were such that

$$\frac{q_J - R}{1 + e} \leqslant a \leqslant \frac{Q_J + R}{1 - e} \tag{1}$$

where q_J, Q_J, and R are Jupiter's perihelion, aphelion and radius of action, and a and e are the objects' semiaxes and eccentricities. This condition is necessary, but not sufficient, for a close encounter with Jupiter. The quotation marks around "comets" and "asteroids" imply that the distributions of the initial eccentricities and inclinations of the two populations have extremes similar to those of asteroids and comets. Actually, our "asteroids" have semiaxes different from those of the real asteroids (which cannot have close encounters with Jupiter!); they look more like short-period comets, while our "comets" can be compared to intermediate-period, direct comets. Moreover the flat shapes of the distributions in e and i of the three samples are dissimilar from those of the real populations.

The main conclusions of the first investigation, described in greater detail in Carusi and Pozzi (1978b) can be summarized as follows:

1. The efficiency of a single close encounter with Jupiter, shown by the variations of the orbital parameters of the minor objects, is generally very poor.

2. The distributions of orbital parameter variations are symmetrical around 0, but their shapes are non-Gaussian (with the noticeable exception of $\Delta\Omega$); great variations are more frequent than they should be for casual events.

3. The number of *peculiar* events such as collisions with Jupiter, ejections on hyperbolic orbits and temporary satellite captures is low: not more than a few percent of the total sample.

Even if a single close encounter is generally less effective than one would expect, it can be very efficient if we deal with peculiar orbits. In the three above-mentioned populations the planet-crossing objects were isolated and an analysis was carried out (Carusi *et al.* 1978; Carusi and Valsecchi 1979b) on the variations and rejuvenation rates of these populations. More details on these computations are reported in Sec. I.

A class of orbits that is strongly affected by an encounter with Jupiter is formed by those nearly tangent to that of the planet. To investigate in greater detail the behavior of these objects, 100 fictitious orbits were generated and processed (Carusi *et al.* 1979; see also Carusi and Valsecchi 1979a, and Sec. II of this chapter).

In all these investigations the computations were carried out using a solar system in which only the sun, Jupiter and the minor object were taken into account. The present extension, described in Sec. III, IV and V, of the earlier work includes the effect due to the Galilean satellites and to the outer planets Saturn, Uranus and Neptune.

Carusi *et al.* (1979) noted that the past evolutions of some observed short-period comets seem to be very similar to those of our fictitious bodies. With the aim of extending this analysis, the past evolution of comet P/Gehrels 3 has been recomputed (a first computation can be found in Rickmann 1979) taking into account the presence of all the outer planets and of the Galilean satellites for the encounter of 1970. The results of these computations are discussed in Sec. VI, while in Sec. VII a review is made of the whole study and some general conclusions are drawn.

TABLE I

Planet-Crossers[a]

	Mercury	Venus	Earth	Mars	Saturn	Uranus	Neptune	Pluto
Geometrical								
A	5	5	12	25	143	77	57	64
B	2	5	9	22	104	35	21	19
C	5	27	66	209	1533	821	439	424
Physical								
A	0	4	10	13	201	100	61	1
B	0	2	6	4	141	36	32	0
C	0	1	0	2	385	82	70	0

[a]A, planet-crossers only before, B after, and C before and after Jupiter encounter.

I. INFLUENCE OF A CLOSE ENCOUNTER WITH JUPITER ON A PLANET-CROSSING POPULATION

The data made available by the first mentioned investigation (Carusi and Pozzi 1978b) permitted the computation of the effects of a close encounter with Jupiter on a population of planet-crossing objects (Carusi et al. 1978; Carusi and Valsecchi 1979b). This problem can be very important in the search for the source regions of meteorites, and in studies of the movements across the solar system of objects that could be responsible for the cratering of the inner planets and of some Jovian satellites (see, e.g., Wetherill 1977, and the references therein).

In the three initial populations (ASTRID, COMET and RANDUN; for a total of 2830 small bodies) the objects crossing the orbits of one or more planets before and/or after the encounter with Jupiter were selected. For "planet crossing object" two definitions were used: a *geometrical* one, i.e. objects with perihelion closer to the sun than the aphelion of a given planet, and a *physical* one, i.e. objects crossing a suitably defined sphere of action of that planet. This second definition is the most appropriate for discussing, for instance, problems of planetary cratering and sources of meteorites. In fact, the requirement for a body to be a *physical* planet crosser in order to have the possibility of hitting a planet does not appear to be too severe if, as in our case, this body can have close encounters with Jupiter. The encounter, if not prevented by any mechanism, will take place within a period of time not long enough to allow slow mechanisms of orbital evolution (e.g. precession of nodes) to be effective, and the orientation of the final orbit has in general no relation to the initial one.

The results are summarized in the second part of Table I. It is easy to see that the effect of a close encounter with Jupiter on a population of *physical* planet crossers is a strong depletion and a simultaneous almost complete rejuvenation. The reason for this fact can be easily understood if we note that the general trend of the final inclinations shown by RANDUN, ASTRID and COMET was a depletion of high inclinations as well as of very low ones. This implies a low probability, for the final orbits, of intersecting one of the tori generated by the revolutions around the sun of the inner planets and their spheres of action, since the radii of the latter are at most 1.5×10^6 km (in the case of the earth), and the inclinations of the inner planets are very small.

The number of objects ejected from the solar system on hyperbolic orbits is 19 out of 2820 and that of the collisions with Jupiter is 9 out of 2820. All but one of the B and C objects reported in Table I may have a further close approach to the planet. It might be concluded that the encounters with Jupiter tend to subtract material both from the inner and the outer regions of the solar system.

II. INVESTIGATION OF TEMPORARY SATELLITE CAPTURES

Among the special cases of the general sample (RANDUN, ASTRID and COMET), i.e. hyperbolic final orbits, collisions with Jupiter and temporary satellite captures (objects which temporarily had elliptical jovicentric orbits), only the last satellite objects showed a common feature in some of their orbital parameters. Events of this kind, already found by Everhart in 1973, have great importance because of the possibility of permanent captures of satellites. The process has been hypothesized by many authors particularly for the outer retrograde satellites of Jupiter; it has not yet been demonstrated.

Two important points must be made on the objects that experienced a temporary satellite capture: the initial orbits were of low inclination and nearly tangent to that of the planet, and the majority of final orbits had changed the type of tangency, i.e. the aphelia became perihelia and viceversa.

These conclusions have suggested that we should generate a new population (CAPTURE) of 100 objects whose orbital parameters are chosen as follows:

1. The angular parameters i, ω, and Ω are the same for all the objects ($i=0°.68$, $\omega=327°.82$, $\Omega=150°.36$) and are those of ASTRID 985, which had remained close to Jupiter for more than 100 years.
2. The eccentricities vary from 0.01 to 0.50, not at random but increasing by steps of 0.01.
3. The semimajor axes are chosen at random, but with the condition that the resulting orbit should be nearly tangent to that of Jupiter, i.e.:

$$Q_J - R \leqslant q \leqslant Q_J + R$$

$$q_J - R \leqslant Q \leqslant q_J + R$$

where q,Q,q_J and Q_J are the perihelia and aphelia of the object and of Jupiter respectively, and R is Jupiter's radius of action (10^8 km).

For each value of the eccentricity two orbits are generated, one tangent at its perihelion (outer tangency) and the other tangent at aphelion (inner tangency). The initial positions of Jupiter and each object are chosen so that they will come together at the point of minimum distance, if there are no perturbations due to Jupiter. The computations covered the time spent by the object inside a sphere of 4×10^8 km radius around Jupiter. The method of computation is the one used in Carusi and Pozzi (1978b). In Figs. 1 and 2 the initial and final positions of these objects in the semiaxis-eccentricity diagram are shown.

We make a few comments on the results of these numerical experiments. First we note that the frequency of binding to Jupiter for this kind of objects is very high (56 %; dots in Fig. 1) as compared with the same frequency for the sample of 2820 objects ($\sim$1 %). We have not found any permanent captures, but if this event is really possible our investigation puts some constraints on the orbital characteristics before the capture. In this respect it may be noted that many objects have been captured during a fly-by, without making a complete orbit. On the other hand, other objects have orbited one or even two times around Jupiter without being bound to it. Therefore, even if an object closes an orbit it does not mean that it is a temporary satellite. Comparing our satellite captures with those computed by using a restricted 3-body problem (see, e.g., Bailey 1971, 1972; Heppenheimer, 1975; Heppenheimer and Porco 1977; Horedt 1972a,b, 1974, 1976), it must be noted that a large fraction of the CAPTURE population does not pass near the Lagrangian points L_2 and L_3, and even if the binding to Jupiter occurs near these points the unbinding can take place anywhere.

Another important comment is that 67 % of the final orbits lie in the a-e diagram on the opposite band with respect to the initial ones. It was noticed in Carusi et al. (1979) that this kind of transition from an inner tangency to an outer one, or vice-versa, has been recognized also in the case of some observed comets (Kazimirchak-Polonskaya 1972); as an example, in Figs. 3 and 4 the evolutions on the a-e diagram for comet P/Whipple and for CAPTURE 49 are shown. The differences are due mainly to the fact that the orbit of P/Whipple has been integrated for a time interval of 400 yr, but the transition from the upper to the lower curve occurred during the close encounter in 1851-1852 (Kazimirchak-Polonskaya 1967).

A third comment concerns the variation of eccentricities. As a consequence of the encounter, the high eccentricities ($0.2\leqslant e\leqslant0.5$) are decreased and low eccentricities ($e\leqslant0.05$) are increased. This phenomenon, already shown in Carusi and Pozzi (1978b) for low eccentricities, produces a clustering of e between 0.05 and 0.2, as shown by the histograms in Fig. 2. All but two of the objects with initial eccentricities belonging to this range

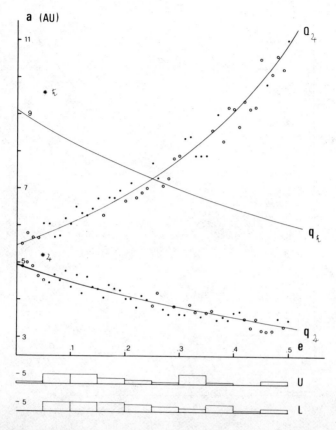

Fig. 1. Initial positions in the semiaxis-eccentricity diagram of the CAPTURE objects. The upper-curve objects are numbered with odd numbers from 1 to 99 in order of increasing e. The lower-curve objects are even and range from 2 to 100. Dots refer to objects which experience a temporary satellite capture. The two histograms (U, upper curve; L, lower curve) indicate the number of captures.

underwent a temporary satellite capture and, among them, almost all the longest and most complex interactions with Jupiter are included. In many cases a temporary unbinding from the sun occurred; for the reasons explained in Carusi *et al.* (1979) these events can occur only during an inner conjunction on a retrograde planetocentric orbit, or during an outer conjunction on a direct planetocentric orbit. In these configurations we have maxima of heliocentric energy, while the minima occur in outer retrograde and inner direct conjunctions. A further observation is that in the points of minimum or maximum semiaxis (and also of the energy) some objects tend to be near their osculating heliocentric aphelia and perihelia respectively. Configurations of this kind are similar to the so-called "mirror

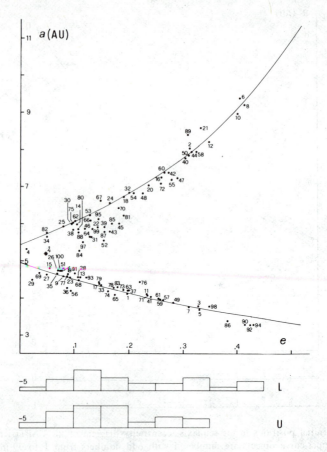

Fig. 2. Final positions of the CAPTURE objects. Histograms refer to the number of lower- and upper-curve objects in each class of final eccentricity.

configurations" (Roy and Ovenden 1955) that imply the time specularity of the trajectories before and after the instant in which the configuration occurs. Actually the mirror theorem requires that all the velocity vectors be perpendicular to all the radii vectors, and this requirement is not completely matched in our cases because Jupiter is not in its aphelion or perihelion and the lines of nodes are not aligned. If a mirror configuration were completely attained, a permanent capture could not occur because the minor object would have to unbind again from Jupiter. This supports the usually accepted idea that in a 3-body problem without dissipative mechanisms the occurrence of a permanent capture is impossible.

The need for a dissipative medium has also been suggested, in a related context, to explain the magnitude-frequency relation of the outer satellites of

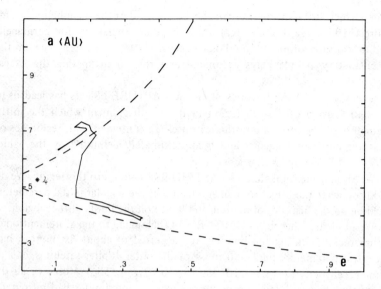

Fig. 3. Evolution of the orbit of comet P/Whipple in the *a-e* diagram. (Data from Kazimirchak-Polonskaya, 1967.)

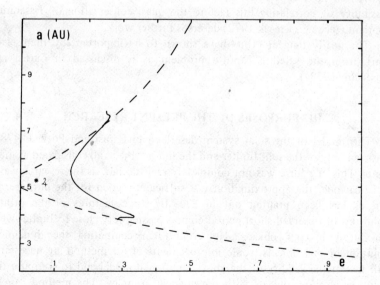

Fig. 4. Evolution of the orbit of the object CAPTURE 49 in the *a-e* diagram.

Jupiter which differs from those of the Trojans and of the main-belt asteroids (Gehrels 1977; see his chapter). Also in the mechanism for permanent satellite capture proposed by Pollack et al. (1979), the presence of the protoplanetary nebula plays a fundamental role in dissipating the excess energy.

The analysis of the histories of all the CAPTURE objects has lead us to the identification of a small region in the a-e diagram in which the initial parameters of the most interesting captures are contained. This region lies on the lower curve of tangency and is approximately delimited by the values $4.2 \leqslant a \leqslant 4.7$ AU and $0.12 \leqslant e \leqslant 0.14$.

We should mention that also ASTRID 985 belongs to this region. We do not know what the effect of a small change in the angular initial parameters would be on the histories of binding and hence on the limits of that region.

We note that the object CAPTURE 24 (belonging to the aforementioned capture region), during the binding, for a period of about six months, had jovicentric parameters similar to those of the outer Jupiter satellites, except for the eccentricity ($a \sim 25.5 \times 10^6$ km, $e \sim 0.79$, $i \sim 153°$). Moreover all the objects that have orbited one or more times around Jupiter had retrograde planetocentric motion. This is in agreement with the analytical results of Moulton (1914) and the numerical experiments of Hunter (1967a), that showed a greater stability against escape (for semiaxes typical of the outer irregular satellites of Jupiter) of the retrograde planetocentric motion.

Finally, all but two CAPTURE objects had nearly tangent final orbits, as can be seen in Fig. 2. This is in agreement again with the work by Hunter (1967b) on the escape of satellites from Jupiter (see his Tables I and II); the possibility of correlating this fact to the quasi conservation of Tisserand's invariant (see, e.g., Kresák 1972) deserves further work.

The nearly tangent orbits have shown to be important by themselves, apart from the satellite capture problem as is discussed in Carusi and Valsecchi (1979a).

III. PURPOSES OF THE PRESENT RESEARCH

The model of the solar system described in Carusi and Pozzi (1978a) consisted only of the sun, Jupiter and the minor object, all considered as mass points. This of course was not completely realistic but, as the research was a statistical one, the approximation was sufficiently good for the purpose. In fact, as has been pointed out in Sec. II, some features of the orbital evolutions of observed short-period comets passing very close to Jupiter were reproduced by our objects. However, when computing the dynamical evolution of real objects, some improvements of the method are necessary, both to test the validity of the above-mentioned model and to compute the evolution of those objects with the greatest accuracy. The method can be improved in at least four ways:

(a) include in the computations the four Galilean satellites;
(b) include the presence of the other outer planets;
(c) using more accurate initial data for the ephemerides of planets and satellites;
(d) take into account the oblateness of Jupiter and the tidal forces.

Let us discuss these points in order of increasing feasibility. It seems that, during a close encounter, tidal forces could be disregarded, in the case of Jupiter, because of the short duration of the event and the small tidal force compared with the total force acting on the object. The effect due to the oblateness of Jupiter is slightly greater, but it acts only at small relative distances (a few planetary radii) and for rather high Jovian latitudes. It is, however, less than the effect due to the Galilean satellites especially at distances from the planet greater than several radii. Moreover, the inclusion of the oblateness would require substantial changes in the formulation of Greenspan's algorithm itself, while the effects of the presence of other bodies can be easily taken into account. This point, in any case, may be worth further numerical work.

On the basis of these considerations we have decided that the first step in the refinement of the method should be the inclusion in the computations of the outer planets and of the Galilean satellites. Now, while the ephemerides of the outer planets are reliable enough, for the Galilean satellites the situation is quite different because it is well known (e.g. see Ferraz-Mello 1975; Arlot 1975) that the Sampson tables (Sampson 1910), on which the ephemerides are based, are not in good agreement with the observed positions. Therefore, as already pointed out by Carusi and Pozzi (1978a), the precision of our computations cannot be better than that allowed by the input data. The present research has had as first objective the test of the influence of the Galilean satellites and of the outer planets on satellite captures.

Increasing the number of interacting bodies slows the computations down considerably and we decided to repeat the evolution only of objects that passed close enough to Jupiter. Therefore only 23 objects having closest approach to the planet at a distance less than 2.5×10^6 km were selected. Eleven of the selected objects belonged to the lower curve, with prevalence of low initial eccentricities, while the remaining twelve upper-curve objects preferentially had higher eccentricity. Moreover it was noted that all but one of the lower-curve objects experienced in the previous research a temporary binding, while all but three of the upper-curve ones experienced no capture. This must be compared with the fact that in the CAPTURE sample there was only a slight difference in the number of captures between lower- and upper-curve objects (31 and 25 respectively; see also Fig. 1). The number of transitions in the present restricted sample is at variance with respect to the total sample, especially for the upper-curve objects; of the 18 (out of 50) upper-curve objects which did not change type of tangency, 10 (out of 12) are in the restricted sample. The same numbers for the lower-curve objects are 15 (out of 50) and 5 (out of 11). It is therefore clear that the restricted sample is not representative of the total one; it can be argued that the depth of the encounter is the probable reason for this different behavior. Finally, the objects CAPTURE 26 and 28, selected for the new study, had initial orbits belonging to the capture region mentioned in Sec. II.

IV. EXTENSION OF THE COMPUTING METHOD

The variations of the model reported in the previous section forced us to change some features of the method rather substantially; a brief outline of the differences between the method described in Carusi and Pozzi (1978a) and the one presented here is necessary. First of all, as we wished to take into account the small effects of other bodies on our minor objects, a new double precision version was prepared of the computer routine of gravitational interactions. Secondly, it was necessary to include a routine which allowed us to change the number of bodies involved in the computations, letting them start from a predetermined position as needed. The variation of the number of bodies was necessary in order to speed up the computations when the minor objects were far from Jupiter and the presence of the Galilean satellites was negligible. A third and more important modification concerned the time step scaling, and this was necessary in order to insure a sufficient precision in the orbits of the Galilean satellites, when present.

In the current version of the program the time step Δt is chosen taking into account all the bodies, in the following way: first the matrix R of the mutual distances is computed, and for each pair of objects i, j (i being the bigger of the two objects) Δt_{ij} is calculated as

$$\Delta t_{ij} = K_i r_{ij}^{\frac{3}{2}}$$

where r_{ij} is the distance between bodies i and j, and

$$K_i = 2\pi / \sqrt{GM_i} / N$$

The meaning of N is the following: as $N \, \Delta t_{ij}$ is the period of the orbit with semiaxis r_{ij} (orbit of the j-th object with respect to the i-th primary), N becomes the number of steps by which this orbit must be divided in order to maintain a given precision of the computation. Experimentally we found that with $N=400$ (the value used in this study) the orbital parameters of the minor object are reliable to the 5th digit. In an n-body problem at every time step the minimum value of Δt_{ij} is chosen as the current value for the time step length, allowing in a natural way the introduction of several "big bodies" and avoiding all the problems related to the definition of the so-called "sphere of action" of the planets.

Finally it should be noticed that some numerical trials allowed us to ignore the influence of the minor bodies on the primaries (Sun, Jupiter, Saturn, Uranus, Neptune and, when present, the Galilean satellites), as the variations with respect to the standard case were beyond the twelfth decimal digit, whereas all the interactions of the primaries (including the mutual perturbations of the Galilean satellites) were computed.

V. ANALYSIS OF THE RESULTS

In order to speed up the computations, all the minor objects belonging to the same tangency curve were processed together. As the initial positions did not correspond to the same true anomaly of Jupiter and then to the same instant (see Sec. II), all the starting points were recomputed taking into account the correct differences of the Jupiter anomaly. This means that only one object of each group had the same initial position as in the CAPTURE study, while the evolutions of the others were affected by Jupiter perturbations for a period of time longer than before. Five different runs were performed:

(i)   Sun, Jupiter and the upper-curve group of objects;
(ii) Sun, Jupiter and the lower-curve group;
(iii) Sun, Jupiter, Galilean satellites and the upper-curve group;
(iv) Sun, Jupiter, Galilean satellites and the lower-curve group;
(v)   Sun, Jupiter, Saturn, Uranus, Neptune and the lower-curve group.

The computations with the satellites and that with the outer planets were made separately in order to investigate the different contributions due to these bodies.

In Table II a short summary of the various computations for each object is given. The first three columns refer to the old computations (CAPTURE), the second ones to the new computations without satellites (CAPTURE/NONSAT) and the last ones to the computations including satellites (CAPTURE/SAT). There is a small difference in the number of satellite captures between CAPTURE and CAPTURE/NONSAT, mainly due to long distance perturbations which are also responsible for the different general behavior of some objects. It may be useful to analyze separately the lower-curve objects and the upper-curve ones; the latter are less interesting for two reasons: (1) only three of them have been captured; and (2) only one (CAPTURE 75) showed a difference when computed with the Galilean satellites, as shown in Fig. 5. The jovicentric path of CAPTURE 75 is given in Fig. 6.

The effects of the Jovian satellites are especially evident in the cases of CAPTURE 26 and 28, which belong to the lower-curve group. In Figs. 7 and 8 the time evolutions of the semimajor axes of these bodies are shown, while the jovicentric paths appear in Figs. 9, 10, 11 and 12. These objects, as well as

TABLE II

Summary of the results of three computations

	CAPTURE			CAPTURE/NONSAT			CAPTURE/SAT		
	Capture	Transition	No. of orbits	Capture	Transition	No. of orbits	Capture	Transition	No. of orbits
18	yes	L – U	none	yes	L – U	none	yes	L – U	none
22	yes	L – U	1ret.	yes	L – L	2ret.	yes	L – L	2ret.
26	yes	L – L	4ret.	yes	L – L	4ret.	yes	L – L	3ret.
28	yes	L – L	3ret.	yes	L – U	13ret.	yes	L – U	4ret.
34	yes	L – U	2ret.	yes	L – U	2ret.	yes	L – U	2ret.
36	yes	L – L	1ret.	no	L – L	1ret.	no	L – L	1ret.
38	yes	L – U	2ret.	yes	L – U	2ret.	yes	L – U	2ret.
46	yes	L – U	1ret.	yes	L – U	1ret.	yes	L – U	1ret.
52	no	L – U	1ret.	yes	L – U	2ret.	yes	L – U	2ret.
56	yes	L – L	none	yes	L – L	none	yes	L – L	none
68	yes	L – L	none	yes	L – L	1ret.	yes	L – L	1ret.
31	no	U – U	none	no	U – U	1ret.	no	U – U	1ret.
39	no	U – U	1ret.	no	U – U	1ret.	no	U – U	1ret.
43	no	U – U	1ret.	no	U – U	1ret.	no	U – U	1ret.
47	no	U – U	1ret.	no	U – U	1ret.	no	U – U	1ret.
51	yes	U – L	none	yes	U – L	none	yes	U – L	none
53	no	U – U	2ret.	no	U – L	none	no	U – U	none
55	no	U – U	1ret.	no	U – U	1ret.	no	U – U	1ret.
67	yes	U – U	2ret.	no	U – U	2ret.	no	U – U	2ret.
69	yes	U – L	2ret.	yes	U – L	2ret.	yes	U – L	2ret.
75	no	U – U	2ret.	no	U – U	2ret.	no	U – U	2ret.
81	no	U – U	1ret.	no	U – U	1ret.	no	U – U	1ret.
87	no	U – U	1ret.	no	U – U	1ret.	no	U – U	1ret.

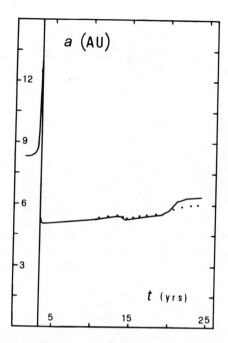

Fig. 5. Time evolution of heliocentric semiaxis of CAPTURE 75. The solid line refers to the computations with and the dotted without the Galilean satellites.

almost all the others, were captured by Jupiter after their exit from the satellite region. CAPTURE 26, without the influence of the satellites, makes four orbits around Jupiter, while the orbits are reduced to three by the influence of the satellites. More striking is the case of CAPTURE 28, which in the present computation has a story completely different and more interesting than in the previous one. In fact, its evolution computed without satellites shows a 62-year capture (the longest continuous one we found until now), during which it makes 13 orbits around Jupiter. The jovicentric orbital parameters of this object are not constant, as can be seen in Fig. 13. In the recomputations with the satellites, CAPTURE 28 unbinds itself from Jupiter after only 4 orbits. In Fig. 13 the time evolution of the jovicentric semiaxis for this case is shown by the solid line. Two remarks can be made on these observations:

1. The only two objects that undergo rather strong perturbations by the Galilean satellites both come from the "capture region" indicated in Sec. II, and they are the only objects that make more than two orbits.
2. Apart from these cases, and the case of CAPTURE 18 and 75 which show small differences in their final orbits, the influence of the Galilean satellites is very small.

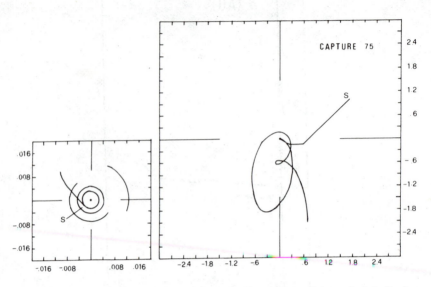

Fig. 6. Jovicentric path of CAPTURE 75 in the rotating frame (Sun on the left; S refers to starting point). In the smaller graph the trajectory within the Galilean region (same reference frame) is shown. Distances are in AU.

Let us finally discuss the influence of the outer planets on the evolution of our objects. The computations, carried out only on the lower-curve objects, have shown negligible differences in the heliocentric and jovicentric parameters with respect to the case without outer planets and satellites. This result justifies the choice, made in the general studies (Carusi and Pozzi 1978*b*; Carusi *et al.* 1979), of disregarding the perturbations due to other planets if only close encounters are concerned. Of course, in the case of longer evolutions as those performed on real comets, the influence of other planets cannot be neglected.

VI. THE EVOLUTION OF THE ORBIT OF COMET P/GEHRELS 3

As we said in Sec. II, some observed comets had one or more close encounters with Jupiter, during which a satellite capture probably occurred. In order to make a more detailed comparison with our fictitious objects, the short-period comet P/Gehrels 3 was selected, for two reasons: it experienced a temporary satellite capture during the encounter of 1970, as shown by Rickmann (1979), and it passed at a distance of only two planetary radii from the surface of Jupiter. This comet is also interesting because it is a potential candidate for evolving to an asteroid-like orbit, as discussed in the

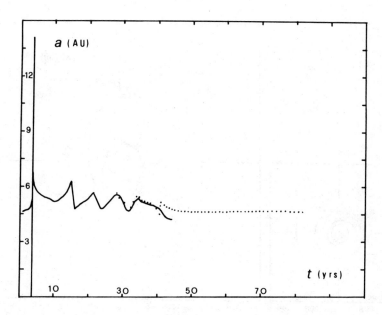

Fig. 7. Time evolution of heliocentric semiaxis of CAPTURE 26. As in Fig. 5, dots refer to the computation without satellites.

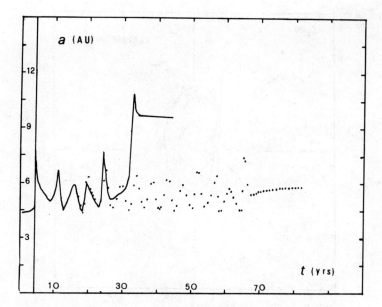

Fig. 8. Time evolution of heliocentric semiaxis of CAPTURE 28.

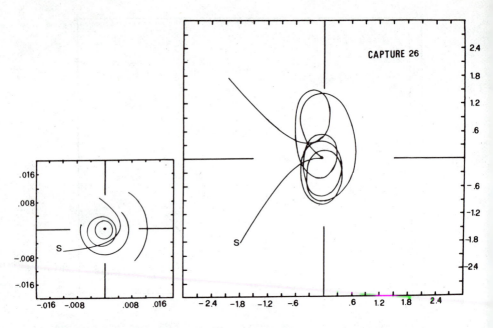

Fig. 9. Jovicentric path, as in Fig. 6, for CAPTURE 26.

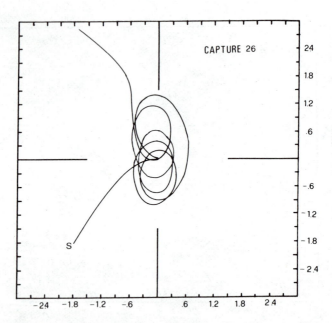

Fig. 10. Jovicentric path of CAPTURE 26 in the computation without satellites.

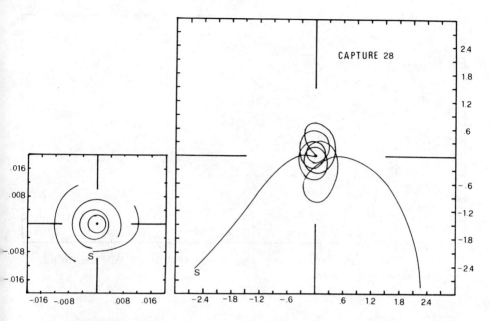

Fig. 11. Jovicentric path, as in Fig. 6, for CAPTURE 28.

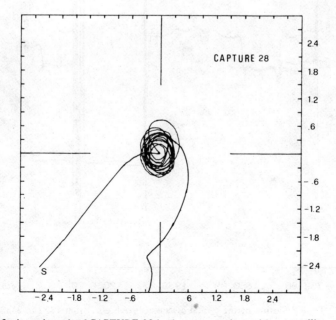

Fig. 12. Jovicentric path of CAPTURE 28 in the computations without satellites.

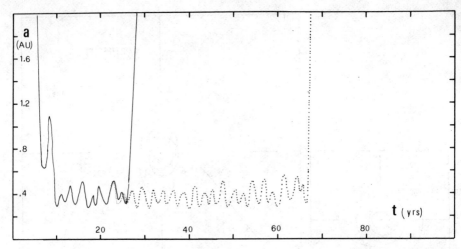

Fig. 13. Time evolution of jovicentric semiaxis of CAPTURE 28. Dots refer to the evolution without satellites.

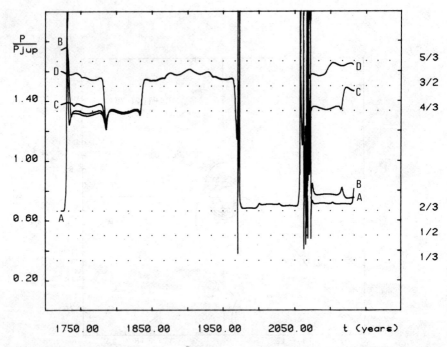

Fig. 14. Time evolution of the period of comet P/Gehrels 3 from 1722 to 2132. Dotted lines indicate the resonances with Jupiter.

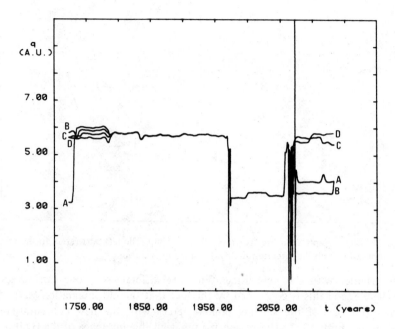

Fig. 15. Time evolution of the perihelion distance of P/Gehrels 3.

chapter by Kresák. Input data of P/Gehrels 3, together with those of all the planets and of the other short-period comets discovered in 1975, were provided by Marsden to Rickmann (1978, personal communication).

A set of four starting points, the same as those in Rickmann (1979), was used to integrate the evolution of the comet from 1974.854 (J. D. 2442360.5) backwards for 252 yr. and forwards for 158 yr, so that the computations cover the period 1722-2132. The method used was the same as described in Sec. IV, and both the outer planets and (only for the 1970 encounter) the Galilean satellites were included. In Figs. 14 and 15 the time evolutions of the period and of the perihelion distance are shown for the four computations.

It is easy to see that the evolution is very similar for the four versions of the comet backwards until an encounter with Jupiter occurred in 1785, and forwards until a very close and long encounter starting in 2055. During this last encounter surely some orbits around Jupiter are made, as is recognizable by the steep peaks and dips in the heliocentric period, which are due to the changes in heliocentric velocity during the revolutions around the planet. As the evolutions divaricate before 1785 and after 2060, nothing can be said about the preceding and subsequent evolution of the comet.

The result of an integration made with the old method (that of our introduction and Secs. I-III), but including the outer planets, is shown in Fig.

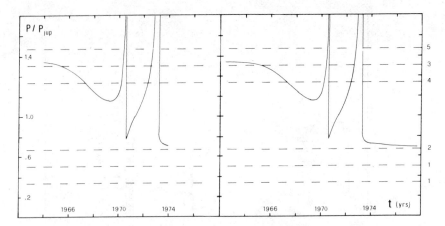

Fig. 16. Same graph as in Fig. 14 from 1960 to 1980. The left part refers to the new computation with the satellites, the right to the old evolution.

16 together with the new integration. The differences as one can see are negligible, and this is confirmed by the fact that the old integration (carried out from one of the starting points) agrees with the new computations backwards until 1832. The reason for the negligible influence of the Galilean satellites can be explained by the relative positions of the bodies when the comet was very close to Jupiter (see the smaller graph in Fig. 17): the comet passed inside the orbits of the four satellites, but with a very high angle between the velocity vectors, and crossed their orbits when the satellites were far from it, as we have verified. The larger graph in Fig. 17 shows the jovicentric path of P/Gehrels 3 during its temporary satellite capture in 1970; it is not very similar to any of the CAPTURE objects.

A comparison of the transfer processes from the supposed "cometary belt" between Jupiter and Saturn (see the chapter by Kresák) to the 2/3 resonance with Jupiter of P/Oterma (Kazimirchak-Polonskaya 1976; Carusi *et al.* 1979) and P/Gehrels 3 shows that the latter will not return (to the same orbit) after two Jupiter revolutions as was the case for P/Oterma in 1961. This can be attributed to the unfavorable 2/3 resonant motion in which P/Gehrels 3 is now moving; the "quasi-mirror" configuration of Oterma in 1950 (halfway between the 1939 and the 1961 encounter) will not occur in 1986 for P/Gehrels 3, so the encounter of 1998 will be less effective and less close than that of 1974 and will not cause a transfer back to the Jupiter-Saturn belt.

VII. GENERAL REMARKS AND CONCLUSIONS

In this final section we will make a few comments on the computations and draw some general conclusions. The first remark concerns the way of starting the Galilean satellites. In the case of the fictitious bodies they were

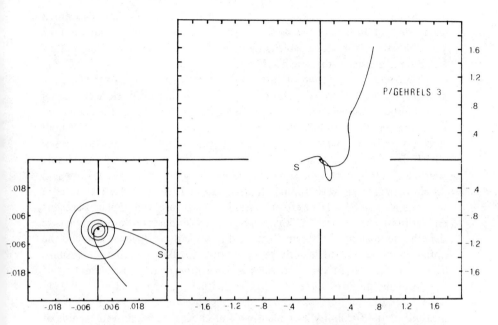

Fig. 17. Jovicentric path of P/Gehrels 3 in the new computations with the satellites. *S* refers to the starting point of our backwards evolution. Distances are in AU.

started as the distance of the closest object from Jupiter became less than 2.5×10^6 km and were stopped when it became greater than 2.5×10^6 km. In the run with the lower-curve objects, the starting jovicentric orbit of Europa was wrong (it had an eccentricity of 0.17). We do not think that this error can invalidate the conclusions of our work; Europa is the smallest of the Galilean satellites and our research is a test, carried out on fictitious objects, of the effect of the presence of satellites around major planets. Therefore a small difference from the real situation is not crucial.

Incidentally, we noted an interesting uncoupling of the orbits of the satellites caused by this error. For comet P/Gehrels 3 the satellites were started when the distance of the comet from Jupiter was 4.97×10^6 km and the starting points were those corresponding to that instant (11:15 AM of 21 August 1970), as derived from Connaissance des Temps. Again the satellites were stopped when the distance from Jupiter became greater than 4.97×10^6 km. In this case the masses of the Galilean satellites were subtracted from that of Jupiter when these bodies were switched on, and were added at the instant of switching off. We had omitted making this correction for the fictitious objects, but an analysis of evolutions of objects that did not suffer change in the computation with the satellites has convinced us that this correction is not very significant in varying the orbital evolution; it only changes the time scale slightly.

In any case we could not avoid small jumps in the heliocentric parameters of Jupiter at the switching the satellites on and off but we think that this effect is an inescapable consequence of the sensitivity of the method based on Greenspan's discrete mechanics.

The general conclusion of this work is that the influence of the Galilean satellites on the dynamical evolutions of objects having a very close encounter with Jupiter is not very relevant. In only 4 cases out of 23 have the evolutions with the satellites shown differences when compared to those without satellites. These differences, however, were important in two cases, CAPTURE 26 and 28, in which the satellite captures were considerably shortened. This conclusion somehow resembles that of the general research on close encounters with Jupiter, that is: low efficiency of the encounter in the general case, but higher effectiveness in special cases (inner planet-crossers and orbits nearly tangent to that of Jupiter). We must, however, emphasize that the number of objects involved in this work is not high enough and the initial sample is not sufficiently representative to allow definite conclusions. Therefore the present study provides an indication of the possible extent of the problem and gives some hints for evaluating the numerical work already done by other authors on comets passing close to Jupiter. It also gives a measure of the flexibility of Greenspan's method, that has proved capable of adapting to problems with a variable number of interacting bodies. Finally we wish to point out that the computations done on comet P/Gehrels 3 agree well with those made on fictitious objects of initial parameters similar to those of the comet (it was an upper-curve object before the encounter of 1970).

Acknowledgments. We would like to thank the Accademia Nazionale dei Lincei for financial support of the computing work, L. Kresák for useful comments on this work and on the method used for the computation of P/Gehrels 3, and H. Rickmann for reviewing the manuscript and providing the starting points mentioned in the text. Special thanks also go to Mrs. Matthews for her kindness and linguistic help.

REFERENCES

Arlot, J. 1975. A comparison of some observations of the Galilean satellites with Sampson's tables. *Celes. Mech.* 12: 39-50.

Bailey, J. M. 1971. Jupiter: Its captured satellites. *Science* 173: 812-813.

Bailey, J. M. 1972. Studies on planetary satellites. Satellite capture in the three-body elliptical problem. *Astron. J.* 77: 177-182.

Belyaev, N. A. 1967. Orbital evolution of comets Neuimin 2 (1916 II), Comas Solà (1927 III), and Schwassmann-Wachmann 2 (1929 I) over the 400 years from 1660 to 2060. *Sov. Astr.-A.J.* 11: 366-373.

Carusi, A. 1979. A new numerical method for the integration of the motion equation for a n-body system: Theory and applications. In *Instabilities in Dynamical Systems*, ed. V. G. Szebehely (Dordrecht: D. Reidel). In press.

Carusi, A., and Pozzi, F. 1978a. A new method for close encounter computation, *Moon and Planets* 19: 65-70.

Carusi, A., and Pozzi, F. 1978b. Planetary close encounters between Jupiter and about 3000 fictitious minor bodies. *Moon and Planets* 19: 71-87.

Carusi, A.; Pozzi, F.; and Valsecchi, G. B. 1978. Planetary close encounters: Changes in a planet crossing population (NASA-TM 79729) 22-24.

Carusi, A.; Pozzi, F.; and Valsecchi, G. B. 1979. Planetary close encounters: An investigation on temporary satellite capture phenomena. In *Dynamics of the Solar System*, ed. R. L. Duncombe (Dordrecht: D. Reidel), pp. 185-189.

Carusi, A., and Valsecchi, G. B. 1979a. Planetary close encounters: Importance of nearly tangent orbits. *Moon and Planets,* (in press).

Carusi, A., and Valsecchi, G. B. 1979b. Effects of a close encounter with Jupiter on different populations of planet crossing objects. *Moon and Planets,* (in press).

Everhart, E. 1972. The origin of short-period comets. *Astrophys. Lett.* 10: 131-135.

Everhart, E. 1973. Horseshoe and Trojan orbits associated with Jupiter and Saturn. *Astron. J.* 78: 316-328.

Everhart, E. 1976. The evolution of comet orbits. In *The Study of Comets,* eds. B. Donn, M. Mumma, W. Jackson, M. A'Hearn and R. Harrington (NASA SP-393, Washington, D. C.: U. S. Government Printing Office), pp. 445-461.

Ferraz-Mello, S. 1975. Problems on the Galilean satellites of Jupiter. *Celes. Mech.* 12: 27-37.

Gehrels, T. 1977. Some interrelations of asteroids, Trojans and satellites. In *Comets, Asteroids, Meteorites,* ed. A. H. Delsemme (Toledo, Ohio: University of Toledo Press), pp. 323-325.

Greenspan, D. 1973. An algebraic, energy conserving formulation of classical molecular and Newtonian *n*-body interaction. *Bull. Amer. Math. Soc.* 79: 423-427.

Heppenheimer, T. A. 1975. On the presumed capture origin of Jupiter's outer satellites. *Icarus* 24: 172-180.

Heppenheimer, T. A., and Porco, C. 1977. New contributions to the problem of capture. *Icarus* 30: 385-401.

Horedt, Gp. 1972a. Capture in the restricted three-body problem. *Acta Astron.* 22: 55-66.

Horedt, Gp. 1972b. Single close encounters in the planetary problem. *Celes. Mech.* 6: 232-241.

Horedt, Gp. 1974. Numerical exploration of the capture problem. *Acta Astron.* 24: 207-213.

Horedt, Gp. 1976. Capture of planetary satellites. *Astron. J.* 81: 675-678.

Hunter, R. B. 1967a. Motions of satellites and asteroids under the influence of Jupiter and the sun. I. Stable and unstable satellite orbits. *Mon. Not. Roy. Astron. Soc.* 136: 245-265.

Hunter, R. B. 1967b. Motions of satellites and asteroids under the influence of Jupiter and the sun. II. Asteroid orbits close to Jupiter. *Mon. Not. Roy. Astron. Soc.* 136: 267-277.

Kazimirchak-Polonskaya, E. I. 1967. Evolution of short-period comet orbits from 1660 to 2060, and the role of the outer planets. *Sov. Astron.-A.J.* 11: 349-365.

Kazimirchak-Polonskaya, E. I. 1972. The major planets as powerful transformers of cometary orbits. In *The Motion, Evolution of Orbits and Origin of Comets,* eds. G. A. Chebotarev, E. I. Kazimirchak-Polonskaya and B. G. Marsden (Dordrecht: D. Reidel), pp. 373-397.

Kazimirchak-Polonskaya, E. I. 1976. Review of investigations performed in USSR on close approaches of comets to Jupiter and the evolution of cometary orbits. In *The Study of Comets,* eds. B. Donn, M. Mumma, W. Jackson, M. A'Hearn and R. Harrington (NASA SP-393, Washington, D. C.: U. S. Government Printing Office), pp. 490-536.

Kresák, L. 1972. On the dividing line between cometary and asteroidal orbits. In *The Motion, Evolution of Orbits and Origin of Comets,* eds. G. A. Chebotarev, E. I. Kazimirchak-Polonskaya and B. G. Marsden (Dordrecht: D. Reidel) pp. 505-514.

LaBudde, R. A., and Greenspan, D. 1976a. Energy and momentum conserving methods of arbitrary order for the numerical integration of equations of motion. I. Motion of a single particle. *Numer. Math.* 25: 323-346.

LaBudde, R. A., and Greenspan, D. 1976b. Energy and momentum conserving methods of arbitrary order for the numerical integration of equations of motion. II. Motion of a system of particles. *Numer. Math.* 26: 1-16.

Moulton, F. R. 1914. On the stability of direct and retrograde satellite orbits. *Mon. Not. Roy. Astron. Soc.* 75: 40-57.

Pollack, J. B.; Burns, J. A.; and Tauber, M. E. 1979. Gas drag in primordial circumplanetary envelopes: A mechanism for satellite capture. *Icarus* 37: 587-611.

Rickmann, H. 1979. Recent dynamical history of the six short-period comets discovered in 1975. In *Dynamics of the Solar System,* ed. R. L. Duncombe (Dordrecht: D. Reidel, in press).

Roy, A. E., and Ovenden, M. W. 1955. On the occurrence of commensurable mean motions in the solar system. II. The mirror theorem. *Mon. Not. Roy. Astron. Soc.* 115: 296-309.

Sampson, R. A. 1910. *Tables of the Four Great Satellites of Jupiter* (London: Wesley).

Sitarski, G. 1968. Approaches of the parabolic comets to the outer planets. *Acta Astron.* 18: 171-195.

Wetherill, G. W. 1977. Evolution of the Earth's planetesimal swarm subsequent to the formation of the earth and moon. *Proc. Lun. Sci. Conf. VIII.* (Oxford: Pergamon Press), pp. 1-16.

DISTANT ASTEROIDS
AND OUTER JOVIAN SATELLITES

J. DEGEWIJ
University of Arizona

and

C. J. VAN HOUTEN
Leiden University

Sixty percent of the sampled objects in the Hilda, Trojan and outer Jovian satellite locations belong to C-type and another 30 % belong to a new group called RD-type (reddish and dark), sometimes referred to simply as D-type. Objects in this group have low albedo values between 2 and 4 % and steep reflection spectra between 0.7 μm and 0.9 μm. Also 944 Hidalgo belongs to this group but shows color variation over its surface. Meteoritic minerals with similar optical reflection spectra are discussed. Trojans with sizes down to 15 km in the cloud preceding Jupiter are ~3.5 times more numerous than those in the following cloud. RD-type Trojans appear more often in the preceding cloud. There is a resemblance of spectrum, albedo and phase relation among the majority of Trojans and the outer Jovian satellites.

Little is known about statistics and physical parameters of asteroids and satellites beyond the main asteroid belt. This is because of the faintness of these objects (V > 16 mag) which makes detection and measurement difficult. The use of the 122-cm Palomar Schmidt telescope aimed towards the center of the Trojan cloud, shows that on a $6°5 \times 6°5$ 103a-O plate exposed for 10 minutes, 300 main-belt asteroids, 20 Trojans and 4 Hildas can

[417]

be found from the apparent motion. The TRIAD orbital elements file (see Bender's tables in Part VII), and also Chapman *et al.* (1978) give orbital elements for 26 Hildas, 12 Trojans preceding, and 10 following Jupiter in its orbit. Assuming an albedo of 0.03, 75 % of these relatively bright Hildas and Trojans have diameters larger than 50 km.

Extensive progress in the knowledge of physical parameters of objects with V>16 mag has been made with an image intensifier and TV acquisition. An inexpensive system for use with the 154-cm Catalina reflector has been developed and built by E. F. Montgomery of the Lunar and Planetary Laboratory. This system made it possible to detect objects down to V~19 mag and to center them in the photometer.

In the first part of this chapter we review the knowledge about distant asteroids like Hildas, Trojans and Hidalgo. In the second part we review the knowledge for the outer Jovian satellites. Finally we discuss whether the distant asteroids show similarities with those in the main belt and whether mixing has taken place between different locations in the solar system. The surface mineralogy will also be discussed.

I. DISTANT ASTEROIDS

Statistics of Trojans

In 1772 Lagrange showed that stable configurations of three masses, m_1, m_2, m_3, are possible if the distances between these masses are equal. A further condition for stability is that $m_1 \geqslant 25\ m_2$ and that m_3 is very small with respect to both m_1 and m_2. If m_1 is taken to be the mass of the sun, m_2 that of Jupiter and m_3 that of an asteroid, then the stability conditions are fulfilled. Stability here means that the mass m_3 can oscillate around the equilibrium point. From the geometry it follows that two such points are possible in Jupiter's orbital plane. They are usually denoted by the symbols L_4 and L_5 and are called libration points or Lagrangian points. Here L_4 is called the preceding Lagrangian point because L_4 preceeds, and L_5 follows, Jupiter in its orbit. The motion of mass points around both locations has given rise to extensive discussion in the literature which will not be reviewed here. It is sufficient to say that the motion around the libration point can be described by a combination of two periods. The longer of the two depends on the amplitude of the oscillation and is on the order of 150–200 yr (Rabe 1961). The shorter period is practically equal to the orbital period of Jupiter, 12 yr.

The theoretical stability of asteroids near the triangular Lagrangian points of the Sun-Jupiter system does not necessarily imply that they are exactly there. The first asteroid near one of these libration points was found by M. Wolf in Heidelberg in 1906. It became customary to name these asteroids after the heroes of Homer's *Iliad;* those found near the point

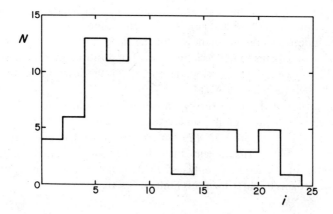

Fig. 1. Distribution of the orbital inclinations of 72 Trojans.

preceding Jupiter were named after the Greek warriors and those near the following point after the defenders of Troy. At the moment 22 Trojan asteroids in both clouds have permanent numbers and have been named. But, in addition, two other groups of orbital elements are available: (1) Trojans found in the Palomar-Leiden Survey (PLS; van Houten *et al.* 1970*b*) – from the Trojans found in this survey 15 had reliable orbital elements and two have now permanent numbers; and (2) Trojans found in a special survey made in 1973 (Minor Planet Circulars 4289–4294); in this survey 34 Trojans were found for which orbital elements were obtained. The material of these three groups, comprised of 69 objects, forms the basis for the following statistical discussion.

Rabe (1967) showed that the eccentricities of the Trojan orbits are dynamically limited; for small-amplitude librations this limit is approximately 0.19, but in the case of large libration amplitudes the limit decreases to 0.08. In accordance with this result none of the orbital eccentricities of the known Trojans exceeds 0.19. The distribution of the inclinations is shown in Fig. 1. There is a possibility that the distribution is bimodal; a group of small and one of large inclinations are separated by a minimum at $i \sim 13°$. But it should be realized that both the PLS and the special survey in 1973 were limited to a region $\pm 6°$ from the ecliptic, so that in this material high-inclination Trojans are severely underrepresented. It seems, therefore, premature to draw definite conclusions about the statistical distribution of the Trojan orbits of high inclination (a detailed discussion is made by Yoder [1979] and in the chapter by Greenberg and Scholl in this book). Plates have been taken in 1975 to extend the survey of the preceding point to higher inclinations.

It was shown in a special survey (van Houten *et al.* 1970*a*), in combination with the PLS result, that the total number of asteroids around the preceding point to a limit of $B(a,0) = 20.9$ or a diameter of 15 km amounts to ~700. This number is uncertain because it is extrapolated from an observed number of 45 and also the concentration of Trojans to the plane of the ecliptic is at present unknown. In the same survey it was shown that the number of Trojans found, as a function of opposition magnitude, shows a linear relation with the same slope as found for the asteroids in the outer zone of the belt ($3.0 < a < 3.5$ AU).

Comparison of the 1973 survey, in which part of the surroundings of L_4 was investigated, with a similar survey for L_5, made in 1971, indicates that the Trojans around L_4 are ~3.5 times more numerous than those around L_5 (van Houten 1980). This means that, to the 20.9-magnitude limit, the surroundings of L_5 contain only ~200 Trojans, bringing the combined number of Trojans brighter than this limit to ~900.

Albedos and Diameters

Two techniques exist for the determination of a geometric albedo: polarimetry and infrared radiometry. The first technique is reviewed in the chapter by Dollfus and Zellner in this book and the second in the chapter by Morrison and Lebofsky. The polarimetric technique has limitations because of the faintness of the objects considered here and their small ranges in phase angle. Also, for dark objects there is no unique interpretation (Zellner et al. 1977b). Zellner et al. observed C-type UBV colors for the Hildas 153 and 361. Zellner (personal communication) found that single polarization measurements for both asteroids are, however, inconsistent with the polarization curve of a typical C-type asteroid in the main belt. Zellner reports further that polarization measurements for Trojan 624 Hektor are consistent with a C-type surface.

Radiometric albedos and diameters for four Trojans were first published by Cruikshank (1977). He used the radiometric information of the 20 μm band which for Trojans is close to the maximum of the Planck curve at 25 μm (120 K). To derive albedos and diameters he used visual estimates of the optical magnitudes. Morrison (1977) recomputed the albedos and diameters in a consistent system and used the absolute visual magnitudes from Gehrels and Gehrels (1979). Gehrels and Tedesco (1979) introduced a more realistic absolute magnitude listing in which the correction to zero degrees phase angle was made by a phase coefficient of 0.039 mag/deg, which is in agreement with the value obtained by van Houten et al. (1970b) for faint and distant asteroids. The absolute magnitudes given in Column 9 of Table I are basically those of Gehrels and Tedesco (1979) with some adjustments for objects observed since.

With the 154-cm Catalina reflector of the Lunar and Planetary Laboratory and the infrared bolometer as described by Low and Rieke (1974), Degewij (1978b) obtained radiometric albedos and diameters at 10.6 μm for four Trojans and three Hildas. The results given in Table II differ

TABLE I

Distant Asteroids and Satellites

	U–B (1)	B–V (2)	n (3)	V–R (4)	V–I (5)	V–I$_J$ (6)	n (7)	S (8)	V(1,0) (mag) (9)	Δm (mag) (10)	n (11)	app (12)	P (hr) (13)	p_V (14)	Diam. (km) (15)	Type (16)	Ref. (17)
944 Hidalgo	0.22	0.75	2					y	10.83	0.5			10.06			RD	
Preceding Trojans																	
588 Achilles	0.23	0.74	1					y	8.97		1	1				RD	
624 Hektor	0.27	0.77	>5	0.44	0.95	1.22	3	y	7.77	1.2			6.92	0.038	190:	RD	a,e
659 Nestor	0.31	0.75	1	0.40	0.80		1		8.86		1	1				C	
911 Agamemnon	0.22	0.77	2					y	8.26	0.46	4	4	6–10	0.035	158	RD	b
1143 Odysseus	0.25	0.79	2						8.55	0.08	2	2				–	
1437 Diomedes	0.24	0.71	4	0.37	0.80	1.06	3		8.40	0.36	9	4	18–24	0.021	191	C	b
1583 Antilochus	0.27	0.76	3	0.42	0.82		1		8.83	0.52	5	3				C	
Following Trojans																	
617 Patroclus	0.24	0.68	5	0.33	0.69		1	y	8.48	0.25	4	2		0.031	152	C	
884 Priamus	0.22	0.71	5					y	8.96	0.34	7	1				RD	
1172 Aeneas	0.27	0.72	4					y	8.53	0.18	4	1		0.039	132	C	e
1173 Anchises	0.29	0.70	3	0.32	0.79		1	y	9.03	0.15	4	2		0.047	95	C	e
1208 Troilus	0.31	0.73	2	0.32	0.68		1	y	9.19	0.28	4	2				C	
1867 Deiphobus	0.25	0.75	3		0.79		1	y	8.69	0.28	5	1				C	
Hildas																	
153 Hilda	0.30	0.68	1		0.90				7.69		1	1				–	
361 Bononia	0.26	0.73	2	0.41			1	y	8.77	0.09	2	2				RD	
958 Asplinda	0.38	0.83	2	0.42	0.93		2		10.70	0.08	2	1		0.18	23	S	
1162 Larissa	0.21	0.65	1					y	9.60:	0.85	2	2	>16	0.14:	42	M	
1212 Francette	0.23	0.69	1					y	7.29		1					C	d
1268 Libya	0.23	0.67	1					y	9.34		1	1				–	

TABLE I (Continued)

Distant Asteroids and Satellites

	U–B (1)	B–V (2)	n (3)	V–R (4)	V–I (5)	$V-I_J$ (6)	n (7)	S (8)	V(1,0) (mag) (9)	Δm (mag) (10)	n (11)	app (12)	P (hr) (13)	p_V (14)	Diam. (km) (15)	Type (16)	Ref. (17)
1345 Potomac	0.29	0.71	3	0.34	0.72		1		9.97	0.09	3	2				C	
1439 Vogtia	0.32	0.75	1						10.64		1	1				–	
1512 Oulu	0.20	0.70	2					y	9.82	0.17	2	2		0.030	83	RD	
1529 Oterma	0.24:	0.76	2					y	10.32	0.06	2	1				RD	
1754 Cunningham	0.26	0.62	1						10.08		1	1				–	
Satellites																	
J6 Himalia	0.30	0.67	5	0.34	0.71		2		8.14	0.12	5		9.5	0.03	170	C	e
J7 Elara	0.28	0.69	5	0.31	0.64		3		10.07	0.51				0.03	80	C	e
J8 Pasiphae	0.34	0.63	2	0.40	0.82		2		10.33:	large?	5	3				C	
S7 Hyperion	0.33	0.77	12	0.61*		0.80	1									–	
S9 Phoebe	0.33	0.70	10						6.91	0.24c			21.2c			C	c
N1 Triton	0.30	0.70	7	0.55*			3		–0.96	0.07						–	

*$V-R_J$ colors given.

y 0.3 – 1.1 μm spectrum is known.

a Dunlap and Gehrels (1969). b Taylor (1971). c Andersson (1974). d Taylor *et al.* (1976). e Cruikshank (1977).

TABLE II

10 μm Radiometric Albedos and Diameters

Object (1)	Date (2)	ΔM_1 (3)	V(1,0) (4)	$N_{10.6}$ (5)	R (AU) (6)	Δ (AU) (7)	α (8)	D (km) (9)	p_V (10)
617 T	9 Jan 78	0.40	8.48	4.52	4.71	4.41	+12	154 ± 5	0.030 ± 1
624 T	9 Jan 78	0.40	7.77	4.52	5.02	4.52	−10	201 ± 8	0.034 ± 2
911 T	9 Jan 78	0.40	8.26	5.46	5.30	4.90	+10	158 ±10	0.035 ± 5
958 Hi	9 Jan 78	0.35	10.70	6.47	3.32	2.36	− 4	23 ± 4	0.179 ±70
1162 Hi	7 Apr 78	0.35	9.60	5.98	3.85	2.98	+ 8	42 ± 5	0.141 ±30
1437 T	9 Jan 78	0.40	8.40	4.60	5.02	4.47	+10	191 ±10	0.021 ± 2
1512 Hi	6-7 Apr 78	0.35	9.82	3.57	3.53	2.53	− 1	83 ± 2	0.030 ± 1

slightly from those published by Degewij (1978b) because of the improvements in absolute magnitudes. Column 3 of Table II gives the transformation from the instrumental system ($\lambda_{eff} = 10.6$ μm) to the standard system ($\lambda_{eff} = 10.0$ μm); details are given by Gradie (1978). Column 5 gives the infrared flux at 10.6 μm and the following columns give respectively, the distance between object and Sun, object and Earth, and the phase angle (negative before and positive after opposition). The diameters and albedos from Cruikshank (1977) and Degewij (1978b) are combined in Columns 14 and 15; Cruikshank's values are adjusted for the absolute magnitudes in Column 9. An equal weight was given to determine an average if the two sources provided data for the same object.

The three Hildas have different albedo types. Both 958 and 1162 have high albedos and 1512 has a low albedo. Three Trojans in the preceding cloud and three in the following cloud all have low albedos between 0.02 and 0.05. It is of interest to look for a systematic difference in albedos for objects in the two clouds. The data in Table I show an average and error in the mean of 0.031 ±0.005 (n=3) for Trojans preceding Jupiter and 0.039 ±0.005 (n=3) for Trojans following Jupiter. The higher number of Trojans, down to a certain magnitude limit, in the cloud preceding Jupiter than the number following Jupiter cannot be explained by systematically different albedos for the objects in the two clouds. If the difference between the average albedos for the three objects in each cloud is real, then the number difference between preceding and following would be even larger.

Reflectance Spectra at Optical and Infrared Wavelengths

The spectral information of solid bodies in the solar system is given in terms of spectral reflectance. This is the ratio of the fluxes of object and Sun, integrated across the spectral response of the filter normalized to unity at 0.55 μm. Extensive work has been done by Chapman et al. (1973), and McCord and Chapman (1975a,b) to derive spectral reflectance curves for 98 bright asteroids. They used 22 narrowband filters in the wavelength domain 0.32 − 1.10 μm and included some Trojans in their program. It was found that both 624 Hektor and 911 Agamemnon yielded unique reflection spectra with a steep rise between 0.7 and 0.9 μm. The spectrum of 1173 Anchises (Chapman 1976) was, however, flat in this domain and showed similarities with a C-type spectrum. Chapman has measured many more spectra (see the chapter by Chapman and Gaffey). Their chapter contains spectral reflectivities for three Trojans preceding Jupiter, five Trojans following Jupiter, five Hildas, and 944 Hidalgo which is the object in a very elongated orbit reaching beyond Jupiter and Saturn. Table I, Column 8 shows a "y" if the spectrum of the object is known.

For many years broadband UBV colors of asteroids have been measured at the University of Arizona. UBV colors for 624 Hektor (Dunlap and Gehrels 1969), 911 Agamemnon and 1437 Diomedes (Taylor 1971), 588 Achilles,

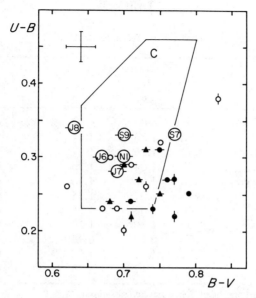

Fig. 2. U-B versus B-V colors for distant asteroids in the 153 Hilda group (○) and the Trojan cloud preceding (●) and following (▲) Jupiter, and some faint satellites. The data are from Table I. Colors at red wavelengths are known for objects marked with bars: vertical bars mean steep reddish RD-type spectra (V-I≥0.9; see Figs. 3 and 4) and horizontal bars mean flat C-type spectra. The color of the sun is at U-B = 0.10 and B-V = 0.63. The domain for C compositional types is drawn as defined by Bowell et al. (1978). The error bar represents the expected error in the mean of the measurements.

1143 Odysseus, and 1583 Antilochus (Zellner et al. 1977a) all in the cloud preceding Jupiter, indicate similar colors near the lower right edge of the C-type region shown in Fig. 2. Also the UBV colors of one Hilda (Taylor et al. 1976) and four other Hildas (Zellner et al. 1977a) are located in this area. Degewij et al. (1978) and Degewij (1978b) give UBV colors for Trojans in the cloud following Jupiter. The broadband technique has the advantage of a shorter integration time than that of narrowband measurements of comparable accuracy. This makes it possible to measure more faint objects per observing night and then it is feasible to repeat the measurements for the same object during a second and, if needed, a third night.

A photometric survey of distant asteroids and satellites was made by Degewij (1978b). The effective wavelengths and half widths were V: 0.55 (0.08) μm, R: 0.65 (0.13) μm, I: 0.83 (0.09) μm, and I_J: 0.86 (0.18) μm. Colors in the Kron-system (Kron et al. 1953) were provided by Weistrop (personal communication) and linear transformations between the instrumental system and the standard system were obtained. The UBVRII$_J$ colors are given in Table I; in Column 4 the "*" means that V−R$_J$ colors in the Johnson system are given. These colors are average values of data

TABLE III

Synthesized Colors from Reflection Spectra[a]

| Asteroid | Reflectances | | | Colors | |
| | V | R | I | V-R | V-I |
	0.55	0.65	0.83		
16 Psyche	0.98	1.06	1.12	0.40	0.88
141 Lumen	1.00	1.00	1.04	0.31	0.77
588 Achilles	1.00	1.10	1.27	0.41	0.99
884 Patroclus	0.98	1.04	1.20	0.38	0.95
911 Agamemnon	0.98	1.06	1.30	0.40	1.04
944 Hidalgo	0.95	1.07	1.25	0.44	1.03
1162 Larissa	1.00	1.03	1.05	0.34	0.78
1172 Aeneas	1.02	1.10	1.10:	0.39	0.81
1212 Francette	1.00	0.93:	0.92:	0.23	0.64
1512 Oulu	1.02	1.12	1.32	0.41	1.01
1529 Oterma	0.98	1.12	1.28	0.46	1.02

[a]Below V, R and I are the effective wavelengths in μm.

obtained on n different nights (columns 3 and 7). Typical standard deviations for objects with V magnitudes between 15 and 16 are 0.02 mag for both UBV and VRII$_J$ colors.

It is possible to determine the scale factor between the measurements for objects observed in common by Chapman and Degewij. With

$$m_{\lambda_1} - m_{\lambda_1} = S_{\lambda_1,\lambda_2} - 2.5 \log (R_{\lambda_1}/R_{\lambda_2}) \qquad (1)$$

we determined S_{λ_1,λ_2}. For the UBV measurements the objects in common were: 361, 617, 588, 624, 884, 911, 1172, 1173, 1208, 1212, 1512 and 1529. For the VRI measurements the objects in common were: 361, 617, 624, 1173 and 1208. The values for S_{λ_1,λ_2} and the standard deviations are: S_{BV} = 0.63 ±0.02 (1172), S_{UV} = 0.82 ±0.05 (1173, 624, 911), S_{VR} = 0.31 ±0.05, and S_{VI} = 0.73 ±0.06 (617). Numbers between parentheses mean that the S value for these objects departs by more than 0.1 mag from the mean and therefore zero weight was given. For both 624 and 911 we notice a discrepancy between a steep UV drop-off as measured by Chapman and a shallower UV drop-off shown by the numerous UBV data. This does not point towards a possibility for spots. The most likely explanation is that the discrepancy is due to calibration errors and limited photon statistics. These scale factor values were used to compute "synthetic" colors from Chapman's spectra for objects without broadband color measurements. The colors are given in Table III which includes also M-type main belt asteroid 16 Psyche and C-type main-belt asteroid 141 Lumen. The VRI colors of Table I and

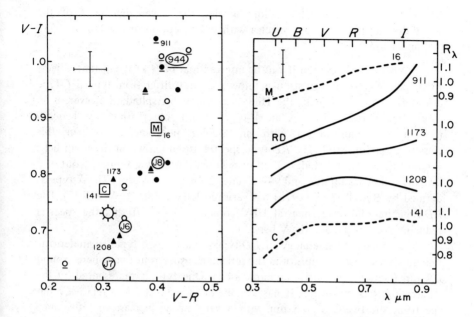

Fig. 3. V-I versus V-R colors for distant asteroids in the 153 Hilda group (○), the Trojan cloud preceding (●) and following (▲) Jupiter, and some faint satellites. The data are from Table I. The underlined symbols represent synthetic colors (see Table III) obtained from spectra measured by Chapman and Gaffey (see their chapter). The color of the sun is at V-I = 0.73 and V-R = 0.31. Reflection spectra for five of the numbered objects are given in Fig. 4. The error bar represents the expected error in the mean of the measurements.

Fig. 4. Reflection spectra for two Trojans 1173 and 1208 in the *C*-type group and Trojan 911 in the reddish *RD*-group. For comparison a typical *C*-type and *M*-type spectrum are given. The spectra from Chapman and Gaffey (see their chapter) are weighted with the broadband color information given in Table I. The error bar represents the expected error in the mean of the measurements.

Table III are plotted in Fig. 3. We draw the following conclusions from Figs. 2, 3 and 4:

1. The distribution of VRI colors indicates two or possibly three groups of spectra. Group 1 contains spectra with a drop-off toward red wavelengths like 1208 Troilus; Group 2 contains *C*-type spectra like 1173 Anchises; and Group 3 contains spectra with a steep rise toward red wavelengths like 911 Agamemnon. This last group with V−I ⩾ 0.9 is designated by *RD*, for reddish and dark distant objects; in the TRIAD file (Part VII of this book) this is simply denoted as Type *D*.

2. The objects with UBV colors in the *C*-domain all have *C*-type VRI colors with the exception of the Hilda asteroid 361 Bononia.

3. The objects with UBV colors outside the lower right part of the C-domain have RD-type spectra with the exception of the Trojans 1583 and 1867.

From this information it can be inferred that 60 % of the distant objects for which spectral information is known, do not differ from typical C-type asteroids in the main belt; the Hilda asteroid 958 Asplinda, however, is an S-type and 1162 Larissa is an M-type. Recent observations by Tedesco (personal communication) with an 8-color photometer confirm this classification for 1162. The RD-type spectra are unknown in the main belt; we see that 3 out of 7 Hildas, 3 out of 6 Trojans preceding Jupiter, 1 out of 5 Trojans following Jupiter, and 944 Hidalgo belong to this type. The R-type as defined by Bowell et al. (1978) has very red UBV colors (U−V ⩾ 1.47). The RD-type has, however, neutral UBV colors (U−V ∼ 1.0) and the spectral reddening occurs beyond the V band.

Regarding the mineralogy of RD-type surfaces, two types of meteoritic minerals show a strong upturn in the reflectance spectrum somewhere around 0.7 μm (Gaffey, personal communication). One type is represented by the Angra dos Reis meteorite. It has an albedo of ∼5.5 % and a relatively flat spectrum shortward of 0.6 μm with a very strong increase in reflectance beyond 0.6 μm (Gaffey 1974). This meteorite consists mostly of a fassaite, a calcic-aluminus pyroxene with iron and titanium. Mao et al. (1977) assigned the strong visual absorption to a complex charge-transfer process involving ferrous iron and titanium. This mineral is a product of high temperature ($\gtrsim 500°$C) igneous and/or metamorphic reactions.

The second type of meteoritic mineral is the iron-rich clay mineral matrix material of the C1 and C2 meteorites, but without the attending opaque phases like magnetite, carbon, etc. This material is strongly absorbant in the blue and visible with an absorption edge between 0.5 and 0.8 μm depending on the chemistry and crystal structure of the specific material. The absorption would arise either from ferrous-ferric alteration products or from condensates of the nebula appropriate for the outer solar system. This is what the C1-C2 matrix material would look like if it were depleted in the opaque carbon and magnetite would look like if it were depleted in the opaque carbon and magnetite grains, although it might be less dark. Gaffey prefers this second interpretation but its biggest stumbling block is how to prevent the incorporation of the opaque phases.

Optical Photometry

Extensive work has been done at the University of Arizona on the variation of the brightness of distant objects with time and phase angle. Lightcurve studies for Trojan 624 Hektor (Dunlap and Gehrels 1969) showed a very elongated (2.75:1) body shape and a rotation period of $6^h 55^m$. The same paper reports moderate amplitudes (0.2–0.4 mag) for two other Trojans; 911 has a fairly short period (6–10 hr) and 1437 has a longer one (18–24 hr). Taylor *et al.* (1976) found from two portions of lightcurves for Hilda asteroid 1212 with an interval of one day that the rotation period must be long, probably >16 hr.

Since insufficient observing time with large telescopes is available to make extensive lightcurve studies for distant objects, the alternative is to obtain as many magnitude points as possible over a range of phase angles. The presence of large lightcurve amplitudes then reveals itself by striking scatter in the resulting phase curves. We have embarked on such a program, especially for the possibility of making a comparison with the phase curves for main belt asteroids of comparable size. This comparison may give an insight into possible differences in surface texture between the objects in both locations. Individual measurements updated to July 1978 are given by Degewij (1978b). More observations have been obtained since and all data are summarized in Table I, Columns 9–12. $B(1,0)$ is the blue absolute magnitude in the UBV photometric system described by Johnson (1965). It is the magnitude at 0.45 μm at one AU to the earth and the sun and zero degrees phase angle as computed from the formulae (Gehrels and Tedesco 1979):

$$B(1,0) = B - 5 \log r - 5 \log \Delta + 0.538 - 0.134 |\alpha|^{0.714} - 7\beta, \text{ for } |\alpha| < 7°$$
$$(2)$$
$$B(1,0) = B - 5 \log r - 5 \log \Delta - |\alpha|\beta \qquad\qquad , \text{ for } |\alpha| \geqslant 7°$$

in which r is the distance between object and Sun in AU, Δ is the distance between object and Earth in AU, α is the phase angle, and the phase coefficient $\beta = 0.039$ mag/deg. In Table I n is the number of different days on which a V(1,0) was obtained, Δm is the difference between the maximum and minimum value of V(1,0), "app" is the number of apparitions for which magnitude determinations were obtained, and $V(1,0) = B(1,0) - (B-V)$.

Degewij (1978b) has found that the deviation of these data from the average phase function for bright asteroids in the main belt (Gehrels 1967) hardly exceeds those of the rms errors (0.03–0.05 mag) in the observations. Exceptions are 884, 911, 1162, 1437, and 1583. For 1437 and 1583, parts of lightcurves have been observed by Gehrels (Dunlap and Gehrels 1969) indicating amplitudes of ~0.4 mag. Firm conclusions about body shapes from only two to nine magnitude observations per object are obviously impossible. However, these distant asteroids do not have lightcurve amplitudes comparable with the maximum amplitude of Hektor's lightcurve (1.2 mag). The Hilda asteroid 1162, however, may be an exception; its diameter is 5 times smaller than that of Hektor. More data for the same asteroids during an opposition 3 yr later will be desirable to determine if a different aspect yields more insight into the true shape.

In the distribution of the collision velocities in the Trojan clouds as compared with the main belt, a great difference has been postulated by Hartmann and Cruikshank (1978). They required low collision velocities between two bodies in the Trojan clouds in order to make partial coalescence without rebound possible. Such a mechanism would explain the elongated shape of Hektor. The slope of the size distribution of the Trojans and that of the outer main belt are similar, however, and no dynamical argument has yet been made to demonstrate that relative velocities are lower among Trojans than among main-belt asteroids. The chapter in this book by Hartmann gives an overview of processes to explain very elongated body shapes.

Hidalgo

944 Hidalgo travels from the inner edge of the asteroid belt, at 2 AU from the sun, to an aphelion distance of 9.6 AU. The orbital eccentricity is 0.66 and the inclination is $42°$. The period of revolution is 14 yr. Kresák (see his chapter in this book) studied in a qualitative way the evolution of objects in elongated orbits and argues that Hidalgo may be an extinct cometary nucleus. Soderblom and Harlan (1976) looked for cometary spectral emission lines, but found none.

During the 1976 opposition, which occurred while Hidalgo was near its perihelion, many physical observations were made. The UBV colors (Degewij *et al.* 1978; Tedesco and Bowell 1979) and VRI colors as derived from the spectral reflectances obtained by Chapman and Gaffey (see their chapter and their Fig. 2) suggest an *RD*-type. There is, however, a variation in UBV colors (Tedesco and Bowell 1979) and presumably also VRI colors during the rotational cycle. Unfortunately no infrared radiometric observations were made so we do not know the albedo with certainty. However, Tedesco and Bowell report a phase coefficient of 0.048 mag/deg which they believe rules out a high albedo, and polarization measurements by B. Zellner (personal communication) do not exclude a dark surface. If we assume an albedo of 0.03, then the diameter of Hidalgo is ∼60 km which appears somewhat large for a cometary nucleus.

It is still impossible to answer the question as to whether Hidalgo is an extinct cometary nucleus as suggested by its extraordinary orbit. Only fragmentary spectrophotometric information is available for cometary nuclei fainter than ∼16 mag. Chapman (personal communication) has obtained a reflection spectrum of Comet Arend-Rigaux between 0.35 and 0.9 μm which shows similarities with an *RD*-type spectrum. A dust cloud may have affected the spectrum, however, and a more detailed study on the relation between the spectral reflectivity and distance to the sun may provide an insight into what an "asymptotic" spectrum of a comet without gaseous activity at great distances should look like (Degewij 1978a). Such a spectrum would be caused mainly by the light reflected from a solid nucleus.

Search for Outgassing Asteroids

Faint objects in the solar system that move in orbits around the sun are called asteroids if no cometary activity is observed. Usually the objects are discovered by direct photography, and an obvious question relevant to the interrelation between comets and asteroids is: do asteroids show weak cometary activity not detectable with direct photography?

A spectrograph is needed to detect weak cometary activity, and to study the physical processes in detail. Degewij observed several faint comets by direct photography using an image intensifier, and a low dispersion (~0.2 μm/mm) spectrograph at the Cassegrain focus of the 228-cm telescope of Steward Observatory. For faint (V $\geqslant$ 15) comets, the coma consists of a dust cloud and a gas cloud. The dust cloud, of unknown composition, contains the majority of material within roughly 10,000 km diameter (~10 arcsec). The gas cloud composed of CN and sometimes C_3 extends much further out and no change in line intensity over the slit length of 20 arcsec for several faint comets was observed. A comparison between direct photography and simultaneous spectroscopy showed that the dust cloud can be detected with both techniques. The gas cloud, however, with its well-defined emission lines, was still detectable with the spectrograph at large (10–20 arsec) distances from the nucleus, whereas the direct photography failed to detect it.

The sensitivity of the equipment for detecting weak gaseous activity initiated a small observing program. The observed asteroids and satellites were: a) Trojan 884 Priamus; b) Hildas 190 Ismene, 499 Venusia, 1162 Larissa, 1512 Oulu, and 1746 Brouwer; c) satellites J6, J7 and S7; and d) asteroids in peculiar orbits: 344 Desiderata, 455 Bruchsalia, 664 Judith, 814 Tauris, 1006 Lagrangea, 1036 Ganymed, 1362 Griqua, 1607 Mavis, 1625 The NORC, 1916 1953RA, and 1977RA. Great care was taken to permit a proper subtraction of the spectroscopic features in the night sky. In most cases a spectrum of the sky near to the object was taken with the same exposure time. All spectra were scanned with the ASTROSCAN at Leiden Observatory and the COMTAL image processing system was used to subtract the night sky spectra from the asteroid and satellite spectra. No gaseous activity was detected with certainty.

II. OUTER JOVIAN SATELLITES

The eight outer satellites of Jupiter show two tight groupings with remarkably similar orbital radii, inclinations and eccentricities for each group. The motions of objects in the first group (J6, J7, J10, J13) are prograde and those in the second group (J8, J9, J11, J12) are retrograde. The semimajor axes for the second group are twice as large as those of the first group.

There is a marked discrepancy between the size distributions of asteroids and outer Jovian satellites. The size distribution for the satellites has a maximum near 40 km. Extensive searches for smaller objects have been made independently by Gehrels and Kowal to a limit of the photographic magnitude as faint as 21.2 (13 km) and only one satellite has been found, J13 (Kowal *et al.* 1975), a member of the prograde group. A suspected faint 14th satellite has been lost again. With the magnitude distribution of the Trojans, about 40 new satellites should have been discovered. Large satellites are deficient also. This topic is discussed by Gehrels (1971, 1977; see his chapter).

Only little is known about albedos. Cruikshank (1977) obtained by infrared radiometry at 20 μm for both J6 and J7 an albedo of 0.03. Degewij *et al.* (1979) showed that polarimetric measurements for J6 are consistent with this low albedo.

UBVRI colors are given in Table I for J6, J7, and J8 (Degewij *et al.* 1979). All three objects have *C*-type spectra and these results show that there apparently is no difference in spectral types between the prograde and retrograde satellite groups.

Also, for the three satellites studied, the variation of the brightness with time and phase angle shows similarities with such information for the distant asteroids. The phase relation of J6 shows a close agreement with that for bright asteroids in the main belt (Degewij *et al.* 1979). The brightness of J6 has been monitored by Andersson (1974). He noticed a change in brightness of 0.2 mag on 16 June 1972. Degewij and Zellner obtained on 28 and 29 November 1976 with the 1.54 m Catalina telescope and the 2.28 m Steward telescope of the University of Arizona two sections of a lightcurve that overlap slightly; it appears that J6 has an asteroid-like lightcurve with two maxima and two minima. The lightcurve period is 9.5 hr and its amplitude is 0.12 mag (Degewij *et al.* 1979).

Photographic lightcurves of the satellites J6 (0.1), J8 (0.4), J9 (0.4), J11 (0.3), and J12 (0.4) were obtained on 15 August 1974 from plates taken by Gehrels with the 1.22 m Palomar Schmidt telescope (Degewij 1978*b.*) The observing period was 6.4 hr. The number between parentheses following the satellite is the standard deviation in magnitudes of the lightcurves of the nearby comparison stars. These high values of the standard deviation are caused by the combination of faintness of the object and a heavy background fog of the plates caused by Jupiter's light scattered in the earth's atmosphere. In similar observing runs on 29 and 30 October 1975, the observing period was 2.5 and 2.9 hr, respectively, for satellites J6 (0.1), J7 (0.2), J8 (0.2), J9 (0.2), J11 (0.2), and J13 (1.4). During these three observing dates only J6 showed some variability; the amplitude was between 0.1 and 0.2 mag. Degewij observed J8 on 28 and 31 January 1979 with the 1.54 m Catalina telescope and found that the V(1,0) magnitudes were similar on both nights with a value of 10.33 (V $\sim$ 16.8). On 30 January the object was not found which may indicate that a large lightcurve amplitude is present.

III. DISCUSSION

With increasing distance from the sun there is a continuous change in the ratio between high- and low-albedo objects (see the chapter by Zellner). Low-albedo objects occur more frequently in the outer zones; a new type of dark surface with reddish RD-type spectra first shows up in the Hilda zone. The seven sampled Hildas show a mixture of both spectral and albedo types, and it is surprising that this distant zone contains with reasonable certainty at least one S-type and one M-type asteroid. Twelve sampled objects in both Trojan clouds all are dark and show a mixture of RD- and C-types equally numerous in the preceding cloud, but only one RD-type out of six objects in the following cloud. This explains the systematic difference in the average B-V color between objects in the two clouds.

With the present limited physical information the following weak constraints can be given for the origin of Hildas and Trojans and their interrelation:

1. A theory of the origin of the Hildas must account for the occurrence of at least two high-albedo type asteroids in this distant location.
2. To explain the 3.5— times higher density of Trojans in the preceding cloud compared with that in the following cloud, van Houten (1979) proposes that proto-Jupiter swept up material from the solar nebula, causing a larger density of bodies preceding proto-Jupiter than following it. A fascinating question is: are the RD-type objects extinct cometary nuclei? In the coming years reflection spectroscopy should be directed to low-activity and distant cometary nuclei, first at optical wavelengths.

The spectral and albedo information shows that there is a similarity between the C-type Trojans and J6, J7 and possibly J8. At infrared wavelengths Cruikshank (personal communication) obtained JHK photometry of J6 and the Trojan 617. The preliminary data show virtually identical reflectances out to 2.2 μm.

There is a resemblance of spectrum, albedo, and phase relation among the majority of the Trojans and the outer Jovian satellites. A dynamical interrelation has been proposed by Bailey (1971, 1972). Greenberg (1976) has discussed recent calculations relating to the possibility that the outer Jovian satellites are captured asteroids. He has shown that Bailey used too many restraints. We are left with two suggestions for the origin of the outer Jovian satellites: a) capture and breakup of an object in a gaseous envelope about Jupiter (Kuiper 1956; Pollack et al. 1979); and b) collision between objects within Jupiter's sphere of influence (Colombo and Franklin 1971). It is not yet possible to distinguish between these mechanisms. The collisional mechanism (b) is less likely because the expected fragmental size distribution is not observed. More physical observations of the smaller satellites are needed to check whether either of these mechanisms applies.

Acknowledgments. This chapter has benefited from discussions with M. J. Gaffey on a possible connection with meteorites. C. R. Chapman kindly made his spectra available before publication. E. Bowell provided all ephemerides, and his suggestions improved the manuscript considerably. Assistance by G. H. Rieke, M. J. Lebofsky and J. Gradie with the use of the infrared radiometer is gratefully acknowledged. This work is supported by the National Aeronautics and Space Administration.

REFERENCES

Andersson, L. E. 1974. A photometric study of Pluto and satellites of the outer planets. Ph. D. dissertation, Indiana University.

Bailey, J. M. 1971. Origin of the outer satellites of Jupiter. *J. Geophys. Res.* 76: 7827-7832.

Bailey, J. M. 1972. Studies on planetary satellites. Satellite capture in the three-body elliptical problem. *Astron. J.* 77: 177-182.

Bowell, E.; Chapman, C. R.; Gradie, J. C.; Morrison, D.; and Zellner, B. 1978. Taxonomy of asteroids. *Icarus* 35: 313-335.

Chapman, C. R. 1976. Interpretation of new asteroid spectrophotometry (abstract). *Bull. Amer. Astron. Soc.* 8: 460.

Chapman, C. R.; McCord, T. B.; and Johnson, T. V. 1973. Asteroid spectral reflectivities. *Astron. J.* 78: 126-140.

Chapman, C. R.; Williams, J. G.; and Hartmann, W. K. 1978. The asteroids. In *Annual Rev. Astron. Astrophys.,* eds. G. Burbidge, D. Layzer and J. Phillips (Palo Alto, Calif.: Annual Reviews, Inc.) 16: 33-75.

Colombo, G., and Franklin, F. 1971. On the formation of the outer satellite groups of Jupiter. *Icarus* 15: 186-189.

Cruikshank, D. P. 1977. Radii and albedos of four Trojan asteroids and Jovian satellites 6 and 7. *Icarus* 30: 224-230.

Degewij, J. 1978*a*. Comet Arend-Rigaux: Not dead yet. *Sky and Telescope* 55: 14.

Degewij, J. 1978*b*. Photometry of faint asteroids and satellites. Ph.D. dissertation, Leiden University.

Degewij, J.; Gradie, J.; and Zellner, B. 1978. Minor planets and related objects. XXV. UBV photometry of 145 faint asteroids. *Astron. J.* 83: 643-650.

Degewij, J.; Andersson, L. E.; and Zellner, B. 1979. Photometric properties of the satellites J6 (Himalia), J7 (Elara), J8 (Pasiphae), S7 (Hyperion), S9 (Phoebe) and N1 (Triton). In preparation.

Dunlap, J. L., and Gehrels, T. 1969. Minor planets. III. Lightcurves of a Trojan asteroid. *Astron. J.* 74: 796-803.

Gaffey, M. J. 1974. A systematic study of the spectral reflectivity characteristics of the meteorite classes with applications to the interpretation of asteroid spectra for mineralogical and petrological information. Ph.D. dissertation, Massachusetts Institute of Technology.

Gehrels, T. 1967. Minor planets. II. Photographic magnitudes. *Astron. J.* 72: 1288-1291.

Gehrels, T. 1971. Physical parameters of asteroids and interrelations with comets. In *From Plasma to Planet,* ed. A. Elvius (Stockholm: Almqvist and Wiksell), pp. 169-178.

Gehrels, T. 1977. Some interrelations of asteroids, Trojans, and satellites. In *Comets, Asteroids, Meteorites,* ed. A. H. Delsemme (Toledo, Ohio: University of Toledo Press), pp. 323-325.

Gehrels, T., and Gehrels, N. 1979. Minor planets and related objects. XXVI. Magnitudes for the numbered asteroids. *Astron. J.* 83: 1660-1674.

Gehrels, T., and Tedesco, E. F. 1979. Minor planets and related objects. XXVIII. Asteroid magnitudes and phase relations. *Astron. J.* 84. In press.

Gradie, J. C. 1978. An astrophysical study of the minor planets in the Eos and Koronis asteroid families. Ph.D. dissertation, University of Arizona.

Greenberg, R. J. 1976. The motions of satellites and asteroids: Natural probes of Jovian gravity. In *Jupiter*, ed. T. Gehrels (Tucson: University of Arizona Press), pp. 122-132.

Hartmann, W. K., and Cruikshank, D. P. 1978. The nature of Trojan asteroid 624 Hektor. *Icarus* 36: 353-366.

Johnson, H. L. 1965. Interstellar extinction in the galaxy. *Astrophys. J.* 141: 923-942.

Kowal, C. T.; Aksnes, K.; Marsden, B. G.; and Roemer, E. 1975. Thirteenth satellite of Jupiter. *Astron. J.* 80: 460-464.

Kron, G. E.; White, H. S.; and Cascoigne, S. C. B. 1953. Red and infrared magnitudes for 138 stars observed as photometric standards. *Astrophys. J.* 118: 502-510.

Kuiper, G. P. 1956. On the origin of the satellites and the Trojans. *Vistas in Astronomy*, ed. A. Beer (Oxford: Pergamon Press), Vol. 2. pp. 1631-1666.

Low, F., and Rieke, G. 1974. The instrumentation and techniques of infrared photometry. In *Methods of Experimental Physics* 12, ed. N. Carleton (New York: Academic Press), pp. 415-462.

Mao, H. K.; Bell, P. M.; and Virgo, D. 1977. Crystal-field spectra of Fassaite from the Angra dos Reis meteorite. *Earth Planet. Sci. Lett.* 35: 352-356.

McCord, T. B., and Chapman, C. R. 1975a. Asteroids: Spectral reflectance and color characteristics. *Astrophys. J.* 195: 553-562.

McCord, T. B., and Chapman, C. R. 1975b. Asteroids: Spectral reflectance and color characteristics. II. *Astrophys. J.* 197: 781-790.

Morrison, D. 1977. Asteroid sizes and albedos. *Icarus* 31: 185-220.

Pollack, J. B.; Burns, J. A.; and Tauber, M. E. 1979. Gas drag in primordial circumplanetary envelopes: A mechanism for satellite capture. *Icarus* 37: 587-611.

Rabe, E. 1961. Determination and survey of periodic Trojan orbits in the restricted problem of three bodies. *Astron. J.* 66: 500-513.

Rabe, E. 1967. Third-order stability of the long-period Trojan librations. *Astron. J.* 72: 10-17.

Soderblom, D. R., and Harlan, E. A. 1976. *IAU Circular* 3007.

Taylor, R. C. 1971. Photometric observations and reductions of lightcurves of asteroids. In *Physical Studies of Minor Planets*, ed. T. Gehrels (NASA SP-267, Washington, D.C.: U.S. Government Printing Office), pp. 117-131.

Taylor, R. C.; Gehrels, T.; and Capen, R. C. 1976. Minor planets and related objects. XXI. Photometry of eight asteroids. *Astron. J.* 81: 778-786.

Tedesco, E. F., and Bowell, E. 1979. Lightcurves and UBV observations of 944 Hidalgo. In preparation.

van Houten, C. J. 1980. Statistics of Trojans. In preparation.

van Houten, C. J.; van Houten-Groeneveld, I.; and Gehrels, T. 1970a. Minor planets and related objects. V. The density of Trojans near the preceding Lagrangian point. *Astron. J.* 75: 659-662.

van Houten, C. J.; van Houten-Groeneveld, I.; Herget, P.; and Gehrels, T. 1970b. The Palomar-Leiden Survey of faint minor planets. *Astron. Astrophys. Suppl.* 2: 339-448.

Yoder, C. F. 1979. Notes on the origin of Trojan asteroids. *Icarus* (special Asteroid issue).

Zellner, B.; Andersson, L.; and Gradie, J. 1977a. UBV photometry of small and distant asteroids. *Icarus* 31: 447-455.

Zellner, B.; Lebertre, T.; and Day, K. 1977b. The asteroid albedo scale. II. Laboratory polarimetry of dark carbon-bearing silicates. *Proc. Lunar Science Conf. VIII* (Oxford: Pergamon Press), pp. 1111-1117.

CHIRON

CHARLES T. KOWAL

Hale Observatories

The discovery, orbit, and origin of the unique minor planet 2060 Chiron are discussed. Dynamical evidence suggests a cometary origin for this object, although its relatively large size and its uniqueness argue against this hypothesis. Several speculations concerning the nature of Chiron are mentioned. An on-going search for distant objects in the solar system is described.

I. DISCOVERY

Minor planet 2060 Chiron was discovered by the author on plates taken during October 18 and 19, 1977 (Kowal *et al.* 1979). Chiron is the only minor planet known which has its perihelion far beyond the orbit of Jupiter. The photographic magnitude of this object at discovery was approximately 19. The object received the provisional designation 1977 UB. The discoverer then named the object *Chiron (kī'rŏn)* after Chiron the Centaur, son of Kronos (Saturn) and grandson of Uranus. The name is appropriate for this object which has an orbit lying between those of Saturn and Uranus. Chiron was discovered during a systematic search for distant objects in the solar system. This "Solar System Survey" is described in Sec. V.

II. ORBIT

After a preliminary orbit was computed by Marsden, old plates were searched for possible pre-discovery photographs of Chiron. Many such photographs were found. Ultimately images of this object were found on plates taken as early as 1895. As a result of this long sequence of plates the orbit of Chiron is quite well determined. The final orbital elements computed by Marsden (1978) are listed as follows:

$$a = 13.6954465 \text{ AU}$$
$$e = 0.3786034$$
$$i = 6°.92293$$
$$\Omega = 208°.71456$$
$$\omega = 339°.10361$$
$$T = 1996 \text{ Feb. } 19.74721 = \text{J.D. } 2450133.24721$$
$$q = 8.5103039 \text{ AU}$$
$$Q = 18.8805891 \text{ AU}$$
$$P = 50.6832 \text{ yr}$$

Although the present period of revolution is 50.7 yr, the mean value over several centuries is about 49 yr. Chiron is therefore roughly in a 3:5 resonance with Saturn. This resonance, however, does not prevent close approaches between Chiron and Saturn. Close approaches to Uranus are also possible, but Saturn is the dominant influence.

Oikawa and Everhart (1979; see Everhart's chapter), and Scholl (1979), have shown that Chiron is in a "chaotic" orbit which will never become stable. Its ultimate fate will be collision with a planet, or permanent ejection from the solar system.

III. PHYSICAL OBSERVATIONS

Little is known about the physical parameters of Chiron. Its absolute magnitude, $B(1,0)$, is +7.0 (Bowell 1979). Narrow band spectrophotometry by J. B. Oke (Matson *et al.* 1979), shows that the reflection spectrum is essentially flat from 3500 to 9000 Å. The single spectrophotometric observation obtained, however, is of rather poor quality at the red end. Unfortunately this spectral information alone does not distinguish between an icy surface with an albedo of 0.5 which would imply a diameter of ~100 km for Chiron and a carbonaceous surface with an albedo of 0.05 which implies a diameter of ~320 km. At least the object does not have a D-type spectrum as some of the Hildas and Trojans have (see the chapter by Degewij and van Houten).

IV. THE NATURE OF CHIRON

In the absence of more detailed compositional information, the origin of Chiron remains a subject of speculation. The simplest theory is that Chiron is a comet which was perturbed into its present orbit. The only real objection to this theory is the relatively large size of Chiron. For example, Chiron was at least six magnitudes brighter than Halley's Comet, when both objects were near the orbit of Uranus. This suggests that the diameter of Chiron is at least 15 times greater than the diameter of the comet, if they have similar albedos. Some estimates of the sizes of cometary nuclei suggest that a few comets could be as large as Chiron (Roemer 1966), but the fact remains that no comet has ever been observed at a heliocentric distance greater than 11.3 AU. Furthermore, if there are very large comets which can be captured into trans-Saturnian orbits, then more objects like Chiron should have been found. The present Palomar "Solar System Survey" is capable of detecting objects two magnitudes fainter than Chiron.

Another possibility for Chiron is that it originated within the asteroid belt, and was ejected from the belt by collisions or by perturbations from Jupiter and Saturn (Smith 1978). A serious objection to this theory is that no intermediate asteroids have been found. One would expect a significant number of asteroids between Jupiter and Saturn if this transport mechanism were possible. Yet, only Hidalgo is known to exist in this region.

Harrington and Van Flandern (1978) have proposed that Chiron and Pluto were originally satellites of Neptune. It is suggested that a massive object passed very close to Neptune and the tidal forces tossed Pluto and Chiron out of the Neptunian system, while moving Triton and Nereid into their present peculiar orbits. In addition to the *ad hoc* nature of such "cataclysmic" hypotheses, this theory suffers from the objection of the well-known Neptune-Pluto resonance, in which Pluto never comes near Neptune. It is also difficult to see how Pluto could have retained a satellite during such a violent encounter.

There are similar theories which suggest that Chiron is an escaped satellite of Saturn or Uranus. The chief advantage of such theories is that they explain the uniqueness of Chiron, by postulating the occurrence of a single event.

Other possible sources for Chiron are the Lagrangian points of Saturn or Uranus. No objects have been found at those points, but searches for Saturnian or Uranian "Trojans" are still incomplete. Wallerstein (1978) has suggested that there may be many asteroidal objects in the outer solar system, with Chiron being the nearest of these objects, and Pluto the largest.

V. THE PALOMAR SOLAR SYSTEM SURVEY

Initially, the plates of the Palomar Solar System Survey were centered on the ecliptic, with each plate covering an area of $6° \times 6°$. Only one quarter of the ecliptic had been photographed when Chiron was discovered. By the beginning of 1979, two-thirds of the ecliptic was covered. It is hoped that this survey will continue for several more years, until a band 30 degrees wide is covered, (15 degrees on either side of the ecliptic). Clearly, there is a possibility that other Chiron-like objects will be found in the future, but thus far Chiron remains unique.

The techniques used in this survey are similar to the methods used by Tombaugh (1961), in his trans-Neptunian planet search. Each month, photographs of the opposition region are obtained on each of two successive nights, using the Palomar 122-cm Schmidt telescope. The plates are immediately examined for comets and fast-moving objects. Later, the plates are blinked for slow-moving objects. Exposure times are 75 min on baked IIIa-J plates with a Wratten 2C filter. The limiting magnitude of the plates is 21-22 (in a non-standard magnitude system between B and V).

VI. FUTURE WORK

The need for further observations of Chiron is obvious. In addition to spectrophotometric and radiometric observations of Chiron itself, distant comets should be observed before they become heated by the sun. Perhaps our only chance of doing this lies with Halley's Comet. We urge that a concerted effort be made to observe this comet while it is still beyond the orbit of Saturn.

REFERENCES

Bowell, E. 1979. A report on recent observations on Chiron. *Icarus* (special Asteroid issue).

Harrington, R. S., and Van Flandern, T. C. 1978. A dynamical study of escaped satellites of Neptune. Preprint.

Kowal, C. T.; Liller, W.; and Marsden, B. G. 1979. The discovery and orbit of (2060) Chiron. In *Dynamics of the Solar System*, ed. R. L. Duncombe (Dordrecht: D. Reidel), in press.

Marsden, B. G. 1978. *Minor Planet Circular No. 4346*.

Matson, D.; Kowal, C. T.; and Oke, J. B. 1979. In preparation.

Oikawa, S., and Everhart, E. 1979. Past and future orbit of 1977 UB, object Chiron. *Astron. J.* 84: 134-139.

Roemer, E. 1966. The dimensions of cometary nuclei. *Mém. Soc. Roy. Sci. Liège* 12: 23.

Scholl, H. 1979. History and evolution of Chiron's orbit. *Icarus* (special Asteroid issue).

Smith, R. C. 1978. Origin of slow-moving object Kowal, *Nature* 272: 229-230.

Tombaugh, C. 1961. The trans-Neptunian planet search. In *Planets and Satellites,* eds. G. P. Kuiper and B. M. Middlehurst (Chicago: Univ. of Chicago Press), pp. 12-30.

Wallerstein, G. 1978. *Sky and Tel.* (letter) 55: 195, 224.

PART IV

Configuration

SATELLITES OF ASTEROIDS

T. C. VAN FLANDERN
U.S. Naval Observatory

E. F. TEDESCO
University of Arizona

and

R. P. BINZEL
Macalester College

The discovery of a probable satellite of a minor planet during the occultation of a star on 7 June 1978 and additional photoelectric events during another such occultation on 11 December 1978 have led to the realization that anomalous sightings during previous occultations of stars by minor planets are possibly also due to satellites. Some features of minor planet lightcurves may be modeled in terms of the rotation of contact binary asteroids, or in terms of eclipsing and shadowing events by orbiting satellites. Calculations show that satellites are gravitationally bound out to distances of ~ 100 times the diameter of the primary. Although the large satellites are probably collisionally stable over the solar system's lifetime, the time scales for tidal evolution are much smaller — typically 10^4–10^7 yr. A dynamical model for a minor planet, and by extension one for comets and fireballs as well, is presented.

I. OCCULTATION RESULTS

The contemporary notion for the existence of satellites of asteroids ("minor satellites") originated as a result of an observation by Maley during the occultation of γ Ceti, a 3.6 mag star, by 6 Hebe on 5 March 1977. Maley, an experienced lunar occultation observer, timed and reported a 0.5 sec event from his location near Victoria, Texas while several observations made in central Mexico and reported to Dunham indicated that the shadow path of Hebe had actually passed some 900 km to the south. Because of Maley's certainty of his observation and the unlikelihood that a terrestrial phenomenon would have occurred in such good time coincidence with the Mexico observations, Dunham and Maley concluded that this observation indicated Hebe apparently had a satellite approximately 20 km in diameter (Dunham and Maley 1977; Dunham 1978*a*).

A year later corroborated observations of a secondary event were made during an occultation of SAO 120774 by 532 Herculina on 7 June 1978. McMahon, another experienced lunar occultation observer, successfully observed a 20.6 sec occultation of the star by Herculina, but he also timed and reported six additional events with durations of 0.5 to 4.0 sec occurring within two minutes before and after the main event (McMahon 1978). Because the light drop in each true occultation was 3.6 magnitudes, McMahon reported these observations with certainty.

The Herculina occultation was also observed visually by Horne at Rosamond, California, and photoelectrically by Bowell and A'Hearn at the Lowell Observatory (Dunham 1978*b*; 1978*c*; Binzel 1978). A secondary event subsequently found in the photoelectric record at Lowell Observatory was in near perfect agreement with the longest secondary event observed by McMahon. These two consistant observations indicated that a secondary body about 50 km across and 1000 km from Herculina (diameter 220 km) had caused this particular event and was thus the first minor satellite seen by two or more independent observers. Although the Herculina events were observed at a zenith distance of $87°.7$, Bowell *et al.* (1978) concluded that the "... secondary occultation, detected only from Flagstaff and Boron, is interpreted as being due to a 46-km-diameter companion to Herculina ..."

The first occultation, of a star by a minor planet (18 Melpomene), for which an organized effort was made to search for satellites occurred on 11 December 1978. Three photoelectric and several visual timings of the occultation by the parent body indicated a 130 km diameter. Three additional photoelectric and one visual record of secondary events are presumed to be due to minor satellites (Dunham *et al.* 1979). (It should be noted, however, that the majority of observers recorded no events.) It presently appears that no two observers were located close enough to have seen the same satellite. This seems, nonetheless, to be strong supporting evidence for the generality of the phenomenon (Dunham 1979*a*,*b*).

In light of the evidence suggesting the existence of minor satellites, investigations of previously observed occultations of stars by minor planets and of early visual observations reveal that minor satellites may have been observed before (Binzel 1978). Anomalous events of this type were reported during earlier occultations involving 2 Pallas on two occasions (Dunham 1977; Binzel 1978; Binzel and Van Flandern 1979), 129 Antigone (Binzel 1978), and 433 Eros (O'Leary et al. 1976; Binzel 1978). In 1926 renowned double star observers van den Bos and Finsen were doing a systematic search for new double stars when they encountered one which they could not identify in the star charts of the Cape Photographic Durchmusterung. Further research revealed that this new "double star" was actually the minor planet Pallas (Finsen 1926). It should be noted, however, that van den Bos also thought he had observed Titan to have the appearance of a non-divided double star. These observers noted that none of these bodies (Eros, Pallas, or Titan) was ever seen clearly divided and that surface markings, rather than binarity, provided a more likely explanation of their observations. If Pallas has a satellite it should be confirmed by speckle interferometry or Space Telescope imaging in the near future.

In addition to the Melpomene event, several secondary events have been observed during four other occultations subsequent to the Herculina event of 7 June 1978 (Dunham 1978c, 1979c,d; Sheffer 1979). Even if only a few of these observations are due to minor satellites it nevertheless suggests that such satellites are both numerous and commonplace.

The evidence that these occultations do indeed represent satellites of minor planets, rather than some other phenomenon, is the following. Intensity drops generally correspond to total occultation of the star's light, rather than partial occultations, as for the rings of Uranus; and only the star disappears, with the asteroid usually remaining visible. These drops are often several magnitudes, insuring that visual timings by experienced observers can be relied upon (c.f. the chapter by Millis and Elliot, in this book). (Observers of lunar grazing occultations, which are similar in most aspects to occultations by minor planets, are 90% reliable in their observations of definite events.) The one large corroborated satellite of Herculina was shown to be co-moving through space with Herculina, virtually ruling out a chance alignment of distant asteroids. Ordinary stars near the ecliptic, when monitored photoelectrically, do not show occultation events if they are not near an asteroid; nor are such secondary events ever seen during lunar occultations. Although most observers have monitored a star for 10–20 minutes before and after the occultation, the furthest minor satellite to date was 4 minutes away, and the vast majority have been within 1.5 min of the main event.

Despite these arguments, Reitsema (1979) has proposed the hypothesis that many secondaries are spurious. Visual observers, it is argued, may be seeing events caused by brightness fluctuations due to atmospheric scintillation and/or turbulence. The point is also made that five of the six secondary events reported by McMahon were not seen at either of the remaining two observing sites. But these correspond to objects too small to occult the star from the other sites. The question really being posed is why did McMahon see six secondary events, while Bowell (in half the observing time) saw only one, and the third observer none? We discuss possible explanations for this in Sec. V. (However, even if all five of these events were spurious that does not lessen the probability of the sixth's being real. The probability of spurious events of similar duration occuring with the required time coincidence at two sites hundreds of miles apart is obviously rather low.) Reitsema notes in addition that photoelectric observations are also subject to error, spurious events in this case being produced by the passage of clouds, birds, airplanes, or other objects, through the field of view. He notes that events can also be generated through having the star drift in and out of the photometer aperture. This, presumably, is more of a problem with the small transportable telescopes often employed in occultation observations than with permanently mounted observatory instruments. If a star drifts out of the aperture, the brightness should fade gradually rather than abruptly as it would in the case of a true occultation. If, on the other hand, the star is quickly removed from and returned to the aperture, as could happen if the telescope were bumped by the observer or shaken by a gust of wind, the tracing should have the same general appearance as an actual event but the brightness level would fall to the level of the sky rather than to the level of the asteroid. These observing conditions have led Zellner (personal communication) to remark that quiet photoelectric records are hard to make while spurious events, on the other hand, are easily produced. An ideal occultation observation would therefore consist of the following: measurements of the brightness levels of the sky, asteroid-plus-sky, and star-plus-sky made before and after the event together with an ∼20 min long tracing centered on the predicted time of the occultation. A tracing with reasonably sharp edges (asteroids, and especially their satellites, are, after all, small irregularly shaped bodies) which falls to the asteroid-plus-sky level is then, in all likelihood, a real event. It is not uncommon for the asteroid brightness and dynamic range of the equipment to be such as to make the difference between sky level and asteroid-plus-sky level indistinguishable.

Reitsema (1979) notes the importance of multiple photoelectric observations of occultation events in interpreting secondary events in terms of minor satellites. Each of the telescopes involved in the experiment he reports on (a near occultation of a star by 13 Egeria on 28 February 1979, observed with two portable and one fixed telescope [the Steward Observatory 21-inch reflector] separated by approximately 6 km in a direction parallel to the occultation track) recorded several spurious events each, most of which were noted as times when the star had drifted out of the photometer aperture. The record from the fixed telescope showed two events which were not identified as guiding errors and which were recorded as periods of constant brightness by the portable telescopes. He therefore concluded that even these events were definitely of, unspecified, local origin. Not mentioned, but also of obvious concern, is the fact that if these spurious events were caused by electrical interference (line surges, RF noise, etc.), then a second telescope operating from the same power line and located nearby might also have recorded these same events. It therefore appears necessary to separate the telescopes designed to record redundant observations by a great enough distance to eliminate this possibility.

The experiment performed at the Lowell Observatory, and described in the chapter by Millis and Elliot, to test the reliability of visual occultation observations under conditions in which the expected magnitude change was unknown, showed that most observers reported more than one spurious event during an observing period lasting from five to ten minutes and that performance was degraded when the eyepiece was in a slightly awkward position, which would often be the case in actual observing situations. This experiment used ten observational astronomers and four non-observers as subjects and, in addition to the conclusions mentioned above, noted that the experienced observers performed better than the non-observers and that brightness changes in excess of 0.75 mag were detected without fail. Their ultimate conclusion was that "... visual occultation observations are unlikely to be consistently reliable unless the star is at least as bright as the asteroid, and then only if the observer is experienced." In this connection it is worth noting that both Maley and McMahon are experienced visual observers and that both of their observations involved stars which were more than 3.5 mag brighter than the occulting asteroid. Maley's star, γ Ceti, was over a magnitude brighter than the artificial star used in the Lowell Observatory experiment. At the present time the evidence is such that we can not say with certainty that satellites of asteroids actually exist; this same evidence is, nevertheless, of sufficient quality to be convincing to some.

There have now been four reports of an additional phenomenon lasting several minutes before and after the main occultation. McMahon (1978) reported that the star often appeared "diffuse" near the main event time for Herculina. Four of six Japanese observers at widely separated locations reported periods of "scintillation" of the starlight for the Metis event. Przybyl reported two periods of "turbid image" during a later occultation by Herculina in August 1978. And Van Flandern and Schmidt reported visual and photoelectric flickering of unusual amplitude shortly after the Melpomene occultation event. This photoelectric record is shown in Fig. 1. Are these phenomena perhaps due to minor satellites that are too small to produce distinct occultations? Or could they be a kind of "coma" around a minor planet, similar to that possessed by comets? In this connection it is worth noting Hartmann's (1972, p. 157) remark that "some observers have reported a faint halo around Irene." A definitive test of the reality of this phenomenon could be obtained by simultaneously monitoring the occulted star and a similar nearby star.

II. EVIDENCE FROM LIGHTCURVES

Shortly after von Oppolzer's (1901) announcement of Eros' variability,

Fig. 1. The top photoelectric trace is a sample from a 30 min observation, obtained by Van Flandern and R. E. Schmidt was the 61-cm U.S. Naval Observatory reflector one minute prior to the 11 December 1978 occultation of SAO 114159 by 18 Melpomene, and is typical of the atmospheric noise both before and after the event. The lower trace at 25 sec after the reappearance exhibits the reported scintillation which lasted approximately one minute. Each trace represents ∼ 3 sec of time. In the lower figure the top reference mark shows the mean unocculted level, while the lower reference marks show the individual wholly occulted level for the secondary and primary stars.

Andre (1901) suggested that the brightness variation was due to the mutual eclipses of two orbiting bodies. Andre was led to this conclusion because of the similarity of Eros' lightcurve to that of eclipsing variable stars of the β Lyrae type. Figure 2 displays the lightcurves of these two objects. Indeed van den Bos and Finsen (1931) reported having observed Eros to have a notched appearance resembling an unresolved double star with a separation of 0.″18 and a magnitude difference of 0.2 leading to revivial of the double planet idea by Lundmark (1932) and Pickering (1932). Similar results were obtained by Heintz (1975) during the 1974-75 apparition. It is unlikely, however, that Eros presently consists of two freely-orbiting components. Radar observations made during the 1974-75 apparition by Jurgens and Goldstein (1976) and Campbell *et al.* (1976) agreed, in general, with the optically derived model for the size and shape of Eros, viz., a cylinder with hemispherical ends having overall dimensions of 12 x 12 x 31 km (Dunlap 1976), or perhaps tidally coalesced spheres. Jurgens and Goldstein's observations rule out an orbiting binary model since they observed maximum

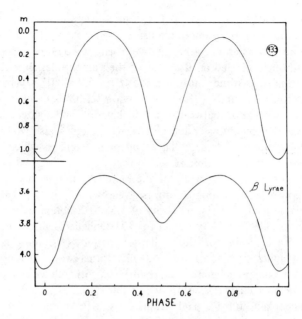

Fig. 2. Comparison of lightcurves between 433 Eros (Dunlap 1976) and eclipsing binary Beta Lyrae (Menzel *et al.* 1970). The striking similarity originally led Andre to postulate Eros as being double. *m* is relative magnitude.

power returned at their center frequency. They were, however, unable to fit their observations with a symmetric model and concluded that the projected rotation axis does not equally divide the projected area. Both radar studies found Eros to be significantly rougher at their respective wavelengths than the moon or the terrestrial planets. It is therefore the extreme shape of Eros which is responsible for the results of the visual double star observers and for its β Lyrae-like lightcurves.

Eros, however, is not the only asteroid to display a β Lyrae-like lightcurve. Chang and Chang (1963) remarked that their lightcurve of 532 Herculina resembled that of β Lyrae. Herculina has been observed at three different oppositions always with a lightcurve amplitude less than 0.2 mag. Eros has a diameter of ∼ 25 km and Herculina one of ∼ 250 km; while Eros is not a binary object, Herculina may be, in addition to having its probable distant satellite. Cook (1971) and Hartmann and Cruikshank (1978; see the chapter by Hartmann) have proposed that 624 Hektor is a contact binary, Cook believing the two components to be orbiting one another while Hartmann and Cruikshank propose a model consisting of a pair of spheres joined together to form a dumbbell shaped object. Photometrically the two models would be essentially indistinguishable if the orbiting pair were in mutually synchronous orbit.

Tedesco and Zappalà (1979) have called attention to a group of 200-km-sized, main belt, large-amplitude lightcurve asteroids (LALA's). If Hektor were a main belt asteroid, rather than a Trojan, it would be a member of this group. Although selected on the basis of their amplitudes all LALA's display rapid rotation (logarithmic mean period, $<P>$ = 5.7 ± 0.07 hr) compared with other asteroids of similar size ($<P>$ = 9.65 ± 1.33 hr). Since there is, in general, no correlation between rotational period and lightcurve amplitude (see the chapter by Burns and Tedesco), we are led to ask whether LALA's could be examples of large, tidally collapsed satellites which have gently collided with their primaries. If so, a lower limit on the final period of the collapsed pair, assuming objects of the same size and density together with conservation of angular momentum, is given by $P = 6.62 \ \rho^{-1/2}$ hr, where ρ is the density in g cm^{-3}. Using a density of 1.5 ± 0.2, obtained from Wijesinghe and Tedesco's (1979) binary model for 171 Ophelia's lightcurve, results in a value of 5.4 ± 0.4 hr for this collapsed period. (Assuming a density of 2.5 ± 1.0 implies a value of 4.5 ± 1.0 hr.) Either way, the suggestion that LALA's are collapsed binaries, or result from other low-velocity collisions (see Hartmann's chapter), seems reasonable. This explanation would also serve to explain Jurgens and Goldstein's (1976) finding that Eros is asymmetric and very rough and may also account for van den Bos and Finsen's (1931, p. 334) description of Eros as appearing "notched." 1620 Geographos, often described as "cigar shaped" due to its two magnitude rotational amplitude, may be an example of an asteroid chain.

It should be noted, however, that both Eros and Geographos are very small objects (diameter < 20 km), and hence may be fragments of larger asteroids. Nevertheless, such elongated shapes are not common among main-belt asteroids (Tedesco and Zappalà 1979), not even among very small main-belt asteroids (Degewij 1978). Is this apparent over-abundance of elongated Earth-approaching asteroids a clue to their origin? If a significant fraction of them are extinct comets, as argued in the chapter by Shoemaker et al., does this mean that multiple bodies are more common among comets than small asteroids?

Another type of lightcurve like an eclipsing star exists in the lightcurves of 44 Nysa and 511 Davida which resemble that of W Ursa Majoris as shown in Fig. 3. It is generally believed that asteroid lightcurves result from the rotation of irregularly shaped bodies about their shortest axes (see the chapter by Burns and Tedesco). Lightcurve structures produced by surface albedo features are probably of minor importance as shown by Degewij *et al.* (1979). We have thus far been unable to find a model which will reproduce Nysa's lightcurves. Knowledge of Nysa's pole orientation, combined with more sophisticated modeling techniques and intensive lightcurve observations of its flat-bottomed minimum, to be obtained in late 1979, should enable us to find such a model.

It is even more difficult, however, to explain the Algol-like lightcurves of

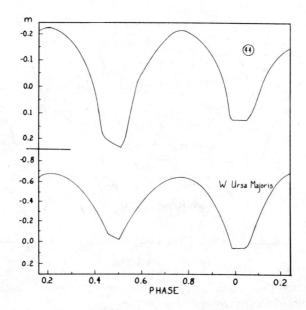

Fig. 3. Illustration of the flat minima of 44 Nysa (Shatzel 1954) and a comparison with contact binary star W Ursa Majoris. The similarity of lightcurves may indicate that 44 Nysa is a contact binary asteroid. *m* is relative magnitude.

asteroids 46 Hestia, 49 Pales and 171 Ophelia, displayed in Fig. 4, observed by Scaltriti and Zapallà (personal communication) and Tedesco (1979*b*). The appearance of these lightcurves, coupled with reports of secondary events observed during occultations, led Tedesco (1979*a*) to propose that the lightcurves of 49 Pales and 171 Ophelia were due to mutual eclipses of binary asteroids.

Minor satellites will produce features in asteroid lightcurves in four distinct ways: occultations, transits, eclipses, and shadow transits. The first two involve direct changes in the projected area, while the last two involve shadowing phenomena which cause changes in the illuminated projected area (c.f. Hartmann's chapter). Wijesinghe and Tedesco (1979) show that a simple binary model provides a reasonable fit to the observed lightcurve of 171 Ophelia. Their results suggest that a typical binary system might involve a roughly spherical primary with a fairly large satellite which would be occulted by and also transit the primary. Under most aspect and phase angle conditions this configuration would give the classical two-maxima, two-minima lightcurve with the maxima being equal, but the minima being equal only if the two bodies have roughly the same albedo. With the exclusion of shadowing events, such a system would be similar to an eclipsing binary star system where the shape of the maxima and minima are dependent upon the relative diameters of and distances between the two stars. Asteroids,

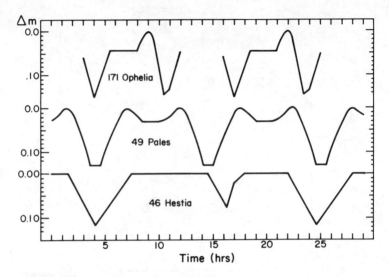

Fig. 4. Algol-like asteroid lightcurves composited from observations by Tedesco (1979*b*) and Scaltriti and Zappalà (personal communication).

smaller satellites or satellites in more distant orbits would tend to produce flat maxima and minima, as shown in Fig. 4, while closer or larger satellites would tend to produce much more rounded extrema. Indeed flat extrema *are* sometimes observed in asteroid lightcurves, the best example being the minima of 44 Nysa (Shatzel 1954) shown in Fig. 3.

Such a model, though possible, is probably not typical of most minor planets. Lightcurves are most certainly due to the rotation of the primary body plus features caused by topography and minor satellites. Minor satellites having diameters ~ 10 to 20 % that of their primary would cause light variations on the order of 0.01−0.04 mag as they undergo occultation and/or eclipse phenomena, and should thus be detectable as dips or other short term variations in the regular asteroid rotation curves. Such items are indeed a commonly observed lightcurve feature (e.g., see Schober 1979). Through tidal forces it is likely that many large minor satellites orbit with a period synchronous to the rotation of their primary. Such objects would produce features in the same portion of the rotational lightcurve through each cycle and would therefore be difficult to distinguish from features caused by surface irregularities.

The minor satellites in nonsynchronous orbits should be readily recognizable, as they would produce anomalous features in different portions of the regular asteroid rotation curve with some definite periodicity. It is possible that analysis of existing lightcurve data will reveal such nonrotationally synchronous anomalies, although definite period determination for the minor satellites may be difficult without continuous observations over several cycles of the lightcurve. To maximize the detection of these minor satellites and also to determine their periods unambiguously, continuous photoelectric observations of any given asteroid should be made over several cycles. Such observations would require a coordinated observing program between at least two observatories on opposite sides of the earth, and would serve to resolve any ambiguities in the asteroid's rotational period as well as reduce any effects of changing phase angle on the shape of the lightcurve. It was just such a cooperative effort which enabled Tedesco et al. (1978) to determine the rotation period of 1580 Betulia and recognize that its lightcurve was extraordinary. (Near opposition Betulia had a lightcurve amplitude of 0.20 mag and displayed a tertiary maximum having an amplitude of ~ 0.03 mag. Away from opposition its rotational amplitude increased to ~ 0.50 mag while the amplitude of the tertiary maximum increased to ~ 0.30 mag.)

Examples of lightcurve effects which may be attributable to satellite events and/or collapsed satellites lying on the surfaces of their primary include the following:

1. Lightcurve maxima sharper than minima as seen for 129 Antigone (Scaltriti and Zappalà 1977).
2. Complex lightcurves as seen for 24 Themis (van Houten-Groeneveld et al. 1979; Tedesco 1979b), 29 Amphitrite (Debehogne et al. 1978; van Houten-Groeneveld et al. 1979) and 51 Nemausa (Chang and Chang 1963; Wamsteker and Sather 1974).
3. Increase in lightcurve amplitude with increasing solar phase angle as seen for 349 Dembowska, 354 Eleanora (Zappalà et al. 1979b), 944 Hidalgo (Tedesco and Bowell 1979), and 1580 Betulia (Tedesco et al. 1978).
4. Asteroids such as 532 Herculina (Chang and Chang 1963; Groeneveld and Kuiper 1954; Harris and Young 1979) whose lightcurves show two maxima and minima per cycle at one apparition but only one of each at another apparition.
5. Triple maxima and minima per rotation cycle as seen in the lightcurves of asteroid 1580 Betulia (Tedesco et al. 1978) and comet d'Arrest (Fay and Wisniewski 1978).
6. Contact binary-like lightcurve as seen for 44 Nysa mentioned above.

Multiple asteroids could be directly detected through Space Telescope imaging, interferometric techniques (such as speckle interferometry), and coordinated, redundant teams of occultation observers. We are currently modeling binary systems for which sufficiently high quality data exist in an attempt to determine whether their behavior can be explained as being due to topography or will require the presence of satellite events. Wijesinghe and Tedesco (1979) and Zappalà et al. (1979a) have shown that some asteroid lightcurves can be modeled in terms of satellite events but whether these same lightcurves can be satisfactorily explained without invoking such events has yet to be established. Additional lightcurve observations at a number of phase angles and during several different apparitions will establish whether these objects are actually binary. Continuous lightcurves should reveal the existence of minor satellites. As shown by Wijesinghe and Tedesco (1979) fitting a binary model to an observed lightcurve yields the mean density for the pair as a byproduct. If these lightcurves are produced by eclipsing systems, lightcurve observations will provide valuable information, unavailable by any other presently known method, on the bulk properties of asteroids.

III. DYNAMICAL CONSIDERATIONS FOR MINOR SATELLITES

Given that an asteroid-satellite pair formed, we wish to know whether such a system would be stable with respect to the three destabilizing forces, viz., (a) gravitational perturbations, (b) collisional events, and (c) tidal evolution. We address each in turn below.

Gravitational Stability.

What is the radius of the sphere of influence for an asteroid, within which it can retain satellites even against solar perturbations? Let M be the mass of Sun, m the mass of asteroid, R the distance of minor planet from Sun, r the distance of minor satellite from asteroid; then, very approximately, the maximum value of r before solar perturbations begin to dominate can be calculated from

$$\frac{m}{r^3} = \frac{M}{R^3} . \tag{1}$$

A derivation of this result is given by Szebehely (1978), whose slightly smaller radius of the sphere of influence is perhaps too conservative since it excludes some outer Jovian satellites. Note that the powers of r and R are 3, not 2, since the solar force is nearly equal on asteroid and satellite, and only differential force can operate to pull the two apart. As a rule of thumb, the radius of the gravitational sphere of influence of a body in the inner solar system (not too near a planet) is roughly 100 times its own diameter.

For an object the size of Herculina, the sphere of influence extends to more than 30,000 km. Hence a minor satellite only 1000 km distant is very stably attached to Herculina. If the minor satellite has a diameter of 1/5 that of Herculina, and a mass (assuming the same density) of 1/125, it follows that *its* sphere of influence with respect to Herculina extends to about 1/5 of the distance to the parent asteroid. Therefore it is possible that one or both of the two occultation events that McMahon observed, close to Herculina's large confirmed satellite, may actually be bodies orbiting the satellite rather than the main asteroid.

Collisional Stability.

An asteroid with a diameter of 1 km has a cross-sectional area of $7.9 \times 10^9 \, \text{cm}^2$. If it moves with the typical 5 km sec^{-1} velocity with respect to other nearby asteroids, it sweeps out a relative volume of $1.2 \times 10^{23} \, \text{cm}^3$ per yr. Half of all asteroids lie within 0.42 AU of the ecliptic between 2.2 and 3.3 AU from the sun. The total volume of space containing half the asteroids is therefore $\sim 4.3 \times 10^{40} \, \text{cm}^3$. From Chebotarev and Shor (1978), we obtain the following relationships:

$$\log m = 26.39 - 0.6 \, g$$
$$\log d = 3.71 - 0.2 \, g \tag{2}$$

where m is mass in grams, d is diameter in kilometers, and g is mean photographic absolute magnitude for an asteroid of density 2.5 g cm^{-3}. Then we consider two cases for the possible number density of the asteroids.

Case I considers that the asteroid population has reached a "steady-state" by collisional evolution. Following Dohnanyi (1969), and adjusting constants, particularly the mean albedo of the asteroids, to conform to the Polomar-Leiden Survey (PLS) (van Houten *et al.* 1970), we adopt

$$\log \left(0.5 \, \frac{dN}{dg} \right) = -3.56 + 0.50g$$
$$dN = -3.91 \times 10^{18} m^{-1.833} dm$$
$$N(m) = 4.69 \times 10^{18} m^{-0.833} \tag{3}$$
$$N(d) = 9.06 \times 10^5 d^{-2.50}$$

where $0.5 \, \dfrac{dN}{dg}$ is the number of asteroids in the entire belt per half magnitude interval in g, dN is the number in the mass range from m to $m+dm$, and N is the total number with mass or diameter larger than or equal to the given value.

Case II simply takes the best empirical fit of the power-law distribution from the PLS data. These values are

$$\log \left(0.5 \, \frac{dN}{dg} \right) = -2.27 + 0.42g$$
$$dN = -2.31 \times 10^{16} m^{-1.70} dm$$
$$N(m) = 3.3 \times 10^{16} m^{-0.70} \tag{4}$$
$$N(d) = 6.83 \times 10^5 d^{-2.1} .$$

Multiplying dN by the cross-sectional area of each asteroid ($0.52 m^{2/3}$) and integrating, we can also derive the total cross-sectional area A of all asteroids combined, in cm^2, for cases I and II:

$$A\ (I)\ =\ 1.23 \times 10^{19} m^{-0.166}$$
$$=\ 3.61 \times 10^{16} d^{-0.50}$$
$$A\ (II)\ =\ 3.63 \times 10^{17} m^{-0.033} \tag{5}$$
$$=\ 1.13 \times 10^{17} d^{-0.10}$$

Dividing the total cross-sectional area for *half* the asteroids into their total volume gives the mean path length to collision for asteroids of diameter d, $2.38 \times 10^{24} d^{0.50}$ cm in Case I, and $7.61 \times 10^{23} d^{0.10}$ cm in Case II. At 5 km sec^{-1} velocity, this implies mean collisional lifetimes (for collisons with bodies of comparable size) of $15.1 \times 10^{10}\ d^{0.50}$ yr or $4.8 \times 10^{10}\ d^{0.10}$ yr, respectively. The orbital velocity of a minor satellite about its parent, assuming a density of 2.5 g cm^{-3}, is given by

$$\Delta v\ =\ 0.12 \rho d / \sqrt{n}\ =\ 0.30\, d / \sqrt{n} \tag{6}$$

in m sec^{-1}, where n is the number of multiples of d in the satellite's mean distance. Apparently we can take n to be typically about 5, as for Herculina's presumed satellite, for which Δv is roughly 31 m sec^{-1} (d=230 km). Then a collision producing a velocity change of order $\sqrt{2}\Delta v$ will suffice in general to de-orbit the satellite, resulting in escape. For v = 5 km sec^{-1}, we calculate that the mass Δm which can de-orbit a satellite of mass m is $\Delta m/m = \Delta v/v = 8.5 \times 10^{-5}\ d \sqrt{n}$ which implies $\Delta d/d_s = 0.044\ d^{1/3}\ n^{-1/6}$ (Δd = diameter of colliding body, d_s = diameter of satellite). Therefore the impact of a 350 m body on a 1 km satellite orbiting 5 diameters from a 230 km parent will in general cause the satellite to escape. Now we see that the mean lifetime between collisions of such satellites with the more numerous bodies of diameter Δd is still on the order of 10^{10} yr. Hence, collisional stability for the lifetime of the solar system is virtually assured for all minor satellites larger than 100 m diameter, unless the number density of asteroids is actually considerably greater than that inferred from the PLS.

Tidal Stability.

Nonrigid gravitating bodies raise tidal bulges on each other, which can result in the conversion of some rotational angular momentum into orbital angular momentum, or vice versa. For a satellite revolving in the same direction as the rotation of its parent, the satellite will decay from orbit, and the parent's rotation will tend to speed up, if the satellite is inside the synchronous orbit (the orbit where parent rotation and satellite revolution periods are the same). Mars' inner satellite Phobos is in such a situation, and will last only perhaps another 10^8 yr. Conversely, a satellite outside the synchronous orbit evolves outward, slowing the parent's rotation, as for the Earth-Moon system. At the same time, the eccentricity of the orbit will tend to diminish (the orbit will become circular) in the case of decay, or will increase if the satellite evolves outward. The inclination of a satellite to its parent's equator always tends to decrease under the influence of solid body tidal forces.

For a satellite of mass Δm revolving about a parent of mass m and density ρ in g cm^{-3}, the radius of the synchronous orbit (in units of the parent's diameter d), which we designate n_s, is given by

$$n_s = 1.87 \, P^{\frac{2}{3}} \, \rho^{\frac{1}{3}} = 2.54 \, P^{\frac{2}{3}} \tag{7}$$

where P is the period in days for both rotation and revolution. Then the ratio of orbital to rotational angular momentum for any satellite is

$$10 \, \frac{\Delta m}{m} \, n^{\frac{1}{2}} \, n_s^{\frac{3}{2}} \approx 19 \, \frac{\Delta m}{m} \tag{8}$$

where n is the mean distance of the satellite in units of the parent's diameter d. (At the synchronous orbit, $n = n_s$.) The approximation on the right is found by taking a density of 2.5 and typical rotation period of 0.4 day (see the chapter by Burns and Tedesco). When most of the angular momentum is in the satellite ($\Delta m/m > 1/19$, or the diameter ratio less than 2.7:1), the parent asteroid's pole orientation and rotation period will be substantially altered from their original values by the tidal interchange of angular momentum; but when the reverse is true, it is the satellite's orbit which is most altered.

If the satellite's revolution period is P and its time rate of change is $\dot{P}$, then

$$\dot{P} = 0.64 \sin 2\epsilon \left(\frac{\Delta m}{m} \right) n^{-5} \tag{9}$$

where ϵ is the angle by which the tidal bulge leads the satellite. This is a modification of the formula derived by Jeffreys (1952). For solid body tides, ϵ is proportional to $(\omega_P - \omega_s)$, where ω_P is the rotational angular velocity of the parent and ω_s is the orbital angular velocity of the satellite. The variation of ϵ with time will be ignored in the following discussion. This formula for $\dot{P}$ enables us to estimate a characteristic time scale $P/\dot{P}$ of tidal evolution for a typical minor satellite if we assume that values of ϵ for asteroids are similar to those estimated from the Phobos-Mars case. (The Earth-Moon values are dominated by friction in the oceans, rather than solid body tides.) For Phobos-Mars, the observed $\dot{P} = -1.4 \times 10^{-2}$, implying $\epsilon = -0°.43$ for $(\omega_P - \omega_s)$

$= -778°/$day. To estimate the order of magnitude of $P/\dot{P}$ for minor satellites, let us take values appropriate to the large satellite of Herculina: $n \sim 5$, $\Delta m/m \sim (\Delta d/d)^3 \sim 0.01$, $P \sim 0.004$ yr, $(\omega_p - \omega_s) \sim +670$ deg/day. Hence $\epsilon \sim + 0°.37$ and $P/\dot{P} \sim 1.5 \times 10^5$ yr, which gives an approximate idea of the time scale for tidal evolution. Considering the uncertainties in these estimates, $P/\dot{P}$ probably lies between 10^4 and 10^7 yr.

If a satellite were decaying toward the surface of its parent, it would reach there in much less than this characteristic time, since $P/\dot{P}$ is proportional to mean distance to the $13/2$ power. Moreover the orbit would have become circular and equatorial in the process, since the tidal forces damp out eccentricity and inclination with roughly the same characteristic time scale. Alternatively, a satellite entirely outside of the synchronous orbit would still tend to become coplanar and equatorial, but would have increasing eccentricity. Indeed, the eccentricity increase would continue until escape of the satellite. This is an interesting feature, because relatively large-mass satellites which are evolving toward escape can be used to set an upper limit to the time since formation of the system, once their current orbits and tidal accelerations are directly measured. It should be possible given enough samples to distinguish between asteroid belt ages of greater than 10^9 yr as in traditional theories, or less than 10^7 yr as required by the theory of origin in a planetary break-up event (Van Flandern 1978; also see Van Flandern 1977 and critical discussion following it). Because the time scale for tidal evolution is proportional to the mass ratio of parent to satellite, it is clear that satellites of small mass (say, 10^{-6} of their parent's mass or less) could not have been appreciably altered by tidal evolution, even over the entire lifetime of the solar system. This condition holds until we consider a satellite size so small that solar radiation pressure becomes an appreciable perturbing force. Clearly dust size grains and smaller have been completely removed by solar radiation.

IV. ORIGINS

Permanent satellite captures require some force operating in addition to gravitation. If two bodies collide at typical relative velocities of 5 km sec^{-1}, much of the debris will likewise be moving at speeds far greater than escape velocity, which is on the order of 100 m sec^{-1} or less. That debris which does not escape may indeed orbit; but the orbit of each particle must necessarily continue to intersect the surface of the parent body, from where it came. Permanent orbits therefore require multiple collisions or some other process to keep the debris in orbit for more than one revolution. Moreover, for stability, the pericenter distance will usually have to be several radii in order to avoid rapid tidal decay. Such collisional processes could be invented, but do not seem very probable for large asteroids.

As suggested by Tedesco (1979a) and Hartmann (see his chapter), such collisional processes are most likely to occur during the collisional destruction of asteroids. This follows because under such conditions there are numerous, relatively densely packed objects moving with low relative velocities away from the impact site (Williams 1975). Indeed, 171 Ophelia, the asteroid having an Algol-like lightcurve modeled by Wijesinghe and Tedesco, is a member of the Themis dynamical family, a family believed to be of common origin from physical observations as well as dynamical arguments (Tedesco 1979b and the chapter by Gradie et al.). Interestingly enough, 24 Themis itself, the largest (210 km) fragment from an approximately 300 km parent body (Tedesco 1979b) has a complex lightcurve indicative of major topographic features. It is possible that these features are due to tidally-decayed satellites, or other fragments which failed to escape during the disruptive event.

Alternatively, the planetary break-up theory proposed by Van Flandern (1978) produces minor satellites in a very straightforward way. As debris flies away from the disintegrating planet in all directions with a great range of velocities, the radius of the gravitational sphere of influence of each body expands with the cube of its increasing distance from the parent planet, limited only by the proximity of another body, or ultimately by the sun. As these radii expand, many other smaller bodies and much debris are trapped within the sphere of influence; for a certain range of relative velocities, escape will then no longer be possible. Hence numerous minor satellites around each minor planet would seem to be an inevitable result of such a planetary break-up process.

If the planetary break-up model is correct, then virtually all asteroids should have satellites and debris in orbit around them due to the recent formation associated with this model. If, on the other hand, the collisional production mechanism is right, then multiple asteroids should be less than ubiquitous. The fraction of asteroids having satellites will then depend not only on the collisional probabilities, but also on the particular breakage mechanisms involved.

A third formation mechanism is that proposed by Davis et al. (see their chapter) in which multiple asteroids are survivors of systems which originated during the asteroid formation era when, according to their scenario, the asteroids moved in orbits with low relative velocities. In this scenario the on the belt population at the time the asteroids' relative velocities had been pumped up to the point where collisional destruction became dominant.

V. MODELS

Asteroids.

From the foregoing, it becomes feasible to hazard the following guesses about a model for a typical asteroid satellite system. Very large satellites

Fig. 5. Davy-Y crater chain on the moon, believed to be of impact origin. (NASA Photograph.)

($\Delta m/m \sim 1$) would be fully evolved tidally; hence these would all by now have escaped, or be in synchronously-locked orbits, or they would be resting on the surface of their primary. The frequency of occurrence of contact binary asteroids should be an indicator of the frequency of such relatively large satellites in the original distribution. In this connection it is interesting to contemplate the fate of several nearly equal masses. After the largest pair come into contact, the next largest would gradually evolve down onto one *end* of the first dumbbell-shaped pair, and so on, forming a chain of attached asteroids. This may be the case for 1620 Geographos, which varies in brightness as if it were six times as long as it is wide. For smaller minor planets, which may have many satellites of size comparable to the primary, such an asteroid chain may be a cause of crater chains seen on the moon's surface (see Fig. 5). Almost all satellites of appreciable size inside the synchronous orbit would by now have impacted on their primaries, affecting the spin rates considerably.

For somewhat smaller mass ratios, the effect of tidal evolution on inclinations should be apparent, with direct revolution in or near the primary's equator tending to dominate. Hence many minor satellites may tend to become coplanar. This may be a partial explanation of why one observer saw six minor satellites, a second saw only one, and a third, none, for the Herculina occultation of June 1978. If the preferred plane were close to being in the line-of-sight, this might happen. Alternatively we would be led to wonder if the periods of revolution of the minor satellites could have tended to come into resonance with one another, and to what extent the satellites are arranged in hierarchies (satellites with satellites). Bodies of much smaller mass would retain their original three dimensional distribution. These may have been responsible for the "diffuse" or "flickering" appearance of the occulted star during the Herculina, Metis and Melpomene occultations, which lasted several minutes before and after the main occultation (see Fig. 1). They may also be the explanation for the faint glow reportedly seen around some asteroids (Hartmann 1972, p. 157).

It is usually assumed that asteroid families are the result of collisional fragmentation. We can now offer an alternative hypothesis. If some asteroids have clouds of satellites around them, then a collision between unrelated parent asteroids might have the effect of immediately removing both parent asteroids from their respective clouds of satellites, with each leaving behind a cloud of objects no longer gravitationally bound and moving at small relative velocities. The result would be an asteroid family, all of whose members would generally be smaller than 50 km. Any such collision which had occurred within the last 10^5 yr would result in a jet stream, since there would then have been insufficient time for the longitude of nodes and perihelia of the various former satellites to complete a single revolution relative to one another, and all orbital elements of the family members would be similar. Such an idea should be verifiable, since the nodes and perihelia can be traced back over 10^5 yr, and should tend to coincide at the epoch of the collision releasing the jet stream members. This idea also predicts that asteroids which have been involved in major collisions, whether fragmented or not, would today be devoid of satellites, since they would in general have been forced to exceed the escape velocity from their satellite systems. If both parents fragment extensively, there might be as many as four populations of family members — two of fragments from each parent, and two of former satellites. This is indeed seen for the Flora family, which was divided into four sub-families by Brouwer (1951). A mechanism similar to this has been proposed by Tedesco (1979c) to explain the anomalous properties of the Flora families.

It requires only a small additional leap of the imagination to realize that, if minor planets could have retained material down to dust or gas molecule sizes, they would be indistinguishable in appearance from comets. They are already believed to have chemical compositions very similar to comets (Millman 1977).

Comets.

Let us examine what we know about comets in the light of this new idea, namely that asteroids have satellites. What if comets also had satellites? Lyttleton (1977) has already summarized many objections to the usual "dirty snowball" model of comets if the source of their comas and tails is supposed to be a central nucleus. Yet the alternative model he offers suffers from problems of insufficient gravitational binding. The following points from Lyttleton's paper are relevant:

Some comets are observed to have more than one nucleus. (Are these satellites?) The coma of a comet *contracts* as the comet approaches the sun and *expands* as it recedes. (So does the radius of the gravitational sphere of influence of a primary with satellites.) Comets are often observed to split, but with small relative velocities. (As the radius of the sphere of influence contracts, some satellites will find themselves outside of it, and therefore on independent solar orbits.) Comet comas are seen at large heliocentric distances where solar radiation pressure is too small to produce them. (In this new model, comets bring their own orbiting dust with them; there is no need to drive it off the surface of a nucleus.) New comets, making their first approach to the sun, lose enough material that they are 2 or 3 magnitudes fainter on their second return; but losses on subsequent returns are barely detectable. (The comet would lose all its material outside of the radius of the minimum sphere of influence, which occurs at perihelion, on its first pass. These losses are caused by gravitational escape. Subsequent passages at the same heliocentric distance would produce losses due to radiation pressure only.) When the earth intersects the orbit of a comet, meteor showers are observed. (These may be former comet satellites which have escaped due to the contraction of the gravitational sphere of influence. In this connection it should be noted that a close approach to Jupiter can accelerate the demise of a comet in two ways: the close approach may itself establish a new minimum to the comet's sphere of influence, causing the loss of much additional material; or Jupiter may lower the comet's perihelion, allowing the sun to produce a new minimum sphere of influence for the comet at its next perihelion passage. In either case, a new cloud of meteors should escape into the comet's orbit.)

Meteors.

As we have noted, satellites may occur in hierarchies; hence satellites may have smaller satellites, etc.; the radius of the sphere of influence is ~ 100 times the diameter of the primary, even down to meter-sized bodies.

There are three pieces of observational evidence from well-observed fireballs, all of whose major fragments were tracked to the ground and recovered, that meteors may likewise be multiple bodies (Revelle, personal communication). The most important evidence is that fireballs "break up" in the atmosphere at heights far too great, considering the strength of the materials. The split has been observed at more than 100 km above the earth's surface. This is easy to understand if these are originally multiple, rather than single, bodies. Also, both the brightness-to-mass ratios and the radar cross-sections of the fireballs are considerably higher than theory permits. Here again we postulate that a small cloud of debris, instead of a single isolated body, is entering the atmosphere, accounting for both phenomena.

VI. SUMMARY

In this chapter we have presented the evidence for the existence of minor satellites (Secs. I and II) and shown that if it is accepted at face value it implies that many asteroids have satellites. We then proceeded to show that the existence of such systems is consistent with known physical laws (Sec. III) and that reasonable methods of formation exist (Sec. IV). Finally we have speculated on what effect the existence of such objects would have on our understanding of asteroids, meteors and comets (Sec. V).

Although the existence of minor satellites is still based on circumstantial evidence it will not remain so for long; the existence, or nonexistence of such objects will be established within the next few years. The discovery of asteroidal satellites will revolutionize our theories about the origin, evolution and dynamics of all minor bodies in the solar system.

REFERENCES

André, Ch. 1901. Sur le systeme formé par la planète double (433) Eros. *Astron. Nach.* 155: 27-30.

Binzel, R. P. 1978. Further support for minor planet multiplicity. *Occul. Newsl.* 1: 152-153.

Binzel, R. P., and Van Flandern, T. C. 1979. Minor planets: The discovery of minor satellites.

Bowell, E.; McMahon, J.; Horne, K.; A'Hearn, M. F.; Dunham, D. W.; Penhallow, W.; Taylor, G. E.; Wasserman, L. H.; and White, N. M. 1978. A possible satellite of Herculina (abstract). *Bull. Amer. Astron. Soc.* 10:594.

Brouwer, D. 1951. Secular variations of the orbital elements of minor planets. *Astron. J.* 56: 9-32.

Campbell, D. B.; Pettengill, G. H.; and Shapiro, I. I. 1976. 70-cm radar observations of 433 Eros. *Icarus* 28: 17-20.

Chang, Y. C., and Chang, C. S. 1963. Photometric investigations of variable asteroids. II. *Acta Astron. Sinica* 11: 139-148.

Chebotarev, G. A., and Shor, V. A. 1978. The structure of the asteroid belt. *Fund. Cosmic Phys.* 3: 87-138.

Cook, A. F. 1971. 624 Hektor: A binary asteroid? In *Physical Studies of Minor Planets*, ed. T. Gehrels (NASA SP-267, Washington, D.C.: U.S. Government Printing Office), pp. 155-164.

Debehogne, H.; Surdej, A.; and Surdej, J. 1978. Photoelectric lightcurves of the minor planets 29 Amphitrite, 121 Hermione, and 185 Eunike. *Astron. Astrophys. Suppl.* 32: 127-133.

Degewij, J. 1978. Photometry of faint asteroids and satellites. Ph.D. dissertation, Leiden University.

Degewij, J.; Tedesco, E. F.; and Zellner, B. 1979. Spots on Asteroids. *Icarus* (special Asteroid issue).

Dohnanyi, J. S. 1969. Collisional model of asteroids and their debris. *J. Geophys. Res.* 74: 2531-2554.

Dunham, D. W. 1977. Observations of planetary occultations. *Occ. Newsl.* I: 98.

Dunham, D. W. 1978a. Occultations of stars by minor planets (abstract). *Bull. Amer. Astron. Soc.* 9: 621.

Dunham, D. W. 1978b. Satellite of minor planet (532) Herculina discovered during occultation. *Occ. Newsl.* I: 151-152.

Dunham, D. W. 1978c. More on Herculina. *Occ. Newsl.* II: 2-3.

Dunham, D. W. 1979*a*. Duplicity of both (18) Melpomene and SAO 114159 discovered during occultation. *Occ. Newsl.* II: 12-16.

Dunham, D. W. 1979*b*. More on Melpomene. *Occ. Newsl.* II: 16.

Dunham, D. W. 1979*c*. Juno larger than expected. Probable graze by distant satellite. *Occ. Newsl.* II: 12.

Dunham, D. W. 1979*d*. Other late 1978 asteroidal occultations. *Occ. Newsl.* II: 24-25.

Dunham, D. W.; Dunham, J. B.; Van Flandern, T. C.; Schmidt, R. E.; Skillman, D. R.; Espenok, F.; A'Hearn, M. F.; Bolster, R. N.; Taylor, G. E.; Klemola, A. R.; Wasserman, L. H.; Williamon, R. M.; and Poss, H. L. 1979. Occultation of SAO 114159 by 18 Melpomene and possible satellite. *Icarus* (special Asteroid issue).

Dunham, D. W., and Maley, P. D. 1977. Possible observations of a satellite of a minor planet. *Occ. Newsl.* 1: 115-117.

Dunlap, J. L. 1976. Lightcurves and the axis of rotation of 433 Eros. *Icarus* 28: 69-78.

Fay, T. D., and Wisniewski, W. 1978. The lightcurves of the nucleus of comet d'Arrest. *Icarus* 34: 1-9.

Finsen, W. S. 1926. In the report on the 106th annual general meeting. *Mon. Not. Roy. Astron. Soc.* 86: 209.

Groeneveld, I., and Kuiper, G. P. 1954. Photometric studies of asteroids. II. *Astrophys. J.* 120: 529-546.

Harris, A., and Young, J. 1979. Photoelectric lightcurves of asteroids 42, 45, 56, 103, 532 and 558. *Icarus* (in press).

Hartmann, W. K. 1972. Moons and Planets: An introduction to planet science. (Belmont, Calif.: Wadsworth Publ. Co., Inc.).

Hartmann, W. K., and Cruikshank, D. P. 1978. The nature of Trojan asteroid 624 Hektor. *Icarus* 36: 353-366.

Heintz, W. D. 1975. Micrometer observations of (433) Eros. *Astrophys. J.* 200: 787.

Jeffreys, H. 1952. *The Earth,* 3rd ed. (Cambridge: Cambridge University Press), p. 240.

Jurgens, R. F., and Goldstein, R. M. 1976. Radar observations at 3.5 and 12.6 cm wavelength of asteroid 433 Eros. *Icarus* 28: 1-16.

Lundmark, K. 1932. The probable mass of Eros. *Lund Obs. Circ.* 7: 169-170.

Lyttleton, R. A. 1977. What is a cometary nucleus? *Quart. J. Roy Astron. Soc.* 18: 213-233.

McMahon, J. H. 1978. The discovery of a satellite of an asteroid. Proc. of Astronomy West '78 Conf., San Luis Obispo, Calif., July 1978. See summary in *Sci. News* 114: (1978).

Menzel, D. H.; Whipple, F. L.; and de Vaucouleurs G. 1970. *Survey of the Universe* (Englewood Cliffs. N.J.: Prentice Hall).

Millman, P. M. 1977. The chemical composition of cometary meteoroids. In *Comets, Asteroids, Meteorites,* ed. A. M. Delsemme (Toledo, Ohio: University of Toledo Press), pp. 127-132.

O'Leary, B.; Marsden, B. G.; Dragon, R.; Hauser, E.; McGrath, M.; Backus, P.; and Robkoff, H. 1976. The occultation of Kappa Geminorum by Eros. *Icarus* 28: 133-146.

Pickering, W. H. 1932. The mass, density, and albedo of Eros. *Pop. Astron.* 40: 10-11.

Reitsema, H. J. 1979. The reliability of minor planet satellite observations. *Science* 205: 185-186.

Scaltriti, F., and Zappalà, V. 1977. Photoelectric photometry of the minor planets 41 Daphne and 129 Antigone. *Astron. Astrophys.* 56: 7-11.

Schober, H.-J. 1979. Rotation period of the minor planet 337 Devosa: An unusual object with triple extrema in the photoelectric lightcurve. *Astron. Astrophys. Suppl.* 35: 337-343.

Shatzel, A. V. 1954. Lightcurve of 44 Nysa. *Astrophys. J.* 120: 547-550.

Sheffer, Y. 1979. More on Juno. *Occ. Newsl.* II: 27.

Szebehely, V. 1978. Stability of artificial and natural satellites. *Celes. Mech.* 18: 383-389.

Tedesco, E. F. 1979*a*. Binary asteroids: Evidence for their existence from lightcurves. *Science* 203: 905-907.

Tedesco, E. F. 1979*b*. A photometric investigation of the colors, shapes, and spin rates of Hirayama family asteroids. Ph.D. dissertation. New Mexico State University.

Tedesco, E. F. 1979c. Was Proto-Flora binary? *Icarus* (special Asteroid issue).

Tedesco, E. F., and Bowell, E. G. 1979. UBV observations of 944 Hidalgo (in preparation).

Tedesco, E.; Drummond, J.; Candy, M.; Birch, P.; Nikoloff, I.; and Zellner, B. 1978. 1580 Betulia: An unusual asteroid with an extraordinary lightcurve. *Icarus* 35: 344-359.

Tedesco, E. F. and Zappalà, V. 1979. Rotational properties of asteroids: Correlations and selection effects. Submitted to *Icarus*.

van den Bos, W. H. and Finsen, W. S. 1931. Physical observations of Eros. *Astron. Nach.* 241: 329-334.

Van Flandern, T. C. 1977. A former major planet of the solar system. In *Comets, Asteroids, Meteorites,* ed. A. H. Delsemme (Toledo, Ohio: University of Toledo Press), pp. 475-481.

Van Flandern, T. C. 1978. A former asteroidal planet as the origin of comets. *Icarus* 36: 51-74.

van Houten, C. J.; van Houten-Groeneveld, I.; Herget, P.; and Gehrels, T. 1970. Palomar-Leiden survey of faint minor planets. *Astron. Astrophys. Suppl.* 2: 339-448.

van Houten-Groeneveld, I.; van Houten, C. J.; and Zappalà, V. 1979. Photoelectric photometry of seven asteroids. *Astron. Astrophys. Suppl.* 35: 223-232.

von Oppolzer, E. 1901. Notiz. Betr. Planet (433) Eros. *Astron. Nach.* 154: 297.

Wamsteker, W., and Sather, R. E. 1974. Minor planets and related objects. XVIII. Five-color photometry of four asteroids. *Astron. J.* 79: 1465-1470.

Wijesinghe, M. P., and Tedesco, E. F. 1979. A test of the plausibility of eclipsing binary asteroids. Submitted to *Icarus*.

Williams, J. G. 1975. Asteroid families. *Bull. Amer. Astron. Soc.* 7: 343.

Zappalà, V.; Scaltriti, F.; Farinella, P.; and Paolicchi, P. 1979a. Asteroidal binary systems: Detection and formation. Submitted to *Moon and Planets*.

Zappalà, V.; van Houten-Groeneveld, I.; and van Houten, C. J. 1979b. Rotation period and phase curve of the asteroids 349 Dembowska and 354 Eleonora. *Astron. Astrophys. Suppl.* 35: 213-221.

DIVERSE PUZZLING ASTEROIDS AND
A POSSIBLE UNIFIED EXPLANATION

WILLIAM K. HARTMANN
Planetary Science Institute

Recent observations have led to unconventional models of certain asteroids, suggesting forms unsuspected a few years ago. Some of these include binary asteroids (e.g., 532 Herculina, 18 Melpomene), very irregular asteroids (1580 Betulia), and very elongated asteroids unlikely to be collisional fragments (624 Hektor). A connection is suggested between this observational work and ongoing theoretical work concerning collisions of large, comparable-sized asteroids ($r/R \gtrsim 0.01$). Such collisions have different consequences from the collisions usually considered ($r/R \lesssim 0.01$). The new work suggests possible sources of elongated and binary asteroids. Quantitative models show that at the high impact velocities, typical of the main belt, dispersal of most fragments occurs while at lower collision speeds only partial dispersal occurs, leaving partially re-assembled brecciated bodies. During such collisions, binary pairs may be produced as a result of the interactions of the adjacent fragments. Low-speed collisions of comparable-sized asteroids, typical of tidally evolving binary pairs, pairs of adjacent fragments, or Trojan pairs, can leave such pairs minimally fractured and held in contact by gravity, creating elongated bodies. Minor collisions could occasionally reseparate such objects or alter tidal evolution of orbiting binary pairs. Thus some elongated asteroids may be products of collisional accretion, rather than of collisional fragmentation as usually thought, while some binary asteroids and brecciated meteorites

[466]

may be produced during collisional fragmentation of large, comparable-sized asteroids.

I. 624 HEKTOR: AN EXAMPLE OF A PUZZLING ASTEROID

Trojan asteroid 624 is by about a factor 2 the largest Trojan, and yet it has the largest lightcurve amplitude of any asteroid of its size. The light variation, ranging up to 3.1:1, is very probably due to elongated shape, not to an albedo asymmetry (Dunlap and Gehrels 1969) especially since the albedos of the two different rotation poles (during 1972 and 1977 corresponding to pole-on views) are nearly the same namely 0.03 ± 0.003 (Hartmann and Cruikshank 1978).[a] The puzzle is that the conventional idea to explain extreme lightcurves, elongated shape associated with origin as a collisional fragment, is especially unsuited to explain Hektor, because Hektor would then be a singularly large and splinter-shaped fragment in a swarm of small round fragments, all isolated in a single cloud. Dunlap and Gehrels (1969) reported a 20-40 % amplitude for the second-biggest Trojan, 911 Agememnon, whose diameter is about 150 km. Degewij (1977, 1978; see the chapter by Degewij and van Houten) has studied lightcurves and/or magnitude-phase relations of eight other Trojans ranging down to 90 km diameter and generally finds light variations less than 20 %, compared to Hektor's 210 %. It would seem a strange collision which, in the isolation of a Lagrangian cloud, would produce one giant splinter (150 x 300 km) and many other spheroids smaller than 150 km.

The original work on Hektor was by Dunlap and Gehrels (1969). Analyzing many years' observations of Hektor, they not only discovered its unusual light amplitude, but also showed that the rotation pole is only about 10° ± 2° out of the ecliptic, so that in various years, one pole, then a broadside, and then the other pole are presented to earth. To match the lightcurve, Dunlap and Gehrels suggested a cylindrical shape 2.6 times as long as it was wide. The origin of such an elongated body was not addressed.

Cook (1971) suggested a different explanation for Hektor's puzzling properties. He asserted that the compressive stresses would be too great for such a cylinder to persist. His estimates need to be corrected to the presently estimated dimensions of Hektor, giving a central compressive pressure of roughly 60 bars. Cook compared the central pressure to a strength of only about 10 bars for the chondrite, Lost City, and concluded that a cylinder this size would not persist, but collapse. However, in the discussion appended to Cook's chapter Hartmann noted that most meteorites are stronger than the Lost City value and that reported ordinary chondrites range from 60 to 3700

---

[a] *Note added in proof.* In April 1979 at Mauna Kea Observatory, Cruikshank and I obtained the first simultaneous in-phase lightcurves in reflected sunlight and thermal infrared, proving that Hektor varies primarily due to elongated shape, not albedo patchiness. This work is not yet published. Degewij (1978) incorrectly indicates that we had established this during our earlier work.

bars in strength. Carbonaceous chondrites are among the weakest meteorites; on the other hand, of 12 sampled Trojans 75 % match carbonaceous chondrite colors while 25 % are dark but do not match these colors (see the chapter by Degewij and van Houten in this book). Strength thus seems a plausible but uncertain criterion for ruling out a cylindrical shape.

Cook (1971) proposed instead the interesting idea that Hektor might be a binary either orbiting or in contact. This suggestion was based partly on librations suspected by Cook from the Dunlap-Gehrels lightcurves. This model was not positively established, and the origin of such a binary was not considered.

Hartmann and Cruikshank (1978) pointed out that in the Jovian Lagrangian cloud, mean encounter velocities (1-2 km sec^{-1}) are substantially below those in the main belt (5 km sec^{-1}) and that there may be a low-velocity tail in the velocity distribution absent in the main belt (an asteroid trapped *at* the Langrangian point would have zero velocity relative to the cloud). Therefore they suggested that a partly-fractured contact binary or elongated object could have formed by low-speed collision of two of the original 150 km-scale Trojans. They suggested that a brightened zone of crushed material analogous to lunar rays, around the collision zone, could explain the 3.1:1 lightcurve, which would be $\leqslant$ 2:1 in Cook's purely geometric contact binary or co-orbiting binary model with two spheres. Cook (1978, personal communication) points out that tidal stresses on two touching spheroids might deform the initial spheres to elongated shapes capable of giving the 3.1:1 lightcurve. John V. Lambert (1979, personal communication) has pointed out that the light ratio would also be affected by the scattering properties of the surface, with the required length-to-width ratio reduced if one changes from a case where brightness $\propto$ surface area, to a case with a Lambertian surface. Thus the exact shape and photometry of the definitely elongated Hektor remain intriguing problems.

II. OTHER PUZZLING ASTEROIDS

Table I lists some current models of various other puzzling asteroids which seem at first glance unrelated to the Hektor problem, but may in fact have a connection. As is now well known, a number of asteroids have unusual properties. Occultations have given marginal evidence of satellite companions, with the best coming from 18 Melpomene, where both the occultation by the 135-km diameter primary and that by the reported 48-km satellite may have revealed that the occulted star itself was a binary (Dunham *et al.* 1979). Some evidence also comes from an occultation by 532 Herculina, a 220-km asteroid found to have a 50-km satellite (Binzel and Van Flandern 1978; see the chapter by Van Flandern *et al.*). Binzel and Van Flandern also list other, much less convincing cases.

A different class of unusual asteroid was pointed out by Tedesco *et al.*

TABLE I

Current, Unusual Asteroid Models

	Dimension	Comment	Reference
1580 Betulia	~6 × 8 km	Very irregular lightcurve; topographic irregularity comparable to radius. Binary or contact binary?	Tedesco, et al. (1978) This chapter
18 Melpomene	135 km with 10 km topography	48-km satellite ~ 750 km away; possible other satellites < 300 m; reseparated pair?	Dunham et al. (1979) This chapter
532 Herculina	220 km	50-km satellite ~ 1000 km away.	Binzel and Van Flandern (1978)
624 Hektor	150 × 300 km	Partially consolidated low-velocity collision product.	Hartmann and Cruikshank (1978)
49 Pales	~140 km[a]	~ 50-km satellite about 450 km away.	Tedesco (1979)
171 Ophelia	~90 km[a]	30-km satellite about 300 km away.	Tedesco (1979)
44 Nysa	~80 km[b]	Lightcurve resembles contact binary.	Tedesco (1979)

[a]Based on diameters summarized by Morrison (1977) for asteroids of equal $B(1,0)$ magnitude as listed in *Ephemerides of Minor Planets 1979*. Satellite data based on size ratio and orbit parameters reported by Tedesco.

[b]Morrison (1977).

(1978), who noted that 1580 Betulia, a 7-km regolith-free Mars-crosser and first *C*-type (carbonaceous?) object found among the Mars-crossers, has a bizarre three-lobed lightcurve. Tedesco *et al.*, interpreted Betulia as an asymmetric "collisional fragment" with a topographic irregularity on one face "of size comparable to the radius." Betulia's is the most asymmetric lightcurve known, in terms of the size of its third lobe.

Following the recognition of probable binary pairs from occultations, Tedesco (1979) pointed out still another class of unusual asteroids with nearly constant lightcurves, but pronounced dips resembling lightcurves of eclipsing binary stars. He proposed that 49 Pales and 171 Ophelia are orbiting binaries and lists nine other suspected binaries or contact binaries (see the chapter by Van Flandern *et al.*).

III. COLLISIONAL HISTORY AS A UNIFYING FACTOR IN EXPLAINING MANY PUZZLING ASTEROIDS

As a result of our work on Hektor, I began a study of the results of collisions of comparable-sized bodies. These collisions are different from the most common and most studied types of collisions, which are those between bodies of markedly different size, and which usually result in simple cratering. Comparable-sized bodies were defined as capable of colliding and disrupting the larger body; usually this implies their radius ratio $r/R \gtrsim 0.01$. This study (Hartmann 1979) described qualitatively, and to first order quantitatively, five conceptually different outcomes of importance:

1. rebound and mutual escape of the two bodies (low velocity);
2. rebound and fallback, producing an unfractured contact binary;
3. with increased energy of collision, disruption of the bodies but insufficient energy to allow escape, so that most fragments reassemble into a brecciated spheroid;
4. disruption with sufficient energy that most bodies escape entirely, leading to total disruption;
5. a transition between (2) and (3), in which substantial fracturing occurs but with insufficient energy to dislodge most material or reassemble the body, so that the result is a brecciated body in the form of an elongated object.

This single framework suggests an origin for all the various types of asteroids described above. To aid the discussion, Fig. 1 presents a set of collision outcomes calculated from the equations given by Hartmann (1979) for collisions of equal-sized bodies. The asteroids in this case are assumed to have properties intermediate between the two sets of examples calculated in that paper. The internal collisional strength, *S,* is that measured for coherent rock-bodies such as basalts or ordinary chondrites ($S = 4 \times 10^6 \, \rho$ ergs cm^{-3} where ρ = density, assumed to be 2.5 g cm^{-3} — see Hartmann [1979] for

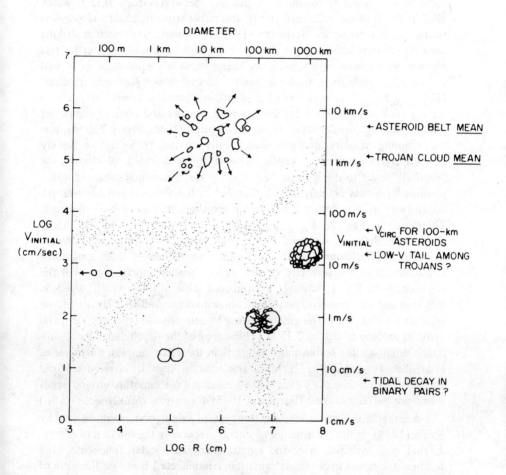

Fig. 1. Regimes of different collision products (indicated by cartoons) for equal-sized colliding asteroids of specified initial radius and approach velocity (at large separation). Stippled bands give indication of transition regimes separating distinct outcomes of rebound, unfractured contact binary, reassembled brecciated spheroid, and total dispersal. An important transition case of a partly fractured contact binary is also illustrated by a cartoon. Approximate velocity regimes of different kinds of collisions in the present solar system are indicated at left. The two colliding asteroids in this calculation are assumed to have interiors with strength and elasticity of basalt or other igneous rock, but surface layers of loose or weakly bonded fragmental material that reduce rebound velocities and effective elasticity during collisions. For further discussion and other calculated models, see Hartmann (1979).

further discussion of those parameters).

The surface layers of the asteroids are asssumed to be granular or weakly consolidated material to some unspecified depth (perhaps $0.1R$), which absorbs much of the collisional energy and rather strongly inhibits rebound as found experimentally by Hartmann (1978) for spheroidal projectiles striking semiinfinite powder/rock layered targets. To make this point clearer for this chapter, an experiment was performed allowing nearly equal-sized spheroidal suspended granitic rocks to swing freely into each other. Rebound velocities ($V_{rebound}/V_{impact}$) were recorded as artificial granular layers were built up on the rock surfaces, first by dusting with powder and then by cementing granular mortar powder layers of various depth on them. Figure 2 shows how the rebound velocity dropped dramatically as the "regolith" or weakly bonded layer increased in depth. Because of the necessity of affixing the "regolith" layer to the granitic "asteroid" in this conceptual experiment, the regolith layer was actually weakly bonded with somewhat variable strength dependent on the drying time and proportions of mortar and moisture. Therefore, in Fig. 2, $V_{rebound}/V_{impact}$ does not drop to the levels of $\leqslant 0.01$ that are observed for impact into semiinfinite dry powders in vacuum (Hartmann 1978), but rather to values ~ 0.2, apparently characteristic of the elasticity of the weakly bonded mortar layer. For the calculations in Fig. 1, however, I assumed a value of 0.03, which, based on this test and the earlier experiments, should correspond to a layer of loose regolith powder and gravel grading into a loosely bonded layer of fragmental material as deep as 0.1 to $0.25\,R$. (The scaling of the depth dependence with size is unknown for bodies much larger than these lab samples, but it seems clear that $V_{rebound}/V_{impact}$ of a fractured or regolith covered asteroid would be well below the values of 0.85 measured for smoothly ground basalt spheres or ~ 0.5 measured [Hartmann 1978] for natural clean igneous rocks.)

Another major uncertainty in this type of calculation, re-emphasized by Kaula (1979), is the partitioning of energy, especially the fraction of energy, k, that goes into fracturing and kinetic energy of ejected fragments. (The fraction 1-k is lost in plastic deformation, heating, etc.) Based on the work of O'Keefe and Ahrens (1977) k is here assumed to be 0.2, although there is a wide range of uncertainty.

The cartoons in Fig. 1 show the five important collisional products mentioned above. Stippled bands, rather than sharp curves, are plotted to divide the various regimes of collisional outcome in Fig. 1, to remind the reader that transitions occur and that the positions of the curves can shift by up to several stippled-band widths if the materials have other properties, such as the crumbly character of weak carbonaceous chondrites (discussed by Hartmann 1979). However, the presence of basaltic achondrites and irons suggests that many larger asteroids have melted and resolidified into igneous rock, so that the parameters assumed here are the most likely values.

The various unusual asteroid types described earlier may all be discussed

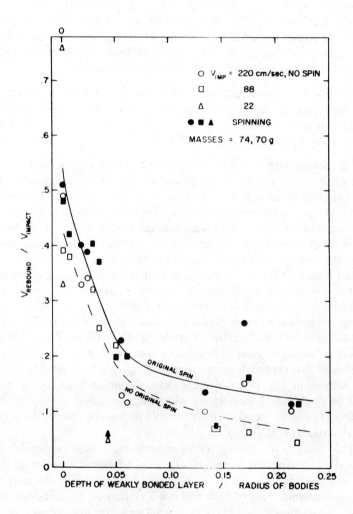

Fig. 2. Results of a "conceptual experiment" designed to show how buildup of a regolith layer consumes energy during collision of comparable-sized asteroids. Each point is an average of results from 6 to 9 collisions. Clean, natural granite spheroidal rocks were suspended and rebounded at about 0.33 to 0.76 V_{impact}. Weakly bonded surface layers were built up by adding first a coating of dry mortar dust, and then by adding thicker layers of moist mortar powder and glue, allowed to dry partially. Rebounds averaged faster for bodies given original spin (ca. 2400 rpm) which often translated into extra rebound energy. No velocity scaling effect was found over one order of magnitude in V_{impact}.

in terms of this diagram. For example, the diagram clearly shows that elongated asteroids are not necessarily collisional fragments; elongated asteroids can also be produced by low-velocity collision of comparable-sized bodies. Approach velocities (the relative orbital velocities at large separation) need to be lower than the mean in the belt or Trojan clouds for this to occur. Two modes of low-velocity ($\lesssim 100$ m sec^{-1}) collisions are:

1. if there is a low-velocity tail in the velocity distribution, particularly in the Trojan clouds where bodies at L_4 or L_5 would have zero velocity relative to the swarm; or
2. by inward tidal decay or co-orbiting pairs.

The surface circular velocity of a 100-km diameter asteroid with $\rho = 2.5$ is only 42 m sec^{-1}, and the radial closing velocity, da/dt, is estimated one or two orders of magnitude less. This would be the effective impact velocity during decay. Hartmann and Cruikshank (1978) suggested that Mode (1) produced Hektor; Binzel and Van Flander (1979) suggest that Mode (2) occurs frequently. Figure 1, as well as calculations for other types of materials in Hartmann (1979), shows that Mode (2) would certainly produce contact binaries. (The reader should note that Fig. 1 is calculated to include free-fall gravitational acceleration as two asteroids approach, which is why larger asteroids undergo more damage for a given V_{initial}, approach velocity at large separation.) Hence Fig. 1 is not directly applicable to the case of inward tidal evolution. Even for asteroids with $D > 100$ km, the collision that terminates tidal evolution would occur with da/dt much less than escape velocity, so that neither disruptive fragmentation nor rebound (lower left and right regimes in Fig. 1) would occur. Virtually unfractured contact binaries could be produced at all sizes for which tidal evolution brings two bodies together. The term "contact binary" should perhaps be reserved for origin by tidal evolution, i.e. Mode (2).

The chance of finding two non-orbiting "field asteroids" (term analogous to "field stars," meaning objects not involved in systems), moving with the low V_{initial}, drops as V_{initial} decreases. Thus the most likely elongated asteroids created by collisional accretion are the ones near the top of the contact binary field in the diagram. This means that the most probable predicted elongated accretional asteroids would range in length (two diameters) from about 10 to 300 km. If asteroids were weaker, these sizes would decrease; for a weak, crumbly carbonaceous chondrite composition, the range would be more like 200 m to 10 km. Tedesco (1979; see the chapter by Burns and Tedesco) reports that an excess of elongated objects exists in a size range around 100-200 km; the excess may be associated with this mechanism.

The above suggests a creation mechanism for orbiting binary asteroids and a third mode for causing low-velocity collisions yielding elongated objects. Figure 1 shows that total disruption is the most common mode of

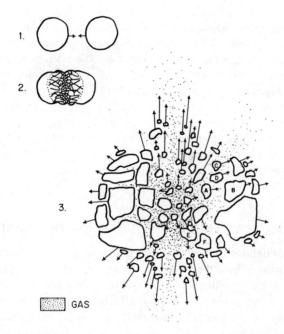

GAS

Fig. 3. Schematic cross-sections through colliding pair of comparable-sized asteroids, showing possible interactions of particles. Velocities average higher near initial impact site. Fragment A catches up to and collides with B; C and D have parallel equal vectors and eventually fall together; E and F could become an orbiting pair; gas from volatilized material could add drag effects. Through these chaotic interactions, binary pairs may be produced during collisions of large, comparable-sized asteroids, which would produce more complex ejecta swarms than the spray from cratering impacts on surfaces.

collision of comparable-sized bodies in the asteroid belt today. Figure 3 gives schematic cross-sections through such a colliding equal-sized pair. Unlike the case in a simple cratering collision, a host of large, asteroid-sized fragments will be produced in a chaotic expanding swarm. After the two bodies approach (Sketch 1) they begin to interpenetrate (2) with plastic deformation and fracturing. As the bodies decelerate, the shock wave (dotted line in Sketch 2) expands, so that peak energy densities are higher close to the impact point and less on the far sides. Brecciation, vaporization of volatiles on ice matrix, for example, and ejecta velocities would all be maximized near the impact point and less on the far sides. We can thus forsee many energy damping mechanisms in this chaos, which would prevent a simple ballistic expansion of all pieces separately away from the impact site. For example in (3), Fragment A hits slower-moving Fragment B. C and D happen to have nearly equal parallel velocity vectors; eventually, they fall together on their way outward. These cases yield a Mode (3) for making elongated bodies. E and F acquire similar but unequal vectors, and go into orbit around each other. A gaseous medium of volatilized matter may accelerate some particles and act as a drag medium on other faster ones. The main point is that many interactions may occur which would lead to pairs and clusters of associated bodies. This conclusion supports Tedesco's (1979) suggestion that binary asteroids might come from multibody interactions in collisions like those that produced Hirayama families. A reviewer of this chapter notes that evolution by impacts, either in the dispersing cloud of fragments or later, increase e and a and can lead to total separation or to periapse collisions that create a contact binary, a mechanism that might be as common as the tidal decay suggested by Binzel and Van Flandern.

IV. EFFECTS OF ASTEROID COMPANIONS AND DEBRIS: SUMMARY

Asteroid satellites that have not orbitally decayed into surface contact with their primaries have the interesting property that they might (especially through continued collisions) have rotation periods and orientations independent of the revolution period and orientation of the system. In cases where the sizes of the pair are comparable (say, $\leqslant 3:1$ in diameter giving $\leqslant 9:1$ in area) the lightcurves could be substantially complicated by occultations and different rotational periodicity. Mars-crossing asteroid 1580 Betulia, interpreted as a single very irregular object to explain a highly irregular lightcurve (Tedesco et al. 1978), might be re-examined in this light. Tedesco (1979, personal communication) indicates that such examination is in progress.

A second effect is that the more common satellites are, the more bias there is against discovery of irregular-shaped asteroids, since an irregular asteroid with a more spherical satellite would have a smoother lightcurve than one without, due to "buffering" by the more constant cross-section area of the satellite. If minor satellites come to be accepted as common, we may need to re-examine our statistics on asteroid shapes, which would tend to be somewhat more irregular than thought.

A similar effect would be on spectrophotometric studies of composition. If pairs of asteroids of unequal composition form orbiting or contact binaries, then spectral signatures will be buffered not only by the conventional effects of cratering or regolith on the primary, but by blending of spectral features from the secondary.

Table II summarizes some of the results given here, in terms of observational consequences on asteroid lightcurves. In addition to the classical explanations of lightcurves in terms of (1) albedo variations or (2) irregular shape due to fragmentation, eight other effects involving collisional accretion or binary formation are seen to be possible.

In summary, the high-amplitude lightcurve of Trojan asteroid 624 Hektor

TABLE II

Asteroid Lightcurve Irregularities

Summary	Case	Possible Explanations
Albedo Effect	1.	Albedo variations (probably a small effect, based on polarimetric results).
Collisional Fragmentation	2.	Irregular shape caused by collisional fragmentation (the conventional explanation; e.g., Tedesco et al. [1978], as applied to 1580 Betulia).
Collisional Accretion — Mode (1)	3.	Dumbbell, or irregular-shaped, compound asteroid caused by direct low-velocity collisional accretion of one field object onto another of comparable size without fragmentation (Hartmann-Cruikshank [1978] explanation of 624 Hektor lightcurve).
Collisional Accretion — Mode (2)	4.	Dumbbell, or irregular-shaped, contact binary caused by tidal evolution of a satellite inward onto surface of primary (very low velocities implying preservation of shape of each body after contact is achieved).
Collisional Accretion — Mode (3)	5.	Dumbbell, or irregular-shaped, compound asteroid caused by fall-together of adjacent fragments during a collision of two large-scale asteroids.
Binary Phenomena	6.	Presence of minor satellite (or multiple asteroid swarms) with satellite rotation period differing from primary rotation period, giving long-term beat phenomena in lightcurves.
	7.	Occultation of minor satellite by primary (two occultation "seasons" during each asteroid revolution around sun).[a]
	8.	Transit of primary by minor satellite (two transit "seasons" during each asteroid revolution around sun.)[a]
	9.	"Satellite" eclipse with minor satellite in primary shadow (two satellite eclipse "seasons" during each asteroid revolution around sun).[a]
	10.	"Solar" eclipse with minor satellite shadow falling on primary (lower amplitude dip than preceding three cases since penumbral shadow smaller than satellite; two solar eclipse "seasons" during each asteroid revolution around sun).[a]

[a]At least one of cases 6-9 would be in progress an estimated 0.2 % of the time for a 50-km satellite 1000 km from a 200 km asteroid orbiting perpendicular to the ecliptic plane. This would approach 10-20% if satellite orbits and primary asteroid heliocentric orbits lie in the ecliptic plane. A higher percentage would be achieved for closer pairs, approaching 100 % for a contact binary due to mutual shadowing, etc.

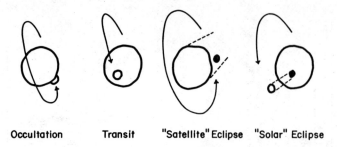

Occultation   Transit   "Satellite" Eclipse   "Solar" Eclipse

Fig. 4. The four types of occultation, transit, and eclipse phenomena that could complicate asteroid lightcurves. Eight sets of events per synodic period occur, since two nodal passages occur. These could be rare and discrete for widely separated pairs with high orbital i; they could be discrete, but occur every revolution period for pairs with widely separated pairs with heliocentric and satellite orbits lying in the ecliptic plane; and they could be non-discrete and continual for closely spaced pairs. In the extreme of contact binaries, they become identical with topographic effects.

reveals a likely candidate for an elongated object, perhaps formed by direct collision (Case 3 in Table II). Hektor seems a particularly poor candidate for being a fragment; it seems less likely to be an orbiting binary or a contact binary formed by tidal evolution because of the absence of an obvious mechanism to form an initially co-orbiting pair of Trojans. Could further dynamical analysis of Trojans reveal a way of so coupling two random large members of a Lagrangian swarm? Then perhaps Hektor is a tidally evolved contact binary, instead of a direct-accretion elongated body.

The irregular lightcurve of Mars-crosser 1580 Betulia, interpreted by Tedesco *et al.* (1978) as a fragmentary object with a large irregularity on one side (Case 2), may be a fallen-together (Case 5) or tidally evolved pair (Case 4) that originated in a major collision in the belt. Alternatively, it may be a binary pair with unequal rotational spins (Case 6), produced in such a collision. Mars-crossers and Earth-crossers are *a priori* good candidates for having undergone recent major collisions, which are needed to put them into Jupiter-perturbed belt orbits, whence they are thrown into short-lived inner solar system orbits.

The recently suspected binary asteroids, if real, may also be pairs coupled during major collisions of comparable-sized bodies in the belt. Further work on their possible unequal rotation periods (Case 6) or occultation/transit/ eclipse phenomena (cases 7-10) could affect our interpretation of asteroid lightcurves, shapes, and histories in general.

Acknowledgments. I thank D. R. Davis for helpful discussions of the eclipse phenomena and other dynamical problems, and also C. R. Chapman, D. P. Cruikshank, R. Greenberg, E. Tedesco, and S. Weidenschilling, for additional helpful discussions. This work was supported by the Planetary Astronomy and Lunar Programs of the National Aeronautics and Space Administration.

REFERENCES

Binzel, R. P., and Van Flandern, T. 1978. The Discovery of minor satellites. Science 203: 903-905.

Cook, A. F. 1971. 624 Hektor: A binary asteroid?. In *Physical Studies of Minor Planets,* ed. T. Gehrels (NASA-SP 267, Washington, D. C.: U. S. Government Printing Office), pp. 155-163.

Degewij, J. 1977. Lightcurve analysis of 170 small asteroids. *Proc. Lunar Sci. Conf. VIII* (Oxford: Pergamon Press), pp. 145-148.

Degewij, J. 1978. Photometry of faint asteroids and satellites, Ph.D. dissertation, Leiden University.

Dunham, D.; Van Flandern, T.; Schmidt, R.; A'Hearn, M.; Skillman, D.; Willamon, R.; Poss, H., and Erickson, G. 1979. More evidence for satellites from an occultation by 18 Melpomene. *Icarus* (special Asteroid issue).

Dunlap, J. L., and Gehrels, T. 1969. Minor planets. III. Lightcurves of a Trojan asteroid. *Astron. J.* 74: 796-803.

Hartmann, W. K. 1978. Planet formation: Mechanism of early growth. *Icarus* 33: 50-61.

Hartmann, W. K. 1979. A special class of planetary collisions: Theory and evidence. *Lunar Sci. X.* The Lunar and Planet. Inst. (in press).

Hartmann, W. K., and Cruikshank, D. P. 1978. The nature of Trojan asteroid 624 Hektor. *Icarus* 36: 353-366.

Kaula, W. M. 1979. Equilibrium velocities at a planetesimal population. *Icarus* (submitted).

Morrison, D. 1977. Sizes and albedos of the larger asteroids. In *Comets, Asteroids, Meteorites,* ed. A. H. Delsemme (Toledo, Ohio: University of Toledo Press), pp. 177-184.

O'Keefe, J., and Ahrens, T. 1977. Impact-induced partitioning, melting, and vaporization on terrestrial planets. *Proc. Lunar Sci. Conf. VIII* (Oxford: Pergamon Press), pp. 3357-3374.

Tedesco, E. F. 1979. Binary asteroids: Evidence for their existence from lightcurves. *Science* 203: 905-907.

Tedesco, E.; Drummond, J.; Candy, M.; Birch, P.; Nikoloff, I.; and Zellner, B. 1978. 1580 Betulia: An unusual asteroid with an extraordinary lightcurve. *Icarus* 35: 344-359.

POLE ORIENTATIONS OF ASTEROIDS

RONALD C. TAYLOR
University of Arizona

Photometric astrometry is a method of determining the pole orientation, sidereal period and sense of rotation of an asteroid. This is done by dividing the number of rotational cycles, corrected to a sidereal frame of reference, into the observed synodic intervals between lightcurve maxima. This method is not explicitly model dependent nor does it use the lightcurve amplitudes. Photometric astrometry, with its subroutines, formulas and limitations is presented. Published pole orientations are reviewed.

Attempts to locate the north pole orientation and sense of rotation of asteroids have led to various methods; a review of several techniques can be found in Vesely (1971). All early methods center around the assumption that lightcurve amplitudes are a function of pole positions. An asteroid lightcurve is the change of the observed brightness versus time caused by changes in its illuminated projected area, and/or albedo variations, as it rotates. Two methods for finding poles have been developed which do not depend on lightcurve amplitudes. The first uses the arrival time of epochs of maximum light; this is called photometric astrometry and is my main topic. The method calculates the pole position, sense of rotation and the sidereal period of rotation. The second method is model dependent and plots unit distance magnitudes, $V(1,\alpha)$, against phase angles α; it was first used to determine the pole orientation of 39 Laetitia (Sather 1976).

Another method of determining the sense of an asteroid's rotation has been outlined by Matson (1971). This infrared technique is further described by Morrison (1977) who presents less than definitive results on the sense of rotation for 6 asteroids. The same can be said of Hansen's (1977) results, also based on infrared measurements. Both papers agree that 1 Ceres clearly spins in a prograde sense.

The purpose of this chapter is to explain photometric astrometry step by step. The defining equations are in the Appendix to this chapter. The results of published pole determinations are presented in Sec. VII.

I. DEFINITION OF PHOTOMETRIC ASTROMETRY

The fundamental concept of photometric astrometry is as follows. The time interval between the epochs of maximum light as seen from the earth is greater than the interval as seen from the stars when the asteroid is increasing in longitude and spins in a prograde sense; the amount of this effect will be referred to in Sec. IV as $\Delta L/360$. Photometric astrometry is the method of converting observed synodic intervals into consistent values for a sidereal period, pole orientation and sense of rotation. The original concept of determining pole orientations of asteroids by this method was developed by Groeneveld and Kuiper (1954) who refer to formulae developed in earlier studies on asteroid Eros. The present form of photometric astrometry is more recognizable in the work of Gehrels (1967). Modifications and refinements have been made over the years and are continuing. Some of the major changes in the original routine are:

1. The phase shifts which are adjustments to the arrival time of epochs of maximum light due to phase angle changes (Gehrels 1967) have evolved to "visual points" (Gehrels *et al.* 1970), and to the currently used "time shifts" (Dunlap *et al.* 1973);

2. It is necessary to add one rotational cycle for each orbital cycle (see the appendix to Taylor 1973);

3. Routines involving the addition of cycles over long intervals (Taylor 1973) will be modified in this chapter.

The basic formula for photometric astrometry is (Taylor 1973)

$$\text{Sidereal period} = \frac{\Delta T_c}{\Delta N \pm \dfrac{\Delta L}{360} + \dfrac{\Delta \delta}{360} + \Delta n} \tag{1}$$

where for each interval between epochs ΔT_c is the time difference in days with each time corrected for light time, ΔN is the integral number of cycles from an arbitrary starting point (see Sec. II), and $\pm$ means that the $+$ is used for direct and − for retrograde spin. The values within the parentheses are corrections to the number of cycles (see Sec. IV) and are a function of pole orientation; L is the longitude of the subearth point on the asteroid; δ is the time shift; and n is the number of cycle corrections.

A pole orientation is selected and sidereal periods are calculated for each time interval between selected epochs of maximum light on the lightcurves. The mean of the sidereal periods and the average residual from the mean period are found. This routine is repeated for various pole orientations and the average residual is determined for each trial; the solution of the pole and its probable error are indicated by the smallest of the average residuals.

An asteroid observed at ecliptic longitudes that are 180° apart will give identical lightcurves if the north pole coordinates are changed by 180° in longitude and kept the same in latitude. This results in pole solutions being found in pairs, 180° apart in longitude; the effect was first pointed out by Russell (1906, p. 18) in his second conclusion with the sentence, "It is always possible (theoretically) to determine the position of the asteroid's equator, (except that the sign of the inclination remains unknown)." The fact that some poles are not exactly 180° apart in longitude or at the same latitude is due to the lack of precision of the solution, and/or to the inclination of the asteroid orbit.

II. INPUT DATA

The ideal data for photometric astrometry are several lightcurves observed during one opposition, from which an estimate of the synodic period can be made, and at least one good lightcurve from each of several different oppositions. When the only available data are lightcurves from one opposition, resolution is insufficient to define a pole unless a broad range of aspects is covered as is the case for Earth-approaching asteroids.

The lightcurves, one per opposition, are first superposed in pairs. Lightcurves that do not superpose with a high degree of coincidence cannot be used in this preliminary analysis because they may be inverted lightcurves (see Sec. VII). The time difference between epochs that coincide within ±5 min (corrected for light time) is used in the determination of the mean synodic period and the number of cycles per interval. Lightcurve maxima are preferred to minima which have sometimes appeared to shift with time (presumably due to shadowing) relative to the maxima.

Rotational cycles over long intervals are counted by using the mean synodic period, which is the average of the instantaneous synodic periods of the asteroid, as seen from the earth, over its entire orbit, including periods from intervals of increasing as well as decreasing longitudes. The mean synodic period is actually determined by dividing each of the time intervals by trial periods until one is found that gives whole-number quotients. Until the mean synodic period is more fully understood it is arbitrarily established that the trial periods be within ± 5 min of the initial estimate of the synodic period and the quotients within ± 0.1 of a whole number. The quotients generated by the mean synodic period are the number of synodic cycles the asteroid has spun during each time interval. This method of counting cycles over long intervals is more accurate and convenient than the equivalent method described in Sec. III of Taylor (1973).

Some intervals between lightcurve maxima may not be considered at first if their lightcurves do not superpose well. The above mean synodic period can, however, next be used to see if any of those intervals are a whole number of cycles apart. From all accepted intervals only those that have a sufficient number of significant digits are used in photometric astrometry. The time intervals and the corresponding whole number of synodic cycles are the values of ΔT_c and ΔN in Eq. (1). The remaining part of the denominator of the formula involves corrections to the number of cycles and these corrections are discussed in Sec. IV.

III. GEOMETRIC PARAMETERS

Figure 1 illustrates the geometric parameters in photometric astrometry. The subscript notation for the coordinate systems has SE for the subearth point, SS for the subsolar point, a for the asteroid, s for the sun, and p for the pole. Table I lists the variables, their primary applications and where they are discussed in this chapter. The parameters needed are the ecliptic coordinates of the asteroid, λ_a and β_a; the ecliptic longitude of the sun, λ_s; the Earth-asteroid distance, ρ; the Earth-Sun distance, R; the solar phase angle, α; and trials of pole orientation expressed in ecliptic coordinates, λ_p and β_p.

The solar phase angle is defined by the Earth-asteroid-Sun angle and in this work is always positive. The aspect A is defined as the Earth-asteroid-pole angle; it is measured from the north pole only and ranges between $0°$ and $180°$. Astrocentric obliquity is defined by the angle γ of the dihedral angle formed by the planes of the phase and aspect angles; γ is measured from $0°$ to $180°$. The three parameters, phase, aspect and obliquity are referred to as the orientation parameters and are used in a check of the final pole solution (see Sec. V).

As a check on the input data the solar-aspect angle E (defined as the Sun-asteroid-pole angle), the aspect A, and phase α should form a spherical triangle. If the sum of any two sides is not greater than the third side then the triangle does not exist because either an error was made in one of the input parameters, or the three vertices of the triangle (the pole, subearth and subsolar points) lie on the arc of a great circle.

IV. ADJUSTMENTS TO THE NUMBER OF CYCLES

As mentioned at the end of Sec. II, there are corrections to the whole number of synodic cycles, as seen from the earth, which need to be applied

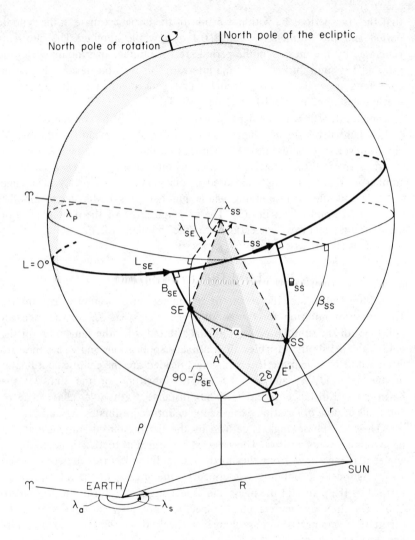

Fig. 1. The geometric parameters of photometric astrometry. *SE* is the subearth point and *SS* the subsolar point on the asteroid surface. The λ, β coordinate system is associated with the north pole of the ecliptic, the L,B system with the north pole of rotation. The true aspect A, obliquity γ, and solar-aspect E are measured from the north pole of rotation and so are the supplements of A', δ', and E'. δ is the time shift and α the solar phase. The subscripts are defined in the text. The formulas associated with this figure are in the Appendix.

when deriving the sidereal period. These corrections will transform the denominator of Eq. (1) into sidereal cycles, as seen from the sun.

$\Delta L/360$ is the fractional part of a cycle that a body would have to rotate in order for the same feature on the surface to be facing the earth at both the

TABLE I

Summary of Geometric Parameters

Variable		Formula in Appendix	Primary Application	Sec. for Disc.
λ_{SE}, β_{SE}	Astrocentric coordinates of subearth point	(2),(3)	Cycle correction n	IV
λ_{SS}, β_{SS}	Astrocentric coordinates of subsolar point	(4),(5)	Cycle correction n	IV
L_{SE}	Pole dependent longitude of subearth point	(1)[a],(6)	Time shift δ Sidereal period	IV I
B_{SE}	Pole dependent latitude of subearth point	(7)	Aspect A Obliquity γ	V V
L_{SS}	Pole dependent longitude of subsolar point	(8)	Time shift δ	IV
B_{SS}	Pole dependent latitude of subsolar point	(9)	Obliquity γ Solar aspect E	V III,V
A	Aspect	(10)		V
γ	Astrocentric obliquity	(11)		V
δ	Time shifts	(1)[a](12)	Sidereal period	I,IV
E	Solar aspect	(13)		III
n	Cycle corrections	(1)[a](14)	Sidereal period	I,IV
E_c	Time of calculated epoch	(15)	Epoch 0-C	V

[a]Eq. (1) in Sec. 1.

beginning and end of interval ΔT. The L formulas are in Eq. (6) in the Appendix.

$\Delta\delta/360$ is one half the fractional part of a cycle that a body would have to rotate in order to move the maximum cross-sectional area from facing the earth to facing the sun. The time shift (δ) formulas are in Eq. (12) in the Appendix. δ agrees well with observed time shifts of epoch arrivals in the model studies of elongated bodies by Dunlap (1972). It is not known, however, how meaningful δ is for near-spherical objects. Another observation is that time shifts do not appear to be a significant correction term when applied to observations at less than 20° phase.

Δn is an adjustment in the number of cycles, one for each orbital rotation. Each time the asteroid completes an orbit about the sun one additional rotational cycle must be added (or subtracted for retrograde spin) to the synodic cycles in order to determine the sidereal cycles. These cycle corrections are a function of the pole orientation and the inclination of the

TABLE II

An Example of Finding Initial Cycle Corrections
For an Orbital Period of 4.5 yr

	Epoch[a]	L_{SE} (deg)	n_i
1	1960.0	52	0
2	1963.0	320	0
3	1964.0	20	1
4	1968.0	190	1
5	1969.7	63	2
6	1973.2	40	3

[a]Epochs 2 and 3 are in the same orbital period; however, the epoch 3 L_{SE} exceeded $360°$, hence $n_i=1$.

orbit. The appendix of Taylor (1973) explains how to calculate corrections; however, the routine has been generalized and the present discussion is to clarify the process of finding the proper correction integers n.

First, one lightcurve epoch per opposition is selected and a derivation is made of the orbital period and of L_{SE} for each epoch at pole orientation $\lambda_p = 0°$ and $\beta_p = 90°$ (therefore $L_{SE} = \lambda_{SE}$). Each epoch has an *initial* cycle correction n_i which is found as follows. The first epoch is assigned $n_i = 0$ (Table II has an example). All succeeding epochs whose L_{SE} are between the value L_{SE} of the first epoch and $360°$ and within one orbital period of the first epoch also have $n_i = 0$. The initial cycle correction changes to $n_i = 1$ for later epochs whose L_{SE} is beyond that of the original $360°$ and time is still within one orbital period of the first epoch. n_i remains 1 until L_{SE} is again $360°$. n_i becomes 2 as L_{SE} again passes over $360°$ during the second orbital period, etc.

These initial cycle corrections n_i only apply for a pole orientation at $\lambda_p = 0°$ and $\beta_p = 90°$. In the Appendix, Eq. (14) lists the relations that define the cycle correction n for any pole orientation. The Δn values are used in Eq. (1) of Sec. I.

V. CHECKS ON THE POLE SOLUTION

Three checks are made on the final solution. First, if two lightcurves are at the same orientation parameters (solar phase, aspect and astrocentric obliquity), one would expect their superpositioning to be nearly perfect. The words "the same orientation parameters" mean $\Delta\alpha \leqslant 10°$, $\Delta A \leqslant 10°$ and $\Delta\gamma \leqslant 10°$.

A second check is to plot lightcurve amplitudes as a function of aspect to see if a direct relationship exists. Dunlap (1971) demonstrated that there is no unique amplitude-aspect relation even for simple models and that the relation is dependent on the orientation parameters, each of which affects the shape and amplitude of the lightcurve. However, the plot should appear reasonably smooth. For simple models the greatest amplitudes should occur near equatorial aspects.

A third check on the final solution is to calculate epochs of maximum light based upon the derived pole and sidereal period and compare them to the observed epochs. With 4 Vesta, for example, the mean residual of the observed and computed epochs is 9 min; appendix Eq. (15) gives the formula for calculating a second epoch E_c from an observed epoch E_o.

If a lightcurve is inverted in relation to the other lightcurves; that is, when maxima become minima and vice versa (as will be explained in Sec. VII), then the residual will be either 1/4 or 3/4 of the period. Such inverted lightcurves are not included in the residual analysis.

VI. LIMITATIONS AND ANOMALIES

It is of critical importance that there be a unique mean synodic period (as defined in Sec. II). If epochs of lightcurve maxima, one per opposition, do not give such a consistent period then it is assumed that one or more of the lightcurves may be inverted (upside-down) or switched (a primary maximum becoming a secondary maximum) relative to the other lightcurves. These problems are further discussed in the next section.

Photometric astrometry cannot yet determine poles of asteroids that are lower in latitude than the subearth points of the observations. We are experiencing this problem with 1685 Toro and 1566 Icarus which have highly inclined orbits. If the pole is lower than the subearth point, then L_{SE} does not sweep out 360° each orbital period and this causes problems with ΔL values of Eq. (1) which we have not yet resolved.

There is a disturbing example where a consistent pole solution is shown to be wrong after acquisition of additional data. For instance, in order to test the validity of the pole and model for 39 Laetitia (Sather 1976), Tedesco (personal communication) obtained a lightcurve of this asteroid in October 1978. According to Sather's pole the 1978 lightcurve has the same orientation parameters as the 1972 lightcurves (Sather 1976), but the lightcurves do not look the same. Sather's pole must be wrong; it was arrived at by a model-dependent magnitude-phase relation and not by photometric astrometry as he was unable to determine cycles over long intervals. We now believe this is because Laetitia has one or more lightcurve inversions and we are currently involved in this study.

One would hope that each newly acquired lightcurve would not mean a new pole solution, or photometric astrometry would not give the true pole orientation of an asteroid and is at best a first estimate. We would prefer to think that each new lightcurve would give a refinement of the pole, and that there is merit in acquiring at least one high-quality lightcurve per opposition.

VII. FUTURE WORK

The most pressing problem in photometric astrometry is twofold: the determination of a unique mean synodic period and the possible existence of lightcurve inversions and switches (defined in Sec. VI). The problems appear to be related. Fortunately the problems have not occurred with all the asteroids we have analyzed. However, in our reductions of 44 Nysa and 39 Laetitia there appears to be a mean synodic period that requires at least one lightcurve to be inverted with respect to the others and/or one or more primary maximum to be switched to secondary maximum.

TABLE III

Published Pole Orientations

Asteroid		λ_p, β_p		Reference	Comment
2	Pallas	228,43		Schroll *et al.* (1976)	
4	Vesta	139,47	333,39	Taylor (1973)	if double period
		151,49	350,40	Taylor (unpublished)	if single period
5	Astraea	148, 9		Taylor (1978)	
6	Hebe	5,50		Gehrels and Taylor (1977)	
7	Iris	11,41		Taylor (1977)	
8	Flora	157,10		Gehrels and Owings (1962)	
9	Metis	156,15		Gehrels and Owings (1962)	
22	Kalliope	215,45		Scaltriti *et al.* (1978)	
39	Laetitia	121,37		Sather (1976)	see Sec. VI
44	Nysa	105,30		Gehrels and Owings (1962)	
433	Eros	16,12		Dunlap (1976)	
		17,10		Scaltriti and Zappala (1976)	
		15, 9		Millis *et al.* (1976)	
511	Davida	122,10		Gehrels and Owings (1962)	
624	Hektor	324,10		Dunlap and Gehrels (1969)	
1566	Icarus	49, 0		Gehrels *et al.* (1970)	see Sec. VI
1580	Betulia	140,20		Tedesco *et al.* (1978)	
1620	Geographos	200,60		Dunlap (1974)	
1685	Toro	200,55		Dunlap *et al.* (1973)	see Sec. VI

How do we know which lightcurves should be inverted and/or switched (each combination resulting in a different pole solution)? What is causing the inversions and/or switches? In answer to the first question, we are attempting to develop an independent method of arriving at the mean synodic period, a method which will indicate which lightcurves are truly inverted and/or switched. The second question requires two answers. The switches may simply be caused by the fact that we observe different hemispheres of the asteroid. The inversions are harder to explain. The asteroid may not be a single body system! We are investigating switches, inversions, and the mean synodic period, with the possibility of a laboratory model program to assist us in understanding these phenomena.

Table III is a listing of pole solutions that have been published since 1970. Poles published prior to 1970 are in Tables I and II of Vesely (1971). Some solutions in Table III were obtained with the magnitude-phase relation method of Sather (1976), others with earlier versions of photometric astrometry and the majority with some type of relation based on amplitude versus either ecliptic longitude or aspect. In addition, there are enough data available to find the pole orientation of perhaps 15 to 20 asteroids by photometric astrometry and we plan to work on that task.

Acknowledgments. I am grateful to T. Gehrels for his encouragement to proceed with photometric astrometry and to E. Tedesco, M. Wijesinghe and C. Vesely for their helpful suggestions.

APPENDIX

Formulas used in photometric astrometry (see Sec. III and Table I).

$$\lambda_{SE} = \lambda_a \pm 180 \text{ such that } 0 \leqslant \lambda_{SE} \leqslant 360. \tag{2}$$

$$\beta_{SE} = -\beta_a. \tag{3}$$

$$\lambda_{SS} = \lambda_s \pm \text{Arccos} \frac{R - \rho\cos\beta_a\cos(\lambda_s - \lambda_a)}{\sqrt{\rho^2\cos^2\beta_a + R^2 - 2\rho R\cos\beta_a\cos(\lambda_s - \lambda_a)}} \tag{4}$$

with the $-$ used only when $0 < \lambda_a - \lambda_s < 180$ or $\lambda_a - \lambda_s < -180$ and the $+$ only when $-180 < \lambda_a - \lambda_s < 0$ or $180 < \lambda_a - \lambda_s$.

$$\beta_{SS} = -\text{Arctan} \frac{\rho\sin\beta_a}{\sqrt{\rho^2\cos\beta_a + R^2 - 2\rho R\cos\beta_a\cos(\lambda_s - \lambda_a)}} \tag{5}$$

or $\beta_{SS} = -\text{Arcsin} \left(\frac{\rho\sin\beta_a}{r} \right)$ if r is given.

The quadrant of the longitude of the subearth point L_{SE} is   (6)
determined by adjusting the initial longitude value L'_{SE} according
to the following relations:

$$L_{SE} = 180 - |L'_{SE}| \text{ if } \sin L'_{SE} \geqslant 0 \text{ and } Q > 0$$
$$L_{SE} = |L'_{SE}| \qquad \text{if } \sin L'_{SE} \geqslant 0 \text{ and } Q < 0$$
$$L_{SE} = 180 + |L'_{SE}| \text{ if } \sin L'_{SE} \leqslant 0 \text{ and } Q < 0$$
$$L_{SE} = 360 - |L'_{SE}| \text{ if } \sin L'_{SE} \leqslant 0 \text{ and } Q < 0$$

$$Q = \frac{\sin\beta_{SE} - \sin\beta_p \sin B_{SE}}{\cos\beta_p \, \cos B_{SE}}$$

$$L'_{SE} = \text{Arcsin} \left(\frac{\cos\beta_{SE} \sin(\lambda_{SE} - \lambda_p)}{\cos B_{SE}} \right) \quad Q < L'_{SE} \leqslant 90.$$

$$B_{SE} = \text{Arcsin} \left[\sin\beta_{SE} \sin\beta_p + \cos\beta_{SE} \cos\beta_p \cos(\lambda_{SE} - \lambda_p) \right] \quad (7)$$

The quadrant of the longitude of the subsolar point L_{SS} is   (8)
determined by adjusting the initial longitude value L'_{SS} according
to the following relations:

$$L_{SS} = 180 - |L'_{SS}| \text{ if } \sin L'_{SS} \geqslant 0 \text{ and } R > 0$$
$$L_{SS} = \qquad |L'_{SS}| \text{ if } \sin L'_{SS} \geqslant 0 \text{ and } R < 0$$
$$L_{SS} = 180 + |L'_{SS}| \text{ if } \sin L'_{SS} \leqslant 0 \text{ and } R > 0$$
$$L_{SS} = 360 - |L'_{SS}| \text{ if } \sin L'_{SS} \leqslant 0 \text{ and } R < 0$$

$$R = \frac{\sin\beta_{SS} - \sin\beta_p \sin B_{SS}}{\cos\beta_p \cos B_{SS}}$$

$$L'_{SS} = \text{Arcsin} \left(\frac{\cos\beta_{SS} \sin(\lambda_{SS} - \lambda_p)}{\cos B_{SS}} \right) \quad 0 < L'_{SS} \leqslant 90.$$

$$B_{SS} = \text{Arcsin} \left(\sin\beta_{SS} \sin\beta_p + \cos\beta_{SS} \cos\beta_p \cos(\lambda_{ss} - \lambda_p) \right) \quad (9)$$

$$A = 90 - B_{SE} . \quad (10)$$

$$\gamma = \text{Arccos} \left(\frac{\sin B_{SS} - \sin B_{SE} \cos\alpha}{\cos B_{SE} \sin\alpha} \right) \quad (11)$$

$$\delta = \pm \text{Arctan} \sqrt{\frac{\sin(S-A)\sin(S-E)}{\sin(S-\alpha)\sin S}} \quad \text{with } S = \frac{1}{2}(\alpha + A + E) \tag{12}$$

when δ is $+$ if $0 < L_{SS} - L_{SE} < 180$ or $-360 < L_{SS} - L_{SE} < -180$

δ is $-$ if $180 < L_{SS} - L_{SE} < 360$ or $-180 < L_{SS} - L_{SE} < 0$

δ is also half the difference between L_{SS} and L_{SE}.

$$E = 90 - B_{SS}. \tag{13}$$

If $\lambda_{SE} < 180$ and $\beta_{SE} < 0$ (14)

then if $\qquad 0 \leqslant \lambda_p \leqslant \lambda_{SE}$ and $\qquad 0 \leqslant \beta_p \leqslant 90$ then $n_i = n$

if $\qquad \lambda_{SE} < \lambda_p \leqslant \lambda_{SE} + 180$ and $\quad 0 \leqslant \beta_p \leqslant 90 \qquad n_i - 1 = n$

if $\quad \lambda_{SE} + 180 < \lambda_p < 360$ and $\quad 0 \leqslant \beta_p \leqslant |\beta_{SE}| \qquad n_i = n$

if $\quad \lambda_{SE} + 180 < \lambda_p < 360$ and $\quad |\beta_{SE}| < \beta_p \leqslant 90 \qquad n_i - 1 = n.$

If $\lambda_{SE} > 180$ and $\beta_{SE} < 0$

then if $\qquad 0 \leqslant \lambda_p \leqslant \lambda_{SE} - 180$ and $\quad 0 \leqslant \beta_p \leqslant 90$ then $n_i = n$

if $\quad \lambda_{SE} - 180 < \lambda_p \leqslant \lambda_{SE}$ and $\quad 0 \leqslant \beta_p \leqslant |\beta_{SE}| \qquad n_i + 1 = n$

if $\quad \lambda_{SE} - 180 < \lambda_p \leqslant \lambda_{SE}$ and $\quad |\beta_{SE}| < \beta_p \leqslant 90 \qquad n_i = n$

if $\qquad \lambda_{SE} < \lambda_p < 360$ and $\quad 0 \leqslant \beta_p \leqslant |\beta_{SE}| \qquad n_i = n$

if $\qquad \lambda_{SE} < \lambda_p < 360$ and $\quad |\beta_{SE}| < \beta_p \leqslant 90 \qquad n_i - 1 = n.$

If $\beta_{SE} > 0$

then if $\qquad 0 \leqslant \lambda_p \leqslant \lambda_{SE}$ and $\qquad 0 \leqslant \beta_p \leqslant 90$ then $n_i = n$

if $\qquad \lambda_{SE} < \lambda_p < 360$ and $\qquad 0 \leqslant \beta_p \leqslant \beta_{SE} \qquad n_i = n$

if $\qquad \lambda_{SE} < \lambda_p < 360$ and $\qquad \beta_{SE} < \beta_p \leqslant 90 \qquad n_i - 1 = n.$

$$E_o + P(I \pm K) = E_c \tag{15}$$

where E_o is the time of an observed epoch,

P is the adopted sidereal period,

I is an integral number of sidereal cycles found by taking the $\Delta T_{(c)}$ of Eq. (1), dividing by P, and rounding up if retrograde and down if prograde rotation.

$\pm$ means $+$ if prograde and $-$ if retrograde rotation,

$$K \text{ is } \left[(L_2 + \delta_2) - (L_1 - \delta_1)\right] \div 360; \text{ add } 360 \text{ to } (L_2 + \delta_2) \text{ if it}$$

is less than $(L_1 + \delta_1)$. Eqs. (6) and (12) explain L and δ.

E_c is the calculated arrival time of an epoch.

A more accurate value of E_c is found by taking the mean of several trials of E_0 from different oppositions.

REFERENCES

Dunlap, J.L. 1971. Laboratory work on the shapes of asteroids. In *Physical Studies of Minor Planets*, ed. T. Gehrels (NASA SP-267, Washington, D.C.: U.S. Government Printing Office), pp. 147-154.

Dunlap, J.L. 1972. Laboratory work on the shapes of asteroids. M.S. Thesis, University of Arizona.

Dunlap, J.L. 1974. Minor planets and related objects. XV. Asteroid (1620) Geographos. *Astron. J.* 79: 324-332.

Dunlap, J.L. 1976. Lightcurves and the axis of rotation of 433 Eros. *Icarus* 28: 69-78.

Dunlap, J.L., and Gehrels, T. 1969. Lightcurves of a Trojan asteroid. *Astron. J.* 74: 796-803.

Dunlap, J.L.; Gehrels, T.; and Howes, M.L. 1973. Minor planets and related objects. IX. Photometry and polarimetry of (1685) Toro. *Astron. J.* 78: 491-501.

Gehrels, T. 1967. Minor planets. I. The rotation of Vesta. *Astron. J.* 72: 929-938.

Gehrels, T. 1970. Photometry of asteroids. In *Surfaces and Interiors of Planets and Satellites*, ed. A. Dollfus (London: Academic Press), pp. 319-376.

Gehrels, T., and Owings, D. 1962. Photometric studies of Asteroids. IX. Additional lightcurves. *Astrophys. J.* 135: 906-924.

Gehrels, T.; Roemer, E.; Taylor, R.C.; and Zellner, B.H. 1970. Asteroid (1566) Icarus. *Astron. J.* 75: 186-195.

Gehrels, T., and Taylor, R.C. 1977. Minor planets and related objects. XXII. Phase functions for (6) Hebe. *Astron. J.* 82: 229-237.

Groeneveld, I., and Kuiper, G.P. 1954. Photometric studies of asteroids. I. *Ap. J.* 120: 200-220.

Hansen, O. 1977. On the prograde rotation of asteroids. *Icarus* 32: 458-460.

Matson, D.L. 1971. Infrared observations of asteroids. In *Physical Studies of Minor Planets*, ed. T. Gehrels (NASA SP-267, Washington, D.C.: U.S. Government Printing Office), pp. 45-50.

Millis, R.L.; Bowell, E.; and Thompson, D.T. 1976. UBV photometry of asteroid 433 Eros. *Icarus* 28: 53-68.

Morrison, D. 1977. Asteroid sizes and albedos. *Icarus* 31: 185-220.

Russell, H.N. 1906. On the light variations of asteroids and satellites. *Ap. J.* 24: 1-18.

Sather, R.E. 1976. Minor planets and related objects. XIX. Shape and pole orientation of (39) Laetitia. *Astron. J.* 81: 61-73.

Scaltriti, F., and Zappala, V. 1976. Photometric lightcurves and pole determination of 433 Eros. *Icarus* 28: 29-36.

Scaltriti, F.; Zappala, V.; and Stanzel, R. 1978. Lightcurves, phase function and pole of the asteroid 22 Kalliope. *Icarus* 34: 93-98.

Schroll, A.; Haupt, H.F.; and Maitzen, H.M. 1976. Rotation and photometric characteristics of Pallas. *Icarus* 27: 147-156.

Taylor, R.C. 1973. Minor planets and related objects. XIV. Asteroid (4) Vesta. *Astron. J.* 78: 1131-1139.

Taylor, R. C. 1977. Minor planets and related objects. XXIII. Photometry of asteroid (7) Iris. *Astron. J.* 82: 441-444.

Taylor, R. C. 1978. Minor planets and related objects. XXIV. Photometric observations for (5) Astraea. *Astron. J.* 83: 201-204.

Tedesco, E.; Drummond, J.; Candy, M.; Birch. P.; Nikoloff. I.; and Zellner, B. 1978. 1580 Betulia: An unusual asteroid with an extraordinary lightcurve. *Icarus* 35: 340-355.

Vesely, C.D. 1971. Summary on orientations of rotation axes. In *Physical Studies of Minor Planets,* ed. T. Gehrels (NASA SP-267, Washington, D.C.: U.S. Government Printing Office), pp. 133-140.

ASTEROID LIGHTCURVES:
RESULTS FOR ROTATIONS AND SHAPES

JOSEPH A. BURNS
Cornell University

and

EDWARD F. TEDESCO
University of Arizona

Some processes fundamental to the development of asteroid rotations are considered, using physical arguments. The processes include: (a) collisions which change spin frequency and direction; (b) fragmentation, which determines the efficacy of collisional spin-up; (c) presence of a limiting spin rate, at least for objects without strength; and (d) damping by internal dissipation of any wobble induced through collisions. Experimental and theoretical studies of asteroid shapes are summarized. A discussion concerning the maximum elevation difference, which can be sustained over the solar system's age by a viscoelastic body is presented.

There is no statistically significant size dependence for rotation period within either the S or C taxonomic type. There is a change in asteroid characteristics near 175 km; objects with $D < 175$ km have $P \approx 11$ hr while larger objects have $P \approx 8$ hr. The observed asteroids do not spin at rates that exceed, or even closely approach, the rotational break-up frequency for fluid bodies. The spins are shown in three independent ways to point in random directions, suggesting that collisions are important. Internal energy dissipation apparently is efficient enough to damp any wobble due to off-axis collisions.

Fluctuations observed in the brightnesses of asteroids provide important, otherwise unobtainable, information on the minor planets. To illustrate this for an ideal case, imagine that the brightness of an asteroid could be continuously monitored over the course of many years. Then variations of several kinds would be seen. First, fluctuations about the mean brightness would be observed as areas with differing reflectivity or unusual elevation (whether they be craters or bumps) appear and disappear over the limb of the spinning body. These usually minor changes are superposed on normally larger variations, having many-hour periods, which are due to gross changes in projected cross-sectional area or, much less often, in surface properties. From these longer-period variations we find the asteroid's rotation rate as well as, in principle, some measure of its varying surface properties. In addition, if the asteroid is considered to be a triaxial ellipsoid, an estimate of the object's shape can be obtained from the total amplitude of the lightcurve. It is possible that other variations, having periods like that of the spin, but due instead to the free precession, could be observed; however, so far these have never been seen, implying that internal damping must be present.

If we were to watch our asteroid for a yet longer time, say on the order of a year or more, we would notice a modulation in the shape of its lightcurve as its aspect (i.e., our line of sight relative to its rotational pole position) changed. Ideally from these variations it should be possible to deduce the precise shape of the asteroid. As an added benefit, the direction of the rotation axis could also be ascertained (see the chapter by Taylor). Waiting even longer, i.e., thousands of years, an alteration in the rotation pole would be detected. This could result from the body's forced precession because of the solar torque, from collisions with other asteroids, or from the damping of any nutation; if discernible, these would, respectively, provide information on the asteroid's internal density structure, on the prevalence of collisions and on the object's anelasticity.

Not surprisingly, with such a prospective wealth of information available, there has been an accelerating growth in the investigation of asteroid lightcurves. The systematic early work carried out by Kuiper and his students resulted in the first tabulation of asteroid rotation rates by Alfvén (1964). A much lengthier compilation (Gehrels 1970) contained in an article describing photometric techniques and results for asteroids followed. McAdoo and Burns (1973) then attempted to interpret Gehrels' data that had been extended by Taylor (1971). As described below, many other tabulations and interpretations came after these, culminating in the papers by Harris and Burns (1979) and Tedesco and Zappalà (1979). The latter report and this chapter are based on the information of the TRIAD file (The Tucson Revised Index of Asteroid Data; see Tedesco in Part VII of this book). Since TRIAD is to be kept current, it will be the prime resource for future studies on asteroid rotation rates and shapes.

This chapter is organized in the following way. First, we outline the pertinent dynamical processes and then describe in greater depth the possible causes of brightness variations. Next a short discussion is presented on how one retrieves the rotation period and shape given a particular set of data; theoretical and experimental studies on shapes are also mentioned. A summary of past studies is provided with particular emphasis laid on the important issue of selection effects. We then plot the results (spin frequency versus size, lightcurve amplitude versus size, and spin frequency versus lightcurve amplitude), including correlations with taxonomic type, and consider their implications. Throughout we suggest possible directions for future studies and mention some speculative topics.

I. DYNAMICS

Celestial bodies spin as a consequence of several processes, whose relative importance may be ascertained only by investigating the rotations of many objects. Asteroids are particularly suited for such a study since a much larger data set is available for them and since this information extends over objects whose orbits are quite varied and whose masses differ by nearly nine orders of magnitude.

To start with, celestial objects spin owing to the manner in which they accumulated. This happens since in general the incoming bodies strike off-center and thus bring in angular momentum H. Not all accumulated bodies deliver positive H but, given appropriate circumstances, the spin rate will build in the mean, due to the random walk nature of collisions. Many workers (e.g. Guili 1968a; Harris 1977) believe that the relatively fast and prograde rotations of planets result from the overwhelming influence of colliding bodies which were on nearly circular orbits; this mechanism might be less effective for the asteroids since their zone may have been fairly disordered in early times. Others (Safronov, personal communication, 1979) maintain that the prograde spin of most planets is a natural outcome of the accumulation process for any object in orbit about another primary; in this view, all asteroids would preferentially spin prograde, regardless of the primordial orbits in their region. Only observations of the spin directions of these objects will tell which model is correct.

In the case of the planets, following the epoch of their growth, collisions little affect their initial rotations because the planets rarely, if at all, undergo significant impacts; they are, and always have been, by far the most massive objects in their own zones and their orbits do not often intersect others. The same is not true for the minor planets. Because of the irregularity of asteroid orbits, as represented by the moderately large mean inclination and eccentricity they have today, and because of the large number of comparable-sized asteroids, any particular minor planet occasionally meets another object having a mass greater than say 10^{-3} of its own. For example, a 100 km asteroid suffers such an impact every 10^8 yr or so (Chapman and Davis 1975; see the chapter by Davis *et al.* in this book); of course, owing to the asteroid size distribution, smaller minor planets undergo even more frequent collisions of this relative importance. From simple arguments (see below, and Burns 1971), such collisions have the capacity to cause an appreciable modification of the asteroid's spin. Whether they do modify spin rates, or not, depends on whether the angular momentum they carry is transferred; some models (chapter by Davis *et al.*) would suggest that only the largest objects, $D > 200$ km (which are gravitationally bound), and the smallest, $D \sim 1$ km (which receive little angular momentum transfer) can be appreciably affected by single collisions in today's solar system. On the other hand, if collisions do transfer angular momentum, the primordial spins of asteroids, for at least all but the largest, may no longer be seen, being obscured by the effects of subsequent impacts. Large impacts, owing to the random walk nature of the accumulated angular momentum, cause rotation rates in the mean to increase. Counteracting this is a drag term (Harris 1979*a*) that can systematically slow the rotation; it arises mainly due to collisions with very small particles. Such particles add no net angular momentum to the asteroid but, instead, will share the asteroid's angular momentum, thereby decreasing the spin rate.

Besides collisions, the only other force that has been seriously suggested for changing the spin state of asteroids is radiation pressure. Icke (1973) proposes that, since the absorption coefficient can be a function of surface temperature, the sunlight striking the afternoon side of a rotating object can produce a torque different from that impinging on the morning side. Icke demonstrates that for a particular functional form this "ponderomotive" force can drive asteroids to a stable rotation rate, independent of size, in which state the radiation torque is zero. He associates this stable rate with the fact that most asteroids have spin periods within a factor or two of ~ 8 hr (Burns 1975). Nevertheless we do not give too much weight to this process because, (a) it has a questionable physical basis; (b) it only marginally satisfies the data, requiring several billion years to drive asteroids to the stable state even if no other disrupting process, like collisions, interferes; and (c) it makes two predictions that are not borne out by the vastly larger data set now available, namely that rotation rate should be proportional to the semimajor axis of the asteroid, and that a class of very fast rotators should exist.

Öpik (1958, 1977) instead claims that the crudely constant spin rate of asteroids must mean that shielding effects are operating because otherwise magnetic damping would cause small asteroids, if they possessed iron cores, to slow down.

The fact that primordial collisions produce prograde rotations for the planets has been illustrated in numerical experiments by Giuli (1978*a,b*) and Kiladze (1971), among others. Analytic theories modeling the angular momentum delivered by collisions during the planetary accumulation process have been attempted by Safronov (1971, 1972; the second reference contains summaries of other Soviet work notably by Artem'ev and Kiladze), Kiladze (1977) and Harris (1977, 1978). These models usually fall short by an order of magnitude in explaining planetary rotation rates but they do have the correct sign. When applied to the accretion of the asteroids, the models predict even slower rates.

Given that the asteroids also had some primordial rotation produced during their accumulation, we can easily demonstrate that subsequent collisions can alter it significantly. A crude measure of the relative angular momentum available for transfer to a target mass M in a collision with a "bullet" of mass m having relative velocity v_c is $(m/M)(v_c/v_s)$, where v_s is the velocity of M's surface. For typical asteroid parameters ($v_c \sim 5$ km sec^{-1}, rotation period $P \sim 8$ hr) the velocity ratio in this expression is $10^6 r^{-1}$, where r is the target's radius in km. Thus even relatively small bullets are capable of delivering large quantities of angular momentum, and thereby of appreciably modifying the spin of an asteroid. This model ignores the angular momentum added to, or lost by, the target due to the debris ejected at the crater site (cf. Burns and Safronov 1973; Hawkes and Jones 1978; Harris 1979a).

This simple calculation also illustrates one of the limitations of the collisional model. It is immediately obvious that as the colliding mass ratio (m/M) grows, impacts become increasingly important since they can transfer enormous quantities of angular momentum. However, beyond that qualitative statement, little can be said about the rotational consequences of major collisions since the complication of fragmentation enters. Experiments demonstrate that at hypervelocities the colliding body becomes more and more capable of disintegrating the target once $m/M \gtrsim 10^{-3}$; a better attempt to characterize breakup is available in the Davis $et\ al.$ chapter. Once disruption occurs, no adequate model is available to specify the resulting rotations of the individual fragments following the event.

Theoretical models for the evolution of the mean rotation rate of bodies in a collisionally evolving system have been developed by Napier and Dodd (1974), Dohnanyi (1976) and Harris (1979a; see also the discussion in the chapter by Davis $et\ al.$). Harris (1979a) proposes that the spin rate ω changes as a function of r according to the finite difference relation

$$\frac{\Delta\omega}{\Delta r} = -\frac{(5\zeta/4)^2}{(3-q)\alpha}\left(\frac{m_1}{m}\right)^2 \frac{v^2}{r^3\omega} + \frac{5}{3(2-q)\alpha}\left(\frac{m_1}{m}\right)\frac{\omega}{r} \qquad (1)$$

in which ζ is the ratio of the angular momentum carried off by the debris to that brought in by the impacting body, $q \approx 11/6$ is the exponent in the asteroid differential number density, m_1 is the largest mass to have collided with m (in other words, the equation is valid only until fragmentation occurs) and α is the average fractional decrease in asteroid radius per disruption. The first term on the right represents the fluctuation of the asteroid's angular momentum from the expected zero value; this is due to the random walk manner in which angular momentum is delivered during collisions. The second term represents the fact that the collisional flux contributes no additional angular momentum (at least on the average) but does bring in mass, which must share in the original object's angular momentum. This sharing is easily understood for totally inelastic collisions and, due to the spherically symmetric character of most ejecta patterns, sharing also happens when material is ejected.

An approximate solution to Eq. (1), valid for large asteroids where gravity dominates over strength, has been found. Above some radius r_1 it is constant and equal to

$$\omega_0 = \frac{9}{8} \omega_s \zeta \gamma \left(\frac{2-q}{3-q}\right)^{1/2} \tag{2}$$

where $\omega_s = (4\pi\rho G/3)^{1/2}$ is the surface orbit frequency about the asteroid, ρ the density, and γ is the ratio of the disruptive energy to the total gravitational binding energy. At sizes smaller than r_1, $\omega \sim r^{-1}$. The radius at which the transition from the constant ω_0 occurs is

$$r_1 = \left(\frac{5}{3}\right) \left[\frac{2}{(2-q)\alpha\gamma}\right]^{1/2} \frac{\sigma}{\rho v \omega_s} \tag{3}$$

with σ a yield stress. For strong materials $r_1 \approx 2$ km. If Harris' model is correct, we should expect to find a nearly constant spin rate as a function of size for large objects of particular interior properties. Given all other quantities to be equal, the spin rate should be proportional to $\rho^{1/2}$, according to Eq. (2). This dependence comes about because denser objects, with their stronger gravity fields, can sustain larger impacts before breakup. The turn-up radius to faster rotation speeds should identify asteroid strengths.

It is the second term, a drag, in Eq. (1) which permits the rotations to remain constant for large asteroids. This result obviates the need of introducing structurally weak asteroids (Napier and Dodd 1974; Burns 1975; Dohnanyi 1976; Degewij and Gehrels 1976), magnetic damping (Öpik 1958, 1977), or "ponderomotive" braking (Icke 1973) as explanations for the relatively narrow spread in observed rotation periods.

While it is a major step forward in understanding the development of asteroid rotations, the Harris theory (1979a) still has limitations. First, of necessity, it is a statistical theory, treating all collisions alike. This means that, whether the projectile hits the asteroids full-face or strikes a glancing blow, all collisions are handled the same insofar as their propensity for transferring angular momentum or for causing fragmentation is concerned; this may be a crucial simplification. Furthermore, for lack of any better assumption, the spin of objects following a catastrophic breakup is taken to be identical to that before fragmentation. A perhaps more important objection to the theory is that the asteroids may not have yet reached equilibrium (chapter by Davis et al.) so that the theory may not be applicable at all. This contention itself can be challenged since the test of Davis *et al.* for equilibrium requires the scaling of collision data over many orders of magnitude as well as the selection of Harris' γ parameter, which, as Davis *et al.*'s Fig. 1 shows, is quite sensitive to the model chosen.

There are two other consequences implied by any collisional model. First, since the random component of angular momentum delivered through collisions is comparable to, or larger than, the original value, at least for objects with $D < 200$ km, the rotational poles – which in essence point in the direction of **H** – should be randomly oriented in space. (The data available on rotational poles confirm this; see the chapter by Taylor.) Second, the rotational rates should display a Maxwellian scatter about some mean rate, as demonstrated to be true in the histograms of Harris and Burns (1979).

If post-accumulation collisions are indeed a major agent in producing asteroid rotations, then some explanation must be given for why asteroids are found in pure spin, i.e., why the **H**, and ω vectors and one of the body's principal axes of inertia are all aligned. This alignment of rotation is implicit in the fact that asteroid lightcurves are not observed to exibit beat phenomena; this absence, which is only known in the grossest sense because serious investigations for precession have not been made, means that in general the minor planets are *not* freely precessing objects. The manner in which a hypothetical wobbling (or tumbling) would modify the lightcurves is demonstrated by the computer diagrams of Sher (1971) for several specific examples. Without the contradiction of the observations, one might otherwise have expected these vectors to be misaligned because any collision, except one passing directly through the center of mass, will change the asteroid's **H** vector instantaneously without affecting the body's orientation. Thus, even if a body axis and **H** were aligned initially, after the collision they no longer would be. Hence the periodic nature of lightcurves implies either that collisions have not occurred or that a damping process is active (Prendergast 1958; Burns 1971). The only damping mechanism with sufficient strength is that due to the internal friction or imperfect elasticity of the material making up the asteroid (Prendergast 1958; Burns and Safronov 1973; McAdoo and Burns 1974). It is a property of all real materials that energy is lost if they undergo a cyclic stress-strain history. Since the minimum energy state for a fixed angular momentum has **H** lying along the axis of maximum momentum of inertia (Lamy and Burns 1972), internal energy loss causes alignment. This mechanism gives a characteristic time (Burns and Safronov 1973) for the alignment of the body axes with **H** of

$$\tau \sim \mu Q/(\rho k r^2 \omega^3) \tag{4}$$

where μ, Q and ρ are, respectively, the object's rigidity, anelasticity factor and mass density; k is a constant depending on body shape equal to about 10^{-1}–10^{-2}. Typical time scales are 10^5 yr for large objects ($r \sim 10^2$ km) and

10^7 yr for small ($r \sim 1$ km), rapidly-spinning, irregular asteroids.

A model incorporating the misaligning effect of collisions with the above damping has been developed by Burns and Safronov (1973) who find that the mean nutation angle $\bar{\beta}^2 \sim r^{-2.6} \omega^{-2.5}$. Typical values of $\bar{\beta}$ are negligibly small unless the size of the asteroid is less than a few km; however, by that size the model has failed because the original asteroid's fragmentation becomes inevitable. From the analytic form for $\bar{\beta}^2$, the candidate most likely to be seen in free wobble (i.e., in the undamped state) should be small, spinning slowly, strong (M taxonomic type?), and not especially irregular; on the other hand, significant nonsphericity is helpful in making the wobble apparent from the lightcurve (cf. Sher 1971).

Bounds on the interior models of asteroids would be available if this wobble could be detected because the wobble time scale (Burns 1971) for an axially symmetric body is

$$\tau_W = (P/\Delta)(C/A) \tag{5}$$

where P is the spin period and $\Delta = (C-A)/C$ is the relative difference in the moments of inertia C and A. Since the latter quantity is of the same order as the relative difference in the axes of the spheroid, it may be of order 1 or 10^{-1}; thus the wobble period is not much different from the spin period. For the more likely case when the body's inertia ellipsoid is triaxial (as it would be for a body of unspecified shape), a free nutation and precession will in general be present. The motion, and therefore the interpretation, in such a case would be considerably more complicated but if decipherable, then would define one more moment of inertia. Only for a single object (321 Florentina; see Gehrels 1970, p. 358) has precession been suspected as possibly accounting for day-to-day variations; however, the observations of this asteroid are of poor quality. Systematic searches (cf. Dunlap 1974) for precession have, however, been carried out on only a few sets of lightcurves.

There is one last variation of the rotation axis that should be noted. Asteroids are surely not spherical and hence they are subject to a gravitational torque due to the sun. The body's response to this torque is a forced precession of its H vector with a time scale (Burns 1971)

$$\tau_f = T^2/P\Delta \tag{6}$$

where T is the orbit period. A discovery of this precession would also bound interior models and, when coupled with a free wobble measurement, would give explicit values for the moments of inertia. This forced motion of the rotation axis has been invoked to account for apparent changes in the rotational pole position of 433 Eros (Chen et al. 1976); however, since the time scale of the precession is so long ($\sim 10^3 - 10^4$ yr), a more likely explanation is simply that Eros' rotational pole has been determined badly over the years (cf. Vesely 1971).

There is a limiting spin rate, independent of asteroid size, beyond which rate particles would leave the surface of a rotating sphere of mean density $\bar{\rho}$:

$$\omega_s > (4\pi\bar{\rho}G/3)^{\frac{1}{2}} \tag{7}$$

corresponding to a spin period of a few hours. Burns (1975) has extended this result and found that particles no longer would be bound gravitationally to the tips of a slightly ellipsoidal object when

$$\omega_s > (4\pi\bar{\rho}G/15)^{\frac{1}{2}} [11-6\log^{-1}(0.4\Delta m)]^{\frac{1}{2}} \tag{8}$$

where Δm is the observed brightness variation due to the ellipsoidal shape. Objects spinning faster than this would have portions of their interiors in tension; there are only a few possible candidates (e.g., 1566 Icarus if $\rho > 2.8$ g cm^{-3}, and 321 Florentina if $\rho > 2.28$ g cm^{-3}).

II. BRIGHTNESS VARIATIONS ASSOCIATED WITH ROTATION

It was not until a full century after the discovery of the first asteroid that variations (with periods on the order of hours) in the brightness of an asteroid were detected by Oppolzer (1901) for Eros. Immediately it was realized that such changes could be due to rotation and that they could be generated in several independent ways: (a) via eclipses and occultations of two orbiting bodies (see the chapter by Van Flandern *et al.*); (b) by surface albedo variegation; and (c) by irregular shape. And of course several of these may operate simultaneously.

The eclipse model, originally proposed by Andre (1901), permits one to estimate the shapes and relative albedos of the components from the nature of the lightcurve (cf. Wijesinghe and Tedesco 1979). It only requires that one or both components be elongated and/or have different albedos in cases where the lightcurve amplitude exceeds 0.75 mag. Most asteroid lightcurves however do not seem to be those of binary systems although there are possible exceptions (Tedesco 1979*a*) to this statement. At any rate, both because of the character of most lightcurves and because it is difficult to envision ways commonly to produce binary systems, the eclipse model has not been widely applied.

If an asteroid were spherical and albedo variations (spottedness) were the sole cause of its brightness variation, then its lightcurve would be simply periodic, having the period of the body's rotation. In fact in the extreme example of a body composed of two hemispheres of different albedos, such as is approximately true for Saturn's satellite Iapetus, the lightcurve would be a singly periodic sinusoid. Vesta's lightcurve is the only one believed to be singly periodic and so its low-amplitude lightcurves (0.08 − 0.14 mag) are apparently produced by albedo differences (Degewij and Zellner 1978; Gradie *et al.* 1978; Degewij *et al.* 1979; see the chapter by Dollfus and Zellner).

The lightcurve developed during the rotation of a smooth triaxial figure of revolution which lacks any albedo features would also be singly periodic because, as already described, the rotational motion would be a pure spin about the shortest axis (Burns 1971). In this case however the period of the lightcurve would be one half the rotation period.

For the more general situation where the body is aspherical and there is a difference in the physical properties (whether gross shape, albedo, texture, or something else) over the surface, a doubly periodic lightcurve results. In such curves the general features of alternate oscillations are often nearly the same, being produced by changes in the cross-section viewed, but the details usually differ since they are the result of local surface variations. The vast majority of observed asteroid lightcurves have this character and are, at least, doubly periodic, thus leading one to believe that they are produced by the rotation of such irregularly shaped, possibly somewhat spotted, bodies. However we must remember, along with Russell (1906), that in general one cannot distinguish between the rotation of a spotted sphere and an irregular shape of uniform albedo from observations of brightness changes alone. Nevertheless, the fact that asteroid surfaces, as inferred from polarization, color and spectrophotometric observations during a rotation, are remarkably uniform (Degewij *et al.* 1979; chapter by Bowell and Lumme in this book) immediately suggests that most brightness changes reflect changes in projected cross-section. Accepting this, it is possible to estimate the contribution due to albedo spots by attributing the slightly unequal brightness levels of the alternate maxima or minima of the doubly-periodic lightcurve to average albedo differences on opposite sides of the asteroid. Since these brightness differences seldom exceed a few hundredths of a magnitude, we conclude that albedo variations play only a minor role in producing major lightcurve features.

This conclusion that albedo variations over asteroid surfaces are small further implies that gross compositional differences are not seen from one region of the surface to the next. It is still not clear whether this is due to the fact that such differences do not exist anywhere in the asteroid or whether it is the consequence of obscuration by impact-generated debris which rapidly covers an originally nonuniform surface (Degewij *et al.* 1979; chapter by Housen *et al.*). However, studies of Hirayama family members (Gradie and Zellner 1977; Gradie 1978; chapter by Gradie *et al.*; Tedesco 1979*b*) suggest that the parent bodies of many of these objects were probably quite homogeneous, while the rest are usually not too heterogeneous. This again supports the idea that asteroid lightcurves are primarily due to shape. Thus throughout the remainder of this review we will consider that the major lightcurve features are produced by changes in the cross-section while smaller lightcurve structure may come from any number of causes (craters or bumps, spottedness, textural variations, etc.) yet to be revealed.

On longer time scales (months and years) brightness variations, due now to changes either in solar phase angle or in aspect angle, also exist. The two long-term effects can be separated, since during a typical apparition of a main-belt asteroid the phase angle varies from about $0°$ to $20°$ whereas in contrast the aspect is fairly constant. Since phase effects are considered in the chapter by Bowell and Lumme, we do not discuss them further except to note that in any case where a variation in the rotational amplitude occurs during an apparition, the amplitude is invariably greater at larger phase angles (Gehrels 1956; Gehrels and Taylor 1977; Zappalà et al. 1979, Tedesco et al. 1978; Tedesco and Bowell 1979). This is an apparent confirmation of Veverka's (1971) contention that for irregular asteroids the phase coefficient of the lightcurve minima will be greater than that of the maxima.

Changes in aspect most often produce larger long-term variations than do phase effects. Pole positions (see the chapter by Taylor) may be determined from the character of such variations after several aspects have been viewed, but the reliability of these positions remains questionable. Given only observations at one opposition, no information on aspect angle can be obtained except in the case of Earth-approaching asteroids where the aspect can change appreciably during a single opposition. Furthermore, since aspect is such an important parameter in defining a lightcurve, not knowing it complicates the interpretation of shape from lightcurve amplitudes measured during a single opposition. Moreover, the shape determined from lightcurves of one apparition will always underestimate the body's true departure from sphericity. Even when observations from several oppositions are available, shape determinations are not straightforward and in fact still tend to underestimate the body's elongation. Only with an accurate knowledge of the pole coordinates is there any hope of a good determination of shape (Dunlap 1971). Nevertheless, as discussed below, we, along with previous workers, still use Δm as an approximate indicator of shape.

From a knowledge of the pole coordinates one can determine the sidereal period and thereby, in principle at least, ascertain the sense of rotation (see Taylor's chapter). Radiometric observations (Morrison 1977; Hansen 1977; the chapter by Morrison and Lebofsky) have also been used to infer the sense of rotation for several of the largest asteroids, as first suggested by Matson (1971). Since knowing the sense of rotation for many large asteroids may help distinguish a primordial origin from a subsequent collisional origin for rotation, such studies are encouraged.

A discussion of various methods used in determining asteroid rotation periods from photoelectric lightcurve observations is given by Taylor (1971), while reduction techniques used in analyzing photographic lightcurves are presented by Lagerkvist (1977) and Degewij (1978). There is little ambiguity involved when ascertaining rotation periods provided the observations are of good quality (i.e., high signal-to-noise ratio) and cover an appreciable fraction of a rotation cycle. For this reason periods are more accurately known for asteroids which spin rapidly and have moderate-to-large ($\gtrsim$ 0.2 mag) rotational amplitudes; this is especially true of photographic determinations. For reasons given below large amplitude lightcurves generally appear smooth while small amplitude lightcurves frequently display numerous small-scale features. In almost all cases, however, the lightcurves show two distinct pairs of maxima and minima before repeating the lightcurve cycle. When the light-curve is more complicated than this we consider the peaks or valleys to be extreme only when the amplitude of one or more of these features becomes nearly as large as that of the lightcurve upon which it is superposed; one then speaks of a lightcurve having three or more pairs of extrema.

Lightcurve observers have generally been aware of, and accounted for, the presence of lightcurve features in determining rotation periods (cf. Taylor 1971). Most of the entries of high quality (i.e., QUAL $\geqslant$ 2 in Part VII, Table VI) should therefore be quite reliable. Periods assigned qualities of 1, however, are determined from a single lightcurve covering less than a complete cycle. The assumption that every lightcurve is doubly periodic was then used to estimate a rotation period. Nearly all rotation periods based on photographic observations were also determined under this assumption. Should there be three or more maxima per rotation cycle the period obtained would be shorter than the true period.

Since the vast majority of lightcurve observations are made at phase angles < 15° on asteroids > 50 km in diameter we feel reasonably certain that the double periodic assumption is valid in the overwhelming majority of cases. Nevertheless, one should be forewarned that such may not be the case for small, not very irregularly shaped, asteroids.

III. CONSIDERATIONS OF ASTEROID SHAPE

Even with the progress expected in observational astronomy following the implementation of the Space Telescope in the 1980's, for observers most asteroids will remain fuzzy dots of light, having little shape. Moreover, although speckle interferometry has reached the point of being able to define sizes for the very largest asteroids (Worden and Stein 1979, and Worden's chapter in this book), it too will not help provide information on shape for years, if at all. Hence, because we are not yet capable of actually imaging a minor planet, indications of asteroid shape must come from indirect means. At present, just one such method, based on the amplitudes of lightcurves, is used but it is crude enough that others may become important in the 1980's. These methods will be mentioned first before we consider more traditional lightcurve studies of shape.

Radar reflections (see the chapter by Pettengill and Jurgens in this book) can be returned from many minor planets, especially those that closely approach the earth, as demonstrated for Icarus (Goldstein 1969), Toro (Goldstein *et al.* 1973), and Eros (Jurgens and Goldstein 1976; Campbell *et al.* 1976). The Doppler spread of such returned signals is determined by the maximum width of the spinning object off its rotation axis; given a strong enough signal with frequency and time resolution, one would be able ideally to ascertain the asteroid's shape. More often one merely knows the Doppler spread in the echo spectra from which the object's radius can be determined *only* if the asteroid's rotational period and pole are independently available. Four Earth- and Mars-crossing asteroids (433 Eros, 1566 Icarus, 1580 Betulia and 1685 Toro) and 1 Ceres have been observed with radar (chapter by Pettengill and Jurgens) but the measurements are not good enough yet for the determination of shapes.

The accurate timing of a stellar occultation (see the chapter by Millis and Elliot) by an asteroid furnishes a chord length across that asteroid. If enough stations observe the occultation so that many chords are known, the contour of the asteroid can be estimated; in essence one is viewing the asteroid's silhouette projected onto Earth's surface by the occulted star's light. This technique was first applied to 433 Eros by O'Leary *et al.* (1976). Other occultations have been by 2 Pallas (Wasserman *et al.* 1979), 18 Melpomene (Dunham *et al.* 1979) and 532 Herculina (Bowell *et al.* 1978*b*). While this method is very valuable in calibrating other schemes of measuring asteroid sizes, it is of limited usefulness in defining asteroid shapes because too many chords need to be found and because stellar occultations by asteroids are too infrequently predicted (O'Leary 1972; chapter by Millis and Elliot). Lunar occultations of asteroids, while more often visible at a given observatory than a stellar occultation by an asteroid, are worth even less for determining asteroid figures since only the remaining reflecting area is measured during an asteroid's obscuration by the lunar limb.

Most estimates of asteroid shapes come from inverting lightcurves. The basis of the technique is easily understood by considering a triaxial ellipsoid (a,b,c) of uniform surface properties rotating normal to the line of sight at zero phase angle; a is the semi-axis of the long side, b that of the intermediate and c the short length. For such a body the projected area would vary from πac to πbc as the asteroid rotates; thus for a constant phase function, a brightness variation (Gehrels 1970)

$$\Delta m = 2.5 \log(a/b) \qquad (9)$$

would be produced. If the rotation axis is tilted relative to the line of sight, a value less than this will be obtained and, as the aspect changes during a synodic period, Δm will vary (see figures in Surdej and Surdej 1978); this would permit c to be estimated. The picture is much less clear for a surface with an unknown phase function *and* an unknown shape but, given enough information, bounds can be placed on the scattering properties of the asteroid's surface (Surdej and Surdej 1978) in addition to its shape and pole. These prizes motivate observations of asteroids over several oppositions.

Most attempts at deducing an asteroid's shape from its lightcurves adopt a particular figure, e.g., a cylinder with hemispherical caps on the ends (Dunlap and Gehrels 1969; Dunlap 1974) or a triaxial ellipsoid (Sather 1976; Zellner 1976), and then vary the model dimensions to match the observations. More complicated versions of the same approach give 1580 Betulia as a cratered elongated asteroid (Tedesco *et al.* 1978) and 624 Hektor as a contact binary (Hartmann and Cruikshank 1978, and Hartmann's chapter). These are all interesting exercises but beg the question of shape since lightcurves do *not* uniquely specify the object's form. These studies do, however, illustrate the limitations of Eq. (9), which is found often to exaggerate nonsphericity. The laboratory experiments of Dunlap (1971), who used a photometer to observe rotating styrofoam models covered with plasticene and then dusted with powdered rock, demonstrate the same thing. In addition Dunlap noticed that his model lightcurves were usually much smoother than those observed for the asteroids even when the models had deviations up to 20% from the average dimension; surface irregularities and rough texture seemed to ease the mismatch. Dunlap also noted that no single amplitude-aspect function existed; this called into question one technique which had been used earlier to determine rotational pole positions. Since our understanding of asteroid shapes is so primitive, experiments such as these deserve further elaboration.

The smaller structure seen in most lightcurves can be directly associated with surface features of a given size, regardless of whether those features are produced by nonuniform reflective properties or by some lumpiness of the surface. Two methods (Goguen *et al.* 1976) present themselves. In the first, the deviation Δm in magnitude from some mean lightcurve is taken to be caused by a plus (or minus) change in the albedo of the feature by its value (i.e. the feature totally absorbs light, perhaps because no material is there, or it reflects twice as effectively as the mean surface). Then the linear scale of the feature is

$$L \sim r[\pi(10^{-\Delta m^*/2.5} - 1)]^{\frac{1}{2}} \qquad (10)$$

This expression would also be valid if the change in reflecting area were caused by the appearance or disappearance of a satellite (see the chapter by Van Flandern *et al.*). On the other hand, we could instead use the width Δt of the feature on the time plot to give

$$L' \sim 2\pi r \Delta t/P. \qquad (11)$$

This last result is not entirely accurate for a satellite-caused brightness variation unless the satellite is close by. If $L \approx L'$, the size of the feature has been found.

In passing we note that Eq. (10) implies that the relative roughness of an asteroid is given only by Δm^*, the absolute size of the lightcurve feature. This should aways be kept in mind because one can have a misleading qualitative impression that large amplitude lightcurves (e.g. 433 Eros, 624 Hektor, 1245 Calvina, 1620 Geographos) are remarkably smooth while small amplitude lightcurves (e.g. 29 Amphitrite, 51 Nemausa) seem more frequently to be quite complex. This impression is generated by the plotting scale being much smaller for quasi-spherical objects so that often the signatures of topographic features having the same relative lumpiness are more noticeable on a small amplitude lightcurve. Similarly, if albedo features were important contributors to asteroid brightness variations, we would also expect them to be more easily detected on the lightcurves of nearly spherical objects.

In preparation for discussing plots of Δm versus size for several asteroid groups we first summarize experimental and theoretical investigations into asteroid shape. Fujiwara et al. (1978) examined the shape distribution for fragments produced in laboratory high-velocity impacts between cylindrical polycarbonate projectiles and cubic basalt block targets. These catastrophic experiments were held at 2.6 and 3.7 km sec^{-1} between objects 0.37 g and about 1500 g, respectively, and each experiment produced hundreds of fragments large enough to be measured. Relative sizes of the semiaxes (a,b,c) of a typical fragment were found to be $2:\sqrt{2}:1$ with approximately Gaussian distributions about these values. Markedly elongated fragments were not found and equidimensional objects were also rare. Hartmann (personal communication, 1977) has carried out comparable experiments at substantially slower speeds (26–50 m sec^{-1}) and obtained similar results. It is possible, however, that these studies have little to do with the shape of the detritus from collisions between objects twenty orders of magnitude or so more massive, particularly for the larger asteroids where gravitational binding energy starts to dominate material strength; nevertheless they are all we have. Since experiments for which scaling considerations will not be important are unlikely ever to be performed, theoretical investigations of expected shapes are needed.

More than the patterns of catastrophic fragmentation determine current asteroid shapes, since virtually every colliding body will chip away at the asteroid in its own distinctive way. The largest of the craters so formed can account for the shapes of those few asteroids whose lightcurves contain three maxima and minima (cf. Tedesco et al. 1978) for the collision to be any bigger would split the asteroid apart. Yet other collisions produce a pockmarked surface on all scales like those of Phobos and Deimos (see the chapter by Veverka and Thomas). It has even been put forth (Degewij 1977, 1978) that tinier yet collisions may account for the fact that very small asteroids are markedly spherical.

Too high a rotation speed can also modify an asteroid's shape because particles at the tips of the longest axes of an asteroid feel the weakest gravity but have the largest centrifugal accelerations when attached. If the rate given by Eq. (8) is exceeded, particles at the tips will leave the surface and thereby result in a more spherical asteroid. Burns (1975) has shown that only if $\rho \gtrsim$ 2.80 g cm^{-3} for 1566 Icarus (a U object) or $\rho \gtrsim$ 2.28 g cm^{-3} for 321 Florentina (probably an S object; see Harris and Burns 1979) will all loose surface particles remain attached. Intriguingly, 1566 is unusually spherical for an Earth-approaching object; has it lost its ends?

Theoretical work on shape so far has concerned what irregularities could be sustained by objects of different size and composition over the age of the solar system (Cook 1971; Johnson and McGetchin 1973). In addition to this research Dermott (1979) computes the rotational distortion $(a-c)/c$ of the largest asteroids about their spin axes to be of order 10%, assuming the objects are able to adopt hydrostatic figures; the precise value depends upon the interior density structure chosen. Even if Dermott's model is valid and the asteroids are undisturbed by collisions, it would be difficult to measure such shape with enough accuracy to distinguish internal structure by any technique other than spacecraft imaging.

Johnson and McGetchin (1973) compute the compressive stress at the base of a topographic element of height h resting on the surface of a non-rotating, spherical planet. By comparing this to the ultimate strength of the (assumed incompressible) material comprising the planet, they are able to bound the possible elevation differences that the object's strength can support to

$$H \leqslant 3\sigma/(4\pi G\rho^2 r). \tag{12}$$

If Ceres is made up of a relatively weak substance, such as carbonaceous chondritic material or ice (?), for which σ might be 10^6-10^7 dyne cm^{-2} (cf. Pollack et al. 1979), the elastically supported relief would be less than a few, or a few tens, of km. From Eq. (9) the corresponding brightness variation, if Ceres had a uniform albedo, would be between 0.04 and 0.004 mag; the observed amplitude is 0.04 mag. This result is of interest following the polarization observations of Ceres by Degewij and Zellner (1978; see the chapter by Dollfus and Zellner) which establish that Ceres' brightness variations can be attributed entirely to topographic features and/or irregular shape. As Eq. (12) shows, smaller asteroids, because of their weaker gravity, are permitted to have bigger features in an absolute sense; this provides much larger relief in a relative sense. For virtually all asteroids except the few largest, h/r can be easily greater than 10^{-1} regardless of the strength of asteroids.

The limit, Eq. (12), on relief is valid only as long as the material responds rigidly. However, over long times and especially at elevated temperatures, geomaterials are known to deform plastically; i.e., they "flow" in order to relieve the imposed stress. Johnson and McGetchin (1973) quote Darwin's expression for the relaxation time of a bulge of wavelength $\sim 2r/n$

$$\tau_R = 19\nu n/2gr, \tag{13}$$

where ν is the kinematic viscosity and g is surface gravity. Viscosity is a decreasing exponential function of temperature for most materials and so there may be an intimate connection between surface topography and the asteroid's thermal history (see the chapter by Sonett and Reynolds). Presuming that the asteroids have rock-like viscosities and that they have not been heated near their melting points, where viscosity drops sharply, Eq. (13) tells us that topography of all scales will not be appreciably reduced by plastic flow. On the other hand, the fact that $\overline{\Delta m} = 0.106 \pm 0.017$ mag for the seven largest asteroids ($r > 150$ km) may be evidence that these objects, at one time, at least, have had internal temperatures high enough that τ_R was short causing surface features to collapse. An equally valid interpretation for the lack of notable asphericity on these large asteroids would be that their surface layers (at least) are a loose and therefore weak aggregation of material which, as it is jostled about by impacts, fills a smooth equipotential surface (Chapman, personal communication).

Remembering that for the most part terrestrial relief is isostatically compensated rather than elastically supported, one caveat must be given about the above discussion. If some of the asteroids are differentiated, as suggested for Vesta, they may display larger elevation differences than the above formula would indicate.

Cook (1971) has calculated the unbalanced stress at the center of a rotating Jacobi ellipsoid. For large, highly elongated objects, like 624 Hektor, the computed stress can exceed the strengths of candidate materials; this has been used to argue against the initial model of Dunlap and Gehrels (1969) and has motivated Cook's binary model as well as the coalesced barbell version by Hartmann and Cruikshank 1978; see Hartmann's chapter).

IV. THE DATA

In the eight years since the publication of the last asteroid book (Gehrels 1971) the number of catalogued asteroids having lightcurve observations has increased more than five-fold. This has been due primarily to the work of six groups: Debehogne, Surdej and Surdej at the European Southern Observatory (Chile), Harris at the Jet Propulsion Laboratory, Lagerkvist at Upsalla University (Sweden), Scaltriti and Zappalà at Torino Observatory (Italy), Schober at Graz Observatory (Austria), and Tedesco at New Mexico State University.

The data set in Part VII; Table VI is essentially that used by Tedesco and Zappalà (1979), i.e., the TRIAD lightcurve file (Bender et al. 1978; Part VII of this book). Harris and Burns (1979) as well as Tedesco and Zappalà summarize previous studies of asteroid rotational properties.

All data sets of asteroid rotational properties are subject to major selection effects. These include the biases of other TRIAD data such as the bias toward observing the higher albedo asteroids found in the inner regions of the asteroid belt and the selection of small objects from special asteroid classes (e.g., Apollo-Amors and Hirayama family members) together with the tendency to observe small objects using photographic photometry. The inclusion of photographic lightcurve data in the sample to be analyzed introduces a bias of its own since over half the objects observed photographically have lightcurve amplitudes below the technique's detection threshold. This means that rotational information on asteroids having low-amplitude lightcurves is excluded by these observations. See Harris and Burns (1979), and Tedesco and Zappalà (1979) for further discussion of selection effects.

The influence of these selection effects, as well as the presence of considerable scatter in the rotations and shapes as a natural outcome of collisions, mean that simple conclusions are not easily derived from the data. This is especially true because, as we shall find, the data are only subtly changed throughout the parameter space we study. Nevertheless, it is clear what selection effects are present and, with effort, their influence can be significantly reduced. Another problem is that, although the data base has been rapidly expanding, there has not been a concomitant improvement in statistics because variables are continually entering the analysis. For example, studies at the end of the 1970's have sought the way rotation rate is influenced by taxonomic type, shape and size (Harris and Burns 1979; Tedesco and Zappalà 1979; Renschen 1978), by family membership (Tedesco 1979b), and by orbital position (Tedesco and Zappalà 1979) whereas earlier studies (e.g. Alfvèn 1964) treated only size effects. However, if the 1980's are as productive as the 1970's have been, asteroid rotational properties will become sufficiently well known to permit a much more detailed and definitive analysis than that presented here. As a result we shall learn much about the interior properties of asteroids and about the collisional history of these once mysterious bodies.

V. RESULTS AND IMPLICATIONS

This section will review the papers by Harris and Burns (1979) and by Tedesco and Zappalà (1979), especially trying to weigh the relative limitations of the two works, since they occasionally come to different conclusions. These papers adequately summarize previous studies but make use of data sets much larger and better defined than those of earlier researchers.

Rotation Rate versus Size

Asteroid rotation properties have been tabulated by Alfvèn (1964), Gehrels (1970), Taylor (1971), McAdoo and Burns (1973), Icke (1973),

Dohnanyi (1976), Lagerkvist (1978), Schober (1978), Harris and Burns (1979) in addition to Tesesco and Zappalà (1979). One way of representing the rotational characteristics of the minor planets is to display spin angular momentum verus asteroid mass; the latest version of this is by Burns (1975), who attempted to include the effect of shape on both quantities. Early work plotted rotation periods P versus absolute magnitudes B(1,0) but more dynamically meaningful parameters are now available: rotation frequency f replaces P, since angular velocity is closer than spin period to the angular momentum which is transmitted in collisions, and diameter D supplants B(1,0) so that effects can be determined as functions of size; Harris and Burns (1979) along with Tedesco and Zappalà (1979) employ these parameters.

Figure 1 shows a linear function of log f (rotation frequency in revolutions per hr) and log D (diameter in km) fit to the available data. This amounts to determining the logarithmic mean diameter and logarithmic mean frequency of the data as well as the line's slope; a negative slope means smaller objects rotate faster. Values for various subsets of these data are prooontod in Tahle I We see that for the entire sample f increases with decreasing size, confirming a trend first noticed by McAdoo and Burns (1973). However, upon separating the sample by taxonomic type (Bowell *et al.* 1978*a*), Harris and Burns (1979) discover that the C asteroids are larger and spin more slowly ($P{\sim}11$ hr) than the non-C's ($P \sim 9$ hr). This tendency, along with selection effects, apparently explains the result of McAdoo and Burns (1973). The data within a specific taxon (S,C) is too localized in diameter or too limited in number to see whether there might be size effects for a particular type. The difference in the spin rates of C objects versus S objects is taken to be caused by the densities of the two taxa varying in the ratio 2:3 in light of the collisional theory of Harris (1979*a*), or by the S objects being stronger. This difference in spin rate between taxa, however, may be caused by size differences instead (see below). M objects (Harris 1979*b*) spin faster than the S asteroids but not by a statistically meaningful amount. If the several taxa do indeed spin at distinct rates, this would be the first evidence that these classifications refer not only to the nature of surface layers but also the properties of the deep interiors; this would lend credence to the idea that most asteroids are compositionally uniform, as suggested by the uniformity of surface properties on individual objects (cf. Degewij *et al.* 1979) and among some family members (chapter by Gradie *et al.*).

Harris and Burns (1979) find, after removing C and S objects, that the remainder of their sample also seems to have faster spin rates at smaller sizes (see Table I). In part this may be due to these objects having a variety of internal compositions which means that they would behave, for example, like a combination of C and S bodies. However, selection effects may be even more important; most of the very small members, those with $D < 15$ km, are Apollo/Amor objects, which many scientists believe to be disparate from main-belt asteroids. Moreover, data on many other small minor planets

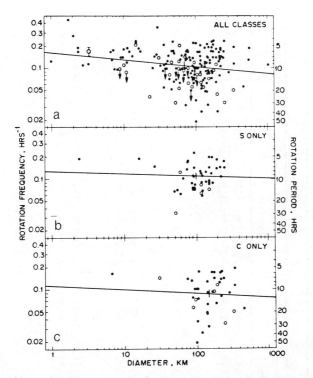

Fig. 1. Asteroid rotation frequency in rev/hr versus diameter in km. The results for 182 asteroids with known rotational properties in early 1979 are plotted. The solid dots are secure values; open dots are asteroids with uncertain rotations. Points with descending arrows indicate those asteroids for which only lower limits on their rotation periods were known. The plotted lines are linear least squares solutions for log f versus log D where all points in the plot, including lower limit points, but extended to their probable values, have been used. Similar solutions for secure values only are listed in Table I. The small tick mark on the plotted solution indicates the mean log D of the asteroids for that solution. The large open circle with error bar is the mean rotation rate (with formal uncertainty) of the ten asteroids in this sample whose orbits cross those of Earth or Mars plotted at the mean log D of those objects. All asteroids (top figure); Class S asteroids only (middle); Class C asteroids only (bottom). (Figure from Harris and Burns, 1979.)

($10 < D < 50$ km) came from photographic photometry, a technique that, due to selection effects, favors finding short periods, as can be demonstrated. The removal of these points lessens the statistical significance of the slope. Hence the quandary becomes clear; to get a good handle on size effects Harris and Burns needed to go to points that are suspect. This should be remedied in the 1980's. It is well within the capability of current telescopes to find rotation periods of small main-belt objects and their study should begin.

Some further points should be made about the rotation of small objects. First, Harris and Burns argued that one can see from Fig. 1 that, regardless of

TABLE I

Least Squares Solutions for Rotation Rate versus Size

| Data Included | | | Sample Size | Solution | | | | |
Composition	Restrictions	Fig.		mean log D	D(km)	mean log f	P(hr)	Slope
All	None	1a	182	1.843	69.7	-0.972 ± 0.017	9.38	-0.098 ± 0.032
	No reliab 1, No >		143	1.845	70.0	-0.932 ± 0.018	8.55	-0.094 ± 0.033
Earth- or Mars-crossers only	--	Φ in 1a	10	0.514	3.3	-0.733 ± 0.058	5.41	-0.126 ± 0.171
S only	None	1b	47	1.988	97.3	-0.972 ± 0.025	9.38	-0.038 ± 0.076
	No reliab 1, No >		41	1.992	98.2	-0.945 ± 0.024	8.81	-0.054 ± 0.069
C only	None	1c	40	2.148	140.6	-1.064 ± 0.039	11.59	-0.056 ± 0.128
	No reliab 1, No >		32	2.158	143.9	-1.045 ± 0.046	11.09	-0.011 ± 0.150
M only[a]	None		11			-0.900 ± 0.050	8.14	
All but C and S	None		95	1.642	43.9	-0.933 ± 0.024	8.57	-0.079 ± 0.042

[a]From Harris (1979b)

the fact that the plotted line is suspect because of selection effects, the mean value for Apollo/Amor objects essentially lies on it; of course, to some extent, the slope is determined by these inner solar system objects. Second, the rotation rates of very small asteroids have been determined by photographic techniques. Degewij (1977, 1978) used a photographic survey of unnumbered asteroids ($D \approx$ a few km) taken with the 122-cm Palomar Schmidt telescope on plates spaced a few minutes apart. Rotation periods for only 27 asteroids out of 130 were found; their mean period was 5.6 hr. This might be a biased value due to the following selection effects: (i) the observing runs were fairly short (4 hr and 6.4 hr); (ii) changes of at least 0.2 mag need to be present before a period could be ascertained; and (iii) the periods of nonsinusoidal, or even multiple maxima, lightcurves are not well determined by the technique.

In the other recent major study of asteroid rotation properties, Tedesco and Zappalà (1979), in an effort to consider as unbiased a sample as possible, presented Fig. 2 which contains only nonfamily main-belt objects; all points come from photoelectric determinations. With this sample the mean rotation rates for the two types are: for C (46 objects), $\bar{f} = 0.100 \pm 0.008$ rev/hr ($P = 10.0$ hr) and for S (49 objects), 0.106 ± 0.007 rev/hr (9.4 hr) where the quoted uncertainties are standard deviations of the means. Since these rates differ by less than a standard deviation ($\sim 0.6\,\sigma$), they are taken as indistinguishable. Tedesco and Zappalà (1979) then look for size effects by separating the sample into two size bins. They find that objects above a diameter of 175 km have a mean rotation rate (for 34 objects) of 0.141 ± 0.010 rev/hr ($P = 7.1$ hr) while for those of smaller sizes (100 objects): $\bar{f} = 0.104 \pm 0.005$ rev/hr (9.6 hr), a difference of 3.3σ. Tedesco and Zappalà argue that their choice of the dividing size has physical significance because it lies near the region in which there are also dramatic changes in asteroid shapes as well as in the size-frequency distribution (Zellner and Bowell 1977; see Zellner's chapter). Both these changes could imply that the large, rapidly spinning objects are primitive, at least in the sense that their spin rates have not been substantially altered since their formation (cf. the chapter by Davis *et al.*). As in the result of Morris and Burns, there is a certain physical plausibility in this.

In order to separate differences in rotation rates due to size effects from those caused by the influence of taxonomic types, it would be extremely valuable to have photoelectric observations of small ($D < 50$ km) main-belt asteroids. An improvement in the statistics of M-type objects would also help because we would then have reliable data on a third taxonomic type which might permit us to separate size effects from compositional effects.

Harris and Burns (1979) measure the dispersion of asteroid rotation rates about their mean curve. They find it to be in good agreement with a three-dimensional Maxwellian distribution (Fig. 3). As they argue, this suggests that the rotation axes are randomly oriented in space, which is also indicated by

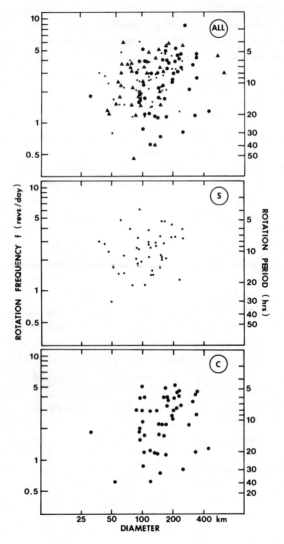

Fig. 2. Rotation frequency versus diameter for all (top figure), S-type only (middle), and C-type only (bottom). • indicates S-type; ●, C-type; △ and ▲, all other types. Only photoelectrically observed main-belt asteroids not belonging to any of the first nine Hirayama families are plotted. (Figure from Tedesco and Zappalà, 1979).

the rotational pole data (see the chapter by Taylor), and that the system has evolved collisionally. It appears for both samples (Figs. 1 and 2) that the rotation rates of C asteroids exhibit more scatter than do the S types; this is seen as well in Fig. 3. This difference is not statistically certain and, if real, has unknown implications.

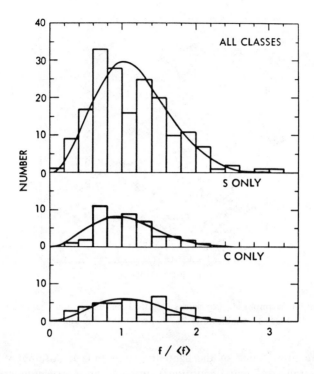

Fig. 3. Histograms of relative rotation frequency for each of the three data sets illustrated in Fig. 1. Each asteroid's rotation rate is measured relative to the logarithmic mean rotation rate, as obtained from a least squares solution for the entire data set, at the asteroid's diameter. The curve in each histogram is the three-dimensional Maxwellian distribution properly normalized and scaled for the data points with the observed mean log f. (Figure from Harris and Burns, 1979.)

The randomness of rotational pole directions is demonstrated by another technique in Fig. 4. Here we have plotted Δm versus ecliptic longitude and it is seen that a scatter diagram results. An earlier version of this plot by Gehrels (1970) seemed to have some structure which now, with more complete data, has disappeared. The relative paucity of points for longitudes between about 180° and 270° results from the fact that asteroids have oppositions at these longitudes during summers in the northern hemisphere. Fewer lightcurve observations are made during these periods due to a combination of factors, including the asteroids' southerly declinations, the short nights, and, at least in the southwestern United States, the generally cloudy summer weather.

Shape versus Size and Taxonomic Type

Asteroid shapes have received less attention than rotation rates, partly since the causes and implications of shape variations are not well understood. Figure 5 is taken from Harris and Burns (1979) and gives $\overline{\Delta m}$ versus log D

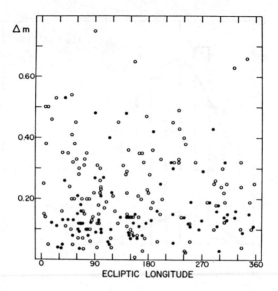

Fig. 4. Lightcurve amplitude Δm versus ecliptic longitude at time of observation (cf. Gehrels 1970, Fig. 14.) ● indicate $D \geqslant 200$ km; ○, $D < 200$ km.

($\overline{\Delta m}$ is the mean value of the lightcurve amplitude if different values are obtained during different oppositions). We note an increasing nonsphercity with smaller size, according to the plotted linear fit, which is a least squares fit to a linear function of the two variables. The individual taxa also show this property but the slopes then are determined by a few points and therefore are statistically insignificant.

Once again, however, Harris and Burns (1979) were confronted by the inadequacy of their data due to selection effects; the small-size end of the diagram relies on Apollo/Amor objects, which perhaps should not be plotted along with main-belt asteroids, and on points determined by photographic photometry. The photographic method needs at least a Δm of 0.2 mag in order to detect variations and large amplitude lightcurves are therefore favored. Harris and Burns (1979) sought the effect of these biases and removed both photographic photometric and Apollo/Amor points; the slope remained but was statistically much less significant. They also evaluated a mean $\overline{\Delta m}$ for five subsets of the data broken into five size ranges. The results for these subsets are shown with error bars in Fig. 5 (top) and demonstrate that the slope is developed well before the Apollo/Amor objects are reached; on the other hand, there would not be much slope if only the data for the three largest size bins were included.

Figure 6, taken from Tedesco and Zappalà (1979) gives $\overline{\Delta m}$ versus log D for the photoelectrically observed, main-belt, nonfamily asteroids in the TRIAD lightcurve file. These authors compared the mean $\overline{\Delta m}$'s for the 26 C

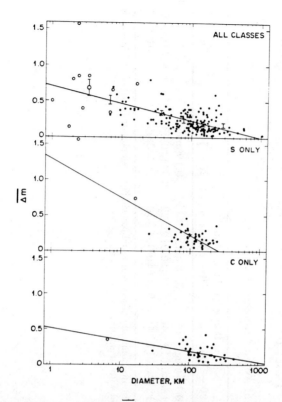

Fig. 5. Mean lightcurve amplitude $\overline{\Delta m}$ versus diameter in km for the asteroids with known variability plotted in Fig. 1. Open circles are Earth- or Mars-crossing asteroids; the large open circle with error bar is the mean (with formal uncertainty) for the ten such asteroids in the sample, plotted at the mean log D of those objects. The solid lines are least squares fits for the plotted data in each figure. The smaller error bars indicate the mean $\overline{\Delta m}$ values for five subsets of the data: $D < 20$ km; 20 km $< D < 50$ km; 50 km $< D < 100$ km; 100 km $< D < 200$ km; and $D > 200$ km. (Figure from Harris and Burns, 1979.)

types and 30 S types having diameters between 75 and 175 km, obtaining values of 0.181 ± 0.024 and 0.193 ± 0.017 mag, respectively. They concluded that the mean rotational amplitudes, and hence the mean shapes, of C- and S-type asteroids are essentially indistinguishable, in agreement with studies by Harris and Burns (1979) and Bowell (1977).

Since there is no dependence of $\overline{\Delta m}$ on taxonomic type, Tedesco and Zappalà divided their sample into the same five diameter bins used by Bowell (1977) and by Harris and Burns (1979). Table II, taken from Tedesco and Zappalà, presents their results and compares them with previous results. They concluded that all three studies are in agreement down to diameters of ~50 km.

TABLE II

The Size Dependence of Asteroid Rotational Amplitudes

Source	n	200	n	100 – 200	n	50 – 100	n	20 – 50	n	20
Bowell (1977)		0.18 mag		0.16 mag		0.18 mag		0.19 mag		—
Harris and Burns (1979)	23	0.182±0.032	54	0.184±0.013	40	0.238±0.019	22	0.341±0.034	25	0.526±0.059
Tedesco and Zappalà (1979)	24	0.188±0.028	60	0.181±0.011	40	0.206±0.015	10	0.362±0.057	(36)	(0.28)[a]

Column group header: Diameter Range (km)

[a]Excludes Earth-approaching asteroids and assumes that for every photographically observed asteroid for which the amplitude is sufficiently large for a rotation period to be estimated, there exists an additional asteroid with an amplitude of 0.15 mag.

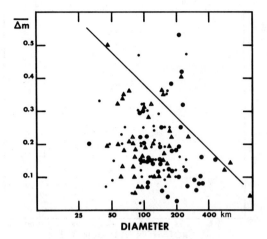

Fig. 6. Mean lightcurve amplitude $\overline{\Delta m}$ versus diameter D for the same sample as plotted in Fig. 2. · indicates S types; •, C types; and ▲, all others. (Figure from Tedesco and Zappalà, 1979.)

The mean $\overline{\Delta m}$'s of about 0.2 or so found by Harris and Burns and by Tedesco and Zappalà for $D > 50$ km are comparable to those found by Bowell (1977) who in a systematic UBV survey of some 400 asteroids detects statistical variations in asteroid brightnesses upon repeating the observations, usually within a few days. He believes that these changes reflect lightcurve variations. Bowell's data suggest that shapes are not dependent on size, at least for objects larger than about 25 km. Degewij *et al.* (1978) have detected only one highly variable asteroid out of twenty small main-belt asteroids observed during an hour's run and in this sense support Bowell's belief that small asteroids are not unusually irregular. Also in conflict with the lightcurve results are the results from photographic lightcurve surveys. The Palomar study (Degewij and Gehrels 1976; Degewij 1977, 1978) of small unnumbered objects found only 27 out of 130 objects to exhibit brightness variations larger than 0.2 mag.

The reasons are not clear for the discrepancy between the various results on lightcurve amplitudes. If real, for we are not yet fully convinced that it is, then it may tell us something about the gross shapes of intermediate size $(10 \lesssim D \lesssim 50$ km) asteroids. Taken at face value these results suggest that asteroids in this size region (10 to 50 km) have more irregular shapes than either larger or smaller asteroids, perhaps because larger asteroids are able to pull themselves into near-spherical shape while smaller ones are polished smooth (Degewij's, 1977, "pebble in a stream" model). In his view Earth-approaching asteroids are, presumably, able to retain their extreme shapes because they are no longer in the "stream."

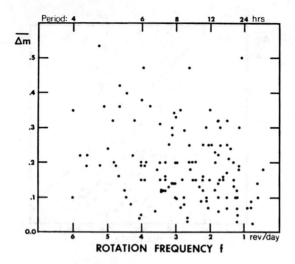

Fig. 7. Mean lightcurve amplitude $\overline{\Delta m}$ versus rotation frequency f for the same sample as plotted in Fig. 2 (Tedesco and Zappalá, 1979).

It is unfortunate that we run out of photoelectric lightcurve data right in the size range where the problem starts to become interesting. It is, however, entirely possible that the problem itself arises for precisely this reason. Better data is needed to resolve this issue.

Shape versus Spin Rate

Figure 7 perhaps shows a tendency within this scatter diagram for minor planets with higher rotation frequencies to be somewhat more irregular. Harris and Burns (1979) found a similar slight slope for their sample but concluded it was of marginal statistical importance. A purported correlation between $\overline{\Delta m}$ and f was also pointed out by McAdoo and Burns (1973) from very incomplete data. The result would suggest that whatever produced $\overline{\Delta m}$ also spun asteroids faster (collisions?) or that some common selection effect was acting for the smaller sizes. The result is the reverse of what would be expected if objects of various shapes were given the same angular momentum density and then allowed to align (cf. Lamy and Burns 1972; Burns 1975).

VI. CONCLUSION

We have seen that the precise manner in which the brightnesses of asteroids fluctuate may help to elucidate conditions in the asteroid belt, in particular the nature of collisions and their importance. The character of asteroid lightcurves may also indicate some fundamental physical properties of minor planets, for example, their densities, strengths and anelasticities. However, as yet we have not understood all that the current observations are telling us. Thus, we need more data but need also improved physical insight so as to unravel the complex physical processes that determine the rotations and shapes of asteroids.

Acknowledgments. We thank C. Chapman, J. Degewij, R. Greenberg, A. Harris and an anonymous referee for valuable criticism of a preliminary version of this manuscript. D. Chu carried out some calculations under extreme circumstances. C. Vesely's help in preparing Fig. 4 is gratefully acknowledged. JAB acknowledges the kind hospitality of l'Observatoire de Paris, Section d'Astrophysique, Meudon, where the penultimate version of his part of the chapter was prepared.

Note added in proof: The tabulation of lightcurve parameters (see Tedesco, Part VII, Table VI) and the listing of diameters and types (Bowell *et al.*, Part VII, Table VII) contained in this book permit another answer to the question of how rotational frequencies depend on asteroid size and/or type. We have treated this observational material in the manners of Harris and Burns (1979) and of Tedesco and Zappalà (1979) but looked at more subsets of the data. Compared to the sample used by Harris and Burns, the current TRIAD file is about 30% larger; the present table also changes the preferred type of about 25% of the objects while it modifies significantly (i.e., by more than 25%) the diameters of about 15 objects and revises the quality rating of the data for various asteroids. Moreover, here we do not attempt, as Harris and Burns did, to estimate the most likely spin rates for those asteroids whose rotational periods are known only as lower limits, but instead simply use the listed periods. The data set investigated by Tedesco and Zappalà is the same as that tabulated here except that they have utilized an earlier compilation of asteroid diameters; furthermore, Tedesco and Zappalà do not include any rotation rates determined from photographic measurements nor do they include in their analysis asteroids other than main-belt, non-family members.

Based on our statistical results, some general statements can be made about the current sample. Asteroids with well-determined data typically tend to rotate 10% faster than other asteroids and to be somewhat larger. The restriction to just photoelectric data does not change any conclusions from those based on the entire sample. With all subsets of the data main-belt, non-family objects usually spin about 90% as fast as the remaining asteroids. Little difference occurs when one considers only objects with unequivocally identified taxa versus objects whose taxon is the first given by Bowell *et al.* (Part VII, Table VII).

When the entire sample is treated in the manner of Harris and Burns we find, for unequivocally determined C types (59 objects, $\overline{\log f} = -1.041 \pm 0.032$) $\overline{P} = 10.99$ hr at $\overline{\log D} = 2.163 \pm 0.032$ (D = 145.5 km) whereas for S asteroids (70 objects, $\overline{\log f} = -1.046 \pm 0.034$) $\overline{P} = 11.12$ hr at $\overline{\log D} = 1.809 \pm 0.057$ (D = 64.4 km). Looking at only high-quality data, i.e., QUAL $\geqslant 2$, S objects ($\overline{P}$ = 9.18 hr) seem to rotate more rapidly than the C's (P = 9.48 hr) as found by Harris and Burns but this is not statistically secure. Before one too quickly jumps from the results for C and S objects to the conclusion that taxonomic type makes no difference on rotation rate, note the very rapid spins of (89) objects which are neither C nor S: $\overline{\log f} = -0.939 \pm 0.028$ ($\overline{P}$ = 8.69 hr) for $\overline{\log D} = 1.572 \pm 0.062$ (37.3 km); the possible distinctiveness of these objects is even more clearly seen when only (33) high-quality data are considered ($\overline{P}$ = 6.27 hr at $\overline{D}$ = 67.4 km). This increase in the spin rate for non-C/S objects is also seen when only main-belt, non-family objects are viewed. Insofar as the relationship size versus rotation rate is concerned, the tendency for small asteroids having high-quality data to spin faster than large ones is probably still present in the entire data set (slope = -0.061 ± 0.042) but the reverse is true if only main-belt, non-family objects are viewed (slope = 0.100 ± 0.074), unless for the entire sample one first removes objects larger than 225 km (then, slope = -0.114 ± 0.050). When looking at specific taxa, we observe little change in spin rate with size, ignoring the fact that very large objects seem to spin faster; this is perhaps not true for the non-C/S sample, when the effect of small, rapidly spinning non-main-belt objects is predominant.

As the book's TRIAD file is the same as that used by Tedesco and Zappalà, it is not surprising that we arrive at results virtually indistinguishable from theirs for main-belt, non-family asteroids, even though we included those observed photographically: $\bar{f} = 0.109 \pm 0.007$ rev/hr (61 C objects; $\bar{P} = 9.17$ hr) and $\bar{f} = 0.106 \pm 0.007$ rev/hr (53 S objects; $\bar{P} = 9.43$ hr) whereas $\bar{f} = 0.133 \pm 0.009$ rev/hr (36 objects with $D > 175$ km; $\bar{P} = 7.52$ hr) and $\bar{f} = 0.107 \pm 0.005$ (127 objects with $D < 175$ km; $\bar{P} = 9.35$ hr). Similar numbers arise when all asteroids in the TRIAD file are included. The dividing size of 175 km chosen by Tedesco and Zappalà does not seem to be critical; comparable results are found if the separation occurs at either 150 km or 225 km. In fact the result of different rates for different sizes can be seen to be principally a consequence of relatively fast rotations at large sizes while there are slow rates at intermediate sizes: for example, for all objects, $D > 150$ km, $\bar{f} = 0.131 \pm 0.008$ rev/hr (55 objects; $\bar{P} = 7.63$ hr); with $100 < D < 150$ km, $\bar{f} = 0.106 \pm 0.008$ (47 objects); and with $D < 100$ km, $\bar{f} = 0.116 \pm 0.005$ rev/hr (143 objects; $\bar{P} = 8.62$ hr). In agreement with Tedesco and Zappalà, large objects ($D > 225$ km) may rotate unusually fast ($\bar{f} = 0.143 \pm 0.011$ rev/hr; 22 objects; $\bar{P} = 6.99$ hr). These results are visible with effort in Figs. 1 and 2. A difference in spin rates also appears when objects of a given taxa are considered; for example, taking the separation size at 150 km, the $\bar{f}$ of the larger objects, whether S or C, exceeds that of the smaller asteroids by $0.02 - 0.03$, depending on the sample, whereas the typical deviation in each mean rate is only 0.01. The distinction between the rotation rates of large and small asteroids is less apparent if one considers only high-quality data. In agreement with the above treatment, the mean rotation rate of main-belt, non-family objects which are not classified S or C is greater than that of members of either of these taxa: $\bar{f} = 0.126 \pm 0.007$ (66 objects; $\bar{P} = 7.64$ hr); this rapid spin rate is faster yet if only high-quality points are included: $\bar{f} = 0.164 \pm 0.012$ rev/hr (43 objects; $\bar{P} = 6.10$ hr).

So a first analysis of the data contained in the June 1979 TRIAD file suggests that asteroid rotation rates are a function of size and that they may depend on taxonomic type. It is still true, however, that to see these rotational properties, we have often been forced to go to questionable data or to small subsets in the sample. Thus our plea for more observations remains so that the trends we currently see can be firmly established and ultimately understood.

REFERENCES

Alfvén, H. 1964. On the origin of the asteroids. *Icarus* 3: 52-56.

Andre, Ch. 1901. Sur le système formé par la planète double (433) Eros. *Astron. Nach.* 155: 27-30.

Bender, D.; Bowell, E.; Chapman, C.; Gaffey, M.; Gehrels, T.; Zellner, B.; Morrison, D.; and Tedesco, E. 1978. The Tucson Revised Index of Asteroid Data. *Icarus* 33: 630-631.

Bowell, E. 1977. UBV photometric survey of asteroids (abstract). *Bull. Amer. Astron. Soc.* 9: 459.

Bowell, E.; Chapman, C.R.; Gradie, J.C.; Morrison, D.; and Zellner, B. 1978a. Taxonomy of asteroids. *Icarus* 35: 313-335.

Bowell, E.; McMahon, J.; Horne, K.; A'Hearn, M.F.; Dunham, D.W.; Penhallow, W.; Taylor, G.E.; Wasserman, L.W.; and White, N.M. 1978b. A possible satellite of Herculina (abstract). *Bull. Amer. Astron. Soc.* 10: 594.

Burns, J.A. 1971. The alignment of asteroid rotation. In *Physical Studies of Minor Planets*, ed. T. Gehrels (NASA SP-267, Washington, D.C.: U.S. Government Printing Office), pp. 257-262.

Burns, J.A. 1975. The angular momentum of solar system bodies: Implications for asteroid strengths. *Icarus* 25: 545-554.

Burns, J.A., and Safronov, V.S. 1973. Asteroid nutation angles. *Mon. Not. Roy. Astron. Soc.* 165: 403-411.

Campbell, D.B., Pettengill, G.H.; and Shapiro, I.I. 1976. 70-cm radar observations of 433 Eros. *Icarus* 28: 17-20.

Chen, D.-H.; Wu, Z.-X.; and Yang, X.-Y. 1976. The motion of Eros' rotation axis in space. *Acta Astron. Sinica.* 17: 176-184.

Cook, A.F. 1971. 624 Hektor: A binary asteroid? In *Physical Studies of Minor Planets*, ed. T. Gehrels (NASA SP-267, Washington, D.C.: U.S. Government Printing Office), pp. 155-163.

Degewij, J. 1977. Lightcurve analyses for 170 small asteroids. *Proc. Lunar Sci. Conf. VIII* (Oxford: Pergamon Press), pp. 145-148.

Degewij, J. 1978. *Photometry of Faint Asteroids and Satellites* Ph.D. dissertation, University of Leiden.

Degewij, J., and Gehrels, T. 1976. Spin and strength of small asteroids (abstract). *Bull. Amer. Astron. Soc.* 8: 459.

Degewij, J.; Gradie, J.; and Zellner, B. 1978. Minor planets and related objects. XXV. *UBV* photometry of 145 faint asteroids. *Astron. J.* 83: 643-650.

Degewij, J.; Tedesco, E. F.; and Zellner, B, 1979. Albedo and color contrasts on asteroid surfaces. *Icarus* (special Asteroid issue).

Degewij, J. and Zellner, B. 1978. Asteroid surface variegation. *Lunar Science IX*. The Lunar and Planet. Inst., pp. 235-237.

Dermott, S. F. 1979. Shapes and gravitational moments of satellites and asteroids. *Icarus* 37: 575-586.

Dohnanyi, J. S. 1976. Sources of interplanetary dust: Asteroids. In *Interplanetary Dust and Zodiacal Light,* eds. H. Elsässer and H. Fechtig (Berlin: Springer-Verlag), pp. 187-205.

Dunlap, J. L. 1971. Laboratory work on the shape of asteroids. In *Physical Studies of Minor Planets* ed. T. Gehrels (NASA SP-267, Washington, D.C.: U.S. Government Printing Office), pp. 145-154.

Dunlap, J. L. 1974. Minor planets and related objects. XV. Asteroid (1620) Geographos. *Astron. J.* 79: 324-332.

Dunlap, J. L.; and Gehrels, T. 1969. Minor planets III. Lightcurves of a Trojan asteroid. *Astron. J.* 74: 796-803.

Dunham, D.; Dunham, J. B.; Van Flandern, T. C.; Schmidt, R. E.; Skillman, D. R.; Espenok, F.; A'Hearn, M. F.; Bolster, R. N.; Taylor, G. E.; Klemola, A. R.; Wasserman, L. H.; Williamson, R. M.; and Poss, H. L. 1979. Occultation of SAO 114159 by 18 Melpomene and possible satellite. *Icarus* (in press).

Fay, T. D., and Wisniewski, W. 1978. The lightcurve of the nucleus of Comet d'Arrest. *Icarus* 34: 1-9.

Fujiwara, A.; Kamimoto, G.; and Tsukamoto, A. 1978. Expected shape distribution of asteroids obtained from laboratory impact experiments. *Nature* 272: 602-603.

Gehrels, T. 1956. Photometric studies of asteroids. V. The lightcurve and phase function of 20 Massalia. *Astrophys. J.* 123: 331-336.

Gehrels, T. 1970. Photometry of asteroids. In *Surfaces and Interiors of Planets and Satellites*, ed. A. Dollfus (London: Academic Press), pp. 319-376.

Gehrels, T., ed. 1971. *Physical Studies of Minor Planets* (NASA SP-267, Washington, D.C.: U.S. Government Printing Office).

Gehrels, T., and Taylor, R. C. 1977. Minor planets and related objects. XXII. Phase functions for (6) Hebe. *Astron. J.* 82: 229-237.

Giuli, R. T. 1968a. On the rotation of the earth produced by gravitational accretion of particles. *Icarus* 8: 301-323.

Giuli, R. T. 1968b. Gravitational accretion of small masses attracted from large distances as a mechanism for planetary rotation. *Icarus* 9: 186-190.

Goguen, J.; Veverka, J.; Elliot, J. L.; and Church, C. 1976. The lightcurve and rotation period of asteroid 139 Juewa. *Icarus* 29: 137-142.

Goldstein, R. M. 1969. Radar observations of Icarus. *Icarus* 10: 430-431.

Goldstein, R.M.; Holdridge, D. B.; and Lieske, J. H. 1973. Minor planets and related objects. XII. Radar observations of (1685) Toro. *Astron. J.* 78: 508-509.

Gradie, J. C. 1978. An astrophysical study of the minor planets in the Eos and Koronis asteroid families. Ph.D. dissertation, University of Arizona.

Gradie, J.; Tedesco, E. F.; and Zellner, B. 1978. Rotational variations in the optical polarization and reflection spectrum of Vesta (abstract). *Bull. Amer. Astron. Soc.* 10: 595.

Gradie, J., and Zellner, B. 1977. Asteroid families: Observational evidence for common origins. *Science* 197: 254-255.

Hansen, O. 1977. On the prograde rotation of asteroids. *Icarus* 32: 458-460.

Harris, A. W. 1977. An analytic theory for the origin of planetary rotation. *Icarus* 31: 168-174.

Harris, A. W. 1978. Dynamics of planetesimal formation and planetary accretion. In *The Origin of the Solar System,* ed. S. F. Dermott (Chichester; John Wiley), pp. 469-492.

Harris, A.W. 1979*a*. Asteroid rotation. II. A theory for the collisional evolution of rotation rates. *Icarus* (in press).

Harris, A. W. 1979*b*. Are M-type asteroids metal cores? (abstract) *Lunar Science X.* The Lunar and Planet. Inst., pp. 500-502.

Harris, A. W., and Burns, J. A. 1979. Asteroid rotation. I. Tabulation and analysis of rates, pole positions and shapes. *Icarus* (in press).

Hartmann, W. K., and Cruikshank, D. P. 1978. The nature of Trojan asteroid 624 Hektor. *Icarus* 36: 353-366.

Hawkes, R. L., and Jones, J. 1978. The effect of rotation on the initial radius of meteor trains. *Mon. Not. Roy. Astron. Soc.* 185: 727-734.

Icke, V. 1973. Distribution of the angular velocities of the asteroids. *Astron. Astrophys.* 28: 441-445.

Johnson, T. V., and McGetchin, T. R. 1973. Topography on satellite surfaces and the shape of asteroids. *Icarus* 18: 612-620.

Jurgens, R. F., and Goldstein, R. M. 1976. Radar observations at 3.5 and 12.6 cm wavelength of asteroid 433 Eros. *Icarus* 28: 1-15.

Kiladze, R. 1971. Rotation of the planets as a result of accretion. *Astron. Vestnik* 5: 159-166. (In Russian.)

Kiladze, R. 1977. On the role of swarms of circumplanetary particles in developing spin. *Bull. Abastumani Obs.* 48: 191-212. (In Russian.)

Lagerkvist, C.-I. 1977. Photographic photometry of main-belt asteroids. *Uppsala Astron. Obs. Rept.* No. 9.

Lamy, P. L.; and Burns, J. A. 1972. Geometrical approach to torque free motion of a rigid body having internal energy dissipation. *Amer. J. Phys.* 40: 441-445.

Matson, D. L. 1971. Infrared observations of asteroids. In *Physical Studies of Minor Planets,* ed. T. Gehrels (NASA SP-267, Washington, D.C.: U.S. Government Printing Office), pp. 45-50.

McAdoo, D. C., and Burns, J. A. 1973. Further evidence for collisions among asteroids. *Icarus* 18: 285-293.

McAdoo, D. C., and Burns, J. A. 1974. Approximate axial alignment times for spinning bodies. *Icarus* 21: 86-93.

Morrison, D. 1977. Asteroid sizes and albedos. *Icarus* 31: 185-220.

Napier, W. Mc., and Dodd, R. J. 1974. On the origin of the asteroids. *Mon. Not. Roy. Astron. Soc.* 166: 469-293.

O'Leary, B. T. 1972. Frequencies of occultations of stars by planets, satellites and asteroids. *Science* 175: 1108-1112.

O'Leary, B. T.; Marsden, B. G.; Dragon, R.; Hanser, E.; McGrath, M.; Backus, P.; and Roboff, H. 1976. The occultation of κ Geminorum by Eros. *Icarus* 28: 133-146.

Öpik, E. J. 1958. Magnetic fields and rotation (stars, satellites, meteors and planets). *Irish Astron. J.* 5: 69-70.

Öpik, E. J. 1977. Magnetic damping of rotation. *Irish Astron. J.* 13: 14-21.

Oppolzer, E. von 1901. Notiz. betr. planet (433) Eros. *Astron. Nach.* 154: 297.

Pollack, J. B.; Burns, J. A.; and Tauber, M. E. 1979. Gas drag in primordial circumplanetary nebulae: A mechanism for satellite capture. *Icarus* 37: 587-611.

Prendergast, K. H. 1958. The effects of imperfect elasticity in problems of celestial mechanics. *Astron. J.* 63: 412-415.

Renschen, C. P. 1978. On the spin rate of the S-type asteroids. *Astron. Nachr.* 299: 103-105.

Russell, H. N. 1906. On the light variations of asteroids and satellites. *Astrophys. J.* 24: 1-18.

Safronov, V. S. 1971. Rotation of giant planets while accreting gas. *Solar Sys. Res.* 5: 139-144.

Safronov, V. S. 1972. *Evolution of the Protoplanetary Cloud and Formation of the Earth and Planets.* (Jerusalem: Israel Program of Tech. Translation).

Sather, R. E. 1976. Minor planets and related objects. XIX. Shape and pole orientation of (39) Laetitia. *Astron. J.* 81: 67-73.

Schober, H.-J. 1978. Photoelectric lightcurve and the period of rotation of the asteroid 200 Dynamene: A further object with low spin rate. *Astron. Astrophys. Suppl.* 31: 175-178.

Sher, D. 1971. On the variation in light of tumbling bodies. *Astrophys. Space Sci.* 11: 222-231.

Surdej, A., and Surdej, J. 1978. Asteroid lightcurves simulated by the rotation of a three-axes ellipsoid model. *Astron. Astrophys.* 66: 31-36.

Taylor, R. C. 1971. Photometric observations and reductions of lightcurves of asteroids. In *Physical Studies of Minor Planets,* ed. T. Gehrels (NASA SP-267, Washington, D.C.: U.S. Government Printing Office), pp. 117-131.

Tedesco, E. F. 1979*a*. Binary asteroids: Evidence for their existence from lightcurves. *Science* 203: 905-907.

Tedesco, E. F. 1979*b*. A photometric investigation of the colors, shapes, and spin rates of Hirayama family asteroids. Ph.D. dissertation, New Mexico State University.

Tedesco, E. F., and Bowell E. 1979. UBV photometry and lightcurves of asteroid 944 Hidalgo. To be submitted to *Astron. J.*

Tedesco, E. F., and Zappalà, V. 1979. Asteroid rotation properties: Correlations and selection effects. Submitted to *Icarus.*

Tedesco, E.; Drummond, J.; Candy, M.; Birch, P.; Nikoloff, I.; and Zellner, B. 1978. 1580 Betulia: An unusual asteroid with an extraordinary lightcurve. *Icarus* 35: 344-359.

Vesely, C. D. 1971. Summary on orientations of rotation axes. In *Physical Studies of Minor Planets,* ed. T. Gehrels (NASA SP-267, Washington, D.C.: U.S. Government Printing Office), pp. 133-140.

Veverka, J. 1971. The physical meaning of phase coefficients. In *Physical Studies of Minor Planets,* ed. T. Gehrels (NASA SP-267, Washington, D.C.: U.S. Government Printing Office), pp. 79-90.

Wasserman, L. W.; Millis, R. L.; Franz, O. G.; Bowell, E. F.; White, N. M.; Giclas, H. L.; Martin, L. J.; Elliot, J. L.; Dunham, E.; Mink, D.; Baron, R.; Honeycutf, R. T.; Henden, A. A.; Kephart, J. E.; A'Hearn, M. F.; Reitsema, H. J.; Radick, R.; and Taylor, G. E. 1979. The diameter of Pallas from its occultation of SAO 85009. *Astron. J.* 84: 259-268.

Whipple, F. L. 1978. On the nature and origin of comets and their contribution to planets. *Moon and Planets* 19: 305-315.

Wijesinghe, M. P., and Tedesco, E. F. 1979. A test of the plausibility of eclipsing binary asteroids. *Icarus* (special Asteroid issue).

Worden, S. P., and Stein, M. K. 1979. Angular diameter of the asteroids Vesta and Pallas determined from speckle observations. *Astron. J.* 84: 140-142.

Zappalà, V.; van Houten-Groeneveld, I.; and van Houten, C. J. 1979. Rotation period and phase curve of the asteroids 349 Dembowska and 354 Eleonora. *Astron. Astrophys. Suppl.* 35: 213-221.

Zellner, B. 1976. Physical properties of asteroid 433 Eros. *Icarus* 28: 149-153.

Zellner, B., and Bowell, E. 1977. Asteroid compositional types and their distributions. In *Comets, Asteroids, Meteorites,* ed. A. H. Delsemme (Toledo, Ohio: University of Toledo Press), pp. 185-198.

COLLISIONAL EVOLUTION OF ASTEROIDS: POPULATIONS, ROTATIONS, AND VELOCITIES

D. R. DAVIS, C. R. CHAPMAN
R. GREENBERG, S. J. WEIDENSCHILLING
Planetary Science Institute

and

A. W. HARRIS
Jet Propulsion Laboratory

The collisional evolution of various initial populations of asteroids is simulated numerically and compared with the present asteroid size-frequency distribution to find those populations which collisionally relax to the present belt. Both orbital and size distributions are treated, as well as the simultaneous evolution of two collisionally interacting populations with different physical properties. If the initial belt distribution was a power law, the initial belt population at the time when the present high-collision speed was established was probably only modestly larger than the present population. However, other distributions allow a more massive early belt. The rotational evolution due to collisions of asteroids with power-law distributions is also examined and compared with observations, leading to conclusions generally in agreement with those of size evolution. The high-collision speed in the present belt is likely due to Jupiter. Gravitational stirring by massive Jupiter-scattered planetesimals or secular resonances sweeping through the belt are the most probable mechanisms.

The dominant process affecting asteroids during modern epochs has been their gradual grinding down due to collisions. Average orbits are moderately eccentric and inclined, resulting in typical relative velocities of 5 km sec^{-1}. Asteroids are sufficiently numerous and their cross-sections sufficiently large that most have suffered major collisions during solar system history. Depending on the magnitude of the collision, effects may be relatively superficial (cratering and regolith formation) or extremely destructive (comminution, catastrophic fragmentation, and dispersal of fragments). Regolith formation on asteroids is treated in the chapter by Housen *et al.* in this book. We consider in the present chapter only those collisions that individually or collectively modify the gross character of asteroids, especially their size distribution, rotations, and their bulk geological properties.

Collisions affect asteroids in many ways. Energetic collisions serve both to destroy large asteroids and create smaller ones. They expose, on asteroid surfaces, materials originally buried at great depth. Collisions modify asteroid spins and may be largely responsible for asteroid shapes. Hirayama families are thought to be the result of past major inter-asteroidal collisions. Dynamical processes proposed for depleting Kirkwood gaps and for transporting asteroid fragments into Earth-crossing orbits to fall as meteorites usually invoke collisions. If collisions serve to fragment and disrupt the largest asteroids on time scales comparable with the age of the solar system, then the present asteroid population itself may be a collisional remnant of a vastly greater early population (Chapman and Davis 1975). On the other hand, if the efficiency of disruption by collision is relatively low, some traits of asteroidal sizes and properties may still reflect early processes of accretion.

A key question about the history of asteroids is how and when the asteroid velocities were pumped up to 5 km sec^{-1} from the necessarily much lower velocities required for the original accretion of the asteroids. At the end of this chapter, we discuss the mechanisms that may have been responsible for these high velocities and thus have been responsible for the reason why an asteroidal planet failed to accrete.

Piotrowski (1953) first showed that catastrophic collisions among asteroids may be frequent on the geological time scale. In an analytic study of asteroid collisions, Dohnanyi (1969) concluded that the observed asteroid size-frequency distribution is essentially in equilibrium at all sizes except the largest ones; creation of new asteroids as the collisional debris of larger bodies balances collisional destruction. With a cumulative power law of the form

$$N(m) = Am^{-\beta+1} \tag{1}$$

where $N(m)$ is the number of asteroids with mass $> m$ and A is a constant, the theoretical equilibrium solution has a population index $\beta \approx 11/6 = 1.833$, which agreed well with $\beta = 1.839$ that Dohnanyi found from a least squares fit to the McDonald Survey (Kuiper *et al.* 1958). (Throughout this paper we refer to incremental diameter indices b which are related to β as $b = 3\beta - 2$.) If asteroids were today a collisionally relaxed population evolving only slowly with time, their size distribution would be independent of their original distribution — hence one could not learn about the initial asteroid population from purely collisional models.

Subsequent to Dohnanyi's work, a wealth of new data has become available regarding physical properties of asteroids. Most asteroids fall into one of two classes, either C or S, each of which has its own size frequency distribution as shown in the chapter by Zellner. In particular, neither population can be represented by a simple power law with a constant population index. Most asteroids are larger than previously believed, implying larger collisional cross-sections. Furthermore, the gravitational binding energy neglected in previous collisional studies is increasingly important as body size increases; it dominates material strength for asteroids larger than ~ 50 km diameter.

Chapman and Davis (1975) constructed numerical models of asteroid collisional evolution which included improved models of collisional physics and new data on asteroid size frequency distributions. They found the asteroids to be a collisionally evolved population consistent with a wide range of initial distributions except those having more than a few bodies larger than Vesta. Asteroids larger than ~ 500 km diameter are especially difficult to disrupt at impact speeds of ~ 5 km sec^{-1}. Based on a hypothesis that large S-type asteroids are collisional remnants of bodies large enough to have differentiated and the fact that only one body has survived intact (Vesta), they estimated the initial asteroid belt mass at 300 times that of the current belt or about 20 % of Earth's mass.

In this chapter, we present a new formulation of the problem of asteroid collisional evolution and describe implications for the asteroid size frequency distribution and rotational rates. Several conclusions are:

1. Collisional lifetimes, the mean time between collisions that remove at least 50 % of a body's mass, of most asteroids larger than 50 km diameter are comparable to or greater than the age of the solar system. Hence, present collision lifetimes do not distinguish whether such asteroids are original condensations or fragments of larger bodies.

2. Most hypothetical initial C asteroid populations, except those with numerous bodies larger than 300 km diameter, collisionally evolve to the present belt in 4.5 Gyr. The major part of the asteroid population is in collisional equilibrium.

3. Most asteroids larger than 100 km diameter probably are fractured throughout much of their volume; i.e., they have developed a mega-regolith.

4. It is predicted that the rotation rate for large, gravitationally bound

asteroids should be the equilibrium value between infrequent large collisions tending to spin up asteroids and numerous small collisions which dampen rotation. The prediction agrees with the observed mean rotation rate only if most (50 – 100 %) of the collisional kinetic energy is converted into kinetic energy of ejecta, an unexpectedly large percentage. Results from rotational studies confirm those from collision evolution modeling; namely, that for power-law distributions with a population index (incremental diameter) < 4, the initial population could not have been appreciably more numerous than the present belt population.

5. The current asteroidal encounter speed may be due to, (a) gravitational stirring by massive Jupiter-scattered planetesimals, or (b) gravitational resonances with Jupiter which swept through the asteroid zone during early stages of solar system formation (Chapman *et al.* 1978; Heppenheimer 1979).

6. The large mean eccentricities and inclinations of asteroids, resulting in an average encounter speed of 5 km sec^{-1}, do not result from collisions between reasonable populations of early asteroids and Jupiter-scattered planetesimals (JSP). Except for unusual distributions, the high encounter speed (3.5 – 20 km sec^{-1}) between asteroids and JSP will collisionally destroy the asteroid population before there is a significant change in the mean orbital elements of asteroids.

I. COLLISION DYNAMICS FOR NUMERICAL MODELS

If the function F describes the complete distribution of a population, i.e. size, physical properties and orbital distribution, then the evolution of a population is found by calculating $F(t)$ given $F(0)$, the initial distribution. The evolutions of the size distribution and the orbital distribution are quite interdependent in general. In order to treat arbitrary population distributions and to incorporate detailed physical collisional models, a numerical approach is the best one for determining $F(t)$. We have developed two computer simulations addressing different aspects of the general problem of asteroid collisional evolution. The first program (P1) models the simultaneous evolution of both the mass and orbital distributions of a single population while the second (P2) computes the simultaneous mass evolution of two interacting populations assuming constant mean collision velocities. A description of the collisional physics relevant to asteroid collisions is included here along with an overview of the numerical simulation. A more complete description of the P1 simulation may be found in Greenberg *et al.* (1978). The model of collisional physics incorporated into the two simulations is essentially identical.

The outcome of a high speed collision between two bodies depends upon many factors: body sizes, collision speed and impact parameter, energy partitioning, material properties, and physical states of the bodies.

We define *fragmentation* or fracturing as the process of crushing part or all of the body, and *disruption* as the process of fragmenting and dispersing a body. Hence a target could be fractured, but not disrupted, provided gravity were the dominant binding mechanism of the body. Catastrophic disruption of a solid body requires that sufficient collisional kinetic energy be available to (a) fracture the material bonds and (b) impart sufficient kinetic energy to these fragments so that they can escape their mutual gravitational attraction. Let γ denote diameter ratio D_t/D_p, where D_t is the target diameter and D_p the minimum projectile diameter required to catastrophically disrupt a body of diameter D_t by a head-on collision at velocity v. When material strength is the dominant cohesive mechanism, laboratory experiments (Gault and Wedekind 1969; Fujiwara *et al.* 1977; Hartmann 1978) indicate that the size of the largest fragment and the distribution of fragments depends on target material properties and the kinetic energy density imparted to the target material. From experiments Fujiwara *et al.* (1977) have derived a relation for the fractional mass of the largest fragment from hypervelocity impacts into basalt targets, which we adopt in a form suitable for scaling to properties of other materials. This relation may be inverted to find γ_s (i.e. the γ value with strength dominant):

$$\gamma_s = \left[\rho_i/\rho_t \left(\frac{1.35 v^2 f_{\ell}^{0.8}}{S} - 1 \right) \right]^{\frac{1}{3}} \qquad (2)$$

where S is the material impact strength, ρ_i and ρ_t are the projectile and target densities respectively, and f_{ℓ} is the mass fraction of the initial target body contained in the largest fragment. γ_s is treated as a constant in our models, which neglects a possible decrease in effective strength for the body due to partial fracturing. When gravitational binding dominates over material strength, additional kinetic energy (i.e., a larger projectile) is needed to disperse the fragments and disrupt the body; hence γ is smaller than γ_s and varies with target size. To produce a largest fragment f_{ℓ} in this case, there must be sufficient kinetic energy in the ejecta that the fraction $(1 - f_{\ell})$ of the initial mass has a velocity exceeding v_e, the escape speed of the target body. The power-law velocity distribution similar to that discussed by Greenberg *et al.* (1978) for crater ejecta is used to represent the distribution of ejecta speeds, $f = v/v_c)^{-k}$, where f is the fraction of ejecta mass moving with speed $> v$ while v_c is a parameter depending on the total ejecta kinetic energy and k is the velocity distribution index. The value of k must be > 2 for the total kinetic energy in the distribution to be finite when integrated over all speeds, hence only k values > 2 are considered. If the entire volume of the target is fractured, then the fractional mass of the target core, f_{ℓ}, is

$$f_{\ell} = 1 - (v_e/v_c)^{-k}, \text{ when } v_e \geqslant v_c \text{ and}$$

$$f_{\ell} = 0 \text{ when } v_e < v_c.$$

(3)

The diameter ratio for catastrophic disruption when gravity is dominant, γ_g, found by equating the energy needed to remove the ejecta to the fraction of the impact energy partitioned into ejecta velocities during the collision, is given by

$$\gamma_g = \left[\rho_i/\rho_t \left(\frac{(k-2)f_{KE}}{k(1-f_{\ell})^{\frac{2}{k}}} (\frac{v}{v_e})^2 - 1 \right) \right]^{\frac{1}{3}}$$

(4)

where f_{KE} is the fraction of the collisional energy that is partitioned into kinetic energy of the ejecta. For catastrophic disruption, f_{ℓ} is taken to be 0.5, i.e., the largest fragment contains 50 % of the original target mass. The 0.5 value is not particularly critical since experiments demonstrate that there is a sharp transition in largest fragment size with energy density for bodies for which the material strength is dominant. When gravity dominates, the variation of largest fragment size with energy depends on k. The projectile size required to produce a largest fragment f_{ℓ} is independent of f_{ℓ} for large k but varies as $(1-f_{\ell})^{-1/3}$ as $k \to 2$. Figure 1 illustrates the variation of γ_g with target size for several collision speeds.

The distribution of fragments from any collision is calculated using the above model. If material strength is dominant, a power-law distribution is used with the largest fragment found by Eq. (2), and the population index is calculated from the mass of the largest fragment and the total ejecta mass. If gravity is dominant, the "core" of the original target is calculated from Eq. (3), while the size distribution of the escaping fraction of ejecta is found from the above relations for the strength dominant case. Typical resulting distributions are illustrated in Figs. 2 and 3, for barely catastrophic and super-catastrophic collisions when gravity or material strength dominates.

For asteroids with dominant gravitational binding, there are many collisions that are energetic enough to fracture much of the target volume, but do not impart enough energy to disperse a significant fraction of the target's mass. Depending on the projectile size distribution, resultant bodies might be fractured throughout much of their volume by numerous impacts of this type before there is a collision sufficiently energetic to disrupt the body. In absence of mechanisms such as intense heating to reconstitute the solid body, it seems plausible that many large asteroids remain gravitationally bound "rubble piles" of megaregolith.

In our numerical simulations the asteroid size distribution is modeled at larger sizes (typically 10 to 10^3 km diameter) by a series of discrete diameter bins each spanning a factor of 2 in mass; smaller asteroids (< 10 km) are

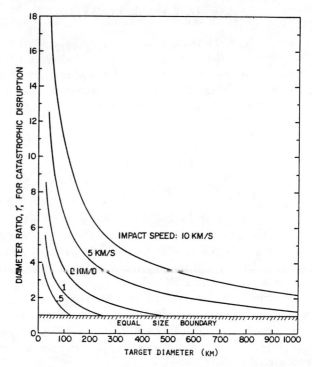

Fig. 1. Diameter ratio (target diameter/projectile diameter) necessary to barely destroy catastrophically a body as a function of target diameter for several impact speeds. Material parameters for C asteroids (Table I) are used. These curves illustrate the γ variation for the case where gravity is dominant; γ increases down to the size at which material strength becomes important. Below this size γ is constant $= \gamma_s$, which for the physical properties of C asteroids has values > 50.

represented by a power-law distribution of the form $dN = CD^{-b}\,dD$, where dN is the number of bodies with diameters between D and $D + dD$, and b is the incremental diameter population index which is attached to the smallest diameter bin. The collisional evolution is found by summing the negative changes in each diameter bin due to any collisional destruction by impacting bodies in other bins and in the tail, and the positive changes due to creation of new bodies by fragmentation of larger ones. The P1 program considers all possible bin interactions while P2 considers only interaction of a target bin with bins having diameter $> D_t/\gamma$. Hence, in P2 cratering erosion is neglected; the effect of this assumption is addressed later.

The P1 model computes orbital evolution and changes in encounter speed due to gravitational stirring as well as collisional damping (c.f. Greenberg *et al.* 1978). Orbital eccentricity and inclination are computed for

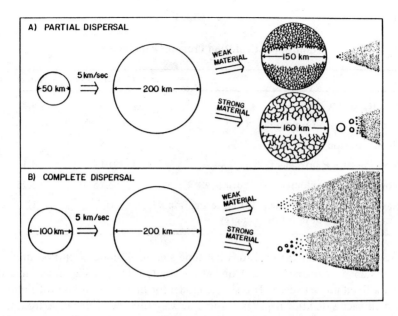

Fig. 2. Collisional outcome of barely catastrophic collisions (A) and supercatastrophic collisions (B) for bodies composed of weak material ($S = 6 \times 10^4$ ergs cm^{-3}) and bodies with impact strength comparable to basalt (3×10^7 erg cm^{-3}). Gravity is the dominant binding mechanism for 200 km bodies made of either strong or weak material.

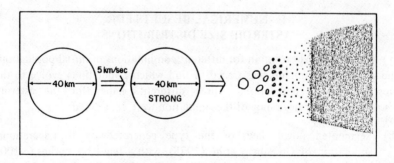

Fig. 3. Outcome for supercatastrophic collision with strength dominant. For smaller collisional kinetic energies, the size of the largest fragment increases.

each diameter bin and the encounter speed between any pair of bins is determined by the associated orbital elements. Program P2 uses an input encounter speed which is used to compute collision probabilities based on a particle-in-a-box model. In both programs, a normal distribution of impact speeds (with the variances being a program input) about the mean may be used with a random number generator to model the actual impact speed

TABLE I

Collisional Evolution Program Parameters

	C Asteroids	S Asteroids
Density (g cm^{-3}), ρ	2.5	3.5
Impact strength (erg cm^{-3}), S	6×10^4	3×10^7
Mean collision speed (km sec^{-1})	5	5
Fraction of KE into ejecta KE, f_{KE}	0.1	0.1
Slope of ejecta velocity distribution, k	2.25	2.25
Mass excavation coefficient for cratering impacts, grams per erg of collisional energy	10^{-8}	10^{-9}

between two diameter bins for each time step. Collisional evolution models vary <25 % at diameter <250 km between cases using variable impact speeds and a fixed impact speed. This simple model for finding collision probabilities and impact velocities neglects systematic variations with distance from the sun. Ip (1977a) calculated the probability of catastrophic destruction in the present belt as a function of orbit elements and found a variation of nearly an order of magnitude for the ranges 2.0 AU $< a <$ 3.2 AU; $0.05 < e < 0.3$; $0°.05 < i < 30°$. Mean values of collision probability therefore differ by ~ 3 from extreme values in the present belt.

II. NUMERICAL RESULTS FOR ASTEROID SIZE DISTRIBUTIONS

The collisional evolution for different combinations of initial populations and physical parameters was traced to find which populations evolve to the present asteroid populations. Three different shapes for the asteroid distribution at the beginning of the simulation were considered:
(a) power law;
(b) segmented power law, of the type generated by the accretional simulation of Greenberg *et al.* (1978), with a few large bodies (~ 1000 km) but most of the mass of the distribution in small bodies; and
(c) "Gaussian" about a mean size as proposed by Anders (1965) and Hartmann and Hartmann (1968).
The parameters given in Table I were used in all experiments unless otherwise noted.

Values of impact strength, adopted from Greenberg *et al.* (1978), are experimentally determined values appropriate for solid rock and basalts for S asteroids, while for C asteroids, the value chosen is representative of weakly bonded material such as consolidated dirt clumps investigated by Hartmann (1978). The adopted value for f_{KE} (0.1) is typical of values computed by

O'Keefe and Ahrens (1977) for anorthosite and iron projectiles impacting semiinfinite anorthosite targets at speeds of 5–45 km sec^{-1}. Our choice of f_{KE} is lower by a factor of 5 than that of Greenberg *et al.* (1978); however, they were treating a much lower velocity regime for which a larger value of f_{KE} may be more appropriate. The slope of the ejecta velocity distribution is derived from Gault *et al.* (1963), while cratering mass excavation coefficients are those adopted by Greenberg *et al.* for weak (*C*) and rocky (*S*) material.

Single component system (Program P1)

Before discussing results from various initial populations, we consider the effect of changing physical parameters and assumptions of the model. The collisional evolution of an initial belt having a power-law distribution with largest bodies 1000 km in diameter and having a total mass of 0.1 $M_\oplus$ is shown for several cases in Fig. 4, in order to illustrate different physical parameters and to evaluate the contribution of cratering erosion on the population evolution.

Inclusion of cratering erosion in addition to catastrophic disruption increases the rate of evolution of the population somewhat, but is apparently not the dominant process shaping the evolution. For small bodies with negligible gravity, cratering erosion dominates the collisional evolution for steep populations ($b > 4.0$); however, since lifetimes are shorter for smaller bodies, an initially steep distribution rapidly relaxes to a shallower distribution with an evolution dominated by catastrophic disruption. We have studied steep initial populations with indices up to $b = 6$, and find that collisional destruction at the small sizes reduces the small size population index to $b \approx 3$ within $\lesssim 10^8$ yr for *C* asteroids.

The critical parameters affecting collisional evolution by catastrophic disruption are, f_{KE} the fraction of collisional energy going into ejecta kinetic energy, k the slope of the ejecta mass-velocity distribution and to a lesser degree S the impact strength. The first two quantities determine the collisional energy required to disperse a body dominated by gravity. In this case there must be sufficient kinetic energy of the ejecta such that at least half the target mass is accelerated to a velocity greater than escape speed. For a given ejecta velocity distribution, the total ejecta energy required for dispersal is determined by the size of the target. Hence, the smaller f_{KE}, the smaller γ_g and the larger the projectile must be to supply the energy required for dispersal (for fixed collision speed).

On the other hand, for fixed f_{KE} and γ_g, the outcome of the collision depends on k, which in our models can only vary between 2 and ∞. The case $k = \infty$ means that all ejecta have the same speed v*, so the minimum requirement for dispersal is that v* = v_e. As k decreased toward 2, the distribution flattens and more energy is carried off by the high-velocity end of the distribution. Consequently more collisional energy is required (smaller

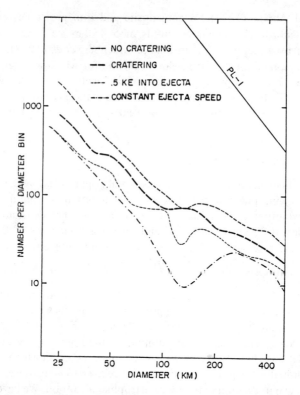

Fig. 4. Collisional evolution of an initial power-law distribution (PL-1) over 4.5 Gyr comparing the effects of neglecting cratering erosion, increasing the fraction of collisional energy that goes into ejecta energy, or using a fixed velocity for ejecta rarther than a distribution. Nominal parameters (Table I) are employed otherwise.

γ_g) in order to disperse at least 50 % of the target mass. The effect of these parameters on collisional evolution is also illustrated in Fig. 4, which compares the evolution of the same initial population but with 50 % of the collision energy partitioned into ejecta kinetic energy in one case, and a nearly constant value of ejecta speed for the second case. The choice of parameters has an important effect on the collisional evolution; better information regarding these quantities from both collisional experiments and theoretical modeling is important for evolution studies.

Unless mean collision velocities have been significantly larger in the past than the current 5 km sec^{-1}, our modeling implies there could not have been many more large (500–1000 km) asteroids in the past than presently exist. As shown in Fig. 4, and discussed by Chapman and Davis (1975), if there were a large number of such bodies initially then there could have been significant collisional depletion, but the residual number of such bodies would be much larger than exists today in the asteroid belt. The effect on asteroids of other

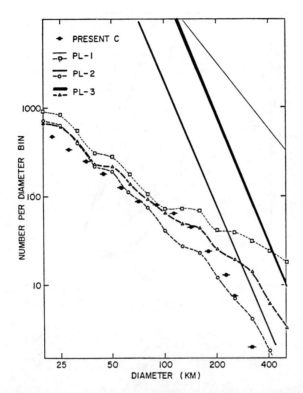

Fig. 5. Collisional evolution over 4.5 Gyr using nominal C parameters for 3 initial power-law distributions. PL-1 containing 0.1 $M_\oplus$ initially, PL-2 with 0.05 $M_\oplus$, and PL-3 with 0.5 $M_\oplus$. The present C population is from the bias-corrected statistics of Zellner and Bowell (1977).

populations of bodies, such as Jupiter-scattered planetesimals, is addressed in a later section, but the result that large asteroids were only modestly more populous in early times remains unaltered. The difficulty in collisionally reducing a large initial population of big asteroids to the present population is further illustrated in the following sections which describe the evolution of several populations in each of the previously defined distributions.

Pure Power-Law Distribution. The evolution of three initial power-law distributions is shown in Fig. 5. These populations cover a range of initial mass from 0.05 $M_\oplus$ to 0.5 $M_\oplus$ distributed with slopes of $b = 3.5$ and 5.75. All populations evolve to near the present belt at diameters smaller than ~100 km, while the number of large diameter bodies essentially reflects the initial distribution. For power-law distributions, the initial belt mass could not have been much less than 0.05 $M_\oplus$. Small initial masses, requiring shallower slopes than case PL-2 would result in collisional depletion for intermediate size

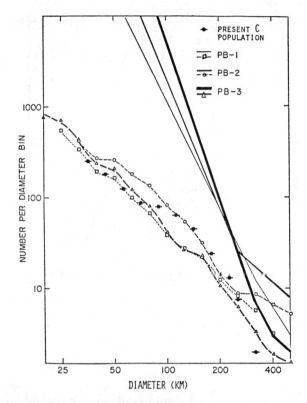

Fig. 6. Evolutionary outcomes for initial distributions of the type found by Greenberg *et al.* (1978) at the end of their simulation of intermediate-state planet growth by accretion. These distributions have a few large bodies (500-1000 km diameter), but most of the mass of the population is in smaller bodies. Population PB-1 has an initial mass 0.01 $M_\oplus$, PB-2 contains 0.05 $M_\oplus$, while PB-3 has an initial mass of nearly 1 $M_\oplus$. The present *C* population is as in Fig. 5.

asteroids. An upper bound is limited only by the steepness of the population index, but even extreme values suggest that the initial belt mass could not be much larger than ~0.1 $M_\oplus$. In all cases, the small diameter population evolves to a slope of 2.4, which is essentially the equilibrium value for these models. The initially steep population indices typically relax to a value near 2.4 in ~ 2 × 10^8 yr.

Non-Power Law Distribution. Greenberg *et al.* (1978), in a simulation of planetesimal accretion, found population distributions with a few large bodies but with most of the mass in small bodies at the end of the early stages of growth. This may be a plausible initial population distribution for asteroids. The evolution of several such initial distributions is illustrated in Fig. 6, covering a mass range of 0.01 $M_\oplus$ (PB-1) to nearly an Earth mass (PB-3). The smallest initial belt results in excessive depletion at sizes around 100 km,

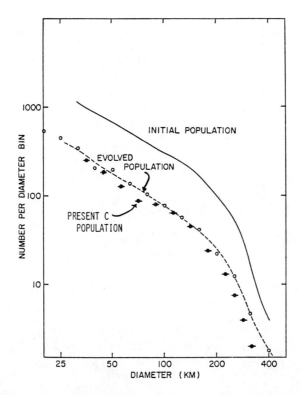

Fig. 7. Evolution of a small initial belt containing 3 times the present belt mass. The present C population is as in Fig. 5.

while the 0.05 $M_\oplus$ Case PB–2 results in too many bodies larger than 400 km diameter. The massive initial belt, PB–3, with only a few large bodies, does evolve to a final distribution close to the present C population. The collisional models indicate that no strong constraint can be placed on the mass of the initial belt for initial population distributions of this type; both massive or small initial belts are consistent with the present population.

Non-power law initial populations that are only modestly larger than the present belt, are consistent with the observed population as illustrated by Fig. 7. This case started with an initial belt having a mass three times that of the present belt with most of the additional mass in bodies < 400 km diameter.

Gaussian Initial Distributions. Not all initial distributions that have been suggested for asteroids evolve to the present distribution. Notable exceptions are some initial "Gaussian" distributions. The evolution of two such initial populations is illustrated in Fig. 8; the population G–1 is a small initial belt while G–2 has a mass of 100 times the present belt and both populations

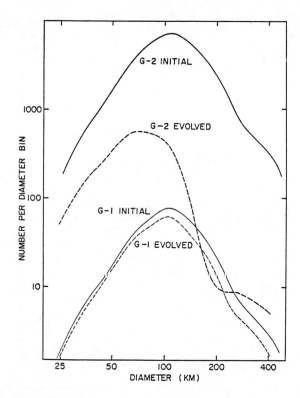

Fig. 8. Collisional outcomes after 4.5 Gyr of initial "Gaussian" populations having a number peak at 100 km. Population G-1 has an initial mass about that of the current belt, while G-2 is 100 times as massive.

peak at 100 km diameter. The evolution over 4.5 Gyr shows little change in the shape of the population, although the amplitude and peak number of the distribution change significantly in the G−2 case.

Such populations retain their shape because the energy required to fragment most of these asteroids is determined by gravitational binding. The debris from such bodies is usually thoroughly comminuted (Fig. 2) and forms a collisional tail of very small particles separated from the main part of the distribution (off the small diameter edge of Fig. 7); only the "cores" from barely catastrophic collisions could fill the gap, but these are relatively infrequent occurrences. Of course, if the energy density required to fracture asteroids were much larger than currently believed, i.e. if material strength dominates the binding of 100 km asteroids, then a more continuous distribution of fragments would be created. More sophisticated models for the collisional physics, including distributions of impact energy over the target volume and off center collisions may also modify the above result.

Two-component collisional evolution (Program P2)

The simultaneous collisional evolution of the two populations of asteroids having different strengths was modeled in order to represent the two main asteroid classes, C and S. (All non-C asteroids were assumed to have the same physical parameters as S asteroids.) Different initial populations were explored to see if the interaction of populations with differing physical properties would further constrain the range of initial populations. The parameters of Table I were used in these models while a mean collision speed of 5 km sec^{-1} was adopted for each population colliding with itself and for collisions between the two populations.

Analysis of several initial power-law distributions for both components suggested that the C/S ratio could not have differed significantly in the early solar system from its present value. This is because each class can smash the other with essentially the same efficiency, particularly when gravity dominates. Hence the same reduction due to collisions will apply to both populations; if collisions have reduced 100 km C asteroids by a factor 5, the S population will have been reduced by a similar factor. This result suggests that the increasing C/S ratio with semimajor axis across the belt (c.f. Zellner's chapter) is not due to preferential destruction of C asteroids in inner regions of the belt where velocities may be somewhat higher than in the outer part. The C/S variation is probably a consequence of the primordial processes that originally formed asteroids.

The two-component model was also used to study the population evolution of asteroids bombarded by Jupiter-scattered planetesimals (JSP). Several investigators (Weidenschilling 1975; Ip 1977b; Safronov 1972) have suggested that such a population could have reduced the mass in the asteroid zone through collisions. The effect of JSP on asteroids depends on the total mass and distribution of bombarding bodies and asteroids.

The evolution of an initial ~ 1 $M_\oplus$ asteroid belt with largest bodies of 1000 km diameter bombarded by a 3 $M_\oplus$ JSP population at 7.5 km sec^{-1} is shown in Fig. 9. The JSP population is removed by Jupiter after 10^8 yr and the residual asteroid population is allowed to collisionally evolve at 5 km sec^{-1} over the next 4.5×10^9 yr. The evolution in this particular case does not evolve to the present belt; there are excessively large asteroids due to the difficulty of smashing such bodies. Jupiter-scattered bodies certainly could have collisionally depleted a massive early asteroid belt, just as such a population would collisionally relax by self-colliding at 5 km sec^{-1}.

Collisional lifetimes in the present belt for both C and non-C asteroids are given in Fig. 10 for two sets of collisional parameters. The longer lifetimes for each population are for $f_{KE} = 0.1$, $k = 9/4$, while the shorter values result from using $f_{KE} = 0.5$, $k = 20$. The latter values are extremely favorable for collisional destruction and they constitute a minimum bound on present lifetimes for collisional destruction. The upper curves are lifetimes based on

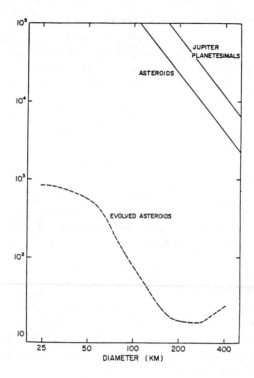

Fig. 9. Evolution after 4.5 Gyr of an initial power-law asteroid population containing 1 $M_\oplus$ colliding with a 3 $M_\oplus$ population of Jupiter-scattered planetesimals for 10^8 yr and subsequently collisionally evolving only with itself.

our nominal collisional parameters; however actual values for f_{KE} and k are poorly known and, as can be seen from Fig. 10, asteroid lifetimes are somewhat sensitive to these parameters. The largest asteroids are essentially indestructible by collisions with other asteroids at 5 km sec^{-1}. For example, at 5 km sec^{-1}, Ceres could be broken up only by a collision with another Ceres-sized body; but there are none in the present belt. The only existing asteroid capable of destroying Ceres would be one with a diameter > 500 km impacting at ⩾ 10 km sec^{-1}, namely Pallas.

    While most asteroids are being destroyed and collisionally eroded, we point out that Ceres is very near the size at which net accretion could result in the dynamical environment of the present belt. If Ceres had been significantly larger by the time asteroid velocities were pumped up to their present values, then the "missing planet" likely would not be missing (provided there was sufficient mass in the belt at that time).

    Hirayama families are treated in the chapter by Gradie *et al.*, so we

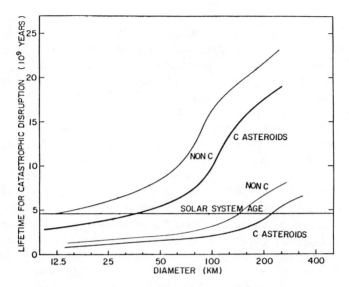

Fig. 10. Collisional lifetimes in the present belt are shown for *C* and non-*C* asteroids as a function of asteroid diameter for two sets of collisional parameters. The upper curves are for nominal collisional parameters (Table I), while the shorter lifetimes (lower curves) are for "efficient" collisional parameters, $f_{KE} = 0.5$ and $k = 20$; other parameters are the same as in Table I.

restrict ourselves to a few comments on these presumed collisional remnants. There is a diversity of size-frequency distributions among the families; however, many of the distributions are consistent with our models for various types of collisional events. For instance, Williams family 189 Flora could result from a large cratering impact; Williams family 132 Concordia resembles barely catastrophic disruption; and family 3 Koronis looks like a completely catastrophic collision. Ip (1979) has also pointed out interpretations of Hirayama families in terms of collisional event types. Since collisions are the dominant process that removes family members and collisional lifetimes are size-dependent, the size-frequency distributions of families may contain information on the ages of the families.

III. COLLISIONAL INFLUENCE ON ASTEROID ROTATION RATES

The subject of collisional evolution of asteroid rotations has been discussed in detail by Harris (1979). In this section, we summarize the principal results and limitations of that work. The reader is referred to the above paper for a more complete and quantitative development.

The importance of collisions in establishing the present rotational motion of asteroids has been suggested in a qualitative way by several authors

(Safronov 1972; McAdoo and Burns 1973; Napier and Dodd 1974). Safronov and Zvjagina (1969) derived expressions for the acquisition of angular momentum by a planet resulting from many randomly oriented collisions during the accretion process. Burns and Safronov (1973) used a similar analysis to study the accumulation of transverse angular momentum by asteroids. Dohnanyi (1976) applied a similar random walk analysis to the problem of asteroid rotation, but failed to recognize the decelerating effect of collisions on an already spinning object. Hence he concluded that collisions spin the asteroids ever faster to the limit of rotational bursting. The observed distribution of rotation rates does not conform to such a model (Harris and Burns 1979; see also the chapter by Burns and Tedesco in this book).

The Harris (1979) model of collisional evolution of rotation rate follows the derivation of Safronov and Zvjagina (1969) for the increase of random rotational motion by a growing planet. The following simplified dimensional derivation illustrates the essence of the model. The increment of angular momentum produced by a completely inelastic collision between a projectile mass m at velocity v and a target body of mass M and diameter D is $mv\ell$, where ℓ is the impact parameter. For randomly directed impact trajectories, the expected value of ℓ is $D/2\sqrt{2}$. The above increment of angular momentum is randomly oriented with respect to the pre-collision angular momentum of M; hence the increment must be added quadratically:

$$\langle\, dh^2 \,\rangle \,=\, \frac{1}{2}\, m^2 v^2\, (D/2)^2 . \tag{5}$$

The differential of the angular momentum, h^2, can be expanded as follows:

$$\begin{aligned}
d(h^2) &= d\left[\frac{2}{5}\, M\left(\frac{D}{2}\right)^2 \omega\right]^2 \\
&= \frac{4}{25}\left(\frac{3}{4\pi\rho}\right)^{\frac{4}{3}}\left(\frac{10}{3}\,\omega^2 M^{\frac{7}{3}}\, dM + 2M^{-\frac{10}{3}}\,\omega d\omega\right)
\end{aligned} \tag{6}$$

where ω is the rotation frequency of the asteroid. The constant 2/5 is taken as appropriate for a homogeneous sphere. The two expressions (5) and (6) can be equated, with the mass increment dM replaced by m as appropriate for an inelastic collision, and solved for $d\omega$ as

$$d\omega \,=\, \frac{25}{4}\left(\frac{m}{M}\right)^2 \frac{v^2}{D^2\omega} \,-\, \frac{5}{3}\,\frac{m}{M}\,\omega . \tag{7}$$

From the above expression it is clear that a succession of collisions tends to increase the mean rotation rate as a random walk (first term), but at a lesser rate with increasing ω due to the quadratic addition. On the other hand, the rotation rate is decreased due to the addition of mass, which must be given a share of the already existing angular momentum (second term). Strictly speaking, the added mass shares the total post-impact angular momentum; hence the above relation is valid only if the imparted angular momentum is small compared to the initial angular momentum. This decrease can be thought of as a drag on the asteroid and is indeed greater the faster the spin of the body. It is thus evident that the mean rotation rate must seek an equilibrium value, rather than increasing indefinitely.

So far, the rotational effect of impacts of bodies of a particular size has been considered, while a real asteroid is hit by a variety of different size objects. The Harris (1979) theory adopts a power-law size distribution of bodies, and integrates the angular velocity change over the size distribution to a largest impacting mass, m_1. The resulting equation for $d\omega$ has the same form as Eq. (7), but also has coefficients in each term depending on the slope of the distribution; for reasonable distributions, $b \sim 2.5 - 3.5$, the coefficients are of $0(1)$. Also, the theory is restricted to distributions with $b < 4$. If m_1 is taken as the largest mass that does not catastrophically disrupt the target body, then Eq. (7) is the expected change in angular velocity of a body up to the time it is collisionally destroyed.

In the above development, it has been assumed that collisions are inelastic. This, of course, is unrealistic for hypervelocity impacts on asteroidal-sized bodies which typically are erosive rather than accretional events. However, in the absence of preferential impact directions, impacts should be uniformly distributed over the surface. Ejecta from such high speed events escape nearly uniformly in all directions with respect to the rotating asteroid surface, even for rather oblique angles of incidence. Hence they carry with them angular momentum, but do not change the rotation frequency. The drag effect, therefore, is still expected and Eq. (7) is applicable, even though mass is lost rather than gained in high velocity collisions. It should be noted that this model may be only a crude approximation to the true situation for large, oblique impacts which are the most efficient type of collisions for momentum transfer. The mass loss dM due to all collisions prior to disruption has an associated diameter change which we express as

$$dD = -\alpha D . \qquad (8)$$

By our definition of catastrophic disruption, that the largest fragment contains less than half the mass of the original body, $\alpha > 1 - 2^{-1/3} \approx 0.2$. The value of α is at most 1.0, corresponding to total pulverization. (It will turn out that the solutions are not strongly sensitive to the value of α; thus the above crude model is sufficient.) By dividing Eq. (7) by (8), a differential equation is obtained:

$$\frac{d\omega}{dD} = -\frac{25}{4}\left(\frac{m}{M}\right)^2 \frac{v^2}{\alpha D^3 \omega} + \frac{5}{3}\frac{m}{M}\frac{\omega}{\alpha D}. \tag{9}$$

It is clear from the form of Eq. (9) that the largest collisions are of greatest importance in momentum transfer. This is true up to the point of catastrophic breakup. We assume, for lack of better insight, that still larger collisions are no more effective than barely sub-catastrophic collisions in producing rotational impulses to the fragments; the excess impulse becomes linear momentum of the scattered fragments.

For asteroids for which gravity is the principal binding force, the mass ratio m/M necessary to disrupt a body is found as

$$\frac{m}{M} \approx \frac{12}{5}\frac{GM}{f_{KE}Dv^2}. \tag{10}$$

The parameter f_{KE} is equivalent to γ^{-1} in Harris (1979) and, for the gravity dominant case, f_{KE} can be interpreted as the ratio of target gravitational binding energy to collisional kinetic energy for barely catastrophic disruption. For m/M given by Eq. (10), Eq. (9) becomes

$$\frac{d\omega}{dD} = -\frac{9\omega_0^4}{16\alpha v^2 f_{KE}^2}\frac{D}{\omega} + \frac{\omega_0^2}{2f_{KE}\alpha v^2}D\omega \tag{11}$$

where $\omega_0 = (4/3\ \pi G\rho)^{1/2}$ which is the surface orbit frequency about the asteroid. Harris (1979) finds the asymptotic solution to the above equation — valid for $D \gg D_0$ which is the diameter at which material strength equals the gravitational binding strength — to be a constant rotation rate. Setting $d\omega/dD = 0$, we obtain

$$\omega = \frac{3\omega_0}{2\sqrt{2}\sqrt{f_{KE}}}. \tag{12}$$

In the more complete theory, the numerical constant $3/(2\sqrt{2})$ is replaced by a value ~ 0.2. As ω_0 corresponds to a rotation period ~ 2 hr, the theoretical equilibrium rotation period agrees with observations only if a large fraction $(0.5 - 1.0)$ of the collisional kinetic energy is converted into ejecta kinetic energy.

At the opposite end of the size spectrum, only material strength is important in resisting disruptive collisions. The minimum mass ratio to disrupt small asteroids may be found from Eq. (2) as

$$\frac{m}{M} \approx \frac{2S}{\rho v^2}. \tag{13}$$

With this value in Eq. (9), it is shown by Harris (1979) that the following is an asymptotic solution for the case $D \ll D_0$:

$$\omega = \frac{5S}{\sqrt{\alpha}\rho v D} \left(1 + \frac{10S}{3\alpha\rho v^2}\right)^{-\frac{1}{2}}. \tag{14}$$

For $S \approx 3 \times 10^7$ ergs cm^{-3}, v = 5 km sec^{-1}, $\rho = 3$ g cm^{-3}, and $\sqrt{\alpha} = 1/2$, the above relation yields

$$\omega \approx \frac{2 \cdot 10^{-3}}{D} \text{ sec}^{-1} \tag{15}$$

corresponding to a period $P \approx 0.9\,D$ hr, when D is in km.

The asymptotic rotation rates for the pure-gravity binding case and the pure-strength binding case can be achieved if the rotational relaxation time is less than the mean lifetime against catastrophic disruption. As noted earlier, the largest bodies impart the greatest angular momentum; so, approximating the total collisional angular momentum by that due to the largest impacting body, the condition for rotational relaxation is that the collisional angular momentum exceed the pre-existing angular momentum or

$$\frac{mv\,(D/2)/\sqrt{2}}{\frac{2}{5}\,M\,(D/2)^2\,\omega} \gtrsim 1. \tag{16}$$

For the gravity binding case, we use m/M from Eq. (10) which gives the following lower limit for D:

$$D \gtrsim \frac{2\sqrt{2}\,\omega v f_{KE}}{3\,\omega_0^2}. \tag{17}$$

For the present values of v ≈ 5 km sec^{-1} and $\omega \approx 2 \times 10^{-4}$ sec^{-1}, the above expression yields $D \gtrsim 10^3\,f_{KE}$ for rotational relaxation to occur before catastrophic disruption occurs. Using m/M from Eq. (13) for the strength dominant case, the corresponding diameter limit becomes

$$D \lesssim \frac{10}{\sqrt{2}}\,\frac{S}{\rho v \omega} \sim 2 \text{ km} \tag{18}$$

when parameter values following Eq. (14) are used. There may be an increase in the observed rotation rates for the smallest asteroids (see the chapter by Burns and Tedesco).

The rotational theory predicts that large asteroids and small asteroids will collisionally relax to the values given by Eqs. (12) and (14) respectively before they are collisionally destroyed, assuming that the asteroid population can be modeled with a power-law size-frequency distribution. Current data (see Zellner's chapter), however, show that a power law is only marginally adequate to represent the asteroid population.

With these and other previously noted caveats in mind, we now consider some implications of the rotational evolution theory. For large asteroids the value of the equilibrium rotation rate depends critically on the choice of model parameters, particularly f_{KE}. If $f_{KE} \sim 0.1$, as was adopted for collisional models, the predicted equilibrium rotation period for asteroids larger than 100 km diameter is $\sim$3 hr, much shorter than is observed. There are three possible explanations for this disparity:

(a) that f_{KE} is greater than our preferred value of 0.1,
(b) that equilibrium rotation periods have not yet been reached for gravitationally bound bodies, or
(c) this model is inadequate to explain asteroidal rotation evolution.

The first possibility cannot be ruled out, as our nominal value of f_{KE} derives from studies of crater impacts into an infinite half-space and it is conceivable that for real finite volumes, less collisional energy is dissipated as heat and a greater fraction is converted into ejecta kinetic energy. It is not clear that the large values of f_{KE} (>0.5) necessary to produce agreement between theory and observations can be achieved in nature. Further studies are certainly needed in this area.

The second possibility implies that the number density of asteroids could not have been significantly greater in the past than it is today. Otherwise, the time scale for rotational relaxation would have been less than the age of the solar system and large asteroids would have maintained rotational equilibrium as they collisionally relaxed. This conclusion is consistent with the result from Sec. II that, for power-law distributions, only modestly bigger initial belts evolve to the present belt unless the population index is steep ($b > 4$); however, the rotational theory is valid only for $b < 4$.

The third possibility requires that more sophisticated models capable of treating non-power law distributions should be applied to investigate the origin of asteroid rotations. Also, most small asteroids are multigeneration fragments and the initial rotation rates of such fragments may be different from that of the parent body; this possibility is not treated in the Harris model. At this time, we can only note that the rotation studies to date do not significantly restrict the range of initial asteroid belt populations relative to what has been found from collisional studies. New insights based on a simultaneous treatment of both collisional and rotational evolution may be more informative than the sum of the individual investigations.

For example, if it turns out that $f_{KE} \sim 0.9$ is an appropriate value for high-speed collisions between asteroids, then the equilibrium period is about 10 hr; but from Eq. (17) only the largest asteroids could collisionally relax before they are destroyed. The rotation periods of most asteroids would then be considered either primordial, if there were little collisional evolution, or considered to be established by the distribution of rotation rates among fragments of catastrophic collisions. If $f_{KE} \sim 0.1$ is really appropriate for asteroidal collisions and future studies of non-power law distributions confirm our results based on Eqs. (12) and (17), then one could conclude that there has been only modest collisional evolution of asteroids since their velocities were pumped up, and that the initial belt mass was only $\sim 2-3$ times the present mass. Consequently, mass depletion in the asteroid zone occurred prior to, or simultaneous with, the pumping up of collision speeds.

The frequency of asteroidal satellites and binary asteroids could provide a constraint on the belt population at the time collisional destruction became dominant. Plausible mechanisms have not yet been proposed in order to produce asteroid satellites as frequent products of high-speed collisions, which is the dominant process affecting asteroids over much of their history. In the absence of collisional formation mechanisms, we might assume that binary asteroids and asteroidal satellites, if they exist, are products of the asteroid formation era, produced before collisionally destructive high encounter speeds were developed. Thus, those that survive today are the bodies that chanced to have both members escape destruction, an especially severe requirement for smaller bodies. Therefore, asteroids that have satellites could be interpreted as first-generation objects that survived from the early age of the belt and if these are common today, then the initial population could have been only modestly larger than the present belt population. If satellites are not common, then either the belt was substantially bigger initially or such objects were never common. Clearly, confirmation of the existence and frequency of asteroidal satellites and an understanding of how they might have formed are important questions relating to the origin and evolution of asteroids.

IV. EARLY ASTEROIDAL EVOLUTION:
THE ORIGIN OF THE HIGH VELOCITIES

Currently asteroids are moving in orbits of moderate eccentricity and inclination that produce a mean collision speed ~ 5 km sec^{-1}. These velocities are sufficiently high that virtually all inter-asteroidal collisions result in erosion or fragmentation rather than accretion, with the exception of Ceres which is nearly big enough to be accreting at the mean asteroid impact speed. However, the relative collision speed must have been smaller in the past in order for asteroids to have accreted from still smaller sized planetesimals. Thus, some process must have converted the original low velocities to the high velocities that dominate asteroidal evolution today. Discovering that process is fundamental to understanding the asteroids, for, as many previous researchers have speculated, it might be just that process which changed asteroidal planetesimals from net accretion to net fragmentation that prevented the formation of an asteroidal planet. Additionally, the present collisional regime may be capable of removing much original mass from the asteroid zone, thus accounting for the small mass present in the asteroidal belt today compared with the masses of other planets. The mass in the asteroid zone must have been higher during accretion, for the mass surface density in the present belt would require intervals $\sim 10^9 - 10^{11}$ yr to grow bodies 1000 km in diameter, based on growth times of Safronov (1972).

Several approaches can be taken to solving this problem. First, can the asteroids have stirred each other up by mutual gravitational interactions? The present population is clearly incapable of doing so. Even if all asteroids were capable of gravitationally interacting with each other, which they are not, the amount of stirring cannot exceed the order of the escape velocity of the largest body in the population, Ceres, which is about 0.5 km sec^{-1}. Impact velocities, calculated using $\theta \sim 3-5$ in equilibrium relations by Safronov (1972), have values ~ 0.25 km sec^{-1} for the present belt, a factor of 20 too low.

Perhaps some precursor population to the present belt, of which the present population is a remnant, could have stirred itself. Such self-stirring requires a population containing a single body substantially larger than Ceres, or even the moon, or perhaps a population with numerous large (but not quite so extremely large) bodies. The case of a single large body cannot work because even Ceres is nearly large enough to accrete in the present collisional environment and we have already seen how difficult it is to collisionally erode or destroy bodies significantly larger than Ceres. In short, there is no ready mode to remove such a large body, once it was formed. A large number of moderately large asteroids (e.g. Ceres-sized) can somewhat enhance the stirring effect over that of a single body, but the stirring efficiency drops rapidly as the relative velocities exceed the escape velocity of the bodies. Moreover, there remains the difficulty of accounting for the removal of these bodies. To test these speculations, we numerically traced the evolution of a massive ($1 \, M_\oplus$) initial belt with a low (200 m sec^{-1}) mean collision speed and with the largest bodies of Ceres' size (1000 km diameter). Both accretion of the largest bodies and velocity pumping were noted; however the simulation indicated that a planet would be formed before collisional destruction of remaining small bodies became dominant.

Another approach to the problem is to appeal to an external source. Because of the proximity of the asteroids to Jupiter, many scientists have invoked Jupiter's influence to explain the low mass of the asteroids and their high velocities. Jupiter does exert important gravitational influences within the asteroid belt, but as described in the chapter by Safronov, these effects are located at discrete resonances, and it is difficult to see how the population as a whole can have been substantially affected by Jupiter's direct action in the present dynamical environment of asteroids.

However, in the early stages of solar system formation, gas drag coupled with secular resonances with Jupiter could have pumped up eccentricities and inclinations in the inner solar system (Chapman et al. 1978; Ward 1978; Heppenheimer 1979). Such a mechanism, sweeping through the asteroidal zone as the nebula dissipated, could explain the large mean encounter speed.

Jupiter can affect the asteroids more substantially through indirect effects. For instance, Weidenschilling (1975) and Safronov (1972) have proposed that the bombardment of asteroidal planetesimals by Jupiter-zone planetesimals, scattered by that planet into the asteroidal region, could account for depletion of mass and enhanced velocities of asteroids. However, the depletion of mass through collisional fragmentation is easier to understand than any appreciable effect on velocities. From conservation of momentum it is clear that the orbital velocity of an asteroid can be changed by as much as a large fraction of the relative velocity only by collision with a body having the same mass as the asteroid.

Alternatively, it takes collisions with n^2 bodies of $1/n$-th the mass of the asteroid, impacting from random directions, to augment the asteroid's velocity by n times the relative velocity. But, at 1 km sec^{-1} a 50 km asteroid is fragmented and dispersed by collision with a body having a mass ratio m/M larger than $\sim 1/30$; at 5 km sec^{-1}, the limiting mass ratio is $\sim 1/1000$. For a body as large as Vesta the limiting mass ratio is $1/10$ at 5 km sec^{-1}. In other words, asteroids are too weak to be moved intact by collisions. Impacts with n^2 bodies of sufficiently small mass to be below the catastrophic fragmentation limit are implausible since, for reasonable size distributions, the probability of impact with larger bodies approaches unity. Small asteroids could be accelerated to higher relative velocities through multi-generational collisions (i.e., fragments of fragments of fragments). But, some of the largest asteroids have the largest eccentricities and inclinations; as pointed out by Whipple et al. (1972), the high inclination of 580 km diameter Pallas sets a strong constraint on the origin of asteroids.

One appealing explanation of the high velocities in the asteroid belt is gravitational stirring of the early population by one or more massive Jupiter-scattered planetesimals. As a rough criterion for the stirring to be more effective than collisional destruction, the escape velocity of the perturbing body should exceed the mean encounter velocity (several km sec^{-1}); this implies a mass on the order of one Earth mass. Each asteroid would be expected to experience several encounters within a distance of 10^{-2} AU, resulting in accumulated relative velocities of a few km sec^{-1}. It is reasonable to assume that such a large body would be accompanied by a population of smaller bodies. They might, depending on their population index, cause catastrophic fragmentation of much of the original asteroidal population, while the large body was increasing the velocities of the survivors. Alternatively, the fragmentation may have been due to mutual collisions among the asteroids after their relative velocities were enhanced.

Numerical simulations of the effect of a few Earth-sized JSP on a hypothetical early asteroid belt were carried out using the P1 program. At the time the JSP had been deflected into asteroid-crossing orbits by Jupiter, the asteroid belt was assumed to contain 0.1 $M_\oplus$ with 500 km diameter bodies being the largest asteroids, while the mean collision speed was 200 m sec^{-1}, low enough so that most asteroids were accreting. The mean collision speed between asteroids increased to about 5 km sec^{-1} within 10^6–10^7 yr depending on the number and size of JSP as well as the characteristics of their orbits. For example, 5 one-$M_\oplus$ JSP moving in orbits having a mean encounter speed of ~ 6 km sec^{-1} with the asteroids, pumped up the collision speed to 5 km sec^{-1} in 3×10^6 yr. Since the time scale for increasing asteroid velocities is comparable to the time scale for elimination of JSP by Jupiter (Weidenschilling 1975), this mechanism provides a plausible explanation for the origin of the long mean impact speed among asteroids. The collision evolution subsequent to removal of the JSP was similar to that described in Sec. II.

The existence of such large Jupiter-scattered bodies depends on the mode of formation of the outer planets, but it is a plausible consequence of some models. If Jupiter and Saturn formed by hydrodynamic accretion of gas onto solid protoplanetary cores, the required core masses are at least several times that of Earth (Harris 1978). If such cores existed, it is unlikely that they were the only large solid bodies which formed in those zones. Estimates of planetesimal size distributions are highly model-dependent (c.f. Safronov 1972, 1979; Wetherill 1976; Greenberg 1979); a second-largest body in the zone of Jupiter (or Saturn) on the order of one Earth mass cannot be ruled out. Alternatively, one or more large bodies may have grown in dynamical isolation from the Jupiter and Saturn cores until the latter became large enough to accrete gas. The time scale for gas accretion is much shorter than that for accretion of solids. The increase in mass of Jupiter and Saturn would render unstable the orbits of the remaining bodies. Their increased velocities, and possibly depletion of gas in their vicinity, would prevent their further growth.

Features of their subsequent orbital evolution can be deduced from the work of Everhart (1973) and Ip (1977*b*). Everhart used numerical integration to show that orbits between Jupiter and Saturn are generally unstable on a time scale of 10^5 yr. Most such starting orbits led to perihelia < 2.6 AU for some time before ejection from the solar system. In Ip's Monte Carlo simulations most objects that had started in Saturn- and Uranus-crossing orbits also evolved into Jupiter-crossers before being ejected from the solar system (the most probable fate) or having collided with a planet. Both approaches indicated that only a small fraction of such objects would achieve perihelia $\leqslant 1$ AU before ejection, as suggested by Weidenschilling (1975) from estimates of the ejection probability. Therefore the orbits of the terrestrial planets should have been unaffected by encounters with a massive body (though one might explain Mars' eccentricity by a rare moderately close encounter).

There is additional empirical evidence for the existence of large planetesimals in the outer solar system. Safronov (1965) estimated the mass of the largest body impacting on each planet, on the assumption that planetary obliquities are due to a random component of angular momentum from infalling planetesimals. If the axial tilts of Saturn and Uranus are due to such impacts, the bodies involved were on the order of one Earth mass.

The advantages of this JSP scenario for the early evolution of asteroidal velocities are:

1. There is no problem analogous to that of getting rid of very large proto-asteroids, since the scattered body (or bodies) is ejected by Jupiter.
2. The stirring is accomplished by gravitational encounters rather than by collisions; the latter would surely fragment the asteroids more effectively than stirring them.
3. Large asteroids, such as Pallas, are stirred as effectively as small bodies.
4. The reduction of mass in the asteroidal zone may be accomplished by an accompanying population of smaller JSP or by collisions among the asteroids at their enhanced velocities, depending on the original mass distributions of both components.
5. The existence of one or more Earth-sized JSP is a plausible accompaniment to the formation of the outer planets (although its necessity remains to be demonstrated) and the time scale for their evolution is reasonable.

Possible limitations of this scenario include: (1) There is no ready explanation of the Kirkwood gaps; they must be formed by a different process than the velocity pumping mechanism, and (2) the degree of gravitational stirring depends on the random orbital evolution of the few largest Jupiter-scattered bodies, and therefore on the statistics of small numbers. More rigorous calculations need to be carried out to show that such an event, together with subsequent collisional evolution among the asteroids themselves, can produce the observed distributions of eccentricities and inclinations, as well as their mean values.

The orientation of rotation axes of large asteroids does provide a test of collisional versus gravitational stirring mechanisms. If such bodies were somehow strong enough or lucky enough to survive collisions with Jupiter-scattered planetesimals, or were fragments of still larger primordial bodies, their spin axis directions should have been randomized. However, gravitational stirring, whether by massive planetesimals or secular resonances, would not affect their rotational states. There is some evidence, although marginal, for a tendency for the largest asteroids to have prograde rotation

(Morrison 1977; see the chapter by Taylor). If confirmed, this would indicate that their rotational states were established during accretion and that their velocities were established by gravitational perturbations.

Acknowledgments. We benefited greatly from several stimulating discussions with V. S. Safronov during his recent visits to Pasadena and Tucson. At the Planetary Science Institute, the invaluable assistance of J. Wacker and A. Hostetler in program development and data generation is gratefully acknowledged, as is the careful and patient typing and editing of the manuscript by J. Dingell. This work was supported at the Planetary Science Institute and at Jet Propulsion Laboratory by the National Aeronautics and Space Administration.

REFERENCES

Anders, E. 1965. Fragmentation history of asteroids. *Icarus* 4: 399.

Durno, J. A., and Safronov, V. S. 1973. Asteroid nutation angles. *Mon. Not. Roy. Astron. Soc.* 165: 403-411.

Chapman, C. R., and Davis, C. R. 1975. Asteroid collisional evolution: Evidence for a much larger early population. *Science* 190: 553-556.

Chapman, C. R.; Williams, J. G.; and Hartmann, W. K. 1978. The asteroids. *Ann. Rev. Astron. Astrophys.*, eds. G. Burbidge, D. Layzer and J. Phillips (Palo Alto, Calif.: Annual Reviews, Inc.), Vol. 16, p. 33.

Dohnanyi, J. W. 1969. Collisional model of asteroids and their debris. *J. Geophys. Res.* 74: 2531-2554.

Dohnanyi, J. S. 1976. Sources of interplanetary dust: Asteroids. In *Interplanetary Dust and Zodiacal Light,* eds. H. Elsässer and H. Fechtig (Berlin: Springer-Verlag), pp. 187-205.

Everhart, E. 1973. Horseshoe and Trojan orbits associated with Jupiter and Saturn. *Astron. J.* 78: 316-328.

Fujiwara, A.; Kamimoto, G.; and Tsukamoto, A. 1977. Destruction of basaltic bodies by high velocity impact. *Icarus* 31: 277-288.

Gault, D. E.; Shoemaker, E. M.; and Moore, H. J. 1963. Spray ejected from the lunar surface by meteoroid impact. NASA TN D-1767.

Gault, D. E., and Wedekind, J. A. 1969. The destruction of tektites by micrometeoroid impact. *J. Geophys. Res.* 74: 6780-6794.

Greenberg, R. 1979. Growth of large, late-stage planetesimals. *Icarus* (in press).

Greenberg, R.; Wacker, J.; Hartmann, W. K.; and Chapman, C. R. 1978. Planetesimals to planets: Numerical simulation of collisional evolution. *Icarus* 35: 1-26.

Harris, A. W. 1978. The formation of the outer planets. *Lunar Science IX Abstracts.* The Lunar and Planet. Inst., pp. 459-461.

Harris, A. W. 1979. Comments on asteroid rotation. II. A theory for the collisional evolution of rotation rates. *Icarus* (in press).

Harris, A. W., and Burns, J. A. 1979. Asteroid Rotation. I. Tabulation and analysis of data on rates, shapes and rotation poles. *Icarus* (in press).

Hartmann, W. K. 1978. Planet formation: Mechanism of early growth. *Icarus* 33: 50-61.

Hartmann, W. K., and Hartmann, A. C. 1968. Asteroid collisions and evolution of asteroidal mass distribution and meteoritic flux. *Icarus* 8: 361.

Heppenheimer, T. A. 1979. Secular resonances and the origin of eccentricities of Mars and the asteroids. *Lunar Science X Abstracts.* The Lunar and Planet. Inst., pp. 531-533.

Ip, W. H. 1977a. On the orbital dependence of the asteroidal collision process. *Icarus* 32: 378-380.

Ip, W. H. 1977*b*. On the early scattering processes of the outer planets. In *Comets, Asteroids, Meteorites,* ed. A. H. Delsemme (Toledo, Ohio: University of Toledo Press), pp. 485-490.

Ip, W. H. 1979. On three types of fragmentation processes observed in the asteroid belt. *Icarus* (special Asteroid issue).

Kuiper, G. P.; Fujita, Y.; Gehrels, T.; Groeneveld, I. 1958. Survey of asteroids. *Astrophys. J. Suppl.* 3: 289-428.

McAdoo, D. C., and Burns, J. A. 1973. Further evidence for collisions among asteroids. *Icarus* 18: 285-293.

Morrison, D. 1977. Asteroid sizes and albedos. *Icarus* 31: 185-220.

Napier, W. McD., and Dodd, R. J. 1974. On the origin of asteroids. *Mon. Not. Roy. Astron. Soc.* 166: 469-489.

O'Keefe, J. D., and Ahrens, T. J. 1977. Impact induced energy partitioning, melting and vaporization on terrestrial planets. *Proc. Lunar Sci. Conf. VIII* (Oxford: Pergamon Press), pp. 3357-3374.

Piotrowski, S. I. 1953. The collisions of asteroids. *Acta Astron. Ser. A* 6: 115-138.

Safronov, V. S. 1965. Sizes of the largest bodies landing on the planets during their formation. *Sov. A. J.* 42: 1270.

Safronov, V. S. 1972. Evolution of protoplanetary clouds and formation of the earth and planets. NASA TT F-677.

Safronov, V. S. 1979. On the relative velocities and accumulation of preplanetary bodies. Submitted to *Icarus.*

Safronov, V. S. and Zvjagina, E. V. 1969. Relative sizes of the largest bodies during the accumulation of planets. *Icarus* 10: 109-115.

Ward, W. R. 1978. Solar nebula dispersal and the stability of the planetary system (abstract). *Bull. Amer. Astron. Soc.* 2: 591.

Weidenschilling, S. J. 1974. A model for accretion of the terrestrial planets. *Icarus* 22: 426-435.

Weidenschilling, S. J. 1975. Mass loss from the region of Mars and the asteroid belt. *Icarus* 26: 361-366.

Wetherill, G. W. 1976. The role of large bodies in the formation of the earth and moon. *Proc. Lunar Sci. Conf. VII* (Oxford: Pergamon Press), pp. 3245-3257.

Whipple, F. L.; Lecar, M.; and Franklin, F. A. 1972. The strange case of Pallas. In *L'origine due Système Solaire,* ed. H. Reeves (Paris: C. N. R. S.) pp. 312-319.

Zellner, B., and Bowell, E. 1977. Asteroid compositional types and their distributions. In *Comets, Asteroids, Meteorites,* ed. A. H. Delsemme (Toledo, Ohio: University of Toledo), pp. 185-197.

CHRONOLOGY OF ASTEROID COLLISIONS AS RECORDED IN METEORITES

D. D. BOGARD
NASA-Johnson Space Center

Because meteorites are derived from asteroidal objects, the chronology of meteorites give age information on the histories of the asteroidal parents. With a few possible exceptions, the radiometric chronology of less than 4.4 Gyr determined in meteorites can be attributable to collisions among meteorite parent bodies. Three types of collisional chronologies which have been determined are: (1) Cosmic ray exposure ages, which generally date the time of the last fragmentation of the parent object when the meteorite was reduced to ∼ meter size, and which generally fall over the wide range of 1-1000 Myr; (2) The collisional shock ages shown by many meteorites, which are believed to date the times of major collisions among parent bodies in the asteroid belt in the time interval of ∼ 30–700 Myr; (3) The brecciation ages of several meteorites which are believed to date the times of their collisional formation from regoliths of larger parent bodies over the period of ∼ 1.4–4.4 Gyr. The collisional chronology of most meteorites is consistent with their derivation from a relatively few parent objects, and with an origin of most types from the asteroid belt.

The various meteorite types which fall on Earth almost certainly had an origin from belt asteroids in orbit between Mars and Jupiter or from asteroid-like objects in Earth-crossing orbits (Arnold 1965; Chapman 1976; Anders 1975, 1978; Wetherill 1974, 1976, 1977; see also chapter by Wasson and Wetherill). It is commonly believed that collisions among belt asteroids cause extensive fragmentation of these objects, and that some fragments are gravitationally perturbed by Jupiter into highly eccentric orbits which may eventually evolve into Earth-crossing orbits. On the other hand, some meteorites probably derive from Apollo asteroids which some investigators believe to be extinct comets rather than fragments of belt asteroids (Wetherill 1976). The ordinary chondrites and the irons, which comprise about 80% and 6% respectively of all fallen meteorites, show evidence of slow cooling rates such as would exist deep in parent objects of at least 20-100 km diameter (Wood 1967; Goldstein and Short 1967; see also Wood's chapter). Differences in the relative abundances of oxygen isotopes indicate that several major classes of stone meteorites (e.g., carbonaceous chondrites, H and L chondrites and achondrites) originated from different parent objects (Clayton *et al.* 1976). Similarly, groupings of the relative abundances of Ga, Ge, Ni and Ir among various classes of iron meteorites suggest that several parent bodies may be represented (Wasson 1974, and references therein). These observations imply that many ordinary chondrites and iron meteorites were derived from the interior of a few large ($\gtrsim$ 50 km) parent bodies by extensive collisional fragmentation. On the other hand, meteorites such as basaltic achondrites and brecciated chondrites must have originated on or near the surface of their parent bodies and could have been ejected into space by smaller collisions which did not catastrophically disrupt the parent object.

Meteorites exhibit a variety of evidence of the collisional history of their parent bodies, and the times of occurrence of many of these events can be determined. Such collisional chronology is pertinent to the spatial distribution, the collisional history and the nature of meteorite parent bodies. For convenience, the chronologies of asteroid collisions as recorded in meteorites are divided here into three categories: (1) the cosmic ray exposure age characteristic of each meteorite which generally dates the time of the last fragmentation of the parent object when the meteorite was reduced to a size no larger than a few meters in diameter; (2) the collisional shock age shown by many meteorites, which is believed to date the time of major parent body collisions in the asteroid belt; and (3) the brecciation and/or formation age of some meteorites which may be the result of collisions on the surfaces of relatively large parent bodies.

I. COSMIC RAY EXPOSURE AGES

Collisional events causing catastrophic disruption of parent objects $\gtrsim$ 30 meters diameter or ejection of deeply buried material from yet larger parent bodies can produce fragments whose diameters are on the order of meters. When this occurs, high energy cosmic ray particles can penetrate the fragments and induce nuclear reactions. The rate at which a specific nuclide is formed from these reactions depends upon the flux and energy spectrum of cosmic rays and the probability of a nuclear interaction producing that product. Most estimates of the production rate of any stable nuclide, S are made from measurements in meteorites of a radioactive nuclide, R and estimates or laboratory determinations of the R/S production ratio by high energy particles (Anders 1963). Some radioactive nuclides commonly determined are ^{22}Na, ^{26}Al, ^{81}Kr and ^{40}K. Stable nuclides commonly determined are those which occur in the lowest natural abundance in meteorites, and most often are various isotopes of the noble gases, He, Ne, Ar, Kr and Xe. If the concentration of a stable cosmic ray-produced species measured in a meteorite is divided by its production rate, the resulting time is the period between initiation of the nuclear reactions and the fall of the meteorite on Earth. This cosmic ray exposure age generally dates the time of the last fragmentation of the parent object when the meteorite was reduced to a size on the order of meters. Through a large body of empirical evidence the exposure age of individual meteorites can usually be determined to better than a factor of two and often to within ±15%.

Cosmic ray exposure ages of nearly all stony meteorites are in the range of 0.1-60 Myr (0.1-60 × 10^6 yr), whereas iron meteorites commonly have exposure ages in the range of 100-1000 Myr. Histograms of exposure ages for meteorites representing several different classes of stones and irons are shown in Figs. 1 and 2. Although these data do not include all meteorites of a given class, and although the histograms may not be accurate in detail, the major trends are undoubtedly correct. Exposure ages of meteorites in some classes show a tendency to cluster (e.g., Zähringer 1968; Ganapathy and Anders 1969; Mazor et al. 1970; Herzog and Cressy 1977; Voshage 1978). The H chondrites show a strong clustering at 3-6 Myr. Meteorites in this age cluster comprise 15-20% of all meteorite falls. Five out of nine specimens of the uncommon diogenite class show essentially the same age of ∼ 14 Myr, and three others have similar ages of ∼ 24 Myr. Among the various classes of iron meteorites significant age clusters occur at ∼ 650 Myr for the III A and III B groups and at ∼ 400 Myr for the IV A group. A few iron and stone meteorites also show evidence of a two-stage cosmic ray exposure at different subsurface depths, which presumably indicates more than one collisional break-up event for these meteorites.

Clustering of exposure ages suggest that a single collisional event produced all the meteorites in each cluster. Consequently, the many hundreds of meteorites for which exposure ages have been determined represent a considerably smaller number of collisional events, probably involving relatively few parent bodies. Conceivably, all meteorites of a given class, say the L chondrites, may have been derived as ejecta from a limited number of major impacts over the past ∼ 50 Myr on the surface of a single parent body. This scenario requires that the parent object presently be located in an orbit from which ejected material could readily reach the earth, and suggests an orbit which crosses the orbits of the earth and the asteroid belt (e.g., Wetherill 1974). This argument particularly applies to the Farmington chondrite (Levin et al. 1976) which has an exposure age of only 0.025 Myr.

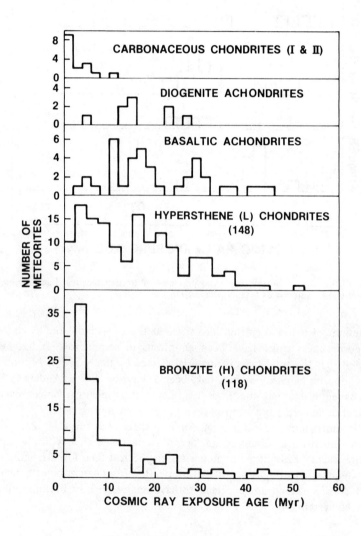

Fig. 1. Histograms of cosmic ray exposure ages of stony meteorites. Data sources are: ordinary chondrites, Zähringer 1968; basaltic achondrites, Ganapathy and Anders 1969; diogenites, Herzog and Cressy 1977; carbonaceous chondrites, Mazor *et al.* 1970.

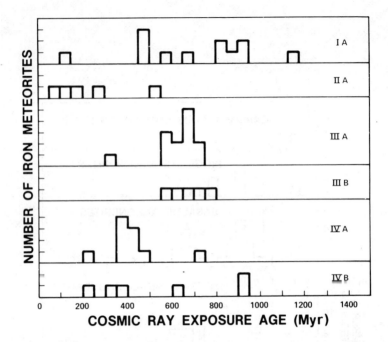

Fig. 2. Histograms of cosmic ray exposure ages of iron meteorites. Each increment represents one meteorite. (After Voshage 1978.)

The amount of material ejected into space at the times indicated by clusters of exposure ages must have been enormous in some cases. It has been estimated that the total mass of material ejected by the ~ 5 Myr collision on the H chondrite parent body may have been as large as 10^{16} g (Anders 1978), which is equivalent to the mass contained in a spherical object over a kilometer in diameter or an impact crater of even greater diameter.

The much longer exposure ages of iron meteorites suggest a somewhat different history, as discussed in the next section. The typically short exposure ages of carbonaceous chondrites may reflect their low strengths and greater tendency to fragment, or their derivation from parent objects such as extinct comets, which are expected to have relatively short lifetimes in the inner solar system (Wetherill 1974).

II. COLLISIONAL SHOCK AGES

The silicate and metal phases of a large number of stone and iron meteorites show petrological and textural evidence of having experienced shock-produced pressure waves as a result of collisions among parent bodies. Strong shock in meteorites can produce blackening of the specimen, transformation of olivine into a fine-grained, polycrystalline state and of feldspar into glass, and preferred orientation of crystals (e.g., Heymann 1967 and references therein). The reheating which accompanies shock can cause recrystallization of the deformed metal into new phases on the Fe-Ni phase diagram (e.g., Wood 1967; Taylor and Heymann 1971) and loss of ^{4}He and ^{40}Ar previously formed by the radioactive decay of U and K. Shock effects in meteorites can be detected for shock pressures as low as ~ 130 kbar. However, several chondrites have been shocked to pressures of > 500 kbar and heated to temperatures of $> 1000°$C (e.g., Smith and Goldstein 1977; Taylor and Heymann 1971). A sawed surface of the heavily shocked Rose City H chondrite (Fig. 3) shows partial melting, separation and apparent flow of the metal from the silicate, and unmelted clasts. These features suggest relatively rapid mixing and cooling of material with appreciably different temperatures. The shock reheating temperature of parts of Rose City may have been as high as 1200°C (Begemann and Wlotzka 1969).

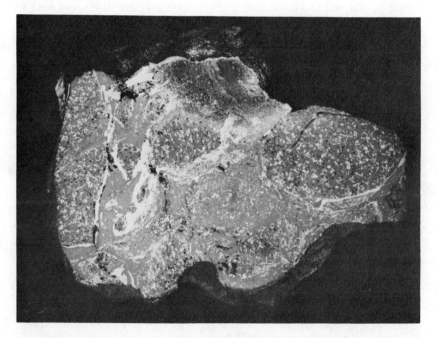

Fig. 3. A sawed interior surface of the strongly shock-reheated Rose City chondrite showing partial melting and segregation of metal from silicate. Note that two clasts appear unmelted but are rimmed with metal, which suggests that the clasts were much cooler when they were mixed with the partially melted host. Scale is approximately 1:1. (Photograph courtesy of Center for Meteorite Studies, Arizona State University.)

Nearly all iron meteorites in groups III A and III B have been shocked to pressures $\geqslant$ 130 kbar, and this may have occurred in a single catastrophic disruption of their parent body $\sim$ 650 Myr ago (e.g., Jain and Lipschutz 1971; Voshage 1978). A very large number of fragments must have been formed in order that a significant number ($>$ 55) of these fragments arrived on Earth as meteorites 650 Myr later. Similarly, the majority of Group IV A irons show evidence of shock pressures of $\geqslant$ 130 kbar and may have been produced when their parent body was catastrophically disrupted $\sim$ 400 Myr ago. The same events which produced the shock effects in III A, III B, and IV A irons presumably initiated their cosmic ray exposure as well. On the other hand, other classes of irons do not show large proportions of shocked specimens, nor do they tend to show clustering of their exposure ages.

An appreciable fraction of ordinary chondrites, primarily the L chondrites, have been shocked. Many of these meteorites have lost a substantial portion of their radiogenic ^{4}He and ^{40}Ar as a result of shock heating. Unlike many shocked irons, shocked L chondrites do not show a clustering of exposure ages. A general correlation does exist among the apparent U, Th-^{4}He ages and the relative classifications of many ordinary chondrites according to degree of shock and degree of reheating (Wänke 1966; Taylor and Heymann 1969). Figure 4 shows that U-He ages tend to decrease with increasing shock classification (A through D) for L chondrites. The U-He, gas-retention ages of several shocked L chondrites lie in the interval of 300-700 Myr, and an appreciable fraction of these have nearly concordant K-Ar ages. These observations led to the suggestion that the parent body of L chondrites was involved in a collisional event $\sim$ 500 Myr ago (e.g., Heymann 1967). This major collision may have disrupted a rather

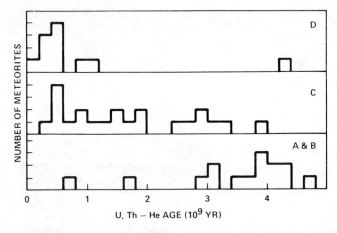

Fig. 4. Histograms of U, Th-He ages of ordinary chondrites as a function of increasing levels of collisional shock. Relative shock classifications are: A-light, B-moderate, C-intense, and D-very intense. (After Taylor and Heymann 1969.)

large parent object to satisfy both the geochemical arguments for a deep-seated origin of these meteorites (e.g., Wood 1967) and the observation that a large fraction of L chondrites have been shocked. Furthermore, cooling rates determined for several severely shocked chondrites indicate that some of the fragments resulting from this major collision must have resided in objects of >100 meters in size (Smith and Goldstein 1977).

A more precise way to determine the times of shock events in stone meteorites is based on the relatively new ^{39}Ar-^{40}Ar technique, a type of K-Ar dating. This technique measures the amount of ^{40}Ar and the apparent K-Ar age for various phases in a meteorite which degas Ar at different temperatures, and in many cases can obtain the ages of events that produced only partial gas loss (Turner 1969). In ^{39}Ar-^{40}Ar dating a sample is irradiated with fast neutrons to convert a portion of the ^{39}K to ^{39}Ar, and the ratio of ^{39}Ar to ^{40}Ar is measured as the sample is degassed in the laboratory in a series of increasing temperature steps. Chondrites release their ^{39}Ar in two temperature intervals which probably represent two mineral phases of different chemical composition. In most chondrites the lower temperature phase contains the greater concentration of K, and its K-Ar age is more easily reset by shock heating. The ^{39}Ar-^{40}Ar release profiles for the shocked Wethersfield L chondrite and the Wellman H chondrite are shown in Fig. 5. Wellman was not significantly reheated by the shock event and its K-Ar age has not been noticeably reset from a characteristic formation age of ~ 4500 Myr. Wethersfield was moderately reheated (metal reheating Class III, out of a maximum of V), and those K lattice sites which degas at lower temperature have been essentially entirely reset to an age of 510 ± 20 Myr. Those K lattice

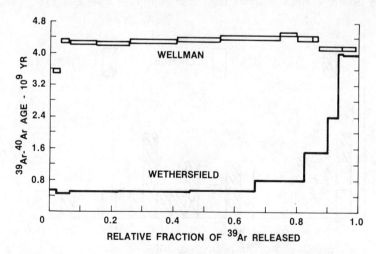

Fig. 5. ^{39}Ar-^{40}Ar ages as a function of fraction of ^{39}Ar (K) released for the Wellman and Wethersfield chondrites. A shock degassing age of ~ 520 Myr is indicated for Wethersfield. Data are from Bogard et al. 1976, and unpublished data of the author.

sites which degas at higher temperature have been only partially reset. As might be expected, the degree of reheating, not shock, is the critical parameter in determining the fraction of ^{40}Ar loss (Bogard *et al.* 1976). Although loss of ^{40}Ar is very sensitive to shock reheating, the reliability of a derived ^{39}Ar-^{40}Ar degassing age depends upon interpretation of the ^{39}Ar-^{40}Ar release spectrum which in some cases can be highly complex. Thus, not all shock-heated chondrites yield ^{39}Ar-^{40}Ar data which are interpretable as a dateable event.

Several shocked L chondrites have also been investigated by the ^{87}Rb-^{87}Sr dating technique (Gopalan and Wetherill 1971). Although whole rock analyses of most specimens did not reveal any unambiguous resetting of the Rb-Sr chronometer, density separates from one meteorite yielded an isochron consistent with a shock age of ~ 170-350 Myr.

Shock degassing ages determined by ^{39}Ar-^{40}Ar analyses of several meteorites representing all three classes of ordinary chondrites are shown in Fig. 6. Four additional L chondrites have been reported to show degassing ages of 300-400 Myr and ~ 500 Myr (Turner and Cadogan 1973; Cadogan and Turner 1975). Three possible clusters in these shock ages appear to exist at ~ 40 Myr, ~ 300 Myr and ~ 500 Myr. The ^{39}Ar-^{40}Ar ages for most shocked chondrites are similar to the cosmic ray exposure ages of iron meteorites, but considerably greater than the cosmic ray exposure ages of stone meteorites. Thus, the shock-degassing ages of chondrites in general appear to be due to more intense and older collisional events compared to those which initiated cosmic ray exposure. This observation is analogous to many lunar highland rocks whose radiometric ages actually represent resetting of the isotope chronometers by large-scale impacts ~ 3.8-4.2 Gyr (~ 3.8-4.2

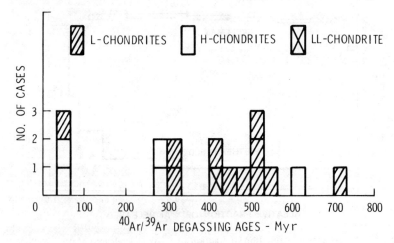

Fig. 6. Histogram of ^{39}Ar-^{40}Ar shock-degassing ages of ordinary chondrites. (After Bogard *et al.* 1976.)

$\times 10^9$ yr) ago. These lunar rocks show cosmic ray exposure ages which are considerably younger than their metamorphic ages, even in those cases where the exposure ages are believed to have been initiated by the formation of craters hundreds of meters in diameter. As seen from Figs. 6 and 2, several collisions spanning hundreds of millions of years are required to explain the shock ages deduced for stone and iron meteorites. Furthermore, since geochemical data (e.g., Clayton *et al.* 1976; Wasson 1974) suggest that certain meteorite types did not derive from a common parent body, several parent bodies were probably involved in these collisions. It is plausible that all L chondrites were derived from a single parent body which experienced major collisions ~ 500, ~ 300, and ~ 40 Myr ago.

Shock effects not associated with brecciation are rare in achondritic meteorites. Recent investigations of the Shergotty achondrite, which does show obvious evidence of shock, gave a well-defined ^{39}Ar-^{40}Ar age for a plagioclase separate of 250 Myr (Bogard *et al.* 1979) and a ^{87}Rb-^{87}Sr isochron age of 165 Myr (Nyquist *et al.* 1979). These ages were interpreted as upper and lower limits to the time of collisional heating of Shergotty, whereas the Sm-Nd age of 620 Myr was interpreted as the probable formation time. From considerations of diffusion rates and model cooling rates, it was concluded that the post-shock burial depth of Shergotty prior to its exposure to cosmic rays ~ 2 Myr ago was at least 100 meters. Because Shergotty, like other achondrites, is believed to have originated by mineral differentiation processes within a melt (Stolper and McSween 1979), the parent body of Shergotty is presumably large in order to have generated such a melt. Furthermore, in a few important geochemical respects this parent body was similar to the earth (Stolper and McSween 1979). Although the reason for the difference between the K-Ar and Rb-Sr ages for Shergotty is not well understood, it is significant that the Rb-Sr chronometer was reset by shock at least as completely as the K-Ar chronometer.

III. RADIOMETRIC DATING TECHNIQUES AND INTERPRETATIONS

At this point, it is appropriate to make a few general comments on techniques and interpretations of radiometric dating as applied to meteorites. The K-Ar, Rb-Sr, U-Pb, U-He, and Sm-Nd techniques are all based on the natural decay of a radioactive parent nuclide (e.g., K) to a stable daughter nuclide (e.g., Ar). The rate at which a given parent decays into its daughter depends only on the parent concentration, and unlike particle-induced nuclear reactions, is independent of meteorite size, composition, location, etc. In principle the time since a particular event (e.g., meteorite formation from a melt) may be simply and precisely calculated from the amount of parent which has decayed in that time period. The amount of decayed parent, however, can only be determined by measuring the accumulated daughter. In practice it is often difficult to determine that amount of daughter which formed in the given time interval because of the presence of daughter which formed in an earlier interval and primordial daughter which was incorporated into the parent body when it formed. For example, only a few percent of the ^{87}Sr present in most meteorites was formed by decay of ^{87}Rb over the lifetime of the solar system. To determine the amount of daughter which decayed over the period of interest, it is often necessary to analyze several phases of a meteorite in order to determine isochron ages. Discussion of isochron ages can be found in reference books on chronology (e.g., Faul 1966).

How precisely the time elapsed since a given event can be measured also depends upon how thoroughly that event was recorded in the dating systems. Formation of a meteorite by melting or condensation from a gas phase would generally cause complete equilibration of the radiometric systems, and the time of that event would be strongly recorded. On the other hand, moderate metamorphic events such as shock heating may cause only partial equilibration of the radiometric systems, and thus the record of that event will be more difficult to determine. Various dating systems react differently to moderate metamorphic events in proportion to the ease with which the parent and daughter nuclides can equilibrate by solid state diffusion. Thus Ar, being a gas, is often more sensitive to such events than are the refractory elements Sm and Nd (e.g., Nyquist et al. 1979). Similarly, various phases within a meteorite may show different degrees of equilibration for the same dating system and thus show different apparent "ages" which may not accurately date the event. Such ages are often upper limits to the time of the metamorphic event. For a dated event to be "reliable" generally requires that the radiometric system used show major equilibration for some phase of the meteorite, for example, an isochron with minimum scatter or a $^{39}Ar/^{40}Ar$ plateau over a significant proportion of the Ar release. Close agreement of radiometric ages by more than one technique also lends support to interpretation of that age as an identifiable event. Examples are the similar ^{39}Ar-^{40}Ar and Rb-Sr shock ages for Shergotty (discussed above) and the identical K-Ar, Rb-Sr, and Sm-Nd ages for the nakhlite achondrites (discussed in the next section).

IV. BRECCIATION AND YOUNG FORMATION AGES

Many meteorites representing several classes of chondrites and achondrites are indurated mixtures of materials called breccias formed on or near the surfaces of their parent bodies. These breccias form by repeated cycles of comminution, mixing and compaction of a regolith as a result of multiple impacts into the surface of the parent body. Recent models (Housen et al. 1979; and their chapter, this book) suggest that regoliths may be very thin on small (<10 km) rocky asteroids, but may be hundreds of meters thick on large (>100 km) rocky asteroids. These regoliths are probably poorly mixed. An appreciable number of meteorite breccias contain solar wind gases and solar flare tracks, which were implanted into fine-grained material dispersed on the surface of the parent regolith before formation of the meteorite (e.g., MacDougall *et al.* 1973; Rajan 1974). Whereas most meteorite breccias are mixtures of the same or similar material, several contain mm- to cm-sized fragments which are foreign to the host meteorite (Wilkening 1977, and references therein). These inclusions of one meteorite type in another (carbonaceous chondrite inclusions appear to be the most common) have been found in several classes of meteorites, and probably resulted from impact of one type of parent body into the regolith of another. All of these characteristics of meteorite breccias appear consistent with an origin from the regoliths developed on belt asteroids. The chronologies of these breccias and their inclusions, therefore, pertain to the times of dynamic regolith evolution on asteroids.

Achondrites probably were derived by differentiation of molten material produced during early heating of their parent body(s). Several meteorites representing more than one class of achondrites have given Rb-Sr, U, Th-Pb, or Sm-Nd isochron ages of ~ 4.5-4.6 Gyr (e.g., Birck and Allegre 1978; Wasserburg *et al.* 1977; Lugmair *et al.* 1975; Minster and Allegre 1976). Material with the composition of basaltic achondrites probably occurs on or near the surface of its parent body(s), and the asteroid Vesta has been suggested as a possible source for these meteorites (McCord *et al.* 1970; chapter by Drake in this book). Thus, it is not surprising that most basaltic achondrites are brecciated and that several show disturbance of their isotope chronometers, probably as a result of impact metamorphism. Table I summarizes the results of a large number of chronological investigations of achondrites for which ages significantly less than 4.5 Gyr are indicated. Brief remarks are also made as to the reliability of the ages. Unfortunately, in many cases neither the age nor the nature of the event which modified the age can be precisely determined. Unless a heat-producing event such as impact is particularly severe, the isotope chronometers may be only partially reset, as discussed in the preceeding section. Furthermore, the possible existence in brecciated meteorites of clasts or other phases which have different radiometric ages because of different regolith histories before meteorite compaction, means that no single age exists for such a whole rock sample. Thus, with brecciated meteorites it is particularly important to perform measurements on individual clasts or phases. Using Table I as a guide, we now discuss some of the major features of brecciation/formation ages in meteorites and their implications for asteroid collisions.

Several basaltic achondrites (eucrites and howardites) and one mesosiderite (stony-iron) indicate partial or complete resetting of their ^{39}Ar-^{40}Ar or Rb-Sr isochron ages during the time period of ~ 3.3-4.3 Gyr ago. The ages determined for plagioclase (4.42 Gyr) and glass phases (4.24 Gyr) in Bununu have been interpreted as the times of igneous formation and shock lithification, respectively. For Malvern the ~ 3.6-3.7 age of the glass

TABLE I

Meteorite Ages Significantly Less than 4.5 Gyr
Possibly Attributable to Impacts onto Asteroid Surfaces

Meteorite	Type	Apparent Age Gyr	Technique[a]	Remarks	Reference
Achondrites:					
Bununu glass	howardite	4.24	K-Ar	fair plateau	Rajan et al. 1975
Bununu plagioclase		4.42	K-Ar	fair plateau	Rajan et al. 1975
Malvern glass	howardite	3.7	K-Ar	fair plateau	Rajan et al. 1975; Kirsten, Horn 1975
Malvern clast		3.6	K-Ar	fair plateau	Kirsten, Horn 1975
Malvern plagioclase		2.1 – 4.2	K-Ar	no plateau	Rajan et al. 1975
Bholghati	howardite	~3.4	K-Ar	data scatter	Leich, Moniot 1976
Kapoeta clast A	howardite	3.48	K-Ar	data scatter	Huneke et al. 1977; Rajan et al. 1979
clast B		3.89	Rb-Sr	2-phase isochron	Huneke et al. 1977
		4.55	K-Ar	good plateau	Huneke et al. 1977; Rajan et al. 1979
clast C		3.63	Rb-Sr	good isochron	Huneke et al. 1977
		4.6	K-Ar	fair plateau	Huneke et al. 1977; Rajan et al. 1979
		4.54	Rb-Sr	fair isochron	Huneke et al. 1977
Stannern	eucrite	3.5 – 3.9	K-Ar	data scatter	Huneke et al. 1977
		~3.3	Rb-Sr	data scatter	Birck, Allegre 1978
Pasamonte	eucrite	~4.1	K-Ar	most radiogenic phases scatter	Podosek, Huneke 1973
		2.5 – 4.55	Rb-Sr	no plateau	Birck, Allegre 1978

[a] Rb-Sr and Sm-Nd ages are by internal isochrons; K-Ar ages are by the $^{39}Ar-^{40}Ar$ technique, with one exception see text).

TABLE I (Continued)

Meteorite	Type	Apparent Age Gyr	Technique[a]	Remarks	Reference
Petersburg	eucrite	2.2 – 4.4	K-Ar	no plateau	Podosek, Huneke 1973
Haraiya	eucrite	2.5 – 3.5	K-Ar	data scatter	Bogard et al. 1976a
Shergotty	special	0.25	K-Ar	good plateau for plagioclase	Bogard et al. 1979
		0.16	Rb-Sr	good isochron	Nyquist et al. 1979
		0.62	Sm-Nd	fair isochron	Nyquist et al. 1979
Nakhla	nakhlite	1.3	K-Ar	fair plateau	Podosek 1973
		1.24	Rb-Sr	good isochron	Papanastassiou and Wasserburg 1974
		1.37	Rb-Sr	good isochron	Gale et al. 1975
		1.27	Sm-Nd	good isochron	Nakamura et al. 1977
Lafayette	nakhlite	1.33	K-Ar	good plateau	Podosek 1973
Governador Valadares	nakhlite	1.32	K-Ar	good plateau	Bogard, Husain 1977a
		1.33	Rb-Sr	good isochron	Wooden et al. 1979
Estherville silicate	mesoiderite	4.35	Rb-Sr	fair isochron	Murthy et al. 1977
		3.6 – 4.2	K-Ar	no plateau	Bogard et al. 1976
Chondrites:					
Plainview clast A	H	3.63	K-Ar	fair plateau	Bogard, Husain 1977b
clast B		<3.4	K-Ar	no plateau	Bogard, Husain 1977b
matrix		3.9 – 4.4	K-Ar	scatter	Schultz, Signer 1977
St. Mesmin clast 1	LL	1.4	K-Ar	classical K-Ar technique	Cadogan, Turner 1975
clast 2		4.42	K-Ar	?	Cadogan, Turner 1975
clast 3		4.57	K-Ar	?	Cadogan, Turner 1975

and of a clast is interpreted as the shock-lithification age, which only partially resets the plagioclase. Three different clasts from Kapoeta gave three different Rb-Sr ages and two different K-Ar ages. Because Kapoeta contains abundant gas implanted by the solar wind and because this gas would be lost during impact heating at least as readily as would radiogenic ^{40}Ar, the shock-lithification age for Kapoeta can be no older than ~ 3.5 Gyr, the K-Ar age of the youngest clast. All of these results show that well-developed regoliths existed on the parent body(s) of basaltic achondrites and were subjected to energetic impact events at least as recently as ~ 3.5 Gyr ago and possibly as recently as ~ 2 Gyr. As discussed in the preceeding section, the resetting of the isotope chronometers in Shergotty, which is not a breccia, occurred by collisional heating after formation of the meteorite by crystal differentiation.

The three achondrites classed as nakhlites (the only specimens known for this rare class) are not breccias, but they show essentially concordant ages of ~ 1.3 Gyr by three independent isotope dating techniques. Interpretations of the Rb-Sr and Sm-Nd data preclude the possibility that these meteorites are 4.5 Gyr-old material which was remelted 1.3 Gyr ago, but rather require that igneous crystal fractionation occurred more recently than ~ 3 Gyr ago. Papanastassiou and Wasserburg (1974) interpreted their Rb-Sr results on Nakhla as the time of intense metamorphism. All other nakhlite investigations reported in Table I, however, were interpreted as the time of formation of the meteorites by crystal differentiation. This interpretation is supported by petrological studies (Reid and Bunch 1975). The similarity of the exposure ages (~ 10 Myr) for the three nakhlites suggests that they were derived from their parent body in a single collision (Bogard and Husain 1977a), It would appear to be particularly difficult to produce a large melt on an asteroid ~ 1.3 Gyr ago by residual accretion heat, radionuclide heating, or other commonly suggested mechanisms, and the possibility that the nakhlites formed as a result of melt produced during an intense impact on their parent body must be considered. It is not known whether large impacts can produce sufficient melt for crystal differentiation to occur, but if this was the formation mechanism for the nakhlites, they may represent a low probability phenomenon and may not be representative of their parent body.

Precise radiometric ages determined for an appreciable number of chondrites show formation times of ~ 4.5-4.6 Gyr. The conception that many chondrites may also be regolith-derived breccias (e.g., Wänke 1965; Fodor and Keil 1976; Wilkening 1977) has spurred efforts at determining the chronologies of individual inclusions. An investigation of seven inclusions in the brecciated Weston H chondrite for the relative amounts of cosmic ray-produced noble gases and particle tracks revealed that one inclusion had experienced a 20% longer irradiation than the others (Schultz et al. 1972). These data indicated that the entire meteorite had experienced an ~ 23 Myr exposure age, but that one clast had been previously irradiated for an additional ~ 4 Myr at a depth of ≥ 40 cm. Presumably, the additional irradiation for this one clast occurred in the regolith of the parent body before formation of the meteorite breccia, but the other six clasts were too deeply buried in that regolith to be irradiated. A study of 12 inclusions in the Djermaia H chondrite (Lorin and Pellas 1979) revealed differences in the irradiation conditions experienced by these inclusions before meteorite compaction, and suggested a pre-compaction irradiation time of ~ 15 Myr in a parent body regolith. In another study of the St. Mesmin brecciated LL chondrite (Schultz and Signer 1977) several inclusions gave classical K-Ar ages of ~ 4.4 Gyr, but one inclusion gave a K-Ar age of 1.4 Gyr. Another clast suggested an excess irradiation by cosmic rays of ~ 1.5 Myr compared to all other samples analyzed. ^{39}Ar-^{40}Ar dating of two additional inclusions in St. Mesmin gave distinctly different ages of 4.57 and 4.42 Gyr (Cadogan and Turner 1975). Particle track ages for five carbonaceous (C2) chondrites, which are also known to contain evidence of a regolith history, suggested compaction ages of 4.2-4.4 Gyr (MacDougall and Kothari 1976). Finally, a study of the Plainview brecciated H chondrite gave an ^{39}Ar-^{40}Ar age of 3.6 Gyr for an impact-remelted inclusion and an ill-defined age of 3.8-4.4 Gyr for the host meteorite (Bogard and Husain 1977b). Plainview contains solar wind-implanted gases and inclusions of both carbonaceous and non-carbonaceous material (Wilkening and Clayton 1974; Fodor and Keil 1976). Plainview, like Kapoeta therefore, could not have been appreciably heated since its compaction, which had to be more recent than ~ 3.6 Gyr ago. The compaction times of breccias like Plainview, St. Mesmin, and Kapoeta probably occurred shortly after the ages of their youngest inclusions, at which time they were left deeply buried in the parent body regolith. This is concluded from the observation that cosmic ray exposure ages of these meteorites are shorter by many orders of magnitude compared to ages given in Table I.

V. IMPLICATIONS FOR METEORITE ORIGINS

An appreciable number of chondrites, achondrites, and iron meteorites show isotope formation ages of ~ 4.5-4.6 Gyr, which represent the times of parent body accretion, metamorphism, or early differentiation. With a few possible exceptions, that isotope chronology of less than ~ 4.4 Gyr determined in meteorites can be attributable to collisions among meteorite parent bodies. Four different types of chronologies determined in meteorites are summarized in Fig. 7. A variety of geochemical considerations indicate that the achondrites, irons, and stony irons were derived from parent bodies which had undergone extensive melting and differentiation. Brecciated meteorites appear to have originated from well-developed, but poorly mixed regoliths formed by the impact of asteroidal debris into the surfaces of large parent objects. Chronologies determined for several brecciated achondrites

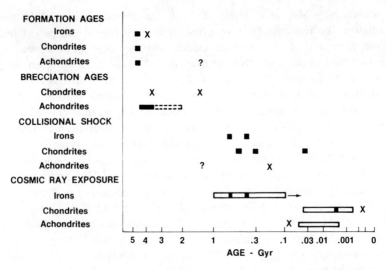

Fig. 7. A summary of four types of event ages determined for stone and iron meteorites. (The age scale is plotted as the fourth root for visual ease.) With the exception of one iron, formation ages cluster in the narrow interval of ~4.4-4.6 Gyr. Brecciation ages of several achondrites and two chondrites appear to occur over the 1.4-4-4 Gyr interval, with achondrites mainly indicating ages of ~ 3.5-4.4 Gyr. Collisional shock ages for most irons cluster at ~400 and ~650 Myr and for several chondrites may cluster at ~ 500, ~ 300, and ~ 30 Myr. Cosmic ray exposure ages for most irons are ~1000-100 Myr and lower, with clusters of ages at ~ 400 and ~650 Myr. Exposure ages of most stones are 50-0.5 Myr, with a significant cluster at ~ 5 Myr, and show oldest and youngest values of ~ 90 and ~ 0.025 Myr. Rectangles represent the range of age determinations on more than one meteorite, whereas an X represents a single meteorite. The question marks represent the nakhalites, whose chronology may be either a formation age, a collisional shock age, or both.

and chondrites indicate that breccia formation was a commonplace process in the time period of ~ 4.4-3.4 Gyr ago, and continued to occur at least as recently as 1.4 Gyr ago. Chronological suggestions of a greater frequency of regolith formation and mixing ~ 4 Gyr ago compared to more recent times is consistent with concepts that large asteroids were more numerous ~ 4 Gyr ago (Chapman and Davis 1975) and with intense bombardment of the moon during this period. Thus, both differentiated meteorites and brecciated meteorites appear more consistent with an asteroidal origin than a cometary origin. Asteroids are known to be objects of varying silicate and metal composition (see chapter by Gaffey and McCord in this book) which have experienced extensive collisional fragmentation (Chapman 1976). In contrast, very little is actually known of the physical and chemical nature of solid materials in the nucleus of comets (O'Dell et al. 1974). Because of the evidence that ordinary chondrites derived from a small number of large parent bodies of similar composition, ordinary chondrites (but not carbonaceous chondrites) are unlikely to have originated from both belt asteroids and extinct comets. Therefore, the existence of brecciated ordinary chondrites like Plainview, which shows several characteristics of existence in a parent body regolith ~ 3.6 Gry ago, suggests that all ordinary chondrites derived from belt asteroids rather than comets.

Many ordinary chondrites and irons and a very few achondrites show evidence of shock caused by collisions of their parent bodies $\sim 10^8 - 10^9$ yr ago. In the case of the iron meteorites these shock events appear to have initiated their exposure to cosmic rays, but for the stony meteorites the shock ages appear to be considerably older than the cosmic ray exposure ages. Typical lifetimes of objects in Earth-crossing orbits are estimated to be on the order of 10^7 yr, whereas much longer typical lifetimes of $\sim 10^8 - 10^9$ are expected for asteroids which are in Mars-crossing orbits (Arnold 1965; Wetherill 1974). The shock-exposure ages of most iron meteorites are consistent with their formation by major collisions in the asteroid belt followed by a relatively long period in which their orbits evolved to intersect that of the earth. Similarly, the shock ages of chondrites and the Shergotty achondrite are consistent with major collisions of their parent bodies in the asteroid belt, followed by evolution of large ($> 10^2$ meter) fragments into Earth-crossing orbits (e.g., Zimmerman and Wetherill 1973). The cosmic ray exposure ages of stony meteorites are consistent with more recent collisions of parent objects already in Earth-crossing orbits.

The majority of meteorites are ordinary chondrites, whereas the vast majority of measured asteroids are of spectral class C or S and are generally believed to be similar to carbonaceous chondrites and stony-irons, respectively. Discounting observational bias because of size or albedo, this suggests that asteroids with compositions of ordinary chondrites and achondrites are few and that some preferential means may exist to bring such objects to Earth. The majority of measured Earth-crossing and Earth-approaching asteroids do show spectral types other than C, and more than one may be similar to ordinary chondrites. The evidence from a variety of geochemical considerations and chronological data indicates that the majority of meteorites derived from relatively few parent objects which had experienced a variety of collisions. For example, a single original parent body may have produced numerous fragments through a long collisional history, and one or more such fragments may be the source of all L chondrites which comprise $\sim 31\%$ of all meteorites falling on Earth. Thus, geochemical and chronological observations are not in conflict with the spectral observations that the majority of meteorites currently falling on Earth are not generally representative of the population of larger asteroids. Hundreds of meteorites have recently been discovered lying on the continental ice of Antarctica, and many of these have considerably longer terrestrial ages than non-Antarctic meteorites (e.g., Yanai *et al.* 1978; Fireman *et al.* 1979). The Antarctic meteorites show essentially the same proportion of major meteorite classes as do recent falls (i.e., ordinary chondrites strongly predominate), which implies that this proportion has not appreciably changed over the past $\sim 10^4$ yr. The relative proportion of major meteorite types, however, may have been significantly different in the more distant past at times comparable to the cosmic ray exposure ages of stony meteorites.

REFERENCES

Anders, E. 1963. Meteorite ages. In *The Moon, Meteorites and Comets*, eds. B. Middlehurst and G. Kuiper (Chicago: Univ. of Chicago Press), pp. 402-495.

Anders, E. 1975. Do stony meteorites come from comets? *Icarus* 24: 363-371.

Anders, E. 1978. Most stony meteorites come from the asteroid belt. In *Asteroids: An Exploration Assessment*, eds. D. Morrison and W.C. Wells, NASA Conf. Publ. 2053, 57-74.

Arnold, J.R. 1965. The origin of meteorites as small bodies. II. The model. *Astrophys. J.* 141: 1536-1556.

Begemann, F., and Wlotzka, F. 1969. Shock induced thermal metamorphism and mechanical deformations in the Ramsdorf chondrite. *Geochim. Cosmochim. Acta* 33: 1351-1370.

Birck, J. L., and Allegre, C. J. 1978. Chronology and chemical history of the parent body of basaltic achondrites studied by the ^{87}Rb ^{87}Sr method. *Earth Planet. Sci. Lett.* 39: 37-51.

Bogard, D. D.; Hirsch, W. C.; and Husain, L. 1976a. ^{40}Ar-^{39}Ar dating of achondritic meteorites (abstract). *Meteoritics* 11: 251-252.

Bogard, D. D., and Husain, L. 1977a. A new 1.3 aeon-young achondrite. *Geophys. Res. Lett.* 4: 69-71.

Bogard, D. D., and Husain, L. 1977b. ^{40}Ar-^{39}Ar dating of Plainview brecciated chondrite: Evidence of regolith formation since 3.7 aeons (abstract). *Bull. Amer. Astron. Soc.* 9: 458.

Bogard, D. D.; Husain, L.; and Nyquist, L. E. 1979. ^{87}Ar-^{39}Ar age of the Shergotty achondrite and implications for its post-shock thermal history. *Geochim. Cosmochim. Acta* 43 (in press).

Bogard, D. D.; Husain, L.; and Wright, R. J. 1976b. ^{40}Ar-^{39}Ar dating of collisional events in chondrite parent bodies. *J. Geophys. Res.* 81: 5664-5678.

Cadogan, P. H., and Turner, G. 1975. Luna 16 and Luna 20 revisited (abstract). *Meteoritics* 10: 375-376.

Chapman, C. R. 1976. Asteroids as meteorite parent-bodies: The astronomical perspective. *Geochim. Cosmochim. Acta* 40: 701-719.

Chapman, C. R., and Davis, D. R. 1975. Asteroid collisional evolution: Evidence for a much larger early population. *Science* 190: 553-555.

Clayton, R. N.; Onuma N.; and Mayeda, T. K. 1976. A classification of meteorites based on oxygen isotopes. *Earth Planet. Sci. Lett.* 30: 10-18.

Faul, H. 1966. *Ages of Rocks, Planets, and Stars.* (New York: McGraw Hill).

Fireman, E. L.; Rancitelli, L.A.; and Kirsten, T. 1979. Terrestrial ages of four Allan Hills meteorites: Consequences for Antarctic ice. *Science* 203: 453-455.

Fodor, R. V., and Keil, K. 1976. Carbonaceous and non-carbonaceous lithic fragments in the Plainview, Texas, chondrite: Origin and history. *Geochim. Cosmochim. Acta* 40: 177-190.

Gale, N.; Arden, J.; and Hutchison, R. 1975. The chronology of the Nakhla achondritic meteorite. *Earth Planet. Sci. Lett.* 26: 195-206.

Ganapathy, R., and Anders, E. 1969. Ages of calcium-rich achondrites. II. Howardites, nakhlites, and the Angra dos Reis angrite. *Geochim. Cosmochim. Acta* 33: 775-788.

Goldstein, J. I., and Short, J. M. 1967. The iron meteorites, their thermal history and parent bodies. *Geochim. Cosmochim. Acta* 31: 1733-1770.

Gopalan, K., and Wetherill, G. W. 1971. Rb-Sr studies on black hypersthene chondrites: Effects of shock and reheating. *J. Geophys. Res.* 76: 8484-8492.

Herzog, G. F., and Cressy, P. J. 1977. Diogenite exposure ages. *Geochim. Cosmochim. Acta* 41: 127-134.

Heymann, D. 1967. On the origin of hypersthene chondrites: Ages and shock effects of black chondrites. *Icarus* 6: 189-221.

Housen, K. R.; Wilkening, L. L.; Chapman, C. R.; and Greenberg, R. 1979. Asteroidal regoliths. *Icarus* (in press).

Huneke, J. C.; Smith, S. P.; Rajan, R. S.; Papanastassiou, D. A.; and Wasserburg, G. J. 1977. Comparison of the chronology of the Kapoeta parent planet and the moon (abstract). *Lunar Science VIII.* The Lunar and Planetary Institute, pp. 484-486.

Jain, A. V., and Lipschutz, M. E. 1971. Shock history of iron meteorites and their parent bodies: A review. *Chem. Erde* 30: 199-215.

Kirsten, T., and Horn, P. 1975. ^{39}Ar-^{40}Ar dating of basalts and rock breccias from Apollo 17 and the Malvern achondrite. *Proc. Soviet-American Conf. on Cosmochemistry of the Moon, and Planets,* Nauka, Moscow, pp. 386-401.

Levin, B. J.; Simonenko, A. N.; and Anders, E. 1976. Farmington meteorite: A fragment of an Apollo asteroid? *Icarus* 28: 307-324.

Leich, D. A., and Moniot, R. 1976. Rare gas chronology of enstatite achondrites and the Bholghati howardite (abstract). *Lunar Science VII.* The Lunar and Planetary Institute, pp. 479-481.

Lorin, J. C., and Pellas, P. 1979. Pre-irradiation history of Djermala (H) chondritic breccia. *Icarus* (special Asteroid issue).

Lugmair, G. W.; Scheinin, N. B.; and Marti, K. 1975. Search for extinct ^{146}Sm. 1. The isotopic abundance of ^{142}Nd in the Juvinas meteorite. *Earth Planet. Sci. Lett.* 27: 79-84.

MacDougall, D.; Rajan, R. S.; Hutcheon, I. D.; and Price, P. B. 1973. Irradiation history and accretionary processes in lunar and meteoritic breccias. *Proc. Lunar Sci. Conf. IV* (Oxford: Pergamon Press), pp. 2319-2336.

MacDougall, J. D., and Kothari, B. K. 1976. Formation chronology for C2 meteorites. *Earth Planet. Sci. Lett.* 33: 36-44.

Mazor, E.; Heymann, D.; and Anders E. 1970. Noble gases in carbonaceous chondrites. *Geochim. Cosmochim. Acta* 34: 781-824.

McCord, T. B.; Adams, J. B.; and Johnson, T. V. 1970. Asteroid Vesta: Spectral reflectivity and compositional implications. *Science* 168: 1445-1447.

Minster, J. F., and Allegre, C. J. 1976. ^{87}Rb-^{87}Sr history of the Norton county enstatite achondrite. *Earth Planet. Sci. Lett.* 32: 191-198.

Murthy, V. R.; Coscio, M. R.; and Sabelin, T. 1977. Rb-Sr internal isochron and the initial ^{87}Sr/^{86}Sr for the Esterville mesosiderite. *Proc. Lunar Sci. Conf. VIII* (Oxford: Pergamon Press), pp. 177-186.

Nakamura, N.; Unruh, D. M.; and Tatsumoto, M. 1977. Nakhla: Further evidence for a young crystallization age (abstract). *Meteoritics* 12: 324-325.

Nyquist, L.; Wooden, J.; Bansal, B.; Wiesman, H.; McKay, G.; and Bogard, D. 1979. Rb-Sr age of the Shergotty achondrite: Implications for metamorphic resetting of isochron ages. *Geochim. Cosmochim. Acta* 43 (in press).

O'Dell, C. R.; Huebner, W. F.; Delsemme, A. H.; Donn, B.; and Whipple, F. L. 1974. Panel discussion on the nucleus of comets. In *The Study of Comets,* eds. B. Donn and M. Mumma (NASA SP-393, Washington, D.C.: U.S. Government Printing Office), pp. 588-637.

Papanastassiou, D. A., and Wasserburg, G. J. 1974. Evidence for late formation and young metamorphism in the achondrite Nakhla. *Geophys. Res. Lett.* 1: 23-26.

Podosek, F. A. 1973. Thermal history of the nakhlites by the ^{40}Ar-^{39}Ar method. *Earth Planet. Sci. Lett.* 19: 135-144.

Podosek, F. A., and Huneke, J. C. 1973. ^{40}Ar-^{39}Ar chronology of four calcium-rich achondrites. *Geochim. Cosmochim. Acta* 37: 667-684.

Rajan, R. S. 1974. On the irradiation history and origin of gas-rich meteorites. *Geochim. Cosmochim. Acta* 38: 778-788.

Rajan, R.; Huneke, J.; Smith, S.; and Wasserburg, G. 1975. ^{40}Ar-^{39}Ar chronology of isolated phases from Bununu and Malvern howardites. *Earth Planet. Sci. Lett.* 27: 181-190.

Rajan, R. S.; Huneke, J. C.; Smith, S. P.; and Wasserburg, G. J. 1979. ^{40}Ar-^{39}Ar chronology of lithic clasts from the Kopoeta howardite. *Geochim. Cosmochim. Acta* 42 (in press).

Reid, A. M., and Bunch, T. E. 1975. The nakhlites. Part II. Where, when, and how. *Meteoritics* 10: 317-324.

Schultz, L., and Signer, P. 1977. Noble gases in St. Mesmin chondrite: Implications to the irradiation history of a brecciated meteorite. *Earth Planet. Sci. Lett.* 36: 363-371.

Schultz, L.; Signer, P.; Lorin, J. C.; and Pellas, P. 1972. Complex irradiation history of the Weston chondrite. *Earth Planet. Sci. Lett.* 15: 403-410.

Smith, B. A., and Goldstein, J. I. 1977. The metallic microstructure and thermal histories of severely reheated chondrites. *Geochim. Cosmochim. Acta* 41: 1061-1072.

Stolper, E., and McSween, H. Y. 1979. Petrology and origin of the Shergottite meteorites. *Geochim. Cosmochim. Acta* 43 (in press).

Taylor, G. J., and Heymann, D. 1969. Shock, reheating, and the gas retention ages of chondrites. *Earth Planet. Sci. Lett.* 7: 151-161.

Taylor, G. J., and Heymann, D. 1971. Postshock thermal histories of reheated chondrites. *J. Geophys. Res.* 76: 1879-1893.

Turner, G. 1969. Thermal histories of meteorites by the ^{39}Ar-^{40}Ar method. In *Meteorite Research*, ed. P. Millman (Netherlands: D. Reidel), pp. 407-417.

Turner, G., and Cadogan, P. H. 1973. ^{40}Ar-^{39}Ar chronology of chondrites (abstract). *Meteoritics* 8: 447-448.

Voshage, H. 1978. Investigations on cosmic-ray-produced nuclides in iron meteorites. 2. New results on ^{41}K/^{40}K ^{4}He/^{21}Ne exposure ages and the interpretation of age distributions. *Earth Planet. Sci. Lett.* 40: 83-90.

Wänke, H. 1965. Der Sonnenwind als Quelle der Uredelgase in Steinmeteoriten. *Z. Naturforsch.* 20a: 946-949.

Wänke, H. 1966. Meteoritenalter and Verwandte Probleme der Kosmochemie. *Fortschritte Chem. Forschung* 7: 322-408.

Wasserburg, G. J.; Tera, F.; Papanastassiou, D. A.; and Huneke, J. C. 1977. Isotopic and chemical investigations on Angra dos Reis. *Earth Planet Sci. Lett.* 35: 294-316.

Wasson, J. T. 1974. *Meteorites* (New York: Springer-Verlag), pp. 29-38.

Wetherill, G. W. 1974. Solar system sources of meteorites and large meteoroids. *Ann. Rev. Earth Planet. Sci.* 2: 203-331.

Wetherill, G. W. 1976. Where do the meteorites come from? A re-evaluation of the earth-crossing Apollo objects as sources of chondritic meteorites. *Geochim. Cosmochim. Acta* 40: 1297-1318.

Wetherill, G. W. 1977. Fragmentation of asteroids and delivery of fragments to earth. In *Comets, Asteroids, Meteorites,* ed. A. H. Delsemme (Toledo, Ohio: University of Toledo Press), pp. 283-291.

Wilkening, L. L. 1977. Meteorites in meteorites. Evidence for mixing among the asteroids. In *Comets, Asteroids, Meteorites,* ed. A. H. Delsemme (Toledo, Ohio: Univ. of Toledo Press), pp. 389-396.

Wilkening, L. L., and Clayton, R. N. 1974. Foreign inclusions in stony meteorites. II. Rare gases and oxygen isotopes in a carbonaceous chondritic xenolith in the Plainview gas-rich chondrite. *Geochim. Cosmochim. Acta* 38: 937-946.

Wood, J. A. 1967. Chondrites: Their metallic minerals, thermal histories, and parent planets. *Icarus* 6: 1-49.

Wooden, J.; Nyquist, L.; Bogard, D.; Bansal, B.; Wiesman, H.; Shih, C.; and McKay, G. 1979. Radiometric ages for the achondrites Chervony Kut, Governador Valadares, and Allan Hills 77005 (abstract). *Lunar Science X.* The Lunar and Planetary Institute, pp. 1379-1381.

Yanai, K.; Cassidy, W. A.; Funaki, M.; and Glass, B. P. 1978. Meteorite recoveries in Antarctica during field season 1977-1978. *Proc. Lunar Sci. Conf. IX.* (Oxford: Pergamon Press), pp. 977-987.

Zähringer, J. 1968. Rare gases in stony meteorites. *Geochim. Cosmochim. Acta* 32: 209-237.

Zimmerman, P. D., and Wetherill, G. W. 1973. Asteroidal source of meteorites. *Science* 182: 51-53.

THE NATURE AND EFFECTS
OF
IMPACT CRATERING ON SMALL BODIES

MARK J. CINTALA
JAMES W. HEAD
Brown University

and

LIONEL WILSON
University of Lancaster

An impact event on a planet-size target body produces a hole, heats a portion of the target material (some of which remains in and near the crater), fragments (compresses) the coherent (porous) target material below and around the final crater, redistributes material across the surface of the target, adds particles to the surficial debris layer, and serves as a transient source of seismic energy. While asteroids and small satellites will undergo the same effects, the relative magnitudes of the various processes could be drastically different. The comparatively weak gravity fields of most small bodies in the solar system will allow permanent escape of much crater ejecta. Since low-velocity ejecta from a coherent target are in general the coarsest fraction, the smallest asteroids will probably possess very blocky regoliths, if any at all. Fine-grained regoliths and/or the surfaces of weakly consolidated asteroids should undergo net compression due to impacts. The small dimensions of most asteroids will not attenuate impact-generated stress waves efficiently, leading to strong seismic effects relative to the low gravities. Compressional and shear waves reflected from free surfaces should mobilize regolith; stronger waves could accelerate weakly-bound or surficial debris beyond the escape velocity. Increasing the size of the impact events relative to that of the body will lead to violent spallation and, finally, destruction of the target.

Without qualification, collisions between asteroids and other forms of solar system debris comprise the most significant macroscopic phenomena affecting the surfaces and interiors of the minor planets. The effects of these collisions (i.e., those involving direct physical contact between the two bodies, thus eliminating from this discussion the "gravitational collisions" of classical mechanics) can range from the formation of a crater on the surface of the larger body to the traumatic termination of both bodies' existence as members of the asteroid population (e.g. Chapman 1974; Chapman and Davis 1975). While the full spectrum of collisions will be touched upon here, the majority of the discussion will concern relatively high-energy, non-catastrophic collisions, the craters that result from these events, and the effects they should have on the surfaces and the interiors of target bodies.

I. THE NATURE OF IMPACT CRATERING

An impact event produces a crater, not as the result of an explosion but rather through the propagation of a shock wave into the target; the cavity is then formed during and through the release of pressure behind the shock front by rarefaction (Gault and Heitowit 1963; Gault *et al.* 1968). While some aspects of impact and explosion cratering mechanisms are significantly different during the earliest portions of their respective events (Gault and Heitowit 1963; Gault *et al.* 1968), the processes subsequent to the initial transfer of energy from the meteorite or explosion source to the target are very similar (Kreyenhagen and Schuster 1977).

A brief, qualitative description of the mechanics associated with an impact cratering event follows. For more detailed descriptions of the various processes involved, the reader is referred to the more extensive treatments of Gault *et al.* (1968), Gault (1974), Guest and Greeley (1977), and the various papers in the volume edited by Roddy *et al.* (1977).

II. THE FORMATION OF AN IMPACT CRATER

In the very earliest stages of the event, as the projectile makes contact with the target, two shock waves are formed; one travels into the target, while the other moves back into the projectile. The combination of high shock pressures (typically on the order of hundreds of kilobars to megabars for the events considered important here) and free surfaces yields violent decompression and high-velocity ejection of molten and vaporized material, giving rise to a hydrodynamic process generally referred to as "jetting" (Gault *et al.* 1968). This jetted mass is rapidly accelerated to velocities well in excess of the impact velocity, and is composed of material contributed from both the target and the projectile. By the time the shock wave reaches the trailing end of the projectile, the majority of the transferral of energy to the target is complete. The time elapsed from initial contact to this stage in the event is on the order of the time taken for the shock wave to traverse the length of the projectile. For a basalt meteoroid of one meter in diameter impacting a basaltic target at 5 km sec^{-1}, this will occur within $\sim 10^{-4}$ sec.

The shock front in the target is now roughly hemispherical in outline, traveling radially away from the impact area. Two factors will cause the wave to attenuate in strength and velocity: geometric effects and irreversible processes, both of which result from the conservation of energy. As the shock front engulfs an increasing volume of material, it must decrease in intensity in order to keep constant the total energy that is available to the system. At the same time, the wave increases the entropy of the material behind it, thus removing energy that might otherwise have been utilized in propagating the shock (Gault and Heitowit 1963). Irreversible processes (such as target heating, phase changes in the constituent minerals, etc.) will cause the most rapid stress decay to occur near the impact site (Gault and Heitowit 1963; O'Keefe and Ahrens 1976, 1977a; Kieffer and Simonds 1979), while geometric attenuation dominates in the later stages of shock propagation (Gault and Heitowit 1963). Ultimately, the shock will decay to a simple elastic wave as the rarefaction overtakes it (Duvall 1968).

The energy imparted to the material behind the shock wave assumes two forms: kinetic, which appears as particle velocity radial to the shock source, and internal. The internal energy manifests itself as irreversible heating of the target material and as work done in decompressing the target during the rarefaction phase. The progressive addition of decompression-generated velocity vectors to the original radial velocity acquired by the target particles imparts a net upward deflection of the particle trajectories. The resultant path taken by a typical target particle is at first directed radially outward from the shock source, but gradually curves upward and outward, finally being ejected from the growing cavity at a velocity and angle dependent on the initial impact conditions, the target and projectile compositions, and the pre-impact position of the particle in question. It should be mentioned that the excavation process takes place at a much more leisurely rate, lasting $\sim$ 10^4-10^5 times longer than the initial compression stage (Gault *et al.* 1968; Gault and Wedekind 1977). Informative diagrams tracing the relationship between shock propagation, decompression, and ejection can be found in Gault *et al.* (1968), Oberbeck (1975), Maxwell (1977), Kreyenhagen and Schuster (1977), and Orphal (1977a,b).

Finally, a portion of the target is driven away from the impact site with very small, horizontal velocity components. Provided the material is sufficiently resistant to compression, the decompression imparts a net inward and upward motion to the material at the bottom of the transient crater (e.g. Ullrich *et al.* 1977). The radially-inward motion of this rebounding material and the fixed volume into which it can fit promotes fracturing and bulking. The final result is a central uplift, or central peak, in the interior of the crater.

III. THE EJECTA

Relative to normal human experience, the formation of an impact crater cavity is an extremely rapid process. The bulk of the ejecta, however, is still in flight after the cavity stops growing (Gault *et al.* 1968; Oberbeck 1975). With the qualification that a large fraction of the most highly shocked target material never leaves the crater (Dence 1968; Stöffler *et al.* 1975; Dence *et al.* 1977; Grieve *et al.* 1977; Hawke and Head 1977 *a,b;* Gault and Wedekind 1978; Simonds *et al.* 1978), there is a strong correlation between the shock pressure felt by a particle of ejecta and the velocity with which it leaves the crater (Oberbeck 1975; Stöffler *et al.* 1975; O'Keefe and Ahrens 1976). In general, the highest velocity ejecta from an event in a homogeneous target originate near the point of impact (the jetted phases and very earliest ejecta following the demise of the projectile), while that with the lowest velocity come from near the rim region (e.g. Stöffler *et al.* 1975; Oberbeck 1975; Oberbeck and Morrison 1976).

The fate of an individual ejecta fragment will depend ultimately on the velocity and angle of its ejection, and on the strength of the ambient gravity field. The highest velocity material could escape the target body completely (e.g. O'Keefe and Ahrens 1977*b;* Cintala *et al.* 1978) or be deposited as high-energy secondary projectiles. Secondary craters, each with its own separate sequence of excavation and ejecta emplacement, will form as a result of impact of ejecta from the primary crater at velocities and ranges dictated by the type of target material and gravity strength (Oberbeck *et al.* 1974; Gault *et al.* 1975). Lower velocity ejecta, of course, will impact close to the crater rim; its potential to create secondary craters will decrease with impact velocity (which is equal to its ejection velocity), until net deposition is greater than net excavation (Oberbeck *et al.* 1974; Morrison and Oberbeck 1975). At the same time, the decreasing area to be covered by the ejecta at smaller ranges will favor the development of a continuous deposit of ejecta as opposed to isolated patches of ejecta and secondary craters at greater distances. The range at which the continuous blanket begins to break up into a more discontinuous deposit is a function of the gravity field which governs the ballistic transport of the fragments and the size of the crater supplying the ejecta (e.g. Gault *et al.* 1975; Morrison and Oberbeck 1978).

Since there is a direct relationship between shock stress and the degree of metamorphism suffered by the affected material (Kieffer 1971; Stöffler 1971; Schaal and Hörz 1977), it follows that there should be a correlation between the range at which an ejected fragment will impact and the shock imprint that it carries — again with the understanding that a substantial portion of the most highly shocked mass remains in or near the crater (Howard and Wilshire 1975; Stöffler *et al.* 1975; Hawke and Head 1977 *a,b*). Decreasing shock effects in the ejecta, then, will be observed in a radial traverse inward toward the crater, until exterior deposits of impact melt begin to appear (Howard and Wilshire 1975; Hawke and Head 1977*a;* Gault and Wedekind 1978). Complicating factors, such as target inhomogeneities and varying ejection angles, arise in actual large-scale events, however, which tend to confuse the modes of emplacement of crater ejecta (Oberbeck 1975; Stöffler *et al.* 1975); lunar sample analysis, for example, is often made difficult by such seemingly malicious vicissitudes of nature.

In a coherent target, higher peak shock stresses imply smaller particle sizes due to fracturing (Gault *et al.* 1963, Öpik 1971; Oberbeck 1975). In general, since ejection angles appear to be relatively constant during the excavation stage of a cratering event (Oberbeck and Morrison 1976; Ivanov 1976), the smallest fragments should be ejected at the highest velocities and travel the greatest distances; the larger fragments, having experienced lower peak pressures, will travel shorter distances from the crater (Oberbeck 1975; Cintala *et al.* 1978). Thus, the coarsest fragments should be found near the crater rim, with the average size decreasing as a function of increasing radial distance. Indeed, this is the generally observed pattern (Moore *et al.* 1969, unpublished report cited in Oberbeck 1975). Although complicating factors certainly arise in natural events, a simplified view such as this can often lead to a better understanding of impact cratering processes.

IV. THE RESULTING CRATER

On the moon, Mercury, and Mars, craters which have not undergone extensive post-formational modification generally take the shape of inverted truncated cones. Small, flat floors in these craters are probably the result of minor amounts of wall slumping as the cavity adjusts to a stable configuration with respect to gravitational forces (Pike 1977; Wood and Andersson 1978). It has been suggested that the transient cavity — i.e. the crater at any given time during the formational event, before modification processes take effect — can be described as a paraboloid of revolution (Dence 1973; Grieve *et al.* 1977). This contention, while justifiable on theoretical grounds and supported by field observations of terrestrial craters (Dence 1973), remains to be proven: the otherwise insignificant wall failure and the rebound effects mentioned earlier tend to mask the transient cavity geometry at the end of the event.

While compression due to the large stresses associated with an impact event is accompanied by rebound in coherent targets, it can be "frozen" to a large extent into the final crater shape in weakly consolidated targets through

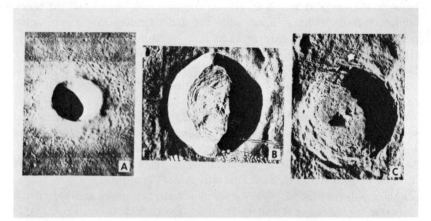

Fig. 1. Three lunar craters, illustrating the growing complexity of interior morphology with increasing crater diameter. (A) Mösting C, 3.8 km in diameter; (B) Dawes, ∼18 km in its maximum dimension; (C) Tycho, averaging ∼85 km in diameter.

irreversible compaction, especially in small-scale craters (centimeters in size) (Braslau 1970; Stöffler *et al.* 1975; Croft 1978). Although the same effects should obtain to some degree in larger craters, the decreasing importance of target strength in events of larger magnitude (see Sec. V below) and the attendant modification processes will tend to obscure the distinction between the two types of target *vis à vis* interior geometry.

Truly large craters (those with diameters measured in tens or hundreds of kilometers) exhibit much more complex interiors. Central peaks become more prominent, grading from single mountains to multiple peaks (Baldwin 1963, chapter 20; Pike 1968, 1977; Wood 1968; Smith and Sanchez 1973; Hale and Head 1979). Wall failure becomes increasingly more important; indeed, crater diameters can be enlarged by more than 25 % through the formation of massive terraces (Fig. 1; Pike 1974*a;* Settle and Head 1979). A major challenge to planetary geologists lies in the reconstruction of large craters to their pre-modification geometries. As might be expected from the formidable complexity of these landforms, results have thus far been varied (Pike 1974*a;* Moore *et al.* 1974; Head *et al.* 1975; Settle and Head 1979).

When crater rim-to-floor depth (R_i) is plotted against diameter (D_r) in a log-log system, a distinctive form emerges as shown in Fig. 2. The small, relatively unmodified craters fall along a line with a slope of 1, while the larger craters plot along a shallower curve. This is true of craters on the moon (Pike 1968, 1974*b*), Mercury (Gault *et al.* 1975; Malin and Dzurisin 1978), and Mars (Cintala and Mouginis-Mark 1979). Craters on Phobos and Deimos, on the other hand, essentially fall along the smaller crater branch of the lunar distribution (Thomas 1978). The lack of a similar kink in the Phobos and Deimos distribution is not surprising since the two satellites do not present environments favorable to the target rebound and gravitationally-driven

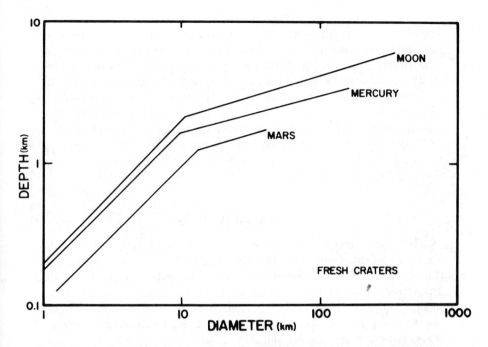

Fig. 2. Depth/diameter fits to fresh crater data from the moon, Mercury, and Mars. The fit for craters on Phobos and Deimos would essentially plot along the steeper branch of the lunar curve. Comparison between the lunar curves and the craters in Fig. 1 illustrates the general relationship between morphology and depth/diameter behavior.

modification mechanisms effective on the terrestrial planets (Quaide *et al.* 1965; Gault *et al.* 1975; Cintala 1977; Pike 1977; Veverka and Duxbury 1977; Cintala *et al.* 1978; Settle and Head 1979).

Gault and Wedekind (1979) have derived an equation to represent the R_i/D_r ratio of a crater formed in any given gravity field, namely,

$$R_i/D_r = 0.193 + 0.05 D_r^{0.12} \left(\frac{g}{g_\oplus} \right)^{0.157} \tag{1}$$

where g and $g_\oplus$ are the target body and terrestrial gravitational accelerations, respectively. The first term on the right hand side is the ratio of apparent depth (R_a, the depth of the crater measured from the original ground surface) to D_r, which has been found to be independent of g in small experimentally produced craters (Gault and Wedekind 1977, 1979). In an attempt to take into account the effects of modeled impact velocity (Hartmann 1977), the expression

$$R_i/D_r = 0.308 V_i^{-0.279} + 0.129 \left(\frac{g}{g_\oplus} \right)^{0.208} V_i^{-0.301} \tag{2}$$

has been derived (Cintala 1979), where V_i is the impact velocity in km sec^{-1}. The first term on the right hand side gives the dependence of R_a/D_r on calculated V_i (Hartmann 1977), while the second term expresses the R_e/D_r ratio (where R_e is the rim height) in terms of g and V_i. This last equation illustrates the decrease in cratering efficiency with increasing V_i, as reflected in the smaller R_a/D_r ratios. It also shows that the rim height increases with increasing g, an effect at least partly due to the confinement of ejecta to regions closer to the crater by higher gravitational accelerations (Gault *et al.* 1975; Cintala 1979). More detailed discussions of crater morphometry can be found in Pike (1968, 1977), Moore *et al.* (1974), Malin and Dzurisin (1977, 1978), Wood *et al.* (1977), and Wood and Andersson (1978).

V. CRATER SCALING

The energy required to form a crater is expended against two resisting agents: the inherent strength of the target and lithostatic forces due to gravity (Chabai 1965; Gault and Wedekind 1977; O'Keefe and Ahrens 1979). Theoretical studies indicate that the sizes of the resulting craters formed by small events (where "small" is a relative term, dependent upon the actual target strength) are governed by the target's strength (resulting in "strength scaling" of crater dimensions), and that

$$D_a \propto E_c^{\frac{1}{3}} \tag{3}$$

where D_a is the apparent crater diameter (i.e., the diameter measured at the original target surface) and E_c is the energy expended in forming the crater (Chabai 1965; Gault and Wedekind 1977). As might be expected intuitively, on the other hand, very large events are affected to a much lesser degree by the target's resistance to deformation and/or fracture; for this class of events ("gravity scaling")

$$D_a \propto \left(\frac{E_c}{g}\right)^{\frac{1}{4}} \tag{4}$$

where, again, g is the local gravitational acceleration (Chabai 1965; Gault and Wedekind 1977). In these large events, more energy is diverted toward overcoming the large "artificial strength" of gravitational forces at depth and toward doing work against the gravitational field in transporting the ejecta out of the crater cavity.

Gault and Wedekind (1977), in choosing between strength and gravity scaling for a specific cratering event, recommend the use of the dimensionless fraction, $s/\rho g D_a$ where s is a measure of strength which depends on the type of target material (Holsapple and Schmidt 1979; O'Keefe and Ahrens 1979) and ρ is the target density. In particular, if $s/\rho g D_a \gg 1$, strength scaling holds, and if $s/\rho g D_a \ll 1$, gravity scaling applies. In cases where $s/\rho g D_a$ is on the order of unity, neither relationship is strictly applicable, and the exponent in the scaling law takes a value between 1/3 and 1/4 (Chabai 1965). Expressions (3) and (4) indicate that gravity scaling generally results in smaller craters than strength scaling for large events in normal geologic (or asteroidal) materials.

VI. THE EFFECTS OF IMPACT CRATERING ON SMALL BODIES

In considering the impact cratering process on asteroids and small satellites, special consideration must be given to the small dimensions and low-gravity environments that they present as targets. An impact event which would be small by planetary standards could be devastating to an asteroid-sized body. While the strong gravitational fields associated with larger planets virtually rule out existence-terminating collisions, very little work per unit mass, in comparison, would be required to convert an asteroid into component pieces of the general meteoroid population. This is, admittedly, an extreme case, but the certainty that such catastrophic collisions have occurred (see the chapter by Davis *et al* in this book) serves as a reminder of the very different effects that impact cratering can have on such small bodies.

It is difficult to separate the nature of the impact cratering process from its effects on small bodies, since one sizeable event could affect a major fraction of the surface area and volume of the target. The following section will discuss the case of cratering on small bodies in the context of a process and its immediate effects on the target. Section IX will consider the long-term effects which this process might have on an asteroid or small satellite.

VII. EFFECTS ON THE TARGET BODY
DURING THE CRATERING EVENT

A. The Crater

With corrections for target size and gravity, all of the cratering phenomena described earlier should hold true for impacts into small bodies. The initial contact and excavation stages should remain the same with no major differences due to the weak ambient gravity fields. Deviations from a similar size event in a target which is planar (or nearly so) will occur, however, if the final crater diameter were a significant fraction of the target body's radius (Cintala *et al.* 1978, 1979). If the gravities of both the hypothetical asteroid and planar targets are taken to be very weak, strength scaling will govern the sizes of the resulting craters. This implies that the crater diameters will be a function of target strength and, therefore, that the stress levels will be essentially the same at the rims of the final craters (e.g., Ivanov 1976). Since the shock stress is a function of distance from the shock source, the crater on the small body will be larger, because the free surface is

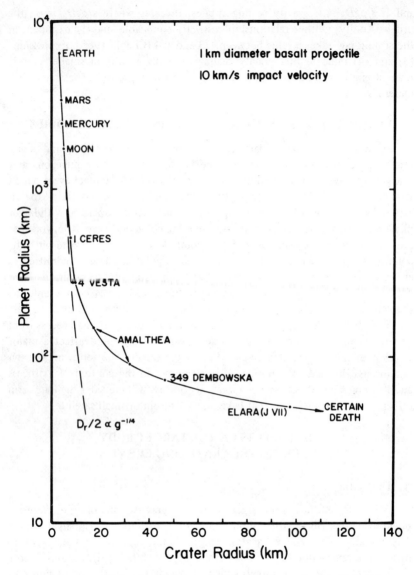

Fig. 3. Crater radii as a function of (spherical) target body curvature. While this curve was constructed for pure gravity scaling, similar trends should hold for strength scaling. The double points for 4 Vesta and Amalthea (JV) represent the possible ranges of radii. The points for 343 Dembowska and Elara are schematic only; events of this magnitude would probably disrupt both bodies.

closer to the shock source for any given surface distance from the impact point (see Fig. 3). It has been shown through theoretical calculations for explosive events (e.g. Orphal 1977*a,b;* Maxwell 1977) that the final crater

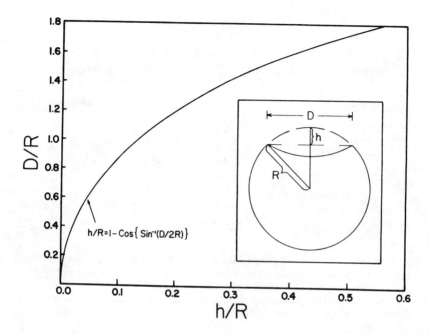

Fig. 4. The effect of target curvature on crater geometry. While the final crater will be shallower than one on a planar surface, a large volume of material *above* a rim-to-rim-chord will have been ejected. This curve was calculated for a spherical target body.

depth is attained before the cavity stops growing laterally. It would appear, therefore, that the final crater depth as measured from the point of impact (i.e., the apparent crater depth, R_a; see Sec. IV) will be the same for both cases. Since all of the target surface above the final crater profile will be ejected, the crater on the asteroid as viewed after the event will be shallower; reconstruction of the target to its true initial shape, however, will reveal the total volume of material actually removed by the cratering event (see Fig. 4). This effect should hold for a set of conditions defined by the magnitude of the event with respect to the size and properties of the target body; for an impact large relative to the target, more complicated geometries will result (Cintala *et al.* 1979).

B. The Ejecta

The ejecta can be divided into two types with respect to their velocity: those which escape the body and those which remain within its gravitational sphere of influence. It has been calculated that the material which is ejected at the highest velocities also comes from the hottest portion of the shocked mass (Gault and Heitowit 1963; O'Keefe and Ahrens 1976). Some of this hot

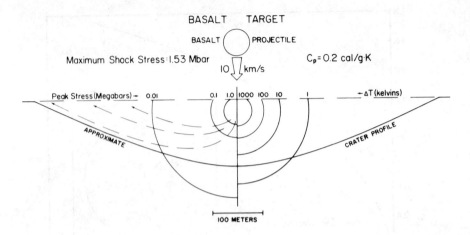

Fig. 5. Peak shock stress and temperature change contours for a 50 m diameter basalt projectile impacting a basalt target. Thin dashed arrows schematically represent the paths followed by material which eventually will be ejected from the growing cavity. C_p is the specific heat at constant pressure used in the calculations, which were made following the method of Gault and Heitowit (1963).

target material, if it is hot enough, will remain within or in the vicinity of the crater as "impact melt". The relative fraction of this melt depends on factors such as angle of impact (Howard and Wilshire 1975; Hawke and Head 1977*a,b;* Gault and Wedekind 1978), local topography (Hawke 1976; Hawke and Head 1977 *a,b*), and crater wall slumping (Hawke and Head 1977*a,b*) which, as was pointed out earlier, will probably be of minimal importance, except perhaps on the largest asteroids. In any event, the fraction of escaping impact-vaporized and melted material will depend on the target body's gravity field strength and on the impact variables, which include target and projectile constitution (O'Keefe and Ahrens 1977*b*).

Farther from the center of impact, the amount of shock heating experienced is much lower (Fig. 5) and shock-induced stress gradients decrease. At the same time, ejection velocities decrease as the position of the final crater rim is approached. These three points lead to the conclusion that material ejected farther from the impact point will occur as cool, low-velocity blocks, which should be emplaced at equally low velocities. If the gravitational acceleration of the target body were very low, only these blocks would be traveling at velocities low enough to remain on the surface of the target (Cintala *et al.* 1978). As larger target bodies are considered, of course, more highly shocked material will be retained. Unfortunately, exact quantities cannot be presented, since the mass/velocity relationship for large crater ejecta is poorly understood. Solutions to this problem through a numerical model can be found in O'Keefe and Ahrens (1977*b*); results for experimental impacts extrapolated to lunar conditions are presented in Gault *et al* (1963) and Schneider (1975).

The effects of non-escaping ejecta on the surface of the target body are highly dependent on the nature of the surface layer and the gravitational acceleration. Low-velocity impact into rock might disintegrate an already-fractured projectile and chip the target, but not much more. The presence of a fine-grained regolith, on the other hand, would promote secondary cratering; low-velocity impact into porous media will result in craters due predominantly to compression (Clark and McCarty 1963; Hartmann 1978; Head and Cintala 1979), while higher velocity secondary projectiles (restricted by escape velocity constraints to higher gravity environments, hence to larger bodies) will excavate and compress the substrate (Oberbeck 1975; Oberbeck *et al* 1974; Stöffler *et al* 1975).

The distance that a particle will travel in a given gravity field depends upon the radius of curvature of the target body, the ejection angle and velocity. For a given impact into a given target structure and composition, material will travel farther in a lower gravity field than in one with a higher gravitational acceleration. If strength scaling were to predominate during the event (see Sec. V), the crater volume would be essentially independent of the gravitational acceleration; the extent of the continuous ejecta blanket should thus be greater in the lower gravity case (Gault *et al* 1975; Morrison and Oberbeck 1978). This is accomplished at the expense of the thickness of the rim deposit of ejecta; since it will travel further and will be smeared over greater areas, the material which would land near the final crater rim on the moon would impact farther from the crater's vicinity under asteroidal conditions. The result would be a thinner rim deposit but a continuous blanket of more uniform thickness along a radial traverse than that expected for higher gravity cases (Morrison and Oberbeck 1978).

Caution must be exercised, however, in extrapolating to the extreme asteroidal environments. One need only consider the fact that the limiting case of zero gravity will give *no* deposit, continuous or otherwise, in order to realize that the widespread nature of crater ejecta at these low gravitational accelerations could limit the extent of the continuous deposits on small asteroids. A constant volume of ejecta spread over a wide area will result in a thinner blanket overall; should the average particle size exceed the average thickness, a discontinuous deposit would result. Calculations describing these phenomena will be a necessity if the dynamics of ejecta emplacement in low-gravity environments are to be utilized in understanding the surface evolution of asteroids and small satellites.

Mention should be made of ejecta which neither escape the gravitational influence of the target body nor return directly to the surface. In the event of a highly irregular target body (433 Eros, 1620 Geographos, etc.), the gravitational equipotential lines would be distorted from those of a sphere. There is no reason to expect that material ejected at relatively high velocities could not go into orbit about the target body. The nature of these orbits would be very complex, as they would be influenced by a periodic forcing function, namely, the rotating asymmetric gravity field of the asteroid. These orbits, if they were to occur, should be subject to instabilities due to resonances with the periodic field variations. Knowledge of the lifetimes of such orbits would be of importance in determining if a debris cloud could surround such asteroids for any considerable length of time – a potential point to consider in planning close fly-by asteroid missions, especially at high relative velocities.

C. Subsurface Effects

Impact craters in coherent targets are accompanied by a subsurface lens of disrupted target material which is not ejected (Baldwin 1963; Beals *et al.* 1963; Shoemaker 1963; Dence *et al.* 1968, 1977; Dvorak and Phillips 1977; Orphal 1979). On the asteroids, this phenomenon will range from concentric and radial fractures surrounding centimeter-scale craters (Moore *et al.* 1962; Curran *et al.* 1977) to nondispersive fragmentation of the target body (Thomas *et al.* 1978; Hartmann's chapter in this book), which, in effect, makes the entire asteroid synonymous with the "breccia lens". Not only would this *in situ* crushing and fracturing extend the depth of any fragmental surface layer – a process akin to the formation of planetary "megaregoliths" (Hartmann 1973; Head 1976) – but it would act as a more subtle agent in contributing material to the ballistically-emplaced portion of the fragmental layer as well (see Sec. VIII below).

A somewhat different effect should obtain on fragmental or weakly consolidated asteroids, such as carbonaceous chondritic bodies. As mentioned in Sec. IV, irreversible compaction of the target could occur if sufficient porosity were to exist (Braslau 1970; Stöffler *et al.* 1975). Provided compaction has not already been accomplished by gravitational forces, an impact event will tend to indurate and remove porosity from the target, thereby increasing its density and acoustic impedance (see Sec. VIII below). If this effect were to be coupled with the compaction associated with secondary cratering (see Sec. VII,B), it would appear that the outermost layers of a porous target, at least, would eventually become more consolidated with time (Cintala *et al.* 1978), except for a relatively thin surficial layer of fine-grained ejecta. Stress waves from craters, those both nearby and "over the horizon," will tend to act against this compaction mechanism; the final results will depend ultimately upon the relative rates at which these processes are operative.

VIII. STRESS WAVE-SURFACE INTERACTIONS

A stress wave incident upon an interface between two media will reflect from the interface and, depending upon the properties of the medium on the other side of the interface, will have some fraction of its energy transmitted across the boundary. Analytic solutions to most cases with simple geometries are obtained easily through energy and momentum considerations (see Rinehart 1968). Briefly, the behavior of a wave incident upon an interface between two dissimilar substances will depend on the acoustic impedances, ρC, of the two media, where ρ is the density and C is the propagation velocity of the wave through the material in question. (The reader is referred to Rinehart [1968] and Kolsky [1963] for a general treatment of stress wave behavior in solids.) The case most relevant to this discussion is that of the incidence of a compressional (longitudinal) wave upon a free surface; this would occur on an asteroid when a shock or elastic wave generated by an impact event reaches the opposite side of the target body (normal incidence for a spherical body) or, for that matter, any free surface (oblique incidence, in general). If the wave were propagated along a normal to the free surface, it would be reflected directly back into the asteroid as a tensile wave of equal amplitude. Oblique incidence of a similar wave would also produce a tensile wave (reflected at an angle equal to the angle of incidence of the compressional wave), but it would also be accompanied by a reflected shear (transverse) wave at an angle different from that of the reflected tensile wave. Since the acoustic impedance can be taken as the constant of proportionality relating the amplitude of the stress wave to the particle velocity acquired by the medium during passage of the wave, it follows that higher amplitude waves will cause higher particle velocities. Provided that the resultant intensity of the waves is larger than the tensile strength of the medium, fragmentation will result; energy not expended in disrupting the material will manifest itself as residual kinetic energy of motion of the fragments.

The tensile strength of fragmental material (such as a regolith, breccia lenses beneath craters, etc.) is very low to nonexistent. Most of the energy associated with a compressional wave incident upon a layer of regolith will impart kinetic energy to the particles in the layer. In a low-gravity environment, very low stresses would be required to loft the constituent particles of a regolith (e.g., Head and Cintala 1979). For example. if the regolith on an asteroid had a density of 1.9 g cm^{-3} and a compressional wave velocity of 10^4 cm sec^{-1} (values determined for the regolith at the Apollo 14 landing site; Mitchell et al. 1971; Kovach et al. 1971), a net upward velocity of 10^3 cm sec^{-1} would result from the normal incidence and resulting reflection of a 9.5×10^6 dyne cm^{-2} compressional wave (10^6 dynes cm^{-2} = 1 bar = 10^5 Pa). This is an extremely weak wave in the context of impact cratering. Stronger waves, both normal and oblique, would cause correspondingly higher "spallation velocities," some of which could easily exceed the escape velocities of the vast majority of the asteroids. While the detailed treatment of this process is beyond the scope of this discussion, it is noted that the masses and velocities of spalled fragments are dependent on the shape, duration and intensity of the stress wave (Rinehart 1975, pp. 203-220).

Very high amplitude stress waves, such as weakly attenuated shock fronts, would give rise to large-scale spallation of the underlying substrate. In general the first area to spall from the target body will be at the antipode (Gault and Wedekind 1969; Schultz and Gault 1975*a,b;* Fujiwara *et al.* 1977; Hughes *et al.* 1977), but deviations from symmetric figures (e.g., large craters) could cause stress concentrations sufficiently intense to cause local disruption and spallation (Rinehart 1968). Increasing stress-wave energies (or weaker waves with longer wavelengths) will lead to more widespread disruption and spallation; ultimately total destruction of the target body will result (Gault and Wedekind 1969; Davis and Chapman 1977; Fujiwara *et al.* 1977; Fujiwara 1978; the chapter by Davis *et al.* in this book). The fate of the fragments will depend upon a number of parameters; detailed treatments of this topic can be found in the chapters by Hartmann and by Davis *et al.*

IX. LONG TERM EFFECTS OF IMPACT CRATERING ON SMALL BODIES

When integrated over time, the effects of impact cratering on asteroids and small satellites define, to a large extent, the geologic histories of their surfaces. It is not the purpose of this qualitative review to construct detailed scenarios of asteroid surface evolution. Instead, a brief summary of the nature and effects of impact cratering on these bodies is presented in the context of processes affecting their evolution.

Considering the sizes of most asteroids and calculated rates of cratering-induced mass loss as a function of escape velocity (O'Keefe and Ahrens 1977*b*), small asteroids must have undergone considerable mass reduction due to escape of crater ejecta (c.f. Chapman 1976). The growth of a fragmental layer on the surface or, alternatively, an inherently weak asteroid (e.g., one with carbonaceous chondritic composition and structure) will reduce the rate of such mass loss due to less efficient cratering and lower ejection velocities (Braslau 1970; Stöffler *et al.* 1975; Chapman 1978; chapter by Housen *et al* in this book). While ejection at velocities greater than escape velocity and violent spallation accompanying large impacts will militate against its retention, the formation of a debris layer, if it were to occur, would enhance the ability of the body to foster a thicker regolith. If a significant regolith were not to be realized, however, even the *in situ* products created by impact events over time could result in surface properties compatible with interpretations of probable surface textures on the basis of Earth-based observations (Cintala *et al.* 1978).

The formation of large craters (i.e., those penetrating any existing fragmental layer) will tend to fragment a target body to progressively greater depths. This would decrease the effective internal strength of the target, possibly to the point of allowing some semblance of gravitationally-induced sphericity at sizes smaller than those expected for pristine material (Hartmann in discussion following Cook 1971, p. 162. At the same time, the efficiency with which the body transmits stress waves will decrease, thus giving less violent effects at free surfaces due to reflected stress waves. On the other hand, the decrease in effective strength will make the body more susceptible to catastrophic disruption by smaller impact events (see the chapter by Davis et al.).

The lower the gravitational acceleration of the target body, the smaller the event is which will produce regolith movement due to stress wave interactions with the free surface. Continual bombardment of a small body covered by a fragmental layer should cause net downslope migration of the debris over time due to "seismic shaking;" this effect, a variation of which has been proposed earlier (Arnold, in discussion following Chapman 1978), would expose more coherent substrates, if they were to exist, in the form of topographic highs as the regolith "ponded" in craters and other topographic lows. This effect might account for the observations that material has collected within craters and hollows on Deimos (Thomas 1978; chapter by Veverka and Thomas). Impact-induced compression of Phobos' regolith, if it were to have occurred over the moon's surface (Cintala et al. 1978), coupled with its greater dimensions (thus allowing more effective attenuation of stress waves; Rinehart 1960; Schultz and Gault 1975a,b), would decrease the effectiveness of this process on the larger Martian satellite.

The interaction of stress waves with free surfaces constitutes a mechanism by which regolith could be mixed or "gardened" without the formation of a nearby crater. Regolith could be lofted and emplaced, becoming mixed in the process, by a cratering event on the other side of the target body; the only necessary condition lies in the generation of a stress wave of sufficient intensity to mobilize the surficial debris layer. Numerous craters large enough to generate such waves should form throughout the history of the target body. Since each (sufficiently large) event will affect a large portion of the surface of the target body the necessity for very high rates of blanketing by crater ejecta should be lessened in order to support an asteroid-regolith origin for gas-rich stony meteorites (Anders 1975, 1978).

Finally, impact-induced heating of volatile-rich and, perhaps, the more anhydrous asteroids might encourage otherwise unlikely reactions in the surface materials (Cintala et al. 1978). The final steady-state products ultimately will depend on a large number of factors, the most important of which is the original composition of the target material. For example, continual bombardment of carbonaceous chondritic surfaces should eventually drive off most volatiles, thus changing the oxidation states of the uppermost layers. Such metamorphism in impact-melted material and recondensation of the more refractory volatilized elements (R. Huguenin and R. Scott 1979, personal communication) could induce unusual and/or unexpected spectral characteristics, especially on the larger asteroids which retain more of the heated ejecta. Caution should be exercised, therefore, in interpreting asteroid spectral data, especially when comparing them to relatively unshocked meteorite specimens.

Acknowledgments. This work was supported by a grant from the National Aeronautics and Space Administration which is gratefully acknowledged. Helpful comments and criticisms were provided by D. Davis, K. Housen and P. Mouginis-Mark. Thanks go to these reviewers and also to S. Bosworth, N. Christy, D. Haas, and L. Ranalli for high quality work in the preparation of the manuscript.

REFERENCES

Anders, E. 1975. Do stony meteorites come from comets? *Icarus* 24: 363-371.

Anders, E. 1978. Most stony meteorites come from the asteroid belt. In *Asteroids: An Exploration Assessment,* eds. D. Morrison and W. C. Wells, NASA Conf. Publ. 2053, pp, 57-75.

Baldwin, R. B. 1963. *The Measure of the Moon* (Chicago: Univ. of Chiago Press).

Beals, C. S.; Innes, M. J. S.; and Rottenberg, J. A. 1963. Fossil meteorite craters. In *The Moon, Meteorites, and Comets,*
University of Chicago Press), pp. 235-284.

Braslau, D. 1970. Partitioning of energy in hypervelocity impact against loose sand targets. *J. Geophys. Res.* 75: 3987-3999.

Chabai, A. J. 1965. On scaling dimensions of craters produced by buried explosives. *J. Geophys. Res.* 70: 5075-5098.

Chapman, C. R. 1974. Asteroid size distribution: Implications for the origin of stony-iron and iron meteorites. *Geophys. Res. Lett.* 1: 341-344.

Chapman, C. R. 1976. Asteroids and meteorite parent bodies: The astronomical perspective. *Geochim. Cosmochim. Acta* 40: 701-719.

Chapman, C. R. 1978. Asteroid collision, craters, regoliths, and lifetimes. In *Asteroids: An Exploration Assessment,* eds. D. Morrison and W. C. Wells, NASA Conf. Publ. 2053, pp. 145-160.

Chapman, C. R., and Davis, C. R. 1975. Asteroid collisional evolution: Evidence for a much larger population. *Science* 190: 553-556.

Cintala, M. J. 1977. Martian fresh crater morphology and morphometry – A pre-Viking view. In *Impact and Explosion Cratering,* eds. D. J. Roddy, R. O. Pepin and R. B. Merrill (New York: Pergamon Press), pp. 575-591.

Cintala, M. J. 1979. Small fresh crater morphometry: A preliminary assessment of the effects of gravitational acceleration and impact velocity (abstract). *Reports of Planet. Geol. Program,* 1978-1979. NASA TM 80339, pp. 176-178.

Cintala, M. J.; Head, J. W.; and Veverka, J. 1978. Characteristics of the cratering process on small satellites and asteroids. *Proc. Lunar Sci. Conf. IX* (Oxford: Pergamon Press), pp. 3803-3833.

Cintala, M. J., and Mouginis-Mark, P. J. 1979. New depth/diameter data for fresh Martian craters and some interplanetary comparisons (abstract). *Reports of Planet. Geol. Program,* 1978-1979. NASA TM 80339, pp. 182-184.

Cintala, M. J.; Wilson, L.; and Gash, P. J. S. 1979. The effects of target curvature on impact cratering events (in preparation).

Clark, L. V., and McCarty, J. L. 1963. The effect of vacuum on the penetration characteristics of projectiles into fine particles. NASA TN D-1519.

Cook, A. F. 1971. 624 Hektor: A binary asteroid? In *Physical Studies of Minor Planets,* ed. T. Gehrels (NASA SP-267, Washington, D. C.: U. S. Government Printing Office), pp. 155-163.

Croft, S. K. 1978. Large crater volumes: Interpretation by models of impact cratering and upper crustal structure. *Proc. Lunar Sci. Conf. IX* (Oxford: Pergamon Press), pp. 3711-3733.

Curran, D. R.; Schockey, D. A.; Seaman, L.; and Austin, M. 1977. Mechanisms and models of cratering in earth media. In *Impact and Explosion Cratering*, eds. D. J. Roddy, R. O. Pepin, and R. B. Merrill (New York: Pergamon Press), pp. 1057-1087.

Davis, D. R., and Chapman, C. R. 1977. The collisional evolution of asteroid compositional classes (abstract). *Lunar Science VIII*, The Lunar and Planetary Institute, pp. 224-226.

Dence, M. R. 1968. Shock zoning at Canadian craters: Petrography and structural implications. In *Shock Metamorphism of Natural Materials*, eds. B. French and N. Short (Baltimore: Mono Book Corp.), pp. 169-184.

Dence, M. R. 1973. Dimensional analysis of impact structures (abstract). *Meteoritics* 8: 343-344.

Dence, M. R.; Grieve, R. A. F.; and Robertson, P. B. 1977. Terrestrial impact structures: Principal characteristics and energy considerations. In *Impact and Explosion Cratering*, eds. D. J. Roddy, R. O. Pepin and R. B. Merrill (New York: Pergamon Press), pp. 247-275.

Dence, M.; Innes, M.; and Robertson, P. 1968. Recent geological and geophysical studies of Canadian craters. In *Shock Metamorphism of Natural Materials*, eds. B. French and N. Short (Baltimore: Mono Book Corp.), pp. 339-362.

Duvall, G. E. 1968. Shock waves in solids. In *Shock Metamorphism of Natural Materials*, eds. B. French and N. Short (Baltimore: Mono Book Corp.), pp. 19-29.

Dvorak, J., and Phillips, R. J. 1977. The nature of the gravity anomalies associated with large young lunar craters. *Geophys. Res. Lett.* 4: 380-382.

Fujiwara, A.; Kamimoto, G.; and Tsukamoto, A. 1977. Destruction of basaltic bodies by high-velocity impact. *Icarus* 31: 277-288.

Gault, D. E. 1974. Impact cratering. In *A Primer in Lunar Geology*, eds. R. Greeley and P. Schultz, NASA – Ames Research Center, pp. 137-175.

Gault, D. E.; Guest, J. E.; Murray, J. B.; Dzurisin, D.; and Malin, M. 1975. Some comparisons of impact craters on Mercury and the moon. *Jour. Geophys. Res.* 80: 2444-2460.

Gault, D. E., and Heitowit, E. D. 1963. The partition of energy for hypervelocity impact craters in rock. *Proc. 6th Hypervelocity Impact Symp.*, Naval Research Lab, U. S. Govt. Res. Rept., pp. 413-456.

Gault, D. E.; Quaide, W. L.; and Oberbeck, V. R. 1968. Impact cratering mechanics and structures. In *Shock Metamorphism of Natural Materials*, eds. B. French and N. Short (Baltimore: Mono Book Corp.), pp. 87-99.

Gault, D. E.; Shoemaker, E. M.; and Moore, H. J. 1963. Spray ejected from the lunar surface. NASA TN D-1767.

Gault, D. E., and Wedekind, J. A. 1969. The destruction of tektites by micrometeoroid impact. *Jour. Geophys. Res.* 74: 6780-6794.

Gault, D. E., and Wedekind, J. A. 1977. Experimental hypervelocity impact into quartz and sand-II, effects of gravitational acceleration. In *Impact and Explosion Cratering*, eds. D. J. Roddy, R. O. Pepin and R. B. Merrill (New York: Pergamon Press), pp. 1231-1244.

Gault, D. E., and Wedekind, J. A. 1978. Experimental studies of oblique impact. *Proc. Lunar Sci. Conf. IX* (Oxford: Pergamon Press), pp. 3843-3875.

Gault, D. E., and Wedekind, J. A. 1979. Experimental results for effects of gravity on impact crater morphology (abstract). *2nd Internat. Colloq. on Mars*, NASA Conf. Publ. 2072, p. 29.

Grieve, R. A. F.; Dence, M. R.; and Robertson, P. B. 1977. Cratering processes: As interpreted from the occurrence of impact melts. In *Impact and Explosion Cratering*, eds. D. J. Roddy, R. O. Pepin and R. B. Merrill (New York: Pergamon Press), pp. 791-814.

Guest, J. E., and Greeley, R. 1977. *Geology on the Moon* (London: Wykeham Publications Ltd.).

Hale, W., and Head, J. W. 1979. Central peaks in lunar craters: Morphology and morphometry (abstract). *Lunar Science X*, The Lunar and Planet. Inst., pp. 491-493.

Hartmann, W. K. 1973. Ancient lunar megaregolith and subsurface structure. *Icarus* 18: 634-639.

Hartmann, W. K. 1977. Relative crater production rates on planets. *Icarus* 31: 260-276.

Hartmann, W. K. 1978. Planet formation: Mechanism of early growth. *Icarus* 33: 50-61.

Hawke, B. R. 1976. Ponded material on the north rim of King Crater: Influence of pre-event topography on the distribution of impact melt. *EOS* 57: 275.

Hawke, B. R., and Head, J. W. 1977a. Impact melt in lunar crater interiors (abstract). *Lunar Science VIII*, The Lunar and Planet. Inst., pp. 415-417.

Hawke, B. R., and Head, J. W. 1977b. Impact melt on lunar crater rims. In *Impact and Explosion Cratering*, eds. D. J. Roddy, R. O. Pepin and R. B. Merrill (New York: Pergamon), pp. 815-841.

Head, J. W. 1976. The significance of substrate characteristics in determining morphology and morphometry of lunar craters. *Proc. Lunar Sci. Conf. VII* (Oxford: Pergamon Press), pp. 2913-2929.

Head, J. W., and Cintala, M. J. 1979. Grooves on Phobos: Evidence for possible secondary cratering origin (abstract). *Reports of Planet. Geol. Program*, 1978-1979. NASA TM 80339, pp. 19-21.

Head, J. W.; Settle, M.; and Stein, R. S. 1975. Volume of material ejected from major lunar basins and implications for the depth of excavation of lunar samples. *Proc. Lunar Sci. Conf. VI* (Oxford: Pergamon Press), pp. 2805-2829.

Holsapple, K. A., and Schmidt, R. M. 1979. A material strength model for apparent crater volume (abstract). *Lunar Science X*, The Lunar and Planet. Inst., pp. 558-560.

Howard, K. A., and Wilshire, H. G. 1975. Flows of impact melt at lunar craters. *J. Res. U. S. Geol. Survey* 3: 237-251.

Hughes, H. G.; App, F. N.; and McGetchin, T. R. 1977. Global seismic effects of basin-forming impacts. *Phys. Earth. Planet. Interiors* 15: 251-263.

Ivanov, B. A. 1976. The effect of gravity on crater formation: Thickness of ejecta and concentric basins. *Proc. Lunar Sci. Conf. VIII* (Oxford: Pergamon Press), pp. 2947-2965.

Kieffer, S. W. 1971. Shock metamorphism of the Coconino sandstone at Meteor Crater, Arizona. *Jour. Geophys. Res.* 76: 5449-5473.

Kieffer, S. W., and Simonds, C. H. 1979. The role of volatiles in the cratering process (abstract). *Lunar Science X*, The Lunar and Planet. Inst., pp. 661-663.

Kolsky, H. 1963. *Stress Waves in Solids* (New York: Dover Publ.).

Kovack, R. L.; Watkins, J. S.; and Landers, T. 1971. Active seismic experiment. In *Apollo 14 Prelim. Sci. Report* (NASA SP-272, Washington, D. C.: U. S. Government Printing Office), pp. 163-174.

Kreyenhagen, K. N., and Schuster, S. H. 1977. Review and comparison of hypervelocity impact and explosion cratering calculations. In *Impact and Explosion Cratering*, eds. D. J. Roddy, R. O. Pepin, and R. B. Merrill (New York: Pergamon Press), pp. 985-1002.

Malin, M., and Dzurisin, D. 1977. Landform degradation on Mercury, the moon, and Mars: Evidence from crater depth/diameter relationships. *J. Geophys. Res.* 82: 376-388.

Malin, M., and Dzurisin, D. 1978. Modification of fresh crater landforms: Evidence from the moon and Mercury. *J. Geophys. Res.* 83: 233-243.

Maxwell, D. E. 1977. Simple Z model of cratering, ejection, and the overturned flap. In *Impact and Explosion Cratering*, eds. D. J. Roddy, R. O. Pepin, and R. B. Merrill (New York: Pergamon Press), pp. 1003-1008.

Mitchell, J. K.; Bromwell, L. G.; Carrier, W. D.; Coates, N. C.; and Scott, R. F. 1971. Soil mechanics experiment. In *Apollo 14 Prelim. Sci. Report* (NASA SP-272, Washington, D. C.: U. S. Government Printing Office), pp. 87-108.

Moore, H. J.; Gault, D. E.; and Lugn, R. V. 1962. Experimental hypervelocity impact craters in rock. In *Proc. 5th Hypervelocity Impact Symp.*, U.S. Govt. Res. Rept., pp. 625-643.

Moore, H. J.; Hodges, C. A.; and Scott, D. H. 1974. Multiringed basins – illustrated by Orientale and associated features. *Proc. Lunar Sci. Conf. V* (Oxford: Pergamon Press), pp. 71-100.

Morrison, R. H., and Oberbeck, V. R. 1975. Geomorphology of crater and basin deposits

– emplacement of the Fra Mauro Formation. *Proc. Lunar Sci. Conf. VI* (Oxford: Pergamon Press), pp. 2503-2530.

Morrison, R. H., and Oberbeck, V. R. 1978. A composition and thickness model for lunar impact crater and basin deposits. *Proc. Lunar Sci. Conf. IX* (Oxford: Pergamon Press), pp. 3763-3785.

Oberbeck, V. R. 1975. The role of ballistic erosion and sedimentation in lunar stratigraphy. *Rev. Geophys. Space Phys.* 13: 337-362.

Oberbeck, V. R., and Morrison, R. H. 1976. Candidate areas for *in situ* ancient lunar materials. *Proc. Lunar Sci. Conf. VII* (Oxford: Pergamon Press), pp. 2983-3005.

Oberbeck, V. R.; Morrison, R. H.; Hörz, F.; Quaide, W. L.: and Gault, D. E. 1974. Smooth plains and continuous deposits of craters and basins. *Proc. Lunar Sci. Conf. V* (Oxford: Pergamon Press), pp. 111-136.

O'Keefe, J. D., and Ahrens, T. J. 1976. Impact ejecta on the moon. *Proc. Lunar Sci. Conf. VII* (Oxford: Pergamon Press), pp. 3007-3025.

O'Keefe, J. D., and Ahrens, T. J. 1977a. Impact-induced energy partitioning, melting, and vaporization on terrestrial planets. *Proc. Lunar Sci. Conf. VIII* (Oxford: Pergamon Press), pp. 3357-3374.

O'Keefe, J. D., and Ahrens, T. J. 1977b. Meteorite impact ejecta: Dependence of mass and energy lost on planetary escape velocity. *Science* 198: 1249-1251.

O'Keefe, J. D., and Ahrens, T. J. 1979. The effect of gravity on impact crater excavation time and maximum depth; comparison with experiment (abstract). *Lunar Science X, The Lunar and Planetary Institute*, pp. 934-936.

Öpik, E. J. 1971. Cratering and the moon's surface. In *Advances in Astronomy and Astrophysics*, ed. Z. Kopal (New York: Academic Press), pp. 108-337.

Orphal, D. L. 1977a. Calculations of explosion cratering-I: The shallow-buried nuclear detonation JOHNIE BOY. In *Impact and Explosion Cratering*, eds. D. J. Roddy, R. O. Pepin and R. B. Merrill (New York: Pergamon Press), pp. 897-906.

Orphal, D. L. 1977b. Calculations of explosion cratering-II: Cratering mechanics and phenomenology. In *Impact and Explosion Cratering*, eds. D. J. Roddy, R. O. Pepin and R. B. Merrill (New York: Pergamon Press), pp. 907-917.

Orphal, D. L. 1979. Depth, thickness, and volume of the breccia lens for simple explosion and impact craters (abstract). *Lunar Science X, The Lunar and Planetary Institute*, pp. 949-951.

Pike, R. J. 1968. Meteoritic origin and consequent endogenic modification of large lunar craters: A study in analytical geomorphology. Ph.D. dissertation, University of Michigan.

Pike, R. J. 1974a. Ejecta from large craters on the moon: Comments on the geometric model of McGetchin *et al. Earth Planet. Sci. Lett.* 23: 265-274.

Pike, R. J. 1974b. Depth/diameter relationships of fresh lunar craters: Revision from spacecraft data. *Geophys. Res. Lett.*. 1: 291-294.

Pike. R. J. 1977. Size-dependence in the shape of fresh impact craters on the moon. In *Impact and Explosion Cratering*, eds. D. J. Roddy, R. O. Pepin and R. B. Merrill (New York: Pergamon Press), pp. 489-509.

Quaide, W. L.; Gault, D. E.; and Schmidt, R. A. 1965. Gravitative effects on lunar impact structures. *Annals N. Y. Acad. Sci.* 123: 563-572.

Rinehart, J. S. 1960. On fractures caused by explosions and impacts. *Quarterly Colo. School Mines* 55: 1-155.

Rinehart, J. S. 1968. Intense destructive stresses resulting from stress wave interactions. In *Shock Metamorphism of Natural Materials*, eds. B. French and N. Short (Baltimore: Mono Book Corp.), pp. 31-42.

Rinehart, J. S. 1975. *Stress Transients in Solids* (Sante Fe: Hyperdynamics).

Roddy, D. J.; Pepin, R. O.; and Merrill, R. B., eds. 1977. *Impact and Explosion Cratering* (New York: Pergamon Press).

Schaal, R. B., and Hörz, F. 1977. Shock metamorphism of lunar and terrestrial basalts. *Proc. Lunar Sci. Conf. VIII* (Oxford: Pergamon Press), pp. 1697-1729.

Schneider, E. 1975. Impact ejecta exceeding lunar escape velocity. *The Moon* 13: 173-184.

Schultz, P. H, and Gault, D. E. 1975a. Seismic effects from major basin formations on the moon and Mercury. *The Moon* 12: 159-177.

Schultz, P. H., and Gault, D. E. 1975*b*. Seismically induced modification of lunar surface features. *Proc. Lunar Sci. Conf. VI* (Oxford: Pergamon Press), pp. 2845-2862.

Settle, M., and Head, J. W. 1979. The role of rim slumping in the modification of lunar impact craters. *J. Geophys. Res.* (in press).

Shoemaker, E. M. 1963. Impact mechanics at Meteor Crater, Arizona. In *The Moon, Meteorites, and Comets*, eds. B. M. Middlehurst and G. P. Kuiper (Chicago: Univ. of Chicago Press), pp. 301-336.

Simonds, C. H.; Floran, R. J.; McGee, P. E.; Phinney, W. C.; and Warner, J. L. 1978. Petrogenesis of melt rocks, Manicouagan impact structure, Quebec. *J. Geophys. Res.* 83: 2773-2788.

Smith, E. I., and Sanchez, A. G. 1973. Fresh lunar craters: Morphology as a function of diameter, a possible criterion for crater origin. *Mod. Geol.* 4: 51-59.

Stöffler, D. 1971. Progressive metamorphism and classification of shocked and brecciated crystalline rocks at impact craters. *J. Geophys. Res.* 76: 5541-5551.

Stöffler, D.; Gault, D. W.; Wedekind, J.; and Polkowski, G. 1975. Experimental hypervelocity impact into quartz sand: Distribution and shock metamorphism of ejecta. *J. Geophys. Res.* 80: 4062-4077.

Thomas, P. C. 1978. The morphology of Phobos and Deimos. Ph.D. dissertation, CRSR 693, Cornell University.

Thomas, P. C.; Veverka, J.; and Duxbury, T. 1978. Origin of the grooves on Phobos. *Nature* 273: 282-284.

Ullrich, G. W.; Roddy, D. J.; and Simmons, G. 1977. Numerical simulations of a 20-ton TNT detonation on the Earth's surface and implications concerning the mechanics of central uplift formation. In *Impact and Explosion Cratering*, eds. D. J. Roddy, R. O. Pepin, and R. B. Merrill (New York: Pergamon Press), pp. 959-982.

Veverka, J., and Duxbury, T. C. 1977. Voking observations of Phobos and Deimos: Preliminary results. *J. Geophys. Res.* 82: 4213-4223.

Wood, C. A. 1968. Statistics of central peaks in lunar craters. *Comm. Lunar Planet. Lab.* 7: 157-160.

Wood, C. A., and Andersson, L. 1978. New morphometric data for fresh lunar craters. *Proc. Lunar Sci. Conf. IX* (Oxford: Pergamon Press), pp. 3669-3689.

Wood, C. A.; Head, J. W.; and Cintala, M. J. 1977. Crater degradation on Mercury and the moon: Clues to surface evolution. *Proc. Lunar Sci. Conf. VIII* (Oxford: Pergamon Press), pp. 3503-3520.

REGOLITH DEVELOPMENT AND EVOLUTION
ON ASTEROIDS AND THE MOON

K. R. HOUSEN, L. L. WILKENING
University of Arizona

and

C. R. CHAPMAN, R. J. GREENBERG
Planetary Science Institute

Early descriptions of regoliths on small bodies were devised to account for observations of asteroids (Chapman 1971, 1976) and the gas-rich meteorites (Anders 1975). Lack of agreement between these approaches prompted Housen et al. (1978, 1979) to examine the problem in detail. The resulting model predicted that moderate-sized (100-300 km) asteroids should evolve regoliths up to a few kilometers deep which could be source regions of gas-rich meteorites. Smaller objects should have regoliths ranging from dust coatings to meters-thick layers depending on the strength of the object. Our earlier model could not treat large asteroids, $\gtrsim$ 300 km in diameter. The model, now modified to treat larger-sized objects, predicts regolith depths, on asteroids larger than $\sim$ 300 km, which decrease with increasing size. A regolith depth of $\sim$ 7 m is predicted for the lunar maria in reasonable agreement with the observed depths of $\sim$ 5 m.

Studies of the terrestrial planets have clearly demonstrated that in the absence of geological processes driven by internal heat sources and in the absence of an atmosphere, the most important geological process is impact cratering. Because of their small sizes, asteroids lose heat rapidly and cannot retain atmospheres. Hence, impact processes must dominate their surface evolution. Cratering on large bodies such as the moon results in the formation of regolith, a layer of fragmental, unconsolidated, rocky material overlying more coherent bedrock. A question that has concerned people interested in asteroids is whether regolith would form on objects with such a low gravity that much crater ejecta is lost entirely from the asteroid. What would be the properties of any regoliths formed under such conditions? And can the properties of any meteorites be related to the predicted properties of asteroidal regoliths?

The particulate or dusty nature of asteroidal surfaces has been inferred from astronomical observations, especially polarimetry (Dollfus 1971; see the chapter by Dollfus and Zellner). But polarimetry cannot distinguish between an extensive fragmental layer and a dusty coating on a rocky surface. Radar backscatter data and thermal radiometry for a few asteroids provide information about surface materials at centimeter scales (see the chapters by Pettengill and Jurgens and by Dickel). All observations are consistent with the presence of regoliths on asteroid surfaces, but they do not provide insight into their extent, depth, or evolution.

Studies of meteorites suggest that many if not all meteorites come from asteroids. Characteristics of certain types of meteorites, especially those rich in implanted solar wind gases (gas-rich), strongly suggest that they formed in regoliths (see the chapter by Wasson and Wetherill). Hence, these meteorites provide a body of observations against which theories of the formation and evolution of asteroidal regoliths can be tested. The moon is also a convenient "end member" test case for studies of regolith evolution.

Under the present circumstances the only avenue to the understanding of asteroidal regoliths is a theoretical approach. The purpose of this chapter is to review the various theoretical descriptions of the origin, evolution, and nature of asteroidal regoliths, to compare theory with observations, especially those made on meteorites and to describe our recent progress in the area of regolith modeling.

I. REGOLITH EVOLUTION ON PLANETARY OBJECTS – PREVIOUS WORK

Asteroids

The earliest theories of asteroidal regoliths were developed in the context of understanding the texture of the uppermost microns of asteroid surfaces as inferred from astronomical observations, especially polarimetry (Dollfus 1971). Chapman (1971, 1976) developed an early model for asteroidal regoliths that emphasized the loss of most ejecta due to the "sandblasting" effect of numerous small impacts into rocky surfaces; he concluded that asteroidal regoliths are very thin, except for those on the largest bodies, and that exposure of brecciated meteorites to cosmic ray tracks and the solar wind must have occurred during an early epoch of asteroidal accretion, prior to the development of the net erosive collisional processes effective today.

Langevin and Arnold (1977) recognized that modifications might be made to the lunar regolith evolution models they reviewed so that the asteroidal case could be treated. But until 1978 the only other discussions of asteroidal regoliths have been in the context of interpreting the features of brecciated, gas-rich meteorites. Following examination. of the first lunar samples, it was hypothesized that irradiation of such meteorites might have occurred in a surficial regolith similar to that on the moon. The evidence for the existence of regoliths on meteorite parent bodies accumulated gradually over roughly a century. The nineteenth century descriptions of various stony meteorites made reference to their brecciated structures (see Wahl 1952 for a brief review). However, according to Wahl the failure of meteorite authority Cohen in 1903 to impute genetic significance to brecciation textures led to neglect of this aspect of meteorite studies for the first half of this century. Wahl attempted to revive interest in the importance of the brecciation structures for understanding meteorite origins, but his paper seems to have been ignored. Wahl concluded that most stony meteorites were mixtures of fragments, "accumulation breccias," and they necessarily formed as "tuffs" which were welded mixtures of different-appearing fragments from chemically similar sources or a single parent body.

A separate line of evidence bearing on the regolith origin of meteorites developed in the 1960's through the study of noble gases in meteorites. A crucial idea was introduced by Wänke (Suess et al. 1964; Wänke 1965) who suggested that high concentrations of helium and neon (solar-type gases) in gas-rich meteorites were due to implantation of solar wind particles in these meteorites. Suess et al. (1964) noted that high concentrations of solar-type gases occurred in meteorites with brecciated textures. Eberhardt et al. (1966) showed that the solar-type gases were present on the surfaces of mineral grains. This was further evidence for low-energy implantation by the solar wind. The proof of the implantation hypothesis was the observation of nuclear particle tracks created by solar cosmic rays in individual mineral grains in the same meteorites which contained the solar-type gases (Lal and Rajan 1969; Pellas et al. 1969). However, the connection to regolith on a meteorite-parent-body regolith had not yet been made since it was concluded by Lal and Rajan and Pellas et al. that the mineral grains had been irradiated in a dispersed state in space. The arrival of lunar samples a few months later provided the final clue. As described by Wilkening (1970, 1971) the characteristics of nuclear particle tracks in lunar soils and rocks left no doubt that the characteristics of the brecciated, solar gas-rich meteorites could be formed on the surface of a parent body as opposed to free space. Additional evidence such as the discovery of microcraters on glassy spheres in a solar gas-rich meteorite (Brownlee and Rajan 1973), the presence of agglutinates (Rajan et al. 1974), and foreign clasts in gas-rich meteorites (Wilkening 1973;

Wilkening and Clayton 1974), and many additional studies of tracks and noble gases in meteoritic and lunar breccias (Macdougall *et al.* 1974; Rajan 1974) made impregnable the case for the existence of regoliths at some stage in the evolution of meteorite parent bodies. Debates have concerned the *location* of these regoliths (e.g. on the surfaces of comets or asteroids) and the *epoch* (i.e. in modern-day regoliths or during the accretionary history of the parent bodies).

An early treatment of these problems was that of Anders (1975, updated 1978) who noted the evidence that brecciated, gas-rich meteorites have shorter cosmic ray exposure ages and lower gas contents than do lunar soils. He argued that the amount of implanted solar gases should be inversely proportional to the square of the distance from the sun and directly proportional to the mean residence time of a grain at the surface. Anders ascribed the differences between the gas contents and exposure ages of meteorites and lunar soils to differences in cratering rates. In fact, he argued that mean residence times are inversely proportional to mean cratering rates. Thus the data implied to Anders that the meteorites have come from an environment in which the cratering rate is 1 to 3 orders of magnitude greater than in the vicinity of the moon and at a location a few times farther from the sun. Langevin and Maurette (1976) came to a similar conclusion after applying their Monte Carlo model, for lunar regolith evolution, to the gas-rich meteorite Kapoeta.

Matson *et al.* (1977) have taken a unique approach to the consideration of asteroidal regoliths. They address the question of whether lunar-regolith-like processes occur on asteroids by examining asteroid spectral data for the tell-tale signs of optical "maturation," involving inverse correlation between redness and albedo. Finding no such evidence, Matson *et al.* address several possible theoretical explanations for the differences between asteroids and the moon. They conclude that asteroidal regoliths are thinner and coarser than the lunar regolith and that they are created by impacts at velocities too low to produce much glass. Neither the evidence addressed by Matson *et al.* nor the other remote-sensing data about asteroid surfaces is really capable of addressing the all-important question of how deep asteroid regoliths may be. For now we must rely on theoretical approches and the indirect implications of meteorites.

The Moon

In recent years, many models of the lunar regolith have been developed. Rather than attempting to review this field we will only briefly mention the kinds of models that exist. For a detailed review of the subject, the reader is referred to the paper of Langevin and Arnold (1977).

The lunar models can be broadly classified into two types depending on which properties of the regolith they attempt to explain.

1. Models of the macroscopic properties (e.g. depth) of the regolith. To date no purely analytical (i.e. non-Monte Carlo) models exist which consider the formation of regolith from bedrock. Monte Carlo simulations of regolith formation in portions of the lunar maria have yielded distributions of surface elevation consistent with observation and have illustrated the dominant role played by large craters in regolith growth (Oberbeck *et al.* 1973). These simulations are limited to rather small areas ($\lesssim 250$ km^2) because of the large amount of computer storage required.

2. Models of the microscopic properties of the regolith. Most of the efforts have been concentrated in this area in an attempt to model the mixing history of the regolith and the histories of individual regolith grains, and to explain such characteristics as grain size distributions, nuclear particle track distributions, and rare gas contents. Early, analytical modeling illustrated how turnover rates rapidly decrease with increasing depth below the surface and pointed out the existence of a ~ 1 mm thick zone of extensive mixing at the surface (Gault *et al.* 1974). Much more detailed Monte Carlo methods have been successfully used to explain the exposure histories of individual regolith grains and the layering of surface material.

Models of asteroidal regoliths require, for the most part, the same input data that are used in lunar models, e.g. one must specify a mass distribution of impacting objects, an impact velocity and the size of crater produced by an impact of given energy. Moreover, one must also account for some processes that lunar models do not. For instance, in contrast with the moon, most asteroids lose some fraction of their impact ejecta. Also, since asteroid models must apply to bodies of various sizes and internal compositions, one must consider how the physics of the cratering process changes from asteroid to asteroid. Fortunately, these added complications do not make the situation unmanageable.

In our study of asteroidal regoliths we have taken an analytical approach. Both the macroscopic and microscopic properties of asteroid regoliths are considered. Consequently we have modeled a wide range of impact processes varying from large-scale events, which dominate regolith growth and erosion, to small-scale impacts, which govern the gardening history of the regolith. A Monte Carlo approach could not adequately consider the effects of the entire size spectrum of impact events over the whole asteroid surface. The methods we have used are distinctly different from any which have been applied to the lunar regolith. In the following sections we describe our efforts in modeling the evolution of regoliths on asteroidal bodies.

II. REGOLITH EVOLUTION OF SMALL BODIES –
RECENT EFFORTS

The ideal theoretical model of a planetary regolith would give such parameters as regolith thickness and "gardening" rate and depth as both functions of time and functions of the fraction of the surface over which they apply. The thickness at a given time, for example, should be obtained in terms of a surface probability distribution in order to interpret local measurements of the thickness, and to determine how many local measurements are needed to give a truly representative sample.

Such an ideal model has yet to be constructed. Rather, models to date give a single value of each parameter at any given time. The value represents some sort of average over the surface distribution. Interpretation of any model requires an understanding of what portion of the surface this average represents. Whether explicitly stated or not, each model gives the average over some limited portion of the planetary surface.

Let us consider why averaging should not be done over the entire planetary surface. Regolith evolution is governed by the impact of a population of bodies with a range of sizes. Crater counts (e.g. on the moon) or direct asteroid population studies (of Zellner; see his chapter) show that the impacting population is dominated in number and in cross-section by small bodies, but that it is dominated in mass by a significant number of larger bodies whose impacts produce anomalously large craters that are not typical of the relatively uniform properties of those portions of the surface saturated with smaller impacts. The probability distribution for a parameter describing the regolith (e.g. thickness) might be expected to have a spike at values corresponding to the uniformly saturated regions, but with significant probability of other values due to the anomalous craters. The average over the entire surface would likely represent neither the uniform nor the anomalous region very well. On the other hand, averaging over only that portion of the surface which has been smoothly saturated by smaller impacts (the "typical region") gives a value which is at least characteristic of the spike, even if it cannot describe the whole distribution function. This concept of a typical region is central to our considerations of asteroidal regoliths in this chapter.

For the lunar regolith, a model of the typical region can be constructed by considering the effects, up to a given time, of that portion of the population (the smaller bodies) which saturates the surface. Any effects of the larger bodies remain local to their anomalous impact sites. For asteroids, such an approach is not justified. While the large craters are still distinct from the typical terrain, the low gravity permits global distribution of the ejecta from even these anomalous craters. Hence, the large impacts cannot be ignored; their ejecta thicken the regolith in the typical region. Also, as time goes on, ever larger craters saturate the surface and must be considered as part of the typical region. A model for regolith development in the typical region of an asteroidal surface is summarized below.

That model applies to only those asteroids that are small enough and/or strong enough that ejecta are distributed nearly uniformly over the globe. For larger asteroids ($\gtrsim$ 10 km for sandy bodies with correspondingly low-velocity ejecta and $\gtrsim 300$ km for rocky bodies with higher ejecta velocities) regolith evolution is intermediate between the small asteroid and the lunar cases. The large anomalous craters may still distribute some ejecta widely, but a significant fraction of the ejecta is deposited in a localized annulus around each crater. As time goes on, the annuli around the larger craters may begin to saturate the surface before the craters do themselves. Hence, the ejecta near a crater must be incorporated into the modeled typical region before the crater itself is. This concept has now been included in our model; the computational methods and results for larger asteroids will be described in the next section. Furthermore, as a test of its validity, the new model has been applied to the moon and gives results consistent with observations.

The model of Housen *et al.* (1978, 1979) describes regolith evolution in the typical region for asteroids, which are small enough so that their crater ejecta are widely spread over the surface. We present a qualitative discussion of this model to provide a physical understanding of regolith evolution and to set the stage for a generalization to large bodies where ejecta are not widespread. Since it is impact cratering which generates regolith, let us consider how various sizes of craters build up or deplete the regolith layer. The impacting crater population is often assumed to be of the form, $N \sim D^{\alpha}$, where N is the number (per unit area per unit time) of craters of diameter D or larger and α is a constant. Our limited knowledge of cratering in the asteroid belt suggests a value of α in the range $-3 < \alpha < -2$, to the extent that the entire diameter range of craters can be described by a single value of α. It is easily shown that for $\alpha < -2$, the number of craters drops off sufficiently fast with increasing diameter so that "small" craters occupy more surface area than do large ones, i.e., the total crater area contained in some diameter interval, centered on a diameter D_1, is greater than the area contained in an interval of the same width centered on diameter D_2 where $D_2 > D_1$. Also, if $\alpha > -3$ then the number of craters decreases so slowly with diameter that most of the volume is contained in "large" craters. Thus, when $-3 < \alpha < -2$, small craters are the first ones to saturate the surface while the large craters excavate the most surface material.

Now, since the smaller craters saturate first, we can define a quantity, $D_s(t)$, to be the diameter of the largest craters which have saturated the surface by the time t. Clearly $D_s(t)$ depends on the precise definition of saturation, which we discuss later. The crater population can now be considered in two parts.

1. Those craters smaller than $D_s(t)$ saturate the surface and hence compose the typical region. They deplete the regolith because some fraction of their ejecta escapes the asteroid. However, they also create new regolith if they are large enough to puncture through the existing debris layer to comminute underlying "pristine" material.

2. The craters larger than $D_s(t)$, up to a maximum diameter D_r of a crater

that catastrophically ruptures the asteroid, are widely dispersed and lie outside the typical region. However, they generate a substantial amount of ejecta and the portion of ejecta which is not lost to space is spread widely over the surface; hence these craters deposit regolith onto the typical region. The effects of all craters are summarized in Fig. 1*a*.

Consider now the temporal evolution of the regolith depth. Early in the evolution, only the very smallest craters are sufficiently numerous to have saturated the surface. These small craters remove very little regolith. Most sizes of craters are larger than D_s, and hence deposit a lot of material into the typical region causing the regolith depth initially to increase. With passing time, D_s increases (i.e. D_s moves to the right in Fig. 1*a*) so that ever larger craters are incorporated into the typical region resulting in more depletion of the regolith layer. Simultaneously, the diameter range, D_s to D_r, of craters depositing material into the typical region, shrinks. These two effects slow down the regolith buildup and eventually cause the regolith depth to decrease. Finally, at some point in time, a crater of diameter D_r or larger forms and ruptures the asteroid. By rupture we do not mean that the asteroid necessarily ceases to exist, for the fragments may fall back together again. We merely stop our analysis of regolith evolution when a sufficiently violent impact occurs to change the asteroid from a more or less coherent body surrounded by a regolith layer into a fragmented ball of debris. Before moving on to actual results of calculations, we mention briefly another important aspect of regolith modeling.

When modeling the regolith origin of brecciated solar gas-rich meteorites, it is necessary to be able to predict various quantities (e.g. solar flare and cosmic ray track densities) observed in meteorites. These quantities depend on how long regolith grains are exposed at or near the surface of the asteroid. Thus, one needs to know how extensively the regolith is churned or "gardened" by craters. The amount of gardening is found by comparing the formation rate of ejecta blankets of a specific depth to the formation rate of craters which excavate to the same depth. These burial and excavation rates for some large asteroids are shown in the next section. An alternative method is to consider how often a "point" (whose position is fixed with respect to the center of the asteroid) in the regolith is excavated. An expression for the number of excavations as a function of time is given in Housen *et al.* (1979). Both of these methods are used to determine the degree of regolith gardening.

The calculations of Housen *et al.* (1978, 1979) are performed for asteroids of two compositional strengths. Asteroids of diameter 1 km to 300 km, whose internal strength is roughly that of basalt, are modeled. Ejecta on larger bodies are not widespread and, hence, must be treated by methods given in the next section. The smaller of these rocky bodies (diameter $\lesssim 10$ km) maintain only thin dusty coatings because of their very small gravity fields. Regolith depth increases with the size of the asteroid, because larger

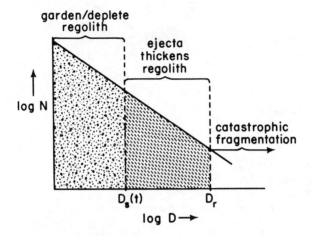

(a) Small asteroids

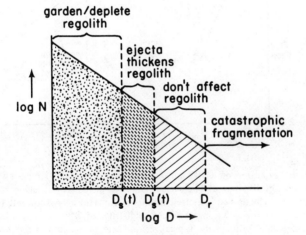

(b) Large asteroids and the Moon

Fig. 1. The population of craters that form on an asteroid's surface is shown, where N is the number of craters of diameter D or larger per unit time and area. At time t only craters smaller than some diameter $D_s(t)$ saturate the surface and so compose the typical region. These craters garden and deplete (i.e. eject to space) regolith in the typical region. In Fig. 1a it is shown that on small asteroids crater ejecta are globally distributed, so the regolith is thickened by all craters outside the typical region (i.e. $D_s < D < D_r$, where D_r represents the smallest crater that can rupture the asteroid). Figure 1b gives a diagram for large bodies showing that ejecta are not widespread, so only those craters whose ejecta blankets saturate the surface [$D_s'(t)$ represents the largest of these] add regolith to the typical region.

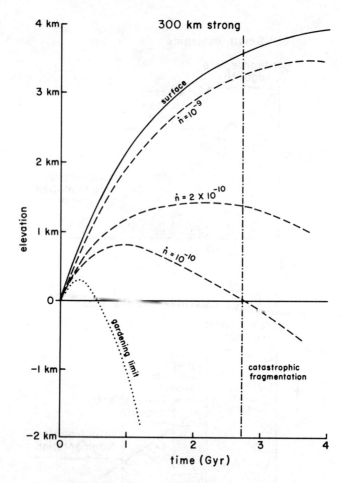

Fig. 2. Regolith evolution on a 300 km diameter asteroid of high strength where crater ejecta are globally distributed. The surface elevation in the typical region is shown as a function of time. The dashed lines are contours of constant excavation rate (yr⁻¹). The gardening limit, below which material rests undistributed, is dictated by the size of the largest crater which forms in the typical region. A regolith depth of roughly 3.5 km is developed before the asteroid is catastrophically fragmented by a large impact. After fragmentation, the asteroid may continue on as gravitationally bound debris.

bodies retain more impact ejecta. Thus the 300 km body accumulates ~ 3.5 km of regolith (Fig. 2). We also study "weak" objects, whose internal strength is comparable to that of very loosely bonded regolith. In such weak targets, experiments show that crater ejecta velocities are low, so ejecta are not widespread on bodies larger than ~ 10 km. Weak asteroids of diameter 1 km to 10 km develop centimeters to meters of regolith. For all of these asteroids, regolith gardening is found to be minimal because the formation rate of ejecta blankets exceeds the excavation rate by craters. Grains in the

TABLE I

Comparison of Gas-rich Meteorites and Lunar Soils and Breccias

	Gas-rich Meteorites	Ref.	Lunar Samples[a]	Ref.
Impact Glass (volume %)	rarely > 1 %	f	as much as 50 %	i
Agglutinates (volume %)	rare	g	up to 60 %	h
Glassy Spherules (volume %)	rare[b]	c	up to 10 %	i
Micrometeorite Craters	rare	c,e	ubiquitous	f
Helium $- 4(\text{cm}^3 \text{STP/g})$	$10^4 - 10^5$	f	$\gtrsim 10^7$	f
Track-rich Grains	1-20 %	e	20-100 %	d

[a]Lunar soils and unmetamorphosed breccias.

[b]Cases employing only achondrites in the comparison.

[c]Brownlee and Rajan 1973.

[d]Crozaz and Dust 1977.

[e]Goswami et al. 1976.

[f]Data collated from many different sources.

[g]Kerridge and Kieffer 1977; Rajan et al. 1974.

[h]Heiken 1975.

[i]Warner 1972.

regolith are excavated very few times, if at all. This is in sharp contrast to the lunar regolith where reworking of material is the dominant process. The relative immaturity of gas-rich meteorites compared with lunar breccias (see Table I and II) is explained quantitatively by the high rate of blanketing compared with excavation for the cases mentioned above. For example, Housen et al. (1978, 1979) concluded that moderate-sized asteroids 100-300 km in diameter would provide a suitable environment in which gas-rich meteorites could be formed. In this environment blanketing by widespread ejecta occurs at a rate sufficient to provide 1 m shielding of buried surface materials from cosmic rays within 2 Myr while also permitting adequate exposure durations of $10^2 - 10^4$ yr for lower energy solar-flare particles and solar wind gases. In order to account for the large fraction of meteorites that are gas-rich, Housen et al. required that several generations of surficial regolith be incorporated into deeper megaregoliths prior to parent-body disruption and delivery of meteorite fragments to Earth.

Another model of regolith evolution on small bodies has been outlined by Duraud et al. (1979) and Dran et al. (1979). In contrast to the model of Housen et al. (1978, 1979) the results of their modeling calculations suggest that regoliths which achieve an equilibrium thickness of 2 m on small objects (diameter ~ 20 km) are the locations in which gas-rich meteorites form. They

TABLE II

Radiation Affecting Meteorites

	Energy (MeV/n)	Range (cm)	Ages (y)
Galactic Cosmic Rays	peaks near 100 extends to $> 10^{15}$	$1 - 2 \times 10^2$	$\sim 10^6$
Solar Cosmic Rays (Heavy nuclei)	1-10 sharply decreasing for higher energy	10^{-2}	$10^3 - 10^4$
Solar Wind	10^{-3}	10^{-5}	$> 10^2$

envision deposition of ejecta to take place as a "steady rain," providing the opportunity for space weathering in the superficial regolith. These small asteroids erode at a rate of ~ 10 m/Gyr limiting the integrated exposure of regolith to energetic particles. However, this rate appears to be much too small to restrict to 10^6 yr the upper limit on the exposure to galactic cosmic rays in gas-rich meteorites. We also question the assumptions made by this group regarding the distribution of ejecta during the cratering processes. A complete description of this model had not been published at the time of this writing.

III. GENERALIZATION OF OUR MODEL TO LARGE ASTEROIDS

The Housen *et al.* model has been restricted so far to asteroids sufficiently small that their ejecta blankets surround the entire bodies. Unfortunately, this leaves out a significant number of asteroids. For example, most asteroids are known to be *C*-type (see Zellner's chapter in this book) which may imply fairly low compositional strengths. Even bodies composed of inherently strong material may have been sufficiently fractured and comminuted so as to respond to impacts as weak material. As previously stated, the Housen *et al.* assumption of widespread ejecta does not apply to weak bodies larger than ~ 10 km diameter. Also omitted from consideration are strong bodies larger than ~ 300 km, which include such important potential meteorite parent bodies as Vesta. It is also important, in order to test the validity of the general approach of Housen *et al.* to be able to treat the lunar regolith case, for which we have abundant experimental evidence including returned core samples and knowledge of the crater populations and regolith depths.

As ejecta are restricted more and more to the proximity of a crater on larger bodies, it is no longer possible to consider a uniform layer of ejecta surrounding the body. Rather, ejecta blankets vary in thickness from a maximum near the crater rim dwindling away to practically zero far from the crater. The formation and superposition of these ejecta blankets results in a spectrum of surface elevations. However, our model gives a single number, an "average" surface elevation or regolith depth. Clearly this average is of little use if the surface elevation varies greatly over the surface. In order for the average to be meaningful, we retain the concept of a typical region over which most of the ejecta is considered to be approximately of uniform depth. We can assure ourselves of a roughly uniform surface elevation by requiring that ejecta blankets overlap one another extensively, i.e. saturate the surface, before they are included in the typical region. This is analogous to our earlier treatment of erosion by small craters on asteroids. Craters were not included in the typical region until they had overlapped one another extensively. This method allows us to characterize the regolith depth on large bodies without actually specifying the shape of ejecta blankets. Our only requirement is that blanket thickness does not vary wildly with distance from the crater rim so that when saturation of ejecta has occurred, the blankets combine to form a relatively uniform layer of debris.

How do we know when ejecta blankets have saturated the surface? If ejecta blankets had sharp, well-defined boundaries, it would be easy to determine when enough overlap had occurred. In order to make possible a precise definition of ejecta saturation, we construct boundaries for the blankets, in the form of annuli surrounding craters. A boundary does not imply that a blanket cuts off at the edge of an annulus. Rather the annuli are drawn around craters merely to help us decide when saturation has occurred. Specifically, ejecta blankets are said to saturate the surface only if their associated annuli are sufficiently numerous to meet our mathematical criterion of saturation, defined below. Once ejecta blankets (annuli) have saturated, we compute the surface elevation under the assumption that all ejecta is spread in a uniform layer. It is in this sense that our model gives the average surface elevation.

How big should the annuli be? The size chosen determines how well our average surface elevation characterizes the true distribution of elevations. This point is illustrated schematically in Fig. 3 where a portion of the surface is shown both in plan view and cross-section for three different sizes of annuli. So as not to clutter the drawings, only one size of crater is drawn. In each case enough craters are drawn so that the annuli just saturate. Figure 3a shows that annuli which are too big saturate before the blankets do, resulting in highly non-uniform ejecta deposits which are not described well by the average surface elevation computed in the model. Very small annuli (Fig. 3c) saturate long after the blankets do (note, for clarity the circles representing the craters were omitted from Fig. 3c). Hence there are blankets, from craters larger than those shown in the plan view, which overlap enough to produce uniform effects but which are omitted from the typical region because their annuli have not yet saturated. The model therefore underestimates the true regolith depth. Annuli which more realistically

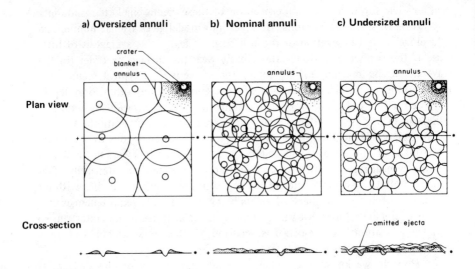

Fig. 3. Annuli are drawn around craters to help determine when ejecta blankets have saturated the surface (i.e. when ejecta should be included in the typical region) Oversized annuli saturate too early resulting in a distribution of surface elevations which is poorly approximated by our average elevation. Undersized annuli saturate long after blankets do. Hence there are craters (larger than those shown in Fig. 6) whose uniform ejecta deposit is wrongfully ignored because their annuli are not saturated. Annuli that better represent the extent of ejecta blankets (nominal case) saturate when a nearly uniform debris layer is produced by overlap of blankets.

represent the extent of ejecta blankets (Fig. 3b) saturate when the blankets have overlapped sufficiently to produce a relatively uniform debris layer. We are now faced with the problem of finding this nominal annulus width: a width small enough to give meaning to our average surface elevation, but large enough so that the regolith depth is not underestimated. Not only must we specify the width of an annulus for a given crater, but we must also specify how the annulus size varies with crater diameter. We adopt two models for the radial width (i.e. outer radius minus inner radius) of an annulus: (1) the width is a constant a_1 independent of crater diameter, as is implied by some experimental and theoretical studies of ejecta velocity distributions, and (2) the width is a constant, a_2, times the crater diameter, as suggested by some studies of ejecta blanket topographic profiles. These two models, which are hereafter referred to as Type 1 and Type 2 ejecta blankets respectively, produce two distinct expressions for the surface elevation in the typical region as a function of time. The method used to compute nominal values for a_1 and a_2 is described below.

The definition of crater saturation, developed by Housen *et al.* (1978, 1979), is used here and extended to include saturation of ejecta blankets (annuli). If f is the fraction of total asteroidal surface area which is taken up by the typical region, then one can show

$$f = e^{-\lambda} \tag{1}$$

where λ is either the integrated area (expressed in units of the asteroid surface area) of all craters produced in some diameter interval, or in the case of ejecta blankets the integrated area of all annuli produced by craters in some diameter interval, during time t. That is,

$$\lambda = \int_{\Delta D} (\text{area}) \, t \, dN(D) \tag{2}$$

where $dN(D)$ is the number of craters produced on the asteroid in the diameter increment dD, per unit area per unit time, and ΔD is the diameter interval we integrate over. By area in Eq. (2) we mean the area of a crater or of an annulus. For example, we compute the diameter, $D_s(t)$, of the largest crater that saturates the surface at time t by integrating in Eq. (2) over all atypical craters,

$$\lambda = \int_{D_s}^{D_r} (\pi D^2/4) t \, dN(D) \tag{3}$$

where D_r is the diameter of the largest crater that can form without rupturing the asteroid. In words, Eq. (1) tells us that f is the fraction of surface area exterior to those atypical craters over whose areas we integrate in Eq. (3); that is f is the fractional area occupied by the typical region. An expression for $D_s(t)$ is obtained by specifying the form of the cratering flux, $dN(D)$. We adopt a power law of the form

$$dN(D) = -K\alpha D^{\alpha-1} \, dD \tag{4}$$

where K and α are constants determined from observational data on asteroids and the meteoroid complex combined with scaling laws that relate impact kinetic energy to resulting crater diameter. Substituting Eq. (3) into Eq. (1) and solving for D_s yields

$$D_s(t) = \left[D_r^{\alpha+2} + C/t \right]^{\frac{1}{(\alpha+2)}} \tag{5}$$

where

$$C = -4(\alpha + 2)\ln(f)/\pi K \alpha . \tag{5a}$$

Similarly, in the case of ejecta blankets, we let $D_{s,i}'(t)$ be the diameter of the largest craters whose ejecta annuli have saturated the surface at time t.

The subscript i (i = 1,2) indicates the ejecta blanket model used. In Eq. (2) we integrate from $D'_{s,i}$ (t) to D_r and the "area" is the area of an annulus for either Type 1 or 2 ejecta blankets. From Eq. (1) and an equation for λ analogous to Eq. (3) we find, in the case of Type 1 blankets,

$$(D_r^{\alpha+1} - D'^{\alpha+1}_{s,1})/(\alpha + 1) + a_1(D_r^{\alpha} - D'^{\alpha}_{s,1})/\alpha + C_1/t = 0 \qquad (6)$$

where

$$C_1 = -\ln(f)/\pi K a_1 \alpha . \qquad (6a)$$

Eq. (6) is solved numerically to find $D'_{s,1}$ (t). An analytical expression is possible for Type 2 blankets,

$$D'_{s,2}(t) = \left[D_r^{\alpha+2} + C_2/t\right]^{\frac{1}{(\alpha+2)}} \qquad (7)$$

where

$$C_2 = -(\alpha + 2)\ln(f)/\pi K \alpha\, a_2(a_2 + 1). \qquad (7a)$$

Notice that together, $D_s(t)$ and $D'_{s,i}$ (t) split the crater population into three parts (Fig. 1b).

1. Craters smaller than $D_s(t)$ remove regolith from the typical region.
2. The craters larger than $D_s(t)$ but smaller than $D'_{s,i}$ (t) do not saturate the surface and so do not reside in the typical region. However, the ejecta blankets of these craters are saturated and so contribute to regolith buildup.
3. Craters larger than $D'_{s,i}$ (t) do not affect regolith evolution in the typical region because neither the craters themselves nor their ejecta annuli saturate the surface.

Even before performing the calculations, we might guess, by comparing Fig. 1a to 1b, that regolith buildup will be reduced on large asteroids compared to small bodies, because $D'_{s,i}(t)$ reduces the number of large craters which deposit ejecta into the typical region. We must remember, of course, that large asteroids retain more impact ejecta than do small ones; it is not yet clear how regolith thickness should vary with asteroid size. We next use Eqs. (5), (6) and (7) to derive an expression for the surface elevation in the typical region.

The surface elevation, R_i (i = 1,2 for Type 1 and 2 blankets), of the typical region, with respect to the center of the asteroid, changes during each modeled time step due to (1) *erosional* ejection of material from the asteroid by the saturating sub-D_s cratering impacts; (2) *deposition* of ejecta that fail to escape the asteroid from sub-$D'_{s,i}$ impacts, whose ejecta blankets saturate the surface; and (3) *migration* of the typical region into previously atypical regions during each time step, which requires a retrospective accounting for the erosion and deposition that had occurred earlier. We consider each of these processes separately.

(1) *Erosion.* Craters are assumed to have the shape of "spherical caps" with depth/diameter ratio μ. The volume, ϕ, of material excavated from a crater of diameter D is

$$\phi = \pi\mu(3 + 4\mu^2)D^3/24. \tag{8}$$

Multiplication of ϕ by the crater flux and integration over diameters smaller than D_s yields

$$(dR_i/dt)_{erosion} = -\int_0^{D_s(t)} \phi(D)dN. \tag{9}$$

(2) *Deposition.* If γ is the fraction of ejecta that has sufficient velocity to escape the asteroid, then the rate of change of elevation due to fallback of ejecta from all sub-$D'_{s,i}$ craters is

$$(dR_i/dt)_{deposition} = \int_0^{D'_{s,i}(t)} \phi(D)(1-\gamma)dN. \tag{10}$$

(3) *Migration of the typical region.* As previously atypical craters or annuli are incorporated into the typical region, the prior erosion or deposition of material from these areas must be taken into account. In a time increment, dt, previously atypical craters in the diameter interval $D_s(t)$ to $D_s(t + dt)$ saturate the surface and previously atypical ejecta blankets associated with craters in the interval $D'_{s,i}(t)$ to $D'_{s,i}(t + dt)$ saturate the surface. These craters have been accumulating from time 0 to t. Thus the incremental change in R_i due to migration of the typical region is

$$(dR_i)_{migration} = -t\int_{D_s(t)}^{D_s(t+dt)} \phi(D)dN + t\int_{D'_{s,i}(t)}^{D'_{s,i}(t+dt)} \phi(D)(1-\gamma)dN \tag{11}$$

or

$$(dR_i/dt)_{migration} = K\alpha D_s^{\alpha-1} \phi(D_s)t dD_s/dt$$
$$- K\alpha(1-\gamma)D'^{\alpha-1}_{s,i} \phi(D'_{s,i})t\, dD'_{s,i}/dt. \tag{12}$$

The expression for the total rate of change of R_i is found by summing Eqs. (9), (10) and (12),

$$(dR_i/dt)_{total} = C_0 \left[D_s^{\alpha+3}(t) - (1-\gamma)D'^{\alpha+3}_{s,i}(t) + (\alpha+3)tD_s^{\alpha+2}(t)dD_s(t)/dt \right.$$

$$\left. -(1-\gamma)(\alpha+3)tD'^{\alpha+2}_{s,i}(t)dD'_{s,i}(t)/dt \right] \tag{13}$$

where

$$C_0 = \pi K \alpha \mu (3 + 4\mu^2)/24(\alpha+3). \tag{13a}$$

Equation (13) is a general expression which is numerically integrated to find the surface elevation in the typical region as a function of time. Note that $dD_s(t)/dt$ is obtained from Eq. (5) and $dD'_{s,i}(t)/dt$ is obtained from Eqs. (6) or (7).

In order for the above equations to be useful, one must of course specify the values of the parameters involved. We will not elaborate on the selection or justification of the numerical values used or their physical significance in the model unless pertinent to subsequent discussions. Unless otherwise stated, we adopt the values given in the paper of Housen *et al.* (1979), to which the reader is referred for a more detailed treatment. Many of the selected parameters are based on laboratory cratering experiments into weak materials (sand) and strong materials (basalt).

As will be discussed in the next section, the nature of regolith development is greatly affected by the size of crater produced in any given impact. In general, we assume the relationships between impact kinetic energy, E, and resulting crater diameter, D, to be of the form

$$D \propto E^{\frac{1}{3}} \quad \text{(strength scaling)} \tag{14}$$

or

$$D \propto E^{0.29} g^{-\frac{1}{6}} \quad \text{(gravity scaling)} \tag{15}$$

where g is the gravitational acceleration at the asteroid surface. As demonstrated by Gault and Wedekind (1977), strength scaling applies when target strength dominates over gravity, i.e., when $s/\rho g D \gg 1$, where s is the target strength and ρ is the target density. Gravity scaling applies when $s/\rho g D \ll 1$. Actually the theoretical limiting values for the gravity scaling exponents of E and g are $1/4$ and $-1/4$ respectively, but the adopted values of 0.29 and $-1/6$ are obtained from experiments in real materials. One can use the ratio $s/\rho g D$ to determine whether Eq. (14) or (15) is appropriate for a particular asteroid. In Fig. 4 we show the regions where $s/\rho g D$ is large or small (note that s and ρ are fixed for each of the two types of asteroids we model, strong or weak). The main effect of the two scaling laws on regolith evolution is that gravity scaling tends to produce relatively few large craters and more small ones, compared with strength scaling.

In order to specify the nominal values for annuli widths (i.e. a_1 and a_2) we employ experimentally determined velocity distributions of crater ejecta

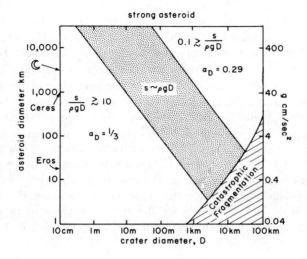

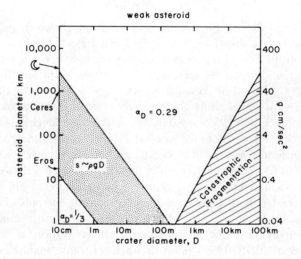

Fig. 4. Domains of applicability of two cratering laws. The appropriate law depends on the ratio of target strength s, to gravitational strength $\rho g D$ (ρ is the target density, g the gravity, and D the crater diameter). A large ratio suggests "strength scaling" where $D \propto$ K.E.$^{1/3}$ (K.E. is the impact kinetic energy). A small ratio suggests "gravity scaling" where $D \propto$ K.E.$^{0.29}$. The stippled area indicates a zone of transition between the two cratering laws; here the ratio is neither large nor small. (This figure was constructed from the work of Gault and Wedekind 1977.)

(c.f. Housen *et al.* 1979 and references therein). Using these velocity data and an ejection angle of $45°$, the nominal value of a_1 is computed so that 90% of the ejected mass resides within the annulus. For Type 2 blankets, the ejecta annulus width, for a crater of diameter D, is a_2D. We choose a_2 such that $a_2 = a_1 D'_{s,2}$ where $D'_{s,2}$ is evaluated at the end of the evolution. This is done to insure that the largest craters (i.e. those near $D_{s,2}$ at the end of the evolution), which dominate regolith growth, do not have annuli whose widths are inconsistent with the value computed from the ejecta velocity data. Our choice of the 90th percentile in the ejecta distribution is somewhat arbitrary. However, as stated earlier (Fig. 3), we must avoid making the annuli too large (e.g. using 100 %) or too small ($\lesssim$ a few tens of percent?). We note in the next section that our ejecta blanket model works reasonably well when we model regolith evolution for the lunar maria. Realizing the uncertainty involved in the choice of 90 %, we illustrate, in our calculations, the changes in regolith evolution when the annulus width is varied from the nominal value.

IV. RESULTS

Model calculations for two large asteroids are shown in Figs. 5 and 6. Figures 5*a* and 6*a* show, for both Type 1 and 2 ejecta blankets, the surface elevation in the typical region as a function of time. For each ejecta blanket type, we plot two curves representing elevations when annuli widths are varied by ± 25 % from nominal values. Figures 5*b* and 6*b* illustrate the relative rates at which ejecta blankets bury regolith grains versus the rate of excavation by craters. We show only one burial curve for each ejecta blanket type (which corresponds to the nominal annulus width) because the variation of ± 25 % results in very little variation in the burial curves. The curves in Figs. 5*b* and 6*b* are cumulative in the sense that the depth shown represents burial by blankets of a given thickness or greater and excavation to a given depth or greater. Both Figs. 5 and 6 represent strong asteroids whose internal strength is comparable to that of basalt. We do not present results for large weak asteroids since the adopted low ejecta velocities for weak targets (Housen *et al.* 1978, 1979) result in a blanket whose annular area is much smaller than the area of the crater itself. Thus craters saturate the surface before their ejecta blankets do, and so are incorporated into the typical region while their ejecta remain in the atypical region. We do not consider this to be a physically realistic situation. Probably the effective strength of all large bodies is reasonably strong.

The evolution for a 1000-km diameter asteroid is shown in Fig. 5. Gravity scaling is used, because the craters which affect the regolith depth the most (i.e. craters near $D'_{s,i}$ at the end of the evolution) are roughly 100 km in diameter. Figure 4 suggests, for craters of this size, that gravity scaling is

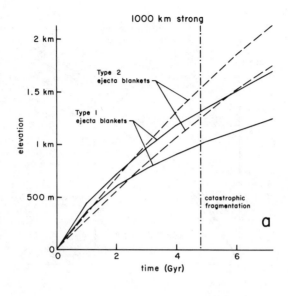

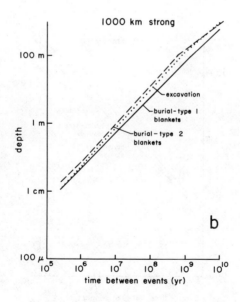

Fig. 5. Regolith evolution on a 1000-km diameter asteroid of high strength for the two ejecta blanket models defined in the text. Figure 5*a* shows the surface elevation in the typical region as a function of time. For each ejecta blanket type, the upper and lower curves correspond to a blanket size 25 % greater and 25 % less than the nominal value computed from ejecta velocity data. Figure 5*b* shows the thickness of ejecta blankets (of depth ⩾ to that shown) which form and the depth to which an excavation occurs as a function of the time interval between these events.

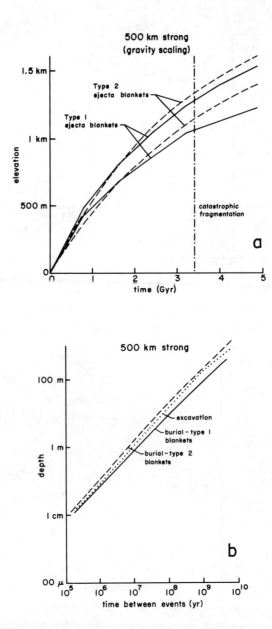

Fig. 6. A 500 km diameter asteroid of high strength. The amount of regolith developed is roughly equal to that on the 1000 km body because even though more ejecta escape here (5 % as opposed to 1 % for the 1000 km asteroid) the ejecta blankets on this asteroid are bigger so more craters can deposit material into the typical region. Just as for the 1000 km object, excavation rates are comparable to burial rates. Regolith grains should be excavated few times over the lifetime of the asteroid.

appropriate. The nominal ejecta blanket size is chosen to be 75 km on the basis of experimental data on ejecta velocities. For each ejecta blanket type, the lower of the pair of curves corresponds to an annulus width of 55 km (nominal value minus 25 %). As is expected, less widespread ejecta means less regolith is developed in the typical region because it takes longer for ejecta blankets to saturate the surface and deposit their material onto the typical region. Roughly 1.2 km of regolith is accumulated before the asteroid is catastrophically ruptured by an energetic impact. Figure 5b shows that the rates of burial and excavation are comparable so that some gardening of the regolith is expected. If we consider a point on the surface at some time (whose position is fixed with respect to the center of the asteroid), then the rate of excavation experienced by the point decreases with time because of burial by ejecta. By integrating the excavation rate it is found that regolith grains may be excavated a few times over the lifetime of the asteroid.

The case of a 500 km asteroid is shown in Fig. 6. Once again, the largest craters that deposit material onto the typical region are large enough (~ 110 km) to warrant the use of gravity scaling (Fig. 4). Even though somewhat more ejecta are lost from this asteroid than from the 1000 km body (5 % as opposed to 1 %), roughly comparable regolith depths are developed during the same period of time. This is because of the more widespread blankets on this smaller object. Here the nominal annulus width is 150 km. Reference to Fig. 6b shows that again some gardening of the regolith is expected.

V. DISCUSSION OF RESULTS

Asteroids

We may gain general insight into regolith evolution on asteroids by comparing the results for large asteroids with the calculations for smaller ones. A summary of regolith depths for various sizes of strong asteroids is given in Table III. For the smallest bodies, nearly all impact ejecta escape, so only thin dusty coatings are expected to exist. With increasing asteroid diameter, more ejecta are retained, hence, more regolith is developed. This trend is exhibited by all asteroids which are small enough such that crater ejecta are more or less globablly distributed (diameters $\lesssim 300$ km). For even larger bodies, three effects are important: (1) still more ejecta are retained; (2) a transition is made from strength scaling to gravity scaling; and (3) ejecta blankets become less widespread. The latter two effects, which tend to decrease regolith depth, dominate over the retention of more ejecta, as illustrated in Table III. Gravity scaling produces a steeper crater flux, i.e. a more negative value of α in Eq. (4) than does strength scaling. This results in relatively fewer large craters and more small ones. Since for the adopted impacting population most of the ejecta mass is from the larger craters, the reduction in large craters reduces the amount of regolith developed. This,

TABLE III

Regolith Depths Accumulated on Strong Asteroids before Rupture

Asteroid Diameter (km)	Fragmentation Time (Gyr)	Regolith Depth
10	0.48	< 1 mm
100	1.52	200 m
300	2.63	3.5 km
500	3.39	1.2 km
1000	4.80	1.2 km

together with the decreased ejecta blanket areas, causes the regolith depth to decrease on the large asteroids.

For still larger bodies even less regolith would develop. Notice that once a transition from strength scaling to gravity scaling has been made, crater diameters decrease with increasing gravity. As ejecta blanketing decreases and more numerous small craters are produced, we find that regolith stirring and gardening increase for large asteroids. The surface does not gain a protective shield of deposited ejecta as fast as on smaller bodies.

Comparison with the Moon

In order to check our model for large bodies and to help delineate the differences between regolith evolution on asteroids and the moon, we have used our model to compute the regolith depth expected to develop on the lunar maria. To model the lunar environment we reduce the mass flux of impacting projectiles by a factor of 10^3 and increase the impact velocity from 5 km sec^{-1} to 20 km sec^{-1} (appropriate to the change from 3 AU to 1 AU). Although there is evidence that the mass flux is more nearly constant with heliocentric distance for small particles, the flux of large objects (e.g. observable asteroids) which dominate regolith production is lower by the factor of 10^3 at 1 AU. But by underestimating the lunar flux of smaller particles, the amount of gardening of the lunar regolith is underestimated by our method and, hence, is not discussed. The craters produced by the adopted flux at 1 AU in a diameter interval about 10 km agree well with crater counts on typical lunar maria. Using a blanket size of 10 km (the nominal value is $\sim$ 20 km) we find that $\sim$ 7 m of regolith is accumulated in 3.5 Gyr ($\sim$12 m accumulates for 100 km blankets). The calculated depths are slightly higher than the observed value ($\sim$ 5 m), but the agreement is satisfactory given the precision of the input parameters; therefore, we have confidence in our model for large asteroids.

In general, asteroids that do not lose most of their ejecta develop thicker regoliths than those on the lunar maria. The two main effects, illustrated in

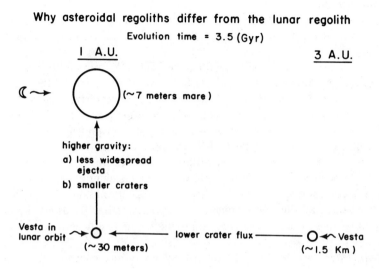

Fig. 7. A schematic illustration of the processes which make asteroidal regoliths differ from the lunar regolith. An asteroid at 3 AU develops more regolith, in the same period of time, than does a similar sized body at 1 AU because of the higher flux of large craters at 3 AU. For two bodies of different size, both at 1 AU (or 3 AU), the larger body develops less regolith because the higher gravity makes less widespread ejecta blankets and smaller craters.

Fig. 7, are differences in crater flux, and size of body.

(1) *Differences in crater flux.* Consider for example, a Vesta-sized object at 3 AU. Roughly 1.5 km of regolith is developed in 3.5 Gyr. If this body were moved to 1 AU, it would accumulate only ∼ 30 m of regolith in 3.5 Gyr due solely to the reduced flux at 1 AU. Obviously, for equal evolution times, if the cratering flux is reduced then the amount of regolith is also reduced. The smaller mass flux and higher impact velocity at 1 AU result in a crater production rate which is a factor of 100 below that at 3 AU. The reason that the regolith depth only decreases by a factor of 50, rather than by a factor of 100, is because regolith growth is not linear in time.

(2) *Size of the body.* Comparing the Vesta-sized asteroid at 1 AU with the moon, we see that the moon develops the lesser amount of regolith. The higher gravity field of the moon causes ejecta to be less widespread. Thus the ejecta blanket annuli, associated with craters of some given diameter, take longer to saturate on the moon than on the asteroid, i.e. ejecta take longer to be incorporated into the typical region on the moon. Thus, at any given time, less regolith has been deposited into the typical region on the moon. The differing gravity fields on the two objects has another effect. The most important craters (i.e. just smaller than $D'_{s,i}$ at 3.5 Gyr) which contribute to regolith buildup are roughly a few kilometers in diameter for the moon and ∼

20 km for the Vesta-sized body at 1 AU. Reference to Fig. 4 shows that gravity scaling applies to both bodies. This means that impacts of the same energy produce a smaller crater on the moon, so that again less regolith is developed.

In summary, the deepest regoliths occur on rocky asteroids ~ 300 km in diameter. Smaller and larger objects both evolve thinner regoliths; smaller objects lose too much of their ejecta and larger objects have smaller craters and more localized ejecta blankets due to their higher gravity fields.

Our conclusions are based on the assumptions of present-day velocities and collision rates in the asteroid belt and on present-day solar activity. For these parameters, there are no quantitative estimates which pertain to earlier epochs in solar system evolution. However, our assumption of present-day values is conservative in the sense that the increased collision rates and lower relative velocities, which are thought to have characterized the asteroid belt in very early times, would make regolith production more efficient.

Acknowledgments. We appreciate the feedback on our previous and current modeling efforts which we have received from many of our colleagues, especially E. Anders, D. E. Gault, W. K. Hartmann, P. Pellas, G. W. Wetherill and A. Woronow. We wish to thank J. Wasson whose constructive criticism resulted in many improvements in our manuscript. Thanks also go to Lynn Lane and Reea Rodriguez for the extra effort they invested in producing the manuscript. This work was supported in part by a NASA grant to the University of Arizona and a NASW contract with the Planetary Science Institute.

REFERENCES

Anders, E. 1975. Do stony meteorites come from comets? *Icarus* 24: 363-371.

Anders, E. 1978. Most stony meteorites come from the asteroid belt. In *Asteroids: An Exploration Assessment,* eds. D. Morrision and W. C. Wells, NASA Conf. Publ. 2053, pp. 57-76.

Brownlee, D. E., and Rajan, R. S. 1973. Micrometeorite craters discovered on chondrule-like objects from Kapoeta meteorite. *Science* 182: 1341-1344.

Chapman, C. R. 1971. Surface properties of asteroids. Ph.D. dissertation, Massachusetts Institute of Technology.

Chapman, C. R. 1976. Asteroids as meteorite parent bodies: The astronomical perspective. *Geochim. Cosmochim. Acta* 40: 701-719.

Crozaz, G., and Dust, S. 1977. Irradiation history of lunar cores and the regolith. *Proc. Lunar Sci. Conf. VIII* (Oxford: Pergamon Press), pp. 3001-3016.

Dollfus, A. 1971. Physical studies of asteroids by polarization of the light. In *Physical Studies of Minor Planets,* ed. T. Gehrels (NASA SP-267, Washington, D. C.: U. S. Government Printing Office), pp. 95-116.

Dran, J. C.; Duraud, J. P.; Langevin, Y.; and Maurette, M. 1979. The predicted irradiation record of asteroidal regoliths and the origin of gas-rich meteorites. *Lunar Science X,* The Lunar and Planet. Inst., pp. 309-311.

Duraud, J. P.; Langevin, Y.; and Maurette, M. 1979. An analytical model for the regolith evolution of small bodies in the solar system. *Lunar Science X,* The Lunar and Planet. Inst., pp. 323-325.

Eberhardt, P.; Geiss, J.; and Grogler, N. 1966. Distribution of rare gases in the pyroxene and feldspar of the Khor Temiki Meteorite. *Earth and Planet. Sci. Lett.* 1: 7-12.

Gault, D. E.; Hörz, F.; Brownlee, D. E.; and Hartung, J. B. 1974. Mixing of the lunar regolith. *Proc. Lunar Sci. Conf. V* (Oxford: Pergamon Press), pp. 2365-2386.

Gault, D. E., and Wedekind, J. A. 1977. Experimental hypervelocity impact into quartz sand. II. Effects of gravitational acceleration. In *Impact and Explosion Cratering,* eds. D. J. Roddy, R. O. Pepin and R. B. Merrill (New York: Pergamon Press), pp. 1231-1244.

Goswami, J. N.; Hutcheon, I. D.; and Macdougall, J. D. 1976. Microcraters and solar flare tracks in crystals from carbonaceous chondrites and lunar breccias. *Proc. Lunar Sci. Conf. VII* (Oxford: Pergamon Press), pp. 543-562.

Heiken, G. 1975. Petrology of lunar soils. *Rev. Geophys. Space Phys.* 13: 567-587.

Housen, K. R.; Wilkening, L. L.; Chapman, C. R.; and Greenberg, R. 1979. Asteroidal regoliths. *Icarus* (in press).

Housen, K. R.; Wilkening, L. L.; Greenberg, R.; and Chapman, C. R. 1978. Regolith evolution on small bodies. *Lunar Science IX,* The Lunar and Planet. Inst., pp. 546-548.

Kerridge, J. F., and Kieffer, S. W. 1977. A constraint on impact theories of chondrule formation.

Lal, D., and Rajan, R. S. 1969. Observations on space irradiation of individual crystals of gas-rich meteorites. *Nature* 223: 269-271.

Langevin, Y., and Arnold, J. R. 1977. The evolution of the lunar regolith. *Ann. Rev. Earth Planet. Sci.* 5: 449-489.

Langevin, Y., and Maurette, M. 1976. A Monte Carlo simulation of galactic cosmic ray effects in the lunar regolith. *Proc. Lunar Sci. Conf. VII* (Oxford: Pergamon Press), pp. 75-91.

Macdougall, D.; Rajan, R. S.; and Price, P. B. 1974. Gas-rich meteorites: Possible evidence for origin on a regolith. *Science* 183: 73-74.

Matson, D. L.; Johnson, I. V.; and Veeder, G. J. 1977. Soil maturity and planetary regoliths: The moon, Mercury, and the asteroids. *Proc. Lunar Sci. Conf. VIII* (Oxford: Pergamon Press), pp. 1001-1011.

Oberbeck, V. R.; Quaide, W. L.; Mahan, M.; and Paulson, J. 1973. Monte Carlo calculations of lunar regolith thickness distributions. *Icarus* 19: 87-107.

Pellas, P.; Poupeau, G.; Lorin, J. C.; Reeves, H.; and Audouze, J. 1969. Primitive low-energy particle irradiation of meteoritic crystals. *Nature* 223: 272-274.

Rajan, R. S. 1974. On the irradiation history and origin of gas-rich meteorites. *Geochim. Cosmochim. Acta* 38: 777-788.

Rajan, R. S.; Brownlee, D. E.; Heiken, G. H.; and McKay, D. S. 1974. Glassy agglutinate-like objects in the Bununu howardite. *Meteoritics* 9: 394-397.

Suess, H. E.; Wänke, H.; and Wlotzka, F. 1964. On the origin of gas-rich meteorites. *Geochim. Cosmochim. Acta* 28: 595-607.

Wahl, W. 1952. The brecciated stony meteorites and meteorites containing foreign fragments. *Geochim. Cosmochim. Acta* 2: 91-117.

Wänke, H. 1965. Der Sonnenwind as Quelle der Uredelgase in Steinmeteoriten. *Zs.f. Naturf.* 20a: 946-949.

Warner, J. L. 1972. Metamorphism of Apollo 14 breccias. *Geochim. Cosmochim. Acta* 1: 623-643.

Wilkening, L. L. 1970. On the early history of meteorites. Evidence from glasses, from fossil particle tracks and from the noble gases. Ph.D. dissertation, University of California at San Diego.

Wilkening, L. L. 1971. Particle track studies and the origin of gas-rich meteorites. *Center for Meteorite Studies* (monograph), Tempe, Ariz.: Arizona State University.

Wilkening, L. L. 1973. Foreign inclusions in stony meteorites. I. Carbonaceous chondritic xenoliths in the Kapoeta howardite. *Geochim. Cosmochim. Acta* 37: 1985-1989.

Wilkening, L. L., and Clayton, R. N. 1974. Foreign inclusions in stony meteorites. II. Rare gases and oxygen isotopes in a carbonaceous chondritic xenolith in the Plainview gas-rich chondrite. *Geochim. Cosmochim. Acta* 38: 937-945.

PHOBOS AND DEIMOS:
A PREVIEW OF WHAT ASTEROIDS ARE LIKE?

J. VEVERKA and P. THOMAS
Cornell University

Phobos and Deimos are small, very dark asteroid-like satellites. While both are irregular, their shapes can be approximated by triaxial ellipsoids. Phobos is about 27 x 21 x 19 km across; Deimos is about 15 x 12 x 11 km. The synchronous spin periods of the satellites are comparable with those of asteroids – 7^h39^m for Phobos and 30^h17^m for Deimos. Both are heavily cratered and completely covered with a regolith whose microstructure appears to be lunar-like. While the disk-integrated photometric properties of the two satellites are similar, high-resolution Viking Orbiter images have shown that the surfaces have significantly different morphologies. On Deimos, but not on Phobos, one sees brighter albedo patches and craters with conspicuous fill. The surface of the inner satellite is covered by sets of extensive, pitted, linear depressions which have no counterparts on Deimos. The Viking images of Phobos and Deimos provide valuable clues to what the surfaces of small asteroids are like. For instance, the data indicate that even tiny bodies, with negligible surface gravities, can retain hundreds of meters of regolith, and display high surface densities of ejecta blocks. The data suggest also that craters on small asteroids will be similar in morphology to those on the moon, with raised rims and depth-diameter ratios comparable to those of small lunar craters. The Phobos/Deimos experience proves that the surfaces of two small bodies can be nearly identical in terms of many disk-integrated properties but yet be strikingly different in morphology. It is noteworthy that extensive and rather uniformly distributed albedo markings occur on Deimos, the presence of which would not be suspected from disk-integrated *measurements. All in all, asteroid surfaces are probably much more varied and interesting than suggested by available disk-integrated observations.*

While we no longer think of asteroids as simple points of light, still, by and large, most of our data about minor planets refer to the average physical characteristics of their surfaces. This situation is unavoidable: Earth-based observations of asteroids are restricted to disk-integrated measurements by the minute apparent diameters of these bodies. Yet, with the first spacecraft mission to an asteroid only ten or fifteen years away, we must strive to relinquish some of the simplistic ideas about asteroids and their surfaces that have resulted from this restricted perspective. Almost certainly the surfaces of asteroids will prove much more varied and interesting than we suspect from available disk-integrated observations.

While we have yet to obtain our first detailed look at the surface of an asteroid, spacecraft investigations of Phobos and Deimos, the two tiny satellites of Mars, probably provide a good preview of what asteroid surfaces are like.

Although it has been suggested that Phobos and Deimos may be captured asteroids, or at least that they are objects which formed in the general region of the asteroid belt and were captured by Mars (Burns 1978; Hunten 1978; Veverka 1978 etc.), the validity of this view is not essential to the argument in this chapter. The suggestion is based on the low mean density of the satellites (Table I; Duxbury 1979, personal communication) and on their spectral reflectance curves (e.g. Pang and Rhodes 1979). These data indicate that Phobos and Deimos are made of some low-density carbonaceous material of the sort that some models (Lewis 1974) claim formed only in the asteroid belt and beyond. Burns (1978), Hunten (1978) and others have discussed specific scenarios for the capture of Phobos and Deimos very early in the evolution of Mars. Burns, following a discussion by Pollack *et al.* (1978), invokes capture due to the drag from a "circumplanetary envelope" or "preplanetary nebula," while Hunten relies on drag from a "protoatmosphere."

Whatever the provenance of Phobos and Deimos, their physical characteristics — sizes, shapes, rotation periods, surface gravities — are similar to those of small asteroids (Table I). If Phobos were still in the asteroid belt it would be classified as a *C* object in the TRIAD scheme: Deimos would fall close to the boundary between *C* and *U* objects (Table II; Fig. 1). Both on a p_V versus (U-V) and on a (U-B) versus (B-V) graph, Phobos plots as a *C* object (Figs. 1*a, b*). Strictly speaking Deimos plots as a *U* object (Figs. 1*a,b,c*), but the error bars are large enough for it to be a *C*. Note that Fig. 1*a* definitely excludes the possibility that Deimos in an *E* object, while its low mean density (Table I) excludes the possibility that it is an *M* object. The view that Deimos should be considered an extreme *C* object is consistent with the fact that its spectral reflectance curve is generally similar to that of Phobos (Pang and Rhodes 1979).

TABLE I

Physical Characteristics of Phobos and Deimos

	Phobos	Deimos	Notes[a]
Mean Radius: R(km)	10.5	6.5	A
Axes (radii): a x b x c(km)	13.5x10.5x9.0	7.5x6.0x5.0	B
a/b	~1.3	~1.3	
a/c	~1.5	~1.5	
Spin Period: P(hr)	7.65	30.29	C
Lightcurve Amplitude: Δm	~0.2	~0.2	D
Mean Density: $\bar{\rho}$ (gcm^{-3})	1.9 ± 0.5	1.5 ± 0.5	E
Surface Gravity: g(cm sec^{-2})	~0.3 −0.6	~0.3	F
Escape Velocity: V_{esc} (m sec^{-1})	~15	~10	F

[a]Notes: A: Volume = $\frac{4}{3}\pi R^3$; B: Approximating shape by a triaxial ellipsoid (Duxbury 1979); C: Synchronous with orbital period; D: Estimated from shape (Noland and Veverka 1976a); E: From Duxbury (1979, personal communication); F: From Goguen and Burns (1978); Housen and Davis (1978).

TABLE II

Photometric Characteristics of Phobos and Deimos

	Phobos	Deimos	Notes[a]
Mean Opposition Magnitude: V_0	11.4	12.45	A
(B-V)	0.65 ± 0.05	0.65 ± 0.05	B
(U-B)	0.28 ± 0.1	0.18 ± 0.1	B
p_V	0.05 ± 0.01	0.06 ± 0.01	B
P_{min}(%)	—	-1.5 ± 0.1	C
Phase Coefficient: β(mag/deg)	0.032	0.030	D
Intrinsic Phase Coefficient: β_i (mag/deg)	0.019	0.017	D

[a]Notes: A: Veverka (1977); B: From French (1979, personal communication); C: Zellner (1972); D: Noland and Veverka (1976b; 1977a,b).

It has been stressed by Chapman (1978) and others that the present environment of Phobos and Deimos is sufficiently different from that of a typical asteroid that the surface morphologies and detailed regolith properties of Phobos and Deimos may be quite different from those of asteroids of a comparable size. The main point of this argument is that the satellites are in the gravitational field of Mars and that the trajectories of ejecta are

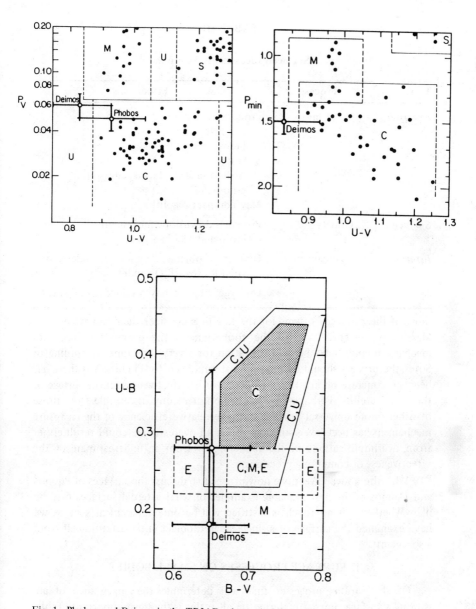

Fig. 1. Phobos and Deimos in the TRIAD scheme (c.f. Bowell *et al.* 1978).

determined mostly by Mars' gravity field and not by the field of the satellites themselves. Soter (1971) and others have discussed a mechanism by which the satellites can recapture ejecta, and it has been argued that the surfaces of Phobos and Deimos could have been modified significantly by this process in terms of their morphology, regolith properties, crater densities, etc. Since

TABLE III

Surface Processes on Asteroids

Type	Process	Effect
External	Cratering	Produces regolith blocks craters spallation scars, facets, edges etc. grooves (?) May compact regolith.
Surface	Gravity	Produces downslope movement of loose material.
Internal	"Volcanism"	On some C objects, may produce release of volatiles (eg. H_2O vapor). On larger objects may produce lava flows.

most of the proposed origins for the satellites consider them to have been in Mars' orbit essentially since the formation of the planet, the recapture mechanism could have been in operation for a very long time. According to Soter the process should have been most important for Phobos. Yet there has been no concrete observational evidence of its effectiveness on the surface of the inner satellite. Perhaps the effects are not readily discernible from those of other, more universal processes, or the relative efficiency of the recapture mechanism has been overestimated. Such an overestimate could result either from overimplifications in the modeling or from an underestimate of the effectiveness of competing processes.

With the above caveat, we now proceed to discuss the surfaces of Phobos and Deimos as the best available examples of what asteroid surfaces may be like. Whether this approach is justified will become apparent as soon as we have examined the surface of a single small C object in the asteroid belt from a spacecraft.

I. SURFACE PROCESSES ON SMALL BODIES

Of the various processes that will determine the appearance of an asteroid's surface, impact cratering on all scales is the most important (Table III). Impacts will comminute the outer layers of a body into regolith, produce local topography (craters, large blocks of ejcta, etc.) and determine the overall shape of the object.

1.1 Shape of Small Bodies Determined by Cratering

The present shapes of Phobos and Deimos clearly result from a long

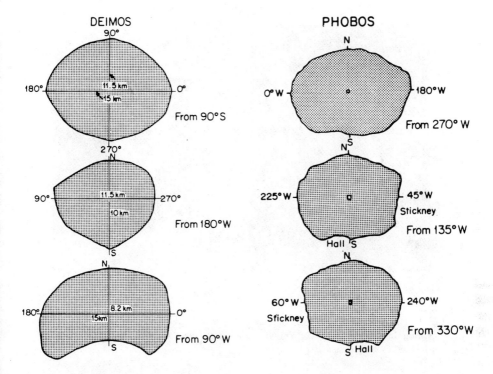

Fig. 2. Silhouettes of Deimos.   Fig. 3. Silhouettes of Phobos.

history of cratering on a variety of scales. While both objects are *approximately* triaxial in form (Duxbury 1974), their precise shapes are actually much more complicated, as can be seen from the silhouette sketches in Figs. 2 and 3. Thomas (1978) has shown that the shape of Deimos can be thought of as a series of irregular facets joined by pronounced ridges. These facets and ridges suggest that present-day Deimos has been fashioned by impacts from a larger object (c.f. experiments by Gault and Wedekind 1969). The facets, edges, and the saddle-like topography approximately centered on the south pole of Deimos (Fig. 2) do not have the morphology of craters, but rather those of spallation scars (c.f. Gault and Wedekind 1969). In fact, the largest distinctly recognizable crater on Deimos today is only 2.3 km in diameter.

On Phobos, well-defined craters range up to 10 km in diameter (Stickney). The three largest Stickney, Hall (5 km) and Roche (5 km) have modified the satellite's shape conspicuously (Fig. 3). High phase angle images such as that shown in Fig. 4*b* clearly demonstrate that the surface topography, limb profiles, cross-sections or silhouettes of small bodies are largely determined by the distribution of large and medium-sized craters. It is

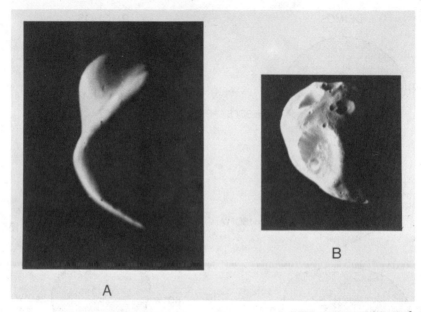

Fig. 4. The irregular shapes of Phobos and Deimos. (a) Deimos at a phase angle of 122°
(Viking Image 464A03). (b) Phobos at a phase angle of 87°. The large crater is
Stickney. (Viking Image 203A15.)

less evident, but equally true that these characteristics are also determined by
a superposition of large craters of all ages including some very degraded ones
(Fig. 5). It was noted by Thomas (1978) that the degraded rims of those very
old craters often line up into more or less continuous ridges which can be
traced for tens of kilometers along the satellite's surface (Fig. 5).

The low phase angle lightcurves of Phobos and Deimos are probably
roughly sinusoidal with an amplitude of about 0.2 mag (Noland and Veverka
1976a) and one expects that they could be matched fairly well by using
ellipsoidal models of the satellites. The true surfaces, however, are much more
complicated (Figs. 2, 3, 4), and under certain viewing and lighting conditions,
especially at large phase angles, simple approximations to the shapes can be
quite misleading (Fig. 4a). One can also see that at large phase angles the
lightcurves of small asteroids will probably be very complicated and
extremely difficult to interpret in terms of any simple model.

1.2 The Morphology of Craters on Small Bodies

The possible morphology of impact craters on low-gravity objects was
the subject of much speculation before the first Mariner 9 images of Phobos
and Deimos became available in late 1971. Some investigators doubted that
such craters would have rims, while a few even speculated that it would be

Fig. 5. Ridges and old, degraded craters on Phobos. A small ray-crater is indicated by the letter (R). (Viking Image 246A62.)

Fig. 6. Variations in crater morphology on Phobos.

difficult to recognize anything resembling craters at all. Needless to say the craters were there in profusion and at least superficially, they resembled those on much larger bodies such as the moon.

Craters of varying degrees of freshness (or degradation can be identified in the Viking images of Phobos and Deimos (Figs. 6 and 7). The fresher craters have conspicuously raised rims (Fig. 6*f*; Crater #1 in Fig. 7*a*) suggesting that circum-crater uplift is important in producing raised rims. Thomas (1978) found that the depth-diameter relation is similar to the lunar one, being approximately

$$d = 0.2D \qquad (1)$$

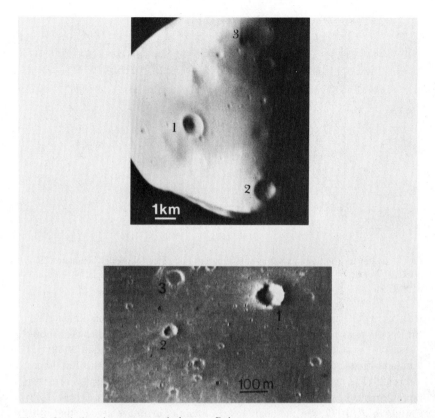

Fig. 7. Variations in crater morphology on Deimos.

for craters between 200 meters and 10 km in diameter. Here d is the crater depth and D is the crater diameter.

A detailed examination of the morphology of craters on asteroids would permit crucial tests of various theories of crater formation. For example, Hartmann (1972, 1973) has presented evidence that the morphology of large craters depends significantly on surface gravity. On the moon he describes the following sequence of crater morphology as a function of increasing crater diameter: bowl-shaped craters, craters with one central peak, craters with a cluster of central peaks, craters with a single inner ring, and finally, craters with multiple central rings. By comparing craters on the moon, earth and Mars, Hartmann concluded that the critical diameter D_c at which a transition of crater morphology occurs is fundamentally a function of g (surface gravity measured in cm sec^{-2}) and that, approximately:

$$D_c \propto g^x \qquad (2)$$

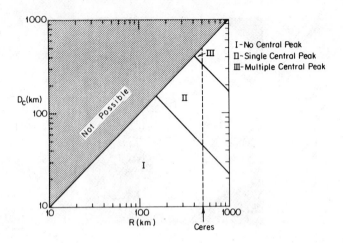

Fig. 8. Crater morphology as a function of parent body size following Hartmann (1972, 1973). See text for details. R is the radius of the parent body; D_c is the crater diameter.

with $x \sim -1$. Using this relationship and the observed transition diameters on the moon reported by Hartmann one can predict that central peaks should not occur in craters on any asteroid having a radius smaller than about 150 km, and that multiple central peaks could occur only on the few largest asteroids (Fig. 8). Whether or not Hartmann's gravity scaling is correct, it represents a testable proposition. Head and his coworkers (Head 1976; Cintala *et al.* 1977; see the chapter by Cintala *et al.*) have argued that other parameters, such as target strength and impact velocity, are much more important in determining crater morphology than is surface gravity. Again, asteroid surfaces provide a means of testing this contention; by looking at two asteroids of different composition but similar surface gravity, and located in the same part of the asteroid belt (similar impact velocities) one could test these two divergent viewpoints.

While our observations of Phobos and Deimos do not provide any critical test of Hartmann's hypothesis (Fig. 8), they are consistent with it, since no definite central peaks have been identified in any of the craters on the two satellites.

In terms of interior morphology, relatively fresh craters on Phobos range from those that are bowl-shaped (Fig. 6a) to those with hummocky floors (Fig. 6e, 6f). An interesting question is whether there is any evidence for a systematic transition of crater morphology with increasing crater diameter that could be atributed to the effects of a surface which is inhomogeneous with depth. Oberbeck and Quaide (1967) and Quaide and Oberbeck (1968) showed that in certain lunar mare regions craters with $D \geqslant 50$ m tend to be bowl-shaped while craters with $D \geqslant 100$ m tend to have concentric floors.

Using laboratory simulations they demonstrated that this observation could be interpreted as evidence for a layer of loose regolith some 10 m in depth overlying a more compact subsurface. Their experiments indicated that the transition from bowl-shaped to flat-floored craters occurs when the ratio R of the crater diameter D to the regolith depth δ reaches 6, and that the transition from flat-floored craters to those with concentric floors occurs when $R = D/\delta$ exceeds 11. If one assumes $d = 0.2\,D$ then the critical value of d/δ for the first transition is ~ 1, as one would expect.

One can use these results to predict that on a typical asteroid having a diameter of 100 km, which according to the model calculations of Housen et al. (1979; see their chapter) might have ~ 100 m of regolith, most of the craters smaller than 500 meters in diameter should be bowl-shaped.

It is difficult to use this method to obtain any definitive estimate of the regolith depth on either Phobos or Deimos. On Deimos the details of crater morphology are commonly obscured by infilling (c.f. Sec. 1.4). On Phobos, craters with hummocky floors which could be interpreted as the equivalents of flat-floored or concentric geometries do occur, but no consistent transition in crater morphology with size can be identified for those craters which are abundant enough to provide a statistically significant sample – say $D \lesssim 1$ km. From this we conclude that generally on Phobos $\delta \gtrsim 200$ m, an estimate consistent with the discussion in Sec. 1.4.

1.3 Characteristics of Crater Ejecta on Phobos and Deimos

In spite of their very low surface gravities (Table I), Phobos and Deimos appear to have accumulated significant amounts of regolith (Sec. 1.4). While conspicuous ejecta patterns are rare – as might be expected, since even on the moon these are associated with only the freshest craters – a variety of related features is observed. Some of these prove that a few crater ejecta are not only retained, but are retained in the vicinity of the parent crater.

First, on Phobos, there are a few small craters which show associated bright rays in their immediate vicinity (c.f. Fig. 5). Larger craters display bright rims at very small phase angles (Fig. 9). Similar bright rims are observed on the moon in the case of the fresher craters which have texturally rough rims consisting of blocky ejecta. These bright rims prove that some ejecta are retained locally.

Second, dark halo craters are seen on both Phobos and Deimos (Fig. 10), again proving that crater ejecta are partly retained directly by the satellites, without the help of any esoteric recapture mechanisms.

Finally, large blocks, typically 5-10 m wide and 2-5 m high, occur on both satellites (Fig. 10), often near the rims of large craters, strongly suggesting localized retention of coarse ejecta.

These observations demonstrate that ejecta are partly retained directly on bodies even as small as Phobos and Deimos, and provide the means of

Fig. 9. Possible slump in a crater on Phobos (Viking Image 343A13).

building up substantial regoliths. Some of the ejecta are localized in the vicinity of the parent crater. Apparently debris from impacts need not be spread uniformly over the entire body as has been suggested by a few modelers.

No clusters of secondary craters have been identified on either satellite (Thomas *et al.* 1979*a*). Perhaps the escape velocities are too low: nonescaping ejecta must hit the surface at less than 10–20 m sec^{-1}, which may not be sufficiently high to produce recognizable secondaries. If this hypothesis is correct, then there should be a minimum size for a rocky asteroid below which the surface will not show typical secondaries. The asteroid's escape velocity must exceed the critical impact velocity needed to produce secondaries.

1.4 The Nature of the Regolith on Phobos and Deimos

A variety of remote sensing evidence proves that regoliths exist on Phobos and Deimos and that they are similar in texture to the lunar regolith. The evidence consists of photometric, polarimetric and thermal infrared measurements (Veverka 1978), and is supported by direct inspection of high-resolution Viking images (Thomas 1978).

There is evidence that the regolith is layered, at least locally. Layering appears to be exposed in the walls of some craters on Phobos (Veverka 1978) at depths of 100-200m. Some interpretations of dark halo craters also require a layered regolith. On Deimos one sees irrefutable evidence of such layering in the superposition of thin brighter albedo streamers (Fig. 11) over darker crater fill (Thomas et al. 1979*b*).

Regolith depth on Phobos has been estimated from the morphology of grooves and from their characteristic widths to average ~100-200 m (Thomas

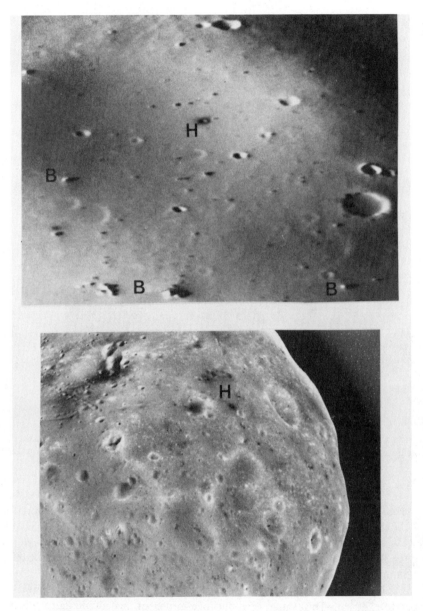

Fig. 10. Dark halo craters (H) on Phobos and Deimos. Top: Deimos (Viking Image 423B63). Bottom: Phobos (Viking Image 252A61). Note the numerous blocks (B) in the Deimos picture.

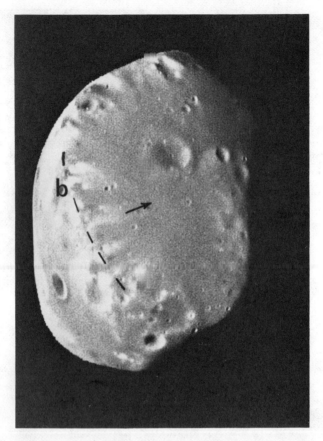

Fig. 11. Bright albedo streamers on Deimos trending downslope (arrow) from a ridge
(---). (Viking Image 428B22.)

et al. 1979*a*), an estimate consistent with that based on the morphology of
craters given in Sec. 1.2.

Thomas (1978) has shown that many craters on Deimos are filled with
10–20 m of sediment, probably ejecta (Fig. 7), providing an estimate of the
minimum average depth of regolith on this satellite.

It is possible that the average depth of regolith on both satellites is
~100–200 m. This value is about one order of magnitude larger than
predicted for asteroids of a similar size by the calculations of Housen *et al.*
(1979; see their chapter). Whether this difference can be ascribed to the fact
that the two satellites exist in the gravitational well of Mars, or whether it
points to a deficiency in the regolith modeling codes is not clear at the
present time.

There is evidence that on both satellites the regolith is thinnest near local highs, ridges and crater rims, and is thickest in the local lows. This effect is most conspicuous on Deimos, where the downslope movement of brighter albedo material is very obvious (Fig. 11; Thomas *et al.* 1979*b*).

While arguments can be advanced that Phobos and Deimos retain regoliths anomolously well due to their location in the gravity field of Mars, one should remember that at the present time it is difficult to prove this assertion rigorously or to demonstrate conclusively that asteroids of a similar size in the asteroid belt do not have well-developed regoliths. Such regoliths will be stratigraphically complicated due to both the original superposition of ejecta and to subsequent modification by downslope movement.

1.5 Downslope Movement on Phobos and Deimos

Even though surface gravity on Deimos is only 10^{-3} g, there is abundant evidence of downslope movement. Thomas *et al.* (1979*b*) showed that both the brighter albedo material in the streamers (Fig. 11) and the darker crater fill (Fig. 4*b*) are subject to such movement, the precise mechanism of which is unclear. Is it initiated by thermal creep, seismic vibrations associated with impacts, or by more esoteric processes such as electrostatic leviation?

Evidence of downslope movement on Phobos is much rarer, partly because the two materials which are most subject to it do not occur on the inner satellite (Sec. 2); however, a few examples of what appears to be slumping of crater walls can be found (Fig. 9). Veverka and Duxbury (1977) argued that since the density of small craters on the floors and inner walls of large craters such as Roche (Fig. 6*b*) is as high as in the surrounding areas, slumping of crater walls on Phobos cannot be effective in general. A similar argument can be made in terms of the preservation of grooves in such craters.

Gravity slumping of features as firm as crater walls is not effective on bodies as small as Phobos and Deimos, but gravity can displace deposits of very loose surface material downslope. Why such materials are abundant on Deimos but not on Phobos is a problem which is discussed, but not resolved, in Sec. 3.

1.6 Evidence of Internal Processes

Active internal process on objects as small as Phobos and Deimos should not be expected. The only surface features on either object that may have been modified partially by quasi-internal processes are the grooves on Phobos. Thomas *et al.* (1979*a*) and Thomas and Veverka (1979) argue that the grooves owe their origin to the nearly catastrophic impact which produced Stickney. They suggest that this large impact not only fractured Phobos (at least near its surface) but that enough local heating ensued to release steam along the fractures. This steam would have been derived from the dehydration of the low-density carbonaceous material of which Phobos is believed to consist. The suggestion is that this steam would eject regolith overlying the fractures and produce the characteristic morphology of the Phobos grooves including a beaded or pitted appearance and possible raised rims along some segments.

Thomas and Veverka (1979) show that the largest crater on Deimos, 2.3 km in diameter, was not sufficient to fracture Deimos in a similar manner, thus accounting for the absence of grooves on the outer satellite. However, they suggest that grooves can be common on the surfaces of a significant fraction of smaller asteroids.

II. ALBEDO MARKINGS

Based on available disk-integrated data there is a tendency to consider the surfaces of most asteroids as uniformly dull in terms of color and albedo variations. However, it should be remembered that such observations can reveal only the grossest variations such as those on the Galilean satellites (Veverka 1977). More subtle, and more uniformly distributed variations can be disguised by the hemispheric averaging implicit in such observations. Phobos, and especially Deimos, provide good examples of this sort of information loss. While disk-integrated photometry of both satellites is consistent with the concept of uniformly dull surfaces (Noland and Veverka 1976b), when the disks are resolved the actual situation is much more interesting. True, both objects are uniformly dull in terms of color variations (at least at levels exceeding variations of 5 %), but they are quite interesting in terms of brightness and albedo variations; this remark applies especially to Deimos.

For convenience, albedo and brightness variations on the two bodies can be divided into five categories: (1) brighter albedo markings on Deimos; (2) bright crater- and groove-rims on Phobos; (3) dark halo craters on Phobos and Deimos; (4) dark patches (outside of craters) on Phobos; and (5) dark markings on crater floors.

2.1 Brighter Albedo Markings on Deimos

Brighter albedo markings occur extensively on Deimos as diffuse patches and as long, tapered streamers which trend downhill (Thomas et al. 1979b). They are prominent at all phase angles (Fig. 11) and have been known since Mariner 9 in 1971-2. Noland and Veverka (1977a) analyzed the Mariner 9 photometric data and concluded that these features appear brighter because their albedo is about 30 % higher than that of the background. Their phase function, for phase angles $20° \leqslant \alpha \leqslant 80°$, was found to be similar to that of the background material, suggesting a similar texture. Recent Viking data indicate that at $\alpha < 20°$ the phase functions may begin to differ and that the bright material may have a steeper opposition effect (phase functions are explained in the chapter by Bowell and Lumme).

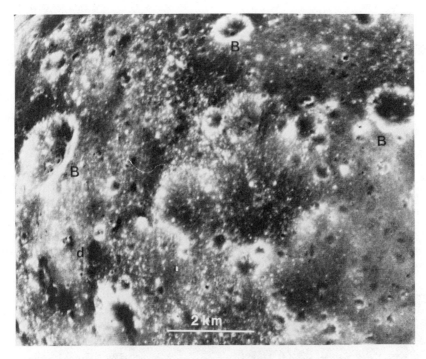

Fig. 12. A low phase angle view of Phobos showing bright crater rims (B) and low-albedo patches (d). (Viking Image 250A14.)

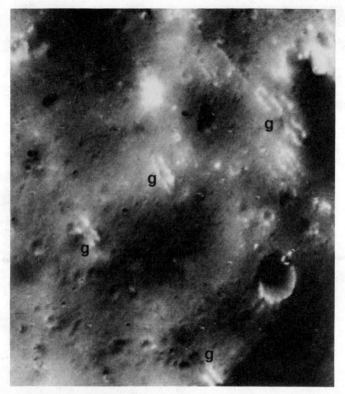

Fig. 13. A low phase angle view of Phobos showing bright groove rims (g) where the grooves cross local highs such as crater rims and ridges. (Viking Image 250A15.)

The brighter markings have an absolute geometric albedo of ~ 0.08 in visible light, compared to an average value of 0.06 for Deimos as a whole. Comparable bright patches and streamers do not occur on Phobos (Sec. 3).

2.2 Bright Crater- and Groove-Rims on Phobos

At low phase angles the rims of many craters on Phobos appear some 10–20 % brighter than their surroundings (Fig. 12). This phenomenon is almost certainly a texture effect since it disappears at larger phase angles (Fig. 6).

There is a marked tendency for these bright rims to be most conspicuous along ridges, ramparts of large craters, and other topographic highs. A closely related phenomenon is shown in Fig. 13. At low phase angles the sides of grooves become highlighted in a manner analogous to that of the crater rims. This enhancement is restricted to those groove segments which cut across local topographic highs, such as ridges and the rims of large craters.

2.3 Dark Halo Craters on Phobos and Deimos

Halo craters occur on both satellites and have already been discussed in Sec. 1.3 (Fig. 10). Distinct dark halos are seen in association with small craters $D \leqslant 50-100$ m, but more extensive and diffuse dark areas, some associated with craters, are seen in low phase angle images of Phobos.

2.4 Dark Patches (Outside of Craters) on Phobos

Low phase angle images of Phobos show many dark patches, only some of which have associated craters within them (Fig. 12). These markings disappear at larger phase angles and must represent local differences in opposition effect, and hence in surface texture.

2.5 Dark Markings on Crater Floors

The photometric properties of these markings (Fig. 14b) which are conspicuous all over Phobos in images taken at large phase angles ($\geqslant 70°$), have been studied by Goguen *et al.* (1978). At small phase angles these markings are only slightly darker than their surroundings but at large phase angles ($\sim 90°$) their contrast may reach 100 %. This strong variation of contrast with phase angle suggests that the markings appear dark at large phase angles because of a rougher texture which produces more shadows. The observed phase dependence can be fitted by a simple model of a pitted surface for which the depth/diameter ratio of the pits is $\sim 1/2$. Such a rough texture is suggestive of a vesicular material and Goguen *et al.* (1978) propose that the material could be solidified impact melt. Such markings tend to be associated with the fresher, less degraded craters on Phobos and are seen wherever high resolution coverage of crater floors at high phase angles exists.

Similar markings are either absent or rare on Deimos. The only possible candidate is shown in Fig. 14a. The absence of such markings on the outer satellite can be attributed to the pervasive fill which obscures the floors of most craters (Fig. 7).

III. DIFFERENCES BETWEEN PHOBOŠ AND DEIMOS: IMPLICATIONS FOR ASTEROIDS

The Viking investigations of Phobos and Deimos prove that the surfaces of two small asteroid-like bodies can be nearly identical in terms of many disk-integrated properties, but be strikingly different in surface morphology.

Both satellites have dark, gray surfaces with nearly identical geometric albedos and closely similar spectral reflectance curves. Yet in detail, the surfaces are very different. The surface of the inner satellite is covered by sets of extensive, pitted, linear depressions, the grooves, which have no counterpart on Deimos. On the other hand, on Deimos, but not on Phobos, there are extensive patches of brighter albedo material and conspicuous fill within most

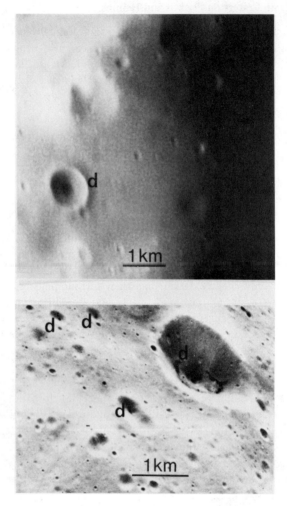

Fig. 14. Dark material in craters (d). Top: Deimos (Viking Image 391B45). Bottom: Phobos (Viking Image 248A01).

craters. These significant differences, and specifically the existence of the bright patches on Deimos, would not be discernable from disk-integrated measurements. Here we have a strong suggestion that asteroid surfaces are probably much more varied and more interesting than suggested by available disk-integrated observations.

The absence of grooves on Deimos is easy to rationalize (Thomas and Veverka 1979), but the striking difference in the amount of fine debris on the two surfaces is not. Before discussing a few suggested explanations for this remarkable difference, it is interesting to note specifically that this gross difference in surface morphology has very little effect on the phase coefficients of the two surfaces.

Noland and Veverka (1976b; 1977a,b) found that for phase angles $20° \leqslant \alpha \leqslant 80°$ the phase coefficient β of Phobos is slightly larger than that of Deimos: 0.032 mag/deg and 0.030 mag/deg, respectively; they attributed the higher value for Phobos to this satellite's rougher surface (this was *before* the presence of grooves on Phobos was realized). A more accurate comparison can be made in terms of β_i, the intrinsic phase coefficients determined from the brightness variation with phase angle of a representative patch of surface — not from the variation of the disk-integrated light. Thus the intrinsic phase coefficient, unlike the disk-integrated value, does not depend on the actual shape of the body. Values of β_i are 0.019, 0.018, 0.017 mag/deg for Phobos, the moon and Deimos, respectively (Noland and Veverka 1977a,b). Since for the purposes of this discussion the albedos of all three surfaces are similar (i.e. low) the difference in β_i reflects differences in effective surface roughness. The data suggest that in terms of shadows, the surface of Phobos is rougher than that of the moon (probably because of the grooves) while that of Deimos is smoother (probably because most craters are filled in to appreciable degrees). But note that the changes in β_i associated with these drastic changes in surface morphology are very small. It is important to realize that in interpreting the phase coefficients of irregular objects such as asteroids, or Phobos and Deimos, the intrinsic phase coefficient β_i is a more reliable parameter than the usual disk-integrated phase coefficient β which, unlike β_i, does depend on the actual shape of the object. Thus, differences in β can in part be due to differences in shape. β_i does not depend on the shape of the object, but evidently cannot be derived from disk-integrated data.

While the difference in the amount of loose material on the surfaces of Phobos and Deimos is evident, the reason why Deimos appears to retain so much more debris is not obvious.

A favorite suggestion has to do with the different location of the two satellites in the gravity field of Mars, the implication being that the proximity of Phobos to the planet somehow makes it more difficult to retain ejecta on its surface. However, three-body calculations of escape velocities at different points on Phobos and Deimos (Housen and Davis 1978; Goguen and Burns 1978) do not show any such effect. Even with help from Mars, at no point on Phobos do escape velocities become smaller than those on Deimos. If the gravity field of Mars plays a role, the effect must be more subtle.

Another possible explanation involves a marked difference in the composition of Phobos and Deimos. Ejection velocities during hypervelocity impacts depend strongly on the strength of the target material (Stöffler *et al.* 1975); if Deimos consisted of a much weaker material than Phobos, typical ejection velocities could be lower. However, the similar photometric properties of the two satellites and their nearly identical mean densities are not consistent with any gross compositional difference.

A more esoteric explanation has been mentioned by Whipple (1978, personal communication), namely that the Mars system has been accumulating interplanetary dust and that most of it is swept up by Deimos before much of it has a chance to spiral in to the orbit of Phobos. However, as Whipple himself has noted, the implied concentrations of interplanetary dust required for such an explanation are unrealistic.

Shoemaker (1978, personal communication) has stressed the stochastic aspect of cratering events and has suggested that most of the blanketing material on Deimos could be the result of a single highly unusual cratering event.

It could also be claimed that Phobos was blanketed once much like Deimos is today, but that the nearly catastrophic cratering event which produced Stickney and the grooves (Thomas *et al.* 1979a; Thomas and Veverka 1979) knocked off most of this layer — a sort of regolith spallation. This idea seems to be inconsistent with the old age of Stickney and the grooves — some 3×10^9 yr according to Thomas *et al.* (1979a). Simple calculations show that, since that time, a significant regolith should have been rebuilt, unless the rebuilding of regoliths on regolith-free bodies is less efficient than expected under conditions obtaining during the past 3×10^9 yr (Hartmann 1978). Although the question remains to be resolved, it is conceivable that Mars really is not a culprit in this drama and that two small bodies, very similar in size, can retain certain types of ejecta to remarkably different degrees and hence end up with strikingly different surface morphologies. Spacecraft exploration of the asteroid belt should resolve this and many other important questions.

Acknowledgment. This work was supported by a National Aeronautics and Space Administration grant under the Planetary Geology Program.

REFERENCES

Bowell, E.; Chapman, C. R.; Gradie, J. C.; Morrison, D.; and Zellner, B. 1978. Taxonomy of asteroids. *Icarus* 35: 313-335.

Burns, J. 1978. The dynamical evolution and origin of the Martian moons. In *Vistas in Astronomy,* eds. A. Beer and P. Beer (Oxford: Pergamon Press), Vol. 22, pp. 193-210.

Chapman, C. R. 1978. Asteroid collisions, craters, regoliths and lifetimes. In *Asteroids: An Exploration Assessment,* eds D. Morrison and W.C. Wells, NASA Conf. Publ. 2053, pp. 145-160.

Cintala, M.; Wood, C.; and Head, J. 1977. The effects of target characteristics on fresh crater morphology: Preliminary results for the moon and Mercury. *Proc. Lunar Sci. Conf. VIII* (Oxford: Pergamon Press), pp. 3409-3426.

Duxbury, T. C. 1974. Phobos: Control network analysis. *Icarus* 23: 290-299.

Gault, D. E., and Wedekind, J. 1969. The destruction of tektites by micrometeoroid impact. *J. Geophys. Res.* 74: 6780-6794.

Goguen, J.; Veverka, J.; Thomas, P.; and Duxbury, T. 1978. Phobos: Photometry and origin of dark markings on crater floors. *Geophys. Res. Lett.* 5: 981-984.

Hartmann, W. K. 1972. Interplanet variations in scale of crater morthology: Earth, Mars, Moon. *Icarus* 17: 707-713.

Hartmann, W. K. 1973. Martian cratering, 4, Mariner 9 initial analysis of cratering chronology. *J. Geophys. Res.* 78: 4096-4116.

Hartmann, W. K. 1978. Planet formation: Mechanism of early growth. *Icarus* 33: 50-61.

Head, J. W. 1976. The significance of substrate characteristics in determining morphology and morphometry of lunar craters. *Proc. Lunar Sci. Conf. VII* (Oxford: Pergamon Press), pp. 2913-2929.

Housen, K. R.; Wilkening, L. L.; Chapman, C. R.; and Greenberg, R. 1979. Asteroidal regoliths. *Icarus* (in press).

Hunten, D. M. 1978. Capture of Phobos and Deimos by protoatmospheric drag. *Icarus* 37: 113-123.

Lewis, J. 1974. The temperature gradient in the solar nebula. *Science* 186: 440-443.

Noland, M., and Veverka, J. (1976*a*). Predicted lightcurves of Phobos and Deimos. *Icarus* 28: 401-404.

Noland, M., and Veverka, J. 1976*b*. The photometric function of Phobos and Deimos. I. Disk-integrated results. *Icarus* 28: 405-414.

Noland, M., and Veverka, J. 1977*a*. The photometric functions of Phobos and Deimos. II. Surface photometry of Deimos. *Icarus* 30: 200-211.

Noland, M., and Veverka, J. 1977*b*. The photometric functions of Phobos and Deimos. III. Surface photometry of Phobos. *Icarus* 30: 212-223.

Oberbeck, V. R., and Quaide, W. L. 1967. Estimated thickness of a fragmental surface layer of Oceanus Procellarum. *J. Geophys. Res.* 72: 4697-4704.

Pang, K., and Rhoads, J. W. 1979. Deimos: Spectral evidence for a carbonaceous chondrite composition. *Science* (in press).

Pollack, J. B.; Burns, J.; and Tauber, M. E. 1978. Gas drag in primordial circumplanetary envelopes: A mechanism for satellite capture. *Icarus* 37: 157-611.

Quaide, W.L., and Oberbeck, V. R. 1968. Thickness determinations of the lunar surface layer from lunar impact craters. *J. Geophys. Res.* 73: 5247-5270.

Soter, S. 1971. The dust belts of Mars. *Center Radiophys. Space Res. Rept.* 462, Cornell University.

Stöffler, D.; Gault, D. E.; Wedekind, J.; and Polkowski, G. 1975. Experimental hypervelocity impact into quartz sand: Distribution and shock metamorphism of ejecta. *J. Geophys. Res.* 80: 4062.

Thomas, P. 1978. The morphology of Phobos and Deimos. Ph.D. dissertation, Cornell University.

Thomas, P., and Veverka, J. 1979. Grooves on asteroids: A prediction. *Icarus* (special Asteroid issue).

Thomas, P.; Veverka, J.; Bloom, A.; and Duxbury, T. C. 1979*a*. Grooves on Phobos: Their distribution, morphology and possible origin. *J. Geophys. Res.* (in press).

Thomas, P.; Veverka, J.; and Duxbury, T. C. 1979*b*. Deimos: Downslope movement at 10^{-3} g. Submitted to *Icarus*.

Veverka, J. 1977. Photometry of satellite surfaces. In *Planetary Satellites*, ed. J. Burns (Tucson: University of Arizona Press), pp. 171-209.

Veverka, J. 1978. The surfaces of Phobos and Deimos. In *Vistas in Astronomy*, eds. A. Beer and P. Beer (Oxford: Pergamon Press), Vol. 22, pp. 163-192.

Veverka, J., and Duxbury, T. 1977. Viking observations of Phobos and Deimos: Preliminary results. *J. Geophys. Res.* 82: 4213-4223.

Zellner, B. H. 1972. Minor planets and related objects. VIII. Deimos. *Astron. J.* 77: 103-185.

PART V

Composition

REFLECTANCE SPECTRA FOR 277 ASTEROIDS

CLARK R. CHAPMAN
Planetary Science Institute

and

MICHAEL J. GAFFEY
University of Hawaii

Visible and near-infrared reflectance spectra are presented for 277 asteroids. These represent the results of an 8-yr program and constitute all spectra available in 1979. There are approximately 80 recognizably different spectral types among the asteroids surveyed. Relatively few of these can be explained as mixtures of C- and S-type materials on individual asteroids, although Eos family members may be an exception.

Incident sunlight is transmitted through mineral grains on an asteroidal surface prior to being reflected toward Earth. Thereby it acquires spectral information diagnostic of surface mineralogy (see the chapter by Gaffey and McCord in this book). Bobrovnikoff (1929) was the first astronomer to consider that asteroids might not be gray reflectors of sunlight. He made microphotometric tracings of photographic spectra of 12 asteroids and deduced real differences among them. He noted especially the high ultraviolet reflectance of Pallas and the generally redder colors of most other asteroids. (He even correctly deduced the rotation period of Vesta from its color variations alone — it was nearly fifty years later before modern observers reached the same conclusion for essentially the same reasons.) Subsequent efforts at photographic colorimetry of asteroids were beset by the limitations of the technique combined and asteroidal lightcurve variations (Chapman *et al.* 1971).

Photoelectric photometry, mainly in broadband UBV filters, commenced in the mid-1950's and accelerated during the 1970's yielding colors for over 700 objects to date (see the chapter by Bowell and Lumme). In the late 1960's it became possible to obtain precision photometry at better spectral resolution and over an extended range of wavelengths using narrowband filters. Since the first narrowband reflectance spectrum for an asteroid was published by McCord *et al.* (1970), application of this technique has been one of the chief reasons for the increasing interest in asteroids. Preliminary spectra for 11 asteroids using about two dozen filters covering the range 0.3 to 1.1 μm (Chapman *et al.* 1971) already revealed important differences among asteroid spectra, including the presence or absence of near-infrared absorption features. During the next four years, continued spectral reconnaissance observations raised the total of measured spectra to 98 (Chapman *et al.* 1973*a,b*, McCord and Chapman 1975*a,b;* and Pieters *et al* 1976). The present chapter presents the results of the continuation of the narrowband spectrophotometry program.

In the late 1970's, asteroid spectrophotometry has branched in two additional directions: (a) a program of eight-color photometry has been initiated that can be applied efficiently to virtually all numbered asteroids and can reveal some of the important spectral traits noted in the two-dozen-filter spectrophotometry; (b) advances in detector sensitivity and optical instrumentation are starting to permit better exploitation of the wavelength range between 1.1 μm and the onset of thermal radiation near 4 μm; a combination of broadband, narrowband, and Fourier spectroscopic techniques has already been applied usefully to some of the brighter asteroids (see the chapter by Larson and Veeder).

So far, the largest set of asteroidal reflectance data taken with sufficient spectral resolution to be diagnostic of mineralogy remains the narrowband program of Chapman, McCord, and their associates. The first 98 asteroidal reflectance spectra, published between 1973 and 1976, led to important early insights about the physical nature of asteroids. The spectra demonstrated that asteroids have a wide variety of surface compositions. Interpretation of the spectra in terms of mineralogy, including comparisons with meteorites, led to the understanding that most asteroids are composed of the same suites of minerals as are many distinct meteorite types, including silicate-rich assemblages, metal-rich assemblages, and carbonaceous assemblages (Chapman and Salisbury 1973; Johnson and Fanale 1973; Gaffey and McCord 1978). Assemblages similar to ordinary chondrites were found on Earth-approaching asteroids 433 Eros and 1685 Toro, but were not found on main-belt asteroids. Together with albedo-sensitive data, asteroidal spectrophotometric parameters led to the widely used *C-S-M* taxonomy introduced by Chapman *et al.* (1975), and extended by Bowell *et al.* (1978). Bias-corrected data led to the recognition of the predominance of *C*-type asteroids in the main belt and the decreasing proportion of *S*-type asteroids with semimajor axis (see Zellner's chapter).

During the past five years, continued reconnaissance spectrophotometry has been obtained at Kitt Peak National Observatory, Mauna Kea Observatory and Lowell Observatory. Although some of these data have been reported in a preliminary fashion, we have postponed final publication until completion of a recalibration of our standard stars. This chapter constitutes the first publication of the newly reduced spectra, including re-averages and recalibrations of published spectra. 277 asteroidal spectra are presented; these have been entered in the TRIAD data file given in Part VII. Another file in TRIAD contains spectrophotometric parameters derived from the spectra; the parameters are defined in the introduction to the spectral parameter file.

I. OBSERVING PROGRAM AND DATA REDUCTION

Asteroid brightnesses are measured in approximately 25 filters, usually spanning the wavelength range 0.32 to 1.08 μm, using the McCord double-beam photometer. Several different photomultipliers are used. The most useful detector has been a high-quantum-efficiency gallium-indium-arsenide detector, sensitive from the ultraviolet through about 1.0 μm. When that detector is not available, observations are often obtained with two detectors: S-20 for 0.32 to 0.83 μm and S-1 for > 0.8 μm. In the latter case, asteroid brightnesses sometimes vary between observations of the two wavelength ranges; in general, we attempt to use the absolute stellar calibrations to scale and attach the two parts of the spectrum, but in the event of an apparent mismatch due to lightcurve effects, we simply shift the infrared to match the values in the couple of overlapping filters.

During the observing sessions, standard stars are observed both to determine extinction corrections in each filter and as color standards. The stars, many of them *A*- or *B*-type Oke/Hayes standards (see Chapman *et al.* 1973*a*) and others of solar type, are typically 5th magnitude. In an extensive measurement program at Hawaii, Owensby *et al.* (1979) have calibrated these stars against each other and tied them to α Lyrae. We have adopted, with small changes, the α Lyrae calibration of Nygard (1975), which relies substantially on a lunar calibration against returned soils. The α Lyrae/Sun ratios employed here are given in Column 2 of Table I in the Appendix.

We note that the present calibration differs from that used for reducing previous asteroid data by amounts exceeding 5% in a few filters. The average calibration difference listed in Column 3 of Table I, has been used to recalibrate all previously published data. Compared with the previous calibration we now believe that:

1. Asteroid spectra are straighter through the visible (i.e., the BEND

parameter is reduced).

2. The 0.65 μm absorption feature previously common in spectra is present only rarely.

3. The 0.9 μm pyroxene absorption is generally deeper and centered at shorter wavelength than previously believed.

The errors in the present calibration cannot be quantitatively estimated but should be smaller than the differences between the old and new calibrations shown in Column 3 of Table I. To reflect this uncertainty the adopted error bars are restricted to values of $\geqslant 0.03$.

In reducing the data to counts/sec for the object being observed, C_{obj}, a variety of possible instrumental effects must be evaluated and corrections made when appropriate. The data reduction procedure can be defined in the form of the following equation:

$$C_{\text{obj}} = CC(C_{B1}) \times C_{B1} - B1B2 \times CC(C_{B2}) \times C_{B2} \tag{1}$$

where $CC(C_{Bn})$ is a coincidence correction and depends on the measured count rate in each channel, C_{B1}, C_{B2}. $B1B2$ (beam ratio) is the ratio of intensity of the same source measured in both channels. The measured counts in each channel are composed of several discrete contributions:

$$C_{B1} = C_{\text{obj}} + C_{\text{sky}} + C_{\text{dark}}$$

$$C_{B2} = C_{\text{sky}} + C_{\text{dark}} \, . \tag{2}$$

The final form of the equation is:

$$C_{\text{obj}} = \left[CC(C_{B1}) \times (C_{\text{obj}} + C_{\text{sky}} + C_{\text{dark}}) \right] - \left[B1B2 \times CC(C_{B2}) \times (C_{\text{sky}} + C_{\text{dark}}) \right] . \tag{3}$$

The *dark counts*, C_{dark}, are intrinsic detector noise and depend on the thermal condition of the detector (temperature and thermal stability), the light levels to which the detector was exposed previously (seconds to hours, depending on the light level), the operating voltage of the tube, and the amplifier or discriminator setting. The dark count level is checked throughout the night by counting with the detector completely shielded from light.

Coincidence effects arise as a result of a finite upper limit to the counting rate for the system, so that dual events too close together are counted as a single event. The net effect is that as a number of events C_E increases toward and/or past the counting rate limit of the system C_L, the measured counts C_M asymptotically approach C_L. For an idealized detector, the relationship should be:

$$C_M = C_L \times \left[1.0 - \exp(-C_A/C_L)\right] . \tag{4}$$

We have found a small departure from the idealized relationship in the sense that C_L can be thought of as increasing slightly with increasing count rate. The saturation level C_L is calculated for each night from observations of pairs of relevant standard stars. This value is then used to correct the measured counts for standard stars to the actual number of events, $CC(C_{Bn})$. However, for the low count levels encountered for asteroids, the effects of coincidence are always negligible.

Beam inequality arises when the optical paths, mirrors and alignment of the two beams are not identical. Orientation of the chopping and stationary mirrors and detector placement are the major factors which contribute to beam inequality. Observations which have swapped "object" and "sky" are ratioed to normal observations of the same point (e.g., star) or diffuse (e.g., bright sky) source.

In order to minimize any nonlinearity effects of detector sensitivity and due to uncertainties in the coincidence levels, it is often necessary to obscure the brighter standard stars with neutral density filters. However, since such filters are not exactly spectrally neutral over the spectral range we measure, a correction must be made for the counts of any object observed with a neutral density filter. This is done by ratioing a faint star observed with a neutral density filter to itself without such a filter. For each run this value is compared with a smooth average for that neutral density filter from previous observations.

Once correct object counts (both asteroids and stars) are obtained, the extinction coefficient for each night or portions thereof is determined using the standard stars. The effective counts of the standard star can then be calculated at the same airmass as the asteroid observations and a ratio (asteroid/star) calculated. This ratio is multiplied by a previously established calibration of this star with respect to the sun in order to produce the reflectance curve of the asteroid:

$$\frac{\text{asteroid}}{\text{sun}} = \frac{\text{asteroid}}{\text{star}} \times \frac{\text{star}}{\text{sun}} . \tag{5}$$

The standard deviation of the mean, σ, is utilized as an indicator of the general confidence in the spectra. However, since there is also a possible systematic error in the calibrations of the standard stars (Owensby *et al.* 1979), adopted error bars are never less than 0.03 even though the σ may be less. In order that the reader may estimate the quality of any particular data set, two tables are provided. Table II (see Appendix) lists all observing nights that recently have been incorporated into the spectral data sets and the parameters of the data reduction. Table III in the Appendix lists the nights on which data for each asteroid were obtained, in order of asteroid number, and the relative weights assigned to each run. Separate reductions for each night and complete documentation of the data reduction are archived at the Planetary Science Institute. In many cases these have been compared with independent reductions of the same data performed at the University of Hawaii.

In recent years, the asteroids have been selected for observation in an attempt to survey a wide variety of orbital and physical properties. The sample thus has certain biases of which users of the data should be aware. In giving priority to the numbered asteroids for observation, added weight has been given for the following traits:

(1) brighter apparent magnitude;

(2) membership in several of Williams' (1971) Hirayama families;

(3) potential sources of meteorites (primarily as listed in Tables 4 and 5 of McCord and Chapman 1975b);

(4) unusual semimajor axes (either inside or exterior to the main belt);

(5) unusual UBV colors.

Several other factors have received some weight. In addition, a fraction of the asteroids have been selected at random, provided they are brighter than the 15th to 17th limiting magnitude (depending on telescope aperture).

The asteroid spectra (see Appendix to this chapter) have all been normalized to unity at 0.57 μm, using a weighted average of reflectances in several filters in the 0.5-0.63 μm range. The spectra are arranged in order by asteroid number, using alternating symbols. The vertical tick marks are separated by 0.2 in reflectance. The abcissa is wavelength in μm. Spectral parameters calculated from these spectra are listed in the TRIAD file (see Part VII).

II. DISCUSSION OF THE NEW SPECTRA

The spectra have not been available sufficiently long for a definitive analysis, but we can describe some of the more interesting new spectra of individual asteroids or classes of asteroids. One of the most significant is that of 496 Gryphia, which is a 15- to 20-km diameter member of the Flora family near the inner edge of the belt. This spectrum more nearly resembles laboratory measurements of ordinary chondrites (most nearly types L4 or H4) than that of any other main-belt asteroid, excluding 349 Dembowska. (Dembowska has a visible spectrum similar to that of LL6 chondrites, but its 2-μm infrared spectrum has been interpreted as indicating either a high metal content [Veeder et al. 1978] or high olivine content [see the Larson and Veeder chapter] that may place it outside permissible ranges for ordinary chondrites.) It is increasingly apparent that sizable parent bodies for ordinary chondrites are rare or absent in the main asteroid belt, unless our interpretations of these spectra are grossly in error (see the chapter by Gaffey and McCord).

We have now observed a substantial sample of known asteroids beyond 3.8 AU. Although they exhibit a range of spectral types, they are generally unlike the common main-belt types. In general, spectra of these distant objects (including Hildas, Thule, Trojans, and Hidalgo) tend to be relatively reddish while at the same time being relatively bright in the ultraviolet. Thus, they have very low or negative values of the BEND parameter. The exceptionally high infrared reflectances of some of these objects (e.g., Hilda 1512 Oulu; Trojans 624 Hektor, 588 Achilles, and 911 Agamemnon; and 944 Hidalgo) are especially noteworthy since those with measured albedos are very dark. More discussion of these asteroids, based in part on these spectra, is given in the chapter by Degewij and van Houten.

Several asteroids exhibit in extreme form the unusual spectral traits of previously observed 354 Eleonora very red, straight, ultraviolet-visible slope and a sharp inflection to diminishing reflectances into the infrared. Examples include 197 Arete, 246 Asporina, and 446 Aeternitas. At the opposite extreme, 785 Zwetana — already known to have exceptional UBV colors — appears to have the flattest, most featureless spectrum of any observed asteroid.

We show here the first spectra of one important taxonomic class of asteroids namely the E-types that had been defined previously on the basis of UBV colors and albedo data (Zellner 1975). 44 Nysa and 64 Angelina both have slightly reddish, straight spectra, similar to M-type spectra, but they are distinguished as E-types based on their albedo. The third E-type listed by Bowell *et al.* (1978), is 434 Hungaria, but its newly measured spectrum is probably incompatible with the archetypal E and is now classified as U (unclassifiable). In fact Hungaria's spectrum has some similarity to that of Vesta; but there is a discrepancy between our spectrum of Hungaria and the available UBV colors, so further measurements are required.

Figure 1 illustrates average spectral curves for asteroids typed as C, S, and M in TRIAD (Part VII of this book). The averages exclude asteroids for which an unambiguous type is not available and also exclude points having error bars in excess of ± 0.08; the spectra have not been weighted. Given the diversity of spectra within the C, S, and M groups, one might expect filter-to-filter deviations in the average spectra to reveal calibration errors; by this criterion, such filter-to-filter systematic errors are quite small. The small absorption feature evident in the M-type average spectrum is possibly an artifact of the small number of known M's for which infrared data have been obtained.

A preliminary attempt has been made to group the 277 spectra into significantly different spectral groups in order to assess the variety of spectral types and the number of distinct mineralogical assemblages that are revealed within the errors of the technique. This is an attempt to update the 34 groupings of McCord and Chapman (1975a,b) analyzed by Chapman (1976). As a first step, a computer program was written to identify all asteroids

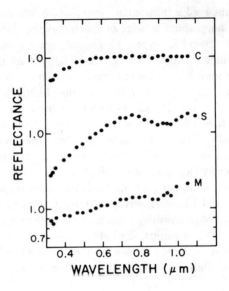

Fig. 1   Average spectra for asteroids typed as *S, C,* and *M.*

having spectral parameters (from the TRIAD file) falling within a modest range of those for each other asteroid; crude limits were also placed on albedo and UBV colors.

Using these results as a guide, one of us (C.R.C.) obtained the groups reported in Table IV (see Appendix) by overlaying the various similar spectra on each other and making comparisons. In this preliminary sorting it is not certain that all of the groups are statistically different from each other, but most are, implying that the spectrophotometric program has recognized about 80 different assemblages among the 255 asteroids for which the error bars are sufficiently low to permit meaningful comparisons. (Many of the separate groups reported for incomplete *S* types are consistent with one or more of the complete *S*-type spectra, as shown in parentheses, but they cannot be related uniquely since the infrared data are missing.) The assignment of an *individual* asteroid to a group is often ambiguous, either because its error bars may be relatively large or because its spectrum lies near an arbitrary boundary between two or more spectral groups that in reality are part of a continuum. Each group is numbered according to the lowest numbered asteroid in the group; that asteroid is not always the most representative member.

Figure 2 illustrates the average spectra for the groups containing more than a single member. Generally, any interpretation of mineralogical assemblages from the spectrum of one asteroid (as reported in the chapter by Gaffey and McCord) should be applicable to any other asteroid in the same group. Asteroids in different groups should have surfaces composed of

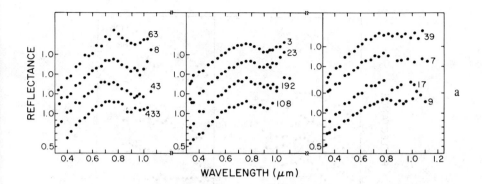

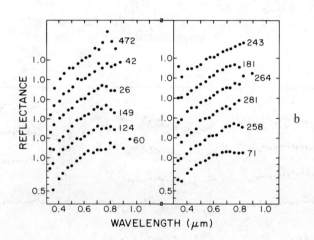

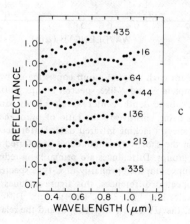

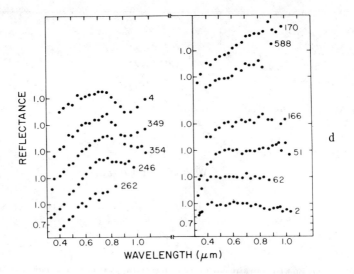

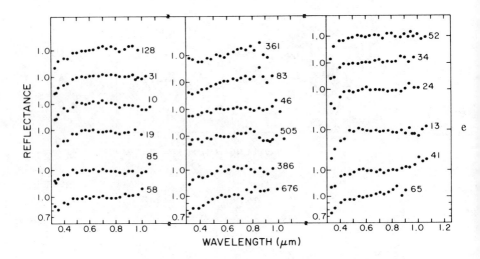

Fig. 2. Average spectra for each spectral group consisting of two or more asteroids are plotted. Fig. 2a: Complete S and S-like spectral groups. Differences are primarily in overall redness, BEND, band depth, center wavelength of band, and band width (the latter two parameters being sensitive to the ratio of pyroxene to olivine). Fig. 2b: S- and S-like groups for spectra lacking infrared data. Differences are primarily in overall redness, BEND, and in the short-wavelength edges of infrared absorption features. Fig. 2c: M-like and E-like groups. Differences are primarily in redness and departures from linearity. Fig. 2d: Groups composed primarily of U-type spectra or of R-like spectra. In addition to obvious spectral differences, this figure includes groups characterized by both low and high albedos. Fig. 2e: C-like groups. Differences are primarily in ultraviolet reflectance, the UV elbow wavelength, and the relative infrared reflectance.

different surfaces materials, or at least slightly different proportions of the same materials, or with different physical properties (e.g., grain size distributions).

Some of the variety of spectra may be revealed by sorting them according to two spectral parameters; an example is Fig. 3. Here, small versions of 274 spectra are plotted according to R/B and BEND. Spectra near the periphery of this plot are labeled by asteroid number; they are among the more unusual spectra.

A question of general interest is the degree to which some (or even many) of the spectra may in fact represent averages of one (or more) highly disparate types. The most commonly hypothesized mixture is of C and S types, as might result from a low-velocity collision involving two disparate asteroids or even from observational averaging of a double asteroid composed of a C and an S. Figure 4 shows a range of mixtures of $C:S$ from 1:6 to 6:1. For a typical $C:S$ albedo ratio of 1:4, this range corresponds to an area-weighted mixture of $C:S$ from 2:3 to 24:1. (It should be noted that if C and S material were mixed on a microscopic scale, photons reflected from the two types of material might have interacted with both, resulting in nonlinearities not represented by the simple mixtures in Fig. 4.) A qualitative comparison of the spectral mixtures with the most populous spectral groups shown in Fig. 3 reveals that such mixtures are probably uncommon in the asteroid belt.

Of particular interest is whether the large relatively homogeneous Hirayama families with UBV colors roughly intermediate between C and S (especially the Eos family; see the chapter by Gradie et $al.$) could be composed of individual asteroids having macroscopic mixtures of C and S material. A fairly close spectral match can, in fact, be obtained by averaging 31-group spectra (typical albedo 0.04) with 7-group spectra (typical albedo 0.16) in the ratio of 4:5, corresponding to an areal average of 16:5. This spectral mixture is compared with the average spectrum for 7 Eos family members in Fig. 5. A possible slight discrepancy arises from the fact that the spectral mixture would have an albedo of about 0.07, which compares with the typical Eos family albedos of 0.09.

III. CONCLUSIONS

The 277 asteroid spectra shown here represent most of the results of eight years of reconnaissance spectrophotometry of asteroids. (Data remain to be reduced for observing runs in July/August 1974 and May 1975.) Our asteroid spectrophotometric program is progressing now at a more modest pace and observational targets are no longer being selected in a reconnaissance mode; a new 8-filter observing program being conducted by other observers will fulfill this function for the remaining numbered asteroids. Instead, we are concentrating on asteroids known from other data to be unusual or those

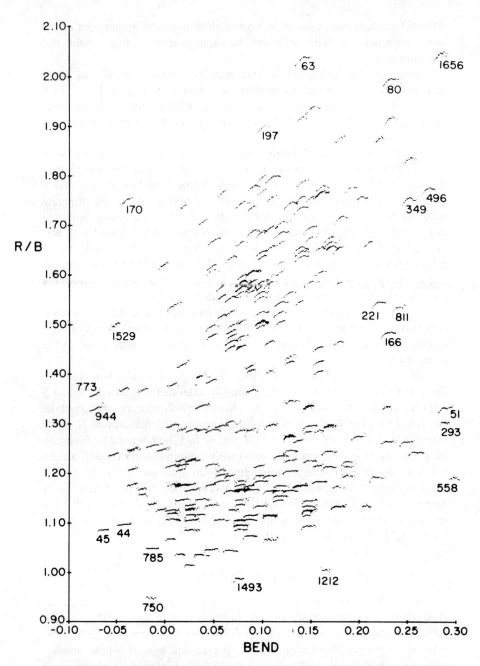

Fig. 3. Spectra are plotted as a function of spectral parameters R/B and BEND. The vertical and horizontal scales of the spectra are in the same proportion as in Figs. 1 and 2. Three spectra are outside of the bounds of the plot: 246 and 446 are plotted off the top of the graph and 1512 off the left.

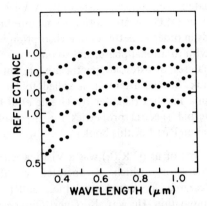

Fig. 4. The four spectra are averages (not weighted for albedo) of *C*- and *S*-type spectra in the following proportions, top to bottom: 6 to 1, 6 to 3, 3 to 4, and 1 to 6. Spectra of 7 to 9 actual asteroids were used rather than the averages of all *C* and *S* asteroids; but the asteroids selected have individual spectra very similar to the type averages shown in Fig. 1.

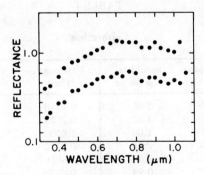

Fig. 5. The upper curve is an average spectrum for the seven Eos family members (as defined in the chapter by Gradie *et al.*) having good quality spectra. The spectrum below it, down-shifted by 0.3, is an average of five *S*-type spectra (for asteroids 7, 15, and 68, with the first two given double weight) and four *C*-type spectra (34, 141, 145, and 511). The *S* group used is interpeted as being olivine-rich. To within the appropriate errors, the *C* + *S* average is indistinguishable from the average Eos family spectrum.

dynamically associated with unusual asteroids. A new observing program initiated by one of us (M.G.) is attempting to acquire higher precision data for specific asteroids in order to better define absorption band characteristics, spectral variations with rotational phase, and so on.

Preliminary analyses of asteroid spectra are presented in several other chapters in this book. Special attention is called to the analysis of spectra grouped by Hirayama families (chapter by Gradie *et al.*), and discussion of implications of asteroid spectral properties for the evolution of asteroids (chapter by Chapman in Part I of this book).

Acknowledgments. One of us (C.R.C.) was a Visiting Astronomer at the Kitt Peak National Observatory which is operated by the Association of Universities for Research in Astronomy, Inc., under contract with the National Science Foundation. He was also Guest Observer at the Mauna Kea Observatory, and wishes to thank the Institute for Astronomy at the University of Hawaii for the allocation of observing time at that observatory. One observing run was conducted jointly with E. Bowell at the Lowell Observatory. T. McCord made available his spectrophotometer and associated electronics for this program. Substantial help in preparing for the observing runs and acquisition of the data was provided by P. Owensby, M. Rognstad, M. Brookes, and J. Gettys. Other observers included J. Bosel, D. Davis, W. Ezell, A. Goldberg, R. Greenberg, and J. Wacker. This research is supported by the NASA Planetary Astronomy Program.

APPENDIX

TABLE I

Calibration

Wavelength (μm)	Adopted α Lyr/Sun	Recalibration factor	Wavelength (μm)	Adopted α Lyr/Sun	Recalibration factor
0.3345	2.260	1.00	0.7340	0.632	1.02
0.3400	– –	1.00	0.7645	0.597	1.02
0.3530	1.936	1.00	0.8000	0.559	1.00
0.3780	2.308	0.96	0.8310	0.527	0.99
0.4040	2.722	1.05	0.8660	0.517	0.93
0.4300	2.071	0.99	0.8990	0.515	0.95
0.4675	1.564	1.02	0.9330	0.513	1.02
0.5020	1.321	1.01	0.9500	– –	1.00
0.5365	1.138	1.00	0.9700	0.506	0.97
0.5675	1.000	1.00	1.0010	0.471	0.98
0.6040	0.869	1.02	1.0330	0.476	0.99
0.6310	0.807	1.06	1.0640	0.460	1.04
0.6680	0.714	1.05	1.0990	0.416	1.00
0.7010	0.694	1.05			

TABLE II

Observing Nights

UT Date	Telescope	Detector	Dark Count per sec	$B2/B1$ Beam Inequality	Quality of sky and extinction[b]	Standard Star
2/12/75	KPNO 50″	GA	40.0	0.87	E	θ Crt
2/13/75	KPNO 50″	GA	27.5	1.00	F	η Hya
2/13/75	KPNO 50″	GA	27.5	1.00	F	θ Crt
2/13/75	KPNO 50″	S–1	30.0	1.00	F	θ Crt
2/13/75	KPNO 50″	S–1	30.0	1.00	F	η Hya
8/11/75	MKO 88″	S–20	7.0	0.92	E	58 Aql
8/11/75	MKO 88″	S–20	7.0	0.92	VG	ξ² Cet
8/12/75	MKO 88″	S–20	6.0-7.5	0.92	E	58 Aql
8/12/75	MKO 88″	S–20	6.0-7.5	0.92	VG	ξ² Cet
8/13/75	MKO 88″	S–1	1.0-2.5	0.93	G	58 Aql
8/13/75	MKO 88″	S–1	0.38	0.93	F	θ Vir
8/13/75	MKO 88″	S–20	5.6-7.0	0.93	F	ξ² Cet
8/14/75	MKO 88″	S–20	6.0	0.94	E	58 Aql
8/14/75	MKO 88″	S–20	6.0	0.94	E	ξ² Cet
8/15/75	MKO 88″	S–20	6.0	0.93	G	58 Aql
8/15/75	MKO 88″	S–1	1.0-1.6	0.93	G	ξ² Cet
10/10/75	KPNO 50″	GA	0.7-1.4	0.75	G	ξ² Cet
10/11/75	KPNO 50″	GA	0.4	0.85	E	29 Psc
10/11/75	KPNO 50″	S–1	2.4	0.89	G	29 Psc
10/12/75	KPNO 50″	GA	1.2	1.08	G	ξ² Cet
10/12/75	KPNO 50″	S–1	0.0	1.08	G	29 Psc
10/13/75	KPNO 50″	GA	0.35	1.01	G	29 Psc

TABLE II (Continued)

Observing Nights

UT Date	Telescope	Detector	Dark Count per sec	$B2/B1$ Beam Inequality	Quality of sky and extinction[b]	Standard Star
10/13/75	KPNO 50"	GA	0.35	1.02	VG	ξ^2 Cet
10/13/75	KPNO 50"	S-1	5.00	1.01	G	ξ^2 Cet
10/13/75	KPNO 50"	S-1	3.00	1.01	G	29 Psc
5/7/76	KPNO 36"	S-20	50.0-70.0	0.97	VG	θ Vir
5/7/76	KPNO 36"	S-20	70.0-80.0	0.97	VG	58 Aql
5/7/76	KPNO 36"	S-1	0.4-0.6	0.97	F	θ Vir
5/7/76	KPNO 36"	S-1	0.3-0.5	0.97	F	58 Aql
5/22/76	LO 72"	GA	10.0-4.0	0.97	G	58 Aql
5/23/76	LO 72"	GA	4.5	0.99	E	58 Aql
5/24/76	LO 72"	GA	0.0	1.00	G	58 Aql
9/16/76	KPNO 84"	GA	3.3	1.09	G	29 Psc
9/17/76	KPNO 84"	GA	3.0-4.0	1.02	G	29 Psc
9/17/76	KPNO 84"	GA	2.6-2.9	1.02	G	10 Tau
9/18/76	KPNO 84"	GA	2.8-4.5	C 92-0.95	E	29 Psc
9/18/76	KPNO 84"	GA	2.6-3.2	0.91	G	10 Tau
9/19/76	KPNO 84"	GA	3.2-5.4	0.96	E	29 Psc
9/19/76	KPNO 84"	GA	2.9-3.8	0.96	G	10 Tau
9/28/76	KPNO 84"	S-20	0.0	1.00	G	10 Tau
11/16/76	KPNO 50"	S-1	1.0-11.0	0.97	G	10 Tau
11/17/76	KPNO 50"	GA	0.0	1.00	E	10 Tau
11/17/76	KPNO 50"	S-1	0.0	1.00	E	10 Tau
11/18/76	KPNO 50"	GA	0.0	1.00	E	10 Tau

TABLE II (Continued)

Observing Nights

UT Date	Telescope	Detector	Dark Count per sec	B2/B1 Beam Inequality	Quality of sky and extinction[b]	Standard Star
11/18/76	KPNO 50"	S–1	0.0	1.00	E	10 Tau
6/9/77	KPNO 50"	GA	6.5-8.0	1.05	E	58 Aql
6/10/77	KPNO 50"	GA	4.0-9.5	1.05	E	58 Aql
6/10/77	KPNO 50"	GA	8.0-9.5	1.06	VG	θ Vir
6/11/77	KPNO 50"	GA	3.2-4.0	1.06	E	58 Aql
6/12/77	KPNO 50"	GA	5.0-5.15	1.055	E	θ Vir
6/12/77	KPNO 50"	GA	3.5-5.15	1.055	G	58 Aql
6/13/77	KPNO 50"	GA	5.2-6.4	1.06	E	θ Vir
6/13/77	KPNO 50"	GA	4.4-6.0	1.06	G	58 Aql
6/14/77	KPNO 50"	GA	5.7-6.1	1.05	VG	θ Vir
6/14/77	KPNO 50"	GA	5.3-5.7	1.05	E	58 Aql
6/15/77	KPNO 50"	GA	5.6	1.02	G	θ Vir
6/15/77	KPNO 50"	S–1	2.7-3.5	1.05	G	θ Vir
6/15/77	KPNO 50"	S–1	2.7-2.9	1.05	E	58 Aql
6/15/77	KPNO 50"	GA	4.7-6.7	1.05	E	58 Aql
6/16/77	KPNO 50"	GA	5.6	1.05	E	θ Vir
6/16/77	KPNO 50"	GA	4.8-5.6	1.05	E	58 Aql
6/16/77	KPNO 50"	S–1	2.6-3.8	1.05	VG	58 Aql
11/5/77	MKO 88"	GA[a]	0.4-0.7	0.99	E	10 Tau
11/6/77	MKO 88"	GA[a]	0.36-0.46	0.99	VG	10 Tau
11/7/77	MKO 88"	GA[a]	0.4-0.55	0.98	E	10 Tau
4/3/78	KPNO 84"	GA[a]	0.7	0.88	G	θ Vir

TABLE II (Continued)

Observing Nights

UT Date	Telescope	Detector	Dark Count per sec	B2/B1 Beam Inequality	Quality of sky and extinction[b]	Standard Star
4/4/78	KPNO 84″	GA[a]	0.55-0.6	0.90	F	ε Hya
4/4/78	KPNO 84″	GA[a]	0.6-0.7	0.90	E	θ Vir
4/5/78	KPNO 84″	GA[a]	0.65	0.90	E	θ Vir
4/5/78	KPNO 84″	GA[a]	0.7-0.75	0.90	G	ε Hya
4/6/78	KPNO 84″	GA[a]	0.55	0.89	VG	ε Hya
4/6/78	KPNO 84″	GA[a]	0.55-0.58	0.89	VG	θ Vir

[a] A 1.0 neutral density filter was used for standard star calibration.

[b] E = excellent; VG = very good; G = good; F = fair.

TABLE III

Asteroid Averaging Summary

Asteroid No.	Run[a]	Weight	Asteroid No.	Run[a]	Weight
1	10/13/75	0.20	24	R	
	6/15/77	0.50	25	R	
	R	1.00	26	R	
	6/12/77	0.50	27	R	
	(10/13/75)[b]	0.30	28	R	
2	R	1.00	29	R	
	8/15/75	0.30	30	R	
3	R		31	R	
4	R		32	R	
5	R		34	R	
6	R	1.00	36	2/13/75	0.30
	2/13/75	0.05	37	8/12/75	1.00
	(2/13/75)	0.05		(8/15/75)[b]	0.30
7	R		39	6/10/77	0.70
8	(5/7/76)	0.10		6/10/77	0.30
	R	1.00		R	1.00
	5/7/76	0.30	40	R	
9	R		41	5/7/76	0.50
10	R			(5/7/76)	0.15
11	R			(5/7/76)	0.15
12	10/11/75	1.00		5/7/76	0.50
	R	1.00	42	8/13/75	1.00
	10/10/75	1.20		10/10/75	0.80
	(10/11/75)[b]	0.20		(10/10/75)[b]	0.10
13	R		43	11/17/76	1.00
14	5/7/76	0.50		8/13/75	0.60
	(5/7/76)[b]	0.20		11/16/76	0.10
	R	1.00		(11/16/76)[b]	0.20
15	R		44	10/11/75	0.70
16	R	1.00		8/11/75	1.00
	10/10/75	0.40		(8/15/75)[b]	0.40
17	R			(10/11/75)[b]	0.10
18	R		45	8/11/75	1.00
19	R			10/12/75	0.30
20	11/17/76	1.00		(10/12/75)	0.05
	(11/17/76)[b]	0.40	46	6/9/77	1.00
	11/17/76	0.20		(6/15/77)[b]	0.40
	8/12/75	1.00		6/15/77	0.70
	8/12/75	0.50		6/12/77	0.80
	10/13/75	0.10	48	R	
	(10/13/75)	0.20	51	R	
21	R		52	10/11/75	0.20
22	9/18/76	1.00		R	1.00
	R	0.60		(10/11/75)	0.05
23	R		53	R	

TABLE III (Continued)

Asteroid Averaging Summary

Asteroid No.	Run[a]	Weight	Asteroid No.	Run[a]	Weight
54	10/10/75	1.00		6/15/77	0.50
58	R		106	11/17/76	1.00
60	R			11/16/76	0.05
62	9/18/76	1.00		(11/16/76)	0.20
	9/16/76	0.50	108	R	
63	(5/7/76)[b]	0.05	110	11/17/76	0.80
	R	1.00		8/13/75	0.40
	5/7/76	0.30		9/17/76	0.70
64	5/24/76	1.00		11/16/76	0.30
	2/13/75	0.50		(11/16/76)[b]	0.40
	2/13/75	0.10	113	10/11/75	1.00
	(2/13/75)	0.40	115	R	
65	5/7/76	1.00	116	2/13/75	0.50
	(5/7/76)	0.10		2/13/75	0.10
66	10/10/75	1.00		(2/13/75)[b]	0.30
68	R		119	R	
69	11/5/77	1.00	121	9/16/76	1.00
	R	0.50	122	9/16/76	1.00
71	8/14/75	1.00	124	10/11/75	1.00
78	11/17/76	1.00	128	11/5/77	1.00
	11/16/76	0.05	129	5/7/76	1.00
	(11/16/76)[b]	0.20		(5/7/76)[b]	0.10
79	R		130	R	0.50
80	R			8/15/75	1.00
82	R		136	4/5/78	1.00
83	9/16/76	1.00	139	R	
84	R		140	R	
85	R		141	R	
87	6/14/77	1.00	144	8/12/75	1.00
	6/16/77	1.00		8/13/75	0.50
	(6/16/77)[b]	0.20	145	R	
88	R		149	6/13/77	1.00
89	R		150	6/9/77	1.00
90	R			6/16/77	0.70
92	8/11/75	1.00		(6/16/77)[b]	0.30
	9/17/76	0.80	156	(6/15/77)[b]	0.30
93	R			6/15/77	1.00
94	6/10/77	1.00		6/12/77	1.00
97	R	0.60	158	9/19/76	1.00
	8/15/75	0.80	163	R	1.00
	(10/13/75)	0.05		9/17/76	1.00
	10/13/75	0.40	164	10/13/75	0.03
105	6/13/77	0.05		10/10/75	1.00
	6/13/77	1.00	166	R	
	(6/15/77)	0.50	167	5/23/76	1.00

TABLE III (Continued)

Asteroid Averaging Summary

Asteroid No.	Run[a]	Weight	Asteroid No.	Run[a]	Weight
169	6/12/77	1.00	326	R	
170	R		335	R	
175	11/18/76	1.00	337	R	
	(11/18/76)[b]	0.15	338	9/18/76	1.00
176	R	1.00	339	11/18/76	1.00
	6/13/77	0.40	340	6/14/77	1.00
181	R		341	11/6/77	1.00
185	8/14/75	1.00	344	11/17/76	1.00
192	R			(11/17/76)[b]	0.10
194	R		345	5/22/76	1.00
196	R		347	4/4/78	1.00
197	4/6/78	1.00	349	R	
198	9/17/76	1.00	354	R	
200	R		356	R	
208	2/13/75	1.00	361	9/28/76	1.00
210	R		363	5/24/76	1.00
213	R		365	10/13/75	1.00
216	4/5/78	1.00	372	6/16/77	1.00
	8/14/75	1.00	374	9/17/76	1.00
217	6/14/77	1.00	375	9/16/76	1.00
	(6/15/77)[b]	0.20	386	8/14/75	1.00
	6/15/77	1.00		8/15/75	0.50
220	8/11/75	1.00	389	9/18/76	1.00
221	R		391	8/15/75	1.00
230	R		402	R	
236	4/6/78	1.00	403	9/18/76	1.00
243	8/15/75	1.00	409	R	
246	4/3/78	1.00	413	6/14/77	1.00
258	9/18/76	1.00	415	2/13/75	0.20
262	4/4/78	1.00		(2/13/75)	0.40
264	10/13/75	1.00	416	9/19/76	1.00
268	9/19/76	1.00	419	8/13/75	1.00
279	8/11/75	0.60	423	9/17/76	1.00
	8/15/75	0.40	426	8/14/75	1.00
281	8/15/75	1.00	433	R	
293	6/11/77	1.00	434	5/23/76	1.00
308	10/11/75	1.00	435	10/13/75	1.00
313	11/17/76	0.90	439	8/12/75	1.00
	11/16/76	0.05	441	6/11/77	1.00
	(11/16/76)[b]	0.10	446	R	
323	8/15/75	1.00	453	8/12/75	1.00
324	R	1.00	462	R	
	2/13/75	0.10	468	11/6/77	1.00
	(2/13/75)[b]	0.05	471	9/18/76	1.00
325	4/6/78	1.00	472	10/13/75	1.00

TABLE III (Continued)

Asteroid Averaging Summary

Asteroid No.	Run[a]	Weight	Asteroid No.	Run[a]	Weight
481	R		750	11/18/76	1.00
488	11/7/77	1.00	758	11/6/77	1.00
490	9/18/76	1.00	760	4/3/78	1.00
496	11/5/77	1.00	770	4/5/78	1.00
505	R		772	10/11/75	1.00
510	8/12/75	1.00	773	8/13/75	1.00
511	R	1.00	781	4/5/78	1.00
	2/13/75	0.20	782	9/19/76	1.00
513	9/28/76	1.00		9/17/76	0.30
526	6/13/77	1.00	783	4/6/78	1.00
532	R		785	5/23/76	1.00
554	R		790	6/14/77	1.00
558	5/23/76	1.00	801	8/14/75	1.00
560	8/14/75	1.00	811	4/4/78	1.00
562	4/4/78	1.00	839	8/13/75	1.00
563	R		846	11/7/77	1.00
574	9/19/76	1.00	858	6/10/77	1.00
579	6/10/77	1.00	884	8/11/75	1.00
582	8/15/75	1.00	887	11/5/77	1.00
584	R			11/5/77	0.90
588	4/6/78	0.05		4/6/78	0.30
	4/6/78	0.05		R	1.00
599	9/17/76	1.00	895	5/24/76	1.00
613	11/18/76	1.00	909	11/6/77	1.00
617	11/7/77	1.00	911	R	
624	4/5/78	1.00	925	4/6/78	1.00
	R	0.40	944	11/17/76	1.00
628	4/4/78	1.00		(11/17/76)[b]	0.10
639	8/14/75	1.00		9/18/76	0.60
648	10/12/75	1.00		9/28/76	0.60
654	R		969	10/10/75	1.00
660	10/13/75	1.00	976	9/19/76	1.00
674	R		1015	6/12/77	1.00
676	8/14/75	1.00	1019	8/12/75	1.00
695	6/11/77	1.00	1025	4/3/78	1.00
	6/16/77	0.40	1036	5/23/76	1.00
	(6/16/77)[b]	0.10		4/5/78	0.20
696	11/7/77	1.00		4/6/78	0.10
704	R		1055	11/7/77	1.00
712	11/17/76	1.00	1058	6/9/77	1.00
	(11/17/76)	0.50	1075	6/10/77	0.70
714	R			6/10/77	0.30
739	R		1088	4/3/78	1.00
741	6/11/77	1.00	1103	4/6/78	1.00
747	2/13/75	0.20	1162	4/6/78	1.00
	(2/13/75)[b]	0.20	1172	8/11/75	1.00

TABLE III (Continued)

Asteroid Averaging Summary

Asteroid No.	Run[a]	Weight	Asteroid No.	Run[a]	Weight
1173	8/14/75	1.00		5/23/76	1.00
1199	6/12/77	0.50		5/23/76	1.00
	6/12/77	0.50		5/23/76	1.00
1208	11/7/77	1.00		5/23/76	1.50
1212	8/15/75	1.00	1595	6/13/77	1.00
1263	6/16/77	1.00	1620	9/18/76	0.90
1284	9/19/76	1.00		9/28/76	1.00
1317	10/11/75	1.00	1636	8/14/75	0.01
1330	6/16/77	1.00		8/14/75	1.00
1364	11/7/77	1.00	1645	9/28/76	1.00
1449	4/6/78	1.00	1650	10/12/75	1.00
1493	9/18/76	1.00	1656	4/4/78	1.00
1512	4/4/78	1.00	1685	R	
1529	11/17/76	1.00	1717	8/13/75	1.00
1566	6/14/77	1.00	1727	4/5/78	1.00
	6/15/77	1.00	1830	11/18/76	1.00
1580	5/22/76	0.30			

[a]R means "recalibrated" from McCord/Chapman or other previously published average.

[b]Infrared data scaled to visible data in overlap region due to suspected lightcurve effects exceeding 3 %. Parentheses around dates indicate infrared data only.

TABLE IV

Spectral Groups

C-like groups

1172	19: 163, 654
361: 617, 1173	676: 909, 976
24: 36, 66, 88	210
194	46: 185, 426
41: 156, 344	83: 220, 347, 510, 790
52: 78	505: 704
34: 121, 164, 375, 423, 1015	1
10: 468	13: 48, 106
31: 54, 139, 141, 511, 554, 712, 846	811
150	293
65: 94, 268, 363, 365, 490, 613	345
58: 105	45
128: 313, 488, 696	140
386: 409, 439, 481, 1025	356
144	85: 324, 1580

TABLE IV (Continued)

Spectral Groups

S-like groups (complete spectra)	S-like groups (incomplete spectra)
25	243: 1284
115	42 (1449): 858
12	264: 599
1449	181 (115): 391, 639
196	158
433: 584	258 (9, 116): 389
43: 1685	124 (3, 7, 12, 116): 198, 374, 416, 471,
28	513, 839, 925, 1727
39: 119, 714	26 (7, 8, 17, 43, 108, 192): 113, 169,
7: 15, 68	340, 1058, 1620, 1830
192: 760	60: 122
887	1036 (28, 43, 887)
63: 197	149 (17, 43, 433, 1055, 1087): 453
116: 167	
17: 462	71: 339, 402, 403, 579, 1075
1055	582 (108, 196)
8: 27	281 (7, 23, 1449): 323
108: 341	695 (43)
23: 40, 79, 674	472: 574, 1636
9: 29, 30, 230, 236	
89: 563	
3: 5, 6, 11, 14, 18, 20, 32, 37, 82, 532	

M- or E-like groups

16: 21, 22, 87, 129, 217, 325, 413, 739, 741, 1103
64: 69, 92, 97, 216, 338, 801, 1645
136: 279
435: 441, 773, 884, 944
335: 526, 1330, 1493
44: 895
213: 419, 750, 785

R-like and U groups

4: 208	758	166: 1162, 1595
349: 1088	221	51: 176
496	558	62: 90, 175
354: 1019	434	2: 372
246: 446	170: 308, 624, 628, 1529, 1717	110
80	588: 1199, 1512	130
262: 770, 782, 1656	911	

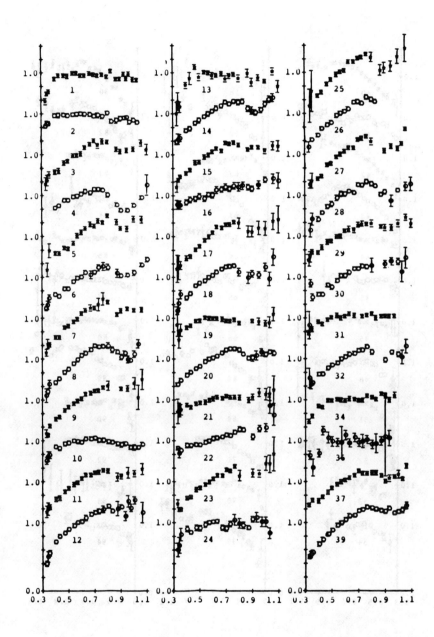

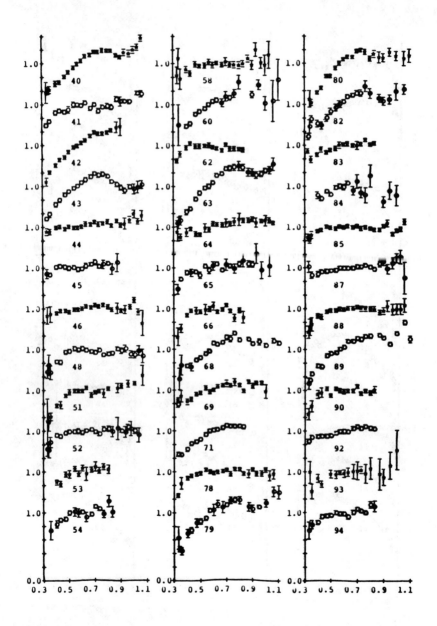

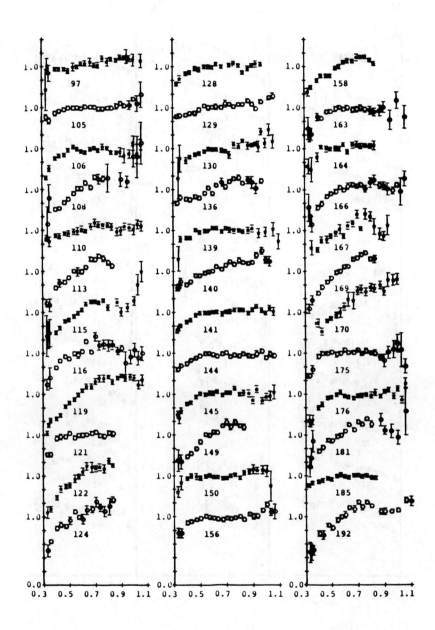

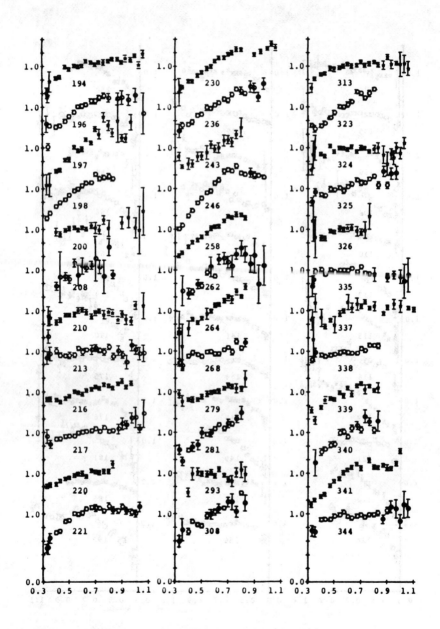

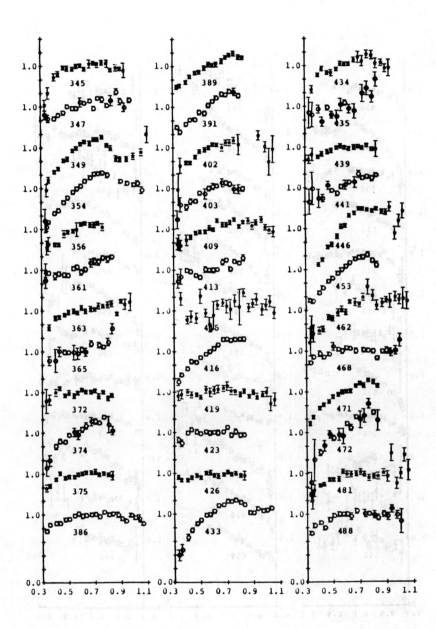

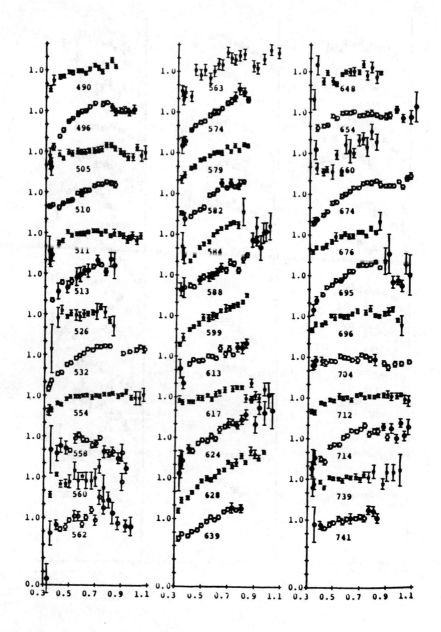

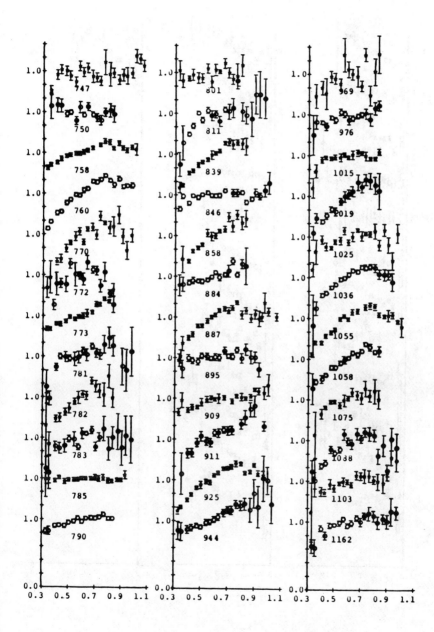

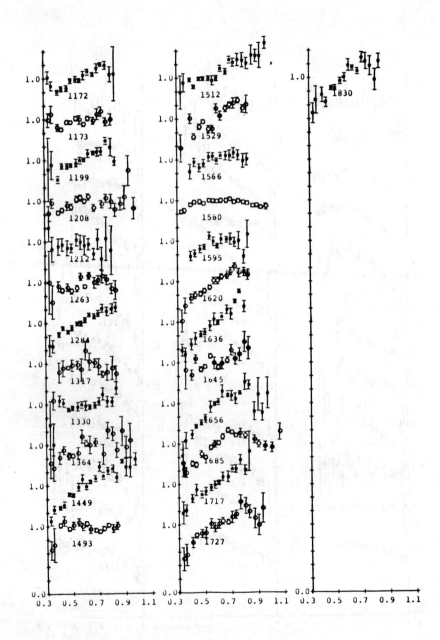

REFERENCES

Bobrovnikoff, N. T. 1929. The spectrum of minor planets. *Lick Obs. Bull.* XIV: 18-27

Bowell, E.; Chapman, C. R.; Gradie, J. C.; Morrison, D.; and Zellner, B. 1978. Taxonomy of asteroids. *Icarus* 35: 313-335.

Chapman, C. R. 1976. Asteroids as meteorite parent bodies. The astronomical perspective. *Geochim. Cosmochim. Acta* 40: 701-719.

Chapman, C. R.; Johnson, T. V.; and McCord, T. B. 1971. A review of spectrophotometric studies of asteroids. In *Physical Studies of Minor Planets,* ed. T. Gehrels (NASA SP-267, Washington, D. C.: U. S. Government Printing Office), pp. 55-65.

Chapman, C. R.; McCord, T. B.; and Johnson, T. V. 1973a. Asteroid spectral reflectivities. *Astron. J.* 78: 126-140.

Chapman, C. R.; McCord, T. B.; and Pieters, C. 1973b. Minor planets and related objects. X. Spectrophotometric study of the composition of (1685) Toro. *Astron. J.* 78: 502-505.

Chapman, C. R.; Morrison, D.; and Zellner, B. 1975. Surface properties of asteroids: A synthesis of polarimetry, radiometry, and spectrophotometry. *Icarus* 25: 104-130.

Chapman, C. R.; and Salisbury, J. W. 1973. Comparison of meteorite and asteroid spectral reflectivities. *Icarus* 19: 507-522.

Gaffey, M. J., and McCord, T. B. 1978. Asteroid surface materials: Mineralogical characterizations from reflectance spectra. *Space Sci. Rev.* 21: 555-628.

Johnson, T. V., and Fanale, F. P. 1973. Optical properties of carbonaceous chondrites and their relationship to asteroids. *J. Geophys. Res.* 78: 8507-8518.

McCord, T. B.; Adams, J. B.; and Johnson, T. V. 1970. Asteroid Vesta: Spectral reflectivity and compositional implications. *Science* 168: 1445-1447.

McCord, T. B., and Chapman, C. R. 1975b. Asteroids: Spectral reflectance and color characteristics. *Astrophys. J.* 197: 781-790.

McCord, T. B., and Chapman, C. R. 1975b. Asteroids, spectral reflectance and color characteristics. II. *Astrophys. J.* 197: 781-790.

Nygard, S. 1975. Alpha Lyrae/Sun flux ratios for use in standard star calibrations: Results of three techniques. M. S. thesis, Massachusetts Institute of Technology.

Owensby, P. D.; Pieters, C.; and Gaffey, M. J. 1979. Standard star calibration for use in planetary reflectance studies. In preparation.

Pieters, C.; Gaffey, M. J.; Chapman, C. R.; and McCord, T. B. 1976. Spectrophotometry (0.33 to 1.07 μm) of 433 Eros and compositional implications. *Icarus* 28: 105-115.

Veeder, G. J.; Matson, D. L.; and Smith, J. C. 1978. Visual and infrared photometry of asteroids. *Astron. J.* 83: 651-663.

Williams, J. G. 1971. Proper elements, families, and belt boundaries. In *Physical Studies of Minor Planets,* ed. T. Gehrels (NASA SP-267, Washington, D. C.: U. S. Government Printing Office), pp. 177-181.

Zellner, B. 1975. 44 Nysa: An iron-depleted asteroid. *Astrophys. J.* 198: L45-L47.

MINERALOGICAL AND PETROLOGICAL
CHARACTERIZATIONS OF ASTEROID
SURFACE MATERIALS

MICHAEL J. GAFFEY and THOMAS B. McCORD
University of Hawaii

Gaffey and McCord (1978) have reviewed the methodology and results of asteroid surface material characterizations carried out through 1977; these results are reviewed and updated in this chapter. Modifications of previous interpretations can arise from the acquisition of new or improved data (spectral resolution, coverage and/or precision) or from improvements in interpretive methodology. Since interpretations are likely to be continually improved, special emphasis is placed on the methodology utilized in making mineralogical interpretations from spectra. In particular, recent advances in the interpretive approach are stressed. It is hoped that the student would derive from this discussion some understanding of the basis for, as well as the limitations of, interpreting spectra and thus be able to evaluate more fully the specific results of past and future studies of asteroid surface materials.

In the eight years since the first asteroid book (Gehrels 1971) there has been a rapid expansion of the state-of-the-art for characterizing asteroid surface materials in terms of their mineralogical properties. The observational data base has increased both in the range of spectra coverage and resolution, and in the number of objects observed. The growth and present state of this data set are reviewed elsewhere in this book (chapter by Chapman and Gaffey).

Similarly, the techniques and calibrations necessary for interpreting these spectral data in order to obtain mineralogical information have undergone a quantum jump in sophistication. In 1970 only UBV color data were available for a range of meteorites, which are the most reasonable comparision materials for asteroid surfaces (Hapke 1971). Spectral reflectance measurements of laboratory samples with broader spectral coverage and higher spectral resolution were beginning to become available for certain meteorites such as the basaltic achondrites (McCord et al. 1970). During the early 1970's a number of spectral reflectance studies were carried out on meteorites including Johnson and Fanale (1973), Chapman and Salisbury (1973), Salisbury et al. (1975), and the most complete study in Gaffey (1976). During the same interval, work was being carried out to relate specific spectral properties (e.g. absorption band position) to mineral compositions (Adams 1974, 1975; Adams and Goullaud 1978; Bell and Mao 1973; Burns 1970, 1974).

As the data base of meteorite and other cosmically interesting spectra accumulated and as the understanding of the functional relationships between spectral and mineralogical properties improved, efforts were made to interpret the concurrently expanding set of asteroidal spectral variations. Care should be exercised in reading the early papers since subsequent improvements in techniques or observational data may have significantly modified some of their conclusions. It is useful to review the previous interpretive work both in terms of understanding the evolution of interpretations of asteroid surface materials and in terms of placing each set of results in context with previous and subsequent work.

McCord et al. (1970) measured the 0.3–1.1 μm spectrum of the asteroid 4 Vesta and identified an absorption feature near 0.9 μm as due to the mineral pyroxene. They suggested that the surface material was similar to basaltic achondritic meteorites. Hapke (1971) compared the UBV colors of a number of asteroids to a variety of lunar, meteoritic, and terrestrial rocks and rock powders. He concluded that the surface material of these asteroids could be matched by powders similar to a range of the comparison materials but not by metallic surfaces. Chapman and Salisbury (1973) concluded that some matches between 0.3– 1.1 μm spectra of asteroids and meteorites were found for several meteorite types including enstatite chondrites, a basaltic achondrite, an optically unusual ordinary chondrite and, possibly, a carbonaceous chondrite. Johnson and Fanale (1973) with similar data and laboratory mixtures showed that the albedo and spectral characteristics of some asteroids are similar to C1 and C2 carbonaceous chondrites and others to iron meteorites. Both papers noted the problem of defining precisely what constituted a "match," and both raised the question of subtle spectral modification of asteroid surface materials by in situ space weathering processes. Salisbury and Hunt (1974) raised the question of the effects of terrestrial weathering on meteorite specimens and the validity of matches between the spectra of such specimens and the asteroids. McCord and Gaffey (1974) utilized absorption features and general spectral properties to characterize 14 asteroids and identified mineral assemblages similar to carbonaceous chondrite, stony-iron, iron, basaltic achondrite, and silicate-metal meteorites. At that time it was possible to establish the general identity of the spectrally important minerals in an assemblage, but very difficult to establish their relative abundances.

Chapman *et al.* (1975) developed an asteroidal classification system which utilized spectral, albedo, and polarization parameters to define two major groups which included most of these bodies. The first group was characterized as having low albedos ($\leqslant 0.09$), strong negative polarizations at small phase angles ($\geqslant 1.1\%$) and relatively flat, featureless spectral reflectance curves in the 0.3–1.1 μm spectral region. These parameters were similar to those for carbonaceous chondrites and these asteroids were designated as "carbonaceous" or *C* type. The second group was characterized as having higher albedos ($\geqslant 0.09$), weaker negative polarizations (0.4–1.0%) and reddish, sometimes featured spectral curves. These parameters were comparable to those for most of the meteorites which contain relatively abundant silicate minerals, so this group was designated "silicaceous or stony-iron" or *S* type. A small minority ($\sim 10\%$) of the asteroids could not be classified in this system and were designated "unclassified" or *U* type (see Zellner's chapter).

This simple classification scheme can be quite useful since it often seems to separate these two major types of objects, and the observational parameters on which the system is based can be measured for objects fainter than those for which complete spectra can be obtained. The choice of terminology is unfortunate, however, since it implies a specific definition of surface materials in meteoritic terms, which was not intended. Any "flat-black" spectral curve would be designated *C* type *whether or not the surface material would be characterized as carbonaceous by any other criteria.* Thus asteroid surface materials similar to such diverse meteoritic assemblages as ureilites, black chondrites and the carbon-poor C4 meteorite Karoonda could, by one or another of the observational criteria, be classified as *C* type, leading persons not familiar with the classification criteria to conclude that such materials were carbonaceous.

The problem is complicated by the terminology for the meteorites themselves. As Mason (1971) noted, the term "carbonaceous" immediately implies the presence of carbon or carbonaceous compounds as a distinctive component. However, a number of the carbonaceous chondrites, especially types C3 and C4, contain less carbon than many non- 'carbonaceous' meteorites. Indeed, the current working definition of the carbonaceous chondrites ($SiO_2/MgO < 1.5$ [by weight] ; Van Schmus and Wood 1967) has nothing to do with carbon, although the abundance of carbon and other volatiles (e.g. water) are important criteria in further subdividing this chemical group. It is essential to realize that the term "carbonaceous asteroid" is not necessarily equivalent to "carbonaceous meteorite" (as broad as that latter classification is). "Carbonaceous" as presently used is a *spectral/albedo/polarization classification designation* with respect to asteroids and a *compositional designation* with respect to meteorites. A similar objection can be raised with respect to the "silicaceous" terminology since it implies a degree of specificity not present in the classification criteria. It must also be noted that the criticism is with respect to the terminology and the confusion which it engenders and not with respect to identification of these two major groups, defined by the parameters chosen (e.g. albedo, polarization, etc.), which appear to be significant.

Thus, while the C and S classification of asteroids cannot be viewed as descriptions of mineralogy or petrology, it does provide valid characterizations with respect to the chosen parameters. Since the groups appear in each of the data sets used (albedo, polarization, color) a single measurement such as UBV color can generally be used to classify the asteroid (Zellner et al. 1975; Zellner et al. 1977a; Zellner and Bowell 1977; Morrision 1977a, b). This approach can also be utilized to identify anomalous objects (Zellner 1975; Zellner et al. 1977b) or to establish possible genetic relationships between members of asteroid dynamical families (Gradie and Zellner 1977). Chapman (1976) utilized the basic C-S classification system but identified subdivisions based on additional spectral criteria (R/B Slope, Bend and Band Depth; McCord and Chapman 1975a, b; see Chapman and Gaffey, in Part VII, for the definitions) which are mineralogically significant.

Johnson et al. (1975) measured the infrared reflectance of three asteroids through the broad bandpass J, H and K filters (1.25, 1.65 and 2.2 μm) and concluded that these were consistent with the infrared reflectance of suggested meteoritic materials (see the chapter by Larson and Veeder). Matson et al. (1977a, b) utilized infrared H and K reflectances to infer that space weathering or soil maturation processes were relatively inactive on asteroid surfaces in contrast to that on surfaces on the moon and Mercury. Veeder et al. (1978) restated this conclusion and supported the interpretation that metallic NiFe was the most plausible candidate as a major phase in the surface material of most S-type asteroids.

A very favorable apparition in early 1975 permitted the measurement of a variety of spectral data sets for the Earth-approaching asteroid 433 Eros. Pieters et al. (1976) measured the 0.33–1.07 μm spectral reflectance of Eros through 25 narrow-bandpass filters. This curve was interpreted to indicate an assemblage of olivine, pyroxene, and metal, with metal abundance equal to or greater than that in the H-type chondrites. Veeder et al. (1976) measured the spectrum of Eros through 11 filters from 0.65–2.2 μm and concluded that their spectral data indicated a mixture of olivine and pyroxene with a metal-like phase. Wisniewski (1976) concluded from a higher resolution spectrum (0.5–1.0 μm), that this surface was best matched by a mixture of iron or stony-iron material with ordinary chondritic material (i.e. iron + pyroxene + olivine), but suggested that olivine is absent or rare. Larson et al. (1976) measured the 0.9–2.7 μm spectral reflectance curve for Eros and identified NiFe and pyroxene, but found no evidence of olivine or feldspar. The dispute over the olivine content arose because of an incomplete understanding of the spectral contribution of olivine in a mixture and because of slight differences in the observational spectra near 1 μm. The uncertainty in the metal abundance was due to incomplete quantitative understanding of the spectral contribution of metal in a mixture with silicates.

Lebofsky (1978) showed evidence for the presence of an H_2O-related absorption feature near 3 μm in the reflectance spectrum of the asteroid 1 Ceres. Larson *et al.* (1979) combined their observational data with that from several other sources to provide a 0.4–3.6 μm reflectance spectrum of Ceres and Pallas. They concluded from these data that the surface materials of these two asteroids were consistent with mixtures of opaques and hydrated silicates, such as are found in type C1 and C2 meteorites.

Gaffey and McCord (1978) have presented the most complete summary of asteroid surface material characterizations and interpretive methodology. They provided mineralogical characterizations for sixty-five asteroids and concluded that most asteroid surfaces were composed of assemblages of meteoritic minerals but that no genetic links had been established between any asteroid and a meteorite specimen or type. Most main-belt asteroids exhibited surfaces analogous, but not necessarily identical, to type I or II carbonaceous chondrites. A significant number of objects exhibited surface materials spectrally dominated by a metallic NiFe component. They concluded that a wide variety of assemblages was present on these objects and that general classification groups such as *C* or *S* were composed of a variety of diverse types which became evident upon more detailed investigation of objects in each group. Gaffey (1978) concluded that unsampled meteorite-types were present in the asteroid population and in particular that the surface materials of the asteroids Ceres and Pallas, while similar in certain respects to type I and II carbonaceous chondrites (e.g. hydrated silicates, opaque-rich), were unlike any known type of meteorite.

I. INTERPRETIVE METHODOLOGY

Ever since the recognition of meteorites as extraterrestrial matter and the discovery of the minor planets, it has been suggested that the former are derived from the latter (e.g. Olbers 1803). More recently, efforts have focused on the relative contributions of various orbit modifying mechanisms for delivering asteroidal debris (meteorites) to the earth (e.g. Kozai 1962; Anders 1964; Arnold 1965; Williams 1969; Zimmerman and Wetherill 1973; Peterson 1976), or on the relative contribution to the meteoritic flux of different (orbital) groups of asteroids or asteroid precursors such as comets (e.g. Wetherill and Williams 1968; Wetherill 1974, 1976; Anders 1971, 1975); these topics are most recently and completely reviewed in several chapters in this book.

Thus most investigators interested in determining the composition of asteroids from observational data have concluded that the meteorites represent the best available comparison material (e.g. Watson 1938; McCord *et al.* 1970; Hapke 1971; Johnson and Fanale 1973; Chapman and Salisbury 1973; Larson *et al.* 1979). Attempts to match directly laboratory meteorite colors or spectra with observational data have met with only limited success for several reasons. The most serious uncertainty in any spectral matching approach lies in the problem of deciding what constitutes a "match." In an empirical curve matching program, there is no basis for deciding whether a deviation from an exact match is mineralogically significant. Hapke (1971) pointed out that a large variation in the UBV colors of a material can be produced by varying the particle size distribution. Salisbury and Hunt (1974) discussed the uncertainty introduced by terrestrial weathering of meteorites. Chapman and Salisbury (1973) and Johnson and Fanale (1973) discussed space weathering (surface bombardment processes, etc.) as a possible source of mismatch. The problem is compounded by the virtual certainty that the selection of meteoritic material arriving at the earth's surface is both biased and incomplete to a significant but unknown degree with respect to the distribution of asteroidal materials (Chapman and Salisbury 1973; Johnson and Fanale 1973; McCord and Gaffey 1974; McCrosky *et al.* 1971; Ceplecha and McCrosky 1976). Despite these uncertainties, the color or spectrum matching approach does provide a good survey technique and can be usefully applied if care is taken to recognize its limitations.

The alternate approach to interpreting spectral data involves the recognition and quantitative characterization of spectral features which are diagnostic of specific minerals or mineral types. These spectral feature parameters (e.g. absorption band position, width, intensity and symmetry) are relatively insensitive to variations in the physical properties (e.g. particle sizes) of a mineral assemblage. Also, since the interpretation depends on the identification of specific mineral phases in an assemblage rather than on the assemblage as a discrete entity, the necessary calibration need only include the range of possible mineral species and an understanding of the effects on the spectral parameters of mixing phases together. One does not need to have measured all possible combinations of mineral phase mixtures and physical property variations to provide a complete comparison set. The former task is very large but achievable, the latter task is probably transfinite.

The interpretation of mineralogically diagnostic spectral features does place more stringent demands on both the observer and the interpreter. It requires an understanding of physical processes which act to produce the spectral features and also requires observational spectra with sufficient spectral precision, resolution and wavelength coverage to quantitatively characterize the appropriate spectral parameters. As shall be seen below, the degree of sophistication of the interpretive calibration varies with both the mineral species and the type of assemblage. Likewise the spectral coverage, resolution and precision differ widely for the data on different asteroids. It is worth noting that, within reason, any increase in either the interpretive capability or the observational data quality can bring about a commensurate improvement in the sophistication of the surface material characterization.

Since the early 1960's a major effort has been underway to define the physical processes which govern the interaction of light with the common rock-forming silicate minerals (e.g. White and Keester 1966, 1967; R. G. Burns 1970b; Bell *et al.* 1975; Hazen *et al.* 1977). This work based on crystal field theory, ligand field theory and molecular orbital theory, has been summarized for mineralogical systems by Burns (1970a).

Elements of the first transition series (Ti, V, Cr, Mn, Fe, Co, Ni, Cu) and their petrologically important cations (especially Fe^{2+}, Ti^{3+}) have an outer (valence) unfilled d-shell in their electron distribution. When such a cation is located in a crystallographic site, surrounded by anions, certain of the outer electron orbitals experience strong repulsions and undergo splitting to higher energy. The orbitals undergoing the least electronic repulsion become the groundstate orbitals in which the electrons tend to reside. The energy difference between the groundstate and any excited state is termed the *crystal field splitting energy* and is a direct function of the particular cation and the crystal site in which it resides. A photon whose energy corresponds to the splitting energy of a particular cation can be absorbed. This gives rise to specific absorption features in the reflectance spectra of transition-metal silicates such as olivine, pyroxene and feldspar. Adams (1975) has shown that the positions of these bands are distinctive for each type of mineral; Adams (1974) has calibrated the precise positions of the two pyroxene bands with respect to the mineral chemistry (iron and calcium content) of the specimens. Adams and Goullaud (1978) have studied the 1.1–1.3 μm Fe^{2+} absorption feature present in the spectra of plagioclase feldspars. The center of this band shifts systematically toward longer wavelength with increasing anorthite content (calcic feldspar) up to about An_{65}. The strength (absorbance) of the band increases with increasing iron content.

In addition to these crystal field or electronic absorption features, several other types of mineralogical absorption features are of increasing importance in the characterization of asteroid surface materials as observational data further into the infrared (toward 5 μm) and into the ultraviolet (toward 1000 Å) become available. *Vibrational features* arise as a result of photon excitation of molecular groups at their fundamental vibrational (bending, stretching and rotational) frequencies as well as at overtones and harmonics of these frequencies. Such vibrational features are present in the spectra of most materials (e.g. salts, silicates, carbonates, ices, etc.; Aronson and Emslie 1974; Liese 1975; Sill 1973). However, with regard to asteroid surfaces, the hydrated minerals and hydrocarbon compounds, found in certain types of meteorites, have been given the most attention (Larson *et al.* 1979; see the chapter by Larson and Veeder).

The water molecule has three fundamental vibrational modes which affect the infrared spectral region: ν_1, the symmetric OH stretch; ν_2, the

H-O-H bend and ν_3, the asymmetric OH stretch. In an H_2O vapor phase the modes correspond to 2.73 μm, 6.269 μm and 2.66 μm respectively, while those in ice are shifted to 3.105 μm, 6.06 μm and 2.94 μm. Hydrated mineral phases have a strong absorption feature in the region 2.9–3.3 μm and two major overtone or combination features near 1.4 μm ($2\nu_3$) and 1.9 μm ($\nu_2 + \nu_3$). The existence of both the 1.4 μm and 1.9 μm features indicates undissociated water molecules in the mineral while the 1.4 μm alone indicated OH groups such as hydroxyls. The exact wavelength position of these features is a function of the crystal structure and site in which the molecule resides. Therefore, such wavelength positions should be mineralogically diagnostic. In practice, only a few cases have been investigated. However, Hunt and Salisbury (1970, 1971) have presented the reflectance spectra of a variety of minerals which contain water-related spectral features. Larson et al. (1979) present spectra showing the 3 μm water feature in several terrestrial and meteoritic minerals.

Charge transfer, exciton (electron-hole pair) formation and *valence-conduction band* transitions dominate the ultraviolet and vacuum ultraviolet spectral region in geologic materials. Nitsan and Shankland (1976) and Hapke et al. (1978) have considered this portion of the spectrum and have concluded that mineralogical information is present. The capability of instruments on Earth-orbiting satellites (e.g. International Ultraviolet Explorer (IUE)) has increased the importance of determining what mineralogical information can be obtained by studying the ultraviolet spectra of minor planets.

The interpretation of the spectral characteristics of mineral mixtures in order to establish mineral abundance as well as composition is a more complex problem than identifying the presence of minerals. For example, Adams and McCord (1970), Nash and Conel (1974) and Gaffey (1974, 1976) have noted that the relative abundance of a mineral phase does not in general correlate linearly with its apparent spectral abundance. Johnson and Fanale (1973), and Nash and Conel (1974) used laboratory mixtures to show that an opaque phase (carbon and magnetite, respectively) would dominate the reflectance spectrum of a mixture if dispersed, even when present in small amounts. Gaffey (1976) discussed how the relatively more optically dense pyroxene phase dominates the spectrum of a pyroxene-olivine-feldspar assemblage.

Three general classes of minerals are important in controlling the optical properties of meteoritic mineral assemblages: (1) nickel-iron minerals, (2) silicate phases (olivine, pyroxene, feldspar, and clay minerals or "layer-lattice" silicates), and (3) opaque phases (carbon, carbon compounds, magnetite). Of these minerals, the Fe^{2+} silicates exhibit the best understood and most easily interpreted diagnostic spectral features.

The apparent contribution of a specific mineral phase to the reflectance spectrum of a mixture is a function of the wavelength-dependent optical properties and of the relative abundance and distribution of that phase. The mineral phase with the greatest optical density at a particular wavelength tends to dominate, out of proportion to its actual abundance, the reflectance spectrum of a mixture at that wavelength. This can be shown by comparison of the spectra of pyroxene, olivine and a pyroxene-olivine mixture. The pyroxene spectral reflectance curve (Fig. 1a) is characterized by two relatively narrow, symmetric absorption features centered near 0.90 μm and 1.9 μm, while the olivine spectrum (Fig. 1c) is characterized by a broad, asymmetric feature centered near 1.0 μm. The optical density of the pyroxene phase in the 1 μm region is nearly an order of magnitude greater than that of the olivine phase with an equivalent iron content. A mineral assemblage containing approximately equal amounts of olivine and pyroxene (Fig. 1b) produces a spectral curve which is essentially that of the pyroxene phase, with the long-wavelength edge of the pyroxene feature depressed by the weaker absorption of the olivine phase. The broadening and asymmetry of the resulting feature is a function of the relative abundance of olivine and pyroxene. A preliminary calibration of this functional relationship is shown in Fig. 2.

Computer deconvolution techniques can be utilized to resolve a spectrum into its superimposed component spectra. This technique can be used to remove the continuum and strong absorptions to reveal weak absorption features such as that of feldspar (Fig. 3). The relative intensity of the pyroxene and feldspar absorption features can be correlated with the relative abundance of these mineral phases for equilibrium or quasi-equilibrium pyroxene/plagioclase assemblages (McFadden and Gaffey 1978). The relationship between relative absorption band intensity and mineral abundance for pyroxene/plagioclase assemblages is shown in Fig. 4. The slope of this relationship can be calculated *a priori* from independent information concerning the abundance and molar absorption coefficients of Fe^{2+} in the two mineral species. A great deal of additional work needs to be carried out in order to define completely calibrations for the range of possible mixtures of these types of silicate minerals.

The metallic nickel-iron (NiFe) phases are significant or dominant mineralogical constituents in a variety of meteoritic assemblages. The 0.35–2.5 μm reflectance spectra of these minerals (Johnson and Fanale 1973; Gaffey 1974, 1976) exhibit no discrete diagnostic electronic absorption features. The overall shape of the reflectance spectrum is distinctive for the general group of NiFe minerals containing less than 25% nickel. A typical NiFe spectral curve (Figure 1d-right) is characterized by a nearly linearly increasing continuum reflectance at wavenumbers greater than 10,000–12,500 cm^{-1} ($\lambda < 0.8$–1.0 μm). At lower wavenumbers (<1.0 cm^{-1} or $\lambda > 1.0$ μm), the spectrum exhibits an upward (positive) curvature with decreasing frequency. The degree of positive infrared curvature seems to be inversely proportional to the Ni content of the alloy (Gaffey 1976).

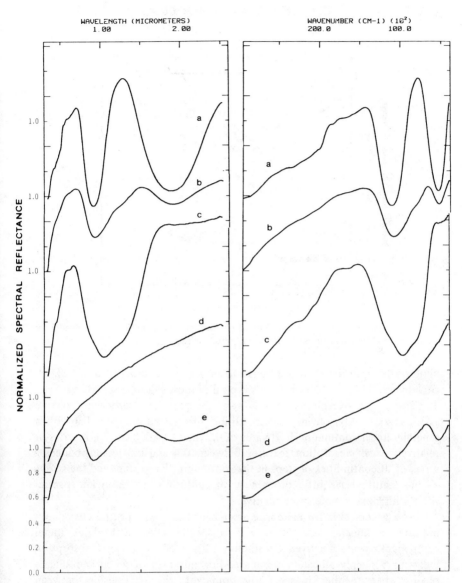

Fig. 1. Normalized spectral reflectance curves of meteoritic minerals and mineral assemblages versus wavelength (left) and wavenumber (right): (a) pyroxene, 32%; (b) L6 chondrite (olivine + pyroxene + metal), 31%; (c) olivine, 36%; (d) nickel-iron metal, 22%; (e) L4 chondrite, 21%: Spherical albedos at the normalization wavelength (0.56 μm) are indicated for each sample.

By contrast, an assemblage primarily made up of silicate minerals such as pyroxene, olivine and feldspar (Fig. 1*b*) has a distinctly nonlinear spectral reflectance curve in the region shortwards of the 0.9 μm absorption feature, (Fig. 1*b*-right). This strong blue and ultraviolet decrease in reflectance is the result of increasingly efficient charge transfer absorption for higher energy

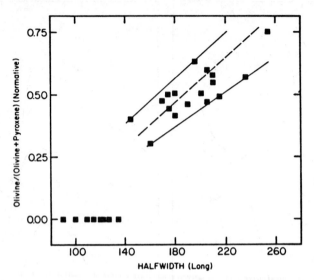

Fig. 2. Correlation of half width at half height the long-wavelength wing of Band I for olivine + pyroxene mixtures versus olivine/pyroxene ratio.

photons between the various cations and anions in the silicate structure. Such curvature and blue-UV absorption is typical of most silicate assemblages.

The major exceptions to this rule are pure or nearly pure olivine assemblages ($>$90% olivine) which exhibit spectral curves (Fig. 1c) with a generally linear continuum shortwards of 0.7 μm ($>$ 14,000 cm^{-1}). A series of relatively weak absorption features between 0.4 and 0.6 μm introduces a series of discontinuities or steps in the continuum. These steps and the strong olivine feature leave little probability of confusion of the olivine spectral curve with a linear NiFe spectral curve.

As a general rule for meteorite assemblages, a linear spectral continuum indicates a mineral assemblage with a spectrally dominant NiFe metal component, while a curved continuum with a blue-ultraviolet absorption indicates a mainly silicate assemblage. For mixtures of metal and silicate phases where neither is spectrally dominant, the relationship between apparent spectral abundance of the metal phase and its actual mineralogical abundance is a complex function of abundance, grain size, distribution and oxidation state. The complexity of this relationship can be illustrated by an example as follows.

The metal abundance for L-type chondritic assemblages is approximately 10%, but the apparent spectral contribution of the metal phase, judged by the linearity of the continuum is significantly different between an L6 assemblage (Fig. 1b-right) and L4 assemblage (Fig. 1e-right). Measurements of metal grain size (Dodd 1976) show an increase in median grain size with increasing degree of metamorphism (judged by the compositional homogeneity of the silicate

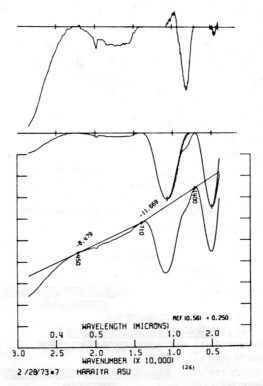

Fig. 3. Computer deconvolution of spectral curve of pyroxene-plagioclase mixture, involving establishment of empirical continuum (A) for original curve (B), continuum minus spectrum (C), spectral features are mirrored about center of feature and subtracted to isolate weak feldspar band (D).

grains) by a factor of at least 2 or 3 from L4 to L6. This would represent about an order of magnitude difference in the surface area of metal grains. This is, to first order, spectrally equivalent to a larger metal abundance with the same grain size distribution. The special behavior of the L4 assemblage is further complicated by the possible presence of a spectral blocking (opaque) phase (Gaffey 1976). Such a component would tend to suppress the higher albedo portions of the spectrum and reduce the blue-ultraviolet falloff as well as the absorption band depth.

The importance of understanding the spectral properties of the opaque-rich (or low albedo) assemblages is evident when one considers that about three quarters of all asteroids have surface materials with geometric albedos of 6.5% or less. When investigators have considered candidate materials, from among the météorites, to match these dark asteroid spectra, the carbonaceous chondrites have received most of the attention. However, even a cursory examination of the albedos for the range of studied meteorites (Fig. 5) reveals a diverse variety of dark (spherical albedo $\leq 9\%$) meteorite assemblages

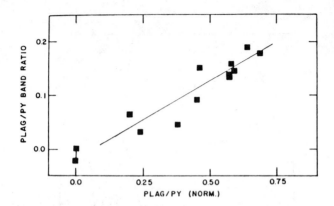

Fig. 4. Correlation of plagioclase/pyroxene absorption band depth with plagioclase/-pyroxene mineral abundance.

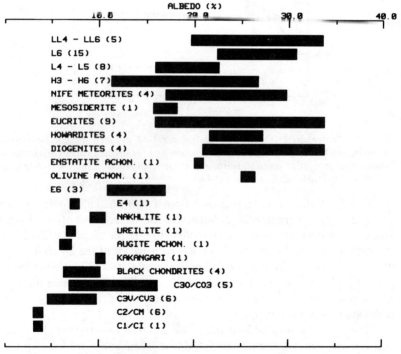

Fig. 5. Spherical albedos of meteorites.

including carbonaceous chondrites, black chondrites, K-chondrites (Kakan-gari), enstatite chondrites, augite achondrites and ureilites (olivine-pigeonite-carbon achondrites). Since albedo is a sensitive function of physical properties, such as particle size (Adams and Filice 1967) and surface microstructure, any meteoritic assemblage with a low albedo must be considered as a potential candidate for some C asteroid surface materials until it can be ruled for any specific body on the basis of criteria other than albedo. However, as might be expected from the diversity of assemblages represented by the dark meteorites, they exhibit a corresponding diversity of spectral features which are correlated with the specific mineralogy of each type. A detailed discussion of the relationships between mineralogy and petrology for the low albedo assemblages can be found in Gaffey (1979). These results will be discussed here in an abridged form.

The relatively restricted group of opaque-rich assemblages represented by the carbonaceous chondrites is far from a simple system, either mineralogically or spectrally. For a much more complete discussion of the mineralogical, petrological and chemical properties alluded to here, the reader is referred to Nagy (1975) and a series of papers by McSween (1977a-d) and McSween and Richardson (1977), and the references contained therein.

Type I Carbonaceous Chondrites

The Type I, C1 or CI carbonaceous chondrites are composed primarily of of a relatively iron-rich (Fe/Si $\sim$1.1) clay mineral or layer lattice silicate which apparently consists of an intimate mixture of several structural types. Major accessory minerals include magnetite ($\sim$5–15 wt%), soluble magnesium salts ($\sim$5–15 wt%) and carbon, both as organic compounds and graphite ($\sim$2–5 wt%). Water ($\sim$10 wt%) is present in the clay minerals and as water of hydration in the magnesium salts. Two CI meteorites have been studied spectrally: Orgueil (Johnson and Fanale 1973; Gaffey 1974, 1976; Salisbury *et al.* 1975; Larson *et al.* 1979) and Alais (Gaffey 1974, 1976). Gaffey (1976) concluded that the spectral reflectance curve of Alais indicated that it had been strongly altered by terrestrial processes. Three different samples of Orgueil were measured. The Johnson and Fanale (1973) results are in good agreement, except for minor slope differences, with the Gaffey (1976) and Larson *et al.* (1979) results for a shared sample. The Salisbury *et al.* (1975) spectrum appears to exhibit the same characteristics as the (presumably) terrestrially weathered Alais sample. However, since the CI meteorites are quite heterogeneous breccias, one cannot, at this time, rule out the possibility that the Alais and the Salisbury *et al.* (1975) Orgueil spectra represent samples of a different clast-type from these breccias.

Two major features can be seen in the reflectance spectrum of Orgueil (Fig. 6a). A very strong H_2O-related band is located near 3 μm (see the chapter by Larson and Veeder in this book) and a strong charge transfer (presumably Fe^{2+}–Fe^{3+} and/or Fe^{3+}) feature or set of overlapping features shortwards of 0.56 μm. Several weaker ($\sim$2–6%) features are also present, including an Fe^{3+} feature ($\sim$0.88–0.90 μm) and two H_2O-related features ($\sim$1.38 μm, $\sim$1.94 μm). The water-related features, especially the 3 μm

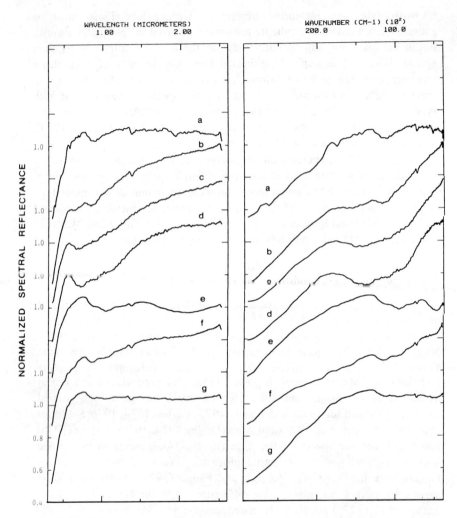

Fig. 6. Normalized spectral reflectance curves of the dark meteorites (spherical albedos at 0.56 μm): (a) CI, Orgueil – 4.7%; (b) CM, Cold Bokkevelt (<75 μm) and Murray, average – 4.9%; (c) CM, Meghei and Murchison, average – 5.1%; (d) CM, Nogoya – 5.3%; (e) CO3, average – 11.1%; (f) CV3, Allende and Mokoia, average – 8.5%; (g) CV3, Vigarano, Grosnaja (<75 μm) and Leoville, average – 9.4%; (h) C4-5, Karoonda – 9.6%; (i) Kakangari – 10.2%; (j) Novo-Urei – 8.4%; (k) Black chondrites, average – 9.4%; (l) E4, Abee – 7.7% and (m) E6, average – 17.4%. (Part 2 of Fig. 6 is found on the following page.)

feature, are produced by the water intrinsic to the clay mineral matrix material. The iron-related features are characteristic of both the high iron content of these meteoritic clay minerals and their relatively oxidized condition ($Fe^{2+} \approx Fe^{3+}$).

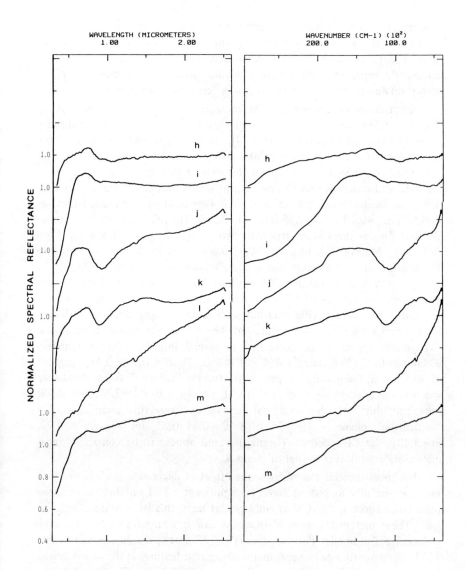

WAVELENGTH (MICROMETERS)
1.00   2.00

WAVENUMBER (CM-1) (10^2)
200.0   100.0

NORMALIZED SPECTRAL REFLECTANCE

Type II Carbonaceous Chondrites

The Type II, C2 or CM carbonaceous chondrites are composed of two major phases: an iron-rich (Fe/Si ~1.8) clay mineral matrix (~50–80 wt%) and mafic inclusions (~15–40 wt%) which are predominantly low-iron olivine (~Fa_1, range Fa_{0-60}) with lesser amounts of pyroxene (~Fs_1, Fs_{0-15}). Soluble sodium and magnesium salts (~10 wt%), carbon (organic and graphite, ~1–2 wt%) and magnetite (~0.5 wt%) constitute the main accessory phases. The matrix material, a hydrated layer lattice silicate with abundant oxidized iron, resembles that of the CI meteorites in a general way. However, these two matrix materials have quite different oxygen isotope ratios (Clayton *et al.* 1976; Clayton and Mayeda 1978) which precludes any genetic relationship. The presence of this matrix material in both CI and CM meteorites apparently represents the action of similar processes (i.e. low-temperature alteration and hydration) acting on two different parent materials.

Spectral measurements of CM meteorites have been carried out by Watson (1938), Johnson and Fanale (1973), Gaffey (1974, 1976), Salisbury *et al.* (1975) and Larson *et al.* (1979). All spectra of CM meteorites share, with those of the CI meteorites, the strong iron charge transfer feature(s) shortwards of 0.56 μm and should also exhibit the very strong 3 μm water feature. Additional weaker (a few percent relative intensity) features include a 1.4 μm water band and at least four (apparently) iron-related features (~0.62 μm, ~0.70 μm, ~0.85 μm, ~0.92 μm). The relative intensity of these features divides the CM spectra into two groups. In group A spectra (Cold Bokkevelt, Murray; Fig. 6b), the ~0.62 μm and ~0.85 μm feature dominate this portion of the spectrum and a water feature is present at ~1.38 μm. In the group B spectra (Meghei, Murchison; Fig. 6c), the ~0.70 μm feature dominates with no 1.4 or 1.9 μm water bands present. Nogoya appears to be a member of group B (Fig. 6d) but with an even more intense ~0.70 μm feature and a ~0.92 μm feature present. The spectral differences between the two groups appear to be correlated to several interdependent parameters including: $Fe^{2+}/(Fe^{2+}+Fe^{3+}) - A = 0.347-0.378$, $B = 0.396-0.413$; percent of Fe^{2+} in a "serpentine"-type silicate (versus "chlorite" from Mossbauer spectroscopy; Vdovykin *et al.* 1975) $- A = 6-18\%$; $B = 25-35\%$; and Mg/Si (wt), probably reflecting an actual increase in the relative abundance of a "serpentine" phase $- A = 0.75-0.78$, $B = 0.81-0.83$, for Nogoya = 0.90. Spectrally the CI spectrum (Orgueil) would appear to be controlled by a matrix material similar to that of group A.

For most mineral assemblages, the effect of decreasing particle size is to increase the albedo but to have no significant effect on their normalized reflectance spectra. For CM assemblages, at least, this latter statement is not true. These meteorites show a strong increase in near-infrared ($\lambda \gtrsim 1.2$ μm) reflectance for smaller particle sizes (PS $\lesssim$ 75 μm) (see Johnson and Fanale 1973). Apparently one or more major absorption features in the near-infrared is weaker (than those in the 0.6–0.9 μm region) in the finer particle size fraction, resulting in a significant "reddening" of the reflectance spectrum ($\Delta Ref/\Delta\lambda > 0$). Two possible mechanisms could produce this effect. First, the absorption features in the near-infrared may have a significantly lower (but still quite high) optical density than those in the 0.6–0.9 μm spectral region, such that in the coarser size fractions, all spectral regions are nearly saturated (Δ absorbance/Δ path-length $\rightarrow$ 0) while in the finer fraction, those in the 0.6–0.9 μm region are nearly saturated but those in the near-infrared are much less saturated. The second alternative is that the significant disparity between the grain sizes (GS) in the matrix ($<$ 1 μm) and those of the chondrules and inclusions (~100 μm, 10–500 μm), could cause the latter phases to be spectrally enriched in the smaller size fraction. That is, the effective spectral abundance of a phase is a function of the mean photon path length in that phase, which in turn is a function of the mean distance between scattering boundaries in the material. For particles which are homogenous single crystals, this mean distance is related to the *particle size,* while for particles which consist of aggregates of grains, the mean scattering boundary distance is related to *grain size.* Thus for the CM meteorites the number of matrix scattering elements remains essentially unchanged (i.e., PS>>GS) while for the coarse crystalline inclusions and chondrules, a decrease in particle size rapidly increases the number of scattering elements $[N\sim(PS)^{-3}]$, so that their effective spectral contribution would rapidly increase also. Since these inclusions are primarily olivine, a rapid rise in the infrared reflectances would be expected. Thus, the relatively high infrared reflectance of C2-type asteroid surface materials should be interpreted with care. Moreover, the importance of studying spectral/particle size effects in these assemblages is clear, and future spectral studies should keep this in mind.

Type III Carbonaceous Chondrites

The Type III or C3 chondrites fall into two major groups: C3O (Ornans subtype) or CO and C3V (Vigarano) or CV. It must be noted that since each of these three major classifications (i.e. Type n, Cn, or C (I,M,O,V) – utilizes somewhat different criteria, not all meteorites, grouped together in one system, will automatically be in a single group in the other systems. For example, Mokoia has been classified as Type III, and as both C2V and C3V, and shares certain characteristics of both types 2 and 3 (see below).

CO meteorites are composed mainly of a fine-grained olivine matrix ($\sim$Fa$_{50}$) interstitial between abundant heterogenous (mean $\sim$Fa$_{11}$–Fa$_{34}$, range Fa$_{0-60}$) olivines (matrix/chondrules $\sim$0.5). Low calcium clinopyroxene ($\sim$Fs$_5$) and magnetite ($\sim$4%) form the major accessory phases. The water and carbon contents are low ($\sim$0.1 – 1.4 wt% and $\sim$0.2 – 0.6 wt%, respectively) which are consistent with the lower abundance (relative to CV types) of the matrix which, though essentially anhydrous, is the primary location of these volatile phases. McSween (1977*b*) suggests that the Ornans-type is a metamorphic sequence with increasing mean fayalite content of the olivine inclusions and decreasing olivine heterogeneity in the following order: Kainsaz, Felix, Ornans, Lance, Isna and Warrenton.

CV meteorites are chemically similar to CO meteorites but are distinguished on the basis of the morphology and larger size of chondrules and on the abundance of the dark matrix material (matrix/chondrules $\sim$ 2). Low calcium clinopyroxene and magnetite ($\sim$1 – 9 wt%) are the main accessory phases. Water and carbon are present ($\sim$0.1 – 4 wt% and $\sim$0.2 – 2 wt%, respectively) in higher abundances than in the CO type. The mafic minerals have lower mean iron contents ($\sim Fa_{10}$, $\sim Fs_6$) than those of the CO type, but are similarily quite heterogenous. Matrix content (matrix/chondrules + inclusion) forms three distinct groups (McSween 1977c): $1.15 \pm .03$ – Kaba, Grosnaja, Bali; $0.62 \pm .08$ – Leoville, Vigarano, Mokoia, Allende; and 0.24 – Coolidge.

The strongest spectral feature in the C3 spectra is the intense charge transfer feature or features in the blue and ultraviolet portion of the spectrum. The effective edge of this feature in the visible or near-infrared (i.e. where its absorbance is overwhelmed by other absorption features at longer wavelengths) is an excellent discriminator between the C2- and C3-type assemblages. This edge is defined as where the slope in the spectrum ($\Delta Ref/\Delta \lambda$) equals zero or is significantly discontinuous. In the C2 spectra this edge is near ~ 0.54–0.56 μm at significantly shorter wavelengths than that of the C3 assemblages (CO ~ 0.64–0.73 μm, CV ~ 0.75–0.78 μm). (It is important to note that these band edges do not appear to be affected by particle size variations.) Several of the C3 spectra exhibit weak inflections at shorter wavelengths, including: Kainsaz (the least metamorphosed CO specimen) ~ 0.55 and 0.63 μm; Mokoia (CV2) ~ 0.52 and 0.63 μm; Allende ~ 0.50 and 0.66 μm; and Vigarano ~ 0.50 and 0.59 μm. The weakness of these inflections should rule out any ambiguity. A shift of this edge toward shorter wavelengths, for materials with this range of albedos, would imply the absence or strong depletion of the species primarily responsible for the blue-ultraviolet features (Fe^{3+}).

In the 3 μm spectral region, the presence or absence of a strong water band also serves to discriminate between the hydrous C2 and anhydrous C3 assemblages (Larson et al. 1979; see the chapter by Larson and Veeder). The potential for utilizing the 3 μm feature to discriminate between the structural/chemical types of the C2 matrix material is currently being investigated by Feierberg and coworkers.

The CO spectra all show absorption features for olivine and pyroxene (Fig. 6e). With increasing degree of metamorphism (McSween 1977b) these bands are deeper and better defined: Felix, Ornans and Warrenton have 6%, 9% and 15% absorption features at ~ 1.03 μm (olivine). There is a persistent H_2O feature at ~ 1.38 μm which decreases in absorbance with increasing grade. In the spectrum of Kainsaz, the H_2O feature is strongest and the olivine-pyroxene 1 μm feature is distorted by a strong band near 0.88 μm. This feature could arise either from iron oxide in the matrix or from the inclusion of a small amount of group A CM matrix material within the matrix of this meteorite. That Kainsaz is spectrally the most "primitive" (i.e. CM-like) of the CO meteorites is quite in agreement with the McSween (1977b) metamorphic sequence.

The CV chondrites do not show, as clearly as the CO spectra, the olivine and pyroxene absorption features. This agrees with the lower mean iron content in these phases. They generally have lower albedos, consistent with their higher content of the dark, fine-grained olivine matrix. Spectrally, the CV meteorite split into two groups. The first group (A) includes Allende and Mokoia (Fig. 6f) and the second group (B) includes Vigarano, Grosnaja and Leoville (Fig. 6g). Group A spectra exhibit a weak ($\sim$4%) iron (magnetite, iron oxide?) feature at $\sim$0.86 μm and a water band at $\sim$1.38 μm. Olivine is evidenced only by the presence of a relatively weak inflection between 1.0 μm and 1.1 μm, while a pyroxene feature is evident near 2 μm as a weak depression below a linear continuum. The intensity (depth below a linear continuum) of the olivine and pyroxene features in Allende is about twice that in Mokoia. The infrared reflectance increases with increasing wavelength, reaching a normalized reflectance of $\sim$1.3 at 2.5 μm.

Group B samples appear to have slightly higher albedos, than those of group A, and a flatter infrared reflectance curve (normalized reflectance at 2.5 μm is $\sim$1.0–1.1). Weak olivine ($\sim$1.05 μm) and pyroxene ($\sim$2 μm) features are seen in the spectra of Grosnaja and Vigarano. The $\sim$0.85 μm iron band is less prominent than in the group A spectra, which would support the idea that it arises from absorption by small amounts of CM type B matrix material. A weak water feature is present at $\sim$1.38 μm.

In the petrologic affiliations classification of McSween (1977c), the spectra type A and B meteorites fall into different groups: Type A (Allende and Mokoia) constitute the *oxidized, low-matrix, many-opaques-in-chondrules* group; Type B meteorite, Grosnaja is a member of the *oxidized, high-matrix, few-opaques-in-chondrules* group and Type B meteorites Vigarano and Leoville are members of the reduced group. Further work investigating this apparent relationship is needed.

C4 and C5 Carbonaceous Chondrites

The C4,5 carbonaceous chondrites are represented by Coolidge and Karoonda, the former exhibiting a reflectance spectrum which appears to show the effects of significant terrestrial weathering. Karoonda consists primarily of olivine (Fa_{34}) with accessory magnetite ($\sim$8 wt%), plagioclase, pigeonite, and pentlandite (a NiFe sulfide). Carbon and water are present in only trace amounts. Petrologically it appears to be strongly recrystallized and metamorphosed, as evidenced by the extreme homogeneity of the olivine (Van Schmus 1969). Spectrally (Fig. 6h) it exhibits a strong ($\sim$10%), well-defined olivine absorption feature ($\sim$1.06 μm). A weaker feature is present as an inflection near 0.9 μm, which appears to be a combination of an Fe^{3+} feature with a side lobe in the olivine feature. It is possible that pyroxene is also contributing to this feature, but no evidence of a 2 μm pyroxene band is seen. A very weak H_2O band appears to be present near 1.38 μm and the blue-ultraviolet absorption edge (see above) is at $\sim$0.75 μm.

Kakangari

Kakangari does not fall easily into one of the more conventional classifications, and as such is treated separately. It is relatively reduced (oxidation state intermediate between E and H chondrites), chemically similar to ordinary chondrites but has an oxygen isotope pattern similar to C2 carbonaceous chondrites (Graham and Hutchison 1974; Clayton *et al.* 1976; McSween and Richardson 1977). Petrologically it consists of abundant chondrules of olivine ($\sim Fa_5$) and pyroxene ($\sim Fs_7$), ranging in size from 0.3–3 mm, set in a fine-grained, enstatite groundmass matrix ($\sim Fs_0$) which includes abundant fine opaque inclusions of iron sulfide and nickel-iron. Estimated abundances are: olivine + pyroxene, ~ 40 wt%, pyroxene > olivine; groundmass, ~ 30 wt%; FeS ~ 15 wt% and NiFe ~ 9 wt%. Spectrally, Kakangari (Fig. 6*i*) exhibits a strong blue-ultraviolet absorption feature with a band edge at ~ 0.61 μm. At least three weak, "iron" features are present (~ 0.5 μm, ~ 0.67 μm, ~ 0.85 μm) but which mineral they are associated with (FeS_N?) is unclear. Apparently the weak, ~ 0.91 μm feature is due to pyroxene but no clear 2 μm feature is discernible. Until the nature of the iron features is better understood, investigations of the relationships between the spectral and mineralogic properties cannot be pursued effectively.

Ureilites

Ureilites are a class of feldspar-free achondrites consisting mainly of olivine with smaller amounts of clinopyroxene (pigeonite), NiFe (kamacite), troilite and carbon (present as a disequilibrium assemblage of graphite, diamond and organic material). These meteorites are enriched in magnesium and depleted in iron and aluminum relative to chondrites. Their carbon content (1.5–2.9 wt%) is higher than most meteorites including many Type 3 carbonaceous chondrites. Clayton *et al.* (1976) showed that the ureilites formed a unique group on a $\partial^{17}O-\partial^{18}O$ plot and that they were related to, but not directly derived from, the anhydrous phases of the C2 and C3 meteorites. Current models of origin involve the injection of a carbon-rich material into a differentiated ultramafic parent material (e.g. Wasson *et al.* 1976; Boynton *et al.* 1976; Higuchi *et al.* 1976; and Berkley *et al.* 1976).

Two subtypes of the ureilites are recognized: Type I (type specimen – Novo Urei) and Type II (type specimen – Goalpara). The Type I ureilites are distinguished by (a) coarser-grained olivines (up to 4 mm) versus very fine-grained aggregates of olivine, (b) twinning of the clinopyroxene, (c) a net-like distribution of metallic NiFe versus concentration in plate-like grains, and (d) smaller diamond-graphite aggregates ($\geqslant 0.3$ mm versus $\geqslant 0.9$ mm) (Vdovykin 1970). This class of meteorites has been reviewed in detail by Vdovykin (1970) with additional data in Vdovykin (1976).

Of the ureilites only Novo Urei has been studied spectrally (Gaffey 1974, 1976). This meteorite consists of olivine ($\sim Fa_{24}$, ~ 65 wt%), pigeonite ($\sim Fs_{24}$, ~ 20 wt%), kamacite (NiFe, ~ 6 wt%), carbon (~ 2 wt%) and troilite (FeS, ~ 2 wt%). The spectral reflectance curve of Novo Urei (Fig. 6*j*) exhibits a strong ($\sim 15\%$) 1 μm and a well-defined 2 μm pyroxene feature ($\lambda \sim 0.925$

μm and ~2.1 μm). The 1 μm feature is broadened toward longer wavelengths and the infrared reflectance is increased (normalized reflectance ~1.25 at 2.5 μm) by the olivine component. The width of the long wavelength half of the 1 μm feature is ~0.150 μm, which would correspond to an olivine/(olivine + pyroxene) ratio of 1/3 (Fig. 2) which is significantly less than that expected for an assemblage with an actual ratio of ~0.75 (~0.245 μm). This is not surprising in view of the shock history of the ureilites. Under conditions of moderate to severe shock, olivine develops a microfracture texture (dislocations and mosaicing; Carter *et al.* 1968; Ashworth and Barber 1975) much more readily than pyroxene. The significantly decreased mean distance between scattering boundaries for the olivine relative to pyroxene reduces the effective spectral abundance of the former.

Several other distinctive spectral features are also present. The blue-UV absorption edge is moderately strong with the absorption edge at ~0.60 μm. The reflectance plateau between ~0.60 and 0.75 μm is characteristic of olivine-rich assemblages. A weak inflection between ~1.2 and ~1.5 μm is due to the long-wavelength sidelobe of the olivine feature (see Fig. 1c), feldspar being very rare. A weak ~1.37 μm water band is present.

No specimen of the Goalpara type of the ureilites has been studied spectrally. However, taking into consideration the mineralogic and petrologic differences between the two types, the systematic variation of their spectral reflectance properties, with respect to the Novo-Urei type, can be estimated. The (apparently) higher shock history of the Goalpara type has produced more scattering boundaries within the olivine crystals which should decrease the effective spectral contribution of the olivine phase. The coarser graphite-diamond aggregates, by virtue of decreased surface area, should decrease the effect of the opaque carbon phases. Thus the reflectance spectra of Goalpara type assemblages should have higher albedos and higher apparent pyroxene/olivine abundances.

Black Chondrites

Black Chondrites are ordinary chondrites, predominately L-type, which have undergone a severe shock event (Heymann 1967). The shock darkening and mosaicing of their olivine (ol/ol+py $\gtrsim$ 0.5), has lowered their albedos and decreased the relative spectral contribution of the olivine phase. The spectra of these meteorites (Fig. 6k) all show the 1 and 2 μm pyroxene bands, the intensity of which varies inversely with the albedo of the sample (and presumably with the shock). The apparent spectral abundance of the olivine, proportional to the width of the long-wavelength half of the 1 μm feature, also varies in a similar manner. The short-wavelength position of the 1 μm feature (~0.90 μm), consistent with the low calcium content of ordinary chondritic pyroxenes, serves to distinguish these from ureilites whose relatively calcic pyroxenes produce a band minimum at longer wavelengths (~0.925 μm). (Note that this should not be utilized as the sole discriminatory parameter, since an increased spectral contribution from olivine will lead to a similar shift.) The long-wavelength position of the blue-ultraviolet absorption edge for the black chondrites ($\gtrsim 0.70$ μm) serves to discriminate between these assemblages and the CM assemblages, while their pyroxene-dominated spectral curves are distinct from the olivine-dominated curves of the CO assemblages. The spectral reflectance curves of the black chondrites are increasingly reddened ($\Delta\text{Ref}/\Delta\lambda > 0$) with decreasing albedo, which presumably represents the effect of creating small metallic iron inclusions by *in situ* shock metamorphism and reduction at the high shock levels.

Thus the spectral reflectance curves of the black chondrites exhibit a degree of uniqueness which parallels their discrete mineralogy and petrology with respect to other types of dark meteorites. There should be little difficulty in distinguishing and characterizing these distinct mineral assemblages. However, distinguishing between the different types of black chondrites (e.g. H, L, LL), which differ primarily in the relative abundance of olivine, pyroxene and silicates, would be a much more difficult task. As noted above, the apparent relative abundance of minerals in a spectrum can be significantly altered by shock effects. For example, the olivine-rich, metal-poor LL-type assemblages can be made to spectrally mimic the olivine-poor, metal-rich H-type assemblages. Although albedo could serve to indicate the degree of shock, the intrinsic spread in ordinary chondrite albedos with petrologic grade would probably not permit its use.

Enstatite Chondrites

Enstatite chondrites represent a very reduced form of chondritic matter. They are primarily assemblages of nearly iron-free silicates (enstatite and plagioclase, $\sim$40–60 wt% and $\sim$5–10 wt%, respectively), NiFe (kamacite, $\sim$17–28 wt%), and sulfides (troilite-FeS and niningerite – [Mg, Fe] S, $\sim$7–15 wt%) (Mason 1966; Reid and Cohen 1967; Keil 1968). The spectral reflectance curve (Fig. 6*l*) of the E4 chondrite, Abee (Type I; Keil 1968) exhibits no strong absorption features and, in a wavelength plot, has a very nearly linear increase in reflectance into the near-infrared (normalized reflectance $\sim$1.7 at 2.5 μm). A weak, $\sim$2%, iron-related feature is present near $\sim$0.86 μm which could arise either from the trace abundance of Fe^{2+} in the enstatite, or from a terrestrially produced Fe^{3+} alteration product or, more likely, from a combination of both. The weak $\sim$1.37 μm water feature is almost certainly of terrestrial origin. The shape of this reflectance curve is similar to that of metallic NiFe. It seems likely that the series of inflections in the curve near 0.5–0.6 μm arises from Fe^{2+} or Fe^{3+} transitions but they are not understood at this time. Gaffey and McCord (1978) suggested that the low albedo might arise from thin dark oxide (wustite) coatings on the metal grains. It appears equally likely that the low albedo is simply an effect of multiple ($\sim$2) reflections between the metal grains in the enstatite matrix.

The spectral reflectance curve of the E6 (Type II) assemblages (Fig. 6m) has a higher albedo and a significantly flatter infrared portion (normalized reflectance ~1.2–1.3 at 2.5 μm). A strong blue-ultraviolet absorption extends to about 0.6 μm and two weak features at ~0.65 and ~0.85 μm are present. These weak features could arise either from the trace amounts of iron in the pyroxene or from Fe^{3+} transition in an alteration product, or again, both. Strictly speaking the E6 assemblages are not "dark materials;" however, they do provide an indication of the direction of variation in spectral properties with metamorphism.

Dark Meteorites and Dark Asteroids: Conclusions

The low-albedo meteorites constitute a diverse variety of both chondritic and achondritic mineral assemblages. Their spectral reflectance curves exhibit spectral features related to their characteristic mineralogy, although many relationships remain to be completely understood. It is evident from a consideration of the observational data of low-albedo asteroids, predominantly (but not exclusively) the C-type, that these objects also exhibit a diverse range of spectra. The interpretation of these spectra is now becoming feasible and should be pursued with vigor.

II. TWO CASE HISTORIES

It is instructive to consider in detail the evolution in the interpretive process and surface material characterization for two extensively studied asteroids: 4 Vesta and 1 Ceres. In particular it is important to understand the relative significance of different spectral data sets in the characterization of these two objects with quite different surface materials.

Vesta

Vesta (Fig. 7) has been the most closely and persistently studied asteroid and continues to be of great interest to both the astronomical and meteoritic communities. The following summary illustrates the increasing sophistication of both the data base and the interpretation of these data.

1. McCord *et al.* (1970) measured the reflectance spectrum of Vesta with moderate spectral resolution and coverage (~0.40–1.08 μm, 24 filters). They identified a deep absorption band (~0.92 μm) which they interpreted as diagnostic of a pigeonite (pyroxene with moderate calcium content). The spectrum was matched to that of a eucritic basaltic achondrite (pyroxene + plagioclase feldspar). A second pyroxene band was predicted near 2.0 μm.

2. Chapman (1972) obtained a spectral curve of Vesta with the absorption feature centered near 0.95 μm which was interpreted to indicate a more calcium- or iron-rich pigeonite.

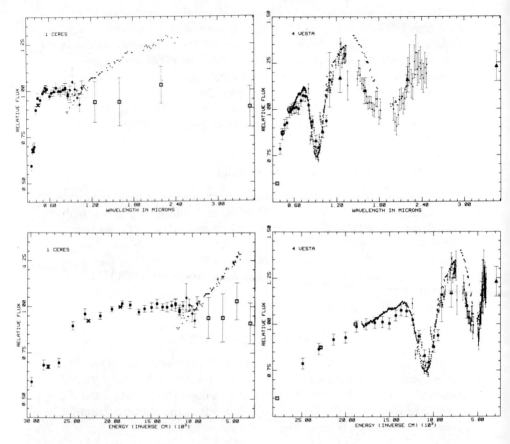

Fig. 7. Spectral reflectance data for two asteroids from various investigators: (a) 4
Vesta and (b) 1 Ceres versus wavelength (top) and versus wavenumber (bottom). UBV
data from TRIAD. JHKL data from Veeder *et al.* (1978). Visible to near-infrared data
from Chapman and Gaffey (see their chapter). Infrared interferometry (small points)
from Larson and Veeder (see their chapter). Infrared spectrum of Vesta (small points
with error bars) from McFadden and McCord (1978). Near-infrared spectrum of Vesta
(small symbols) from McFadden *et al.* (1977).

3. Chapman and Salisbury (1973) compared this spectrum to a range of
   meteorite spectra and concluded that it was best matched by a
   laboratory spectrum of the howarditic basaltic achondrite, Kapoeta.

4. Veeder *et al.* (1975) measured a high-resolution (~50Å) 0.6–1.1 μm
   reflectance spectrum of Vesta, determined the absorption band position
   to be 0.92±0.02 μm and interpreted this to represent a calcic pyroxene
   or eucritic basaltic achondrite.

5. Johnson *et al.* (1975) measured the broad bandpass reflectance of Vesta
   at 1.65 and 2.20 μm (H and K filters) and concluded that these data
   were consistent with values expected for a basaltic achondritic surface

material. They emphasized the need for higher resolution spectra beyond 1.0 μm.

6. Larson and Fink (1975) measured the 1.1–2.5 μm reflectance of Vesta relative to the moon. They identified the predicted second pyroxene band and confirmed the existence of pyroxene in the surface material. They indicated that no absorption bands for olivine, feldspar or ices were seen in the spectrum. However, their band position (2.06±0.01 μm) was artificially shifted toward longer wavelengths because of the slope in the lunar spectrum used as the standard.

7. McFadden et al. (1977) measured the high-resolution (20–40Å) 0.5–1.06 μm spectrum and determined the band position to be 0.924±0.004 μm. This band position was interpreted to indicate the presence of a pyroxene with a 10–12 mole % calcium content. They suggested that the symmetry of the absorption feature indicated little or no olivine.

8. Larson (1977) presented the 1.0–2.5 μm reflectance curve of Vesta relative to the sun. The band minimum (2.00±0.05 μm) was within the field of eucrite meteorites, although it overlapped with the howardite field.

9. Feierberg et al. (1978) discussed a new 0.8–2.5 μm spectrum of Vesta from which they concluded that the infrared spectrum of this asteroid was most consistent with a Ca-rich achondritic composition, similar to the howardites. They also noted a spectral feature at 1.2 μm which they concluded was evidence of a major plagioclase feldspar component in the surface material.

10. McFadden and McCord (1978) discussed their own 0.3–2.5 μm reflectance spectrum of Vesta and identified the pyroxene and plagioclase absorption features. They concluded that the relative ratio plagioclase/pyroxene is 0.66±0.2.

Ceres

Ceres (Fig. 7) is the largest asteroid and constitutes more than a third of the total mass of the minor planets. Ceres is an atypical member of the dark C-type asteroids which dominate the main-belt population. Also the spectral and albedo properties of the Martian moons, Phobos and Diemos are essentially identical to those of Ceres. These facts have contributed to a strong and continuing interest in this object. The results of this work are summarized as follows.

1. Johnson and Fanale (1973) compared the reflectance spectrum of Ceres to the laboratory spectra of several carbonaceous chondrites and a "carbonaceous" construct. They noted that the spectrum of this asteroid did not match that of any known meteorite or ordinary mineral assemblage. They concluded, however, that their construct (a

mixture of powdered carbon and the clay mineral, montmorillonite) did provide a good match.

2. Chapman and Salisbury (1973) attempted to match the spectrum of Ceres to the laboratory spectra of a number of meteorites and concluded that among their selection, no meteorites could provide a good match. They noted, however, that certain opaque-rich basalts did provide a reasonable match.

3. McCord and Gaffey (1974) concluded that the spectral properties of Ceres indicated a surface material dominated by an opaque phase (e.g. carbon) and thus probably represented some type of carbonaceous surface.

4. Chapman et al. (1975) stressed that Ceres, as well as several other of the largest asteroids, have spectral albedo curves which set them apart from the more typical C-type asteroids. They proposed three possible surface materials: (i) basalts rich in opaques, (ii) a different kind (perhaps ultraprimitive) of carbonaceous meteoritic material, and (iii) a metamorphosed carbonaceous-type material, such as the C4 meteorite, Karoonda. They concluded that the relatively low density of Ceres ($\rho \sim 2.2$–3.1 g cm^{-3}) would mitigate against a basaltic object ($\rho \sim 3.5$ g cm^{-3}). They preferred the third alternative.

5. Chapman (1976) distinguished Ceres from most other C-type asteroids and designated it and several other asteroids with similar spectra a "C-asterisk". He suggested that the surface material was similar to Karoonda or possibly an opaque-rich basalt.

6. Morrison (1976) compared the densities of Ceres and Vesta and concluded that the relatively low density of the former implied a high bulk volatile content (e.g. H_2O) similar to Type I or II carbonaceous chondrites.

7. Gaffey and McCord (1978) considered the spectral albedo properties of Ceres at length and concluded that the surface material must consist of an opaque phase mixed with a relatively transparent (silicate) phase such as olivine or an iron-poor clay mineral. They ruled out any material with a (spectrally) abundant Fe^{3+} phase such as is typical of the C1 or C2 meteorites. Their preferred alternative was an olivine-magnetite assemblage similar to Karoonda or opaque-rich basalts, but considered that an unknown type of opaque-rich, iron-poor clay mineral assemblage (analogous to but chemically distinct from C1 or C2 meteorite assemblages) could not be ruled out.

8. Lebofsky (1978) observed Ceres in the 3 μm spectral region and detected an H_2O-related absorption feature which indicated the presence of hydrated minerals in the surface material of this body.

9. Larson et al. (1979) observed Ceres in the 1.0–2.6 μm spectral region and combined these data with existing data to provide 0.4–3.6 μm spectral coverage. They concluded that the surface material of Ceres

was consistent with mixtures of opaques and hydrated silicates, such as are found in C1 and C2 meteorites.

10. Gaffey (1978) considered the extended spectral albedo data for Ceres and concluded that the surface material of this body was unlike any known meteorite type. He suggested two types of mineral assemblages as possible candidates: (i) a hydrated iron-poor silicate mixed with abundant opaques, or (ii) an opaque-rich hydrated assemblage with silicates containing normal chondritic iron contents but predominantly in the bivalent state (Fe^{2+}). He proposed that the former assemblage could be a result of either low-temperature accretion from an iron-depleted nebular region or of an efficient *in situ* leaching of iron from the silicate and its concentration into large (opaque) oxide grains (magnetite). The latter assemblage requires the establishment of a state of chemical disequilibrium (the H_2O normally promotes the oxidation of Fe^{2+} to Fe^{3+}). This might be accomplished by a heating and reduction episode but it is difficult to envision how the hydrated phases could retain their water through such an episode. However, the reaction kinematics for C1-C2 type assemblages are not known well enough to rule out this alternative.

The evolution of the observational data base is parallel for Vesta and Ceres, resulting from improvements in instrumental capabilities. The evolution of an interpretation for the surface material of each has diverged significantly. In the case of Vesta, the initial work recognized a mineralogically diagnostic spectral feature, the 0.9 μm pyroxene absorption band. Subsequent work has concentrated on more precise characterization of this feature, and the verification and characterization of additional predicted features related to pyroxene and other expected mineral phases. There exists no current dispute that its surface material contains abundant pyroxene and feldspar and that it is similar to either eucrite or howardite basaltic achondrites (see Drake's chapter).

The interpretation of the spectral albedo of Ceres is still a matter of active controversy. There is general agreement among investigators that the surface material is an assemblage of an opaque phase mixed with hydrated, presumably silicate, minerals. The dispute arises when meteoritic equivalents are considered, and to a large extent derives from incorrect or ambiguous use of meteorite terminology. Since carbonaceous chondrites have generally lower albedos than most other chondrites, and since they contain hydrated minerals, many investigators have assumed that the low albedo, hydrated surface material of Ceres must be carbonaceous chondrite material.

However, as was noted in our introduction, the carbonaceous chondrites are distinguished from other chondrites on the basis of chemical criteria which do not require the presence of either carbon or water or a low albedo. There is no meteoritic evidence to indicate that the lack of hydrated, opaque-rich members of the other chondritic groups or of altered material is anything other than a random selection effect in material arriving at the earth's surface during recent time.

Moreover, the conclusions by a number of investigators that the visible spectrum of Ceres cannot be matched to any studied carbonaceous chondrite specimen has not been challenged so much as discounted by the proponents of a C1 or C2 type surface material. An understanding of the spectral behavior of carbonaceous chondritic material at wavelengths below 0.6 μm and its relationship to iron oxidation state and distribution and to possible modifying processes will be needed to completely resolve this controversy. In the meantime, careful use of the adopted meteorite terminology would go far to alleviate the communication failure aspects of this dispute.

Mineralogical characterizations of asteroid surface materials

Gaffey and McCord (1978) have discussed in some detail their interpretation of surface mineralogy for 65 asteroids. Since that work was published, additional spectral data have become available and the interpretations of some objects have been modified or refined. A current summary of these results is given in Table I.

It is useful to restate the major conclusions of that study as a basis for further discussion:

1. Asteroid surface materials are generally composed of assemblages of meteoritic minerals.
2. Mineral assemblages analogous to most meteorite types, with the significant exception of the ordinary chondrites, have been found on main-belt asteroids.
3. The main-belt population is dominated by materials similar to C1 or C2 assemblages (i.e. iron-rich, clay minerals with abundant opaques) although a number of spectrally distinct subtypes of unknown implication exist.
4. A significant minority of the main-belt asteroids have undergone an intense heating episode which permitted melting and differentiation to take place.
5. Co-existing with this population is the majority which appears to have undergone no significant thermal event at any time.
6. All asteroids larger than ~450 km in diameter have undergone significant heating; this conclusion now appears premature.
7. The "compositional" distributions in population of Earth-approaching and Earth-crossing asteroids appears to be quite different from that of the main belt, including a significant if not dominant component of ordinary chondrite-like assemblages and including very few of the C2-like (i.e. low albedo) assemblages. (The statistics for the Apollo-Amor population are poor and observational bias against dark C-type bodies may reduce this discrepancy.)

TABLE I

Asteroid Surface Materials: Characterizations*

Asteroid	Spectral Type	Mineral Assemblage[a]	Meteoritic Analog[b]	CMZ Type[c]
1 Ceres	F	HLFe, Opq	None	C
2 Pallas	F	HLFe, Opq	None	U
3 Juno	RA-1	NiFe ~ (Ol~Px)	Ol-Px Stony-Iron	S
4 Vesta	A	Cpx	Eucrite	U
6 Hebe	RA-2	NiFe>Cpx	Mesosiderite	S
7 Iris	RA-1	NiFe, Ol, Px	Ol-Px Stony-Iron	S
8 Flora	RA-2	NiFe≳Cpx	Mesosiderite	S
9 Metis	RF	NiFe, (Sil(E))	E. Chon, Iron	S
10 Hygiea	TB	Phy, Opq(C)	C1-C2 (C1-CM)	C
11 Parthenope	RF	NiFe, (Sil (E))	E. Chon, Iron	S
14 Irene	RA-3	NiFe, Px	Px Stony-Iron	S
15 Eunomia	RA-1	NiFe~(Ol>>Px)	Ol-Px Stony-Iron	S
16 Psyche	RR	NiFe, Sil(E)	E. Chon, Iron	M
17 Thetis	RA-2	NiFe, Cpx	Mesosiderite	S
18 Melpomene	TE	Sil(O), Opq(C)	C3	S
19 Fortuna	TA	Phy, Opq(C)	C1-C2 (CI-CM)	C
25 Phocaea	RA-2	NiFe, Px, Cpx	Px Stony-Iron	S
27 Euterpe	RA-2	NiFe, Px, Cpx	Px Stony-Iron	S
28 Bellona	TE	Sil(O), Opq(C)	C3	S
30 Urania	RF(?)	---	---	S
39 Laetitia	RA-1	NiFe ~ (Ol~Px)	Ol-Px Stony-Iron	S
40 Harmonia	RA-2	NiFe > Px	Mesosiderite	S
48 Doris	TA	Phy, Opq(C)	C1-C2 (CI-CM)	U
51 Nemausa	TC	Phy, Opq(C)	C1-C2 (CI-CM)	U
52 Europa	TA	Phy, Opq(C)	C1-C2 (CI-CM)	C
58 Concordia	TABC	Phy, Opq(C)	C1-C2 (CI-CM)	C
63 Ausonia	RA-3	NiFe, Px	Px Stony-Iron	S
79 Eurynome	RA-2	NiFe ~ Cpx	Mesosiderite	S
80 Sappho	TD	Sil(O), Opq(C)	C3	U
82 Alkmene	TE	Sil(O), Opq(C)	C3	S
85 Io	F	f	f	C
88 Thisbe	TB	Phy, Opq(C)	C1-C2 (CI-CM)	C
130 Elektra	TABC	Phy, Opq(C)	C1-C2 (CI-CM)	U
139 Juewa	TB	Phy, Opq(C)	C1-C2 (CI-CM)	C
140 Siwa	RR	NiFe, Sil(E)	E. Chon, Iron	C
141 Lumen	TA	Phy, Opq(C)	C1-C2 (CI-CM)	C
145 Adeona	TA	Phy, Opq(C)	C1-C2 (CI-CM)	C
163 Erigone	TA	Phy, Opq(C)	C1-C2 (CI-CM)	C
166 Rhodope	TC	Phy, Opq(C)	C1-C2 (CI-CM)	U
176 Iduna	TA	Phy, Opq(C)	C1-C2 (CI-CM)	U
192 Nausikaa	RA-2	NiFe~(Px>Ol)	Px-Ol Stony-Iron	S
194 Prokne	TC	Phy, Opq(C)	C1-C2 (CI-CM)	C

TABLE I (Continued)

Asteroid Surface Materials: Characterizations*

Asteroid	Spectral Type	Mineral Assemblage[a]	Meteoritic Analog[b]	CMZ Type[c]
210 Isabella	TABC	Phy, Opq(C)	C1-C2 (CI-CM)	CMEU
213 Lilaea	F	f	f	C
221 Eos	TD	Sil(O), Opq(C)	C3	U
230 Athamantis	RF	NiFe, (Sil(E))	E. Chon, Iron	S
324 Bamberga	TABC	Phy, Opq(C)	C1-C2 (CI-CM)	C
335 Roberta	F	f	f	EU
349 Dembowska	A	Ol, (NiFe)	Ol. Achondrite	R
354 Eleonora	RA-1	NiFe~Ol	Pallasite	U
433 Eros[d]	— —	Px~Ol, NiFe	H Chondrite	S
462 Eriphyla	RF (?)	— — —	— — —	U
481 Emita	TABC	Phy, Opq(C)	C1-C2 (CI-CM)	C
505 Cava	TA	Phy, Opq(C)	C1-C2 (CI-CM)	U
511 Davida	TB	Phy, Opq(C)	C1-C2 (CI-CM)	C
532 Herculina	TE	Sil(O), Opq(C)	C3	S
554 Peraga	TA	Phy, Opq(C)	C1-C2 (CI-CM)	C
654 Zelinda	TC	Phy, Opq(C)	C1-C2 (CI-CM)	U
674 Rachele	RF (?)	— — —	— — —	S
704 Interamnia	F	f	f	U
887 Alinda	TD	Sil(O), Opq(C)	C3	S
1685 Toro[e]	— —	Px, Ol	L Chondrite (?)	S

*Table modified from Gaffey and McCord 1978.

[a]Mineral assemblage of asteroid surface material determined from interpretation of reflectance spectra: NiFe (nickel-iron metal); Ol (olivine); Px (pyroxene, generally low calcium orthopyroxene); Cpx (clinopyroxene, calcic pyroxene); Sil(O) (mafic silicate, most probably olivine); Sil(E) (spectrally neutral silicate, most probably iron-free pyroxene [enstatite], or iron-free olivine [forsterite]); Phy (phyllosilicate, layer lattice silicate, meteoritic clay mineral, generally hydrated, unleached with abundant subequal Fe^{2+} and Fe^{3+} cations); HLFe (hydrated silicate spectrally low in Fe^{3+} such as a leached clay mineral); Opq (opaque phase, either carbon or magnetite indicated on basis of meteoritic associations).

Mathematical symbols ($>$, greater than; $\gg$, much greater than; $\sim$ approximately equal) are used to indicate relative abundance of mineral phases. In cases where abundance is undetermined, order is of decreasing apparent abundance.
Asteroidal spectra which are ambiguous between "TDE" and "RF" are not characterized mineralogically.

[b]Meteoritic analogues are examples of meteorite types with similar mineralogy but genetic links are not established. For example, objects designated as analogous to mesosiderites could be a mechanical metal-basaltic achondritic mixture.
[c]Asteroidal color-albedo-polarization classification from the TRIAD file (Part VII of this book).
[d]Pieters et al. 1976.
[e]Chapman et al. 1973.
[f]Type "F" spectra without 3 μm data could be either a Ceres-like assemblage or a Karoonda-like assemblage (olivine + abundant magnetite).

At the present time all of these conclusions still appear valid with the possible (probable?) exception of No. 6. As noted in the previous section, the discovery of a 3 μm water band in the spectrum of 1 Ceres essentially ruled out the possibility that it had an opaque-rich basalt surface. However, since the process that did lead to the formation of the present surface material is unknown, the existence of a moderate thermal event is neither supported nor ruled out.

The most significant modification of these conclusions arises from a reconsideration of the spectra of the low-albedo asteroids, including most or all of those classified as C, many of the U objects and some of the lower albedo members of the S and M groups. Gaffey and McCord (1978) distinguished several spectral subtypes within this population including their $T1$-$T5$ and F groups. Consideration of the nearly 300 spectra now available (see the chapter by Chapman and Gaffey) in light of the discussion of diagnostic spectral features for the dark meteorites (above), makes it clear that this diversity was significantly underestimated. There appear to be in excess of 15 significant subgroups among the dark asteroids. For those with spectra having sufficiently high signal-to-noise levels, detailed mineralogic interpretations are now possible.

REFERENCES

Adams, J.B. 1974. Visible and near-infrared diffuse reflectance spectra of pyroxenes as applied to remote sensing of solid objects in the solar system. *J. Geophys. Res.* 79: 4829-4836.

Adams, J.B. 1975. Interpretation of visible and near-infrared diffuse reflectance spectra of pyroxenes and other rock-forming minerals. In *Infrared and Raman Spectroscopy of Lunar and Terrestrial Minerals,* ed. C. Karr (New York: Academic Press), pp. 91-116.

Adams, J.B., and Filice, A.L. 1967. Spectral reflectance 0.4 μm to 2.0 μm of silicate rock powders. *J. Geophys. Res.* 72: 5705-5715.

Adams, J.B., and Goullaud, L.H. 1978. Plagioclase feldspars: Visible and near-infrared diffuse reflectance spectra as applied to remote sensing. *Proc. Lunar Sci. Conf. IX,* (Oxford: Pergamon Press), pp. 2901-2909.

Adams, J.B., and McCord, T.B. 1970. Remote sensing of lunar surface mineralogy: Implications from visible and near-infrared reflectivity of Apollo 11 samples. *Proc. Lunar Sci. Conf. 1st* (Cambridge, Mass: M.I.T. Press), Vol. 3, pp. 1937-1945.

Anders, E. 1964. Origin, age and composition of meteorites. *Space Sci. Rev.* 3: 583-714.

Anders, E. 1971. Interrelations of meteorites, asteroids and comets. In *Physical Studies of Minor Planets,* ed. T. Gehrels (NASA SP-267, Washington, D.C.; U.S. Government Printing Office), pp. 429-446.

Anders, E. 1975. Do stony meteorites come from comets? *Icarus* 24: 363-371.

Arnold, J.R. 1965. The origin of meteorites as small bodies. III. General considerations. *Astrophys. J.* 141: 1548-1556.

Aronson, J.R., and Emslie, A.G. 1975. Applications of infrared spectroscopy and radiative transfer to earth sciences. In *Infrared and Raman Spectroscopy of Lunar and Terrestrial Minerals,* ed. C. Karr (New York: Academic Press), pp. 143-164.

Ashworth, J.R., and Barber, D.J. 1975. Electron petrography of shock deformed olivine in stony meteorites. *Earth Planet. Sci. Lett.* 27: 43-50.

Bell, P.M., and Mao, H.K. 1973. Optical and chemical analysis of iron in Luna 20 plagioclase. *Geochim. Comochim. Acta* 37: 755-759.

Bell, P.M.; Mao, H.K.; and Rossman, G.R. 1975. Absorption spectroscopy of ionic and molecular units in crystals and glasses. In *Infrared and Raman Spectroscopy of Lunar and Terrestrial Minerals,* ed., C. Karr (New York: Academic Press), pp. 1-38.

Berkley, J.L.; Brown, H.G.; Keil, K.; Carter, N.L.; Mercier, J.C.C.; and Huss, G. 1976. The Kenna ureilite: An ultramafic rock with evidence for igneous, metamorphic and shock origin. *Geochim. Cosmochim. Acta* 40: 1429-1437.

Boynton, W.V.; Starzyk, P.M.; and Schmitt, R.A. 1976. Chemical evidence for the genesis of the ureilites, the achondrite Chassigny and the nakhlites. *Geochim. Cosmochim. Acta* 40: 1439-1447.

Burns, R.G. 1970*a*. *Mineralogical Applications of Crystal Field Theory,* (New York: Cambridge University Press).

Burns, R.G. 1970*b*. Crystal field spectra and evidence of cation ordering in olivine minerals. *Amer. Mineral.* 55: 1608-1632.

Burns, R.G. 1974. The polarized spectra of iron in silicates: Olivine. A discussion of neglected contributions from Fe^{2+} ions in M(1) sites. *Amer. Mineral.* 59: 625-629.

Carter, N.L.; Raleigh, C.B.; and DeCarli, P.S. 1968. Deformation of olivine in stony meteorites. *J. Geophys. Res.* 73: 5439-5461.

Ceplecha, Z., and McCrosky, R.E. 1976. Fireball end heights: A diagnostic for the structure of meteoric material. *J. Geophys. Res.* 81: 6257-6275.

Chapman, C.R. 1972. *Surface properties of asteroids.* Ph.D. dissertation, Massachusetts Institute of Technology.

Chapman, C.R. 1976. Asteroids as meteorite parent bodies: The astronomical perspective. *Geochim. Cosmochim. Acta* 40: 701-719.

Chapman, C.R.; Morrison, D.; and Zellner, B. 1975. Surface properties of asteroids: A synthesis of polarimetry, radiometry and spectrophotometry. *Icarus* 25: 104-130.

Chapman, C.R., and Salisbury, J.W. 1973. Comparisons of meteorite and asteroid spectral reflectivities. *Icarus* 19: 507-522.

Clayton, R.N., and Mayeda, T.K. 1978. Multiple parent bodies of polymict brecciated meteorites. *Geochim. Cosmochim. Acta* 42: 325-327.

Clayton, R.N.; Onuma, N.; and Mayeda, T.K. 1976. A classification of meteorites based on oxygen isotopes. *Earth Planet. Sci. Lett.* 30: 10-18.

Dodd, R.T. 1976. Iron-silicate fractionation within ordinary chondrite groups. *Earth Planet. Sci. Lett.* 28: 479-484.

Feierberg, M.; Larson, H.; Smith, H.; and Fink, U. 1978. Spectroscopic evidence for at least two achondritic parent bodies (abstract). *Bull. Amer. Astron. Soc.* 10: 595.

Gaffey, M.J. 1974. A systematic study of the spectral reflectivity characteristics of the meteorite classes with applications to the interpretation of asteroid spectra for mineralogical and petrological information. Ph.D. dissertation, Massachusetts Institute of Technology.

Gaffey, M.J. 1976. Spectral reflectance characteristics of the meteorite classes. *J. Geophys. Res.* 81: 905-920.

Gaffey, M.J. 1978. Mineralogical characterizations of asteroid surface materials: Evidence for unsampled meteorite types. (abstract) *Meteoritics* 13: 471-473.

Gaffey, M.J. 1979. Optical and spectral properties of the low albedo meteorites: Applications to the interpretation of the spectra of dark asteroids (in preparation).

Gaffey, M.J., and McCord, T.B. 1978. Asteroid surface materials: Mineralogical characterizations from reflectance spectra. *Space Sci. Rev.* 21: 555-628.

Gehrels, T., ed. 1971. *Physical Studies of Minor Planets* (NASA SP-267, Washington, D.C.: U.S. Government Printing Office).

Gradie, J., and Zellner, B. 1977. Asteroid families: Observational evidence for common origins. *Science* 197: 254-255.

Graham, A.L., and Hutchison, R. 1974. Is Kakangari a unique chondrite? *Nature* 251: 128-129.

Hapke, B. 1971. Inferences from optical properties concerning the surface texture and compositions of asteroids. In *Physical Studies of Minor Planets,* ed. T. Gehrels, (NASA SP-267, Washington, D.C.: U.S. Government Printing Office), pp. 67-77.

Hapke, B.W.; Partlow, W.D.; Wagner, J.K.; and Cohen, A.J. 1978. Reflectance measurements of lunar materials in the vacuum ultraviolet. *Proc. Lunar Sci. Conf.*

IX (Oxford: Pergamon Press), pp. 2935-2947.

Hazen, R.M.; Mao, H.K.; and Bell, P.M. 1977. Effects of compositional variation on absorption spectra of lunar olivines. *Proc. lunar Sci. Conf. VIII,* (Oxford: Pergamon Press), pp. 1081-1090.

Heymann, D. 1967. On the origin of hypersthene chondrites: Ages and shock effects of black chondrites. *Icarus* 6: 189-221.

Higuchi, H.; Morgan, J.W.; Ganapathy, R.; and Anders, E. 1976. Chemical fractionations in meteorites − X. Ureilites. *Geochim. Cosmochim. Acta* 40: 1563-1571.

Hunt, G.R., and Salisbury, J.W. 1970. Visible and near-infrared spectra of minerals and rocks: I. Silicate minerals. *Mod. Geol.* 1: 283-300.

Hunt, G.R., and Salisbury, J.W. 1971. Visible and near-infrared spectra of minerals and rocks: II. Carbonates. *Mod. Geol.* 2: 23-30.

Johnson, T.V., and Fanale, F.P. 1973. Optical properties of carbonaceous chondrites and their relationship to asteroids. *J. Geophys. Res.* 78: 8507-8518.

Johnson, T.V.; Matson, D.L.; Veeder, G.J.; and Loer, S.J. 1975. Asteroids: Infrared photometry at 1.25, 1.65 and 2.2 μm. *Astrophys. J.* 197: 527-531.

Keil, K. 1968. Mineralogical and chemical relationships among enstatite chondrites. *J. Geophys. Res.* 73: 6945-6976.

Kozai, Y. 1962. Secular perturbations of asteroids with high inclination and eccentricity. *Astron. J.* 67: 591-598.

Larson, H.P. 1977. Asteroid surface compositions from infrared spectroscopic observations: Results and prospects. In *Comets, Asteroids, Meteorites,* ed. A.H. Delsemme (Toledo, Ohio: University of Toledo Press), pp. 219-228.

Larson, H.P.; Feierberg, M.; Fink, U.; and Smith, H.A. 1979. Remote spectroscopic identification of carbonaceous chondrite mineralogies: Applications to Ceres and Pallas. *Icarus* (in press).

Larson, H.P., and Fink, U. 1975. Infrared spectral observations of asteroid (4) Vesta. *Icarus* 26: 420-427.

Larson, H.P.; Fink, U.; Treffers, R.R.; and Gautier, T.N. 1976. The infrared spectrum of asteroid 433 Eros. *Icarus* 28: 95-103.

Lebofsky, L.A. 1978. Asteroid 1 Ceres: Evidence for water of hydration. *Mon. Not. Roy. Astron. Soc. 182,* Short comm. pp. 17p-21p.

Liese, H.C. 1975. Selected terrestrial minerals and their infrared absorption spectral data (4000-300 cm^{-1}). In *Infrared and Raman Spectroscopy of Lunar and Terrestrial Minerals,* ed., C. Karr (New York: Academic Press), pp. 197-229.

Mason, B. 1966. The enstatite chondrites. *Geochim. Cosmochim. Acta* 30: 23-39.

Mason, B. 1971. The carbonaceous chondrites − A selective review. *Meteoritics* 6: 59-70.

Matson, D.L.; Johnson, T.V.; and Veeder, G.J. 1977a. Soil maturity and planetary regoliths: The Moon, Mercury and the asteroids. *Proc. Lunar Sci. Conf. VIII* (Oxford: Pergamon Press), pp. 1001-1011.

Matson, D.L.; Johnson, T.V.; and Veeder, G.J. (1977b). Asteroid reflectivities at 1.65 and 2.2 microns. In *Comets, Asteroids, Meteorites,* ed. A.H. Delsemme (Toledo, Ohio: University of Toledo Press), pp. 229-241.

McCord, T.B.; Adams, J.B.; and Johnson, T.V. 1970. Asteroid Vesta: Spectral reflectivity and compositional implications. *Science* 168: 1445-1447.

McCord, T.B., and Chapman, C.R. 1975a. Asteroids: Spectral reflectance and color characteristics. I. *Astrophys. J.* 195: 553-562.

McCord, T.B., and Chapman, C.R. 1975b. Asteroids: Spectral reflectance and color characteristics. II. *Astrophys. J.* 197: 781-790.

McCord, T.B., and Gaffey, M.J. 1974. Asteroids: Surface composition from reflection spectroscopy. *Science* 186: 352-355.

McCrosky, R.E.; Posen, A.; Schwartz, G.; and Shao, C.-Y. 1971. Lost City meteorite − Its recovery and a comparison with other fireballs. *J. Geophys. Res.* 76: 4090-4108.

McFadden, L.A., and Gaffey, M.J. 1978. Calibration of quantitative mineral abundances determined from meteorite reflection spectra and applications to solar system objects (abstract). *Meteoritics* 13: 556-557.

McFadden, L.A., and McCord, T.B. 1978. Prospecting for plagioclase on Vesta (abstract). *Bull. Amer. Astron. Soc.* 10: 601.

McFadden L.; McCord, T.B.; and Pieters, C. 1977. Vesta: The first pyroxene band from new spectroscopic measurements. *Icarus* 31: 439-446.

McSween, H.Y., Jr. 1977*a*. On the nature and origin of isolated olivine grains in carbonaceous chondrites. *Geochim. Cosmochim. Acta* 41: 411-418.

McSween, H.Y., Jr. 1977*b*. Carbonaceous chondrites of the Ornans type: A metamorphic sequence. *Geochim. Cosmochim. Acta* 41: 477-491.

McSween, H.Y., Jr. 1977*c*. Petrographic variations among carbonaceous chondrites of the Vigarano type. *Geochim. Cosmochim. Acta* 41: 1777-1790.

McSween, H.Y., Jr. 1977*d*. Chemical and petrographic constraints on the origin of chondrules and inclusions in carbonaceous chondrites. *Geochim. Cosmochim. Acta* 41: 1843-1860.

McSween, H.Y., and Richardson, S.M. 1977. The composition of carbonaceous chondrite matrix. *Geochim. Cosmochim. Acta* 41: 1145-1161.

Morrison, D. 1976. The densities and bulk compositions of Ceres and Vesta. *Geophys. Res. Lett.* 3: 701-704.

Morrison, D. 1977*a*. Asteroid sizes and albedos. *Icarus* 31: 185-220.

Morrison, D. 1977*b*. Radiometric diameters of 84 asteroids from observations in 1974-1976. *Astrophys. J.* 214: 667-677.

Nagy, B. 1975. *Carbonaceous Meteorites,* (New York: Elsevier Publ.).

Nash, D.B., and Conel, J.F. 1974. Spectral reflectance systematics for mixtures of powdered hypersthene, labradorite and ilmenite. *J. Geophys. Res.* 79: 1615-1621.

Nitsan, U., and Shankland, T.J. 1976. Optical properties and electronic structure of mantle silicates. *Geophys. J. Royal Astron. Soc.* 45: 39-87.

Olbers, W. 1803. Uber die von Himmel gafallenen Steine. *Ann. Physik* 14: 38-45.

Peterson, C. 1976. A source mechanism for meteorites controlled by the Yarkovsky effect. *Icarus* 29: 91-111.

Pieters, C.; Gaffey, M.J.; Chapman, C.R.; and McCord, T.B. 1976. Spectrophotometry (0.33 to 1.07 μm) of 433 Eros and compositional implications. *Icarus* 28: 105-115.

Reid, A.M., and Cohen, A.J. 1967. Some characteristics of enstatite from enstatite achondrites. *Geochim. Cosmochim. Acta* 31: 661-672.

Salisbury, J.W., and Hunt, G.R. 1974. Meteorite spectra and weathering. *J. Geophys. Res.* 79: 4439-4441.

Salisbury, J.W.; Hunt, G.R.; and Lenhoff, C.J. 1975. Visible and near-infrared spectra: X. Stony meteorites. *Mod. Geol.* 5: 115-126.

Sill, G.T. 1973. Reflection spectra of solids of planetary interest. *Comm. Lunar Planet. Lab.* 10: 1-7.

Van Schmus, W.R. 1969. Mineralogy, petrology and classification of Types 3 and 4 carbonaceous chondrites. In *Meteorite Research,* ed. P. Millman (Dordrecht: D. Reidel), pp. 480-491.

Van Schmus, W.R., and Wood, J.A. 1967. A chemical-petrologic classification for the chondritic meteorites. *Geochim. Cosmochim. Acta* 31: 747-765.

Vdovykin, G.P. 1970. Ureilites. *Space Sci. Rev.* 10: 483-510.

Vdovykin, G.P. 1976. The Haverö meteorite. *Space Sci. Rev.* 18: 749-776.

Vdovykin, G.P.; Grachev, V.I.; Malysheva, T.V.; and Saratova, L.M. 1975. A Mossbauer study of the forms of iron in carbonaceous meteorites. Part I. The iron phases in carbonaceous chondrites and ureilites. *Geochim. Inter.* 12: No. 6, 152-164.

Veeder, G.J.; Johnson, T.V.; and Matson, D.L. 1975. Narrowband spectrophotometry of Vesta (abstract). *Bull. Amer. Astron. Soc.* 7: 377.

Veeder, G.J.; Matson, D.L.; Bergstralh, J.T.; and Johnson, T.V. 1976. Photometry of 433 Eros from 0.65 to 2.2 μm. *Icarus* 28: 79-85.

Veeder, G.J.; Matson, D.L.; and Smith, J.C. 1978. Visual and infrared photometry of asteroids. *Astron. J.* 83: 651-663.

Wasson, J.T.; Chou, C-L.; Bild, R.W.; and Baedecker, P.A. 1976. Classification of and elemental fractionation among ureilites. *Geochim. Cosmochim. Acta* 40: 1449-1458.

Watson, F.G. 1938. Reflectivity and color of meteorites. *Proc. Nat. Acad. Sci.* 24: 532-537.

Wetherill, G.W. 1974. Solar system sources of meteorites and large meteoroids. In *Ann.*

Rev. Earth Planet. Sci., ed. F.A. Donath (Palo Alto, Calif.: Annual Reviews, Inc.) Vol. 2, pp. 303-331.

Wetherill, G.W. 1976. Where do the meteorites come from? A re-evaluation of the Earth-crossing Apollo objects as sources of chondritic meteorites. *Geochim. Cosmochim. Acta* 40: 1297-1317.

Wetherill, G.W., and Williams, J.G. 1968. Evaluation of the Apollo asteroids as sources of stone meteorites. *J. Geophys. Res.* 73: 635-648.

White, W.B.; and Keester, K.L. 1966. Optical absorption spectra of iron in rock forming silicates. *Amer. Mineral.* 51: 774-791.

White, W.B. and Keester, K.L. 1967. Selection rules and assignments for the spectra of ferrous iron in pyroxene. *Amer. Mineral.* 52: 1508-1514.

Williams, J.G. 1969. Secular perturbations in the solar system. Ph.D. dissertation, University of California at Los Angeles.

Wisniewski, W.Z. 1976. Spectrophotometry and UBVRI photometry of Eros. *Icarus* 28: 87-90.

Zellner, B. 1975. 44 Nysa: An iron-depleted asteroid. *Astrophys. J.* 198: L45-L47.

Zellner, B.; Andersson, L.; and Gradie, J. 1977*a*. UBV photometry of small and distant asteroids. *Icarus* 31: 447-455.

Zellner, B., and Bowell, E. 1977. Asteroid compositional types and their distributions. In *Comets, Asteroids, Meteorites,* ed. A.H. Delsemme (Toledo, Ohio: University of Toledo Press), pp. 185-197.

Zellner, B.; Leake, M.; and Morrison, D. 1977*b*. The E asteroids and the origin of the enstatite achondrites. *Geochim. Cosmochim. Acta.* 41: 1759-1767.

Zellner, B.; Wisniewski, W.Z.; Andersson, L.; and Bowell, E. 1975. Minor planets and related objects. XVIII. UBV photometry and surface composition. *Astron. J.* 80: 986-995.

Zimmerman, P.D., and Wetherill, G.W. 1973. Asteroidal source of meteorites. *Science* 182: 51-53.

INFRARED SPECTRAL REFLECTANCES
OF ASTEROID SURFACES

HAROLD P. LARSON
University of Arizona

and

GLENN J. VEEDER
Jet Propulsion Laboratory

This review compares the types of compositional information produced by three complementary techniques used in infrared observations of asteroid surfaces: broadband JHKL photometry, narrow band photometry, and multiplex spectroscopy. The high information content of these infrared observations permits definitive interpretations of asteroid surface compositions in terms of the major meteoritic minerals (olivine, pyroxene, plagioclase feldspar, hydrous silicates, and metallic Ni-Fe). These studies emphasize the individuality of asteroid surface compositions, the inadequacy of simple comparisons with spectra of meteorites, and the need to coordinate spectral measurements of all types to optimize diagnostic capabilities.

The importance of infrared spectral measurements to compositional analyses of planetary and stellar atmospheres is well established. More recently, observational and interpretive capabilities have permitted identifications of minerals on asteroid surfaces and more stringent characterizations of the assemblages in which they are dispersed. This emerging diagnostic capability requires combining the data produced by the complementary observational techniques of visible spectrophotometry and infrared photometry and spectroscopy. The composite spectral reflectances created in these syntheses often reveal, by inspection alone, compositional information that might remain hidden in the individual sets of data. Complete spectroscopic and photometric coverage from the near-ultraviolet at 0.4 μm to the infrared at 3 μm exists for asteroids nos. 1, 2, 4, 5, 8, 12, 18, 29, 39, 324, 349 and 433. Their composite spectral reflectances reveal mineralogically significant differences. This awareness emphasizes the need to study many asteroids as unique objects rather than as equivalent members of broadly classified groups.

Comparing these asteroid data with spectra of meteorites and terrestrial minerals is an important part of the interpretive program. While close matches are possible in some cases, Vesta and eucrites being a classic example, the high information content of the composite spectral reflectances permits more sophisticated analyses in terms of individual mineral components whether or not close meteoritic analogs can be found. These analyses contribute to such fundamental questions as: the source of primitive (carbonaceous chondrite) meteoritic matter, the source of differentiated (achondritic) meteoritic matter, the nature of highly reddened surfaces, and the very low albedos of some asteroids. The following section briefly reviews the observational methods employed in infrared studies of asteroid surfaces. Representative data are included to illustrate each technique. Section II reviews the compositional information that has been deduced from the composite spectral reflectances of several intensively studied asteroids. The final section indicates probable future directions for remote mineralogical analyses of asteroid surfaces.

I. INFRARED OBSERVATIONAL TECHNIQUES

Broadband JHKL photometry

These observations produce absolute photometric magnitudes that are measures of the radiant flux received at telescopes from astronomical sources. The magnitudes are fundamental spectral parameters useful in themselves for classifying objects. They also permit limited degrees of compositional analysis. As functions of wavelength, the J (1.25 μm), H (1.6 μm), K (2.2 μm), and L (3.4 μm) magnitudes constitute a coarse spectral grid that may reveal some of the broad spectral signatures of rock-forming minerals, although assignments based upon this evidence alone are not necessarily unique.

Table I contains broadband photometric data, reduced to relative reflectances, for thirty-one asteroids. A relative reflectance, R_λ, is the ratio of an observed photometric magnitude, m_λ, to that at some reference wavelength. In this chapter all relative reflectances are equal to unity at 0.56

TABLE I

Relative Reflectances from Broadband Photometric Observations of Asteroids[a]

Asteroid	$R\lambda$ (1.25μm)	$R\lambda$ (1.6μm)	$R\lambda$ (2.2μm)	$R\lambda$ (3.45μm)	Type[b]	Footnote
1 Ceres	0.94 ± 0.11	0.94 ± 0.13	1.03 ± 0.10	0.91 ± 0.11	U	c,d
2 Pallas	0.90 ± 0.09	0.79 ± 0.08	0.83 ± 0.08	0.89 ± 0.09	U	c,d
3 Juno		1.19 ± 0.10	1.30 ± 0.10		S	c
4 Vesta	1.17 ± 0.08	1.10 ± 0.07	1.16 ± 0.07	1.23 ± 0.08	U	c,d
5 Astraea	1.15 ± 0.16	1.25 ± 0.10	1.30 ± 0.15		S	c,e
6 Hebe	1.15 ± 0.15	1.21 ± 0.09	1.30 ± 0.12		S	c,e
7 Iris	1.16 ± 0.10	1.31 ± 0.09	1.44 ± 0.09		S	c,e
8 Flora	1.25 ± 0.12	1.42 ± 0.10	1.55 ± 0.11		S	c,f
9 Metis		1.27 ± 0.09	1.45 ± 0.10		S	c
10 Hygiea	0.81 ± 0.11	0.96 ± 0.07	1.06 ± 0.10	1.55 ± 0.11	C	c,d
12 Victoria		1.51 ± 0.12	1.72 ± 0.13		S	c
14 Irene		1.39 ± 0.08	1.52 ± 0.10		S	c
15 Eunomia	1.21 ± 0.09	1.24 ± 0.08	1.38 ± 0.08		S	c,f
16 Psyche		1.14 ± 0.09	1.37 ± 0.10		M	c
19 Fortuna		0.97 ± 0.11	1.10 ± 0.11	1.08 ± 0.12	C	c,d
20 Massalia		1.28 ± 0.09	1.39 ± 0.10		S	c
22 Kalliope		1.14 ± 0.12	1.46 ± 0.14		M	c
27 Euterpe		1.26 ± 0.11	1.37 ± 0.10		S	c
39 Laetitia		1.38 ± 0.08	1.49 ± 0.08		S	c
40 Harmonia		1.38 ± 0.11	1.46 ± 0.11		S	c
43 Ariadne		1.44 ± 0.12	1.47 ± 0.19		S	c
44 Nysa		1.03 ± 0.12	1.15 ± 0.11		E	c

TABLE I (Continued)

Relative Reflectances from Broadband Photometric Observations of Asteroids[a]

Asteroid	$R\lambda$ (1.25μm)	$R\lambda$ (1.6μm)	$R\lambda$ (2.2μm)	$R\lambda$ (3.45μm)	Type[b]	Footnote
51 Nemausa	1.05 ± 0.11	1.24 ± 0.13	1.28 ± 0.10	1.06 ± 0.11	U	c,d
63 Ausonia		1.50 ± 0.10	1.70 ± 0.12		S	c
129 Antigone	1.02 ± 0.10	1.28 ± 0.07	1.45 ± 0.08		M	c,e
192 Nausikaa		1.47 ± 0.07	1.62 ± 0.10		S	c
230 Athamantis		1.33 ± 0.15	1.45 ± 0.14		S	c
349 Dembowska		1.38 ± 0.10	1.57 ± 0.14	1.79 ± 0.16	R	c,d
354 Eleonora		1.48 ± 0.14	1.69 ± 0.16		U	c
433 Eros	1.3 ± 0.14	1.5 ± 0.1	1.7 ± 0.1		S	e,g
511 Davida	0.85 ± 0.11	1.03 ± 0.08	1.19 ± 0.08		C	c,e

[a] Entries are relative reflectances scaled to unity at 0.56 μm.

[b] Classification as revised by Bowell et al. (1978).

[c] $R\lambda$ (1.6μm) and $R\lambda$ (2.2μm) from Veeder et al. (1978).

[d] $R\lambda$ (1.25μm) and $R\lambda$ (3.45μm) computed from unpublished observations courtesy of L. Lebofsky.

[e] $R\lambda$ (1.25μm) computed from Chapman and Morrison (1976); c.f., Matson et al. (1978).

[f] $R\lambda$ (1.25μm) computed from Leake et al. (1978); c.f., Matson et al. (1978).

[g] $R\lambda$ (1.6μm) and $R\lambda$ (2.2μm) from Veeder et al. (1976).

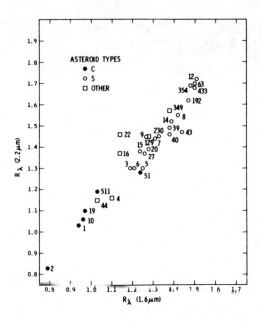

Fig. 1. Color plot comparing the reflectance measurements of asteroids in the H(1.6 μm) and K(2.2 μm) photometric bands relative to that at 0.56 μm. The increasing reflectivity with wavelength (reddening) is well developed and appears related to the metallic component of asteroid surfaces. The asteroid types (*C,S*, etc. defined by Chapman *et al.* (1975) are correlated with the degree of reddening. (Figure reproduced from Veeder *et al.* 1978.)

μm. If $R_\lambda > 1$ for $\lambda > 0.56$ μm, the asteroid's surface is described as having a reddened photometric color. These relative reflectances are useful in distinguishing between asteroid compositional classes (Veeder *et al.* 1978). Color plots such as in Fig. 1 compare relative reflectances at various wavelengths. Their systematic rather than random distribution suggests that the generally reddened nature of asteroid surfaces at infrared wavelengths may be due to a specific surface mineral. Candidate materials include metal (Ni-Fe) and lunar-like glass. The metallic component is preferred by Matson *et al.* (1977) for their extensive interpretation of photometric observations of asteroids.

When examined in greater detail, the color plot in Fig. 1 suggests correlations between asteroid infrared reflectances and taxonomic schemes (Bowell *et al.* 1978) used to classify asteroids. Type *C* (carbonaceous) objects, for example, have flat infrared reflectance curves, and these objects cluster at the bottom of the color plot in Fig. 1. The infrared relative reflectances of *S* (silicaceous) objects, on the other hand, increase with wavelength and these objects form most of the linear progression in Fig. 1. The compositional implications resulting from this type of analysis depend upon statistical associations involving measurements of a large population of asteroids. For the faintest asteroids that can be observed with current sensitivities, these photometric data are the only infrared reflectance measurements possible. Laboratory comparison measurements of minerals and higher spectral resolution observations of the brighter asteroids are being used to strengthen the conclusions drawn from these photometric studies.

Narrow band photometry

The spectral passbands of the JHKL photometric system were chosen with respect to transmission windows in the earth's atmosphere which are defined by saturated water vapor absorptions. The filter passbands centered in these transmission windows are not optimally placed with respect to absorptions in minerals. The 2 μm pyroxene absorption, for example, is centered in the 1.9 μm terrestrial H_2O band and this important mineral cannot be convincingly revealed in broadband measurements. On the other hand, narrow band filters ($\Delta\lambda \sim 0.1$ μm) judiciously located with respect to a characteristic absorption feature of a mineral permit much higher sensitivity for detecting that constituent. This is the basis of an observational technique that has special value in the thermal infrared ($\lambda > 3$ μm) where high background radiation levels restrict the sensitivity of broadband measurements. Lebofsky (1978) used this method to search for the water of hydration band at 3 μm on asteroid surfaces. The presence of this band may indicate an unaltered nebula condensate, or a hydrothermally altered surface. Figure 2 illustrates the location of Lebofsky's narrow band filters with respect to the water of hydration band measured in a Type C2 primitive meteorite. The nearly featureless spectrum of an anhydrous Type C30 carbonaceous chondrite is included in Fig. 2 for comparison. These narrow band photometric measurements can clearly distinguish between the two very different mineral assemblages. Also, these narrow band measurements are tied into the standard JHKL system for ease in combining the sets of data.

Table II contains typical observational data by Lebofsky (unpublished) using a narrow band photometric system. A prominent decrease in the relative reflectances in the 3 μm region, compared to those in the J and K bands, may indicate the presence of a hydrated mineral. The water of hydration band is definitely present on Ceres (Lebofsky 1978) and, though weaker, also on Pallas.

Multiplex spectroscopy

Maximizing the information return of infrared observations of asteroids requires continuous spectral coverage of a broad spectral region (0.8 − 2.5 μm, for example) at sufficient resolution to reveal many absorptions of the

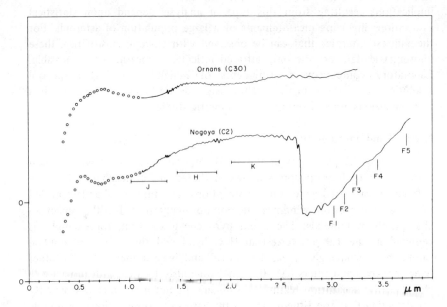

Fig. 2. Illustration of an observational technique for distinguishing between a primitive mineral assemblage containing hydrated silicates such as a Type C2 carbonaceous chondrite and one containing high-temperature silicates (C30 type). Narrow band filters F1-F5 ($\Delta\lambda \sim 0.1$) are spaced throughout the spectral region around 3 μm containing the hydrated H_2O absorption. When combined with broadband JHK measurements, a coarse spectral grid is produced that can reveal this important diagnostic spectral feature of primitive mineral assemblages. (Laboratory data adapted from Larson *et al.* 1979.)

important rock-forming minerals. This requires spectroscopic instrumentation, but the faintness of most asteroids prevents the use of scanning spectrometers of the filter wheel or grating type. Although Vesta, the brightest asteroid, has been observed in the near-infrared using these classical techniques (McFadden and McCord 1978), the signal-to-noise ratio, spectral resolution and wavelength precision are not competitive with modern spectroscopic capabilities.

The Fourier transform is a method whereby a general function which can be represented as a sum of sines and cosines is decomposed into these component sines and cosines. In the equation

$$B(s) = \int_{-\infty}^{\infty} A(x)e^{-2\pi ixs}\,dx = \int_{-\infty}^{\infty} A(x)\cos(2\pi xs)dx - \int_{-\infty}^{\infty} A(x)\sin(2\pi xs)dx \tag{1}$$

the function $B(s)$ is the Fourier transform of the function $A(x)$. $B(s)$ can be the intensity as a function of frequency, or spectrum, of light received from an astronomical source. A Fourier transform spectrometer divides the light

TABLE II

Relative Reflectances from Narrowband Photometric Observations of Asteroids[a]

Asteroid	R_λ (J,1.25 μm)	R_λ (K,2.2 μm)	R_λ (F1,3.01 μm)	R_λ (F2,3.11 μm)	R_λ (F3,3.25 μm)	R_λ (F4,3.43 μm)	R_λ (F5,3.73 μm)
1 Ceres	0.94 ± 0.01	1.03	0.71 ± 0.02	0.72 ± 0.02	–	0.93 ± 0.04	1.18 ± 0.10
2 Pallas	0.90 ± 0.03	0.83	0.68 ± 0.03	0.76 ± 0.04	0.76 ± 0.02	0.77 ± 0.04	0.79 ± 0.03
4 Vesta	–	1.16	1.25 ± 0.04	1.30 ± 0.03	–	1.18 ± 0.03	1.32 ± 0.07
19 Fortuna	–	1.10	–	0.91 ± 0.03	0.98 ± 0.02	1.18 ± 0.13	–
349 Dembowska	–	1.57	–	–	1.70 ± 0.04	–	1.33 ± 0.15

[a]Unpublished observations courtesy of L. Lebofsky. Entries are relative reflectances scaled to unity at 0.56 μm.

from the source into two beams and inputs them into a classical Michelson interferometer. Translation of one of the mirrors in the interferometer causes the two beams to interfere with one another. Each element of the spectrum produces a sinusoid of a certain frequency, whose amplitude is determined by the intensity of that spectral element. The sum of these sinusoids produces an output beam with intensity varying as a function of path difference. This function corresponds to $A(x)$ in the above equation and is the quantity measured by the spectrometer. The Fourier transform of $A(x)$ is then calculated by a computer, the result being $B(s)$, or the spectrum of the source.

There are two basic advantages of Fourier transform spectrometers over conventional scanning spectrometers of the grating or filter wheel types.

1. An interferometer receives information about the entire spectral range during an entire scan, while the conventional instrument receives information only in a narrow band at a given time. Thus, the sensitivity of the Fourier spectrometer is greater by a factor proportional to the number of resolution elements in the spectrum. This is called the multiplex advantage.

2. For a conventional instrument, the resolution is dependent on the size of the filter wheel, or for a grating instrument, on the width of the slit. For a given input aperture size, a conventional spectrometer must be optimized for a certain resolution which can not easily be varied. However, the resolution of the interferometer is not strongly limited by the size of the input aperture. Thus, the resolution used in observations with a Fourier spectrometer can be chosen according to the width of the spectral features one wishes to measure. This is called the throughput advantage.

The high efficiency and greater flexibility of multiplex (Fourier) methods permit infrared spectroscopic observations of numerous asteroids. Instrumentation developed at the Lunar and Planetary Laboratory at the University of Arizona (Larson and Fink 1975a) has produced many infrared spectra of asteroids as well as spectra of many satellites of Jupiter and Saturn (Larson and Fink 1977). Asteroids as faint as visual magnitude 9-10 have been observed in reasonable integration times (<6 hr) with modest aperture (1.5 m) telescopes. The spectral resolution is 25 cm^{-1} (resolving power of 200 at 2 μm), more than adequate for mineralogical analyses. Table III lists the asteroids for which infrared spectroscopic observations have been obtained by Fourier methods. Six of the infrared spectra are displayed in Fig. 3. All differ in compositionally significant ways. They can be approximately characterized in terms of meteoritic analogs as follows: Vesta and Dembowska, strong silicate absorptions similar to achondrites; Eros and Victoria, highly reddened stony-iron mixtures; Ceres and Pallas, primitive (C1 or C2) carbonaceous chondrites. Detailed analyses of some of the infrared

TABLE III

Infrared Spectroscopic Observations of Asteroids at the
Lunar and Planetary Laboratory of the University of Arizona

Asteroid	Date	Telescope Aperture (m)	Visual Magnitude	Integration Time (hr)
1 Ceres	Jan 76	2.3	7.5	3.0
1 Ceres	May 78	1.5	7.7	4.2
2 Pallas	Dec 76	4.0	8.0	2.0
4 Vesta	May 74	1.5	6.6	13.7
4 Vesta	May 78	1.5	5.6	1.8
5 Astraea	Mar 79	1.5	9.4	3.6
8 Flora	Dec 78	1.5	9.4	4.6
12 Victoria	Jun 78	1.5	9.4	6.5
18 Melpomene	Dec 78	1.5	9.0	4.6
29 Amphitrite	Sep 78	1.5	9.1	3.2
39 Laetitia	Oct 78	1.5	9.4	3.2
324 Bamberga	Oct 78	1.5	9.0	4.6
349 Dembowska	Dec 77	2.3	9.8	5.6
433 Eros	Jan 75	2.3	7.8	2.3

spectra are available (Vesta: Larson and Fink 1975*b;* Eros: Larson *et al.*
1976; Ceres and Pallas: Larson *et al.* 1979; Dembowska and Vesta: Feierberg
et al. 1979).

II. COMPOSITE ASTEROID SPECTRAL REFLECTANCES AND COMPOSITIONAL IMPLICATIONS

Infrared observations of asteroid Ceres illustrate the contributions of the
experimental methods discussed above. The composite reflection spectrum of
Ceres in Fig. 4 combines three independent sets of data: visible
spectrophotometry (Chapman *et al.* 1973), infrared spectroscopy (Larson *et
al.* 1979), and infrared photometry (Lebofsky 1978). These data were
combined by scaling them in their regions of overlap. The narrowband
photometric data points above 3 μm were attached to the infrared spectrum
by fitting their accompanying JHK measurements to the spectrum's
continuum level. The extensive spectral overlap of the infrared spectrum with
the visible spectrophotometry permits scaling these data sets. This composite
thus defines the relative spectral reflectance of Ceres over the 0.4-3.5 μm
region. Spectral features that can be directly related to surface minerals
include the ultraviolet falloff, the broad inflection at 1 μm, the slightly
reddened infrared reflectance, and the very obvious absorption near 3.1 μm.
These features were collectively used by Larson *et al.* (1979) to conclude that

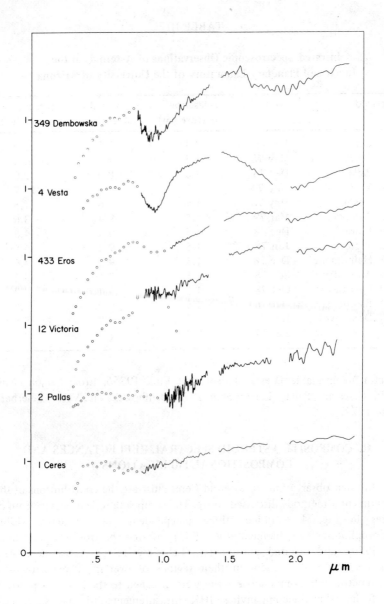

Fig. 3. Composite spectral reflectances of asteroids assembled from visible spectrophotometry (Chapman *et al.* 1973) and infrared spectroscopy. All of the infrared spectra (resolution ∼25 cm^{-1}) were produced by Fourier methods at the Lunar and Planetary Laboratory. Pyroxene minerals produce the obvious absorptions at 0.95 and 1.9 μm on asteroids 4, 12, 349, and 433. In addition, olivine is evident on Dembowska in the 1.1-1.4 μm region, and plagioclase feldspar is detectable on Vesta in the 1.3 μm region when compared with meteoritic analogs (see Feierberg *et al.* 1979 for details).

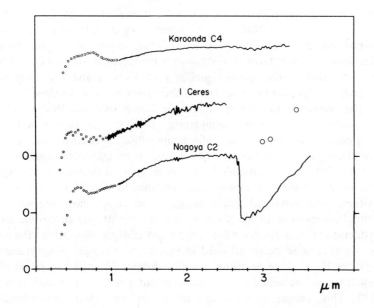

Fig. 4. Comparison of the composite reflection spectrum of Ceres with the two types of carbonaceous chondrites frequently associated with its surface composition. (Figure reproduced from Larson *et al.* 1979.)

the surface of Ceres is a hydrated mineral assemblage that, if restricted to comparisons with meteorites, most closely resembles Type C2 carbonaceous chondrites. This association was frequently suggested in previous analyses of the individual sets of data, but other interpretations invoking rather different mineralogies were also proposed (see Larson *et al.* 1979, for the chronology and references to previous work on the surface composition of Ceres). Figure 4 contains laboratory comparison spectra of the two kinds of meteoritic matter most frequently associated with Ceres' surface: primitive carbonaceous chondrites (C1 or C2), and higher grade Type C4 (Karoonda) material. By simple inspection of the spectra in Fig. 4, however, the C4 association is fundamentally incompatible with the composite reflection spectrum of Ceres. The comparison of Ceres' surface with primitive meteoritic matter is most consistent with the spectral data. This new observational constraint is due primarily to infrared observations, particularly the 3 μm narrow band photometry.

The same procedure is being used to initiate interpretations of other recently acquired asteroid spectra. The six spectra in Fig. 3 have each been joined to published visible spectrophotometry to reveal more clearly important diagnostic features in the 1 μm region. One surprising result of the synthesis is the similarity of Eros and Victoria. Two apparently erroneous spectrophotometric data points near 1.1 μm implied a very deep infrared absorption on Victoria that was not confirmed by the infrared spectroscopic observations. Previous interpretations of the seemingly unique spectral reflectance of Eros favored a stony-iron composition that had no exact meteoritic analog. The spectral similarity of Victoria and Eros may now change the perspective in which each of these asteroids must be viewed.

The composite spectral reflectances of Dembowska and Vesta in Fig. 3 also provide an interesting comparison. Both spectra have pronounced absorptions characteristic of high-temperature pyroxene and olivine assemblages. Vesta's surface composition has been associated with eucrites since the first spectrophotometric data were reported (McCord et al. 1970). Subsequent infrared observations have been used to search for compositional variations with rotational phase angle, to determine more precisely the chemical composition of Vesta's pyroxene component, and to provide more convincing evidence for its subtle plagioclase feldspar absorption. The very high signal-to-noise ratios achieved in recent spectroscopic observations of Vesta (see Fig. 3) have permitted a critical review of Vesta's association with achondrites and its candidacy as a eucrite parent body source (Feierberg et al. 1979). This unique asteroid appears to be an intact, geochemically differentiated body whose surface composition is intermediate to that of eucrites and howardites. Compositional arguments favor Vesta as the eucrite parent body source (see the chapter by Drake in this book), but no dynamical mechanism has yet been found to deliver fragments of Vesta to the earth on the time scale (10^6-10^7 yr) required by the cosmic ray exposure ages of eucrites (Wetherill 1974).

The infrared reflection spectrum of Dembowska in Fig. 3 displays a combination of olivine and pyroxene absorptions that has no meteorite counterpart, although recent finds in Antarctica may eventually provide one. Figure 5 compares Dembowska's composite spectral reflectance with several meteorite spectra that Feierberg et al. (1979) used to help establish this asteroid's surface composition. A 10:1 mixture of an olivine similar to that in Chassigny and a pyroxene similar to that in diogenites would closely resemble Dembowska. This describes an olivine-rich, magmatically differentiated assemblage, a geochemically significant missing group in our terrestrial collections of achondrite meteorites. This interpretation was suggested as a possibility by Gaffey and McCord (1978) using just the 0.4-1.1 μm reflectance data, but uncertainty over the pyroxene component prevented a definite conclusion. The very obvious pyroxene band at 2 μm in Dembowska's spectrum in Fig. 5, however, resolves this question. A previous interpretation of the same visible spectrophotometry by McCord and Chapman (1975) in terms of an ordinary chondrite (LL6- or L6-type) composition is less consistent with the infrared data. Matson et al. (1978) inferred an olivine plus metal composition, but their broadband infrared data did not resolve the pyroxene component. It is interesting to note that if a high albedo and

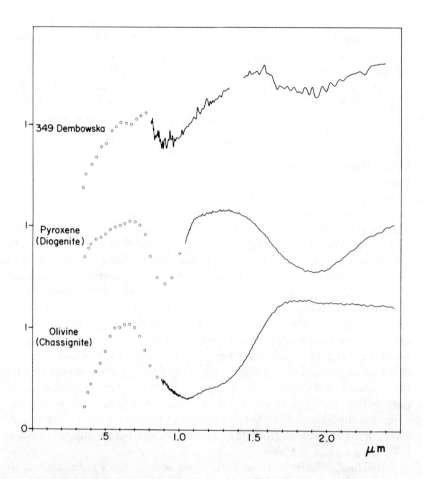

Fig. 5. Comparison of the spectral reflectance of Dembowska with pyroxene and olivine meteoritic minerals. This asteroid's surface is most consistent with an olivine-rich achondritic composition with no known meteoritic analog. (Figure adapted from Feierberg *et al.* 1979.)

spectrally featureless mineral, such as Ni-Fe, were added to Dembowska's olivine-rich achondrite composition, the resulting spectral reflectance might resemble more closely Eros and Victoria than an LL6- or L6-type chondrite.

Feierberg *et al.* (1979) conclude that their analysis of the olivine-rich achondritic assemblage on Dembowska's surface implies geochemical differentiation, as on Vesta. For Dembowska, however, compositional arguments based upon cosmochemical models require that its crust must have been stripped away to reveal the observed olivine-rich mantle (see the chapter by Drake). These crustal fragments may have contributed to terrestrial eucrite flux, but Drake considers this possibility improbable.

These examples illustrate a significant shift in the interpretation of asteroid infrared reflectances, from simple comparisons with meteorite spectra to the firm identification of specific minerals. Figure 6 summarizes the spectral reflectances of the important meteoritic minerals. Almost all of their prominent absorptions occur at wavelengths of 1 μm or longer. The pyroxene, olivine and feldspar signatures are best studied with Fourier methods because of their unique combination of high efficiency, wide spectral bandpass, and high spectral resolution in the 0.8-2.5 μm region. The hydrous silicates are adequately revealed with narrowband photometry, and the reddening effect of metallic Ni-Fe can be discerned in JHKL photometry. Thus infrared spectral measurements of asteroids provide optimum opportunities for detailed mineralogical interpretations.

Table IV summarizes the surface compositions of the asteroids whose infrared spectra have been observed by Fourier methods. There are significant differences between interpretations based upon visible reflectance data alone and those using all available spectral measurements. The differences are due primarily to the unique infrared spectral signatures of pyroxenes at 2 μm and hydrous silicates at 3.1 μm, both important meteoritic minerals indicative of rather different compositions. Fewer meteoritic associations are given for the infrared-based interpretations, which suggests that simple matching of spectra of meteorites to asteroids is not a very productive interpretative procedure. Moreover, Table IV shows that when meteoritic analogs are proposed in each data base for an asteroid, as often as not they disagree. Those associated with the infrared reflectances are probably closer to the asteroid's actual composition. Recall from Fig. 4, for example, the obvious incompatibility of Ceres and the frequently proposed Type C4 carbonaceous chondrite material, when both are viewed from the broader perspective of infrared spectral reflectance data.

The final column in Table IV lists the asteroid types in the taxonomic system of Chapman et al. (1975) updated by Bowell et al. (1978). Although the system is not built upon compositional analyses, its creators did define mineralogically suggestive classes such as C (carbonaceous), S (stony-iron), M (metal), E (enstatite) and R (red). Table IV permits the first comparison of these classifications with mineralogical analyses based upon composite spectral reflectances. With the exception of Eros, all objects identified as S-type in Table IV are inner main-belt asteroids with diameters in the 125-200 km range. The S asteroids in Table IV have composite infrared reflectances with no close meteoritic counterparts, but preliminary analyses indicate pyroxene, olivine, and abundant metal (i.e., stony-irons) as their primary constituents. Thus the association of S-type asteroids with metal-silicate assemblages is supported by infrared reflectance data.

Interestingly, those asteroids designated U (unclassified) in Table IV are actually among the best understood from the point of view of their constituent minerals. The unique spectrum of Dembowska should also qualify

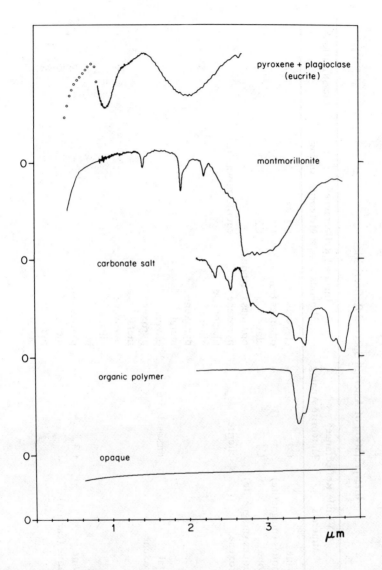

Fig. 6. Typical reflection spectra of the major meteoritic minerals (for an olivine spectrum see Fig. 5). The unique infrared spectral signatures of olivine, pyroxene, plagioclase and hydrated silicate minerals permit their detection on asteroid surfaces. Opaque minerals are implicitly present through low albedo. (Figure reproduced from Larson *et al.* 1979.)

TABLE IV

Interpretation of Asteroid Spectral Reflectivities

Asteroid	Visible Reflectance[a]		Infrared Reflectance[b]		Classification[c]
	Minerals	Meteorite Analog	Minerals	Meteorite Analog	
1 Ceres	olivine opaques (magnetite)	C4	clays opaques (carbon)	C2	U
2 Pallas	olivine opaques (magnetite)	C4	clays opaques (carbon)	C2	U
4 Vesta	pyroxene	eucrite	pyroxene plagioclase	eucrite/howardite	U
5 Astraea			olivine pyroxene metal		S
8 Flora	metal pyroxene	mesosiderite	olivine pyroxene metal		S
12 Victoria			pyroxene olivine metal		S
18 Melpomene	olivine opaque (carbon)	C3	pyroxene metal		S
29 Amphitrite			pyroxene metal		S

TABLE IV (Continued)

Interpretation of Asteroid Spectral Reflectivities

Asteroid	Visible Reflectance[a]		Infrared Reflectance[b]		Classification[c]
	Minerals	Meteorite Analog	Minerals	Meteorite Analog	
39 Laetitia	metal olivine pyroxene	stony-iron olivine-pyroxene	metal olivine pyroxene		S
324 Bamberga	clays	C1,C2	olivine opaques	ureilite	C
349 Dembowska	olivine metal	achondrite	olivine pyroxene	achondrite	R
433 Eros	pyroxene olivine metal	H chondrite	pyroxene olivine metal		S

[a] As summarized in Gaffey and McCord (1978).
[b] Provisional summary from the dissertation research of Feierberg.
[c] As summarized in Bowell et al. (1978).

it for a U classification. Its present designation as an R-type (extreme reddening) is certainly deserved based upon its composite spectral behavior, but its spectrum, and by implication its composition as discussed above, is unusual in other significant respects, too. Ceres closely resembles a primitive mineral assemblage (see comparison with a type C2 carbonaceous chondrite in Fig. 4), but its classification has changed from C (Chapman *et al.* 1975) to C^* (Chapman 1976) to U (Bowell *et al.* 1978) and finally back to C (see the chapter by Zellner in this book). The comparatively featureless spectral characteristics of carbonaceous chondrite assemblages at visible and near-infrared wavelengths may account in part for this indecision in classifying Ceres, but its infrared spectral reflectance leaves little doubt as to some of its constituent minerals, particularly hydrous silicates. For these U-type asteroids, then, the taxonomic system is not mineralogically informative, and it may be creating unnecessary confusion regarding current awareness of their compositions.

Only one C-type asteroid (Bamberga) appears in Table IV. Provisional interpretation of Bamberga's composite spectral reflectivity (0.4-3.5 μm) requires a mixture of olivine (broad 1 μm absorption) and abundant opaques (very low albedo). No hydrous silicate component has been detected in narrowband photometric observations (Lebofsky 1979, personal communication). The meteoritic analog proposed in Table IV for Bamberga's surface is a ureilite. This association is not intended to be used literally. Rather, it indicates that a mixture of carbon and olivine seems to be the most reasonable starting point for interpreting the asteroid's composite spectral reflectance. Bamberga's classification as a C-type (carbonaceous) asteroid is consistent with the spectroscopic data, but this taxonomic system was unable to single out such a demonstrably carbonaceous surface as that of Ceres. This should reinforce statements by the authors of this taxonomic system (Bowell *et al.* 1978) that the classifications are not based upon mineralogical interpretations, in spite of the suggestive labels used.

The initial comparisons of taxonomic classifications with the observed compositions of asteroids reveal some suggestive correlations, but it is clear that only the higher information content of composite spectral reflectances leads to convincing mineralogical characterizations of these surfaces. The taxonomic classifications are better suited to identifying compositional trends in large populations of asteroids, and in this complementary role they add substantially to understanding the origin and evolution of asteroids.

III. FUTURE DIRECTIONS

The infrared reflectances reviewed here are results of new observational programs employing unique instrumental capabilities. The spectral diversity of these asteroidal data emphasizes the need to study intensively many asteroids as unique objects. The broad spectral bandpass and high resolution of spectroscopic observations are essential for preparing the composite reflectances in Figs. 3 and 4. From the log of observations in Table II, groundbased opportunities can be extrapolated to include, in the near future, asteroids as faint as visual magnitude 11 to 12. This assumes additional improvements in detector sensitivity, continued use of multiplex methods, use of larger telescopes, and careful attention to the compromise between spectral resolution, integration time and signal-to-noise ratio that must be established during observations. This capability will permit adequate study of the S-type asteroids since the silicate absorptions of this class (pyroxene, olivine) lie below 2.5 μm. For C-type asteroids, however, there is a need to explore spectroscopically the wavelength region from 2.5 to 4.0 μm where the characteristic CH and OH bands occur in organic polymers and hydrous silicates. As Larson et al. (1979) demonstrate, these absorptions are blended to such a degree that continous spectral coverage at high resolution and high signal-to-noise is required to distinguish between them. The spectroscopic measurements are difficult for even the brightest asteroids due to obscuring molecular absorptions in our atmosphere and to higher thermal background radiation levels at these longer wavelengths. One consequence of the background flux is that detector performance becomes limited by background noise, and the multiplex advantage of a Fourier spectrometer is lost. The ideal solution to both of these observational constraints is an earth-orbiting, cryogenically-cooled telescope such as the SIRTF (Shuttle Infra Red Telescope Facility). Predictions of spectroscopic capabilities in the 3 μm region indicate that asteroids as faint as visual magnitude 10 or 11 can be observed (signal-to-noise ratio ~50, resolving power ~100, integration time ~10 min). This will eventually permit spectroscopic coverage of a significant number of C-type objects in the 3 μm region, but current work will continue to depend upon groundbased narrow band photometric observations.

Acknowledgments. The authors would like to thank S. Ridgway, L. Lebofsky, and two other anonymous referees for their suggestions, which were helpful in the revision of this chapter. Thanks also go to M. Feierberg who helped in the final preparation of the manuscript.

REFERENCES

Bowell, E.; Chapman, C. R.; Gradie, J. C.; Morrison, D.; and Zellner, B. 1978. Taxonomy of asteroids. *Icarus* 35: 313-335.

Chapman, C. R. 1976. Asteroids as meteorite parent bodies: The astronomical perspective. *Geochim. Cosmochim. Acta* 40: 701-719.

Chapman, C. R.; McCord, T. B.; and Johnson, T. V. 1973. Asteroidal spectral reflectivities. *Astron. J.* 78: 126-140.

Chapman, C. R., and Morrison, D. 1976. J, H, K photometry of 433 Eros and other asteroids. *Icarus* 28: 91-94.

Chapman, C. R.; Morrison, D.; and Zellner, B. 1975. Surface properties of asteroids: A synthesis of polarimetry, radiometry and spectrophotometry. *Icarus* 25: 104-130.

Feierberg, M.; Larson, H. P.; Fink, U.; and Smith, H. A. 1979. Spectroscopic evidence for two achondrite parent bodies: 349 Dembowska and 4 Vesta. Submitted to *Geochim. Cosmochim. Acta.*

Gaffey, M. J., and McCord, T. B. 1978. Asteroid surface materials: Mineralogical characterizations from reflectance spectra. *Space Science Rev.* 21: 555-628.

Larson, H. P.; Feierberg, M.; Fink, U.; and Smith, H. A. 1979. Remote spectroscopic identification of carbonaceous chondrite mineralogies: Applications to Ceres and Pallas. *Icarus* (in press).

Larson, H. P., and Fink, U. 1975a. An infrared Fourier spectrometer for laboratory use and for astronomical studies from aircraft and groundbased telescopes. *Appl. Optics* 14: 2085-2095.

Larson, H. P., Fink, U. 1975b. Infrared spectral observations of asteroid 4 Vesta. *Icarus* 26: 420-427.

Larson, H. P., and Fink, U. 1977. The application of Fourier transform spectroscopy to the remote identification of solids in the solar system. *Appl. Spectros.* 31: 386-402.

Larson, H. P.; Fink, U.; Treffers, R. R.; and Gautier, T. N. 1976. The infrared spectrum of asteroid 433 Eros. *Icarus* 28: 95-103.

Leake, M.; Gradie, J.; and Morrison, D. 1978. Infrared (JHK) photometry of meteorites and asteroids. *Meteorites* 13: 101-120.

Lebofsky, L. A. 1978. Asteroid 1 Ceres: Evidence for water of hydration. *Mon. Not. Roy. Astron. Soc.* 182: 17-21.

Matson, D. L.; Johnson, T. V.; and Veeder, G. J. 1977. Soil maturity and planetary regoliths: The Moon, Mercury and the asteroids. *Proc. Lunar Sci. Conf. VIII* (Oxford, Pergamon Press), pp. 1001-1011.

Matson, D. L.; Veeder, G. J.; and Lebofsky, L. A. 1978. Infrared observations of asteroids from earth and space. In *Asteroids: An Exploration Assessment*, eds. D. Morrison and W. C. Well. NASA Conf. Publ. 2053, pp. 127-144.

McCord, T. B.; Adams, J. B.; and Johnson, T. V. 1970. Asteroid Vesta: Spectral reflectivity and compositional implications. *Science* 168: 1445-1447.

McCord, T. B., and Chapman, C. R. 1975. Asteroids: Spectral reflectance and color characteristics. II. *Astrophys. J.* 197: 781-790.

McFadden, L. A., and McCord, T. B. 1978. Prospecting for plagioclase on Vesta (abstract). *Bull. Amer. Astron. Soc.* 10: 601.

Veeder, G. J.; Matson, D. L.; Bergstralh, J. T.; and Johnson, T. V. 1976. Photometry of 433 Eros from 0.65 to 2.2 μm. *Icarus* 28: 79-85.

Veeder, G. J.; Matson, D. L.; and Smith, J. C. 1978. Visual and infrared photometry of asteroids. *Astron. J.* 83: 651-663.

Wetherill, G. W. 1974. Solar system sources of meteorites and large meteoroids. *Ann. Rev. Earth Planet. Sci.* 2: 303-331.

AQUEOUS ACTIVITY ON ASTEROIDS: EVIDENCE FROM CARBONACEOUS METEORITES

J. F. KERRIDGE
University of California at Los Angeles

and

T. E. BUNCH
NASA Ames Research Center

Carbonaceous chondrites of groups CI and CM were formed by impact brecciation and aqueous alteration of earlier generations of mineral phases within the surface regions of two or more parent bodies. Those parent bodies were probably asteroids, rather than comets, although a problem still exists in delivering such material safely to Earth. Aqueous activity may have been widespread on asteroids.

Despite the common perception of carbonaceous chondrites as pristine nebular material, these meteorites actually display abundant evidence for significant geochemical processing on their parent body or, more probably, bodies. It is the purpose of this chapter to review evidence for such processing and thence to infer conditions which prevailed upon the parent bodies (see also McSween 1979). These conditions may then be used to constrain theories concerning the source of carbonaceous chondrites. In fact, it will be shown that this evidence favors an asteroidal rather than cometary source, although these arguments should be considered together with other lines of evidence, summarized by Wasson and Wetherill in this book and elsewhere by

Anders (1975, 1978) and Wetherill (1974, 1978).

Attention will be focused here upon the most volatile-rich meteorites, of groups CI and CM; it is not yet clear to what extent these findings may be applied to the less volatile-rich groups CO and CV (including the widely studied Allende chondrite), although preliminary results indicate that these meteorites also show evidence for secondary alteration (Bunch and Chang 1979a), as well as mild thermal metamorphism (McSween 1977). We shall begin with CI chondrites, describing salient petrographic features and pertinent mineralogical details, followed by a similar description of CM chondrites. The probable relationship between these two groups of meteorites will be considered and a qualitative scenario describing their geochemical evolution will be outlined. Finally, this scenario will be compared with our scanty knowledge concerning conditions on asteroidal and cometary surfaces.

I. OBSERVATIONS OF CI CHONDRITES

Petrography

The most extensively studied CI chondrite, Orgueil, was described by Boström and Fredriksson (1966) as "a bituminous clay with a clastic texture." This texture is readily apparent in polished sections thinner than about 20 μm, with individual clasts generally delineated from neighboring clasts and/or interstitial material by differences in opacity, color, microtexture or, less commonly, mineralogy. Most clasts have broadly similar composition (Boström and Fredriksson 1966) despite substantial chemical heterogeneity on a micron scale (Kerridge 1976). The major mineral constituent (60 to 65%) of both clasts and interstitial material consists of ill-defined phyllosilicates (hydrated Mg,Fe-silicates), within which are embedded a variety of minerals, some of which are described below. Clasts are sometimes bounded by, more rarely transected by, veins formed from sulfates or, less commonly, carbonates, of magnesium and calcium (DuFresne and Anders 1962; Boström and Fredriksson 1966; Richardson 1978). Pronounced lamellar textures, indicative of depositional or flow structures, are visible in some clasts (Fig. 1) and within interstitial material (see also Nagy 1966, 1975).

The clastic texture of CI chondrites is strongly reminiscent of breccias produced in the lunar regolith or in the regolith of other meteorite parent bodies such as those of the gas-rich chondrites and achondrites. In those cases, it is incontrovertible that meteoroid impact has been responsible for production of clasts by fragmentation, and for induration of the breccia by means of impact lithification (Chao et al. 1971; Kurat et al. 1974). In the case of CI chondrites it is difficult to visualise any process other than impact capable of producing the population of clasts. However, it seems likely that induration was effected largely by mineralization, rather than by impact-related processes.

Fig. 1. Clast in the CI chondrite Orgueil, composed mostly of ill-defined hydrated silicates, showing pronounced lamellar texture. Width of field of view is 1 mm.

Richardson (1978) showed from textural relations among vein-filling minerals that three generations of vein formation may be discerned in Orgueil, characterized by deposition of carbonate (calcium-, magnesium-rich), calcium sulfate and magnesium sulfate, respectively. Transitions between generations were apparently controlled by variations in availability of CO_2 and leachable calcium. The epoch of vein formation apparently coincided, at least approximately, with that of impact brecciation.

A further constraint on the temporal relationship between mineralization and brecciation may be derived from the observation of fresh, unaltered xenocrysts of olivine, $(Mg,Fe)_2SiO_4$, and pyroxene, $(Mg,Fe)SiO_3$, within the matrix of CI chondrites (Reid *et al.* 1970; Kerridge and Macdougall 1976). The pristine surfaces of these isolated mafic grains would not have survived the environment which produced the carbonates and sulfates by leaching of calcium and magnesium from the matrix in which the mafics now reside. In addition it seems clear that the mafics were introduced during the period of impact-induced regolith turnover and brecciation. Finally, preservation of some veins showing no evidence of impact-related disturbance leads to reconstruction of a sequence in which impact brecciation coincided with pervasive leaching of matrix and crystallization of carbonates and early generations of sulfates, followed by reduction in severity of leaching and intensity of brecciation, with progressively larger lithic units retaining their competence, and final vein deposition taking place in effectively undisturbed regolith. Isolated olivines and pyroxenes buried within intact clasts survived; others presumably did not.

Two points are noted. First, unlike lunar and other meteoritic breccias, CI meteorites reveal no evidence of hypervelocity shock effects such as impact-derived glass; even fracturing of grains is rare. It follows that the impacts responsible for brecciation must have been mild, presumably the result of very low projectile velocities. Second, leaching and mineralization apparently did not move significant quantities of material over distances sufficient to perturb bulk meteorite analyses, i.e., on a centimeter scale. In fact, bulk contents of calcium and magnesium in CI chondrites provide a reasonable match with solar abundances (Anders 1971; Holweger 1977) and conform closely to chondritic distribution patterns for these elements (Larimer and Anders 1970) indicating that CI parental material was neither depleted nor enriched in these elements.

Mineralogy

Although the entire mineral suite of CI chondrites (except for the trace mafics olivine and pyroxene) is consistent with formation by secondary alteration processes at low temperature, some mineral species are less informative than others. Thus the phyllosilicate material broadly resembles an ultra-fine grained terrestrial clay, too chemically heterogeneous and structurally disordered for specific identification (Kerridge 1967), although minor amounts of material characterizable as serpentine, $(Mg,Fe)_3Si_2O_5(OH)_4$, and montmorillonite, $(Al,Mg,Fe)_4Si_8O_{20}(OH)_4$, are present (Bass 1971). The range of conditions capable of producing such material is considerable; it does not seem possible, for example, to specify whether hydration of the silicates took place *via* the liquid or vapor phase, although evidence that cation exchange affected the phyllosilicate composition (Kerridge 1977) suggests the former. For more positive information, however, we must turn to the sulfates, carbonates and magnetite.

Sulfates. Veins of magnesium sulfate, with minor amounts of calcium sulfate, are a prominent feature of CI chondrites in hand specimen and thin section. The state of hydration presently observed is a function of local humidity and reveals nothing about the state of hydration as originally crystallized. Richardson (1978) noted that although hexahydrite, $MgSO_4 \cdot 6H_2O$, may be observed to be pseudomorphous after epsomite, $MgSO_4 \cdot 7H_2O$, the reverse is never found, and thus concluded that epsomite was the dominant preterrestrial form. Although the vein fillings are strongly suggestive of deposition from a liquid phase, the possibility must be considered that they were formed by reaction between a gas, containing sulfur dioxide and oxygen, and the walls of fissures through which the gas

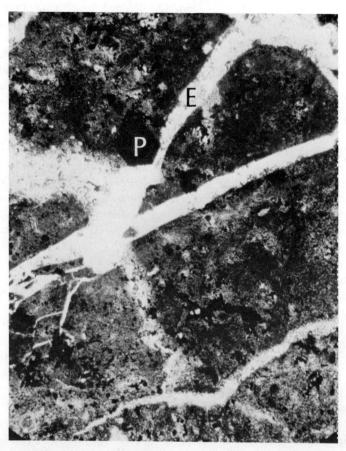

Fig. 2. Euhedral hexagonal pyrrhotite, Fe_7S_8, grain (P) protruding into vein of epsomite, $MgSO_4 \cdot 7H_2O$, (E) in Orgueil. Such a mineral association shows clearly that the sulfate was formed elsewhere and was transported into the vein by liquid water. Width of field of view is 0.3 mm. (Micrograph by courtesy of S. M. Richardson.)

was diffusing. However, this mechanism may be eliminated by means of an important observation made by Richardson (personal communication) and illustrated in Fig. 2. This shows a euhedral pyrrhotite, Fe_7S_8, grain protruding into a vein of epsomite. If this sulfate had been formed *in situ* by reaction involving a gas containing oxidized sulfur, the sulfide grain would also have reacted with the gas and would have decomposed. The mineral association in Fig. 2 requires that the sulfate was formed elsewhere and was transported into its present location in solution.

Carbonates. Of the three carbonates found in CI chondrites (dolomite; a pure calcium carbonate — possibly aragonite; and a ferroan magnesite with highly variable iron contents — generally referred to as breunnerite), the dolomite, $CaMgCO_3$, is probably the most useful diagnostic agent. From the degree of crystal perfection in such material, DuFresne and Anders (1962) inferred crystallization times in contact with liquid water of at least a thousand years. This is an aspect of the mineral chemistry of CI chondrites which would repay further detailed study by means of accurate measurements of the strengths of ordered reflections in x ray diffraction patterns.

Magnetite. The unusual morphologies exhibited by CI magnetite, Fe_3O_4, grains have provoked speculation that this material formed by condensation in the primitive solar nebula (e.g., Jedwab 1971; Herndon and Wilkening 1978). This idea received considerable support from an iodine-xenon retention age determined by Lewis and Anders (1975) for a magnetic separate, consisting largely of magnetite, from the Orgueil meteorite. They found the largest ^{129}Xe anomaly correlated with ^{127}I (i.e., the oldest age) then measured for any material and concluded that CI magnetite did, in fact, form in the solar nebula. Since then, older ages have been measured in a number of other meteorites (Drozd and Podosek 1976; Niemeyer 1979) and also it has been shown that the unusual morphologies were probably produced by crystallization from an aqueous medium (Kerridge *et al.* 1979c).

This last conclusion depends upon detailed comparisons between the most common morphologies among CI magnetites and terrestrial forms. A very close resemblance was demonstrated between the abundant "framboidal" magnetite (Jedwab 1965) and pyrite framboids found in marine sediments, which Sweeney and Kaplan (1973) have shown are developed only in hydrous conditions. Similarly, the spherulitic texture, commonly found in CI magnetites (Boström and Fredriksson 1966), has been shown to require crystallization from an amorphous medium under very specific conditions. Crystallization of the nonequilibrium fibrous form which constitutes a spherulite requires a delicate balance to exist between growth rate of the fiber and the rate of diffusion of impurities away from the growing crystal face (Keith and Padden 1963, 1964; Lofgren 1971). In general these conditions may be satisfied by growth from a melt or glass, but the prevalent low-temperature nature of the CI mineralogy precludes such growth media and seems to require, again, crystallization from an aqueous fluid, in this case most probably an iron-rich gel. Finally, the somewhat less common "plaquettes" as seen in Fig. 3 (Jedwab 1967, 1971), which have excited comparison with the platy morphologies predicted by Donn and Sears (1963) to characterize many nebular condensates, have been shown to have growth habits which are entirely inconsistent with crystallization from a gas and actually require their crystallization to have been controlled, probably epitaxially, by the microenvironment in which they grew (Kerridge *et al.* 1979c). This environment was probably supplied by the crystal lattice of the

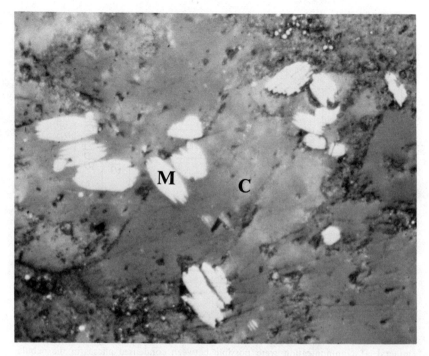

Fig. 3. Magnetite, Fe_3O_4, "plaquettes" (M) within carbonate (C) in Orgueil. The form of these plaquettes was apparently controlled epitaxially by the carbonate crystal structure. Width of field of view is 0.1 mm.

carbonate grains in which these plaquettes frequently occur; either the magnetites nucleated and grew upon the growth faces of the carbonates or else they formed by exsolution from crystallized carbonates. In either case it seems clear that plaquette formation was controlled by the conditions of carbonate crystallization.

It follows that not only did the magnetite forms probably result from crystallization in the presence of liquid water, but this episode of mineralization apparently occurred very early in solar system history, as evidenced by the ancient iodine-xenon age (Lewis and Anders 1975).

Summary

Petrographic features of CI chondrites indicate formation during an epoch of low-energy impact brecciation, roughly contemporaneous with a period of aqueous activity on the parent planetesimal. Similarly, several morphological, textural and crystallographic characteristics of sulfates, carbonates and magnetites appear to demand an origin by crystallization from liquid water. Composition and structure of the major matrix component, the

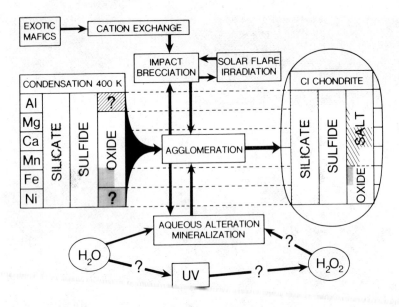

Fig. 4. Schematic summary of the processes needed to convert hypothetical low temperature (<400 K) nebular condensates into the observed CI chondrites. The shaded areas correspond to the proportion of each element in various types of host mineral. If communication were possible between condensed phases in the nebula, some oxidized iron could have entered the silicates, but evidence for this process is currently lacking.

ill-defined phyllosilicates, are also consistent with such an origin (Kerridge 1967, 1976, 1977). Composition and structure of sulfide phases also indicate an origin by secondary alteration although conditions cannot be specified at present (Macdougall and Kerridge 1977; Kerridge *et al.* 1979b). Formation of CI chondrites by alteration of low-temperature condensate material is illustrated schematically in Fig. 4.

II. OBSERVATIONS OF CM CHONDRITES

Petrography and mineralogy

The CM chondrites are examples of regolith breccias that, in addition to containing different CM clasts, chondrules, mineral fragments and "high-temperature" inclusions such as are common in CO and CV chondrites, show evidence for hydrothermal aqueous alteration. Matrices of these meteorites were converted to a secondary mineral assemblage consisting mostly of phyllosilicates; calcite, $CaCO_3$; magnetite; hydrated iron oxides; pentlandite, $(Fe,Ni)_9S_8$; and "poorly characterized phases" (*PCP*). Trace amounts of gypsum, $CaSO_4 \cdot 2H_2O$, and whewellite, $CaC_2O_4 \cdot H_2O$, have been

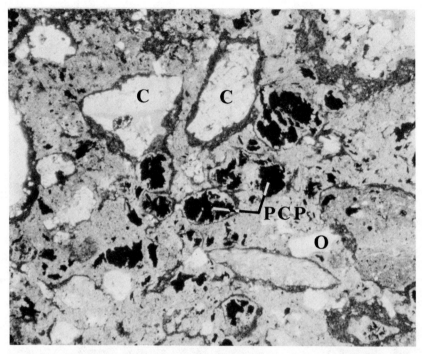

Fig. 5. Matrix in Murray CM chondrite. Light grains are mostly calcite, $CaCO_3$, (C), a few are relict olivine, $(Mg,Fe)_2SiO_4$, (O). Dark grains are "Poorly Characterized Phase," (PCP). Aphanitic (very fine grained) clasts, mostly comprising ill-defined hydrated silicates, are surrounded by darker material. Width of field of view is 0.46 mm.

reported in the Murchison chondrite (Fuchs *et al.* 1973). Ultra-thin section observations of the matrices indicate an interlocking arrangement of phyllosilicates of different color and composition in which all other phases and inclusions are embedded (Fig. 5). Calcite typically occurs as small (<20 μm) single grains or as grain aggregates that show well-defined growth twinning. Most calcite is bounded by *PCP*, which available data indicate is a mixture of Fe,Ni sulfides and sulfates, together with high concentrations of carbon and organic polymers (Bunch and Chang 1979*b*). *PCP* either protrudes into calcite, has formed within fractures, or occurs between grain boundaries of calcite aggregates (Fig. 6), all of which suggest overlapping or post-calcite formation. A similar paragenetic relationship may be inferred from phyllosilicate intergrowths with calcite. *PCP* is also common in isolated patches throughout matrices or as fine-grained aggregates associated, and contemporaneously formed, with a gray Mg-rich phyllosilicate. Magnetite is present either as irregularly shaped, rather large grains (up to 40 μm) or more commonly as globular clusters. Hydrated iron oxides occur as small grains (<10 μm) associated with phyllosilicates in altered outer portions of

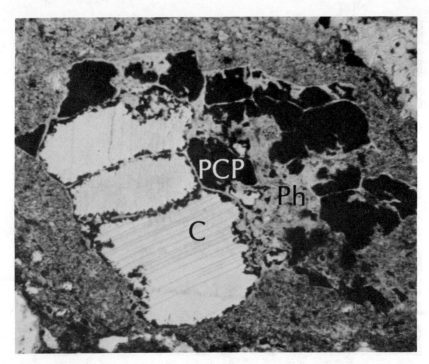

Fig. 6. "Poorly Characterized Phase," PCP surrounding and partly protruding into calcite (C) along grain boundaries and fractures. Light gray matrix in PCP is an iron-rich phyllosilicate (Ph), different in composition from that which surrounds the entire assemblage. Murray CM chondrite. Width of field of view is 0.18 mm.

chondrules or ferro-magnesian aggregates and as products of pseudomorphic replacement (Fig. 7).

Heterogeneous alteration is typified by the presence of completely altered chondrules, clasts, mineral fragments and glass spherules, reminiscent in textural appearance of terrestrial altered volcanic tuffs, mixed together with material showing only partial alteration. The admixture of altered and relatively unaltered phases implies complicated cycles of alteration episodes or a continuum of alteration and regolith turnover with infall of fresh material.

Examination of four CM chondrites (Murchison, Murray, Cold Bokkeveld and Nogoya) suggests variable degrees of alteration. Whereas the first three contain small to moderate amounts of altered material, Nogoya is nearly completely altered. Moreover, each secondary mineral has variable compositional limits and the crystallinity of phyllosilicates in Nogoya, as shown by x ray diffraction, is superior to that in other CM phyllosilicates (Bunch and Chang 1979b; C. B. Moore, personal communication). Nogoya matrix is a clearly defined assemblage of Fe-rich and Mg-rich phyllosilicates,

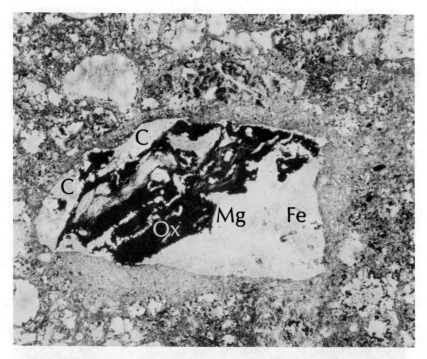

Fig. 7. Hydrated iron oxides (Ox), magnesium-rich phyllosilicates (Mg), iron-rich phyllosilicates (Fe) and calcite (C) in pseudomorphic replacement texture, possibly after olivine. Width of field of view is 0.46 mm.

calcite and secondary sulfides. The most striking feature of Nogoya that best exemplifies aqueous alteration, is the pseudomorphic replacement of ferro-magnesian silicates and opaque minerals in chondrules and aggregates by phillosilicates, calcite and iron oxides (Fig. 8). Although the overall texture and mineral assemblage of Nogoya implies a straightforward aqueous alteration scenario, observations of calcite suggest somewhat erratic changes in environmental conditions. Earliest formed calcite, which is twinned and commonly shows lattice strain, has been partially recrystallized in many cases into strain- and twin-free fine-grained calcite aggregates (Fig. 8).

Evidence for aqueous alteration in CM chondrites is less pronounced than in CI's. Late-stage veining of CI chondrites, which is an obvious example of aqueous transport, is limited in CM's to isolated occurrences of cross-cutting sulfide-*PCP* veins and, rarely, iron oxide veins. Flow orientation of matrix components in CM chondrites is suggestive of aqueous activity. Alternatively, this textural feature can also be explained as an artefact of pre-alteration plastic flow. In either case, the apparent flow orientation in CM chondrites, and also CV chondrites (King *et al.* 1978; Bunch and Chang 1979*a*), strongly

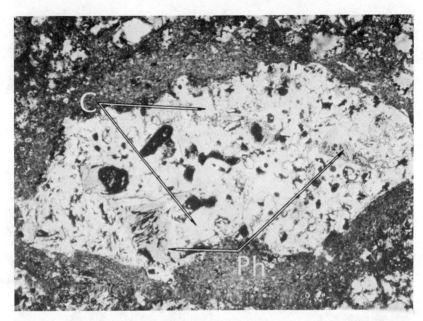

Fig. 8(a)

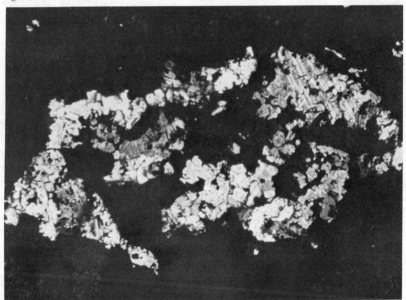

Fig. 8(b)

Fig. 8. (a) Aggregate inclusion in Nogoya pseudomorphed by calcite (C) and phyllosilicate (Ph). Plane light.
(b) Same area viewed with cross polarizers. Phyllosilicates appear as black field. Width of field of view is 0.46 mm.

suggests mechanical events which could only be achieved upon a parent body. Aqueous alteration has also been observed in at least two CV chondrites (Bunch and Chang 1979a); although alteration is not as extensively developed in those examples as in the meteorites described above, there is clear evidence for hydrothermal aqueous alteration.

CM chondrites also show evidence for shock effects due to hypervelocity impact, presumably dating, at least in part, from the period of regolith turnover. These effects include planar features in olivines (Bunch, unpublished work) and asterated x ray diffraction patterns from some olivines (DuFresne and Anders 1962).

Summary

Petrographic features of CM chondrites indicate formation as regolith breccias with some evidence for hypervelocity shock effects. Mineral associations and textures in matrix material are characteristic of aqueous alteration on the parent body, with common occurrence of pseudomorphic replacement structures and evidence for depositional or flow textures.

RELATIONSHIP BETWEEN CI AND CM CHONDRITES

It will have been apparent from the foregoing that there are strong similarities between CI chondrites and the matrices of CM's; however, they are not identical. This is clear from their oxygen isotopic compositions (Clayton and Mayeda 1977) which occupy different regions within a three-isotope plot (Fig. 9). In addition, although their mineralogies are apparently very similar, several subtle but real differences emerge on close scrutiny. Thus the bizarre morphologies characteristic of CI magnetites are entirely absent from CM meteorites (Jedwab 1968), except for very rare occurrences within CI xenoliths occasionally found in CM meteorites (Kerridge, unpublished work). Also, compositions of iron-nickel sulfides from CI and CM chondrites scarcely overlap each other when plotted on an Fe-Ni-S ternary diagram (Kerridge et al. 1979a,b). Finally, although analyses of individual olivine grains from both types of meteorite show closely similar distributions of iron content within both the major homogeneous grains and minor zoned grains, CI chondrites show a positive correlation between iron and calcium contents, whereas CM chondrites show no correlation at all in homogeneous grains and only an occasional correlation in zoned grains (Kerridge and Macdougall 1976; Fredriksson and Keil 1964; Fuchs et al. 1973).

From observations such as these, some preliminary conclusions may be drawn, though it should be emphasized that the detailed relationship between CI's and CM's remains to be evaluated. It seems clear, for example, that both types of meteorite originated within planetesimal regoliths similarly affected by meteoroid impact and aqueous activity, though the former effect seems to

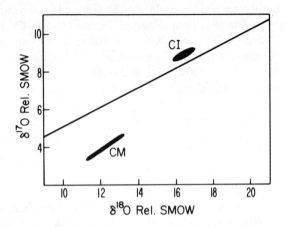

Fig. 9. CI and CM chondrites have clearly differing oxygen isotopic compositions, as illustrated in this plot of ratios $^{17}O/^{16}O$ versus $^{18}O/^{16}O$ expressed as *per mil* deviations from Standard Mean Ocean Water (SMOW). Chemical or physical isotopic fractionation causes compositions to move along the solid line, which is also the locus of all naturally occurring terrestrial compositions, and are clearly incapable of converting one type of carbonaceous chondrite into the other. (Illustration by courtesy of R. N. Clayton.)

have been enhanced and the latter reduced in severity in the CM regolith compared with that of the CI's. These differences, and the oxygen isotope data, indicate that CI and CM regoliths were on two, or more, parent bodies rather than in different parts of the same body. However, the isotopic differences are sufficiently small and the elemental abundances are sufficiently similar, that it seems unlikely that the parent bodies were widely separated in space. These observations also suggest that both types of meteorite formed in the same kind of parent body, i.e., either both come from asteroids or both come from comets. (A comet is here considered to be any object that exhibits cometary activity at some time during its existence, regardless of whether or not it ends up in an asteroidal-type orbit.)

III. CI-CM FORMATION CONDITIONS: A SCENARIO

The region of the parent body in which a CI or CM chondrite was formed must satisfy the following requirements. First, it must have been accessible to meteoroid impact in order to generate the observed brecciated structure. This clearly implies a location on the surface of the parent body, a conclusion supported by the evidence for solar flare irradiation of some individual mineral grains within both CI and CM meteorites (Kerridge and Macdougall 1976; Goswami *et al.* 1976). In addition, a small proportion of grains from CM chondrites reveal hypervelocity impact craters on their surfaces, as observed on grains from the lunar regolith (Goswami *et al.* 1976). Although elaborate models may be constructed, interpreting these irradiation and impact phenomena in terms of exposure while dispersed in space, the weight of the evidence points to an origin within a regolith environment (Goswami *et al.* 1976).

Second, the formation region must have been able to retain liquid water for a geologically significant time period, conceptually on the order of $\geqslant 10^3$ yr (DuFresne and Anders 1962). Although such a condition is easy to satisfy on an object of planetary size, such a large size is ruled out for CI chondrites because the low-impact velocities implied by the lack of observable shock features are incompatible with the gravitational fields of objects larger than a few tens of kilometers in radius. Such a low gravitational field, however, makes it difficult for liquid water to have been anything other than a transient phenomenon on the parent body surface. A possible solution to this impasse was advanced by DuFresne and Anders (1962) who showed that internal heating, for example by radionuclide decay, of an object initially containing chemically bound water, would produce a narrow zone of liquid water within the object. This zone would migrate towards the surface as heating progressed, culminating in an epoch in which the liquid zone would immediately underly a surficial icy layer analagous to permafrost on Earth. This layer would be self-healing as a result of the ratio of the heats of fusion and evaporation of water, so that the lifetime of the near-surface aqueous layer would have been governed by the rate of sublimation of ice. DuFresne and Anders (1962) showed that an object 100 km in radius (on the large side for a CI parent body) in a typical asteroidal orbit would have retained liquid water for about 200 My, which is certainly far longer than needed to explain the observations described above. However, these calculations neglected the effect of regolith turnover by meteoroid impact, so that their estimated lifetimes should probably be reduced somewhat. Nonetheless, it seems clear that periods of aqueous activity could have been sustained in internally heated objects characterized by radii and orbital parameters typical of asteroids. During the time that the active zone overlapped the zone of impact-induced turnover, the carbonaceous chondrites acquired their final form.

The evidence considered above has related to the place of formation of carbonaceous chondrites; what do we know about the time of formation? Most of the currently available information from radiochronology about carbonaceous meteorites consists of whole-rock model ages which are not instructive in this regard. Gray *et al.* (1973) showed that the rubidium-strontium systematics of the Allende CV chondrite, though including some very ancient material, had been disturbed at a relatively recent, though indeterminate, date precluding construction of an internal isochron. A similarly complex, multistage evolution was concluded for the uranium-thorium-lead systematics of the Orgueil CI chondrite by Tatsumoto *et al.* (1976). It appears that only two lines of evidence bear upon the time when the CI and CM chondrites were formed. First, the iodine-xenon retention interval for the magnetic separate from Orgueil already mentioned (Lewis and Anders 1975) indicates that the period of aqueous activity on the CI parent body was one of the earliest events yet dated in the solar system. The lack of an absolute time marker in this dating scheme prevents us from calculating how soon after nebular condensation this episode of alteration took place. Lewis and Anders (1975) also obtained an identical isochron, to within 2×10^5 yr, for a magnetic separate from the Murchison CM chondrite, suggesting that aqueous activity on the CM parent body coincided with that on the CI body, which is not intuitively unreasonable if the aqueous zone in each body were caused by internal heating from decay of a short-lived radionuclide, such as ^{26}Al (Lee *et al.* 1977).

Second, measurement of fission tracks in isolated olivine grains, originating from ^{244}Pu and ^{238}U in surrounding matrix, permits a compaction age to be calculated if the primordial Pu/U ratio is known. This approach has yielded ages of 4.2 to 4.4 Gyr for CM chondrites (Kothari and Macdougall 1976) and about 4.3 Gyr for Orgueil (Macdougall 1977). The simplest interpretation of these ages is that they represent the epoch of impact brecciation, although, in view of the apparent simultaneity of brecciation and mineralization, this interpretation is not readily consistent with the iodine-xenon ages discussed above. In fact, as discussed by Kothari and Macdougall (1976), interpretation as a straightforward compaction age is dependent upon a number of assumptions, in addition to the Pu/U ratio mentioned above. The fission track record may be disturbed, for example, by temporary residence in a matrix with a different actinide content from the present matrix, or by thermal erasure of tracks.

Available evidence therefore suggests that mineralization and brecciation of CI and CM chondrites occurred very early in solar system history. However, the possibility cannot be ruled out that these ancient ages are misleading, as a result, say, of inheriting by the magnetite of radiogenic xenon as well as iodine during a recent mineralization and accumulation of fission tracks by olivine in some earlier environment. Clearly, more radiochronological data for carbonaceous chondrites are required.

If the observations described above are taken at face value, CI and CM chondrites were formed in the near surface regions of bodies which were large enough to generate and retain a liquid water zone and which were sufficiently close to the sun to accumulate solar flare particle tracks at a time very soon after condensation of the solar system.

When attempting to relate this scenario to conditions on asteroidal or cometary surfaces, we must recognize that our knowledge of such objects is still very limited. However, formation of regolith on asteroids is generally accepted (e.g., Housen *et al.* 1979) and there appears to be convincing spectroscopic evidence for hydrated minerals on several asteroids (Matson *et al.* 1978; L. Lebofsky, personal communication). Thus there currently seems to be no observational evidence which precludes formation of carbonaceous chondrites within some asteroidal surfaces.

Considering cometary surfaces, on the other hand, the situation is much more complex. It is worth noting that, although the surface temperature of a comet could exceed $0°C$ in the inner solar system, any water produced by such solar heating would be promptly lost to space, ruling out significant aqueous alteration of the kind observed in carbonaceous chondrites. Thus, both types of putative parent body apparently require production of water by internal heating. Overt loss of material, both gas and solid, from the surfaces of active comets makes it highly unlikely that a suitable regolith could remain in place long enough to produce the observed geochemical and petrographic features of carbonaceous chondrites. However, this constraint would not necessarily apply to the surfaces of either possible protocometary cores or extinct cometary nuclei.

In the former case, the regolith episode must have taken place after condensation and significant agglomeration of ice, in order to make the nebula transparent to the solar flare particles which irradiated CI and CM olivines. It is also necessary that the nebular temperature dropped to the condensation point of ice without alteration of the olivines, which presumably condensed earlier. In addition, the protocomet must have been in a part of the solar system characterized by significant meteoroid bombardment, though at low to moderate velocities. This observation, and the solar flare track record, imply a heliocentric distance which was not too great, <8 AU according to Anders (1975).

In the case of an extinct comet, the age data cited above indicate that its active phase occurred shortly after condensation of the solar system. Thus, the object must then have been in an orbit frequently penetrating well into the inner solar system. Such an orbit would have had a lifetime of only about 10^6 to 10^7 yr against ejection by Jupiter (Wetherill 1974). In order to be able to supply contemporary meteorites, therefore, perturbation into a longer-lived orbit would have been necessary.

It is difficult to assign probabilities to such scenarios, mostly because models of cometary origin are currently so imprecise that the plausibility of the various *ad hoc* constructs outlined above cannot be assessed. In light of present knowledge, the only *ad hoc* assumption involved in an asteroidal scenario is the existence of a dynamical mechanism capable of nonviolently transporting main-belt material into earth-crossing orbit. We are encouraged by the progress being made in this area (Wetherill 1974, 1978; Wetherill and Williams 1977). We therefore favor an asteroidal source for carbonaceous chondrites, but agree with Wasson and Wetherill (this book) that a cometary source cannot currently be ruled out. In view of the similarities in visible reflectance spectra between C-type asteroids and carbonaceous chondrites, particularly of group CM, and the dominance of C-type asteroids in the main belt (see the chapter by Zellner in this book), it would appear as though aqueous alteration of asteroidal surfaces may be commonplace. However, the infrared reflectance spectral data (summarized by Larson and Veeder in this book) suggest that such a conclusion may be premature. The correspondence, among existing asteroid spectra, between "visible C-type" and "infrared CM ($C2$)-type" is far from perfect and only for the latter have hydroxyl absorption bands been demonstrated so far (Larson and Veeder, in this book).

An additional complication in the interpretation of hydroxyl absorption spectra in terms of asteroidal aqueous activity arises from the fact that the water which was responsible for alteration of the carbonaceous chondrite material must have been accreted by the parent body in some form, and hydrated silicates seem at least as plausible in this respect as condensed ice. It may well be that the putative precursor material would be spectroscopically indistinguishable from the products of aqueous alteration on a planetesimal. (In fact, the canonical interpretation of C-type spectra among asteroidal spectroscopists is in terms of such "primordial" material.) Thus asteroidal spectral data are unlikely to *confirm* the aqueous alteration hypothesis, which will probably continue to depend upon the indirect association of observed alteration textures and mineralogies in chondrites with an origin for these objects within the asteroidal region.

Acknowledgments. We thank R. N. Clayton, T. Gehrels, I. R. Kaplan, J. D. Macdougall, H. Y. McSween, C. B. Moore and S. M. Richardson for advice, assistance, observations and/or meteorite samples. One of us (JFK) acknowledges support from the National Aeronautics and Space Administration.

REFERENCES

Anders, E. 1971. How well do we know "Cosmic" abundances? *Geochim. Cosmochim. Acta* 35: 516-522.
Anders, E. 1975. Do stony meteorites come from comets? *Icarus* 24: 363-371.
Anders, E. 1978. Most stony meteorites come from the asteroid belt. In *Asteroids: An exploration assessment,* NASA Conf. Publ. 2053: pp. 57-75.
Bass, M. N. 1971. Montmorillonite and serpentine in Orgueil meteorite. *Geochim. Cosmochim. Acta* 35: 139-148.
Boström, K., and Fredriksson, K. 1966. Surface conditions of the Orgueil meteorite parent body as indicated by mineral associations. *Smithsonian Misc. Publ.* 151: 3.
Bunch, T. E., and Chang, S. 1979a. Thermal metamorphism (shock?) and hydrothermal alteration in C3V meteorites. *Lunar Science X,* The Lunar and Planetary Institute, pp. 164-166.
Bunch, T. E., and Chang, S. 1979b. Carbonaceous chondrites II: Carbonaceous chondrite phyllosilicates and light element geochemistry as indicators of parent body processes and surface conditions. *Geochim. Cosmochim. Acta* (in press).
Chao, E. C. T.; Boreman, J. A.; and Desborough, G. A. 1971. The petrology of unshocked and shocked Apollo 11 and Apollo 12 microbreccias. *Proc. Lunar Sci. Conf. II* (Cambridge: MIT Press), pp. 797-816.
Clayton, R. N., and Mayeda, T. K. 1977. Anomalous anomalies in carbonaceous chondrites. *Lunar Science VIII,* The Lunar Science Institute, pp. 193-195.

Donn, B., and Sears, G. W. 1963. Planets and comets: Role of crystal growth in their formation. *Science* 140: 1208-1211.

Drozd, R. J., and Podosek, F. A. 1976. Primordial [129]Xe in meteorites. *Earth Planet. Sci. Lett.* 31: 15-30.

DuFresne, E. R., and Anders, E. 1962. On the chemical evolution of the carbonaceous chondrites. *Geochim. Cosmochim. Acta* 26: 1085-1114.

Fredriksson, K., and Keil, K. 1964. The iron, magnesium, calcium and nickel distribution in the Murray carbonaceous chondrite. *Meteoritics* 2: 201-217.

Fuchs, L. H.; Olsen, E.; and Jensen, K. J. 1973. Mineralogy, mineral-chemistry and composition of the Murchison (C2) meteorite. *Smithsonian Contr. Earth Sci. No. 10.*

Goswami, J. N.; Hutcheon, I. D.; and Macdougall, J. D. 1976. Microcraters and solar flare tracks in crystals from carbonaceous chondrites and lunar breccias. *Proc. Lunar Sci. Conf. VII* (Oxford: Pergamon Press), pp. 543-562.

Gray, C. M.; Papanastassiou, D. A.; and Wasserburg, G. J. 1973. The identification of early condensates from the solar nebula. *Icarus* 20: 213-239.

Herndon, J. M., and Wilkening, L. L. 1978. Conclusions derived from the evidence on accretion in meteorites. In *Protostars and Planets,* ed. T. Gehrels (Tucson: University of Arizona Press), pp. 502-515.

Holweger, H. 1977. The solar Na/Ca and S/Ca ratios: A close comparison with carbonaceous chondrites. *Earth Planet. Sci. Lett.* 34: 152-154.

Housen, K. R.; Wilkening, L. L.; Chapman, C. R.; and Greenberg, R. 1979. Asteroidal regoliths. *Icarus* (submitted).

Jedwab, J. 1965. Structures framboidales dans la météorite d'Orgueil. *C. R. Acad. Sci. Paris* 261: 2923-2925.

Jedwab, J. 1967. La magnétite en plaquettes des météorites carbonées d'Alais, Ivuna et Orgueil. *Earth Planet. Sci. Lett.* 2: 440-444.

Jedwab, J. 1968. Variations morphologiques de la magnetite des météorites carbonées d'Alais, Ivuna et Orgueil. In *Origin and Distribution of the Elements,* ed. L. H. (Oxford: Pergamon), pp. 467-478.

Jedwab, J. 1971. La magnetite de la météorite d'Orgueil vue au microscope électronique à balayage. *Icarus* 15: 319-340.

Keith, H. D., and Padden, F. J. 1963. A phenomenological theory of spherulitic crystallization. *J. Appl. Phys.* 34: 2409.

Keith, H. D., and Padden, F. J. 1964. Spherulitic crystallization from the melt. *J. Appl. Phys.* 35: 1270.

Kerridge, J. F. 1967. Mineralogy and genesis of the carbonaceous meteorites. In *Mantles of the Earth and Terrestrial Planets,* ed. S. K. Runcorn (London: Wiley), pp. 35-47.

Kerridge, J. F. 1976. Major element composition of phyllosilicates in the Orgueil carbonaceous meteorite. *Earth Planet. Sci. Lett.* 29: 194-200.

Kerridge, J. F. 1977. Correlation between nickel and sulfur abundances in Orgueil phyllosilicates. *Geochim. Cosmochim. Acta* 41: 1163-1164.

Kerridge, J. F., and Macdougall, J. D. 1976. Mafic silicates in the Orgueil carbonaceous meteorite. *Earth Planet. Sci. Lett.* 29: 341-348.

Kerridge, J. F.; Macdougall, J. D.; and Carlson, J. 1979a. Iron-nickel sulfides in the Murchison meteorite and their relationship to phase Q1. *Earth Planet. Sci. Lett.* (in press).

Kerridge, J. F.; Macdougall, J. D.; and Marti, K. 1979b. Clues to the origin of sulfide minerals in CI chondrites. *Earth Planet. Sci. Lett.* (in press).

Kerridge, J. F.; Mackay, A. L.; and Boynton, W. V. 1979c. Magnetite in CI carbonaceous meteorites: Origin by aqueous activity on a planetesimal surface. *Science* (submitted).

King, T. V. V.; Butler, J. C.; and King, E. A. 1978. Petrofabric study of the Allende meteorite. *Meteoritics:* 517-518.

Kothari, B. K., and Macdougall, J. D. 1976. Formation chronology for C2 meteorites. *Earth Planet. Sci. Lett.* 33: 36-44.

Kurat, G.; Keil, K.; and Prinz, M. 1974. Rock 14318: A polymict breccia with chondritic texture. *Geochim. Cosmochim. Acta* 38: 1133-1146.

Larimer, J. W., and Anders, E. 1970. Chemical fractionations in meteorites. III. Major element fractionations in chondrites. *Geochim. Cosmochim. Acta* 34: 367-387.

Lee, T.; Papanastassiou, D. A.; and Wasserburg, G. J. 1977. Aluminum-26 in the early solar system: Fossil or fuel? *Astrophys. J.* 211: L107-L110.

Lewis, R. S., and Anders, E. 1975. Condensation time of the solar nebula from extinct [129]I in primitive meteorites. *Proc. Nat. Acad. Sci.* 72: 268-273.

Lofgren, G. 1971. Spherulitic textures in glassy and crystalline rocks. *J. Geophys. Res.* 76: 5635-5648.

Macdougall, J. D. 1977. Time of compaction of Orgueil. *Meteoritics* 12: 301-302.

Macdougall, J. D., and Kerridge, J. F. 1977. Cubanite: A new sulfide phase in CI meteorites. *Science* 197: 561-562.

Matson, D. L.; Veeder, G. J.; and Lebofsky, L. A. 1978. Infrared observations of asteroids from Earth and space. In *Asteroids: An exploration assessment,* eds. D. Morrison and W. C. Wells, NASA Conf. Publ. 2053, pp. 127-144.

McSween, H. Y. 1977. Petrographic variations among carbonaceous chondrites of the Vigarano type. *Geochim. Cosmochim. Acta* 41: 1777-1790.

McSween, H. Y. 1979. Are carbonaceous chondrites primitive or processed? A review. *Rev. Geophys. Space Phys.* (in press).

Nagy, B. 1966. Investigations of the Orgueil carbonaceous meteorite. *Geol. För. Stock. Förhandl.* 88: 235-272.

Nagy, B. 1975. Carbonaceous Meteorites. (Amsterdam: Elsevier).

Niemeyer, S. 1979. I-Xe dating of silicate and troilite from IAB iron meteorites. *Geochim. Cosmochim. Acta* (in press).

Reid, A. M.; Bass, M. N.; Fujita, H.; Kerridge, J. F.; and Fredriksson, K. 1970. Olivine and pyroxene in the Orgueil meteorite. *Geochim. Cosmochim. Acta* 34: 1253-1255.

Richardson, S. M. 1978. Vein formation in the CI carbonaceous chondrites. *Meteoritics* 13: 141-159.

Sweeney, R. E., and Kaplan, I. R. 1973. Pyrite framboid formation: Laboratory synthesis and marine sediments. *Econ. Geol.* 68: 618-634.

Tatsumoto, M.; Unruh, D. M.; and Desborough, G. A. 1976. U-Th-Pb and Rb-Sr systematics of Allende and U-Th-Pb systematics of Orgueil. *Geochim. Cosmochim. Acta* 40: 617-634.

Wetherill, G. W. 1974. Solar system sources of meteorites and large meteoroids. *Ann. Rev. Earth Planet. Sci.* 2: 303-331.

Wetherill, G. W. 1978. Dynamical evidence regarding the relationship between asteroids and meteorites. In *Asteroids: an exploration assessment,* eds. D. Morrison and W. C. Wells, NASA Conf. Publ 2053, pp. 17-35.

Wetherill, G. W., and Williams, J. G. 1977. Origin of differentiated meteorites. In *Proc. 2nd. Int. Conf. on Origin and Abundance of the Elements* (Paris: Pergamon Press, in press).

GEOCHEMICAL EVOLUTION OF THE EUCRITE PARENT BODY: POSSIBLE NATURE AND EVOLUTION OF ASTEROID 4 VESTA?

MICHAEL J. DRAKE
University of Arizona

The calculated composition of the eucrite parent body suggests that it was composed of ~ 90% of a feldspathic peridotite mantle (± core) and ≤10% of basaltic achondrite crust. The absence in our museums of meteorites of appropriate mineralogy and bulk composition representative of the mantle suggest that the eucrite parent body remains intact. Using compositional and dynamical discriminants all planetary bodies except the asteroids are eliminated from candidacy. In the survey of bright asteroids with radii >25 km, only Vesta displays the requisite optical properties. Although hypothetical anhydrous asteroids in highly eccentric orbits and asteroids with radii of less than 25 km remain in principle as candidates, we conclude tentatively that Vesta is the eucrite parent body. Until infrared reflectance spectra are obtained for the Shergottites, the arguments presented in favor of Vesta being the eucrite parent body apply equally well to the Shergottites. Regardless of the outcome of such measurements, a paradox exists unless the eucrites and Shergottites are derived from a single, isotopically and compositionally heterogeneous parent body. Dynamical objections to the identification of Vesta as the parent planet of basaltic meteorites are answered with an ad hoc two-stage evolutionary model.

It has been known since the work of McCord *et al.* (1970) that asteroid 4 Vesta is spectroscopically unique. These authors noted the similarity between the spectrum of Vesta and those of certain basaltic achondrite meteorites, specifically the eucrites. They proposed, therefore, that the surface of Vesta is covered with basalts, a conclusion which has been reinforced in subsequent studies (Larson and Fink 1975; Chapman 1976; Feierberg *et al.* 1979; see the chapters by Gaffey and McCord and by Larson and Veeder). Although we believe that some, although probably not all, meteorites are ultimately derived from the main belt, we cannot *prove directly* at this time that any meteoritic basalts come from Vesta. The candidacy of Vesta as the eucrite parent body remains ambiguous. It may be possible through additional infrared reflectance measurements of basaltic meteorites to rule out all meteoritic basalts except the eucrites as being consistent with the surface rocks of Vesta, in which case the eucrites may be considered at least as analogues of Vestan basalts. Thus if we can decipher the petrogenesis of the eucrites and their parent body, we may be able to infer the geological history of Vesta.

I. PROPERTIES OF EUCRITES

Petrology

Eucritic meteorites (plagioclase-pigeonite achondrites) appear to have originated as extraterrestrial basaltic melts (Duke and Silver 1967). Most eucrites are now brecciated but a few, e.g., Ibitira, preserve igneous textures which have been modified only by subsequent minor shock and thermal metamorphic events (Wilkening and Anders 1975; Steele and Smith 1976). Most eucrites also share a common mineralogy, consisting of approximately 40% by weight plagioclase (An 80-95), and 60% Ca-poor clinopyroxene (Wo 5-15, En 25-30, Fs 50-60) which generally is associated with a Ca-rich clinopyroxene. The FeO/(FeO+MgO) molar ratio in the pyroxene is generally 0.57-0.65. There are some exceptions to this pattern, however. In particular, a few unusual eucrites such as Moore County and Serra de Magé are unbrecciated and coarse grained, and appear to be cumulates. In these eucrites the pyroxene phase has inverted to hypersthene, and exsolution lamellae of a high-Ca pyroxene are visible in thin section. There is a suggestion of planar orientation of plagioclase in both Moore County (Hess and Henderson 1949) and Serra de Magé (Duke 1963) which could be attributed to crystal accumulation in a magma chamber.

Geochemistry

We shall restrict our summary of the geochemistry of the eucrites to the coherent group of refractory and lithophile rare earth elements (REE) and to W, an element which is an indicator of the fractionation of metal from silicates.

Rare earth patterns for many eucrites have been reported over the past 15 years by a number of authors using various techniques. Several representative patterns are shown in Fig. 1. Note that a number of REE

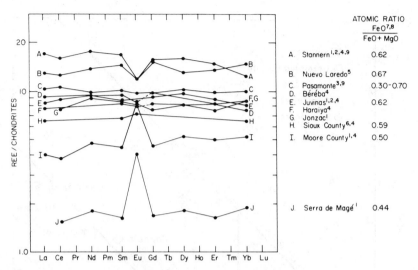

Fig. 1. Rare earth abundance patterns for the eucrites, normalized to chondritic abundances. Sources for the data are indicated by superscript: [1]Schnetzler and Philpotts (1969); [2]Schmitt et al. (1963); [3]Schmitt et al. (1964); [4]Jérome (1970); [5]Gast and Hubbard (1970); [6]Ma and Schmitt (1976); [7]Duke and Silver (1967); [8]Takeda et al. (1976b); [9]Gast et al. (1970). When more than one reference is listed for a eucrite, the average was plotted; in such cases that standard deviation was generally in the range 5-10%. The small positive Eu anomaly in Sioux County may be due to sampling problems (excess of plagioclase) in the sample analyzed by Jérome (1970).

patterns cluster at 8-10 x CC1 with negligibly small negative Eu anomalies, the type example being Juvinas. Sioux County appears as an isolated pattern at approximately 6.5 x CC1. Stannern and Nuevo Laredo have higher REE abundance patterns than the Juvinas group, but Stannern has a comparable Fe/(Fe + Mg) ratio while that of Nuevo Laredo is higher. Moore County and Serra de Magé show low trivalent REE patterns and significant positive Eu anomalies, which are consistent with the hypothesis that these two eucrites are cumulates.

Tungsten abundances in the eucrites are illustrated in Fig. 2. The eucrites (and other basaltic achondrites) are depleted in W relative to chondrites by a factor of approximately 19. This depletion is generally attributed to metal/silicate fractionation (e.g., Rammensee and Wänke 1977), either during nebular condensation or subsequent planetary differentiation.

Ages

The ages of several eucrites have been determined from precise mineral isochrons using the Rb/Sr, Sm/Nd, and Pb/Pb dating techniques. The most thoroughly studied eucrite, Juvinas, yields an age of 4.60 ± 0.07 AE using the Rb/Sr method (Birck et al. 1975; Allegre et al. 1975; Birck and Allegre 1978)

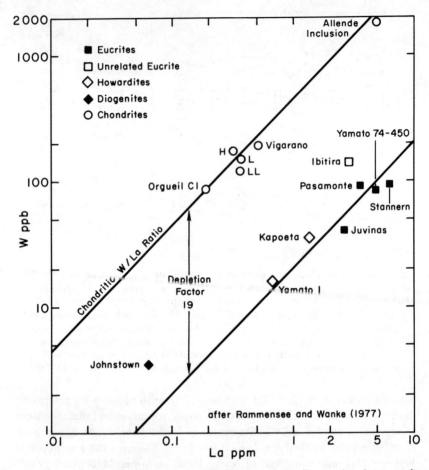

Fig. 2. W versus La for a variety of meteorites. The correlation lines are based on a much larger data set summarized in Rammensee and Wänke (1977).

and 4.56 ± 0.08 using the Sm/Nd method (Lugmair *et al.* 1975). Ibitira has an Rb/Sr mineral isochron age of 4.53 ± 0.1 AE (Birck *et al.* 1975; Birck and Allegre 1978). Pasamonte yields mineral isochron ages of 4.56 ± 0.14 AE using the Sm/Nd method and 4.57 ± 0.01 AE using the Pb-Pb (207/204 versus 206/204) method, but Rb/Sr systematics indicate a more recent disturbance of this system (Nakamura *et al.* 1976; Birck and Allegre 1978). Other eucrites such as Béréba, Sioux County, and Stannern yield younger "ages" (Birck *et al.* 1975; Birck and Allegre 1978). The ages close to 4.6 AE are interpreted as crystallization or formation ages while younger ages are interpreted as subsequent disturbance of older materials. The concordance of most basaltic achondrites on a whole rock isochron with the same primitive initial $^{87}Sr/^{86}Sr$ ratio (BABI), which is amongst the most primitive Sr observed in the meteorites (Papanastassiou and Wasserburg 1969; Gray *et al.* 1973), is consistent with the inference that the eucrites formed in a single melting event at ~ 4.6 AE.

II. EVOLUTION OF EUCRITES AND THEIR PARENT BODY

Experimental Petrology

The time of the event which produced the eucrites appears to be well established at ~ 4.6 AE. The nature of the event, however, has been the subject of speculation for many years. The differentiated meteorites provide a record of inhomogeneous source regions, variable degrees of partial melting and fractional crystallization, multiple genesis of magma from a single source region, and remelting of earlier-formed igneous rocks. These observations provide evidence that early intense heat sources were both spatially and temporally variable. A comprehensive discussion is given by Mittlefehldt (1979).

The eucrites appear to reflect relatively simple igneous differentation eucrites, however. There are two main classes of models: (1) the eucrites were produced by extensive fractional cyrstallization of more magnesian liquids (Mason 1962; Schnetzler and Philpotts 1969; McCarthy *et al.* 1973; Shimizu and Allegre 1975; Takeda *et al.* 1976a); and (2) the eucrites were produced by partial melting of a polymineralic source region (Stolper 1975, 1977).

Stolper demonstrated that the former hypothesis is inconsistent with experimentally-determined phase equilibria and the composition of the eucrites (Fig. 3). See Consolmagno and Drake (1977) for an abbreviated discussion. Stolper's experiments demonstrated that many eucrites (e.g. Juvinas) are saturated with five phases (olivine-Fo_{65}, pigeonite-Wo_5En_{65}, plagioclase-An_{94}, Cr-rich spinel, and metal) at temperatures close to their liquidi, and plot close to a pseudoperitectic point in appropriate phase space at appropriate redox potentials. Other eucrites plot away from the pseudoperitectic point in positions consistent with the hypothesis that they represent cumulates. The essence of Stolper's (1975) model is that (1) eucrites that appear to represent primary liquids, such as Juvinas, Stannern, Ibitira, Sioux County, are produced by varying fractions of partial melting of an olivine-pigeonite-plagioclase-spinel-metal mantle; (2) unusual eucrites such as Moore County, Serra de Magé may represent partial cumulates from similar melts; and (3) eucrites that appear to be derivative liquids because they exhibit higher $Fe/(Fe + Mg)$ ratios, such as Nuevo Laredo, represent residual liquids after separation of cumulus pyroxene and plagioclase. Although this model has been extended and has become more complex in Stolper (1977), the basic hypothesis remains unchanged.

Geochemical Modeling

Once the *process* by which the ecuritic basalts formed is understood, we

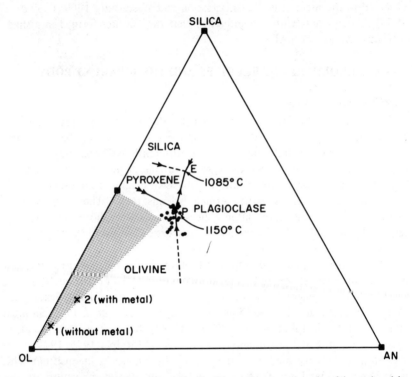

Fig. 3. Schematic "pseudo-ternary" liquidus diagram for basaltic achondrite compositions. Dots represent compositions of non-cumulate eucrites. Points 1 and 2 are the bulk compositions of the silicate portion of the eucrite parent body computed for the cases of no metal and 30% metal, respectively, in the planet. The shaded area is the range of permissible bulk compositions estimated by Stolper. (After Stolper 1975, 1977.)

may quantitatively calculate the amounts of the various minerals present in the source region to be melted. We take advantage of the fact that each mineral accepts trace elements (e.g., REE) into its structure in a characteristically individual way (Fig. 4). Upon melting a mineral, the silicate liquid produced contains a trace element pattern which is approximately complementary to that in the mineral. The trace element pattern in a silicate liquid produced by melting a polymineralic source region is a fingerprint or a hieroglyphic which may be deciphered to yield the composition of the source region. The mathematics and assumptions involved have been discussed by Gast (1968), Shaw (1970, 1977), Weill *et al.* (1974), Hertogen and Gijbels (1976) and Drake (1979).

Applying trace element modeling to the eucrites, Consolmagno and Drake (1977) concluded that the eucrite parent body has approximately average solar system (chondritic) composition for all lithophile elements

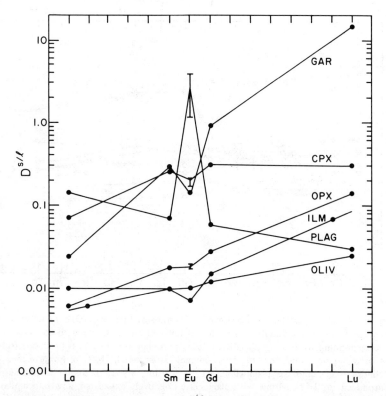

Fig. 4. Solid/melt partition coefficients $(D^{s/\ell})$ for several important minerals, GAR = garnet (Arth 1976), CPX = Ca-rich clinopyroxene (Grutzeck *et al.* 1974), OPX = orthopyroxene (Weill and McKay 1975), ILM = ilmenite (McKay and Weill 1976), PLAG = plagioclase feldspar (Drake and Weill 1975), OLIV = olivine (Schnetzler and Philpotts 1970).

except the most volatile, and that the bulk of the eucrites correspond to ∼ 10% partial melting of an olivine-rich source region with plagioclase essentially exhausted at that fraction of melting (Fig. 5). Fukuoka *et al.* (1977) reach similar conclusions, although they differ in the conclusions that the eucrite parent body was chondritic absolute with respect to the REE and, hence, in the fraction of melting to which most eucrites correspond.

The close correspondence of the crystallization age of the eucrites (∼ 4.5 AE) with the age of the solar system appear to preclude extensive chemical differentiation of the eucrite parent body prior to the event which produced eucritic basalts. The eucrites may be interpreted, therefore, as having been produced in a single partial melting event. The equivalent of this event may have manifested itself on larger bodies such as the moon and earth (which retain internal heat more efficiently) in a global complete melting or magma ocean epoch (Goleš and Seymour 1979). If this interpretation is correct, and

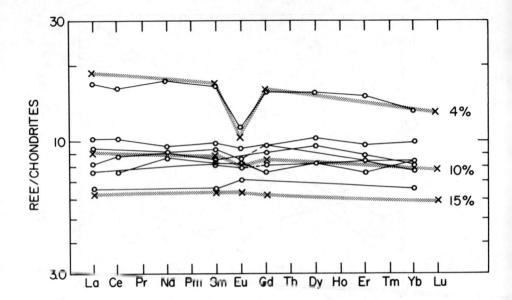

Fig. 5. Model REE patterns (x) compared with measured patterns (O) for eucrites which
appear to represent liquids derived by partial melting. A liquid with a composition
corresponding to 40% plagioclase and 60% pyroxene (see Fig. 1) is produced from a
source region consisting of 50% olivine, 30% metal, 10% orthopyroxene, 5%
clinopyroxene and 5% plagioclase. At 4% melting, a pattern similar to Stannern is
produced. At 10% melting, with plagioclase effectively exhausted, a pattern similar to
the main group of eucrites (e.g., Juvinas) is produced. At 15% melting, with all
plagioclase and most of the pyroxene exhausted, a pattern similar to Sioux County is
produced. A source region consisting initially of 85% olivine, 5% orthopyroxene, 5%
clinopyroxene and 5% plagioclase gives virtually identical results; the amount of metal
in the system cannot be resolved by REE patterns. Variations in the relative
proportions of phases in the source region can be made to provide an even closer fit to
some aspects of the REE patterns; e.g., 4% plagioclase results in a better match to the
Eu anomaly of Stannern than 5% plagioclase, but the inherent precision of the model
calculations is inadequate to warrant such "fine-tuning." (After Consolmagno and
Drake 1977.)

if the eucrite parent body is the product of homogeneous accretion, our
calculations have yielded not only the composition of the source region of
the eucrites, but the composition of the eucrite parent planet as a whole.
There is remarkable agreement between this bulk composition and bulk
compositions calculated by independent methods (see Table I and Fig. 6).
The compositions calculated by Consolmagno and Drake (1977) are plotted
in Fig. 3.

The amount of metal in the eucrite parent body is uncertain. Figure 2
shows that the basaltic achondrites are depleted in the siderophile element W

TABLE I

Bulk Compositions of the Eucrite Parent Body

	Consolmagno and Drake (1977)		Hertogen et al. (1977)	Morgan et al. (1978)	Dreibus et al. (1977)
	no metal	30% metal			
Na_2O	0.04	0.06	0.05	0.05	0.07
MgO	29.7	28.0	29.4	28.5	32.5
Al_2O_3	1.8	2.55	2.4	2.5	3.0
SiO_2	39.0	41.3	41.2	39.8	46.1
CaO	1.2	1.85	2.0	2.06	2.7
FeO	28.3	26.3	24.0	26.6	14.6
TOTAL	100.04	100.06	99.05	99.51	98.77

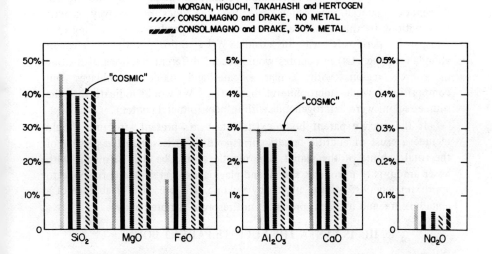

Fig. 6. Bar graph comparing estimates of the bulk composition of the eucrite parent body. The figures from Dreibus et al. (1977), Hertogen et al. (1977) and Morgan et al. (1977) were computed by a technique which utilized elements that are correlated in the condensation sequence. The bulk composition labeled "no metal" represents the estimate by Consolmagno and Drake (1977) for a source region 85% olivine (Fo_{65}), 10% pigeonite (Wo_5En_{65}), and 5% plagioclase (An_{94}). The bulk composition labeled "30% metal" represents the silicate portion only of a source region containing 30% metal, 50% olivine, 15% pigeonite, and 5% plagioclase. Horizontal lines represent cosmic abundances in oxide weight percent of SiO_2, MgO, Al_2O_3, CaO, and FeO where FeO is computed by substracting the amount of Fe sufficient to give the parent body a 10% FeO content, an average of the estimtes of Hertogen et al. (1977) and Morgan et al. (1977). (After Consolmagno and Drake 1977.)

relative to cosmic values by a factor of approximately 19. This depletion appears to be the result of planetary processes rather than nebular processes (Rammansee and Wänke 1977; Newsom and Drake 1979). If the factor of 19 depletion correlation line conveys genetic information, W must have been in the oxidized state in order to be covariant with La. Thus the depletion of the source region in W must have occurred prior to reaching the silicate solidus temperature of $\sim 1150°C$, possibly due to metal separation to form a core. This proposal raises a problem in that core formation must have occurred very rapidly, because of the ~ 4.6 AE internal mineral isochron ages of some eucrites. Separation of metal in the solid state would be far too slow on an asteroidal-sized body. A possible solution would be separation of a Fe-FeS eutectic melt at $\sim 1000°C$ (or lower temperatures if significant amounts of Ni are present). Such a process could occur rapidly ($\leqslant 10^8$ yr) but is problematical in that it has not been demonstrated that the required amount of S is available. Indeed, the general depletion of the eucrites in volatile elements might argue against the presence of significant quantities of S. Alternatively the factor of 19 depletion correlation line may be viewed as a mixing line generated by a continuum of mixtures of primarily eucritic and diogenetic material, in which case the line does not convey genetic information. In this view the eucrites could be produced by melting of a feldspathic peridotite with a cosmic W/La ratio. Lithophile element abundances (e.g., La) in eucrites would record different fractions of melting of source regions with similar silicate and oxide assemblages and compositions, while siderophile elements (e.g., W) would indicate that the source regions were heterogeneous with respect to metal content.

If the eucrite parent body exists intact, its present structure should include a crust of eucritic basalts corresponding to less than ten percent of the total volume of the planet. The mantle should be rich in olivine with lesser amounts of pyroxene, spinel, and plagioclase, and should correspond to approximately 90% of the planet by volume. The presence or absence of a metallic core, and its size if present is not resolved as noted above.

III. IS VESTA THE EUCRITE PARENT BODY?

Thus far we have side-stepped the question of whether or not Vesta is the parent planet of the eucrites, preferring instead to use the eucrites as analogues for the evolution of Vesta basalts. At this point it is appropriate to present evidence for and against Vesta being the eucrite parent body. Arguments supporting the hypothesis that Vesta is the eucrite parent body are based largely on geochemical and observational considerations. Arguments against Vesta being the eucrite parent body are based largely on dynamical considerations.

Evidence in favor

First we must seek an answer to the question: does the eucrite parent body still exist? If evidence existed that the planet had been disrupted at some point in the past, then Vesta could not be the eucrite parent body.

As noted in the previous section, the consensus that the eucrites correspond to relatively small degrees of partial melting (Consolmagno and Drake 1977; Fukuoka *et al.* 1977; Stolper 1977) implies that the bulk of the eucrite parent body should consist of an olivine-rich mantle ($\sim Fo_{65}$) with or without a metal/sulfide core, covered by a veneer of basalts. Thus if the eucrite parent body had been totally disrupted, we would expect to find vastly more samples of the mantle in our meteorite collections than samples of crust. Yet there is not one suitable candidate for the mantle of the eucrite parent body in our collections. Even with the doubling of the number of known meteorites with the collection of meteorites from Antarctica, this conclusion has not changed. Unless a mechanism can be developed in which crust and mantle evolve dynamically into different orbits such that only the crustal fragments impact the earth, we must conclude that *the eucrite parent body is intact* and that the eucrites have been chipped from the crustal veneer of basalts by small impacts.

How big is the eucrite parent body? Using considerations of rates of crystal settling from magmas, and buoyancy of magmas, Walker *et al.* (1978) concluded that the apparent ambient gravitational field was consistent with a planet of radius 10-100 km. Examination of their assumptions indicates that 10 km is an *extreme* lower limit. This radius range is that of the asteroids and would seem to rule out the larger planets. We can eliminate the larger planets on more direct grounds, however.

The composition of the eucrites indicates that they evolved on a volatile-poor, anhydrous body. Mercury would fit the description, but is eliminated on dynamical and compositional grounds. Spectrophotometric observations of Mercury's surface place an upper limit of 6% of FeO by weight (Adams and McCord 1977). In contrast the eucrites typically contain approximately 18 wt% FeO. Venus may be eliminated because it is volatile-rich and surface rocks are subjected to greenschist facies metamorphism, in contrast to the eucrites. In addition there are dynamical grounds for elimination (Wetherill 1974). We know that lunar samples differ significantly from the eucrites in a wide range of properties. Mars is eliminated on dynamical grounds and because it is volatile-rich relative to the eucrites. Phobos and Deimos resemble the carbonaceous chondrites rather than the eucrites. The Jovian planets and satellites, and comets are all eliminated because they are volatile-rich. We are left to search the asteroids for a suitable object. The survey of bright asteroids is complete down to a radius of ~ 25 km (see chapter by Zellner). Of all surveyed asteroids, only the unique asteroid 4 Vesta has spectral and optical properties that match the eucrites. Thus Consolmagno and Drake (1977) concluded on the basis of geochemical and observational evidence that Vesta is the eucrite parent body.

Evidence against

The seemingly plausible conclusion that Vesta is the eucrite parent body has been challenged on dynamical grounds (Wetherill 1978). Cosmic ray exposure ages for the eucrites are $10^6 - 10^7$ yr (Heymann et al. 1969); these ages probably simply reflect their dynamical and/or collisional lifetimes of $10^7 - 10^8$ yr (Wetherill 1974, 1976). Cosmic rays penetrate rocky bodies to a depth of approximately one meter. Thus if the eucrites were ejected from the eucrite parent body as fragments of one meter or less, the cosmic ray exposure ages would record their true lifetimes in space.

The problem is that the time scale required to perturb a fragment from an asteroid such as Vesta (a main-belt asteroid far from a convenient resonance with Jupiter) into an Earth-crossing orbit appears to be $10^8 - 10^9$ yr (Wetherill 1977). A high relative velocity (> 5 km sec^{-1}) collision of a projectile with Vesta $10^6 - 10^7$ yr ago, with insertion of resultant excavated crust into an Earth-crossing orbit, appears to be precluded by the lack of evidence for extreme shock in the eucrites.

One possible solution to these conflicting conclusions (Hostetler and Drake 1978) would involve ejection from Vesta of fragments that were large compared to the approximately one-meter penetration depth of cosmic rays, at a time significantly older (e.g., 3-4 $\times$ 10^9 yr) than cosmic ray exposure ages. These fragments could then be removed by low-probability mechanisms (e.g., multiple low-energy collisions and/or interactions with Jupiter resonances after low-energy collisions) into Earth-crossing orbits on time scales more consistent with dynamical requirements. Once in Earth-crossing orbits these fragments would be subject to probable dynamical and/or collisional lifetimes on the order of 10^7 yr (Wetherill 1974, 1976). Thus one would predict that these objects would be disrupted, exposing fresh surfaces to space and restarting the cosmic ray exposure clocks for most of the material. This scenario is admittedly *ad hoc,* but finds at least circumstantial support in the two-stage irradiation history of Serra de Magé (Carver and Anders 1970). Its credibility would also be enhanced if a small Earth-crossing asteroid were found to have a Vesta-like spectrum.

IV. OTHER POSSIBLE PARENT BODIES?

In the previous sections I have presented plausibility arguments that Vesta is the eucrite parent body. This position cannnot be rigorously defended, however. For example, we cannot rigorously exclude the possibility that a Vesta-like main-belt asteroid with a radius of less than 25 km exists, may be close to a resonance with Jupiter, and thus be a more convenient source of the eucrites. We note, however, that such bodies are only barely allowed by the conservative estimate of the lower limit on the radius of the eucrite parent body (Walker et al. 1978). We also cannot rigorously exclude the possibility that a Vesta-like asteroid with a highly eccentric, cometary-like orbit may exist, and may not have been detected because of its present great distance from the earth. We note, however, that objects with such orbits tend to be comets or extinct comets which are not suitable parent bodies for the eucrites because of their volatile-rich nature. Main-belt asteroid 349 Dembowska (like Vesta, far from a convenient resonance with Jupiter), has an infrared spectrum indicative of low-Fe olivine and low-Fe, low-Ca pyroxene with an olivine/pyroxene ratio of ~ 10:1 (Feierberg et al. 1979). Feierberg et al. note that a spectrum of the mantle of the eucritic parent body would be similar to that of Dembowska. Could Dembowska be the eucrite parent body stripped of its crust of basalts? Again the possibility cannot be rigorously excluded although circumstantial evidence argues against the hypothesis. The absence in our collections of meteorites representative of the mantle of the eucrite parent body requires that (1) either impacts of Dembowska have conveniently stripped only the basaltic crust away during the age of the solar system, or (2) a dynamical mechanism must be found which selectively inserts only basaltic crustal fragments into Earth-intersecting orbits. Both proposals appear to be implausible. Thus we conclude that Dembowska is probably not the eucrite parent body, although it may ultimately prove to be the source of rare achondritic meteorite types like the new finds in Antarctica.

V. VESTA AS THE SHERGOTTITE PARENT BODY?

The shergottites (Shergotty and Zagami) are composed primarily of low-Ca and high-Ca clinopyroxenes, and maskelynite. Their infrared reflectance spectra have not been measured, but are expected to be similar to Vesta and the eucrites. The 1-μm and 2-μm band centers may be shifted towards longer wavelengths (see Adams 1974) because of the presence of high-Ca clinopyroxene in approximately equal proportion to low-Ca clinopyroxene (Stolper and McSween 1979). At least three events are evident in the history of the shergottites: (1) a primary igneous event of unknown age; (2) a shock event at ~ 165 Myr which reset Rb/Sr isotopic systematics, partially outgassed Ar isotopes, but probably did not result in peak temperatures and pressures above 400°C and 300 kbar respectively (Nyquist et al. 1978); and (3) exposure to space at ~ 2.5 Myr as indicated by cosmic ray exposure ages (Heymann et al. 1968).

Could the shergottites be derived from Vesta? Direct derivation by insertion into Earth-crossing orbit due to impact is excluded because relative velocities of > 5 km sec^{-1} are required. A relative velocity of this magnitude would result in melting and shock effects more severe than observed in the shergottites (see Wetherill 1977). Incomplete degassing of Ar isotopes also appears to exclude melting at ~ 165 Myr (Nyquist et al. 1978). A multistage insertion into Earth-intersecting orbit cannot be excluded, however, as discussed for the eucrites. For example, an object ejected from Vesta could reach the 1:2, 2:5 and 1:3 Kirkwood gaps as a result of impacts of relative velocities ~ 2.5 km sec^{-1}, ~ 1.5 km sec^{-1}, and $\lesssim 1$ km sec^{-1} respectively under most favorable circumstances (Weidenschilling, personal communication). The probable time scales for evolution into an Earth-intersecting orbit from these resonances are uncertain, but may be as short as 10^5 yr for the 1:2 resonance (Zimmerman and Wetherill 1973). Wetherill (1977) has shown that probable time scales for objects close to the ν_6 secular resonance to evolve into Earth intersecting orbits are $10^8 - 10^9$ yr. In this case ejection velocities from Vesta may be too large to be consistent with observed petrographic features, however. The $\simeq 165$ Myr event is consistent with these time scales and could represent the event which excavated the parental fragment of the shergottites from Vesta. The 2.5 Myr exposure ages may result from fragmentation during a subsequent minor impact event.

Resolution of the candidacy of Vesta on the shergottite parent body awaits measurement of the infrared reflectance spectra of the shergottites. The most recent measurement of the infrared reflectance spectrum of Vesta by Feierberg et al. (1979; see the chapter by Larson and Veeder), in which the 1-μm and 2-μm bands were simultaneously measured for the first time, places the band centers at 0.92 ± 0.01 μm and 1.97 ± 0.02 μm, between the fields occupied by the eucrites and the howardites. If the band centers for the shergottites are shifted towards longer wavelengths relative to the eucrites as predicted, the candidacy of Vesta as the shergottite parent body will be compromised. Note, however, that the arguments presented in favor of the present existence of the eucrite parent body apply equally well to the shergottites. Thus a paradox exists unless the eucrites and shergottites come from a single isotopically and compositionally heterogeneous parent body.

VI. SUMMARY

The similarity of the surface of Vesta to the eucritic meteorites as inferred from spectral and other optical properties suggests that eucrites may be analogs for Vestan basalts. The eucrite parent body accreted with an approximately cosmic complement of non-volatile components, but was severely depleted in volatile components. Following its assumed homogeneous accretion, the eucrite parent body was composed of feldspathic peridotite consisting of 50-85% olivine, of composition $\sim$ Fo$_{65}$, 5-15% of low-Ca pyroxenes with a Fe/(Fe+Mg) ratio similar to olivine, approximately 5% of Na-poor plagioclase, an unknown ($\sim$ 10%?) amount of metal, and minor amounts of other phases such as sulfides (e.g., troilite) and oxides (e.g., spinel). Soon after or contemporaneously with accretion, the silicate solidus temperature ($\sim 1150°$C) was reached within the interior of the planet. This stage may have been immediately preceded by separation of metal to form a core. Basalts erupted on the surface for a relatively short time period. The planet then became quiescent except for external events such as impacts.

The question of whether the eucrites are actually *derived* from Vesta, rather than simply being *analogs* of Vestan basalts is considered. Arguments based on the calculated composition of the eucrite parent body and the relative abundances of meteorite types in our museums suggest that the eucrite parent body is still intact. Dynamical and compositional considerations eliminate all observed solar system bodies except main-belt asteroids as candidates for the eucrite parent body. Estimates of the size of the eucrite parent body indicate a *minimum* radius of 10 km, with larger radii being more probable. The survey of bright asteroids is complete down to a radius of 25 km. Of these objects, only asteroid 4 Vesta has spectral and other optical properties consistent with eucritic basalts. Thus geochemical and observational arguments point to Vesta as the eucrite parent body.

This conclusion is based on plausibility arguments and the process of elimination, and cannot rigorously withstand all criticism. Dynamical objections involving the difficulty in delivering fragments from Vesta to Earth on a time scale of $\sim 10^7$-10^8 yr consistent with cosmic ray exposure ages have been raised. An *ad hoc* scenario in which large objects were ejected from Vesta much earlier than 10^7-10^8 yr ago and were subsequently fragmented into $\leqslant 1$ meter-sized meteoroids has been proposed to circumvent the objection. We cannot rigorously exclude the possibility that the eucrite parent body is a Vesta-like asteroid of radius less than 25 km, that the eucrite parent body has a highly eccentric orbit and has not been detected because it has been out of observable range since the advent of modern astronomical instrumentation, or that Dembowska is the eucrite parent body. Each hypothesis appears to be improbable, however. Finally, most of the arguments presented in favor of Vesta as the eucrite parent body apply equally well to the shergottites. Possible resolution of this ambiguity may await measurement of the infrared reflectance spectra of the shergottites. The nature of Vesta may remain unresolved until Vesta is targeted as part of an asteroid mission. If Vesta is not the eucrite parent body we will be forced to consider other, at present less plausible, hypotheses for the source of the eucrites. If Vesta is the eucrite parent body, a paradox exists concerning the location of the shergottite parent body. Regardless of the outcome, dynamicists will have a powerful impetus to investigate the delivery of meteorites from unfavorably located asteroids to the earth.

Acknowledgements. This paper is a synthesis of an ongoing research program involving several present and former students. The contributions of G. Consolmagno, C. Hostetler, and H. Newsom are particularly noted. Discussions with E. Anders, D. Burnett, C. Chapman, G. Wetherill, S. Weidenschilling and L. Wilkening have been illuminating. A critical comment in a review by E. Stolper led to the treatment of the shergottite question. This work was supported by the National Aeronautics and Space

Administration.

REFERENCES

Adams, J. B. 1974. Visible and near-infrared diffuse reflectance spectra of pyroxenes as applied to remote sensing of solid objects in the solar system. *J. Geophys. Res.* 79: 4829-4836.

Adams, J. B., and McCord, T. B. 1977. Mercury: Evidence for an anorthositic crust from reflectance spectra (abstract). *Bull. Amer. Astron. Soc.* 9: 457.

Allegre, C. J.; Birck, J. L.; Fourcade, S.; and Semet, M. P. 1975. Rubidium-87/strontium-87 age of Juvinas basaltic achondrite and early igneous activity in the solar system. *Science* 187: 436-438.

Arth, J. G. 1976. Behavior of trace elements during magmatic processes – a summary of theoretical models and their applications. *J. Res. U.S. Geol. Survey* 4: 41-47.

Birck, J. L., and Allegre, C. J. 1978. Chronology and chemical history of the parent body of basaltic achondrites studied by the ^{87}Rb-^{87}Sr method. *Earth Planet. Sci. Lett.* 39: 37-51.

Birck, J. L.; Minster, J. F.; and Allegre, C. J. 1975. ^{87}Rb-^{87}Sr chronology of achondrites. *Meteoritics* 10: 364-365.

Carver, E. A., and Anders, E. 1970. Serra de Magé: A meteorite with an unusual history. *Earth Planet. Sci. Lett.* 8: 214-220.

Chapman, C. R. 1976. Asteroids as meteorite parent bodies: The astronomical perspective. *Geochim. Cosmochim. Acta* 40: 701-719.

Consolmagno, G. J., and Drake, M. J. 1977. Composition and evolution of the eucrite parent body: Evidence from rare earth elements. *Geochim. Cosmochim. Acta* 41: 1271-1282.

Drake, M. J. 1979. Trace elements as quantitative probes of differentiation processes in planetary interiors. *Rev. Geophys. Space Phys.* (in press).

Drake, M. J., and Weill, D. F. 1975. Partition of Sr, Ba, Ca, Y, Eu^{+2}, Eu^{+3} and other REE between plagioclase feldspar and magmatic liquid: An experimental study. *Geochim. Cosmochim. Acta* 38: 689-712.

Dreibus, G.; Kruse, H.; Spettel, B.; and Wänke, H. 1977. The bulk composition of the moon and the eucrite parent body. *Proc. Lunar Sci. Conf. VIII* (Oxford: Pergamon Press), pp. 211-227.

Duke, M. B. 1963. Petrology of basaltic achondrites. Ph.D. dissertation, California Institute of Technology.

Duke, M. B., and Silver, L. T. 1967. Petrology of eucrites, howardites and mesosiderites. *Geochim. Cosmochim. Acta* 31: 1637-1665.

Feierberg, M. H.; Larson, H.; Fink, U.; and Smith, H. 1979. Spectroscopic evidence for at least two achondrite parent bodies: Asteroids 349 Dembowska and 4 Vesta. *Geochim. Cosmochim. Acta* (submitted).

Fukuoka, T.; Boynton, W. V.; Ma, M.-S. and Schmitt, R. A. 1977. Genesis of howardites, diogenites, and eucrites. *Proc. Lunar Sci. Conf. VIII* (Oxford: Pergamon Press), pp. 187-210.

Gast, P. W. 1968. Trace element fractionation and the origin of tholeiitic and alkaline magma types. *Geochim. Cosmochim. Acta* 32: 1057-1086.

Gast, P. W., and Hubbard, N. J. 1970. Rare earth abundances in soil and rocks from the Ocean of Storms. *Earth Planet. Sci. Lett.* 10: 94-101.

Gast, P. W.; Hubbard, N. J.; and Wiesmann, H. 1970. Chemical composition and petrogenesis of basalts from Tranquility Base. *Proc. Apollo 11 Lunar Sci. Conf.*, ed. A. A. Levinson (New York: Pergamon Press), 1143-1163.

Goleš, G. G., and Seymour, R. S. 1979. Terrestrial magmatic ocean: Evidence for its former existence and implications for Earth's mantle. *Science* (submitted).

Gray, C. M.; Papanastassiou, D. A.; and Wasserburg, G. J. 1973. The identification of early condensates from the solar nebula. *Icarus* 20: 213-239.

Grutzeck, M.; Kridelbaugh, S.; and Weill, D. 1974. The distribution of Sr and REE between diopside and silicate liquid. *Geophys. Res. Lett.* 1: 273-275.

Hertogen, J., and Gijbels, R. 1976. Calculation of trace element fractionation during partial melting. *Geochim. Cosmochim. Acta* 40: 313-322.

Hertogen, J.; Vizgirda, J.; and Anders, E. 1977. Composition of the parent body of eucritic meteorites (abstract). *Bull. Amer. Astron. Soc.* 9: 459.

Hess, H. H., and Henderson, E. P. 1949. The Moore County meteorite: A further study with comment on its primordial environment. *Amer. Mineral.* 34: 494-507.

Heymann, D.; Mazor, E.; and Anders, E. 1968. Ages of the calcium-rich achondrites. I. Eucrites. *Geochim. Cosmochim. Acta* 32: 1241-1268.

Heymann, D.; Mazor, E.; and Anders, E. 1969. Ages of the Ca-rich achondrites. In *Meteorite Research,* ed. P. M. Millman (Dordrecht: D. Reidel), pp. 444-457.

Hostetler, C. J., and Drake, M. J. 1978. Quench temperatures of Moore County and other eucrites: Residence time on eucrite parent body. *Geochim. Cosmochim. Acta* 42: 517-522.

Jérome, D. Y. 1970. Composition and origin of some achondritic meteorites. Ph.D. dissertation, University of Oregon.

Larson, H., and Fink, U. 1975. Infrared and spectral observations of Asteroid 4 Vesta. *Icarus* 26: 420-427.

Lugmair, G. W.; Scheinin, N. B.; and Marti, K. 1975. Search for extinct ^{146}Sm. 1. The isotopic abundance of ^{142}Nd in the Juvinas meteorite. *Earth Planet. Sci. Lett.* 27: 79-84.

Ma, M.-S., and Schmitt, R. A. 1976. Possible source materials for eucritic achondrites based on multi-linear regression analysis of trace element data. *Meteoritics* 11: 324-325.

Mason, B. 1962. *Meteorites* (New York: Wiley).

McCarthy, T. S.; Erlank, A. J.; and Willis, J. P. 1973. On the origin of eucrites and diogenites. *Earth Planet. Sci. Lett.* 18: 433-442.

McCord, T. B.; Adams, J. B.; and Johnson, T. V. 1970. Asteroid Vesta: Spectral reflectivity and compositional implications. *Science* 168: 1445-1447.

McKay, G. A., and Weill, D. F. 1976. Petrogenesis of KREEP. *Proc. Lunar Sci. Conf. VII* (Oxford: Pergamon Press), pp. 2427-2447.

Mittlefehldt, D. W. 1979. The nature of asteroidal differentiation processes: Implications for primordial heat sources. *Proc. Lunar Sci. Conf. X* (Oxford: Pergamon Press, submitted).

Morgan, J. W.; Higuchi, H.; Takahashi, H.; and Hertogen, J. 1978. A "chondritic" eucrite parent body: Inference from trace elements. *Geochim. Cosmochim. Acta* 42: 27-38.

Nakamura, N.; Tatsumoto, M.; and Unruh, D. M. 1976. Rb-Sr, Sm-Nd, and U-Th-Pb systematics of the Pasamonte meteorite. *Meteoritics* 11: 339-340.

Newsom, H. E., and Drake, M. J. 1979. Metal depletion in the eucrites: Evidence for a core or for a heterogeneous mantle in the eucrite parent body. *Lunar Science X.* The Lunar and Planetary Institute, pp. 910-912.

Nyquist, L. E.; Wooden, J.; Bansal, B.; and Weismann, H. 1978. A shocking Rb/Sr age for the Shergotty achondrite. *Lunar Science IX.* The Lunar and Planetary Institute, pp. 820-822.

Papanastassiou, D. A., and Wasserburg, G. J. 1969. Initial strontium isotopic abundances and the resolution of small time differences in the formation of planetary objects. *Earth Planet. Sci. Lett.* 5: 361-376.

Rammensee, W., and Wänke, H. 1977. On the partition coefficient of tungsten between metal and silicate and its bearing on the origin of the moon. *Proc. Lunar Sci. Conf. VIII* (Oxford: Pergamon Press), pp. 399-409.

Schmitt, R. A.; Smith, R. H.; Lasch, J. E.; Mosen, A. L.; Olehy, D. A.; and Vasilevskis, J. 1963. Abundances of the fourteen rare-earth elements, scandium, and yttrium in meteoritic and terrestrial matter. *Geochim. Cosmochim. Acta* 27: 577-622.

Schmitt, R. A.; Smith, R. H.; and Olehy, D. A. 1964. Rare-earth, yttrium, and scandium abundances in meteoritic and terrestrial matter. II. *Geochim. Cosmochim. Acta* 28: 67-86.

Schnetzler, C. C., and Philpotts, J. A. 1969. Genesis of the calcium-rich achondrites in light of rare-earth and barium concentrations. In *Meteorite Research,* ed. P. M. Millman, (Dordrecht: D. Reidel), pp. 206-216.

Schnetzler, C. C., and Philpotts, J. A. 1970. Partition coefficients of rare-earth elements between igneous matrix material and rock-forming mineral phenocrysts. II. *Geochim. Cosmochim. Acta* 34: 331-340.

Shaw, D. M. 1970 Trace element fractionation during anatexis. *Geochim. Cosmochim. Acta* 34: 237-243.

Shaw, D. M. 1977. Trace element behavior during anatexis. In *Magma Genesis, Bull. 96, State of Oregon Dept. of Geol. and Mineral Industries*, pp. 189-213.

Shimizu, N., and Allegre, C. J. 1975. The genetic relationships among diogenites, howardites and eucrites. *Meteoritics* 10: 488.

Steele, I. M., and Smith, J. V. 1976. Ibitira eucrite: Mineralogy and comparison with lunar samples and other eucrites. *Lunar Science VII*, The Lunar Science Institute, pp. 833-835.

Stolper, E. 1975. Petrogenesis of eucrite, howardite and diogenite meteorites. *Nature* 258: 220-222.

Stolper, E. 1977. Experimental petrology of eucritic meteorites. *Geochim. Cosmochim. Acta* 41: 587-611.

Stolper, E., and McSween, H. Y., Jr. 1979. Petrology and origin of the Shergottite meteorites. *Geochim. Cosmochim. Acta* (in press).

Takeda, H., Ishii, T.; and Miyamoto, M. 1976a. Characterization of crust formation on a parent body of achondrites and the moon by pyroxene crystallography and chemistry. *Lunar Science VII*, The Lunar Science Institute, pp. 846-848.

Takeda, H.; Miyamoto, M.; and Duke, M. B. 1976b. Pasamonte pyroxenes, a eucritic analogue of lunar mare pyroxenes. *Meteoritics* 11: 372-373.

Walker, D.; Stolper, E. M.; and Hays, J. F. 1978. A numerical treatment of melt/solid segregation: Size of eucrite parent body and stability of the terrestrial low velocity zone. *Jour. Geophys. Res.* 83: 6005-6013.

Weill, D. F., and McKay, G. A. 1975. The partitioning of Mg, Fe, Sr, Ce, Sm, Eu, and Yb in lunar igneous systems and a possible origin of KREEP by equilibrium partial melting. *Proc. Lunar Sci. Conf. VI.* (Oxford: Pergamon Press), pp. 1143-1158.

Weill, D. F.; McKay, G. A.; Kridelbaugh, S. J.; and Grutzeck, M. 1974. Modeling the evolution of Sm and Eu abundances during lunar igneous differentiation. *Proc. Lunar Sci. Conf. V.* (Oxford: Pergamon Press), pp. 1337-1352.

Wetherill, G. W. 1974. Solar system sources of meteorites and large meteoroids. *Ann. Rev. Earth Planet. Sci.* 2: 303-331.

Wetherill, G. W. 1976. Where do meteorites come from? A re-evaluation of the Earth-crossing Apollo objects as sources of chondritic meteorites. *Geochim. Cosmochim. Acta* 40: 1297-1318.

Wetherill, G. W. 1977. Fragmentation of asteroids and delivery of fragments to Earth. In *Comets, Asteroids, and Meteorites* ed. A. H. Delsemme, (Toledo, Ohio: University of Toledo Press), pp. 283-291.

Wetherill, G. W. 1978. Dynamical evidence regarding the relationship between asteroids and meteorites. In *Asteroids: An exploration assessment*, ed. D. Morrison and W. Wells, NASA Conf. Publ. 2053, pp. 17-35.

Wilkening, L. L., and Anders, E. 1975. Some studies of an unusual eucrite: Ibitira. *Geochim. Cosmochim. Acta* 39: 1205-1210.

Zimmerman, P. D., and Wetherill, G. W. 1973. Asteroidal source of meteorites. *Science* 182: 51-53.

ASTEROID TAXONOMY
AND
THE DISTRIBUTION OF THE COMPOSITIONAL TYPES

B. ZELLNER
University of Arizona

Physical observations of minor planets documented in the TRIAD computer file are used to classify 752 objects into the broad compositional types C, S, M, E, R, and U (unclassifiable) according to the prescriptions adopted by Bowell et al. (1978). Diameters are computed from the photometric magnitude using radiometric and/or polarimetric data where available, or else from albedos characteristic of the indicated type.

An analysis of the observational selection effects leads to tabulation of the actual number of asteroids, as a function of type and diameter, in each of 15 orbital element zones. For the whole main belt the population is 75% of type C, 15% of Type S, and 10% of other types, with no belt-wide dependence of the mixing ratios on diameter. In some zones the logarithmic diameter-frequency relations are decidedly nonlinear. The relative frequency of S-type objects decreases smoothly outward through the main belt, with exponential scale length 0.5 AU. The rarer types show a more chaotic, but generally flatter, distribution over distance. Characteristic type distributions, contrasting with the background population, are found for the Eos, Koronis, Nysa and Themis families.

Individual asteroids tend to be rather uniform in their surface optical properties, yet they differ remarkably one from another; when closely examined, each object may be unique. Nevertheless it is possible to classify most of them, on the basis of remotely sensed optical properties, into a few broadly defined types. A type classification implies an estimate of geometric albedo, which together with the photometric magnitude implies a diameter which should be reliable at least in a statistical sense. Thus we can use the thermal-radiometric, polarimetric, spectrophotometric, and UBV data now available for more than a third of the numbered asteroid population, together with an analysis of the observational selection effects, to derive the distribution of the types over diameter, orbital elements, and family membership.

The fact that most of the brighter asteroids can be divided into two principal optical types, now designated C and S, was first noted by Chapman (1973), Zellner (1973), and explicitly by Chapman et al. (1975), who also made the first attempt to derive bias-corrected frequency distributions of the types. As the data base grew with the vigorous observational programs of succeeding years, further such analyses were made by Chapman (1976), Morrison (1977), Zellner and Bowell (1977), and Degewij (1978) among others. From close examination of the data available in late 1977 for 523 objects, Bowell et al. (1978) refined the taxonomic system and adopted the classification algorithm used here.

In the system of Bowell et al., five types C, S, M, E, and R are recognized on the basis of seven observational parameters as described below. An additional type designation U (for unclassifiable) is employed for objects which are known to belong to none of the recognized types. The system uses a few broad groups, chiefly those naturally defined by bimodalities or hiatuses in one or more parameters, rather than a large number of subsets which could be distinguished in the same data for the better observed asteroids. While it is expected that the classification scheme reflects significant similarities and differences in surface composition, the scheme itself is entirely empirical and divorced from any mineralogical or meteoritical interpretations.

An alternative taxonomic system, developed by Gaffey and McCord (1977, 1978 and this book), emphasizes interpretations in terms of mineralogical assemblages and is applied to narrowband spectrophotometric data only.

In this chapter I use the physical data listed in the TRIAD (Tucson Revised Index of Asteroid Data) tables (Part VII of this book), together with the classification criteria of Bowell et al., to generate type classifications of 752 objects. The data base and the classification and diameter algorithms are discussed in the following section. A bias model for correction of observational selection effects is developed in Sec. II, and in Sec. III bias-free distributions of the types over diameter and distance are derived. My attention is essentially confined to the main belt; the cis-Martian asteroids are discussed in this book by Shoemaker et al., the Hildas and more distant objects are covered by Degewij and van Houten and the family groups are discussed in more detail in the chapter by Gradie *et al.*

I. THE TAXONOMIC SYSTEM

Table I gives a general description of the optical properties associated with each asteroid type, probable mineralogical interpretations, and possible meteoritic analogues. The potential identifications between classified asteroids and known types of meteorites must be taken with extreme caution. Generally it is not possible to observe an asteroid and to find a meteoritic analogue with any degree of confidence. In most cases the argument must proceed in the opposite direction. For example, we know that the enstatite achondrites have to come from somewhere, and the only solar system objects known to have the right optical properties are the rare E-type asteroids (Zellner *et al.* 1977).

The C asteroids provide fairly good optical equivalents for the carbonaceous chondrites, but some of the minor planets appear to be darker than any known meteorite. Optically it is very difficult to distinguish between pure metal (nickel-iron meteorites) and pure metal plus spectrally-neutral silicates (enstatite chondrites); either or both may be present among the M asteroids. The nature of the S objects is still an open question. Superficially they resemble ordinary chondrites, but according to spectroscopic interpretations the free metal content is on the order of 50% (McCord and Gaffey 1974), and stony-irons provide the only possible meteoritic equivalents. That is not to say, of course, that they could not have a mineralogy unknown to meteoritics and a thermal history very different from that implied by the stony-iron identification. Anders (1978) has summarized compelling arguments that the common iron-poor ordinary chondrites come from the asteroid belt, but few if any source bodies can be identified at present. In my opinion, the only really convincing identification that can be made at present is that between Vesta and some of the basaltic achondrites. Classification, mineralogical interpretation, and meteoritic identification of minor planets are three distinct steps, and only the first is of concern for the remainder of this chapter.

Table II lists the numerical type discriminants in seven-dimensional classification space. The geometric albedo p_v is obtained from thermal-radiometric or polarimetric data, or the weighted average of the two (see chapters by Morrison and Lebofsky and by Dollfus and Zellner in this book). The other parameters are the maximum depth P_{min} of the negative polarization branch, the spectrophotometric parameters R/B, BEND, and DEPTH, and the broadband color indices B − V and U − B from UBV photometry. R/B is a measure of the overall redness of the visible part of the reflection spectrum, BEND is a measure of its curvature, and DEPTH is a measure of the strength of the Fe^{2+} absorption feature near 0.95 μm wavelength (precise definitions are given by McCord and Chapman 1975; see

TABLE I

Description of the Asteroid Types

Type	Albedo	Spectrum	Mineralogy	Meteoritic Analogues[a]
C	low	relatively flat, weak features	silicates plus opaques (carbon)	carbonaceous chondrites
S	moderate	reddish Fe^{2+} absorptions	silicates plus metal	stony irons (H chondrites?)
M	moderate	slightly reddish, featureless	metal, or metal plus neutral silicates	nickel-irons enstatite chondrites
E	high	flat featureless	neutral silicates	enstatite achondrites
R	moderate to high	red, strong features	Fe^{2+} silicates	various or unknown (ordinary chondrites?)
U	various	unusual	various	various or unknown (certain achondrites)

[a] See text.

TABLE II

Definition of Classes

Parameter	C	S	M	E	R
Albedo p_v	$\leqslant 0.065$	$0.065-0.23$	$0.065-0.23$	$\geqslant 0.23$	$\geqslant 0.16$
$P_{min}(\%)$	$1.20-2.15$	$0.58-0.96$	$0.86-1.35$	$\leqslant 0.40$	$\leqslant 0.70$
R/B	$1.00-1.40$	$1.34-2.07$	$1.06-1.34$	$0.9-1.35$	$\geqslant 1.70$
BEND	$0.00-0.21$	$0.00-0.20$	$\leqslant 0.0-1$	$\leqslant 0.10$	$\geqslant 0.20$
DEPTH	$0.95-1.00$	$0.80-1.00$	$0.90-1.00$	$0.90-1.00$	$\leqslant 0.90$
$B-V$	$\geqslant 0.64$[a]	—[b]	$0.67-0.77$	$0.60-0.79$	—[c]
$U-B$[d]	$0.23-0.46$[a]	$\geqslant 0.34$[c]	$0.17-0.28$	$0.22-0.28$	—[c]

[a] Additionally $4.60(B\text{-}V)-3.17 \leqslant (U\text{-}B) \leqslant (B\text{-}V)-0.27$. Type U allowed 0^m02 inside limits when only UBV photometry is available.

[b] Additionally $B\text{-}V \geqslant (U\text{-}B)/7.0+0.74$; $1.70(B\text{-}V)-1.12 \leqslant (U\text{-}B) \leqslant (B\text{-}V)-0.33$; $(U\text{-}V) \leqslant 1.47$. Type U allowed 0^m02 inside limits, except for the last, when only UBV photometry is available.

[c] $(U\text{-}V) \geqslant 1.47$.

[d] Type U always allowed for $U\text{-}B \leqslant 0.28$ when only UBV photometry is available.

the chapter by McCord and Gaffey). The spectrophotometric data base has been greatly augmented since the work of Bowell *et al.* (1978), and the fundamental reduction to the solar spectrum has been recalibrated by Chapman and Gaffey (see their chapter in this book). Table II is unchanged from Bowell *et al.* except for minor modifications in the BEND criteria necessitated by the spectrophotometric recalibration.

Figure 1 provides an example of the classification data, for the geometric albedo and U – V color. Only high-weight data are plotted, and in these parameters (as in others) the type domains could be substantially narrowed if all the data were of such quality. There appears to be a genuine gap between types *C* and *S* or *M* in the best albedo data, the gap being occupied only by Eos family members and a few objects otherwise known to be unclassifiable.

The computer algorithm initially assumes that all five classes are allowed, and then proceeds to eliminate one or more types on the basis of the available data. Type *U* is then generated if all five classes are excluded. Multiple classifications cannot survive when all seven observational parameters are available, but ambiguities often do arise in the case of incomplete data. A geometric albedo between 0.065 and 0.16, for example, allows both types *S*

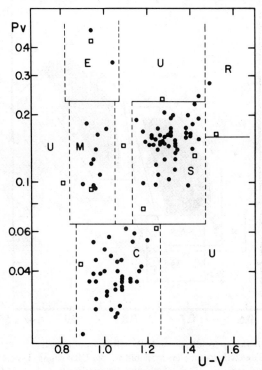

Fig. 1. Domains of the asteroid types in albedo and U-V color index. Data are from the TRIAD tables (Part VII of this book); only data of observational weight two or higher are plotted. Open squares represent objects independently known to be unclassifiable.

and *M;* UBV or spectrophotometric data are needed to resolve the uncertainty. Ambiguous types most frequently arise in the case of UBV photometry alone. As illustrated in Fig. 2, there is a substantial region of overlap between types *C, M,* and *E* in the UBV colors. This color domain also contains objects independently known to be unclassifiable, and we always add *U* to the type when U-B < 0.28 and only UBV data are available. The designation *U* is also added for putative *E* classifications in the absence of albedo-sensitive data, and to Type *R* in the absence of UBV colors or spectrophotometry. Multiple types are listed in the order *CSMERU,* or roughly from the most common to the least common types.

It should be kept firmly in mind that the classification procedure does not *generate* types but *excludes* them. Type designation *CMUE,* for example, really means "not *S* or *R*" and nothing more. Even an unambiguous classification such as *C* really means only "not *S, M, E,* or *R*". It does not deny the possibility that the *C* class could be subdivided, nor that the object could ultimately be reclassified into a new, presently unrecognized type.

The TRIAD tables list classification results for 752 minor planets, from

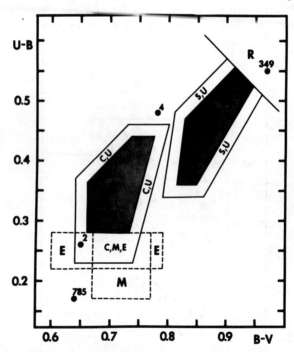

Fig. 2. Domains of the asteroid types in UBV colors, from Bowell *et al.* (1978). In cases where the only diagnostic parameters available are the UBV colors, Type U is generated outside the shaded areas for *C* and *S* (resulting in classifications *CU* or *SU*); when U-V > 1.47 (resulting classification *RU*); and when U-B ≤ 0.28 (classifications *MU, EU, CEU, MEU,* or *CMEU*). Numerical coefficients for the type boundaries are given in Table II.

the listed spectrophotometry of 277 objects, thermal radiometry of 195, optical polarimetry of 108, and useful UBV data (both color indices available with nonzero weight) for 690 objects. The improvement of the spectrophotometry has had a salutary effect on the quality of the classifications, especially in marginal cases. Quite a few objects that were previously thought to be exceptional (e.g., 69 Hesperia, 85 Io, 887 Alinda, and 1685 Toro) are now recognized as belonging to familiar types. Ceres was formerly classified U but is now (marginally) allowed as a C object; it remains unusual, however, and should not be thought of as a prototype for the C class.

The type classifications are almost never ambiguous and can be taken with a high degree of confidence when both albedo-sensitive and spectrum-sensitive data of any kind are available. Still, however, we are subject to a limitation common to any classification exercise based on inhomogeneous data: Objects are likely to be assigned to familiar types when only a few parameters are available, but subsequently recognized as unusual after more complete study. The U object 80 Sappho, for example, qualifies as an S-type in six parameters but fails in the parameter BEND. Since BEND is available for only 37% of our sample, there are certain to be additional such cases, even among the apparently secure types.

The TRIAD tables also list adopted diameters for all classified objects. The diameters are computed according to

$$2 \log d = 6.244 - 0.4 \ [B(1,0) - (B-V)] \ - \log p_v \qquad (1)$$

where d is the diameter in kilometers. The absolute magnitude $B(1,0)$ is taken from the TRIAD magnitude file (maintained by Gehrels, see Part VII), and $B-V$ from the TRIAD color file (Bowell, also in Part VII) or else assumed according to the type if no color has been observed. Diameters listed with weight 3 are generated from measured polarimetric or radiometric albedos, or their weighted average, if the total weight is $\geqslant 2$. Diameters listed with weight 2 are generated on the basis of C or CU classifications with adopted albedo 0.037, or classifications S, SU, SM, or M (but not MU) with adopted albedo 0.16. Such diameters are quite adequate for statistical purposes and may be more reliable than a single noisy radiometric or polarimetric determination. Diameters of weight 1 are based on the following albedo assumptions, which give only a best guess at the size: Multiple classifications that begin with C are assigned $p_v = 0.037$; those beginning with S or M, $p_v = 0.16$, and all other combinations (including unambiguous U), $p_v = 0.10$.

The algorithms for type and diameter are followed strictly, even in cases for which they are likely to lead to peculiar distortions. Well-observed Trojans, for example, are usually unclassifiable and are invariably found to be quite dark in radiometric albedo (see chapter by Degewij and van Houten in this book). In the absence of albedo data, however, the U classification leads to adopted albedo 0.10 and hence an artificially small diameter. Degewij and van Houten introduced a new type *"D"* for some of the Trojans, but the spectroscopic data necessary to identify an object as Type D in the absence of albedo information are usually not available. Similar cases can be found among the Hirayama families; objects 171, 268, and 379 in the Themis family, for example, are quite likely to be C-types like the rest of the Themis members. Each, however, fails the C classification on the basis of a single noisy or marginal parameter, and thus is assigned the default value $p_v = 0.10$.

II. THE BIAS MODEL

The set of classified asteroids is heavily biased in favor of bright objects, i.e., those which are large, nearby, and/or of high albedo. The limits of sampling can be pushed to smaller diameters and into more distant zones by further observations, but the bias against low-albedo objects will always remain. Additional biases arise from special attempts to sample objects in certain Hirayama families and other interesting regions.

Table III lists 15 zones which are believed adequately to separate out the observational biases with respect to orbital elements. The Eos, Koronis, Nysa, and Themis zones represent true dynamical families, whereas the Flora zone is defined in terms of osculating elements only and contains several identifiable families. We cannot say, *a priori*, which zones have been most thoroughly inventoried or which most poorly, or what the bias functions are within each; that must be determined by a comparison between the number of objects classified and the number known to be present in each zone.

The bias analysis is illustrated in Figs. 3 through 5 for the three principal zones in the main belt. The fundamental assumption is that, for each classified asteroid with bias factor N, there are $N - 1$ additional objects of identical type and diameter in the same zone. The bias factor is computed, in half-magnitude bins of mean opposition blue magnitude, as the ratio of the number of asteroids present in that zone to the number sampled. Implicit assumptions are that the selection bias is a function (not necessarily monotonic) of the apparent magnitude only, and that the bias factor cannot depend on the taxonomic type (i.e., the type is unknown before the classification observation is made). The method suffers from counting noise in poorly sampled or poorly populated zones, but it is nevertheless the best analysis that can be made from the data.

A further complication is introduced by the incompleteness of the numbered or catalogued population beginning about apparent magnitude 15. The incompleteness arises for S objects of diameter less than about 10 km in Zone I but for C objects as large as 60 km in Zone IV. As a correction for the incompleteness I have, in each zone, attached a line of constant slope to the magnitude-frequency relation for the numbered asteroids. The line is attached at the magnitude bin centered at 15.75, and the slope is the mean value found in the Palomar-Leiden Survey (log N proportional to 0.39 times the apparent

TABLE III

Orbital Element Zones

Zone	Description	Criteria[a]			Number in TRIAD[b]	Number Classified[b]
A	Apollo-Amor	$q = a(1-e) \leqslant 1.65$			55	19
HU	Hungarias	$1.82 \leqslant a \leqslant 2.00$	$e \leqslant 0.15$	$i \geqslant 16°$	17	7
PH	Phocaeas	$2.25 \leqslant a < 2.50$	$e \leqslant 0.35$	$i \geqslant 17°$	42	10
FL	Floras	$2.06 \leqslant a \leqslant 2.295$	$0.08 \leqslant e \leqslant 0.20$	$i \leqslant 10°$	201	37
NY	Nysa family	$a = 2.43$	$e = 0.17$	$i = 3°$	13	8
I	Main-belt I	$2.06 \leqslant a \leqslant 2.50$	$e \leqslant 0.35$	$i \leqslant 30°$	283	89
II	Main-belt II	$2.50 < a \leqslant 2.82$	$e \leqslant 0.35$	$i \leqslant 30°$	578	244
Eos	Eos family	$a = 3.01$	$e = 0.17$	$i = 10°$	86	39
Ko	Koronis family	$a = 2.85$	$e = 0.05$	$i = 2°$	50	18
Th	Themis family	$a = 3.13$	$e = 0.15$	$i = 1°$	83	23
III	Main-belt III	$2.82 < a \leqslant 3.27$	$e \leqslant 0.35$	$i \leqslant 30°$	557	192
IV	Main-belt IV	$3.27 < a \leqslant 3.65$	$e \leqslant 0.35$	$i \leqslant 30°$	46	26
HI	Hildas	$3.80 \leqslant a \leqslant 4.20$	$e \leqslant 0.35$	$i \leqslant 30°$	28	12
T	Trojans	$5.06 \leqslant a \leqslant 5.30$	— No test —		21	12
Z	Exceptional	— None of the above —			18	4

[a] Tests for zone membership are made in the order indicated, using the osculating elements semimajor axis a, eccentricity e, and inclination i. For the Eos, Koronis, Nysa and Themis families, membership is from lists provided by Tedesco; the approximate mean of proper elements is indicated.

[b] Numbered asteroids only.

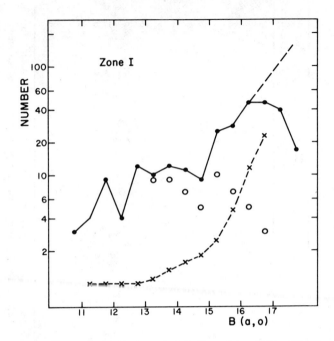

Fig. 3. Bias diagram for asteroids in main-belt Zone I (see Table III). The solid line represents the count of numbered asteroids in half-magnitude bins, and open circles the count of classified asteroids. The ratio of the two is the bias factor, given by the lower dashed line. The upper dashed line is the Palomar-Leiden extrapolation of the numbered asteroid population.

magnitude, where N is counted in half-magnitude bins; van Houten *et al.* 1970). For magnitudes fainter than 16.0 the bias analysis then uses either the actual count of numbered asteroids or else the PLS extrapolation, whichever is larger. The procedure is rather arbitrary and certainly subject to considerable doubt; however it appears to be the best that one can do. In the tables that follow, all incompleteness-corrected results are given within parentheses.

Table IV lists bias factors obtained in 12 zones. No attempt is made to model the bias effects for Apollo-Aten-Amor objects, since the real biases are more a result of the discovery circumstances than of selection of targets for classification observations. Similarly no attempt is made to perform a bias analysis for the Trojans, for technical reasons noted above. Of the four classified asteroids which belong to none of the specified bias zones (i.e., which fall into Zone Z), 279 Thule and 944 Hidalgo are of unusual type, while 1252 Celestia is a 20-km S object in a high-inclination orbit near 2.7 AU heliocentric distance. The fourth object is 2 Pallas, and we should not forget the enigma posed by that very large, very unusual minor planet in an orbit of almost uniquely high inclination (see the end of Cameron's chapter).

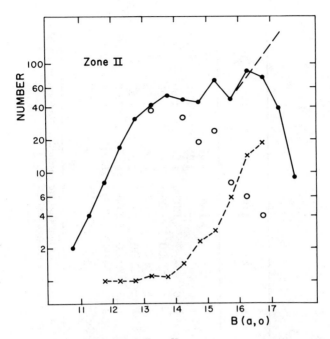

Fig. 4. As in Fig. 3, but for main-belt Zone II.

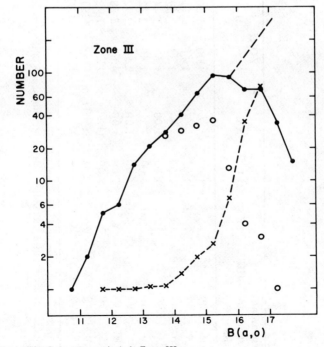

Fig. 5. As in Fig. 3, but for main-belt Zone III.

TABLE IV

Bias Factors[a]

Apparent Magnitude Range	HU	PH	FL	NY	ZONE							
					I	II	Eos	KO	TH	III	IV	HI
12.0–12.49	—	—	1.00	—	1.00	1.00	1.00	—	1.00	1.00	1.00	—
12.5–12.99	—	1.00	1.00	—	1.11	1.11	1.00	—	—	1.05	1.00	—
13.0–13.49	1.00	1.00	1.00	—	1.33	1.09	1.00	—	1.00	1.08	1.00	—
13.5–13.99	—	1.50	1.00	1.00	1.57	1.44	—	1.00	1.00	1.41	1.00	1.33
14.0–14.49	—	2.00	1.00	—	1.80	2.32	1.25	1.00	1.00	2.00	1.75	1.00
14.5–14.99	1.50	2.00	4.00	1.00	2.50	2.92	1.38	1.00	2.00	2.58	1.57	—
15.0–15.49	2.33	X	7.00	1.00	4.67	5.88	1.25	1.00	1.00	6.92	2.00	3.00
15.5–15.99	—	X	5.00	(1.50)	11.5	14.2	2.27	2.00	3.00	(35)	(6.0)	2.50
16.0–16.49	X	X	(86)	(5.00)	(23)	(29)	X	4.00	(7.7)	(74)	1.20	(1.75)
16.5–16.99	(26)	X	X	X	X	(1.82)	X	X	(18)	(3.46)	(16.0)	(12)

[a]Symbol (—) represents zones and magnitude bins that are vacant (unpopulated), and symbol (X) bins that are unsampled. Numbers given within parentheses are computed with corrections for incompleteness of the numbered asteroid population (see text).

Given a bias factor for each classified asteroid, we can now compute type mixing ratios, diameter-frequency relations, etc., for the entire belt population or any subset thereof, so long as consistent diameter limits are used within each zone or combination of zones. The question arises, however, of what to do about the ambiguous types. They cannot be rejected from the analysis, since the validity of the bias model depends entirely on the assumption that the type (ambiguous or otherwise) is unknown before the object is observed and hence included in the data set. For the following analyses I have assumed that all multiple types beginning with *"C"* are indeed of type *C,* and that all beginning with *"S"* are indeed of Type *S.* Types *M, E, R, U,* and all combinations thereof are combined into "other types". Since *S* objects are well recognized by UBV photometry alone, which accounts for a good fraction of the data set, mixing ratios are most confidently specified in the form *S*/total.

III. RESULTS

Tables V, VI, and VII list the diameter-frequency distributions, both observed and bias-corrected, in each zone for types *C, S,* and other types. The main belt is found to be 95% sampled for diameters > 100 km, 76% sampled for diameters > 63 km, and 48% sampled for all diameters > 32 km. For $d > 32$ km the sampling varies from 100% down to about 25%, depending on albedo and distance.

Table VIII lists the 30 largest asteroids. The bias analysis tell us that this list is essentially complete, i.e., only one object with distance < 3.65 AU and diameter potentially > 200 km (for albedo 0.03) is missing. (The unsampled object is 154 Bertha in Zone III.) It remains possible that one or two of the ambiguous types in the list are really of higher albedo and hence smaller diameter than assumed. It is also possible that a few of the largest asteroids classified *U* but lacking albedo information (e.g., 48 Doris) are really very dark and have diameters in the 200 km range.

Figures 6 and 7 illustrate some diameter-frequency relations that can be plotted from the tables. Diameter and distance bins are sometimes grouped for better counting statistics in the plots, but no data smoothing of any kind has been applied in either the tables or the figures. For those accustomed to power-law diameter-frequency relations, some of the curves should be astonishing. In Zone I (including Floras) there are just as many *S*-type objects of diameter 100 km as of diameter 20 km. The counting statistics are poor (5 − 6 objects) in each bin, but the aggregate effect is indisputable; the sampling is essentially complete to 20-km diameter and the bias correction makes hardly any contribution. In Zone II there is a genuine depletion of *C*-type objects at diameters near 60 km, or alternatively an excess of 100-km objects. Figure 8, for asteroids neither *C* nor *S* in the whole main belt, yields the closest approximation to a power-law distribution that can be found in the data.

TABLE V

Diameter-Frequency Results for Asteroids of Type C

Diameter Range (km)	PH		I		II[b]		Zone[a] EOS		III		TH		IV	
	Obs.	Corr.	Obs.	Corr.	Obs.	Corr.	Obs.	Corr.	Obs.	Corr.	Obs.	Corr.	Obs.	Corr.
15.9–19.9	1	2.0	1	(23)	0	X	0	X	0	X	0	X	0	X
20.0–25.0	1	2.0	2	(34)	2	(211)	0	X	0	X	0	X	0	X
25.1–31.6	0	0	4	18.7	1	(29)	0	X	1	(346)	0	X	0	X
31.7–39.8	1	1.5	5	12.5	5	40	1	1.25	1	(74)	2	(36)	1	(16)
39.9–50.0	0	0	3	5.2	7	26.4	5	0	4	(179)	3	(18)	0	X
50.1–63.0	0	0	4	6.3	8	20.4	0	6.6	9	45	3	9.0	0	X
63.1–79.4	1	1	4	5.3	13	20.6	0	0	23	56	2	2	1	2.0
79.5–99.9	0	0	3	3.3	23	27.5	0	0	19	46	1	1	4	6.3
100–125	0	0	6	6	29	32.2	0	0	17	24.7	3	3	4	7.0
126–158	0	0	0	0	19	11	0	0	19	21.1	3	3	4	4
159–199	0	0	1	1	4	4	0	0	12	12.4	0	0	1	1
200–250	0	0	0	0	1	1	0	0	7	7	1	1	2	2
251–316	0	0	0	0	0	0	0	0	3	3	0	0	2	2
317–398	0	0	0	0	0	0	0	0	1	1	0	0	0	0
399–500	0	0	0	0	0	0	0	0	1	1	0	0	0	0

[a] Although the Flora region is sampled down to about 40 km for C-type asteroids, only a single example is found, object 336 Lacadiera of diameter 69 km and bias factor unity. In the Nysa family a single asteroid, 877 Walkure with bias factor unity, is provisionally classified *CMEU*.

[b] Asteroid 1 Ceres, a peculiar C-type of diameter 1025 km, is also located in Zone II.

TABLE VI

Diameter-Frequency Results for Asteroids of Type S[a]

Diameter[b] Range (km)	PH		FL		Zone[c]									
					I		II		EOS		KO		III	
	Obs.	Corr.	Obs.	Corr.	Obs.	Corr.	Obs.	Corr.	Obs.	Corr.	Obs.	Corr.	Obs.	Corr.
12.6–15.8	0	X	5	26.0	5	23.7	1	(29)	0	X	0	X	0	X
15.9–19.9	1	2.0	3	6.0	1	2.5	0	0	2	4.5	1	4.0	0	X
20.0–25.0	1	2.0	1	1	1	1.8	6	23.4	3	6.8	3	6.0	0	X
25.1–31.6	0	0	2	2	0	0	11	29.1	7	9.9	2	2	4	23.3
31.7–39.8	1	1.5	0	0	4	5.8	9	16.5	1	1.4	3	3	4	9.7
39.9–50.0	0	0	5	6.0	5	6.0	13	15.2	4	5.3	1	1	2	4.6
50.1–63.0	0	0	4	4.1	4	4.1	7	8.1	1	1.25	0	0	7	9.8
63.1–79.4	1	1	4	4.1	4	4.1	12	12.7	1	1	0	0	5	6.1
79.5–99.9	0	0	6	6.0	6	6.0	7	7	1	1	0	0	8	8.3
100–125	0	0	4	4	4	4	6	6	0	0	0	0	1	1
126–158	0	0	3	3	3	3	3	3	0	0	0	0	1	1
159–199	0	0	1	1	2	2	2	2	0	0	1	1	1	1
200–250	0	0	0	0	2	2	2	2	0	0	0	0	0	0
251–316	0	0	0	0	0	0	1	1	0	0	0	0	0	0

[a] Entries in parentheses represent results corrected for incompleteness of the numbered population.

[b] Diameter bins are taken each unit of 0.2 in the common logarithm of the diameter.

[c] No S-type asteroids are identified in the Nysa or Themis families. A single S-type object, 692 Hippodamia of diameter 47 km, is identified in Zone IV; the associated bias factor is 1.57.

TABLE VII

Diameter-Frequency Results for Asteroids of Unusual Type

Diameter Range (km)	PH		FL		NY		I		II		Zone EOS		KO		TH		TH		IIIa[a]		IV	
	Obs.	Corr.	Obs.	Corr.	Obs.	Corr.	Obs.	Corr.	Obs.	Corr.	Obs.	Corr.	Obs.	Corr.	Obs.	Corr.	Obs.	Corr.	Obs.	Corr.	Obs.	Corr.
12.6–15.8	0	X	2	7.0	0	3.0	1	X	1	(29)	0	X	0	X	0	X	0	X	0	X	0	X
15.9–19.9	0	X	2	4.0	2	2	4	7.2	4	57	0	X	0	X	0	X	0	X	0	X	0	X
20.0–25.0	0	X	0	2	1	0	2	1.3	2	11.8	1	2.3	0	X	0	X	1	X	1	(35)	0	X
25.1–31.6	1	2.0	0	0	1	0	4	1.6	4	14.6	4	8.1	3	10.0	2	10.7	3	10.7	3	(84)	1	(10)
31.7–39.8	0	0	1	0	1	1	3	1.6	3	7.6	3	4.8	2	3	2	2	5	2	5	30	1	(6)
39.9–50.0	0	0	0	2	2	0	5	2.1	2	6.8	2	2.6	1	1	1	0	7	0	7	16.9	0	X
50.1–63.0	0	0	1	1	1	0	7	1.1	1	8.3	1	1.2	1	0	0	1	3	1	3	5.7	3	5.1
63.1–79.4	0	1	0	0	1	1	8	1	8	8.9	1	0	1	0	1	1	4	1	4	5.3	1	1
79.5–99.9	0	0	0	0	1	0	0	1	0	0	1	1	2	0	2	2	2	2	2	2.5	0	0
100–125	0	0	0	0	1	0	2	1	2	2	0	0	0	0	4	0	4	0	4	4.1	0	0
126–158	0	0	0	0	1	0	2	1	2	2.1	0	0	0	0	4	0	4	0	4	4.4	0	0
159–199	0	0	0	0	0	0	0	0	0	0	0	0	0	0	1	0	1	0	2	2	0	0
200–250	0	0	0	0	0	0	0	0	0	1	0	0	0	0	0	0	0	0	1	1	0	0

[a] Zone III also contains the unclassifiable object 704 Interamnia, of diameter 338 km and bias factor unity.

TABLE VIII

The Largest Asteroids

Asteroid	Type	Diam. (km)	Asteroid	Type	Diam. (km)
1 Ceres	C	1025	24 Themis	C	249
2 Pallas	U	583	3 Juno	S	249
4 Vesta	U	555	16 Psyche	M	249
10 Hygeia	C	443	13 Egeria	C	245
704 Interamnia	U	338	216 Kleopatra	CMEU	236?
511 Davida	C	335	165 Loreley	C	228
65 Cybele	C	311	19 Fortuna	C	226
52 Europa	C	291	7 Iris	S	222
451 Patientia	C	281	532 Herculina	S	219
31 Euphrosyne	C	270	250 Bettina	CMEU	211?
15 Eunomia	S	261	702 Alauda	CU	217
324 Bamberga	C	256	747 Winchester	C	208
107 Camilla	C	252	423 Diotima	C	209
87 Sylvia	CMEU	251?	386 Siegena	C	203
45 Eugenia	U	250	375 Ursula	C	200

Table IX lists the type mixing ratios as a function of distance, zones II and III being subdivided for this purpose. The diameter cutoff varies with distance, but is fixed within each zone. The S/total ratio as a function of distance, excluding the families, is plotted in Fig. 9. As noted by Zellner and Bowell (1977), the proportion of S objects drops smoothly outward through the main belt, the distribution being well described by an exponential decay with scale length 0.53 AU.

Figure 10 illustrates the fraction of asteroids not of S- or C-type as a function of distance. The data points scatter widely, and the distribution is either flat or gently sloped. Several people have called attention to an apparent excess of unusual asteroids in the region of several Kirkwood gaps near 2.9 AU. The effect is genuine, but it is of dubious statistical significance. In Zone IIIA we expect about four objects of unusual type with diameters >50 km; the observed number is seven with bias factors adding up to eight.

Apparently the high-inclination Phocaea group contains a broad representation of types, while the low-inclination Floras tend to have a relatively high proportion of R, M, and unclassifiable objects and are almost entirely lacking in C asteroids. It is difficult to be quantitative, however, since the sampling is poor for C asteroids smaller than 40 km and there are very few larger objects of any kind in either group. The Phocaeas are better compared with Zone I, for which the mean semimajor axis is nearly the same.

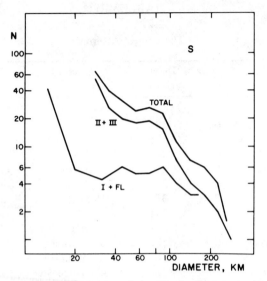

Fig. 6. Bias-corrected diameter-frequency relationship for S-type asteroids in Zone I plus Floras, in combined zones II and III, and for the whole main belt (including the families and other groups).

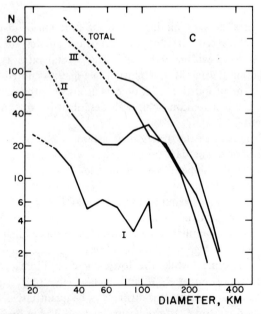

Fig. 7. Bias-corrected diameter-frequency relationships for C-type asteroids in zones I, II, and III, and for the whole main belt. Dashed lines represent results that are corrected for incompleteness of the numbered asteroid population.

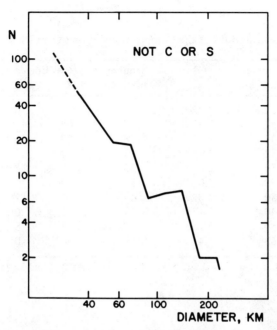

Fig. 8. Bias-corrected diameter-frequency relation for asteroids of types other than *S* or *C,* for the whole main belt.

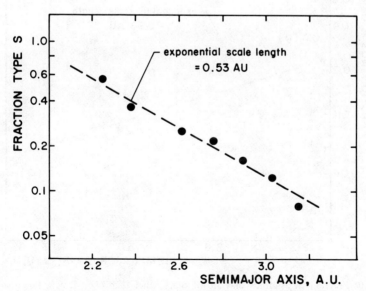

Fig. 9. Fraction of asteroids that are of Type *S,* as a function of semimajor axis, from Table IX. The dashed line is a least-squares fit and has slope corresponding to an exponential scale length of 0.53 AU.

TABLE IX

Distributions of the Compositional Types Over Distance

Zone[a]	Diameter Cutoff km	$\langle a \rangle$	Number Observed	Bias-Corrected Percentage		
				C	S	Other
FL	32	2.250	9	11:	56:	33:
I	32	2.381	89	47	43	10
II A	32	2.614	118	60	25	25
II B	32	2.759	93	70	22	8
III A	50	2.894	30	65	16	19
EOS	32	3.017	21	29	36	29
III B	50	3.029	43	80	12	8
TH	40	3.135	17	90	0:	10
III C	50	3.153	87	84	8	8
IV	63	3.445	19	96	0:	4:

[a]Zone II is subdivided at $a = 2.705$ AU, and Zone IV at $a = 2.96$ and 3.075 AU.

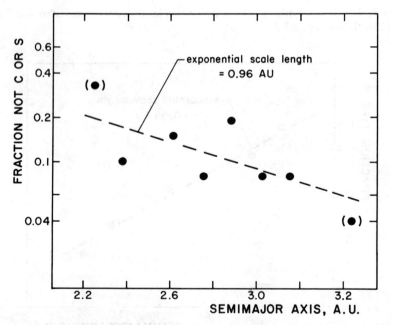

Fig. 10. Fraction of asteroids that are of types other than C or S, as a function of semimajor axis. The statistics are poor for the nearest and most distant zones. The dashed line is a weighted least-squares fit, and has slope corresponding to an exponential scale length of 0.96 AU.

For diameters > 40 km Zone I contains about 30 asteroids of Type S, 27 of Type C, and 7 of other types, while the Phocaeas contain one of each. Within limitations of the present counting statistics, it may be that both Floras and Phocaeas have type abundances that are "normal" for their distances.

Some of the families do show abundances that are strikingly at variance with the background population. Among the eight classified objects in the Nysa family, one is provisionally classified $CMEU$ and the others are all apparently of the rare E-type. There are no S asteroids among 23 objects classified in the Themis family, and no C objects among 18 in the Koronis family. The physical content of these families and of others is discussed in the chapter by Gradie *et al.* In some cases, however, the bias analysis urges caution. In the Koronis family, C asteroids are less than 50% sampled down to the diameter of the largest S and U objects present.

The E asteroids are genuinely quite rare. Only two are positively identified, namely 44 Nysa of diameter 68 km and 64 Angelina (60 km). The bias analysis allows no unsampled objects of diameter > 50 km and albedo > 0.30 in the entire main belt. Among the ambiguously classified objects there are nine potential E asteroids of d > 50 km. If as many as three of them are indeed of Type E (against all expectations), the class still accounts for only 1% of the main-belt population. The R objects may be even more uncommon. They are less thoroughly inventoried because of their (comparatively) lower albedos, but they are unambiguously recognizable in UBV or spectrophotometric data alone, and there are almost certainly only three (349 Dembowska, 446 Aeternitas and 863 Benkoela) of diameter > 50 km. Bowell *et al.* (1978) have noted the extreme rarity of objects with optical properties like Vesta. We who look for source bodies for the high-albedo types of meteorites have not many candidates to choose from.

IV. DISCUSSION

For the fainter asteroids the classification data are generally of poorer quality, and several of the classification parameters are likely to be missing. Thus we may feel some concern over the statistical quality of the type assignments, and of the resulting bias analysis, as we compare nearer and larger asteroids with smaller and more distant ones. At least three factors are involved: First, noise in the classification data for the fainter objects tends to depopulate the C- and S-types and thereby leads to an excess of objects designated U. Secondly, as noted above, well-observed asteroids are more likely to be recognized as unusual than the ones observed in only one or two parameters. Finally, there are the uncertainties associated with the incompleteness correction. The first and second perturbations work in opposite directions, while the third is unpredictable.

That serious problems of this nature are unlikely can be demonstrated in several ways. Consider Fig. 11 for the diameter-frequency distribution of S

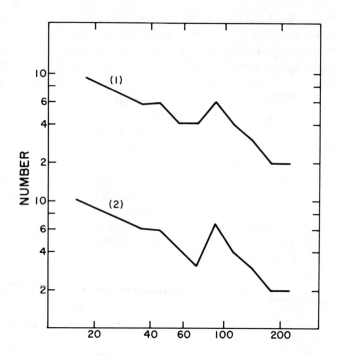

Fig. 11. Bias-corrected diameter-frequency relation for *S*-type asteroids in Zone I. The upper line is from the standard bias model in Table VI. The bottom line is from a separate analysis excluding objects classified on the basis of a single kind of data with unit observational weight.

asteroids in Zone I. The upper curve is plotted from Table VI, while the lower curve is from a completely different bias model, based on a limited data set for which all objects classified on the basis of weight-one data of any single kind are excluded. The second model suffers from enhanced counting noise, but there are no significant differences in the results.

Also consider Table X, which gives type abundance ratios for the whole main belt as a function of diameter. There is a genuine excess of unusual types among the very largest asteroids, but the number of unusual objects in question is only five. There also appears to be a slight excess of unusual objects at the smallest diameters; the effect may be genuine or it may be an artifact of noise in the classification data. Nevertheless we may feel rather confident in describing the main-belt population as 75% of Type *C*, 15% of Type *S*, and 10% of other types, with no belt-wide dependence of the abundance ratios on diameter.

Concerning future observations, we should keep in mind all the uncertainties noted above and remember that we have confident, unambiguous classifications for hardly more than 200 asteroids. The

TABLE X

Type Frequencies for the Whole Main Belt

Diameter	Number	Percentage of Type		
		C	S	Other
> 200 Km	30	70	13	17
> 158 Km	58	76	14	10
> 100 Km	195	75	15	10
> 63 Km	463	73	17	10
> 32 Km	1100:	72	15	13

inhomogeneity of the data is particularly exasperating, and limits the use of powerful statistical techniques such as cluster analysis (e.g., Pike 1978). The new eight-color survey (Gradie *et al.* 1978) is intended to provide homogeneous spectroscopic data in the $0.33 - 1.05$ μm range for a thousand minor planets. The IRAS satellite (see the chapter by Morrison and Niehof) will provide thermal-radiometric data for a very large number of minor planets but the results will be poorly interpretable unless the program is supported by vigorous groundbased work.

The outstanding problem for understanding the asteroid population is its mixed complexion: objects of grossly different surface mineralogy are found side by side. Either (1) the asteroids formed that way; or (2) the types evolved one into another by processes that are not well-behaved functions of diameter and heliocentric distance; or (3) the distinct types were formed in various parts of the solar system and subsequently relocated in the main belt; or else (4) we are being deceived by the presence or absence of superficial coatings on what are really very similar bodies. The fourth possibility can be rejected on several grounds, one of which is the evidence from the families. The third mechanism may provide the correct explanation for some of the rarer types of asteroids. Looking at the remarkably smooth curve displayed in Fig. 9, however, I have difficulty believing that the relocation, if any, of the dominant C- and S-types was by purely gravitational means. Surely the foreign objects would tend to pile up at a few discrete resonance points. For the same reason, I no longer believe that the general decrease of S-types with distance can be attributed to essentially stochastic effects such as the collisional break-up of a few large parent bodies. The smooth exponential decay is instead suggestive of diffusion under viscous drag forces. Perhaps the C objects were formed at greater distances and diffused inward during the early phases of the solar system. Still it is hard to see how they could cross the Kirkwood gaps unless the process was very rapid or unless Jupiter was not yet formed. We are not far from a good description of the main-belt population; we may yet be some distance, however, from an understanding of its origins.

Acknowledgments. This work is supported by the National Aeronautics and Space Administration.

REFERENCES

Anders, E. 1978. Most stony meteorites come from the asteroid belt. In *Asteroids: An Exploration Assessment,* eds. D. Morrison and W. C. Wells, NASA Conf. Publ. 2053, pp 57-74.

Bowell, E.; Chapman, C. R.; Gradie, J. C.; Morrison, D.; and Zellner, B. 1978. Taxonomy of Asteroids. *Icarus* 35: 313-335.

Chapman, C. R. 1973. Mineralogy of the asteroid belt and relationships to meteorites (abstract). *Bull. Amer. Astron. Soc.* 5: 388.

Chapman, C. R. 1976. Asteroids as meteorite parent-bodies: The astronomical perspective. *Geochim. Cosmochim. Acta* 40: 701-719.

Chapman, C. R.; Morrison, D.; and Zellner, B. 1975. Surface properties of asteroids: A synthesis of polarimetry, radiometry, and spectrophotometry. *Icarus* 25: 104-130.

Degewij, J. 1978. Photometry of faint asteroids and satellites. Ph. D. dissertation, Leiden University.

Gaffey, M. J., and McCord, T. B. 1977. Asteroid surface materials: Mineralogical characterizations and cosmochemical implications. *Proc. Lunar Sci. Conf. VIII,* (Oxford: Pergamon Press), pp. 113-143.

Gaffey, M. J., and McCord, T. B. 1978. Asteroid surface materials: Mineralogical characterizations from reflectance spectra. *Space Sci. Rev.* 21: 555: 628.

Gradie, J.; Tedesco, E.; and Zellner, B. 1978. A photometric system for faint asteroids (abstract). *Bull. Amer. Astron. Soc.* 10: 594.

McCord, T. B., and Chapman, C. R. 1975. Asteroids: Spectral reflectance and color characteristics. *Astrophys. J.* 195: 553-562.

McCord, T. B., and Gaffey, M. J. 1974. Asteroids: Surface compositions from reflection spectroscopy. *Science* 186: 352-355.

Morrison, D. 1977. Asteroid sizes and albedos. *Icarus* 31: 185-220.

Pike, R. J. 1978. Statistical classification of asteroids from surface properties. *Lunar Science IX.* The Lunar and Planet. Inst., pp. 901-903.

van Houten, C. J.; van Houten-Groeneveld, I.; Herget, P.; and Gehrels, T. 1970. The Palomar-Leiden survey of faint minor planets. *Astron. Astrophys. Suppl.* 2: 339-448.

Zellner, B. 1973. Polarimetric albedos of asteroids (abstract). *Bull Amer. Astron. Soc.* 5: 388.

Zellner, B., and Bowell, E. 1977. Asteroid compositional types and their distributions. In *Comets, Asteroids, Meteorites: Interrelations, Evolution, and Origins,* ed. A. H. Delsemme (Toledo, Ohio: Univ. Toledo Press), pp. 185-197.

Zellner, B.; Leake, M.; Morrison, D.; and Williams, J. G. 1977. The E asteroids and the origin of the enstatite achondrites. *Geochim. Cosmochim. Acta* 41: 1759-1767.

PART VI

Evolution

NON-EQUILIBRIUM EFFECTS ON THE CHEMISTRY
OF NEBULAR CONDENSATES:
IMPLICATIONS FOR THE PLANETS AND ASTEROIDS

MILTON BLANDER
Argonne National Laboratory

Kinetic effects, for example nucleation constraints and slow reactions, should have been important in nebular condensation. Consideration of these effects leads to the prediction of pressure-dependent compositions and physical properties of nebular condensates which is consistent with (1) the differences between different classes of chondritic meteorites, (2) some of the differences between planets, and (3) the presence of oxidized iron on the moon and in the eucrite parent body (presumably an asteroid) despite the low abundances of volatiles. Diffusion effects appear to be important for understanding oxygen isotope anomalies in refractory inclusions in Allende. The consideration of kinetic effects leads to more information concerning nebular processes than if equilibrium is assumed.

Chondritic meteorites appear to be composed of relatively primitive materials which condensed from a nebula and which last crystallized about 4.6 billion years ago (see, also, the chapter by Kerridge and Bunch). Consequently, they should provide our most important clues on the origins of condensed matter in the solar system. A large number of observations have been made of compositional, textural and mineralogical disequilibrium in chondrites. For example, we have: (a) compositional gradients within individual crystals in Type 3 chondrites; (b) disequilibrium between chondrules and the matrix of ordinary and carbonaceous chondrites; (c) silicon dissolved in metal in enstatite chondrites; (d) nonequilibrated oxygen isotopes in refractory inclusions in carbonaceous chondrites; (e) polymict meteorites in which different portions of a single rock and even different chondrules differ in composition, mineralogy and texture; (f) glass; and (g) assemblages which are out of equilibrium – such as $MgS+FeSiO_3$ dissolved in enstatite ($MgSiO_3$); diopside ($CaMgSi_2O_6$)+gehlenite ($Ca_2Al_2SiO_7$); and Fe_3O_4 + low Fe^{++} silicates. Even such an abbreviated list of observations of departures from equilibrium provides support for the idea that nonequilibrium effects have been important in the formation of meteorites. Despite these observations, there have been few systematic studies of the influence of kinetic factors on the formation of chondrites. In this chapter we will examine some of the kinetic factors that one would expect to be important in the formation and accretion of primordial condensates from a nebula. We will show that the influence of these factors on composition, mineralogy and texture can be great and that most of the major properties of chondrites can be explained if these factors were important. The implications for the planets and asteroids follow from the conclusion that the chemical and physical properties of condensates from a gas of a given composition are a function of the pressure. Consequently, one would expect different materials to have condensed in different parts of a nebula, with departures from equilibrium tending to be greater at lower nebular pressure.

I. KINETIC INFLUENCES ON NEBULAR CONDENSATES

From known physico-chemical concepts one may deduce some of the kinetic factors which may have been important in the formation of nebular condensates. Perhaps the most important influences on condensation processes are related to barriers to the nucleation of certain phases. Nucleation constraints on condensation and on crystallization from a liquid lead to the formation of metastable phases and to departures of composition, mineralogy and texture from those expected at equilibrium. Kinetic constraints may also arise because of slow reactions of solid phases. Such reactions should be especially important for chain and network silicates, e.g., pyroxenes. Slow diffusion in, or between, mineral phases can also help to maintain disequilibrium already present in chondritic precursors. Below we will discuss some of these kinetic factors. Misconceptions concerning nucleation exist because it is a relatively unfamiliar subject to most physical scientists. Consequently, the discussion will emphasize nucleation constraints and especially those aspects of nucleation kinetics which are little understood.

Blander and Katz (1967) and Blander and Abdel-Gawad (1969) were the first to consider the influence of nucleation constraints on nebular condensation and accretion. From nucleation theory one predicts that all vapors supersaturate before they can condense, with the supersaturation being greatest for materials having high surface tensions (surface free energies for solids) and which tend to nucleate homogeneously. The fundamental equation for homogeneous nucleation of a pure substance such as iron is (Becker and Döring 1935; Katz and Blander 1973):

$$J = 4.5 \times 10^{33} \left(\frac{p}{T}\right)\left(\frac{\sigma M}{d}\right)^{\frac{1}{2}} \exp\left[\frac{-17.56M^2\sigma^3}{d^2T^3(\ln S)^2}\right] \qquad (1)$$

where J is the number of nuclei formed per cm^3 per second; p is the partial pressure of condensing species (atm); σ is the surface tension (ergs cm^{-2}); M is the molecular weight (gm/mole); T is temperature; d is the density (gm cm^{-3}); S which is always greater than unity, is the critical supersaturation ratio (p/p_e) for some value of J and p_e; and p_e is the equilibrium vapor pressure of the condensing species. Iron has an extremely high surface tension ($\sim$1800 ergs cm^{-2} for the liquid and higher for the solid) and from Eq. (1) one calculates high supersaturations for iron (or iron-nickel alloys) even for the long time scales one considers for the process of condensation in a nebula. This arises from the fact that the exponential term in Eq. (1) is large ($\sim e^{-50}$ for cosmic time scales) so that large changes in J and in the time scale are made with only a relatively small change in supersaturation ratio. For example, a change in the critical supersaturation ratio from 1000 to 100 corresponds to a change in J and in the time scale, of about 27 orders of magnitude and a change from 1000 to 250 corresponds to a change in J, and time scale, of $\sim 10^{12}$. Thus nucleation barriers should be important no matter what the time scale may be. The relatively small shift in supersaturation ratios, S, for very large changes in time scale, J, means that any given large supersaturation will occur at somewhat lower temperatures, or nebular pressures, when the time scale is shifted from, let us say years to billions of years.

A generally held misconception concerning the nucleation of condensates is that nuclear and cosmic radiation and the consequent ionization should remove nucleation barriers completely. This misconception arises because of general familiarity with cloud chambers, that are filled with water vapor, for the detection of radiation ionization. However, it is known that even for a dipolar material such as water, ions merely lower the nucleation barrier and do not eliminate it. For nonpolar materials it has been shown experimentally that the influence of radiation is negligibly small (Katz 1975, personal communication). Consequently, the influence of ionizing radiation on the condensation of iron is probably negligible.

A real influence on nucleation barriers arises because of the tendency for the co-condensation of oxides, including silicates, with metal. Preformed dust can provide sites for the heterogeneous nucleation of metals. For heterogeneous nucleation, the prefactor in Eq. (1) is smaller and the exponent is multiplied by a quantity $\phi(\theta)$ which is a function of the wetting angle. Because the surface tension of iron ($\geqslant$1800 ergs cm^{-2}) is very much higher than those for oxides (300-600 ergs cm^{-2}), the function ϕ can be shown to differ little from unity and, because of the smaller prefactor for heterogeneous nucleation, the rate of heterogeneous nucleation can be shown to differ little from, or even be smaller than, the rate of homogeneous nucleation (Dunning 1969; Anders 1969). Consequently, the presence of "dust," if its surface free energy is not large, will not catalyze the nucleation of metallic iron, or iron-nickel alloys.

Superimposed on these considerations is an even more important thermodynamic effect which leads to the coating and isolation of the metal by oxides or silicates. From the thermodynamics of surfaces one deduces that low surface tension materials, e.g. silicates, should coat high surface tension materials, e.g. iron-nickel alloys. Consequently, since oxides including silicates tend to co-condense with iron-nickel alloys in the nebula, even if metal is nucleated in the nebula, it will become coated. The metal surface will then be isolated from the nebula and will not be a nucleation site for further condensation of metal. Thus a "solar-type" nebula is an ideal place in which to expect large nucleation barriers to the condensation of metals.

If for the sake of simplicity we look only at five of the major nebular elements, H, O, Si, Mg and Fe, we can gain further insights into yet other kinetic constraints. The possible reactions of these five elements are:

$$Fe(g) \rightarrow Fe(s) \tag{2}$$

$$2Mg + SiO + 3H_2O \rightarrow Mg_2SiO_4(s) + 3H_2 \tag{3}$$

$$Mg + SiO + 2H_2O \rightarrow MgSiO_3(s) + 2H_2 \tag{4}$$

$$Mg_2SiO_4(s) + SiO + H_2O \rightarrow 2MgSiO_3(s) + H_2 \tag{5}$$

$$Fe + H_2O \rightarrow FeO(s) + H_2 \tag{6}$$

$$3Fe + 4H_2O \rightarrow Fe_3O_4(s) + 4H_2 \tag{7}$$

$$2Fe + SiO + 3H_2O \rightarrow Fe_2SiO_4 \text{ (pure solid or solid soln.)}$$

$$+ 3H_2 \tag{8}$$

$$2Fe(s) + 2MgSiO_3(s) + 2H_2O \rightarrow Fe_2SiO_4(ss) + Mg_2SiO_4(ss) + 2H_2 \tag{9}$$

At equilibrium, in a nebula in the pressure range usually considered (10^{-6}-1 atm), reactions (2) and (3) occur at relatively high temperatures followed by Reaction (5). After essentially all Mg and SiO has condensed or reacted according to these reactions, Reaction (9) will occur to form oxidized iron in solid solution. Fe_3O_4 can form only at very low temperatures. Reactions (5) and (9) are solid-solid reactions involving the formation or decomposition of enstatite which is a chain silicate (pyroxene). One expects such reactions to be very sluggish even at moderately high temperatures. We have examined similar reactions in E(5,6) chondrites such as, for example:

$$\text{Si(in metal)} + 2\text{FeSiO}_3 \text{(in solid soln. in enstatite)} + 3\text{Mg}_2\text{SiO}_4$$

$$\rightarrow 2\text{Fe(in Fe-Ni alloy)} + 6\text{MgSiO}_3 \ .$$

From known thermodynamic data on silicon alloys (Sakao and Elliott 1975) and silicates, and compositional data on enstatite chondrites, we calculate that such reactions froze in at $\sim$ 1300-1400 K. Since from known experiments the slow step can be shown to be related to the formation or decomposition of the pyroxene, this range of temperatures is probably a reasonable guess for the freezing in of reactions (5) and (9). For our calculations we have chosen 1350 K.

We have performed calculations on the condensation of Mg, Si, Fe, and O from a nebular gas with relative compositions given by Cameron (1973) over a wide range of nebular pressures using a computer program of Gordon and McBride (1971). We find similarities between our condensation results and different chondritic types if we impose constraints deduced from nucleation theory on Reaction (2) and a constraint at 1350 K on reactions (5) and (9). We find that enstatite chondrite-like materials appear to form at high nebular pressures (10^{-1} -10^{-2} atm), ordinary chondrites near 10^{-3} atm, and carbonaceous chondrites at low nebular pressures ($\leqslant 10^{-4}$ atm). Schematically, the condensates in these pressure ranges are given in Table I. We see that at high pressures (labeled enstatite chondrites) the condensation of metal is blocked, forsterite is condensed and conversion to enstatite begins before metallic iron starts to condense. Because all this happens at temperatures >1350 K the assemblage can appear to be at or close to equilibrium with the nebula. Near 10^{-3} atm (ordinary chondrites) the condensation of metallic iron is blocked, forsterite is condensed and only partly converted to enstatite before Reaction (5) is frozen in. When iron becomes sufficiently supersaturated it reacts with the SiO in the gas according to Reaction (8) to form Fe_2SiO_4 in some form. As the nebula cools further any iron still in the nebular gas will ultimately condense as metal. At low pressures ($\leqslant 10^{-4}$ atm) the condensation of metal and the conversion of forsterite to enstatite are essentially completely blocked. The iron then will tend to react following reactions (7) and (8) to form Fe_2SiO_4 or Fe_3O_4 at temperatures which are much higher than one would calculate at equilibrium. The possibility of the formation of FeO by Reaction (6), as indicated in our calculations, has never been considered. Under conditions where metal ultimately nucleates, FeO will disproportionate according to the reaction

$$4\text{FeO(s)} \rightleftarrows \text{Fe} + \text{Fe}_3\text{O}_4\text{(s)} \ . \tag{10}$$

Fuchs (Blander and Fuchs 1975) has observed many associations currently consisting of magnetite with a lesser amount of metal on one surface of the magnetite, which could have been FeO originally. However, such associations

TABLE I

Schematic Representation of Calculated Condensation Sequence

T (K)	Enstatite Chondrites		Ordinary Chondrites		Carbonaceous Chondrites	
	Blocked	Formed	Blocked	Formed	Blocked	Formed
1600	Fe	Mg_2SiO_4	Fe	Mg_2SiO_4	Fe	Mg_2SiO_4
		$MgSiO_3$	$MgSiO_3$	$MgSiO_4$	$MgSiO_3$	
		Fe		"Fe_2SiO_4"		$\begin{cases} Fe_2SiO_4 \\ FeO \\ Fe_3O_4 \\ Fe \end{cases}$
				Fe		
900						

could have been formed by other mechanisms. In any case the mineralogy calculated in Table I corresponds to that found in the classes of chondrites indicated. Thus, with the imposition of only two kinetic constraints we can rationalize the formation of different chondrite classes. The most important result is that the *composition of condensates is a function of nebular pressure.*

Yet another result of nucleation constraints must be considered. Surface free energies of solids are greater than those for liquids so that the barriers for the nucleation of solids are much greater than those for liquids. Consequently one would expect the formation of metastable supercooled liquid condensates, in a manner analogous to the formation of liquid water in clouds down to -40°C. Metastable silicates were investigated by Blander *et al.* (1976) and appeared to be capable of remaining metastable down to temperatures as low as 1300 K.

This lowest temperature for chondrule crystallization is close to that where reactions to form enstatite freeze in. The apparent freezing in of reactions which form pyroxenes might reflect the blockage of the full condensation of SiO_2 into a metastable liquid after that liquid has crystallized. If most solid reactions are blocked but dissolution of condensates into liquids is not blocked, then the crystallization of different and independent droplets at different temperatures will lead to the different mineralogies and the different Mg/Si ratios found in the various chondrules in the same chondrite. A similar blockage of solid-solid reactions could have been the reason for compositional differences found among calcium-aluminum-rich inclusions (CAI's) in the Allende meteorite. Thus the variety of compositions and textures of chondrules and CAI's could be readily produced by this mechanism.

Deeply supercooled metastable droplets which crystallize at different temperatures should produce a variety of textures and forms which resemble meteoritic chondrules. Thus chondrules were probably crystallized from metastable liquids and the large variety of observed chondrule textures probably reflects, at least in part, differences in the degree of supercooling at the time crystallization is initiated (Blander *et al.* 1976).

Thus far we have not considered refractory elements such as Ca and Al which are among the first condensates expected from a solar nebula (Lord 1965). The first detailed calculation of equilibrium refractory condensates was made by Grossman (1972) using abundances given by Cameron (1968). Blander and Fuchs (1975) have recalculated these equilibrium condensates at 10^{-3} atm to take into account phases not considered by Grossman. Cameron's 1968 abundances were used for direct comparison with Grossman's results; the change to more recent abundances does not lead to significant differences. The results are given in Table II. The temperatures given are temperatures of first formation. The major influence of moderate changes in nebular pressure is to shift the temperatures with a general

TABLE II

Condensation from a Nebula at 10^{-3} Atm

Phase	T
Corundum (Al_2O_3)	1775
Hibonite ($CaAl_{12}O_{19}$)	1765[a]
$CaAl_4O_7$	1735[a]
Perovskite ($CaTiO_3$)	1647
Gehlenite ($Ca_2Al_2SiO_7$)	1625
Spinel ($MgAl_2O_4$)	1513
Metallic iron-nickel (Fe-Ni)	~1473
Diopside ($CaMgSi_2O_6$)	1450

[a]Phases not considered by Grossman (1972).

tendency for the oxides, including silicates, to form in the same order. The uncertainty in the data on hibonite is great enough so that it could be the first phase that tends to form.

If one imposes nucleation constraints on the formation of solids and, as predicted by nucleation considerations, if one forms instead a metastable liquid phase, that liquid phase tends to form at about 1740 K and consists almost completely of a CaO-Al_2O_3 solution close to the hibonite composition. As the temperature drops, the CaO and SiO_2 contents increase followed by a significant increase in the MgO content of the melt. It has been shown experimentally that metastable liquids of such compositions can persist and form glasses readily on a laboratory time scale (Blander et al. 1976; Blander and Fuchs 1975). Consequently one would expect that such metastable liquids can form and supercool considerably in the nebula. The textures and mineral relationships of materials crystallized from such metastable liquids should differ considerably from those of equilibrium condensates.

Refractory condensates (CAI's) have been observed in carbonaceous meteorites, as well as in some ordinary chondrites. A careful investigation has shown that the CAI's in the Allende meteorite (C3) appear to have formed a liquid or a glass (Blander and Fuchs 1975) and could not be an agglomeration of solid condensates. Some of the evidence for this is:

1. the spherical and ovoid shapes of some CAI's;
2. the highly refractory minerals (hibonite, perovskite) rimming some CAI's;
3. textures characteristic of crystallization from a melt;
4. minerals which are completely enclosed by more refractory minerals;
5. glass, and fine grained grossular stringers;
6. mineral relationships generally as expected from a liquid to solid

transformation;

7. spinel sprinkled about in all minerals both more and less refractory than spinel, i.e., spinel appears to have crystallized before most of the major minerals.

Despite statements and interpretations of ambiguous observations in the literature to the contrary (e.g., Grossman *et al.* 1975), the evidence favors a liquid origin for CAI's in C3 chondrites. At lower pressures, i.e. where we believe C1 or C2 chondrite precursors formed, similar materials might condense to crystalline solids directly.

If another important element, sulfur, is considered, further and independent evidence is obtained which correlates each class of chondrites with a particular range of pressure. At equilibrium, metallic iron-nickel alloys condense before they are converted to FeS at about 680 K by the reaction

$$Fe + H_2S \rightarrow FeS + H_2 . \tag{11}$$

If the metal phase does not nucleate and Fe supersaturates in the nebula, then Reaction (11) occurs at much higher temperatures than at equilibrium (Blander 1971); how much higher is a function of pressure. Above $10^{-2.8}$ atm, where we believe enstatite chondrites form, the temperature is above the eutectic temperature of the Fe-FeS system and direct condensation to a liquid is possible. Below $10^{-2.8}$ atm where e.g., ordinary chondrites form, only solids form. Many of the differences in mineralogy and chemistry between the enstatite and ordinary chondrites are consistent with this predicted difference in the mode of condensation. For example, the activity coefficients of the sulfides of chalcophile elements should be much smaller in liquid FeS than in solid and one would expect the co-condensation of other chalcophiles (Zn, Cd, etc.) with the liquid. This prediction is consistent with the fact that chalcophile elements are generally more abundant in enstatite chondrites than in ordinary chondrites.

If the process of accretion of CAI's and/or chondrules with later condensates occurs simultaneously with condensation, most of the properties of chondrites can be explained, such as the correlation of volatiles and of mineral equilibration with "metamorphic" grade, as well as the properties of polymict chondrites.

The major conclusion is that if one takes kinetic constraints into account, chondrites appear to have formed in a far simpler manner and with fewer *ad hoc* assumptions than is postulated in other theories and that the chemistry of nebular condensates depends on pressure.

II. OXYGEN ISOTOPE ANOMALIES

Recent observations of isotope anomalies in CAI's (e.g., Clayton *et al.* 1977; Wasserburg *et al.* 1977; Clayton and Mayeda 1977) have led to speculations concerning the mechanical mixing of solids from the solar nebula with materials injected into the nebula from a supernova (Clayton *et al* 1977). If correct, these speculations would have important implications concerning planets and asteroids. However, such a hypothesis is not consistent with the evidence that the CAI's have been liquid. To explain the observations of oxygen isotope anomalies of once molten CAI's in which separated minerals from one CAI have different anomalies, the entire CAI must have been strongly anomalous. Consequently, if the source of the anomaly was a supernova, one must find a mechanism for transporting not just a few grains but whole CAI's, many of which are millimeters in size, from a supernova into the solar nebula. This appears unlikely.

The measurements of Clayton *et al.* (1977) are plotted in Fig. 1. These include measurements of individual CAI's in C2, C3 and C4 chondrites as well as of separated minerals of individual CAI's in Allende. Some minerals, such as pyroxenes, were shown to have much larger anomalies than melilites. A line through the experimental points has a slope of about 0.94. The original model of Clayton *et al.* involved the mixing of solids containing essentially

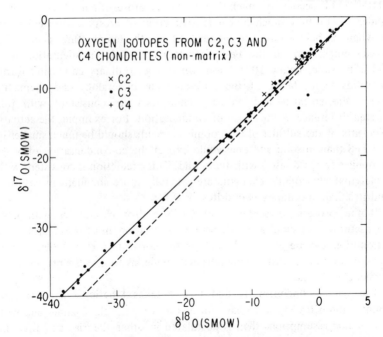

Fig. 1. Oxygen isotope anomalies in carbonaceous chondrites. The points are from Clayton *et al.* (1977); SMOW ≡ Standard Mean Ocean Water. The dashed line represents a model for the mechanical mixing of ^{16}O-rich solids with nebular materials having "normal" oxygen isotopic composition. The solid line was calculated from Eq. (12) with δ^{17} and δ^{18} normalized to "normal" isotopic composition i.e., in "normal" material, $\delta^{17} = \delta^{18} = 0$, in Eq. (12).

only ^{16}O with solids containing normal abundances of ^{17}O and ^{18}O. However, such a model would lead to a slope of unity for a plot such as in Fig. 1. The data are not consistent with this model, which predicts anomalies along the dashed line in Fig. 1.

A simple diffusion model (Blander and Fuchs 1975) leads to an explanation of the oxygen isotope anomalies. If an ^{16}O-rich CAI which had already crystallized was transferred to a region containing normal abundances of ^{17}O and ^{18}O, each of the mineral phases in a given inclusion would tend to equilibrate with the normal oxygen by diffusion. The equilibration is a function of the diffusion coefficient divided by the square of the crystal size (D/a^2). There are two influences on the diffusion coefficients. One is the isotope effect such that ^{17}O diffuses faster than ^{18}O with a maximum ratio of $(18/17)^{1/2} = 1.03$. The ratio can be less than but never greater than this value. The second influence depends on differences in diffusion rates between various minerals. For example, one expects low diffusion coefficients for oxygen in chain silicates such as pyroxenes. From this model one obtains a relation between δ^{17} and δ^{18} for the maximum isotope effect

$$\delta^{17} = 0.825(\delta^{18})^{1.03} \tag{12}$$

where δ^{17} and δ^{18} are properly normalized. The solid line in Fig. 1 was calculated from Eq. (12) and appears to fit the data very well. If the isotope effect on oxygen diffusion is less than the maximum, the fit could be improved even further. The diffusion model can be seen to be consistent with both a liquid origin and with the observed oxygen isotope anomalies. This conclusion was also supported by the observations of Wasserburg et al. (1977).

If a source of anomalous CAI solids in a supernova is unlikely, then what are their origins? One possibility requires a layered, or otherwise inhomogeneous, nebula with condensation of CAI precursors at high temperature in an outer ^{16}O-rich region, followed by transfer (e.g., by convective and/or gravitational forces) to an inner region containing "normal" oxygen isotopes. Future observations on isotope anomalies should help to further test for such possibilities and may help to unravel details of the structure and compositions of the nebular sources of the planets and asteroids.

III. IMPLICATIONS FOR THE PLANETS AND ASTEROIDS

The most important implications of our work for the planets and asteroids arise from the dependence of the chemistry of condensates on nebular pressure. The differences between the high metal content of the innermost planets, Mercury, Venus and Earth, and the lesser metal content of the moon and Mars, with small cores, if any, are consistent with their origins at different nebular pressures, with the metal-rich planets forming at higher pressures ($>10^{-3}$) than the metal-poor bodies, Mars and the moon.

In addition we can correlate the apparently mysterious combination of a low volatile content, the paucity of metal and the presence of oxidized iron on the moon and the eucrite parent body with our model (see the chapter by Drake in this book; Stolper *et al.* 1979). Such a combination cannot form at equilibrium with a nebula. With the imposition of nucleation constraints on the formation of metallic iron, oxidized iron forms at temperatures above those where volatiles, such as Na, for example, are essentially fully condensed.

Within the context of the model presented, one can not only make predictions concerning the possible compositions of condensates expected for different models for the formation and accretion of planets and asteroids, but one can also relate the chemistry of such bodies to a particular source region in the nebula. For example, if the model is correct, the moon and the eucrite parent body (presumably an asteroid, [see Drake's chapter]) are accreted from relatively high-temperature nebular condensates (>900 K at 10^{-3} atm) from a relatively low-pressure source ($\leqslant 10^{-3}$ atm). Mercury is largely an accretion of high-pressure condensates which might, at most, have a thin veneer of low-pressure condensates.

We gain much more information from studies of meteorites, planets and asteroids if nonequilibrium effects were important in the formation of the precursors of these bodies than if condensates were formed at equilibrium. This is true because these bodies preserve the thermodynamic tendencies of the source nebula as well as kinetic information. Although nonequilibrium effects make it much harder to study materials formed in the nebula, these effects should help us learn far more concerning our primordial origins.

Acknowledgments. This chapter was completed while the author was a Regents Lecturer at the University of California at Los Angeles. The hospitality of the University and especially of J. Wasson is gratefully acknowledged. This work was supported by the National Aeronautics and Space Administration.

REFERENCES

Andres, R. P. 1969. Homogeneous nucleation in a vapor. In *Nucleation,* ed. A. C. Zettlemoyer (New York: Marcel Dekker Inc.), pp. 69-108.
Becker, R., and Döring, W. 1935. Kinetic treatment of nuclear formation in supersaturated vapors. *Ann. Physik* 24: 719-752.
Blander, M. 1971. The constrained equilibrium theory: Sulphide phases in meteorites. *Geochim. Cosmochim. Acta* 35: 61-76.
Blander, M., and Abdel-Gawad, M. 1969. The origin of meteorites and the constrained equilibrium condensation theory. *Geochim. Cosmochim. Acta* 33: 701-716.
Blander, M., and Fuchs, L. H. 1975. Calcium-aluminum rich inclusions in the Allende meteorite: Evidence for a liquid origin. *Geochim. Cosmochim. Acta* 39: 1605-1619.

Blander, M., and Katz, J. L. 1967. Condensation of primordial dust. *Geochim. Cosmochim. Acta* 31: 1025-1034.

Blander, M.; Planner, H.; Keil, K.; Nelson, L. S.; and Richardson, N. L. 1976. The origin of chondrules: Experimental investigation of metastable liquids in the system Mg_2SiO_4-SiO_2. *Geochim. Cosmochim. Acta* 40: 889-896.

Cameron, A. G. W. 1968. A new table of abundances of the elements in the solar system. In *Origin and Distribution of the Elements,* ed. L. H. Ahrens (London: Pergamon Press). pp. 125-143.

Cameron, A. G. W. 1973. Abundances of the elements in the solar system. *Space Sci. Rev.* 15: 121-146.

Clayton, R. N., and Mayeda, T. K. 1977. Correlated oxygen and magnesium isotopes in Allende inclusions. I. Oxygen. *Geophys. Res. Lett.* 4: 295-298.

Clayton, R. N.; Onuma, N.; Grossman, L.; and Mayeda, T. K. 1977. Distribution of the presolar component in Allende and other carbonaceous chondrites. *Earth Planet. Sci. Lett.* 34: 209-224.

Dunning, W. J. 1969. General and theoretical introduction. In *Nucleation,* ed. A. C. Zettlemoyer (New York: Marcel Dekker Inc.), pp. 1-67.

Gordon, S., and McBride, B. J. 1971. Computer program for calculation of complex chemical equilibrium compositions, rocket performance, incident and reflected shocks, and Chapman-Jouget detonations. (NASA SP-273, Washington, D. C.: U.S. Government Printing Office).

Grossman, L. 1972. Condensation in the primitive solar nebula. *Geochim. Cosmochim. Acta* 36: 597-619.

Grossman, L.; Fruland, R. M.; and McKay, D. S. 1975. Scanning electron microsopy of a pink inclusion from the Allende meteorite. *Geophys. Res. Lett.* 2: 37-40.

Katz, J. L., and Blander, M. 1973. Condensation and boiling: Corrections to homogeneous nucleation theory for non-ideal gases. *J. Colloid Interf. Sci.* 42: 496.

Lord, III, H. C. 1965. Molecular equilibria and condensation in a solar nebula and cool stellar atmospheres. *Icarus* 4: 279-288.

Sakao, H., and Elliott, J. F. 1975. Thermodynamics of dilute bcc Fe-Si alloys. *Met. Trans.* 6A: 1849-1851.

Stolper, E.; McSween, Jr., H.; Hays, J. F. 1979. A petrogenetic model of the relationships among achondritic meteorites. *Geochim. Cosmochim. Acta* 43: 589-602.

Wasserburg, G. J.; Lee, T.; and Papanastassiou, D. A. 1977. Correlated O and Mg isotopic anomalies in Allende inclusions. II. Magnesium. *Geophys. Res. Lett.* 4: 299-302.

PRIMORDIAL HEATING OF ASTEROIDAL
PARENT BODIES

C. P. SONETT
University of Arizona

and

R. T. REYNOLDS
NASA Ames Research Center

Most meteorites show evidence of thermal processing either because of metamorphic changes or as a result of melting and differentiation. Proposed mechanisms for supplying this energy generally rely upon short-lived radioisotopes or electrical induction, though accretion is sometimes mentioned, and more exotic models have been discussed. Interest in isotopic heating has been heightened by the discovery of ^{26}Al in Allende inclusions and also by the proposal that a lunar core and dynamo resulted from the radioactive decay of superheavy elements during the early solar system. Electrical induction as a heat source can be scaled to a broad range of solar system conditions, but corroborative evidence for these conditions is inconclusive. The accretion mechanism is probably not viable for the asteroidal and meteorite parent bodies, because the high kinetic energy requirement is inconsistent with the formation of the objects and their regoliths in the presence of a weak gravitational field.

The study of the thermal evolution of asteroid parent bodies is subject to all the difficulties and uncertainties attendant upon studies of planetary thermal histories. There are at least three major additional impediments to the

[822]

understanding of the origin and evolution of these bodies. Firstly, not only must a formative process be theoretically constructed to account for their existence (as is usual for any planetary body), but additional destructive processes and scenarios must also be hypothesized. This is necessary because the objects currently in existence are presumably only remnants of the original population. Secondly, although a great deal of quantitatively precise and accurate information is available regarding both meteorites and asteroids (all of which must somehow eventually be included within any comprehensive theoretical framework), the correspondence between the data and knowledge of conditions on any individual parent body is typically ambiguous, nonunique, and very model dependent. This would generally be true even if it were not for the following and third difficulty. In contrast to a single (albeit quite complex) evolutionary history of a given planetary body, here one deals with a bewildering multiplicity of objects having great differences possible in their physical parameters, spatial distributions and in the development of these quantities with time.

Today it is generally agreed among workers in the field of meteorites and its related cosmogony that the material of which the solar system is formed condensed over a wide range of temperatures (Grossman and Larimer 1974). Much of the siliceous and siderophilic materials have been melted and differentiated at high temperatures, but the presence of water, hydrated minerals, and of Pb and other nonaqueous volatiles in the meteorites shows the existence of a wide temperature range which was undoubtedly at least partially due to time variations of solar luminosity and to varying distance from the primordial sun. For a broad review see Anders (1971a,b).

Most of our information regarding the very early solar system comes from study of the physics and chemistry of meteorites. From a petrologic standpoint some of the outstanding features, not found in earth rocks, are the very existence of the irons, the stony-irons, and the ordinary chondrites. All of these objects show evidence of having been processed through high temperatures; it is known that the temperature extremes could not be due to entry into the earth's atmosphere, even though the neighboring gas temperature can rise to many thousands of degrees C. This is because the time for heat to flow into the meteorite is long compared to the flight time through the atmosphere (Lovering *et al.* 1960). More accurately, according to Lovering *et al.*, the rate of ablation stays in step with the inward flow of heat. Thus, for example, the $100°C$ isotherm is at a constant depth. The fiery beginning of meteorites was recognized from the earliest times when Chladni (see Wood 1968) first observed the igneous character of meteorites (Kerridge argues that even carbonaceous chondrites exhibit secondary thermal events; c.f. his chapter in this book).

It is a remarkable fact that the one or more major thermal events which took place in the early pre-history of the solar system after accumulation of parent objects remain(s) expressed in meteorites. Even the ordinary chondrites, which do not show obvious evidence of differentiation do exhibit metamorphism. Because meteorites are small on a planetary scale, they cannot retain heat for long. Therefore the thermal processing which many of them underwent must have taken place in parent objects of a size capable of accumulating and storing thermal energy.

Among the ordinary chondrites, at least among types 3-6, (Van Schmus and Wood 1967) an increasing degree of metamorphism is noted. The irons, stony-irons, (pallasites and mesosiderites), and the basaltic achondrites all appear to have been completely melted and differentiated. The one exception to this is the IA-IB group of irons which Wasson (1972) argues is of non-igneous origin because of the presence of undeformed olivine grains. (Metamorphism refers to chemical and/or physical modification *below* the solidus, while differentiation means separation during melt, i.e. above the solidus.) Another and highly controversial viewpoint holds that at least some of the irons arose from direct condensation from the vapor phase into the crystalline form (e.g. Bloch and Muller 1971). But it is difficult to understand, for example, how the giant Widmanstätten patterns such as those found in the Cape York meteorite could arise from the dissociation of Fe and/or Ni carbonyl. Since this does not seem to be the way that the irons evolved, and because parent objects are still required for the basaltic meteorites and stony-irons, this hypothesis does not aid in understanding their genesis.

In spite of the high temperatures commonly thought to be associated with the iron, stony-iron and basaltic meteorites, as noted earlier not all meteorites show the effects of strong heating. The carbonaceous chondrites, which contain magnetite, imply a peak temperature of about $400°C$, since this mineral becomes unstable at higher temperature. Also any simple model of parent body heating does not account for the formation of the chondrules. At least some of these appear to have cooled rapidly, on a time scale of minutes, some $10^{10} - 10^{13}$ times faster than the thermal relaxation time for an asteroidal or meteoritic parent object. This suggests a completely different means of heating these objects or their parent material.

As noted before, the attainment of temperature sufficient for metamorphism or differentiation requires that thermal energy be accumulated and stored for an extended period of time, a process which cannot take place in small, meteoritic-sized objects. This implies storage of the meteorites within parent objects. It is not known with certainty whether the asteroids and the meteorites are derived from common parents, but much of the research today is based at least tacitly upon this assumption. Asteroidal spectrophotometry suggests a range of surface types corresponding to the meteorites, but a final answer to the relationship of asteroids and meteorites probably will have to await the direct recovery of asteroidal material by space flight.

The earliest view on the source of energy required for thermal modification were based upon the long-lived radionuclides, i.e. those with half lives measured in large fractions or multiples of aeons. The "great debate" of the 1950's involved various protagonists and antagonists and centered on the age of meteorites and their thermal processing. At this time Kuiper (1954) proposed that the solar system originated some 5×10^9 yr ago; this included the idea that thermally insulated objects (those with extremely long thermal relaxation times) were formed then, and that episodic melting, yielding the basaltic meteorites, took place some 5×10^8 yr later due to the presence of the long-lived isotopes ^{40}K, ^{23}Th, ^{235}U and ^{238}U, the half life of the latter being on the order of the age of the solar system itself. With the first radiometric determination of the great antiquity of meteorites (Patterson 1955; Schumacher 1956) it became evident that the long-lived nuclides could not provide sufficient heat in a reasonable time period between the accumulation of the body and its subsequent cooling to a temperature low enough to set radiometric "clocks". An additional, now historical, difficulty concerning Kuiper's (1954) proposal was the existence of chondrites which appeared thermally unaltered (Urey 1956). Today we know these to have been, in most cases, altered thermally, but their metamorphism was apparently not generally recognized at the time.

As a means of estimating the ages of the elements, Brown (1947) suggested the presence of ubiquitous relict daughters of now extinct (fossil) radioisotopes throughout the solar system. In recognition of the difficulties with long-lived nuclides, Urey (1955) proposed such fossil nuclides as heat sources; in particular ^{26}Al was taken to be the most promising. Since then ^{26}Al has been the premier candidate fossil nuclide both because of its potential abundance and its ease of formation from ^{26}Mg. (The discovery of relict ^{26}Mg in Allende has greatly heightened interest in the fossil nuclide hypothesis.) Urey later abandoned the ^{26}Al hypothesis because its expected uniform distribution throughout the solar system (the principle of the homogeneous solar nebula) would have melted the moon very early, in conflict with his ideas on the genesis of that body (Urey 1958).

Both ^{244}Pu (Kuroda et al. 1966) and ^{129}I (Reynolds 1960) have been inferred in meteorites from the trace xenon isotopes which are produced by the fission of ^{244}Pu and the decay of ^{129}I. Additionally ^{244}Pu fission tracks are observed in lunar whitlockite (Drozd et al. 1972). Although both isotopes are important in radiometric relative age determinations, the concentrations appear far too low to have provided a significant heat source in the early solar system.

A competing source of energy for the various meteoritic thermal episodes (and also perhaps for the moon and Mercury) is provided by electromagnetic induction, a process which can provide a powerful heating source under certain sets of conditions which could plausibly have existed in the protoplanetary nebula. A high level of inductive heating would result from either or both the interplanetary electric and magnetic fields if they were enhanced as would have been the case if the early sun were rapidly rotating and if the solar plasma flow at the time was increased by some 10^8 times over the present value. These conditions are representative of an early solar evolutionary phase similar to the present conditions for T Tauri stars (Sonett and Herbert 1977).

Several exotic schemes have been proposed for providing the energy for thermal processing of meteorites. These include the heat of reaction of condensed radicals (still a possibility for energizing cometary outbursts) (Urey and Donn 1956) and the collapse of large spherical masses (Urey 1963; Bainbridge 1962). All schemes for heating have strong links to the physics and chemistry of the early solar system. The significance of heating by radioisotopes requires their production and incorporation into parent objects prior to their decay to meaninglessly low levels. The high Z elements could have been produced in supernova explosions which might also have served as the collapse trigger for the solar system. An alternative means of production (but not r-process elements) would be some version of the Fowler, Greenstein and Hoyle "giant solar flare" model for the early sun (Fowler et $al.$ 1962). Although certain aspects of this model, such as a neutron sea, are not supported by present data, the idea of intense solar activity is consistent with present views on the solar atmosphere.

I. HEAT FLOW

To a considerable degree our understanding of the interior of planets and of constraints upon the evolution of the early solar system depends upon the details of the heat transfer process within the planets and the radiation to space from their surfaces. The basic law governing the conductive flow of heat expresses the proportionality between the heat flux vector $\mathbf{F}$ and the temperature gradient ∇T, i.e.

$$\mathbf{F} = -k\nabla T \tag{1}$$

where the constant of proportionality k is called the thermal conductivity. For anisotropic media, k is a symmetric second order tensor, but for planetary calculations the medium is assumed isotropic and is taken to be a scalar, and in practice $k = k(T)$.

The conservation theorem for heat flow (divergence theorem) is given by

$$\mathbf{F} \cdot \mathbf{n}\, dS = (\nabla \cdot \mathbf{F})dV \tag{2}$$

where dS is the unit of surface area of the surface S enclosing a volume V, and $\mathbf{n}$ is the unit normal to the surface, positive outwards. If Q is the rate at which heat is being generated

$$Q dV = (\rho c\, \frac{\partial T}{\partial t} + \nabla \cdot \mathbf{F})dV \tag{3}$$

so that from Eq. (1) the differential question for heat flow is given by

$$\rho c = \frac{\partial T}{\partial t} = \nabla \cdot (k \nabla T) + Q \tag{4}$$

where ρ is the mass density and c the heat capacity here taken as c_p, the heat capacity at constant pressure. Commonly $k/\rho c_p = \kappa$ is called the thermal diffusivity.

For planets the initial condition at $t = 0$ is a prescribed distribution of temperature throughout the object, while the boundary condition equation,

$$k \frac{\partial T}{\partial t} + \sigma T^4 = 0 \tag{5}$$

is also specified. Solutions to Eq. (4) are given by Allan and Jacobs (1956), Carslaw and Jaeger (1959), Kopal (1969), Urey (1962) and many others. Application to the calculation of heat flow for parent objects is made by Fricker *et al.* (1970) as well as by Allan and Jacobs.

Although heat transport is usually assumed to depend upon lattice conductivity, corrections for radiative transport at very high temperatures are sometimes necessary. The total effective coefficient of thermal conductivity K is then of the form

$$K = k + \frac{16 n^2 \sigma T^3}{3\epsilon} \tag{6}$$

where n is the refractive index, σ the Stefan-Boltzmann constant, and ϵ the opacity. Minear and Hubbard (unpublished) discuss in detail the effects of porosity upon estimates of heat flow. When material is porous, the lattice conductivity is reduced since contact between grains becomes poorer, and greater dependence exists upon the radiative losses (see the chapter by Wood in this book).

The loss of heat from a parent object competes with the accumulation of heat from the decay of radioisotopes (or from other sources). This is illustrated in the work of Allan and Jacobs (1956) who solve the heat flow equation, yielding

$$T_c = \frac{2J}{\nu} \sum_{m=1}^{\infty} e^{-\lambda t} - e^{-m^2 \nu t} \frac{(-1)^m}{(\frac{\lambda}{\nu} - m^2)} \tag{7}$$

where $T_c = T_c(t)$ is the central temperature and t is time, J is the initial *rate* of production of heat normalized by the product ρc, where ρ is the mass density and c the specific heat, λ is the decay constant; $\nu = \pi^2 \kappa / a^2$ where κ is

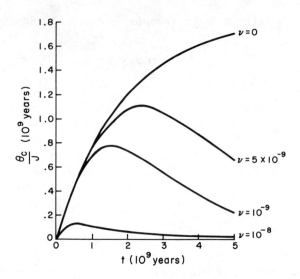

Fig. 1. The central temperature of planetary objects versus time, with $\nu = \pi^2 \kappa/a^2$ as the generalized parameter specifying the radius a and the thermal diffusivity κ. (Figure from Allan and Jacobs, 1956.)

the thermal diffusivity. A similar equation for the mean temperature is given by Allan and Jacobs. Equation (6) shows that the diffusivity and the heat production enter through exponentials. Calculation of the cooling of bodies of varying radius specified in normalized form by ν are shown in Fig. 1 for a single radioactive species with decay constant $\lambda = 5.5 \times 10^{-10}$ yr^{-1}. The effect of decreasing radius (and therefore higher heat radiation rate to space) is seen in these curves, where the lower cases correspond to the smaller radii. Increasing the diffusivity has the same effect as decreasing the radius, as noted in the expression $\nu = \pi^2 \kappa/a^2$.

The occurrence of convection in parent objects (provided that the stability constraints on convection are met) would be to further decrease the ability of the body to store heat, i.e. the effective rate of loss of heat would be increased. Thus Urey's early conviction that long-lived radioisotopes could not supply sufficient heat at a sufficient rate to both raise temperatures to melting and replace the heat loss from the surface, is confirmed.

Fricker et al. (1970) have also discussed at length the problem of cooling and the thermal histories of small, i.e. parent, objects. They include the effects of melting and differentiation, including movement of radioisotopes due to their instability to fit into the lattice structure of the iron phase. The initial temperature distribution, as expected, plays a central role in the subsequent evolution and is modeled by the assumption of various quantities of short-lived radioisotopes or electrical heating.

Detailed tabulations of the thermal conductivity of rocky matter are given in Clark (1966). For many purposes a value of 0.2 cal/g °C for thermal conductivity is assumed. To this should be added a heat of fusion of about 100 cal/g if the material is carried through the melting point. From the standpoint of pressure this is generally ignorable for asteroidal-sized objects. For example, for Ceres (radius = 513 km; see TRIAD tables in this book) the central pressure is about 5.9 kbars (1 kbar (kilobar) = 10^3 bars = 10^9 dynes cm^{-2}). At this pressure the change in melting point is only about 20 °C and need not be accounted for.

Thermal conduction has been considered to be the major heat transfer process, except in a molten region with a superadiabatic temperature gradient where fluid convection dominates. But it is now well established, for silicate bodies of the size of the moon or larger, that solid-state convection can be the dominant mechanism for the transport of energy within planetary interiors. Consolmagno and Lewis (1978) and Reynolds and Cassen (1979) have shown that solid-state convection ought to be important at relatively small-length scales (tens of kilometers) for bodies composed of ice.

The two parameters, whose values determine the stability of a body to solid-state convection, are the viscosity of the material and the size of the convecting region. Small silicate bodies are thus less likely to exhibit thermal instability than the objects discussed above, since the viscosity of silicates is much higher than that of ice and the thickness of the potential convecting layers is greatly restricted in size.

A simple calculation can illustrate this point. The relation expressing the stability of a self-gravitating sphere of fluid with uniform heat generation is given by the Rayleigh number R_a (see Schubart et al. 1969) where

$$R_a = \frac{\alpha \rho G \, Q r^6}{k \kappa \eta} \tag{8}$$

and the condition for instability requires that the Rayleigh number so calculated exceed a critical value $R_{a/cr}$ for a given configuration. For this case the critical Rayleigh number has been determined to be 5785. In Eq. (8) ρ is the density, α the coefficient of thermal expansion, r the radius, κ the thermal diffusivity, k the thermal conductivity, Q the heat production/unit volume, G the gravitational constant and η the kinematic viscosity.

In order to estimate the minimum radius of a sphere which will just become unstable $(R_a/R_{a/cr} - 1)$ we calculate the stability for an equilibrium temperature conduction profile which just reaches melting at the center of the body. Such a profile has a central temperature given by $T_c = T_s + Qr/6\kappa$ (Carslaw and Jaeger 1959), where T_c is the central temperature and T_s the surface temperature.

Using nominal values for the material parameters, $\alpha = 3 \times 10^{-5}$ cm^{-3}, $\kappa = 4 \times 10^5$ erg cm^{-1} sec^{-1} deg^{-1} ,$\rho = 3$ g cm^{-3}, $\eta = 10^{22} cm^2 sec^{-1}$ and for

$T = (T_c - T_s) = 1200$ °C and $R_{a]cr} = 5785$, the radius at which the sphere becomes just unstable is 855 km. This result is relatively insensitive to variations of the material parameters since they enter into the calculations only as the fourth root of their numerical values. Although increased heat production rates can yield still steeper thermal profiles in the outer layers of a small body, the interior in this case will necessarily have a lower thermal gradient and for very high heat production rates melting will occur, reducing the thickness of any potential solid-state convecting layer.

The stability criterion is only weakly dependent upon the details of the shape of the temperature profile for a fixed ∇T (Cassen and Reynolds 1974). These considerations lead to the conclusion that solid-state convection would not significantly influence the thermal evolution of silicate bodies smaller than about 500 km in radius.

II. THE SOURCE OF HEAT

The homogeneous form of Eq. (4) permits us to determine the cooling time for a source-free planet, but usually the term Q representing the rate of generation of heat is nonzero, so that $Q \neq 0$ for $r < a$ where a is the planetary radius. The limitations of accretional heating as a planetary heat source have been noted by several authors (e.g. Safronov 1969; Sonett et al. 1975). Accumulation of thermal energy by this means requires both high impact velocity and the blanketing of impact sites by some means so that heat is stored rather than reradiated. The former takes place either by fast accretion where the blanketing rate is higher than the loss rate, or by burial of the incoming object and its energy to a depth sufficient to assure its long-term retention. The importance of accretional energy as a means of producing metamorphism or melting is difficult to evaluate because of the uncertain time scale for accretion. The velocity distribution and the form of the accretion function have been studied by Weidenschilling (1974, 1976). Generally for objects as small as the moon, accretion as a source of strong heating, e.g. the magma ocean, is marginal except for very fast formation, and this is not in accord with most estimates of the formation time for the large objects which range upwards to 10^8 yr.

Recent ideas on accretion at hypervelocity impact suggest that shock waves propagating to the deep interior can transport thermal energy to great depths, thus avoiding the problem of heat leaks and blanketing. These ideas are currently being intensively explored and it is too early to return a verdict (Kaula 1979). But for the smaller objects such as asteroidal and meteoritic parent objects, accretional heating in any form is likely to be an inefficient mechanism. Regoliths probably exist on these objects; they cannot form in the face of very high energy impacts (Housen et al. 1979; see their chapter). Such impacts would remove more material than they add since the gravity is weak. Thus the two requirements are in direct conflict. Even if accretional

TABLE I[a]

Some Potentially Significant Short-lived Radionuclides

Nuclides	Half Life (yr)	Decay Mode	Available Decay Energy (Mev)	Estimated Abundance (atoms/10^6 atoms Si)	Initial Heat Output (cal g^{-1} yr^{-1})
^{10}Be	2.5×10^6	β^-	0.25	4.5	5.3×10^{-5}
^{26}Al	7.2×10^5	β^+; EC	3.18	100	5.6×10^{-2}
^{36}Cl	3×10^5	β^+; EC	0.32	50	6.6×10^{-3}
^{60}Fe	$\sim 3 \times 10^5$	β^-	2.88	100	~ 0.12
^{129}I	1.7×10^7	β^-	0.11	0.5	3.9×10^{-7}
^{237}Np	2.2×10^6	α; β^-	43.9	0.03	7.2×10^{-5}
^{244}Pu	7.6×10^7	α; β^-	15.8	0.02	5.1×10^{-7}
^{247}Cm	$>4 \times 10^7$	α; β^-	16.8	0.004	$<2.1 \times 10^{-7}$

[a]Table from Fish et al. 1960.

heating were efficient, the bulk of the deposited heat is at large radii (except for deeply penetrating shock waves), since the gravitational energy released by impact increases quadratically with radius. The view that the lunar magma ocean formed from accretional energy sources is based upon this observation. (For an alternate mechanism see Sonett et al. 1975.)

III. RADIOISOTOPIC HEATING

Since the recognition of the existence of significant post-accumulation thermal events in the early solar system, and the possible role of radionuclides, ^{26}Al has been the most studied nuclide because of its expected high abundances, high decay energy and reasonable half life. Other nuclides have been suggested, particularly by Fish et al. (1960) who used the nuclear tables of Strominger et al. (1958) for identifying isotopes with half lives between 10^5 and 10^8 yr. This time period is optimum, both from the standpoint of producing sufficiently fast heating rates to obtain melting, and also decaying slowly enough to remain as an effective heating source after the time required for the accumulation of the body. The Lederer et al. (1967) tables list a large number of potential isotopes. These need to be examined from the standpoint of both energy and abundance. The listing of candidate isotopes with short half lives from Fish et al. (1960) is shown in Table I. Of these isotopes the actinides are lithophilic, as is ^{26}Al and probably ^{10}Be. Only for ^{129}I and ^{36}Cl is there the possibility of siderophilic behavior. Of course ^{60}Fe is included among the siderophiles. These isotopes might be of significance for a fossil dynamo, e.g. in the moon (Runcorn 1978a). However the lifetimes are probably too short.

To raise the temperature of silicate material by about 1500 °C requires $\sim$ 1800 cal g^{-1} and to melt the material will require about 100 cal g^{-1}. The heating rate necessary to produce this amount of energy in 10^7 yr is 1.8×10^{-4} cal g^{-1} yr^{-1}; this is 5000 times the present-day chondritic heat production rate and does not take into account the heat lost from the surface. Since the magnitude of the heating is proportional to the mass, and the total heat flow from the body is proportional to the surface area, the ratio of heat produced to heat lost scales as $1/4$ for a homogeneous body. It requires a higher heating rate to heat a small body to the same temperature than for a large one. Of the nuclides listed by Fish *et al.* (1960), only ^{60}Fe, ^{36}Cl and ^{26}Al appear to be contenders for planetary heating, based upon their heat outputs and abundances. But as the pre-accumulation time is increased, the initial concentration given in Table I must be revised upwards to make up the pre-accumulation losses. The abundance estimate ^{26}Al/^{27}Al $= 6 \times 10^{-5}$ from Lee *et al.* (1976) corresponds to 5.3×10^{-6} for ^{26}Al/Si, compared to 10^{-4} given in Fish *et al.* (1960). The thermal output from the Lee *et al.* estimate of concentration is 3×10^{-3} cal g^{-1} yr^{-1}. Storage in the solar nebula for, say, ten half lives, i.e., $\tau = 7.2 \times 10^6$ yr increases the initial concentration required to attain the above heat intensity at accumulation time by the factor $e^{10} = 2.2 \times 10^4$. This would be a somewhat implausible concentration of ^{26}Al in view of the estimated total Al abundance in the solar system.

From a general standpoint all short-lived nuclides must have been produced either in nuclear reactions related to the early sun, external to the solar system in a nearby supernova, or else they originated in a more generalized stellar scenario but within a time period sufficiently close to the accumulation time of the planets and minor objects so that the decay during the intervening time would not have adversely affected the nuclide concentration. This could be a serious constraint if a long isolation time exists between the generation of the isotope and the accumulation of objects. On the other hand, late additions to the solar system of extrasolar nuclides appear to have taken place (Clayton *et al.* 1973) and whether the source of anomalous ^{26}Al is solar or extrasolar is uncertain.

In considering the very large concentration of ^{26}Al found in Allende, it should be noted that the ordinary chondrites are metamorphosed, but not melted. A lower bound on concentration is inferred from meteorites which have differentiated, since any amount larger than that required for melting would suffice, but the metamorphosed meteorites define an upper bound, which depends upon the initial temperature (Pellas and Storzer 1977). A variation in ^{26}Al is required to reconcile the concentration differences that are implied. This might take place via an inhomogeneous distribution of ^{26}Al. The differing thermal histories of meteorites suggest variable solar nebula concentrations.

With the discovery of oxygen isotopes in Allende inclusions (Clayton *et al.* 1973), it has become less certain that the time intervals deduced from I/Xe and Rb/Sr isotopic ratios can be simply interpreted. These differences are important in attempting to assess the radioisotope concentration at the time of accumulation of meteorite parent objects. $^{129}I/^{129}Xe$ formation intervals have been interpreted to extend to 14×10^6 yr (Podosek 1972). Were ^{26}Al incorporated uniformly in meteorites with a formation interval of 14×10^6 yr, the concentration would be reduced by a factor of 2.8×10^8 and ^{26}Al would be an insignificant heat source. $^{244}Pu/U$ ratios suggest even greater formation intervals (St. Severin-Petersburg, 146×10^6 yr and St. Severin-Lafayette, 350×10^6 yr).

IV. ELECTRICAL HEATING

A nonradiogenic mechanism for the source of thermal energy for the heating of asteroidal parent objects is based upon ohmic losses incurred by electrical currents generated in the asteroids (Sonett *et al.* 1970). Two models have been considered, one based upon the time variable component of the interplanetary magnetic field and leading to a system of eddy currents within objects, the other based upon the currents generated in the asteroids by the interplanetary electric field. The two modes correspond to transverse electric (TE) and transverse magnetic (TM) in the convention of electromagnetic theory (Stratton 1941) and they can operate simultaneously. The idea draws upon the earlier solar system model of Hoyle (1960) who, along with Alfvén (see Alfvén and Arrhenius 1976) earlier, proposed a rapidly rotating Sun connected to an interplanetary plasma via a magnetic field which supplied the mechanism for transfer of solar angular momentum to the nebula. Electrical heating also requires the presence of an intense flow of plasma from the sun, corresponding approximately to the flow hypothesized to take place in a T Tauri-like pre-main sequence phase of the sun. The rapid solar spin conjectured for the early sun is in accord with observation of stellar rotation where stars of spectral type later than F5V show a discontinuously slower spin than earlier types (Kraft 1972), inferring the presence of a stellar wind. This inference is supported by the observation of chromospheric spectral lines in such objects. The mode, which requires a high surface conductivity for an object, usually implies a high background temperature. One source for such a high background temperature would be a high-opacity protoplanetary solar nebula. But a temperature higher than about 400 °K is inconsistent with the evidence of the carbonaceous chondrites. Most rocky materials give low conductivities at these temperatures although Brecher *et al.* (1975) and Briggs (1976) have measured the low temperature conductivity of carbonaceous chondrites, finding values sufficiently high as to void the high-temperature requirement.

TE-mode heating is circumscribed by the electrical skin depth which in turn is a strong function of time. Because of this dependence, the temperature is a nonmonotonic function of asteroidal radius with a peak attained near the surface for many model cases. This mode can also explain the formation of a lunar and Mercurian magma ocean because of the tendency to heat outer layers preferentially. Models involving combinations of TE and TM induction have been calculated for the moon (Sonett *et al.* 1975) and a model has been computed to study induction heating while the moon was being formed (Herbert *et al.* 1977). Such "zero age" models are complicated and whether their application in full complexity to the asteroids is warranted has not yet been determined.

Calculations of asteroidal heating due solely to the TM mode have been reported by Sonett (1971). The results show a strong dependence upon the background temperature because of the aforementioned dependence of the current density upon electrical conductivity, which in turn is dependent upon the temperatures. The calculations can be improved by better determinations of the rock conductivities, which are still poorly known. However, a general result is that the heating rate, and thus peak temperature, varies strongly inversely with radius. This is due to the decrease in current density and attendant reaction magnetic field as r decreases. Thus the back pressure of the field upon the solar wind is less and the induction exceeds that for larger objects before the back pressure of the magnetic field saturates the induction process. The eventual cutoff in heating at small radius is due to radiative loss to space from the surface.

Because of the strong dependence of induction upon electrical conductivity, in turn coupled to heating via temperature, a thermal runaway is implied. This is a natural consequence of the limitations of the calculations which do not include an accurate assessment of the quenching of the induction for large values of current. A significant side effect of the mutual dependence of both modes upon the electrical conductivity is their coupling. This is potentially of great significance to model building because the two modes are sensitive, for example, to radius in quite different ways; thus combining the modes can result in thermal profiles distinct from those issuing from TE or TM excitation separately.

Herbert and Sonett (1978) have investigated application of the TM mode heating mechanism to the asteroid belt. The heating rate varies inversely as the square of the solar distance and both this and dependence upon asteroid radius are seen. The predicted thermal histories are qualitatively consistent with the interpretations deduced from surface spectrophotometry data.

A potentially important aspect of the induction mechanism is the link with radioisotope heating. Although initially the two mechanisms are decoupled and add heat arithmetically, as time progresses the electrical induction (TM) is strongly quenched. The reason is that the addition of isotopic heating causes the asteroid interior to heat more rapidly, leading to a more rapid increase in the internal electrical conductivity. Under these conditions most of the heat is deposited in the outer regions of the body where the ohmic resistance is larger, and this heat is easily radiated to space. Thus, although the heating rate can remain high, heat is not easily retained.

V. SOLAR HEATING

One classical view of the protosolar contraction consists of Hayashi contraction (Ezer and Cameron 1962; Hayashi 1966) at approximately constant temperature with decreasing luminosity as the radius decreases, until a minimum is reached. The contraction proceeds until the Sun's core achieves a radiative balance after which the sun follows a track towards the main sequence governed by radiative equilibrium. These comments and those that follow are based upon early ideas of solar evolution. Some newer calculations of the pre-main sequence track do not display the strongly superluminous Hayashi phase. Leaving open the distinct possibility that the superluminous phase is archaic, it should be noted that study of the sun's pre-main sequence behavior is still the object of investigation and the final verdict is not in. Because of this and the tendency in the literature to discuss heating by a superluminous Sun, the discussion to follow is included. As noted below even if a superluminous phase did exist, heating of this type was probably not important.

The sun is superluminous for about 2×10^6 yr, but the superluminosity is not large for a good deal of this time. For example, the superluminosity is greater than 10 $L_{\odot}$ for 7×10^4 yr and superluminous by 100 $L_{\odot}$ for only $2-6 \times 10^3$ yr. This time is short, and the superluminous phase must take place after accumulation of the asteroidal parent objects if it is to provide a heat source as suggested by some workers, e.g. Wasson (1974) (see Fig. 2). (It is, however, possible to envisage a scenario where external solar heating takes place while the asteroidal parent objects are accumulating. This has apparently not been studied; it would require rather rapid accumulation which seems implausible.)

Asteroidal parent objects would be expected to spin, a result of retaining relict regular momentum of the parent matter from which they are formed. Therefore the solar insolation will be seen by the rotating body as a heat source of the form (Condon 1958),

$$T(r,t) = T_0 R^{-\frac{r}{d}} \cos[\omega t - (r/d)] \qquad (9)$$

where T_0 is the temperature difference between the illuminated and dark sides of the object, r is the depth at which the temperature T is measured, d is the thermal skin depth at which $d = 2\kappa/\omega$ where κ is the thermal diffusivity, and ω the angular spin rate in radians/time. For a primordial spin period of 10 hr (taken from the spins of planets which are thought to be not tidally degraded), $\omega = 1.7 \times 10^{-4}$ rad sec^{-1}. For typical rocky matter $\kappa \sim 10^{-6}$ m^2 sec^{-1} and the skin depth is only 2.2 m. Equation (9) shows that at π skin

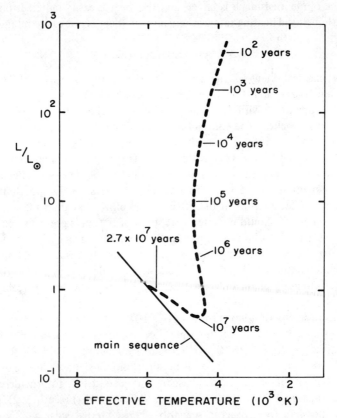

Fig. 2. A typical pre-main sequence evolutionary track for a star of one solar mass showing the fully convective phase to $\sim 10^5\text{-}10^6$ yr followed by the conversion to nuclear burning at $\sim 10^7$ yr and eventually the connection to the main sequence (solid line). (Adapted from Ezer and Cameron, 1962.)

depth the phase of the thermal wave is inverted and the amplitude depressed to $0.043\ T_0$. Thus only a negligibly thin surface layer will be heated. In the extreme case of an object facing the sun, the synodic spin period is zero and the surface temperature is determined by the steady-state insolation which is given by $L_\odot/4\pi R^2$ where $L_\odot$ is the solar luminosity and R is the distance from the center of the sun to the point of observation. For a sphere illuminated by the sun, the insolation is balanced by the radiative output given by the Stefan-Boltzmann law so that

$$T = [L\ \cos\theta/8\pi\epsilon\sigma R^2]^{\frac{1}{4}} \tag{10}$$

where θ is the angle between the unit normal to the surface and the solar direction, ϵ the emmissivity, σ the Stefan-Boltzmann constant and R the solar distance. A more accurate average temperature can be calculated (Colburn *et al.* 1972) but is probably not warrranted here.

The superluminous sun has been briefly reviewed by Wasson (1974) who concludes that electrical induction and superluminosity are the principal, though somewhat unsatisfactory, candidates for asteroidal parent object heating. (This antedates the discovery of ^{26}Al in Allende inclusions.) The superluminous sun hypothesis depends strongly upon the time scale for the period of superluminosity as well as the dynamical spin of the object being illuminated, and its distance from the sun.

VI. SUPERHEAVY ELEMENTS

Interest in superheavy elements geochemically and cosmochemically has increased since the possibility of the use of minerals as detectors of relict superheavy fission products or as carriers of still extant radioisotopes (Flerov *et al.* 1976; Stoughton *et al.* 1973; Gentry *et al.* 1976) has been proposed. Secondly, superheavy elements could provide a potent source of internal heating very early in solar system evolution (Runcorn *et al.* 1977; Runcorn 1978*b*, and 1979, personal communication) and, if the estimated half-lives are correct, supply the energy required to drive planetary dynamos (Runcorn 1978*a*); this also requires that the superheavy elements be siderophile so that they can be incorporated into the Fe core of the planet.

Superheavy elements are those which belong to the conceptual "island" of stability centered on $Z = 114$ and neutron number $N = 184$. The stability is relative, falling off rapidly due to alpha decay and spontaneous fission to either side of $N = 184$ for fission and with increasing Z for alpha decay (Seaborg 1979).

There is doubt as to the accuracy of the proposed values of the half lives for fission and for alpha decay. Runcorn *et al.* (1977) suggest a half life of $10^{-8}–10^{9}$ years, but according to Randrup (1974) this degree of stability against fission (lifetime of 10^{8} yr) is achieved only at $Z = 114$, $N = 184$. Xe isotopes are generated by superheavy fission. In particular, several workers have commented upon the excess (^{131}Xe–^{136}Xe) first seen in Renazzo (see, e.g. Anders *et al.* [1975] and Lewis *et al.* [1975]). Estimates of the time required for the accumulation of planets and parent objects range from a few times 10^{3} yr upwards to some 10^{8} yr depending upon the authority cited. The isotopic evidence certainly favors the longer time estimates. If so, then superheavies of short half life cannot be thermally significant, because the abundances so required for a long accretion time would be totally unrealistic. It should also be noted that kinetic energies of fission fragments (172 Mev for $Z = 92$) are far higher than those from alpha decay (4 Mev for $Z = 92$) (Seaborg *et al.* 1979). Based upon the calculations of Randrup (1974), fission is generally a far more potent source of energy, except for a small region of the stability diagram, where the half life against alpha decay is more than 34 times less than for fission (this is for $Z = 114$ for which the fission kinetic energy is 235 Mev and the alpha kinetic energy is 7 Mev).

The existence of superheavy elements in biotite mica is reported by Gentry *et al.* (1976) from examination of x ray spectra whose excitation is induced by low-energy protons. However Bosch *et al.* (1976) argue that the reported spectra are misinterpreted. In a provocative set of papers, Flerov (1974) and Flerov *et al.* (1976) report the results of neutron multiplicity experiments from presumed fissions of superheavy elements in the meteorites Efremovka, Allende and Saratov (see also Popeko *et al.* 1974; Stoughton *et al.* 1973). Zvara (1977) reports the possible separation of superheavies from meteoritic matter; the volatiles yielded a neutron multiplicity result suggesting 0.02 decay/day-kg of meteorite which is nearly the values reported by Flerov *et al.* (1977).

An alternate search for superheavies is reported by Flerov *et al.* (1976) and Perelygin *et al.* (1977) who look for the distribution of nuclear damage tracks in pallasite olivines as a means of determining a possible contribution of energetic superheavy nuclei to the galactic cosmic ray background. The results so far are indeterminate. The idea that superheavies could have contributed to, or were wholly responsible for, the heating episodes in meteorites is considered by Runcorn (personal communication). A difficulty with this hypothesis, in addition to those discussed above, is the lack of evidence for fission relicts in concentrations sufficient to have been thermally significant.

The difficulties with the detection of superheavy elements (SHE) is reinforced by noting Pitzer's (1975) consideration of relativistic shell corrections which may seriously modify the expected chemistry and volatility of SHE. Furthermore, Nozette and Boynton (1979) show that an upper bound for the SHE/U ratio in the iron meteorite Santa Clara is 10^{-3}. If this low value is not due to volatilization and if Santa Clara is representative, this estimate would probably rule out SHE as a significant heat source.

VII. CHONDRULES

Chondrules represent, thermally, a special class of object since they are some 10^8 times smaller than the parent objects and their thermal relaxation times are thus much shorter. Their ubiquity and the evidence that they were processed thermally show that even on the scale of 0.1–1 cm, solar system objects were subject to heating events.

Chondrules, or at least the droplet chondrules, are thought to have formed by collisions of pre-chondrular matter at velocities greater than ~ 3 km sec^{-1} (Kieffer 1975). However there are problems with the idea of collision induced heating because the high collision velocity can lead to fragmentation as well as droplet formation (Kerridge, personal communication). Whipple (1966) and also Cameron (1966) proposed that lightning in the primitive solar nebula had melted the pre-chondrular matter. This idea is consistent with the appearance of melts (fulgamites) on the earth, and lightning is a pervasive phenomenon in turbulent atmospheres. There are, however, many difficulties with the hypotheses in detail, especially regarding the passage of the required current through pre-chondrular material.

An alternate source of heat arises from consideration of the interplanetary magnetic field, for which there is considerable evidence of occurrence at a very early time (e.g., meteoritic magnetization, the likely damping of high solar spin rates by magnetic breaking and inductive heating of parent objects). With the presence of an interplanetary field, reconnection of adjacent regions of magnetic fields of alternating polarity can take place supplying a high heating rate by means of the acceleration of electrons to relativistic energies; these then can deposit heat within pre-chondrular clumps (Sonett 1979). Although reconnection is considered an exotic plasma process, evidence for it is widespread in the solar system ranging from solar flares to the earth's magnetosphere.

VIII. METAMORPHIC TEMPERATURES

Metamorphism is defined as the petrologic and chemical change induced in a rock by exposure to elevated temperatures that do not exceed the solidus temperature. Such changes are commonly observed in the ordinary chondrites, those silicate-rich meteorites containing round rocky inclusions (chondrules) within a matrix of chemically similar but usually physically distinct material. The slow disappearance of chondrules by metamorphism proceeds sequentially through chondrite types 3–6. The metamorphism of the ordinary chondrites is the result of increased temperatures which endure for an extended period of time.

There are a number of indicators of the peak temperature reached in the neighborhood of a meteorite. These include gas retention, nuclear track annealing, thermoluminescence, the distribution of Fe and Mg between augite and orthopyroxene and, generallly, the degree of volatile loss. Track annealing was first discussed by Fleischer et al. (1968) for the meteorite Toluca; the inferred peak temperature was 550 °K. Pellas and Storzer (1977) have used ^{244}Pu track annealing as a chronometer in St. Severin and Shaw. Their work defines cooling rates rather than peak temperatures. Gas retention and thermoluminescence have been used to confirm peak temperatures of 400 °K. Type 3 chondrites display a volatile pattern consistent with $T_{max} <$ 450 °K. In general $450 < T_{max} < 650$ is observed for the Type 3 chondrites (Wasson 1974).

The other extremum of metamorphic temperature is some 1100 °K for Type 6 chondrites using the augite-orthopyroxene distribution (Van Schmus and Koffman 1967). Other studies, including rare gas release procedures, generally find $950 < T_{max} < 1150$ °K for Type 6 chondrites (e.g. Bunch and Olsen 1974).

In contra-distinction to the chondrites with these metamorphic temperatures, the irons, stony-irons (pallasites and mesosiderites) and the basaltic achondrites all come from complete melts which differentiated. Depending upon the water content the melting temperature of the basalts can range upwards from 1000 °K, while for the irons the melting temperature is about 1500 °K depending upon the Ni alloy content. Thus a wide range of temperatures is required to explain the thermal episodes. This can be at least partially caused by the depth dependence of temperature in meteorite parent objects.

IX. PARENT OBJECTS AND COOLING RATES

As noted earlier, because of the relatively small sizes of recovered meteorites, it has long been recognized that plausible heat sources could not supply sufficient heat for metamorphism or melting, since the rate of heat loss to space is excessive. Smaller bodies lose heat faster since the surface-to-volume ratio varies as $1/r$ for spherical objects in this manner. One exception to these remarks is extrinsic heating by a superluminous Sun, discussed in Sec. V. Thus it follows that burial in parent objects is the only way for meteorites to have been exposed to high temperatures; for any extended length of time the high temperature results from the accumulation and storage of heat from a source with a low rate of energy production.

Various other lines of evidence such as cosmic ray exposure ages, show that meteorites are derived from larger objects within which they were shielded for considerable time.

Information on the parent interiors of such parent bodies is obtained from experiments on the cooling rates of meteorites, primarily though not exclusively the irons (c.f. Wasson 1971). Metallographic rates of cooling of octahedrites show variations from 1 to 10 °C/10^6 yr, suggesting depths of burial of 50–260 km (Wood 1964). The kamacite-taenite Widmanstätten structures are dramatic evidence of the transition from the paramagnetic taenite phase of Fe/Ni to the ferromagnetic kamacite. The actual transition may, however, be more complex than simple precipitation of the kamacite phase as the temperature is lowered, accompanied by diffusion of Ni from the taenite to the kamacite, because an intermediate state (martensite) probably exists before the final transition to kamacite. The lifetime of this metastable state is uncertain and probably dependent upon events such as secondary heating, collisions, etc. Wood (1964) has been able to estimate the degree of "supercooling" prior to the precipitation of the Widmanstätten pattern. Cooling rates and supercooling have also been extensively studied by Goldstein and Ogilvie (1965) and by Goldstein and Short (1967). Figure 3 from Goldstein and Short (1967) shows some 27 meteorites, pallasites and irons, all having cooling rates between 0.5 and 40 °C/10^6 yr. Curiously the pallasites have cooling rates one to two orders lower than the irons. For a

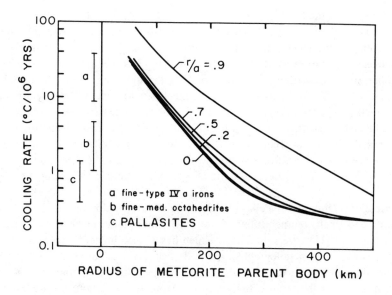

Fig. 3. Cooling rates for various classes of meteorite versus radius and at varying depth in the body. (Adapted from Goldstein and Short, 1967.)

review of the metallographic cooling rates, their comparison to argon ages and fission track ages see the chapter by Wood.

Since diffusion in the Widmanstätten structure is studied in a restricted temperature range, we are, for example, restricted to a knowledge of the thermal histories above 300 °C. Because pallasites have substantially lower cooling rates than the irons (Goldstein and Short 1967), either the conventional models of parent bodies are in error, or these objects came from different parent bodies, for it is commonly held that the pallasites originated in a boundary layer between a metal core and crustal (silicate) matter, while the irons are core material. The argument for a range of parent objects is supported by the Ge and Ga correlation and by cooling rates suggesting several bodies. Wasson (1972) has argued that Ge and Ga are correlated with a reducing environment (higher temperature) implying a range of parent core sizes.

As was noted earlier, the very large Fe-Ni crystals often found probably reflect an origin in cooling from a solic mass (core). The cooling rates determined from study of the Fe/Ni phase properties indicate cooling within an insulating shell.

The rate of cooling of a meteorite establishes a model depth in the parent object within which the meteorite cooled through a specific temperature range. For the irons, this range ends at $\sim$ 350–400 °C because of the inhibition of further diffusion of Ni through the host taenite. The requirement that cooling took place from elevated temperatures is additional, though perhaps unnecessary, evidence for the presence of a heat source which earlier had raised the temperature of the parent object to a value sufficient for metamorphism (chondrites) and melting and differentiation (irons, stony-irons and basaltic achondrites). These studies cannot reveal any information, in the case of the metallographic technique, regarding temperatures higher than some 600–700 °C; neither this method nor the track retention technique of Pellas (1972) can give information on the duration of extended high temperatures or on the chronology preceding the peak temperature. However, the latter, which has a much higher temperature limit than does the metallographic method, can give no information on the chronology between the time of the peak temperature T_{peak} and the onset of kamacite growth because the transition from the high-temperature gamma phase to the joint $\alpha - \gamma$ phase is probably preceded by transition to the diffusion-free or martensite phase (Wood 1964). This phase is metastable but for a generally unknown period of time which can probably be strongly influenced both by trace contaminants and by mechanical stress such as collisional shock.

Currently a large number of iron meteorites have been investigated for their cooling rates in the restricted temperature range below about 900 °C where the phase diagram for Fe/Ni separates into two phase boundaries (Wood 1964; Goldstein and Ogilvie 1965; Goldstein and Short 1967). Studies on the irons have been extended to the pallasites by Goldstein and Short (1967), with resulting cooling rates of 0.4–1.0 °C/10^6 yr. The problem of the cooling rates of the stones (chondrites) has been attacked by Pellas and co-workers (Pellas and Storzer 1977). (This reference contains a compact but important discussion of ^{244}Pu tracks used for cooling rates of *chondrites.*) They examine the annealing rate of nuclear tracks from the fission of ^{244}Pu in whitlockite compared to the ^{238}U track record. Earlier studies have been done on the relative rate of release of fission xenon and annealing of tracks. These results (primarily for St. Severin and Shaw) show cooling rates of 1 °C/10^6 yr.

Cooling rates are useful primarily for establishing model dependent sizes for the hypothesized parent objects. For the irons, the time interval from T_{peak} to the onset of nucleation of the two phase system (kamacite-taenite) cannot be obtained because of the uncertain time interval which the Fe/Ni metal spends in the joint $\alpha - \gamma$ field as metakamacite (martensite). Even the possibility of multiple exposures to heating at elevated temperature can take place without a subsequent fossil record. Though such events are entirely possible for the inductions scenario, it is unlikely that this would take place for nuclide heating, since we know of no way to repeat a heating cycle without the injection of new radioisotopes.

These comments show that the determination of cooling rates of meteorites and their parents cannot establish the chronology of the parent bodies (or of the solar system more generally) through the heating cycle. Sufficiently low temperatures need to be observed such that the metallographic history in the case of the irons and the track history in the case of the chondrites, can be established.

An interesting result of the chondrite cooling rate, which establishes the parent object radius as greater than about 100 km, is that the $^{26}Al/^{27}Al$ ratio of 6×10^{-5} (Lee et $al.$ 1976) reported for Allende is almost certainly too high for the chondrites; they would then have been completely melted and this is not the case. Pellas and Storzer (1977) suggest an initial ratio less than 10^{-5}.

X. THE SCALING PROBLEM

There exists evidence for an early melting of at least the outer parts of the moon (magma ocean) (e.g. Lugmair et $al.$ 1975) and less certain evidence for early differentiation of the planet Mercury (Murray et $al.$ 1974, 1975). Further, there seems to be no reason for excluding the possibility that the other terrestrial planets were also subject to ancient thermal activity, though for these the evidence would likely have been lost long ago because of obliteration of the record by erosion of the geological features. Nevertheless, with the evidence from the moon and from Mercury it appears likely that the metamorphism of the meteorite parent objects was synchronized with an extended period of thermal activity within at least the inner solar system at a very early period of time. These statements are based partially upon the fact that the whole solar system was evolving at the same time as the meteorite parent objects. Comparisons of the evolution of all bodies within a consistent framework could provide crucial information on the origin and evolution of each individual object. In considering all the solar system planetary and asteroidal-like objects, it is apparent that not all hypothesized heat sources could be efficient for all objects. In particular, accretional heating will not provide a significant heat input to bodies less with than a 500–1000 km radius. (See Wetherill, 1976 for a new mechanism of heating by impact).

We cannot yet hope to uncover a scenario which can account for primordial heating of all these bodies because, in addition to weathering on planets with atmospheres, long-term heating from long-lived radionuclides will also mask the very early history, especially when melting and convection are taken into account. But in at least some of the bodies of the solar system a record has been maintained; the moon, Mercury and the meteorites, and possibly the asteroids being the key examples where an early "fossilized" record is preserved. Mittlefehldt (1979) has shown evidence of multiple episodes of early heating of meteorite parent bodies. The data consist of fractionation patterns for the rare earths which are most easily explained as originating in successively depleted magma sources. If his interpretation is correct, then, at least for the objects sampled by the meteorites, episodic heating is most easily accounted for by a process such as an intermittant T Tauri inductive phase rather than by radionuclides whose heating is most likely monotonically diminished with time. T Tauri stars are known to be highly variable.

Regarding tidal heating it is our opinion that such heating would likely not be important even if present. Its existence would have to result from spin re-alignment after a collision in which the body was rotating about other than the axis of maximum moment of inertia, or alternatively from tidal interaction in a binary system. Neither seems energetically plausible.

Acknowledgments. We thank our colleagues for an intermittently continuous dialogue without which the assembly of the material of the text would have been prohibitive. We also thank G. Garcia for unending patience in manuscript preparation and F. Jakopin for bibliographical work. One of us (CPS) was supported by a grant from the planetary geophysics and geochemistry program of the National Aeronautics and Space Administration.

REFERENCES

Alfvén, H., and Arrhenius, G. 1976. *Evolution of the Solar System,* (NASA SP-345, Washington, D. C.: U. S. Government Printing Office).

Allan, D. W., and Jacobs, J. A. 1956. The melting of asteroids and the origin of meteorites. *Geochim. Cosmochim. Acta* 9: 256-272.

Anders, E. 1971a. Conditions in the early solar system. In *Nobel Symposium 21, From Plasma to Planet,* ed. A. Elvius, (New York: Wiley), pp. 123-156.

Anders, E. 1971b. Meteorites and the early solar system. *Ann. Rev. Astron. Astrophys.* 9: 1-34.

Anders, E.; Higuchi, H.; Gros, J.; Takahashi, H.; and Morgan, J. W. 1975. Extinct superheavy element in the Allende meteorite. *Science* 190: 1262-1271.

Bainbridge, J. 1962. Gas imperfections and physical conditions in gaseous spheres of lunar mass. *Astrophys. J.* 136: 202-210.

Bloch, M. R., and Müller, O. 1971. An alternative model for the formation of iron meteorites. *Earth Planet. Sci. Lett.* 12: 134-136.

Bosch, F.; El-Goresy, A.; Krätschmer, W.; Martin, B.; Povh, B.; Nobiling, R.; Traxel, K.; and Schwalm, D. 1976. Comment on the reported evidence for primordial superheavy elements. *Phys. Rev. Lett.* 37: 1515-1517.

Brecher, A.; Briggs, P. L.; and Simmons, G. 1975. The low temperature electrical properties of carbonaceous chondrites. *Earth Planet. Sci. Lett.* 28: 37-45.

Briggs, P. L. 1976. Solar wind heating of asteroids. Master's thesis, Massachusetts Institute of Technology.

Brown, H. 1947. An experimental method for the estimation of the age of the elements. *Phys. Rev.* 72: 348.

Bunch, T. A., and Olsen, E. 1974. Restudy of pyroxene-pyroxene equilibrium temperatures for ordinary chondrite meteorites. *Contr. Mineral. Petrol.* 43: 83-90.

Cameron, A. G. W. 1966. The accumulation of chondritic matter. *Earth. Planet. Sci. Lett.* 1: 93-96.

Carslaw, H. S., and Jaeger, J. E. 1959. *Conduction of heat in solids*, (London, Oxford University Press).

Cassen, P. M., and Reynolds, R. T. 1974. Convection in the moon: Effect of variable viscosity. *J. Geophys. Res.* 79: 2937-2944.

Clark, S. P. 1966. Thermal conductivity. In *Handbook of Physical Constants*, ed. S. P. Clark (Geol. Soc. Am. Mem. 97), pp. 459-482.

Clayton, R. N.; Grossman, L.; and Mayeda, T. K. 1973. A component of primitive nuclear composition in carbonaceous meteorites. *Science* 182: 485-488.

Colburn, D. S.; Sonett, C. P.; and Schwartz, K. 1972. Unipolar interaction of Mercury with the solar wind: The steady state bow shock problem. *Earth Planet. Sci. Lett.* 14: 325-337.

Condon, E. U. 1958. Heat transfer. In *Handbook of Physics*, eds. E. U. Condon and H. Odishaw (New York: McGraw-Hill), Chapt. 5, pp. 61-72.

Consolmagno, G. J., and Lewis, J. S. 1978. The evolution of icy satellite interiors and surfaces. *Icarus* 34: 280-293.

Drozd, R. J.; Hohenburg, C. M.; and Ragan, D. 1972. Fission Xenon from extinct ^{244}Pu in 14301. *Earth Planet. Sci. Lett.* 15: 338-346.

Ezer, D., and Cameron, A. G. W. 1962. A study of solar evolution. *Canadian J. Phys.* 43: 1497-1517.

Fish, R. A.; Goles, G. G.; and Anders, E. 1960. The record in the meteorites. III. On the development of meteorites in asteroidal bodies. *Astrophys. J.* 132: 243-258.

Fleischer, R. L.; Price, P. B.; and Walker, R. M. 1968. Identification of ^{244}Pu fission tracks and the cooling of the parent body of the Toluca meteorite. *Geochim. Cosmochim. Acta* 34: 21-31.

Flerov, G. N. 1974. Search for superheavy elements. In *Proc. Int'l. Conf. on Reactions between Complex Nuclei*, Vol. 2, eds. R. D. Robinson, F. K. McGowan, and J. B. Ball (New York: Elsevier), pp. 459-481.

Flerov, G. N.; Ter-Akop'yan, G. M.; Popeko, A. G.; Fefilov, B. V.; and Subbotin, V. G. 1977. Observation of a new spontaneously fissile nuclide in certain meteorites. *Sov. J. Nucl. Phys.* 26: 237-239.

Flerov, G. N.; Zholud, T. P.; Otgonsuren, O.; Perelygin, V. P.; and Wiik, H. B. 1976. On search for tracks of heavy and superheavy cosmic-ray nuclei in crystals from Pallasites. *Geochim. Cosmochim. Acta* 40: 305-307.

Fowler, W. A.; Greenstein, J. L.; and Hoyle, F. 1961. Deuteronomy. Synthesis of Deuterons and the light nuclei during the early history of the solar system. *Amer. J. Phys.* 29: 393-403.

Fowler, W. A.; Greenstein, J. L.; and Houle, F. 1962. Nucleosynthesis during the early history of the solar system. *Geophys. J. Roy. Astron. Soc.* 6: 148-219.

Fricker, P. E.; Goldstein, J. I.; and Summers, A. L. 1970. Cooling rates and thermal histories of iron and stony-iron meteorites. *Geochim. Cosmochim. Acta* 34: 475-491.

Gentry, R. V.; Cahill, T. A.; Fletcher, N. R.; Kaufmann, H. C.; Medsker, L. R.; Nelson, J. W.; and Flocchini, R. G. 1976. Evidence for primordial superheavy elements. *Phys. Rev. Lett.* 37: 11-15.

Goldstein, J. I., and Ogilvie, R. E. 1965. The growth of the Widmanstätten pattern in metallic meteorites. *Geochim. Cosmochim. Acta* 29: 893-920.

Goldstein, J. I., and Short, J. M. 1967. Cooling rates of 27 iron and stony-iron meteorites. *Geochim. Cosmochim. Acta* 31: 1001-1023.

Grossman, L.; and Larimer, J. W. 1974. Early chemical history of the solar system. *Rev. Geophys. Spa. Phys.* 12: 71-101.

Hayashi, C. 1966. Evolution of protostars. *Ann. Rev. Astron. Astrophys.* eds. G. Burbidge, D. Layzer and J. Phillips (Palo Alto, Calif.: Annual Reviews, Inc.), Vol. 4, pp. 171-192.

Herbert, F.; Sonett, C. P.; and Wiskerchen, M. J. 1977. Model "zero-age" lunar thermal profiles resulting from electrical induction. *J. Geophys. Res.* 82: 2054-2060.

Herbert, F., and Sonett, C. P. 1978. Primordial metamorphism of asteroids via electrical induction in a T Tauri-like solar wind. *Astrophys. Space Sci.* 55: 227-239.

Housen, K. R.; Wilkening, L. L.; Chapman, C. R.; and Greenberg, R. 1979. Asteroidal regoliths. *Icarus* (in press).

Hoyle, F. K. 1960. On the origin of the solar nebula. *Quart. J. Roy. Astron. Soc.* 1: 28-55.

Kaula, W. M. 1979. Thermal evolution of earth and Moon growing by planetesimal impacts. *J. Geophys. Res.* (in press).

Kieffer, S. W. 1975. Droplet chondrules. *Science* 189: 333-340.

Kopal, Z. 1969. Thermal history of the moon. In *The Moon,* ed. Z. Kopal (Dordrecht: D. Reidel), pp. 91-109.

Kraft, R. 1972. Evidence for changes in angular velocity of the surface regions of the sun and stars. In *Solar Wind,* eds. C. P. Sonett, P. J. Coleman, and J. M. Wilcox (NASA SP-308, Washington, D. C.: U. S. Government Printing Office), p. 276.

Kuiper, G. P. 1954. On the origin of the lunar surface features. *Proc. Nat. Acad. Sci.* 40: 1097-1112.

Kuroda, P. K.; Rome, M. W.; Clark, R. S.; and Ganapathy, R. 1966. Galactic and solar nucleosynthesis. *Nature* 212: 241-243.

Lederer, C. M.; Hollander, J. M.; and Perlman, I. eds. 1967. *Table of Isotopes,* (New York: Wiley).

Lee, T.; Papanastassiou, D. A.; and Wasserburg, G. J. 1976. Demonstration of ^{26}Mg excess in Allende and evidence for ^{26}Al. *Geophys. Res. Lett.* 3: 109-112.

Lewis, R. S.; Scrinivasan, B,; and Anders, E. 1975. Host phase of a strange Xenon component in Allende. *Science* 190: 1251-1262.

Lovering, J. F.; Parry, L. G.; and Jaeger, J. C. 1960. Temperature and mass loss in iron meteorites during albation in the Earth's atmosphere. *Geochim. Cosmochim. Acta* 19: 156-167.

Lugmair, G. W.; Scheimin, N. B.; and Marti, K. 1975. Sm-Nd age and history of Apollo 17 basalt 75075; Evidence for early differentiation of the lunar exterior. *Lunar Science VI.* The Lunar and Planet. Inst., pp. 1419-1429.

Mittlefehldt, D. W. 1979. The nature of asteroidal differentiation processes: Implications for primordial heat sources. *Proc. Lunar Sci. Conf. X* (Oxford: Pergamon Press). In press.

Murray, B. C.; Belton, M. J. S.; Danielson, G. E.; Davies, M. E.; Gault, D. E.; Hapke, B.; O'Leary, B.; Strom, R. G.; Suomi, V.; and Trask, N. 1974. Mercury's surface: Preliminary description and interpretation from Mariner 10 pictures. *Science* 185: 169-179.

Murray, B. C.; Strom, R. G.; Trask, N. J.; and Gault, D. E. 1975. Surface history of Mercury: Implications for terrestrial planets. *J. Geophys. Res.* 80: 2508-2514.

Nozette, S., and Boynton, W. V. 1979. A study of rare earth elemental abundances in iron meteorites. Submitted to The Meteoritical Society Heidelberg Meeting; abstract to appear in *Meteoritics.*

Patterson, C. C. 1955. Pb^{207}/Pb^{206} ages of stone meteorites. *Geochim. Cosmochim. Acta* 7: 151-153.

Pellas, P. 1972. Irradiation history of grain aggregates in ordinary chondrites. Possible clues to the advanced stages of accretion. In *From Plasma to Planet,* ed. A. Elvius (Stockholm: Almqvist and Wiksell), pp. 65-90.

Pellas, P., and Storzer, D. 1977. On the early thermal history of chondritic asteroids derived by 244-Plutonium fission track thermometry. In *Comets, Asteroids, Meteorites,* ed. A. H. Delsemme (Toledo, Ohio: University of Toledo Press), pp. 355-363.

Perelygin, V. P.; Stetsenko, S. G.; Pellas, P.; Lhagvasuren, D.; Ogonsuren, O.; and Jakupi, 1977. Long-term averaged abundances of VVH cosmic ray nuclei from studies of olivines from Maryjahahti meteorite. *Nucl. Track Detect.* 1: 199-205.

Pitzer, K. S. 1975. Are elements 112, 114, and 118 relatively inert gases? *J. Chem. Phys.* 63: 1032-1034.

Podosek, F. A. 1972. Gas retention chronology of Petersburg and other meteorites. *Geochim. Cosmochim. Acta* 36: 755-772.

Popeko, A. G.; Sobelev, N. K.; Ter-Akop'yan, G.M.; and Goncharov, G. N. 1974. Search for superheavy elements in meteorites. *Phys. Lett.* 52B: 417-420.

Randrup, J.; Larson, S. E.; Möller, P.; Sobiczewski, A.; and Lukasiak, A. 1974. Theoretical estimates of spontaneous fission half-lives for superheavy elements based on the modified oscillator model. *Phys. Scripta* 10A: 60-64.

Reynolds, J. H. 1960. I-Xe dating of meteorites. *J. Geophys. Res.* 65: 3843-3846.

Reynolds, R. T., and Cassen, P. M. 1979. On the internal structure of the major satellites of the outer planets. *J. Geophys. Res. Lett.* (in press).

Runcorn, S. K. 1978a. An ancient lunar core dyamo. *Science* 199: 771-773.

Runcorn, S. K. 1978b. On the possible existence of superheavy elements in the primeval moon. *Earth Planet. Sci. Lett.* 39: 193-198.

Runcorn, S. K.; Libby, L.; and Libby, W. 1977. Primeval melting of the Moon. *Nature* 270: 676-681.

Safronov, V. S. 1969. *Evolution of the Proto-planetary Cloud and Formation of the Earth and Planets,* (Transl. in English, NASA TTF-677, Washington, D. C.: U. S. Government Printing Office).

Schubert, G.; Turcotte, D. L.; and Oxburgh, E. R. 1969. Stability of planetary interiors. *Geophys. J. Roy. Astron. Soc.* 18: 441-460.

Schumacher, E. 1956. Age determination of stone meteorites by the Rubidiumstrontium method. *Helv. Phys. Acta* 39: 531-538.

Seaborg, G. T.; Loveland, W.; and Morrissey, D. J. 1979. Superheavy elements: A crossroads. *Science* 203: 711-717.

Sonett, C. P. 1971. The relationship of meteorite parent body thermal histories and electromagnetic heating by a pre-main sequence T Tauri Sun. In *Physical Studies of the Minor Planets,* ed. T. Gehrels (NASA SP-267, Washington, D. C.: U. S. Government Printing Office), pp. 239-245.

Sonett, C. P. 1979. On the origin of chondrules. Submitted to *Geophys. Res. Lett.*

Sonett, C. P.; Colburn, D. S.; and Schwartz, K. 1975. Formation of the lunar crust: An electrical source of heat. *Icarus* 24: 231-255.

Sonett, C. P.; Colburn, D. S.; Schwartz, K.; and Kiel, K. 1970. The melting of asteroidal-sized bodies by unipolar dynamo induction from a primordial T Tauri sun. *Astrophys. Space Phys.* 7: 446-488.

Sonett, C. P., and Herbert, F. 1977. Pre-main sequence heating of planetoids. In *Comets, Asteroids, Meteorites,* ed. A. H. Delsemme (Toledo, Ohio: University of Toledo Press), pp. 429-437.

Stoughton, R. W.; Halperin, J.; Drury, J. S.; Perey, F. G.; Macklin, R. L.; Gentry, R. V.; Moore, C. B.; Noakes, J. E.; Milton, R. M.; McCarthy, J. H.; and Sherwood, D. W. 1973. A search for naturally occurring superheavy elements. *Nature Phys. Sci.* 246: 26-28.

Stratton, J. 1941. Electromagnetic theory, (New York: McGraw-Hill).

Strominger, D.; Hollander, J. M.; and Seaborg, G. T.; 1958. Table of isotopes. *Rev. Mod. Phys.* 30: 585.

Urey, H. C. 1955. The cosmic abundances of potassium, uranium, and thorium, and the heat balances of the Earth, Moon, and Mars. *Proc. Nat. Acad. Sci.* 41: 127-144.

Urey, H. C. 1956. Diamonds, meteorites, and the origin of the solar system. *Astrophys. J.* 124: 623-637.

Urey, H. C. 1958. The early history of the solar system. *Proc. Chem. Soc.* pp. 67-78.

Urey, H. C. 1962. The origin of the moon and its relationship to the origin of the solar system. In *The Moon, Proc. Symp. 141 IAU,* (New York: Academic Press), pp. 133-148.

Urey, H. C. 1963. The origin and evolution of the solar system. In *Space Science,* ed. D. P. LeGalley (New York: Wiley), pp. 123-168.

Urey, H. C.; and Donn, B. 1956. Chemical heating for meteorites. *Astrophys. J.* 124: 307-310.

Van Schmus, W. R., and Koffman, D. M. 1967. Equilibration temperatures of iron and magnesium in chondritic meteorites. *Science* 155: 1009-1011.

Van Schmus, W. R., and Wood, J. A. 1967. A chemical-petrologic classification for the chondritic meteorites. *Geochim. Cosmochim. Acta* 31: 747-765.

Wasson, J. T. 1971. An equation for the determination of iron-meteorite cooling rates. *Meteoritics* 6: 139-148.

Wasson, J. T. 1972. Parent body models for the formation of iron meteorites. In *Proc. 24th Int. Geo. Congr.* 15: 161-168.

Wasson, J. T. 1974. *Meteorites* (New York: Springer-Verlag).
Weidenschilling, S. J. 1974. A model for accretion of the terrestrial planets. *Icarus* 22: 426-435.
Weidenschilling, S. J. 1976. Accretion of the terrestrial planets. II. *Icarus* 27: 161-170.
Wetherill, G. 1976. The role of large bodies in the formation of the earth and moon. In *Proc. Lunar Sci. Conf. VII* (Oxford: Pergamon Press), pp. 3245-3257.
Whipple, F. L. 1966. Chondrules: Suggestion concerning the origin. *Science* 153: 54-56.
Wood, J. A. 1964. The cooling rates and parent planets of several iron meteorites. *Icarus* 3: 429-459.
Wood, J. A. 1968. Meteorites and the origin of planets. (New York: McGraw-Hill).
Zvara, I. 1977. Experiments on chemical concentration of a new spontaneously fissile nuclide from material of the Allende meteorite. *Sov. J. Nucl. Phys.* 26: 240-243.

REVIEW OF THE METALLOGRAPHIC COOLING RATES OF METEORITES AND A NEW MODEL FOR THE PLANETESIMALS IN WHICH THEY FORMED

JOHN A. WOOD

Harvard-Smithsonian Center for Astrophysics

The cooling rates of meteorites through $\sim900°\text{-}650°K$, as read from their metal alloy compositions, are reviewed. Metallographic cooling rates are compared with the cooling rates that appear to be required by the K/Ar and $^{40}Ar/^{39}Ar$ ages of five meteorite classes, and discrepancies are found in all cases. Either (1) the metallographic cooling rates (and also ^{244}Pu fission cooling rates) are systematically in error, being too slow by a factor of ~6; or (2) the traditional thermal model for parent meteorite planets (having constant dimension and uniform physical properties) is oversimplified and the Ar closure temperatures for chondrites derived by Turner et al. are too low.

An alternative parent planet model is proposed and numerically modeled, in which accretion of thermally insulating particulate matter, heat generation by ^{26}Al decay, melting or sintering of the particulate matter into conductive rock, and establishment of the properties of the meteorites occurred concurrently. Meteorite chronologies are somewhat easier to understand in this context, since the initially small, hot (thus sintered and conductive) bodies would have cooled rapidly to isotopic closure, but later cooling might have been much slower as a result of continued accretion of insulating particulate matter.

Most of the meteorites we have access to have been thermally processed. In some cases the systems they were once part of were actually melted, whereupon phase separation produced zones of pure metal liquid, pure silicate melt, and more complex mixtures of liquid metal and solid silicate minerals; solidification of such zones produced the irons, achondrites, and many of the stony-iron meteorites. Other meteorites, especially the chondrites, have been heated severely enough to promote textural recrystallization and internal chemical equilibration, but not enough to melt. These episodes of heating are believed to have occurred in planetesimals of asteroidal dimension in the early solar system.

The planetesimals must have been small, because they cooled rapidly enough to allow their substance (now the meteorites) to begin accumulating radiogenic gases, in some cases including the decay products of short-lived radionuclides, soon after the solar system formed. The heat source in the planetesimals is not known. Solar system levels of U, Th, and K are not high enough to produce major amounts of radiogenic heat in small planetesimals, and the amount of accretional energy deposited during the assembly of asteroidal-sized bodies would also be trivial. Two potential heat sources have been contemplated: radiogenic heating from the decay of short-lived ^{26}Al ($t_{1/2} \sim 0.7 \times 10^6$ yr; Urey 1955; Fish et al. 1960), and electrical heating by dynamo induction from a pre-main-sequence T Tauri "solar wind" (Sonett et al. 1968; see the chapter by Sonett and Reynolds).

The discovery of Al-correlated ^{26}Mg, the product of ^{26}Al decay, in inclusions of the Allende chondrite (Lee et al. 1976), appears to have enhanced the plausibility of the first of these mechanisms. The level of ^{26}Al originally present in Allende inclusions (^{26}Al/^{27}Al $\sim 6 \times 10^{-5}$; Lee et al. 1976) would have been sufficient to cause melting in rocky bodies larger than ~ 7 km radius (Lee et al. 1977; Herndon and Herndon 1977), though there is no assurance that all early solar system material contained Al with this percentage of ^{26}Al.

Electrical inductive heating of early planetary material remains a possibility that is difficult to test or constrain. The gaseous protoplanet concept of Cameron (1978; see his chapter in this book) also offers a heating mechanism; the hypothetical protoplanets are heated by self-gravitational compression, which would make condensed matter that accumulated in them correspondingly hot. There is also a real possibility that none of the mechanisms named above was responsible for the heating of objects in the early solar system.

I. METALLOGRAPHIC COOLING RATES

Most thermally processed meteorites contain metallic nickel-iron with a 6.5-20 % Ni content. Metal in this compositional range which has cooled slowly from high temperatures has imprinted in it a record of the rate at which it cooled through the temperature range $\sim 900°$–$\sim 650°$K. Under equilibrium conditions two alloy phases (α, or kamacite; and γ, or taenite) are stable in this temperature range, but the compositions and proportions of the phases vary with temperature. As temperature declines the amount of taenite in an equilibrium system diminishes, but its Ni content increases. Nickel has to be moved into cooling taenite crystals via solid-state diffusion to accomplish this. Diffusion coefficients decrease exponentially with temperature, so a point is reached during cooling of real systems when the interiors of taenite crystals can no longer be supplied with the Ni that equilibrium dictates. A Ni diffusion gradient is frozen into each taenite crystal, the character of which is an expression of the cooling rate: the slower the system cooled, the lower the temperature at which solid-state diffusion ceased to be able to move Ni, and so the higher the Ni content of taenite interiors of a given size were when diffusion did cease. Given a knowledge of the Fe-Ni phase diagram and the temperature dependency of the diffusion coefficients, the process can be computer-modeled. The reader is referred to Wood (1964) and Goldstein and Ogilvie (1965) for more comprehensive discussions of the metallurgical evolution involved and the estimation of cooling rates.

A number of workers have determined metallographic cooling rates for meteorites; their results are assembled in Table I and summarized in Fig. 1. ("Anomalous" iron meteorites, those not included by Scott and Wasson [1975] in discrete chemical subgroups, have been omitted.) Several different short-cuts to establishing the relationship between alloy compositions and cooling rates have been employed. The "technique" column of Table I indicates which approach was used, and supplies a literature reference that explains it more fully. Techniques V through Z are arranged in order of decreasing reliability, according to my opinion. The most recently developed (and highly ranked) techniques allow explicitly for the effect of phosphorus on the system. Unfortunately the efforts of Willis and Wasson (1978a) and Moren and Goldstein (1978) to introduce this refinement have led to divergent results, i.e. the former find a narrow range of cooling rates for the Group IVA irons and the latter a broad range. It is difficult for an outsider to evaluate the conflicting arguments applied by these workers in a complex and still somewhat uncertain field. I have ducked the issue by including the cooling rates of both groups (Vw and Vg). In all other cases where cooling rates have been determined by several workers, only the value I consider most reliable is reported.

Where the cooling rates of meteorites have been estimated by other techniques, they have tended to confirm the metallographic cooling rates. Kulpecz and Hewins (1978) computer-modeled growth of schreibersite crystals in metal systems and found that cooling rates of $0.01 - 0.1°K/10^6$ yr would be required to produce the schreibersite in the Emery mesosiderite, as compared to $0.1°K/10^6$ yr in Table I. (However, schreibersite growth, like the final Ni content of taenite crystals, is diffusion controlled, so these two cooling-rate estimates are not completely independent.) Crozaz (1979) concluded from the absence of ^{244}Pu tracks in whitlockite of the Estherville mesosiderite, and the low density of U-fission tracks relative to the amount of U present, that this meteorite must have cooled unusually slowly, consistent

TABLE I

Survey of Metallographic Cooling Rates of Meteorites

IRONS

	Cooling rate, °K/10⁶ yr	Ref- erence	Tech- nique		Cooling rate, °K/10⁶ yr	Ref- erence	Tech- nique
Group IA				**Group IB**			
Ashfork	2	A	Z	Colfax	1	A	X
Balfour Downs	1.5	A	Z	Four Corners	1.9	D	Z
Bischtübe	2	A	Z	Persimmon Creek	5	E	Z
Bogou	3	A	Z	Pitts	0.8	A	Z
Bohumilitz	2	A	Z	Woodbine	2.7	D	Z
Cañon Diablo	3	B	W	**Group IC**			
Cranbourne	3	A	Z	Arispe	8	C	Y
Deport	1	A	Z	Bendego	~9	F	Z
Leeds	2	A	Z	Mt. Dooling	~6	F	Z
Lexington County	2	A	Z	Santa Rosa	~1000	F	Z
Mazipil	2.5	A	Z	St. Francois County	~2	F	Z
Odessa	4	C	X	Union County	~6	F	Z
Ogallala	1	A	Z				
Rifle	2	A	Z	**Group IIC**			
Toluca	1.6	D	W	Fallinoo	200	A	Z
Yenberrie	3	A	Z	Eumerina	100	A	Z
Youndegin	3	B	W	Ferryville	250	A	Z
				Salt River	200	A	Z
				Wiley	15	D	W

TABLE I (Continued)

Survey of Metallographic Cooling Rates of Meteorites

	Cooling rate, °K/10^6 yr	Reference	Technique		Cooling rate, °K/10^6 yr	Reference	Technique
Group IID				**Group IIIA (continued)**			
Carbo	1	D	W	Canyon City	5	A	Z
Elbogen	0.8	A	Z	Cape York	3	A	Z
N'Kandhla	1	A	Z	Carthage	2	A	Z
Rodeo	2.5	A	Z	Casimiro de Abreu	1.5	A	Z
Wallapai	1.1	D	W	Casas Grandes	4	A	Z
				Characs	4	A	Z
Group IIE				Chilkoot	4	A	Z
Arlington	8	A	Z	Costilla Peak	4	A	Z
Kodaikanal	100	A	Z	Cumpas	2.2	A	Z
				Descubridora	4	A	Z
Group IIIA				Drum Mtns.	2	A	Z
Aggie Creek	1.5	A	Z	Franceville	3	A	Z
Apoala (pseudo)	4	A	Z	Gundaring	4	A	Z
Bagdad	5	A	Z	Harriman (Om)	5	A	Z
Bartlett	1.5	A	Z	Henbury	8	A	Z
Basedow Range	7	A	Z	Kenton County	8	A	Z
Billings	3	A	Z	Kyancutta	2	A	Z
Boxhole	8	A	Z	Lénártó	1	A	Z
Cacaria	6	A	Z	Livingston (Montana)	8	A	Z
Canton	5	A	Z	Loreto	3.5	A	Z

TABLE I (Continued)

Survey of Metallographic Cooling Rates of Meteorites

	Cooling rate, $^\circ K/10^6$ yr	Reference	Technique		Cooling rate. $^\circ K/10^6$ yr	Reference	Technique
Group IIIA (continued)				Group IIIA (continued)			
Madoc	10	A	Z	Tamarugal	2	A	Z
Mereditas	4.5	A	Z	Thunda	2	A	Z
Milly Milly	3.5	A	Z	Trenton	2.5	A	Z
Morito	7	A	Z	Willamette	6	A	Z
Nejed	70	A	Z	Williamston	10	A	Z
Norfolk	6	A	Z				
Norfork	5	A	Z	Group IIIB			
Plymouth	2	A	Z	Acargas	1.5	A	Z
Point of Rocks (iron)	2	A	Z	Baquedano	2	A	Z
Providence	2	A	Z	Bear Creek	1.2	A	Z
Puente del Zacate	3	A	Z	Bella Roca	1.5	A	Z
Roebourne	2.5	A	Z	Breece	1.2	A	Z
Ruff's Mtn.	1.5	A	Z	Campbellsville	1	A	Z
Sacramento Mtns.	5	A	Z	Chupaderos	1.2	A	Z
Samelia	3	A	Z	Cleveland	1.5	A	Z
San Angelo	7.5	A	Z	Cuernavaca	1.3	A	Z
Santa Apolonia	40	C	X	El Capitan	1.8	A	Z
Spearman	4	D	W	Grant	5.1	D	W
Susuman	5	A	Z	Joe Wright Mtn.	1.5	A	Z

TABLE I (Continued)

Survey of Metallographic Cooling Rates of Meteorites

	Cooling rate, °K/10⁶ yr	Reference	Technique		Cooling rate, °K/10⁶ yr	Reference	Technique
Group IIIB (continued)				**Group IIID**			
Knowles	1.5	A	Z	Dayton	6.5	D	W
Mt. Edith	2	A	Z	Föllinge	2	G	Z
Orange River (iron)	1	A	Z	Tazewell	2.2	D	W
Owens Valley	2	A	Z				
Roper River	1.2	A	Z	**Group IIIE**			
Sams Valley	1	A	Z	Coopertown	1.2	A	Z
Sanderson	1.2	A	Z	Staunton	1	A	Z
Tambo Quemado	1.2	A	Z	Willow Creek	1	A	Z
Tieraco Creek	1.5	A	Z				
Verkhne Dnieprovsk	1.2	A	Z	**Group IIIF**			
Wonyulgunna	2	A	Z	Abancay	15	A	Z
				Clark County	8	A	Z
Group IIIC				Moonbi	12	A	Z
Anoka	4	C	X	St. Genevieve County	20	A	Z
Carlton	4.2	D	W				
Edmonton (Kentucky)	1.9	D	W	**Group IVA**			
Havana	5	G	Z	Altonah	20	D	W
Mungindi	4	G	Z	Bishop Canyon	65	A	Z
				Bodaibo	60	A	Z

TABLE I (Continued)

Survey of Metallographic Cooling Rates of Meteorites

	Cooling rate, °K/10⁶ yr	Reference	Technique
Group IVA (continued)			
Bristol	20	D	W
Boogaldi	10	A	Z
Charlotte	25	H	Vg
Chinautla	6	H	Vg
Duchesne	4	H	Vg
Gibeon	35	H	Vg
	28	I	Vw
Harriman (Of)	? 14	J	Vw
	20	H	Vg
Hill City	25	J	Vw
	6	H	Vg
Huizopa	40	D	W
Jamestown	90	A	Z
La Grange	? 200	H	Vg
Mantos Blancos	15	J	Vw
	4	H	Vg
Maria Elena (1935)	70	A	Z
Mart	7	A	Z
Muonionalusta	25	H	Vg
New Westville	25	J	Vw

	Cooling rate, °K/10⁶ yr	Reference	Technique
Group IVA (continued)			
New Westville	3	H	Vg
Obernkirchen	100	A	Z
Otchinjau	50	A	Z
Pará de Minas	25	J	Vw
	25	H	Vg
Putnam County	60	A	Z
Seneca Township	12	A	Z
Signal Mtn.	65	H	Vg
Social Circle	70	A	Z
Western Arkansas	65	A	Z
Wood's Mtn.	30	A	Z
Yanhuitlan	80	A	Z
Group IVB			
Cape of Good Hope	7	A	Z
Hoba	9	A	Z
Iquique	80	A	Z
Klondike	20	A	Z
Kokomo	2	A	Z

TABLE I (Continued)

Survey of Metallographic Cooling Rates of Meteorites

	Cooling rate, °K/10⁶ yr	Reference	Technique		Cooling rate, °K/10⁶ yr	Reference	Technique
Group IVB (continued)				**Pallasites (continued)**			
Tawallah Valley	20	A	Z	Glorietta Mtn.	0.5	D	W
Tlacotepec	8	A	Z	Huckitta	0.8	K	Y
Weaver Mtns.	20	A	Z	Ilimaes	0.8	K	Y
				Imilac	0.5	D	W
STONY IRONS				Itzawisis	0.5	K	Y
Pallasites				Krasnojarsk	0.8	K	Y
Admire	1	K	Y	Lipovsky	1	K	Y
Ahumada	1	K	Y	Marburg	0.8	K	Y
Albin	0.5	D	W	Marjalahti	0.5	K	Y
Anderson	0.5	K	Y	Mt. Vernon	0.8	D	W
Antofagasta	1.2	K	Y	Newport	1	D	W
Argonia	1	K	Y	Ollague	0.8	K	Y
Brahin	0.8	K	Y	Pojoaque	0.5	K	Y
Brenham	0.5	D	W	Rawlinna	1	K	Y
Calderilla	0.8	K	Y	Salta	0.5	K	Y
Eagle Station	2	K	Y	Santa Rosalia	0.5	K	Y
Esquel	1	K	Y	Somervell County	0.8	K	Y
Finmarken	0.8	K	Y	South Bend	0.8	K	Y
Giroux	1.3	D	W	Springwater	0.4	D	W

TABLE I (Continued)

Survey of Metallographic Cooling Rates of Meteorites

	Cooling rate, °K/10⁶ yr	Reference	Technique
Pallasites (continued)			
Thiel Mtns.	1.5	K	Y
Mesosiderites			
Crab Orchard	0.15	L	X
Emery	0.1	R	X
Estherville	0.1	L	X
Hainholz	0.1	L	X
Lowicz	0.1	L	X
Mincy	0.1	L	X
Morristown	0.1	L	X
Patwar	0.1	L	X
Udei Station	10	L	X
Vaca Muerta	0.1	L	X
Veramin	0.1	L	X
Siderophyre			
Steinbach	6.5	D	W

	Cooling rate, °K/10⁶ yr	Reference	Technique
Lodranite			
Lodran	~ 20	M	Y
STONES			
Ordinary chondrites			
Tieschitz, H3	1	O	X
Bath, H4	25	Q	X
Serra, H4	10	P	X
Welman, H4 or H5	3	Q	X
Cee Vee, H5	15	Q	X
Ehole, H5	2	O	X
Forest City, H5	10	O	X
Leighton, H5	100	Q	X
Malotas, H5	4	Q	X
Pantar, H5	5	O	X
Sutton, H5	2	Q	X
Mt. Browne, H6	10	Q	X
Salles, H6	5	O	X
Mezö-Madaras, L3	1	O	X
Bjurböle, L4	1	Q	X

TABLE I (Continued)

Survey of Metallographic Cooling Rates of Meteorites

	Cooling rate, °K/10⁶ yr	Reference	Technique
Ordinary chondrites (continued)			
Adelie Land, L5	1	Q	X
Elenovka, L5,6	2	Q	X
Bruderheim, L6	6	N	X
Kandahar, L6	3	Q	X
Mocs, L6	10	O	X
Tillaberi, L6	130	P	X
Waconda, L6	2	Q	X

	Cooling rate, °K/10⁶ yr	Reference	Technique
Chainpur, LL3	0.2	O	X
Olivenza, LL5	4	S	X
Ensisheim, LL6	1	S	X
Lake Labyrinth, LL6	50	S	X
St. Severin, LL6	1	S	X
Carbonaceous chondrites			
Felix, C3(0)	0.3	O	X
Lancé, C3(0)	1	O	X

REFERENCES

A Goldstein (1969).
B Short and Anderson (1965).
C Wood (1964, 1967a).
D Goldstein and Short (1967).
E Wasson (1970).
F Scott (1977).
G Wasson and Schaudy (1971).
H Moren and Goldstein (1978).
I Willis and Wasson (1978b).
J Willis and Wasson (1978a).
K Buseck and Goldstein (1969).
L Powell (1969).
M Bild and Wasson (1976).
N Wood (1967b).
O Wood (1967a).
P J.M. Malezieux, quoted by Pellas and Storzer (1977).
Q Taylor and Heymann (1971).
R Nehru et al. (1978).
S Taylor (1976).

TECHNIQUES

Vg As X, but using Fe-Ni-P phase diagram and diffusion coefficients; Moren and Goldstein (1978).

Vw As X, but using Fe-Ni-P phase diagram and diffusion coefficients; Willis and Wasson (1978a).

W Detailed fitting of computer-modeled Ni diffusion profile to observed profile in taenite; Fe-Ni system; Goldstein and Short (1967).

X Central Ni content of taenite vs. taenite half-width; Fe-Ni system; Wood (1964, 1967a).

Y Determination of γ composition immediately adjacent to α interface; Short and Goldstein (1967).

Z Kamacite band width vs. bulk Ni content of metal phase; Short and Goldstein (1967).

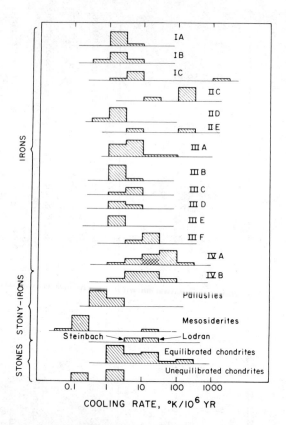

Fig. 1. Histogram of metallographic cooling rates of meteorites, from Table I. The vertical scale varies among meteorite classes, to even out large height differences between classes. In the Group IVA iron meteorite histogram, hatching with a positive slope represents the cooling rates of Willis and Wasson (1978*a*,*b*); hatching with negative slope comprises the cooling rates of Moren and Goldstein (1978) and others.

with the remarkably slow metallographic cooling rates of the mesosiderites generally. Pellas and Storzer (1977) object that cooling rates of ordinary chondrites deduced from fission-track retention are generally lower than metallographic cooling rates, but for the most part the discrepancies are small and can be interpreted as due to a positive second derivative in the chondrite cooling histories, since most of the track data are applicable to a lower temperature range (600–300°K) than that in which diffusion ceased in taenite.

The cooling rates for irons and pallasites appear to have met with wide acceptance, at least as approximate values. Some meteoriticists have reservations about the chondrite and mesosiderite cooling rates, partly because the alloy phases are dispersed in these meteorites, not in physical contact with one another, and partly because the mesosiderite cooling rates are so slow as to be difficult to understand in the traditional parent meteorite planet context.

II. COOLING IN ROCKY PARENT METEORITE PLANETS

Presumably the cooling rates of Table I reflect the time constants for diffusive heat loss from the parent meteorite planets, which can be crudely estimated. Assume the meteorites cooled through roughly $1000°K$ from their peak temperatures, and that their parent planets were made of rocky material of thermal conductivity 6.6×10^5 J/cm yr $°K$, heat capacity 1.2 J/g $°K$, and mass density 3.6 g/cm^3 (therefore having a thermal diffusivity, $K/\rho C_p$, of 1.5×10^5 cm^2/yr. These values are appropriate for ordinary chondrite material. Then cooling rates of $1°$ and $10°K/10^6$ yr, representative of most of the meteorites in Table I, correspond to cooling times of $\sim10^9$ yr and $\sim10^8$ yr respectively, and therefore (multiplying each by the thermal diffusivity) to characteristic dimensions of ~120 and ~40 km. These are, very roughly, the size objects made of ordinary chondrite material would have to be to lose heat at rates corresponding to $1°$ and $10°K/10^6$ yr. Fortunately they correspond to the dimensions of asteroids, the only reasonable source for meteorites most of us can conceive of. More exact studies of the thermal evolution of rocky asteroids have been made by numerically integrating the heat flow equation (Wood 1964, 1967a; Goldstein and Short 1967; Fricker *et al.* 1970). The relationship between total size of a rocky parent body, position in it, and the cooling rate through $770°K$ is shown in Fig. 2 (Wood 1967a). It is worth noting that arbitrarily low cooling rates cannot be obtained in planetary interiors.

III. METEORITE COOLING RATES VERSUS RADIOMETRIC AGES

The small computation made in the last section reminds us that the meteorite cooling rates correspond to quite substantial cooling times relative to the age of the solar system, which means there should be an observable correlation between the radiometric ages of the meteorites and their cooling rates. All other things being equal, the slower the cooling rate the more time must have elapsed before a meteorite became cool enough to immobilize isotopic exchanges or noble gas losses, so the younger the meteorite should appear. In practice it is hard to draw the comparison, primarily because so much uncertainty attaches to the isotopic closure temperatures (those temperatures beneath which radiogenic daughters cease to be outgassed from or exchanged between various host minerals) and secondarily because assumptions must be made about the high-temperature, pre-kamacite-plus-taenite phase of thermal evolution of meteorite systems. When the attempt is made, unfortunately, substantial inconsistencies are found between the radiometric and metallographic time scales for most meteorite classes.

All ages in the discussion to follow are relative to the time scale defined

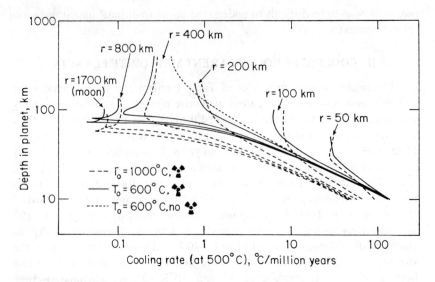

Fig. 2. Cooling rates through 770°K (500°C) at a range of depths, in rocky parent planets of 50-1700 km overall radius (Wood 1967a). Two uniform initial temperatures, and the presence and absence of chondritic levels of long-lived radioactivities, are assumed. Several curves reverse slope at minimum cooling rates of ~0.1°K/10⁶ yr. In the lower branches of these curves (at shallow depths), cooling is monotonic from the assumed initial temperature; in the upper branches the planetary material first heats up from radioactive decay, then at a much later time cools as shown.

by the revised ^{87}Rb and ^{40}K decay constants recommended by Steiger and Jäger (1977). In this time scale the age of the solar system is 4.52, not 4.6×10^9 yr.

Equilibrated Ordinary Chondrites

Turner et al. (1978) find a mean ^{40}Ar/^{39}Ar age for ordinary chondrites of $4.44 \pm 0.03 \times 10^9$ yr, meaning that Ar closure occurred $50\text{-}110 \times 10^6$ yr after the nominal origin of the meteorites. Minster and Allègre (1979) find a whole-rock Rb/Sr isochron for H-group ordinary chondrites of $4.52 \pm 0.05 \times 10^9$, signifying closure to Sr $<\sim 50 \times 10^6$ yr after meteorite formation.

Onuma et al. (1972) find temperatures of oxygen isotope equilibration of $1220° \pm 100°$K for types 5 and 6 ordinary chondrites, and ~960°K for Bjurböle (Type 4). These are lower limits on the temperatures of metamorphism of these meteorites. Turner et al. deduce Ar closure temperatures of ~500°K for ordinary chondrites. The Sr closure temperature is not known, but a value of ~1000°K, based on experimental measurements of the diffusion coefficient of Sr in diopside (Sneeringer and Hart 1978) is

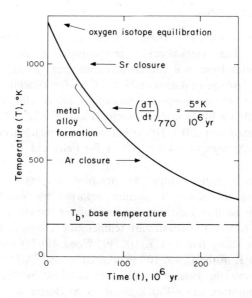

Fig. 3. Cooling of a system from metamorphic temperatures toward a base temperature of 170°K, according to Eq. (1). The example shown is applicable to many of the equilibrated chondrites.

probably not totally unrealistic.

Metallographic cooling rates can be projected back in order to estimate cooling times by integrating the simple relationship

$$dT/dt = k(T - T_b) \tag{1}$$

where T = temperature; T_b = the temperature toward which the system cools, taken here to be 170°K, a representative surface temperature in the asteroid belt; t = time; and k is a constant, fixed by any given metallographic cooling rate and the T at which it applies, $\sim$770°K. If we assume the ordinary chondrite systems began cooling at 1220°K and cooled through 770°K at 1°-5°K/10^6 yr (Fig. 3), the cooling times to Sr and Ar closure are found from Eq. (1) to be $\sim$150 and $\sim$700 $\times$ 10^6 yr for 1°K/10^6 yr cooling; $\sim$30 and $\sim$140 $\times$ 10^6 yr for 5°K/10^6 yr cooling. Thus the faster-cooling chondrites reach isotopic closure temperatures in times that are consistent with the radiometric ages of chondrites, but there is a clear discrepancy between these ages and 1°K/10^6 yr cooling. (Unfortunately there is practically no overlap between the list of chondrites in Table I and those dated by Turner *et al.*, making comparisons for individual chondrites impossible.) The effect of possible errors in metallographic cooling rates will be considered separately.

Unequilibrated Chondrites

Though unequilibrated, these meteorites have experienced mild metamorphism. The metamorphic temperature is not known, and undoubtedly varies from one chondrite to another. Turner *et al.* (1978) obtain a $^{40}Ar/^{39}Ar$ age of 4.45 ± 0.05 × 10^9 yr for Tieschitz, equivalent to 20-120 × 10^6 yr between meteorite formation and Ar closure. Minster and Allègre (1979) find a Rb/Sr internal isochron age of 4.53 ± 0.06 × 10^9 yr for the same meteorite (<50 × 10^6 yr before Sr closure). Mazor *et al.* (1970) find younger K/Ar ages of ~4.0 × 10^9 yr for Felix and ~3.7 × 10^9 yr for Lancé.

The highest metamorphic temperature recorded in Tieschitz, corresponding to the lowest-Ni taenite reported by Wood (1967a), is ~730°K. It may be that Tieschitz was never above the closure temperature for Sr. If 730°K was the maximum temperature attained, the estimated metallographic cooling rate (~1°K/10^6 yr; Wood 1967a) would carry the system to ~610°K in the 120 × 10^9 yr allowed by its $^{40}Ar/^{39}Ar$ age. This is still ~100°K above the closure temperature obtained by Turner *et al.* for equilibrated chondrites, and ~300°K above an Ar closure temperature these authors derived specifically for Tieschitz. It is questionable, however, if a closure temperature based on gas release from a bulk sample of Tieschitz has physical meaning that can be used to predict gas release in the parent planet, because K in unequilibrated chondrites is undoubtedly sited in several different phases (including glass) of highly variable dimension. For the same reason, the closure temperature for recrystallized chondrites probably cannot be applied to Tieschitz. Thus it cannot be said with certainty that the radiometric ages and metallographic cooling rates of unequilibrated chondrites are inconsistent.

A qualitatively puzzling fact about these meteorites is the apparent discrepancy between their little-metamorphosed state, which suggests a position near the surfaces of their parent bodies, and their relatively slow cooling rates (0.2°-1°K/10^6 yr, as compared to 1°-10°K/10^6 yr in most equilibrated chondrites), which appear to require a position of deep burial.

Irons

The silicate inclusions in Group I irons have been radiometrically dated. Bogard *et al.* (1968) report K/Ar ages of 4.55 ± 0.1 × 10^9 yr for silicate inclusions in Toluca (1.6°K/10^6 yr) and Four Corners (1.9°K/10^6 yr) respectively; that is, times to Ar closure of <70 × 10^6 yr and <120 × 10^6 yr. Wasserburg and Burnett (1969) found that silicate inclusions in Group IA irons Pine River, Linwood, Toluca, Odessa, and Copiapo define a Sr/Rb isochron age of 4.45 × 10^9 yr, in substantial agreement with the K/Ar ages; however, this age is uncertain by several times 10^8 yr.

Wasson (1970) argues that the Group I irons have not experienced bulk melting. If we assume they formed by metamorphism at $\sim 1270°K$, the metallographic cooling rates indicate that Toluca and Four Corners would have taken ~ 450 and $\sim 380 \times 10^6$ yr, respectively, to cool to Turner et al.'s closure temperature for ordinary chondrites ($\sim 500°K$), much in excess of the time allowed by K/Ar dating. In the time available, the two systems could have cooled to no less than $\sim 1080°$ and $\sim 920°K$, respectively.

However, Turner et al.'s closure temperature for recrystallized chondrites is not strictly applicable, because as these authors show the closure temperature varies with the radius, a, and activation energy for Ar diffusion, E, of the K host minerals, and with the cooling rate, c, of the system, as

$$E/RT_c - 2 \ln T_c = \text{const.} - \ln c - 2 \ln a \qquad (2)$$

where T_c is the closure temperature and R the gas constant.

There is remarkably little information in the literature about the textures and dimensions of K-bearing minerals in the inclusions of Group I irons, and essentially none about the particular inclusions that were radiometrically dated (above references). El Goresy (1965) and Henderson (1965) show photomicrographs of 50-100 μm anorthite grains in the nodules of Toluca; Buchwald (1975) alludes to silicate grains of 30-100 μm diameter in Four Corners. The host K phase in equilibrated chondrites is oligoclase; its dimension is ~ 50 μm in Type 6 chondrites, smaller in lower-grade chondrites (Van Schmus and Wood 1967).

The differences between host mineral grain sizes and cooling rates appear to be negligibly small; a factor of two difference in grain size would lead to a change in closure temperature of only $\sim 10°K$ (assuming $E \sim 70$ kcal/mole, a value found for the equilibrated chondrites with highest closure temperatures by Turner et al. [1978]). The nature of the K host mineral differs in the two cases, being oligoclase (sodic feldspar) in equilibrated chondrites and anorthite (calcic feldspar) in the irons studied. This must lead to differences in the values of E for Ar diffusion. The degree of the difference is not known, but it is very unlikely to be large enough to cause Ar closure at $\sim 1000°K$, as the K/Ar ages of Group I irons (above) appear to require. For this to be the case, E in anorthite would have to be ~ 143 kcal/mole, a value that appears impossibly high and unreasonably different from the value in oligoclase, which is crystallographically similar enough to anorthite to permit unlimited miscibility of the two minerals.

If the Group I irons formed by melting and differentiation, which seems more likely to me than the model of selective accretion and metamorphism proposed by Wasson (1970), the cooling time is longer and the problem even worse. Thus there appears to be a serious discrepency between the radiometric ages and cooling times of Group I irons.

Pallasites

Megrue (1968) found a K/Ar age of $4.22 \pm 0.1 \times 10^9$ yr for Krasnojarsk, Marjalahti, and Springwater. Taking $0.6°K/10^6$ yr to be a characteristic cooling rate for these objects in the metallographic range, they can have cooled from the temperature of solidification ($\sim1770°K$) only as far as $\sim1350°K$ in the $\sim300 \times 10^6$ yr allowed by the K/Ar ages. Even if we assume that the pallasites' small content of K (~2 ppm; Megrue 1968) is dissolved in their very large (~1 cm) olivine crystals. Eq. (2) shows that the activation energy for Ar diffusion in the latter would have to be ~171 kcal/mole to bring about Ar closure at $1350°K$. Though it is unsafe to extrapolate from Turner et al.'s estimated activation energy in a completely different mineral, this seems an improbably high value, and the radiometric and metallographic time scales of pallasites appear to be inconsistent.

An apparent discrepancy in the planetary siting of pallasites noted by Wood (1978), though unrelated to chronology, is that the pallasitic structure could not have survived as a cumulate layer of olivine crystals immersed in molten metal at the core-mantle interface of a parent meteorite planet if the latter was larger than ~10 km radius, because the gravitational acceleration in larger planets would have caused layers overlying the pallasite cumulate to exert forces that squeezed the (somewhat plastic) olivine crystals together into a compact metal-free layer. Yet the slow cooling rates of pallasites appear to require residence in a rocky planet of >200 km radius (Fig. 2).

Mesosiderites

These stony-irons have fairly old radiometric ages, though not as old as the other meteorite classes discussed. Murthy (1978) reports a $4.24 \pm 0.03 \times 10^9$ yr Rb/Sr isochron age for Estherville, and a $^{40}Ar/^{39}Ar$ age of $\sim3.6 \times 10^9$ yr. K/Ar ages obtained by earlier workers tend to agree with this value, within broad error limits: Crab Orchard, 2.5-4.7×10^9 yr, and Mincy, 2.7-3.9×10^9 yr (Begemann et al. 1976); Lowicz, 3.0-3.4×10^9 yr (Kirsten et al. 1963).

Mesosiderites display the slowest metallographic cooling rates of any meteorite class, $\sim0.1°K/10^6$ yr. (Indeed, Cobleigh et al. [1970], noting that the cooling time of mesosiderites appears to exceed the age of the solar system, proposed a program to visit museums and feel mesosiderites for residual warmth.) Most of the mesosiderites appear not to have been melted, but the compositions and homogeneity of orthopyroxene-clinopyroxene exsolution structures in them point to major periods of metamorphism at $>1250°K$ (Nehru et al. 1978). Application of Eq. (1) shows that cooling from $1250°K$ at a rate which would equal $0.1°K/10^6$ yr at $770°K$ would have brought Estherville only to $\sim1100°K$ in the $\sim0.9 \times 10^9$ yr allowed by its $^{40}Ar/^{39}Ar$ age. Though the feldspar grain sizes in mesosiderites are highly nonuniform, a large proportion of the feldspar in the matrix of Estherville is roughly 50 μm in diameter (my observation), comparable to the dimensions of plagioclase in Type 6 chondrites. From Eq. (2) we find that E for Ar diffusion in Estherville anorthite would have to be ~165 kcal/mole to cause Ar closure to occur at $1100°K$. As in the case of the irons and pallasites, this seems an unacceptably high activation energy. It appears impossible to reconcile the cooling rate of Estherville, and presumably mesosiderites in general, with their radiometric ages.

It is also worth noting that the textures and polymict clast population of mesosiderites strongly suggest they were once regolith breccias or impact melt deposits, i.e., emplaced on planetary surfaces; yet their cooling rates, the slowest of all meteorite classes, would appear to require the deepest burial. This is the same paradox noted earlier for unequilibrated chondrites.

In summary, it appears that the cooling times of all meteorite types, with the possible exception of the faster-cooling equilibrated chondrites, are more or less inconsistent with the radiometric ages of the meteorites, in the sense that cooling rates based on metallographic data are too slow for the meteorites to reach the estimated closure temperatures in the times allowed by radiometric ages.

The degree of uncertainty in the metallographic cooling rates has not been discussed. Wood (1967a) estimates that the cooling rates are accurate to within a factor of 2.5; Goldstein and Short (1967) put the error factor at 2.0. Moren and Goldstein (1978) and Willis and Wasson (1978a,b), by differently modeling the evolution of the same Group IVA irons, obtain cooling rates that differ by as much as a factor of five (Table I). It is interesting to ask what the error in metallographic cooling rates would have to be, to bring cooling times into line with radiometric (especially Ar retention) ages. Estimates were made of the required cooling rates for the five meteorite classes discussed above, taking Turner et al.'s closure temperatures at face value and assuming that differences in the activation energy for Ar diffusion among K host minerals are negligible.

The results are summarized in Table II. Metallographic cooling rates are too slow by a factor of about 6 to be consistent with the Ar retention ages. The cooling rate discrepancy is very similar for all five of the meteorite classes. In fact, the uniformity of the discrepancy is remarkable, considering the range of cooling rates and radiometric ages spanned, and the variability of diffusion geometry and K hosts. This suggests rather strongly that there is a systematic error in the published absolute values of metallographic cooling rates, such that the cooling rates should be adjusted by a factor of six upward. There is no apparent reason why the entire body of metallographic cooling rate data should be in error by such a large factor, however, and the price one pays for solving the problem in this way is the creation of a new and serious discrepancy with the cooling rates derived from ^{244}Pu fission track thermometry, which are ~$1°K/10^6$ yr in the temperature range $600°-300°K$ (Pellas and Storzer 1977).

TABLE II

Summary of discrepancies between metallographic cooling rates and radiometric ages of meteorite groups

Meteorite group	Metallographic cooling rate, $^\circ K/10^6$ yr	Time to Ar closure	$\left(\dfrac{dT}{dt}\right)_{770}$ needed to account for K/Ar or $^{40}Ar/^{39}Ar$ age, $^\circ K/10^6$ yr[a]	Discrepancy (ratio of cooling rates)
Equilibrated chondrites	1-5	$50\text{-}110 \times 10^6$ yr	6.6-14	~4.3[b]
Tieschitz (unequilibrated ordinary chondrite)	~1	$\leqslant 120 \times 10^6$ yr	$\geqslant 6.9$	$\geqslant 6.9$
Group I irons	~1.7	$~100 \times 10^6$ yr	~7.2	~4.2
Pallasites	~0.6	$~300 \times 10^6$ yr	~3.2	~5.3
Estherville (mesosiderite)	~0.1	$~900 \times 10^6$ yr	~0.8	~8.0

[a] Assuming cooling according to Eq. (1), Ar closure temperatures of Turner et al. (1978)
[b] Ratio of geometric means of ranges.

IV. AN ALTERNATIVE PARENT METEORITE PLANET MODEL

The alternative to adjusting metallographic (and fission track) cooling rates is to question the realism of previous thermal modeling of the parent meteorite planets, and the accuracy of the Ar closure temperatures of Turner *et al.* (1978).

Thermal evolution calculations (references in Sec. II) have traditionally assumed that the parent planets came into existence instantly as several-hundred-km objects of uniform density and thermal conductivity; the effect of the initial heat source has been approximated by assuming a uniformly high initial temperature throughout the objects. The thermal model of Fig. 2 as well as the simple model of Eq. (1) and Fig. 3 make this assumption. In fact the parent planets must have grown by accretion and their temperatures must have risen to peak values over a finite period of time. At any given time they must have consisted of materials having a range of physical properties, with high-density highly-conductive rock and low-density insulating particulate matter as end members. Melting or metamorphism would have tended to transform particulate matter to rock; collisions would have had the reverse effect. The most widely held views of events in the early solar system hold that all these processes – accretion, heat generation, collisional destruction, and establishment of the properties of the meteorites – happened on similar time scales. If so, the thermal evolution of the parent meteorite planets would have been determined by a complex interplay of these effects. A thermal evolution study that attempts to take all of them into account may seem distastefully model-dependent; but the alternative, studies that ignore them, may lead to totally wrong conclusions.

An accreting planet might be expected to cool at first rapidly, then (as its thermal inertia increases) much more slowly. Qualitatively this works in the right direction to resolve the discrepancies noted; relatively rapid cooling is needed from temperatures of melting or metamorphism to the isotopic closure temperatures, but much slower cooling seems to have occurred in the temperature ranges of metal alloy formation and fission track retention. The trouble is that the Ar closure temperatures of Turner *et al.* lie *beneath* the temperature range of alloy formation, which prevents cooling to closure from being any more rapid than the time of alloy formation.

These closure temperatures must be considered approximate, since they do not take into account variability of the grain-size distribution or mineralogy of K hosts. However, a detailed examination of closure temperatures is beyond the scope of this chapter. In the following sections I will assume the possibility that the effective Ar closure temperature was within or above the temperature range of alloy formation; and I report on the results of a thermal modeling study that includes the effects of planetary growth by accretion of insulating dust (either primary condensate particles from the solar nebula or fragmental debris from earlier objects demolished by collisions), and which takes into account the large difference in thermal

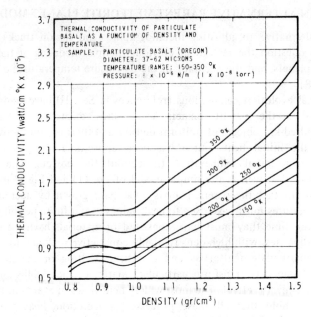

Fig. 4. Experimentally determined thermal conductivities of basalt powders, as a function of density (compaction) and temperature (Fountain and West 1970).

conductivity between solid rock and unconsolidated dust or soil and acknowledges that high internal temperatures transform dust into dense rock by melting or sintering it.

V. INSULATING PROPERTIES OF PARTICULATE AGGREGATES

The insulating effect of a layer of unconsolidated particulate matter on a planet has been referred to repeatedly (e.g., Urey *et al.* 1971; Lee *et al.* 1977) whenever it has seemed desirable to attain or hold a high temperature in a small space, but the consequences of such an insulating layer have never been fully explored. The thermal conductivity of basalt powder in a vacuum as a function of temperature and density (compaction) is shown in Fig. 4 (Fountain and West 1970). The dependence on temperature is as $A + BT^3$, where the cubic term is attributable to radiative transfer betweeen particles. Thermal conductivity probably bottoms out at a value not much smaller than the lowest conductivity shown. (The thermal conductivity of a porous aggregate cannot be made arbitrarily small by continuing to increase porosity, because a point is reached where conductive transfer through contacts between grains becomes small compared to radiative transfer; further increases in porosity only enhance radiative transfer, and therefore net conductivity.) The thermal conductivity of lunar soil measured in the laboratory agrees closely with the basalt powder data (0.6-1.5×10^{-5} W/cm

°K at 160°K, 1.3 g/cm^3, Cremers and Hsia 1973). *In situ* measurements of regolith conductivity by the Apollo heat flow experiment yielded values at a density of ~2.0 gcm^3 that are somewhat higher than a projection of the curves in Fig. 4 would predict (~10^{-4} W/cm °K at 300°K; Langseth *et al.* 1976), but by less than a factor of two. These soil conductivities are two or three orders of magnitude smaller than the thermal conductivities of corresponding solid rock or meteorite materials.

VI. SINTERING OF PARTICULATE AGGREGATES

The following discussion will assume that particulate aggregates having the insulating properties just described played a major role in the assembly of the meteorite parent planets, but that this soil or dust had the property of *sintering* into dense, conductive rocky material if its temperature ever exceeded some critical sintering temperature. Metal and dielectric powders are known to sinter at temperatures that are a substantial fraction of their melting temperatures; the process is believed to depend upon volume diffusion of lattice vacancies away from the voids between particles, leading to closure of the voids, increase in the areas of contact between particles (and therefore an increase in thermal conductivity), and shrinkage (densification) of the aggregate. Rates of volume diffusion are exponentially dependent upon temperature. Obviously, however, temperature is not the only variable that promotes sintering. Time at a particular temperature, static pressure, shock pressure, powder composition, and gases in the powder voids would also affect the process. In the case of powders that consist of primitive condensates, which are more or less disequilibrium mineral assemblages, the process would be driven not only by the excess surface energy of the finely divided powder but also by the tendency of the minerals to reconstitute themselves as a more stable assemblage, with lower thermodynamic free energy, as they sintered.

Clearly the process is very complex and infinitely variable under different circumstances, making it impossible to model realistically. For purposes of discussion in this chapter, and for use in the numerical model that was examined, I have simply assumed the existence of a fixed sintering temperature, at or above which soil or dust instantly and irreversibly transforms to rock, independently of all the other variables just acknowledged. This is admittedly a gross simplification, but still a major improvement over the approximation used in previous studies, which was to ignore the presence and effect of particulate layers altogether.

VII. THE HEAT SOURCE

I have considered the initial heating of the parent meteorite objects to be due to ^{26}Al, not electrical induction, because (a) the presense of ^{26}Al in the early solar system has been proven; (b) its effect is, or can be assumed to be, independent of the size, temperature, electrical conductivity, and heliocentric distance of a planetesimal; and (c) the diminution of its effect with time is straightforward and easily incorporated in a numerical model. The model to be examined serves to illustrate effects of the interplay between a heat source and an accretional epoch having similar time scales, and the results can be applied at least qualitatively to the evolution of electrically-heated as well as ^{26}Al-heated planetesimals.

The heating effects of ^{40}K, U, and Th are also included in the model; they play an important role in the thermal evolution of small planetesimals. The thermal effects of ^{244}Pu and ^{129}I in the early solar system are negligible.

VIII. A QUALITATIVE CONSIDERATION OF THE THERMAL EVOLUTION OF A ROCK/DUST PLANETESIMAL: THE CONSTANT-MASS CASE

In assessing the effects of an insulating dust layer on the thermal evolution of a planetesimal, it is instructive to consider two extreme possibilities. The first is an already-accreted dust sphere of several km dimension, having the properties postulated above, and liberally endowed with ^{26}Al. As decay of ^{26}Al heats the interior of the planetesimal, an isotherm corresponding to the sintering temperature of the dust expands outward, transforming everything inside it into conductive dense rock. The planetesimal is mantled by a layer of dust which grows thinner and thinner. Where does the encroachment stop? This happens when the residual dust layer becomes thin enough to accomodate a heat flow just equal to the flux of heat, from current and past ^{26}Al decay, that the rocky interior of the planetesimal is trying to unload. The rate of heat flow through a layer is equal to the product of the conductivity of the layer and the thermal gradient in it. Temperatures at the boundaries of the dust layer are fixed in this situation; the surface of the layer is at the blackbody temperature of the planetesimal, and the lower boundary is at the sintering temperature. If the lower boundary should exceed this temperature it sinters, moving the boundary upward to where the sintering temperature still prevails, and thinning the layer. The thinner the layer, the steeper the thermal gradient in it is (since the temperature contrast between top and bottom is fixed), and so the greater the heat flow through it. If the latter equals or exceeds the heat flux trying to escape the interior of the planetesimal, the rock/dust interface stabilizes; if not, temperature at the base of the layer mounts, more sintering occurs, and the layer grows thinner.

Basically the system acts to match the thermal time constant of the dust layer to the time constant of the rocky zone inside it. Since the dust layer is 10^2-10^3 times less conductive than the rock, it must be 10^2-10^3 times thinner to achieve a match; i.e., a few meters or tens of meters thick for a planetesimal a few km in dimension.

Such a thin layer is not adequate to maintain warmth inside a small

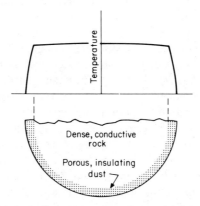

Fig. 5. Quasi-steady-state temperature distribution in a planetesimal composed of rock covered by a layer of insulating dust; schematic.

planetesimal once decay has removed the ^{26}Al heat source, however, no more adequate than a several-km rocky body (to which it is thermally matched) would be. Consequently the internal temperature drops, along an exponential curve that expresses the heat production of remaining ^{26}Al, to the blackbody temperature of the planetesimal in question. Cooling through the temperature range of metal alloy formation would be at a rate of the order of $100°K/10^6$ yr. That such a body may have been able to achieve the melting temperature internally is not primarily due to the insulating property of the dust, since this is largely removed before the melting temperature is reached, but simply to the thermal inertia of several km of rock relative to the rapid rate at which ^{26}Al can deposit energy during its first half-life of decay.

IX. THERMAL EVOLUTION OF A PLANETESIMAL THAT CONTINUES TO ACCRETE DUST: THE THERMAL QUASI-STEADY-STATE

The other extreme possibility to consider is a planetesimal containing no ^{26}Al, in a thermal quasi-steady-state such that its insulating dust layer, with outer surface at the blackbody temperature and inner surface at the sintering temperature, is thick enough to allow heat flow just equal to the small amount of heat being generated by long-lived radioactivities. Temperatures in the rocky interior would be all but isothermal, since this region is so much more conductive than the dust layer (Fig. 5). Temperatures would also be almost constant in time (extremely low cooling rates). Anticipating results of the modeling study to be described, the dust layer would have to be kilometers rather than meters thick to sustain this quasi-steady-state. Continued accretion of dust onto a planetesimal after the effect of ^{26}Al heating had peaked out would cause the situation described in the previous section to tend toward this section's quasi-steady-state.

There is some particular accretion history that would lead to exactly this postulated quasi-steady-state. Of course we cannot assume nature to be obliging enough to provide such a postulated condition; real accretion rates onto early solar system planetesimals were undoubtedly greater or less than this idealized accretion rate. If accretion rates were *less* than the idealized value, cooling rates through the temperature range of metal alloy formation would have been intermediate to the rates for a constant mass object ($\sim 100°K/10^6$ yr) and the quasi-steady-state situation ($\sim 0°K/10^6$ yr, assuming the sintering temperature and the alloy-formation temperature range were similar). This is just what is observed in most meteorites. If accretion rates were *greater* than the idealized value, and a dust layer too thick to pass the internal heat being generated was created, the system would respond by heating up internally, sintering additional dust at the base of the layer, and restoring (indeed, maintaining) the layer at the steady-state thickness. Thus for a *range* of accretion rates, not just a particular idealized accretion rate, planetesimals would arrive at a thermal configuration similar to the quasi-steady-state postulated in this section.

The reason why temperature in such an idealized situation is not strictly constant is that the long-lived radionuclides which maintain temperature in the system are decaying and thus providing heat at a slowly diminishing rate. When heat generation has declined by 10 %, the thermal gradient in the insulating dust layer must decline by ~ 10 % to reduce the heat flow through the layer to the rate of heat generation. If accretion has ceased, this means the temperature at the base of the layer must decrease to ~ 90 % of its initial value. The cooling rate of such a system due to decay of long-lived radionuclides can be estimated; it is simply the product of the decay constant of the dominant heat-producing radionuclide and the temperature range through which the system has to cool to reach the blackbody temperature. After ^{26}Al has decayed to insignificance, the dominant radionuclide in a chondritic system would be ^{40}K, for which $\lambda = 4.962 \times 10^{-10}/yr$. For assumed sintering and blackbody temperatures of $\sim 800°K$ and $\sim 170°K$, respectively, the system cooling rate works out to $\sim 0.3°K/10^6$ yr. This constitutes an effective lower limit on long-duration cooling rates for planetesimals: a slower cooling rate could be produced only by continuing accretion indefinitely, to permit a growth in dust layer thickness that would diminish heat flow through it without decreasing the temperature at its base. Accretion in the early solar system is generally held to have been largely completed in the first $10^7 - 10^8$ yr.

Figure 2 shows a lower limit to planetary cooling rates for essentially the same reason; it is all but impossible to contrive a planetary context where the cooling rate is less than that dictated by ^{40}K decay, other than briefly.

Cooling rates in the mesosiderites and some unequilibrated chondrites are close to this limiting value reached in the quasi-steady-state situation.

To rationalize the apparent thermal histories of the mesosiderites and

TABLE III

Peak temperatures and cooling parameters of planetesimals with initial radius 5 km, which continue to accrete dust at exponentially diminishing rates

Entry:	characteristic time of accretion (= e-folding time of accretion rate)			
Initial accretion rate, cm/yr	1×10^6	3×10^6	1×10^7	3×10^7
	A 1408°K	B 1408°K	C 1408°K	D 1408°K
maximum temperature attained cooling rate, °C/10^6 yr	75°/Myr	16°/Myr	4.8°/Myr	3.0°/Myr
radius	4.0 km	4.8 km	6.4 km	7.5 km
dust thickness	87 m	0.4 km	1.4 km	2.1 km
time	4.0 Myr	5.7 Myr	7.3 Myr	7.9 Myr
0.1	E 1479°K	F 1488°K	G 1493°K	H 1494°K[b]
	50°/Myr	7.1°/Myr	0.6°/Myr	bottoms out
	4.9 km	7.3 km	14 km	15.4 km
	0.13 km	1.0 km	5 km	3.2 km
0.3	4.7 Myr	7.4 Myr	11 Myr	10.1 Myr
				(803 °K)

(when planetesimal cools through 800°K)

TABLE III (Continued)

Peak temperatures and cooling parameters of planetesimals with initial radius 5 km, which continue to accrete dust at exponentially diminishing rates

Entry: maximum temperature attained
cooling rate, °C/10⁶ yr
radius
dust thickness } when planetesimal cools through 800°K
time

Initial accretion rate, cm/yr	characteristic time of accretion (= e-folding time of accretion rate)			
	1×10^6	3×10^6	1×10^7	3×10^7
1.0	I 1826°K	J melts	K melts[b]	L melts[a]
	36°/Myr	1.4°/Myr	bottoms out	< 0.53°/Myr
	7.8 km	16.2 km	40.1 km	>69.9 km
	0.18 km	2.1 km	1.9 km	> 2.9 km
	6.3 Myr	14.2 Myr	30.0 Myr	>25.7 Myr
			(817°K)	(833°K)
3.0	M melts	N melts[a]		
	33°/Myr	< 8.3°/Myr		
	16.2 km	>38.1 km		
	0.11 km	> 0.3 km		
	10.9 Myr	>29 Myr		
		(891°K)		

[a] These runs had not cooled through 800°K in 1024 sec of computer time. Parameters and final temperatures when runs terminated are shown.
[b] These runs cool through the minimum temperatures shown, then begin to warm up. Program cannot accommodate a reversal in cooling rate, terminates.

perhaps also the slower cooling chondrites, it appears necessary that temperatures in the parent planets tended to asymptotically approach some "base temperature" that was higher than the assumed planetary surface temperature, ~170°K. This would lead to relatively rapid initial cooling from metamorphic temperatures, then much slower cooling rates as the base temperature was approached. If the base temperature was close to the temperature range of meteorite alloy formation, this could explain very slow metallographic cooling rates and yet allow relatively rapid cooling from metamorphic temperatures to temperatures of isotopic closure (as long as these were higher than the temperatures of alloy formation).

One possible explanation for a base temperature >170°K is that the parent body surface temperatures were in fact much higher in the early solar system, at a time when the sun was in a super-luminous phase of its evolution. Unfortunately the sun would be required to maintain its exaggerated luminosity for half the age of the solar system to account for the cooling record in the mesosiderites, a possibility that is not admitted by stellar evolution studies or the early geologic and paleontologic record of the earth.

It was anticipated that if a planetesimal evolved toward the thermal quasi-steady-state condition postulated in this section, the sintering temperature of the accreting particulate matter that formed an insulating blanket around the hot planetesimal core would effectively constitute a base temperature toward which the core would cool asymptotically. The model calculations to be described confirmed this. For most of the computations made, a sintering temperature of 800°K was assumed, in order to produce very slow cooling in the temperature range of alloy formation.

X. NUMERICAL MODELING OF ACCRETING, SINTERING, ^{26}AL-BEARING PLANETESIMALS

To test the processes discussed above quantitatively, a crude computer program was set up which follows the thermal evolution of small bodies that accrete and sinter. The program assumes that low-conductivity, low-density (1.3 g/cm^3) dust sinters at a specific temperature, irreversibly, into high-conductivity, high-density (3.6 g/cm^3) rock. The conductivity profile used, as a function of temperature, is shown in Fig. 6. The planetary material was assumed to have the chondritic abundance of Al and ^{26}Al/^{27}Al of 5×10^{-5}, the initial isotopic composition of Al in an Allende inclusion found by Lee *et al.* (1977). It also contains the chondritic abundances of U, Th, and K.

In addition to assessing thermal evolution, the program makes volume adjustments in the planetesimal as its substance is sintered, and redistributes all radionuclides (Al as well as U, Th, and K) to the top of the melted zone. The planetesimal can grow during the course of the computer run; dust is added at an exponentially declining rate. Entered as input are, (a) an assumed initial size for the planetesimal, perhaps representing Goldreich-Ward

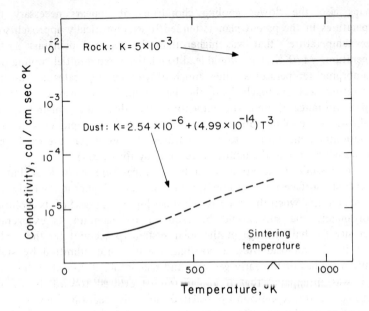

Fig. 6. Conductivities of dust and rock used in the modeling computations of this article. The solid curve for dust represents the experimental measurement of Fountain and West (1970) for dust of density 1.2 g cm^{-3}; the dashed curve is an extrapolation according to the $A + BT^3$ relationship fitted to their data.

(1973) coalescence; (b) an initial accretion rate; and (c) a characteristic accretion time, the time in which the accretion rate falls to $1/e$ of its initial rate. Details of the program are discussed in the Appendix to this chapter. The program involves assumptions and approximations, and should be viewed critically. Principal findings of the modeling study are as follows.

1. A planetesimal initially composed of dust, 20 km in radius, which does not accrete further, will melt and sinter into the configuration shown in Fig. 7 (left). Volume decrease on sintering/melting of the substance of the planetesimal shrinks its radius from 20 to 14.3 km. The residual layer of unsintered dust is only 6.2 m thick. Although the planet is almost totally sintered, the increased conductivity does not allow heat to be shed as fast as it is being generated by ^{26}Al in the first half-life after creation of the planetesimal, so extensive interior melting occurs. Cooling thereafter is fast, $\sim 200°$K/10^6 yr through the temperature range in which metal alloy compositions were established. At least 40 % as much ^{26}Al as the amount assumed in this case (see above) is needed to cause melting in a planetesimal of 20 km radius.

2. Table III shows a matrix of results of planetesimal models that assume a starting nucleus of nominal Goldreich-Ward dimension (5 km radius) and accrete additional dust onto the nucleus at various rates. Model J, for which

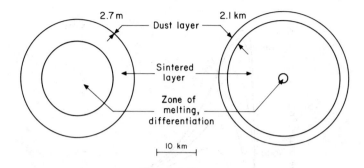

Fig. 7. Cross-sections of two planetesimals after ^{26}Al heating, accretion, and sintering cease. Left: a planetesimal, initially all dust (20-km radius), which does not accrete further, sinters and melts to this configuration. Right: outcome of Model *J*, Table II.

the initial accretion rate is 1 cm/yr and the characteristic time of accretion is 3×10^6 yr, demonstrates that a small planetesimal can melt and cool to the temperature range of alloy formation rapidly (14.2×10^6 yr), yet cool through this range at only $1.4°K/10^6$ yr. The detailed thermal history of this model appears in Fig. 8; its final configuration is shown in Fig. 7 (right).

    3. To cause melting to occur, the initial accretion rate onto the Goldreich-Ward nucleus must be >1 cm/yr. This value is critically dependent on the size assumed for the nucleus. For nuclei larger than 5 km radius, smaller accretion rates suffice to cause melting. Accretion rates insufficient to promote melting may still be adequate to produce the metamorphic effects observed in ordinary chondrites and mesosiderites.

    4. Damping the cooling rate of a small planetesimal, once melting has occurred, to values near those read from the metal phases of meteorites, requires that the accretion be sustained over a period of a few million years; in the accretion model assumed, a characteristic time of accretion $>3 \times 10^6$ yr is needed. This value is not strongly dependent on the dimension of the nucleus. Accretion need not continue beyond the time when ^{26}Al ceases to be an important heat source.

    5. It appears that models with initial accretion rates >1 cm/yr *and* characteristic accretion times $>3 \times 10^6$ will all produce melting and slow cooling through 800°K. Planetesimals cannot be modeled in this regime, however, because the computer runs take excessive amounts of time, and because the run of the interior temperature tends to "bottom out" — that is, insulation by the dust layer becomes so effective that the cooling rate diminishes to zero and then continuing radioactive heat generation begins to increase internal temperatures again. This should lead to the self-governing quasi-steady-state situation discussed in the previous section, wherein the sintered zone expands outward and the internal temperature remains nearly

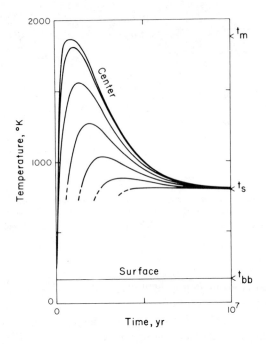

Fig. 8. Thermal evolution of Model *J*, Table II, for the first ten million years. Temperature curves at 2-km intervals above the planetesimal center are shown. t_m, t_s and t_{bb} are the assumed temperatures of melting, sintering, and blackbody radiation equilibrium.

constant, but an artifact of the program prevents it from dealing with a system that reverses the sign of its thermal gradient.

6. To obtain very slow cooling rates like this in the vicinity of the sintering temperature requires, in general, the presence of several km of insulating dust. This can be read from Fig. 9. For example, the heat generated by chondritic U, Th, and K decay in a sintered zone of 10 km radius amounts to 1.3×10^{-8} cal/sec per cm^2 of surface area of the zone. If this zone surface is to be held at 800 K, and just 1.3×10^{-8} cal/cm^2 sec is to be allowed to escape through the dust layer, then Fig. 9 shows that the layer must be 5 km thick.

7. The sizes of melted zones produced in the models of Table III are very small relative to the final dimensions of the planetesimals (Fig. 7, right). Changing the accretion parameters is ineffective in increasing the scale of the melted zone. The most effective way to increase the proportion of melted material is by increasing the size of the initial planetesimal nucleus, or by assuming its almost instantaneous growth during a preliminary period of extra rapid accretion.

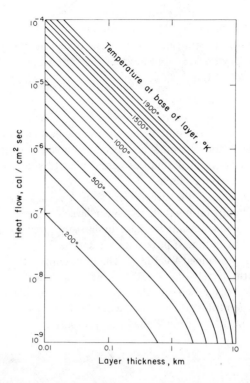

Fig. 9. Steady-state heat flow through a dust layer having the conductivity profile of Fig. 6, as a function of layer thickness and base temperature (a range of sintering or base temperatures is considered). Heating by decay of long-lived radioactivity causes the downturns of curves at lower right. A typical steady-state temperature distribution is shown in Fig. 10.

XI. CONCLUSIONS

The model thermal histories in Table II can be related to the cooling histories of meteorites as follows.

Iron Meteorites

Models J and M melt, cool rapidly at first, and then cool through 800°K (a temperature representative of the range in which metal alloys are formed) at rates characteristic of the metallographic cooling rates read in iron meteorites. The very high cooling rates of the Group IIC irons appear to require that their parent body accreted little or no additional dust after melting and differentiation. This could be modeled with a faster accretion time than those shown in the table, or with a zero initial accretion rate, though in the latter case an initial radius $>\sim 5$ km would have to be assumed to produce melting. Presumably, differences in accretion rates among the various parent planets reflect different orbital parameters. A parent body that was perturbed into an orbit inclined enough to keep it out of the dust-rich midplane of the planetary system most of the time would accrete relatively little dust, and cool rapidly.

Ordinary Chondrites

The parent planets of ordinary chondrites may have had nuclei too small, or their initial accretion rates may have been too small, to melt at all; or they may come from late-accreted zones of parent planets where melting affected only the earliest-accreted material.

Note that in Fig. 8 the deeper levels of a planet cool more rapidly through a given temperature than shallower levels. This is a reversal of the depth-cooling rate relationship we have understood previously (Fig. 2) and stems from the distribution of peak temperatures in Fig. 8 versus the uniform initial temperatures assumed for Fig. 2. The peak temperature distribution in Fig. 8 is, in turn, largely due to the protracted instead of instantaneous accretion of the planetesimal. The cooling rates of equilibrated versus unequilibrated chondrites are better understood in terms of a depth-cooling rate relationship like that of Fig. 8 than the relationships of Fig. 2.

Pallasites

Presumably these solidified as small ($<\sim 10$ km radius) objects, thereby preserving their characteristic structures, then grew or joined larger objects in time to cool relatively slowly through the alloy-formation temperature regime.

Mesosiderites

It is necessary to postulate that one or more early-formed planetesimals melted, differentiated, and erupted surface lava flows which were then brecciated and mingled with iron meteorite debris, after which the system accreted several km more of material early enough that it still contained an amount of ^{26}Al sufficient to raise the temperature of the mesosiderite layer to $\sim 1000°$C. The heat production had to be fast enough to produce this temperature rise in a rocky, conductive body, without much assistance from insulating dust, since $1000°$C is much in excess of the postulated sintering temperature. That such a high-temperature pulse can be retained for a short time in a small conductive body is demonstrated by the thermal evolution model of Fig. 8, where the thickness of the dust layer was only ~ 5 m at the time when internal temperatures peaked. Subsequently the parent object(s) had to accrete enough additional dust, and over a long enough period, to put it in a category similar to models H and K of Table III. These "bottoms out" models lead to approximations of the thermal quasi-steady-state condition discussed in Sec. IX, in which cooling is chiefly due to the decay of long-lived radioactivities and is very slow ($\sim 0.3^\circ K/10^6$ yr).

The range of observed taenite compositions in mesosiderites corresponds to the temperature range $\sim 770^\circ - \sim 570^\circ K$. At $0.3^\circ K/10^6$ yr, the mesosiderite systems spent $\sim 700 \times 10^6$ yr in this temperature range. If, as the postulated model makes possible, the mesosiderites cooled rapidly to $\sim 770^\circ K$, and if the effective Ar closure temperature for mesosiderites (taking into account coarse as well as fine feldspar grains) is not much less than $\sim 570^\circ K$, then the relatively old ages of these meteorites can be accounted for. (Use of $0.3^\circ K/10^6$ yr as the cooling rate for mesosiderites entails acceptance of a factor of three error in their reported cooling rates, of course; this is half of the possible 6x error pointed out in Sec. V.)

Thus the examined model can be used to rationalize the properties of the various meteorite classes. However, the model rests on a number of assumptions, and constrains the magnitude of several early solar system processes. Whether these assumptions and constraints correspond to reality or not remains to be demonstrated.

Sintering Temperature

The sintering temperature assumed for these calculations, $800^\circ K$, is rather lower than might seem intuitively correct, but it must be remembered that our intuition is founded on the time scale of laboratory experiments and industrial processes, not the millions of years in which planetesimals accrete and [26]Al decays; and also that here "sintering" refers only to enough textural alteration of a particulate aggregate to increase its conductivity to rock-like values, not necessarily enough to thoroughly recrystallize it into a high-grade chondrite, for example. By this criterion the unequilibrated chondrites are sintered, since they are nearly as dense and conductive as igneous rocks; for that matter, the conductivity of C1 chondritic material is more nearly rock-like than soil-like.

Thermal Conductivity of Thick Dust Layers

The value of thermal conductivity assumed for accreted dust (Fig. 6) is fairly conservative, but it is debatable whether or not such a low thermal conductivity would be applicable throughout the extent of a kilometers-thick dust layer. The lunar regolith is compacted to a density of ~ 2.0 g/cm^3 only a few cm beneath the surface, and has a thermal conductivity an order of magnitude greater than the value used in the present modeling study (Langseth et al. 1976). Such high-conductivity material would not produce the thermal histories the meteorites appear to require. My reasoning was that a very young dust layer in an almost-zero gravity field would not tend to compact as the lunar regolith has. Meteoroid impacts in the dust layer would also have an effect on net density, however, which I have not been able to assess.

Accretion Parameters

The e-folding times for accretion rates that produce the desired thermal behavior in planetesimals (Table III), $3\text{-}30 \times 10^6$ yr, are hearteningly similar to the accretion times of planets based on considerations of orbital dynamics (e.g., Safronov 1972). On the other hand Greenberg et al. (1978) obtain much shorter times, $\sim 10^4$ yr, for the assembly of planets of asteroidal dimension, from a numerical simulation of collisional evolution in the early solar system. However, this time scale is dependent upon a relatively high assumed surface density of solid matter available for accretion in the solar system, 8 g/cm², which is a mean value equal to the mass of the terrestrial planets divided by the surface area out to 2.6 AU. Presumably the 10^4 yr time scale applies to accretion in the vicinity of Earth. If the much lower surface density that probably obtained locally in what is now the asteroid belt is used in Greenberg et al.'s numerical simulation, much longer accretion times are obtained (C.R. Chapman, personal communication). It is unclear whether the processes simulated by Greenberg et al. and by me are reconcilable.

I am unable to judge the plausibility of the initial accretion rates required (0.3–3 cm/yr). Both this parameter and the e-folding time of accretion can, to some extent, be traded off for planetesimal nuclei of dimensions other than the 5-km radius assumed.

Al Content of Early Planetary Material

If the level of ^{26}Al in Allende inclusions when they solidified (^{26}Al/^{27}Al $\sim 6 \times 10^{-5}$; Lee et al. 1976) applied generally to rocky bodies containing the chondritic abundance of Al, it would have been sufficient to cause melting in bodies larger than ~ 7 km radius (Lee et al. 1977; Herndon and Herndon 1977). The amount of energy generated by total decay of this much ^{26}Al, if all conserved as heat, is ~ 5 times that needed to initiate melting. However, much of the Al in chondritic material appears in different forms than Ca,Al-rich inclusions, and it is not clear that this non-inclusion Al was similarly sweetened with ^{26}Al.

It is significant that Schramm et al. (1970) found ^{26}Al/^{27}Al $< 0.026 \times 10^{-5}$ in Juvinas and Pasamonte, eucrites that are indisputably the products of melting; the ratio is even smaller in Moore County. This upper limit is only ~ 2 % of the ^{26}Al (assuming chondritic Al) needed to initiate melting. For these meteorites to have been melted by ^{26}Al requires either that at least six half-lives of ^{26}Al ($\geqslant 4.3 \times 10^6$ yr) elapsed between the time of melting and the time of eruption and solidification; or that the eucrites after solidification were metamorphosed severely enough to homogenize differences in the isotopic composition of Mg between phases. These seem like extreme requirements, but in fact Walker *et al.*'s (1978) model for melt segregation in the eucrite parent body operates on a time scale of millions of years, so it is possible that ^{26}Al in eucritic lavas decayed below detectability before the lavas solidified.

Inefficiency of the Process

As noted earlier, in most of the models tested the volume melted and differentiated was quite small compared with the amount of overlying sintered and unsintered material that served to modulate its thermal history (e.g. Fig. 7, right). Thus each iron or stony-iron meteorite in our collections is expected to represent only the "kernel" from a much larger mass of "waste material" that has been lost to us. I do not know if this constitutes an objection to the model. Iron and stony-iron meteorites survive in space about 50 times longer than stones, according to their cosmic ray exposure ages, so we would not expect to find among the meteorites representative proportions of differentiated cores and the sintered layers that overlaid them.

Acknowledgment. I am grateful to P. Pellas for constructive criticism of the original draft of this chapter. The research was supported in part by a grant from the National Aeronautics and Space Administration.

APPENDIX

Programming the Thermal Evolution of an Accreting, Sintering, ^{26}Al-Heated Planetesimal

The basic numerical computation of the thermal evolution of planetesimals was carried out by the traditional finite-difference method. The redistribution of radioactive heat sources (in this case including ^{26}Al as well as K, U, and Th) by melting has also become traditional by now. Allowance for accretional growth of a planetesimal at least in a step-wise fashion, by addition of new rows of points at the surface edge of the computational array, was straightforward. The planetesimal was considered to consist of rock (3.6 g/cm^3) in all parts of the planetesimal that had ever exceeded the sintering temperature, t_{sint}, and particulate matter, 1.3 g/cm^3, elsewhere. After every computational step the presence of the t_{sint} isotherm was located, and points in the array were changed from dust to rock if necessary. Since such an adjustment decreases the volume of the material that changes hands, the temperature distribution in the dust overlying the sintered region had to be shifted downward in the computational array by a rather tedious process. In some cases such shrinkage led to elimination of a row of points at the surface edge of the computational array.

The program used the values of thermal conductivity shown in Fig. 6 for dust and rock. The program and I found it very difficult to deal with this dichotomy of rock and dust conductivities. In the first place, the program produced spurious results that apparently resulted from trying to apply the finite difference method of integration across a discontinuity of conductivity as drastic as that of Fig. 6. An experiment with a conductivity profile that varied somewhat more gradually did not seem to help. In the second place, sintering is capable of reducing the residual dust layer to a very small fraction of the radius of the planetesimal, as noted in the text; a thickness much smaller than the smallest radial distance increments that can practicably be used in the computational array.

The approach that was finally settled upon was to treat thermal evolution in the rocky zone by the finite difference method; assume that the dust layer was always in the thermal steady-state and transported heat at a rate consistent with this; compare the heat output from the rocky zone with the capacity for heat flow of the dust layer; and make the dust layer thinner (by additional sintering), if necessary, to match the two. The dust layer was effectively placed outside the computational grid, and finite difference calculations were not extended into it

Numerical integration yielded the relationship between thickness of a dust layer having the conductivity properties of Fig. 6, the temperature at its base (surface temperature was assumed equal to $170°K$, characteristic of blackbody temperatures in the asteroid belt), and the rate of steady-state heat flow through it. This is shown in Fig. 9, which was used to match layer thickness to required heat throughput as just mentioned.

The procedure outlined has the following advantages.

1. It eliminates the need to carry the finite difference calculations across a conductivity discontinuity.
2. It can deal with dust layers of arbitrarily small thickness, since the thickness is represented by a real number rather than by an integral multiple of the radial distance increment.
3. For the same reason, it allows accretion to proceed smoothly in many small increments, instead of periodically whenever enough accreted material has built up to warrant addition of a new row of points to the computational array. (When the latter mode of accretion is used, the sudden addition of one radial-distance-increment of highly insulating material can introduce such a disturbance in the internal thermal evolution of a planetesimal as to obscure more general trends.)
4. Use of this rather sensitive device at the surface of the planetesimal, where the most complicated things are happening, permits a relatively coarse computational mesh to be used in the sintered zone, saving computational time.

However, the procedure also has some serious disadvantages and pitfalls.

1. The data of Fig. 9 hold only for flab-slab geometry. To allow for

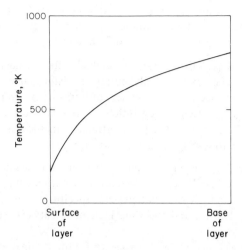

Fig. 10. The steady-state temperature distribution in a dust layer with boundaries at the blackbody temperature and at an 800°K sintering temperature; thermal conductivity as in Fig. 6.

spherical geometry would entail the redetermination of a portion of Fig. 9 at each computational step. The flab-slab geometry is a tolerable approximation if the dust layer thickness is small compared to the planetesimal radius, but is unrealistic in early stages of the thermal history when a planetesimal is composed mostly or entirely of dust.

2. Radioactive decay within the dust layer complicates application of the method. Decay of long-lived radioactivities in very thick layers produces the down-turn of curves at the lower right of Fig. 9. In even thicker layers, radioactivity can produce temperatures in the layer that are greater than the (supposedly maximum) temperature at the base of the layer. Because of this complication, and because of the rapid temporal variation of heat production by ^{26}Al, the effect of this nuclide was omitted from the dust layer altogether. For this reason as well as (1.), use of the steady-state dust layer in the earliest period of planetesimal evolution is impractical.

3. The temperature distribution in the dust layer is assumed always to be the steady-state configuration (Fig. 10), in spite of the fact that additions of cold dust at the surface and subtractions (by sintering) of hot dust from the base would introduce transient deviations from a steady-state temperature distribution and heat flow rate.

4. The procedure outlined does not explicitly conserve energy. For example, if rapid accretion builds up the dust layer thickness in a short time, the steady-state temperature distribution, with temperatures increasing monotonically inward through the layer, is assumed to exist immediately, with no concern as to where the calories came from that

warmed the cold dust to these temperatures. In reality, the addition of a large amount of cold dust would have the effect of quenching high temperatures in a planetesimal interior, at least temporarily.

Because of these difficulties, it seemed most prudent to use the steady-state dust layer method only during the later stages of planetesimal evolution, after the volume and thickness of dust had grown relatively small. In the first stages of thermal evolution, it was considered preferable to tolerate the idiosyncracies of the finite difference method as applied to both rock and dust. The following procedure was used.

A. The planetesimal begins as pure dust, at the assumed blackbody temperature of $170°K$. The initial radius, accretion parameters, sintering temperature, and ^{26}Al content are specified. The radial distance increment is 100 m, and the time increment between computations is 187.5 yr.

B. The finite difference method is used to follow thermal evolution. When the temperature at a computational point exceeds t_{sint}, that point transforms to rock, and all overlying material is adjusted for the shrinkage involved. When the rock/dust interface has reached 0.804 of the radius of the planet (encompassing 75 % of its mass), the mode of operation changes to that discussed above, where only the rocky interior of the planetesimal is treated by the finite difference method, and steady-state heat flow is assumed in the dust layer. Specifically, after each computational step the temperature gradient at the outermost levels of the rocky zone is used to calculate heat flow out of the zone, ΔQ_r. Figure 9 is used to determine the rate of heat flow through the current dust layer, ΔQ_d. If $\Delta Q_r > \Delta Q_d$, the dust thickness is diminished (sintered) until the two heat flow rates match. If $\Delta Q_r < \Delta Q_d$, temperature at the rock/dust interface is lowered to a value that causes the two heat flow rates to match.

C. Twenty computational steps after the mode of computation is changed, the scale of the computational array is coarsened to 200 m and 750 yr to conserve machine time. The thermal evolution models are typically run for 20×10^6 yr.

REFERENCES

Begemann, F.; Weber, H.W.; Vilcsek, E.; and Hintenberger, H. 1976. Rare gases and ^{36}Cl in stony-iron meteorites: Cosmogenic elemental production rates, exposure ages, diffusion losses and thermal histories. *Geochim. Cosmochim. Acta* 40: 353-368.

Bild, R.W., and Wasson, J.T. 1976. The Lodran Meteorite and its relationship to the ureilites. *Mineral. Mag.* 40: 721-735.

Bogard, D.; Burnett, D.; Eberhardt, P.; and Wasserburg, G.J. 1968. ^{40}Ar-^{40}K ages of silicate inclusions in iron meteorites. *Earth Planet. Sci. Lett.* 3: 275-283.

Buchwald, V.F. 1975. *Handbook of Iron Meteorites* (Berkeley: Univ. of California Press).

Buseck, P.R., and Goldstein, J.I. 1969. Olivine compositions and cooling rates of pallasitic meteorites. *Geol. Soc. America Bull.* 80: 2141-2158.

Cameron, A.G.W. 1978. Physics of the primitive solar nebula and of giant gaseous protoplanets. In *Protostars and Planets,* ed. T. Gehrels (Tucson: University of Arizona Press), pp. 453-487.

Chapman, C.R. 1976. Asteroids as meteorite parent-bodies: The astronomical perspective. *Geochim. Cosmochim. Acta* 40: 701-719.

Cobleigh, T.J.; Holz, J.; and Goldstone, J. 1970. Paper rejected for presentation at the 33rd Annual Meeting of the Meteoritical Society.

Cremers, C.J., and Hsia, H.S. 1973. Thermal conductivity and diffusivity of Apollo 15 fines at low density. *Proc. Lunar Sci. Conf. IV* (Oxford: Pergamon Press), pp. 2459-2464.

Crozaz, G. 1979. Thermal history of the mesosiderite Estherville revisited (abstract). *Meteoritics* 12: 200-201.

El Goresy, A. 1965. Mineralbestand und Strukturen der Graphit und Sulfideinschlüsse in Eisenmeteoriten. *Geochim. Cosmochim. Acta* 29: 1131-1151.

Fish, R.A.; Goles, G.G.; and Anders, E. 1960. The record in the meteorites. III. On the development of meteorites in asteroidal bodies. *Astrophys. J.* 132: 243-258.

Fountain, J.A., and West, E.A. 1970. Thermal conductivity of particulate basalt as a function of density in simulated lunar and martian environments. *J. Geophys. Res.* 75: 4063-4069.

Fricker, P.E.; Goldstein, J.I.; and Summers, A.L. 1970. Cooling rates and thermal histories of iron and stony-iron meteorites. *Geochim. Cosmochim. Acta* 34: 475-491.

Goldreich, P., and Ward, W.R. 1973. The formation of planetesimals. *Astrophys. J.* 183: 1051-1061.

Goldstein, J.I. 1969. The classification of iron meteorites. In *Meteorite Research,* ed. P.M. Millman (Dordrecht: D. Reidel), pp. 721-737.

Goldstein, J.I., and Ogilvie, R.E. 1965. The growth of the Widmanstätten pattern in metallic meteorites. *Geochim. Cosmochim. Acta* 29: 893-920.

Goldstein, J.I., and Short, J.M. 1967. Cooling rates of 27 iron and stony-iron meteorites. *Geochim. Cosmochim. Acta* 31: 1001-1023.

Greenberg, R.; Wacker, J.F.; Hartmann, W.K.; and Chapman, C.R. 1978. Planetesimals to planets: Numerical simulation of collisional evolution. *Icarus* 35: 1-26.

Henderson, F.B., III. 1965. Mineralogy of sulfide nodules in some iron meteorites. *Publ. of the Department of Geological Sciences,* Harvard University.

Herndon, J.M., and Herndon, M.A. 1977. Aluminum-26 as a planetoid heat source in the early solar system. *Meteoritics* 12: 459-465.

Kirsten, T.; Krankowsky, D.; and Zähringer, J. 1963. Edelgas und Kaliumbestimmungen an einer grösseren Zahl von Steinmeteoriten. *Geochim. Cosmochim. Acta* 27: 13-42.

Kulpecz, A.A., and Hewins, R.H. 1978. Cooling rate based on schreibersite growth for the Emery mesosiderite. *Geochim. Cosmochim. Acta* 42: 1495-1500.

Langseth, M.G.; Keihm, S.J.; and Peters, K. 1976. Revised lunar heat-flow values. *Proc. Lunar Sci. Conf. VII* (Oxford: Pergamon Press), pp. 3143-3171.

Lee, T.; Papanastassiou, D.A.; and Wasserburg, G.J. 1976. Correction. Demonstration of ^{26}Mg excess in Allende and evidence for ^{26}Al. *Geophys. Res. Lett.* 3: 109-112.

Lee, T.; Papanastassiou, D.A.; and Wasserburg, G.J. 1977. Aluminum-26 in the early solar system: Fossil or fuel? *Astrophys. J.* 211: L107-L110.

Mazor, E.; Heymann, D.; and Anders, E. 1970. Noble gases in carbonaceous chondrites. *Geochim. Cosmochim. Acta* 34: 781-824.

Megrue, G.H. 1968. Rare gas chronology of hypersthene achondrites and pallasites. *J. Geophys. Res.* 73: 2027-2033.

Minster, J.-F.; and Allègre, C.J. 1979. ^{87}Rb-^{87}Sr chronology of H chondrites: Constraint and speculations on the early evolution of their parent body. *Earth Planet. Sci. Lett.* 42: 333-347.

Moren, A.E., and Goldstein, J.I. 1978. The effect of P on the cooling rates of the Group IVA iron meteorites. *Earth Planet. Sci. Lett.* 40: 151-161.

Murthy, V.R. 1978. Rb-Sr and ^{40}Ar-^{39}Ar systematics of the Estherville mesosiderite. *Lunar Science IX.* The Lunar and Planet. Inst., pp. 781-783.

Nehru, C.E.; Hewins, R.H.; Garcia, D.J.; Harlow, G.E.; and Prinz, M. 1978. Mineralogy and petrology of the Emery mesosiderite. *Lunar Science IX.* The Lunar and Planet. Inst., pp. 799-801.

Onuma, N.; Clayton, R.N.; and Mayeda, T.K. 1972. Oxygen isotope temperatures of "equilibrated" ordinary chondrites. *Geochim. Cosmochim. Acta* 36: 157-168.

Pellas, P., and Storzer, D. 1977. On the early thermal history of chondritic asteroids derived by 244-plutonium fission track thermometry. In *Comets, Asteroids, Meteorites,* ed. A.H. Delsemme (Toledo, Ohio: University of Toledo Press), pp. 355-363.

Powell, B.N. 1969. Petrology and chemistry of mesosiderites. I. Textures and composition of nickel-iron. *Geochim. Cosmochim. Acta* 33: 789-810.

Safronov, V.S. 1972. Accumulation of the planets. In *On the Origin of the Solar System,* ed. H. Reeves (Paris: CNRS), pp. 89-113.

Schramm, D.N.; Tera, F.; and Wasserburg, G.J. 1970. The isotopic abundance of ^{26}Mg and limits on ^{26}Al in the early solar system. *Earth Planet. Sci. Lett.* 10: 44-59.

Scott, E.R.D. 1977. Composition, mineralogy, and origin of Group IC iron meteorites. *Earth Planet. Sci. Lett.* 37: 273-283.

Scott, E.R.D., and Wasson, J.T. 1975. Classification and properties of iron meteorites. *Rev. Geophys. Space Phys.* 13: 527-546.

Short, J.M., and Andersen, C.A. 1965. Electron microprobe analyses of the Widmanstätten structure of nine iron meteorites. *J. Geophys. Res.* 70: 3745-3759.

Short, J.M., and Goldstein, J.I. 1967. Rapid method of determining cooling rates of iron and stony-iron meteorites. *Science* 156: 59-61.

Sneeringer, M., and Hart, S. 1978. Sr diffusion in diopside (abstract). *EOS* 59: 402.

Sonett, C.P.; Colburn, D.S.; and Schwartz, K. 1968. Electrical heating of meteorite parent bodies and planets by dynamo induction from a premain sequence T Tauri "solar wind." *Nature* 219: 924-926.

Steiger, R.H., and Jäger, E. 1977. Subcommision on geochronology: Convention of the use of decay constants in geo- and cosmochronology. *Earth Planet Sci. Lett.* 36: 359-362.

Taylor, G.J. 1976. Cooling rates of LL-chondrites. *Meteoritics* 11: 374-375.

Taylor, G.J., and Heymann, D. 1971. The formation of clear taenite in ordinary chondrites. *Geochim. Cosmochim. Acta* 35: 175-188.

Turner, G.; Enright, M.C.; and Cadogan, P.H. 1978. The early history of chondrite parent bodies inferred from ^{40}Ar-^{39}Ar ages. *Proc. Lunar Planet. Sci. Conf. IX* (Oxford: Pergamon Press), pp. 989-1025.

Urey, H.C. 1955. The cosmic abundances of potassium, uranium, and thorium and the heat balances of the earth, the moon and Mars. *Proc. Nat. Acad. Sci.* 41: 127-144.

Urey, H.C.; Marti, K.; Hawkins, J.W.; and Liu, M.K. 1971. Model history of the lunar surface. *Proc. Lunar Sci. Conf. II* (Oxford: Pergamon Press), pp. 987-998.

Van Schmus, W.R., and Wood, J.A. 1967. A chemical-petrologic classification for the chondritic meteorites. *Geochim. Cosmochim. Acta* 31: 747-765.

Walker, D.; Stolper, E.M.; and Hays, J.F. 1978. A numerical treatment of melt-solid segregation: Size of the eucrite parent body. *J. Geophys. Res.* 83: 6005-6013.

Wasserburg, G.J., and Burnett, D.S. 1969. The status of isotopic age determinations on iron and stone meteorites. In *Meteorite Research,* ed. P.M. Millman (Dordrecht: D. Reidel), pp. 467-479.

Wasson, J.T. 1970. The chemical classification of iron meteorites. IV. Irons with Ge concentrations greater than 190 ppm and other meteorites associated with Group I. *Icarus* 12: 407-423.

Wasson, J.T., and Schaudy, R. 1971. The chemical classification of iron meteorites. V. Groups IIIC and IIID and other irons with germanium concentrations between 1 and 25 ppm. *Icarus* 14: 59-70.

Willis, J. and Wasson, J.T. 1978a. Cooling rates of Group IVA iron meteorites. *Earth Planet. Sci. Lett.* 40: 141-150.

Willis, J., and Wasson, J.T. 1978b. A core origin for Group IVA iron meteorites: A reply to Moren and Goldstein. *Earth Planet. Sci. Lett.* 40: 162-167.

Wood, J.A. 1964. The cooling rates and parent planets of several iron meteorites. *Icarus* 3: 429-459.

Wood, J.A. 1967a. Chondrites: Their metallic minerals, thermal histories, and parent planets. *Icarus* 6: 1-49.

Wood, J.A. 1967*b*. Criticism of paper by H. E. Suess and H. Wänke, 'Metamorphosis and equilibration in chondrites.' *J. Geophys. Res.* 72: 6379-6383.

Wood, J.A. 1978. Nature and evolution of the meteorite parent bodies: Evidence from petrology and mineralogy. In *Asteroids: An Exploration Assessment,* ed. D. Morrison and W. C. Wells (NASA Conference Publ. 2053), pp. 45-55.

ORIGIN OF IRON METEORITES

EDWARD R. D. SCOTT
Carnegie Institution of Washington

Iron meteorites divide into 12 groups with 5-150 members and ~50 grouplets with 1-4 members. Each group and grouplet formed by melting of some chondrite-like material in a separate location. Cosmochemical models for the chemical trends within these groups except IAB and IIICD suggest that trends result from fractional crystallization of molten cores. Cooling rates at 800 K are usually 1-10 K/Myr, and thermal models for asteroidal chondritic bodies require burial depths of ~100 km to produce these cooling rates. But it is unlikely that 50 asteroids 100 km in size have been broken up to provide iron meteorites.

A possible explanation is that irons melted and solidified in km-sized bodies which were subsequently accreted into much larger bodies in which they cooled slowly through 800 K. Such accretion must have occurred early when collision velocities were low. Except in IAB and IIICD, which never formed a core, and IIE which did, metal and silicate were not intimately mixed on a meter scale during this accretion.

Such a two-stage origin is compatible with dynamical calculations of Wetherill and Williams which suggest that most irons and other differentiated meteorites could be produced by collisions on a few, favorably located S-type asteroids with diameters between 50 and 200 km like 6 Hebe and 8 Flora. But it does not explain satisfactorily the correlation of cooling rate and composition within groups found by some workers.

Iron meteorites are traditionally believed to have formed in cores in asteroidal parent bodies, in part because of analogies with the earth's structure. Some authors, however, have begun to favor an origin for the irons as raisins isolated in a silicate matrix in these bodies. There is still general agreement that the irons come from asteroids.

The aim of this chapter is to review information about the origin of iron meteorites, both their sources in the solar system and the processes which formed them. Emphasis is placed on information derived from studies of the composition and structure of irons. A previous review by Wasson (1972) concluded that one group of irons were raisins but the largest group formed a core of an asteroid. Sears (1978b) has questioned whether any irons formed in cores. For an introduction to the mineralogy, structure and composition of iron meteorites the reader is referred to other reviews, e.g., Buchwald (1975, Vol. 1; 1977), Goldstein and Axon (1973) and Wasson (1974). The classification of irons is reviewed in considerable detail by Scott and Wasson (1975), but their origin and formation history is not discussed by these authors.

Frequent comparisons are made in this chapter to properties of metal in other differentiated meteorites, like the pallasites and mesosiderites, and in the chondrites, which almost by definition have not been differentiated by planetary igneous processes. Although there are gaps in our knowledge about the sources and formation history of these meteorites, obviously they can provide some clues to the origin of irons. Wasson (1974) and Larimer (1978) discuss the classification, properties and origins of chondrites and differentiated meteorites. It has sometimes been assumed that the melting processes which produced differentiated meteorites destroyed any record of earlier processes, but this is not the case.

I. CLASSIFICATION

Approximately 600 iron meteorites are known and about 500 of these have been analyzed for Ni, Ga and Ge and classified according to the scheme devised by Wasson and associates (Wasson 1974; Scott and Wasson 1975, 1976). These elements are the most useful for classifying irons (for reasons explained later) but other chemical or mineralogical parameters may be used (Scott 1972; Buchwald 1975). On Ge-Ni (Fig. 1) or Ga-Ni plots the analyses fall into well-defined clusters. Ten groups have relatively narrow concentration ranges compared to the total observed range. Nickel, Ga and Ge vary by factors of less than 1.4, 1.2 and 2.3 in each of these groups, c.f. factors of > 6, 10^3 and 10^4 for all irons. Two groups, IAB and IIICD, have much wider concentration ranges and there is other geochemical and mineralogical evidence that they have a different origin from the other groups. The Ni-rich tail of IAB (called IB) and IIICD, which are marked in Fig. 1 with dashed lines, account for only 4% of all irons. The concentration

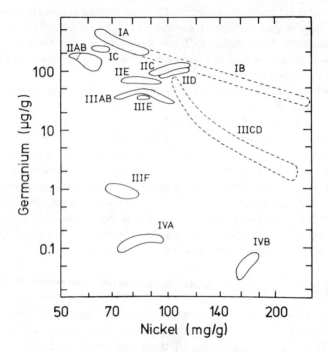

Fig. 1. Logarithmic plot of Ni against Ge concentrations showing the 12 fields where
86% of all analyzed iron meteorites plot. Group populations are shown in Fig. 2. All
the irons in a single group formed in the same parent body but several groups may have
been in a single body. Ge and Ni are the most useful elements for classifying irons but
many other chemical or mineralogical parameters can be used. Groups IAB and IIICD
have large fields in this figure and have a different origin from the remainder.

ranges and analytical errors for individual irons are under 5% for each of the
elements Ni, Ga and Ge. (A possible exception is Ni in IAB irons, judging
from the variation found in Canyon Diablo samples, 7-9% Ni.)

The populations and properties of the 12 groups are shown in Table I.
Some 14% of all irons do not belong to these groups and are related to 3 or
fewer irons; they are called anomalous. (The minimum population for a group
was arbitrarily set at 5.) Although no one has ever suggested that these irons
have an origin which is any different from those in the groups, the word
'anomalous' might imply an unusual history for these meteorites. Since there
is evidence that they have a similar origin to irons in the groups (Scott 1979),
some other term such as "grouplets" is a useful alternative name for the
anomalous irons. Figure 2 shows a histogram of mean Ge concentrations and
populations in all the groups and grouplets (except for IB, IIICD and two
anomalous irons). The 69 anomalous irons divide into 5 doublets, 2 triplets, 1
quadruplet and 49 unique irons. Further analytical and mineralogical studies
of these irons will probably reveal some more genetic relationships. Cosmic

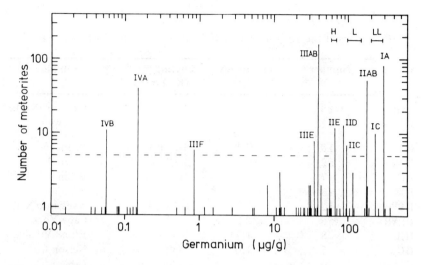

Fig. 2. Histogram showing mean Ge concentration and population of the groups and the 14% of irons which lie outside these groups. These so-called 'anomalous' irons have similar properties to the groups and probably formed in the same way. The minimum population for a group is arbitrarily set at 5. Most iron groups have Ge concentrations close to those in the H, L and LL chondrite groups.

ray exposure ages, which are discussed below, show that in at least two groups, the members were once in a single parent body. In each of the groups the irons can be arranged in a chemical sequence in which trace element concentrations vary smoothly. Since the chemical trends cannot be extrapolated from one group to the next, it is clear that each group formed in a separate chemical system. Whether this means that no two groups ever shared the same parent body is not clear.

Originally only 4 groups were recognized; those were labeled I to IV in order of decreasing Ge concentration. But subsequent studies revealed the 12 shown in Fig. 1. Note that IVA is no more closely related to IVB than it is to IIIF. At various times subgroups have been identified in some of the 12 groups but the uniformity of chemical trends within the groups suggests that these are not chemically significant divisions. Thus for group IIIAB, on all interelement graphs except those involving Ga and Ge, the chemical trends in IIIA (the low-Ni end of the group) exactly match those in IIIB when extrapolated to higher Ni concentrations. Similarly in group IVA the existence of subgroups is not supported by the well-defined chemical trends (Scott and Wasson 1975). In IVB, which only has 11 members, there is a hiatus between low and high Ni members. If this is assumed to be a result of poor sampling, the IVB trends are just like those in nearly all other groups and probably result from planetary, igneous processes (Scott 1978a). The contrary view (Larimer and Rambaldi 1978) that two IVB subgroups formed

TABLE I

Properties of the 12 Iron Meteorite Groups with 5 or More Members

Group	Number	Frequency (%)	Cooling rate[d] (K Myr^{-1})	Example
IAB[a]	88	18.3	1-5	Canyon Diablo
IIICD[a]	12	2.4	1-5	Tazewell
IC[c]	10	2.1	3->100	Bendegó
IIAB	52	10.8	2-10	Coahuila
IIC	7	1.4	100-500	Ballinoo
IID	13	2.7	1-2	Needles
IIE[b,c]	12	2.5	0.2-400	Weekeroo Sta.
IIIAB	156	32.3	1-10	Cape York
IIIE	8	1.7	0.5-2	Rhine Villa
IIIF	5	1.0	5-20	Nelson Co.
IVA	40	8.3	3-200	Gibeon
IVB	11	2.3	5-200	Hoba
Others	67	13.9	0.3->1000	Mbosi

[a]Meteorites contain chondritic silicates.

[b]Meteorites contain differentiated silicates.

[c]Wide cooling rate range is not correlated with metal composition.

[d]Cooling rates derived by Goldstein and Short technique (1967), except for IIAB (Randich and Goldstein 1978).

at different nebular pressures and temperatures by unidentified processes is much less attractive. Wood (1978) proposes that groups IVA and IIIAB should both be subdivided into no less than 3 groups on the basis of chemistry and cooling rates. To ignore the uniformity of chemical trends within these two groups seems ill advised.

The chemical classification of irons has been criticized because so many irons (14%) do not belong to the 12 named groups. By contrast only a few of the ~1000 chondrites lie outside the 8 groups, H, L, LL, CI, CM, CO, CV and E: Pontlyfni, Winona and Mt. Morris (Wisconsin) (which are all related to IAB irons); Tierra Blanca, Kakangari; Acapulco, Antarctic meteorite ALHA 77081; chondritic inclusions in Cumberland Falls. But the much greater abundance of anomalous irons compared to anomalous chondrites *is* entirely consistent with the populations of the named groups of irons and chondrites. Figure 3 is a logarithmic graph of cumulative group frequency against group population (Rajan and Scott, unpublished) showing data for 485 irons (finds and falls) (Scott and Wasson 1975) and 603 chondrite falls (Wasson 1974). (The proportion of anomalous irons among the 452 finds is only slightly lower than among the 31 falls [14%, c.f., 19%], but the proportions of rare chondrites, like LL and E, are much lower in the finds than the falls. Thus only data for chondrite falls are shown.)

The lines shown are trends drawn through data for groups having more than 5 members. If these trends are extrapolated (dashed lines) they predict numbers of anomalous irons (12%) and anomalous chondrites (1.8%) which are not too different from those observed (14% and 0.3%, respectively). The graph does predict that some more pairings will be discovered amongst the unique irons. This treatment assumes that the irons and chondrites both form homogeneous populations. This is probably a good assumption for the irons but for the chondrites a separate origin for the C group (e.g., in comets) cannot be excluded. However, if data for C chondrites were omitted the graph would predict even fewer anomalous chondrites. There are many factors which could affect the population of a group (e.g., size and number (?) of parent bodies, location, bombardment rate, strength and collision lifetime of meteorites, and unknown dynamic factors), and some of these may be responsible for the different gradients of irons and chondrites in Fig. 3. In the irons there is some tendency for the smaller groups and grouplets to have faster cooling rates which could suggest that size of the body is an important factor. An entirely different explanation for the different gradients in Fig. 3 is considered at the end of this chapter.

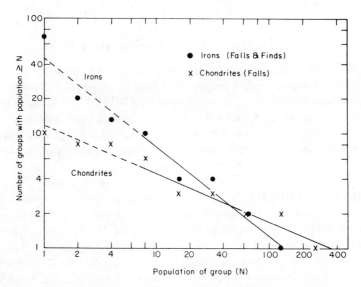

Fig. 3. Logarithmic graph of cumulative group frequency against group population for irons and chondrites. Lines through data for the groups are extrapolated to include grouplets. For both irons and chondrites the observed number of grouplets ($N < 5$) is close to that predicted by the extrapolations of data for the groups ($N \geqslant 5$). Irons are samples of 12 groups and about 50 grouplets whereas chondrites come from only 8 groups and less than four grouplets. (From Rajan and Scott, unpublished.)

Important clues to the origins of the groups are provided in their mineralogy (Buchwald 1975, 1977). Especially useful are the abundance and composition of silicate inclusions and sulfide nodules. Silicates would quickly float out of molten metal so their abundance in groups IAB and IIE shows that metal in these groups must have frozen soon after metal and silicates were mixed. Silicates in groups IAB and IIICD are similar to those in chondrites but in IIE they have probably been igneously differentiated. Related to the IAB irons are three chondrites with similar elemental and isotopic compositions to the IAB silicates (Bild 1977; Davis *et al.* 1977). Ages and isotopic data from these silicate inclusions are not discussed in this chapter (see references in Scott and Wasson 1975; Clayton and Mayeda 1978; Niemeyer 1979). Rb-Sr formation ages are 4.6 ± 0.1 Gyr except for the IIE iron, Kodaikanal, which gives an age of 3.8 ± 0.1 Gyr. The recent discovery of evidence of ^{107}Pd in a IVB iron (Kelly and Wasserburg 1978) shows that the time interval between the last element production and the beginning of the fractionation of metal from silicate was comparable to the half life of ^{107}Pd, 6 Myr.

In the following sections, information in the cosmic ray exposure ages, cooling rates and chemical compositions of the irons is examined for clues to their parent bodies and formation history. Comparisons are made between metal in irons and metal in other types of meteorites to provide additional constraints.

II. COSMIC RAY EXPOSURE AGES

The cosmic ray exposure age measures the time for which the meteorite existed as a sub-meter object in space. The most extensive results for irons are those obtained by Voshage (1967, 1978) using the ^{40}K-^{41}K method (Fig. 4). All but one of the ages for 19 IIIAB irons are within experimental error of 650 Myr, and 8 of 10 group IVA irons have an age of 400 Myr. This strongly suggests that in each of these groups nearly all the irons were produced in a single collision. Thus the members of each group must have resided in one parent body prior to this collision. Except for 3 groups where there are insufficient data (IIC, IIE and IIIF) the remaining groups do not seem to have been involved in such catastrophic collisions. Groups IAB and IIAB, which have 12 and 9 analyzed members, respectively (Fig. 4), were both involved in at least 4 collisions between 100 and 1200 Myr ago.

The ^{40}K-^{41}K method does not provide accurate results for irons with ages ⩽ 100 Myr. Wasson (1974) estimates that around 10% of all irons have ages under 100 Myr. By contrast all chondrites have ages under 100 Myr. This may be because irons and chondrites come from different parts of the solar system, or it may be due to the relatively short collisional lifetime for sub-meter silicate bodies (⩽ 100 Myr). This latter factor certainly accounts for the absence of any achondrites with ages ⩾ 100 Myr which must have

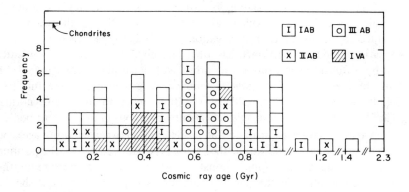

Fig. 4. Cosmic ray exposure ages of 76 iron meteorites determined by the ^{40}K-^{41}K method (Voshage 1967; 1978). Two groups, IIIAB and IVA, show well defined peaks at 650 and 400 Myr respectively. Nearly 40% of all irons were produced by collisions at these two times. The remaining groups like IIAB and IAB show a wide range of exposure ages and were evidently produced in numerous collisions during the last 1000 Myr.

once surrounded the irons in their parent bodies. The relative paucity of exposure ages above 1000 Myr may be because this is the collisional lifetime for irons (Wetherill and Williams 1979).

The frequency of collisions that produce iron meteorites will be dependent on factors like the cratering rate, size of parent bodies, mechanical strength of metal (or metal-silicate mixture) and whether or not the metal is in a core or raisins. Voshage (1978) interprets his data with models in which IIIAB and IVA both formed in cores whereas group IAB, which shows a quasi-continuous age distribution, formed as raisins. He notes that IIA irons are single crystals of Fe,Ni and their lower mechanical strength probably accounts for the spread of low ages in this group (50-500 Myr). The same explanation cannot account for the age spread of 250-900 Myr in group IVB, which probably formed in the same way as IIIAB and IVA (see below), as IVB irons are not known to be weaker.

The two groups with clustered exposure ages have a preponderance of heavily shocked members. Jain and Lipschutz (1971) find that nearly all IIIAB irons and ~60% of IVA members experienced shock pressures above 130 kbar. Other groups have abundances of shock-melted sulfides like those in these two groups, but tend to lack shock-hatched kamacite (Buchwald 1975). These other groups may have been annealed after deformation to remove deformation effects from the kamacite but not the sulfide. Thus it is not known for certain that the 650 and 400 Myr events were larger than any other meteorite-producing collisions, but they certainly produced nearly 40% of all our iron meteorites.

Evidence for solar heating of irons during cosmic ray exposure is provided by a few irons which have low $^3He/^4He$ ratios. Both isotopes are produced predominantly by spallation; nearly half the ^3He comes from spallation produced ^3H, which has a half life of 12.3 yr. Tritium is evidently lost by diffusion from irons having orbits with small perihelia (≤ 0.4 AU ?). Most of these irons also show metallographic evidence for reheating (see Schultz et al. 1971; Buchwald 1971). Other reheated irons with normal $^3He/^4He$ ratios were presumably reheated by collisions in their parent bodies.

Attempts have been made to find correlations between exposure ages and the compositional trends within the groups. As discussed below, the position of a meteorite in the compositional sequence may be a measure of the meteorite's location in the parent body. For group IIICD, 5 irons show a good inverse correlation of Ni concentration and exposure age, but correlations are not visible in groups IIAB and IVB. There is no good model for IIICD compositional trends, although an origin in raisins seems probable in view of the many similarities with group IAB. The age-composition relationship would be consistent with gradual collisional stripping of a parent body in which the Ni content of metal raisins increased with depth. The absence of IIICD irons containing 7-10% Ni may be a result of collisional destruction as these would have been produced over 1000 Myr ago.

III. COOLING RATES

Much valuable information about meteorite parent bodies can be derived from estimates of the cooling rates experienced by the meteorites. Two techniques, one based on metallography, the other on ^{244}Pu fission tracks, give broadly similar results; most meteorites cooled through the range 900-400 K at the rate of 1-100 K/Myr. Burial depth and size of the parent body cannot be uniquely determined from this information. However, thermal models for differentiated asteroidal bodies of chondritic composition suggest that burial depths for the iron meteorites were typically 10-200 km (Wood 1967; Fricker et al. 1970). Fricker et al. find, for example, that a cooling rate of 10 K/Myr could be achieved at the center of a 100 km radius body or at burial depths of 40 and 15 km in bodies with radii 200 and 300 km, respectively.

The heating source which melted metal in these asteroids is not known for certain; ^{26}Al or electromagnetic induction are the two most likely candidates (see Sonett and Herbert 1977 and the chapter by Sonett and Reynolds). In both cases the source of heat would have been removed when the meteorites were cooling through 900-400 K. (The concentration of long-lived radionuclides like ^{40}K and ^{235}U will not significantly influence cooling rates in these small bodies until the temperature falls to within $\sim$200 K of the surface temperature.) Under conductive cooling the cooling rate will be proportional to $T-T_0$ where T_0 is the surface temperature. Since the black-body temperature at 2.5 AU is 176 K, cooling rates will decrease by a factor of 2 in the temperature intervals 1000 to 590 and 590 to 380 K.

Thermal models for meteorite parent bodies are reviewed in the chapter by Wood. He considers models in which the insulating material controlling the cooling rate is not achondritic silicate but fine dust. If the thermal conductivity of the dust is 100 times lower than that of achondritic silicate, a dust layer of 1-km thickness would provide the same insulation as a 10-km silicate layer. Thus the burial depths and planetesimal sizes calculated by Fricker *et al.* (1970) may be too large. A very different thermal model in which cooling is controlled by the rate of decrease in the luminosity of the early sun is briefly considered by Wasson (1972). In this discussion it is assumed that meteorites were buried at depths of 10-200 km.

Most meteorite cooling rates have been derived from a study of the composition of the metal phases. The ^{244}Pu fission technique has only recently been applied to irons and will be considered briefly at the end of this section. The metallographic technique has been developed with such success that there is probably more information on the cooling rates of iron meteorites than on any other geological samples. The method relies on the compositional changes that occur during the transformation of face-centred cubic Fe,Ni (taenite) to body-centred Fe,Ni (kamacite) on cooling through the temperature range 900-700 K. Below this temperature range, diffusion rates are too sluggish for any measurable changes to occur. Experimentally determined values for the equilibrium phase compositions, diffusion coefficients and the bulk composition of meteoritic metal allow computer modeling of the transformation. The cooling rate is derived by comparing the observed dimensions and compositional gradients of kamacite and taenite crystals with those generated by the computer model for various cooling rates (Wood 1964; Goldstein and Short 1967). One other parameter that has to be considered is undercooling below the equilibrium temperature for nucleation of kamacite. Goldstein and Short find that this is typically 100 K. Assuming that this is correct for all irons, the cooling rate can be estimated simply from the bulk Ni concentration of the iron and the width of the kamacite bands using curves of Goldstein and Short, or an equation calculated from these curves by Wasson (1971).

Cooling rates derived by the Goldstein-Short technique from kamacite widths are shown in Table I. They are normalized to a temperature of 770 K and are probably accurate to a factor of < 5. Cooling rates in groups IIC, IIIF, IVA and IVB are nearly all in the range 10-500 K/Myr, whereas the remaining groups and most grouplets generally have slower cooling rates in the range 1-10 K/Myr. The total range is 0.3 to 10^3 K/Myr. There are a few irons (< 10) with cooling rates up to 10^8 times faster but these may have been reheated. Mean cooling rates of groups and grouplets are inversely correlated with the mean Ge concentration (Scott 1978*a*). This may mean that metal grains which inefficiently condensed, or subsequently lost, volatiles like Ge tended to form cores in small bodies or near-surface raisins in larger bodies.

The range of cooling rates in a *single* group provides important information about the distribution of members in the parent body at the time of cooling. Because of the high thermal conductivity of metal, irons in a single core should have indistinguishable cooling rates. The early work of Goldstein and Short (1967) suggested that groups IAB and IIIB (the Ni-rich end of IIIAB) had uniform cooling rates and may have formed in a core, whereas group IVA and other groups had wide ranges and formed as raisins in their respective parent bodies. However, geochemical data which is discussed below suggests to some workers that groups IIIAB and IVA both formed in cores. To help resolve these conflicts between metallographic cooling rates and geochemical models, new studies of cooling rates in group IVA have recently been made. Group IVA was chosen because the Goldstein-Short cooling rates vary from 5-200 K/Myr (systematically increasing with decreasing Ni) and it has low concentrations of P, a minor element which has appreciable effects on phase relations and diffusion coefficients in the binary Fe-Ni system.

Willis and Wasson (1978*a,b*) in their investigation of group IVA cooling rates did not use kamacite bandwidths, as calculated bandwidths are sensitive to uncertainties in undercooling and the distance between kamacite nucleation sites. Instead they adapted Wood's technique (1964, 1967) in which the central Ni content of taenite grains is plotted as a function of crystal size. By looking only at crystals smaller than 10 μm in size, which equilibrated at lower temperatures, they minimized effects due to undercooling uncertainties. However, the difficulty with this approach is that the phase equilibria and diffusion coefficients are measured at high temperatures (above 800-900 K) and have to be extrapolated to the temperatures at which small grains are still equilibrating ($\sim$700 K). Small crystals are also more sensitive to the effects of shock reheating (Axon 1979). These authors used an iterative technique, matching computer profiles of Ni in kamacite to try and obtain more accurate values for the equilibrium Ni concentrations in kamacite at these low temperatures. They also used a psuedo-binary phase diagram to compensate for the higher concentration of P in high Ni irons. Their results for six IVA irons (Fig. 5), show a uniform cooling rate of 20 K/Myr, contrary to earlier data of Goldstein and Short (1967).

Moren and Goldstein (1978) using almost identical techniques to those of Willis and Wasson (1978*a*) reinvestigated the cooling rates in group IVA irons, but their results confirmed Goldstein and Short's (1967) work. They therefore denied the claim of Willis and Wasson that IVA irons could have cooled through 700 K in a core. An important difference between these workers was their choice of published diffusion data; their measurements of Ni concentrations in taenite and their techniques of computer modeling were in general agreement. Willis and Wasson had shown that the systematic

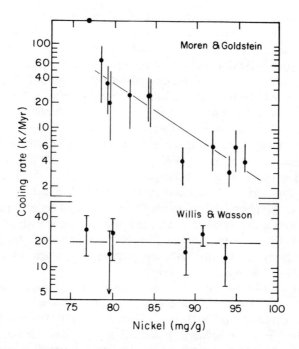

Fig. 5. Cooling rates plotted against Ni for group IVA irons. According to Moren and Goldstein (1979), there is a systematic decrease in cooling rate with increasing Ni, which means that group IVA irons could not have cooled through 900 K in a core, due to the high thermal conductivity of metal. Willis and Wasson (1978a) find uniform cooling rates and favor a core origin for this group.

variation of cooling rate with Ni content could be eliminated largely by using different diffusion coefficients, but Moren and Goldstein considered their adversaries' choice of diffusion data to be ill judged.

Results of another study by Moren and Goldstein (1979) using a more sophisticated ternary Fe-Ni-P model for kamacite growth are also shown in Fig. 5; their results agree closely with their earlier work. The vertical bars in Fig. 5 through their data represent what they consider to be the maximum allowable changes in cooling rates as a result of input errors. Since the errors on their diffusion data included the values favored by Willis and Wasson, the results in Fig. 5 directly contradict those of Willis and Wasson. However, uncertainty remains in the magnitude of the errors in the equilibrium phase data for kamacite. The iterative techniques used to refine these data do *not* converge, and it is possible that the uncertainties are greater than the value of ±0.1% Ni assumed by Moren and Goldstein. (The experimental errors at 770 K are ±1% Ni.) Because of great difficulties in assessing the error in the cooling rates, these authors note (personal communication 1979) that they are unable to provide a clear statement of the accuracy of their cooling rates.

Clearly new experimental determinations of phase equilibria and diffusion coefficients at low temperatures and better methods of estimating errors are needed to settle this issue. At present the cooling-rate data tend to suggest that IVA irons could not have cooled through 700 K in a core.

One possible method of checking whether the range of IVA cooling rates is genuine would be to analyze a suite of Canyon Diablo samples. There are local variations on a scale of 0.1 to 1 m in the Ni concentration of this IA iron, 7-9% Ni. If Moren and Goldstein's ternary model (1979) were to give the same cooling rates for all Canyon Diablo samples, this would tend to confirm their results for group IVA.

Another metallographic technique has been used to obtain cooling rates for 7 IIA irons, which do not contain taenite (Randich and Goldstein 1978). The growth of phosphide in kamacite was modeled by a ternary Fe-Ni-P model, which is closely analogous to that used for the growth of kamacite in taenite. Results range from 0.8 to 10 K/Myr, but in view of the precision of around a factor of ±3, the data are probably consistent with a uniform cooling rate of 2 K/Myr.

Wood (1967) determined cooling rates for chondrites in the range 900-700 K from a study of the composition of their metal grains. His values and those obtained later by Taylor and Heymann (1971) range from 0.2 to 100 K/Myr; most lie between 1 and 10 K/Myr. Mesosiderites, which are breccias composed of metal and igneously differentiated silicates, give the lowest cooling rates of 0.1 K/Myr (Powell 1969), according to Wood's technique. Estimates of the cooling rate in one mesosiderite by Kulpecz and Hewins (1978) using Randich and Goldstein's (1978) model for phosphide growth in kamacite give a value between 0.1 to 0.01 K/Myr. This value may be too low because of the presence of taenite, which is absent in IIA irons. Metal in pallasites, which are mixtures of metal and olivine, cooled at around 1 K/Myr according to Buseck and Goldstein (1969). With the exception of reheated meteorites, all these meteorites have metallographic cooling rates of 0.1 to 100 K/Myr, like most of the irons.

There is an unresolved problem in reconciling cooling rates of 0.1 to 1 K/Myr in the range 900-700 K with some dating techniques which give very old ages. According to Turner et al. (1978), many chondrites cooled to 510 ±120 K, the Ar retention temperature, 4.48 ±0.03 Gyr ago. These meteorites must therefore have cooled through the range 1000-500 K in less than say 200 Myr, giving a minimum average cooling rate of 2.5 K/Myr. There are only two chondrites for which Wood calculated cooling rates below 1 K/Myr and these were not dated by Turner et al. For the mesosiderites which give the slowest cooling rates of 0.1 K/Myr, there are no ^{40}Ar-^{39}Ar ages available.

Finally, brief mention must be made of the ^{244}Pu fission track method of estimating cooling rates developed by Pellas and Storzer (1977). By counting the tracks in phosphates and in the surface of olivine, pyroxene and feldspar crystals which were adjacent to phosphates, it is possible to estimate the concentrations of ^{244}Pu in phosphate at the temperatures at which these minerals begin to retain tracks. For example, if the density of ^{244}Pu fission tracks in the surface of olivine is half that in the surface of feldspar, we can deduce that 82 Myr (the ^{244}Pu half life) have elapsed between the 50% track retention temperatures of feldspar and olivine. These temperatures are 600 and 470 K respectively, giving a mean cooling rate of 1.6 K/Myr. The ^{244}Pu content of phosphate at 920 K can also be estimated from the concentration of fission Xe; this temperature is the 50% retention temperature for Xe in phosphate (Pellas and Storzer 1977).

Fission track estimates for 12 chondrites gave surprisingly uniform cooling rates of 0.7 to 1.6 K/Myr in the range 300-600 K (Pellas and Storzer 1979), in good general agreement with the metallographic technique. But two of these chondrites, Shaw and Tillaberi, give metallographic cooling rates > 100 K/Myr. This discrepancy may arise because the two techniques are measuring the cooling rates over different temperature ranges (see Scott and Rajan 1979). Turner et al. (1978) suggest that partial loss of tracks by annealing might lead to an underestimate in cooling rates. However, any loss of tracks by annealing should be revealed by a reduction in track length. One limitation of the fission track method is that because of the errors in measuring track densities, cooling rates faster than 80 K/Myr cannot be distinguished. Preliminary data for chondritic inclusions in two IAB irons (Benkheiri et al. 1979) give cooling rates of $\sim$ 1.5 K/Myr. These results compare very favorably with the typical cooling rates of 1-5 K/Myr determined for IAB irons with the Goldstein-Short technique.

IV. COSMOCHEMICAL MODELS

Studies of the compositions of individual minerals in iron meteorites provide details about temperatures (see above) and pressures (see Anders 1964) in the parent bodies. The latter did not exceed a few kbar and show that parent bodies were less than 1000 km in radius. The bulk compositions of irons not only identify genetically related groups but provide a record of their formation history. This information is less direct than that discussed above as cosmochemical models must be devised and then tested with chemical data. The record in iron meteorites is more complicated than in chondrites as the irons have been extensively affected by planetary igneous processes, which the chondrites escaped. However, the redeeming feature of the irons is that their chemical fractionation patterns are generally simple. It has proved possible to identify those chemical features due to condensation and accretion processes in the nebula, and those features due to planetary melting processes. The detailed mechanisms of the planetary and nebula processes are not well established, although various promising models are available.

A comparison of the concentrations of siderophile (metal-loving)

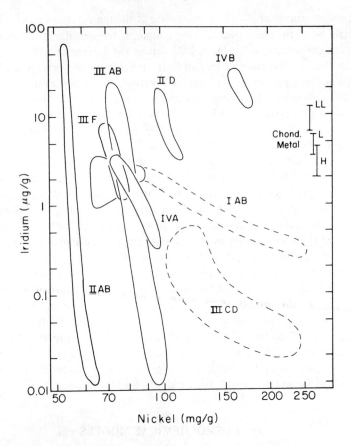

Fig. 6. Logarithmic graph of Ir plotted against Ni showing the compositional fields of the main iron meteorite groups. In two groups of irons the total range of Ir concentrations is over a factor of 1000. Metal from a group of equilibrated chondrites, shows an Ir range of less than a factor of two, much less than in an iron meteorite group. The Ir-Ni fractionation in the iron groups occurred during planetary melting or solidification of molten metal, a process the chondrites escaped.

elements in chondritic metal grains with those in iron meteorites shows two major differences. The concentration range in an iron group is nearly always much wider than that in a group of chondrites. Iridium (Fig. 6) and other Pt metals show this trend most effectively. The total range of Ir concentration of metal in a group of equilibrated ordinary chondrites is less than a factor of 2 (e.g., Chou *et al.* 1973; Rambaldi 1976) whereas in most iron groups it exceeds a factor of 10, and in IIAB and IIIAB 1000. The second major difference is that the mean compositions of iron meteorite groups generally show a much wider range than those of metal in all the chondrite groups. This is most obvious for Ga and Ge (Fig. 2) and Ir (Fig. 6).

The broad range of siderophile concentrations in irons is attributed largely to two processes. A primary process, which probably occurred in the nebula during condensation and accretion, established the mean siderophilic ratios in each group. It is likely that the same process was important in establishing the elemental ratios in chondrite groups also. A secondary process occurred as a result of melting and solidification of molten metal in the parent bodies. This process is responsible for the wide concentration range within each group of irons. Since chondrites have never been melted after accretion, metal in chondrite groups is more homogeneous than a typical iron meteorite group.

Primary Fractionation

Fractionation of elements between dust and gas in the solar nebula can be modeled most simply by assuming that chemical and thermal equilibrium prevailed. Following Urey, Larimer (1967) calculated the changes in the composition of dust during condensation, and Larimer and Anders (1967) showed how these changes could account for many chemical variations in chondrites. The achievements and defects of the equilibrium condensation theory in chondrites have been reviewed by Grossman and Larimer (1974) and Arrhenius (1978), respectively.

During slow cooling from temperatures above 1500 K which the equilibrium theory requires, siderophile elements will dissolve into Fe,Ni grains so that their vapor pressure equals their partial pressure in the nebula. A convenient parameter for initial investigations of condensation is the temperature at which 50% of an element has condensed. Fig. 7 shows the 50% condensation temperatures calculated for a variety of elements assuming a total nebular pressure of 10^{-4} atm (Grossman and Larimer 1974; Wai and Wasson 1977; Palme and Wlotzka 1976; Kelly and Larimer 1977). Also shown are the mean elemental abundances in three groups of chondrites, CM2 and CO3 (both carbonaceous) and an ordinary group H5,6, together with mean abundances in 3 groups of irons IIIAB, IIIF and IVB and in one unique iron, Denver City (Mason 1971; Wai and Wasson 1977; Takahashi et al. 1978; Krähenbühl et al. 1973; Laul et al. 1973; Case et al. 1973; Scott 1972; Scott 1978a,b and references listed therein). Elemental concentrations are divided by those in CI chondrites, which are believed to be closest in composition to that of the non-volatile portion of the original solar nebula. Chondrite abundances are normalized using Si and irons using Ni, as is customary, although it can be seen that use of Ni for chondrites also would not affect the plot significantly.

Both irons and chondrites show the same general trends in Fig. 7 namely refractory elements which condense between 1800 and 1400 K have abundances relatively close to CI levels, whereas the volatile elements have abundances which decrease with decreasing condensation temperature. The existence of these trends in chondrites was recognized by Larimer and Anders

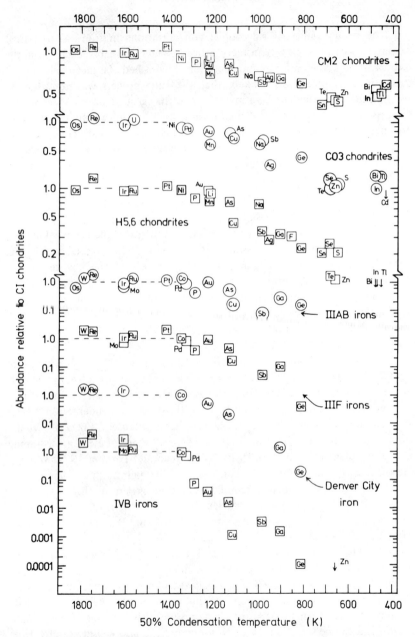

Fig. 7. Elemental abundances in three groups of chondrites (CM2, CO3 and H5,6), three groups of irons (IIIAB, IIIF and IVB) and a unique iron plotted against 50% nebular condensation temperatures calculated for a total pressure of 10^{-4} atm. Concentration data are normalized to CI chondrite concentrations using Si for chondrites and Ni for irons. In both irons and chondrites, refractory elements on the left have abundances $\simeq$ 1, but volatile element abundances decrease with increasing volatility. Note the changes in the vertical scale below CM2 and H5,6 chondrites.

(1967). However, new analytical data and thermodynamic calculations by the authors listed above have greatly reduced the scatter in these trends. That the irons would also show a correlation between abundance and volatility for siderophiles was predicted by Wasson and Wai (1976), and demonstrated by Wai *et al.* (1978) and Scott (1978*b*).

One difference between the concentrations of volatile elements in irons and those in chondrites (Fig. 7) is that some groups and grouplets of irons (e.g., IIIF and IVB) are depleted to a much greater extent. However, *most* irons have Ge concentrations (Fig. 2), for example, which are comparable to those in equilibrated chondritic metal. Assuming that similar nebular mechanisms are responsible for trends in both irons and chondrites, this difference may be due to sampling: the irons come from nearly 70 groups and grouplets whereas the chondrites come from less than 15.

There is little agreement about the mechanisms responsible for the depletions of volatiles shown in Fig. 7. Most of the proposed mechanisms have been devised to explain chondrite abundances and fall into one of the following simplified divisions:

(a) Incomplete Condensation in the Nebula. Three possible causes for incomplete condensation have been identified: accretion of dust grains, kinetic effects and dust-gas fractionation. The first two are broadly similar in that volatile condensation is impeded as the grains fail to maintain equilibrium with the gas during cooling due to increased size of dust aggregates or slower rates of diffusion. Clearly one or both effects must have been very important for some elements. Larimer and Anders and their colleagues have claimed that accretion effects were almost entirely responsible for the depletion of highly volatile elements like Bi, In, Tl in type 5 and 6 ordinary chondrites (e.g., Larimer 1973). But the idea that the kinetic effects were unimportant so that temperatures and pressures in the nebula during accretion can be calculated from the equilibrium condensation theory has been vigorously disputed, e.g., by Blander (1975). Larimer (1967) provides a good discussion of the effects of loss of equilibrium due to limited diffusion at low temperatures, but it is Blander who has championed the importance of kinetic effects although not specifically to explain the general volatility-abundance correlations in Fig. 7 (see e.g., Blander 1971).

The third possible nebular mechanism for loss of volatiles, dust-gas fractionation, envisages volatiles being depleted by such processes as preferential settling of dust grains to the nebular plane or gradual loss of gas from the nebula during condensation. In the model of Wai and Wasson (1977) it is the most important mechanism for depleting the moderately volatile elements. Anders and associates (e.g., Takahashi *et al.* 1978) believe it plays a more minor role.

(b) Chondrule Formation. According to Larimer and Anders (1967), the relatively flat portions on the right side of the volatility-condensation

temperature diagrams (Sb-Cd in CM2 chondrites, Te-Tl in CO3 and Sb-S in H5,6) are a result of mixing a high-temperature fraction lacking volatiles (largely chondrules) with a low-temperature fraction (matrix), which has CI abundances. The proportions are approximately 1:1 in CM2 chondrites and 3:1 in CO3 and H5,6 chondrites. Departures from the plateaus predicted by this model are attributed to incomplete loss of volatiles during chondrule formation or dust-gas fractionation (e.g., Takahashi *et al.* 1978). Wai and Wasson (1977) suggest, however, that volatiles were either not lost from chondrules or immediately recondensed. Anders and his associates consider the latter may have happened and envisage that the volatile depletion may result from preferential settling of chondrules (another type of dust-gas fractionation). Anders (1977) and Wasson (1977) provide a vitriolic debate on the importance of chondrule formation in establishing volatile abundances.

(c) Thermal Metamorphism in Parent Bodies. The strong depletion of highly volatile Bi, In and Tl in type 5 and 6 ordinary chondrites is attributed by Wood, Wasson and Blander (see Wasson 1974) to thermal metamorphism in the parent bodies. Lipschutz and his colleagues believe that open-system metamorphism is responsible for the loss of these and other volatiles in E and C chondrites but not in ordinary chondrites (see Ikramuddin *et al.* 1977). Arguments against this model, the relevance and interpretations of the heating experiments of Lipschutz and his collaborators are given by Takahashi *et al.* (1979). Relevant to the depletion of Bi, In and Tl in ordinary chondrites is the plot of primordial $^{20}Ne/^{36}Ar$ against primordial ^{36}Ar in ordinary chondrite minerals (Alaerts *et al.* 1977). Concentrations of Ne and Ar are correlated with those of In, Bi and Tl. Thus if losses are due to diffusion during metamorphism, a positive correlation would be expected as Ne would be lost more readily than Ar. Alaerts *et al.* find a negative correlation but with their additional data (1979) the correlation has become very much weaker.

This brief review reveals no consensus on mechanisms and models which might explain depletions of volatiles in chondrites. It seems safe to conclude that none of the mechanisms listed above can be definitely excluded. For irons there are far fewer papers dealing with primary fractionation mechanisms, in part because the similarity to chondritic trends has only recently been recognized.

Kelly and Larimer (1977) and Sears (1978a, 1979) use the accretion mechanism alone to explain the volatile abundances in irons. They calculate temperatures and pressures of accretion by matching mean compositions of iron meteorite groups with those for metal grains cooling in equilibrium with the nebula. Kelly and Larimer find many elements in group IVB which have concentrations close to those calculated for metal grains at 1270 K and 10^{-5} atm. Sears considers that groups IAB, IIAB, IIIAB and IVA (but not IVB) accreted at lower temperatures (600-670 K). In his model the volatile depletion is caused by low pressures, 10^{-8} atm for IVA, c.f., 10^{-4} atm for IAB and IIAB. The condensation sequences due to decreasing temperatures and increasing pressures are similar but not identical. Sears' pressure range is much larger than other authors have considered and requires that the irons formed over an enormous range of heliocentric distances, from 1 to > 10 AU (Sears 1979).

At present the accretion model has only been tested for a few elements in a few iron meteorite groups. In view of the large number of variables, pressure, temperature and activity coefficients for most elements, it is by no means certain that the accretion model is correct, but it clearly provides many useful insights. However, it is certainly not the only process that controlled volatile abundances in chondrites, and it is likely that another process affected the irons too.

Wasson and Wai (1976) take the view favored here that the irons and chondrites experienced similar volatile fractionations, and they invoke gas-dust fractionation for both groups of meteorites. No quantitative calculations have been made for such a model, but it should prove possible to derive a condensation sequence for an open system in which gas slowly leaks out. An alternative type of dust-gas fractionation is discussed by Wai and Wasson (1977). They consider that volatiles would begin to condense homogeneously forming aerosols which fail to accrete, but the necessity for homogeneous nucleation at low temperatures is not clear. The enormous variation of Ge abundances among the groups (1 to 10^{-4}) requires a similar variation in dust-gas fractionation which Sears (1979) finds puzzling.

Although these models have had some success in explaining the broad trends among irons, no detailed conclusion can yet be drawn from them about the nature of the parent bodies of iron-meteorite groups. It must also be stressed that the strong arguments of Blander (1975) and Arrhenius (1978) against the concept of equilibrium between gas and dust at low temperatures mean that the concept of accretion temperatures (especially those below $\sim$ 1000 K) must be treated with extreme caution.

The similarity of trends in the irons and chondrites (Fig. 7) is consistent with the idea that the mean relative abundances of siderophiles in iron groups were inherited from the metal grains which separated by planetary melting from some chondrite-like material. The discovery of chondrites with large volatile depletions like those in group IVA would confirm this theory. An alternative idea is that metal grains accreted preferentially to form iron meteorites. Such heterogeneous accretion has been proposed for group IAB irons (Wasson 1972), but it is doubtful whether heterogeneous accretion should be generally invoked.

Secondary Fractionation

Chemical variations *within* iron meteorite groups are very regular. In each

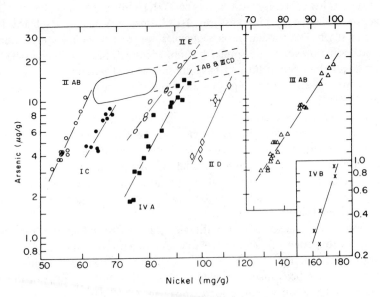

Fig. 8. Logarithmic graph of As plotted against Ni showing analyses of iron meteorites in groups IIAB, IC, IIE, IVA, IID, IIIAB and IVB. All these groups and IIIF, which is not shown, have a positive As-Ni correlation. Groups IAB and IIICD, which have a different origin from the other groups, have flatter slopes and are shown only in outline. The steep trends within groups result from planetary igneous processes, probably fractional crystallization of molten metal. In this process Ni and As would be concentrated in the last solids which crystallize.

group, the irons can be ranked in a sequence in which Ni, Au, As, Co, Pd, P, Mo concentrations tend to increase, while Ir, Os, Ru, Pt, Re, W and Cr decrease (Scott 1972; Yavnel 1975). The inverse correlation of Ni and Ir is shown in Fig. 6. Figure 8 shows as As-Ni plot for iron meteorites in groups IIAB, IC, IIE, IVA, IID, IIIAB and IVB. In each of these groups and IIIF, which is not shown, there is a positive correlation of Ni and As. Groups IAB and IIICD show less steep slopes, and groups IIC and IIIE have insufficient data to define a trend. Nickel data are from Wasson (1974), Scott and Wasson (1976) and As values from Smales *et al.* (1967), Scott (1977*b*, 1978*a*, and unpublished work). The existence of all these correlations, which were not known when the groups (except IC) were defined, illustrates the power of the chemical classification in revealing useful information in the irons.

The origin of secondary fractionation in iron groups (with the possible exceptions of IAB and IIICD) is very probably the distribution of elements between solid and liquid metal during melting or crystallization of metal in planetesimals. Consider first the chemical fractionation that may occur during solidification of a molten core or pool produced by melting and gravitational separation of metal. Elements will be distributed between solid and liquid

Fe,Ni according to the distribution (or partition) coefficient k which is defined as the ratio of the elemental concentration in the solid to that in the liquid, C_S/C_L. If each fraction of liquid in turn solidifies in equilibrium with the liquid, which is kept completely mixed by convection, but does not equilibrate with previously solidified metal, then C_S is given by

$$C_S = k \, C_L^{\circ} \, (1\text{-}f)^{k-1} \tag{1}$$

where C_L° is the original bulk concentration in the pool, and f is the fraction of liquid that has already crystallized. Detailed studies of core solidification in asteroids have not been made. Nevertheless, at present it seems plausible, but is not proven, that fractional crystallization as described by the above equation would occur during solidification of such cores.

Analyses of metal from such a core when plotted on inter-element logarithmic graphs like Figs. 6 and 8 will define a straight line of gradient $(k^A - 1)/(k^B - 1)$, where k^A and k^B are the distribution coefficients for elements A and B. Nickel and other elements with $k < 1$ will preferentially accumulate in the liquid, while elements like Ir with $k > 1$ will become depleted in the liquid. Because there is qualitative agreement between binary-phase diagram predictions and trends in the groups (except for Cr), a fractional crystallization model seemed to be promising to Scott (1972). Groups IAB and IIICD were excluded because they have very different trends and it is unlikely that they ever formed a single molten pool or core. Since then, experimental values for distribution coefficients of some elements in Fe,Ni have been obtained (Bild and Drake 1978; Goldstein and Friel 1978). I believe that these initial studies, while not entirely in agreement, do provide limited support for the crystallization model (Table II). Obvious difficulties are Cr and Ir. Bild and Drake suggest that the differences between experimental k values may result from lack of equilibrium.

The presence of 0.1-1% S in the form of FeS nodules in many irons shows that some liquid must have been trapped in the solidifying metal, as S is almost insoluble in solid metal, although easily dissolved in molten metal. Under these conditions, the fractional crystallization equation becomes

$$C_S = k \, C_L^{\circ} \, (1\text{-}f)^{(k-1)(1-\alpha)} \tag{2}$$

where α is the fraction of trapped liquid. The actual proportion of trapped liquid in the irons is difficult to estimate as the bulk S of the liquid is not known. Note that trapping of liquid does not change the gradient of the line defined by the solids on inter-element logarithmic plots, although it reduces the efficiency of fractionation. However, a few percent S in the liquid may appreciably affect the distribution coefficients. In an ingenious but speculative model to explain the anomalous Cr behavior, Kracher et al. (1977) suggest that a second S-rich immiscible liquid was preferentially

TABLE II

Comparison of Experimental Values for Solid/Liquid
Metal Distribution Coefficients with those Derived from IIIAB
Irons, Assuming they Fractionally Crystallized

| Element | Experimental values | | IIIAB Irons |
	(1)	(2)	(3)
Ni		0.88±.03	0.90
Au	0.18±0.2	0.64±.05	0.50
P		0.44±.1	0.17
Ge		0.77±.05	0.9-1.4
Co	0.97±.02	0.88±.03	0.94
Cr	0.53±.03	0.86±.05	2.1
Pt	0.9±.3	1.27±.05	1.7
Ir		2.1±.05	3.4

1. Bild and Drake, 1978
2. Goldstein and Friel, 1978
3. Scott, 1977, except for Ge from Wai *et al.* (1978)

trapped by the solid. There are no experimental data showing immiscibility in S-rich melts of meteoritic composition.

Kelly and Larimer (1977) suggest that the trends in some groups were produced during melting, and not during solidification. They propose that trends in group IAB, which are different from those in IIIAB, were produced during fractional melting (the inverse of the process described above). The concentration of an element in each fraction of liquid C_L is then given by

$$C_L = k'C_S^o (1-F)^{k'-1} \qquad (3)$$

where k' is the reciprocal of k, F is the fraction of solid that has melted and C_S^o is the initial concentration in the solid. This model for IAB is unappealing for the following reasons: (1) The required distribution coefficients are further from the measured values than those deduced from IIIAB data, e.g., Ni 0.5-0.7, Ir 20, Ge 12. (2) More extreme coefficients are needed to explain trends in group IIICD, which almost certainly has a similar origin to group IAB. (3) The abundance of Ni-poor relative to Ni-rich irons in IAB requires that most IAB irons represent extremely large values for $F \sim 0.8$. Because virtually all the S would be removed into early-formed liquids with low F values, the model cannot explain the high S abundance of IA irons. There is no other plausible model to explain IAB trends. Nebular processes have been invoked because of the consistent depletion of refractory elements in high Ni IAB and IIICD irons (Scott and Bild 1974). But the extent of these depletions is far larger than chondrites show. Clearly models for primary and secondary fractionations in irons cannot be considered entirely satisfactory while the origin of trends in IAB and IIICD is uncertain.

Because of the wide variation in the Goldstein-Short cooling rates in IVA irons (Fig. 5) and the smaller fractionation of Ir compared to that of IIIAB (Fig. 6), Kelly and Larimer (1977) proposed a partial melt origin for this group also. In their IVA model, the solid fractionally melts (as above) but the liquids aggregate into pools. The IVA irons would then have compositions C'_L given by

$$C'_L = C^o_S/F \; [1-(1-F)^{k'}] \; . \tag{4}$$

This model gives curves on log-log plots. Group IVA data do not show trends which are more curved than those of group IIIAB and a fractional melting model would fit equally well. Kelly and Larimer probably preferred the aggregation model as the IVA irons have CI Ir/Ni ratios in Ni-poor members. This is in accord with their primary fractionation models which predict a bulk CI Ir/Ni ratio for IVA, and the aggregation model which gives the final Ni-poor liquid ($F=1$) the bulk composition. (Note that models for primary and secondary fractionations are not chosen independently.) Like the melting model for IAB, the aggregate melting model for IVA requires a lower k value for Ni (0.8) and a higher value for Ir (10) than those listed in Table II.

Kelly and Larimer (1977) discuss, but do not seem to favor, a 2-stage model for group IIIAB trends in which metal pods are produced by partial melting, and then the pods fractionally crystallize. Thus the large IIIAB fractionation of Ir is achieved by two processes. Such a composite origin for IIIAB trends seems unlikely in view of the good fit of data to a single straight line on many plots, e.g., Fig. 8. Melting and crystallization would produce trends on logarithmic graphs with quite different gradients. They conclude, and I agree, that group IIIAB and most of the other groups fractionally crystallized, presumably in cores.

Arguing against partial melting models for group IVA, Willis and Wasson (1978a) note that if the parent body is internally heated, the degree of partial melting will increase with depth. Then deeply buried irons will be poorer in Ni and cool slower, contrary to what Moren and Goldstein (1979) find (Fig. 5). If IVA did experience fractional crystallization in a core, one explanation for the flatter trend in group IVA on the Ir-Ni plot (Fig. 6) is that stirring of the melt was less efficient during solidification of the IVA core.

Although Kelly and Larimer (1977) give a detailed theoretical description of trace element partition during partial melting, there have been no detailed studies of whether or not liquid metal would separate gravitationally from solid metal during asteroidal melting. Further, it is not known how such melts could be prevented from mixing or, in the case of internally heated bodies, forming cores. With external heating, it is easier to envisage how metal raisins, and not cores, could form.

This discussion of primary and secondary fractionation in irons has concentrated on the groups. However, it is possible to show that the grouplets have experienced very similar processes to the groups, even though the trends and mean concentrations of grouplets may not be known (Scott 1978b, 1979). On element-Ga plots the mean concentrations of the groups are positively correlated for volatile elements like Ge, P and Au, as would be expected from Fig. 7. Now the grouplets define very similar positive trends on these diagrams showing that they experienced the same primary fractionation as the groups. The scatter of grouplets from these primary fractionation curves on several element-Ga graphs for elements which show large secondary fractionations is largely explained if the grouplets were igneously differentiated like IIIAB.

Primary fractionation in groups was investigated above by averaging data for each group. The assumption that our sample of the groups is reasonably representative is probably fairly good for large groups like IIIAB. One can argue that the total range of Ir concentrations in IIIAB, for example, is almost as large as that shown by all the irons including the grouplets. So it is unlikely that group IIIAB extends to much higher or lower Ir concentrations. Another method of calculating the mean composition of groups is to assume the correctness of the fractional crystallization model and to use the calculated k values and the lowest Ni concentration in the group (where $f \simeq$ 0). In group IIIAB this gives a fairly similar result to a simple average of the data (Scott 1977a), but in other groups with fewer analyses it may give better means (Wai et al. 1978).

Other Processes

Not discussed so far is loss of Fe and other elements with some lithophilic tendencies by oxidation. If Fe alone is lost, abundance ratios, like those in Fig. 7, will not be affected, but on logarithmic element-Ni plots (Figs. 6 and 8) the data would be shifted diagonally upwards on straight lines of constant element/Ni ratio. Sears (1978a, 1979) argues that oxidation of Fe from metal occurred in the nebula during equilibrium condensation at 500-700 K, and not after the iron meteorite parent bodies had accumulated. Thus he deduces accretion temperatures in this range for groups IAB, IIAB and IIIAB directly from their Ni contents. The variety of degrees of oxidation in chondrites is similarly attributed by Larimer (1968, 1973) to equilibrium nebular condensation processes, but there are difficulties in understanding how the postulated nebular oxidation of Fe operated (see Arrhenius 1978). The broad ranges of iron concentrations within and among olivine and pyroxene crystals in unequilibrated chondrites implies that these silicates never equilibrated with the nebula.

Even fairly strong siderophiles in irons could be fractionated from each other by metal-silicate equilibration if the proportion of metal in the chondritic starting material was low. Metal-silicate distribution coefficients for the siderophiles discussed above probably vary from 20 to 2000. Thus as long as chondritic metal contents were not below 5%, siderophile element ratios in the irons would not be changed by more than a factor of two because of oxidation. Very much larger effects are found in metal from C chondrites and achondrites, which typically have metal contents $\leqslant 1\%$.

Not mentioned in the discussion of chondrite compositions were the loss of refractories and effects of metal-silicate fractionation (see Grossman and Larimer 1974). On the logarithmic scales in Fig. 7 these effects are relatively minor. However, chondrites do not all contain CI abundances of refractory siderophiles, and the processes which affected abundances in chondrites (e.g., Rambaldi 1976) almost certainly affected the irons too. It is not known how refractory siderophile ratios were changed but clearly one cannot assume that irons had exact CI refractory abundances prior to igneous secondary fractionation.

The discovery of refractory-rich inclusions in Allende suggested to Wänke *et al.* (1974) that the large secondary fractionations of refractories like Ir within iron groups (Fig. 6) might result from some nebula condensation process. It is now known that grains with large refractory fractionations exist in these inclusions (El Goresy *et al.* 1978). However, it seems unlikely that refractory concentrations in irons were affected greatly by nebular processes as all bulk compositions of all chondrites are relatively close to CI levels (0.5 to 2). Iron groups IAB and IIICD might be exceptions to this rule, as planetary fractionation models do not explain their refractory depletions convincingly.

In the following section there is a brief comparison of the compositions of irons and metal in stony-iron meteorites and a discussion of their interrelationships. To test whether metal in stony irons might be derived from iron meteorites, their refractory element abundances are compared.

V. OTHER METAL-RICH METEORITES

The two main classes of stony-iron meteorites both have roughly equal proportions of metal and silicates and they probably formed by violent mixing of metal and silicate from different sources. For the pallasites, the mixing occurred at great depth probably at the core-mantle interfaces of planetesimals, whereas the mesosiderites are surface breccias (e.g., Anders 1964; Floran 1978). The mesosiderites contain mineral and differentiated rock fragments of great diversity, some of which are similar to known achondrites such as diogenites and eucrites. In the pallasites, however, the silicate is almost entirely coarsely crystalline olivine. In both cases the distribution and amount of metal is very variable. The shorter collision lifetimes of metal-free silicates suggest that metal-rich regions of silicate-metal mixtures may be preferentially delivered to Earth.

The compositions of metal in these stony-irons are in agreement with the origins outlined above. Pallasites come from at least two different parent bodies. Most ($\sim75\%$) belong to a so-called main group which has concentrations of refractory siderophile elements with $k > 1$, like Ir, which are considerably below those in chondritic metal. The general levels of elemental concentrations are consistent with an origin during an igneous differentiation process like that responsible for secondary fractionation in irons. In fact there is a good match between the mean composition of main-group metal and that predicted for the liquid after 80% of a IIIAB melt has fractionally crystallized (Scott 1977a). Thus these pallasites probably formed during final stages of solidification of the IIIAB core. There is no obvious group of irons with which to associate the second group of pallasites (Eagle Station types) although there are possible candidates among the grouplets.

Metal in 17 mesosiderites has strikingly uniform concentrations of Ir (2-8 μg/g) close to those in chondrites (Wasson et $al.$ 1974). Now a random selection of 10 irons would have Ir contents that vary by a factor of 10^2-10^3. Although data for the other elements which show large secondary fractionations are lacking, it seems unlikely that metal in mesosiderites ever participated in the secondary igneous fractionation experienced by the iron groups. Gallium and Ge concentrations lie in the well-populated IIIAB region but cannot be definitely associated with any group of irons. It seems probable, therefore, that metal in mesosiderites had a very different history from that of the iron meteorites. Possibly this metal was produced by impact melting close to the surface of planetesimals.

There are other metal-rich meteorites with well-analyzed metal. Bencubbin, Weatherford and Mt. Egerton have chondritic levels of Ir (2-3 μg/g) (Kallemeyn et $al.$ 1978; Scott and Wasson 1976). Bencubbin has been identified as an impact-produced breccia. Its metal has low concentrations of volatile siderophiles like Ga and Ge which show a correlation between abundance and condensation temperature as in Fig. 7. The only known meteorites containing silicate and appreciable metal with non-chondritic levels of Ir are those in iron group IB.

The absence of typical irons from surface breccias like mesosiderites and howardites is consistent with the idea that irons formed in large pools or cores which were deeply buried. In general, iron meteorites do not seem to have been mixed with silicates after secondary fractionation. Group IIE is an exception which is discussed in the next section.

VI. PARENT BODIES

Previous sections have provided information about parent bodies of the various iron meteorite groups during three different periods: (1) melting and solidification of metal ~4.6 Gyr ago; (2) slow cooling from 900-700 K about 1-1000 Myr later; and (3) during breakup into sub-meter sized meteorites 100-1000 Myr ago. In at least two groups the irons have stuck together through all three stages, but their parent bodies may have been altered during this time. Two other sources of information are provided by dynamical calculations on the origins of irons, and physical measurements of asteroids.

Wetherill and Williams (1979) calculate the meteorite production rate on Earth from large asteroids with low inclinations in the inner portion of the asteroid belt. As a result of resonant Jupiter and Saturn secular perturbations and close encounters with Mars, the transit times of collision ejecta to Earth are consistent with the 100-1000 Myr exposure ages of irons. Order-of-magnitude calculations suggest that several favorably located S-type asteroids with diameters of 50-200 km, including 6 Hebe, 8 Flora and 18 Melpomene, could alone provide the observed flux of iron meteorites.

According to physical measurements of asteroids (see chapter by Gaffey and McCord) the most likely sources of iron meteorites are the M-type asteroids which have perhaps 10-100% metal on their surfaces and S asteroids which could have around 25-80% metal. The grains of metal may be km- or mm-sized in dimensions. Of 560 main-belt asteroids with diameters over 50 km, 5% are M-type and 16% S-type.

If each of the 60 groups and grouplets of irons comes from a separate parent body, most of these would have diameters above 50 km according to the Goldstein-Short cooling rates and conventional thermal models. There may just have been enough asteroids to provide a 50-km sized parent body for each of the 60 odd groups and grouplets but it is unlikely that we are receiving deeply buried samples from all of them. Further, only a few of them are near locations which are thought to produce meteorites. This argues against the idea that the irons formed cores in 50-km sized bodies. There is another supporting argument which depends on the correctness of the fractional crystallization model and the probability that the core would have crystallized radially. As the group IIIAB sample must represent a nearly complete sample of this core, the 650-Myr collision must have largely destroyed an appreciable portion of a 50-km asteroid. As such events are considered to be very rare, it is very unlikely that several large asteroids were destroyed in this way. Even if the IIIAB irons were distributed as raisins the same arguments suggest that they could not have been distributed throughout a 50-km object (Fig. 9).

To resolve this discrepancy between mean cooling rates of groups and cosmochemical models, we can consider if the body in which the metal melted and solidified in a core was a small body which was later incorporated into a 50-km sized body. If several such small differentiated bodies were incorporated into the large one, we could be getting many groups from one body (see D in Fig. 9), satisfying the dynamical requirements discussed above. The minimum size for group IIIAB is 0.3 km (Wasson 1974, p. 126). Presumably such events, if they occurred, happened early when collisional velocities were low so that an appreciable fraction of a km-sized IIIAB body could be accreted onto a larger body. Except in IIE, these events did not mix metal and silicate intimately. The absence of metal having non-chondritic Ir

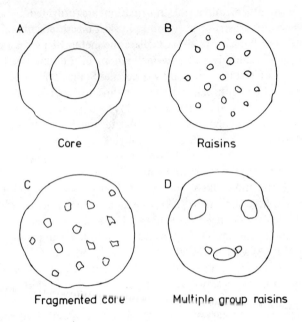

Fig. 9. Schematic models for iron meteorite parent bodies. Chemical modeling of secondary fractionation trends within groups suggests that most iron groups once formed in cores in their respective parent bodies like Model *A*. In an alternative raisin model (*B*), secondary trends are produced by partial melting instead of fractional crystallization. Raisins seem less plausible unless external heating caused melting. Two arguments tend to suggest that meteorite parent bodies like models *A* and *B* were not 100 km in size, as suggested by cooling rates. One solution may be that bodies like *A* and *B* were km-sized and were later accreted into bodies like *D* which were 100 km in size and contained several or many groups of irons. Irons in two groups, IC and IIE, may have been distributed as raisins (Model *C*) after fragmentation of the cores in which they once resided.

levels in mesosiderites and other meteoritic breccias suggests that iron meteorite fragments were not widely dispersed by such collisions.

There is some evidence that is independent of whether fractional crystallization or partial melting occurred, which shows that some iron meteorite parent bodies were broken up between secondary fractionation and slow cooling at 700 K. In group IC and IIE (Scott 1977*b*) there are chemical trends just like those in group IIIAB, but the very wide ranges of the Goldstein-Short cooling rates in IC and IIE are not correlated with the chemical trends as they are in IVA, for example. Since both igneous models for secondary fractionation would probably produce a radial variation in metal composition, the parent bodies must have been fragmented and the irons reaccreted with burial depths unrelated to their composition (see C in Fig. 9).

It seems unlikely that reaccretion could bury meteorites at depths which were correlated with their composition. Thus the cooling rates in IVA of Moren and Goldstein (1979), if correct, present a problem for this model. If cooling rates in group IVA are all within experimental error of 2 or 3 values, as Wood (1978) proposes, an apparent correlation between composition and cooling rates could be attributed to chance. Although Moren and Goldstein's cooling rates (Fig. 5) do cluster into two groups, IVA bandwidths and Ni contents are positively correlated and provide little evidence for clustering of Goldstein-Short cooling rates. Of course, even an apparently uniform cooling rate (Willis and Wasson 1978a) does not preclude IVA fragments from being dispersed in rock over a narrow range of burial depths.

The idea that meteorites may contain a record of conditions in more than one generation of parent bodies was championed by Urey (1959) and has recently been stressed by Levin (1977). To account for the survival of textures in pallasites, Wood (1978) proposes that molten metal and olivine were mixed in small planetesimals ($<$ 10 km radius), which were subsequently incorporated into larger bodies. He also suggests that irons may have formed cores in small bodies which then became raisins in large bodies. For mesosiderites, deep burial inside large bodies after formation in surface breccias is needed to explain very slow metallographic cooling rates.

If both irons and chondrites were melted and metamorphosed respectively in small bodies which later accreted into large bodies, this might help to explain the different gradients in Fig. 3. Chondrites which have been metamorphosed to varying degrees in different planetesimals may still be classified in the same group, whereas irons which melted in different planetesimals might be classed in separate groups. Late accretion of the chondritic planetesimals could prevent melting if [26]Al was the heat source.

Acknowledgments. I thank J. W. Larimer and D. W. Sears for helpful reviews of this chapter and many other colleagues for invaluable discussions, especially V. F. Buchwald, A. M. Davis, J. I. Goldstein, R. S. Rajan and J. T. Wasson.

REFERENCES

Alaerts, L.; Lewis, R. S.; and Anders, E. 1977. Primordial noble gases in chondrites: The abundance pattern was established in the solar nebula. *Science* 198: 927-930.

Alaerts, L.; Lewis, R. S.; and Anders, E. 1979. Isotopic anomalies of noble gases in chondrites – III. LL chondrites. *Geochim. Cosmochim. Acta* 43. In press.

Anders, E. 1964. Origin, age and composition of meteorites. *Space Sci. Rev.* 3: 583-714.

Anders, E. 1977. Critique of "Nebular condensation of moderately volatile elements and their abundances in ordinary chondrites" by Chien M. Wai and John T. Wasson. *Earth Planet. Sci. Lett.* 36: 14-20.

Arrhenius, G. 1978. Chemical aspects of the formation of the solar system. In *The Origin of the Solar System*, ed. S. F. Dermott (New York: Wiley), pp. 521-581.

Axon, H. J. 1979. The preterrestrial history of group IVA iron meteorites. *Earth Planet. Sci. Lett.* 42: 237-238.

Benkheiri, Y.; Pellas, P.; and Storzer, D. 1979. A cooling history of Copiapo (IA) iron: Preliminary results. *Icarus* (special Asteroid issue).

Bild, R. W. 1977. Silicate inclusions in group IAB irons and a relation to the anomalous stones Winona and Mt. Morris (Wis.). *Geochim. Cosmochim. Acta* 41: 1439-1456.

Bild, R. W., and Drake, M. J. 1978. Experimental investigations of trace element fractionation in iron meteorites: I. Early results. *Proc. Lunar Planet. Sci. Conf. IX* (Oxford: Pergamon Press), pp. 1407-1421.

Blander, M. 1971. The constrained equilibrium theory: Sulphide phases in meteorites. *Geochim. Cosmochim. Acta* 35: 61-76.

Blander, M. 1975. Critical comments on a proposed cosmothermometer. *Geochim. Cosmochim. Acta* 39: 1315-1320.

Buchwald, V. F. 1971. Tritium loss resulting from cosmic annealing compared with the microstructure and microhardness of six iron meteorites. *Chem. Erde* 30: 33-57.

Buchwald, V. F. 1975. *Handbook of Iron Meteorites* (Berkeley: University of California Press).

Buchwald, V. F. 1977. The mineralogy of iron meteorites. *Phil. Trans. Roy. Soc. London A* 286: 453-491.

Buseck, P. R., and Goldstein, J. I. 1969. Olivine compositions and cooling rates of pallasitic meteorites. *Bull. Geol. Soc. Amer.* 80: 2141-2158.

Case, D. R.; Laul, J. C.; Pelly, I. C.; Wechter, M. A.; Schmidt-Bleek, F.; and Lipschutz, M. E. 1973. Abundance patterns of thirteen trace elements in primitive carbonaceous and unequilibrated chondrites. *Geochim. Cosmochim. Acta* 37: 19-34.

Chou, C.-L.; Baedecker, P. A.; and Wasson, J. T. 1973. Distribution of Ni, Ga, Ge and Ir between metal and silicate portions of H-group silicates. *Geochim. Cosmochim. Acta* 37: 2159-2171.

Clayton, R. N., and Mayeda, T. K. 1978. Genetic relations between iron and stony iron meteorites. *Earth Planet. Sci. Lett.* 40: 168-174.

Davis, A. M.; Ganapathy, R.; and Grossman, L. 1977. Pontlyfni: A differentiated meteorite related to the group IAB irons. *Earth Planet. Sci. Lett.* 35: 19-24.

El Goresy, A.; Nagel, K.; and Ramdohr, P. 1978. Fremdlinge and their noble relatives. *Proc. Lunar Planet. Sci. Conf. IX* (Oxford: Pergamon Press), pp. 1279-1303.

Floran, R. J. 1978. Silicate petrography, classification, and origin of the mesosiderites: Review and new observations. *Proc. Lunar Planet. Sci. Conf. IX* (Oxford: Pergamon Press), pp. 1053-1081.

Fricker, P. E.; Goldstein, J. I.; and Summers, A. L. 1970. Cooling rate and thermal histories of iron and stony-iron meteorites. *Geochim. Cosmochim. Acta* 34: 475-491.

Goldstein, J. I., and Axon, H. J. 1973. The Widmanstätten figure in iron meteorites. *Naturwissenschaften* 60: 313-321.

Goldstein, J. I., and Friel, J. J. 1978. Fractional crystallization of iron meteorites, an experimental study. *Proc. Lunar Planet. Sci. Conf. IX* (Oxford: Pergamon Press), pp. 1423-1435.

Goldstein, J. I., and Short, N. M. 1967. The iron meteorites, their thermal history and parent bodies. *Geochim. Cosmochim. Acta* 31: 1733-1770.

Grossman, L., and Larimer, J. W. 1974. Early chemical history of the solar system. *Rev. Geophys. Space Phys.* 12: 71-101.

Ikrammudin, M.; Matza, S.; and Lipschutz, M. E. 1977. Thermal metamorphism of primtive meteorites – V. Ten trace elements in Tieschitz H3 chondrite heated at $400\text{-}1000°C$. *Geochim. Cosmochim. Acta* 41: 1247-1256.

Jain, A. V., and Lipschutz, M. E. 1971. Shock history of iron meteorites and their parent bodies: A review, 1967-1971. *Chem. Erde* 30: 199-215.

Kallemeyn, G. W.; Boynton, W. V.; Willis, J.; and Wasson, J. T. 1978. Formation of the Bencubbin polymict meteoritic breccia. *Geochim. Cosmochim. Acta* 42: 507-515.

Kelly, W. R., and Larimer, J. W. 1977. Chemical fractionations in meteorites – VIII. Iron meteorites and the cosmochemical history of the metal phase. *Geochim. Cosmochim. Acta* 41: 93-111.

Kelly, W. R., and Wasserburg, G. J. 1978. Evidence for the existence of ^{107}Pd in the early solar system. *Geophys. Res. Lett.* 5: 1079-1082.

Kracher, A.; Kurat, G.; and Buchwald, V. F. 1977. Cape York: The extraordinary mineralogy of an ordinary iron meteorite and its implication for the genesis of IIIAB irons. *Geochem. J.* 11: 207-217.

Krähenbühl, U.; Morgan, J. W.; Ganapathy, R.; and Anders, E. 1973. Abundance of 17 trace elements in carbonaceous chondrites. *Geochim. Cosmochim. Acta* 37: 1353-1370.

Kulpecz, A. A., and Hewins, R. H. 1978. Cooling rate based on schreibersite growth for the Emery mesosiderite. *Geochim. Cosmochim. Acta* 42: 1495-1500.

Larimer, J. W. 1967. Chemical fractionations in meteorites – I. Condensation for the elements. *Geochim. Cosmochim. Acta* 31: 1215-1238.

Larimer, J. W. 1968. Experimental studies on the system Fe-MgO-SiO$_2$-O$_2$ and their bearing on the petrology of chondritic meteorites. *Geochim. Cosmochim. Acta* 32: 1187-1207.

Larimer, J. W. 1967. Chemical fractionations in meteorites – I. Condensation of the and cosmobarometry. *Geochim. Cosmochim. Acta* 37: 1603-1623.

Larimer, J. W. 1978. Meteorites: Relics from the early solar system. In *Origin of the Solar System,* ed. S. F. Dermott (New York: Wiley), pp. 347-393.

Larimer, J. W., and Anders, E. 1967. Chemical fractionations in meteorites – II. Abundance patterns and their interpretation. *Geochim. Cosmochim. Acta* 31: 1239-1270.

Larimer, J. W., and Rambaldi, E. R. 1978. Trace element chemistry of iron meteorites. *Meteoritics* 13: 537.

Laul, J. C.; Ganapathy, R.; Anders, E.; and Morgan, J. W. 1973. Chemical fractionations in meteorites – VI. Accretion temperatures of H-,LL-, and E- chondrites, from abundance of volatile trace elements. *Geochim. Cosmochim. Acta* 37: 329-357.

Levin, B. J. 1977. Relationship between meteorites, asteroids and comets. In *Comets, Asteroids, Meteorites,* ed. A. H. Delsemme (Toledo, Ohio: Univ. of Toledo Press), pp. 307-311.

Mason, B. 1971. *Handbook of Elemental Abundances in Meteorites* (New York: Gordon and Breach).

Moren, A. E., and Goldstein, J. I. 1978. Cooling rate variations of group IVA iron meteorites. *Earth Planet. Sci. Lett.* 40: 151-161.

Moren, A. E., and Goldstein, J. I. 1979. Cooling rates of group IVA iron meteorites determined from a ternary Fe-Ni-P model. *Earth Planet. Sci. Lett.* (in press).

Niemeyer, S. 1979. I-Xe dating of silicate and troilite from IAB iron meteorites. *Geochim. Cosmochim. Acta* (in press).

Palme, H., and Wlotzka, F. 1976. A metal particle from a Ca,Al-rich inclusion from the meteorite Allende, and the condensation of refractory siderophile elements. *Earth Planet. Sci. Lett.* 33: 45-60.

Pellas, P., and Storzer, D. 1977. On the early thermal history of chondritic asteroids derived from 244-Plutonium fission track thermometry. In *Comets, Asteroids, Meteorites,* ed. A. H. Delsemme (Toledo, Ohio: Univ. of Toledo Press), pp. 355-363.

Pellas, P., and Storzer, D. 1979. Pu-244 fission track thermometry and its application to meteorites. "Comptes Rendus de Journées de Planétologie" (Paris, Editions du CNRS), in press.

Powell, P. N. 1969. Petrology and chemistry of mesosiderites – I. Textures and composition of nickel-iron. *Geochim. Cosmochim. Acta* 33: 789-810.

Rambaldi, E. 1976. Trace element content of metals from L-group chondrites. *Earth Planet. Sci. Lett.* 31: 224-238.

Randich, E., and Goldstein, J. I. 1978. Cooling rates of seven hexahedrites. *Geochim. Cosmochim. Acta* 42: 221-234.

Schultz, L.; Funk, H.; Nyquist, L.; and Signer, P. 1971. Helium, neon and argon in separated phases of iron meteorites. *Geochim. Cosmochim. Acta* 35: 77-88.

Scott, E. R. D. 1972. Chemical fractionation in iron meteorites and its interpretation. *Geochim. Cosmochim. Acta* 36: 1205-1236.

Scott, E. R. D. 1977a. Geochemical relationships between some pallasites and iron meteorites. *Mineral. Mag.* 41: 265-272.

Scott, E. R. D. 1977b. Composition, mineralogy and origin of group IC iron meteorites. *Earth Planet. Sci. Lett.* 37: 273-284.

Scott, E. R. D. 1978a. Iron meteorites with low Ga and Ge concentrations – composition, structure and genetic relationships. *Geochim. Cosmochim. Acta* 42: 1243-1551.

Scott, E. R. D. 1978b. Primary fractionation of elements among iron meteorites. *Geochim. Cosmochim. Acta* 42: 1447-1458.

Scott, E. R. D. 1979. Origin of anomalous iron meteorites. *Mineral. Mag.* (in press).

Scott, E. R. D., and Bild, R. W. 1974. Structure and formation of the San Cristobal iron meteorite, other IB irons and group IIICD. *Geochim. Cosmochim. Acta* 38: 1379-1391.

Scott, E. R. D., and Rajan, R. S. 1979. Thermal history of the Shaw chondrite. *Proc. Lunar Planet. Sci. Conf. X.* Submitted.

Scott, E. R. D., and Wasson, J. T. 1975. Classification and properties of iron meteorites. *Rev. Geophys. Space Phys.* 13: 527-546.

Scott, E. R. D., and Wasson, J. T. 1976. Chemical classification of iron meteorites – VIII. Groups IC, IIE, IIIF and 97 other irons. *Geochim. Cosmochim. Acta* 40: 103-115.

Sears, D. W. 1978a. Condensation and the composition of iron meteorites. *Earth Planet. Sci. Lett.* 41: 128-138.

Sears, D. W. 1978b. *The Nature and Origin of Meteorites* (Bristol: Adam Hilger).

Sears, D. W. 1979. Did iron meteorites form in the asteroid belt? – Evidence from thermodynamic models. *Icarus* (special Asteroid issue).

Smales, A. A.; Mapper, D.; and Fouché, K. 1967. The distribution of some trace elements in iron meteorites as determined by neutron activation. *Geochim. Cosmochim. Acta* 31: 673-720.

Sonett, C. P., and Herbert, F. L. 1977. Pre-main sequence heating of planetoids. In *Comets, Asteroids, Meteorites,* ed. A. H. Delsemme (Toledo, Ohio: Univ. of Toledo Press), pp. 429-437.

Takahashi, H.; Gros., J.; Higuchi, H.; Morgan, J. W.; and Anders, E. 1979. Volatile elements in chondrites: Metamorphism or nebular fractionation? *Geochim. Cosmochim. Acta* (in press).

Takahashi, H.; Janssens, M.-J.; Morgan, J. W.; and Anders, E. 1978. Further studies of trace elements in C3 chondrites. *Geochim. Cosmochim. Acta* 42: 97-106.

Turner, G.; Enright, M. C.; and Cadogan, P. H. 1978. The early history of chondrite parent bodies inferred from ^{40}Ar-^{39}Ar ages. *Proc. Lunar Planet. Sci. Conf. IX* (Oxford: Pergamon Press), pp. 989-1025.

Urey, H. C. 1959. Primary and secondary objects. *J. Geophys. Res.* 64: 1721-1737.

Voshage, H. 1967. Bestrahlungsalter und Herkunft der Eisenmeteorite. *Z. Naturforsch. Ser. A.* 22: 477-506.

Voshage, H. 1978. Investigations on cosmic-ray produced nuclides in iron meteorites. 2. New results on $^{41}K/^{40}K$-$^{4}He/^{21}Ne$ exposure ages and the interpretation of age distributions. *Earth Planet. Sci. Lett.* 40: 83-90.

Wai, C. M., and Wasson, J. T. 1977. Nebular condensation of moderately volatile elements and their abundances in ordinary chondrites. *Earth Planet. Sci. Lett.* 36: 1-13.

Wai, C. M.; Wasson, J. T.; Willis, J.; and Kracher, A. 1978. Nebular condensation of moderately volatile elements, their abundances in iron meteorites, and the quantization of Ga and Ge abundances. *Lunar and Planetary Science IX*, The Lunar and Planet. Inst., pp. 1193-1195.

Wänke, H.; Baddenhausen, H.; Palme, H.; and Spettel, B. 1974. On the chemistry of the Allende inclusions and their origin as high temperature inclusions. *Earth Planet. Sci. Lett.* 23: 1-7.

Wasson, J. T. 1971. An equation for the determination of iron-meteorite cooling rates. *Meteoritics* 6: 139-147.

Wasson, J. T. 1972. Parent-body models for the formation of iron meteorites. *Proc. 24th Intern. Geol. Cong.* 15: 161-168.

Wasson, J. T. 1974. *Meteorites – Classification and Properties* (New York: Springer Verlag).

Wasson, J. T. 1977. Reply to Edward Anders: A discussion of alternative models for explaining the distribution of moderately volatile elements in ordinary chondrites. *Earth Planet. Sci. Lett.* 36: 21-28.

Wasson, J. T.; Schaudy, R.; Bild, R. W.; and Chou, C.-L. 1974. Mesosiderites – I. Compositions of their metallic portions and possible relationship to other metal-rich meteorite groups. *Geochim. Cosmochim. Acta* 38: 135-149.

Wasson, J. T., and Wai, C. M. 1976. Explanations for the low Ga and Ge concentrations in some iron meteorites. *Nature* 261: 114-116.

Wetherill, G. W., and Williams, J. G. 1979. Origin of differentiated meteorites. In *Proceedings of the 2nd International Conference on Origin and Abundance of Elements,* ed. H. de la Roche (Oxford: Pergamon Press, in press).

Willis, J., and Wasson, J. T. 1978a. Cooling rates of group IVA iron meteorites. *Earth Planet. Sci. Lett.* 40: 141-150.

Willis, J., and Wasson, J. T. 1978b. A core origin for group IVA iron meteorites: A reply to Moren and Goldstein. *Earth Planet. Sci. Lett.* 40: 162-167.

Wood, J. A. 1964. The cooling rates and parent planets of several iron meteorites. *Icarus* 3: 429-459.

Wood, J. A. 1967. Chondrites: Their metallic minerals, thermal histories, and parent planets. *Icarus* 6: 1-49.

Wood, J. A. 1978. Nature and evolution of the meteorite parent bodies: Evidence from petrology and metallurgy. In *Asteroids: An Exploration Assessment,* eds. D. Morrison and W. C. Wells, NASA Conf. Publ. 2053, pp. 45-55.

Yavnel, A. A. 1975. Element distribution trends in iron meteorites. *Geochem. Internat.* 12: 113-123.

DYNAMICAL, CHEMICAL AND ISOTOPIC EVIDENCE REGARDING THE FORMATION LOCATIONS OF ASTEROIDS AND METEORITES

JOHN T. WASSON
University of California at Los Angeles

and

GEORGE W. WETHERILL
Carnegie Institution of Washington

The study of meteorites is closely linked to the study of asteroids. Even though some meteorites may conceivably be cometary material derived from Apollo-Amor objects it is essentially certain that all the present terrestrial influx of meteorites is derived from objects which can be termed "asteroidal" from the observational point of view. At the present time asteroidal bodies are for the most part found in the main asteroid belt. However, the history of the evolving population of this region is complex and poorly understood. At the earliest stage of solar system history the density of matter in the asteroid belt was probably essentially as large as that in the vicinity of Earth or Jupiter, but later it was greatly reduced by processes associated with the intense bombardment accompanying the formation of the giant planets. Since these earliest times this region has been the principal stable storage place in the solar system, wherein stray bodies from elsewhere can be retained for billions of years. Thus the indigenous residual population of the asteroid belt has been, and very likely still is being, augmented by the addition of bodies formed at a wide range of heliocentric distances. Even if this additional material does not represent the major mass of the asteroid belt, the mode of its implantation tends to place it in asteroidal orbits inherently slightly less stable than orbits of indigenous asteroids, from whence it can be preferentially extracted into Earth-crossing orbits. Semiquantitative mechanisms have been described for placing residual planetesimals from the vicinity of Earth and Venus into the inner part of the asteroidal belt. Extinct comets are today evolving into asteroidal Apollo-Amor objects with orbits largely in the asteroidal belt.

There is chemical and isotopic evidence from meteorites relevant to the question of the original place of formation of these asteroidal parent bodies. The degree of oxidation may be a useful criterion for distinguishing bodies which formed near or interior to 1 AU (enstatite meteorites, chondritic clasts in IAB irons) from those which formed at greater distances. Oxygen isotope data suggest that the parent bodies of the CM, CO and CV chondrites formed in one or two distinct regions of the solar system, probably at greater heliocentric distances than other chondrite groups. The near absence of foreign xenolithic clasts other than CM-like fragments in ordinary chondrite breccias is interpreted as indicating that the parent bodies of the different groups of ordinary chondrites were in nearly circular orbits that were not mutually intersecting although they did intersect those of the CM-like chondrites.

This chemical and petrographic evidence must be reconciled with dynamic theories for the injection and removal of material from the asteroid belt, with spectrophotometric data on asteroids, and with observations of bright meteors. It appears plausible that most differentiated meteorites may be residual Earth-Venus material stored in the inner belt as S-asteroids or as uncharacterized fragments of the crusts and mantles of such bodies (although rare achondrites may conceivably be derived from Mars), and that carbonaceous meteorites are derived from both the asteroid belt and from comets via Apollo objects. Evidence for the site of origin of the remaining chondritic material is at present conflicting. Degrees of oxidation suggest that some of this material was formed within a 1-AU radius but it is difficult to understand how it was placed in the asteroid belt without being fragmented and mixed with the presumably differentiated planetesimals that became the parent bodies of the differentiated meteorites. The near absence of spectrophotometric evidence of main-belt ordinary chondrite asteroids is puzzling. Apollo objects appear to be adequate sources for ordinary chondrites, but it is quantitatively difficult to place a sufficient number of fragments of the belt asteroid into Apollo orbits. The alternative of deriving ordinary chondrites from comets via Apollos conflicts with the compositional evidence for an origin of their parent bodies in the inner solar system. Investigations directed to the ultimate resolution of the problems are suggested.

Most of our detailed data relevant to events and processes during the earliest history of the solar system has come from the study of meteorites. Interpretation of these data is closely related to asteroid studies. Measured cosmic ray exposure ages of meteorites demonstrate that the objects falling to Earth today were produced late in solar system history by fragmentation and cratering of larger parent bodies. The following discussion will lead to the conclusion that bodies which at least at present fit the observational definition of an asteroid represent the only plausible sources for the great majority of the meteorites in our collections.

Meteorites differ widely in their chemical and physical nature, and it is most plausible that many of these differences result from corresponding differences in the compositions and histories of their parent bodies. These differences lead to distinction of two broad categories of meteorites. The *chondrites* are primitive meteorites having nearly solar abundance ratios of the nonvolatile elements; their compositions and many of their textural features were established by processes occurring prior to the formation of their parent bodies. In contrast, the *differentiated meteorites* were largely formed by melting and igneous differentiation processes in the interiors of parent bodies.

Meteorites can be further classified into *groups,* on the basis of similarities in numerous properties. It will be assumed that each chondrite group (Table I) formed in a separate parent body. Some evidence supporting this assumption will be presented. Thus 10 parent bodies appear to be necessary to account for the chondrite groups listed, and roughly another 10 parent bodies to account for additional chondrite "grouplets" having only one to four members. Grouplets, especially among the irons, are too numerous to list in Table I.

There are 7 groups of differentiated silicate-rich meteorites ("achondrites") and 13 groups of iron meteorites (Table I). In addition, there are perhaps 10 grouplets of achondrites and 40-50 grouplets of irons. It is unlikely that any two iron groups or grouplets formed in the same parent body, but some or all achondrites may have formed in the parent bodies responsible for iron meteorite groups or grouplets. Thus ~60 parent bodies are required to account for the differentiated meteorites (Scott 1979). It is improbable but not impossible that chondrites and differentiated meteorites originated in the same parent body. The closest known link is between the IIE irons and the H chondrites; H group metal resembles IIE metal in its content of Ga and Ge (Chou *et al.* 1973; Rambaldi 1977; Scott and Wasson 1976), two elements whose degree of condensation-accretion may have varied with distance from the sun, and some IIE irons contain silicate inclusions having O-isotope composition unresolvable from H chondrite values (Clayton and Mayeda 1978).

The identification of the parent bodies among solar system bodies has been a principal goal of meteoritic studies. Candidates include comets, asteroidal bodies of various kinds, the surfaces of planets and their satellites, and possibly undiscovered classes of bodies, as suggested by the recently discovered Saturn- and Uranus-crossing object, Chiron. Even an interstellar contribution cannot be entirely ruled out, although orbits of photographic meteors show that any interstellar component must be small ($\leq 0.1\%$). Evidence for and against the association of particular meteoritic classes with

TABLE I

Meteorite Classification: Groups of Meteorites Having 5 or More Members

Clan	Group	Synonym[a]	Example	Falls	Finds	Freq. (%)
Chondrites						
refractory-rich	CV c.	C3 carbonaceous c.	Allende	8	3	1.1
	CO c.		Ornans	6	1	0.85
minichondrule	CM c.	C2 carbonaceous c.	Murchison	14	0	2.0
volatile-rich	CI c.	C1 carbonaceous c.	Orgueil	5	0	0.71
ordinary	LL c.	amphoterite c.	St. Mesmin	51	16	7.2
	L c.	hypersthene c.	Bruderheim	278	192	39.3
	H c.	bronzite c.	Ochansk	229	230	32.3
IAB-inclusion	IAB c.	—	Copiapo	6	see below	0.85
enstatite[b]	EL c.	enstatite c.	Indarch	5	3	0.71
	EH c.		Khairpur	...	...	...
—	other chondrites	—	Kakangari	2	2	0.28
Differentiated Meteorites						
igneous	EUCrites[d]	basaltic or Ca-rich	Juvinas	20	4	2.8
	HOWardites	or pyroxene-plag. ac.	Kapoeta	18	1	2.5
	DIOgenites	hypersthene ac.	Johnstown	8	0	1.1
	MESosiderites	stony-irons	Estherville	6	14	0.85
	PALlasites		Krasnojarsk	2	33	0.28
	UREilites	olivine-pigeonite ac.	Novo Urei	3	3	0.42
enstatite[b]	AUBrites	enstatite ac.	Norton County	8	1	1.1
—	other differentiated silicate-rich meteorites	—	Shergotty	7	6	0.99

TABLE I (continued)

Meteorite Classification: Groups of Meteorites Having 5 or More Members

Clan	Group	Synonym[a]	Example	Falls	Finds	Freq. (%)
— —	IAB irons	none relevant[e]	Canyon Diablo	7	83	0.85[f]
— —	IC irons	—	Bendego	0	10	0.09
— —	IIAB irons	—	Coahuila	5	47	0.49
— —	IIC irons	—	Ballinoo	0	7	0.06
— —	IID irons	—	Needles	2	11	0.12
c	IIE	—	Weekeroo Stat.	1	11	0.11
— —	IIF	—	Monahans	0	5	0.05
c	IIIAB	—	Cape York	6	151	1.46
— —	IIICD	—	Tazewell	2	10	0.11
c	IIIE	—	Rhine Villa	0	8	0.08
— —	IIIF	—	Nelson County	0	5	0.05
— —	IVA	—	Gibeon	2	38	0.38
— —	IVB	—	Hoba	0	11	0.10
— —	other irons	—	Mbosi	6	62	0.62

[a]Synonyms occuring commonly in the current literature. c. = chondrite, ac. = achondrite, plag. = plagioclase.

[b]The enstatite clan chondrites and the aubrites are very closely related.

[c]Clan relationships involving irons are still poorly understood. The only known close relationships are between IIIAB and IIIE irons (and these possibly also with the pallasites) and between the IIE irons and the ordinary chondrites

[d]First three letters are symbol of the group.

[e]The structural classification of irons into hexahedrites, octahedrites and ataxites does not lead to genetically related groups.

[f]Fall frequencies of iron meteorites are calculated by arbitrarily allocating the 32 observed falls confirmed in Buchwald (1975), to the frequencies of all irons classified by Scott and Wasson (1975).

various candidate objects, and evidence regarding the original solar system location of the parent bodies are the principal topics of this review.

I. THE FORMATION REGIONS OF METEORITE PARENT BODIES

Bodies ranging in size from 1 to 1000 km were probably formed prior to the formation of the planets, and represent an intermediate stage of planetary formation. It may be expected that bodies in this size range formed at all solar distances from inside the orbit of Mercury to beyond that of Neptune. Most of these objects have been swept up by the planets or ejected from the solar system by close encounters with the major planets. In order to be preserved for the entire 4.5-Gyr history of the solar system, it is necessary that the parent bodies have been stored in orbits that are stable on this time scale. Most of those remaining in the inner solar system are now in the asteroid belt between 1.8 and 4.0 AU from the sun. Long-term storage places also exist in the outermost solar system, particularly in the distant Oort cloud, from which comets are derived at present. However, it is possible that some small bodies are still stored in stable inner solar system orbits (c.f., Weissman and Wetherill 1973), possibly in resonances resembling that in which Toro is currently trapped (Danielsson and Ip 1972; Williams and Wetherill 1973), but, unlike that of Toro, not destabilized by Mars' perturbations. In addition, dynamic mechanisms are known which will transfer bodies from short-lived orbits into more stable ones. For example, bodies may be removed from the inner solar system and trapped in metastable Mars-crossing orbits (Wetherill 1978); alternatively, following collisions their fragments could be trapped directly in orbits in the main belt.

It is also possible that bodies formed initially in the outer solar system can be transferred into orbits between Mars and Jupiter having long-term stability. The chief mechanism (discussed in Sec. III) involves the operation of nongravitational forces to reduce the aphelion of short-period comets to values sufficiently small to avoid major perturbations by Jupiter. Thus the present location of a parent body in the solar system does not necessarily correspond to the heliocentric distance at which it was originally formed.

The H_2O-rich composition of Uranus implies that temperatures during the formation of the solar system reached the ice condensation temperature beyond ~15 AU, and it is possible that ice condensation occurred inside the orbit of Saturn. As a result, it is plausible to assume that planetesimals in the major planet region resembled comets. However it is also possible that a variety of objects formed in this region. The formation of planetesimals by dust-layer gravitational instabilities (Edgeworth 1949; Gurevich and Lebedinski 1950; Safronov 1960, 1969; Goldreich and Ward 1973) could have occurred sequentially during the cooling of the solar nebula, with earlier objects being mainly chondritic and later objects mainly icy. Later accumulation processes could have produced hybrids of various kinds, most of which would be called comets at least during their first several passages through the solar system.

Although not required for many of our arguments, it will simplify the following discussion if we first make 3 plausible assumptions:

1. The fractionations observed between groups of chondrites primarily resulted from differences in nebular conditions as a function of distance from the sun, and, in some cases, differences resulting from sequential formation of planetesimals as a function of time at a fixed distance from the sun. Compositional hiatus between groups reflect incomplete sampling of the originally continuous distribution.

2. The wide range in properties between the extremes of the known chondrite distribution (from the highly reduced enstatite chondrites to the highly oxidized carbonaceous chondrites) encompasses most of the range initially present in the solar system. This assumption is based on the observed very wide systematic variation in degree of oxidation of the chondrites and on dynamic calculations indicating that a small but significant sample of bodies formed from the orbit of Mercury to that of Neptune could have been stored in orbits that provide meteorites at the present time. Additional populations of materials, never compacted enough to form tough meteoroids capable of surviving atmospheric passage, may also exist.

3. Most or all differentiated meteorites formed by the igneous differentiation of parent bodies having more or less chondritic compositions, thus at least for the less volatile elements it is reasonable to try to infer the properties of the precursor chondritic materials from those of the differentiated meteorites.

There are three chief properties of chondrites that will be considered to have varied in roughly systematic fashion with distance from the sun: O-isotope composition, refractory[a] element abundance, and degree of oxidation. Certain other properties, such as volatile element and siderophile abundances, sometimes show significant fractionations between groups that appear to have formed at about the same distance from the sun, and will not be discussed in this chapter.

Figure 1 illustrates the variation in refractory abundance and degree of oxidation (roughly proportional to the $FeO_x/(FeO_x+MgO)$ ratio, where FeO_x is the amount of Fe bound to O; in most cases oxidized Fe is FeO, but Fe_3O_4 is present in carbonaceous chondrites) among the 10 groups of chondrites. Because $FeO_x/(FeO_x+MgO)$ ratios in the unequilibrated EH chondrites are much higher than those that would be in equilibrium with the observed Si contents of the Fe-Ni metal, the EL chondrite $FeO_x/(FeO_x+MgO)$ value is assumed to hold for both enstatite chondrite groups. One observes a rough

[a]Refractory during nebula condensation processes; Ca, Al, Sc, rare earths are examples.

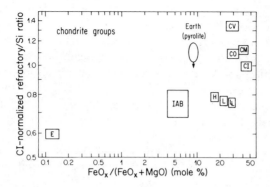

Fig. 1. The $FeO_x/(FeO_x+MgO)$ ratio provides a rough measure of the degree of oxidation of chondrites that contain Fe-Ni; this ratio is expected to increase with increasing distance from the sun (see text for discussion). The abundance relative to Si of elements such as Ca, Al and Sc that are refractory during nebular condensation processes tends to correlate with $FeO_x/(FeO_x+MgO)$. A plausible hypothesis is that the observed trend from lower left to upper right reflects the formation of these chondrite groups at increasing distances from the sun. If the pyrolite upper-mantle composition is a reasonable estimate for the bulk Earth composition, its position suggests that the IAB and H-group chondrites are the group that formed nearest 1 AU.

correlation between refractory abundance and degree of oxidation.

Also plotted in Fig. 1 is pyrolite, the estimate of Ringwood and Kesson (1977) of the composition of the terrestrial upper mantle, which is taken as a rough approximation of the whole earth composition. The $FeO_x/(FeO_x+MgO)$ value of pyrolite (0.08) is probably a reasonably precise (to within $\sim\pm20\%$) whole earth estimate. Crustal formation cannot have led to significant alteration of the mantle $FeO_x/(FeO_x+MgO)$ ratio. Although the Fe/Mg ratio in the crust is much higher than that in the mantle, concentrations of both elements are lower in the crust than in the mantle, and the mass of the crust is only $\sim0.5\%$ that of the mantle. A more difficult question is whether or not the mantle as a whole has suffered igneous differentiation; fractional crystallization of an initially molten mantle could have resulted in the lower mantle being significantly more magnesian than the upper mantle. However, the apparent existence of deep mantle "plumes" consisting of low-melting materials is inconsistent with an early fractionation of the entire mantle. It appears probable that the upper mantle $FeO_x/(FeO_x+MgO)$ ratio is close to the whole mantle value.

The refractory abundance of pyrolite is more approximate, and is mainly based on carbonaceous chondritic abundances. Some upper mantle rocks show even lower Ca and Al abundances (Bickle *et al.* 1976) and igneous evolution of the mantle would tend to enrich Ca and (especially) Al upwards. Thus we suggest that the pyrolite composition should be considered an upper limit on the refractory abundance in the earth.

If the variations in chondrite refractories and oxidation state primarily reflect formation at differing distances from the sun, the data in Fig. 1 suggest that the IAB and H chondrites are those that formed nearest to 1 AU. Here a caveat is in order. Because of its large gravitational field the earth accreted materials not present in chondritic planetesimals formed near 1 AU: (a) Primordial material formed at 1 AU but too fine-grained to settle to the nebula midplane and agglomerate through dust-layer instabilities; and (b) material formed elsewhere in the solar system (e.g., cometary matter) and captured by the earth after the period of chondrite formation was largely complete. It seems likely that only a small fraction of the Earth's mass resulted from these two sources, but there is presently insufficient evidence to confirm this view.

There are reasons to believe that the trend from enstatite to carbonaceous chondrites in Fig. 1 represents increasing distance from the sun. At any particular time, average nebular temperature and pressure probably increased radially towards the sun. Higher pressures and temperatures favor more rapid grain growth, and thus earlier settling of grains to the nebular midplane. Following the settling of grains the mean collapse time for the dust layer instability mechanism also increases with distance from the sun. Thus chondritic materials formed nearer the sun should have higher nebular equilibration temperatures, and as pointed out by Latimer (1950), the higher the equilibration temperature, the smaller the fraction of Fe bound to O.

In Fig. 2 are illustrated the O-isotope compositions determined by R. N. Clayton in the chondrite groups, one chondrite grouplet, and in the igneous clan of silicate-rich differentiated stones. Three lines are shown on the diagram: (a) Fractionation of isotopes in an initially well-mixed system results in arrays having slopes of 0.52 on $\delta^{17}O$-$\delta^{18}O$ diagrams; the terrestrial fractionation line is such an array. (b) Fractionation of isotopes in the matrix of CM chondrites has produced a roughly linear array with slope near 0.5 about 2 $^o/oo$ lower in $\delta^{17}O$ than the terrestrial line. (c) The anhydrous minerals in CM, CO and CV chondrites form a linear array having a slope of 0.94; as noted by Clayton et al. (1973), this line appears to result from the mixing of two components, one of which may lie fairly near the terrestrial line, the other at least as ^{16}O-rich as the most extreme measured sample ($\delta^{18}O$ ~50 $^o/oo$).

The chondrite groups fall into two distinctly different regions in Fig. 2. Most groups fall near the terrestrial fractionation line between 3 and 6$^o/oo$ $\delta^{18}O$. Mean CO and CV compositions fall along the CM, CO, CV anhydrous minerals mixing line, the CM chondrites between mean CO and mean CM matrix, and mean CI compositions just above the terrestrial fractionation line at $\delta^{18}O \simeq 17$.

These data could be produced by mixing and/or fractionating two hypothetical presolar components, a common one that included all the gaseous O (H_2O, CO) lying near the terrestrial fractionation line, and a rare

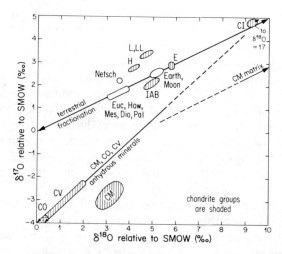

Fig. 2. A plot of $^{17}O/^{16}O$ (expressed as $\delta^{17}O$) versus $^{18}O/^{16}O$ (expressed as $\delta^{18}O$) shows that 6 chondrite groups (EH, EL, IAB, H, L, LL) and the igneous clan of different silicate-rich meteorites have O-isotope compositions that plot very near the composition estimated for the bulk Earth and bulk Moon, suggesting an approach to complete isotopic mixing in the solar nebula. In contrast three of the carbonaceous groups have highly unequilibrated O isotopes and compositions that fall far away from the terrestrial composition. It seems probable that the 6 groups closely similar to the earth formed in the hotter, inner portion whereas the highly unequilibrated groups formed in the cooler outer portion of the solar system. This is consistent with the general sequence inferred from refractory abundance and degree of oxidation. The O-isotope composition of the CI chondrites is near the terrestrial fractionation line, but their $\delta^{18}O$ value is far from the terrestrial value, and closest to that in the matrix of CM chondrites.

one similar to the most ^{16}O-rich samples along the CM, CO, CV anhydrous mineral mixing line. A plausible way to preserve the rare ^{16}O-rich component is in the form of incompletely evaporated presolar grains. In the hotter, inner portion of the solar nebula maximum nebular temperatures were high, and most presolar grains probably evaporated, thus the O-isotope composition in chondrites formed in the inner solar system should plot near the terrestrial fractionation line. The preservation in CM, CO, and CV chondrites of materials having high ^{16}O contents implies an origin in cooler outer portion of the solar nebula. The CI chondrites plot near the terrestrial fractionation line but differ by ~11 °/oo in $\delta^{18}O$ from the groups that cluster near the Earth's composition. The absence of a preserved ^{17}O anomaly in this group may be associated with the fact that they consist entirely of fine-grained materials, whereas the isotopic anomalies are preserved in the large-grained fraction of the CM, CO and CV groups. The fine-grained material may have condensed from an isotopically well-mixed gas or initial heterogeneities may have disappeared as a result of isotopic exchange between grains and gas.

Thus the O-isotope data seem consistent with the general conclusions based on degree of oxidation; IAB and E chondrites formed nearer to the sun and the 4 carbonaceous chondrite groups farther from the sun.

An estimate of the bulk Earth O-isotope composition based on the analysis of ultramafic rocks that have not been altered by reaction with ground water is shown in Fig. 2. The O-isotope compositions of olivine-rich lunar rocks (a dunite, the igneous glass samples from Apollo 15 and 17 sites) are in the same range as terrestrial ultramafics. The chondrite group nearest to the earth in composition consists of the chondritic blocks in the IAB iron meteorites. Although IAB plots slightly below the terrestrial fractionation line, a hybrid of IAB and the H-group (the two groups having $FeO_x/(FeO_x+MgO)$ ratios on either side of the terrestrial value) could plot directly on this line within the uncertainty of the bulk Earth value. In addition, the E chondrites (both EH and EL) and the igneous clan of differentiated silicate-rich meteorites fall on or near the terrestrial line displaced from the earth 0.5 $^o/oo$ higher and 1.5 $^o/oo$ lower, respectively. It appears probable that all these groups formed in the inner solar system, and reasonably likely that some actually formed inside 1 AU.

II. PROPERTIES OF METEORITE BRECCIAS

Because the accretion of a substantial fraction of extra-parent body materials as sizable (several mm or larger) rock fragments requires low relative velocities between accretor and accretee, the meteoritic breccias offer unique information about the populations of materials in low-velocity crossing orbits at the time and place where the breccia formed.

Breccias are divided into three end types. *Monomict* breccias consist of only one rock type, with no significant variation among large fragments (clasts) or between clasts and the fine-grained matrix. These breccias originate from the crushing and relithification of a single rock, and thus do not offer information about accretion. In *polymict* breccias clasts show distinctly different mineralogical and chemical composition; the matrix composition is generally that expected from mixing the different clast types. The term *genomict* was coined to cover a common intermediate kind of breccia; most clasts are very similar in composition (thus many of these were earlier designated monomict), but differ in texture and in detailed composition; e.g., the mean olivine composition may be 5-10 % different from clast to clast. Genomict breccias also contain rare foreign clasts, and thus some were earlier designated polymict.

Some genomict and polymict breccias have high contents of solar-type (little interelement fractionation) rare gases. Various experiments have shown these gases to be surface correlated, e.g., concentrations are far higher in the fine-grained matrix than in clasts. In most cases the rare-gas distribution and the brecciated texture are very similar to those observed in lunar regolith breccias, and it seems reasonable to believe that these meteoritic breccias formed under similar circumstances, including incorporation of the solar rare gas as a result of implantation of solar wind (Wänke 1965). The presence of microcraters and glassy agglutinates strengthens this conclusion (Rajan 1974). In order for the solar wind to flow it is necessary that the residual solar nebula have been dissipated. In order that the solar wind strike mineral grains these must have been directly on the surface of the parent body or possibly dispersed in space, but any dispersal could not have led to enough opacity to shield out the solar wind. The product of solar wind flux and the duration of the irradiation must have been sufficient to account for the observed rare-gas concentrations.

Studies of lunar material show that no meteorites are from the moon, and the most plausible loci for the present-day formation of gas-rich meteorites are the surfaces of large (e.g., ≥ 100 km) bodies in the asteroid region. If so, the observed high abundances of gas-rich meteorites indicate that this regolith cannot be a surficial layer (depth $\lesssim 1$ % the radius of the present body), as asteroid collision calculations show that the mass yield from asteroidal fragmentation is dominated by deep and even totally destructive impacts (Wetherill 1967; Dohnanyi 1969). Whether or not a body with the low surface gravity of an asteroid can be expected to possess the required deep regolith is not clear. Many workers have argued against anything but production of a very surficial regolith in the present-day asteroid belt, whereas Anders (1975) has concluded that it is possible that almost the entire asteroid consists of regolithic material developed by collisions in the present asteroid belt. Very extensive regoliths with similar characteristics may also have developed during the accretional history of the meteorite parent bodies (Wasson 1972). Understanding the relative importance of deep versus shallow and primordial versus more-recent types of regolith in producing the effects seen in gas-rich brecciated meteorites is an unresolved problem of major significance. Combined theoretical and experimental work directed toward a detailed understanding of the probable nature of asteroidal regoliths of these two kinds is badly needed. A significant start in this direction has been made (Housen et al. 1979; Chapman 1978; see the Housen et al. chapter in this book). Until our knowledge of this problem is greatly advanced, it cannot be said whether or not the effects observed are compatible with meteoritic regoliths of either kind.

Many ordinary (H, L, LL) chondrites are genomict, and a large fraction of these show textural and rare-gas properties indicating that they are regolith breccias. Roughly 10, 2.5 and 8%, respectively, of the H, L and LL chondrites contain resolvable amounts of solar-type rare gases (Wasson 1974). The fractions that are petrographically genomict are 25, 10, and 62%, respectively (Binns 1967). These fractions seem too large to be understandable in terms of a thin regolith layer on the surface of a moderately large (radius $\gtrsim 100$ km) parent body. In Table II are listed the foreign clasts observed in 15 of these

TABLE II

Characterized Foreign Chondritic Clasts in Meteorite Breccias

	Host Class	Clast Class	Note	Reference
Host Chondrite				
Abbott	H5	CM	a	Fodor et al. (1976)
Allende	CV3	CM	a	Clarke et al. (1971); Fruland et al. (1978)
Bencubbin	Chon	LL	a	Kallemeyn et al. (1978)
		CM	a	Kallemeyn et al. (1978)
		Chon	d	McCall (1968); R. Binns (1977, personal communication)
Cynthiana	L4	C	b, e	Van Schmus (1967)
Dimmitt	H3,4	C	b, e	R. V. Fodor (1979, personal communication)
Holyoke	H4	C	b	Ramdohr (1972)
Krymka	LL3	CM	g	Lewis et al. (1979)
Lancé	CO3	CM	a	Kurat (1975)
Leighton	H5	CM	a, e	Wilkening (1976)
Leoville	CV3	CM	a, e	Keil et al. (1968)
Mezö-Madaras	L3	CM	a	Van Schmus (1967)
Murchison	CM2	C	b	Fuchs et al. (1973)
Plainview	H5	CM	a	Wilkening and Clayton (1974)
		CM	a	Fodor and Keil (1976)
Pultusk	H5	C	b, e	Wilkening (1976)
Rio Negro	L3	C	b	Fodor et al. (1977)
St. Mesmin	LL6	H	b	Pellas (1972); Dodd (1974)

TABLE II (continued)

Characterized Foreign Chondritic Clasts in Meteorite Breccias

	Host Class	Clast Class	Note	Reference
Sharps	H3	CM	a	Fredriksson et al. (1969)
Tieschitz	H3	C	b	Kurat (1970)
Tysnes Island	H4	C	b, e	Wilkening (1976)
Weatherford	Chond	CV	a	Mason and Nelen (1968)
Weston	H4	C	b	Noonan et al. (1976)
Host Achondrite				
Cumberland Falls	Aub	Chon	c	Binns (1969)
Jodzie	How	CM	a	Bunch (1975)
Kapoeta	How	CM	a	Wilkening (1973)

aSome features anomalous (e.g., different concentrations of some volatiles).

bGroup characterization uncertain: CM one possibility.

cUngrouped chondrite having degree of reduction similar to IAB but distinctly different O-isotope composition.

dThe host Bencubbin is a reduced chondrite unrelated to the major groups; its texture appears to have been destroyed by plastic flow as a result of shock-induced shear; the "chon" clast may be compositionally identical to the host, but characterization is incomplete. Weatherford is closely related to Bencubbin; the available data are consistent with a chondritic host.

eNo published petrographic description.

fCarbonaceous chondrite intermediate in properties between CV and CO.

gLewis et al. (1979) suggest that this inclusion is CI but we find their reported volatile pattern more in keeping with a CM description.

breccias; all except one consist of material that resembles CM chondrites (C-rich, small chondrules, high-matrix content). The one exception is an H-group clast in the St. Mesmin LL chondrite. The properties of H, L and LL chondrites are so similar (Table I and Figs. 1, 2) that they must have formed in neighboring solar system locations. How did it happen that most, probably >99 %, of their clasts belong to the same group as the host (see, e.g., Fodor and Keil 1978), and most of the remainder consist of CM chondrite material? What do these statistics imply regarding the time of formation — early in solar system history or more recently in the asteroid belt (collisions are too infrequent to generate regoliths greater than ~10m in thickness outside the asteroid belt)? The large fraction of the H and LL ordinary chondrites that are genomict seems to require such breccias to comprise large fractions of their parent bodies. Since most of the material from asteroids results from deep impacts, sometimes including destruction of the parent body, it seems highly improbable that the high abundance of breccias results from the selective sampling of a lunar-type régolith. During this inferred accretionary period the regions where the different ordinary chondrite parent bodies were forming were separated enough in heliocentric distance to prevent the accretion of other classes of ordinary chondrite materials. If we are correct that these are accretionary breccias then the CM-like material was introduced from an external source, probably a source located farther from the sun.

Evidence favoring some recent regolith formation are the 1.4-Gyr K-Ar and ^{39}Ar-^{40}Ar ages reported for the H clast in St. Mesmin (Schultz and Signer 1977; Turner, 1978, private communication) and the 3.6-Gyr age of a Plainview clast (see Bogard's chapter in this book). The St. Mesmin result is complicated by 4.5-Gyr Rb-Sr ages on microporphyry inclusions (Minster 1979). These inclusions were melted and intruded the meteorite after some brecciation had occurred (Dodd and Jarosewich 1976). Whether or not evidence for at least two epochs of regolith formation in a single meteorite is a probable occurrence requires more detailed study of asteroid regolith formation processes. It has also been suggested that the low ^{39}Ar-^{40}Ar ages found in certain ordinary chondrites by Turner (1969) and Bogard et al. (1976) are evidence for a recent regolith history. However the chondrites with low ages show strong shock effects most probably associated with a major impact on their parent body; none of them is solar-gas-rich. Except for St. Mesmin and Plainview, there is no evidence for concluding that these low ages reflect regolith processes.

Wasson (1972) proposed that the heating and recrystallization that produced the different clast textures in genomict chondrites occurred while the materials were in planetesimals, and that gas-incorporation and formation of the regolithic structures occurred during the accumulation of these planetesimals to form larger parent bodies. This offered a simple explanation for the fact that gas-rich meteorites were equally distributed among petrologic types 3-5 and that the cosmic ray age distribution in gas-rich H chondrites is the same as that in the whole population. This remains a viable model, but we cannot exclude other models in which the heating occurs in a single parent body.

The polymict howardites are solar-gas-rich and clearly of regolithic derivation. Their dominant components are basaltic and ultramafic pyroxenitic materials that probably originated in the same parent body. The only foreign (extra-parent-body) clasts reported are CM chondrites (Table II); Chou et al. (1976) and Hertogen et al. (1978) have shown that the siderophile patterns in howardites indicate that the dominant chondritic component closely resembles CM chondrites. Siderophile concentrations indicate that CM material accounts for ~3 % of the matrix of these howardites. Since the amount present as recognizable clasts exceeds 0.5 % (Wilkening 1973), it follows that the CM material was accreted at velocities low enough to allow $\gtrsim 20$ % of the accreting projectile to survive as clasts. If, as seems likely, the CM-like material originated external to the region where the howardite parent body was located, how were the eccentricites of the orbits of the CM materials reduced to the low values consistent with low-velocity accretion? A suggestion is that collisions with local material reduced the eccentricities; if the CM planetesimals were cometary, the energy released by such collisions might mainly heat up the presumably more compressible icy fraction and leave the silicates relatively intact. Rb-Sr isochron ages of clasts in the Kapoeta howardite range from 4.5 to 3.7 Gyr (Papanastassiou et al. 1974) indicating that regolithic activity declined to a low level at the end of the early intense bombardment recorded by lunar highlands samples. However, this lower age limit may simply reflect incomplete sampling. About 10 % of the howardites are heavily shocked, but it is not known whether this occurred during the regolithic period or at the time the meteoroid was ejected from the parent body.

A large fraction of the aubrites (achondrites consisting almost entirely of enstatite, $MgSiO_3$) are solar-gas-rich. They contain metal that is probably a foreign component, and a large fraction of one (Cumberland Falls) is a reduced chondrite not closely related in chemical or O-isotope composition to any of the chondrite groups. Three aubrites (Bishopville, Norton County, Pena Blanca Spring) have been dated reasonably precisely, and their ages are >4.5 Gyr (Bogard et al. 1967; Podosek 1971; Minster and Allegre 1976). There is no evidence for regolithic stirring in more recent times.

The two (of 4) carbonaceous chondrite groups having high-volatile contents also contain solar gases and solar-flare particle tracks. "Compaction ages," the time when any stirring ceased, based on [244]Pu tracks produced in neighboring crystals, are in the range 4.2-4.4 Gyr (Macdougall and Kothari 1976; Macdougall 1977). A key question is whether these reflect the end of low-impact-velocity accretion or the end of a high-impact-velocity regolithic phase. If these meteorites formed >8 AU from the sun, calculations by Safronov (1969) indicate that the accretionary phase could have lasted ca. 0.2

Gyr, consistent with the compaction ages. A regolith phase similar to that experienced by the shocked and reheated lunar, H-group, howardite and aubrite breccias can be ruled out for the CM and CI chondrites, since high-velocity impacts would have led to extensive dehydration of the layer lattice silicates (as commonly observed in the CM-like clasts listed in Table II).

Anders (1975) attempted to determine the distance from the sun at which solar-gas-rich meteorites formed by relating their maximum solar-gas contents to their cosmic ray ages (used as upper limits on their regolith ages) and comparing the observed distribution with that found in mature lunar soils. In order that this treatment be valid it is necessary to assume that the meteorite breccias formed by processes closely similar to those currently prevailing on the moon but with the solar wind and projectile flux scaled to that expected in the asteroid belt, i.e., that the mean flux and velocity of bombarding meteoroids and the solar wind flux have not varied with time, and that a negligible fraction of the ejecta escaped the parent body (of unknown size). We doubt that these assumptions are even correct for those gas-rich breccias that formed in recent times, let alone those forming during the period 3.6–4.5 Gyr ago. Nonetheless, the chemical and isotopic evidence presented earlier support Anders' conclusion that the ordinary chondrites and howardites probably formed within 4 AU of the sun. However, these chemical arguments do not constrain the origin of CM and CI chondrites to the region inside 4 AU; as discussed in more detail below, it appears equally probable that their parent bodies originated in the outer solar system.

In summary, the chemical and isotopic evidence is most consistent with the formation of the chondrites in two distinct locations: LL, L, H, IAB, EH and EL chondrites in the inner solar system, CI, CM, CO and CV chondrites in the outer solar system. The heliocentric distance at which the transition from inner to outer occurs cannot yet be defined; it is almost certainly as large as 2 AU, and may be several times greater. The inner solar system groups are distinguished from the outer groups in terms of their lower refractory abundances, lower degrees of oxidation, and oxygen-isotope compositions close to that of the earth and moon. The sequencing of formation locations among the inner or outer solar system groups is not clear, since the various properties do not vary in concert. The speculation by Wasson (1977) that, in order of increasing heliocentric distance, the inner solar system sequence of clans is enstatite-IAB-ordinary (see Table I) still appears to be the best working hypothesis.

III. DYNAMICAL ARGUMENTS FOR AND AGAINST PARTICULAR CLASSES OF PARENT BODIES CONSIDERED IN THE LIGHT OF OTHER EVIDENCE

The preceding discussion concerns the original site of formation of the meteorite parent bodies. This may or may not correspond to their present location in the solar system, as dynamic processes exist which can transfer objects from one heliocentric distance to another. Therefore if a class of objects in the present solar system is hypothesized to be the source of meteorites which, on the basis of the preceding discussion have the characteristics of having originated elsewhere, then it is necessary to explain how and when this orbital transfer occurred. Furthermore, in every case there must exist a quantitatively adequate mechanism for transferring meteoritic fragments from the present location of their parent body to the earth. In this Section these dynamical aspects of the problem of the origin of meteorites will be discussed for the various candidate sources.

Asteroids

Asteroids are bodies ranging up to 1000 km in diameter which are almost entirely confined to the wide region between Mars and Jupiter. They exhibit no coma of volatile compounds, and most likely consist of mixtures of silicates and metal. Spectrophotometric studies (McCord *et al.* 1970; Chapman 1976; see the chapters by Chapman and Gaffey and by Gaffey and McCord) show large differences in composition among the asteroids; although these reflection spectra only result from the outermost 100 μm of the asteroid surface, the inferred composition probably applies to the entire regolith, and may, as commonly assumed, apply to the entire planet. The asteroids primarily fall into two classes: the most abundant *C*-type appears to consist of carbonaceous material, and the *S*-type of mixtures of silicates and metal. There is a distinct hiatus in albedos between these two classes (Morrison 1977). Unlike the comets, direct association of photographic meteoroids with an asteroidal source is not possible, as the orbits of belt asteroids do not intersect the orbit of the earth. However, there are mechanisms by which the orbits of asteroidal bodies can evolve into Earth-intersecting orbits.

In addition to the asteroids proper, confined to the region between Mars and Jupiter, there is a significant number of objects which are asteroidal in appearance, but whose orbits are not confined to the asteroid belt. The orbit of 944 Hidalgo (see the chapter by Degewij and van Houten) extends beyond Saturn and the Earth-approaching Apollo-Amor objects have perihelia within 1.3 AU (see the chapter by Shoemaker *et al.*). Although these can be operationally defined as asteroids, it is probable that many, and possibly most, of these objects are genetically more closely related to short-period comets. Because their short dynamic lifetimes demand constant resupply, the Apollo-Amor objects will be discussed separately from main-belt asteroids.

Asteroids are strong *prima facie* candidates for meteorite sources because collisions among the asteroids must provide a quantity of small debris (10^{13}-10^{15} g annually), more than adequate to supply the present flux of Earth-impacting matter, *provided* that there exist mechanisms able to place a sufficient fraction ($\sim 10^{-4}$) of this material into Earth-crossing orbits on the short time scale ($\sim 10^6$-10^7 yr) defined by the cosmic ray exposure history of stony meteorites. It is also necessary that the shock damage associated with this transfer mechanism usually be limited to that associated with low shock pressures (10-100 kbar) recorded in stony meteorites. Iron meteorites commonly record higher shock pressures (130-300 kbar) (Lipschutz 1968).

Difficulties in finding such mechanisms have been a problem in the past. Suitable mechanisms must be primarily gravitational in nature, as collisional shock associated with more than small (<0.1 AU) changes in semimajor axis is probably excessive. Gravitational perturbations of Mars-crossing and Mars-grazing asteroidal fragments by Mars have been suggested as such a gravitational mechanism (Arnold 1965a,b; Anders 1971) but until recently it appeared that, except for iron meteorites, this mechanism required transit times too long ($\sim$500 Myr) to be reconciled with cosmic ray exposure histories. Calculations of collisional lifetimes of meter-size meteorites with aphelia in the asteroid belt, using best estimates for the destructability of silicate material and the abundance of smaller asteroidal debris, yield lifetimes of only 10-40 Myr. This result corresponds to the short lifetime extreme of the results presented earlier (Wetherill 1967) when these parameters were even more uncertain than they are at present. Regardless of whether the observed exposure ages (commonly 2-20 Myr) were primarily dynamically or collisionally limited, they presented a severe difficulty for an asteroidal origin.

Production of relatively low-velocity (<200 m sec^{-1}) fragments in proximity to various regions in the asteroid belt in which the motion of fragments is in resonance with the motion of the giant planets has now been semiquantitatively shown to be an adequate source. Fairly large ($\sim$100 m) fragments can be produced by collisions in the vicinity of the Kirkwood 2:1 gap at 3.28 AU, in which the orbital period is commensurable with the period of Jupiter. The resulting resonant motion will at times cause these fragments to be in highly eccentric orbits with aphelia beyond 4 AU and perihelia $\sim$2 AU. These orbits will be stabilized by librational relationships which preclude close encounters to Jupiter. However, statistically probable collisions of these $\sim$100-m bodies with smaller asteroidal debris produces low-velocity meteorite-size fragments some of which escape the libration region. Strong Jupiter perturbations cause these to random walk into Earth-crossing on a time scale short ($\sim 10^6$ yr) relative to the cosmic ray ages of stony meteorites (Zimmerman and Wetherill 1973). This chain of events has been criticized on the grounds that requiring two collisions, close approaches to Jupiter, etc., renders it too complex and by inference too *ad hoc* to be taken seriously. Such reasoning is fallacious, as there is no reason to suppose that nature provides meteorites to Earth by mechanisms which are simple for us to describe to one another in preference to those which are probable. The mechanisms described are real phenomena of significant and estimable probability which cannot fail to occur.

The principal problem is a quantitative one, as best estimates of the meteorite yield on Earth from this source are 10^7-10^8 g per year, uncertain by at least an order of magnitude. Thus it is not clear if a major or only a minor part of the Earth's meteorites are produced in this way. Scholl and Froeschlé (1977) have presented evidence that the mechanism described above may be more effective for the 5:2 Kirkwood gap than for the 2:1 case. Large asteroids in proximity to these Kirkwood gaps are listed in Tables III and IV.

Williams (1973) proposed that asteroid collision fragments ejected at fairly low velocities ($\sim$300 m sec^{-1}) in the vicinity of the secular resonance ν_6 (Fig. 3) would experience large enough excursions in eccentricity to permit their becoming Earth-crossing. A more efficient variant of this mechanism has been described by Wetherill (1974, 1977) and Wetherill and Williams (1979). This is a "synergistic" mechanism by which the nonlinear interaction of secular resonance with the coupled system of the larger planets and Mars' perturbations can perturb a rather large yield ($\sim$10^{10} g yr^{-1}) of meteorite-size asteroidal fragments into Earth-crossing. The typical time required for this material to impact the earth is $\sim$5 $\times$ 10^8 yr, but a significant

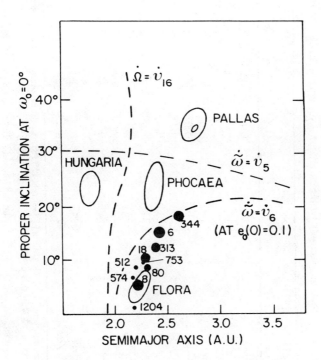

Fig. 3. Observed distribution of large asteroids in the inner portion of the asteroid belt and in the vicinity of the ν_6 resonance. The large open "ovals" approximately define the limits of the Hungaria, Flora, Phocaea, and Pallas regions of the asteroid belt.

TABLE III

Large Asteroids with Semimajor Axes within 0.1 AU of the 2:1 Kirkwood Gap (3.28 AU)

Asteroid	a (AU)	e	i (deg)	$B(1,0)$[a]	Class[b]	Diameter[b] (km)
106 Dione	3.17	0.18	5	8.8	C	139
511 Davida	3.18	0.18	16	7.4	C	323
154 Bertha	3.18	0.10	21	8.5	C	191
92 Undina	3.19	0.07	10	7.9	C	244
702 Alauda	3.19	0.03	21	8.3	C	205
758 Mancunia	3.20	0.13	6	9.6	C	119
175 Andromache	3.21	0.20	3	9.6	C	113
530 Turandot	3.21	0.20	8	10.3	C	81
381 Myrrha	3.21	0.12	13	9.7	C	126
108 Hecuba	3.22	0.09	4	9.7	S	61
122 Gerda	3.22	0.06	2	9.2	C	139
895 Helio	3.22	0.14	26	9.5	C?	
745 Mauritia	3.24	0.07	14	11.0	?	
903 Nealley	3.24	0.05	12	10.9	?	

[a]Absolute magnitudes from Gehrels and Gehrels (1978). Some of these data differ somewhat from those in the TRIAD published by these same workers (see Part VII in this book).

[b]Classifications and diameters from Morrison (1977) and Zellner and Bowell (1977), Fowell et al., (1978), and Bowell (1978, personal communication).

TABLE IV

Large Asteroids with Semimajor Axes Within 0.1 AU of 5:2 Kirkwood Gap (2.82 AU)[a]

Asteroid	a (AU)	e	i (deg)	$B(1,0)$	Class	Diameter (km)
146 Lucina	2.72	0.07	13	9.2	C	141
45 Eugenia	2.72	0.08	7	8.3	C	226
410 Chloris	2.72	0.24	11	9.5	C	134
156 Xanthippe	2.73	0.23	10	9.8	C	104
140 Siwa	2.73	0.21	3	9.6	C	103
110 Lydia	2.73	0.08	6	8.7	C	170
200 Dynamene	2.74	0.13	7	9.5	C	123
185 Eunike	2.74	0.13	23	8.7	C	169
247 Eukrate	2.74	0.24	25	9.3	C	142
387 Acquitania	2.74	0.24	18	8.4	S	112
173 Ino	2.74	0.21	14	9.1	C	142
308 Polyxo	2.75	0.04	4	9.3	U	138
128 Nemesis	2.75	0.12	6	8.8	C	164
71 Niobe	2.76	0.17	23	8.3	S	115
93 Minerva	2.76	0.14	9	8.7	C	168
356 Liguria	2.76	0.24	8	9.3	C	150
41 Daphne	2.76	0.27	16	8.1	C	204
1 Ceres	2.77	0.07	11	4.5	C	1003
88 Thisbe	2.77	0.17	5	8.1	C	210
39 Laetitia	2.77	0.11	10	7.4	S	163
2 Pallas	2.77	0.23	35	5.2	U	608
148 Gallia	2.77	0.19	25	8.5	S	106

TABLE IV (continued)

Large Asteroids with Semimajor Axes Within 0.1 AU of 5:2 Kirkwood Gap (2.82 AU)[a]

Asteroid	a (AU)	e	i (deg)	$B(1,0)$	Class	Diameter (km)
532 Herculina	2.77	0.17	16	8.0	S	150
393 Lampetia	2.77	0.33	15	9.2	C	129
28 Bellona	2.78	0.15	9	8.2	S	126
68 Leto	2.78	0.18	8	8.2	S	126
139 Juewa	2.78	0.17	11	9.2	C	163
446 Aeternitas	2.79	0.07	11	10.2	R	40
216 Kleopatra	2.79	0.25	13	8.1	M	128
354 Eleonora	2.80	0.12	18	7.5	S	153
346 Hermentaria	2.80	0.10	9	8.9	S	84
236 Honoria	2.80	0.19	8	9.5	S	65
441 Bathilde	2.81	0.08	8	9.5	M	66
804 Hispania	2.84	0.14	15	8.9	C	141
385 Ilmatar	2.85	0.13	14	8.8	S	96
81 Terpsichore	2.85	0.21	8	9.6	C	112
129 Antigone	2.87	0.21	12	7.9	M	115
47 Aglaja	2.88	0.14	5	9.2	C	158
471 Papagena	2.89	0.24	15	7.9	S	143
386 Siegena	2.90	0.17	20	8.4	C	191
238 Hypatia	2.91	0.09	12	9.2	C	154
22 Kalliope	2.91	0.10	14	7.3	M	177
16 Psyche	2.92	0.13	3	6.9	M	250
674 Rachele	2.92	0.20	14	8.5	S	102
349 Dembowska	2.92	0.09	8	7.2	R	144

[a]Data from the same sources as for Table III.

fraction (~1 %) can impact within 50 Myr. In these papers the mechanism is proposed as the most probable source of some groups of iron meteorites, which typically have exposure ages of 10^8 to 10^9 yr and of some minor portion of the differentiated silicate meteorites, e.g., the basaltic achondrites. The asteroids which supply most of this material are those near the $\hat{\nu}_6$ resonance. This includes the populous Flora region with semimajor axes ~2.25 AU, low inclinations, and with eccentricities which permit the parent objects to come within ~0.05 to 0.1 AU of Mars' aphelion for favorable combinations of the long-period "secular" variations in the orbits of both the asteroids and Mars (Fig. 3).

Most of these asteroids are *S*-type, believed to consist of a mixture of metallic iron, pyroxene, and olivine, but nevertheless different from ordinary chondrites in composition. This identification is consistent with derivation of differentiated meteorites from this region. The other known mechanisms for removing meteorites from the asteroid belt produce exposure ages much shorter than those typically associated with iron meteorites. As a result it seems likely that either *S*-type or less abundant *M* ("metallic") asteroids in the vicinity of the ν_6 resonance represent the principal sources of iron meteorites and of some of their complementary silicate differentiates. If the *S* asteroid spectra can be reconciled with those of ordinary chondrites, this region could also be a source of chondritic meteorites. If *S* asteroids are of chondritic composition, asteroids near the 5:2 Kirkwood gap (Table IV) could be the dominant sources of ordinary chondrites.

In addition to their oxidation state and oxygen isotope composition, other chemical and petrological arguments are relevant to the question of whether or not meteorite sources are asteroidal. Even the undifferentiated chondritic meteorites have had a complex chemical history. The volatile-poor ordinary chondrites have in many cases been heated and metamorphosed at temperatures as high as ~900°C (Van Schmus and Koffman 1967; Bunch and Olsen 1974). It has proven difficult to determine the heat source that could have heated objects to metamorphic temperatures at solar distances ⩾2.4 AU. However, there is strong evidence for an asteroidal origin for basaltic materials, and it has been *demonstrated* that the basaltic meteorites experienced a melting event ~4.5 × 10^9 yr ago. It does not seem extreme to suppose that other bodies presently in the asteroid belt underwent the less severe heating required to explain ordinary chondrite textures and mineralogy. Although it is possible to imagine mechanisms for heating cometary interiors, there is no direct evidence that they have been hot.

Another class of chemical arguments is based upon a presumably known relationship between the temperatures at which meteoritic materials condensed from the nebula (as deduced from their mineralogy and severe heating required to explain ordinary chondrite textures and mineralogy. Although it is possible to imagine mechanisms for heating cometary interiors, there is no direct evidence that they have been hot.

and time be well understood, in contrast to the present diffuse state of knowledge concerning the processes and conditions of star and planetary system origin. Insofar as astrophysical calculations of the initial temperature of the solar nebula are valid, they indicate that temperatures were well below the condensation temperature of many meteoritic minerals, except at unreasonably small heliocentric distances (Black and Bodenheimer 1976; Tscharnuter 1978). Arguments for an extended high-temperature nebula rest on cosmochemical evidence (Wasson 1978), not on theoretical astrophysics. Condensation theories fail to explain how the distinctly different C and S classes of asteroids are formed at similar (and overlapping) heliocentric distances. If one assumes that most meteorites come from parent bodies in the asteroid belt, the detailed variations in oxygen isotopic composition among the different meteorite classes (Clayton *et al.* 1976; Clayton and Mayeda 1978) are difficult to explain in terms of condensation at similar heliocentric distances. These phenomena appear to require that materials originally formed at significantly different heliocentric distance were subsequently mixed by conceivable but poorly characterized gravitational and nongravitational mechanisms. This considerably complicates the identification of an asteroidal belt origin based on distance inferred from nebula evolutionary models.

In addition to the one large C-type asteroid located near the ν_6 resonant surface (313 Chaldaea), others are in proximity to the Kirkwood gaps (Tables III and IV). Assuming these asteroids are indeed similar to carbonaceous chondrites, they are strong candidate sources for such meteorites.

The large asteroid 4 Vesta has frequently been proposed as the source of the basaltic achondrite meteorites (specifically, the eucrites); see Drake's chapter. Its reflectance spectrum is in accord with this identification. However, its perihelion is so far from Mars' aphelion and its semimajor axis so far from resonant values that there is no dynamical reason to expect a significant yield of meteorites from this asteroid. From the dynamic point of view, it seems more likely that differentiated silicate meteorites are derived from the silicate materials stripped off large S asteroids such as 6 Hebe and 8 Flora and Apollo-Amors derived from these bodies. Consolmagno and Drake (1977) have questioned the validity of this inference in view of the near absence among differentiated meteorites of the peridotitic residues of basalt formation, and have proposed Vesta as a basaltic achondrite source for which only the surficial basalt and slightly deeper pyroxene-rich (diogenite) layers are exposed. If there were compelling evidence for Vesta having such a fragmentation history, this argument should be given serious weight. On the other hand, there is no *a priori* reason that the earth should sample all parts of a fragmented parent body with equal efficiency.

Hostetler and Drake (1978) propose that the dynamical difficulties of obtaining meteorites from Vesta can be removed by assuming that ~100 m fragments are removed from Vesta by cratering, and that these fragments are subsequently perturbed into Earth-crossing orbits. Meteorites are subsequently produced by further fragmentation of these large fragments. Insofar as this suggestion makes use of known dynamical processes this simply represents an Apollo object source, with the Apollo object produced by the "synergistic mechanism," proposed both as a source of meteorites (Wetherill 1974; Wetherill and Williams 1979) and of Apollo objects (Wetherill 1976). The difficulties of obtaining meteorites from Vesta remain the same: the very low yield of relatively unshocked material which ultimately becomes Earth-crossing expected from Vesta, in contrast with material derived from much more favorably situated objects such as 6 Hebe and 8 Flora and probably a number of other differentiated objects that have been reduced by fragmentation to sizes too small to have been studied telescopically. Unless new perturbation mechanisms are discovered which remove this selection effect, the Hostetler-Drake proposal does not solve the problem. In view of the dynamical problems associated with this identification, it seems premature to definitely conclude that Vesta is the eucrite parent body.

If ordinary chondrites do come from asteroids, it is becoming increasingly puzzling why almost none of the asteroids match their reflectance spectra. The sampling is now sufficiently complete that recognition of only an occasional asteroid having the spectrum expected of ordinary chondrite material (e.g., 496 Gryphia) still leaves a discrepancy. Another main-belt asteroid (the Apollo-Amor objects will be discussed separately) previously held to be of ordinary chondritic composition by spectrophotmetric observers, 349 Dembowska, is now believed on the basis of longer wavelength infrared data to be composed of differentiated mantle-type materials (Feierberg et al. 1979; see the chapter by Larson and Veeder). Various qualitative explanations of this discrepancy have been proposed: poor sampling of asteroids, observation of only a surficial layer of the asteroid, changes in reflectance caused by solar wind sputtering or micro-particle bombardment, contamination of surface layers by foreign material. Reasons for rejecting these hypotheses can be given. Nevertheless quantitative understanding of spectrophotometric data is not sufficiently advanced to rule out the possibility that asteroids like Dembowska or the S-type asteroids could be sources of ordinary chondrites.

Comets

Typical active comets are small ($\sim$1 to 10 km) objects containing more or less equal quantities of volatile compounds of H, C, N, and O and more refractory compounds, e.g., silicates and metal. The size distribution of comets is poorly known. Delsemme (1978) and Kresák (1978) have inferred that the comet size distribution differs from that of asteroids in that there is a marked deficiency of very small comets. There is some evidence that very large comets exist, e.g., the comet of 1729 had an estimated magnitude of

~4, even though its perihelion was at 4 AU. Although the calculation is obviously model dependent, it seems likely that this comet has a diameter of more than 100 km.

The success of the "dirty snowball" model of a comet (Whipple 1950) has probably accidentally led to a misconception in the minds of some workers, particularly those in related fields of study. This is that comets are principally composed of ice, and can be visualized more or less as a glacier or snowbank. In fact, the material emitted by a comet contains as much dirt as snow (Delsemme 1977). Furthermore, only the smaller, nonvolatile particles can be "blown off" with the volatiles, and the presence of larger bodies (e.g., $\gtrsim 25$ cm) in comets decreases the fraction of ice. As the ice is volatilized, the fraction of rocky material will increase and will tend to accumulate as residual material. Fireball studies show that the more massive nonvolatile cometary material is abundant and this conclusion is strengthened by the evidence for nearly-extinct and extinct cometary nuclei. Thus at least half, and possibly 90 % of even an active comet may be nonvolatile dust and rocky matter. As discussed above, it is possible that a comet may be more like a breccia than an iceberg, and that pieces of ice are as likely to be clasts as matrix in this breccia, a consequence of H_2O and (possibly) CO_2 being as solid as anything else at the low temperatures which prevailed in the region in which comets were formed. At present, most (and probably all) comets are derived from the outermost regions of the solar system, the Oort cloud at a distance of 10^4-10^5 AU (0.1 to 1 lightyear) (Marsden 1977). They become observable only when they are gravitationally perturbed by passing stars into the inner solar system. The volatilization of their H, C, N and O compounds produces a *coma* ~10^4 km in diameter, and an ionized tail (up to ~10^8 km in length), which render the comets visible.

Comets are definitely associated with much of the interplanetary flux of bodies impacting the earth, including those in the mass range under discussion (10^2 to 10^7 g). This has been established by photographic studies of the orbits of these bodies as they enter the atmosphere (Ceplecha and McCrosky 1976), and comparison of these orbits with those of known comets. Positive identification with particular comets is possible in many cases. In many additional cases similarity of both orbits and physical properties (as indicated by their ablation or fragmentation in the atmosphere) to objects associated with known comets demonstrates their cometary association. However, only three meteorite falls (the undifferentiated ordinary chondrites Pribram, Lost City and Innisfree) are contained among these photographic meteoroids. As will be discussed further subsequently, their orbits and physical properties, although well determined, do not define whether or not they are of cometary origin.

It is practically certain that no meteorites in our collections have been derived from historically observed active comets. The shortest known cosmic ray exposure age is that of the ordinary chondrite Farmington (19 kyr) and this is unique. In contrast, the volatile content of comets is insufficient to continue the observed volatile loss for more than 10 kyr. Furthermore, it is unlikely that most meteorites are derived from comets during the active stage of their history, as they are usually too massive to be swept along by the outflowing gases. Although meteorites could be freed from comets during more violent cometary outbursts, or following disruption while passing close to the sun, it seems most likely that cometary meteorites, if they exist, are derived from nonvolatile residues of comets that survive the active lifetime of a short-period comet. This nonvolatile material could be a core that was originally mantled with volatile ices, a loosely aggregated collection of meteoritic fragments originally scattered through the icy material, or ice-poor portions of a regolithic breccia. If carbonaceous chondrites are of cometary origin, the solar-flare track data (Macdougall and Kothari 1977; Goswami and Lal 1979) seem most consistent with large scale (>1 m) initial segregation of silicates and ices prior to comet formation. Thus the comet may be heterogeneous on this scale.

There is good, if not compelling, evidence that nonvolatile residues of comets exist. There is a gradation in activity between highly volatile comets newly arrived from the Oort cloud, and the short-period comets. This trend continues down to apparently severely volatile-depleted short-period comets, such as Encke (Sekanina 1971) and barely active comets such as Arend-Rigaux and Neujmin I. The natural end-members of this series are the nonvolatile Apollo-Amor objects, and it has frequently been proposed that some or all of these bodies are extinct comets (Öpik 1963, 1966; Anders and Arnold 1965; Wetherill and Williams 1968; Wetherill 1976, 1979).

One short-period comet (Encke) is presently in an orbit with aphelion at 4.1 AU, well within the orbit of Jupiter. An orbit of this kind is relatively stable with respect to gravitational perturbations, in contrast to Jupiter-crossing bodies which will be ejected from the solar system in $\lesssim 10^5$ yr. Sekanina (1971) has shown how the "jet effect" (nongravitational forces which are the reaction to the comet's emitted dust and gas) could have reduced Encke's aphelion to its present value during the last ~ 3000 yr. At present these forces are small, implying relatively little emission of gas, which is compatible with Encke being nearly extinct. On this line of reasoning, it can be predicted that during the next few hundred years, Encke will become an Apollo object. Nor does it appear to be alone. A number of meteor streams exhibit physical characteristics very much the same as the Taurid meteors, generally considered to be fragments of Encke (Ceplecha and McCrosky 1976; Ceplecha 1977). It is most plausible that these streams have been recently derived from unobserved extinct comets, because the time required for the streams to evolve into orbits as small as, for example, that of the Geminids (aphelion = 2.6 AU) is $\sim 10^6$ yr (Wetherill 1976), whereas a stream will only remain coherent for $\sim 10^4$ yr.

The recent discovery of the Apollo object 1978SB with an orbit very similar to that of Encke may raise the question of whether the Taurids should be associated with this body, rather than with Comet Encke. Meteor astronomers attach much significance to the similarity of $\tilde{\omega} = \omega + \Omega$ for the Taurids and Encke. On the other hand, the value of ω for these meteors is highly constrained by the requirement that the node of the meteors be near 1 AU in order that they impact the earth. If $\tilde{\omega}$ remained strictly constant during the secular perturbations which produced the difference between the present values of ω and Ω for Encke and the Taurids, then the stream would be observed only for a very restricted range of Ω, which is not the case. The similarity of orbits of 1978SB and Comet Encke also leads to the obvious speculation that they are in some way genetically related. This is supported by the observation that 1978SB has a low albedo (see the chapter by Shoemaker *et al.*), and thus may be carbonaceous in composition. However no adequate dynamical mechanism has been proposed for producing the observed dispersion in ω and Ω for these two bodies within the 10^3-10^4-yr period during which Encke has been in its present orbit.

Fragments of extinct comets are not confined to small meteors. Large objects, kilograms in mass, are associated with these streams. There is even evidence that very large ($\sim$100 ton) bodies are sometimes found in streams (Table V).

So it seems very likely that extinct comets exist and that large meteoroids derived from them impact the earth. To a large extent the orbits of these meteoroids will be similar to those derived from the asteroid belt by the mechanisms discussed in the previous section. Are these meteoroids ever meteorites, i.e., can they survive passage through the atmosphere and be recovered from the ground? No direct evidence exists, but there is circumstantial evidence that this may be the case. Ceplecha and McCrosky (1976) have shown that meteoroids in the 10^{-2} to 10^{-7} g mass range differ considerably in their physical strength and ability to penetrate deeply into the atmosphere. These authors classify fireballs into three groups, based on the variation of their end height with photometric mass and velocity. The classification of fireballs has been reinvestigated (ReVelle and Wetherill 1978; Wetherill *et al.* 1978). Although the latter classification differs conceptually from that of Ceplecha and McCroskey, for the most part the same fireballs are grouped together. The following discussion will be given in terms of the Ceplecha-McCrosky approach, and the consequences of the more recent work will be appended.

Class III, the weakest of all, is associated with a number of well-established cometary meteor streams, and is nearly certain to be of cometary origin. Class II is significantly stronger. Some objects of this class are also associated with cometary streams (e.g., Taurids and Encke). Ceplecha *et al.* (1977) have obtained a spectrum of one of these bodies which had a terminal mass of 70 g, and hence survived passage through the atmosphere.

TABLE V

χ Orionid Fireballs

	Magnitude (max)	Initial Mass (photometric)	a (AU)	e	i (deg)	ϖ	Ω	Date	End Height (km)
N. χ Orionids			2.22	0.79	2	179	258	12/4 to 12/15	
EN041274	−21	10^5 kg	1.98±.18	0.76	3.5	174	252	12/4/74	56
EN031267	−11.3	14 kg	2.20	0.79	3.9	173	250	12/3/67	55
(?)PN39115B	−10.9	1.1 kg	2.33	0.83	2.1	197	269	12/21/75	62
PN42388	−9.3	1.6 kg	2.22	0.79	2.8	176	255	12/7/74	59
PN39833	−8.7	1.3 kg	2.38	0.83	3.5	185	256	12/9/67	63
D614764	−2.8		2.20	0.77	2.9	175	259	12/11/61	54
S. χ Orionids			2.18	0.78	7	180	79	12/7 to 12/14	
PN39469A	−9.7	1 kg	2.33	0.78	5	172	78	12/10/66	56
H2349	−3.7		2.22	0.73	5	166	80	12/13/50	56
H1544	−1.6		2.23	0.75	5	167	77	12/10/47	57

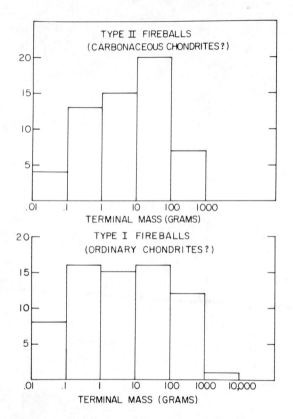

Fig. 4. Observed distribution of the terminal masses of Prairie Network fireballs (Ceplecha and McCrosky 1976). It is seen that significant terminal masses are found for both Type I and Type II fireballs. Most fireballs belong to these classes.

The spectrum shows strong CN bands and therefore contains carbonaceous matter. It seems most plausible to associate this body with some type of carbonaceous chondrite.

Class I fireballs are strongest. Many fireballs (~1/3) fall into this class which includes the three photographed by fireball networks which were recovered as meteorites, and proved to be ordinary, noncarbonaceous chondrites. There is every reason to believe that any of them could have reached the ground if they had been large enough, or had entered the atmosphere at sufficiently low velocity. Their terminal mass distribution (Fig. 4) indicates that it is common for both Type I and Type II meteoroids to have finite terminal masses. Except for the problem that asteroids do not appear to be of ordinary chondritic composition, there is no particular reason why most of these fireballs could not be of asteroidal origin, accelerated into Earth-crossing by one of the gentle resonance gravitational mechanisms discussed in the previous section. If so, this would oppose the present consensus that essentially all large and small meteors are derived from comets.

However, there also seem to be Type I and Type II bodies in *prima facie* cometary orbits, e.g., with aphelia beyond Jupiter or in retrograde motion. Many of these are of high atmospheric entry velocity, >25 km sec^{-1}, and scaling to the velocities of the Type I bodies actually recovered as meteorites may have caused them to be erroneously assigned to Type I. But this is not always the case. Six ($\sim$3 %) of the Prairie Network fireballs (McCrosky *et al.* 1977) are low-velocity (<20 km sec^{-1}) Type I or II bodies with aphelia, Q, beyond Jupiter's perihelion (see Table VI). These have low terminal masses consistent with their small initial masses. It is very unlikely that this is asteroidal material which impacted the earth while in the process of being ejected from the solar system, as their number is a factor of >100 larger than the number calculated from studies of the orbital evolution of such material.

The Ceplecha-McCrosky classification was based on an empirical statistical correlation between observed fireball end heights and their photometric masses and pre-atmospheric velocities. This approach explictly deviates from classical single-body meteor theory. ReVelle and Wetherill (1978) and Wetherill *et al.* (1978) took a different approach which retains the formalism of single-body theory. They noted that the three ordinary chondrites which were photographed by meteor networks (Pribram, Lost City, Innisfree) were among the brightest objects observed by these networks. For any reasonable power-law size distribution of chondritic fragments, it follows that there should be many more smaller chondritic fragments represented by fainter fireballs. In this way it is estimated that about 30 % of the fireballs photographed by the Prairie Network are ordinary chondrites, and the problem is that of identifying which they are. ReVelle (1979) noted that the deceleration and end heights of the Lost City and Innisfree fireballs could be described quite well by the single-body theory provided a "dynamic mass" lower than the actual mass was used. This is in contrast to the photometric mass, which appears to err in the opposite direction of being too high (ReVelle and Rajan 1979). It is assumed that the chondritic fragments should share this characteristic of obeying single-body theory with a dynamic mass determined from deceleration data. Deceleration data were used to calculate dynamic masses for observed fireballs, and the single-body theory was used to scale the observed end height for differences in velocity, entry angle and dynamic mass. In order to reduce the extrapolation in the scaling of velocity, only meteors with pre-atmospheric velocities less than 18 km sec^{-1} were used in this initial study. When this was done it turned out that a large fraction of those fireballs which survived deceleration to less than 8 km sec^{-1} had scaled end heights of 21 ± 1 km, within 1 km of those of Lost City and Innisfree (Fig. 5). A smaller fraction of the surviving fireballs had end heights up to 4 km higher. The majority of the fireballs, including almost all those with final velocities higher than 8 km sec^{-1} had higher-scaled end

TABLE VI

"Strong" Jupiter-Crossing Fireballs with Low Entry Velocity

Number	a (AU)	e	i (deg)	Q (AU)	V_{ENTRY} (km sec^{-1})	$M_{INITIAL}$[a] (g)	M_T[b] (g)	M_T^*[c] (g)	Type	Scaled End Height (km)
PN39057	4.2	0.76	0.1	7.3	14.7	1400	40	80	I	22.3
PN39820B	3.1	0.69	11	5.3	16.7	1100	19	8	I	25.5
PN39972	5.5	0.84	3	10.4	18.1	170	1	0.7	I	23.3
PN42357C	3.0	0.67	12	5.0	16.0	360	7	20	I	27.7
PN42312	3.0	0.69	14	5.1	17.5	430	2	6	I	24.4
PN41282	4.5	0.80	2	8.1	17.5	1900	1	20	II	—d

[a]$M_{INITIAL}$ is the initial photometric mass given by Ceplecha and McCrosky (1976).

[b]M_T is the terminal mass calculated by the formal procedure of Ceplecha and McCrosky (1976).

[c]M_T^* is the estimated photometric mass near the end point at ~8 km sec^{-1};

[d]Data inadequate.

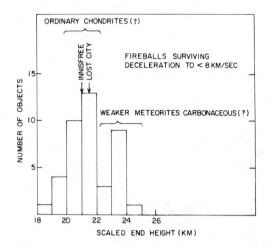

Fig. 5. A large fraction of the photographic network fireball meteoroids that survived deceleration to <8 km sec^{-1} and thus could have survived atmospheric passage have scaled end heights (see text for definition) of 21±1 km. The recovered ordinary chondrites Innisfree and Lost City have scaled end heights within this range. Other meteoroids in the same peak are inferred to have comparable strengths, and to largely consist of ordinary chondrite material. The weaker the material, the higher its scaled end height. The bodies having slightly greater end heights, i.e., ~23.5 km, are inferred to represent slightly weaker material such as CM or CI carbonaceous chondrites.

heights ranging up to 45 km and possibly higher.

Since the number of bodies which group with Lost City and Innisfree in the 21-km group is about the number expected for reasonable power-law distribution (Fig. 6), and they are all smaller than Lost City and Innisfree, it is plausible to suppose these are the small ordinary chondrites which should be present in the fireball population. Inasmuch as the physical properties of achondrites and some carbonaceous chondrites are similar to ordinary chondrites, a few of these less abundant recoverable meteorites may also be represented in the 21-km group. Those fireballs which failed to survive deceleration in the atmosphere and which had scaled end heights above 30 km are unlikely to be represented in meteorite collections because of their inability to survive atmospheric penetration. This leaves the 13 surviving objects with end heights in the 22 to 26 km range shown in Fig. 5. These are likely to be potentially recoverable meteorites for which the single-body theory is not applicable, presumably due to excessive fragmentation deep in the atmosphere. It is suggested that these are weaker objects, e.g., CI carbonaceous chondrites or more friable members of other meteorite classes.

Although the physical model used in this work is quite different from that of Ceplecha and McCrosky, both their Class I fireballs and the 21-km group of this newer work share the property of having very low end heights

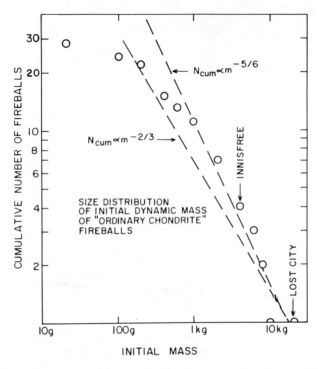

Fig. 6. The size distribution of the "initial dynamic mass" of fireball meteoroids having scaled end heights $\leqslant 22$ km sec^{-1} follow size distributions $N_{cum} \propto m^{-\alpha}$ where α is between 2/3 and 5/6. Such a power-law distribution is predicted by fragmentation models. The falloff at masses <1 kg is a selection bias resulting from sensitivity limits of the observing network. The distribution is consistent with essentially all these objects being ordinary chondrites.

for their size, entry velocity and entry angle. Therefore the 21-km group primarily contains Type I fireballs. However, the trans-Jupiter fireballs classified as Type I in Table VI and for which adequate deceleration data exist do not fall in the 21-km group. Four of the 5 objects for which data are available have scaled end heights in the 22- to 26-km group, although one of these was not included in Fig. 5 since its last measured velocity was 9 km sec^{-1}. The scaled end heights of these objects are given in the last column of Table VI. As suggested above, it is plausible to suppose that these are carbonaceous bodies of cometary origin with appreciable physical strength. However, until "ground truth" is established by the identification of recovered meteorites with objects having these scaled end heights, this identification must be quite tentative. Therefore, although the conclusion is limited by the present availability of data, a comet-meteorite association is suggested regardless of whether the Ceplecha-McCrosky classification or that of ReVelle *et al.* is used. If it could be shown that identifiable meteorites are associated with these large-aphelion orbits of cometary affinity, it would be plausible to associate meteorites of the same strength class in more common orbits with extinct comets.

Apollo-Amor objects

Apollo-Amor objects are small bodies (typically ~1 km diameter, but ranging up to ~30 km) with perihelia less than a rather arbitrary value of 1.3 AU, and usually with aphelia in the asteroid belt. These orbits are dynamically unstable on the time scale of the solar system, and like the meteorites they must be derived from sources elsewhere in the solar system, probably including both comets and asteroids (Anders and Arnold 1965; Wetherill 1976). Their Earth-crossing or near Earth-crossing nature identifies them as prime candidate meteoroid and meteorite sources (Anders 1964; Levin et al. 1976; Wetherill 1976). However, up to the present no clear-cut orbital identifications have been made, although tentative identification of a few Apollos with known small meteoroid streams has been proposed (Sekanina 1973). Spectral measurements show that all but three Apollo-Amor objects studied using these techniques resemble S-type asteroids more than C-type (Gaffey and McCord 1978), but statistics are biased in favor of the S-type objects because of their higher albedos. In any case, it is of interest to note that they are not all of the same composition and that this range of compositions includes high albedo, silicate objects resembling differentiated and undifferentiated silicate meteorites.

The fact that Apollo objects are in Earth-crossing orbits and are exposed to asteroidal collisions near aphelion implies that at least some Earth-impacting meteoroids must be derived from these bodies. Although Amor objects are not Earth-crossing at present, it has been shown (Wetherill 1979) that evolution of Apollos into Amors and vice-versa is so rapid that many Amors must be former or future Apollos. Williams has shown that a number of Amors can make close encounters to the earth on the even shorter 10^4-10^5 yr time scale associated with secular perturbation (see the chapter by Shoemaker et al.; Wetherill and Williams 1968). With regard to their role as meteorite sources, the only questions are of yield and mechanical strength. Calculations of the yield show that this could be large enough to supply the entire flux of chondritic meteorites ($\sim 10^8$ g yr^{-1}) and is at least high enough to supply ~1 % of this material. No definite information regarding their strength is available.

If some fireballs could be associated with known Apollo objects, strength information could be obtained from the end heights of the meteoroids. However, it is not obvious that such associations can be found, as the exposure ages of stone meteorites are comparable to the time scale for major orbital evolution of Apollo objects. However, in the case of meteorites such as Farmington with very short exposure ages, this could prove possible (Levin et al. 1976).

One problem with identifying the Apollo-Amor objects with ordinary chondrites is that the chondrite radiants (Astopovich 1939; Simonenko 1975) and times of fall (Wetherill 1968, 1969) are at least at first sight not in agreement with dynamical calculations of the expected distribution of these quantities. This question needs to be examined in the light of more recent work on the aerodynamic selection effects accompanying the entry of fireballs into the atmosphere (ReVelle 1976) and the orbital evolution of Apollo-Amors (Wetherill 1978; the chapter by Shoemaker et al.).

As discussed above, Apollo-Amors are not permanent residents of the inner solar system, but are derived from asteroidal or cometary sources, or more likely, both. Thus they can be thought of as big meteoroids which can either fragment into small meteoroids or before fragmentation impact the earth, forming craters 1-100 km in diameter. Some of these bodies are probably the extinct comets discussed earlier, whereas others can be derived from the inner asteroid belt (Levin et al. 1976; Wetherill 1976, 1978). The resulting orbits are similar in either case (Wetherill 1978). Dynamical considerations suggest that the cometary component should predominate. Interpretation of spectral observations leads to an ambiguity. A large number of Apollos appear to have reflectance characteristics similar to ordinary chondritic material, though recent discoveries have increased the population of C-type objects (see the chapter by Shoemaker et al.). Most workers would interpret these spectra to indicate an asteroidal origin but the spectrophometric work showing that ordinary chondritic material is rare or absent in the asteroid belt conflicts with such intuitive interpretations.

One argument against the derivation of ordinary chondrites from comets via Apollo objects has been that an Apollo object of cometary origin could not have suffered the massive collision 500 Myr ago suggested by the apparent clustering in gas-retention ages (Anders 1964; Heymann 1967; Turner 1969). Recent theoretical studies of the evolution of Apollo-Amor objects of either cometary or asteroidal origins show that when secular resonance is included, an asteroidal parent body cannot necessarily be inferred from these ages even if they are interpreted literally (Wetherill 1978). A significant fraction (~15 %) of Apollos is transferred into the Amor and Mars-crossing region for times as long as 2 Gyr, then returned to Earth-crossing. Thus an Apollo object of cometary origin can have been in the asteroid belt 500 Myr ago, been cratered and shocked by collisions, then returned to an Apollo-type orbit <100 Myr ago. However, a typical 1-km Apollo object would seem unlikely to develop a full-fledged lunar-style regolith, insofar as this is required to explain the results on gas-rich chondrites.

It might have been thought that the gross difference between typical cometary and asteroidal orbits would make it relatively easy to distinguish between cometary and asteroidal sources once the orbits of meteorites are known. This is not the case. In order for asteroidal material to impact the earth as a meteorite, it is necessary that it be placed into a more eccentric orbit with perihelion within the orbit of the earth. On the other hand, comets or cometary residua will have short dynamical lifetimes in the inner solar system unless their orbits evolve into orbits with aphelia inside Jupiter's orbit, i.e., become similar to the orbit of Encke's comet (aphelion = 4.1 AU) and subsequently evolve into Apollo objects. Thus asteroidal and cometary meteorites will have similar orbits, with Earth-crossing perihelia, and aphelia in or near the asteroid belt. The distinction will be further blurred as a consequence of perturbations by Earth and Venus which will tend to "equilibrate" the distribution of Earth-crossing orbits. It is known from radiant and time-of-fall statistics (Wetherill 1971) that at least ordinary chondritic meteorites must evolve from initial Earth-crossing orbits with perihelia near Earth and aphelia near Jupiter. Both asteroidal and cometary sources can have this general characteristic. However, more subtle differences exist which in the future may be helpful in identification of candidate sources.

Planetary surfaces

It is generally thought that planetary surfaces are not likely source regions for meteorites of any kind, because acceleration to the escape velocity from a planetary body even as small as the moon is expected to cause shock effects greater than those observed. In spite of this difficulty, a lunar origin was proposed for carbonaceous chondrites (Urey 1965), H-group ordinary chondrites (Hintenberger et al. 1964), eucrites and howardites (Duke and Silver 1967), and it was suggested that L-group ordinary chondrites were from Mars (Wänke 1966). The samples returned from the moon by the Apollo and Luna programs showed that lunar material is not present in our meteorite collections. This reinforced arguments indicating that such material should show more extreme shock effects than those observed in most stony meteorites, and it seemed likely that if we were not receiving meteorites from the nearby Moon, the probability of their coming from larger and more distant planets was nil (e.g., Wetherill 1974).

However, this conclusion may prove hasty. Comparison of chemical and isotopic data on tektites with lunar and terrestrial material has persuaded almost all workers that these impactites are of terrestrial origin. Nevertheless, the aerodynamic evidence that the Australian tektites were accelerated to above the earth's atmosphere (i.e., to 7-11 km sec^{-1}) has not been seriously challenged (Baker 1958; Chapman and Larson 1963). Thus it appears that it is possible for an object to survive acceleration to nearly escape velocity even from a planet with as large a gravity field and as massive an atmosphere as the Earth's. To be sure, the tektites were melted in the process, but the tektite data do show that, provided suitable target materials are available, some objects can be ejected from smaller planets with smaller atmospheres, e.g., the moon and Mars.

Several workers (Urey 1957; Lin 1966) have proposed that terrestrial acceleration of tektites was a result of volatilization of the nucleus of a comet upon impact with the earth. In these theories it is hypothesized that the massive quantities of cometary water relased by the impact aerodynamically accelerate the tektite up to the escape velocity, at the same time preserving relatively low velocities between the tektite and the gas surrounding it. Terrestrial cratering by comets certainly occurs, probably with a frequency ~10 % that associated with Apollo impact, i.e., about one continental cometary crater >20 km in diameter every 10 Myr. At typical cometary impact velocities the kinetic energy of the impacting body is sufficient to volatilize several hundred times its own mass. Therefore the quantity of water released by a comet impacting the ocean could be much greater than that contained within the nucleus of the comet itself. Little is known about the partition of energy for impacts into an aqueous target. However, Mars is more similar to the earth than the moon inasmuch as the Viking data have been interpreted showing that the Martian subsurface contains water in the form of permafrost up to a depth of ~1 km. When combined with its relatively small atmosphere and gravity field, conditions should be more favorable for ejection without melting from Mars than from the earth In contrast, the absence of a water-rich layer on the moon may be the reason we have no meteorites from the moon.

Although detailed studies have not been made, preliminary work shows that an appreciable quantity of Mars ejecta will become Earth-crossing on the short time scale associated with the exposure ages of stony meteorites. It is not clear if collisions of this ejecta with asteroidal material will be sufficiently frequent to eliminate unobserved longer exposure ages.

Ejecta from Mars would be expected to be differentiated and heavily shocked, perhaps severely enough to reset radiometric clocks. The differentiated stony meteorite Shergotty (like its sister meteorite Zagami) has been heavily altered by shock; in particular, the abundant plagioclase has been converted to glass. A recent Rb-Sr investigation by Nyquist et $al.$ (1979) yielded a Rb-Sr isochron age of 165 Myr; a ^{39}Ar-^{40}Ar age by Bogard (see his chapter in this book) gave a poorly defined plateau of ~250 Myr. The Rb-Sr age is interpreted as the time when the shock alteration occurred, the higher ^{39}Ar-^{40}Ar age presumably reflecting incomplete Ar outgassing during the shock event. If any meteorite came from Mars, Shergotty seems the best candidate. The chief problem with this hypothesis seems to be its short cosmic ray age of 2 Myr. If Shergotty was removed from Mars 165 Myr ago, the ejected mass must have had dimensions of ≳5 m until ~2 Myr ago. Nyquist et $al.$ infer a diameter of ~100 m in order to retain enough heat to permit diffusive resetting of the Rb-Sr and argon ages at 165 Myr. Here the tektite analogy may break down; we have no evidence that such large masses can be ejected from the surface of a planet as massive as Mars.

As noted by Stolper et $al.$ (1979), Nakhla and two siblings (Lafayette,

Governador Valdares) and Chassigny and a sister (Brachina) may be related to Shergotty. The O-isotope data of Shergotty and the Nakhla twin Lafayette are consistent with such a relationship (Clayton *et al.* 1976), and rare-earth data suggest a genetic link between Nakhla and Chassigny (Boynton *et al.* 1976). Radiometric ages of the Nakhla trio are ~1.3 Gyr (Podosek 1973; Papanastassiou and Wasserburg 1974; Gale *et al.* 1975; Bogard and Husain 1977), and a K-Ar age of 1.39 Gyr for Chassigny was reported by Lancet and Lancet (1971). These low ages and the presence of hydrated silicates (Bunch and Reid 1975; Floran *et al.* 1978) in these meteorites would be consistent with an origin on Mars. Chassigny has been shocked to pressures of 150-200 kbar, while evidence for shock in the Nakhla trio is absent. Papanastassiou and Wasserburg (1974) attribute some scatter in the Nakhla Rb-Sr mineral isochron to late metamorphism, whereas Gale *et al.* (1975) hold that the ~1.3 Gyr event was of igneous origin. As discussed for Shergotty, the relatively low cosmic ray ages of the Nakhla trio of 10 Myr (Ganapathy and Anders 1969) and 9 Myr for Chassigny (Lancet and Lancet 1971) may present problems for an origin on Mars.

Although it is obvious from the foregoing discussion that there are serious difficulties which must be surmounted before a Martian origin for these meteorites appears likely, the possibility that we have rocks from Mars in our meteorite collections is important enough to warrant giving some attention to this consideration.

IV. SUMMARY AND SUGGESTIONS FOR FUTURE WORK

Evidence of various sorts indicates that most differentiated meteorites are presently derived from the asteroidal belt, either directly or through the intermediary of Apollo-Amor objects. Some, possibly most, of these differentiated asteroids in such orbits may have formed 0.4-1.5 AU from the sun and been introduced into their present orbits by the inverse of the resonance process proposed for their present-day removal from such orbits. Relatively small shifts in the position of the secular resonances resulting from the growth of the planets could leave some of these implanted in stable orbits outside the present resonances. More distant *S*-asteroids and possibly even Vesta conceivably were implanted by still larger shifts in the secular resonance. Many of the objects near these resonances have *S*-type spectra, which most authorities associate with differentiated meteoritic material.

The origin of the most abundant meteorite falls, the three groups comprising the ordinary chondrite clan, has not been defined. Although some probably are associated with the Apollo asteroids having similar spectra, the exact region of the solar system where such asteroids formed cannot yet be specified. If the asteroid belt is the source of ordinary chondrite material, then surface processes have significantly altered the characteristic reflectance spectra. On the other hand, the cosmochemical arguments given in Sec. II suggesting that ordinary chondrites originated in the inner solar system are inconsistent with a cometary origin.

Spectrophotometric studies show that there is opaque material with low albedo in the asteroid belt; it appears that this is carbonaceous. Some of these dark asteroids are adjacent to resonances that can gently accelerate their low velocity collision ejecta into Earth-crossing. Unless this material is too weak to survive atmospheric passage (i.e., is associated with the Type III fireballs), there should also be carbonaceous meteorites of asteroidal origin.

It is quite possible that some of our meteorites are cometary, and if so, they were probably derived from extinct comets that are now in Apollo-Amor type orbits. These are likely to be undifferentiated meteorites. If the "inner-solar-system" chondrites (Sec. II) were in orbits where H_2O could condense one must explain the absence of evidence of low-temperature minerals such as Fe_3O_4 or hydrated silicates. However, the high volatile contents of the CI and CM chondrites indicate relatively low nebula equilibration temperatures, and suggest that they could originate in comets. The other (CO and CV) groups of carbonaceous meteorites could be asteroidal, but their close petrographic and compositional ties to the CI and CM chondrites suggest similar formation locations for all four groups.

One might think that the abundance of observational, theoretical and experimental evidence relevant to the problem of identification of meteorite sources should permit more clear-cut identifications. The problem is that the evidence does not lead to an internally consistent solution. In general, the smaller the fraction of the available evidence examined, the firmer the conclusions regarding the identification of meteorite sources.

It is possible that the major ambiguities will not be resolved until samples are obtained from asteroids and comets. However, at the present stage, it would appear useful to attempt to resolve which line(s) of evidence is leading us astray. Some of the most important topics of investigation are the following:

1. Perhaps the most straightforward problem would be to resolve the question of whether or not the distribution of chondrite radiants and time of falls is compatible with derivation of most of these bodies from Apollo-Amor objects. This will require selection of a plausible range of fragment-size distributions making use of available or new hypervelocity impact data. This could then be combined with bias-corrected Apollo-Amor statistics (smoothed by theoretical steady-state considerations) and an improved physical theory for meteorite entry, perhaps along the lines of ReVelle (1976). Comparison of the theoretical radiant and time-of-fall distribution with that observed should then permit us to know whether or not the discrepancy is as serious as appears at first glance.

2. Spectrophotometric measurements of asteroids have led to the

conclusion that ordinary chondrites are rare or absent in the main asteroid belt. This leads to the rather ironic situation that while compositional data argue that the ordinary chondrites come from the inner solar system, there do not seem to be suitable asteroids in the most accessible storage place for inner solar system material — the asteroid belt. On the other hand, carbonaceous material appears to be abundant in the asteroid belt, even though an outer solar system formation region seems consistent with the chemical and isotopic properties of carbonaceous chondrites. There are large asteroids adjacent to the 5:2 Kirkwood gap which probably could supply meteorites with the required radiant and time-of-fall distributions of ordinary chondrites. Could these be ordinary chondritic bodies, the spectral signature of which has been obscured by surface alteration processes? Plausible arguments against this possibility have been advanced, but do not seem sufficiently definitive. Further laboratory simulation coupled with theoretical studies of the basic physical processes involved are needed.

3. On a sufficiently short time scale, i.e. 10^2-10^4 yr, the orbital evolution of planet-crossing bodies is deterministic and can be handled by classical methods of celestial mechanics. However, on longer time scales multiple close planetary encounters occur and minor differences in initial orbits result in grossly different final orbits. Under these circumstances the system is best modeled statistically. Although, like a roulette wheel, it is still in principle deterministic, the information required to make deterministic predictions is not available. Nevertheless, in both cases, valid inferences of a probabilistic nature can be made. Discussion of the long-range orbital evolution of planet-crossing bodies has been entirely dependent on these stochastic methods (Öpik 1951, 1977; Arnold 1965a,b; Wetherill 1968, 1977). However, there are a number of assumptions and approximations made in these methods which have never been critically examined using the full body of celestial mechanical understanding. Some recent work (Cox et al. 1978; Cox 1978) represents a start in this direction. It would be trivial to show that the stochastic methods are not rigorous, and trite to say "they should be used with great caution." What is needed is a constructively motivated, critical study of these techniques, directed toward placing them on a better theoretical foundation. This could allow us to have more confidence in interpreting second-order differences between observed and theoretical orbit distributions and to produce more quantitative estimates of expected yields from various sources.

4. A principal basis for the inference that there is meteoritic material of cometary origin is obtained from photographic fireball networks, particularly the Prairie Network (McCrosky et al. 1977). However, the efforts of these networks have primarily been directed toward meteorite recovery, and the calculation of orbits are often deliberately biased

against the most clear-cut fireballs of cometary origin — the shower meteoroids. Meteoroids identified as belonging to the major showers were not reduced in the Prairie Network investigations, and Canadian Network data is not reduced at all unless a meteorite fall is suspected. There are no continuing fireball studies in the United States at present. The inferences tentatively made previously strongly suggest that serious treatment of fireball data may force revision of our present concepts of the physical nature of comets, but this cannot happen unless people work in the field.

5. Many of the arguments used to identify meteorites with their sources are based on regolithic analogs. However, there is very little understanding of how regolithic properties may be expected to vary as a function of solar system history, of heliocentric distance or of mass and composition on the body on which they occur. An important contribution of this kind has been made (Housen *et al.* 1979*a* and their chapter; Chapman 1978). Until we understand much more quantitatively when asteroidal or cometary regoliths formed and can precisely model their abundance of charged particle tracks, microcraters, agglutinates, etc., we do not really know if meteoritic evidence favors or disfavors particular regolithic identifications.

6. There is at present no theory adequate to explain even qualitatively the origin of the principal features of the asteroid belt, e.g., its small mass content, relative velocity distribution, Kirkwood gaps and mixed chemical composition. Development of a theory of this kind will require a much more quantitative understanding of the origin of stars and planetary systems in general, and the sun and planets of our solar system in particular. There has been renewed interest in these problems during the last few years, but the goal is still distant.

7. The compositional hiatus between the chondritic classes need better definition, and a thorough search of breccias, clasts, weathered finds, etc. for meteorites having compositions within these hiatus should be made. The more detailed our understanding of the compositional variations, the greater the probability that the pattern can be understood in terms of a model that applies to materials formed at all distances from the sun, as opposed to the present models that mainly account for the properties of single groups or clans that probably originated within a narrow range of solar distances.

One can be hopeful that investigations along the lines suggested above would help considerably in constructing an internally consistent framework in which to view the problem of identification of meteorites with their sources. It is unlikely, however, that such investigations would lead to the qualitatively distinctive revelations which have followed actual spacecraft missions to the moon and planets. "Ground truth" and sample return may be expected to be the ultimate answer to the identification of meteorites with their sources, and to the realization of the geological context in which these small bits of primordial material should be viewed.

Acknowledgments. We thank C. R. Chapman for a detailed review and L. L. Wilkening and an anonymous referee for useful comments. Partial support of this research has been provided by a grant from the National Science Foundation.

REFERENCES

Anders, E. 1964. Origin, age, and composition of meteorites. *Space Sci. Rev.* 3: 583-714.

Anders, E. 1971. Interrelations among meteorites, asteroids, and comets. In *Physical Studies of the Minor Planets,* ed. T. Gehrels (NASA SP-267, Washington, D. C.: U. S. Government Printing Office), pp. 429-446.

Anders, E. 1975. Do stony meteorites come from comets? *Icarus* 24: 363-371.

Anders, E., and Arnold, J. R. 1965. Age of craters of Mars. *Science* 149: 1494-1496.

Arnold, J. R. 1965. The origin of meteorites as small bodies. II. The model. *Astrophys. J.* 141: 1536-1547.

Arnold, J. R. 1965. The origin of meteorites as small bodies. III. General considerations. *Astrophys. J.* 141: 1548-1556.

Astopovich, I. S. 1939. Some results of the study of 66 orbits of meteorites. *Astron. J. USSR* 16: 15-45.

Baker, G. 1958. The role of aerodynamical phenomena in shaping and sculpturing Australian tektites. *Amer. J. of Sci.* 256: 369-383.

Bickle, M. J.; Hawkesworth, C. J.; Martin, A.; Nisbet, E. G.; and O'Nions, R. K. 1976. Mantle composition derived from the chemistry of ultramafic lavas. *Nature* 263: 577-580.

Binns, R. A. 1967. Structure and evolution of noncarbonaceous chondritic meteorites. *Earth Planet. Sci. Lett.* 2: 23-28.

Binns, R. A. 1969. A chondritic inclusion of unique type in the Cumberland Falls meteorite. In *Meteorite Research,* ed. P. M. Millman (Dordrecht: D. Reidel), 696-704.

Black, D. C., and Bodenheimer, P. 1976. Evolution of rotating interstellar clouds. II. The collapse of protostars of 1, 2 and 5 $M_\odot$. *Astrophys. J.* 206: 138-139.

Bogard, D. D.; Burnett, D. S.; Eberhardt, P.; and Wasserburg, G. J. 1967. ^{87}Rb-^{87}Sr isochron and ^{40}K-^{40}Ar ages of the Norton County achondrite. *Earth Planet. Sci. Lett.* 3: 179-189.

Bogard, D. D., and Husain, L. 1977. A new 1.3 aeon-young chondrite. *Geophys. Res. Lett.* 4: 69-71.

Bogard, D. D.; Husain, L.; and Wright, R. J. 1976. ^{40}Ar-^{39}Ar dating of collisional events in chondrite parent bodies. *J. Geophys. Res.* 81: 5664-5678.

Bowell, E.; Chapman, C. R.; Gradie, J. C.; Morrison, D.; and Zellner, B. 1978. Taxonomy of asteroids. *Icarus* 35: 313-335.

Boynton, W. V.; Starzyk, P. M., and Schmitt, R. A. 1976. Elemental composition and genesis of the ureilites, the achondrite Chassigny and the nakhlites. *Geochim. Cosmochim. Acta* 40: 1439-1448.

Buchwald, V. F. 1975. *Handbook of Iron Meteorites* (Berkeley: University of California Press).

Bunch, T. E. 1975. Petrography and petrology of basaltic achondrite polymict breccias (howardites). *Proc. Lunar Sci. Conf. VI* (Oxford: Pergamon Press), pp. 469-492.

Bunch, T. E., and Olsen, E. 1974. Restudy of pyroxene-pyroxene equilibration temperatures for ordinary chondrite meteorites. *Contrib. Mineral. Petrol.* 43: 83-90.

Bunch, T. E., and Reid, A. M. 1975. The nakhlites. Part I: Petrography and mineral chemistry. *Meteoritics* 10: 303-315.

Ceplecha, Z. 1977. Meteoroid populations and orbits. In *Comets, Asteroids, Meteorites,* ed. A. H. Delsemme (Toledo, Ohio: University of Toledo Press), pp. 143-152.

Ceplecha, Z.; Bocek, J.; and Jezkova, M. 1977. *Bull. Astron. Inst. Czech.* (to be submitted).

Ceplecha, Z., and McCrosky, R. E. 1976. Fireball end heights: A diagnostic for the structure of meteoric matter. *J. Geophys. Res.* 81: 6257-6275.

Chapman, C. R. 1976. Asteroids as meteorite parent-bodies: The astronomical perspective. *Geochim. Cosmochim. Acta* 40: 701-719.

Chapman, C. R. 1978. Asteroid collisions, craters, regoliths and lifetimes. In *Asteroids: An Exploration Assessment,* eds. D. Morrison and W. C. Wells, NASA Conf. Publ. 2053, pp. 145-157.

Chapman, C. R., and Larson, H. K. 1963. On the lunar origin of tektites. *J. Geophys. Res.* 68: 4305-4358.

Chou, C.-L.; Baedecker, P. A.; and Wasson, J. T. 1973. Distribution of Ni, Ga, Ge, and Ir between metal and silicate portions of H-group chondrites. *Geochim. Cosmochim. Acta* 37: 2159-2171.

Chou, C.-L.; Baedecker, P. A.; and Wasson, J. T. 1976. Allende inclusions: Volatile-element distribution and evidence for incomplete volatilization of presolar solids. *Geochim. Cosmochim. Acta* 40: 85-94.

Clarke, R. S.; Jarosewich, E.; Mason, B.; Nelen, J.; Gomez, M.; and Hyde, J. R. 1971. The Allende, Mexico, meteorite shower. *Smithson. Contr. Earth Sci.* 5: 1-53.

Clayton, R. N.; Grossman, L.; and Mayeda, T. K. 1973. A component of primitive nuclear composition in carbonaceous meteorites. *Science* 182: 485-488.

Clayton, R. N., and Mayeda, T. K. 1978. Genetic relations between iron and stony meteorites. *Earth Planet. Sci. Lett.* 40: 168-174.

Clayton, R. N., Onuma, N.; and Mayeda, T. K. 1976. A classification of meteorites based on oxygen isotopes. *Earth Planet. Sci. Lett.* 30: 10-18.

Consolmagno, G. J., and Drake, M. J. 1977. Composition and evolution of the eucrite parent body: Evidence from rare earth elements. *Geochim. Cosmochim. Acta* 41: 1271-1282.

Cox, L. P. 1978. Numerical simulation of the final stages of terrestrial planet formation. Ph.D. dissertation, Massachusetts Inst. of Technology.

Cox, L. P.; Lewis, J. S.; and Lecar, M. 1978. A model for close encounters in the planetary problem. *Icarus* 34: 415-427.

Danielsson, L., and Ip, W.-H. 1972. Capture resonance of the asteroid 1685 Toro by the earth. *Science* 176: 906-907.

Delsemme, A. H. 1977. The pristine nature of comets. In *Comets, Asteroids, Meteorites,* ed. A. H. Delsemme (Toledo, Ohio: University of Toledo Press), pp. 3-13.

Delsemme, A. H. 1978. Empirical data on the origin of new comets. In *Dynamics of the Solar System,* ed. R. L. Duncombe (Dordrecht: D. Reidel, in press).

Dodd, R. T. 1974. Petrology of the St. Mesmin chondrite. *Contr. Mineral. and Petrol.* 46: 129-145.

Dodd, R. T., and and Jarosewich, E. 1976. Olivine microporphyry in the St. Mesmin chondrite. *Meteorites* 11: 1-27.

Dohnanyi, J. S. 1969. Collisional model of asteroids. *J. Geophys. Res.* 74: 2531-2554.

Duke, M. B., and Silver, L. T. 1967. Petrology of eucrites, howardites and mesosiderites. *Geochim. Cosmochim. Acta* 31: 1637-1665.

Edgeworth, K. E. 1949. The origin and evolution of the solar system. *Mon. Not. Roy. Astron. Soc.* 109: 600-609.

Feierberg, M.; Larson, H.; Fink, U.; and Smith, H. 1979. Spectroscopic evidence for two achondrite parent bodies: 349 Dembowska and 4 Vesta. In preparation.

Floran, R. J.; Prinz, M.; Hlava, P. F.; Keil, K.; Nehru, C. E.; and Hinthorne, J. R. 1978. The Chassigny meteorite: A cumulate dunite with hydrous amphibole-bearing melt inclusions. *Geochim. Cosmochim. Acta* 42: 1213-1229.

Fodor, R. V., and Keil, K. 1976. Carbonaceous and noncarbonaceous lithic fragments in the Plainview, Texas, chondrite: Origin and history. *Geochim. Cosmochim. Acta* 40: 177-189.

Fodor, R. V., and Keil, K. 1978. *Catalog of Lithic Fragments in LL Group Chondrites.* Univ. New Mexico Inst. Met. Spec. Publ. 19.

Fodor, R. V.; Keil, K.; and Gomes, C. B. 1977. Studies of Brazilian meteorites. IV.

Origin of a dark-colored, unequilibrated lithic fragment in the Rio Negro chondrite. *Rev. Brasil. Geol.* 7: 45-57.

Fodor, R. V.; Keil, K.; Wilkening, L. L.; Bogard, D. D.; and Gibson, E. K. 1976. Origin and history of a meteorite parent-body regolith breccia: Carbonaceous and noncarbonaceous lithic fragments in the Abott, New Mexico, chondrite. In *Tectonics and Mineral Resources of Southwestern North America,* Geol. Soc. Spec. Publ. Vol. 6, pp. 206-218.

Fredriksson, K.; Jarosewich, E.; and Nelen, J. 1969. The Sharps chondrite – new evidence on the origin of chondrules and chondrites. In *Meteoritic Research,* ed. P. M. Millman (Dordrecht: D. Reidel), pp. 155-165.

Fruland, R. M.; King, E. A.; and McKay, D. S. 1978. Allende dark inclusions. *Proc. Lunar Sci. Conf. IX* (Oxford: Pergamon Press), pp. 1305-1329.

Fuchs, L. H.; Olsen, E.; and Jensen, K. J. 1973. Mineralogy, mineral-chemistry, and composition of the Murchison (C2) meteorite. *Smithson. Contrib. Earth Sci.* 10: 1-39.

Gaffey, M. J., and McCord, T. B. 1978. Asteroid surface materials: Mineralogical characterizations from reflectance spectra. *Space Sci. Rev.* 21: 555-628.

Gale, N. H.; Arden, J. W.; and Hutchison, R. 1975. The chronology of the Nakhla achondritic meteorite. *Earth Planet. Sci. Lett.* 26: 195-206.

Ganapathy, R., and Anders, E. 1969. Ages of calcium-rich achondrites. II. Howardites, nakhlites, and the Angra dos Reis angrite. *Geochim. Cosmochim. Acta* 33: 775-787.

Gehrels, T., and Gehrels, N. 1978. Minor planets and related objects. XXVI. Magnitudes for the numbered asteroids. *Astron. J.* 83: 1660-1674.

Goldreich, P., and Ward, W. R. 1973. The formation of planetesimals. *Astrophys. J.* 183: 1051-1061.

Goswami, J. N., and Lal, D. 1979. Formation of the parent bodies of the carbonaceous chondrites. *Icarus* (special Asteroid issue).

Gurevich, L. E., and Lebedinsky, A. I. 1950. Formation of the planets (in Russian). *Izv. Akad. Nauk. USSR Phys.* 6, 14: 765-799.

Hertogen, J.; Janssens, M.-J.; Palme, H.; and Anders, A. 1978. Late nebular condensates and other materials collected by the meteorite parent bodies. *Proc. Lunar Sci. Conf. IX* (Oxford: Pergamon Press), pp. 497-499.

Heymann, D. 1967. On the origin of hypersthene chondrites: Ages and shock effects of black chondrites. *Icarus* 6: 189-221.

Hintenberger, H.; König, H.; Schultz, L.; and Wänke, H. 1964. Radiogene, spallogene und primodiale Edelgase in Steinmeteoriten. *Z. Naturforsch.* 19: 327-341.

Hostetler, C. J., and Drake, M. J. 1978. Quench temperatures of Moore County and other eucrites: Residence time on eucrite parent body. *Geochim. Cosmochim. Acta* 42: 517-522.

Housen, K. R.; Wilkening, L. L.; Chapman, C. R.; and Greenberg, R. 1979. Asteroidal regoliths. *Icarus* (in press).

Kallemeyn, G. W.; Boynton, W. V.; Willis, J.; and Wasson, J. T. 1978. Formation of the Bencubbin polymict meteorite breccia. *Geochim. Cosmochim. Acta* 42: 507-515.

Keil, K.; Huss, G. I.; and Wiik, H. B. 1968. The Leoville, Kansas, meteorite: A polymict breccia of carbonaceous chondrites and a chondrite (abstract). In *Meteorite Research,* ed. P. M. Millman (Dordrecht: D. Reidel), p. 217.

Kresák, L. 1978. The comet and asteroid population of the earth's environment. *Bull. Astron. Inst. Czech.* 29: 114-125.

Kurat, G. 1970. Zur Genese des kohligen Materials im Meteoriten von Tieschitz. *Earth Planet. Sci. Lett.* 7: 317-324.

Kurat, G. 1975. Der kohlige Chondrit Lancé: Eine petrologische Analyse der komplexen Genese eines Chondriten. *Mineral Petrog. Mitt.* 22: 38-78.

Lancet, M. S., and Lancet, K. 1971. Cosmic-ray and gas-retention ages of the Chassigny meteorite. *Meteoritics* 6: 81-85.

Larimer, J. W., and Anders, E. 1967. Chemical fractionations in meteorites. II. Abundance patterns and their interpretation. *Geochim. Cosmochim. Acta* 31: 1239-2170.

Latimer, W. M. 1950. Astrochemical problems in the formation of the earth. *Science* 112: 101-104.

Levin, B. J.; Simonenko, A. N.; and Anders, E. 1976. Farmington meteorite: A fragment of an Apollo asteroid? *Icarus* 28: 307-324.

Lewis, R. S.; Alaerts, L.; Hertogen, J.; Janssens, M-J.; Palme, H.; and Anders, E. 1979. A carbonaceous inclusion from the Krymka LL-chondrite: Noble gases and trace elements. *Geochim. Cosmochim. Acta* (submitted).

Lin, S. C. 1966. Cometary impact and the origin of tektites. *J. Geophys. Res.* 10: 2427-2437.

Lipschutz, M. E. 1968. Shock effects in meteorites. In *Shock Metamorphism of Natural Materials,* eds. B. M. French and N. M. Short (Baltimore: Mono Book Corp.), pp. 571-583.

Macdougall, J. D. 1977. Time of compaction of Orgueil (abstract). *Meteoritics* 12: 301-302.

Macdougall, J. D., and Kothari, B. K. 1977. Formation chronology for C2 meteorites. *Earth Planet. Sci. Lett.* 33: 36-44.

Marsden, B. G. 1977. Orbital data on the existence of Oort's cloud of comets. In *Comets, Asteroids, Meteorites,* ed. A. H. Delsemme (Toledo, Ohio: University of Toledo Press), pp. 79-86.

Mason, B., and Nelen, J. 1968. The Weatherford meteorite. *Geochim. Cosmochim. Acta* 32: 661-664.

McCall, G. J. H. 1968. The Bencubbin meteorite: Further details, including microscopic character of host material and two chondrite enclaves. *Mineral. Mag.* 36: 726-739.

McCord, T. B.; Adams, J. B.; and Johnson, T. V. 1970. Asteroid Vesta: Spectral reflectivity and compositional implications. *Science* 168: 1445-1447.

McCrosky, R. F.; Shao, C. Y.; and Posen, A. 1977. Prairie network fireball data. I. Summary and orbits. *Center for Astrophysics Preprint* No. 665. (submitted to *Meteoritika,* in Russian).

Minster, J. F. 1979. The ^{87}Rb^{87}Sr chronology of the chondrites and the formation age of the solar system (in French). Ph.D. dissertation, Univ. of Paris, VII, pp. 218-225.

Minster, J. F., and Allègre, C. J. 1976. ^{87}Rb-^{87}Sr history of the Norton County enstatite achondrite. *Earth Planet. Sci. Lett.* 32: 191-198.

Morrison, D. 1977. Asteroid sizes and albedos. *Icarus* 31: 185-220.

Noonan, A. F.; Nelen, J.; and Fredriksson, K. 1976. Mineralogy and chemistry of xenoliths in the Weston chondrite-ordinary and carbonaceous (abstract). *Meteoritics* 11: 344-346.

Nyquist, L.; Wooden, J.; Bansal, B.; Wiesmann, H.; McKay, G.; and Bogard, D. 1979. Rb-Sr age of the Shergotty achondrite and implications for metamorphic resetting of isochron ages. *Geochim. Cosmochim. Acta* 43 (in press).

Öpik, E. J. 1951. Collision probabilities with the planets and distribution of interplanetary matter. *Proc. Roy. Irish Acad.* 54A: 165-199.

Öpik, E. J. 1963. Survival of comet nuclei and the asteroids. *Advan. Astron. Astrophys.* 2: 219-262.

Öpik, E. J. 1966. The stray bodies in the solar system. Part II. The cometary origin of meteorites. *Advan. Astron. Astrophys.* 4: 301-336.

Öpik, E. J. 1977. *Interplanetary Encounters* (New York: Elsevier Scientific Publishing Co.).

Papanastassiou, D. A.; Rajan, R. S.; Huneke, J. C.; and Wasserburg, G. J. 1974. Rb-Sr ages and lunar analogues in a basaltic achondrite: Implications for early solar system chronologies (abstract). *Lunar Science V,* The Lunar and Planet. Inst., pp. 583-585.

Papanastassiou, D. A., and Wasserburg, G. J. 1974. Evidence for late formation and young metamorphism in the achondrite Nakhla. *Geophys. Res. Lett.* 1: 23-26.

Pellas, P. 1972. Irradiation history of grain aggregates in ordinary chondrites. Possible clues to the advanced stages of accretion. In *From Plasma to Planet,* ed. A. Elvius (Stockholm: Almqvist and Wiksell), pp. 65-90.

Podosek, F. A. 1971. Neutron-activation potassium-argon dating in meteorites. *Geochim. Cosmochim. Acta* 35: 157-173.

Podosek, F. A. 1973. Thermal history of the nakhlites by the ^{40}Ar-^{39}Ar method. *Earth Planet. Sci. Lett.* 19: 135-144.

Rajan, R. S. 1974. On the irradiation history and origin of gas-rich meteorites. *Geochim. Cosmochim. Acta* 38: 777-788.

Rambaldi, E. R. 1977. Trace element content of metals from H- and LL-group chondrites. *Earth Planet. Sci. Lett.* 36: 347-358.

Ramdohr, P. 1972. *The Opaque Minerals in Stony Meteorites.* (Berlin: Akademie-Verlag).

ReVelle, D. O. 1976. Dynamics and thermodynamics of large meteor entry: A quasi-simple ablation model. *Herzberg Institute of Astrophysics, Nat. Res. Council Canada,* SR-76-1.

ReVelle, D. O. 1979. A quasi-simple ablation model for large meteorite entry: Theory versus observations. *J. Atmos. Terr. Phys.* (in press).

ReVelle, D. O., and Rajan, R. S. 1979. On the luminous efficiency of meteoritic fireballs. *J. Geophys. Res.* (in press).

ReVelle, D. O., and Wetherill, G. W. 1978. Which fireballs are meteorites? A study of the large meteor deceleration data (abstract). *Meteoritics* 13: 612-613.

Ringwood, A. E., and Kesson, S. E. 1977. Composition and origin of the moon. *Proc. Lunar Sci. Conf. VIII* (Oxford: Pergamon Press), pp. 371-398.

Safronov, V. S. 1960. Formation and evolution of protoplanetary dust condensations (in Russian). *Voprosy Kosmogonii* 7: 121.

Safronov, V. S. 1969. *Evolution of the Protoplanetary Cloud and the Formation of the Earth and Planets.* (Moscow: Nauka; in Russian). Engl. translation, NASA TT F-677. NTIS, Springfield, Va., 1972.

Scholl, H., and Froeschlé, C. 1977. The Kirkwood gaps as an asteroidal source of meteorites. In *Comets, Asteroids, Meteorites,* ed. A. H. Delsemme (Toledo, Ohio: University of Toledo Press), pp. 293-295.

Schultz, L., and Signer, P. 1977. Noble gases in the St. Mesmin chondrite: Implications to the irradiation history of a brecciated meteorite. *Earth Planet. Sci. Lett.* 36: 363-371.

Scott, E. R. D., and Wasson, J. T. 1975. Classification and properties of iron meteorites. *Rev. Geophys. Space Phys.* 13: 527-546.

Scott, E. R. D., and Wasson, J. T. 1976. Chemical classification of iron meteorites. VIII. Groups IC, IIE, IIIF and 97 other irons. *Geochim. Cosmochim. Acta* 40: 103-115.

Sekanina, Z. 1971. A core-mantle model for cometary nuclei and asteroids of possible cometary origin. In *Physical Studies of Minor Planets,* ed. T. Gehrels (NASA SP-267, Washington, D. C.: U.S. Government Printing Office), pp. 423-428.

Sekanina, Z. 1973. Statistical model of meteor streams. III. Stream search among 19303 radio meteors. *Icarus* 18: 253-284.

Simonenko, A. N. 1975. *Orbital Elements of 45 Meteorites. Atlas* (Moscow: Nauka).

Stolper, E.; McSween, H. Jr.; and Hays, J. F. 1979. A petrogenetic model of the relationships among achondritic meteorites *Geochim. Cosmochim. Acta* 43 (in press).

Tscharnuter, W. 1978. Collapse of the presolar nebula. *Moon and Planets* 19: 229-236.

Turner, G. 1969. Thermal histories of meteorites by the Ar^{39}-Ar^{40} method. In *Meteorite Research,* ed. P. M. Millman (Dordrecht: D. Reidel), pp. 407-417.

Urey, H. C. 1957. Origin of tektites. *Nature* 179: 556-557.

Urey, H. C. 1965. Meteorites and the moon. *Science* 147: 1262-1265.

Van Schmus, W. R. 1967. Polymict structure of the Mezö-Madaras chondrite. *Geochim. Cosmochim. Acta* 31: 2027-2042.

Van Schmus, W. R., and Koffman, D. M. 1967. Equilibration temperatures of iron and magnesium in chondritic meteorites. *Science* 155: 1009-1011.

Wänke, H. 1965. Der Sonnenwind als Quelle der Uredelgase in Steinmeteoriten. *Z. Naturforsch.* 20a: 946-949.

Wänke, H. 1966. Meteoritenalter und verwandte Probleme der Kosmochemie. *Fortschr. Chem. Forsch.* 7: 322-408.

Wasson, J. T. 1972. Formation of ordinary chondrites. *Rev. Geophys. Space Phys.* 10: 711-759.

Wasson, J. T. 1974. *Meteorites – Classification and Properties* (New York: Springer).

Wasson, J. T. 1977. Relationship between the composition of solid solar-system matter and distance from the sun. In *Comets, Asteroids, Meteorites,* ed. A. H. Delsemme, (Toledo, Ohio: University of Toledo Press), pp. 551-559.

Wasson, J. T. 1978. Maximum temperatures during the formation of the solar nebula. In

Protostars and Planets, ed. T. Gehrels (Tucson: University of Arizona Press), pp. 488-501.

Weissman, P. R., and Wetherill, G. W. 1973. Periodic Trojan-type orbits in the Earth-Sun system. *Astron. J.* 79: 404-412.

Wetherill, G. W. 1967. Collisions in the asteroid belt. *J. Geophys. Res.* 72: 2429-2444.

Wetherill, G. W. 1968. Stone meteorites: Time of fall and origin. *Science* 159: 79-82.

Wetherill, G. W. 1969. Relationships between orbits and sources of chondritic meteorites. In *Meteorite Research,* ed. P. M. Millman, (Dordrecht: D. Reidel) pp. 573-589.

Wetherill, G. W. 1971. Cometary versus asteroidal origin of chondritic meteorites. In *Physical Studies of Minor Planets,* ed. T. Gehrels (NASA SP-267, Washington, D. C.: U. S. Government Printing Office), pp. 447-460,

Wetherill, G. W. 1974. Solar system sources of meteorites and large meteoroids. *Ann. Rev. Earth Planet. Sci.* 2: 303-331.

Wetherill, G. W. 1976. Where do the meteorites come from? A reevaluation of the Earth-crossing Apollo objects as sources of stone meteorites. *Geochim. Cosmochim. Acta* 40: 1297-1317.

Wetherill, G. W. 1977. Fragmentation of asteroids and delivery of fragments to Earth. In *Comets, Asteroids, Meteorites,* ed. A. H. Delsemme (Toledo, Ohio: University of Toledo Press), pp. 283-291.

Wetherill, G. W. 1978. Steady-state populations of Apollo-Amor objects. *Icarus* (in press).

Wetherill, G, W. 1979. Apollo objects. *Sci. Amer.* 240: 54-65.

Wetherill, G. W.; ReVelle, D. O.; and Rajan, R. S. 1978. Identification of meteorites with bright meteors. *Ann. Rept. Dir. Dept. Terr. Mag.* (Carnegie Institution of Washington) 77: 482-493.

Wetherill, G. W., and Williams, J. G. 1968. Evaluation of the Apollo asteroids as sources of stone meteorites. *J. Geophys. Res.* 73: 635-648.

Wetherill, G. W., and Williams, J. G. 1979. Origin of differentiated meteorites. *Origin and Abundance of the Elements* 2 (in press).

Whipple, F. L. 1950. A comet model. I. The acceleration of Comet Encke. *Astrophys. J.* 111: 375-394.

Wilkening, L. L. 1973. Foreign inclusions in stony meteorites. I. Carbonaceous chondritic xenoliths in the Kapoeta howardite. *Geochim. Cosmochim. Acta* 37: 1985-1989.

Wilkening, L. L., and Clayton, R. N. 1974. Foreign inclusions in stony meteorites. II. Rare gases and oxygen isotopes in a carbonaceous chondritic xenolith in the Plainview gas-rich chondrite. *Geochim. Cosmochim. Acta* 38: 937-945.

Wilkening, L. L. 1976. Carbonaceous chondritic xenoliths and planetary-type noble gases in gas-rich meteorites. *Proc. Lunar Sci. Conf. VII* (Oxford: Pergamon Press), pp. 3549-3559.

Williams, J. G. 1973. Meteorites from the asteroid belt? (abstract). *EOS* 54: 233.

Williams, J. G., and Wetherill, G. W. 1973. Minor planets and related objects. XIII. Long-term orbital evolution of 1685 Toro. *Astron. J.* 78: 510-515.

Zellner, B., and Bowell, E. 1977. Asteroid compositional types and their distributions. In *Comets, Asteroids, Meteorites,* ed. A. H. Delsemme (Toledo, Ohio: University of Toledo Press), pp. 185-197.

Zimmerman, P. D., and Wetherill, G. W. 1973. Asteroidal source of meteorites. *Science* 182: 51-53.

ON THE ORIGIN OF ASTEROIDS

V. S. SAFRONOV
O. J. Schmidt Institute of the Physics of the Earth

The problem of the origin of asteroids should be treated as a part of the more general problem of the origin of planets. Olbers' hypothesis on the formation of asteroids by a disruption of a parent planet agreed with Laplacian cosmogony. But it proved to be in contradiction with more recent data concerning the variation of chemical composition of asteroids with distance from the sun as well as the distribution of their orbits. The theory of the accumulation of the planets from solid material, now widely recognized, is based on a hypothesis formulated by Schmidt more than thirty years ago. According to this theory the origin of the asteroids is caused by an interruption of the accumulation of a planet, its stopping at an intermediate stage due to disturbing actions of bodies in the neighboring zone of Jupiter. Due to the higher spatial density of solids in that zone and due to gravitational instability in that part of the dust disk, the bodies in Jupiter's zone were much larger than those in the asteroidal zone. Their gravitational interactions enhanced the eccentricities of their orbits; when the largest body, Jupiter's embryo, reached the mass $\sim 10^{27}$ g, the bodies began to penetrate into the zone of asteroids. They swept out most of the bodies in this zone and increased the velocities of the remainder. This increase in velocity transformed the accumulation process of the asteroids into the reverse process, namely of their fragmentation by collisions. The present mass distribution of asteroids is evidence in favor of their well-developed collisional evolution. Statistical study of asteroid rotations arrives at the same conclusion. The available data on the direct rotation of the few largest asteroids indicate that these bodies did not undergo catastrophic collisions and that the erosional decrease of their masses was not large.

[975]

The origin of asteroids is closely related to the more general problem of the origin of planets. In this way the problem appeared long before the discovery of the asteroids themselves. The possibility of the existence of a planet between the orbits of Mars and Jupiter was predicted even before the formulation of the law of planetary distances by Titius in 1776 and almost half a century before the discovery of the first asteroid, Ceres. So it was enough for Olbers to discover the second asteroid, Pallas, in order to suggest the hypothesis on the origin of asteroids due to disintegration of a "normal" planet. In the framework of the Laplace cosmogony the hypothesis by Olbers was logically the only possible conclusion because such small bodies could not be formed from gaseous condensations. It contained implicitly a reasonable idea that the characteristic feature of asteroidal evolution was fragmentation by collisions. However, the hypothesis could not explain the cause of disintegration.

Now the problem of asteroid origin is solved on the base of the widespread concept of the formation of planets by accumulation of solid bodies and particles. The idea of the formation of planets from solids was suggested already on the verge of the 20th century by Ligondes (see Poincaré 1911) and by Chamberlin and Moulton, but then was forgotten for many years. It was revived only in 1944 by Schmidt, who used it as a base for a new theory of planetary formation. Schmidt considered the origin of asteroids as a result of an interruption of the accumulation of a planet, the halt at an intermediate stage due to perturbations by massive Jupiter which had time to grow somewhat earlier. The accumulation theory has made considerable progress, many observational data on asteroids have been obtained, many papers on their evolution written, but nevertheless the general idea by Schmidt remains valid as a first reasonable approximation.

I. HYPOTHESES ON THE ORIGIN OF ASTEROIDS

The ever increasing number of asteroids discovered between the orbits of Mars and Jupiter could be considered as support for Olbers hypothesis that they are fragments of a major planet. However, the variety of their orbits has also increased. Even at the end of the last century the explanation of all these orbits by only one breakup of a parent planet was considered questionable and it was supposed that there were several such events. Sultanov (1953) arrived at the same conclusion from the distribution of orbital parameters (semimajor axes, Jacobi constants, angular momenta) of fragments which form by spherically symmetric fragmentation.

Ovenden (1972) has suggested a modified version of Olber's hypothesis. He has assumed that the distances between planets should correspond to a minimum of the perturbation function (a resonance for the whole system) and has computed that such a minimum would take place for the giant planets if a planet with a mass of $\sim 90\ M_\oplus$ existed at a distance of

2.79 AU from the sun and then disappeared (exploded) 16×10^6 yr ago, with only a small part of its fragments remaining in the asteroid belt. The hypothesis was criticized by Napier and Dodd (1973). Any known source of energy is several orders of magnitude less effective than necessary for a disintegration (explosion) of such a massive body. Only a close encounter with Jupiter inside its Roche limit could lead to the disintegration. In this case, however, the system of Galilean satellites would be highly perturbed and, according to Ovenden's own estimates for the restoring of resonances in the system, a time interval of $\sim 2 \times 10^9$ yr would be needed. Nevertheless, Ovenden's hypothesis was acclaimed by Van Flandern (1977) who has calculated the orbits of 60 very-long-period comets backward in time and has concluded that most of them had perihelia at nearly the same point where the comets were $5\text{-}6 \times 10^6$ yr ago (the moment of breakup of the planet). The fallacy of the Ovenden hypothesis has been convincingly shown by Öpik (1977) who has outlined the dramatic consequences of such an explosion for life on Earth.

Kuiper tried to explain the origin of asteroids in the framework of his cosmogonic hypothesis regarding the formation of planets due to gravitational instability in the gaseous component of the protoplanetary cloud. Later he dropped this idea and suggested a hypothesis (Kuiper 1950, 1953) for the formation of small planets by the accumulation of solid material from the cloud. The number of original asteroidal bodies with diameters larger than 50 km according to Kuiper exceeded a hundred. Collisions and fragmentations of these bodies led eventually to the present system of the asteroids.

Alfvén (1964, 1969; see Alfvén and Arrhenius 1970) has suggested another picture of the accumulation of asteroids. He assumed that in a system with collisions, smaller bodies should rotate faster than larger ones (equipartition of the energy of rotation). From the fact that rotational periods of the asteroids are on the average the same for different sizes, Alfvén concluded that collisions did not play an appreciable role in the evolution of asteroids; therefore the accumulation of asteroids has not yet finished and they continue to grow. But Napier and Dodd (1974) point out that this conclusion is in contradiction with the observed mass distribution of asteroids which corresponds to a system with well-advanced collisional evolution.

Alfvén (1969) also suggested "jet streams" which are in some sense similar to meteor streams. By asteroidal streams he means a group of small planets with similar values of "proper elements" a, A, B, π_1 and θ_1. Applied by Hirayama, proper elements are the analog of normal elements (semimajor axis, eccentricity, inclination, longitude of perihelion and longitude of the node) but they remain nearly invariable in perturbed motion. They are derived by integration of equations of motion from which are excluded all periodic terms, as well as secular terms higher than second order, in eccentricity and inclination. The asteroids with similar values of semimajor axes and proper eccentricities and inclinations were defined by Hirayama and later by Brouwer as families of asteroids. The Alfvén streams are smaller groups of bodies with similar orbits. Within the Flora family, for instance, Alfvén detected 3 streams containing 15,13 and 7 members. Arnold (1969) has revised Brouwer's work and found that 42 % of 1735 asteroids considered are members of various families. He confirmed the existence of asteroidal streams detected by Alfvén and found 7 more streams. A definition of asteroid families is given in the chapter by Kozai in this book.

Alfvén's explanation of the origin of the streams and his suggestions on their evolution have not been generally accepted. Inelastic collisions of particles and bodies diminish their relative velocities and make their orbits more similar. Hence, according to Alfvén, a large complex of particles revolving around a central body divides into a number of jet streams. Due to the decrease of the relative velocities of colliding particles, the stream becomes more and more narrow, the accretion becomes more effective than the disintegration and large bodies form. Only when their number becomes very small would the stream disintegrate due to perturbations. But in this picture the role of gravitational perturbations is surely underestimated. At the present density of the asteroid belt, the characteristic time between collisions of bodies is of the order of 10^9 yr, but perturbations would cause a precession of perihelia and nodes of asteroid orbits which would lead to their almost uniform distribution in 10^5-10^6 years (Öpik 1961). Therefore all streams which we observe now should disintegrate and not contract rapidly (on the cosmogonic time scale). They should be relatively young and the most reasonable deduction then would be to relate their formation with a collisional disintegration of asteroids and ejection of fragments with small velocities. But it remains unclear why such young formations are so numerous.

The theory of the formation of planets in torus-like jet streams does not seem reliable. Inelastic collisions of bodies diminish their relative velocities and accordingly diminish the thickness of the disk in which they are contained; but there are no reasons for a concentration of bodies near the orbit of a growing planet. One can expect rather an opposite tendency because the planet sweeps out the bodies colliding with it and scatters over larger volume the bodies which it encounters; as a result the density of material near the planet diminishes. The velocities of bodies are determined by the balance of the energy lost by collisions and the energy gained in encounters (Safronov 1969). They are proportional to the radius of the largest body in the zone and so increase during the accumulation, and they do not decrease. Numerical experiments by Brahic (1977) suggest that even without mutual gravitational perturbations the collisions tend to spread material (disperse jet streams). Ip earlier supported the Alfvén concept of the evolution of jet streams. Now Ip (1978a,b) agrees that the concept was related to the simplified model which did not include perturbations and he attempts to find a place for these streams in the real system of protoplanetary bodies.

A concept of the origin of asteroids is suggested by Cameron (chapter in this book) which is based on his new model (Cameron 1978) of the formation of planets from massive gaseous protoplanets condensed at a very early stage of the evolution of the solar nebula.

II. ACCUMULATION OF SOLID BODIES IN THE ZONE OF ASTEROIDS

An important stage in the early evolution of the protoplanetary cloud is the separation of the dust component from the gas due to the coagulation of dust particles by collisions and to their settling in the central plane of the cloud ($z = 0$). This process essentially depends on the degree of decay of random (turbulent) motions in the gas. At pure laminar rotation (no random motions) a thin dust layer forms in the plane $z = 0$ which becomes gravitationally unstable and disintegrates into numerous dust condensations with masses $\sim 10^{17}$ g in the earth zone and $\sim 10^{22}$ g in Jupiter's zone (Safronov 1969; Goldreich and Ward 1973; Genkin and Safronov 1975). But if there is a source of energy which prevents the damping of random motions, then the dust layer will not reach the high degree of flattening necessary for the instability, and its evolution will be reduced to a growth of solid particles due to their coagulation by collisions. The most unfavorable conditions for the instability occur at small distances from the sun while in the region of giant planets the instability probably takes place. Due to the absence of reliable data on the random motions in the gas, we cannot estimate the position of the boundary between stable and unstable regions in the dust disk.

The asteroid belt has an intermediate position between the terrestrial planets and the giant planets. Nothing can be said about stability of this zone. One cannot exclude the possibility that the much higher density of solids in Jupiter's zone, due to the condensation of abundant volatiles H_2O, NH_3, CH_4, has accelerated the beginning of gravitational instability in the zone, and that the dust condensations have grown large enough to prevent by their perturbations the instability in the neighboring asteroid zone. In this case only a direct growth of particles by coagulation could lead to the formation of larger bodies. The effectiveness of this process is not quite clear. In some papers (see for example Coradini et al. 1977), from a study of the mechanism of the sticking of small particles due to van der Waals and electromagnetic forces, it has been found that the coagulation of particles was possible only at very small relative velocities; therefore, in a cloud with turbulent motions of the gas the particles could not grow larger than $10^{-3} - 10^{-2}$ cm. Silicate particles grow more slowly than metallic ones. In the turbulent gas such small particles cannot settle to the central plane and cannot form a gravitationally unstable dust disk.

There will be no evolution of the cloud into a system of bodies until the gas has dissipated. But how much will remain after the dissipation of the gas (say, at the stage of T Tauri solar wind) also is not clear. This theory of growth was developed for idealized elastic and spherically symmetrical particles. The coagulation of real particles of irregular forms, loose structure, different sizes, and especially of magnetized or charged particles should proceed much easier. Admixture of ices in the asteroid zone should also help the sticking of particles. That is why many scientists believe that particles could grow in size without any limitation. Nevertheless the possibility of turbulent gas motions of unknown scale and intensity brings an uncertainty into our understanding of the earliest stages of the accumulation of solids. In the asteroid zone these motions could be additionally created by large bodies which escaped from Jupiter's zone.

We shall compare now the growth of bodies in the two zones for the cases in the asteroid zone when instability sets in and when it does not.

Gravitational instability in both zones

According to our estimate (Safronov 1969) the masses of dust condensations which form due to the instability at axisymmetric radial perturbations are

$$m_0 \approx \frac{\sigma_p^{\,3}}{\rho^{*2}} \text{ and } \rho^* = \frac{3M_\odot}{4\pi R^3} \tag{1}$$

where σ_p is the surface density of the dust disk and R is the distance from the sun. Due to the condensation of volatiles in Jupiter's zone, σ_p was several times higher than in the asteroid zone. Accordingly the initial masses of condensations in Jupiter's zone were 3–4 orders of magnitude higher than in the zone of the asteroids. The critical density necessary for the instability is $\rho_{cr} = K\rho^* \approx 2\rho^*$ and the critical thickness of the dust disk at the beginning of the instability is

$$H_{cr} = \frac{\sigma_p}{\rho_{cr}} = \frac{4\pi\sigma_p}{3kM_\odot} R^3 \approx \frac{2\sigma_p R^3}{M_\odot} \tag{2}$$

For the zone of Jupiter, H_{cr} is more than ten times higher than in the asteroid zone. Therefore the instability should arise there appreciably earlier. The presence of the gas of much lower density which is not taken into account in these expressions can change the figures. But their large difference in the zones of asteroids and Jupiter should remain.

The duration τ_p of the stage of sedimentation of particles to the equatorial plane through the nonturbulent gas until reaching the critical density ρ_{cr} depends on the sizes of particles. The minimum value of τ_p can be found if we consider the settling down toward the equatorial plane of larger particles coagulating with all the other, nonsettling, particles that they meet on the way. Equating the z-component of the solar gravitation $F_z = -m\Omega^2 z$ to the force exerted by the gas $F_g = cr^2\,dz/dt$ taking into account that $dr = -\rho_p dz/4\delta$ and integrating over z, we find the sedimentation time (Safronov 1969)

$$\tau_p \approx \frac{\sigma_g}{2\sigma_p} \ln\left(\frac{\Delta r}{r}\right) P \tag{3}$$

where $\Delta r \approx \sigma_p/8\delta$ is the increase of the particle's radius ($\Delta r \sim 1$ cm) and $P = 2\pi/\Omega$ is the period of revolution around the sun. For initial sizes of particles $\sim 10^{-3} - 10^{-5}$ cm, one finds $\tau_p \sim 10^4$ yr in the asteroid zone and about half of this value in Jupiter's zone. The maximum value of τ_p is found by the assumption that all particles have initially the same size. They settle down almost with the same velocity without coagulation. Then τ_p is about an order of magnitude greater. The former estimate (3) seems to be more realistic than the latter. Here only relative motion of particles along the z-coordinate is taken into account. Slower rotation of the gas around the sun than Keplerian motion of large bodies creates additional velocities of particles of different sizes in tangential (v_ψ) and in radial (v_R) directions (Whipple 1971; Weidenschilling 1977). Because v_z is proportional to z, we have $v_R > v_z$ for small z, and particles should grow faster than calculated only from v_z (Safronov and Ruzmaikina 1978).

The time scale for the development of gravitational instability of the dust disk and for the formation of condensations is a few tens of periods P of revolution around the sun and therefore much smaller than characteristic times of the other stages of the process.

Dust condensations formed in the rotating disk have acquired their initial rotation in the same direction. Due to this rotation their initial contraction was not considerable and after it led to rotation with Keplerian velocity, the density exceeded only a few times the Roche density ρ_R. Further contraction of condensations took place due to their coalescence by collisions and led to the formation of dense solid bodies ($\rho_f \approx 2$ or 3). The Roche density decreases with the third power of the distance from the sun. Hence at larger distances from the sun the condensation stage was more lengthy (Safronov 1975). In the earth's zone the condensations transformed into bodies after their masses increased on the average by a factor 10^2 and in Jupiter's zone by more than 10^3. The condensations grew faster than the bodies of the same mass due to larger geometrical cross-sections and smaller relative velocities, i.e. due to the larger ratio of gravitational cross-section to the geometrical one. The presence of the gas diminished the relative velocities of the condensations; in the expression for the velocity $V^2 = Gm/\theta r$ the coefficient θ increased up to values 10-15. In addition the gas slowed down the rotation of condensations and caused more rapid contraction.

The duration τ_c of the condensation stage was estimated assuming the power dependence of the angular momentum of condensation K on its mass

$$K = \frac{2}{5} \mu\omega mr^2 \propto m^p .$$

Then   (4)

$$r \propto m^{2p-3} \text{ and } \rho \propto m^{10-6p} .$$

For random directions of relative velocities of coagulating condensations and their axes of rotation, the squared values of their angular momenta are added,

$$K^2 = \Sigma K_i^2 \tag{5}$$

where K_i includes the axial rotation and the relative motion of colliding condensations. These expressions allow us to estimate average sizes and densities of condensations and, with the formula of growth $dm/dt > \pi r_{eff}^2 \rho_p v$, to find the time scale of their transformation into solid bodies. We find

$$\tau_c \approx \frac{2P}{1+2\theta} \left(\frac{m_f}{m_o}\right)^{7-4p} = \frac{2P}{1+2\theta} \left(\frac{\rho_f}{\rho_o}\right)^{\frac{7-4p}{10-6p}} \tag{6}$$

where m_o and ρ_o are mass and density of condensation after its initial contraction and m_f is the final mass of condensation when its density reaches the density of a solid body ρ_f. Assuming for the zone of Jupiter $\sigma_p \sim 40$ g cm^{-2}, $\rho_o \sim 50$ $\rho^* = 5 \times 10^{-8}$ g cm^{-3}, $\theta = 15$, $p = 1.3$, we find $m_o \approx 6 \times 10^{22}$g, $m_f \sim 2 \times 10^{26}$g and $\tau_c \sim 10^6$ yr. The same value of p may be taken for the zone of the outer planets and it then helps in solving the problem of their growth; the stage of condensations in Neptune's zone would last until their mass is $m \sim 0.1\, m_N$ and thus Neptune can grow up to the present mass during about 10^9 yr. For the asteroid zone assuming $\sigma_p \approx 5$ g cm^{-2}, $\rho^* = 6.4 \times 10^{-9}$ g cm^{-3}, $\rho_o = 3.2 \times 10^{-7}$ g cm^{-3}, $\rho_f = 2.5$ g cm^{-3}, $p = 1.3$ we obtain $m_f = 4 \times 10^{21}$ g, $\tau_c = 4/(1+2\theta)\, 10^6$ yr. At the same $\theta = 15$ the value of τ_c is an order of magnitude less than in Jupiter's zone. Planetesimals in the asteroid zone continued to grow until their relative velocities became large enough for the disintegration of bodies by collisions. Such velocities could not be induced by gravitational interactions of planetesimals nor by perturbations from distant planet embyros from Jupiter's zone. But the embryos becoming more massive enhanced the velocities of other bodies in Jupiter's zone and the bodies began to flow into the asteroid zone. They were more massive and their collisions and close encounters with asteroids at velocities 2-4 km sec^{-1} made great disturbances in that zone.

The body which is at Jupiter's distance from the sun, and has a velocity $V = 2$ km sec^{-1} relative to the Keplerian one in the direction opposite the orbital motion, penetrates at the perihelium of its orbit to the middle of the asteroid belt ($R = 2.9$ AU) and at $V = 3$ km sec^{-1} to the inner edge of the belt. In fact the velocities of bodies were distributed roughly according to the Maxwellian law and an appreciable number of bodies had velocities ~ 2-3 km sec^{-1} at 2-3 times the smaller average value of V, say $V \sim 0.8$-1.2 km sec^{-1}. From the expression $V^2 = GM/\theta r$ which gives the average velocity of bodies with an effective mass m of the largest body in the zone, we have $m \propto V^3 \theta^{3/2}$ and find $m \sim (0.8$-2.7$)10^{27}$g for $\theta = 15$. The rate of further growth is determined by the relation $\Delta r \sim (1+2\theta) \sigma_p \Delta t / \rho_f P$, where ρ_f is the density of the body. Due to smaller values of σ_p and θ, the radii of asteroidal bodies grew 5-10 times more slowly than radii of Jupiter zone bodies (JZB). The estimates show that until the moment when JZB began to "shoot" the zone of asteroids, the latter could have grown to radii about 4-7 hundred kilometers (for $\theta = 10$, $r_a \sim 500$ km). However the results depend considerably on the values of θ for both zones which were not constant.

While the number of bodies penetrating the asteroidal zone was still small, the collisions with them were rare, and the majority of asteroids grew further. But the influx of bodies from outside increased and this had two important consequences for the asteroid belt. (1) The more massive JZB newcomers "swept" the asteroids away from the asteroidal zone by collisions. (2) The JZB increased the random velocities of asteroids by close encounters.

Consider first the decrease of the asteroid population due to collisions with massive JZB. Without computer analysis only an order of magnitude estimate is possible. We can compare the frequency of collisions of an asteroid m_a with these stray bodies (of mass m and radius r) and the frequency of collisions of the embryo of Jupiter m_J with the same bodies. Because $m_J \gg m \gg m_a$ the ratio is determined mainly by the collisional cross-sections of m and m_J,

$$\nu_{aJ} = C \frac{r^2 (1 + 2\theta r^2 / r_J^2)}{r_J^2 (1 + 2\theta)}. \tag{7}$$

When relative velocities of JZB are large enough they fill the asteroid's zone with about the same density, and $C \sim 1$. Assuming an inverse power law of the mass distribution of bodies,

$$n(r) = cr^p \tag{8}$$

we can estimate a number N of bodies $m > m_a$ which collide with m_a during the increase of mass m_J from m_{J_0} to $m_{J_s} \sim (10 - 15)m_\oplus$, i.e. to the mass of solids acquired by Jupiter. We have

$$\Delta m_J = \Delta c \int_{r_o}^{\beta r_J} mr^{-p}\, dr \text{ and } \Delta N = \Delta c \int_{r_a}^{\beta r_J} v_{a_J} r^{-p}\, dr \tag{9}$$

where βr_J is the radius of the second largest body ($\beta \sim 0.5 - 0.6$) and r_o is the radius of the smallest body. Integrating these equations we find

$$\frac{\Delta N}{\Delta m_J} = \frac{\int_{r_a}^{\beta r_J} Cr^2(1 + 2\theta r^2/r_J^2)r^p\, dr}{(1 + 2\theta)r_J^2 \int_{r_o}^{\beta r_J} mr^p\, dr} = f(m_J) \tag{10}$$

and finally,

$$N = \int_{m_{J_o}}^{m_{J_s}} f(m_J)\, dm_J . \tag{11}$$

With reliable values of parameters at $p = 3.5$ we obtain $N \sim 1$, i.e. a decrease of the asteroidal population of about a factor of three. The real depopulation was $\sim 10^2$ times ($N \sim 5$). One can assume that the removal of asteroids due to collisions with bodies from Jupiter's zone was accompanied by other mechanisms.

Now consider perturbation of the asteroid due to a close encounter with JZB in the case when the former is moving with the circular Keplerian velocity V_k and the latter is near the perihelion with both orbits lying in the same plane. According to the two-body approximation, a relative velocity vector v rotates through the angle

$$\tan \frac{\psi}{2} = \frac{G(m+m_a)}{Dv^2} \approx \frac{Gm}{Dv^2} \tag{12}$$

where D is a target radius. It can be shown that after such an encounter the asteroid acquires radial and tangential components of the velocity relative to the circular one

$$v_R = 2v \sin \frac{\psi}{2} \cos \frac{\psi}{2} \text{ and } v_\theta = 2v \sin^2 \frac{\psi}{2} \tag{13}$$

and the eccentricity of its orbit becomes

$$e_a = 2v' \sin \frac{\psi}{2}[(1 + 2v' \sin^2 \frac{\psi}{2})^2 + \sin^2 \frac{\psi}{2}(3 + 4v' \sin^2 \frac{\psi}{2})]^{\frac{1}{2}} \tag{14}$$

where $v' = v/V_k = \sqrt{1+e}-1$ and $e = (R_J-R_J)/(R_J+R_a)$ is the eccentricity of the orbit of JZB. Taking for D the minimum possible value $D_m = r\,(1 + v_e^2/v^2)^{1/2}$ (a grazing encounter) we find that the average observed eccentricity of the asteroids $e \sim 0.15$ can be acquired at one encounter with a JZB when its mass exceeds the minimum values 10^{26}g and 2×10^{26}g for the middle and the inner edge of the asteroid belt respectively ($\psi \sim 40°$). We note that the masses of these bodies are several times smaller than the mass of Mars.

The increase in the velocities of asteroids inhibited their growth because the bodies ceased to coalesce at encounters and began to break down; the process of accumulation has changed by the reverse process of disintegration. It would be very interesting to develop a numerical modeling of the interaction of asteroids with JZB in order to evaluate initial physical conditions which could have produced the present asteroidal system.

Gravitational instability in Jupiter's zone but not in the asteroidal zone

Gravitational instability may be prevented only by the chaotic ("thermal") motions along z, created by the gas. If the motions support the system in a quasi-equilibrium state, then the relation $v_p\rho_p = 4\,\sigma_p/P$ holds which permits us to evaluate the rate of growth of bodies by collisions from the formula

$$\frac{dm}{dt} = (1 + 2\theta)\,\pi r^2\,\rho_p v_p\,. \tag{15}$$

For bodies which are less than 1 km in size, θ is small and can be neglected. One finds that the bodies larger than one km, due to mutual gravitational perturbations, have relative velocities exceeding the critical value of 26 cm sec^{-1}, and the gravitational instability cannot set in even without turbulence in the gas. But for the support of the overcritical velocity of bodies somewhat smaller than one km, turbulent gas velocities of ~ 1 km sec^{-1} are needed. With the increase of the masses of bodies the influence of gas on their motion decreases and θ approaches the usual values between 3 and 5.

From Eq. (15) it is found that at the beginning of the penetration of bodies from Jupiter's zone, the radii of the largest bodies in the asteroidal zone at $\theta = 4$ would be about 60 km, i.e. nearly two times smaller than in the preceding case when the gravitational instability was supposed to set in within the asteroidal zone. At this stage the sweeping out of the matter in the zone is negligible, θ is constant, and the body's radius according to Eq. (15) grows proportionally with time. Therefore we always get a relatively smaller radius of bodies ($\Delta r \sim 60$ km) as compared with the previous scheme. Due to the uncertainty of the parameters, there is no preference between the two schemes considered. The early mass distribution of bodies in the asteroidal zone could be quite different in these two variants, but the difference would vanish later in the course of further prolonged evolution.

At sufficiently high velocities the coagulation of bodies changes to their destruction. If a body is impacted by relatively small ones it is eroded by cratering, and a fraction of the ejecta from the craters leave the body at velocities greater than the escape velocity. According to Marcus (1969), the bodies having the strength of basalts can grow by collisions with the present average random velocity of 5 km sec^{-1} in the asteroid belt only if their diameters exceed 800 km. Because the asteroids have a lower strength than that of basalts, one can expect that all of them, including Ceres, are losing mass by mutual collisions. If an asteroid collides with a sufficiently large body, it is destroyed by fragmentation. The minimum mass of such a destructing body was estimated by Dohnanyi (1969) from experimental data by Gault. About the same result is obtained when one considers directly a critical energy density for the destruction (Greenberg *et al.* 1977). Taking for example a reasonable value $\sim 10^7$ erg g^{-1} for this energy density, we find that Ceres might be destroyed by collisions with an asteroid having the diameter of 40 km and a velocity of 5 km sec^{-1}. However, only a very small fraction of the mass of Ceres could obtain the velocity of escape and leave its gravitational field. Only a body with diameter ten times larger could disintegrate Ceres into separate independent fragments.

III. COMPARISON OF THEORETICAL CONSIDERATIONS ON THE ACCUMULATION OF ASTEROIDS WITH THE OBSERVATIONAL DATA

We shall consider here the observational data which at the present stage of knowledge permit us to make a judgement on the accumulation process.

(a) The space distribution of asteroids. This space distribution is rather complicated. The most prominent features are the gaps near the orbital revolution periods of asteroids in commensurabilities with Jupiter's period. Due to the resonance, the orbital eccentricity of an asteroid increases, thus augmenting the probability of its collision with other asteroids.

Another feature is the existence of numerous asteroid families. The most simple explanation of families lies in the collisional fragmentation of large bodies and the carrying away of fragments at low velocities. In general the structure of the assembly of asteroids is the product of a long-enduring evolution of an initially much more numerous population of gravitating bodies.

From the region of the belt nearer to Jupiter the asteroids were ejected by its gravitational perturbations (Lecar and Franklin 1973). There remained only a small number of bodies in stable resonant orbits in the Hilda groups (2:3) and in resonance 3:4. In the main belt, more distant from Jupiter and out of resonance, the number of asteroids continuously decreased due to collisions, so that the collisional time scale has become comparable to the age of the solar system. The structure of the asteroid system as a whole gives evidence in favor of the model of initially multiple embryos, but not of a single one, in the zone of asteroids which had not yet considerably runaway in mass from the other bodies of the zone.

(b) The mass distribution of asteroids. The distribution can be satisfactorily described by an inverse power law n (m) = cm^{-q} (differential function). The power index q for all asteroids except the largest ones is near the value $11/6 \sim 1.8$. For the largest asteroids one should either take a lower $q \sim 5/3$ at the same value of c (Hellyer 1970). or take a lower value of c conserving a constant q (Dohnanyi 1972). Degewij (1978; see Zellner's chapter) has obtained the size distributions separately for C- and S-type asteroids. C asteroids have a steeper slope of the curve at $d > 150$ km. The slope increases also with the distance from the sun.

A theoretical study of the accumulation and fragmentation processes by the methods of the coagulation theory shows that, in the case of the pure accumulation in the whole mass interval with the exception of the largest bodies, the distribution tends to power law with the index $q \sim 1.5$ or 1.6, depending on the coagulation coefficient (Safronov 1969; Zviagina and Safronov 1971). When there is no accumulation, and the bodies are only destroyed or eroded by collisions, the power asymptotic solutions are obtained with $q \sim 11/6$ (Dohnanyi 1969; Bandermann 1972). Hellyer (1970) has obtained $q = 5/3$ for the large bodies and $q = 1.8$ for the remainder. From the study of the more general equation taking into account both the accumulation and the fragmentation (with smoothly changing parameters), the asymptotic power solution with the index $q \sim 1.8$ is also obtained if we exclude the interval of the largest masses (Zviagina *et al.* 1973; Pechernikova *et al.* 1976). These results indicate that after the end of the accumulation stage, the system of asteroids has a far advanced evolution and that the present mass distribution is connected with the fragmentation of asteroidal bodies. The largest masses were probably less influenced by fragmentation.

(c) The rotation of asteroids. The similarity of the rotation periods of asteroids over an enormous interval of masses in some way repeats the same feature of the planetary system. Alfvén (1964) has considered this property as an indication of the noneffectiveness of collisions and fragmentations of asteroids which would lead to a faster rotation of smaller bodies due to a tendency to an equipartition of rotational energies. This conclusion was widely discussed and questioned. According to our model (Safronov 1969), the planets in the course of the accumulation get two components of axial rotation: a regular (direct rotation) and a nonregular one connected with the randomly oriented impact velocities of various bodies. Numerous small bodies and particles contribute mainly to the regular component, because their randomly oriented impulses nearly compensate each other. In an erosional collision the particle gives the body a smaller angular momentum than in an accretional collision with the same velocity. Therefore, in the asteroid belt where the erosion process dominates, the impacting bodies produce a slower rotation than the "normal" direct rotation of planets. If the body acquired a

rapid rotation due to the collision with a larger one, the subsequent collisions with small bodies will tend to slow down this rotation to an average value. Harris has developed a theory for the origin of rotation by collisions in which he has shown that the decelerating effect from the impacts of many particles leads to an equilibrium rotational velocity for all bodies larger than several kilometers, independent of their strength (see the chapter by Burns and Tedesco).

New data on the asteroid rotation have been obtained; data on 143 asteroids are summarized and analyzed by Harris and Burns (1978; see the chapter by Burns and Tedesco). The C-type asteroids (i.e., of the carbonaceous chondrite composition) rotate 20 % more slowly than those not classified as C-type. According to Harris this indicates a lower density by a factor of 1.5 or a lower strength of C-type asteroids as compared to other ones. The rotation axes have a random spatial orientation, and the velocity distribution is a three-dimensional Maxwellian one. This gives us independent evidence that the asteroid belt is a system of interacting bodies with a well-advanced collisional evolution. Meanwhile Morrison (1977; see the chapter by Morrison and Lebofsky) presents new evidence on the direct rotation of the largest asteroids. The simplest explanation of this fact can be the assumption that the largest bodies have not been involved in the collisional evolution. It is easy to estimate (Safronov 1969) that from the collision of two bodies the larger one may get a retrograde rotation only if the mass of the smaller body reaches several percent of its mass. But at the impact velocities of ~ 5 km sec^{-1} the smaller body should completely destroy the larger one even if it has a mass of one half percent of that of the larger body, i.e. several times less than is needed for the origin of retrograde rotation. From this consideration it follows that bodies with retrograde rotation are the fragments of larger bodies disintegrated by collision and that the bodies which did not undergo catastrophic collisions should have a direct rotation.

(d) Chemical composition of asteroids. As is discussed in various other chapters, the asteroids are classified in several compositional types; after the most abundant C-type, the next is the S-type (silicates). The S/C ratio drops monotonously with the distance R from the sun (an exponential scale length is about 0.4 AU; Zellner 1978), continuing smoothly the change in chemical composition of the terrestrial planets with R. This demonstrates that, in spite of the extensive fragmentation of asteroids, there was no complete mixing of bodies in the asteroid belt. This evidence appears to be the most serious argument against the hypothesis of the origin of asteroids from a single destroyed planet of any given mass, or against their origin due to the fragmentation of two colliding bodies.

(e) Relation of asteroids to other small bodies of the solar system. The interrelation of asteroids and meteorites is evident; it is generally accepted that asteroids are the main source of meteorites (Chapman 1977; Wetherill

1977; see the chapter by Wasson and Wetherill). But a direct relationship of certain meteorite types with definite asteroid types is not yet recognized. There may be a relation between asteroids and Trojans (Gehrels 1977; Yoder 1979), and between asteroids and some irregular satellites (see the chapter by Carusi and Valsecchi).

There exists also a relation with comets, but it is less noticeable. On the one hand, a fraction of long-period comets is captured by Jupiter's gravitational field and transformed into short-period comets. Their surface layers gradually evaporate the volatiles, and the comets become extinct and transform into asteroid-type bodies (Levin 1977). On the other hand, the bodies which have originated in the outer part of the asteroid belt and have contained a large amount of volatiles were ejected from these locations by Jupiter and they are partially found in orbits of short-period comets. These bodies have lost the volatiles from the surface but conserved them in the interiors; we may consider them as potential comets. Such considerations have led Genkin (1978) to a hypothesis on the origin of the short-period comets from the comet-like asteroids, whose outer protecting layers are damaged by the impact with small bodies (meteorites), while the volatiles from their interiors are liberated. But the quantitative evaluation of the frequency of the flares of such comets has not yet been made. Finally, of the total mass of primary asteroid bodies ejected by Jupiter from the solar system, a small fraction, $\sim 2 \times 10^{-3}$ (Safronov 1972), was retained in Oort's cloud and some of these enter the inner parts of the solar system as long-period comets. However, they comprise only a thousandth of the total number of incoming comets and they therefore are of no practical interest.

We conclude that, in spite of all the peculiarities of the asteroid system, it can be logically explained in the context of our understanding of the broader problem of planet formation. The system has conserved important features ("imprints") of an early stage of planetary accumulation. Its detailed study will give us valuable information on the origin of planets.

Acknowledgments. I am indebted to R. Greenberg and S. Weidenschilling for a helpful discussion and criticism of this chapter.

REFERENCES

Alfvén, H. 1964. On the origin of the asteroids. *Icarus* 3: 52-56.
Alfvén, H. 1969. Asteroidal jet streams. *Astrophys. Space Sci.* 4: 84-102.
Alfvén, H., and Arrhenius, G. 1970. Structure and evolutionary history of the solar system. *J. Astrophys. Space Sci.* 8: 338-421.
Arnold, J. R. 1969. Asteroid families and "jet streams". *Astron. J.* 74: 1235-1242.
Bandermann, L. W. 1972. Effects of erosion and fragmentation on the mass distribution of colliding particles. *Mon. Not. Roy. Astron. Soc.* 160: 321-338.
Brahic, A. 1977. Systems of colliding bodies in a gravitational field. I. Numerical simulation of the standard model. *Astron. Astrophys.* 54: 895-907.

Cameron, A. G. W. 1978. Physics of the primitive solar nebula and of giant gaseous protoplanets. In *Protostars and Planets*, ed. T. Gehrels (Tucson: University of Arizona Press), pp. 453-487.

Chapman, R. 1977. The evolution of asteroids as meteorite parent bodies. In *Comets, Asteroids, Meteorites*, ed. A. H. Delsemme (Toledo, Ohio: University of Toledo Press), pp. 265-275.

Coradini, A.; Magni, G.; and Federico, C. 1977. Grains accretion processes in a protoplanetary nebula. II. Accretion time and mass limit. *Astrophys. Space Sci.*, 48: 29-87.

Degewij, J. 1978. Photometry of faint asteroids and satellites. Ph.D. dissertation, Leiden University.

Dohnanyi, J. S. 1969. Collisional model of asteroids and their debris. *J. Geophys. Res.* 75: 3468-3493.

Dohnanyi, J. S. 1972. Interplanetary objects in review: Statistics of their masses and dynamics. *Icarus* 17: 1-48.

Gehrels, T. 1977. Some interrelations of asteroids, Trojans and satellites. In *Comets, Asteroids, Meteorites*, ed. A. H. Delsemme (Toledo, Ohio: University of Toledo Press), pp. 323-325.

Genkin, I. L. 1978. A new hypothesis on the origin of short period comets. *Astron. Zirk. Acad. Sci. USSR* 1002: 4-5 (in Russian).

Genkin, I. L., and Safronov, V. S. 1975. The instability of the rotating gravitating systems with radial perturbations. *Astron. Zhurn.* 52: 306-315 (in Russian).

Goldreich, P., and Ward, W. R. 1973. The formation of planetesimals. *Astrophys. J.* 183: 1051-1061.

Greenberg, R.; Davis, D.; Hartmann, W. K.; and Chapman, C. R. 1977. Size distribution of particles in planetary rings. *Icarus* 30: 769-779.

Harris, A. W., and Burns, J. A. 1978. Asteroid rotation. I. Tabulation and analysis of data. *Icarus* (in press).

Hellyer, B. 1970. The fragmentation of the asteroids. *Mon. Not. Roy. Astron. Soc.* 148: 389-390.

Ip, W.-H. 1978a. A note on the problem of jet stream formation. *Astrophys. Space Sci.* 55: 267-269.

Ip, W.-H. 1978b. Model consideration of the bombardment event of the asteroidal belt by the planetesimals scattered from the Jupiter zone. *Icarus* 34: 117-127.

Kuiper, G. P. 1950. The origin of the asteroids. *Astron. J.* 55: 164.

Kuiper, G. P. 1953. Note on the origin of the asteroids. *Proc. Nat. Acad. Sci.* 39: 1159-1161.

Lecar, M., and Franklin, F. A. 1973. On the original distribution of the asteroids. I. *Icarus* 20: 422-436.

Lecar, M., and Franklin, F. A. 1974. On the original distribution of the asteroids. In *The Stability of the Solar System and of Small Stellar Systems*, ed. Y. Kozai (Dordrecht: D. Reidel), pp. 37-55.

Levin, B. J. 1977. Relationship between meteorites, asteroids and comets. In *Comets, Asteroids, Meteorites*, ed. A. H. Delsemme (Toledo, Ohio: University of Toledo Press), pp. 307-312.

Marcus, A. H. 1969. Speculations on mass loss by meteoroid impact and formation of the planets. *Icarus* 11: 76.

Morrison, D. 1977. Asteroid sizes and albedos. *Icarus* 31: 185-220.

Napier, W. M., and Dodd, R. J. 1973. The missing planet. *Nature* 242: 250-251.

Napier, W. M., and Dodd, R. J. 1974. On the origin of the asteroids. *Mon. Not. Roy. Astron. Soc.* 166: 466-489.

Öpik, E. J. 1961. The survival of stray bodies in the solar system. *Contrib. Armagh. Obs.* 34: 185.

Öpik, E. J. 1977. Origin of asteroids and the missing planet. *Irish Astron. J.* 13: 22-39.

Ovenden, M. W. 1972. Bode's law and the missing planet. *Nature* 239: 508-509.

Poincaré, H. 1911. Leçons sur les hypothèses cosmogonique. Paris.

Pechernikova, G. V.; Safronov, V. S.; and Zviagina, E. V. 1976. Mass distribution of protoplanetary bodies. II. Numerical solution of the generalized coagulation equation. *Astron. Zhurn.* 53: 612-619 (in Russian).

Safronov, V. S. 1969. The evolution of the protoplanetary cloud and the formation of the earth and the planets. (Moscow: Nauka Press), transl. from the Russian, NASA TTF-677, 1972.

Safronov, V. S. 1972. Ejection of bodies from the solar system in the course of the accumulation of the giant planets and the formation of the cometary cloud. In *The motion, Evolution of Orbits, and Origin of Comets,* eds. G. A. Chebotarev *et al.* (Dordrecht: D. Reidel), pp. 329-334.

Safronov, V. S. 1975. The duration of the process of the formation of the earth and the planets. In *Cosmochemistry of the Moon and Planets,* ed. A. P. Vinogradov, (Russ.) (transl. in English). NASA SP-370, p. 797, 1977.

Safronov, V. S., and Ruzmaikina, T. V. 1978. On angular momentum transfer and accumulation of solid bodies in the solar nebula. In *Protostars and Planets,* ed. T. Gehrels (Tucson: University of Arizona Press), p. 545-564.

Sultanov, G. F. 1953. Theoretical distributions of the orbital elements of the Olbers' hypothetical planet. *Transactions Sternberg Astron. Inst. Moscow,* 88-89 (in Russian).

Van Flandern, T. C. 1977. A former major planet of the solar system. In *Comets, Asteroids, Meteorites,* ed. A. H. Delsemme (Toledo, Ohio: University of Toledo Press), pp. 475-481.

Weidenschilling, S. J. 1977. Aerodynamics of solid bodies in the solar nebula. *Mon. Not. Roy. Astron. Soc.* 180: 57-70.

Wetherill, G. W. 1977. Fragmentation of asteroids and delivery to fragments to Earth. In *Comets, Asteroids, Meteorites,* ed. A. H. Delsemme, (Toledo, Ohio: University of Toledo Press), pp. 283-291.

Whipple, E. L. 1972. On certain aerodynamic processes for asteroids and comets. *From Plasma to Planets,* ed. A. Elvius, (New York: Wiley), pp. 211-232.

Yoder, C. 1979. Notes on the origin of Trojan asteroids. *Icarus* (special Asteroid issue).

Zellner, B. H. 1978. Geography of the asteroid belt. *Asteroids: An Exploration Assessment,* eds. D. Morrison and W. C. Wells, NASA Conf. Publ. 2053, pp. 99-107.

Zviagina, E. V., and Safronov, V. S. 1971. Mass distribution of protoplanetary bodies. *Astron. Zhurn.* 48: 1023-1032 (in Russian).

Zviagina, E. V.; Pechernikova, G. V.; and Safronov, V. S. 1973. Qualitative solution of the coagulation equation of bodies with the account of fragmentation. *Astron. Zhurn.* 50: 1261-1273 (in Russian).

ON THE ORIGIN OF ASTEROIDS

A. G. W. CAMERON
Harvard-Smithsonian Center for Astrophysics

A general scenario is described for the early history of the solar system. The primitive solar nebula is formed from the infall of gas from a collapsing interstellar cloud fragment. It becomes repeatedly unstable against collapse to form giant gaseous protoplanets. In the course of protoplanet evolution the center of the protoplanet enters a thermodynamic regime in which common rocky minerals become liquids; convection brings solids to the central region where a substantial fraction of them rain out to form a protoplanetary core. In the inner solar system protoplanetary envelopes are tidally stripped away, thus injecting into the solar nebula large quantities of chondrules and inclusions. Late in the development of the solar nebula, after most of the gas has disappeared, turbulence dies out and the small solids settle into a thin layer at midplane of the nebula. Gravitational instabilities in this layer form asteroidal and cometary bodies. Some further consequences of this scenario are discussed.

The events accompanying the formation of the solar system appear to have been very complex, and a large number of physical and chemical processes must be examined in connection with them. Any theory of the origin of the solar system attempts to assemble physical and chemical processes into a scenario with the hope that a logical chain of cause and effect relationships is thereby established, and also the hope that the assemblage of processes does not introduce mutual inconsistencies. The process is actually an iterative one, since the establishment of the scenario raises questions about processes that have been inadequately examined, and in turn the examination of those processes introduces changes.

The formation of the asteroids was but one of many complex events that occurred early in the history of the solar system. The problem of the early history of these small bodies has played practically no role in my thinking about the origin of the solar system. Nevertheless it appears to accommodate their formation in a natural way.

In this chapter I will present the scenario as I currently envisage it. The relevant processes have all been described elsewhere (see e.g. Cameron 1978b). It must be emphasized that the scenario is in a process of constant evolution, and therefore the question of the degree of its accuracy seems to me to be less important than its role in suggesting important problems for future investigation.

I. CHARACTERISTICS OF THE PRIMITIVE SOLAR NEBULA

Star formation commences with the collapse of an interstellar cloud. The conditions required for self-gravity within such a cloud to predominate over pressure forces and bring about the collapse are such that hundreds to thousands of solar masses of interstellar material must be involved. Hence the cloud must fragment extensively, and many stars will be formed as result of a single cloud collapse. At any one time in the galaxy, most interstellar clouds are not in the process of collapse, and hence the collapse process is the exception rather than the rule. There are probably a number of different processes that can lead an interstellar cloud to the threshold for collapse. Most of them are probably fairly violent in a hydrodynamic sense and lead to the rapid compression of the cloud material until self-gravity becomes dominant.

The presence of radioactive ^{26}Al in the early solar system, with a half life of less than one million years (Lee et al. 1976), suggests strongly that a supernova explosion occurred close to the solar nebular at the time of its formation. Cameron and Truran (1977) suggested that the relationship of the supernova to the formation of the solar system was causal rather than accidental. This triggering supernova would have previously existed as a highly luminous O or B star, emitting a prodigious flood of ultraviolet radiation that would strongly ionize and heat the intercloud regions of the interstellar medium, thereby subjecting nearby interstellar clouds to a preliminary degree of compression. When the supernova went off, the gaseous ejecta from the event would sweep past and around these interstellar clouds, subjecting them to an even larger external pressure, and triggering the collapse of those that were situated within ten or twenty parsecs of the explosion.

During the collapse phase of an interstellar cloud, progressive fragmentation is expected to occur as the density rises, and the internal temperature of the collapsing gas is expected to become very low as the result of the increased efficiency of cooling. We focus our attention on one of the ultimate fragments of this cloud, as the material rains down toward a gravitating center which is about to become the primitive solar nebula. A fragment within the gas will have much angular momentum, some inherited from the initial angular velocity of the cloud, but more of it resulting from the large turbulent shearing motions which the cloud should obtain as a result of its violent compression (Cameron 1973). Because of this angular momentum, the gas cannot fall directly in to form a star, but must spread out to form a disk having typically a radius of some tens to hundreds of astronomical units. Rather large hydrodynamic flows will be required to assimilate the chaotic infalling gaseous elements into a smooth disk structure. Even when such a disk structure is achieved, it must be expected that large hydrodynamic flows will continue due to meridional circulation currents (Cameron and Pine 1973; Cameron 1978a). The Reynolds number is very high in such a large disk structure, and therefore it is expected that at least the inner several tens of astronomical units of the disk will be vigorously turbulent as a result of these stirring motions.

The basic theory for the evolution of a viscous disk has been published by Lynden-Bell and Pringle (1974). I have applied this theory to the evolution of the primitive solar nebula (Cameron 1978a), taking the turbulent viscosity about as high as could conceivably be physically realizable. The following summarizes these calculations. During the whole course of the evolution, mass flows inward near the center of the disk to form the sun, and the outer edge of the disk is in the process of continual expansion as mass flows outwards at large distances to help transport energy and angular momentum. It was expected that after a reasonable period of time, on the order of 10^5 yr, the infall of gas from the collapsing interstellar cloud fragment would slacken and mass outflow due to turbulent heating of the coronal region surrounding the disk would begin to predominate. Under these circumstances the outer radius of the primitive solar nebula would continually shrink and the surface density would progressively decrease. The amount of mass in the primitive solar nebula out to the orbit of Neptune was at maximum about 10% of the solar mass, but much more mass lay at larger distances, up to a solar mass of material or more.

The most important lesson learned from these calculations was not the general history of the disk behavior just described, but rather the finding that the disk should have become repeatedly unstable against the formation of large axisymmetric ring instabilities, starting very early in the history of the disk when there was still relatively little matter near the center. Previously Black and Bodenheimer (1976) had found that such a ring instability probably exists in the infalling material right at the center of the primitive solar nebula. Thus the presence of ring-like, or perhaps bar-like, instabilities is expected to be a common situation within the primitive solar nebula. Some three-dimensional hydrodynamic studies of the breakup of these rings upon their gravitational contraction have been carried out, indicating that a breakup into two or three pieces should be commonplace (Tohline 1977; Cook 1977; Norman and Wilson 1978). Such pieces are in mutual unstable orbits, and they will be perturbed into collisions with one another, or in some cases into fairly highly eccentric orbits which may avoid collisions for a period of time. These studies have usually involved more angular momentum than the scenario in Cameron (1978a) but the results can be carried over to the smaller angular momentum case. I have called these pieces and their merged collision products "giant gaseous protoplanets;" their evolution is discussed below. However, it is useful to remark at this point that the initial radius of a giant gaseous protoplanet may be in excess of one astronomical unit, and such large sizes can exist in the inner solar system only because there is relatively little matter there at the time of the formation of the protoplanet, and consequently a large amount of space is available.

The presence of giant gaseous protoplanets in the primitive solar nebula has a profound effect on the evolution of the nebula. Consider first the pair of protoplanets that may be formed out of the initial central ring instability accompanying the collapse process. The rotation of these two protoplanets around their center of mass acts like a bar raising very large tidal perturbations in the surrounding gas of the primitive solar nebula. If there is dissipation and a phase lag in the tidal response of the surrounding gas, then there is an efficient transfer of angular momentum from the orbital motion of the protoplanets into the orbital motion of the surrounding gas, and the protoplanets will rapidly spiral together. Similarly, a protoplanet revolving around the forming sun in the primitive solar nebula will raise tides in the gas which are both internal and external to it. This will result in transfer of orbital angular momentum, first from the inner gas to the motion of the protoplanet, and then from the motion of the protoplanet to the orbital motion of the external gas. Studies by Lin and Pringle (1976), Papaloizou and Pringle (1977), and Lin and Papaloizou (1979) indicate that angular momentum transfer by these tidal interactions is much more efficient than that due to turbulent viscosity in the gas reasonably near the protoplanets. Thus the presence of protoplanets in the primitive solar nebula should have a profound effect upon the evolution of the nebula.

At present there does not exist a good evolutionary model sequence for the primitive solar nebula. If I were to attempt to repeat my previous evolutionary calculations (Cameron 1978a) today, I would make two major modifications. On the one hand, I would reduce the efficiency of the assumed turbulence by about an order of magnitude or so in the assumed value of the turbulent viscosity, since that originally taken was highly optimistic. This effect would reduce the efficiency of the angular momentum transport process, so that the surface density of the disk would tend to build up to a value approximately ten times as high as in the original calculations, in order that the rate of inflow of material to the center of the disk could re-establish its original relationship to the time scale of infall of material from the interstellar cloud fragment. On the other hand, the greater efficiency of angular momentum transport that accompanies the presence of giant gaseous protoplanets in the primitive solar nebula counteracts the above effect, reducing the surface density of the disk. Since these effects operate in opposite directions, it is not clear what the net effect would be relative to the calculations of Cameron (1978a), and this indicates the necessity for a proper understanding of the evolutionary behavior of giant gaseous protoplanets.

II. EVOLUTION OF GIANT GASEOUS PROTOPLANETS

So far, detailed evolutionary studies have been made of giant gaseous protoplanets only under the assumption that they behave like stars and radiate their released heat into a vacuum (DeCampli and Cameron 1979). In the calculations of Cameron (1978a) the amounts of gas which became unstable in the form of a ring were approximately equal to the mass of Jupiter, so that one Jupiter mass has been taken as a nominal reference mass, and the actual masses of interest for giant gaseous protoplanets are probably within a factor of a few of this amount. Masses on the order of that of Jupiter were found to go through their pre-collapse evolutionary phases in about 10^5 yr, plus or minus a factor of a few depending upon uncertainties in the interior opacity. Hydrodynamic collapse occurs when the temperature near the center of the protoplanet becomes high enough ($\sim$ 2700 K) for appreciable amounts of hydrogen molecules to be dissociated into hydrogen atoms, which is an energy-absorbing process for which the only available energy source is the gravitational potential energy that the star releases during a collapse process.

An event which was probably of major importance for the inner solar system was the tidal stripping of the envelopes from giant gaseous protoplanets. There is an inner Lagrangian point between the protoplanet on the one hand and the forming sun and the solar nebula on the other hand, which separates the regions in which these respective bodies have gravitational control over the gas. As dissipation causes gas to move inwards to be accumulated onto the growing sun, the gravitational bonds between the protoplanet and the protosun strengthen, causing the protoplanet to spiral slowly inward. As this happens, the distance from the protoplanet to the inner Lagrangian point gradually decreases. It is expected that if the inner Lagrangian point passes inside the surface of the protoplanet, the surface layer of gas will be stripped away, allowing the lower layers to expand and to be stripped away, and so on until the entire envelope has been stripped. It appears that this process occurs for protoplanets in the inner solar system but not in the outer solar system. This stripping process may be of great significance in connection with the the formation of asteroids.

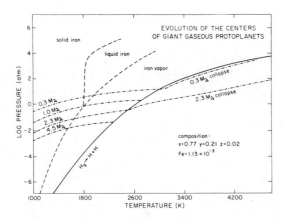

Fig. 1. Thermodynamic diagram showing regions of dissociation of hydrogen molecules, the phases of iron, and the evolutionary tracks followed by the centers of giant gaseous protoplanets of various mass.

It is important that the calculations of the evolution of giant gaseous protoplanets should be improved, since they are not surrounded by a vacuum but by a region of gas and radiation which represents the fluid continuation of the protoplanet envelope into the surrounding primitive solar nebula. I have recently carried out a study (Cameron 1979) which suggests that there is a discontinuity in the thermodynamic conditions between the primitive solar nebula and the protoplanet. Nevertheless, because of the change in the boundary conditions, the detailed studies of evolution by DeCampli and Cameron must be considered to give very uncertain results with respect to the structure of the surface and the evolutionary time scale, but the results of the calculations bearing on the deep interior of the protoplanets should be considerably more trustworthy, since these conditions depend primarily upon energy considerations as expressed in the virial theorem.

The motions in a thermodynamic diagram of the central points in protoplanets of a variety of masses are shown in Fig. 1. This figure is a phase diagram. The solid line shows the points at which molecular hydrogen is 10 % dissociated into atomic hydrogen; the hydrodynamic collapse of a protoplanet begins when the central conditions reach approximately that line. The dashed lines show the positions in the diagram in which iron is gaseous or condensed, and where the condensed phase is either solid or liquid. The dash-dot lines show the evolution of the centers of four protoplanets up to the solid line, with the collapse of two of them being followed thereafter. It may be seen that the tracks for the protoplanets of approximately one Jupiter mass or less pass through the region in which iron is condensed and in the liquid phase. The phase diagram for common high-temperature minerals would look somewhat similar, except that the liquid phase would be present in the thermodynamic diagram in the region extending down to the evolutionary tracks for protoplanets of two Jupiter masses or perhaps somewhat more.

The primitive solar nebula is expected to have condensed solids in it right from the beginning. The energy released by the gravitational collapse does not allow the temperature of the gas in the nebula to be raised high enough to cause complete evaporation of the most refractory constituents of the interstellar grains that enter the solar nebula with the infalling gas, except very close to the axis of the nebula. The modeling which I carried out (Cameron 1978a) indicates that such complete grain evaporation should not take place outside of approximately the orbit of Mercury. This means that the initial condensed constituent of a giant gaseous protoplanet should be in the form of fine grains, with typical grain sizes in the submicron region. The evolutionary calculations of DeCampli and Cameron showed that the giant gaseous protoplanets were most likely convective, so that when thermodynamic conditions arise at the center of the protoplanet whereby iron and common minerals will melt in the small grains carried with the gas, then convection will bring grains into the melting region from throughout the entire volume of the protoplanet.

As long as the grains are solid, there is no assurance that they will stick together upon collision. However, even if such sticking does take place fairly readily, then an optimistic view of the size distribution of the grains when they first enter the protoplanet extends from very small sizes up to the millimeter to centimeter size. However, molten grains should readily coalesce upon collision. Therefore, we are interested in the degree to which molten grains will grow in size by collision in the liquid region near the center of the protoplanet, and subsequently rain out into the deeper interior.

Collisions are brought about as a result of two effects. Very small grains are effectively suspended within the gas, so that they move with the gas as it is subject to turbulent motions arising from the interior convection. It is a characteristic of turbulence that the fluid is sheared, so that any grain suspended in the fluid with respect to the motion of its center will be subject to a fluid flow past its edges. This flow will transport other grains toward glancing collisions with the first one, which will lead to amalgamation if the grains are fluid. Larger grains fall through the fluid with terminal velocities which are proportional to their radii, and hence the larger grains will overtake and amalgamate with the grains of somewhat smaller size. There are a number of complications which modify the effective cross-sectional area of the grains for collision and in some cases lead to the breakup of very large grains. These have been discussed for iron droplets coalescing and falling out within a protoplanet by Slattery (1978). The calculations have been extended to liquid droplets with parameters typical of common minerals in unpublished work of Slattery, DeCampli, and Cameron. A typical result of these calculations is shown in Fig. 2, starting with a flat distribution of droplet sizes

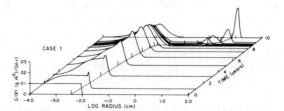

Fig. 2. The rate of change of droplet size for nominal mineral parameters near the center of a giant gaseous protoplanet, as a function of time.

extending from 10^{-4} to 10^{-2} cm. It may be seen that this flat distribution very rapidly steepens into a peaked distribution, with the elimination of particles of very small size due to collisions brought about by shear. The position of the peak in the distribution gradually shifts toward larger sizes, until when the mean radius is ~ 0.5 mm, a bifurcation takes place in which a part of the material very rapidly grows into quite large droplets which fall rapidly through the fluid toward the center of the protoplanet. It should be noted that the time required for this growth to the point of rapid rainout is only about ten years, which is much shorter than the time for convective mixing to circulate solid grains through the central portion of the protoplanet. It is therefore evident that a substantial fraction of the solid grains initially present in the protoplanet can be expected to become part of liquid droplets and to rain out to form a core at the center of the protoplanet on a relatively short time scale.

The fairly high value of the opacity in the interior of the protoplanet which is responsible for the presence of convection is largely due to the suspended grains within the gas. When the smaller grains are eliminated through incorporation in larger liquid droplets, the opacity falls by a fairly significant factor, and convection will cease in the protoplanet when the opacity throughout has fallen by a moderate factor. The initial condensed solid constituent of one Jupiter mass of solar material is approximately equal to one Earth mass, and therefore it is expected that a significant fraction of one Earth mass of material will rain out to form a protoplanetary core.

It is evident that the evolutionary calculations shown in Fig. 1 cannot be trusted past the point where the central conditions pass through the liquid thermodynamic region. When this occurs, on a very rapid time scale, a condensed liquid core will form at the center of the protoplanet and convection will cease. The formation of the core will release a great deal of energy, and it is not even clear whether a liquid thermodynamic region within the protoplanet will remain after this structural readjustment has taken place.

When stripping occurs for one of the inner protoplanets, the bulk of the envelope will be removed from about any liquid core which has formed. The stripping takes place quite rapidly, with the material flowing across the inner Lagrangian point at about the local speed of sound. Consequently, quite large droplets which may still be falling through the gas in the envelope may well be stripped off with the envelope, and will enter the primitive solar nebula. I suggest that this is a major source of meteoritic chondrules and inclusions.

III. FORMATION OF CHONDRULES AND INCLUSIONS

Chondrules and inclusions make up a substantial fraction of the mass of most meteorites, and if asteroids are the source of the meteorites, then we may infer that chondrules and inclusions probably make up a substantial fraction of the mass of the asteroids. This requires that very efficient chondrule and inclusion factories were operating early in solar system history. It appears that most if not all chondrules were at one time at least partially molten, and at least a large number of the inclusions are composed of very refractory minerals and represent the condensation products of a cooling gas of solar composition.

It is difficult if not impossible to find suitable conditions in the primitive solar nebula for the formation of chondrules and inclusions in large numbers. Collisions between small solid objects space may produce molten droplets if the collision velocity is several kilometers per second. However, for such a collision to take place in the presence of the gas of the primitive solar nebula, at least one of the objects must be traveling several kilometers per second through the gas, and it would not travel very far at this velocity before gas drag effects slowed it down. At one time (Cameron 1973) I suggested that chondrules might be formed by collisions of this sort as the result of gas acceleration of small particles in the centimeter size range in which the solid bodies cross from one major turbulent eddy into another and can therefore have very energetic collisions with solid bodies being carried by the gas in the other eddy. However, if the characteristic turbulent velocities of the largest eddies are only a relatively small fraction of the local speed of sound, as I now believe, then this process would be very improbable and could not be regarded as efficient. Once one has formed bodies in the asteroidal-size range, then such bodies can move quite rapidly through the gas with negligible drag deceleration, and collisions of small bodies with the surface regolith may result in some chondrule production. However, it is difficult to believe that meteorites could have a high prevalance of chondrules and also come from asteroids if the asteroids were first required to exist before the chondrules could be produced. Energetic collisions between asteroidal-sized bodies would be highly disruptive to these bodies, and many liquid droplets would be produced, but most of the energy of the collision would not be expended in this way. Therefore, while there may be a certain amount of chondrule production by these processes, it seems unlikely that they can represent the principal chondrule factories which are required.

In the present scenario large amounts of condensed matter form into liquid droplets in a natural manner in the interior of protoplanets. In principle such liquid droplets can be ejected into the primitive solar nebula upon the tidal stripping of a protoplanet envelope. However, there are some significant questions relating to this stripping process. When a central protoplanetary core of liquid material is formed, the entropy of the surrounding gas may be raised sufficiently so that solids sublime directly from the solid to the gas phase. In such a protoplanet, only those chondrules could be ejected which had been swept up by the convective motions earlier in the evolution of the protoplanet, and had not had time to fall out prior to the stripping. Such chondrules would be very small. On the other hand, in such a model, solids will keep settling through the gas until they reach the point where complete evaporation will take place. This can lead to large local enrichments of the hydrogen gas with the evaporation products of common high-temperature minerals. This may lead to a large increase in the oxygen-to-hydrogen ratio and it is possible that the oxygen brought into these regions is isotopically anomalous, being enriched in ^{16}O with which the minerals were originally formed in the supernova environment. Upon envelope stripping, these regions will be subject to adiabatic expansion, whereupon solids will condense from the cooling gas and will typically form a layered structure with a succession of lower temperature condensation products. These are the candidates to become the inclusions in the meteorites.

It is thus possible but not certain that the optimum conditions for production of large numbers of chondrules are different from those which will give a large yield of inclusions. The chondrules and inclusions may come from the stripping of protoplanets which have reached different stages in their evolution. Since the number of protoplanets was probably significantly larger than the number of planets in the solar system today, some of the complications discussed above may not be implausible.

It seems quite possible that from each stripped protoplanet several percent of one Earth mass of condensed material may be formed in this way into chondrules and inclusions released into the primitive solar nebula. This is a considerably larger yield of material than seems plausible by other mechanisms, but it still represents an efficiency of a few percent for conversion of the protoplanetary condensed mass into these small objects. The production of chondrules and inclusions would occur in the inner solar system and these objects would be carried elsewhere in the primitive solar nebula only as the result of diffusion due to the turbulent motions of the gas. Since, in the inner solar system, the gas should have a mean inward drift, the diffusion of chondrules and inclusions upstream in this drift was probably not very efficient. Still, only a very small fraction of the produced material suffices for production of the asteroids.

IV. FORMATION OF ASTEROIDS

The general mechanism which was probably responsible for the formation of the asteroids was the gravitational instability in a thin layer of orbiting solid particles in the general manner described by Safronov (1972; see his chapter) and Goldreich and Ward (1973). In this mechanism, solid materials settled gently through the gas to form a thin disk at the midplane of the primitive solar nebula, and when the surface density of this thin disk of material was great enough, it became gravitationally unstable and local patches of material contracted to form larger bodies. The general expectation for the size of bodies formed in this way is the size characteristic of asteroids and comets, a few kilometers in radius. However, it should be noted that in order for this mechanism to work, turbulence must have had a chance to die out in the primitive solar nebula, since for small particles turbulent gas motions can very easily stir up the distribution of particles and transport them to high levels above the central plane of the solar nebula.

The formation of asteroids is thus tied to the decay of turbulence in the nebula. Let us examine the major causes of turbulence and ask when these will disappear.

One major cause is the infall of material from the collapsing interstellar cloud fragment. In general, the local angular momentum brought in by any parcel of infalling gas will differ from that locally present in the nebula at the point of impact. Hence the gas in the nebula must be continually sorting itself out so that a smooth distribution, monotonically increasing with radius, is continually being re-established within the disk. This process will cease only when the gas inflow ceases. As discussed by Cameron (1978a), there probably arises a time in the evolution of the primitive solar nebula, after about 10^5 years or so, in which the heating of the upper layers of the atmosphere above the disk of the nebula becomes sufficiently great to establish a hydrodynamic expansion, thus terminating the inflow of material. However, this heating is itself a product of vigorous turbulence in the underlying nebula, so the major effect of this process in helping to terminate turbulence is to get rid of excess gas in the primitive solar nebula.

Another major source of turbulence is probably the tendency toward meridional circulation within the primitive solar nebula. This process arises because it is not possible to have surfaces of constant pressure, density, and temperature coincide with surfaces of constant effective potential in a highly flattened rotating system. Thermal imbalances arise which give rise to the meridional circulation currents. These currents can become particularly vigorous for fat disks such as the primitive solar nebula (Cameron 1978a). Because the large-scale flows occur in a system with a very large Reynolds number, they should give rise to fully developed turbulence in the gas. However, the thermal imbalances which provide the driving forces for the meridional circulation depend upon the luminous flux of energy being radiated from the interior of the disk. The effectiveness of this thermal driving term should therefore diminish as the opacity of the gas decreases, which will occur as the surface density of the gas diminishes with time. Gas is lost from the primitive solar nebula as the result both of mass outflow due to hydrodynamic expansion of the upper atmosphere and mass inflow toward the center to add material onto the growing primitive sun. According to this criterion, turbulence will only die out when the surface density in the disk has decreased to a small remnant of its maximum value.

A third major cause of turbulence is the tidal interactions of giant gaseous protoplanets with the primitive solar nebula. This gives rise to a fairly large mass redistribution and hence to large-scale fluid flows which will stir up turbulence. It is clear that this effect will be greatly diminished when tidal stripping of protoplanetary envelopes takes place in the inner solar system. This reduces the mass of the protoplanetary bodies present in the gas by some two orders of magnitude, thereby greatly localizing the effects of tidally-induced turbulence. This means that the formation of the asteroids is unlikely to take place until after protoplanet envelope stripping, so that chondrules and inclusions will be present in the primitive solar nebula. In the outer solar system, the giant planets are at present greatly depleted in hydrogen and helium relative to solar composition. This depletion probably took place as the result of mass loss in planetary winds arising from the expansion of the upper layers of the planetary atmospheres, possibly after the collapse phase of protoplanetary evolution, but this process has not yet been examined in detail. However, it is likely that the mass loss from the giant gaseous protoplanets in the outer solar system was substantial and that therefore a very significant diminution in the tidal forcing function for outer solar nebula turbulence did occur.

These various considerations suggest that the stage was not set for the formation of asteroids until quite late in the evolution of the primitive solar nebula. At that time the scenario suggests that the primitive sun had grown to essentially its full mass, the primitive solar nebula had lost most of its mass, and the inner solar system contained stripped protoplanetary cores. The gas in the inner solar nebula contained chondrules and inclusions ejected from the stripping of giant gaseous protoplanets, as well as large amounts of dust in the form of interstellar grains from which the volatiles had been lost, but which may never have been in the interiors of giant gaseous protoplanets. These grains were intrinsically in the submicron size range, but they may have been clumped together into fluffy accumulates in the millimeter to centimeter size range.

When this material was finally able to settle through gas that has become quiescent to form a thin disk of orbiting solid particles, it is evident that the larger particles would be concentrated there, since they have the largest terminal velocities of descent through the gas. Thus, at the time that the surface density of solids in the thin disk becomes high enough or the thickness small enough for gravitational instability to occur, the chondrules and inclusions and larger clumps of interstellar grains will have settled there. When the instability occurs, the resulting asteroidal bodies should have basically a meteoritic composition. It is not clear whether the gravitational instabilities will be confined to a relatively brief interval of time, consuming most of the available material, or whether they will be stretched out over a considerably longer period of time, and can take place repeatedly in the same region of space. The finer unconsolidated dust will take a considerably longer period to settle through the gas, but eventually it can be expected to accumulate on the surfaces of the asteroidal bodies, forming a thick regolith layer there. This may be the source of the Type I carbonaceous chondrites.

Early collisions among the asteroidal bodies should be accumulative rather than disruptive. This is because neighboring bodies will have been formed in very similar orbits, so that the initial collisions that take place will be very gentle ones. This is probably the way in which the larger asteroids managed to accumulate (see Hartmann's chapter). Later, the random velocity components of the orbital motions of the asteroidal bodies build up, and collisions become destructive. A particular mechanism that can cause this is suggested in the following section.

In the outer solar system, the chondrules and inclusions are not expected to occur, but fine dust should be there in abundance, still containing icy constituents in the outer part of the primitive solar nebula. There may be a reduced efficiency for the formation of asteroidal-sized bodies in the region of the giant planets, but nevertheless I would expect that a significant production of these should occur. Most of these will have been eliminated during the course of time by planetary perturbations. The outer asteroid Chiron is probably a remnant of this class of bodies. In the outer solar nebula, including the region beyond Neptune, the asteroidal-sized bodies which are formed out of the settling icy dust are essentially cometary in composition and should be formed in large numbers. In fact, beyond the region of the outer giant gaseous protoplanets, it is not clear that the primitive solar nebula should have been turbulent, and the formation of these cometary bodies may have occurred at a much earlier stage in the evolution of the nebula than did the formation of the asteroidal bodies. Cameron (1978a) has suggested a mechanism involving a relatively rapid mass loss from the nebula by means of which the outer cometary bodies would have their orbits altered so that they would be ejected into the region of the Oort cloud. The subsequent evolution of the asteroidal bodies must have depended critically upon the availability of heat sources in their interiors. One obvious candidate for this heat source is the radionuclide ^{26}Al. All of the events described here should take place in a time short compared to a half life of this nucleus. Therefore its efficiency as an asteroidal heat source depends upon the general average level of mixture of products of the triggering supernova throughout solar nebula materials. Because there may be significant delays in mixing the products of the supernova explosion into the gas that falls into the solar nebula, it is possible that the asteroids, being late-forming objects, may have higher concentrations of ^{26}Al than do the principal planetary bodies. If this concentration is comparable to that in some Allende inclusions, it is highly probable that this radionuclide would be an effective heat source which could melt the interiors of at least the larger asteroids.

Another potential heat source is associated with the magnetic field transported outward in the early solar wind which existed after the disappearance of the primitive solar nebula. In particular the fluctuations in this field should induce heating in asteroidal interiors via eddy currents (Sonett 1969; see the chapter by Sonett and Reynolds).

As a result of heating the asteroidal interiors, a complex set of mineralogical changes will take place which I will not attempt to describe in detail. Accompanying this there will be an escape of the more volatile substances from the interior of the asteroid through pores and cracks. Some of these substances may recondense near the surface where conditions are cooler. This is probably the way in which boron and mercury have become enriched in carbonaceous chondrites of Type I, with a strong spatial variation in the concentrations of these two elements.

These asteroidal bodies should be formed throughout the inner solar system, not just in the asteroid belt. They will play a very important role in the formation of the terrestrial planets. The liquid cores which are formed at the center of protoplanets, and from which the surrounding envelopes are removed, probably constitute the bulk of the present terrestrial planets, but they do not have a complement of the more volatile elements and minerals. On the other hand, the asteroidal bodies are formed at lower temperatures, and the volatile constituents are present in them, at least initially. The asteroidal bodies will be swept up by the planets in the inner solar system, and the planets will acquire their volatiles in this way. From a geochemical point of view, it may be estimated that $\sim 20\%$ of the earth is acquired in this way, since this amount of meteoritic material is sufficient to bring in the observed amounts of more volatile materials.

V. EXCESS PLANETARY CORES

In the scenario described above the number of axisymmetric ring instabilities which occur in the region of the inner solar system may be comparable to the number of planets, which means that the number of protoplanets formed after the rings break up will be two to three times as great as the number of inner planets. Mutual collisions among them will reduce the number fairly promptly, but it is still likely that there will be an excess of protoplanetary cores following protoplanetary envelope stripping late in the development of the solar nebula.

The angular momentum of the earth-moon system corresponds to the angular momentum contained in the tangential collision of a Mars-sized object with the protoearth. Cameron and Ward (1976) and Ward and Cameron (1978) have discussed how such a collision can lead to a large amount of vaporized rock, which condenses to form an orbiting disk of material about the protoearth, which in turn spreads out beyond the Roche limit and collects to form the moon. This is not part of the present story and I will not discuss it further, but it probably accounts for one of the excess protoplanetary cores.

There is a considerably larger cross-sectional area for a protoplanetary core to pass within the Roche limit of a forming planet like the protoearth than to make a direct collision. This may under some circumstances result in the breakup of the protoplanetary core into a very large number of pieces, each just a few kilometers in radius. Since the protoplanetary cores are formed liquid, they will have differentiated into an iron core plus a rocky mantle, with the liquid rocky mantle subject to convection which will keep its elemental composition fairly well homogenized. If any of these fragments were to survive to the present day in the inner solar system, they will be considered asteroidal bodies with an apparently differentiated composition. I point this out as a possibility that may have happened, but which need not have happened, in case meteoriticists should find this type of asteroidal body required to explain what they see.

Other protoplanetary cores may be perturbed in their orbits to the extent that they will pass close to Jupiter. Under such circumstances, Jupiter will take control over the changes in their orbital elements, and the protoplanetary cores are likely to collide with Jupiter or be ejected from the solar system (see the chapter by Wasson and Wetherill). In the meantime, bodies of such large mass passing through the region of the asteroid belt will produce severe perturbations among the asteroidal bodies there, pumping up their orbital eccentricities and inclinations. For some years it has seemed very strange that one of the largest asteroids, Pallas, should have a very high inclination orbit (Whipple *et al.* 1972), which it is likely to obtain only as the result of a perturbation from a considerably larger body. An excess protoplanetary core must be considered a leading candidate to have done that.

Acknowledgment. This research has been supported in part by a grant from the National Aeronautics and Space Administration.

REFERENCES

Black, D. C., and Bodenheimer, P. 1976. Evolution of rotating interstellar clouds. II. The collapse of protostars of 1, 2, and 5 $M_\odot$. *Astrophys. J.* 206: 138-149.
Cameron, A. G. W. 1973. Accumulation processes in the primitive solar nebula. *Icarus* 18: 407-450.

Cameron, A. G. W. 1978a. Physics of the primitive solar accretion disk. *Moon and Planets* 18: 5-39.

Cameron, A. G. W. 1978b. Physics of the primitive solar nebula and of giant gaseous protoplanets. In *Protostars and Planets,* ed. T. Gehrels (Tucson: University of Arizona Press), pp. 453-487.

Cameron, A. G. W. 1979. The interaction between giant gaseous protoplanets and the primitive solar nebula. *Moon and Planets* (in press).

Cameron, A. G. W., and Pine, M. R. 1973. Numerical models of the primitive solar nebula. *Icarus* 18: 377-406.

Cameron, A. G. W., and Truran, J. W. 1977. The supernova trigger for formation of the solar system. *Icarus* 30: 447-461.

Cameron, A. G. W., and Ward, W. R. 1976. Origin of Moon. In *Lunar Science VII,* The Lunar and Planet. Inst., pp. 120-122.

Cook, T. L. 1977. Three-dimensional dynamics of protostellar evolution. *Los Alamos Publ.* LA-6841-T.

DeCampli, W. M., and Cameron, A. G. W. 1979. Structure and evolution of isolated giant gaseous protoplanets. *Icarus* 38: 367-391.

Goldreich, P., and Ward, W. R. 1973. The formation of planetesimals. *Astrophys. J.* 183: 1051-1061.

Lee, T.; Papanastassiou, D. A.; and Wasserburg, G. J. 1976. Demonstration of ^{26}Mg excess in Allende and evidence for ^{26}Al, *Geophys. Res. Lett.* 3: 109-112.

Lin, D. N. C., and Papaloizou, J. 1979. Tidal torques on accretion disks in binary systems with extreme mass ratios. *Mon. Not. Roy. Astron. Soc.* 186: 799-812.

Lin, D. N. C., and Pringle, J. E. 1976. Numerical simulation of mass transfer and accretion disc flow in binary systems. In *Structure and Evolution of Binary Systems,* eds. P. P. Eggleton, S. Mitton and J. A. J. Whelan (Dordrecht: D. Reidel), pp. 237-252.

Lynden-Bell, D., and Pringle, J. E. 1974. The evolution of viscous discs and the origin of the nebular variables. *Mon. Not. Roy. Astron. Soc.* 168: 603-637.

Norman, M. L., and Wilson, J. R. 1978. The fragmentation of isothermal rings and star formation. *Astrophys. J.* 224: 497-511.

Papaloizou, J., and Pringle, J. E. 1977. Tidal torques on accretion discs in close binary systems. *Mon. Not. Roy. Astron. Soc.* 181: 441-454.

Safronov, V. S. 1972. *Evolution of the Protoplanetary Cloud and Formation of the Earth and Planets,* (Moscow: Nauka Press), transl. from the Russian, NASA TT F-677.

Slattery, W. L. 1978. Protoplanetary core formation by rain-out of iron drops. *Moon and Planets* 19: 443-456.

Sonett, C. P. 1969. Fractionation of iron: A cosmogonic sleuthing tool. II. Heating by electrical induction, *Comm. Astrophys. Space Phys.* 1: 41-48.

Tohline, J. E. 1977. Rotating protostellar collapse in three space dimensions (abstract). *Bull. Amer. Astron. Soc.* 9: 566.

Ward, W. R., and Cameron, A. G. W. 1978. Disk evolution within the Roche limit. In *Lunar Science IX,* The Lunar and Planet. Inst., pp. 1206-1207.

Whipple, F. L.; Lecar, M.; and Franklin, F. A. 1972. The strange case of Pallas. In *On the Origin of the Solar System,* ed. H. Reeves (Paris: CNRS), pp. 312-319.

I. THE TUCSON REVISED INDEX
OF ASTEROID DATA

B. ZELLNER
University of Arizona

In early 1976 a need was felt for a machine-readable compilation of results from the rapidly accumulating physical observations of minor planets. The result is the TRIAD file, maintained at the University of Arizona by a consortium of ten Contributors, and containing all reliable physical parameters for the asteroids. The Contributors select optimized parameters from the available observations, and assign quality codes according to criteria of their own choosing. The file is frequently updated, and is documented by references to the original publications.

The master file is kept at the Lunar and Planetary Laboratory in the form of punched cards as supplied by each Contributor. A magnetic-tape version maintained at the Cyber 175 computer of the University of Arizona is well adapted for large-scale FORTRAN manipulations; it was used for the type classification and bias analysis (see the chapter by Zellner in Part V of this book), and was also used to generate Tables II-VII below. A version maintained on Hewlett-Packard cassette tapes at the Planetary Science Institute in Tucson is better adapted for interactive usage and especially for making two-parameter plots. At this writing, only the historical material in Table VIII has not been reduced to machine-readable form.

The entire substantive content of TRIAD as of June 20, 1979, is listed in the following tables. In some cases data from two or more Contributors are combined into a single table for utility and compactness. Table I lists the Contributors, their areas of responsibility, the number of entries by each, and the locations of the data in the succeeding tables. We are especially appreciative of the continuing efforts by D. Bender to put all orbital elements on a

TABLE I

TRIAD Contributors

Contributor	Data	Number of Entries	Table
D. F. Bender Jet Propulsion Laboratory	Osculating elements, names	2118	II
J. G. Williams Jet Propulsion Laboratory	Proper elements, families	1753	III
M. J. Gaffey University of Hawaii	Spectral reflectance data	277	IV
C. R. Chapman Planetary Science Institute	Spectral reflectance parameters	277	IV
D. Morrison University of Hawaii	Radiometric diameters, albedos	195	V
D. Zellner University of Arizona	Polarimetric parameters	111	V
E. F. Tedesco University of Arizona	Lightcurve parameters	321	VI
T. Gehrels University of Arizona	Magnitudes	2101	VII
E. Bowell Lowell Observatory	UBV colors	744	VII
B. Zellner University of Arizona	Types, adopted diameters	752	VII
F. Pilcher Illinois College	Discovery circumstances	2125	VIII

common, current epoch; of the recalibrated spectral reflectance data for 277 asteroids by C. Chapman and M. Gaffey; and of the exhaustive documentation of lightcurve observations by E. Tedesco. Also we are fortunate to be able to include J. Williams' proper elements and families, which have never before appeared in print.

Errors are inevitable in an undertaking of this magnitude. Because of last-minute submission of some of the data, we were not able to maintain complete consistency among the various tables and the results discussed in various review chapters in this book. Also we have made every attempt to preserve the vital distinction between a zero datum and a blank representing the absence of data, but that distinction may have broken down in a few cases.

It is our intention to maintain the TRIAD file at least for several years to come, and to make listings, card copies, or magnetic-tape versions available to anyone with a genuine professional need. (Address inquiries concerning TRIAD to B. Zellner.) There is also in existence an *apparition file* available in microfiche form, listing opposition dates, distances, phases, magnitudes, and celestial coordinates for all oppositions of the numbered asteroids during the years 1980–1990; address inquiries to D. F. Bender at the Jet Propulsion Laboratory.

Acknowledgments. The maintenance of the TRIAD file at the University of Arizona and its publication in this book is supported by a grant from the National Aeronautics and Space Administration. Individual Contributors are separately funded. C. E. KenKnight deserves much credit for the computer operations associated with TRIAD and for the generation of the tables presented here.

II. OSCULATING ORBITAL ELEMENTS
OF THE ASTEROIDS

DAVID F. BENDER
Jet Propulsion Laboratory

Listed here are osculating orbital elements for 2118 numbered asteroids, of which 17 are considered lost. The columns give asteroid number; name; semi-major axis in AU; eccentricity; inclination; longitude of the ascending node; argument of perihelion; mean anomaly; and Julian date of epoch minus 2,400,000.

This file carries the TRIAD responsibility for correct spelling of asteroid names. The names listed by Pilcher in Table VIII, however, have also been thoroughly researched and should be given some weight in cases of disagreement.

The file was started several years ago with a set of cards supplied to me by B. Marsden who responded to my first request for the data to D. Brouwer. The file has been carefully kept up to date by adding new elements from the *Minor Planet Circulars* as they appear and from the Russian *Ephemerides* each year. It is my aim that all elements should be expressed on a common, current epoch, namely a Julian date of the current year that is divisible by 200. At present this is not yet the case, but in later editions this condition will be approached. In the cases of most of the first 200 asteroids and a few others, the orbital elements have been integrated at the Jet Propulsion Laboratory using perturbations by the earth, Venus, Mars, Jupiter, and Saturn starting from the original epochs (of the *Minor Planet Circulars* or Russian *Ephemerides*). Currently these data have the epoch 2443800.5 or 19 October 1978. It is planned to move data for those cases available to the 1980 epoch of 2444600.5 or 27 December 1980, during the course of the next few months. This program of providing elements at future epochs has been undertaken at the Jet Propulsion Laboratory because of need for accurate ephemerides of asteroids for missions involving planets and asteroids.

In addition the data recently made available by P. Herget have been used to update the file, and most of these data use the 1978 epoch of 2443800.5. Thanks are due to J. G. Williams who has proofread most of the data as they were entered through the years.

[1014]

NUMBER	NAME	A	E	I	NODE	PERI	ANOM	DAYEP
1	CERES	2.76784	.07685	10.598	80.119	73.506	148.321	43800
2	PALLAS	2.77315	.23254	34.800	172.738	309.835	139.535	43800
3	JUNO	2.67127	.25465	13.002	169.930	246.865	291.269	43800
4	VESTA	2.36168	.08967	7.144	103.489	150.618	36.350	43800
5	ASTRAEA	2.57723	.19916	5.349	141.257	355.734	355.177	43800
6	HEBE	2.42419	.20310	14.789	138.541	238.216	185.594	43800
7	IRIS	2.38563	.22989	5.503	259.470	144.293	143.372	43800
8	FLORA	2.20182	.15638	5.888	110.569	284.843	49.267	43800
9	METIS	2.38656	.12248	5.586	68.629	4.918	281.812	43800
10	HYGEIA	3.13822	.11838	3.835	283.158	318.042	41.237	43800
11	PARTHENOPE	2.45169	.10020	4.626	125.116	193.217	83.745	43800
12	VICTORIA	2.33376	.22109	8.374	235.270	68.753	3.806	43800
13	EGERIA	2.57550	.08816	16.498	42.935	80.513	259.769	43800
14	IRENE	2.59773	.16371	9.126	86.397	94.567	261.762	43800
15	EUNOMIA	2.64211	.19776	11.759	292.990	97.374	130.775	43800
16	PSYCHE	2.92155	.13825	3.092	150.133	226.240	251.129	43800
17	THETIS	2.46818	.13816	5.592	125.083	136.016	145.505	43800
18	MELPOMONENE	2.29580	.21752	10.132	150.136	227.375	33.963	43800
19	FORTUNA	2.44190	.15755	1.569	211.599	181.308	334.728	43800
20	MASSALIA	2.40855	.14474	.702	206.236	255.225	177.107	43800
21	LUTETIA	2.43663	.16047	3.073	80.559	249.092	107.648	43800
22	KALLIOPE	2.91090	.10360	13.729	66.025	354.926	340.263	43000
23	THALIA	2.62994	.22985	10.149	66.644	59.521	327.727	43000
24	THEMIS	3.12876	.13314	.761	35.650	112.187	235.879	43800
25	PHOCAEA	2.40148	.25388	21.585	213.766	90.402	209.907	43800
26	PROSERPINA	2.65480	.08875	3.566	45.625	194.522	333.966	43800
27	EUTERPE	2.34728	.17233	1.586	94.387	355.785	239.194	43800
28	BELLONA	2.77658	.15114	9.393	144.165	341.603	194.917	43000
29	AMPHITRITE	2.55409	.07203	6.107	356.003	63.524	311.515	43800
30	URANIA	2.36555	.12672	2.095	307.445	86.048	341.208	43800
31	EUPHROSYNE	3.14788	.22761	26.327	30.725	63.861	353.573	43800
32	POMONA	2.58638	.08507	5.518	220.088	336.999	128.583	43800
33	POLYHYMNIA	2.86084	.34083	1.890	8.289	337.361	218.005	43800
34	CIRCE	2.68710	.10464	5.485	184.283	328.553	56.851	43800
35	LEUKOTHEA	3.00014	.22134	8.029	354.150	210.088	283.743	43800
36	ATALANTE	2.74895	.30102	18.471	358.313	46.458	338.047	43800
37	FIDES	2.64275	.17556	3.080	7.450	60.706	186.958	43800
38	LEDA	2.74018	.15558	6.971	295.617	167.589	31.725	43800
39	LAETITIA	2.76853	.11163	10.381	156.880	207.477	17.072	43800
40	HARMONIA	2.26717	.04700	4.258	93.866	269.187	155.242	43800
41	DAPHNE	2.76689	.26922	15.778	177.913	45.434	199.965	43800
42	ISIS	2.43984	.22697	8.542	84.265	235.532	340.400	43800
43	ARIADNE	2.20313	.16876	3.470	264.513	15.579	358.291	43800
44	NYSA	2.42284	.15113	3.710	131.144	342.038	177.183	43800
45	EUGENIA	2.72035	.08363	6.601	147.572	86.986	28.968	43800
46	HESTIA	2.52484	.17053	2.326	180.769	174.972	37.629	43800
47	AGLAJA	2.87885	.13500	4.989	3.051	313.569	309.081	43800
48	DORIS	3.11310	.06331	6.538	183.465	261.524	282.220	43800
49	PALES	3.09873	.23319	3.180	285.769	112.048	69.556	43800
50	VIRGINIA	2.65094	.28454	2.828	173.310	198.872	18.582	43800
51	NEMAUSA	2.36554	.06614	9.970	175.657	1.086	48.803	43800
52	EUROPA	3.09517	.10909	7.465	128.863	334.815	92.264	43800
53	KALYPSO	2.62103	.20105	5.156	143.513	312.371	223.527	43800
54	ALEXANDRA	2.71099	.19791	11.790	313.220	344.286	350.342	43800
55	PANDORA	2.76100	.14179	7.185	10.313	3.774	60.963	43800
56	MELETE	2.59901	.23363	8.085	193.253	103.126	3.788	43800
57	MNEMOSYNE	3.15380	.11556	15.197	199.011	218.547	45.157	43800
58	CONCORDIA	2.69986	.04469	5.057	160.952	29.550	251.846	43800
59	ELPIS	2.71239	.11918	8.636	169.859	210.627	135.691	43800
60	ECHO	2.39346	.18408	3.593	191.561	269.340	230.763	43800
61	DANAE	2.98385	.16606	18.235	333.621	11.242	323.689	43800
62	ERATO	3.11328	.18545	2.227	125.304	275.631	114.007	43800
63	AUSONIA	2.39527	.12658	5.775	337.721	293.840	167.161	43800
64	ANGELINA	2.68068	.12796	1.315	309.342	178.018	317.632	43800
65	CYBELE	3.42839	.10979	3.553	155.558	111.934	118.134	43800
66	MAJA	2.64674	.17248	3.059	7.620	42.003	231.861	43800
67	ASIA	2.42014	.18813	6.008	202.445	105.307	7.753	43800
68	LETO	2.78200	.18693	7.953	44.087	303.142	359.646	43800
69	HESPERIA	2.97738	.17142	8.550	185.373	287.361	345.610	43800
70	PANOPAEA	2.61403	.18394	11.593	47.687	254.504	220.988	43800
71	NIOBE	2.75564	.17349	23.279	315.825	266.715	314.619	43800
72	FERONIA	2.26590	.12097	5.414	207.741	101.684	52.188	43800
73	KLYTIA	2.66589	.04389	2.365	7.061	55.271	40.148	43800
74	GALATEA	2.78065	.23576	4.043	197.234	172.828	8.075	43800
75	EURYDIKE	2.67118	.30528	4.996	359.282	338.257	223.089	43800
76	FREIA	3.39995	.17439	2.116	204.302	257.648	123.726	43800
77	FRIGGA	2.66864	.13260	2.434	1.273	59.399	179.956	43800
78	DIANA	2.62340	.20313	8.660	333.347	151.060	122.333	43800
79	EURYNOME	2.44402	.19346	4.626	206.563	200.263	13.066	43800
80	SAPPHO	2.29517	.20104	8.656	218.416	138.643	225.068	43800
81	TERPSICHORE	2.85323	.21211	7.844	1.559	48.746	204.996	43800
82	ALKMENE	2.75973	.22381	2.839	25.352	110.750	237.638	43800
83	BEATRIX	2.43091	.08361	4.979	27.345	165.980	345.921	43800
84	KLIO	2.36233	.23535	9.329	327.305	14.036	50.497	43800
85	IO	2.65354	.19305	11.931	203.132	121.698	87.745	43800

NUMBER	NAME	A	E	I	NODE	PERI	ANOM	DAYEP
86	SEMELE	3.10290	.21860	4.804	86.816	303.153	221.464	43800
87	SYLVIA	3.48295	.09268	10.879	73.273	273.424	3.110	43800
88	THISBE	2.76922	.16268	5.235	276.684	34.280	124.178	43800
89	JULIA	2.55110	.18143	16.114	311.159	44.908	157.726	43800
90	ANTIOPE	3.14329	.16779	2.236	70.653	234.875	71.101	43800
91	AEGINA	2.59034	.10461	2.109	10.528	73.585	268.121	43800
92	UNDINA H73	3.20503	.08195	9.884	101.652	244.552	137.219	43800
93	MINERVA	2.75566	.14092	8.549	4.140	273.454	168.250	43800
94	AURORA	3.16252	.08516	8.027	3.057	46.559	247.968	43800
95	ARETHUSA	3.06945	.14945	12.957	243.220	149.823	246.550	43800
96	AEGLE	3.04740	.14208	16.021	321.817	205.259	253.287	43800
97	KLOTHO	2.67019	.29606	11.772	159.822	267.094	207.017	43800
98	IANTHE	2.68829	.18593	15.558	353.835	157.031	55.762	43800
99	DIKE	2.66358	.19617	13.875	41.376	194.089	130.505	43800
100	HEKATE	3.09401	.16054	6.410	127.472	179.663	94.506	43800
101	HELENA	2.58295	.13961	10.182	343.628	345.666	103.345	35800
102	MIRIAM	2.66013	.25329	5.148	210.739	146.110	222.746	44200
103	HERA	2.70201	.08121	5.416	135.959	187.852	311.669	43800
104	KLYMENE	3.13987	.16887	2.853	42.839	24.406	50.560	32760
105	ARTEMIS	2.37220	.17824	21.474	188.004	56.136	137.533	43800
106	DIONE	3.15783	.18265	4.622	62.091	332.012	154.417	43800
107	CAMILLA	3.48698	.07462	9.952	174.015	291.900	257.515	43800
108	HECUBA	3.23214	.06557	4.324	350.718	167.466	31.208	43800
109	FELICITAS	2.69670	.29744	7.921	3.376	54.995	199.674	43800
110	LYDIA	2.73139	.08096	5.975	56.732	282.230	218.940	43800
111	ATE	2.59494	.09971	4.912	305.655	166.943	269.442	44200
112	IPHIGENIA	2.43360	.12912	2.605	323.520	16.737	291.058	44200
113	AMALTHEA	2.37643	.08641	5.042	123.102	79.285	112.423	43800
114	KASSANDRA	2.67806	.13761	4.929	164.023	350.551	332.706	43800
115	THYRA	2.37981	.19253	11.580	308.695	96.071	4.641	43800
116	SIRONA	2.76839	.14064	3.570	63.931	91.460	298.763	43800
117	LOMIA	2.99054	.02694	14.929	349.096	63.972	171.454	38000
118	PEITHO	2.43780	.16106	7.757	47.266	32.531	78.433	43800
119	ALTHAEA	2.58226	.08092	5.784	203.610	170.338	75.033	43800
120	LACHESIS	3.11546	.06362	6.967	341.211	241.530	93.801	43800
121	HERMIONE	3.45956	.13528	7.561	74.558	286.503	108.071	43800
122	GERDA	3.21878	.05580	1.633	178.576	338.425	311.408	43800
123	BRUNHILD	2.69361	.12194	6.414	308.226	123.368	241.947	34000
124	ALKESTE	2.62913	.07917	2.957	187.903	60.424	127.831	44200
125	LIBERATRIX	2.74624	.07801	4.649	168.902	107.990	195.875	43800
126	VELLEDA	2.43885	.10618	2.923	23.083	326.315	333.242	43800
127	JOHANNA	2.75497	.06781	8.256	31.403	91.972	97.157	36000
128	NEMESIS	2.74950	.12544	6.256	76.213	302.834	301.444	43000
129	ANTIGONE	2.86914	.21033	12.232	136.526	107.088	182.242	43800
130	ELEKTRA	3.10944	.21987	22.896	145.338	235.635	177.804	43800
131	VALA	2.43086	.06982	4.961	65.322	156.917	198.855	42000
132	AETHRA	2.61365	.38356	25.072	258.476	254.391	54.484	43800
133	CYRENE	3.06488	.13322	7.232	318.850	293.223	293.019	43800
134	SOPHROSYNE	2.56496	.11517	11.590	346.019	82.372	255.081	44200
135	HERTHA	2.42887	.20400	2.297	343.577	338.400	86.647	43800
136	AUSTRIA	2.28677	.08570	9.568	186.101	132.062	319.529	43800
137	MELIBOEA	3.11102	.22227	13.434	201.952	108.838	263.267	43800
138	TOLOSA	2.44703	.16528	3.207	54.776	258.322	40.173	43800
139	JUEWA	2.78213	.17297	10.914	1.789	165.225	8.481	43800
140	SIWA	2.73115	.21716	3.192	107.027	195.614	70.590	43800
141	LUMEN	2.66699	.21224	11.919	318.509	56.993	48.682	43800
142	POLANA	2.41831	.13553	2.242	290.885	292.844	122.769	43800
143	ADRIA	2.75993	.07313	11.473	333.438	251.087	241.496	34200
144	VIBILIA	2.65382	.23568	4.814	76.224	292.995	237.488	43800
145	ADEONA	2.67287	.14704	12.652	77.321	43.099	1.639	43800
146	LUCINA	2.71859	.06669	13.089	83.795	145.371	45.642	43800
147	PROTOGENEIA	3.13856	.02009	1.922	249.555	133.900	185.782	36000
148	GALLIA	2.77148	.18571	25.318	145.056	251.379	351.956	40000
149	MEDUSA	2.17487	.06537	.939	159.114	250.501	122.608	44200
150	NUWA	2.98246	.12511	2.173	206.958	152.186	315.223	32200
151	ABUNDANTIA	2.59218	.03431	6.448	39.061	127.803	264.969	32200
152	ATALA	3.13308	.08803	12.188	40.814	51.839	207.230	34040
153	HILDA	3.97536	.15357	7.845	228.416	49.171	121.752	32200
154	BERTHA	3.17866	.09840	21.122	36.848	154.154	258.837	43800
155	SCYLLA	2.75880	.27327	11.485	41.700	43.621	57.782	30000
156	XANTHIPPE	2.72838	.22680	9.733	242.182	336.914	181.006	35800
157	DEJANIRA	2.57781	.20074	12.191	62.417	44.630	168.289	35800
158	KORONIS	2.86710	.05608	1.004	278.646	144.635	290.835	43000
159	AEMILIA	3.10533	.10287	6.120	134.011	339.141	238.243	43800
160	UNA	2.72784	.06448	3.842	8.964	49.796	58.083	35800
161	ATHOR	2.37894	.13905	9.052	18.788	293.022	299.846	30600
162	LAURENTIA	3.01633	.18175	6.096	36.068	113.363	230.792	43800
163	ERIGONE	2.36679	.19208	4.806	159.878	297.427	146.428	43800
164	EVA	2.63214	.34705	24.456	76.824	283.468	276.844	43800
165	LORELEY	3.13961	.06771	11.207	302.851	344.933	191.120	43800
166	RHODOPE	2.63580	.21153	12.012	129.363	262.340	80.729	36000
167	URDA	2.85232	.03373	2.197	166.458	123.165	106.612	35000
168	SIBYLLA	3.37795	.05331	4.581	208.278	195.882	170.518	32760
169	ZELIA	2.35758	.13117	5.508	354.634	334.028	315.931	38000
170	MARIA	2.55267	.06504	14.420	301.044	154.025	161.193	40000

NUMBER	NAME	A	E	I	NODE	PERI	ANOM	DAYEP
171	OPHELIA	3.13356	.12864	2.554	101.122	46.752	153.335	36000
172	BAUCIS	2.38129	.11318	10.018	331.722	358.715	178.549	44200
173	INO	2.74536	.20587	14.219	148.084	227.639	64.829	43800
174	PHAEDRA	2.86124	.14123	12.143	327.648	289.946	344.118	40000
175	ANDROMACHE	3.21705	.20849	3.224	22.120	322.153	249.927	38000
176	IDUNA	3.18250	.16955	22.651	200.650	178.808	17.514	44200
177	IRMA	2.77218	.23411	1.423	348.677	35.767	117.518	29320
178	BELISANA	2.45970	.04613	1.907	51.163	210.793	312.708	35800
179	KLYTAEMNESTRA	2.97189	.11327	7.803	252.050	102.348	22.795	44200
180	GARUMNA	2.72321	.15603	.876	313.462	172.205	199.893	35800
181	EUCHARIS	3.12180	.21375	18.738	143.995	315.269	190.776	32200
182	ELSA	2.41629	.18673	2.005	106.940	309.492	60.904	35800
183	ISTRIA	2.79431	.34703	26.449	142.075	263.234	345.389	44200
184	DEJOPEJA	3.17690	.08780	1.149	332.288	215.041	281.030	42000
185	EUNIKE	2.73767	.12878	23.238	153.616	223.279	206.946	43800
186	CELUTA	2.36316	.14943	13.177	14.487	314.874	128.764	43800
187	LAMBERTA	2.72785	.24017	10.630	21.642	195.120	357.556	43400
188	MENIPPE	2.76225	.17741	11.709	241.275	67.774	169.652	40000
189	PHTHIA	2.45144	.03837	5.161	203.585	164.791	343.704	38000
190	ISMENE	3.95263	.17002	6.165	176.566	281.063	315.447	36000
191	KOLGA	2.89362	.08985	11.499	159.723	227.285	284.776	34000
192	NAUSIKAA	2.40343	.24616	6.812	343.100	29.690	190.834	43000
193	AMBROSIA	2.59923	.29819	12.024	350.442	80.110	251.853	35800
194	PROKNE	2.61506	.23917	18.521	159.075	162.476	70.210	43800
195	EURYKLEIA	2.87836	.04146	6.966	7.177	113.643	166.042	40000
196	PHILOMELA	3.11283	.02538	7.272	72.350	232.311	284.014	43800
197	ARETE	2.73827	.16295	8.809	81.482	245.281	262.599	43800
198	AMPELLA	2.45827	.23091	9.274	268.258	88.416	357.959	43000
199	BYBLIS	3.17159	.17506	15.386	89.699	167.450	297.186	38000
200	DYNAMENE	2.73691	.13307	6.904	324.587	83.918	74.562	43000
201	PENELOPE	2.67724	.18135	5.751	156.747	180.044	306.837	44200
202	CHRISEIS	3.07212	.10265	8.839	137.303	358.334	20.769	36000
203	POMPEJA	2.73783	.05901	3.187	348.049	58.226	100.471	38000
204	KALLISTO	2.67110	.17501	8.312	205.607	54.078	303.981	34000
205	MARTHA	2.77642	.03692	10.683	212.077	174.382	239.032	35800
206	HERSILIA	2.74138	.03765	3.777	144.969	300.559	230.487	43800
207	HEDDA	2.28428	.02894	3.810	29.100	191.477	348.813	36000
208	LACRIMOSA	2.89241	.01431	1.758	4.504	128.870	304.133	42000
209	DIDO	3.14480	.06646	7.190	.571	260.457	91.344	44200
210	ISABELLA	2.72180	.12146	5.263	32.515	13.940	5.962	43800
211	ISOLDA	3.03949	.16331	3.874	264.429	171.996	212.311	36000
212	MEDEA	3.11160	.11772	4.276	313.457	104.752	65.919	43800
213	LILAEA	2.75193	.14605	6.809	121.948	161.670	102.265	43800
214	ASCHERA	2.61128	.02939	3.437	342.040	135.700	273.099	44200
215	OENONE	2.76560	.03331	1.700	25.131	318.572	347.500	38640
216	KLEOPATRA	2.79025	.25423	13.152	215.367	178.809	104.370	43400
217	EUDORA	2.87133	.30840	10.458	163.018	153.985	295.276	36000
218	BIANCA	2.66660	.11733	15.212	170.663	61.340	149.405	40000
219	THUSNELDA	2.35399	.22369	10.821	200.723	141.277	116.909	40000
220	STEPHANIA	2.34926	.25712	7.589	258.031	77.569	162.860	38000
221	EOS	3.01218	.10211	10.881	141.857	190.962	347.983	43800
222	LUCIA	3.14694	.13674	2.162	80.207	184.374	108.886	34000
223	ROSA	3.09308	.11801	1.956	48.237	61.038	350.212	34000
224	OCEANA	2.64450	.04404	5.851	353.296	279.206	359.849	34000
225	HENRIETTA	3.34781	.29222	20.750	198.545	100.923	281.478	38000
226	WERINGIA	2.71534	.20250	15.898	135.226	151.523	248.284	35800
227	PHILOSOPHIA	3.13386	.21313	9.156	327.703	263.188	202.728	40000
228	AGATHE	2.20122	.24150	2.543	313.345	17.988	322.231	36000
229	ADELINDA	3.39861	.15843	2.121	29.654	297.914	260.770	38000
230	ATHAMANTIS	2.38256	.06099	9.450	239.563	138.432	270.777	43400
231	VINDOBONA	2.92148	.15480	5.125	351.251	266.702	102.973	38000
232	RUSSIA	2.55040	.17697	6.105	152.520	47.126	280.750	40000
233	ASTEROPE	2.65966	.10029	7.667	222.045	123.909	101.709	40000
234	BARBARA	2.38483	.24562	15.384	144.180	191.113	34.887	44200
235	CAROLINA	2.88090	.06274	9.055	66.259	208.017	147.893	36000
236	HONORIA	2.79817	.18997	7.668	186.302	172.696	71.837	36000
237	COELESTINA	2.76271	.07287	9.758	84.784	197.196	312.509	26000
238	HYPATIA	2.90629	.09117	12.393	184.054	206.927	274.345	40000
239	ADRASTEA	2.97720	.22709	6.152	180.854	207.606	88.086	40000
240	VANADIS	2.66557	.20499	2.102	114.998	299.170	33.627	40000
241	GERMANIA	3.04648	.10587	5.514	270.699	75.652	249.693	43800
242	KRIEMHILD	2.86506	.11717	11.329	207.668	275.290	280.443	36000
243	IDA	2.86029	.04620	1.138	324.046	106.571	113.234	43800
244	SITA	2.17444	.13684	2.834	209.002	164.993	314.372	30360
245	VERA	3.10088	.19778	5.183	61.516	328.826	165.342	36000
246	ASPORINA	2.69504	.10831	15.613	162.530	94.884	315.801	34000
247	EUKRATE	2.74216	.24136	25.016	359.918	54.671	318.853	43800
248	LAMEIA	2.47171	.06051	4.018	246.903	6.788	336.299	38000
249	ILSE	2.37767	.21693	9.643	334.684	40.968	355.890	38000
250	BETTINA	3.14025	.14118	12.892	25.185	69.415	359.575	32760
251	SOPHIA	3.09561	.09475	10.532	156.295	289.197	347.001	38000
252	CLEMENTINA	3.15327	.08477	10.047	202.259	163.742	268.444	40000
253	MATHILDE	2.64680	.26689	6.707	179.655	156.123	6.738	38000
254	AUGUSTA	2.19501	.12177	4.520	28.356	232.146	274.620	36000
255	OPPAVIA	2.74622	.07748	9.485	13.898	153.456	264.668	36000

NUMBER	NAME		A	F	I	NODE	PERI	ANOM	DAYEP
256	WALPURGA		3.00071	.06268	13.323	182.794	46.098	330.748	44200
257	SILESIA		3.12168	.11669	3.653	34.905	30.530	113.352	36000
258	TYCHE		2.61663	.20280	14.275	207.554	154.313	123.414	42000
259	ALETHEIA		3.13566	.12783	10.785	87.821	157.690	288.986	36000
260	HUBERTA		3.44991	.10611	6.336	167.131	173.848	161.742	32200
261	PRYMNO		2.33112	.08988	3.638	96.412	64.108	94.311	42200
262	VALDA		2.55535	.20887	7.738	38.435	22.838	44.176	35800
263	DRESDA		2.88839	.07543	1.295	217.044	158.102	121.523	35800
264	LIBUSSA		2.79925	.13432	10.431	49.727	338.946	90.036	38000
265	ANNA		2.42233	.26288	25.686	335.295	251.634	159.780	40000
266	ALINE		2.80244	.15892	13.392	236.330	148.930	124.005	32200
267	TIRZA		2.77416	.10212	6.025	74.010	192.916	313.650	32200
268	ADOREA		3.09734	.13138	2.435	121.223	62.323	192.971	43000
269	JUSTITIA		2.61348	.21682	5.454	156.835	118.274	79.315	40000
270	ANAHITA		2.19846	.15025	2.366	254.180	79.563	128.114	44200
271	PENTHESILEA		3.00811	.09790	3.549	335.980	56.588	139.016	38000
272	ANTONIA		2.77721	.02997	4.451	37.444	61.446	163.767	40000
273	ATROPOS		2.39419	.16263	20.413	158.655	120.108	165.932	44200
274	PHILAGORIA		3.04711	.11719	3.674	93.119	117.009	176.180	40840
275	SAPIENTIA		2.77290	.16166	4.774	134.201	36.726	146.101	40000
276	ADELHEID		3.11715	.06196	21.640	211.066	273.407	281.560	40000
277	ELVIRA		2.88523	.09206	1.151	231.968	133.259	331.616	40000
278	PAULINA		2.75456	.13521	7.817	62.609	136.919	15.702	25720
279	THULE		4.26093	.03232	2.339	74.584	195.389	85.999	43000
280	PHILIA		2.94444	.10803	7.453	10.498	84.021	238.992	40080
281	LUCRETIA		2.19760	.13191	5.312	31.244	15.543	64.235	36000
282	CLORINDE		2.33903	.08106	9.033	144.625	295.384	186.529	44200
283	EMMA		3.05205	.14436	7.998	304.257	55.400	85.259	43800
284	AMALIA		2.35849	.22144	8.069	233.685	56.802	.622	40400
285	REGINA		3.09382	.20871	17.661	311.975	12.766	2.678	29120
286	ICLEA		3.19195	.04153	17.926	148.927	242.237	227.749	44200
287	NEPHTHYS		2.35413	.02251	10.022	142.368	120.616	149.434	30000
288	GLAUKE		2.76123	.20497	4.334	120.547	82.855	168.476	35800
289	NENETTA		2.87235	.20520	6.678	190.172	187.793	56.220	36000
290	BRUNA		2.33647	.26003	22.295	10.418	104.295	188.701	38080
291	ALICE		2.22196	.09263	1.849	161.348	330.644	169.095	36000
292	LUDOVICA		2.52977	.03050	14.973	43.314	281.103	287.322	36000
293	BRASILIA		2.86158	.10630	15.597	61.720	86.390	349.571	35800
294	FELICIA		3.13340	.24954	6.306	135.835	186.574	330.547	40000
295	THERESIA		2.79704	.16785	2.689	277.005	145.381	35.706	36000
296	PHAETUSA		2.22879	.15963	1.747	121.183	251.815	351.082	37200
297	CAECILIA		3.17418	.12942	7.533	332.640	359.803	279.550	38000
298	BAPTISTINA		2.26367	.09538	6.283	7.909	134.260	293.114	44200
299	THORA		2.43459	.06221	1.605	241.279	150.329	154.016	40000
300	GERALDINA		3.21539	.03526	.749	42.896	344.733	85.781	43800
301	BAVARIA		2.72475	.06514	4.880	142.626	120.679	351.006	34000
302	CLARISSA		2.40569	.11101	3.424	7.671	53.807	245.230	40000
303	JOSEPHINA		3.12374	.06682	6.880	344.021	77.138	36.671	43800
304	OLGA		2.40369	.22066	15.804	158.762	171.061	121.854	44200
305	GORDONIA		3.09642	.19223	4.425	209.451	256.428	124.587	32200
306	UNITAS		2.35739	.15184	7.274	141.552	167.209	293.502	43800
307	NIKE		2.91361	.13905	6.113	101.330	322.210	166.510	32200
308	POLYXO		2.74941	.04009	4.358	181.640	112.207	330.323	43800
309	FRATERNITAS		2.66455	.11475	3.728	356.644	309.206	163.603	40000
310	MARGARITA		2.76310	.11332	3.147	229.860	322.246	138.845	35800
311	CLAUDIA		2.89623	.00545	3.233	81.073	75.026	280.203	38000
312	PIERRETTA		2.78093	.16234	9.034	6.782	260.730	134.990	36000
313	CHALDAEA		2.37611	.18098	11.620	176.382	314.709	149.702	43800
314	ROSALIA		3.14762	.19213	12.564	170.431	194.114	250.125	38000
315	CONSTANTIA		2.24118	.16871	2.425	161.337	172.230	317.996	40000
316	GOBERTA		3.16587	.14960	2.333	123.695	321.342	131.036	36000
317	ROXANE		2.24663	.08556	1.761	151.090	186.212	3.078	36000
318	MAGDALENA		3.20125	.06623	10.636	161.420	307.887	354.247	40000
319	LEONA		3.37678	.23572	10.853	188.702	214.739	101.996	38000
320	KATHARINA		3.01571	.11071	9.358	220.336	149.349	205.876	36000
321	FLORENTINA		2.88545	.04345	2.607	40.418	33.038	60.408	34000
322	PHAEO		2.78118	.24748	7.999	252.740	113.471	206.681	40000
323	BRUCIA		2.38232	.30005	24.237	97.140	290.996	259.633	43800
324	BAMBERGA		2.68524	.33676	11.169	328.067	42.746	283.269	43400
325	HEIDELBERGA		3.18968	.18059	8.563	345.345	73.628	296.422	32200
326	TAMARA		2.31683	.19033	23.743	31.925	237.956	131.563	43800
327	COLUMBIA		2.77776	.06133	7.136	354.715	306.394	152.418	43800
328	GUDRUN		3.11071	.10720	16.119	353.103	101.881	75.021	32200
329	SVEA		2.47665	.02298	15.916	178.034	49.838	39.305	43800
330	ADALBERTA LOST		2.08931		19.980	359.580		175.195	12160
331	ETHERIDGEA		3.02511	.09923	6.077	22.609	335.675	6.658	32200
332	SIRI		2.77178	.09046	2.860	31.868	294.227	37.634	33280
333	BADENIA		3.12836	.16281	3.789	354.499	20.436	276.511	32200
334	CHICAGO		3.86523	.05289	4.665	130.893	147.760	131.814	43000
335	ROBERTA		2.47177	.17878	5.088	148.021	138.908	101.134	43800
336	LACADIERA		2.25183	.09487	5.642	234.964	30.315	144.545	36000
337	DEVOSA		2.38249	.13774	7.865	355.161	98.360	72.441	43800
338	BUDROSA		2.91507	.02210	6.046	287.569	121.853	74.639	43800
339	DOROTHEA		3.01407	.09749	9.937	173.736	162.854	39.978	43000
340	EDUARDA		2.74683	.11695	4.687	27.207	41.448	99.303	38000

NUMBER	NAME	A	E	I	NODE	PERI	ANOM	DAYEP
341	CALIFORNIA	2.19926	.19372	5.673	29.000	292.752	329.831	36000
342	ENDYMION	2.56763	.12992	7.318	232.445	225.897	6.115	44200
343	OSTARA	2.41197	.23042	3.277	38.557	9.021	65.272	40000
344	DESIDERATA	2.59079	.31651	18.481	48.313	235.637	285.181	43800
345	TERCIDINA	2.32561	.06158	9.746	212.305	229.968	53.123	43800
346	HERMENTARIA	2.79532	.10090	8.756	92.142	288.581	20.907	38000
347	PARIANA	2.61149	.16579	11.686	85.720	84.801	116.551	38000
348	MAY	2.97018	.06828	9.754	89.992	11.826	245.287	43800
349	DEMBOWSKA	2.92647	.08967	8.250	32.291	346.710	106.953	43800
350	ORNAMENTA	3.10972	.16181	24.865	90.517	335.046	305.231	32200
351	YRSA	2.76522	.15656	9.229	99.780	29.286	77.219	29760
352	GISELA	2.19421	.14945	3.377	247.004	143.560	331.480	44200
353	RUPERTO-CAROLA	2.73537	.32849	5.645	103.465	317.983	102.009	26000
354	ELEONORA	2.79833	.11354	18.411	140.187	5.548	97.636	43800
355	GABRIELLA	2.53959	.10022	4.296	352.623	103.190	86.081	30480
356	LIGURIA	2.75849	.23721	8.238	354.926	77.631	298.482	43800
357	NININA	3.14397	.09128	15.086	138.322	249.605	298.002	36000
358	APOLLONIA	2.88033	.15050	3.537	172.482	252.040	99.582	35800
359	GEORGIA	2.72814	.15605	6.786	6.262	336.331	78.804	40000
360	CARLOVA	3.00035	.17937	11.683	132.997	288.231	297.417	32640
361	BONONIA	3.94767	.21523	12.658	18.720	70.557	34.459	43800
362	HAVNIA	2.57795	.04650	8.053	27.205	30.456	172.916	40000
363	PADUA	2.74632	.07160	5.957	64.858	294.942	272.452	38000
364	ISARA	2.22109	.14944	6.003	105.210	312.313	53.264	43800
365	CORDUBA	2.80236	.15685	12.815	185.215	214.146	217.694	43800
366	VINCENTINA	3.14218	.06709	10.599	347.321	332.221	108.087	32200
367	AMICITIA	2.21910	.09600	2.944	83.117	54.532	31.137	43800
368	HAIDEA	3.07976	.19831	7.770	227.797	90.827	235.961	35800
369	AERIA	2.64890	.09856	12.710	94.297	268.364	313.623	38000
370	MODESTIA	2.32438	.09143	7.861	290.953	67.678	322.851	36000
371	BOHEMIA	2.72656	.06411	7.386	283.368	341.479	92.110	44200
372	PALMA	3.14684	.26180	23.846	326.900	117.601	354.280	43800
373	MELUSINA	3.12520	.13658	15.381	4.425	345.847	258.404	32200
374	BURGUNDIA	2.77901	.08230	9.001	219.163	24.559	217.661	42000
375	URSULA	3.13358	.09667	15.935	336.412	347.758	170.745	43800
376	GEOMETRIA	2.28817	.17210	5.428	302.113	315.488	235.378	36000
377	CAMPANIA	2.68920	.07780	6.678	210.193	195.422	216.341	38000
378	HOLMIA	2.77565	.13075	6.989	232.893	155.535	293.021	36000
379	HUENNA	3.14540	.17525	1.635	172.614	180.423	293.211	32200
380	FIDUCIA	2.67789	.11401	6.161	95.121	238.291	115.805	40000
381	MYRRHA	3.20729	.11645	12.573	125.135	133.694	213.345	42000
382	DODONA	3.12975	.16494	7.419	315.776	265.765	111.575	34000
383	JANINA	3.12598	.18196	2.660	92.971	320.550	197.015	38000
384	BURDIGALA	2.65059	.14928	5.606	47.845	33.718	118.071	40000
385	ILMATAR	2.84775	.12650	13.570	345.030	188.255	135.566	43400
386	SIEGENA	2.89364	.17265	20.268	166.673	219.100	107.126	43400
387	AGUITANIA	2.73758	.23848	18.103	128.075	156.135	125.228	43800
388	CHARYBDIS	3.00803	.05681	6.465	354.999	327.996	250.319	36000
389	INDUSTRIA	2.60859	.06579	8.126	282.213	266.462	190.075	40000
390	ALMA	2.65146	.13188	12.174	305.233	189.243	344.202	38000
391	INGEBORG	2.32051	.30670	23.124	212.592	146.083	293.438	43800
392	WILHELMINA	2.88466	.14037	14.290	210.529	172.373	75.681	24160
393	LAMPETIA	2.77580	.33404	14.846	213.211	88.838	116.476	43800
394	ARDUINA	2.76173	.22808	6.230	67.319	268.827	55.260	40000
395	DELIA	2.78776	.08303	3.341	260.069	9.249	271.821	35800
396	AEOLIA	2.74234	.15769	2.538	250.477	19.494	31.617	35800
397	VIENNA	2.63389	.24833	12.835	227.927	139.136	246.478	42160
398	ADMETE	2.74023	.22268	9.510	279.917	158.183	280.180	44200
399	PERSEPHONE	3.05004	.07399	13.142	346.682	191.015	235.980	38000
400	DUCROSA	3.13556	.10048	10.595	327.886	230.984	53.670	40000
401	OTTILIA	3.34282	.05494	5.944	36.543	263.463	245.129	40000
402	CHLOE	2.55593	.11813	11.869	129.510	16.544	343.279	40000
403	CYANE	2.81530	.09583	9.132	245.023	250.299	215.406	38000
404	ARSINOE	2.59078	.20312	14.114	92.187	121.103	22.949	43800
405	THIA	2.57862	.25064	11.900	255.039	307.804	354.961	43400
406	ERNA	2.91915	.17712	4.211	316.185	35.802	170.803	38000
407	ARACHNE	2.62394	.06974	7.535	294.994	79.191	87.805	34000
408	FAMA	3.16496	.15047	9.052	298.899	96.193	339.195	38000
409	ASPASIA	2.57708	.06973	11.246	242.194	353.548	211.642	43800
410	CHLORIS	2.72404	.24093	10.947	96.816	171.523	346.734	43800
411	XANTHE	2.93630	.11166	15.311	107.604	179.193	291.058	38000
412	ELISABETHA	2.76177	.04425	13.767	106.983	91.491	196.860	26640
413	EDBURGA	2.58411	.34076	18.830	104.404	250.579	110.438	38000
414	LIRIOPE	3.50720	.07688	9.542	111.890	313.746	119.213	38000
415	PALATIA	2.79219	.30166	8.139	127.024	296.306	318.949	43800
416	VATICANA	2.79187	.21898	12.922	58.088	196.869	229.676	43800
417	SUEVIA	2.79976	.13433	6.589	199.938	344.377	328.495	32200
418	ALEMANNIA	2.59272	.11911	6.813	248.788	124.795	39.978	40000
419	AURELIA	2.59068	.25727	3.950	229.368	42.798	282.576	43800
420	BERTHOLDA	3.41739	.04350	6.689	244.100	193.033	267.758	43800
421	ZAHRINGIA	2.54025	.28362	7.763	187.239	209.398	81.956	43800
422	BEROLINA	2.22779	.21479	5.001	8.979	334.124	110.756	36000
423	DIOTIMA	3.06878	.02938	11.222	69.391	210.937	288.999	44200
424	GRATIA	2.77376	.11053	8.216	99.230	331.686	129.057	38000
425	CORNELIA	2.88623	.05971	4.070	61.408	124.242	358.149	32200

NUMBER	NAME		A	E	I	NODE	PERI	ANOM	DAYEP
426	HIPPO		2.88877	.10156	19.540	311.291	218.355	.671	43800
427	GALENE		2.97766	.11249	5.110	298.131	8.601	261.319	36000
428	MONACHIA		2.30749	.17839	6.205	17.459	14.805	218.451	38000
429	LOTIS		2.60612	.12391	9.512	219.820	167.972	192.100	44200
430	HYBRIS		2.84343	.25764	14.649	249.748	177.705	103.776	34000
431	NEPHELE		3.12269	.18753	1.829	117.071	214.639	338.254	40000
432	PYTHIA		2.36874	.14689	12.128	88.602	173.228	107.807	38000
433	EROS ✦		1.45784	.22265	10.828	303.828	178.478	203.407	42800
434	HUNGARIA		1.94409	.07382	22.510	174.854	123.767	244.010	43800
435	ELLA		2.44900	.15704	1.833	23.022	331.640	114.362	30400
436	PATRICIA		3.19910	.06195	18.607	351.551	15.692	79.381	38000
437	RHODIA		2.38639	.24762	7.361	263.400	60.446	153.843	38000
438	ZEUXO		2.55344	.06605	7.398	48.952	205.158	35.689	38000
439	OHIO		3.12619	.07631	19.165	201.785	237.564	144.754	42000
440	THEODORA		2.21029	.10760	1.596	291.879	178.284	166.257	44200
441	BATHILDE		2.80527	.09199	8.134	253.657	202.328	334.100	43800
442	EICHSFELDIA		2.34500	.07100	6.065	134.727	83.654	37.723	40000
443	PHOTOGRAPHICA		2.21508	.04071	4.232	175.116	349.347	214.560	41960
444	GYPTIS		2.77039	.17603	10.273	195.585	153.740	304.180	43800
445	EDNA		3.17863	.21027	21.419	293.715	74.232	132.580	32200
446	AETERNITAS		2.79735	.12584	10.616	41.969	278.359	42.486	43800
447	VALENTINE		2.98544	.04135	4.808	72.151	311.313	64.736	36000
448	NATALIE		3.15061	.17316	12.673	38.460	288.486	163.473	40000
449	HAMBURGA		2.55414	.16686	3.094	85.703	47.332	226.241	32640
450	BRIGITTA		3.01371	.10168	10.179	14.714	353.695	39.621	38000
451	PATIENTIA		3.06501	.06769	15.202	89.205	338.954	258.501	43800
452	HAMILTONIA	LOST	2.86525	.02970	3.222	93.321	46.410	293.959	15000
453	TEA		2.18340	.10888	5.553	11.380	220.025	62.782	43800
454	MATHESIS		2.62697	.11100	6.305	32.324	175.644	243.211	38000
455	BRUCHSALIA		2.65638	.29459	12.016	76.422	271.782	321.026	43800
456	ABNOBA		2.78332	.18189	14.448	229.500	3.070	49.026	34000
457	ALLEGHENIA		3.09011	.17805	12.948	249.544	130.878	261.838	40640
458	HERCYNIA		2.98991	.24532	12.621	135.558	273.739	310.223	32200
459	SIGNE		2.62060	.21044	10.591	30.569	17.672	323.143	40000
460	SCANIA		2.71716	.10484	4.628	205.265	160.077	48.219	40000
461	SASKIA		3.10660	.15925	1.443	157.042	303.410	83.480	44200
462	ERIPHYLA		2.87330	.08374	3.193	105.156	248.205	32.397	43800
463	LOLA		2.39784	.22008	13.531	36.447	328.497	274.711	38000
464	MEGAIRA		2.80207	.20561	10.155	102.880	257.002	24.284	32200
465	ALEKTO		3.09808	.20152	4.631	303.342	278.639	60.542	38000
466	TISIPHONE		3.36908	.06513	19.083	290.943	247.314	229.763	42000
467	LAURA		2.94579	.10677	6.459	323.140	90.628	119.913	34200
468	LINA		3.15197	.18157	.480	21.105	329.546	171.139	36000
469	ARGENTINA		3.15251	.18133	11.735	334.478	209.054	321.785	36000
470	KILIA		2.40420	.09532	7.237	173.126	45.110	349.579	40000
471	PAPAGENA		2.89235	.22988	14.939	84.125	313.871	112.082	40000
472	ROMA		2.54369	.09292	15.846	127.176	296.608	57.225	40000
473	NOLLI	LOST	2.97909	.25563	27.780	334.148	57.108	89.676	15400
474	PRUDENTIA		2.45471	.21030	8.817	161.732	156.236	227.453	38000
475	OCLLO		2.59465	.37970	18.800	35.011	303.832	8.915	40120
476	HEDWIG		2.64870	.07458	10.936	286.459	359.245	119.582	38000
477	ITALIA		2.41548	.18750	5.300	10.711	321.781	313.422	36000
478	TERGESTE		3.01341	.08889	13.173	234.274	240.735	131.002	36000
479	CAPRERA		2.72122	.21717	8.662	136.605	267.339	61.424	34000
480	HANSA		2.64391	.04465	21.292	237.069	213.754	11.265	40000
481	EMITA		2.74096	.15651	9.847	66.724	348.991	23.890	43800
482	PETRINA		2.99589	.10135	14.474	179.614	86.294	186.751	40000
483	SEPPINA		3.42880	.03527	18.704	174.648	153.966	50.325	38000
484	PITTSBURGHIA		2.66723	.05786	12.506	127.169	189.949	325.212	40000
485	GENUA		2.74828	.19295	13.847	193.857	271.007	87.439	35800
486	CREMONA		2.35180	.16292	11.089	93.928	124.104	190.398	44200
487	VENETIA		2.67028	.08825	10.236	114.779	279.143	308.658	40000
488	KREUSA		3.14771	.17775	11.505	84.821	73.085	348.674	36000
489	COMACINA		3.15004	.04962	12.976	167.592	344.577	110.961	36000
490	VERITAS		3.17457	.08803	9.253	178.633	206.466	196.560	38000
491	CARINA		3.19534	.07848	18.835	175.541	245.826	130.993	38000
492	GISMONDA		3.10768	.18623	1.639	46.516	292.383	351.387	38400
493	GRISELDIS		3.11838	.17355	15.251	358.095	41.090	2.466	40360
494	VIRTUS		2.98632	.05812	7.117	38.686	209.129	348.367	36000
495	EULALIA		2.48782	.13239	2.287	186.620	205.046	341.655	36000
496	GRYPHIA		2.19888	.07922	3.788	207.243	257.888	177.551	44200
497	IVA		2.84990	.30220	4.865	6.617	1.773	24.145	44200
498	TOKIO		2.65005	.22356	9.522	97.219	240.272	267.531	43800
499	VENUSIA		3.96337	.22214	2.082	257.125	180.943	320.830	32800
500	SELINUR		2.61164	.14587	9.793	290.022	73.532	256.812	40000
501	URHIXIDUR		3.15083	.15395	20.960	357.305	349.601	13.884	44200
502	SIGUNE		2.38187	.17959	25.049	132.780	18.085	61.239	38000
503	EVELYN		2.72295	.17760	5.042	69.436	38.812	294.139	32200
504	CORA		2.72073	.21754	12.908	104.611	246.750	42.316	44200
505	CAVA		2.68516	.24478	9.823	90.714	336.329	9.221	43800
506	MARION		3.04488	.14448	16.958	313.369	145.766	141.042	32200
507	LAODICA		3.15304	.10337	9.495	294.526	86.711	359.169	38000
508	PRINCETONIA		3.15965	.02494	13.381	44.943	234.998	209.536	32200
509	IOLANDA		3.00175	.09555	15.395	217.903	155.951	250.928	38000
510	MABELLA		2.60732	.19347	9.531	202.894	89.040	283.840	43800

NUMBER	NAME	A	E	I	NODE	PERI	ANOM	DAYEP
511	DAVIDA	3.18083	.17192	15.897	107.673	338.002	269.337	43800
512	TAURINENSIS	2.18916	.25446	8.759	106.908	248.288	315.927	40000
513	CENTESIMA	3.01236	.08414	9.719	184.631	224.636	279.102	43000
514	ARMIDA	3.04634	.03943	3.869	269.269	106.583	30.721	40000
515	ATHALIA	3.12389	.18006	2.005	122.360	289.470	314.057	16360
516	AMHERSTIA	2.68138	.27164	12.984	329.149	257.049	31.623	40000
517	EDITH	3.13770	.19831	3.210	274.814	140.651	174.716	38000
518	HALAWE	2.53719	.22014	6.747	204.134	117.022	275.169	38000
519	SULVANIA	2.78857	.18580	11.013	44.984	301.172	258.067	39600
520	FRANZISKA	3.00792	.10505	10.983	34.607	20.975	110.427	38000
521	BRIXIA	2.74027	.28268	10.555	90.157	314.349	226.005	34000
522	HELGA	3.62832	.07157	4.420	118.467	241.206	292.071	38000
523	ADA	2.96219	.18361	4.327	261.671	185.560	188.533	36000
524	FIDELIO	2.63735	.12693	8.235	326.833	78.905	102.054	40000
525	ADELAIDE	2.24547	.10213	5.992	203.127	263.376	229.539	43800
526	JENA	3.12461	.12795	2.167	137.645	1.866	264.165	38000
527	EURYANTHE	2.72358	.15211	9.663	120.774	201.453	117.409	34000
528	REZIA	3.39433	.02113	12.686	51.345	56.308	107.696	32760
529	PREZIOSA	3.01943	.09044	11.033	65.667	334.078	90.173	33520
530	TURANDOT	3.19337	.21650	8.556	128.758	203.640	76.623	43000
531	ZERLINA	2.78669	.19642	33.977	197.384	58.003	334.446	43800
532	HERCULINA	2.77350	.17486	16.344	107.457	75.666	64.818	43800
533	SARA	2.98216	.03725	6.525	180.696	31.719	116.810	36000
534	NASSOVIA	2.88517	.05584	3.275	93.939	333.786	212.278	43800
535	MONTAGUE	2.56960	.02435	6.776	84.392	72.917	190.159	44200
536	MERAPI	3.50470	.08161	19.394	60.289	297.970	76.578	32200
537	PAULY	3.05944	.23857	9.920	120.177	184.789	285.069	43800
538	FRIEDERIKE	3.17711	.14821	6.508	141.347	221.782	101.993	44200
539	PAMINA	2.73834	.21320	6.794	274.811	95.312	351.792	40000
540	ROSAMUNDE	2.21920	.09036	5.577	201.804	336.640	291.238	43800
541	DEBORAH	2.81508	.05181	5.995	267.638	359.804	29.027	44200
542	SUSANNA	2.90410	.14218	12.080	153.240	214.960	258.998	38000
543	CHARLOTTE	3.06509	.14581	8.450	295.461	106.819	348.417	44200
544	JETTA	2.59126	.15303	8.337	298.177	340.659	347.115	43800
545	MESSALINA	3.17535	.19260	11.195	334.415	321.193	154.363	40000
546	HERODIAS	2.59670	.11507	14.894	21.662	109.832	326.029	40000
547	PRAXEDIS	2.77659	.23418	16.882	193.303	194.147	104.267	35800
548	KRESSIDA	2.28212	.18572	3.868	108.471	318.809	318.315	25520
549	JESSONDA	2.68254	.26131	3.967	291.686	156.306	57.337	38000
550	SENTA	2.58903	.22174	10.099	270.836	44.424	65.063	38000
551	ORTRUD	2.97047	.12177	.418	7.306	68.278	97.962	36000
552	SIGELINDE	3.16014	.06797	7.639	267.702	340.861	327.939	44200
553	KUNDRY	2.23054	.11012	5.390	72.023	353.825	187.915	44200
554	PERAGA	2.37431	.15305	2.934	295.344	126.539	99.279	43800
555	NORMA	3.15841	.17894	2.645	130.654	351.320	39.710	41760
556	PHYLLIS	2.46678	.09843	5.219	285.973	175.124	55.697	34000
557	VIOLETTA	2.44164	.10312	2.496	293.347	196.100	101.913	34000
558	CARMEN	2.90781	.04197	8.364	143.726	316.832	350.779	42000
559	NANON	2.71068	.06671	9.306	111.978	127.776	212.380	44200
560	DELILA	2.75017	.16205	8.473	105.466	1.994	204.663	36000
561	INGWELDE	3.15990	.13767	1.499	160.169	304.767	139.808	42000
562	SALOME	3.01884	.09852	11.115	70.803	262.315	248.373	40000
563	SULEIKA	2.71474	.23334	10.221	85.167	335.688	302.072	43800
564	DUDU	2.74691	.27498	18.081	70.998	213.234	192.444	38000
565	MARBACHIA	2.44356	.13075	10.938	225.961	288.488	255.815	40000
566	STEREOSKOPIA	3.38545	.10255	4.934	81.076	308.543	118.648	32200
567	ELEUTHERIA	3.13144	.09518	9.265	58.088	141.297	189.109	44200
568	CHERUSKIA	2.88203	.16816	18.372	249.913	171.852	224.754	40000
569	MISA	2.65700	.18050	1.293	302.348	140.926	354.415	38000
570	KYTHERA	3.44442	.09880	1.738	225.975	154.266	323.849	38000
571	DULCINEA	2.40929	.24344	5.260	3.123	25.982	251.067	40000
572	REBEKKA	2.40052	.15722	10.567	194.304	190.881	290.414	40000
573	RECHA	3.01194	.11336	9.834	343.599	30.808	286.388	34000
574	REGINHILD	2.25218	.23922	5.691	336.624	76.092	313.810	43000
575	RENATE	2.55505	.12743	15.079	349.525	333.853	153.620	40000
576	EMANUELA	2.98887	.19232	10.225	300.505	25.920	15.883	32200
577	RHEA	3.11153	.48809	3.960	320.347	354.485	18.566	37200
578	HAPPELIA	2.74890	.19506	6.153	29.575	260.719	261.351	39560
579	SIDONIA	3.01140	.07844	11.029	82.819	226.823	102.773	42000
580	SELENE	3.22614	.09803	3.644	99.280	310.078	345.974	38000
581	TAUNTONIA	3.22354	.01170	21.795	102.523	5.770	171.486	41480
582	OLYMPIA	2.61295	.22237	29.936	155.583	309.450	296.800	40000
583	KLOTILDE	3.18094	.15111	8.215	257.741	257.197	255.743	40000
584	SEMIRAMIS	2.37411	.23306	10.720	281.982	84.462	47.594	43800
585	BILKIS	2.42915	.13328	7.576	180.074	328.604	.528	38000
586	THEKLA	3.04151	.06107	1.608	229.305	247.346	278.670	38000
587	HYPSIPYLE	2.33489	.16728	25.012	324.316	187.848	154.337	40000
588	ACHILLES	5.17383	.14893	10.334	315.877	131.617	83.356	43800
589	CROATIA	3.13004	.05248	10.808	178.154	228.640	202.172	38000
590	TOMYRIS	2.99803	.08075	11.182	106.148	336.266	149.449	44200
591	IRMGARD	2.68040	.20636	12.522	334.451	216.403	320.096	38000
592	BATHSEBA	3.02282	.13503	10.167	168.079	256.124	24.870	43800
593	TITANIA	2.69815	.21742	16.923	76.004	30.163	54.816	40000
594	MIREILLE	2.62762	.35315	32.574	154.720	77.156	346.661	43800
595	POLYXENA	3.21046	.05992	17.882	24.738	251.873	242.784	40000

NUMBER	NAME		A	E	I	NODE	PERI	ANOM	DAYEP
596	SCHEILA		2.93010	.16434	14.651	71.157	173.685	350.177	32200
597	BANDUSIA		2.67205	.14398	12.818	36.765	306.549	6.683	32200
598	OCTAVIA		2.76103	.25089	12.195	91.996	289.798	347.005	40000
599	LUISA		2.77063	.29525	16.651	44.640	292.375	76.351	40000
600	MUSA		2.65991	.05577	10.193	139.272	112.646	110.758	40000
601	NERTHUS		3.13063	.11827	16.124	169.791	156.101	22.402	38000
602	MARIANNA		3.08483	.24797	15.223	332.162	42.069	258.080	43800
603	TIMANDRA	LOST	2.55178	.16446	8.114	344.247	155.229	11.111	17280
604	TEKMESSA		3.16667	.18587	4.428	12.328	31.386	134.154	36000
605	JUVISIA		2.99884	.13812	19.676	342.737	13.157	16.039	42000
606	BRANGANE		2.58682	.21872	8.630	318.540	57.027	287.963	40000
607	JENNY		2.85154	.07679	10.100	285.617	291.553	278.270	32200
608	ADOLFINE		3.02320	.12100	9.371	294.389	71.022	246.393	38000
609	FULVIA		3.09246	.02829	4.167	165.489	107.695	193.380	43800
610	VALESKA		3.07718	.26481	12.785	21.275	357.098	326.586	27200
611	VALERIA		2.98429	.11988	13.448	189.832	252.056	280.258	34000
612	VERONIKA	LOST	3.13887	.26646	20.464	205.812	116.229	22.301	17480
613	GINEVRA		2.92118	.06224	7.679	354.908	64.467	328.949	43000
614	PIA		2.69267	.11084	7.023	217.335	207.761	219.565	38000
615	ROSWITHA		2.63062	.11074	2.777	13.665	245.175	174.118	38000
616	ELLY		2.55259	.05945	15.016	356.005	110.250	38.427	41800
617	PATROCLUS		5.22959	.14089	22.047	43.774	306.263	62.557	43800
618	ELFRIEDE		3.18424	.08771	17.049	111.168	241.049	344.273	36000
619	TRIBERGA		2.51953	.07554	13.715	187.284	176.019	42.831	38000
620	DRAKONIA		2.43470	.13521	7.719	359.899	334.704	327.739	38000
621	WERDANDI		3.11911	.14707	2.323	67.174	35.269	25.839	38000
622	ESTHER		2.41443	.24446	8.634	142.114	255.313	346.614	38000
623	CHIMAERA		2.45917	.11464	14.194	308.122	124.598	11.252	40000
624	HEKTOR		5.15334	.02552	18.258	342.071	179.538	25.297	43800
625	XENIA		2.64499	.22702	12.085	127.464	199.541	265.880	40000
626	NOTHBURGA		2.57337	.24401	25.402	341.542	43.069	192.892	42160
627	CHARIS		2.89883	.06178	6.474	142.404	172.955	88.690	44200
628	CHRISTINE		2.58196	.04390	11.503	111.787	203.018	25.667	44200
629	BERNARDINA		3.11935	.17239	9.338	67.304	33.810	173.668	40760
630	EUPHEMIA		2.62267	.11396	13.886	105.325	37.432	47.847	44200
631	PHILIPPINA		2.79159	.08352	18.916	224.757	276.540	100.553	40000
632	PYRRHA		2.66147	.19374	2.231	357.247	250.178	269.329	38000
633	ZELIMA		3.01701	.08676	10.909	147.576	184.465	133.234	38000
634	UTE		3.04623	.18299	12.315	133.726	218.383	77.560	40000
635	VUNDTIA		3.13409	.09225	11.033	183.648	225.852	204.089	38000
636	ERIKA		2.90884	.17423	7.927	34.724	296.427	43.617	44200
637	CHRYSOTHEMIS		3.15413	.14528	.288	355.462	173.117	232.790	41600
638	MOIRA		2.73568	.15859	7.700	103.567	126.074	345.344	35800
639	LATONA		3.01415	.10961	8.577	280.043	48.522	179.736	43800
640	BRAMBILLA		3.16648	.05863	13.321	235.406	29.441	3.567	38000
641	AGNES		2.21996	.12844	1.715	40.682	17.130	243.676	44200
642	CLARA		3.17149	.14874	8.142	7.474	105.720	168.569	38000
643	SCHEHEREZADE		3.34562	.09428	13.698	253.888	224.228	258.537	38000
644	COSIMA		2.59926	.15374	1.041	109.323	267.613	69.689	38000
645	AGRIPPINA		3.17977	.17641	7.062	.709	85.421	177.239	38000
646	KASTALIA		2.32499	.21262	6.921	302.668	36.907	38.831	41240
647	ADELGUNDE		2.44338	.19063	7.284	254.649	175.019	269.492	40000
648	PIPPA		3.17174	.22555	9.998	292.805	166.992	217.399	36000
649	JOSEFA		2.54950	.27278	12.638	357.164	348.351	127.678	40640
650	AMALASUNTHA		2.45770	.18535	2.553	215.036	177.131	157.920	43800
651	ANTIKLEIA		3.02300	.10054	10.775	38.124	350.665	266.598	44200
652	JUBILATRIX		2.55539	.12532	15.746	85.936	276.713	213.883	38000
653	BERENIKE		3.01212	.04507	11.281	133.368	47.744	72.515	38000
654	ZELINDA		2.29673	.23141	18.143	278.163	213.370	105.342	43800
655	BRISEIS		2.98672	.09074	6.445	130.323	281.108	205.496	36000
656	BEAGLE		3.16832	.11390	.472	186.054	325.154	312.454	32200
657	GUNLOD		2.61088	.11221	10.254	297.797	241.824	57.529	40000
658	ASTERIA		2.85472	.06112	1.515	351.179	59.402	300.406	43800
659	NESTOR		5.26213	.10932	4.514	350.492	336.391	200.534	43800
660	CRESCENTIA		2.53437	.10125	15.268	156.836	105.235	334.729	42000
661	CLOELIA		3.01458	.03848	9.263	335.856	181.371	246.175	44200
662	NEWTONIA		2.55396	.21687	4.127	133.639	166.111	74.896	38000
663	GERLINDE		3.07003	.14941	17.729	233.297	309.119	133.540	38000
664	JUDITH		3.16944	.24351	8.526	176.021	87.151	252.440	38000
665	SABINE		3.16933	.15372	14.656	299.260	314.170	246.928	43800
666	DESDEMONA		2.59270	.24055	7.597	215.176	172.787	341.260	44200
667	DENISE		3.19271	.18465	25.367	152.939	309.839	42.063	38000
668	DORA		2.79543	.23438	6.828	214.581	112.713	14.989	37200
669	KYPRIA		3.01121	.07979	10.774	171.002	112.555	200.873	40000
670	OTTEGEBE		2.80493	.19194	7.531	174.784	193.809	179.708	38000
671	CARNEGIA		3.10297	.06021	8.044	1.090	85.076	266.640	38000
672	ASTARTE		2.55411	.13897	11.127	343.628	309.084	151.136	38000
673	EDDA		2.81540	.01269	2.856	227.648	231.028	303.053	32200
674	RACHELE		2.92148	.19525	13.535	58.256	41.401	306.698	43800
675	LUDMILLA		2.77054	.20226	9.774	263.208	151.713	28.498	43800
676	MELITTA		3.06351	.12211	12.824	150.596	181.172	202.667	38000
677	AALTJE		2.95574	.05024	8.490	273.798	274.969	67.792	30160
678	FREDEGUNDIS		2.57387	.21625	6.099	281.590	119.893	75.065	38000
679	PAX		2.58701	.30917	24.421	112.357	265.520	4.182	43800
680	GENOVEVA		3.13046	.29677	17.842	40.123	240.337	167.248	38000

NUMBER	NAME		A	E	I	NODE	PERI	ANOM	DAYEP
681	GORGO		3.11202	.09458	12.521	178.593	112.446	228.795	38000
682	HAGAR	LOST	2.63210	.15578	11.555	192.887	92.684	348.532	18480
683	LANZIA		3.11523	.05842	18.499	260.030	271.059	19.592	40000
684	HILDBURG		2.43264	.03668	5.511	336.269	291.495	264.550	36000
685	HERMIA		2.23567	.19596	3.643	235.820	78.554	216.467	24120
686	GERSUIND		2.58815	.26615	15.727	243.318	87.313	211.771	43800
687	TINETTE		2.72222	.27191	14.944	334.815	51.324	202.486	36000
688	MELANIE		2.69757	.13818	10.240	171.010	137.808	117.295	40000
689	ZITA		2.31567	.23059	5.742	167.923	187.686	9.648	40480
690	WRATISLAVIA		3.15887	.16737	11.257	254.077	115.029	205.151	34000
691	LEHIGH		3.01418	.11820	13.015	88.450	301.167	97.257	38000
692	HIPPODAMIA		3.35790	.19261	26.140	64.092	53.704	264.344	38000
693	ZERBINETTA		2.94340	.03026	14.205	352.238	284.883	220.642	34000
694	EKARD		2.67038	.32451	15.826	230.304	110.357	325.795	43800
695	BELLA		2.53908	.15872	13.909	275.648	77.688	228.964	43000
696	LEONORA		3.18006	.24678	13.020	299.127	106.633	54.523	43800
697	GALILEA		2.88128	.15464	15.153	15.877	332.605	344.345	32200
698	ERNESTINA		2.86809	.10911	11.547	41.029	96.972	331.626	38000
699	HELA		2.61344	.40760	15.304	243.883	89.356	233.290	30800
700	AURAVICTRIX		2.22931	.10409	6.788	96.680	100.471	88.848	36000
701	ORIOLA		3.01371	.03251	7.100	244.449	309.947	98.852	38000
702	ALAUDA		3.19489	.03176	20.542	289.762	12.289	355.923	43800
703	NOEMI		2.17499	.13736	2.449	213.499	173.820	82.409	38000
704	INTERAMNIA		3.06018	.15318	17.290	280.506	91.948	261.180	43800
705	ERMINIA		2.92371	.05338	25.023	2.877	100.673	84.439	38000
706	HIRUNDO		2.72957	.19288	14.466	325.504	29.365	213.395	38000
707	STEINA		2.18056	.10791	4.268	282.064	89.485	115.670	38000
708	RAPHAELA		2.67108	.08521	3.501	355.267	195.091	356.146	40000
709	FRINGILLA		2.91387	.11197	16.300	324.773	14.758	42.083	33080
710	GERTRUD		3.14302	.11703	1.741	140.160	100.245	56.242	38000
711	MARMULLA		2.23725	.19598	6.087	357.059	298.992	53.406	38000
712	BOLIVIANA		2.57432	.18996	12.815	230.614	180.146	337.401	43000
713	LUSCINIA		3.40959	.14664	10.170	220.343	124.321	341.811	36000
714	ULULA		2.53579	.05721	14.297	233.836	228.393	2.249	43800
715	TRANSVAALIA		2.76794	.08343	13.824	46.248	299.706	248.899	36000
716	BERKELEY		2.81007	.08776	8.509	146.532	50.459	74.735	38000
717	WISIBADA		3.13846	.26224	1.750	346.403	15.488	71.518	36000
718	ERIDA		3.05223	.20383	6.977	39.098	171.519	345.596	36000
719	ALBERT	LOST	2.58391	.54037	10.821	186.093	151.942	10.073	19320
720	BOHLINIA		2.88801	.01402	2.367	35.671	112.792	184.216	44200
721	TABORA		3.55392	.11183	8.386	39.893	349.323	217.270	38000
722	FRIEDA		2.17184	.14520	5.644	45.540	256.286	62.292	38000
723	HAMMONIA		2.99749	.06021	4.978	163.637	241.943	305.377	38000
724	HAPAG	LOST	2.45063	.25398	11.768	204.960	202.683	349.408	19320
725	AMANDA		2.57123	.22242	3.797	69.013	321.110	266.908	24960
726	JOELLA		2.56557	.28540	15.374	242.900	111.331	225.029	24640
727	NIPPONIA		2.56721	.10625	15.001	133.042	273.401	195.907	38000
728	LEONISIS		2.25325	.08810	4.261	82.377	53.949	94.680	42000
729	WATSONIA		2.76013	.09393	18.052	124.445	85.900	42.106	40000
730	ATHANASIA	LOST	2.24363	.17654	4.228	95.573	119.514	357.335	19520
731	SORGA		2.98524	.14335	10.711	47.046	284.819	130.912	32200
732	TJILAKI		2.45690	.04017	11.015	173.399	59.462	74.936	26920
733	MOCIA		3.39107	.07529	20.353	342.211	169.988	267.832	36000
734	BENDA		3.15540	.08659	5.843	3.985	54.149	319.121	40000
735	MARGHANNA		2.72800	.32426	16.768	43.451	308.078	358.006	36000
736	HARVARD		2.20237	.16488	4.374	135.666	199.665	175.246	38000
737	AREQUIPA		2.59215	.24336	12.374	184.542	133.972	6.434	43800
738	ALAGASTA		3.03674	.05455	3.527	132.171	39.668	107.791	40000
739	MANDEVILLE		2.73553	.14460	20.704	136.485	44.072	241.700	44200
740	CANTABIA		3.04866	.11400	10.860	116.631	45.528	102.068	36000
741	BOTOLPHIA		2.72347	.06767	8.416	100.971	59.584	226.531	34000
742	EDISONA		3.01115	.11932	11.227	64.438	283.445	348.525	40000
743	EUGENISIS		2.79502	.05720	4.827	229.625	183.981	188.378	32200
744	AGUNTINA		3.16713	.12291	7.718	142.340	35.485	258.463	44200
745	MAURITIA		3.25955	.06957	13.540	126.694	337.150	238.150	38000
746	MARLU		3.11294	.23906	17.390	3.134	303.797	290.199	32200
747	WINCHESTER		3.00440	.33692	18.187	130.197	274.996	316.630	43800
748	SIMEISA		3.95114	.17556	2.264	266.807	186.296	348.609	30720
749	MALZOVIA		2.24296	.17361	5.386	109.617	127.347	79.190	38640
750	OSKAR		2.44195	.13502	3.947	69.632	71.253	57.193	38000
751	FAINA		2.55046	.15504	15.565	78.628	301.731	253.910	38000
752	SULAMITIS		2.46334	.07127	5.947	84.852	24.487	247.743	36000
753	TIFLIS		2.32946	.22046	10.114	61.541	201.712	226.781	26000
754	MALABAR		2.98558	.05370	24.532	180.296	301.097	159.516	32200
755	QUINTILLA		3.16182	.15662	3.233	177.031	46.698	220.832	38000
756	LILLIANA		3.20156	.14033	20.359	207.765	8.882	57.893	43800
757	PORTLANDIA		2.37290	.10977	8.175	22.401	43.062	250.592	38000
758	MANCUNIA		3.19378	.14416	5.606	105.883	321.138	248.549	43000
759	VINIFERA		2.61655	.20756	19.940	318.200	359.923	338.630	40000
760	MASSINGA		3.16231	.22193	12.761	333.245	190.063	106.907	40000
761	BRENDELIA		2.86222	.06316	2.173	23.999	295.704	136.994	40000
762	PULCOVA		3.15047	.11215	13.124	306.829	178.057	220.095	32200
763	CUPIDO		2.24033	.16655	4.081	289.645	88.098	251.198	44200
764	GEDANIA		3.18094	.11265	10.075	259.698	165.398	260.940	32200
765	MATTIACA		2.54881	.28008	5.574	327.192	70.124	263.301	24160

NUMBER	NAME	A	E	I	NODE	PERI	ANOM	DAYEP
766	MOGUNTIA	3.02242	.09317	10.094	8.242	70.819	76.766	38000
767	BONDIA	3.11379	.18910	2.424	79.770	262.581	20.466	44200
768	STRUVEANA	3.13186	.21635	16.395	39.732	10.693	287.118	36000
769	TATJANA	3.19585	.15516	7.486	40.392	244.280	320.999	36000
770	BALI	2.22059	.15126	4.394	44.448	17.090	160.176	40000
771	LIBERA	2.65047	.24776	14.930	217.916	227.338	90.580	44200
772	TANETE	2.99966	.09896	28.835	64.031	142.494	31.072	32200
773	IRMINTRAUD	2.85678	.08223	16.688	322.629	331.766	133.815	36000
774	ARMOR	3.04301	.17119	5.566	251.037	26.347	201.694	36000
775	LUMIERE	3.01017	.07410	9.296	298.134	165.862	157.687	40000
776	BERBERICIA	2.93427	.16161	18.231	79.785	306.242	48.081	43800
777	GUTEMBERGA	3.21913	.11258	13.085	286.507	236.749	138.177	38000
778	THEOBALDA	3.18046	.26776	13.336	324.590	122.805	259.024	38000
779	NINA	2.66495	.22427	14.595	283.976	47.762	133.515	36000
780	ARMENIA	3.12001	.08355	19.075	144.836	210.713	110.896	44200
781	KARTVELIA	3.23456	.08741	19.134	138.837	154.567	312.253	40000
782	MONTEFIORE	2.17954	.03920	5.263	80.137	81.598	213.682	42000
783	NORA	2.34350	.22837	9.317	141.934	153.750	59.451	43000
784	PICKERINGIA	3.03992	.23769	12.460	16.522	233.665	279.732	36000
785	ZWETANA	2.57568	.20292	12.723	72.338	129.456	150.867	36000
786	BREDICHINA	3.16241	.17128	14.566	89.652	137.552	190.190	38000
787	MOSKVA	2.53921	.12707	14.874	184.070	123.744	53.189	40000
788	HOHENSTEINA	3.12580	.13152	14.378	178.555	39.304	286.824	42000
789	LENA	2.68480	.14787	10.806	232.711	41.676	86.700	40000
790	PRETORIA	3.40269	.15981	20.544	251.789	43.518	157.528	47000
791	ANI	3.13609	.18404	16.292	130.150	201.343	241.767	36000
792	METCALFIA	2.62198	.13163	8.629	265.333	225.571	86.929	38000
793	ARIZONA	2.79594	.12504	15.803	36.331	307.863	54.763	32200
794	IRENAEA	3.15224	.28356	5.402	160.298	132.236	281.806	44200
795	FINI	2.75042	.09887	19.050	17.307	188.324	62.587	40000
796	SARITA	2.63364	.32264	19.006	33.089	328.407	307.325	42000
797	MONTANA	2.53793	.05504	4.464	238.555	351.240	128.856	38000
798	RUTH	3.01452	.04036	9.208	214.745	40.532	237.362	38000
799	GUDULA	2.54141	.02567	5.244	184.383	255.343	208.006	36000
800	KRESSMANNIA	2.19255	.20227	4.266	324.954	346.693	50.599	40120
801	HELWERTHIA	2.60553	.07421	14.100	185.832	335.027	53.775	43800
802	EPYAXA	2.19623	.07917	5.208	7.578	114.989	195.816	40080
803	PICKA	3.20419	.04943	8.624	251.584	82.150	284.326	40000
804	HISPANIA	2.83853	.14076	15.362	347.573	341.897	316.712	43800
805	HORMUTHIA	3.20334	.17818	15.703	166.153	137.138	273.652	41760
806	GYLDENIA	3.21170	.08548	14.114	45.380	92.647	143.880	40000
807	CERASKIA	3.01588	.06751	11.309	132.388	336.498	120.910	40000
808	MERXIA	2.74586	.12694	4.708	181.402	272.246	329.066	35800
809	LUNDIA	2.28272	.19231	7.144	154.308	195.205	89.509	40000
810	ATOSSA	2.17902	.18069	2.599	152.667	194.262	.891	20760
811	NAUHEIMA	2.89658	.07162	3.135	130.802	180.772	275.215	40000
812	ADELE	2.65866	.16675	13.334	7.264	351.249	50.728	40000
813	BAUMEIA	2.22303	.02583	6.306	51.766	315.799	227.612	36000
814	TAURIS	3.15909	.30567	21.807	88.341	298.232	87.456	43800
815	COPPELIA	2.65803	.07496	13.870	57.113	57.653	336.977	38160
816	JULIANA	3.00221	.10746	14.343	128.170	22.120	332.401	36000
817	ANNIKA	2.58839	.18189	11.319	125.755	283.653	269.788	34000
818	KAPTEYNIA	3.16848	.09480	15.647	70.676	302.439	206.647	44200
819	BARNARDIANA	2.19752	.14118	4.899	333.099	305.325	2.182	37960
820	ADRIANA	3.12777	.06216	5.947	118.761	195.400	32.674	36000
821	FANNY	2.77772	.20597	5.366	209.648	31.635	237.380	44200
822	LALAGE	2.25551	.15503	.715	209.830	246.205	357.427	44200
823	SISIGAMBIS	2.22115	.09036	3.642	254.728	217.943	147.999	44200
824	ANASTASIA	2.79241	.13534	8.084	141.973	137.798	211.764	36000
825	TANINA	2.22581	.07477	3.401	101.185	109.743	2.453	38000
826	HENRIKA	2.71354	.20374	7.112	230.808	33.760	304.231	34000
827	WOLFIANA	2.27429	.15673	3.407	173.019	194.105	132.144	24120
828	LINDEMANNIA	3.18598	.05411	1.154	2.662	307.449	98.235	36000
829	ACADEMIA	2.57922	.09802	8.318	352.389	40.102	12.953	38000
830	PETROPOLITANA	3.20039	.09428	3.824	342.663	65.409	49.168	32200
831	STATEIRA	2.21251	.14564	4.835	177.717	224.265	278.384	43800
832	KARIN	2.86371	.08041	.992	255.710	115.645	128.544	34000
833	MONICA	3.00907	.12298	9.808	353.413	34.838	300.946	40000
834	BURNHAMIA	3.15402	.22460	3.954	183.377	90.000	122.409	40000
835	OLIVIA	3.20504	.11826	3.689	310.356	55.640	2.224	40000
836	JOLE	2.18985	.17605	4.839	199.501	179.164	315.823	41160
837	SCHWARZSCHILDA	2.29798	.04046	6.730	199.855	171.574	78.704	38000
838	SERAPHINA	2.87683	.13466	10.395	240.601	114.782	103.110	36000
839	VALBORG	2.61412	.15245	12.573	338.069	337.799	287.266	43800
840	ZENOBIA	3.12799	.10166	9.955	273.833	.555	211.420	34000
841	ARABELLA	2.25497	.06954	3.793	354.549	119.433	349.922	40000
842	KERSTIN	3.22124	.14357	14.626	5.931	351.482	320.128	41880
843	NICOLAIA	2.27913	.20957	7.997	4.542	315.058	23.746	21120
844	LEONTINA	3.18889	.09703	8.874	349.417	342.146	98.629	36000
845	NAEMA	2.93811	.06909	12.642	43.505	290.749	91.696	36000
846	LIPPERTA	3.12796	.18410	.263	262.407	131.754	77.540	43800
847	AGNIA	2.78114	.09642	2.479	270.997	127.663	160.726	43800
848	INNA	3.11027	.16499	1.038	207.755	128.093	229.693	40000
849	ARA	3.17010	.17942	19.552	228.925	59.325	179.486	43800
850	ALTONA	2.99902	.12918	15.479	121.642	129.534	299.124	38000

NUMBER	NAME	A	E	I	NODE	PERI	ANOM	DAYEP
851	ZEISSIA	2.22798	.09040	2.393	140.792	6.532	91.809	44200
852	WLADILENA	2.36314	.27394	23.019	27.269	281.729	120.125	37600
853	NANSENIA	2.31223	.10672	9.217	182.650	58.983	93.766	41960
854	FROSTIA	2.36802	.17363	0.088	190.723	82.238	132.864	37640
855	NEWCOMBIA	2.36236	.17808	10.905	17.004	232.787	188.756	41680
856	BACKLUNDA	2.43633	.11475	14.358	125.463	72.052	297.070	36000
857	GLASENAPPIA	2.19082	.08888	5.304	82.695	238.127	22.361	37960
858	EL DJEZAIR	2.80814	.10590	8.916	67.243	174.134	325.984	38000
859	BOUZAREAH	3.20560	.13412	13.778	37.108	1.098	90.328	32200
860	URSINA	2.79577	.10892	13.328	309.637	19.817	16.091	36000
861	AIDA	3.15011	.08410	8.017	115.424	190.393	269.691	36000
862	FRANZIA	2.80256	.08509	13.913	300.228	118.251	267.455	36000
863	BENKOELA	3.19614	.04949	25.477	116.736	102.966	237.894	42000
864	AASE	2.20790	.19096	5.438	163.218	192.341	77.837	23200
865	ZUBAIDA	2.41611	.19744	13.310	176.864	300.608	249.419	40000
866	FATME	3.12360	.05580	8.647	91.437	271.374	238.460	36000
867	KOVACIA	3.07255	.12698	5.999	47.367	67.695	221.041	38000
868	LOVA	2.70407	.14849	5.825	115.860	284.915	333.996	40000
869	MELLENA	2.69586	.21561	7.819	155.377	104.980	121.254	43000
870	MANTO	2.32173	.26507	6.198	120.939	195.162	347.375	24120
871	AMNERIS	2.22226	.11899	4.251	157.714	65.380	308.048	41760
872	HOLDA	2.73221	.07895	7.361	194.826	18.999	123.928	40000
873	MECHTHILD	2.62704	.14910	5.261	150.218	107.812	230.844	38000
874	ROTRAUT	3.16402	.05965	11.065	191.102	359.960	85.319	40000
875	NYMPHE	2.55354	.14825	14.620	195.992	116.824	113.397	40000
876	SCOTT	3.01343	.10713	11.325	151.257	208.374	176.343	38000
877	WALKURE	2.48584	.15871	4.277	116.149	276.361	200.527	36000
878	MILDRED	2.36333	.23107	2.021	172.697	187.068	8.997	21120
879	RICARDA	2.53100	.15333	13.717	270.109	95.096	180.253	40000
880	HERBA	3.00933	.31692	15.080	264.307	98.093	191.966	34000
881	ATHENE	2.61306	.20777	14.245	277.028	40.330	173.062	43800
882	SWETLANA	3.13290	.26121	5.987	259.118	118.587	37.229	38000
883	MATTERANIA	2.23839	.19841	4.730	285.989	40.403	78.532	24120
884	PRIAMUS	5.16556	.12013	8.910	301.042	332.233	128.825	43800
885	ULRIKE	3.10092	.17755	3.283	149.103	201.898	16.504	39400
886	WASHINGTONIA	3.14850	.28624	16.692	61.048	295.876	224.442	30840
887	ALINDA	2.49963	.55397	9.192	110.428	349.262	76.854	43800
888	PARYSATIS	2.70966	.19312	13.844	124.150	296.921	54.084	38000
889	ERYNIA	2.44566	.20728	8.074	132.461	276.928	132.662	44200
890	WALTRAUT	3.02066	.05945	10.868	160.747	84.643	188.276	44200
891	GUNHILD	2.85959	.02988	13.535	106.099	299.354	178.881	35800
892	SEELIGERIA	3.23488	.08196	21.349	175.684	288.360	317.601	42000
893	LEOPOLDINA	3.05062	.14853	17.037	145.460	223.472	15.194	32200
894	ERDA	3.11268	.12154	12.700	190.666	118.207	339.272	40000
895	HELIO	3.20351	.14825	26.072	264.286	183.547	20.043	43800
896	SPHINX	2.28528	.16326	8.178	254.084	.544	302.429	36600
897	LYSISTRATA	2.54327	.08870	14.232	257.813	21.082	5.626	38000
898	HILDEGARD	2.72536	.37351	10.199	242.300	47.627	7.639	41560
899	JOKASTE	2.90947	.19851	12.395	253.157	125.643	220.882	34000
900	ROSALINDE	2.47302	.16054	11.534	182.768	118.172	157.177	38000
901	BRUNSIA	2.22449	.22058	3.446	265.199	67.023	139.941	38040
902	PROBITAS	2.44680	.17920	6.391	353.155	26.295	234.299	34000
903	NEALLEY	3.24661	.01974	11.666	159.554	248.439	46.336	43800
904	ROCKEFELLIA	2.99485	.08586	15.175	198.065	250.972	241.612	44200
905	UNIVERSITAS	2.21577	.15311	5.327	36.922	342.124	22.883	40000
906	REPSOLDA	2.89352	.08462	11.799	40.355	291.494	81.632	40000
907	RHODA	2.80036	.16048	19.601	43.254	86.428	147.915	40000
908	BUDA	2.47413	.14472	13.355	85.317	24.905	74.666	38000
909	ULLA	3.54533	.09092	18.752	146.968	229.031	254.458	40000
910	ANNELIESE	2.93192	.15076	9.263	50.489	204.398	97.999	34000
911	AGAMEMNON	5.18631	.06723	21.858	337.277	81.697	128.285	43800
912	MARITIMA	3.12537	.18668	18.298	34.491	89.547	25.347	36000
913	OTILA	2.19728	.17076	5.811	95.009	187.152	320.804	22080
914	PALISNA	2.45446	.21371	25.370	255.711	46.930	15.357	37680
915	COSETTE	2.22807	.13922	5.556	9.164	38.649	110.617	38000
916	AMERICA	2.36496	.23599	11.123	329.713	40.522	334.478	38000
917	LYKA	2.38115	.20128	5.140	343.357	358.646	343.060	38120
918	ITHA	2.85364	.18963	12.103	330.787	14.034	327.724	38000
919	ILSEBILL	2.77146	.08463	8.135	230.203	152.455	80.283	34000
920	ROGERIA	2.62150	.10636	11.584	192.705	268.377	28.505	41360
921	JOVITA	3.19287	.15855	16.393	205.370	65.800	290.113	44200
922	SCHLUTIA	2.68785	.19413	7.267	205.705	124.739	129.642	34000
923	HERLUGA	2.61476	.19502	14.476	197.739	199.784	347.879	33080
924	TONI	2.93546	.15764	8.991	150.986	216.758	152.913	32200
925	ALPHONSINA	2.70060	.08067	21.075	299.452	200.554	27.082	42000
926	IMHILDE	2.97890	.18562	16.376	49.640	170.146	16.839	32200
927	RATISBONA	3.22668	.09508	14.498	8.527	139.386	138.565	40000
928	HILDRUN	3.14235	.14175	17.605	129.716	25.881	102.911	39280
929	ALGUNDE	2.23871	.11259	3.908	231.130	21.935	335.231	42160
930	WESTPHALIA	2.43012	.14525	15.322	340.728	329.131	332.971	38000
931	WHITTEMORA	3.16074	.24402	11.283	113.200	305.342	312.724	36000
932	HOOVERIA	2.41987	.09082	8.116	14.915	47.927	313.932	43800
933	SUSI	2.36986	.16388	5.532	141.371	11.027	128.538	33400
934	THURINGIA	2.74763	.21823	14.113	325.900	62.830	339.340	36000
935	CLIVIA	2.21892	.14608	4.043	346.866	56.237	323.052	22600

NUMBER	NAME		A	E	I	NODE	PERI	ANOM	DAYEP
936	KUNIGUNDE		3.15093	.15942	2.384	62.237	255.220	234.424	40000
937	BETHGEA		2.23143	.21767	3.697	243.898	70.776	342.406	33400
938	CHLOSINDE		3.17417	.17112	2.659	119.120	222.267	188.229	38000
939	ISBERGA		2.24735	.17655	2.585	327.233	4.974	214.193	38000
940	KORDULA		3.39986	.14259	6.283	70.010	265.776	354.028	36000
941	MURRAY		2.78426	.19743	6.643	52.334	334.179	6.037	43000
942	ROMILDA		3.16291	.16658	10.586	71.416	324.493	158.696	38000
943	BEGONIA		3.11958	.21206	12.129	114.538	357.503	170.983	40000
944	HIDALGO		5.86079	.65646	42.404	20.966	57.412	346.644	44200
945	BARCELONA		2.63959	.16226	32.882	318.060	160.694	194.090	37600
946	POESIA		3.12353	.13917	1.453	69.820	31.674	252.732	36000
947	MONTEROSA		2.75109	.25146	6.711	48.612	336.322	8.859	34000
948	JUCUNDA		3.03211	.16206	8.666	357.418	159.826	43.686	44200
949	HEL		2.99269	.20250	10.746	322.144	246.669	313.358	36000
950	AHRENSA		2.37101	.16029	23.473	181.860	346.738	332.080	25280
951	GASPRA		2.20977	.17370	4.096	252.999	128.622	353.124	38000
952	CAIA		2.98564	.24918	10.054	18.801	352.886	5.465	40000
953	PAINLEVA		2.78814	.18818	8.674	36.438	259.810	341.528	40000
954	LI		3.13116	.16907	1.124	163.997	143.736	177.725	36000
955	ALSTEDE		2.59410	.28835	10.669	352.461	279.646	347.853	38000
956	ELISA		2.29809	.20410	5.936	193.090	123.406	1.348	24160
957	CAMELIA		2.91921	.08246	14.768	233.029	224.683	9.186	38000
958	ASPLINDA		3.96289	.18841	5.648	343.288	95.640	110.482	44200
959	ARNE		3.20094	.19980	4.447	59.889	322.699	61.083	40000
960	BIRGIT		2.24836	.16582	3.016	249.789	86.066	28.946	22960
961	GUNNIE		2.69219	.09250	10.995	26.512	283.258	123.268	44200
962	ASLOG		2.90483	.10166	2.581	146.111	221.321	25.719	23000
963	IDUBERGA		2.24783	.13750	7.996	62.594	3.527	327.638	22960
964	SUBAMARA		3.04769	.12031	9.083	31.083	8.706	248.362	38000
965	ANGELICA		3.14533	.28750	21.485	40.996	48.425	66.803	44200
966	MUSCHI		2.71766	.13096	14.419	72.544	177.085	140.968	37600
967	HELIONAPE		2.22497	.16912	5.420	82.165	230.885	94.940	40000
968	FETUNIA		2.86679	.13701	11.558	208.758	297.527	249.625	44200
969	LEOCADIA		2.46218	.20701	2.289	290.038	89.901	359.556	22060
970	PRIMULA		2.56251	.26505	5.058	311.643	93.192	11.494	38000
971	ALSATIA		2.64161	.16113	13.791	83.476	4.553	273.317	40000
972	COHNIA		3.06112	.23151	8.357	282.892	91.604	274.270	34000
973	ARALIA		3.22451	.09602	15.756	348.272	94.741	32.394	42000
974	LIOBA		2.53349	.11280	5.469	86.337	301.253	285.507	40000
975	PERSEVERANTIA		2.83448	.03539	2.572	39.082	55.097	22.368	33280
976	BENJAMINA		3.19145	.11520	7.565	245.512	298.439	193.591	43000
977	PHILIPPA		3.11881	.02445	15.178	76.441	65.707	244.145	32200
978	AIDAMINA		3.19053	.23895	21.666	216.207	135.063	286.934	44200
979	ILSEWA		3.14943	.14842	10.041	231.498	106.452	52.084	44200
980	ANACOSTIA		2.73906	.20194	15.920	285.703	69.373	201.106	43800
981	MARTINA		3.10264	.19530	2.077	46.638	294.589	15.939	39400
982	FRANKLINA		3.06916	.23277	13.589	299.876	348.830	171.393	32200
983	GUNILA		3.16882	.08220	14.797	251.471	355.357	176.787	32200
984	GRETIA		2.80425	.19701	9.143	315.057	52.568	280.182	35000
985	ROSINA		2.30015	.27708	4.070	290.582	58.096	213.606	38040
986	AMELIA		3.14530	.18984	14.832	93.749	259.937	97.892	40000
987	WALLIA		3.13480	.23630	8.997	324.202	8.956	173.563	32200
988	APPELLA		3.16175	.21959	1.610	41.924	330.773	71.892	40000
989	SCHWASSMANNIA		2.65914	.25227	14.651	243.726	163.182	66.984	30000
990	YERKES		2.66873	.21604	8.804	353.874	7.782	70.416	44200
991	MCDONALDA		3.13966	.15687	2.099	63.989	244.553	138.233	40000
992	SWASEY		3.02818	.08539	10.818	212.566	342.275	78.900	40000
993	MOULTONA		2.86063	.04960	1.763	184.276	246.140	250.977	40400
994	OTTHILD		2.53081	.11367	15.334	2.496	339.764	153.287	38000
995	STERNBERGA		2.61510	.16786	13.055	221.560	121.453	65.055	44200
996	HILARITAS		3.10206	.12565	.663	348.338	144.231	188.817	44200
997	PRISKA		2.66824	.18322	10.506	247.012	51.113	329.106	44200
998	BODEA		3.12828	.20228	15.570	301.996	67.534	336.521	43800
999	ZACHIA		2.61170	.21726	9.720	215.179	126.215	99.215	38000
1000	PIAZZIA		3.20444	.23423	20.582	324.964	278.028	9.368	40000
1001	GAUSSIA		3.18424	.15614	9.375	259.866	142.623	272.303	38000
1002	OLBERSIA		2.78741	.15285	10.755	344.144	353.053	46.748	35800
1003	LILOFEE		3.16089	.14104	1.819	139.740	310.871	293.239	38000
1004	BELOPOLSKYA		3.38933	.10927	2.956	154.334	225.547	79.383	38000
1005	ARAGO		3.16641	.12300	19.167	349.867	49.957	320.237	38000
1006	LAGRANGEA		3.15300	.34995	11.033	299.254	77.370	5.903	42000
1007	PAWLOWIA		2.70809	.10927	2.549	307.931	74.493	118.734	34000
1008	LAPAZ		3.09089	.07658	8.977	21.046	12.022	44.764	34000
1009	SIRENE	LOST	2.62774	.45380	15.752	229.968	183.390	1.627	23760
1010	MARLENE		2.93158	.10201	3.906	99.035	277.259	234.732	34000
1011	LAODAMIA		2.39357	.34964	5.469	132.595	352.169	29.183	29360
1012	SAREMA		2.48204	.13336	4.049	72.858	22.739	15.788	43800
1013	TOMBECKA		2.68309	.21052	11.885	27.956	97.290	313.105	33280
1014	SEMPHYRA		2.80501	.19856	2.283	252.555	230.097	96.887	38000
1015	CHRISTA		3.20597	.08960	9.416	121.426	263.445	14.768	40000
1016	ANITRA		2.21926	.12801	6.042	8.588	52.606	134.875	42160
1017	JACQUELINE		2.60646	.07390	7.942	118.659	67.333	66.458	44200
1018	ARNOLDA		2.53583	.25313	7.687	359.505	340.971	120.025	34000
1019	STRACKEA		1.91144	.07153	26.978	143.932	121.603	135.559	43800
1020	ARCADIA	LOST	2.78630	.00880	4.277	180.993	52.729	304.449	23880

NUMBER	NAME		A	E	I	NODE	PERI	ANOM	DAYEP
1021	FLAMMARIO		2.73748	.28642	15.813	115.985	284.947	297.736	38000
1022	OLYMPIADA		2.80897	.17282	21.071	112.039	125.302	92.223	38000
1023	THOMANA		3.16280	.11129	10.065	194.397	202.703	324.797	40800
1024	HALE		2.87242	.22383	16.005	59.356	306.218	157.023	31520
1025	RIEMA		1.97921	.03930	26.869	162.937	348.851	198.449	42000
1026	INGRID	LOST	2.25043	.17995	5.390	104.534	211.650	8.672	23640
1027	AESCULAPIA		3.15388	.12806	1.258	29.089	138.057	240.078	44200
1028	LYDINA		3.40920	.11684	9.490	64.797	6.735	108.709	36000
1029	LA PLATA		2.88921	.02695	2.448	30.212	143.582	313.991	36000
1030	VITJA		3.12365	.11589	14.746	188.337	357.194	43.568	40000
1031	ARTICA		3.04720	.05814	17.620	219.024	306.593	50.867	39280
1032	PAFURI		3.13735	.12843	9.472	76.865	183.982	340.047	36000
1033	SIMONA		2.99935	.11917	10.656	188.987	215.651	176.820	44200
1034	MOZARTIA		2.29236	.26337	3.994	305.032	16.777	22.391	24040
1035	AMATA		3.13712	.20238	18.106	1.889	326.267	329.110	38000
1036	GANYMED		2.66575	.53595	26.415	215.757	131.586	65.829	43400
1037	DAVIDWEILLA	LOST	2.18057	.18074	5.379	199.510	177.323	14.951	24080
1038	TUCKIA		3.93568	.23843	9.254	58.251	307.957	321.410	40640
1039	SONNEBERGA		2.67998	.06106	4.547	221.756	326.294	198.829	40000
1040	KLUMPKEA		3.11843	.18592	16.661	279.990	160.342	8.842	44200
1041	ASTA		3.06977	.14821	13.933	60.201	341.839	150.527	40000
1042	AMAZONE		3.20787	.12874	20.830	53.006	299.564	227.730	24280
1043	BEATE		3.09338	.04180	8.929	159.219	172.727	182.863	43800
1044	TEUTONIA		2.57630	.14038	4.266	59.678	226.538	174.109	43000
1045	MICHELA		2.35909	.15806	.257	269.942	163.595	156.791	34000
1046	EDWIN		2.98162	.06737	7.926	10.916	50.235	163.615	39960
1047	GEISHA		2.24146	.19230	5.667	78.164	299.059	142.384	38000
1048	FEODOSIA		2.72893	.18352	15.854	52.904	181.833	275.007	36000
1049	GOTHO		3.09156	.13805	15.148	343.245	36.603	283.698	40000
1050	META		2.62490	.17628	12.526	342.941	64.595	322.458	24400
1051	MEROPE		3.22211	.10243	23.241	181.214	134.647	199.830	38000
1052	BELGICA		2.23649	.14315	4.697	99.267	297.050	72.419	44200
1053	VIGDIS		2.61385	.09757	8.353	18.233	40.427	308.377	25880
1054	FORSYTIA		2.92107	.13855	10.856	85.857	292.923	322.005	44200
1055	TYNKA		2.19790	.20845	5.272	146.851	175.595	87.965	40000
1056	AZALEA		2.23021	.17734	5.430	104.025	211.263	111.249	33280
1057	WANDA		2.89531	.24581	3.514	259.352	111.522	178.941	38000
1058	GRUBBA		2.19642	.18709	3.687	221.744	93.129	154.107	38000
1059	MUSSORGSKIA		2.64171	.18698	10.091	200.484	86.997	2.183	40000
1060	MAGNOLIA		2.23733	.20226	5.930	221.369	83.115	65.080	38000
1061	PAEONIA		3.11521	.22417	2.496	91.319	300.447	210.202	43800
1062	LJUBA		3.00635	.06535	5.617	341.856	99.108	20.226	40000
1063	AQUILEGIA		2.31388	.03931	5.980	95.062	106.003	109.382	40800
1064	AETHUSA		2.54709	.16952	9.428	280.683	19.956	351.733	38000
1065	AMUNDSENIA		2.36056	.29726	8.366	330.643	351.421	5.766	24760
1066	LOBELIA		2.40227	.20777	4.829	344.989	15.767	64.379	40000
1067	LUNARIA		2.86830	.19449	10.526	290.007	113.059	177.469	40000
1068	NOFRETETE		2.90723	.09660	5.499	319.561	264.008	201.525	36000
1069	PLANCKIA		3.13880	.09497	13.512	142.279	33.427	139.387	44200
1070	TUNICA		3.22070	.09371	17.081	165.388	172.007	5.184	41560
1071	BRITA		2.80135	.11077	5.393	52.614	25.158	237.852	40000
1072	MALVA		3.17821	.23227	8.004	36.947	28.680	91.058	38000
1073	GELLIVARA		3.16462	.20633	1.618	39.244	292.016	14.885	40200
1074	BELJAWSKYA		3.16299	.16930	.851	38.768	13.123	328.442	40000
1075	HELINA		3.01246	.11430	11.533	100.983	249.811	5.931	40000
1076	VIOLA		2.47689	.14037	3.307	143.793	302.002	50.893	38000
1077	CAMPANULA		2.39231	.19764	5.418	346.655	11.827	6.227	24760
1078	MENTHA		2.26972	.13857	7.371	93.756	42.941	64.838	34000
1079	MIMOSA		2.87244	.04885	1.187	330.205	104.373	168.515	38000
1080	ORCHIS		2.42095	.25483	4.605	1.790	56.366	265.513	44200
1081	RESEDA		3.09166	.15534	4.234	30.666	7.608	141.484	40000
1082	PIROLA		3.13521	.16677	1.816	148.037	183.187	176.012	38000
1083	SALVIA		2.32810	.18354	5.146	80.803	30.992	78.539	33280
1084	TAMARIWA		2.68844	.13084	3.889	186.746	108.911	278.360	44200
1085	AMARYLLIS		3.17622	.06664	6.649	140.074	128.188	152.848	40000
1086	NATA		3.16378	.04135	8.365	313.541	176.815	314.214	38000
1087	ARABIS		3.01924	.09278	10.081	30.438	23.493	295.786	42280
1088	MITAKA		2.20106	.19596	7.650	54.321	318.651	275.228	37960
1089	TAMA		2.21331	.12805	3.732	71.244	353.649	216.571	37960
1090	SUMIDA		2.35907	.22109	21.508	147.683	337.214	152.133	43000
1091	SPIRAEA		3.41625	.07486	1.183	80.242	8.432	44.631	25200
1092	LILIUM		2.90094	.08342	5.389	308.187	313.871	95.170	35800
1093	FREDA		3.15358	.25410	25.253	56.010	250.194	228.487	40000
1094	SIBERIA		2.54822	.13162	13.918	149.658	306.412	32.901	24560
1095	TULIPA		3.02521	.02054	10.003	179.245	1.714	54.509	30520
1096	REUNERTA		2.60045	.19295	9.500	81.713	246.049	253.212	31200
1097	VICIA		2.63853	.29701	1.518	133.573	175.470	63.752	33560
1098	HAKONE		2.68826	.11664	13.402	329.318	79.070	50.267	34000
1099	FIGNERIA		3.17076	.28348	11.778	23.272	338.782	44.746	44200
1100	ARNICA		2.89762	.06765	1.041	305.559	21.506	294.673	35800
1101	CLEMATIS		3.24136	.06772	21.281	201.905	125.005	303.704	42000
1102	PEPITA		3.07261	.11090	15.797	216.696	117.492	202.966	40000
1103	SEQUOIA		1.93382	.09463	17.899	267.247	77.399	322.417	42000
1104	SYRINGA		2.63098	.34250	6.438	129.347	275.241	4.069	38000
1105	FRAGARIA		3.01256	.09942	10.945	117.235	224.496	262.625	38000

NUMBER	NAME	A	E	I	NODE	PERI	ANOM	DAYEP
1106	CYDONIA	2.59899	.12128	13.080	328.118	229.667	107.625	40000
1107	LICTORIA	3.18770	.12031	7.073	110.752	.263	35.605	38000
1108	DEMETER	2.42620	.25974	24.895	234.215	77.484	74.148	40000
1109	TATA	3.20522	.12559	4.185	268.955	348.786	217.163	42000
1110	JAROSLAWA	2.21816	.24135	5.855	241.726	77.451	27.260	41240
1111	REINMUTHIA	2.99585	.09538	3.883	132.585	231.705	129.125	38000
1112	POLONIA	3.01880	.10651	9.000	303.953	80.474	132.621	32200
1113	KATJA	3.11739	.13411	13.245	325.326	117.547	350.832	36000
1114	LORRAINE	3.08934	.08245	10.741	195.498	208.245	151.389	44200
1115	SABAUDA	3.09939	.17440	15.358	72.536	53.975	51.152	36000
1116	CATRIONIA	2.92149	.22971	16.574	357.355	81.007	101.891	33000
1117	REGINITA	2.24762	.19738	4.332	146.990	149.733	157.888	38000
1118	HANSKYA	3.20406	.06646	14.005	319.181	342.403	89.882	40000
1119	EUBOEA	2.61247	.15435	7.854	57.244	229.707	37.790	41920
1120	CANNONIA	2.21593	.15580	4.046	158.458	218.867	243.022	36000
1121	NATASCHA	2.54678	.16138	6.158	358.451	48.590	50.727	33280
1122	NEITH	2.60718	.25708	4.732	63.166	327.229	305.259	43800
1123	SHAPLEYA	2.22542	.15628	6.418	79.690	316.483	89.350	38000
1124	STROOBANTIA	2.92557	.03517	7.795	22.283	264.349	72.318	41960
1125	CHINA	3.14650	.20251	3.032	96.970	11.337	211.202	43800
1126	OTERO	2.27172	.14686	6.505	.938	135.222	296.262	38000
1127	MIMI	2.59296	.26415	14.776	128.559	281.603	198.900	40000
1128	ASTRID	2.78765	.04632	1.023	59.021	237.236	33.727	40000
1129	NEUIMINA	3.02322	.07987	8.605	269.906	134.880	58.318	38000
1130	NEUJMINA	2.22898	.19731	2.162	216.002	112.988	242.134	40000
1131	PORZIA	2.22869	.28552	3.236	100.813	246.634	10.544	25880
1132	HOLLANDIA	2.68414	.27586	7.252	30.754	267.050	270.580	36600
1133	LUGDUNA	2.18604	.18723	5.374	58.053	305.730	89.378	38000
1134	KEPLER	2.68469	.46640	15.030	6.365	330.498	65.359	43800
1135	COLCHIS	2.66593	.11398	4.553	350.983	2.878	239.832	38000
1136	MERCEDES	2.56412	.25718	8.890	209.675	145.919	55.513	38000
1137	RAISSA	2.42322	.09717	4.328	78.366	276.222	309.832	38000
1138	AIIILA	3.14633	.08081	13.992	284.260	94.671	22.110	34000
1139	ATAMI	1.94727	.25526	13.099	213.191	205.585	44.473	38160
1140	CRIMEA	2.77282	.11337	14.114	72.067	309.215	300.088	42240
1141	BOHMIA	2.27047	.16485	4.275	105.277	274.992	356.966	42000
1142	AETOLIA	3.17114	.10281	2.097	139.287	103.416	212.057	36000
1143	ODYSSEUS	5.22626	.09224	3.142	220.649	234.449	292.334	41960
1144	ODA	3.75542	.09270	9.693	158.420	212.090	30.367	41240
1145	ROBELMONTE	2.42442	.11676	6.233	346.993	264.994	248.935	38000
1146	BIARMIA	3.04888	.24967	17.151	214.665	61.978	72.962	36000
1147	STAVROPOLIS	2.27035	.23199	3.880	265.063	14.689	128.515	40000
1148	RARAHU	3.02248	.10773	10.839	145.414	175.732	199.436	44200
1149	VOLGA	2.90090	.09418	11.728	261.870	114.334	186.295	36000
1150	ACHAIA	2.19079	.20415	2.379	206.509	138.256	91.673	38000
1151	ITHACA	2.40862	.27469	6.557	225.201	122.142	168.201	44200
1152	PAWONA	2.42747	.04271	5.068	331.863	219.312	342.103	40000
1153	WALLENBERGIA	2.19578	.15992	3.335	280.455	27.857	300.398	38000
1154	ASTRONOMIA	3.39813	.04027	4.543	83.838	181.573	131.521	38000
1155	AENNA	2.46325	.15406	6.631	38.860	191.605	89.117	40000
1156	KIRA	2.23676	.04623	1.400	90.713	354.093	293.087	42000
1157	ARABIA	3.19408	.13126	9.550	336.001	318.452	320.458	44200
1158	LUDA	2.56516	.11085	14.901	344.551	56.283	102.833	40000
1159	GRANADA	2.37942	.05823	13.033	347.771	312.357	335.069	36320
1160	ILLYRIA	2.56187	.11753	14.987	3.416	4.075	156.636	40000
1161	THESSALIA	3.16634	.10678	9.382	72.921	299.107	306.840	42000
1162	LARISSA	3.93465	.11243	1.901	39.523	216.863	2.565	44200
1163	SAGA	3.21118	.07056	8.987	127.995	206.682	28.979	40000
1164	KOBOLDA	2.30685	.19592	25.145	156.652	340.232	351.578	40000
1165	IMPRINETTA	3.14067	.19749	12.802	205.100	93.853	267.649	38000
1166	SAKUNTALA	2.54118	.20121	18.799	106.774	188.960	170.317	35800
1167	DUBIAGO	3.42152	.05790	5.698	225.669	45.262	88.037	38000
1168	BRANDIA	2.55062	.22272	12.689	218.604	122.526	304.279	33440
1169	ALWINE	2.31848	.15547	4.045	255.238	175.498	296.633	34000
1170	SIVA	2.32544	.29909	22.264	1.008	58.270	40.938	30400
1171	RUSTHAWELIA	3.14854	.21772	3.053	122.780	285.685	245.308	38000
1172	ANEAS	5.16102	.10249	16.714	246.797	47.649	115.748	43800
1173	ANCHISES	5.28020	.13887	6.925	283.396	38.098	105.701	43800
1174	MARMARA	3.02014	.11442	10.109	1.308	348.844	78.536	38000
1175	MARGO	3.22007	.02879	16.364	238.211	80.777	299.920	27600
1176	LUCIDOR	2.69254	.14153	6.638	272.235	155.110	26.721	44200
1177	GONNESSIA	2.35154	.01443	15.055	252.094	188.879	340.503	44200
1178	IRMELA	2.67872	.18418	6.940	169.999	356.276	263.251	42020
1179	MALLY	2.61660	.17542	8.741	7.570	232.377	318.901	26440
1180	RITA	3.98098	.17296	7.207	88.285	215.481	272.165	40000
1181	LILITH	2.66261	.19743	5.589	260.789	154.550	234.203	40000
1182	ILONA	2.25956	.11781	9.398	336.219	62.166	285.267	38000
1183	JUTTA	2.38408	.12907	2.810	14.995	204.414	103.782	40000
1184	GAEA	2.66805	.07205	11.321	355.635	308.956	31.007	43800
1185	NIKKO	2.23715	.10599	5.704	71.693	1.497	151.686	38000
1186	TURNERA	3.02099	.11390	10.791	43.482	293.477	82.698	36000
1187	AFRA	2.63949	.22243	10.735	328.092	72.456	133.014	28000
1188	GOTHLANDIA	2.19011	.18089	4.826	5.209	6.195	223.582	40000
1189	TERENTIA	2.93353	.11069	9.870	275.643	95.077	136.596	38000
1190	PELAGIA	2.43135	.13115	3.193	26.744	39.720	319.726	26280

NUMBER	NAME		A	E	I	NODE	PERI	ANOM	DAYEP
1191	ALFATERNA		2.89195	.04712	18.453	134.717	57.303	351.000	39120
1192	PRISMA		2.36495	.26175	23.835	1.530	130.007	21.158	26400
1193	AFRICA		2.64611	.12256	14.161	49.970	183.283	355.800	26480
1194	ALETTA		2.91462	.08842	10.882	292.130	242.959	51.323	26480
1195	ORANGIA		2.25743	.20087	7.178	281.097	327.136	182.869	42000
1196	SHEBA		2.65251	.17982	17.684	101.326	260.514	255.388	26480
1197	RHODESIA		2.87932	.23788	12.920	256.608	274.788	256.907	40000
1198	ATLANTIS	LOST	2.24908	.33530	2.724	261.072	81.404	5.443	26600
1199	GELDONIA		3.02389	.02752	8.746	236.253	275.545	181.575	38000
1200	IMPERATRIX		3.05761	.11518	4.605	205.078	49.044	95.589	44200
1201	STRENUA		2.69817	.03853	6.993	203.579	163.931	96.396	40000
1202	MARINA		3.95039	.19689	3.391	50.857	313.045	5.224	38000
1203	NANNA		2.88276	.25098	5.962	224.866	175.241	158.001	40000
1204	RENZIA		2.26280	.29372	1.887	7.945	311.535	23.757	26600
1205	EBELLA		2.53252	.28099	8.934	23.144	346.120	304.537	26400
1206	NUMEROWIA		2.86569	.05533	13.024	324.391	279.829	87.282	44080
1207	OSTENIA		3.01863	.09300	10.396	21.033	43.321	329.915	30320
1208	TROILUS		5.17759	.09275	33.677	47.973	293.849	11.978	43000
1209	PUMMA		3.17402	.12416	6.940	89.739	184.731	27.871	40000
1210	MOROSOVIA		3.01074	.05357	11.243	107.086	162.686	357.129	38000
1211	BRESSOLE		2.92754	.15943	12.778	130.120	209.173	107.477	36000
1212	FRANCETTE		3.95436	.18432	7.591	149.251	351.813	297.636	43800
1213	ALGERIA		3.12568	.14370	13.072	271.850	107.916	242.612	40000
1214	RICHILDE		2.70944	.11887	9.853	286.152	31.121	111.923	38000
1215	BOYER		2.57862	.13395	15.871	123.720	264.980	302.889	38240
1216	ASKANIA		2.23290	.17878	7.601	121.261	144.126	258.842	43800
1217	MAXIMILIANA		2.35255	.15471	5.146	148.326	90.764	254.255	35800
1218	ASTER		2.26335	.10956	3.168	63.823	67.847	25.435	26800
1219	BRITTA		2.21247	.12516	4.420	42.348	22.966	204.504	38000
1220	CROCUS		3.00596	.07523	11.362	114.166	330.617	63.030	26800
1221	AMOR		1.92061	.43605	11.913	171.086	25.889	27.017	38560
1222	TINA		2.79091	.24915	19.714	245.948	57.876	41.755	44200
1223	NECKAR		2.86855	.06178	2.560	40.839	13.336	147.074	40000
1224	FANTASIA		2.30393	.19916	7.869	258.053	128.049	4.402	38000
1225	ARIANE		2.23276	.07532	3.078	12.135	100.041	217.674	41160
1226	GOLIA		2.58289	.11209	9.816	17.322	136.348	211.344	41960
1227	GERANIUM		3.20208	.21233	16.247	2.752	290.794	218.971	38000
1228	SCABIOSA		2.76772	.04128	3.294	308.533	203.395	132.921	38000
1229	TILIA	LOST	3.19535	.16122	.845	200.450	165.558	199.508	34000
1230	RICEIA		2.57220	.18083	10.519	200.803	184.789	191.855	38000
1231	AURICULA		2.66789	.08684	11.517	342.662	243.847	148.231	26640
1232	CORTUSA		3.17850	.14201	10.202	262.688	323.336	315.855	40000
1233	KOBRESIA		2.55431	.05783	5.576	291.410	329.062	336.412	38000
1234	ELYNA		3.01282	.08664	8.536	304.762	87.310	12.820	42000
1235	SCHORRIA		1.91042	.15412	25.003	12.489	43.378	269.153	43800
1236	THAIS		2.43051	.24302	13.176	48.681	305.110	144.724	34000
1237	GENEVIEVE		2.61168	.07785	9.747	57.949	304.501	75.095	36000
1238	PREDAPPIA		2.66719	.14087	12.169	52.043	91.065	19.680	38000
1239	QUETELETA		2.66121	.23294	1.667	72.832	35.541	20.002	44200
1240	CENTENARIA		2.87018	.17356	10.166	324.084	22.851	42.419	38680
1241	DYSONA		3.18680	.10631	23.517	322.073	328.953	38.934	40000
1242	ZAMBESIA		2.73414	.19083	10.191	349.994	52.297	179.252	40000
1243	PAMELA		3.09617	.04517	13.256	246.224	49.047	157.895	36000
1244	DEIRA		2.34297	.09797	8.703	277.269	259.471	143.125	32400
1245	CALVINIA		2.89432	.07776	2.883	151.620	205.402	256.085	43000
1246	CHAKA		2.62118	.30651	16.125	291.149	52.551	178.112	34000
1247	MEMORIA		3.13783	.16742	1.727	162.827	130.891	201.824	38000
1248	JUGURTHA		2.72211	.01678	9.144	79.222	346.547	111.487	44200
1249	RUTHERFORDIA		2.22452	.07646	4.868	258.830	222.588	226.817	40080
1250	GALANTHUS		2.55532	.26785	15.115	291.663	217.965	221.194	40000
1251	HEDERA		2.71642	.15804	6.047	140.759	215.847	47.918	38320
1252	CELESTIA		2.69341	.20354	33.940	140.792	62.990	228.752	38000
1253	FRISIA		3.15675	.21594	1.350	39.610	358.445	99.390	43800
1254	ERFORDIA		3.13426	.03490	7.062	288.216	267.483	196.634	40000
1255	SCHILOWA		3.16044	.15556	8.447	238.922	127.330	96.429	40000
1256	NORMANNIA		3.90646	.07985	4.116	239.158	125.950	13.404	44200
1257	MORA		2.48749	.08263	3.919	213.877	11.542	.609	38000
1258	SICILIA		3.18520	.03531	7.747	300.300	38.497	147.546	40240
1259	OGYALLA		3.10595	.12674	2.390	74.950	154.351	78.594	40000
1260	WALHALLA		2.61435	.03629	8.018	305.346	17.304	170.132	27120
1261	LEGIA		3.15185	.16813	2.439	68.648	95.082	142.879	38000
1262	SNIADECKIA		3.00091	.00916	13.140	124.526	150.967	172.367	40000
1263	VARSAVIA		2.66525	.18903	29.237	158.318	287.203	5.430	38000
1264	LETABA		2.86400	.15288	25.026	235.125	27.456	318.727	36000
1265	SCHWEIKARDA		3.03115	.07274	9.514	314.586	109.074	253.669	42000
1266	TONE		3.36619	.01602	17.240	321.967	313.813	191.862	33880
1267	GEERTRUIDA		2.46334	.18743	4.782	24.847	265.204	153.338	34000
1268	LIBYA		3.93088	.10640	4.413	352.393	139.771	150.303	38000
1269	ROLLANDIA		3.91458	.09055	2.758	134.720	30.784	125.305	40000
1270	DATURA		2.23466	.20721	5.994	97.966	257.413	355.439	26040
1271	ISERGINA		3.12807	.13782	6.668	127.485	269.465	203.052	40000
1272	GEFION		2.78322	.15167	8.441	321.400	2.159	84.037	43800
1273	HELMA		2.39377	.16182	5.404	296.518	48.124	229.195	40000
1274	DELPORTIA		2.22906	.11298	4.410	327.425	242.978	214.748	27000
1275	CIMBRIA		2.67983	.16797	12.869	188.950	194.587	23.044	27000

NUMBER	NAME	A	E	I	NODE	PERI	ANOM	DAYEP
1276	UCCLIA	3.16322	.11222	23.373	114.485	335.195	284.180	43000
1277	DOLORES	2.70028	.23816	6.976	247.309	45.964	273.556	40000
1278	KENYA	2.40369	.26340	10.892	90.437	237.070	115.428	36000
1279	UGANDA	2.36996	.20912	5.723	336.158	295.149	345.936	27200
1280	BAILLAUDA	3.41198	.06269	6.471	294.776	70.393	201.812	38000
1281	JEANNE	2.55930	.20438	7.417	209.938	73.246	98.005	38000
1282	UTOPIA	3.12192	.11151	18.083	324.518	79.682	42.176	40000
1283	KOMSOMOLIA	3.21782	.19737	8.851	157.668	236.733	114.256	40000
1284	LATVIA	2.64361	.17292	10.880	302.755	114.334	288.016	43000
1285	JULIETTA	2.99335	.05168	5.692	318.090	67.973	295.651	44200
1286	BANACHIEWIC	3.02082	.09431	9.727	200.609	106.046	322.034	44200
1287	LORCIA	3.01199	.05894	9.820	202.645	266.216	117.634	43800
1288	SANTA	2.88702	.06591	7.571	299.452	52.871	340.639	38000
1289	KUTAISSI	2.85972	.06250	1.600	193.267	114.604	90.505	40000
1290	ALBERTINE	2.36651	.15404	5.588	307.329	78.089	162.939	40000
1291	PHRYNE	3.01764	.08959	9.081	216.034	112.080	210.652	36000
1292	LUCE	2.54186	.05699	2.165	271.844	236.181	289.790	38000
1293	SONJA	2.22690	.27547	5.358	236.305	98.829	215.755	43800
1294	ANTWERPIA	2.69072	.23312	8.689	81.347	311.349	135.083	36000
1295	DEFLOTTE	3.38628	.12535	2.860	185.368	286.661	211.317	38000
1296	ANDREE	2.41907	.14249	4.109	227.150	234.190	235.057	38000
1297	QUADEA	3.02643	.06994	9.002	296.238	121.638	268.730	42000
1298	NOCTURNA	3.14130	.13687	5.496	300.638	59.161	188.331	36000
1299	MERTONA	2.80131	.18755	7.886	165.667	259.274	208.955	42000
1300	MARCELLE	2.78155	.00923	9.548	82.958	334.856	221.351	40000
1301	YVONNE	2.76658	.27240	33.975	161.453	301.691	198.163	40000
1302	WERRA	3.13683	.16013	2.600	90.151	350.493	183.793	41480
1303	LUTHERA	3.20955	.12822	19.581	71.922	102.427	290.500	43800
1304	AROSA	3.21418	.10164	18.808	87.472	135.313	304.646	43800
1305	PONGOLA	3.01275	.07590	2.330	63.071	147.480	280.815	38000
1306	SCYTHIA	3.14464	.09713	14.950	275.344	137.037	273.066	26200
1307	CIMMERIA	2.25023	.09668	3.940	234.101	205.881	305.134	26240
1308	HALLERIA	2.71006	.00079	5.595	354.558	160.883	163.822	38000
1309	HYPERBORIA	3.21892	.13265	10.218	206.417	254.114	321.472	26000
1310	VILLIGERA	2.39155	.35765	21.048	357.350	87.814	2.241	44200
1311	KNOPFIA	2.42629	.04627	2.817	244.989	243.640	314.974	42000
1312	VASSAR	3.09967	.20844	21.968	129.583	261.316	81.250	38000
1313	BERNA	2.65676	.20821	12.523	298.532	97.933	329.567	40000
1314	PAULA	2.29513	.17473	5.231	264.740	142.679	51.751	34000
1315	BRONISLAWA	3.20604	.08013	7.085	236.444	38.384	81.163	40000
1316	KASAN	2.41013	.32076	23.835	238.362	146.933	67.418	27600
1317	SILVRETTA	3.16899	.25657	20.727	7.082	31.072	256.206	40000
1318	NERINA	2.30796	.20245	24.652	358.025	195.556	270.066	40000
1319	DISA	2.98200	.20818	2.733	258.825	310.357	156.352	30280
1320	IMPALA	2.98405	.23248	19.933	72.131	203.897	49.302	43000
1321	MAJUBA	2.94808	.16432	9.510	318.247	342.795	219.450	40000
1322	COPPERNICUS	2.42577	.22947	23.301	252.855	28.807	53.379	43000
1323	TUGELA	3.19077	.18433	18.735	46.012	126.593	46.290	40000
1324	KNYSNA	2.18469	.16331	4.514	304.576	328.151	352.648	27600
1325	INANDA	2.53859	.26159	7.459	14.071	335.779	104.676	40000
1326	LOSAKA	2.66798	.22345	16.000	101.855	277.703	135.048	38000
1327	NAMAQUA	2.78105	.15938	5.830	57.900	268.216	17.455	43000
1328	DEVOTA	3.50575	.13542	5.722	224.751	168.019	303.598	36000
1329	ELIANE	2.61596	.17508	14.478	131.869	164.355	8.370	40000
1330	SPIRIDONIA	3.17858	.05894	15.932	158.946	13.095	196.422	40000
1331	SOLVEJG	3.10472	.18495	3.080	121.858	183.072	30.189	27360
1332	MARCONIA	3.06177	.13082	2.481	14.252	346.635	250.578	38000
1333	CEVENOLA	2.63333	.13487	14.625	115.090	334.443	116.580	34000
1334	LUNDMARKA	2.91579	.08871	11.446	133.368	129.240	321.773	40000
1335	DEMOULINA	2.24031	.15452	2.536	172.583	197.550	322.676	27600
1336	ZEELANDIA	2.85064	.05965	3.195	97.497	219.414	293.139	36000
1337	GERARDA	2.90959	.09998	17.976	160.446	200.705	248.726	38000
1338	DUPONTA	2.26396	.11265	4.818	325.908	108.549	6.716	27800
1339	DESAGNEAUXA	3.02342	.05474	8.688	291.501	166.206	86.408	36000
1340	YVETTE	3.16981	.14358	.428	346.087	229.275	231.906	38000
1341	EDMEE	2.74256	.07869	13.088	107.618	139.169	280.095	38000
1342	BRABANTIA	2.28917	.20177	20.963	312.765	228.672	199.828	40000
1343	NICOLE	2.56803	.11593	6.056	41.716	232.778	290.113	27920
1344	CAUBETA	2.24807	.12045	5.659	60.121	131.489	16.007	27920
1345	POTOMAC	3.97910	.17798	11.380	137.194	338.239	4.336	41000
1346	GOTHA	2.63043	.17666	13.844	166.519	248.269	188.649	32320
1347	PATRIA	2.57178	.06683	11.905	229.139	198.189	293.380	40000
1348	MICHEL	2.79356	.13576	6.585	87.709	16.068	249.296	40000
1349	BECHUANA	3.01849	.15393	10.001	307.888	304.943	136.061	34000
1350	ROSSELIA	2.85801	.09006	2.933	139.599	237.179	306.361	38000
1351	UZBEKISTANIA	3.19480	.07764	9.722	10.538	39.484	295.149	40000
1352	WAWEL	2.77842	.06440	3.752	186.014	210.399	338.489	44200
1353	MAARTJE	3.01282	.08731	9.185	212.257	97.144	307.453	36000
1354	BOTHA	3.12781	.21298	6.063	31.220	243.811	199.635	29280
1355	MAGOEBA	1.85341	.04475	22.825	224.819	339.617	56.109	40000
1356	NYANZA	3.08239	.05598	7.961	69.976	294.883	5.658	40480
1357	KHAMA	3.19724	.14171	14.002	83.999	289.799	58.139	32880
1358	GAIKA	2.47492	.17181	2.175	21.430	287.031	4.981	38000
1359	PRIESKA	3.11658	.07569	11.104	64.606	335.998	265.122	36000
1360	TARKA	2.63627	.21433	22.812	331.468	286.862	275.617	40000

NUMBER	NAME		A	E	I	NODE	PERI	ANOM	DAYEP
1361	LEUSCHNERIA		3.08236	.12852	21.573	164.799	168.033	210.831	43000
1362	GRIQUA		3.28066	.34192	24.061	121.465	263.368	294.837	30000
1363	HERBERTA		2.90153	.06981	1.096	214.760	109.828	272.952	43800
1364	SAFARA		3.01178	.07110	11.505	64.415	217.076	201.548	38000
1365	HENYEY		2.24870	.12251	5.061	258.892	334.917	96.108	25480
1366	PICCOLO		2.87397	.14038	9.490	24.479	281.147	161.737	38000
1367	1934NA		2.34342	.13144	22.483	270.587	347.335	169.787	40000
1368	NUMIDIA		2.52197	.06246	14.920	18.532	256.998	12.171	34000
1369	OSTANINA		3.10889	.22182	14.269	180.968	126.402	8.122	38000
1370	HELLA	LOST	2.25114	.17038	4.813	306.209	2.114	25.554	28040
1371	RESI		3.21565	.08757	16.517	185.927	95.142	302.249	40000
1372	HAREMARI		2.76738	.14670	16.447	327.603	86.559	34.660	40200
1373	CINCINNATI		3.39717	.32931	38.967	297.284	99.807	174.323	43000
1374	ISORA		2.24999	.27860	5.307	302.576	59.965	249.927	40000
1375	ALFREDA		2.44746	.06879	5.846	52.514	30.634	148.069	40000
1376	MICHELLE		2.22770	.21504	3.549	163.152	155.173	109.899	38000
1377	ROBERBAUXA		2.26018	.09227	6.012	223.238	355.142	2.887	33400
1378	LEONCE		2.37409	.15035	3.602	43.403	201.113	232.191	40000
1379	LOMONOSSOVA		2.52667	.08472	15.535	169.550	31.563	218.842	38000
1380	VOLODIA		3.14185	.11111	10.503	.060	235.428	188.193	43800
1381	DANUBIA		2.48804	.18142	4.682	351.679	29.359	164.184	44200
1382	GERTI		2.22019	.13099	1.569	353.108	245.648	59.676	38000
1383	LIMBURGIA		3.07560	.19115	.014	204.390	152.238	68.875	34000
1384	KNIERTJE		2.67675	.18282	11.835	153.403	273.751	289.812	34000
1385	GELRIA		2.74050	.10728	6.928	114.796	259.637	173.981	44200
1386	STORERIA		2.36386	.28762	11.790	161.247	152.013	316.435	43800
1387	KAMA		2.25835	.20824	5.508	203.375	123.970	11.402	28040
1388	APHRODITE		3.01733	.09329	11.197	54.679	254.751	142.569	40000
1389	ONNIE		2.86596	.01445	2.029	174.631	286.602	178.535	40000
1390	ABASTUMANI		3.43827	.03404	19.948	29.261	6.198	312.661	44200
1391	CARELIA		2.54905	.16288	7.615	104.317	79.839	314.666	29640
1392	PIERRE		2.60850	.20026	12.302	358.621	42.293	336.873	29160
1393	SOFALA		2.43487	.10765	5.841	56.540	184.319	59.969	43800
1394	ALGOA		2.43848	.07803	2.660	178.503	114.532	115.972	40000
1395	ARIBEDA		3.20885	.03924	8.674	245.834	83.541	174.555	40000
1396	OUTENIQUA		2.24815	.16369	4.504	359.501	265.416	326.061	38000
1397	UMTATA		2.68292	.25146	3.528	77.560	204.689	196.175	34000
1398	DONNERA		3.15447	.11063	11.836	297.130	79.980	207.796	40000
1399	TENERIFFA		2.21636	.16535	6.511	161.319	223.144	63.015	42000
1400	TIRELA		3.10919	.24761	15.565	210.413	110.017	5.389	44200
1401	LAVONNE		2.22657	.17925	7.290	277.482	69.866	90.269	38000
1402	ERI		2.68467	.15378	14.292	266.890	9.394	193.895	42000
1403	IDELSONIA		2.71980	.29042	10.144	157.420	190.032	350.770	28400
1404	AJAX		5.20882	.11296	18.089	332.266	58.719	73.066	38400
1405	SIBELIUS		2.25141	.14592	7.038	312.300	94.612	318.910	28400
1406	KOMPPA		2.69707	.09920	12.427	333.622	83.731	30.558	30360
1407	LINDELOF		2.76197	.28477	5.782	269.480	108.218	332.651	40000
1408	TRUSANDA		3.11091	.09802	8.322	202.123	179.744	317.171	38000
1409	ISKO		2.67598	.05539	6.703	177.403	204.294	27.963	44200
1410	MARGRET		3.02179	.10161	10.347	171.194	230.019	71.743	40000
1411	BRAUNA		3.00419	.05331	8.068	285.196	87.384	106.230	38000
1412	LAGRULA		2.21470	.11336	4.727	66.085	12.729	43.122	28560
1413	ROUCARIE		3.02010	.06537	10.214	178.981	298.851	138.228	40720
1414	JEROME		2.78840	.15788	8.822	143.600	.190	324.114	42000
1415	MALAUTRA		2.22317	.08747	3.430	329.190	239.591	230.259	38000
1416	RENAUXA		3.01888	.10269	10.066	353.015	62.423	38.591	38000
1417	WALINSKIA		2.97369	.07329	8.268	96.391	159.429	323.046	38000
1418	FAYETA		2.24184	.20335	7.195	354.984	323.352	259.047	38000
1419	DANZIG		2.29315	.14777	5.722	213.436	231.555	329.438	38680
1420	RADCLIFFE		2.74842	.07675	3.482	261.153	72.566	331.429	38000
1421	ESPERANTO		3.09191	.08432	9.818	43.124	155.265	315.339	40000
1422	STROMGRENIA		2.24744	.16722	2.674	201.278	170.320	282.264	43000
1423	JOSE		2.86038	.08140	2.915	58.375	317.419	324.068	44200
1424	SUNDMANIA		3.18290	.07800	9.209	43.268	313.852	301.926	40000
1425	TUORLA		2.61127	.10146	12.932	186.274	337.337	39.890	28680
1426	1937GF		2.58075	.15940	9.082	335.847	272.382	328.755	28640
1427	RUVUMA		2.74985	.21129	9.374	79.143	239.519	25.887	30680
1428	MOMBASA		2.81373	.13788	17.326	115.829	250.879	141.625	40000
1429	PEMBA		2.54830	.34204	7.704	48.415	294.954	53.289	38000
1430	1937NK		2.56185	.19908	3.310	327.456	351.133	176.115	40000
1431	1937DB		2.61903	.18291	14.006	117.730	222.431	70.340	40000
1432	ETHIOPIA		2.38173	.22579	8.284	123.142	217.216	121.413	40000
1433	GERAMTINA		2.79437	.17211	8.259	321.747	93.421	186.321	40000
1434	MARGOT		3.01832	.06096	10.783	152.967	140.751	293.595	38000
1435	GARLENA		2.64692	.24798	4.026	190.423	265.422	324.743	28400
1436	1936YA		3.14301	.07581	13.877	260.401	35.947	42.002	44200
1437	DIOMEDES		5.08259	.04575	20.607	315.117	130.543	348.974	38400
1438	WENDELINE		3.19192	.20898	2.004	240.951	125.248	328.978	39000
1439	VOGTIA		3.98088	.11461	4.202	36.051	111.821	314.337	38000
1440	ROSTIA		3.16319	.18106	2.296	47.271	349.966	348.754	28800
1441	BOLYAI		2.63202	.23682	13.896	254.768	114.237	9.129	30280
1442	CORVINA		2.87351	.07938	1.245	221.053	127.922	282.848	42200
1443	RUPPINA		2.93725	.05856	1.918	175.010	161.164	7.746	43000
1444	1938 AE		3.15854	.13220	17.750	302.967	315.588	30.265	44200
1445	KONKOLYA		3.11425	.18572	2.303	89.303	271.642	291.808	34000

NUMBER	NAME	A	E	I	NODE	PERI	ANOM	DAYEP
1446	SILLANPAA *	2.24521	.10183	5.262	17.261	195.430	157.466	42020
1447	UTRA	2.53476	.03800	4.835	35.793	63.576	186.468	34000
1448	LINDBLADIA	2.37216	.18659	5.828	45.233	72.504	57.648	30440
1449	1938DO	2.22277	.14125	6.635	110.817	130.447	291.966	28960
1450	RAIMONDA	2.61248	.16867	4.857	74.622	13.907	9.997	44200
1451	GRANO	2.20281	.11803	5.109	174.932	51.275	222.539	44200
1452	1938DZ	3.12686	.18787	14.226	21.518	93.719	171.615	43800
1453	FENNIA	1.89700	.02799	23.675	6.668	254.332	80.083	40840
1454	KALEVALA	2.36490	.14339	5.105	352.794	133.840	64.527	29720
1455	MITCHELLA	2.24616	.12552	7.754	128.100	99.473	228.989	38000
1456	1937NG	3.19401	.22327	10.736	285.835	50.978	136.673	38000
1457	ANKARA	2.69700	.15721	6.076	296.914	294.337	164.258	34000
1458	MINEURA	2.62484	.18181	12.528	181.836	98.434	130.529	40000
1459	MAGNYA	3.15630	.22487	16.907	41.634	330.713	181.932	38000
1460	HALTIA	2.54056	.19180	6.719	74.123	357.316	314.831	42000
1461	JEAN-JACQUES	3.12557	.05213	15.329	104.985	317.007	234.636	40000
1462	ZAMENHOF	3.15107	.10350	1.026	25.947	177.193	281.110	28800
1463	NORDENMARKIA	3.13543	.20845	7.314	331.237	72.766	239.346	40000
1464	ARMISTICIA	3.00425	.04445	11.569	86.736	55.335	94.118	43800
1465	AUTONOMA	3.02388	.17251	9.950	167.627	48.829	204.580	36000
1466	MUNDLERIA	2.37740	.15561	13.134	155.154	73.899	20.153	29040
1467	MASHONA	3.36956	.14303	21.948	326.968	336.675	284.059	40000
1468	ZOMBA	2.19533	.27069	9.956	308.635	22.275	289.640	42000
1469	LINZIA	3.12643	.05586	13.359	189.053	214.574	71.289	40000
1470	CARLA	3.15991	.06208	3.230	359.398	319.917	152.520	38000
1471	TORNIO	2.71644	.11726	13.637	322.481	91.646	330.048	29200
1472	MUONIO	2.23385	.19860	4.574	45.210	317.443	28.280	29200
1473	DUNAS	2.57448	.23762	13.670	216.268	129.288	281.005	43800
1474	BEIRA	2.73293	.49007	26.799	324.878	82.352	37.798	30000
1475	YALTA	2.34950	.16815	4.494	200.290	199.839	146.431	30000
1476	COX	2.28158	.18874	6.311	330.394	349.311	239.359	38000
1477	BONSDORFFIA	3.17558	.28960	15.739	320.826	105.536	340.850	30640
1478	VIHURI	2.66395	.09720	7.906	318.951	159.903	12.787	28920
1479	INKERI	2.67536	.19478	7.326	161.355	77.369	30.814	28960
1480	AUNUS	2.20226	.10996	4.871	63.664	63.658	27.359	28960
1481	TUBINGIA	3.01527	.04611	3.534	354.198	310.278	140.397	40000
1482	SEBASTIANA	2.87248	.03973	2.977	70.878	214.056	6.548	43800
1483	NAKOILA	2.71866	.17717	4.497	71.905	86.562	118.614	44200
1484	POSTREMA	2.73992	.20350	17.262	72.732	125.804	73.154	44200
1485	ISA	3.02663	.11342	8.932	297.922	45.190	222.632	40000
1486	MARYLIN	2.19797	.12460	.078	334.631	347.575	296.387	42000
1487	BODA	3.13772	.11851	2.467	97.663	95.540	26.878	42160
1488	AURA	3.03890	.12123	10.547	355.303	113.273	29.588	29400
1489	1939GC	3.17404	.16861	2.405	155.930	10.232	33.935	29400
1490	LIMPOPO	2.35211	.15433	10.028	254.236	89.471	57.266	38000
1491	BALDUINUS	3.19458	.17463	3.776	314.729	152.835	145.810	40000
1492	OPPOLZER	2.17267	.11681	6.052	137.742	80.340	301.061	30040
1493	SIGRID	2.42957	.20238	2.590	330.644	1.154	13.294	43000
1494	SAVO	2.19052	.13089	2.445	194.992	183.225	155.917	29720
1495	HELSINKI	2.63942	.15335	12.767	13.520	267.149	278.757	30080
1496	TURKU	2.20551	.16246	2.503	294.643	359.068	67.344	29200
1497	TAMPERE	2.89462	.08253	1.063	301.276	24.873	45.364	40000
1498	LAHTI	3.09443	.24438	12.629	266.869	93.376	358.668	29160
1499	PORI	2.67060	.18460	12.186	240.276	72.448	15.414	30600
1500	JYVASKYLA	2.24225	.19024	7.459	19.915	15.712	111.188	30800
1501	BAADE	2.54712	.23436	7.324	16.970	11.259	.649	29200
1502	ARENDA	2.73209	.08646	4.075	204.786	271.988	359.034	36000
1503	KUOPIO	2.62633	.10333	12.378	317.239	174.990	337.473	29240
1504	LAPPEENRANT	2.39884	.15900	11.054	95.031	50.063	33.859	29360
1505	KORANNA	2.65865	.13456	14.460	248.697	340.618	.279	29400
1506	XOSA	2.56668	.26620	12.635	235.008	43.329	236.969	38000
1507	VAASA	2.33134	.24488	9.269	293.229	48.520	32.437	29600
1508	KEMI	2.76543	.42168	28.684	14.485	92.522	106.177	40000
1509	ESCLANGONA	1.86631	.03217	22.317	283.107	267.326	106.941	40000
1510	CHARLOIS	2.66949	.15092	11.886	331.733	163.505	359.339	34000
1511	DALERA	2.35789	.10822	4.074	81.783	94.764	10.972	33360
1512	OULU	3.93325	.16167	6.567	10.690	248.080	119.823	30680
1513	1940EB	2.19279	.09913	3.973	136.084	25.916	33.321	29760
1514	RICOUXA	2.24050	.19971	4.531	145.558	178.214	152.912	40000
1515	1936VG	2.57156	.23409	10.663	49.136	350.574	74.112	43800
1516	HENRY	2.62065	.18686	8.745	126.073	91.553	253.635	44200
1517	BEOGRAD	2.71735	.04311	5.286	64.103	227.155	281.946	34000
1518	ROVANIEMI	2.22546	.14258	6.732	27.823	35.662	338.219	29200
1519	KAJAANI	3.14321	.22643	12.466	16.625	335.313	4.787	37200
1520	IMATRA	3.10721	.09996	15.219	254.263	113.744	47.734	29240
1521	SEINAJOKI	2.84799	.14040	15.063	12.502	48.579	290.682	43000
1522	KOKKOLA	2.36788	.07127	5.361	60.604	29.944	338.852	29240
1523	PIEKSAMAKI	2.24221	.09360	5.151	327.833	185.818	329.681	32920
1524	JOENSUU	3.11360	.11220	12.706	348.102	357.502	105.942	40000
1525	SAVONLINNA	2.69548	.26532	5.901	280.344	62.131	34.181	29600
1526	MIKKELI	2.31498	.18808	6.216	337.280	70.973	354.470	29440
1527	MALMQUISTA	2.22741	.19838	5.193	16.079	303.293	35.368	38000
1528	CONRADA	2.41458	.14389	8.535	139.351	56.958	341.759	38000
1529	OTERMA	3.99759	.19375	9.000	101.496	303.404	63.796	28920
1530	RANTASEPPA	2.24823	.19951	4.418	285.828	83.769	278.656	40000

NUMBER	NAME	A	E	I	NODE	PERI	ANOM	DAYEP
1531	HARTMUT	2.62936	.15212	12.374	279.164	141.562	96.214	36000
1532	INARI	3.00650	.04850	8.773	330.777	122.860	233.864	44200
1533	SAIMAA	3.01912	.03422	10.698	157.009	358.187	188.358	38000
1534	NASI	2.72851	.25323	9.835	62.319	41.403	135.657	43000
1535	PAIJANNE	3.15186	.20487	6.154	266.750	34.881	46.043	29520
1536	PIELINEN	2.20436	.19548	1.518	195.716	169.191	11.704	29560
1537	1940QA	3.05342	.29981	3.777	231.316	144.961	18.501	39800
1538	1940RF LOST	2.36096	.21779	9.469	343.357	11.817	16.338	29960
1539	BORRELLY	3.15368	.18062	1.720	142.983	242.669	11.017	44200
1540	KEVOLA	2.85134	.07971	12.004	52.922	110.803	212.806	36000
1541	ESTONIA	2.76907	.06923	4.898	2.129	187.844	334.924	29320
1542	SCHALEN	3.09512	.11469	2.742	212.146	155.911	281.871	43800
1543	BOURGEOIS	2.62795	.32535	11.170	288.701	22.640	5.537	30200
1544	VINTERHANSENIA	2.37354	.10465	3.342	59.906	355.437	340.472	30280
1545	THERNOE	2.77244	.23820	2.973	52.998	86.156	308.243	30400
1546	1941SG1	3.17307	.13077	16.050	191.624	268.537	198.418	40000
1547	1929CZ	2.64649	.25199	11.692	291.234	153.478	324.022	44200
1548	PALOMAA	2.78856	.08051	16.547	117.368	87.432	7.472	28000
1549	MIKKO	2.23054	.08453	5.555	85.252	4.858	79.635	28600
1550	TITO	2.54827	.30578	8.861	65.318	309.432	37.052	28880
1551	ARGELANDER	2.39398	.06620	3.762	107.046	231.630	70.361	38000
1552	BESSEL	3.00960	.10137	9.887	10.926	36.853	105.831	28960
1553	BAUERSFELDA	2.90764	.09663	3.230	110.849	21.979	286.975	40160
1554	YUGOSLAVIA	2.62051	.20223	12.118	217.031	130.743	198.079	40000
1555	DEJAN	2.69057	.27413	6.077	319.093	45.360	350.991	30240
1556	WINGOLFIA	3.42033	.12651	15.707	92.533	261.115	180.066	40000
1557	ROEHLA	3.00827	.10711	10.334	355.768	357.071	103.001	36000
1558	JARNEFELT	3.22537	.06068	10.508	110.903	285.753	312.375	42280
1559	KUSTAANHEIMO	2.38941	.13659	3.200	327.932	215.462	79.451	43000
1560	STRATTONIA	2.68360	.21375	6.281	289.250	93.575	173.678	44200
1561	FRICKE	3.16943	.15473	4.362	232.459	32.392	220.767	36000
1562	GONDOLATSCH	2.22627	.07815	4.885	129.147	81.892	341.852	44200
1563	NOEL	2.19154	.08535	5.988	53.409	115.215	175.481	40840
1564	SRBIJA	3.14523	.21814	11.066	178.459	227.478	222.229	40000
1565	LEMAITRE	2.39288	.35026	21.426	261.057	115.719	202.114	43000
1566	ICARUS	1.07789	.82658	22.945	87.631	31.040	230.846	40240
1567	ALIKOSKI	3.21294	.08325	17.246	51.362	122.235	266.662	42020
1568	AISLEEN	2.35138	.25357	24.915	145.892	228.312	83.724	43000
1569	EVITA	3.15743	.11498	12.259	99.512	252.021	330.954	43000
1570	BRUNONIA	2.84260	.05787	1.653	190.044	227.124	255.831	43000
1571	1950FJ	3.13771	.12599	14.556	292.809	132.130	6.671	43000
1572	POSNANIA	3.11450	.19947	13.243	6.088	356.022	314.880	43000
1573	VAISALA	2.37041	.23177	24.546	202.042	173.082	134.934	43000
1574	MEYER	3.53359	.04697	14.446	247.291	269.954	267.602	32200
1575	WINIFRED	2.37343	.18036	24.817	206.475	347.331	79.517	43000
1576	FABIOLA	3.12984	.18716	.940	166.985	242.883	349.314	43000
1577	REISS	2.23052	.16619	4.356	123.251	265.581	155.329	43000
1578	KIRKWOOD	3.93566	.23194	.815	73.618	4.602	330.996	42000
1579	HERRICK	3.41176	.15455	8.658	186.261	272.908	75.391	43000
1580	BETULIA	2.19564	.49048	52.041	61.796	159.114	200.934	40000
1581	ABANDERADA	3.16399	.11075	2.538	104.501	95.300	283.061	43000
1582	MARTIR	3.15531	.12457	11.602	93.645	134.266	264.501	43000
1583	ANTILOCHUS	5.27647	.05406	28.303	221.067	186.310	336.126	33600
1584	FUJI	2.37494	.19548	26.661	305.039	188.028	182.587	43000
1585	UNION	2.93116	.30714	26.175	150.022	264.440	236.689	43000
1586	THIELE	2.43057	.10247	4.061	125.481	27.383	311.000	43000
1587	KAHRSTEDT	2.54750	.15016	7.856	357.647	96.185	317.828	43000
1588	DESCAMISADA	3.02917	.06968	11.266	98.706	230.415	213.350	43000
1589	FANATICA	2.41703	.09243	5.255	90.090	288.743	295.768	43000
1590	TSIOLKOVSKAJA	2.23021	.15710	4.351	226.319	51.540	344.595	43000
1591	BAIZE	2.39049	.17815	24.815	90.163	162.398	288.275	43000
1592	MATHIEU	2.76390	.30579	13.516	106.016	175.761	154.509	43000
1593	FAGNES	2.22485	.28080	9.977	119.930	184.342	340.057	33840
1594	DANJON	2.26916	.19547	8.949	69.124	221.816	19.883	44200
1595	1930 ME	2.64494	.10990	4.162	112.013	188.139	156.380	44200
1596	1951EV	2.89166	.12822	13.271	249.518	159.312	171.285	34000
1597	LAUGIER	2.84477	.09009	11.824	158.562	52.691	30.730	43800
1598	PALOQUE	2.33180	.08138	7.529	297.858	300.001	56.758	40400
1599	GIOMUS	3.13775	.12902	6.078	43.502	358.937	2.195	43800
1600	VYSSOTSKY	1.84906	.03767	21.177	60.313	49.761	322.372	32600
1601	PATRY	2.23417	.13004	4.949	74.548	195.332	39.385	33160
1602	INDIANA	2.24466	.10499	4.165	74.871	72.233	315.375	41680
1603	NEVA	2.75380	.09581	8.555	129.833	256.020	141.047	43800
1604	1931 FH	3.02618	.09590	9.391	309.550	34.392	214.430	38000
1605	1936GA=68KP	3.01481	.07425	10.519	174.728	272.877	136.461	28400
1606	JEKHOVSKY	2.69075	.31579	7.654	190.842	141.200	5.656	40000
1607	MAVIS	2.54564	.31169	8.644	123.154	233.267	354.632	38000
1608	1951RZ	2.21402	.16901	3.947	356.920	315.272	291.581	36000
1609	BRENDA	2.58489	.24556	18.688	105.408	227.704	3.689	33960
1610	MIRNAYA	2.20292	.19853	2.203	359.485	14.008	225.337	43000
1611	BEYER	3.19731	.13686	4.246	237.725	78.620	49.451	40840
1612	HIROSE	3.10227	.09279	16.875	320.227	241.932	312.922	33320
1613	SMILEY	2.73774	.25986	7.938	321.651	124.359	306.463	33560
1614	GOLDSCHMIDT	2.99595	.07533	14.079	162.794	345.071	49.121	34120
1615	BARDWELL	3.11744	.18519	1.669	152.963	250.998	56.721	33280

NUMBER	NAME	A	E	I	NODE	PERI	ANOM	DAYEP
1616	FILIPOFF	2.91329	.01548	8.499	48.318	25.026	131.404	40680
1617	ALSCHMITT	3.19100	.13909	13.260	154.978	31.798	143.933	35000
1618	DAWN	2.86924	.03035	3.224	103.012	172.127	16.986	32760
1619	UETA	2.24097	.17583	6.215	61.374	327.191	356.098	34680
1620	GEOGRAPHOS	1.24452	.33535	13.327	336.820	276.437	45.201	43000
1621	DRUZHBA	2.23003	.11868	3.161	181.988	236.770	328.993	24800
1622	CHACORNAC	2.23445	.16353	6.471	4.453	255.005	301.203	34120
1623	VIVIAN	3.13474	.16290	2.491	115.578	320.266	269.253	32800
1624	RABE	3.17421	.11717	1.977	133.808	21.244	234.527	26640
1625	THE NORC	3.16629	.24724	15.476	324.051	276.845	66.743	34640
1626	SADEYA	2.36414	.27424	25.262	279.211	148.452	350.811	42000
1627	IVAR	1.86421	.39667	8.430	132.851	166.963	353.558	36020
1628	STROBEL	3.01296	.06466	19.342	181.130	286.188	91.256	43800
1629	PECKER	2.23902	.15391	9.705	132.405	105.904	267.958	43800
1630	MILET	3.02987	.16763	4.542	55.184	93.997	356.847	34080
1631	KOPFF	2.23529	.21329	7.487	16.982	313.388	42.950	28480
1632	SIEBOHME	2.65619	.13625	5.702	199.788	125.266	255.181	43000
1633	CHIMAY	3.16718	.15233	2.676	114.825	58.481	295.655	43800
1634	1935QP	2.24548	.16254	7.606	90.393	190.518	324.449	43800
1635	BOHRMANN	2.85265	.05930	1.804	184.437	135.420	277.153	40000
1636	PORTER	2.23461	.12778	4.435	168.124	238.296	253.895	40000
1637	SWINGS	3.06794	.05033	14.124	22.279	223.361	125.792	28480
1638	1912OX	2.74821	.18927	.273	201.775	83.373	55.525	35000
1639	BOWER	2.57331	.15038	8.403	324.240	103.853	158.182	43800
1640	1951QA	2.28874	.34260	7.113	355.151	353.788	301.043	43800
1641	1935OJ	3.01736	.10453	9.342	332.024	-.176	45.319	36000
1642	HILL	2.75426	.06685	10.830	339.572	145.359	239.602	33940
1643	BROWN	2.48950	.19864	3.533	288.872	86.581	73.260	40000
1644	RAFITA	2.54829	.15337	6.981	270.841	198.309	332.848	40000
1645	WATERFIELD	3.05761	.11370	1.017	266.711	105.086	331.395	43000
1646	ROSSELAND	2.36129	.11911	8.381	119.793	279.187	109.183	40000
1647	MENELAUS	5.24081	.02586	5.641	239.797	287.372	218.335	42000
1648	SHAJNA	2.23698	.20557	4.557	130.365	133.298	110.189	34320
1649	FABRE	3.02049	.05171	10.807	145.981	15.449	288.266	33700
1650	HECKMANN	2.43583	.16050	2.735	199.996	54.619	122.539	28840
1651	1936HD	2.17983	.06629	5.068	187.550	337.933	52.724	28300
1652	1933UE1	2.25096	.14973	3.191	251.786	11.451	61.137	34640
1653	YAKHONTOVIA	2.61111	.32243	4.078	305.602	86.025	340.906	28800
1654	BOJEVA	3.01528	.09319	10.455	25.671	333.606	5.233	36140
1655	1929WG	2.78187	.23687	9.525	111.879	320.407	3.802	25960
1656	SUOMI	1.87744	.12372	25.066	175.150	287.105	119.337	40000
1657	ROEMERA	2.34816	.23596	23.410	105.059	53.847	105.965	43000
1658	INNES	2.55975	.18400	9.107	95.640	188.663	20.393	34600
1659	PUNKAHARJU	2.78466	.25674	16.545	338.694	34.162	311.052	34500
1660	WOOD	2.39477	.30181	20.563	212.620	276.770	249.978	42020
1661	GRANULE	2.18382	.09096	3.025	262.245	325.453	330.151	20960
1662	HOFFMAN	2.74236	.17301	4.248	331.995	60.230	336.272	23680
1663	VAN DEN BOS	2.24002	.17995	5.367	83.369	273.437	340.564	24760
1664	1929CD	2.33923	.22303	6.121	44.603	105.296	355.729	25640
1665	GABY	2.41366	.20725	10.843	91.753	4.312	63.934	26080
1666	VAN GENT	2.18562	.18191	2.685	263.725	81.421	335.371	26200
1667	PELS	2.18989	.15624	4.620	80.518	195.477	355.457	43800
1668	HANNA	2.80425	.21821	4.722	161.355	187.628	339.652	27280
1669	DAGMAR	3.14439	.10541	.998	19.944	172.252	163.734	27680
1670	MINNAERT	2.90077	.10512	10.532	59.862	16.271	302.848	27720
1671	CHAIKA	2.58713	.25838	3.956	178.048	247.637	333.233	27720
1672	1935BD	3.17154	.28112	1.008	185.885	238.697	52.541	27880
1673	VAN HOUTEN	3.10212	.18145	3.567	209.137	205.555	71.875	33320
1674	GROENEVELD	3.17562	.15374	2.689	96.142	356.768	57.196	28960
1675	SIMONIDA	2.23339	.12528	6.806	30.065	49.354	338.586	34700
1676	1939LC	2.23583	.18608	6.146	54.606	203.187	349.945	29400
1677	TYCHO BRAHE	2.53180	.10731	14.753	338.501	312.455	46.152	29880
1678	HVEEN	3.17054	.09136	10.158	352.879	114.307	5.959	30000
1679	NEVANLISNA	3.12494	.13997	17.972	178.098	86.759	81.480	43000
1680	1942CH	2.72259	.18327	4.261	83.572	155.664	301.871	31040
1681	1948WE	2.69741	.20557	7.218	94.607	359.448	359.204	32960
1682	KAREL	2.23897	.19143	4.035	325.812	8.413	353.317	33140
1683	1950SL	2.73441	.17932	12.497	326.990	344.863	39.298	33520
1684	1951QF	3.09754	.12267	3.644	106.121	148.014	65.142	33880
1685	TORO	1.36767	.43597	9.370	273.961	126.571	96.881	40000
1686	DESITTER	3.16035	.16332	.628	6.502	299.938	321.528	42000
1687	GLARONA	3.15439	.17856	2.643	93.394	321.639	313.647	39020
1688	1951EQ	2.61936	.23796	11.811	246.174	39.618	282.166	33740
1689	FLORIS-JAN	2.44932	.20818	6.358	123.693	262.880	341.913	26260
1690	1948VB	3.03803	.09734	13.022	230.786	157.245	173.797	35580
1691	OORT	3.17135	.16761	1.051	174.414	222.315	356.025	40000
1692	SUBBOTINA	2.78733	.13411	2.404	200.615	108.447	19.056	28400
1693	HERTZSPRUNG	2.80613	.26409	11.988	70.662	233.083	329.197	27960
1694	KAISER	2.39578	.25669	11.094	13.271	355.119	357.794	37190
1695	1941UO	2.78428	.29033	16.571	218.766	138.093	357.914	38660
1696	1939FF	2.26181	.09922	6.052	21.105	163.181	356.057	29340
1697	1940RM	2.37393	.11821	5.676	331.913	90.361	309.864	29900
1698	1934CS	3.15521	.12311	1.528	27.202	119.207	347.016	43800
1699	1941QD	2.21128	.16479	1.973	273.965	49.734	29.856	30280
1700	1940QC	2.36075	.22499	4.537	357.105	13.771	342.741	29880

NUMBER	NAME	A	E	I	NODE	PERI	ANOM	DAYEP
1701	1953NJ	3.17717	.17475	16.370	62.923	251.944	348.218	34560
1702	1924SH	2.85864	.14069	9.986	104.386	238.806	322.513	32760
1703	BARRY	2.21465	.17116	4.523	112.231	211.682	19.986	26240
1704	WACHMANN	2.22292	.08690	.969	259.216	280.454	164.121	43800
1705	1941SL1	2.29938	.24536	7.680	188.737	154.910	20.212	30280
1706	DIECKVOSS	2.12580	.11469	1.872	279.735	338.058	97.639	37940
1707	1932RL	2.21918	.17040	4.053	6.328	41.150	338.146	27000
1708	1929XA	2.92103	.30069	6.056	192.567	247.480	286.033	43800
1709	UKRAINA	2.37814	.21362	7.572	300.006	41.326	43.898	42000
1710	1941UF	2.32267	.26714	8.459	356.815	334.606	54.070	30320
1711	1935BB	3.01231	.11400	11.080	135.686	247.084	91.596	27840
1712	1935KC	3.16469	.15435	19.379	238.840	6.556	8.391	27960
1713	BANCILHON	2.22857	.18455	3.754	61.085	255.383	22.467	33920
1714	SY	2.56443	.15805	7.939	301.197	317.997	23.747	33860
1715	1938GK	2.39829	.24111	11.514	39.015	207.807	359.920	34540
1716	PETER	2.73353	.09426	5.713	244.368	314.171	348.795	27560
1717	1954AC	2.19592	.12906	6.187	340.173	115.311	242.950	43800
1718	1942RX	2.36608	.27599	7.673	203.100	132.376	30.869	34680
1719	JENS	2.65811	.22082	14.293	323.776	56.999	115.274	33360
1720	NIELS	2.18811	.10439	.730	127.231	308.317	161.140	40000
1721	WELLS	3.14422	.06134	16.102	317.931	129.680	313.313	34680
1722	1938EG	2.51407	.04789	5.492	168.502	288.460	56.350	28960
1723	1936FX	3.01390	.03936	10.903	150.453	8.160	58.845	30480
1724	1932DC	2.71128	.06028	12.210	164.417	294.340	72.065	26780
1725	CRAO	2.90293	.09296	3.169	119.011	227.566	350.832	44200
1726	HOFFMEISTER	2.78763	.04259	3.462	231.889	66.825	26.360	27320
1727	1965BA	1.85404	.10215	22.895	132.622	312.506	19.913	38780
1728	GOETHE-LINK	2.56306	.08955	7.219	240.644	63.210	68.858	38700
1729	BERYL	2.23043	.10012	2.448	8.965	261.553	79.417	38300
1730	1936UA	2.78433	.22410	9.506	172.048	223.950	5.020	28480
1731	SMUTS	3.18853	.10077	5.882	152.869	198.520	152.517	40000
1732	HEIKE	3.00870	.11731	10.785	156.248	208.568	162.427	30800
1733	SILKE	2.19297	.08455	4.422	159.307	305.829	33.387	28960
1734	ZHONGOLOVICH	2.77684	.23407	8.352	182.409	184.579	358.803	33920
1735	1948RJ	3.14281	.12377	15.614	10.458	269.188	81.478	32840
1736	FLOIRAC	2.22881	.16895	4.549	159.533	248.146	329.373	39760
1737	SEVERNY	3.00978	.05089	9.402	327.772	223.108	208.347	39420
1738	1930SP	2.18345	.20335	4.876	43.861	283.417	339.777	43800
1739	MEYERMANN	2.26169	.12350	3.397	203.483	80.955	46.336	29500
1740	1939UA	2.46733	.18663	2.023	296.685	76.879	11.532	29560
1741	GICLAS	2.88357	.07070	2.902	55.607	338.070	110.710	36980
1742	SCHAIFERS	2.88881	.09692	2.475	152.542	211.493	357.985	27700
1743	SCHMIDT	2.46847	.14384	6.384	189.454	356.383	178.756	37200
1744	HARRIET	2.22888	.12152	4.414	27.293	154.460	175.487	37200
1745	FERGUSON	2.84500	.05455	3.264	78.715	337.964	305.878	37200
1746	BROUWER	3.97310	.19675	8.399	323.264	54.507	358.923	32640
1747	WRIGHT	1.70905	.11041	21.412	267.923	339.861	6.958	40480
1748	MAUDERLI	3.91912	.23409	3.294	125.725	203.061	27.584	39440
1749	TELAMON	5.25415	.11100	6.068	340.599	109.416	72.033	43800
1750	ECKERT	1.92653	.17305	19.075	273.538	108.290	352.518	33600
1751	HERGET	2.78987	.17478	8.079	240.727	128.525	358.324	35480
1752	1930OK	2.23832	.20005	3.496	237.464	96.666	287.009	26000
1753	1934JM	3.01562	.07580	11.409	59.239	227.326	329.386	27600
1754	CUNNINGHAM	3.99371	.16279	11.995	163.991	120.917	240.443	27600
1755	LORBACH	3.09068	.05021	10.665	157.972	327.220	270.020	28400
1756	GIACOBINI	2.54850	.22833	5.126	291.130	100.233	359.070	37600
1757	1939FC	2.35122	.17611	3.983	39.245	148.459	24.638	40000
1758	1942DK	3.00838	.03455	10.815	113.864	108.041	315.792	40000
1759	KIENLE	2.64920	.31403	4.560	159.034	203.680	33.081	30800
1760	SANDIA	3.16549	.10644	8.349	233.582	329.143	82.188	40000
1761	EDMONDSON	3.16854	.23680	2.469	76.785	53.311	32.695	34000
1762	RUSSELL	2.87589	.07552	2.262	160.825	234.469	12.866	34800
1763	WILLIAMS	2.18937	.20320	4.244	304.516	27.554	76.913	34800
1764	COGSHALL	3.09732	.11981	2.211	152.790	76.290	202.678	34800
1765	WRUBEL	3.15600	.19554	19.984	71.110	261.110	71.066	36000
1766	SLIPHER	2.74905	.08756	5.217	188.757	168.755	22.886	38000
1767	LAMPLAND	3.01950	.09668	9.808	192.307	133.903	55.688	43800
1768	APPENZELLA	2.45051	.18160	3.284	12.568	18.107	26.337	39200
1769	1966QP	2.17849	.14152	1.593	319.226	23.030	77.652	39600
1770	SCHLESINGER	2.45775	.05948	5.315	21.593	39.904	277.277	40000
1771	MAKOVER	3.13336	.16598	11.234	86.590	319.132	79.268	40000
1772	GAGARIN	2.53018	.09944	5.760	88.408	88.685	2.438	40000
1773	RUMPELSTILZ	2.43502	.12989	5.409	74.435	166.925	334.986	40000
1774	KULIKOV	2.87775	.06635	1.847	175.221	248.836	315.370	40000
1775	ZIMMERWALD	2.60578	.18193	12.566	195.994	83.131	245.574	40000
1776	KUIPER	3.10189	.02257	9.452	177.099	306.558	247.158	37200
1777	GEHRELS	2.62553	.01902	3.156	334.891	133.829	257.696	37200
1778	ALFVEN	3.14085	.13198	2.462	106.738	123.680	213.250	43800
1779	1950LZ	2.17531	.16055	.892	254.556	9.946	76.528	37200
1780	KIPPES	3.01712	.05192	8.999	291.392	337.894	153.437	40800
1781	VAN BIESBROECK	2.39499	.10829	6.952	44.391	341.556	331.595	40400
1782	SCHNELLER	3.12385	.14240	1.522	158.028	102.085	335.221	36000
1783	1935 FJ	2.66377	.13234	11.451	190.073	313.036	57.655	28000
1784	1935 MG	2.40505	.13150	1.476	94.919	183.970	9.459	28000
1785	WURM	2.23580	.06810	3.776	283.481	245.566	329.503	30000

NUMBER	NAME	A	E	I	NODE	PERI	ANOM	DAYEP
1786	1948 TL	3.02212	.10309	10.441	16.970	346.119	2.961	32800
1787	1950 SK	3.00247	.05010	8.916	307.416	268.936	149.619	43000
1788	KIESS	3.11721	.15152	.651	161.799	142.195	46.053	34400
1789	DOBROVOLSKY	2.21349	.18838	1.979	101.733	213.889	330.252	39200
1790	VOLKOV	2.23786	.10111	5.116	1.810	146.600	30.356	39600
1791	PATSAYEV	2.74572	.14420	5.368	199.218	72.284	27.033	39600
1792	RENI	2.78064	.27795	8.989	72.925	322.882	81.675	40000
1793	ZOYA	2.22467	.09746	1.500	226.074	321.480	268.363	34800
1794	FINSEN	3.11638	.16830	14.568	221.318	337.058	27.039	40800
1795	WOLTJER	2.78163	.19316	7.537	193.145	75.465	75.106	37200
1796	RIGA	3.34254	.06584	22.735	187.136	354.356	111.889	44200
1797	SCHAUMASSE	2.23655	.02431	3.151	29.709	355.327	9.394	28400
1798	WATTS	2.19859	.12285	6.205	44.179	2.971	141.447	33000
1799	KOUSSEVITZKY	3.02916	.11805	11.475	156.805	190.533	267.000	40800
1800	AGUILAR	2.35730	.13543	5.785	124.106	213.189	25.919	33600
1801	1963UR=52SP	3.01923	.07048	10.974	77.675	11.897	263.957	43800
1802	ZHANG HENG	2.84475	.03806	2.674	142.566	296.943	323.337	38800
1803	ZWICKY	2.34906	.24743	21.588	337.242	252.970	319.405	39600
1804	CHEBOTAREV	2.41008	.02036	3.645	325.729	304.663	291.513	39600
1805	DIRIKIS	3.13094	.12638	2.523	78.705	90.994	50.013	40800
1806	1971LC=27EB	2.23738	.10623	3.839	270.867	192.870	106.084	41000
1807	SLOVAKIA	2.22625	.17779	3.483	236.172	139.559	280.425	41000
1808	BELLEROPHON	2.74716	.17973	2.048	13.491	46.183	322.106	37200
1809	PROMETHEUS	2.92699	.09772	3.264	99.495	229.204	33.012	37200
1810	EPIMETHEUS	2.22356	.09273	4.029	254.061	203.130	278.022	37200
1811	BRUWER	3.13869	.10139	8.467	168.761	126.294	58.805	37200
1812	GILGAMESH	3.00645	.07997	10.241	178.909	263.599	292.199	37220
1813	IMHOTEP	2.68224	.08157	8.107	35.290	165.554	158.062	37200
1814	BACH	2.22584	.13094	4.360	20.484	64.260	358.976	26800
1815	BEETHOVEN	3.14951	.19924	2.728	111.550	351.898	29.042	26800
1816	1936 BD	2.34005	.21839	26.065	153.341	339.103	309.202	28000
1817	1971BG	2.37210	.19127	25.657	88.837	138.869	69.603	29600
1818	BRAHMS	3.16638	.17841	2.975	249.757	73.031	42.854	29600
1819	LAPUTA	3.14825	.22156	23.658	122.840	108.807	8.474	40800
1820	LOHMANN	2.19826	.20999	4.990	145.053	167.846	26.242	33200
1821	1969UN	2.37821	.20286	2.107	297.722	348.239	36.018	33600
1822	WATERMAN	2.17041	.15234	.947	221.315	29.155	22.525	33400
1823	GLIESE	2.22578	.13526	2.892	309.955	295.378	92.749	41200
1824	HAWORTH	2.88545	.03983	1.948	15.536	69.916	97.587	34000
1825	KLARE	2.67731	.11477	4.037	289.186	142.134	250.843	34800
1826	MILLER	2.99898	.07768	9.208	274.844	161.417	262.788	35200
1827	ATKINSON	2.70941	.17833	4.511	220.757	238.695	303.258	38000
1828	KASHIRINA	3.06304	.10628	14.281	184.757	220.175	276.650	39200
1829	DAWSON	2.25119	.12111	6.328	293.326	140.977	301.325	36400
1830	POGSON	2.18823	.05649	3.953	147.171	334.531	80.464	40000
1831	NICHOLSON	2.23911	.12750	5.638	72.319	182.481	326.775	40000
1832	MRKOS	3.20410	.11713	15.023	303.465	90.884	297.236	40400
1833	SHMAKOVA	2.63444	.11337	9.998	158.047	153.093	4.080	40400
1834	1969QP	3.02502	.06903	9.423	268.515	355.133	45.232	40400
1835	GAJDARIYA	2.83243	.08943	.988	297.271	78.523	304.440	40800
1836	KOMAROV	2.78312	.19093	7.014	272.829	9.716	29.712	41200
1837	OSITA	2.20543	.08579	3.843	280.799	313.705	80.481	41200
1838	URSA	3.20786	.03216	22.099	44.222	98.420	245.393	41200
1839	RAGAZZA	2.79926	.16888	10.196	50.919	349.387	343.063	41200
1840	HUS	2.91737	.01927	2.423	40.532	29.409	317.731	41200
1841	MASARYK	3.43227	.08755	2.633	45.311	129.189	216.683	41200
1842	HYNEK	2.26630	.18059	5.347	153.247	124.677	225.288	41400
1843	JARMILA	2.65333	.16892	8.423	266.951	31.006	194.647	41400
1844	SUSILVA	3.01484	.05145	11.796	99.528	71.874	227.014	41600
1845	HELEWALDA	2.96942	.05523	10.701	142.626	326.405	291.178	41600
1846	BENGT	2.33817	.14223	3.188	18.851	74.358	163.606	44600
1847	STOBBE	2.61020	.01724	11.153	106.773	137.901	31.935	44600
1848	1972QN=33QD	2.86967	.04463	1.451	332.309	308.013	64.130	41600
1849	KRESAK	3.05183	.01789	10.779	50.379	110.100	70.982	44600
1850	KOHOUTEK	2.25124	.12499	4.051	68.561	190.142	113.300	44600
1851	LACROUTE	3.11841	.18128	1.670	24.760	344.477	191.264	44600
1852	1955GA=37WH	3.01734	.06374	11.192	95.413	353.196	77.787	44600
1853	1957XE=30YP	3.06673	.04752	15.762	298.643	100.246	153.264	44600
1854	SKVORTSOV	2.54069	.13325	4.914	189.168	273.935	277.286	40000
1855	KOROLEV	2.24770	.08412	3.075	190.773	348.374	179.586	40400
1856	RUZENA	2.23667	.07958	4.740	185.633	54.922	225.726	40800
1857	PARCHOMENKO	2.24328	.13487	4.395	235.822	173.292	312.701	41200
1858	LOBACHEVSKI	2.69728	.07873	1.661	272.155	13.587	55.964	41600
1859	KOVALEVSKAYA	3.20166	.11016	7.735	343.072	253.473	94.503	41600
1860	BARBAROSSA	2.56284	.20797	9.948	132.536	161.504	75.579	42000
1861	KOMENSKY	3.02024	.06302	10.478	23.831	265.778	191.184	41320
1862	APOLLO	1.46969	.55988	6.360	35.566	285.254	9.463	41920
1863	ANTINOUS	2.26040	.60566	18.447	347.899	265.830	4.483	41400
1864	DAEDALUS	1.46104	.61483	22.136	6.260	325.229	6.442	41200
1865	CERBERUS	1.08011	.46694	16.086	212.491	325.000	354.302	41360
1866	SISYPHUS	1.89355	.53979	41.122	63.222	292.858	17.206	41640
1867	DEIPHOBUS	5.20355	.04467	26.805	282.953	358.374	254.601	41200
1868	THERSITES	5.24456	.10762	16.816	197.113	168.673	255.600	44600
1869	PHILOCTETES	5.30619	.06187	3.960	43.639	320.674	238.796	44600
1870	GLAUKOS	5.21005	.03069	6.589	175.824	129.605	257.784	41200

NUMBER	NAME	A	E	I	NODE	PERI	ANOM	DAYEP
1871	ASTYANAX	5.33319	.03366	8.568	145.136	162.396	256.164	41200
1872	HELENOS	5.10565	.04280	14.784	188.840	110.275	264.810	41200
1873	AGENOR	5.24655	.09127	21.870	197.361	353.184	11.084	41200
1874	KACIVELIA	3.12593	.30433	4.875	159.493	193.162	110.283	43000
1875	1969QQ	3.12616	.17619	13.410	193.601	141.644	87.963	43000
1876	NAPOLITANIA	1.96433	.04815	23.107	304.034	243.179	75.456	43000
1877	MARSDEN	3.95609	.21320	17.498	352.741	309.230	333.124	41560
1878	330C=72T07	2.84492	.01506	1.769	187.634	277.269	179.607	44600
1879	35UN=72RS1	2.24547	.14849	1.721	249.789	173.544	322.961	41600
1880	MC CROSKY	2.67434	.07439	4.850	116.635	187.973	315.564	44600
1881	SHAO	3.16449	.11019	9.878	217.867	74.528	83.977	44600
1882	41UJ=72RP1	3.00599	.09357	9.462	201.493	123.409	39.881	41600
1883	42XA=57YM	2.41501	.26087	25.432	74.526	329.871	65.398	44600
1884	1943EB1	2.42307	.26701	21.782	353.382	110.409	39.114	41800
1885	HERERO	2.25096	.24732	5.667	326.087	4.760	208.175	44600
1886	LOWELL	2.62744	.15584	14.906	82.153	215.850	240.590	42000
1887	50TO=700A	3.00606	.11310	9.634	348.829	31.081	151.255	42000
1888	ZU CHONG-ZHI	2.54991	.16170	5.852	244.482	233.279	340.081	38800
1889	PAKHMUTOVA	3.08714	.11963	13.202	55.759	80.545	9.345	40000
1890	KONOSHENKOVA	3.21935	.13288	9.896	70.209	24.616	64.211	40000
1891	GONDOLA	2.70519	.07010	11.524	321.836	9.991	274.429	40000
1892	LUCIENNE	2.46091	.09032	14.015	315.449	93.389	104.559	41800
1893	JAKOBA	2.70836	.05167	10.040	64.405	241.341	71.848	41200
1894	HAFFNER	2.88691	.07238	.904	258.408	117.600	6.191	41200
1895	LARINK	3.17055	.16947	1.834	45.343	57.586	301.273	41200
1896	BEER	2.36812	.22048	2.213	181.877	179.432	7.543	41200
1897	HIND	2.28315	.14176	4.060	63.220	268.082	36.699	41200
1898	COWELL	3.11781	.16926	1.020	163.197	236.171	351.177	41200
1899	CROMMELIN	2.26518	.10590	7.280	51.919	127.235	213.950	41200
1900	KATYUSHA	2.20929	.13544	6.544	281.657	141.864	52.829	41400
1901	MORAVIA	3.25408	.05985	23.873	95.160	120.998	284.471	41400
1902	SHAPOSHNIKOV	3.97485	.22429	12.516	59.481	271.101	275.692	41400
1903	ADZHIMUSHKAJ	3.00232	.04660	10.966	135.312	346.369	93.687	41400
1904	MASSEVITCH	2.74325	.07516	12.841	106.392	259.595	221.085	41400
1905	AMBARTSUMIAN	2.22330	.16268	2.615	201.183	60.441	331.156	41400
1906	NAEF	2.37338	.13396	6.479	354.580	13.755	355.369	41600
1907	RUDNEVA	2.54450	.04718	3.209	151.996	61.017	159.408	41600
1908	POBEDA	2.88988	.03829	4.784	14.492	214.138	146.339	41600
1909	ALEKHIN	2.42223	.22518	1.776	227.300	3.197	95.270	41600
1910	MIKHAILOV	3.04465	.05350	10.341	201.156	323.323	236.650	41600
1911	SCHUBART	3.97901	.16193	1.656	285.379	186.511	307.891	42000
1912	6534PL=68HQ	2.90269	.09581	3.166	76.072	315.315	6.746	44600
1913	SEKANINA	2.87920	.07462	1.577	358.738	35.666	345.966	41600
1914	70EU=30SB1	2.40520	.14986	5.690	120.308	158.896	11.932	43800
1915	QUETZALCOAT	2.52049	.58252	20.550	162.583	347.572	359.689	41720
1916	1953 RA	2.27322	.44959	12.803	340.512	334.751	354.695	42080
1917	CUYO	2.14870	.50509	24.019	187.964	193.866	1.457	40920
1918	1968 UA	3.20068	.11940	9.252	195.791	232.231	232.362	43800
1919	CLEMENCE	1.93590	.09469	19.335	356.571	99.658	324.082	42360
1920	SARMIENTO	1.92994	.10583	22.803	63.625	316.532	358.756	41240
1921	PALA	3.22173	.41636	19.737	352.944	17.552	3.325	42000
1922	ZULU	3.25372	.47501	35.363	225.914	31.661	119.152	44600
1923	64TO2=4011	2.43510	.06488	4.962	352.641	105.652	34.919	44600
1924	69BA=4023P	2.33925	.13152	2.732	350.014	151.581	113.412	44600
1925	34RY=70KH	2.55131	.17915	7.722	113.530	241.701	221.968	42000
1926	35JA=74HB	2.65541	.10882	13.720	93.426	89.708	35.207	42220
1927	36FP=30XN	2.65000	.14913	13.360	27.060	95.246	298.418	42000
1928	38SO=69PA	2.47788	.19817	4.545	180.580	155.706	16.831	42000
1929	39BS=68BH	2.36333	.07541	7.784	65.270	70.427	22.290	40000
1930	LUCIFER	2.90103	.14239	14.036	318.744	340.180	96.407	38800
1931	69QB=57TK	2.54075	.26988	8.169	182.419	161.797	239.666	40000
1932	58DO1=71UB	2.37183	.15900	1.882	188.880	302.383	268.476	41200
1933	TINCHEN	2.35285	.12348	6.878	164.653	213.699	102.880	41400
1934	JEFFERS	2.39092	.29896	23.125	86.403	294.154	67.486	41800
1935	LUCERNA	2.62718	.22585	9.542	199.919	197.783	343.462	42000
1936	LUGARNO	2.67526	.13762	10.260	265.234	254.078	288.326	42000
1937	LUCARNO	2.37782	.15539	12.465	78.770	225.333	115.541	42000
1938	LAUSANNA	2.23638	.15879	3.333	171.419	63.746	353.578	42200
1939	LORETTA	3.14025	.11505	.919	40.177	195.147	95.264	42300
1940	WHIPPLE	3.06027	.06337	6.546	264.111	185.843	18.992	42400
1941	WILD	3.99374	.27810	3.948	60.417	305.050	1.968	41200
1942	JABLUNKA	2.31851	.18409	24.373	346.016	11.050	76.056	41840
1943	ANTEROS	1.43122	.25638	8.700	245.883	337.973	358.625	41800
1944	GUNTHER	2.24039	.23657	5.475	212.283	123.942	143.119	42000
1945	300L=34NL+	2.55592	.17802	4.216	142.571	194.682	194.252	42000
1946	31PH=52PB+	2.29382	.23448	8.159	17.111	338.956	62.416	39600
1947	1935EA	3.15146	.04671	11.891	91.325	116.066	291.460	42000
1948	35GL=35JL+	2.53529	.16743	5.823	23.119	180.542	125.319	44600
1949	36NE=40RK+	2.38360	.22926	4.654	265.014	65.316	19.631	42000
1950	WEMPE	2.17874	.08458	4.225	69.582	52.441	75.949	44600
1951	LICK	1.39041	.06166	39.096	130.186	140.413	323.444	42000
1952	36NO=40CQ+	3.10394	.15095	14.260	78.413	333.362	294.445	44600
1953	29VC=29WD+	3.12731	.16620	2.457	74.267	325.776	105.721	44600
1954	52PH=57QB	2.94294	.30902	14.833	278.417	68.858	221.889	44600
1955	49XN=51EP1+	2.85413	.06470	.997	258.252	156.541	171.527	44600

NUMBER	NAME	A	E	I	NODE	PERI	ANOM	DAYEP
1956	ARTEK	3.20951	.10481	1.430	153.981	321.821	333.555	40800
1957	ANGARA	3.00931	.05981	11.207	50.871	207.345	337.927	40800
1958	65UN=70SB+	3.11421	.15424	10.521	345.606	318.609	16.796	40800
1959	KARBYSHEV	2.31597	.13277	6.196	284.871	30.966	17.038	41600
1960	GUISAN	2.52331	.12802	8.462	22.381	261.232	107.061	42000
1961	DUFOUR	3.20530	.10010	6.646	29.527	54.176	316.838	42000
1962	DUNANT	3.19937	.22197	1.636	15.977	1.311	35.582	42000
1963	BEZOVEC	2.42469	.20745	24.975	106.553	355.389	28.680	42400
1964	33UA12007PL	2.46655	.19190	2.361	238.412	161.987	60.094	44600
1965	56TN=2521PL	2.56764	.10906	2.228	87.864	341.340	88.838	43800
1966	51GJ=2552PL	2.44846	.08887	2.488	125.339	56.120	84.914	43800
1967	MENZEL	2.23315	.13923	3.904	57.486	346.843	15.622	42800
1968	MEHLTRETTER	2.73789	.11326	4.602	71.271	163.067	203.337	44600
1969	35CG=52HW3	3.10225	.14430	3.270	209.772	341.835	121.560	44600
1970	498F=54ER	2.78107	.15795	7.084	312.290	192.201	303.320	44600
1971	HAGIHARA	2.99326	.08447	8.691	300.361	120.363	337.550	41200
1972	YI XING	2.41757	.16978	4.130	46.473	32.092	130.197	42000
1973	680A=69VV1	3.17449	.09833	10.595	183.429	169.458	310.312	40000
1974	1968 OE	3.17235	.08620	10.194	168.305	100.525	21.617	40000
1975	PIKELNER	2.80140	.11835	6.308	170.227	183.516	30.143	42400
1976	KAVERIN	2.38146	.07471	2.375	90.627	118.499	27.888	40800
1977	SHURA	2.78023	.07161	7.768	332.388	308.409	322.141	42000
1978	52WC=71LD	2.19425	.21405	4.346	54.120	267.747	338.225	41200
1979	2006PL71SQ3	2.37403	.10077	6.052	202.378	220.059	147.880	44600
1980	TEZCATLIPOCA	1.70950	.36548	26.839	246.188	115.158	.848	42560
1981	MIDAS	1.77622	.64952	39.848	356.726	267.569	356.984	42680
1982	CLINE	2.30940	.25046	6.837	42.261	278.330	106.352	44200
1983	BOK	2.62147	.09969	9.410	23.471	344.797	298.027	44200
1984	1926TN73SE3	3.01062	.08683	4.775	185.802	125.160	120.245	44200
1985	HOPMANN	3.12250	.14952	17.277	305.803	226.543	43.648	44200
1986	1935SV173SA	3.10094	.19329	2.202	146.518	226.400	45.914	44200
1987	1952RH	2.38419	.22556	23.655	313.907	37.847	139.761	44200
1988	DELURES	2.15347	.10278	4.254	105.990	234.797	244.027	44200
1989	1944DL71SJ2	2.35134	.07628	7.703	81.031	88.176	354.292	44200
1990	1956EE73QM	2.17428	.05097	3.130	193.208	11.630	108.722	44200
1991	1967JL71SU2	2.25010	.20657	5.914	328.172	344.889	205.885	44200
1992	1968DD	2.99280	.04786	10.563	182.565	94.774	95.220	44200
1993	1968DH1	3.05822	.06959	11.458	158.152	149.323	43.050	44200
1994	1939RN61TE	2.67903	.20768	10.207	244.909	87.973	79.811	44200
1995	HAJEK	2.52792	.05806	10.804	46.819	130.564	227.664	44200
1996	ADAMS	2.55930	.13841	15.108	.665	353.688	177.674	44200
1997	LEVERRIER	2.20932	.20691	6.071	352.859	359.967	332.547	44200
1998	TITIUS	2.41846	.06511	7.637	351.680	244.434	321.588	44200
1999	HIRAYAMA	3.11391	.11729	12.514	148.279	346.606	101.533	44200
2000	HERSCHEL	2.38021	.29955	22.757	291.714	129.740	30.810	44200
2001	EINSTEIN	1.93348	.09840	22.687	356.610	217.436	131.530	44200
2002	EULER	2.41651	.06971	8.521	178.433	50.891	343.580	44200
2003	1941BH73AG1	3.05932	.12671	1.882	64.595	64.604	89.029	44200
2004	LEXELL	2.17232	.07929	2.500	4.224	57.502	288.318	44200
2005	HENCKE	2.62154	.16636	12.226	291.015	110.646	126.657	44200
2006	KONSTITUTSIYA	2.32421	.19338	4.925	.701	23.629	257.280	44200
2007	1941SW170QD	2.38403	.11717	3.052	16.835	184.468	312.252	44200
2008	1938SV73SV4	3.21064	.10299	20.689	15.212	210.411	185.545	44200
2009	VOLOSHINA	3.11848	.13645	2.860	107.286	10.693	302.267	44200
2010	CHEBYSHEV	3.09301	.18773	2.438	9.424	27.151	301.525	44200
2011	VETERANIYA	2.38788	.14846	6.185	338.349	3.220	194.195	44200
2012	GUO SHOU-JING	2.32832	.17899	2.911	277.003	35.461	140.387	44200
2013	1936PL71UH4	2.29156	.22458	7.513	96.281	237.249	166.719	44200
2014	VASILEVSKIS	2.40344	.28363	21.415	203.991	81.850	244.423	44200
2015	KACHUEVSKAYA	2.33571	.10401	11.914	344.194	274.107	80.722	44200
2016	HEINEMANN	3.15505	.17330	.929	17.243	341.118	142.659	44200
2017	WESSON	2.25310	.18497	4.858	170.991	135.747	226.089	44200
2018	SCHUSTER	2.18321	.19280	2.555	185.682	157.119	353.533	44200
2019	35SX1=75ND	2.24054	.16621	4.047	251.904	24.077	128.615	44200
2020	36FR=73EP1	3.02473	.06356	11.105	148.768	326.586	170.780	44200
2021	POINCARE	2.30873	.22015	5.484	154.647	163.391	103.948	44200
2022	WEST	2.70574	.11920	5.669	2.404	36.201	247.529	44200
2023	ASAPH	2.87996	.27749	22.311	2.845	357.091	195.140	44200
2024	52UP=38WP	2.32487	.13904	7.313	68.913	290.916	251.527	44200
2025	35EQ=53LG	3.17497	.10048	7.002	330.568	294.611	258.221	44200
2026	COTTRELL	2.44589	.11882	2.466	311.333	209.123	182.749	44200
2027	53VQ1=64VR1	3.02318	.09362	11.027	55.123	351.892	323.946	44200
2028	1968OB1	2.29622	.11307	7.951	242.521	27.087	122.453	44200
2029	69RB=71BX2	2.34983	.12856	5.589	277.807	66.446	305.997	44200
2030	BELYAEV	2.24730	.09335	2.579	169.402	61.660	150.523	44200
2031	BAM	2.23410	.17241	4.750	168.888	212.976	11.788	44200
2032	ETHEL	3.06483	.13281	1.521	29.851	298.883	251.621	44200
2033	53FY=73CA	2.22583	.11136	8.460	321.422	133.995	54.975	44200
2034	41SQ=73EE	2.24644	.18030	8.565	18.810	63.223	69.971	44200
2035	STEARNS	1.88405	.13129	27.755	76.561	200.183	198.825	44200
2036	SHERAGUL	2.24454	.18558	3.974	345.943	305.868	359.608	44200
2037	1973UB	2.30112	.13258	4.257	9.179	345.485	300.619	44200
2038	1973WF	2.43597	.08836	14.765	73.143	182.097	28.627	44200
2039	PAYNE-GAPOSCHKIN	3.16722	.14797	2.528	95.596	46.538	1.484	44200
2040	72XP=74HA	3.10399	.19933	14.660	39.559	85.938	57.424	44200

NUMBER	NAME	A	E	I	NODE	PERI	ANOM	DAYEP
2041	73AP3=2523P	3.15679	.19821	2.982	133.696	277.718	114.893	44200
2042	718Q=4633PL	2.75356	.14906	5.338	17.136	54.877	15.596	44200
2043	36TH=74SW1	3.10149	.11732	3.101	322.157	52.448	185.265	41400
2044	1950VE	2.37978	.34362	24.069	53.561	49.391	350.066	43000
2045	440J=73AE	2.37996	.05470	6.909	10.916	196.504	341.211	42000
2046	34RK=76QS	3.14539	.19027	2.743	74.319	275.353	7.015	42000
2047	1971UA1	1.87213	.00340	25.278	36.157	303.542	343.296	42000
2048	DWORNIK	1.95370	.04271	23.755	157.208	105.210	81.983	42000
2049	1973SH	1.94895	.08429	24.420	199.637	141.309	47.277	42000
2050	FRANCIS	2.32506	.23800	26.598	72.120	170.378	218.608	43000
2051	CHANG	2.84108	.07845	1.344	215.334	174.131	346.489	43000
2052	28TD=76UN	3.00749	.08315	9.507	213.937	201.844	29.979	43000
2053	28RW=76U0	2.80147	.14190	8.522	193.215	235.601	53.693	43200
2054	73FG14097PL	2.96159	.10298	3.789	293.690	182.694	106.195	42000
2055	1974DB	2.31203	.30919	21.517	340.245	243.617	212.449	43000
2056	NANCY	2.21774	.13843	3.922	225.886	144.138	349.243	30600
2057	ROSEMARY	3.08273	.22964	1.442	15.173	19.331	318.427	43400
2058	388H=78A0	3.11348	.15953	2.543	95.119	182.179	351.222	43800
2059	1963UA	2.64330	.53061	11.051	200.691	191.000	32.844	43200
2060	CHIRON	13.69545	.37860	6.923	208.715	339.104	229.072	43400
2061	ANZA	2.26472	.53748	3.735	207.429	155.792	106.257	43800
2062	ATEN	.96627	.18243	18.935	108.069	147.823	194.890	43800
2063	BACCHUS	1.07749	.34931	9.419	32.784	54.889	206.274	43800
2064	THOMSEN	2.17821	.32981	5.702	301.841	1.799	110.234	43800
2065	73YR259RN	2.69919	.23285	6.446	328.237	65.460	3.846	43480
2066	75NZ34LB	2.39482	.12551	3.759	116.903	109.341	16.807	43800
2067	AKSNES	3.94966	.18440	3.066	150.369	302.852	206.087	43800
2068	DANGREEN	2.77279	.09729	12.887	95.865	319.340	294.897	43800
2069	75TT355FT	3.16621	.18043	9.214	48.233	59.109	129.998	43800
2070	74RB264TQ	2.25023	.15431	2.756	.993	349.361	75.872	43800
2071	75XD371QS	2.25158	.15761	3.635	301.920	359.895	59.384	43800
2072	75EL73QE2	2.45095	.16383	4.767	26.074	36.520	66.713	43800
2073	72TU674DK	2.71544	.11337	2.969	84.776	358.970	74.028	43800
2074	SHOEMAKER	1.79980	.08178	30.075	206.772	205.059	207.415	43800
2075	1974VA	2.40469	.24757	27.011	111.813	354.378	343.076	43800
2076	LEVIN	2.27410	.15227	4.991	328.714	74.378	58.971	43800
2077	1974YA	2.32696	.29726	28.100	69.097	343.435	37.828	43800
2078	1975AD	2.36970	.37505	20.135	287.027	98.742	48.605	43800
2079	JACCHIA	2.59720	.08187	13.303	352.782	145.161	245.130	43800
2080	76DG55SH1	2.17691	.06093	3.853	23.530	51.185	10.529	43800
2081	76DH46LA	2.44957	.16454	3.926	66.243	221.013	135.507	43800
2082	74MB=7588PL	2.92189	.16231	3.071	89.998	144.596	333.057	43800
2083	SMITHER	1.87212	.05132	18.454	258.719	228.540	346.610	44000
2084	1935CK	2.39533	.10215	4.835	148.760	247.300	16.636	43800
2085	1965YA	2.69853	.08428	3.885	117.932	293.838	.925	43800
2086	1966BC	2.40124	.11222	6.488	134.823	295.208	192.471	43800
2087	1975YC	2.20571	.05747	1.831	95.228	319.521	338.887	43800
2088	1976DJ	2.20709	.07929	5.541	359.436	90.772	347.992	43800
2089	1977VF	2.53429	.15626	15.422	102.327	287.451	86.567	43800
2090	MIZUHO	3.06493	.14310	11.831	339.767	339.478	256.099	43800
2091	1941HO	3.01605	.05707	11.357	114.534	318.803	205.091	43800
2092	1969UP	2.84777	.03038	3.085	72.176	202.418	79.202	43800
2093	1971HX	2.26897	.16938	6.085	154.535	117.543	41.808	43800
2094	1971TC2	2.23210	.09734	5.026	281.682	250.927	273.190	43800
2095	73SS26036PL	2.64133	.00884	3.595	332.292	10.010	94.193	43800
2096	1939UC	2.44759	.22983	.990	304.676	38.226	99.840	43800
2097	1953PV	3.12227	.26230	4.385	319.083	30.403	180.807	43800
2098	1972QE	2.42307	.12865	6.522	337.496	354.119	229.151	43800
2099	OPIK	2.30314	.36311	26.927	218.435	158.721	105.738	43800
2100	RA-SHALOM	.83201	.43652	15.755	170.336	355.970	232.808	43800
2101	ADONIS	1.87278	.76390	1.370	351.207	41.056	258.448	43800
2102	1975YA	1.29006	.29834	64.014	93.736	61.657	206.240	44200
2103	1960FL30XM	3.15756	.17959	7.663	291.860	240.133	169.048	44200
2104	1963PD55HW	3.19960	.10160	18.411	252.444	291.675	135.173	44200
2105	1976DA	2.33932	.15097	29.307	273.179	155.327	94.651	44200
2106	1936UF62PN	2.70304	.09764	8.029	151.783	247.723	247.782	44200
2107	1941VA48LJ	2.62711	.07733	8.836	221.056	175.614	341.843	44200
2108	1948TR135FN	2.43608	.00542	10.787	327.521	51.938	77.997	44200
2109	1950TH232PA	2.68997	.26131	8.076	161.925	191.483	229.278	44200
2110	1962RD55MA	2.19797	.17755	1.130	139.942	191.951	119.624	44200
2111	1969LG28SO	3.01711	.09489	10.487	167.242	228.788	248.755	44200
2112	1972NP62RA	2.25380	.13713	3.368	243.435	155.431	341.567	44200
2113	1972RJ258AA	2.47375	.09769	6.460	22.985	347.322	302.866	44200
2114	1976HA53GZ	3.18581	.15578	.564	1.050	220.922	222.054	44200
2115	1976UD52HZ3	3.01092	.05481	8.975	241.328	338.794	104.408	44200
2116	1976UM49KP	2.58875	.05822	9.085	170.566	172.913	22.969	44200
2117	1978AC36VJ	2.87003	.07146	2.939	59.371	262.644	24.098	44200
2118	1978PB907CC	2.54765	.21648	6.331	331.224	101.399	22.010	44200

III. PROPER ELEMENTS AND FAMILY MEMBERSHIPS OF THE ASTEROIDS

J. G. WILLIAMS
Jet Propulsion Laboratory

This introduction to the table of proper elements and family identifications is not intended to be detailed or complete but rather to serve as a brief guide to the user. A more comprehensive description, including a discussion of results, is in preparation.

The table of proper elements conforms to the following format:

Column	Parameter
1	Asteroid number.
2	Proper semimajor axis (AU).
3	Proper eccentricity.
4	Proper sine of inclination.
5	Proper longitude of perihelion (deg); equinox and epoch of 1950.0.
6	Proper longitude of ascending node (deg); equinox and epoch of 1950.0.
7	Rate of proper longitude of perihelion (arcsec/yr).
8	Rate of proper longitude of node (arcsec/yr).
9	An index indicating a secular resonance was encountered. The values 1 to 16 identify the resonance number. Numbers 1 through 10 are resonances with the planetary eccentricities and longitudes of perihelion. Numbers 11 through 14 and 16 through 18 are resonances with the planetary inclinations and longitudes of node. A zero value indicates no secular resonance.
10	The closest distance of approach to Mars (AU) along the Sun-Mars-asteroid line. A negative value indicates a Mars crosser.
11	The closest distance of approach to Jupiter (AU) along the Sun-asteroid-Jupiter line.
12	The family identity if a family member.

The proper elements are calculated according to the secular perturbation theory of Williams (1969) and the free oscillations are referred to a zero value for the proper argument of perihelion. The secular theory for the planets is taken from Brouwer and van Woerkom (1950). The closest straight line distances of the asteroid's approach to Mars and Jupiter account for the secular variations of both the asteroid and the planets.

The families with numbers smaller than 100 are reasonably close matches to families found by previous investigators (Brouwer 1951; Arnold 1969; Lindblad and Southworth 1971) and their former numbers have been used. Families with numbers greater than 100 are new to this work. The objects identified as being in family 1A are members of a more compact core in the broader Themis family (No. 1). It should be appreciated that the quality and uniqueness of family identifications varies widely. Difficulties are encountered with families with few members or families which are crowded or overlapping.

The table should be used with the following warnings in mind. The strong secular resonances 5 and 6 prevent the calculation of reasonable proper elements (except for the semimajor axis) so that most of the line has been blanked out. In these cases the proper node or perihelion rate can be bad. For the weaker resonances the proper elements have been calculated, but they will have degraded accuracy compared to normal objects and will occasionally be quite poor. The weaker resonances can cause the failure of the closest approach calculations, and both values are blanked out in such cases. Proper elements should be expected to have lower accuracy for objects near secular resonances or near commensurabilities with Jupiter or for objects in orbits deeply crossing Mars or the earth. Objects in the outer part of the belt will have a poor accuracy for the proper longitude of perihelion rate.

Generally the table contains minor planets with numbers less than 1797 but several categories of objects are excluded due to a lack of a suitable theory: Trojans, Hildas, the commensurate objects 279, 887, 1362, and 1685, the argument of perihelion librator 1373, the Jupiter crosser 944, and objects 330 and 864.

Acknowledgment. This paper presents the results of one phase of research carried out at the Jet Propulsion Laboratory, California Institute of Technology, sponsored by the National Aeronautics and Space Administration.

REFERENCES

Arnold, J. R. 1969. Asteroid families and jet streams. *Astron. J.* 74: 1235–1242.
Brouwer, D. 1951. Secular variations of the orbital elements of minor planets. *Astron. J.* 56: 9–32.
Brouwer, D., and van Woerkom, A. J. J. 1950. The secular variations of the orbital elements of the principal planets. In *Astron. Papers Amer. Ephemeris* 13, Part 2: 85–107.
Lindblad, B. A., and Southworth, R. B. 1971. A study of asteroid families and streams by computer techniques. In *Physical Studies of Minor Planets*, ed. T. Gehrels (NASA SP-267, Washington, D.C.: U.S. Government Printing Office), pp. 337–352.
Williams, J. G., 1969. Secular Perturbations in the Solar System. Ph.D. Dissertation, University of California at Los Angeles.

NUMBER	A	E	SINI	WBAR	ANODE	DWBAR	DNODE	NRES	TOMARS	TOJUP	FAMILY
1	2.767	.097	.169	147.8	78.7	50.6	-58.3		.611	1.918	67	1
2	2.771	.180	.584	156.1	184.3	1.5	-50.1		.013	1.773	129	2
3	2.670	.218	.245	63.6	172.6	40.5	-60.5		.234	1.770		3
4	2.362	.097	.112	228.0	107.1	36.8	-39.5		.308	2.422	169	4
5	2.578	.215	.083	143.5	152.7	47.3	-58.5		.197	1.868		5
6	2.425	.146	.258	39.0	140.2	30.9	-41.2		.141	2.157		6
7	2.386	.210	.115	54.2	264.3	37.8	-46.2		.055	2.127		7
8	2.201	.141	.097	60.0	116.9	31.8	-35.0		.024	2.457	189	8
9	2.386	.125	.083	91.6	65.5	38.6	-41.8		.273	2.321	170	9
10	3.144	.136	.092	224.5	285.1	85.7	-99.5		.865	1.465	110	10
11	2.452	.072	.068	302.6	133.7	41.5	-43.0		.463	2.374		11
12	2.334	.172	.167	293.4	236.8	33.9	-40.6		.081	2.266	171	12
13	2.576	.121	.281	135.1	41.0	32.6	-45.6		.370	2.094		13
14	2.588	.191	.153	181.1	86.0	44.3	-55.2		.259	1.924	150	14
15	2.644	.143	.231	41.2	292.6	39.7	-52.0		.428	1.994	140	15
16	2.922	.100	.045	20.2	171.4	69.1	-73.4		.802	1.810		16
17	2.469	.141	.088	247.3	133.1	41.8	-46.3		.304	2.186		17
18	2.296	.174	.179	29.7	153.2	32.1	-39.0		.023	2.285		18
19	2.442	.131	.039	43.6	233.6	41.7	-44.8		.312	2.243	158	19
20	2.408	.162	.026	114.4	242.5	40.5	-44.9		.210	2.206	162	20
21	2.435	.127	.038	322.4	76.7	41.5	-44.3		.315	2.261	158	21
22	2.910	.109	.222	71.7	63.9	50.7	-65.6		.679	1.718		22
23	2.626	.249	.180	130.9	61.1	44.8	-63.7		.126	1.732		23
24	3.133	.159	.020	155.3	315.1	90.6	-105.5		.794	1.409	1A	24
25	2.400	.183	.417	295.4	214.6	17.0	-38.4		.192	2.071		25
26	2.656	.134	.052	222.1	28.4	51.9	-56.7		.425	1.940		26
27	2.347	.187	.012	102.2	104.6	38.2	-43.6		.094	2.226		27
28	2.776	.176	.153	138.5	151.9	53.4	-66.6		.432	1.735		28
29	2.554	.066	.110	87.8	347.2	44.5	-47.0		.568	2.278		29
30	2.366	.103	.050	49.6	295.9	38.4	-40.4		.309	2.398		30
31	3.156	.099	.469	84.1	23.9	.9	-60.4		.925	1.509		31
32	2.588	.114	.109	189.5	220.7	46.2	-50.5		.468	2.121		32
33	2.865	.300	.039	341.6	334.0	68.9	-104.1		.166	1.308		33
34	2.687	.153	.100	157.7	197.4	51.9	-59.5		.383	1.852		34
35	2.997	.254	.158	200.5	341.2	68.9	-105.0		.366	1.278		35
36	2.747					27.8	-45.2	6				36
37	2.642	.165	.061	80.4	350.4	51.3	-62.3	10			142	37
38	2.740	.163	.141	120.5	293.8	52.4	-63.1		.424	1.795		38
39	2.769	.088	.172	10.3	162.7	50.2	-57.7		.661	1.963	67	39
40	2.267	.019	.064	82.2	99.3	34.6	-34.9		.399	2.702		40
41	2.765	.279	.291	211.9	176.0	40.5	-75.3		.111	1.493		41
42	2.441	.186	.137	314.1	81.4	39.0	-47.0		.155	2.120	157	42
43	2.203	.140	.071	260.7	261.0	32.4	-35.2		.046	2.473	185	43
44	2.422	.177	.054	124.2	145.5	40.7	-46.4		.178	2.158	24	44
45	2.721	.115	.107	217.5	155.8	53.1	-58.6		.563	1.961	133	45
46	2.525	.134	.044	355.8	206.0	45.4	-49.2		.372	2.139		46
47	2.881	.111	.092	302.1	351.5	63.6	-69.9		.729	1.819		47
48	3.112	.064	.115	113.4	194.7	78.1	-84.9		1.067	1.722		48
49	3.090	.193	.085	40.7	287.9	83.1	-105.8		.643	1.352		49
50	2.650	.236	.048	12.7	198.0	52.4	-66.6		.167	1.704		50
51	2.366	.111	.177	170.4	180.6	34.1	-39.0		.259	2.383		51
52	3.097	.119	.113	124.9	135.6	78.4	-90.0		.879	1.569		52
53	2.618	.215	.083	104.0	157.7	49.4	-61.2		.213	1.810		53
54	2.710	.179	.221	289.9	311.7	43.8	-59.6		.383	1.813	138	54
55	2.760	.102	.124	15.4	1.8	53.9	-59.6		.633	1.962		55
56	2.598	.208	.160	287.9	201.1	44.4	-57.4		.219	1.870		56
57	3.153	.095	.271	61.4	202.7	50.1	-77.8		.944	1.547	108	57
58	2.700	.088	.083	189.7	173.5	53.0	-56.1		.623	2.053	132	58
59	2.713	.094	.147	27.9	177.1	49.7	-55.6		.615	2.015		59
60	2.393	.201	.076	111.7	206.3	39.2	-46.3		.092	2.136		60
61	2.984	.122	.319	341.4	330.6	36.4	-65.0		.730	1.686		61
62	3.122	.146	.023	47.6	151.2	88.8	-101.0		.825	1.462	1A	62
63	2.395	.119	.110	254.1	330.5	38.1	-41.7		.289	2.327	165	63
64	2.682	.151	.041	141.3	293.1	53.7	-59.9		.443	1.912		64
65	3.429	.129	.056	244.9	175.6	126.5	-146.6		1.122	1.189		65
66	2.646	.171	.059	41.4	326.1	51.3	-58.8	10	.330	1.869	142	66
67	2.421	.152	.119	238.7	211.0	38.8	-44.1		.226	2.223		67
68	2.782	.144	.132	343.8	39.0	55.0	-64.4		.536	1.827	126	68
69	2.979	.174	.160	124.7	195.4	64.8	-84.0		.607	1.526		69
70	2.616	.144	.198	289.7	43.0	41.8	-51.9		.396	2.017	138	70
71	2.755	.117	.434	222.7	314.3	13.6	-46.0		.642	1.859		71
72	2.266	.077	.105	292.1	212.4	33.7	-35.4		.255	2.564	174	72
73	2.665	.038	.044	136.9	342.8	52.2	-52.8		.709	2.206		73
74	2.780	.199	.077	10.7	214.5	58.9	-72.2		.393	1.683		74
75	2.671	.267	.091	331.3	348.5	52.6	-72.3		.092	1.607		75
76	3.390	.186	.053	105.8	232.3	125.9	-164.2		.890	1.043		76
77	2.668	.109	.048	80.8	338.0	52.5	-55.7		.530	2.025	141	77
78	2.620	.232	.166	132.2	330.3	45.3	-61.5		.169	1.780		78
79	2.444	.175	.090	57.0	216.4	40.8	-47.1		.196	2.134	75	79
80	2.296	.147	.162	7.2	223.4	32.7	-38.0		.095	2.353		80

NUMBER	A	E	SINI	WBAR	ANODE	DWBAR	DNODE	NRES	TOMARS	TOJUP	FAMILY	
81	2.854	.179	.149	57.8	351.9	58.2	-73.5		.496	1.649		81
82	2.765	.246	.051	142.1	5.9	59.6	-78.8		.251	1.576		82
83	2.431	.120	.081	187.7	18.2	40.4	-43.5		.324	2.281		83
84	2.362	.190	.169	340.9	321.9	34.8	-43.0		.068	2.204	171	84
85	2.654	.143	.225	320.5	209.2	40.6	-52.7		.440	1.980	140	85
86	3.108	.176	.067	35.7	85.9	85.8	-105.1		.711	1.385		86
87	3.486	.051	.171	320.0	71.3	98.7	-121.6		1.440	1.390		87
88	2.768	.143	.111	297.5	275.5	55.7	-63.9		.532	1.846		88
89	2.552	.089	.296	37.9	308.4	29.9	-42.2		.289	2.068		89
90	3.148	.150	.024	295.7	43.3	91.7	-105.5		.832	1.419	1A	90
91	2.590	.113	.039	100.1	343.3	48.6	-51.5		.474	2.114		91
92	3.193	.061	.152	311.3	102.4	78.8	-90.0		1.152	1.644	106	92
93	2.755	.138	.158	41.4	135.6	51.3	-61.0	10	.313	1.661	127	93
94	3.158	.068	.145	77.5	355.4	77.1	-87.5		1.100	1.658	106	94
95	3.068	.112	.241	46.7	247.4	54.3	-76.3		.858	1.605	116	95
96	3.051	.164	.298	162.2	317.2	43.9	-77.7		.676	1.469		96
97	2.668	.228	.222	75.6	164.9	42.9	-62.7		.212	1.736		97
98	2.687	.225	.280	156.1	352.6	37.6	-61.2		.217	1.733		98
99	2.664	.215	.231	225.9	36.1	41.6	-60.3		.244	1.779		99
100	3.096	.145	.094	295.0	135.0	81.1	-95.1		.799	1.488	114	100
101	2.584	.104	.182	322.9	338.8	41.5	-48.0		.485	2.159	144	101
102	2.661	.234	.105	1.2	225.4	51.3	-66.5	10	.138	1.663		102
103	2.702	.058	.081	293.2	145.5	53.1	-55.1		.707	2.132	134	103
104	3.149	.141	.044	80.5	16.0	90.6	-103.2		.861	1.447		104
105	2.374	.168	.387	234.7	187.0	19.1	-37.1		.217	2.131		105
106	3.172	.136	.064	39.8	52.3	91.2	-104.3		.890	1.440		106
107	3.488	.084	.171	139.4	182.8	100.2	-128.0		1.318	1.273		107
108	3.218	.123	.086	177.2	337.4	93.6	-107.0		.970	1.427	101	108
109	2.696	.277	.167	64.1	351.9	49.6	-74.4		.076	1.553		109
110	2.733	.047	.090	315.0	49.9	54.3	-56.3		.776	2.139	130	110
111	2.593	.124	.102	125.1	300.4	46.8	-51.6		.446	2.087		111
112	2.434	.090	.056	335.4	309.5	41.1	-42.9		.403	2.350	161	112
113	2.376	.123	.077	190.0	130.1	38.4	-41.3		.268	2.342	170	113
114	2.676	.181	.083	165.4	178.6	52.2	-61.4		.325	1.816		114
115	2.380	.171	.223	66.3	307.7	32.2	-41.5		.083	2.184	163	115
116	2.768	.176	.050	162.0	51.9	58.9	-68.6		.454	1.761		116
117	2.991	.028	.264	135.6	345.4	45.3	-62.8		1.056	1.952		117
118	2.439	.164	.128	92.2	40.8	39.2	-45.6		.210	2.174		118
119	2.581	.049	.108	25.8	212.7	45.7	-47.8		.634	2.294		119
120	3.118	.088	.135	204.6	332.9	76.1	-86.6		.992	1.643		120
121	3.451	.089	.116	1.5	71.3	114.5	-132.3		1.268	1.297		121
122	3.222	.071	.033	183.6	220.0	97.5	-101.3		1.152	1.590		122
123	2.696	.125	.130	84.6	304.7	50.3	-57.0		.490	1.937		123
124	2.630	.080	.056	229.3	207.2	50.2	-52.2		.572	2.127	141	124
125	2.743	.086	.074	250.1	183.0	55.7	-58.9		.678	2.026	132	125
126	2.439	.069	.046	349.2	5.2	41.4	-42.5		.460	2.396		126
127	2.756	.092	.139	152.6	24.8	52.4	-58.4		.636	1.974		127
128	2.750	.088	.091	24.0	73.2	55.5	-59.1		.677	2.015	132	128
129	2.872	.228	.226	41.4	310.0	51.3	-79.0	10	.112	1.258		129
130	3.119					27.8	-76.7	6				130
131	2.431	.098	.073	205.0	60.9	40.6	-42.8		.380	2.339	161	131
132	2.611	.212	.533	143.0	253.6	8.1	-47.2		.089	1.701		132
133	3.065	.148	.150	237.6	315.3	71.4	-88.8		.760	1.510		133
134	2.565	.105	.211	91.6	342.1	38.5	-46.4		.459	2.183		134
135	2.427	.174	.048	315.8	323.6	41.1	-46.5		.191	2.158	160	135
136	2.287	.023	.173	286.2	189.5	31.7	-34.4		.364	2.632		136
137	3.119	.159	.264	296.9	208.2	53.1	-86.2		.719	1.399	113	137
138	2.447	.138	.044	304.4	42.1	41.9	-45.5		.298	2.220	158	138
139	2.785	.200	.195	166.9	357.1	50.2	-68.6		.344	1.642		139
140	2.732	.201	.036	291.7	112.9	57.3	-69.0		.357	1.732		140
141	2.666	.162	.233	18.7	313.8	40.8	-54.9		.390	1.917	140	141
142	2.419	.159	.058	210.9	286.1	40.5	-45.1		.223	2.201	24	142
143	2.761	.094	.212	205.0	328.5	45.8	-56.0		.671	1.984		143
144	2.655	.196	.072	10.5	70.0	51.3	-56.9	10			136	144
145	2.673	.160	.208	128.6	74.2	43.4	-56.1		.410	1.905	138	145
146	2.719	.086	.211	212.0	84.0	43.9	-53.1		.658	2.056		146
147	3.137	.011	.053	195.1	260.5	86.6	-87.6		1.265	1.865		147
148	2.771	.098	.433	22.3	145.0	12.5	-45.0		.711	1.902		148
149	2.175	.079	.025	95.8	183.2	32.0	-32.8		.159	2.635		149
150	2.982	.090	.050	354.1	231.1	73.5	-77.6		.883	1.777		150
151	2.592	.069	.103	171.6	31.3	46.6	-49.1		.592	2.230		151
152	3.140	.074	.204	117.8	35.7	64.5	-81.4		1.053	1.662		152
154	3.184					27.8	.0	6				154
155	2.759	.262	.212	91.1	31.7	48.5	-75.7		.164	1.528		155
156	2.729	.246	.196	214.6	249.5	48.2	-70.7		.198	1.609		156
157	2.579	.209	.210	114.5	56.9	40.0	-54.6		.196	1.900		157
158	2.869	.045	.038	103.7	277.8	65.1	-66.1		.917	2.020	3	158
159	3.106	.117	.091	126.8	143.8	81.6	-91.5		.893	1.565	112	159
160	2.728	.052	.069	95.0	352.7	55.1	-56.6		.760	2.134	134	160
161	2.379	.095	.155	300.3	13.9	35.6	-39.4		.319	2.414		161

NUMBER	A	E	SINI	WBAR	ANODE	DWBAR	DNODE	NRES	TOMARS	TOJUP	FAMILY	
162	3.022	.206	.105	154.6	29.9	75.4	-98.7		.546	1.389		162
163	2.368	.208	.082	107.5	174.0	38.1	-45.6		.053	2.149	166	163
164	2.633	.213	.463	1.8	79.9	15.2	-49.7		.241	1.726		164
165	3.130	.075	.215	264.5	301.5	61.7	-79.6		1.049	1.660		165
166	2.686	.167	.208	39.5	132.3	43.7	-57.3		.393	1.885	138	166
167	2.854	.043	.037	236.7	197.2	64.0	-64.9		.909	2.042	3	167
168	3.379	.025	.089	54.5	221.3	109.4	-114.9		1.439	1.585		168
169	2.358	.093	.096	321.4	346.3	37.2	-39.3		.320	2.434		169
170	2.554	.099	.266	121.2	299.6	33.1	-43.8		.425	2.188	4	170
171	3.134	.161	.024	160.0	101.8	90.6	-106.1		.787	1.401	1A	171
172	2.380	.070	.178	324.7	327.8	34.5	-38.3		.378	2.469		172
173	2.743	.160	.243	16.3	150.2	42.4	-58.9		.455	1.835		173
174	2.861	.133	.235	244.4	324.9	47.3	-63.9		.633	1.769		174
175	3.212	.176	.057	330.8	2.5	98.5	-121.0		.794	1.268		175
176	3.178	.153	.393	16.3	203.6	23.7	-76.4		.768	1.211		176
177	2.770	.198	.040	28.1	314.7	59.6	-71.7		.393	1.700		177
178	2.460	.059	.022	224.4	24.8	42.6	-43.3		.507	2.395		178
179	2.972	.070	.157	350.3	255.8	62.9	-71.1		.926	1.840	121	179
180	2.722	.190	.034	136.4	290.5	56.6	-66.9		.378	1.772		180
181	3.132	.195	.325	114.4	151.8	42.0	-88.7		.613	1.293		181
182	2.416	.175	.019	67.9	123.9	40.9	-46.1		.183	2.164	162	182
183	2.795	.183	.503	40.2	141.1	8.8	-52.8		.366	1.585		183
184	3.183	.113	.038	187.4	304.0	93.8	-102.2		.979	1.499	103	184
185	2.739	.099	.389	356.3	154.7	20.0	-46.0		.681	1.900		185
186	2.362	.075	.222	334.3	12.3	31.3	-36.8		.282	2.409		186
187	2.732	.256	.183	210.0	12.8	49.9	-73.5		.155	1.561		187
188	2.762	.141	.227	295.6	242.6	44.6	-58.8		.533	1.860		188
189	2.450	.011	.098	89.8	211.9	40.5	-41.6		.611	2.526		189
191	2.896	.047	.192	38.0	164.9	53.5	-62.4		.909	1.976		191
192	2.403	.207	.130	16.6	333.5	37.9	-46.5		.072	2.114		192
193	2.600	.265	.252	80.0	345.7	38.2	-61.5		.043	1.741		193
194	2.818	.166	.306	333.7	166.1	31.5	-49.6		.241	1.902		194
195	2.879	.068	.123	154.5	358.5	60.7	-65.6		.848	1.943		195
196	3.114	.039	.108	216.7	68.5	79.1	-83.9		1.151	1.798		196
197	2.739	.130	.137	309.3	78.7	52.2	-60.0		.511	1.881	133	197
198	2.458	.175	.193	.0	268.4	36.6	-46.0		.189	2.131		198
199	3.167	.189	.249	252.6	90.9	60.5	-100.3		.686	1.268		199
200	2.737	.084	.136	70.7	318.8	51.8	-56.9		.608	1.975		200
201	2.678	.140	.094	315.3	168.2	51.6	-57.9		.372	1.856		201
202	3.072	.127	.139	146.0	143.1	72.9	-86.8		.827	1.572		202
203	2.737	.039	.064	84.2	331.5	55.8	-56.9		.809	2.162		203
204	2.671	.177	.156	247.7	210.0	47.9	-59.3		.359	1.858		204
205	2.777	.019	.195	94.9	217.1	47.8	-54.8		.891	2.176		205
206	2.740	.050	.056	135.2	161.2	56.1	-57.4		.777	2.130		206
207	2.284	.064	.058	183.8	20.5	35.3	-36.1		.317	2.586		207
208	2.893	.045	.037	170.8	332.1	66.8	-67.8		.938	1.997	3	208
209	3.148	.076	.132	227.1	351.5	79.1	-88.5		1.059	1.648		209
210	2.722	.095	.085	60.2	20.8	54.1	-57.7		.632	2.019	132	210
211	3.044	.149	.088	90.4	268.2	77.3	-90.3		.745	1.533		211
212	3.113	.090	.094	74.2	307.7	81.4	-88.8		.989	1.642		212
213	2.754	.143	.103	257.8	128.7	55.4	-63.1		.521	1.861		213
214	2.611	.057	.069	143.7	328.0	48.8	-50.2		.639	2.234	143	214
215	2.767	.015	.029	252.3	346.9	58.5	-58.7		.905	2.203		215
216	2.795	.224	.235	40.1	220.8	46.6	-71.0		.310	1.590		216
217	2.869	.276	.188	312.1	175.0	57.5	-91.8		.208	1.363		217
218	2.667	.130	.264	208.6	173.0	36.8	-50.9		.459	2.013	137	218
219	2.354	.164	.198	345.8	206.7	32.9	-40.5		.101	2.251		219
220	2.349	.203	.159	334.0	258.2	34.9	-43.5		.029	2.191	171	220
221	3.012	.071	.174	318.9	147.0	62.8	-73.2		.957	1.796	2	221
222	3.135	.157	.019	247.0	57.7	90.7	-105.3		.801	1.412	1A	222
223	3.089	.136	.027	124.4	10.8	85.0	-95.1		.831	1.525	1	223
224	2.645	.048	.107	233.9	343.5	48.9	-51.1		.690	2.219		224
225	3.382	.150	.421	300.7	203.1	15.0	-88.8		1.004	1.048		225
226	2.712	.172	.268	279.9	140.2	38.5	-56.8		.384	1.855		226
227	3.145	.214	.201	221.9	321.1	70.7	-109.9		.591	1.226		227
228	2.201	.183	.053	330.7	299.5	32.6	-37.0		-.049	2.382		228
229	3.411	.120	.036	319.3	356.5	125.4	-141.2		1.139	1.236		229
230	2.382	.038	.177	65.5	241.5	34.5	-37.8		.456	2.542		230
231	2.919	.166	.105	244.3	339.6	66.3	-79.4		.593	1.620		231
232	2.553	.205	.100	194.2	159.7	45.5	-55.4		.205	1.927	152	232
233	2.660	.064	.147	343.2	228.0	47.3	-51.4		.664	2.168		233
234	2.386	.162	.251	340.0	147.3	30.5	-40.6		.050	2.143		234
235	2.882	.073	.141	242.5	62.1	59.1	-65.3		.840	1.921		235
236	2.800	.149	.134	358.4	195.9	55.9	-66.0		.539	1.796	126	236
237	2.763	.092	.152	247.6	83.1	51.9	-58.4		.622	1.939		237
238	2.907	.103	.218	41.4	185.0	51.3	-65.3	10	.666	1.712		238
239	2.970	.196	.107	32.0	191.0	70.4	-89.4		.536	1.481		239
240	2.664	.180	.020	64.6	136.5	53.3	-61.6		.348	1.853		240
241	3.050	.065	.118	333.6	272.4	72.9	-79.2		1.009	1.782		241
242	2.864	.127	.214	131.7	213.3	50.0	-64.7		.635	1.755		242

NUMBER	A	E	SINI	WBAR	ANODE	DWBAR	DNODE	NRES	TOMARS	TOJUP	FAMILY
243	2.862	.045	.036	122.6	300.6	64.6	-65.6		.909	2.026	3	243
244	2.174	.103	.060	39.8	212.5	31.6	-33.2		.096	2.574		244
245	3.091	.168	.074	33.5	53.5	83.1	-99.7		.729	1.424		245
246	2.695	.100	.269	230.2	165.8	36.7	-50.1		.566	2.064		246
247	2.742	.151	.446	42.9	353.7	13.9	-48.3		.519	1.776		247
248	2.472	.075	.086	225.8	251.2	41.9	-43.7		.475	2.344	156	248
249	2.378	.173	.181	23.5	328.2	34.6	-42.4		.120	2.225		249
250	3.148	.114	.229	112.3	19.9	60.5	-84.6		.930	1.519		250
251	3.094	.095	.176	108.5	163.2	67.8	-81.8		.953	1.641	115	251
252	3.157	.037	.181	359.7	209.0	70.1	-82.5		1.198	1.759		252
253	2.647	.230	.120	334.5	192.9	49.4	-34.1		.184	1.737		253
254	2.195	.116	.070	231.1	22.9	32.1	-34.1		.088	2.531	188	254
255	2.746	.107	.164	167.2	7.9	50.0	-57.6		.593	1.943		255
256	3.001	.100	.233	211.2	187.4	52.7	-70.5		.826	1.696	120	256
257	3.114	.102	.057	79.2	15.5	84.9	-91.3		.954	1.606		257
258	2.616	.159	.255	8.7	213.3	36.7	-50.9		.342	1.995	21	258
259	3.139	.140	.170	237.8	88.1	73.3	-93.9		.840	1.455	109	259
260	3.445	.084	.105	323.4	178.9	116.4	-131.4		1.285	1.322		260
261	2.331	.132	.050	162.3	99.9	37.2	-40.0		.212	2.370	172	261
262	2.555	.197	.130	70.4	29.2	44.2	-54.2		.220	1.950	43	262
263	2.887	.042	.037	23.3	245.1	66.3	-67.3		.943	2.010	3	263
264	2.798	.090	.167	37.3	45.6	52.2	-59.8		.674	1.923	67	264
265	2.419	.175	.485	218.6	329.8	10.1	-37.1		.177	2.077		265
266	2.804	.125	.247	37.9	240.4	43.3	-59.0		.607	1.862		266
267	2.774	.113	.087	247.9	69.0	57.2	-62.3		.631	1.925		267
268	3.097	.170	.025	185.6	139.2	86.7	-102.4		.732	1.413	1	268
269	2.616	.204	.096	255.6	169.4	48.7	-59.6		.244	1.848		269
270	2.198	.092	.053	334.8	250.6	32.5	-33.7		.151	2.585		270
271	3.006	.067	.076	42.9	322.4	73.5	-77.3		.972	1.816		271
272	2.778	.050	.069	151.1	24.1	58.0	-59.6		.815	2.095	38	272
273	2.395	.149	.364	279.8	162.9	21.1	-37.3		.275	2.141		273
274	3.044	.158	.046	205.6	92.8	80.3	-93.2		.725	1.512		274
275	2.771	.201	.071	173.4	142.7	58.7	-71.8		.383	1.688		275
276	3.115	.042	.380	170.1	214.4	19.1	-59.3		1.165	1.687		276
277	2.886	.051	.037	3.7	254.0	66.3	-67.6		.915	1.984	3	277
278	2.755	.172	.125	197.3	58.9	54.3	-65.6		.430	1.779	131	278
280	2.943	.112	.133	116.3	2.6	64.4	-73.6		.771	1.751		280
281	2.188	.134	.084	75.2	26.6	31.6	-34.3		.026	2.484	189	281
282	2.339	.099	.153	108.3	150.4	34.4	-38.0		.270	2.438		282
283	3.046	.105	.162	352.7	300.6	67.3	-79.7		.880	1.661	118	283
284	2.358	.189	.159	277.8	235.0	34.9	-42.8		.063	2.207	171	284
285	3.089	.169	.316	316.4	308.5	41.3	-80.0		.672	1.424		285
286	3.194	.021	.297	140.5	152.7	42.4	-72.0		1.255	1.780		286
287	2.353	.047	.170	183.9	146.2	34.0	-37.0		.404	2.547		287
288	2.760	.242	.066	199.8	126.8	58.8	-77.5		.254	1.590		288
289	2.874	.166	.115	11.2	193.0	62.2	-74.7		.553	1.670		289
290	2.337	.188	.406	127.8	12.0	17.5	-36.7		.145	2.128		290
291	2.222	.141	.036	143.7	183.9	33.5	-36.1		.078	2.467	187	291
292	2.529	.018	.248	179.2	41.1	33.8	-41.3		.628	2.434		292
293	2.863	.120	.263	154.8	59.6	43.0	-61.1		.664	1.819	125	293
294	3.148	.213	.092	312.1	146.2	89.3	-120.7		.616	1.223		294
295	2.797	.152	.068	75.2	278.6	59.9	-67.9		.543	1.794		295
296	2.229	.123	.026	26.4	143.7	33.8	-35.7		.129	2.505	187	296
297	3.171	.110	.144	314.6	326.3	79.7	-94.2		.970	1.514		297
298	2.264	.145	.107	147.4	2.6	33.6	-37.3		.094	2.412	180	298
299	2.434	.041	.043	66.9	252.4	41.3	-41.7		.524	2.468		299
300	3.208	.011	.018	282.8	316.0	95.8	-96.1		1.332	1.796		300
301	2.725	.083	.074	236.6	154.1	54.7	-57.5		.672	2.051	132	301
302	2.406	.105	.061	81.5	352.7	39.8	-42.0		.343	2.346	161	302
303	3.122	.045	.130	92.1	336.4	76.6	-83.6		1.136	1.772		303
304	2.403	.118	.263	339.6	162.8	29.7	-38.8		.108	2.168		304
305	3.089	.198	.098	114.7	220.6	82.0	-106.4		.621	1.337		305
306	2.358	.117	.120	296.5	148.4	36.4	-39.8		.254	2.383	165	306
307	2.908	.125	.087	78.1	103.7	66.1	-73.6		.711	1.751		307
308	2.750	.045	.078	238.1	196.3	55.9	-57.5		.800	2.133	130	308
309	2.665	.116	.073	271.5	342.0	51.7	-55.7		.437	1.941	139	309
310	2.762	.153	.073	190.4	243.5	57.5	-65.3		.511	1.828		310
311	2.898	.041	.037	180.2	73.1	67.1	-68.0		.954	2.005	3	311
312	2.782	.173	.169	249.8	358.6	52.1	-66.1		.409	1.707		312
313	2.376	.226	.214	139.0	184.9	33.0	-45.2		-.039	2.078		313
314	3.151	.144	.210	358.7	175.8	65.5	-91.7		.839	1.420		314
315	2.242	.120	.044	331.4	178.4	34.0	-36.1		.149	2.501	186	315
316	3.175	.134	.024	96.4	149.1	94.2	-105.6		.906	1.442	1	316
317	2.287	.042	.031	331.8	174.8	35.7	-36.0		.375	2.631		317
318	3.199	.075	.178	126.7	168.8	74.0	-86.3		1.107	1.597		318
319	3.392	.184	.198	49.0	192.1	89.0	-141.0		.874	1.036		319
320	3.013	.074	.175	10.9	226.6	62.7	-73.3		.949	1.787	2	320
321	2.886	.046	.038	120.5	14.3	66.2	-67.3		.930	2.000	3	321
322	2.783	.180	.172	4.6	257.4	52.0	-66.9		.388	1.685		322
323	2.383	.195	.438	32.5	101.5	15.3	-38.1		.137	2.068	168	323

NUMBER	A	E	SINI	WBAR	ANODE	DWBAR	DNODE	NRES	TOMARS	TOJUP	FAMILY	
324	2.683	.285	.230	10.9	318.3	43.8	-72.1		.061	1.566		324
325	3.206	.128	.168	67.6	337.9	78.3	-99.4		.936	1.421		325
326	2.318	.165	.412	256.5	28.0	16.2	-34.4		.177	2.211		326
327	2.776	.056	.130	262.5	346.7	54.1	-58.2		.782	2.073		327
328	3.106	.105	.291	120.1	350.6	44.3	-73.4		.915	1.592		328
329	2.476	.113	.284	154.7	182.0	29.8	-40.8		.168	2.090		329
331	3.025	.063	.102	349.6	11.5	72.8	-77.6		.999	1.811		331
332	2.773	.064	.045	306.2	8.9	58.6	-60.2		.773	2.061		332
333	3.124	.130	.075	13.7	336.8	85.3	-96.8		.872	1.505	112	333
334	3.891	.049	.064	233.0	142.4	228.8	-246.3		1.802	1.009		334
335	2.472	.166	.084	275.1	160.3	42.1	-48.0		.245	2.120	154	335
336	2.252	.091	.110	232.3	235.9	33.0	-35.1		.206	2.540	174	336
337	2.383	.153	.143	109.2	350.1	36.5	-42.2		.185	2.270		337
338	2.913	.025	.125	155.6	286.2	62.4	-66.3		1.010	2.036	124	338
339	3.012	.067	.170	318.6	181.3	63.3	-73.1		.970	1.808	2	339
340	2.748	.102	.079	87.1	13.3	56.0	-59.9		.639	1.979	132	340
341	2.199	.129	.092	318.2	25.5	31.9	-34.5		.053	2.489	189	341
342	2.567	.142	.146	109.6	236.5	43.5	-50.5		.374	2.076	149	342
343	2.412	.211	.049	55.6	20.5	40.5	-48.3		.089	2.087		343
344	2.595	.224	.320	265.5	41.1	30.7	-53.6		.015	1.734		344
345	2.325	.092	.181	114.9	215.3	32.6	-36.8		.253	2.449		345
346	2.796	.062	.134	32.8	92.1	55.0	-59.5		.787	2.039		346
347	2.612	.191	.198	170.7	85.1	42.1	-55.5		.269	1.905		347
348	2.970	.079	.151	124.3	89.2	63.7	-72.0		.898	1.814	121	348
349	2.925	.052	.136	24.9	25.3	62.2	-67.6		.937	1.944	124	349
350	3.116	.119	.400	76.6	91.6	19.2	-65.9		.912	1.442		350
351	2.765	.178	.145	138.9	98.0	53.5	-66.5		.417	1.741	128	351
352	2.194	.130	.070	54.4	247.6	32.1	-34.5		.054	2.499	188	352
353	2.736	.307	.084	68.8	111.1	57.4	-86.5		.055	1.443		353
354	2.796	.159	.309	148.2	142.6	35.1	-58.4		.459	1.775		354
355	2.540	.112	.082	114.6	340.7	45.1	-48.4		.439	2.176		355
356	2.757	.173	.103	41.4	303.6	51.3	-64.1	10	.278	1.629		356
357	3.148	.035	.247	47.3	141.5	54.6	-75.1		1.177	1.764		357
358	2.877	.134	.062	76.9	190.9	65.5	-72.7		.664	1.760		358
359	2.729	.117	.119	333.4	357.6	52.8	-59.0		.555	1.941	133	359
360	3.002	.147	.197	73.9	138.1	60.0	-79.1		.703	1.584		360
362	2.579	.042	.134	104.4	20.7	44.3	-47.1		.654	2.316		362
363	2.748	.032	.088	349.6	58.9	55.2	-56.9		.832	2.171	130	363
364	2.221	.154	.096	79.0	111.5	32.5	-36.2		.028	2.426	180	364
365	2.802	.127	.226	46.6	189.5	45.9	-60.0		.607	1.851		365
366	3.142	.047	.193	276.3	341.7	66.7	-80.6		1.147	1.743		366
367	2.219	.147	.040	144.0	86.0	33.3	-36.2		.061	2.456	187	367
368	3.070	.170	.163	308.0	233.2	70.5	-92.6		.690	1.438		368
369	2.649	.055	.204	10.2	94.6	41.9	-48.7		.675	2.220		369
370	2.324	.046	.148	20.0	287.9	34.1	-36.4		.382	2.576		370
371	2.727	.062	.148	240.7	282.8	50.1	-54.8		.710	2.081		371
372	3.146	.156	.453	99.4	332.9	11.7	-70.0		.789	1.287		372
373	3.116	.129	.268	.6	.9	51.3	-79.8	10	.743	1.424		373
374	2.780	.105	.160	41.4	39.8	51.3	-59.6	10	.397	1.694	67	374
375	3.127	.073	.284	309.3	334.1	45.5	-72.3		1.039	1.667	111	375
376	2.289	.168	.111	241.7	298.6	34.4	-39.3		.066	2.337	175	376
377	2.691	.068	.126	66.3	217.8	50.0	-53.8		.657	2.106		377
378	2.777	.091	.136	38.7	239.2	53.8	-59.7		.679	1.974		378
379	3.137	.148	.032	348.1	216.8	90.1	-103.2		.832	1.440	1	379
380	2.678	.087	.089	292.9	94.4	51.6	-54.7		.512	1.990		380
381	3.212	.111	.205	253.3	129.6	69.7	-92.9		.991	1.473		381
382	3.121	.192	.165	216.2	310.6	74.8	-103.7		.655	1.317		382
383	3.135	.148	.025	51.7	90.8	90.2	-103.3		.828	1.439	1A	383
384	2.651	.159	.088	89.0	37.1	50.5	-57.7		.360	1.883		384
385	2.847	.162	.246	171.6	340.2	45.8	-65.7		.532	1.709		385
386	2.896					27.8	-51.7	6				386
387	2.742	.200	.303	273.7	133.4	35.9	-60.3		.306	1.733		387
388	3.006	.042	.119	286.7	345.5	69.4	-74.1		1.044	1.894		388
389	2.608	.098	.162	180.7	281.3	44.0	-49.7		.527	2.141	144	389
390	2.653	.155	.230	140.5	302.1	40.3	-53.5		.402	1.948	140	390
391	2.320	.255	.420	3.3	220.0	17.9	-40.9		-.039	1.984		391
392	2.885	.109	.254	31.5	214.5	44.9	-62.2		.721	1.821	125	392
393	2.775	.241	.318	293.7	215.5	36.1	-67.8		.197	1.577		393
394	2.762	.189	.100	329.7	62.1	56.5	-68.6		.399	1.730		394
395	2.786	.095	.078	247.5	264.2	58.2	-62.1		.695	1.959		395
396	2.742	.167	.063	255.6	258.0	56.7	-65.5		.453	1.814		396
397	2.636	.197	.243	12.9	234.8	39.1	-56.1		.266	1.866		397
398	2.739	.226	.188	86.5	282.2	49.1	-69.0		.257	1.641		398
399	3.053	.112	.236	175.4	341.8	54.7	-75.4		.849	1.623	116	399
400	3.129	.137	.206	194.3	322.1	65.1	-88.7		.834	1.479		400
401	3.342	.045	.395	233.3	25.9	104.1	-110.7		1.335	1.555		401
402	2.556	.152	.195	142.3	132.1	39.8	-49.3		.334	2.069	148	402
403	2.812	.088	.181	41.4	145.2	51.3	-59.8	10	.540	1.772	67	403
404	2.593	.211	.246	204.8	95.1	37.5	-54.7		.188	1.888		404
405	2.584	.258	.252	198.4	261.7	37.5	-59.3		.049	1.781		405

NUMBER	A	E	SINI	WBAR	ANODE	DWBAR	DNODE	NRES	TOMARS	TOJUP	FAMILY	
406	2.916	.142	.092	347.4	306.5	66.6	-76.3		.664	1.691		406
407	2.625	.036	.148	34.7	292.2	45.3	-48.8		.712	2.281		407
408	3.164	.106	.183	52.1	297.6	71.6	-89.9		.971	1.534		408
409	2.576	.093	.213	214.7	244.9	38.7	-46.3		.502	2.202		409
410	2.727	.232	.170	262.3	98.8	50.1	-69.4		.219	1.623		410
411	2.935	.104	.245	272.7	109.0	48.2	-65.6		.773	1.771		411
412	2.763	.071	.222	184.8	107.8	44.3	-54.3		.737	2.051		412
413	2.583					27.8	-67.1	6				413
414	3.503	.070	.145	102.9	114.9	110.9	-132.2		1.385	1.310		414
415	2.788	.277	.144	70.5	136.7	56.7	-84.8		.152	1.460		415
416	2.787	.227	.210	245.8	52.8	49.1	-72.1		.293	1.576		416
417	2.799	.174	.124	184.8	211.0	56.9	-69.0		.469	1.729		417
418	2.593	.084	.137	23.6	252.6	44.8	-48.9		.551	2.197		418
419	2.596	.247	.086	263.1	235.7	48.4	-63.4		.118	1.763	146	419
420	3.418	.044	.135	143.6	250.0	103.5	-116.9		1.402	1.482		420
421	2.538	.258	.141	41.8	193.3	43.3	-59.5		.041	1.820		421
422	2.228	.161	.082	346.4	1.6	33.0	-36.8		.027	2.406	184	422
423	3.068	.052	.178	229.3	66.9	65.3	-76.0		1.069	1.801		423
424	2.774	.099	.124	91.9	101.0	54.7	-60.4		.660	1.958		424
425	2.886	.097	.057	184.4	49.7	65.9	-69.8		.781	1.854		425
426	2.889					27.8	-55.0	6				426
427	2.974	.103	.108	288.9	294.1	68.8	-76.0		.827	1.748		427
428	2.308	.153	.104	44.7	9.3	35.2	-39.5		.126	2.357	175	428
429	2.607	.096	.176	39.9	225.3	42.9	-49.2		.530	2.151	144	429
430	2.841	.252	.272	75.2	252.4	44.4	-78.2		.241	1.472		430
431	3.129	.151	.013	321.5	148.3	90.0	-103.3		.815	1.437	1	431
432	2.370	.139	.198	242.5	89.3	33.3	-39.9		.181	2.299		432
433	1.458	.219	.187	126.2	314.5	15.5	-18.7	13	-.439	3.095		433
434	1.944	.067	.358	285.7	183.1	15.0	-22.6		.094	2.833	190	434
435	2.449	.119	.029	355.3	351.9	42.2	-44.7		.347	2.262	158	435
436	3.197	.038	.326	84.8	347.9	34.4	-69.5		1.125	1.682		436
437	2.386	.198	.155	318.8	262.1	36.2	-44.7		.073	2.163		437
438	2.554	.080	.117	227.7	43.1	44.2	-47.2		.531	2.245		438
439	3.132	.081	.339	116.2	205.1	32.0	-67.7		.885	1.582		439
440	2.210	.151	.039	124.3	278.9	33.1	-36.0		.039	2.451	187	440
441	2.807	.086	.161	121.7	256.2	53.1	-60.3		.716	1.946	67	441
442	2.345	.101	.099	198.1	141.2	36.6	-39.1		.286	2.431		442
443	2.215	.095	.080	157.6	183.1	32.6	-34.2		.162	2.565		443
444	2.771	.140	.184	351.7	203.5	49.6	-61.2		.525	1.836		444
445	3.194	.104	.404	353.1	290.5	16.0	-67.8		1.036	1.425		445
446	2.788	.093	.178	322.4	37.4	50.6	-58.9		.641	1.907	67	446
447	2.986	.015	.066	91.4	64.9	72.5	-73.8		1.110	1.994		447
448	3.144	.130	.223	323.0	35.0	62.2	-87.4		.878	1.473		448
449	2.555	.194	.037	139.6	78.9	47.2	-55.2		.242	1.950	151	449
450	3.015	.065	.174	13.1	8.4	62.8	-73.0		.979	1.811	2	450
451	3.063	.059	.243	104.5	89.2	52.5	-70.7		.998	1.751		451
452	2.865	.063	.036	168.3	90.8	64.9	-66.5		.862	1.973	3	452
453	2.183	.136	.091	199.8	5.0	31.4	-34.2		.010	2.477	189	453
454	2.628	.139	.103	200.4	22.8	48.6	-54.6		.430	2.003		454
455	2.657	.233	.216	347.9	74.8	42.8	-62.4		.195	1.734		455
456	2.784	.203	.265	224.7	232.7	42.0	-65.6		.353	1.676		456
457	3.090	.125	.248	26.9	254.4	54.3	-79.4		.831	1.542	116	457
458	2.994	.188	.227	56.9	139.6	55.8	-83.0		.566	1.463		458
459	2.621	.185	.175	54.9	22.3	44.2	-56.2		.300	1.904		459
460	2.718	.064	.089	5.4	217.4	53.6	-56.0		.710	2.105	134	460
461	3.112	.159	.025	114.3	209.4	88.1	-102.7		.774	1.431	1A	461
462	2.874	.050	.036	345.2	110.5	65.5	-66.6		.908	2.001	3	462
463	2.398	.172	.223	16.4	35.5	32.8	-42.4		.108	2.173	163	463
464	2.802	.152	.166	356.1	102.7	53.2	-65.3		.518	1.769		464
465	3.092	.233	.119	217.2	300.1	81.3	-115.8		.501	1.228		465
466	3.358					27.8	-85.4	6				466
467	2.944	.086	.130	73.6	316.9	64.4	-71.3		.854	1.827		467
468	3.140	.153	.022	348.3	298.9	91.1	-105.1		.819	1.420	1A	468
469	3.166	.200	.226	173.5	326.3	65.7	-105.5		.660	1.235		469
470	2.405	.123	.127	203.2	179.4	37.8	-42.0		.285	2.311		470
471	2.888	.197	.248	47.8	87.2	48.2	-73.8		.451	1.555		471
472	2.542	.103	.264	98.0	130.7	33.0	-43.6		.405	2.190	4	472
473	2.983	.121	.503	10.7	324.6	1.1	-53.8		.626	1.651		473
474	2.454	.173	.151	312.5	169.9	38.8	-46.5		.196	2.138		474
475	2.596	.251	.334	350.9	38.4	30.1	-56.5		-.102	1.630		475
476	2.650	.065	.207	257.3	285.5	41.7	-49.0		.648	2.193		476
477	2.416	.150	.091	328.4	1.6	39.5	-44.2		.237	2.226	75	477
478	3.017	.093	.248	120.6	237.0	50.3	-69.7		.842	1.683	120	478
479	2.721	.214	.153	49.3	140.9	51.3	-67.2	10	.225	1.632	135	479
480	2.644	.012	.376	130.2	238.3	19.8	-39.8		.845	2.237		480
481	2.741	.174	.156	41.4	38.6	51.3	-63.9	10	.340	1.718		481
482	3.000	.087	.253	242.2	183.7	48.6	-67.6		.877	1.753	120	482
483	3.426	.015	.317	203.0	178.5	36.9	-85.2		1.418	1.558		483
484	2.668	.033	.202	277.4	130.4	42.6	-49.1		.751	2.260		484
485	2.750	.183	.264	115.7	198.6	40.5	-60.5		.384	1.774		485

NUMBER	A	E	SINI	WBAR	ANODE	DWBAR	DNODE	NRES	TOMARS	TOJUP	FAMILY	
486	2.352	.184	.187	203.8	96.7	33.5	-41.9		.060	2.215		486
487	2.670	.062	.162	56.6	117.7	46.5	-51.4		.679	2.168		487
488	3.155	.188	.198	159.2	82.0	70.8	-103.9		.687	1.294		488
489	3.152	.079	.220	173.9	172.8	61.7	-81.2		1.054	1.624		489
490	3.175	.052	.159	20.5	186.3	76.1	-86.9		1.164	1.693	106	490
491	3.196	.055	.322	95.0	178.9	35.7	-70.9		1.089	1.646		491
492	3.112	.147	.023	329.2	.7	87.7	-100.0		.813	1.469	1A	492
493	3.123	.126	.277	46.1	352.3	48.4	-78.7		.840	1.513		493
494	2.986	.084	.115	226.3	29.4	68.8	-75.3		.896	1.789		494
495	2.488	.106	.045	44.0	209.3	43.6	-45.9		.414	2.247		495
496	2.199	.128	.077	124.4	212.8	32.1	-34.6		.064	2.499	189	496
497	2.850	.262	.088	8.7	350.6	64.5	-90.7		.259	1.432		497
498	2.650	.183	.156	332.7	95.9	46.8	-58.6		.327	1.866		498
500	2.613	.098	.192	7.5	288.2	41.7	-49.0		.527	2.145	144	500
501	3.155					27.8	-68.1	6				501
502	2.384	.173	.420	151.2	132.5	16.4	-37.1		.206	2.116		502
503	2.723	.187	.076	120.5	60.0	55.3	-66.1		.379	1.774		503
504	2.721	.165	.216	348.6	104.3	44.4	-58.9		.434	1.836	138	504
505	2.685	.239	.163	73.9	93.5	49.6	-67.9		.182	1.664		505
506	3.041	.156	.313	115.3	313.4	40.3	-74.1		.684	1.520		506
507	3.153	.058	.187	41.5	293.1	68.9	-82.9		1.122	1.696		507
508	3.161	.047	.220	210.5	40.7	61.3	-79.3		1.167	1.716		508
509	3.065	.057	.276	25.5	222.1	45.3	-67.6		1.024	1.795		509
510	2.611	.163	.183	280.0	208.1	42.9	-53.4		.349	1.975	148	510
511	3.178	.171	.253	96.5	111.9	58.9	-96.3		.751	1.312		511
512	2.190	.174	.152	13.8	108.7	30.0	-35.3		-.145	2.320		512
513	3.014	.056	.171	68.0	191.7	63.2	-72.5		1.008	1.841	2	513
514	3.047	.011	.088	81.7	271.2	75.6	-78.0		1.180	1.947		514
515	3.120	.154	.019	66.3	151.1	89.0	-102.6		.800	1.440	1A	515
516	2.680	.261	.277	215.9	323.2	38.5	-65.8		.101	1.656		516
517	3.146	.166	.079	62.6	278.8	88.2	-106.7		.773	1.369		517
518	2.537	.186	.135	314.5	212.3	43.0	-52.2		.234	2.001	150	518
519	2.790	.154	.186	352.6	42.3	50.6	-63.9		.454	1.730		519
520	3.006	.082	.184	70.0	29.7	61.1	-72.9		.915	1.770	2	520
521	2.741	.287	.179	41.4	86.4	51.3	-80.2	10	.003	1.399		521
522	3.629	.039	.057	347.1	126.2	158.6	-164.8		1.613	1.298		522
523	2.966	.178	.097	99.2	264.8	70.8	-86.7		.593	1.533		523
524	2.635	.102	.160	60.6	321.6	45.2	-51.4		.540	2.099		524
525	2.245	.143	.117	122.3	208.6	32.7	-36.5		.074	2.426	183	525
526	3.121	.162	.026	147.9	173.0	89.2	-104.6		.772	1.412	1A	526
527	2.726	.125	.151	317.9	123.5	50.2	-58.3		.523	1.906		527
528	3.397	.033	.206	174.5	47.1	78.7	-102.7		1.422	1.543		528
529	3.017	.065	.174	55.9	62.5	63.0	-73.1		.980	1.809	2	529
530	3.210	.174	.128	318.1	133.5	88.4	-114.9		.780	1.278		530
531	2.794	.154	.584	222.0	188.7	-1.3	-47.8		.115	1.848	129	531
532	2.772	.184	.286	176.7	107.1	38.4	-61.2		.399	1.745		532
533	2.981	.077	.114	199.5	191.0	68.4	-74.4		.914	1.816		533
534	2.884	.053	.037	107.4	93.3	66.2	-67.5		.910	1.982	3	534
535	2.569	.059	.101	164.4	83.9	45.5	-47.6		.601	2.277		535
536	3.500	.038	.316	28.0	58.4	37.1	-92.5		1.409	1.409		536
537	3.063	.222	.152	294.8	124.9	73.5	-105.3		.519	1.291		537
538	3.165	.124	.101	.5	149.8	86.2	-99.3		.921	1.481	104	538
539	2.739	.145	.146	41.4	302.4	51.3	-60.8	10	.433	1.803		539
540	2.219	.145	.109	168.2	207.9	32.1	-35.6		.038	2.441	183	540
541	2.815	.068	.124	231.5	269.7	55.9	-61.3		.791	2.004		541
542	2.906	.096	.199	4.3	157.6	53.5	-65.7		.767	1.823		542
543	3.062	.115	.172	56.7	294.9	66.9	-81.9		.859	1.613	118	543
544	2.593	.143	.153	267.0	296.7	43.5	-51.5		.392	2.044	149	544
545	3.189	.163	.214	291.2	332.4	68.2	-99.9		.805	1.320	102	545
546	2.597	.145	.262	142.9	19.7	35.3	-48.6		.355	2.043		546
547	2.773	.203	.293	31.5	196.0	39.0	-63.3		.339	1.696		547
548	2.282	.188	.055	81.9	117.3	35.3	-40.5		.027	2.306		548
549	2.682	.256	.089	97.8	291.3	53.1	-71.6		.138	1.630		549
550	2.589	.186	.198	305.6	268.9	41.1	-53.4		.264	1.948		550
551	2.966	.114	.042	91.7	297.6	73.4	-78.8		.798	1.724		551
552	3.154	.101	.157	232.6	270.5	75.6	-90.4		.984	1.558		552
553	2.231	.123	.083	91.1	72.9	33.1	-35.6		.116	2.490	182	553
554	2.375	.148	.066	77.0	290.4	38.6	-42.4		.208	2.284		554
555	3.169	.189	.031	134.9	156.3	95.6	-119.3		.724	1.277	107	555
556	2.467	.117	.108	118.1	283.1	41.0	-44.8		.360	2.250	155	556
557	2.442	.133	.060	140.0	285.7	41.4	-44.8		.303	2.239	159	557
558	2.908	.058	.133	138.4	150.8	61.4	-66.7		.905	1.946	124	558
559	2.712	.077	.145	219.7	115.0	49.7	-54.8		.659	2.061		559
560	2.751	.174	.129	121.2	106.6	53.7	-65.4		.415	1.774	131	560
561	3.177	.195	.022	109.4	202.7	97.2	-122.5		.708	1.251	107	561
562	3.019	.066	.177	310.8	68.1	62.6	-73.1		.980	1.805	2	562
563	2.713	.231	.160	67.9	87.1	50.2	-68.7		.208	1.635		563
564	2.748	.246	.299	272.1	64.5	37.9	-67.7		.186	1.609		564
565	2.442	.169	.213	156.8	230.1	34.8	-44.4		.176	2.159	163	565
566	3.387	.065	.066	27.4	75.7	115.4	-122.1		1.308	1.445		566

NUMBER	A	E	SINI	WBAR	ANODE	DWBAR	DNODE	NRES	TOMARS	TOJUP	FAMILY	
567	3.138	.123	.148	194.4	55.3	76.3	-92.5		.900	1.508		567
568	2.883	.171	.337	78.3	252.3	33.2	-63.5		.441	1.610		568
569	2.657	.173	.041	98.2	291.6	52.4	-59.8		.348	1.862	142	569
570	3.429	.068	.047	21.5	248.8	124.4	-130.8		1.335	1.394		570
571	2.411	.218	.097	35.2	349.7	39.3	-48.3		.063	2.073	166	571
572	2.400	.127	.189	38.6	198.7	34.6	-40.9		.252	2.309		572
573	3.014	.073	.180	13.0	337.2	61.9	-73.0		.951	1.789	2	573
574	2.252	.229	.115	66.1	328.9	33.1	-41.2		-.118	2.226		574
575	2.555	.077	.262	320.0	346.5	33.4	-43.1		.489	2.248	4	575
576	2.987	.165	.197	315.4	296.2	59.5	-80.6		.635	1.549		576
577	3.117	.139	.110	281.4	322.2	81.0	-95.6		.829	1.488		577
578	2.750	.185	.109	277.1	19.0	55.1	-66.8		.394	1.752		578
579	3.013	.062	.171	282.2	81.1	63.1	-72.8		.986	1.821	2	579
580	3.220	.089	.042	82.7	100.8	97.0	-102.9		1.090	1.536		580
581	3.214					27.8	-87.6	6				581
582	2.611					4.3	-41.2	5				582
583	3.180	.170	.179	156.4	259.5	75.8	-104.8		.770	1.319		583
584	2.374	.170	.216	17.4	281.1	32.5	-41.3		.090	2.203	163	584
585	2.431	.165	.137	153.2	189.2	38.5	-45.1		.199	2.183		585
586	3.041	.086	.046	140.4	249.7	78.6	-82.8		.946	1.730	119	586
587	2.335	.174	.427	154.3	322.0	15.3	-35.3		.162	2.173		587
589	3.133	.023	.185	94.8	184.6	67.5	-79.4		1.219	1.827		589
590	3.001	.078	.174	111.2	107.8	62.2	-72.8		.925	1.788	2	590
591	2.678	.235	.241	184.9	326.4	41.6	-63.6		.192	1.711		591
592	3.023	.106	.177	77.6	174.7	63.5	-77.1		.854	1.681		592
593	2.700	.223	.289	109.4	71.3	36.9	-61.3		.224	1.728		593
594	2.628	.106	.616	216.6	145.3	-6.3	-38.4		.132	2.079		594
595	3.202	.051	.303	246.2	21.1	41.3	-73.4		1.160	1.683		595
596	2.932	.212	.243	41.4	234.9	51.3	-80.5	10	.176	1.214		596
597	2.672	.099	.216	342.6	33.4	41.8	-51.0		.579	2.075		597
598	2.762	.203	.208	28.7	94.1	47.6	-66.5		.342	1.692		598
599	2.772	.209	.305	340.0	45.3	36.7	-63.5		.307	1.678		599
600	2.660	.067	.167	219.7	144.2	45.6	-50.8		.658	2.164		600
601	3.130	.073	.273	318.6	174.9	48.1	-73.5		1.026	1.671	111	601
602	3.087	.187	.292	11.6	324.8	47.5	-86.2		.623	1.355		602
603	2.552	.199	.151	146.3	338.0	42.9	-53.8		.210	1.953	150	603
604	3.151	.162	.082	48.2	354.5	88.2	-106.2		.790	1.377		604
605	3.000					27.8	-60.0	6				605
606	2.587	.176	.174	19.3	312.5	42.8	-53.5		.297	1.969		606
607	2.852	.060	.197	190.9	284.9	50.0	-60.1		.720	1.872		607
608	3.024	.075	.186	1.8	292.2	61.7	-73.5		.955	1.775	2	608
609	3.088	.050	.068	231.0	182.6	81.1	-83.9		1.097	1.790		609
610	3.083	.216	.220	14.9	16.0	63.0	-99.9		.549	1.279		610
611	2.979	.114	.243	93.6	194.8	50.3	-69.4		.748	1.663	120	611
612	3.132	.194	.412	324.2	212.8	23.5	-79.9		.591	1.144		612
613	2.920	.047	.139	100.7	347.2	61.4	-66.9		.949	1.966	124	613
614	2.695	.107	.135	79.5	223.7	49.9	-55.6		.554	2.003		614
615	2.631	.104	.049	246.1	352.1	50.5	-53.4		.493	2.048	141	615
616	2.552	.097	.263	133.9	353.4	33.3	-43.7		.433	2.200	4	616
618	3.188	.035	.274	333.8	112.3	48.6	-74.7		1.197	1.730		618
619	2.520	.037	.241	30.4	191.0	34.2	-41.4		.579	2.402		619
620	2.435	.095	.134	330.3	353.6	38.6	-42.1		.379	2.346		620
621	3.118	.154	.026	113.7	39.4	88.5	-102.2		.796	1.442	1A	621
622	2.414	.212	.152	46.4	147.3	37.5	-47.4		.066	2.091		622
623	2.459	.146	.263	103.0	308.0	31.3	-42.4		.185	2.135		623
625	2.647	.191	.194	320.6	129.9	44.0	-58.0		.300	1.861		625
626	2.574	.182	.448	10.2	335.0	15.0	-44.2		.303	1.879		626
627	2.900	.045	.100	280.0	151.4	63.9	-66.9		.940	1.990		627
628	2.582	.023	.184	253.2	114.2	40.9	-45.6		.699	2.371		628
629	3.130	.177	.148	127.9	83.5	77.9	-102.7		.714	1.348		629
630	2.624	.141	.228	146.7	105.2	39.4	-50.9		.416	2.022	140	630
631	2.791					27.8	-4.9	6				631
632	2.662	.221	.052	236.0	332.4	52.8	-65.6		.221	1.742		632
633	3.017	.059	.177	316.9	152.9	62.5	-72.6		.999	1.827	2	633
634	3.047	.141	.199	347.3	135.9	61.7	-82.0		.756	1.555		634
635	3.140	.055	.193	66.0	189.8	66.8	-80.9		1.117	1.720		635
636	2.910	.138	.135	324.0	28.5	62.2	-73.4		.664	1.714		636
637	3.157	.156	.024	163.6	290.3	92.9	-107.9		.823	1.391	1A	637
638	2.735	.189	.122	221.9	107.1	53.7	-66.3		.364	1.751	131	638
639	3.016	.068	.171	333.9	279.9	63.4	-73.4		.971	1.801	2	639
640	3.163	.088	.247	237.6	238.5	56.3	-80.4		1.021	1.585		640
641	2.220	.131	.021	83.5	22.9	33.5	-35.7		.099	2.492	187	641
642	3.183	.150	.149	132.7	1.7	81.0	-103.1		.844	1.375		642
643	3.352	.080	.261	128.3	255.3	58.6	-94.4		1.201	1.425		643
644	2.599	.128	.005	23.6	196.0	49.5	-53.1		.436	2.054		644
645	3.200	.147	.137	102.1	352.7	84.7	-105.9		.861	1.373		645
646	2.325	.163	.135	338.7	297.3	34.8	-40.3		.108	2.310		646
647	2.444	.191	.144	81.3	257.4	38.8	-47.4		.142	2.105	157	647
648	3.184	.223	.196	112.7	292.2	75.2	-120.1		.586	1.151		648
649	2.551	.233	.218	346.1	353.8	38.7	-55.4		.109	1.875		649

NUMBER	A	E	SINI	WBAR	ANODE	DWBAR	DNODE	NRES	TOMARS	TOJUP	FAMILY	
650	2.458	.161	.056	41.9	231.4	42.2	-47.2		.249	2.152	24	650
651	3.024	.065	.177	41.9	33.1	63.0	-73.5		.986	1.801	2	651
652	2.555	.072	.258	22.9	86.1	33.7	-43.1		.510	2.267	4	652
653	3.014	.079	.182	183.2	136.9	61.7	-73.3		.933	1.772	2	653
654	2.297	.192	.332	133.0	275.0	22.9	-37.1		.077	2.134		654
655	2.989	.064	.098	73.1	137.4	70.6	-74.9		.963	1.844		655
656	3.160	.158	.025	161.5	257.0	93.3	-108.8		.820	1.382	1A	656
657	2.611	.142	.202	175.8	294.6	41.2	-51.2		.403	2.036	138	657
658	2.854	.045	.037	85.6	320.1	64.1	-65.1		.902	2.034	3	658
660	2.535	.088	.262	224.4	160.1	32.9	-42.7		.436	2.234	4	660
661	3.016	.071	.173	166.6	330.6	63.1	-73.5		.960	1.792	2	661
662	2.554	.197	.061	291.6	148.6	46.7	-55.3		.229	1.944	151	662
663	3.062	.162	.342	184.3	239.7	34.7	-74.1		.614	1.449		663
664	3.175	.234	.171	255.9	182.2	80.6	-126.1		.549	1.124		664
665	3.148	.152	.286	249.6	301.2	48.6	-84.8		.787	1.385		665
666	2.594	.209	.142	33.2	223.5	45.4	-57.7		.216	1.869		666
667	3.198	.120	.432	121.4	160.0	11.1	-68.4		.969	1.367		667
668	2.797	.192	.145	319.2	223.9	55.3	-70.7		.411	1.676	128	668
669	3.012	.074	.185	257.6	177.4	61.0	-72.5		.946	1.790	2	669
670	2.803	.154	.127	8.6	183.6	56.7	-67.1		.528	1.778	126	670
671	3.097	.076	.144	132.9	353.8	72.6	-82.4		1.020	1.692		671
672	2.555	.112	.204	278.8	340.1	38.9	-46.6		.434	2.174		672
673	2.815	.041	.065	169.9	240.7	60.5	-61.9		.878	2.083		673
674	2.924	.158	.230	41.4	349.8	51.3	-71.9	10	.473	1.507		674
675	2.770	.190	.191	64.0	267.3	49.7	-66.2		.375	1.698		675
676	3.059	.099	.210	321.8	155.5	59.8	-76.8		.910	1.656	117	676
677	2.956	.079	.170	186.3	273.9	60.0	-69.5		.883	1.828		677
678	2.573	.191	.132	50.2	283.0	44.8	-54.7		.251	1.942	43	678
679	2.586	.186	.449	12.1	111.7	15.1	-45.1		.300	1.856		679
680	3.147	.221	.337	267.2	30.7	41.2	-95.6		.515	1.202		680
681	3.110	.082	.219	270.0	184.5	60.2	-78.4		1.007	1.659		681
682	2.632	.130	.013	260.4	196.8	40.8	-51.2		.454	2.030		682
683	3.116	.095	.343	150.5	260.6	31.4	-67.8		.794	1.534	003	683
684	2.432	.047	.103	216.3	328.1	39.7	-41.2		.502	2.459		684
685	2.236	.146	.079	306.9	238.1	33.3	-36.6		.071	2.435	184	685
686	2.589	.130	.306	337.8	244.8	30.2	-45.6		.243	1.958		686
687	2.723	.207	.293	29.1	327.8	36.7	-60.5		.284	1.745		687
688	2.698	.110	.178	298.1	177.8	46.6	-54.7		.567	2.010		688
689	2.316	.187	.098	358.4	177.5	35.7	-41.6		.055	2.266		689
690	3.148	.121	.225	12.0	257.5	61.6	-86.1		.911	1.498		690
691	3.012	.082	.207	43.9	88.6	57.3	-71.6		.913	1.767		691
692	3.369	.092	.447	106.6	58.0	-.8	-72.8		1.113	1.318		692
693	2.944	.035	.251	211.8	347.8	46.4	-61.4		.997	1.970		693
694	2.670	.210	.321	346.9	236.0	32.0	-56.2		.157	1.730		694
695	2.538	.088	.263	5.0	275.1	32.9	-42.8		.438	2.231	4	695
696	3.182	.171	.272	54.1	300.5	54.0	-94.3		.730	1.303		696
697	2.881	.116	.258	349.8	12.7	44.3	-62.3		.695	1.807	125	697
698	2.869	.132	.196	154.0	37.5	52.8	-66.6		.621	1.752		698
699	2.616	.235	.337	339.3	245.3	29.8	-55.6		-.072	1.623		699
700	2.229	.148	.112	181.2	99.6	32.4	-36.2		.043	2.426	183	700
701	3.013	.070	.143	195.0	249.4	67.0	-74.9		.967	1.798		701
702	3.194	.041	.371	274.6	289.7	20.6	-64.1		1.212	1.597		702
703	2.175	.117	.053	54.0	217.7	31.7	-33.6		.067	2.544	186	703
704	3.062	.081	.324	40.0	281.2	34.7	-64.7		.881	1.693		704
705	2.923	.045	.425	161.0	1.3	9.6	-47.5		.968	1.925		705
706	2.729	.146	.264	352.2	321.0	39.1	-55.5		.475	1.902		706
707	2.180	.075	.081	50.8	276.1	31.4	-32.6		.146	2.615		707
708	2.671	.136	.068	192.0	337.9	52.2	-57.5		.443	1.938		708
709	2.914	.075	.291	332.9	321.7	38.7	-58.7		.835	1.902		709
710	3.135	.154	.021	227.7	183.5	90.6	-104.7		.810	1.421	1A	710
711	2.237	.152	.109	283.7	351.1	32.7	-36.6		.047	2.415	183	711
712	2.576	.173	.235	62.4	234.7	37.6	-50.7		.277	2.006		712
713	3.399	.108	.196	345.4	227.3	85.2	-116.3		1.154	1.285		713
714	2.535	.091	.264	124.8	235.6	32.8	-42.8		.426	2.225	4	714
715	2.768	.042	.228	341.8	43.1	43.6	-53.2		.822	2.127		715
716	2.811	.125	.139	192.7	152.0	56.0	-64.5		.617	1.849	126	716
717	3.146	.216	.047	2.6	315.0	93.4	-123.9		.614	1.219		717
718	3.056	.236	.114	207.7	29.6	78.7	-111.8		.468	1.261		718
719	2.583	.483	.228	338.3	205.5	43.7	-107.9		-.561	1.208		719
720	2.887	.051	.036	172.9	6.8	66.4	-67.7		.917	1.983	3	720
721	3.551	.083	.135	41.1	32.4	121.4	-145.8		1.373	1.213		721
722	2.172	.085	.088	281.1	46.2	31.1	-32.5		.113	2.597		722
723	2.994	.038	.082	88.8	177.3	72.0	-74.5		1.049	1.916		723
724	2.451	.234	.217	55.5	206.8	35.4	-49.7		.018	1.993		724
725	2.573	.195	.050	36.6	61.1	47.9	-56.3		.248	1.922	151	725
726	2.565	.186	.301	9.3	246.9	31.2	-49.0		.125	1.884		726
727	2.568	.087	.250	76.3	136.2	34.9	-44.4		.494	2.227		727
728	2.256	.134	.062	145.7	83.1	34.2	-36.9		.128	2.453	181	728
729	2.760	.113	.303	186.3	125.7	34.0	-52.5		.560	1.937		729
730	2.243	.206	.064	203.2	99.5	33.9	-39.7		-.053	2.297		730

NUMBER	A	E	SINI	WBAR	ANODE	DWBAR	DNODE	NRES	TOMARS	TOJUP	FAMILY	
731	2.988	.100	.180	320.6	42.2	60.9	-73.4		.844	1.736		731
732	2.457	.068	.192	195.4	177.4	36.1	-41.0		.457	2.397		732
733	3.398	.061	.359	184.4	339.1	22.9	-80.0		1.251	1.284		733
734	3.151	.076	.107	87.4	352.8	82.9	-90.4		1.066	1.646		734
735	2.730	.252	.301	357.1	46.0	37.3	-67.1		.160	1.616		735
736	2.202	.106	.073	337.6	144.2	32.3	-34.1		.118	2.546	188	736
737	2.590	.195	.229	314.4	192.3	38.7	-53.5		.239	1.924		737
738	3.035	.096	.046	179.6	147.3	78.2	-83.3		.910	1.706	119	738
739	2.738					27.8	-176.7	6				739
740	3.050	.141	.175	167.7	117.3	66.2	-84.4		.764	1.547		740
741	2.720	.127	.130	181.0	101.3	51.7	-58.8		.444	1.850	133	741
742	3.013	.072	.181	340.9	61.5	61.7	-72.8		.953	1.793	2	742
743	2.793	.043	.099	91.9	237.6	57.3	-59.7		.847	2.097		743
744	3.173	.153	.122	176.5	148.3	84.7	-104.8		.823	1.385		744
745	3.238	.099	.219	147.3	129.1	66.6	-92.2		1.059	1.471		745
746	3.109	.166	.324	309.2	2.8	39.8	-80.6		.680	1.420		746
747	2.998	.245	.363	56.8	135.1	34.5	-84.2		.276	1.275		747
749	2.243	.186	.087	221.0	115.9	33.4	-38.5		-.014	2.337		749
750	2.442	.168	.056	147.3	62.5	41.6	-46.8		.219	2.155	24	750
751	2.552	.114	.256	39.1	79.4	34.2	-44.8		.402	2.171		751
752	2.463	.095	.088	129.4	83.7	41.5	-44.0		.417	2.306	156	752
753	2.330	.211	.161	249.4	56.5	34.1	-42.7		-.015	2.182		753
754	2.988	.053	.416	158.1	183.3	11.4	-51.1		1.020	1.827		754
755	3.175	.178	.057	214.3	197.9	94.1	-116.2		.758	1.304	105	755
756	3.212	.170	.357	197.3	213.4	34.0	-86.6		.682	1.252		756
757	2.373	.109	.139	88.2	16.3	36.1	-39.9		.285	2.389		757
758	3.202	.115	.078	74.6	110.1	92.5	-103.6		.985	1.471	101	758
759	2.618					27.8	-52.6	6				759
760	3.158	.255	.243	166.7	325.5	65.0	-121.5		.459	1.070		760
761	2.863	.047	.037	285.0	353.4	64.7	-65.8		.907	2.021	3	761
762	3.157	.124	.248	139.4	303.8	57.1	-85.1		.895	1.479		762
763	2.241	.130	.084	33.2	284.9	33.4	-36.1		.112	2.467	182	763
764	3.187	.086	.195	79.1	262.0	69.6	-88.0		1.056	1.577		764
765	2.546	.248	.126	43.7	317.6	44.3	-59.1		.077	1.832		765
766	3.021	.081	.180	101.3	1.9	62.6	-74.3		.933	1.757	2	766
767	3.117	.150	.025	336.7	63.2	88.3	-101.3		.808	1.454	1A	767
768	3.142	.135	.271	62.0	35.4	51.3	-83.0	10	.687	1.335		768
769	3.181	.162	.130	271.1	29.7	84.6	-107.6		.798	1.348		769
770	2.221	.157	.067	81.5	38.0	33.0	-36.5		.031	2.425	184	770
771	2.652	.232	.292	90.2	217.7	35.4	-59.1		.157	1.753		771
772	3.001	.090	.467	312.0	153.0	2.7	-52.1	7	.922	1.646		772
773	2.858	.047	.301	260.0	320.4	35.3	-53.6		.848	2.030		773
774	3.050	.170	.117	264.4	254.5	75.3	-93.5		.675	1.466		774
775	3.012	.085	.182	129.6	296.0	61.8	-73.8		.909	1.753	2	775
776	2.933	.119	.299	42.5	81.0	39.0	-63.0		.713	1.753		776
777	3.210	.136	.262	171.5	284.4	56.2	-90.5		.890	1.388		777
778	3.177	.243	.270	103.8	328.4	59.3	-115.8		.501	1.087		778
779	2.667	.158	.278	324.7	281.2	35.5	-52.6		.364	1.925	137	779
780	3.116	.041	.315	.4	147.9	36.7	-66.0		1.092	1.789		780
781	3.223	.067	.319	253.8	142.8	37.6	-74.3		1.091	1.595		781
782	2.180	.104	.083	152.4	83.8	31.4	-33.2		.089	2.558		782
783	2.343	.190	.158	286.1	149.0	34.4	-42.2		.044	2.218	171	783
784	3.107	.228	.239	240.1	5.8	61.7	-105.0		.516	1.216		784
785	2.575	.225	.219	198.8	74.0	39.4	-56.1		.149	1.864		785
786	3.177	.173	.247	221.1	92.6	60.5	-97.3		.748	1.308		786
787	2.540	.063	.264	288.6	187.9	32.7	-42.0		.502	2.289	153	787
788	3.129	.154	.251	207.4	181.2	56.2	-87.5		.769	1.416	113	788
789	2.686	.141	.203	257.1	235.0	44.0	-55.2		.468	1.951	138	789
790	3.406	.169	.391	275.0	249.2	25.0	-100.4		.815	.869		790
791	3.124	.159	.261	320.4	131.3	54.0	-87.0		.733	1.401	113	791
792	2.623	.153	.175	137.9	264.9	44.0	-53.6		.388	1.984		792
793	2.796	.074	.265	346.6	33.4	39.8	-53.9		.739	2.027		793
794	3.139	.277	.106	279.5	181.0	90.8	-145.1		.395	1.043		794
795	2.750	.154	.332	183.0	14.1	30.7	-54.1		.327	1.747		795
796	2.636	.264	.319	11.0	35.6	32.9	-61.2		.019	1.661		796
797	2.537	.078	.094	208.4	244.4	44.5	-46.8		.524	2.266		797
798	3.015	.060	.171	223.7	220.8	63.2	-72.8		.996	1.828	2	798
799	2.542	.024	.089	129.7	175.5	44.8	-45.9		.669	2.397		799
800	2.193	.144	.076	303.7	316.6	32.0	-34.9		.018	2.466	189	800
801	2.605	.118	.250	165.2	190.9	36.4	-47.6		.447	2.116		801
802	2.196	.139	.088	130.9	2.1	31.9	-34.8		.028	2.470	189	802
803	3.204	.041	.169	258.6	254.9	75.9	-88.1		1.225	1.698		803
804	2.839	.101	.269	325.1	344.7	40.8	-57.5		.695	1.906	125	804
805	3.207	.103	.275	221.4	105.5	51.3	-82.8	10	.781	1.296		805
806	3.203	.111	.243	166.2	43.0	59.8	-87.7		.982	1.473		806
807	3.019	.081	.180	136.4	136.6	62.3	-73.9		.932	1.762	2	807
808	2.745	.131	.087	111.1	195.2	55.7	-61.8		.553	1.904	40	808
809	2.283	.143	.121	353.6	160.7	33.8	-37.9		.113	2.396		809
810	2.179	.124	.047	355.5	166.0	31.9	-33.9		.061	2.530	186	810
811	2.897	.062	.040	277.8	148.3	67.1	-68.8		.892	1.946	3	811

NUMBER	A	E	SINI	WBAR	ANODE	DWBAR	DNODE	NRES	TOMARS	TOJUP	FAMILY	
812	2.659	.131	.228	359.1	3.3	40.3	-52.1		.474	2.005	140	812
813	2.223	.038	.099	129.8	51.9	32.4	-33.4		.292	2.680		813
814	3.157	.192	.407	22.2	94.1	24.3	-81.9		.590	1.099		814
815	2.659	.093	.230	131.3	54.3	39.9	-49.7		.575	2.111		815
816	3.002	.139	.235	155.5	129.6	53.3	-75.0		.714	1.588	120	816
817	2.590	.154	.190	62.6	129.9	41.4	-51.3		.355	2.027	148	817
818	3.172	.037	.254	11.5	69.4	53.6	-76.0		1.182	1.732		818
819	2.197	.112	.087	252.8	326.3	31.9	-34.0		.092	2.530		819
820	3.128	.052	.085	267.8	124.2	83.2	-87.2		1.125	1.742		820
821	2.778	.231	.105	232.1	218.7	57.8	-76.2		.292	1.599		821
822	2.256	.181	.025	109.2	237.3	34.7	-39.0		.027	2.350	178	822
823	2.221	.136	.078	127.3	253.4	32.9	-35.7		.077	2.469	185	823
824	2.795	.132	.132	263.8	149.7	55.6	-64.2		.583	1.847	126	824
825	2.226	.111	.051	189.0	109.2	33.4	-35.2		.146	2.528	186	825
826	2.713	.213	.141	41.4	21.2	51.3	-66.8	10	.104	1.538	135	826
827	2.275	.117	.063	17.6	184.6	34.9	-37.0		.184	2.477	179	827
828	3.189	.039	.031	248.1	319.1	93.4	-94.7		1.224	1.725		828
829	2.580	.071	.150	50.7	345.5	43.5	-47.5		.575	2.243		829
830	3.205	.062	.079	85.5	328.5	91.4	-96.5		1.161	1.634		830
831	2.214	.137	.091	65.3	184.5	32.3	-35.3		.057	2.465	182	831
832	2.864	.044	.037	18.7	266.7	64.8	-65.8		.916	2.027	3	832
833	3.010	.083	.179	36.9	346.8	62.1	-73.6		.917	1.763	2	833
834	3.174	.217	.078	262.0	198.4	94.3	-127.9		.627	1.184		834
835	3.208	.067	.084	17.4	302.6	91.2	-97.1		1.146	1.613		835
836	2.191	.142	.093	39.2	205.2	31.6	-34.7		.008	2.461	189	836
837	2.298	.027	.126	100.6	204.2	34.1	-35.5		.405	2.650		837
838	2.898	.078	.201	350.3	244.3	52.6	-63.7		.800	1.870		838
839	2.615	.119	.226	308.7	335.0	39.0	-49.1		.468	2.090		839
840	3.134	.099	.162	255.6	274.6	66.8	-84.6		.969	1.590		840
841	2.255	.109	.065	131.7	344.5	34.2	-36.1		.184	2.510	179	841
842	3.233	.077	.253	6.1	1.8	57.3	-84.4		1.115	1.549		842
843	2.279	.155	.138	319.7	35.18	55.1	80.0		.071	2.363		843
844	3.196	.053	.162	319.3	342.7	76.9	-88.8		1.180	1.668	106	844
845	2.939	.034	.208	294.6	39.4	53.4	-64.0		.983	1.968	122	845
846	3.129	.144	.027	36.8	278.4	89.4	-101.6		.836	1.459	1A	846
847	2.783	.066	.063	57.7	273.3	58.6	-60.6		.775	2.046	37	847
848	3.106	.138	.034	322.5	247.3	86.4	-97.1		.835	1.504	1	848
849	3.151	.124	.351	258.9	229.0	31.4	-73.3		.699	1.396		849
850	2.998	.121	.259	242.3	125.2	48.4	-70.5		.767	1.655	120	850
851	2.228	.140	.040	150.7	158.9	33.6	-36.3		.086	2.463	187	851
852	2.363	.196	.427	307.9	27.5	16.2	-37.7		.127	2.087	168	852
853	2.312	.121	.166	213.8	186.1	33.1	-37.6		.178	2.403		853
854	2.369	.162	.116	258.0	197.7	37.0	-42.4		.156	2.262		854
855	2.362	.177	.190	231.3	10.3	33.7	-41.8		.086	2.221		855
856	2.437	.151	.244	178.7	126.3	32.4	-42.3		.188	2.179		856
857	2.190	.027	.086	300.4	88.0	31.7	-32.3		.271	2.722		857
858	2.808	.132	.139	228.3	63.8	55.8	-64.8		.595	1.834	126	858
859	3.212	.095	.229	57.2	32.0	63.0	-87.5		1.047	1.512		859
860	2.796	.076	.245	315.9	307.0	42.6	-55.1		.746	2.010		860
861	3.144	.082	.120	284.9	118.6	80.6	-89.5		1.037	1.635		861
862	2.803	.075	.259	91.0	299.2	41.0	-54.7		.749	2.012		862
863	3.200	.060	.419	229.3	119.0	6.9	-61.0		1.038	1.662		863
865	2.416	.235	.249	127.8	185.1	32.2	-47.2		-.050	1.992		865
866	3.123	.024	.129	332.9	90.7	76.6	-82.8		1.208	1.838		866
867	3.065	.143	.098	131.9	37.6	78.0	-91.2		.778	1.528		867
868	2.704	.112	.087	52.9	121.4	53.2	-57.8		.559	1.986	132	868
869	2.693	.204	.143	251.8	163.7	50.1	-64.1		.276	1.735	135	869
870	2.322	.226	.094	311.2	125.3	36.1	-44.4		-.032	2.170		870
871	2.222	.147	.076	202.3	166.4	32.9	-36.1		.053	2.444	184	871
872	2.731	.135	.135	205.0	203.2	52.0	-59.9		.473	1.860	133	872
873	2.627	.150	.087	246.1	161.5	49.3	-55.7		.397	1.966		873
874	3.156	.110	.197	191.5	198.0	68.2	-88.1		.950	1.533		874
875	2.555	.083	.263	302.2	199.8	33.3	-43.3		.470	2.230	4	875
876	3.011	.073	.185	353.5	155.3	61.0	-72.4		.947	1.793	2	876
877	2.486	.132	.061	42.2	123.7	43.3	-47.0		.347	2.189		877
878	2.363	.191	.037	2.8	200.6	38.6	-44.5		.097	2.196	164	878
879	2.530	.093	.260	24.5	270.4	33.1	-42.8		.422	2.230	4	879
880	3.002	.217	.331	6.8	267.0	38.9	-81.1		.443	1.382		880
881	2.612	.157	.266	306.8	275.1	35.5	-50.2		.334	1.994	21	881
882	3.132	.216	.140	22.5	266.3	81.1	-115.1		.587	1.228		882
883	2.238	.144	.095	323.3	280.3	33.1	-36.4		.072	2.433	180	883
885	3.097	.146	.046	347.8	170.3	85.2	-97.4		.804	1.486		885
886	3.169	.198	.302	3.2	62.4	47.9	-95.7		.645	1.220		886
888	2.709	.170	.233	74.9	129.0	42.3	-58.1		.404	1.843	140	888
889	2.445	.185	.112	60.7	138.5	39.2	-47.2		.160	2.115	157	889
890	3.023	.077	.183	218.9	166.6	62.1	-73.8		.947	1.768	2	890
891	2.861	.024	.216	114.5	107.3	48.5	-58.4		.957	2.072		891
892	3.229	.022	.365	93.5	179.1	21.6	-65.8		1.279	1.613		892
893	3.052	.100	.283	4.3	147.3	44.8	-70.1		.885	1.665		893
894	3.114	.089	.229	284.5	197.0	58.5	-78.6		.983	1.632		894

NUMBER	A	E	SINI	WBAR	ANODE	DWBAR	DNODE	NRES	TOMARS	TOJUP	FAMILY	
895	3.219	.130	.456	83.9	265.5	6.9	-69.7		.870	1.246		895
896	2.286	.160	.157	234.6	254.6	32.6	-38.3		.053	2.332		896
897	2.544	.075	.261	245.4	258.2	33.1	-42.6		.482	2.263	4	897
898	2.732	.339	.218	282.7	237.3	48.6	-89.2		-.079	1.352		898
899	2.908	.155	.243	27.4	257.5	48.5	-69.4		.591	1.651		899
900	2.473	.117	.209	288.0	188.0	35.7	-42.9		.342	2.258		900
901	2.224	.164	.077	331.5	262.7	33.0	-36.9		.017	2.403	184	901
902	2.447	.145	.116	26.2	343.6	39.9	-45.1		.267	2.207		902
903	3.239	.021	.193	273.3	165.4	72.7	-88.1		1.324	1.733		903
904	2.993	.083	.272	114.2	202.5	44.4	-65.3		.883	1.792		904
905	2.216	.122	.082	38.6	34.0	32.6	-35.0		.101	2.503	182	905
906	2.894	.048	.196	302.3	36.2	53.0	-62.2		.895	1.967		906
907	2.801	.206	.358	135.1	41.8	30.0	-62.1		.134	1.493		907
908	2.474	.171	.220	119.2	83.4	35.3	-45.9		.198	2.120		908
909	3.540	.038	.312	40.9	149.9	38.4	-96.7		1.455	1.378		909
910	2.926	.170	.149	241.6	42.8	62.4	-78.5		.582	1.595		910
912	3.124	.166	.339	132.8	32.5	36.1	-80.0		.648	1.374		912
913	2.197	.135	.091	262.3	98.4	31.8	-34.7		.037	2.477	189	913
914	2.454	.181	.456	284.6	251.2	13.4	-39.3		.214	2.019		914
915	2.228	.134	.094	70.9	1.5	32.7	-35.7		.083	2.464	182	915
916	2.365	.181	.210	19.6	323.8	32.6	-41.7		.059	2.190	163	916
917	2.381	.161	.094	340.3	333.7	38.2	-43.2		.177	2.244	75	917
918	2.865	.158	.221	341.6	325.9	49.8	-67.8		.545	1.670		918
919	2.772	.017	.155	32.5	235.1	51.7	-56.2		.800	2.095		919
920	2.622	.116	.210	119.4	198.1	40.8	-49.9		.480	2.092		920
921	3.173	.152	.303	250.0	207.4	45.4	-85.3		.801	1.377		921
922	2.690	.159	.141	328.3	214.7	49.6	-59.2		.411	1.868		922
923	2.615	.168	.258	46.0	200.6	36.5	-51.6		.314	1.970	21	923
924	2.938	.117	.147	7.9	156.4	62.3	-72.9		.756	1.737		924
925	2.700	.063	.375	152.9	298.1	21.0	-43.2		.745	2.031		925
926	2.983	.205	.272	215.1	47.4	48.1	-81.6		.480	1.428		926
927	3.213	.142	.256	168.9	5.7	58.1	-92.8		.875	1.365		927
928	3.136	.171	.292	151.6	129.9	48.2	-87.1		.709	1.342		928
929	2.239	.118	.080	227.3	233.5	33.4	-35.7		.139	2.497	182	929
930	2.430	.019	.267	256.3	338.2	29.5	-37.1		.335	2.345		930
931	3.168	.205	.188	72.0	119.7	74.9	-112.4		.640	1.223		931
932	2.420	.087	.139	89.2	8.7	37.8	-41.0		.387	2.380		932
933	2.370	.204	.088	155.2	149.9	38.0	-45.3		.064	2.158	166	933
934	2.748	.156	.275	36.6	321.3	38.2	-57.1		.449	1.854		934
935	2.219	.136	.072	65.8	335.9	32.9	-35.6		.076	2.473	185	935
936	3.136	.151	.029	302.5	35.9	90.2	-104.0		.819	1.429	1	936
937	2.231	.166	.082	307.9	245.0	33.1	-37.2		.020	2.394	184	937
938	3.161	.149	.027	34.6	134.2	93.0	-107.0		.847	1.407	1A	938
939	2.247	.129	.050	329.8	312.5	34.1	-36.6		.132	2.473	186	939
940	3.379	.116	.097	331.6	63.8	111.1	-129.1		1.117	1.281		940
941	2.781	.178	.101	25.9	46.3	57.5	-68.9		.448	1.736		941
942	3.160	.133	.167	40.4	70.2	75.1	-95.2		.883	1.454	109	942
943	3.126	.221	.192	122.6	116.1	71.5	-110.8		.554	1.223		943
945	2.636					4.3	-49.7	5				945
946	3.122	.149	.013	119.9	13.0	89.1	-101.8		.817	1.453	1	946
947	2.752	.212	.104	29.1	42.2	56.0	-70.9		.325	1.679		947
948	3.036	.192	.158	163.7	351.5	69.5	-93.6		.594	1.412		948
949	2.998	.212	.226	203.0	315.5	57.1	-88.6		.496	1.388		949
950	2.371	.172	.404	178.1	184.5	17.6	-36.8		.212	2.128		950
951	2.210	.143	.084	40.4	254.0	32.4	-35.4		.042	2.458	189	951
952	2.987	.211	.168	12.9	12.2	65.3	-91.5		.497	1.413		952
953	2.790	.168	.152	282.0	29.1	54.1	-66.8		.470	1.744		953
954	3.139	.147	.024	297.5	224.7	90.6	-103.6		.836	1.439	1A	954
955	2.594	.259	.224	264.5	348.4	40.2	-61.1		.070	1.754		955
956	2.298	.157	.115	310.6	201.0	34.5	-39.2		.100	2.353	175	956
957	2.919	.087	.275	113.1	235.4	41.9	-60.9		.805	1.863		957
959	3.185	.179	.063	28.0	49.6	94.8	-117.5		.763	1.292	105	959
960	2.248	.115	.068	335.3	249.7	33.9	-36.0		.162	2.501	179	960
961	2.693	.065	.186	289.5	21.8	45.3	-51.9		.690	2.141		961
962	2.906	.062	.035	9.9	171.5	67.9	-69.6		.900	1.935	3	962
963	2.248	.152	.128	87.8	60.7	32.4	-36.8		.049	2.398	183	963
964	3.053	.085	.151	50.2	23.8	69.1	-79.2		.952	1.712		964
965	3.154					27.8	-81.3	6				965
966	2.720	.141	.233	237.1	71.3	42.2	-56.0		.497	1.910	140	966
967	2.226	.118	.083	303.2	83.7	32.9	-35.2		.121	2.505	182	967
968	2.868	.148	.219	150.0	216.2	50.0	-67.0		.572	1.689		968
969	2.463	.171	.061	21.7	284.1	42.4	-48.1		.226	2.120	24	969
970	2.562	.239	.117	52.0	307.1	45.4	-59.3		.116	1.834		970
971	2.641	.169	.223	99.6	82.3	40.8	-54.2		.357	1.925	138	971
972	3.062	.175	.186	14.7	282.3	66.3	-91.0		.662	1.434		972
973	3.227	.066	.283	94.5	344.8	47.8	-78.7		1.142	1.583		973
974	2.534	.085	.078	41.8	85.3	44.9	-47.1		.505	2.254		974
975	2.834	.049	.038	141.4	13.1	62.7	-63.7		.874	2.045	3	975
976	3.186	.148	.159	192.4	252.5	79.5	-102.4		.851	1.377		976
977	3.118	.056	.245	180.6	74.9	54.2	-74.2		1.080	1.728		977

NUMBER	A	E	SINI	WBAR	ANODE	DWBAR	DNODE	NRES	TOMARS	TOJUP	FAMILY	
978	3.221	.146	.403	354.0	224.0	20.5	-77.1		.887	1.234		978
979	3.153	.095	.197	332.8	236.8	67.3	-85.4		.998	1.584		979
980	2.741	.117	.306	360.0	284.5	33.3	-51.6		.522	1.937		980
981	3.100	.165	.029	337.0	12.5	86.7	-101.6		.750	1.426	1	981
982	3.072	.221	.258	280.3	300.2	55.5	-96.3		.499	1.276		982
983	3.162	.100	.276	230.1	253.9	49.2	-78.5		.949	1.542	108	983
984	2.803	.125	.178	359.5	309.8	51.7	-62.4		.518,	1.770		984
985	2.300	.227	.095	349.7	284.2	35.3	-43.4		-.058	2.197		985
986	3.142	.129	.251	352.1	93.0	55.8	-84.1		.863	1.479		986
987	3.150	.195	.170	327.0	317.5	76.3	-108.5		.664	1.274		987
988	3.153	.188	.024	15.2	352.3	93.9	-116.4		.714	1.298	107	988
989	2.660	.234	.270	57.9	248.3	38.0	-60.7		.168	1.752		989
990	2.669	.180	.157	2.1	346.6	47.7	-59.1		.347	1.852		990
991	3.145	.137	.024	298.4	31.4	91.0	-102.2		.874	1.465	1	991
992	3.024	.128	.201	194.8	219.5	60.2	-78.2		.781	1.616		992
993	2.861	.046	.036	118.0	218.1	64.6	-65.6		.908	2.025	3	993
994	2.530	.061	.261	351.2	359.6	32.7	-41.7		.498	2.304	153	994
995	2.615	.111	.243	344.4	226.1	37.3	-48.1		.480	2.118		995
996	3.093	.161	.028	143.0	301.4	86.0	-100.2		.756	1.444	1	996
997	2.670	.156	.203	283.8	248.0	43.5	-75.5		.412	1.931	138	997
998	3.126	.125	.305	15.1	299.3	42.4	-76.5		.854	1.518		998
999	2.612	.169	.189	341.0	222.3	42.5	-53.8		.334	1.960	148	999
1000	3.181	.157	.421	246.0	325.7	18.0	-75.0	4	.837	1.247		1000
1001	3.200	.107	.183	58.4	263.4	73.8	-93.3		.997	1.496		1001
1002	2.788	.120	.193	334.6	339.2	49.3	-60.3		.598	1.875		1002
1003	3.150	.151	.022	108.0	183.7	92.0	-106.0		.833	1.414	1A	1003
1004	3.397	.060	.043	14.7	179.3	119.6	-124.2		1.338	1.453		1004
1005	3.164	.089	.343	90.7	346.9	31.5	-70.1		.856	1.504		1005
1006	3.151	.263	.272	19.2	295.7	58.9	-119.5		.408	1.057		1006
1007	2.708	.072	.063	31.9	297.9	54.2	-56.2		.685	2.098	36	1007
1008	3.093	.047	.153	59.5	13.4	70.7	-79.5		1.111	1.789		1008
1009	2.629	.434	.284	58.2	232.6	40.6	-97.1		-.418	1.269		1009
1010	2.931	.067	.048	19.8	100.1	69.4	-71.7		.908	1.895	123	1010
1011	2.394	.383	.084	129.5	145.0	39.8	-67.1		-.356	1.709		1011
1012	2.483	.139	.055	112.0	66.5	43.3	-47.3		.325	2.174		1012
1013	2.684	.217	.220	133.7	25.0	43.5	-62.3		.257	1.746		1013
1014	2.807	.216	.064	131.4	258.1	61.7	-77.6		.367	1.607		1014
1015	3.203	.061	.147	52.1	125.1	80.5	-91.4		1.160	1.634	106	1015
1016	2.219	.137	.103	85.9	2.2	32.2	-35.4		.061	2.462		1016
1017	2.606	.111	.124	181.1	122.4	46.3	-51.1		.490	2.110		1017
1018	2.537	.215	.135	339.0	353.1	43.2	-54.8		.155	1.928		1018
1019	1.912	.065	.422	259.3	147.2	10.6	-20.8		.059	2.879	191	1019
1020	2.787	.045	.076	193.8	195.4	58.2	-59.9		.838	2.096	38	1020
1021	2.738	.227	.286	51.5	120.4	38.5	-64.5		.248	1.665		1021
1022	2.805					27.8	-76.4	6				1022
1023	3.168	.069	.178	42.8	201.3	71.8	-85.8		1.102	1.647		1023
1024	2.866	.170	.271	13.9	59.5	42.9	-66.3		.512	1.678		1024
1025	1.979	.055	.419	171.6	168.2	11.2	-22.0		.141	2.828		1025
1026	2.250	.134	.082	308.7	108.1	33.7	-36.6		.116	2.452	182	1026
1027	3.161	.159	.025	169.0	334.3	93.4	-109.1		.816	1.378	1A	1027
1028	3.402	.101	.150	94.8	59.5	98.9	-122.3		1.190	1.302		1028
1029	2.890	.063	.039	178.6	2.1	66.5	-68.4		.882	1.948	3	1029
1030	3.123	.159	.260	189.8	193.7	54.2	-86.8		.737	1.405	113	1030
1031	3.046	.092	.321	169.8	223.5	35.6	-65.2		.848	1.689		1031
1032	3.131	.150	.146	249.5	73.8	77.2	-96.9		.806	1.431		1032
1033	3.003	.086	.188	57.0	195.3	60.2	-72.7		.898	1.762	2	1033
1034	2.292	.220	.082	317.8	295.6	35.3	-42.6		-.044	2.223		1034
1035	3.140	.146	.319	327.6	1.1	40.5	-79.7		.777	1.447		1035
1036	2.662	.382	.577	355.0	235.6	14.0	-73.0		-.445	1.210		1036
1037	2.181	.146	.103	39.1	203.7	31.1	-34.3		-.023	2.454	189	1037
1039	2.680	.132	.094	194.9	232.9	51.6	-57.4		.394	1.874		1039
1040	3.117	.185	.308	90.9	282.8	44.8	-86.8		.653	1.329		1040
1041	3.072	.109	.224	52.8	58.3	57.6	-77.7		.886	1.615	116	1041
1042	3.225					27.8	-70.2	6				1042
1043	3.093	.028	.147	262.3	167.1	71.4	-79.1		1.170	1.845		1043
1044	2.578	.130	.060	273.0	49.3	47.6	-51.8		.423	2.084	145	1044
1045	2.359	.161	.021	87.9	267.9	38.6	-42.6		.167	2.271	162	1045
1046	2.984	.049	.139	98.4	2.5	65.5	-71.7		.998	1.895		1046
1047	2.241	.157	.087	30.5	79.9	33.3	-37.1		.051	2.406	184	1047
1048	2.731	.202	.261	225.3	49.9	40.7	-62.0		.320	1.743		1048
1049	3.095	.088	.274	23.3	338.9	47.4	-72.4		.949	1.662		1049
1050	2.625	.142	.234	54.8	337.4	38.8	-50.8		.413	2.021	140	1050
1051	3.213	.080	.401	324.8	186.0	14.3	-65.9		1.131	1.493		1051
1052	2.236	.125	.072	57.0	105.4	33.4	-35.9		.122	2.485	188	1052
1053	2.615	.086	.143	78.8	10.5	45.5	-50.0		.570	2.162		1053
1054	2.921	.095	.172	26.6	85.9	58.2	-68.3		.803	1.818		1054
1055	2.198	.145	.089	319.6	154.9	31.9	-35.1		.018	2.455	189	1055
1056	2.230	.128	.084	307.4	108.1	33.0	-35.7		.105	2.481	182	1056
1057	2.893	.205	.091	11.2	265.9	66.1	-83.6		.457	1.541		1057
1058	2.197	.127	.078	307.2	224.9	32.0	-34.5		.060	2.501	189	1058

NUMBER	A	E	SINI	WBAR	ANODE	DWBAR	DNODE	NRES	TOMARS	TOJUP	FAMILY	
1059	2.644	.159	.193	273.7	205.2	43.5	-54.6		.387	1.948		1059
1060	2.237	.149	.120	293.4	224.3	32.4	-36.4		.047	2.415	183	1060
1061	3.121	.182	.024	36.9	88.1	89.9	-108.9		.709	1.354	107	1061
1062	3.007	.065	.109	111.0	332.9	70.8	-76.0		.973	1.822		1062
1063	2.314	.078	.092	180.2	98.1	35.6	-37.3		.314	2.518		1063
1064	2.547	.149	.180	288.3	279.0	40.5	-49.0		.337	2.082	148	1064
1065	2.361	.252	.152	319.1	325.3	35.6	-47.7		-.085	2.061		1065
1066	2.403	.171	.090	3.4	333.6	39.1	-44.7		.174	2.193	75	1066
1067	2.871	.155	.211	41.4	278.6	51.3	-68.3	10	.572	1.688		1067
1068	2.908	.125	.116	211.4	312.3	63.8	-72.6		.701	1.751		1068
1069	3.132	.135	.225	172.4	145.7	61.2	-86.7		.851	1.473		1069
1070	3.219	.067	.286	348.1	169.7	46.8	-77.9		1.133	1.593		1070
1071	2.801	.104	.082	98.4	42.2	58.9	-63.7		.682	1.920		1071
1072	3.173	.209	.140	69.6	25.1	84.1	-119.3		.630	1.213		1072
1073	3.175	.170	.020	321.9	353.9	95.6	-114.5		.791	1.332	1	1073
1074	3.155	.151	.020	65.3	321.8	92.6	-106.6		.837	1.409	1A	1074
1075	3.014	.069	.182	342.3	101.3	61.6	-72.6		.965	1.803	2	1075
1076	2.477	.147	.051	99.9	161.7	43.1	-47.4		.302	2.163	24	1076
1077	2.393	.158	.099	1.4	336.1	38.4	-43.5		.193	2.237	75	1077
1078	2.270	.187	.118	141.1	93.3	33.6	-39.5		-.004	2.307	177	1078
1079	2.874	.047	.036	119.4	304.0	65.5	-66.5		.917	2.009	3	1079
1080	2.420	.241	.092	65.9	348.1	39.9	-50.7		.014	2.005		1080
1081	3.091	.122	.069	43.7	13.8	82.6	-91.9		.872	1.567		1081
1082	3.128	.144	.025	323.7	192.8	89.3	-101.5		.837	1.461	1A	1082
1083	2.329	.212	.076	121.0	77.1	36.7	-43.9		.014	2.193		1083
1084	2.689	.124	.074	274.8	202.7	53.0	-57.8		.515	1.969		1084
1085	3.183	.066	.102	238.4	148.6	86.4	-93.2		1.125	1.645		1085
1086	3.164	.069	.164	145.5	309.4	74.3	-86.7		1.101	1.649	106	1086
1087	3.015	.071	.170	78.2	24.2	63.6	-73.6		.963	1.795	2	1087
1088	2.202	.154	.122	32.2	55.2	31.2	-35.3		-.024	2.414		1088
1089	2.214	.139	.053	87.4	72.0	33.0	-35.6		.069	2.474	186	1089
1090	2.360	.196	.360	135.1	152.3	22.1	-39.2		.118	2.061		1090
1091	3.432	.094	.007	103.4	348.2	128.9	-138.2		1.253	1.304		1091
1092	2.901	.096	.114	239.2	303.5	63.1	-69.6		.783	1.840		1092
1093	3.134	.142	.462	293.8	48.8	8.6	-66.3		.816	1.347		1093
1094	2.548	.148	.238	115.1	154.8	36.1	-47.3		.315	2.096		1094
1095	3.025	.059	.173	180.8	185.9	63.5	-73.4		1.007	1.820	2	1095
1096	2.601	.155	.153	320.8	78.7	44.7	-53.1		.369	2.000	149	1096
1097	2.641	.275	.019	300.3	188.4	52.7	-71.9		.064	1.622		1097
1098	2.689	.090	.247	72.6	326.0	38.9	-50.3		.601	2.094		1098
1099	3.167	.245	.200	5.4	20.9	74.6	-124.1		.502	1.103		1099
1100	2.898	.047	.037	292.6	290.0	67.2	-68.3		.938	1.987	3	1100
1101	3.243	.050	.371	251.4	204.4	20.4	-67.5		1.228	1.523		1101
1102	3.066	.062	.288	325.7	203.9	42.9	-67.0		1.019	1.775		1102
1103	1.934	.073	.360	323.4	271.4	14.8	-22.5		.071	2.832	190	1103
1104	2.630	.319	.119	49.5	136.8	49.7	-77.1		-.077	1.524		1104
1105	3.013	.069	.171	323.8	119.4	63.2	-73.2		.965	1.802	2	1105
1106	2.597	.154	.245	185.7	323.1	37.1	-49.9		.348	2.031	147	1106
1107	3.191	.125	.103	120.8	113.8	88.5	-102.4		.936	1.452	104	1107
1108	2.428	.163	.468	297.0	231.3	11.3	-36.8		.235	2.096		1108
1109	3.210	.134	.095	251.0	272.4	91.9	-107.5		.925	1.402		1109
1110	2.218	.173	.122	314.9	242.6	31.8	-36.7		-.038	2.368		1110
1111	2.994	.062	.052	2.9	145.9	74.1	-76.5		.979	1.848		1111
1112	3.021	.065	.177	39.0	300.8	62.6	-73.1		.985	1.807	2	1112
1113	3.113	.128	.249	106.6	324.0	55.1	-81.6		.840	1.511	116	1113
1114	3.092	.049	.190	61.0	202.2	64.5	-76.9		1.097	1.785		1114
1115	3.101	.164	.269	131.3	68.8	51.3	-85.0	10	.701	1.414	113	1115
1116	2.925	.184	.321	92.4	354.1	37.2	-69.5		.482	1.567		1116
1117	2.248	.161	.074	285.5	158.8	33.8	-37.6		.054	2.395	184	1117
1118	3.210	.037	.257	253.3	316.6	54.2	-78.7		1.220	1.697		1118
1119	2.612	.142	.125	272.3	51.0	46.7	-53.4		.408	2.020		1119
1120	2.216	.121	.073	35.1	167.4	32.8	-35.0		.106	2.507	188	1120
1121	2.546	.139	.114	58.3	348.0	44.2	-49.8		.367	2.103		1121
1122	2.605	.235	.068	36.2	57.2	49.4	-63.0		.148	1.773		1122
1123	2.225	.140	.101	57.3	82.7	32.5	-35.8		.063	2.453		1123
1124	2.927	.044	.132	228.9	14.1	62.7	-67.6		.966	1.967	124	1124
1125	3.147	.224	.032	114.2	95.1	94.8	-127.2		.593	1.194		1125
1126	2.272	.195	.116	141.7	355.4	33.7	-40.1		-.020	2.286	177	1126
1127	2.594	.223	.265	62.0	133.4	36.0	-55.4		.141	1.848		1127
1128	2.789	.047	.012	243.9	338.9	60.2	-60.9		.837	2.090		1128
1129	3.022	.059	.170	68.9	271.4	63.8	-73.4		1.005	1.822	2	1129
1130	2.229	.146	.050	326.0	226.3	33.5	-36.5		.071	2.448	186	1130
1131	2.229	.234	.045	350.0	104.5	33.7	-40.9		-.126	2.253		1131
1132	2.686	.242	.140	294.4	22.0	50.3	-68.7		.148	1.638		1132
1133	2.186	.138	.083	20.2	59.1	31.6	-34.4		.016	2.478	189	1133
1134	2.684	.411	.279	337.6	12.5	42.8	-98.9		-.323	1.245		1134
1135	2.666	.108	.086	5.1	338.7	51.3	-55.0	10	.463	1.959	139	1135
1136	2.565	.214	.169	357.7	218.9	42.5	-55.7		.179	1.899		1136
1137	2.424	.058	.060	357.2	76.4	40.5	-41.5		.471	2.441		1137
1138	3.146	.063	.264	51.2	284.2	50.8	-74.9		.989	1.599	111	1138

NUMBER	A	E	SINI	WBAR	ANODE	DWBAR	DNODE	NRES	TOMARS	TOJUP	FAMILY	
1139	1.947	.220	.245	55.3	202.3	20.9	-28.2		-.221	2.502		1139
1140	2.772	.078	.228	36.8	71.3	44.1	-54.9		.723	2.022		1140
1141	2.271	.132	.065	33.8	111.4	34.7	-37.4		.144	2.445	181	1141
1142	3.179	.120	.026	223.8	172.8	94.1	-103.1		.954	1.482	103	1142
1144	3.755	.047	.159	12.2	165.7	134.6	-172.0		1.688	1.141		1144
1145	2.424	.124	.116	233.5	338.9	39.0	-42.9		.300	2.286		1145
1146	3.047	.207	.327	258.0	212.7	39.8	-83.0		.495	1.367		1146
1147	2.271	.208	.082	267.7	264.1	34.5	-41.1		-.040	2.262		1147
1148	3.016	.089	.175	304.8	151.4	63.0	-74.7		.905	1.740	2	1148
1149	2.898	.063	.224	29.6	263.5	49.4	-61.6		.864	1.915		1149
1150	2.191	.147	.052	350.3	216.6	32.3	-35.1		.020	2.470	186	1150
1151	2.406	.228	.139	347.8	233.8	37.8	-48.3		.022	2.063		1151
1152	2.427	.079	.097	180.2	323.0	39.7	-41.8		.420	2.388	156	1152
1153	2.196	.109	.066	295.6	273.1	32.2	-34.0		.108	2.549	188	1153
1154	3.399	.067	.059	239.9	78.8	118.4	-124.8		1.313	1.426		1154
1155	2.462	.188	.107	219.5	29.6	41.1	-48.7		.176	2.082		1155
1156	2.260	.088	.013	149.4	107.3	34.9	-36.0		.239	2.556		1156
1157	3.195	.118	.184	269.2	333.6	73.7	-94.5		.956	1.466		1157
1158	2.564	.087	.266	72.2	340.7	33.3	-43.7		.467	2.209	4	1158
1159	2.380	.031	.226	201.5	344.4	31.4	-36.6		.410	2.501		1159
1160	2.560	.078	.256	20.7	.1	34.0	-43.5		.503	2.250	4	1160
1161	3.164	.059	.146	22.6	70.0	77.3	-87.3		1.134	1.678	106	1161
1163	3.215	.022	.138	301.8	132.6	82.8	-91.0		1.299	1.757		1163
1164	2.306	.192	.426	148.4	160.4	15.6	-35.5		.095	2.163		1164
1165	3.130	.162	.251	285.7	209.4	56.7	-89.2		.746	1.392	113	1165
1166	2.542					27.8	-75.2	6				1166
1167	3.413	.073	.114	248.1	234.3	109.5	-122.9		1.294	1.387		1167
1168	2.551	.161	.242	342.9	223.7	35.9	-48.2		.293	2.062		1168
1169	2.318	.160	.086	86.8	256.9	36.0	-40.4		.123	2.327		1169
1170	2.326	.212	.140	53.9	354.8	17.8	-38.0		.071	2.087		1170
1171	3.167	.176	.039	56.2	138.3	94.1	-113.7		.765	1.318		1171
1174	3.022	.079	.178	346.7	355.1	62.8	-74.2		.940	1.763	2	1174
1175	3.215	.025	.297	269.2	240.1	42.6	-73.6		1.263	1.747		1175
1176	2.692	.142	.136	79.0	274.1	49.9	-58.0		.452	1.910		1176
1177	3.350	.037	.278	146.5	255.4	50.5	-86.1		1.271	1.518		1177
1178	2.680	.203	.119	160.6	180.1	50.8	-62.6		.207	1.703		1178
1179	2.617	.188	.160	228.7	357.7	45.1	-56.6		.290	1.899		1179
1181	2.664	.187	.117	64.2	265.2	49.9	-60.3		.305	1.824		1181
1182	2.259	.110	.166	73.1	332.0	31.5	-35.3		.131	2.454		1182
1183	2.383	.158	.048	207.8	354.1	39.3	-43.5		.194	2.247	167	1183
1184	2.668	.049	.200	275.4	350.8	42.8	-49.5		.707	2.217		1184
1185	2.237	.124	.088	97.3	71.9	33.2	-35.8		.120	2.482	182	1185
1186	3.021	.071	.179	326.3	38.6	62.5	-73.5		.963	1.787	2	1186
1187	2.640	.177	.217	49.4	322.9	41.3	-55.3		.335	1.903	138	1187
1188	2.191	.140	.079	28.8	358.1	31.8	-34.7		.022	2.474	189	1188
1189	2.931	.065	.194	14.1	276.0	55.3	-65.7		.893	1.892		1189
1190	2.432	.127	.051	83.4	8.8	41.1	-44.1		.311	2.266	159	1190
1191	2.893	.082	.308	164.1	136.4	35.5	-56.9		.767	1.889		1191
1192	2.365	.224	.422	146.4	5.0	17.5	-40.1		.064	2.011		1192
1193	2.646	.144	.233	220.2	46.5	39.6	-52.1		.426	1.989	140	1193
1194	2.914	.127	.216	180.1	289.3	52.1	-68.2		.652	1.689		1194
1195	2.259	.197	.146	232.8	281.2	32.3	-39.1		-.055	2.274		1195
1196	2.653	.103	.297	20.2	101.4	32.2	-47.1		.474	2.053		1196
1197	2.883	.230	.280	172.0	259.0	44.4	-76.9		.354	1.468		1197
1198	2.249	.281	.074	343.7	261.8	34.0	-45.2		-.222	2.123		1198
1199	3.019	.063	.170	174.3	241.3	63.7	-73.4		.989	1.814	2	1199
1200	3.059	.133	.090	235.1	217.8	78.0	-89.1		.808	1.567		1200
1201	2.699	.021	.130	48.3	211.3	59.1	-52.9		.793	2.223		1201
1203	2.884	.223	.117	44.1	235.4	64.0	-85.3		.390	1.500		1203
1204	2.263	.248	.034	315.9	341.6	35.0	-43.5		-.127	2.190		1204
1205	2.532	.244	.148	12.8	17.2	42.5	-57.2		.073	1.865		1205
1206	2.872	.065	.241	212.8	321.3	45.7	-59.0		.847	1.957		1206
1207	3.021	.070	.139	85.9	14.1	62.5	-73.4		.969	1.792	2	1207
1209	3.175	.132	.100	251.9	89.1	87.6	-102.1		.899	1.445	104	1209
1210	3.011	.069	.177	240.2	108.9	62.3	-72.8		.962	1.804	2	1210
1211	2.930	.120	.204	329.5	132.4	54.6	-69.4		.719	1.733		1211
1213	3.132	.077	.252	28.1	273.4	53.6	-76.4		1.013	1.643	111	1213
1214	2.711	.091	.189	303.4	284.6	46.0	-53.9		.631	2.050		1214
1215	2.579	.094	.265	54.3	125.9	33.8	-44.6		.470	2.183	4	1215
1216	2.232	.160	.128	245.4	127.3	32.0	-36.6		.007	2.386	183	1216
1217	2.353	.169	.087	225.0	157.3	37.3	-42.4		.133	2.260		1217
1218	2.263	.152	.043	140.1	58.1	34.8	-38.0		.092	2.406		1218
1219	2.213	.138	.068	88.7	36.7	32.8	-35.5		.066	2.474	185	1219
1220	3.005	.070	.177	121.4	115.7	61.9	-72.3		.954	1.808	2	1220
1221	1.921	.448	.223	199.8	168.5	21.7	-42.1		-.697	2.095		1221
1222	2.790					27.8	-65.7	6				1222
1223	2.869	.043	.037	90.2	14.7	65.1	-66.1		.923	2.025	3	1223
1224	2.304	.168	.155	41.6	260.7	33.4	-39.4		.062	2.306		1224
1225	2.233	.115	.049	129.9	.9	33.7	-35.6		.147	2.515	186	1225
1226	2.584	.140	.170	160.4	13.0	42.6	-50.6		.391	2.063	149	1226

NUMBER	A	E	SINI	WBAR	ANODE	DWBAR	DNODE	NRES	TOMARS	TOJUP	FAMILY	
1227	3.201	.150	.307	294.7	1.5	44.9	-86.9		.825	1.355		1227
1228	2.769	.076	.075	168.8	300.1	57.3	-60.0		.734	2.030	37	1228
1229	3.180	.136	.030	5.2	249.4	94.6	-106.7		.902	1.429	1	1229
1230	2.572	.151	.188	32.5	206.2	40.9	-50.2		.353	2.049	148	1230
1231	2.669	.104	.211	208.3	337.0	42.2	-51.3		.564	2.063		1231
1232	3.181	.159	.207	225.8	267.1	68.7	-98.9		.801	1.356	102	1232
1233	2.555	.068	.115	230.8	288.8	44.3	-47.0		.562	2.272		1233
1234	3.013	.055	.168	50.7	302.1	63.5	-72.5		1.010	1.844	2	1234
1235	1.910	.075	.435	49.6	6.8	9.8	-20.7		.037	2.861	191	1235
1236	2.430	.183	.223	1.7	48.2	33.8	-44.5		.116	2.124	163	1236
1237	2.612	.041	.155	12.5	54.6	44.4	-48.2		.686	2.283		1237
1238	2.667	.159	.210	150.2	49.1	42.7	-55.5		.397	1.927	138	1238
1239	2.664	.241	.014	117.1	34.7	53.8	-68.9		.184	1.697		1239
1240	2.868	.133	.190	337.8	318.8	53.7	-67.1		.627	1.756		1240
1241	3.185	.082	.420	286.6	322.6	9.4	-62.7		1.070	1.538		1241
1242	2.736	.157	.193	49.5	343.0	47.4	-60.0		.464	1.842		1242
1243	3.099	.048	.248	240.4	248.6	52.8	-72.2		1.073	1.756		1243
1244	2.344	.138	.171	169.3	275.6	33.8	-39.3		.171	2.340		1244
1245	2.893	.043	.042	349.7	175.3	66.6	-67.7		.943	2.003	3	1245
1246	2.620	.193	.318	346.1	286.8	30.8	-51.9		.134	1.797		1246
1247	3.138	.160	.031	283.2	205.8	90.7	-106.1		.795	1.402	1	1247
1248	2.722	.028	.141	127.7	77.8	50.2	-53.9		.797	2.175		1248
1249	2.224	.128	.099	134.0	256.9	32.5	-35.3		.088	2.476	182	1249
1250	2.551	.308	.312	146.6	283.9	32.4	-62.0		-.166	1.647		1250
1251	2.717	.116	.093	152.0	148.7	53.6	-58.7		.564	1.966	132	1251
1252	2.696	.095	.579	178.3	131.4	-6.6	-39.5		.363	2.012		1252
1253	3.153	.172	.023	37.6	340.7	93.1	-111.3		.768	1.345	1	1253
1254	3.134	.070	.145	182.9	286.9	75.5	-85.5		1.070	1.675		1254
1255	3.153	.121	.169	10.3	245.0	73.8	-92.1		.915	1.499	109	1255
1256	3.904	.024	.090	307.7	249.9	213.1	-234.0		1.909	1.094		1256
1257	2.488	.108	.080	210.2	224.6	42.8	-45.6		.406	2.246		1257
1258	3.185	.009	.154	316.6	297.7	77.3	-86.4		1.316	1.818		1258
1259	3.100	.165	.024	216.0	55.2	87.1	-102.7		.745	1.425	1	1259
1260	2.614	.017	.155	254.5	301.4	44.5	-47.9		.752	2.344		1260
1261	3.145	.212	.029	170.5	48.0	94.0	-122.6		.631	1.234		1261
1262	3.002	.033	.211	192.5	127.4	55.6	-67.8		1.065	1.911		1262
1263	2.665	.085	.508	102.1	163.4	-.3	-37.8		.640	2.020		1263
1264	2.862	.162	.443	249.4	234.1	15.1	-55.3		.574	1.604		1264
1265	3.012	.026	.184	114.4	311.6	60.8	-70.3		1.097	1.935		1265
1266	3.363	.038	.313	199.4	319.3	39.3	-82.5		1.323	1.559		1266
1267	2.465	.170	.082	279.0	12.8	41.9	-47.9		.230	2.120	154	1267
1270	2.235	.155	.097	3.2	100.3	32.9	-36.7		.041	2.410	180	1270
1271	3.135	.096	.104	44.4	132.5	82.3	-91.2		.987	1.600		1271
1272	2.782	.128	.161	307.4	317.5	52.3	-61.8		.551	1.840	127	1272
1273	2.394	.120	.110	341.9	291.5	38.1	-41.6		.284	2.327	165	1273
1274	2.229	.148	.084	192.5	317.3	33.0	-36.4		.056	2.434	184	1274
1275	2.680	.136	.225	30.2	192.9	41.7	-53.8		.478	1.967	140	1275
1276	3.172	.071	.378	168.8	116.7	20.6	-64.8		1.104	1.526		1276
1277	2.699	.215	.145	285.8	248.3	50.4	-65.9		.238	1.687	135	1277
1278	2.405	.211	.181	323.3	87.9	35.6	-46.0		.054	2.109		1278
1279	2.370	.197	.115	260.0	329.3	37.2	-44.5		.072	2.179		1279
1280	3.413	.018	.134	34.3	292.6	103.0	-114.4		1.493	1.577		1280
1281	2.560	.187	.147	271.2	215.1	43.4	-53.4		.249	1.971	150	1281
1282	3.118	.080	.330	80.9	322.8	34.0	-67.6		.943	1.649		1282
1283	3.202	.164	.150	31.3	162.9	83.0	-108.9		.811	1.309		1283
1284	2.646	.150	.211	71.4	302.4	41.7	-53.3		.416	1.965	138	1284
1285	2.993	.021	.115	74.1	312.0	68.8	-72.6		1.098	1.970		1285
1286	3.023	.074	.178	278.2	207.7	62.8	-73.9		.956	1.777	2	1286
1287	3.012	.075	.180	140.4	209.9	61.9	-73.0		.943	1.785	2	1287
1288	2.885	.028	.150	330.6	296.6	58.1	-63.4		.977	2.048		1288
1289	2.860	.052	.037	270.3	226.8	64.5	-65.7		.890	2.010	3	1289
1290	2.366	.120	.113	38.0	301.9	36.9	-40.4		.256	2.362	165	1290
1291	3.012	.061	.171	309.5	222.3	63.1	-72.7		.990	1.827	2	1291
1292	2.543	.089	.057	156.8	271.6	45.9	-47.9		.503	2.231		1292
1293	2.227	.206	.118	334.6	239.9	32.2	-38.7		-.097	2.293		1293
1294	2.687	.215	.138	33.9	82.7	50.2	-65.1		.232	1.714	135	1294
1295	3.388	.132	.056	122.4	210.1	120.3	-139.5		1.076	1.221		1295
1296	2.418	.159	.088	113.8	233.8	39.7	-44.8		.216	2.201	75	1296
1297	3.021	.058	.176	94.4	294.9	62.8	-72.9		1.005	1.826	2	1297
1298	3.128	.105	.118	355.2	295.9	79.9	-90.7		.946	1.580		1298
1299	2.803	.164	.139	75.8	173.7	55.7	-67.6		.499	1.745	126	1299
1300	2.782	.038	.148	182.2	81.7	52.8	-57.4		.830	2.101		1300
1301	2.767	.163	.579	136.7	174.4	-.0	-48.1		.095	1.837	129	1301
1302	3.122	.162	.024	94.7	83.0	89.3	-104.6		.775	1.413	1A	1302
1303	3.230	.109	.332	180.9	72.2	36.7	-79.1		.940	1.443		1303
1304	3.196	.125	.315	219.7	89.9	41.3	-80.9		.896	1.451		1304
1305	3.014	.107	.026	203.0	37.3	77.4	-82.9		.859	1.690		1305
1306	3.140	.089	.280	76.1	276.1	47.1	-75.1		.980	1.618	108	1306
1307	2.251	.118	.082	103.7	236.0	33.7	-36.1		.154	2.490	182	1307
1308	2.909	.049	.104	180.8	343.7	64.2	-67.6		.937	1.969		1308

NUMBER	A	E	SINI	WBAR	ANODE	DWBAR	DNODE	NRES	TOMARS	TOJUP	FAMILY	
1309	3.227	.132	.194	86.0	211.1	74.2	-99.9		.935	1.389		1309
1310	2.393	.236	.424	318.2	225.2	18.0	-42.2	4	.023	1.939		1310
1311	2.426	.077	.065	153.4	250.6	40.5	-42.0		.428	2.394		1311
1312	3.093					27.8	-68.4	6				1312
1313	2.657	.161	.248	49.1	298.0	38.7	-53.5		.379	1.937		1313
1314	2.295	.163	.107	63.4	266.1	34.7	-39.4		.087	2.345	175	1314
1315	3.212	.082	.140	237.6	242.3	82.6	-94.9		1.099	1.558		1315
1316	2.411	.264	.435	31.3	245.5	18.3	-45.4		-.036	1.858		1316
1317	3.192					27.8	-114.5	6				1317
1318	2.308	.217	.422	190.8	354.4	16.7	-37.5		.033	2.101		1318
1319	2.987	.242	.076	207.0	268.7	76.0	-104.5		.419	1.325		1319
1320	2.986	.221	.325	263.6	67.4	39.9	-81.0		.428	1.385		1320
1321	2.941	.153	.183	290.5	315.3	58.9	-75.6		.639	1.629		1321
1322	2.422	.238	.423	270.7	249.6	18.4	-43.5		.057	1.910		1322
1323	3.204	.176	.330	185.7	47.8	40.8	-90.6		.734	1.265		1323
1324	2.185	.136	.086	249.3	298.2	31.5	-34.3		.017	2.481	189	1324
1325	2.538	.224	.127	349.7	7.9	43.7	-55.9		.136	1.905		1325
1326	2.666	.160	.275	32.2	103.5	35.8	-52.9		.361	1.922	137	1326
1327	2.781	.131	.091	315.0	50.0	57.5	-64.2		.584	1.866	40	1327
1328	3.496	.109	.113	40.9	235.0	122.9	-146.9		1.230	1.182		1328
1329	2.617	.140	.241	284.3	136.0	37.9	-50.0		.406	2.040	140	1329
1330	3.176	.111	.269	172.6	162.9	51.3	-81.7	10	.905	1.471	108	1330
1331	3.104	.171	.036	291.1	137.2	87.3	-104.4		.729	1.403		1331
1332	3.063	.094	.046	354.4	347.1	80.8	-85.8		.940	1.681	119	1332
1333	2.633	.146	.238	105.7	118.0	38.8	-51.4		.406	2.002	140	1333
1334	2.914	.101	.187	241.7	137.8	55.8	-67.4		.770	1.806		1334
1335	2.242	.124	.048	19.5	187.4	34.0	-36.2		.137	2.489	186	1335
1336	2.851	.047	.036	278.0	98.4	63.9	-64.9		.893	2.031	3	1336
1337	2.911	.057	.301	4.2	163.8	36.6	-56.8		.873	1.956		1337
1338	2.264	.126	.090	97.7	318.8	34.0	-36.7		.143	2.460	182	1338
1339	3.021	.063	.171	126.3	289.9	63.5	-73.4		.991	1.812	2	1339
1340	3.183	.167	.028	206.8	295.1	96.4	-114.9		.805	1.330	1	1340
1341	2.742	.089	.211	228.9	109.7	45.2	-54.8		.667	2.018		1341
1342	2.289	.179	.382	180.5	308.3	18.8	-35.0		.128	2.202		1342
1343	2.568	.113	.097	256.7	32.3	45.8	-49.8		.459	2.142		1343
1344	2.248	.163	.089	181.3	58.3	33.5	-37.6		.043	2.385	184	1344
1346	2.627	.148	.247	67.2	169.6	37.7	-50.9		.391	2.010		1346
1347	2.573	.072	.222	108.2	231.8	37.7	-45.1		.551	2.264		1347
1348	2.791	.148	.096	118.1	85.3	58.0	-66.5		.543	1.807		1348
1349	3.013	.159	.205	241.2	306.8	59.5	-81.3		.673	1.538		1349
1350	2.858	.051	.039	23.9	160.4	64.3	-65.5		.889	2.013	3	1350
1351	3.192	.057	.171	92.1	3.7	74.8	-87.9		1.160	1.660	106	1351
1352	2.778	.039	.068	68.7	203.0	58.0	-59.3		.846	2.125	38	1352
1353	3.012	.073	.172	280.9	218.3	63.1	-73.4		.952	1.792	2	1353
1354	3.134	.207	.112	266.2	16.6	85.1	-115.0		.617	1.259		1354
1355	1.853	.070	.403	213.8	230.5	11.5	-20.2		-.003	2.930	191	1355
1356	3.083	.012	.120	349.6	65.8	74.9	-79.7		1.210	1.911		1356
1357	3.190	.094	.229	11.5	83.2	62.1	-85.6		1.033	1.538		1357
1358	2.475	.148	.037	298.1	355.2	43.2	-47.4		.299	2.162		1358
1359	3.120	.041	.175	66.9	61.3	68.8	-79.8		1.148	1.782		1359
1360	2.633	.160	.429	263.5	332.3	16.4	-45.1		.425	1.869		1360
1361	3.084					27.8	-67.1	6				1361
1363	2.903	.047	.034	291.5	247.0	67.7	-68.7		.943	1.982	3	1363
1364	3.012	.073	.183	253.3	60.8	61.4	-72.7		.949	1.790	2	1364
1365	2.249	.141	.104	213.6	258.9	33.1	-36.6		.090	2.434	180	1365
1366	2.875	.116	.167	290.9	18.2	56.5	-67.0		.699	1.806		1366
1367	2.344	.152	.396	253.2	271.2	17.6	-34.9		.241	2.203		1367
1368	2.523	.056	.253	222.5	14.9	33.3	-41.5		.517	2.337	153	1368
1369	3.109	.162	.269	305.1	187.9	51.3	-85.1	10	.668	1.378	113	1369
1370	2.251	.128	.091	297.9	299.3	33.6	-36.4		.127	2.462	182	1370
1371	3.203	.085	.292	248.8	190.0	45.8	-78.2		1.056	1.555		1371
1372	2.767	.119	.304	80.6	325.9	34.2	-53.3		.549	1.913		1372
1374	2.251	.224	.115	9.0	295.6	33.1	-40.8		-.108	2.239		1374
1375	2.448	.079	.090	111.0	46.4	40.8	-42.8		.443	2.363	156	1375
1376	2.228	.165	.063	313.2	177.2	33.4	-37.1		.025	2.403	184	1376
1377	2.260	.125	.118	197.0	226.3	33.1	-36.2		.129	2.455		1377
1378	2.375	.162	.053	230.4	29.5	38.9	-43.3		.175	2.248	167	1378
1379	2.528	.129	.270	180.2	172.1	32.4	-44.1		.309	2.128		1379
1380	3.145	.128	.192	222.6	352.2	68.9	-90.3		.881	1.487		1380
1381	2.491	.178	.087	26.2	338.0	42.9	-49.7		.232	2.068	154	1381
1382	2.220	.147	.029	218.2	325.9	33.4	-36.2		.063	2.456	187	1382
1383	3.083	.148	.023	355.1	276.8	84.8	-96.4		.790	1.496	1	1383
1384	2.677	.160	.209	80.9	158.7	43.2	-56.2		.404	1.913	138	1384
1385	2.741	.064	.104	17.1	119.0	54.0	-57.1		.728	2.085	134	1385
1386	2.364	.228	.205	309.6	169.1	33.1	-45.0		-.051	2.089		1386
1387	2.258	.155	.109	325.1	210.3	33.3	-37.5		.067	2.394	180	1387
1388	3.019	.071	.182	284.6	50.1	61.9	-73.1		.961	1.790	2	1388
1389	2.866	.043	.037	167.3	206.8	64.9	-65.8		.919	2.027	3	1389
1390	3.435					27.8	-85.8	5				1390
1391	2.549	.194	.123	181.6	105.0	44.3	-53.8		.225	1.963	43	1391

NUMBER	A	E	SINI	WBAR	ANODE	DWBAR	DNODE	NRES	TOMARS	TOJUP	FAMILY
1392	2.608	.167	.225	49.2	351.6	39.5	-52.3		.334	1.972	147	1392
1393	2.435	.130	.090	217.4	51.6	40.3	-44.0		.300	2.253		1393
1394	2.439	.064	.050	263.2	197.6	41.4	-42.4		.474	2.408		1394
1395	3.201	.028	.169	296.9	249.8	75.5	-87.1		1.267	1.744		1395
1396	2.248	.153	.081	246.9	349.4	33.7	-37.3		.070	2.411	184	1396
1397	2.685	.249	.044	272.6	64.5	54.7	-71.6		.182	1.657		1397
1398	3.159	.056	.227	12.9	295.7	59.9	-79.1		1.135	1.690		1398
1399	2.216	.136	.118	45.6	166.5	31.7	-35.1		.046	2.455	183	1399
1400	3.114	.160	.308	315.9	216.1	43.2	-81.3		.726	1.421		1400
1401	2.227	.110	.142	359.8	274.5	31.3	-34.4		.099	2.483		1401
1402	2.684	.146	.264	261.2	267.3	37.5	-53.1		.432	1.955	137	1402
1403	2.718	.272	.167	350.6	163.0	50.7	-75.6		.051	1.484		1403
1405	2.252	.139	.133	70.8	308.2	32.4	-36.4		.081	2.422	183	1405
1406	2.696	.081	.228	84.1	329.8	41.4	-51.1		.646	2.101		1406
1407	2.763	.233	.138	20.1	273.3	54.6	-74.2		.258	1.606		1407
1408	3.110	.056	.151	33.6	210.1	72.7	-81.8		1.095	1.745		1408
1409	2.676	.037	.116	50.1	186.9	49.8	-52.3		.735	2.208		1409
1410	3.020	.073	.177	54.1	177.4	62.7	-73.6		.958	1.784	2	1410
1411	3.003	.017	.161	29.0	284.3	63.7	-70.9		1.119	1.966		1411
1412	2.215	.138	.071	101.4	64.8	32.8	-35.5		.066	2.470	185	1412
1413	3.022	.079	.178	143.1	186.6	62.9	-74.2		.939	1.762	2	1413
1414	2.785	.193	.143	151.5	150.4	54.8	-69.8		.398	1.687	128	1414
1415	2.224	.125	.064	188.5	317.4	33.2	-35.5		.110	2.496	188	1415
1416	3.018	.079	.184	75.3	346.8	61.5	-73.4		.936	1.768	2	1416
1417	2.973	.093	.125	236.4	97.1	66.8	-74.5		.856	1.778		1417
1418	2.242	.147	.124	313.7	350.3	32.4	-36.4		.054	2.414	183	1418
1419	2.293	.165	.114	100.5	218.1	34.4	-39.3		.076	2.340	175	1419
1420	2.749	.048	.080	305.8	264.4	55.8	-57.5		.789	2.125	130	1420
1421	3.093	.113	.161	198.3	38.2	70.7	-85.5		.895	1.586		1421
1422	2.247	.130	.055	23.2	213.0	34.1	-36.5		.130	2.471	186	1422
1423	2.860	.044	.037	26.1	40.1	64.5	-65.5		.910	2.030	3	1423
1424	3.187	.025	.149	336.5	37.1	78.4	-87.4		1.268	1.766		1424
1425	2.612	.138	.229	168.1	191.6	38.8	-50.1		.412	2.044	140	1425
1426	2.581	.162	.178	235.2	330.0	42.1	-51.8		.328	2.014	148	1426
1427	2.750	.180	.152	304.2	74.4	52.1	-64.9		.366	1.724		1427
1428	2.810	.079	.287	15.3	116.9	36.9	-53.6		.727	2.002		1428
1429	2.550	.295	.139	342.9	46.2	44.3	-65.3		-.050	1.712		1429
1430	2.559	.175	.069	309.7	315.1	46.6	-53.6		.290	1.995		1430
1431	2.620	.138	.229	334.7	118.5	39.1	-50.5		.421	2.035	140	1431
1432	2.381	.181	.132	338.6	125.7	36.9	-43.9		.115	2.204		1432
1433	2.786	.132	.164	71.6	318.0	52.3	-62.3		.546	1.827	127	1433
1434	3.018	.059	.179	259.9	158.9	62.2	-72.5		.999	1.827	2	1434
1435	2.648	.266	.087	41.4	139.6	51.3	-70.2	10	.049	1.601		1435
1436	3.146	.076	.259	254.6	262.1	52.2	-76.5		.997	1.608	111	1436
1438	3.173	.178	.061	5.9	258.8	93.5	-115.6		.756	1.308	105	1438
1440	3.152	.162	.032	43.6	12.8	92.3	-108.6		.796	1.377	1	1440
1441	2.632	.171	.273	19.8	258.1	35.4	-52.0		.306	1.931	21	1441
1442	2.875	.045	.038	330.7	248.1	65.5	-66.5		.923	2.014	3	1442
1443	2.938	.036	.036	300.2	209.9	70.2	-71.0		1.006	1.977		1443
1444	3.158	.100	.335	235.5	304.5	34.2	-72.0		.902	1.541		1444
1445	3.114	.149	.021	356.0	79.7	88.1	-100.7		.808	1.461	1A	1445
1446	2.246	.136	.086	193.7	9.5	33.5	-36.5		.105	2.450	182	1446
1447	2.536	.057	.076	135.4	25.0	45.0	-46.4		.577	2.321		1447
1448	2.372	.214	.099	126.5	37.1	37.8	-45.9		.039	2.134	166	1448
1449	2.223	.151	.110	218.7	116.0	32.2	-36.0		.029	2.424	183	1449
1450	2.611	.179	.069	97.7	67.9	49.3	-57.3		.308	1.913		1450
1451	2.203	.145	.095	200.6	180.8	31.9	-35.2		.020	2.449	189	1451
1452	3.117	.179	.268	130.9	20.2	53.0	-89.9		.646	1.339		1452
1453	1.897	.041	.416	229.5	1.4	10.8	-20.4		.095	2.942	191	1453
1454	2.365	.177	.094	135.4	344.1	37.6	-43.3		.123	2.228		1454
1455	2.247	.147	.132	204.5	132.6	32.3	-36.5		.056	2.406	183	1455
1456	3.190	.179	.220	331.9	282.0	68.0	-103.6		.748	1.268		1456
1457	2.695	.159	.133	222.1	294.8	50.4	-59.7		.389	1.844		1457
1458	2.627	.152	.230	266.3	186.5	39.4	-51.8		.389	1.991	140	1458
1459	3.148	.144	.278	9.7	40.1	51.3	-84.2	10	.659	1.303		1459
1460	2.541	.189	.100	80.2	70.7	44.8	-53.3		.238	1.983		1460
1461	3.127	.045	.245	119.0	106.2	54.4	-74.2		1.125	1.753		1461
1462	3.145	.134	.024	203.1	323.2	90.9	-101.7		.883	1.473	1	1462
1463	3.147	.159	.153	51.6	324.5	77.5	-99.9		.788	1.386		1463
1464	3.002	.075	.183	160.4	85.9	60.8	-72.1		.934	1.795	2	1464
1465	3.024	.202	.175	207.6	170.9	66.5	-93.5		.551	1.395		1465
1466	2.377	.171	.230	203.7	156.5	31.6	-41.2		.066	2.172	163	1466
1467	3.386	.129	.392	309.6	327.7	19.7	-87.3		1.081	1.115		1467
1468	2.195	.133	.176	345.3	305.0	29.2	-33.7		-.151	2.301		1468
1469	3.124	.032	.234	74.5	194.1	56.6	-74.5		1.173	1.798		1469
1470	3.160	.042	.063	295.2	340.4	88.3	-90.9		1.187	1.743		1470
1471	2.716	.096	.253	79.5	319.7	39.2	-51.7		.609	2.050		1471
1472	2.234	.155	.069	11.9	40.9	33.5	-36.9		.054	2.421	184	1472
1473	2.574	.176	.257	348.8	222.5	35.4	-50.0		.263	1.995		1473
1474	2.735	.154	.601	47.5	320.7	-1.9	-45.9		.039	1.902		1474

NUMBER	A	E	SINI	WBAR	ANODE	DWBAR	DNODE	NRES	TOMARS	TOJUP	FAMILY	
1475	2.349	.148	.087	53.3	209.0	37.1	-41.1		.180	2.314		1475
1476	2.281	.144	.113	313.7	324.2	34.0	-37.9		.114	2.400	180	1476
1477	3.187	.212	.332	75.9	322.4	42.9	-98.2		.596	1.164		1477
1478	2.464	.126	.150	132.9	314.2	39.1	-44.4		.325	2.238		1478
1479	2.676	.199	.135	105.0	9.8	49.6	-62.3		.282	1.784	135	1479
1480	2.202	.165	.075	135.8	61.1	32.3	-36.1		-.015	2.415		1480
1481	3.017	.040	.071	255.5	337.5	74.6	-76.8		1.066	1.887		1481
1482	2.872	.049	.035	236.4	57.1	65.4	-66.5		.909	2.005	3	1482
1483	2.718	.220	.067	166.1	65.2	55.8	-70.0		.285	1.695		1483
1484	2.737	.213	.306	196.2	75.7	35.7	-61.6		.265	1.702		1484
1485	3.026	.080	.175	326.6	295.3	63.5	-74.8		.942	1.756	2	1485
1486	2.198	.072	.010	313.0	239.7	32.8	-33.4		.204	2.637		1486
1487	3.143	.154	.023	195.2	97.8	91.3	-105.6		.817	1.414	1A	1487
1488	3.038	.128	.193	126.3	350.5	61.9	-79.6		.793	1.603		1488
1489	3.189	.191	.036	172.5	188.5	97.9	-123.1		.728	1.249	107	1489
1490	2.353	.091	.192	350.0	254.6	32.9	-37.6		.278	2.425		1490
1491	3.216	.174	.085	122.1	307.8	95.5	-119.1		.798	1.266		1491
1492	2.173	.156	.105	188.7	142.5	30.8	-34.4		-.063	2.422		1492
1493	2.430	.165	.055	327.2	314.6	41.0	-46.1		.213	2.175	24	1493
1494	2.190	.100	.052	43.5	202.1	32.2	-33.6		.122	2.574		1494
1495	2.640	.127	.228	264.5	8.4	39.7	-50.8		.470	2.040	140	1495
1496	2.206	.125	.051	277.3	283.7	32.8	-34.8		.090	2.512	186	1496
1497	2.895	.057	.038	301.4	288.3	67.0	-68.5		.906	1.962	3	1497
1498	3.100	.153	.264	359.3	267.7	52.4	-83.1		.709	1.421	113	1498
1499	2.671	.140	.236	301.0	241.3	40.3	-53.1		.456	1.973	140	1499
1500	2.243	.170	.125	52.6	13.4	32.4	-37.4		.000	2.360		1500
1501	2.547	.205	.126	34.1	6.6	44.0	-54.6		.193	1.938	43	1501
1502	2.732	.106	.083	136.4	218.2	54.8	-59.2		.613	1.982	132	1502
1503	2.627	.135	.230	141.0	313.7	39.2	-50.6		.433	2.034	140	1503
1504	2.399	.196	.184	147.1	93.2	35.3	-44.7		.083	2.148		1504
1505	2.659	.149	.272	218.6	251.5	36.1	-51.7		.395	1.965	137	1505
1506	2.566	.244	.241	266.5	233.6	37.6	-57.0		.079	1.835		1506
1507	2.331	.180	.183	342.9	288.9	33.0	-40.7		.046	2.237		1507
1508	2.768	.154	.587	122.2	19.0	-1.3	-47.3		.096	1.863	129	1508
1509	1.866	.042	.417	208.7	284.1	10.5	-19.8		.062	2.970	191	1509
1510	2.670	.182	.219	142.8	328.3	42.4	-57.1		.345	1.858	138	1510
1511	2.358	.148	.057	174.1	80.7	38.1	-41.7		.193	2.304	167	1511
1513	2.193	.159	.067	157.5	145.1	32.1	-35.5		-.013	2.435		1513
1514	2.241	.152	.075	321.0	155.9	33.6	-37.0		.065	2.421	184	1514
1515	2.566	.212	.171	46.5	45.7	42.4	-55.5		.184	1.902		1515
1516	2.620	.206	.149	211.3	130.0	46.1	-59.0		.243	1.844		1516
1517	2.717	.050	.076	243.6	56.8	54.1	-55.8		.752	2.145	134	1517
1518	2.226	.154	.112	85.6	21.7	32.2	-36.2		.024	2.414	183	1518
1519	3.134	.195	.215	352.7	13.8	66.1	-101.5		.654	1.283		1519
1520	3.108	.057	.283	28.0	255.9	45.0	-69.7		1.078	1.739		1520
1521	2.850	.112	.265	74.3	7.7	42.0	-59.4		.676	1.859	125	1521
1522	2.368	.090	.080	115.7	55.9	38.0	-39.8		.341	2.425		1522
1523	2.242	.147	.096	153.9	319.5	33.2	-36.7		.071	2.425	180	1523
1524	3.107	.090	.226	346.5	343.9	58.9	-78.4		.974	1.636		1524
1525	2.696	.227	.132	41.4	339.2	51.3	-67.3	10	.177	1.653		1525
1526	2.315	.172	.120	63.0	329.4	35.0	-40.5		.081	2.304		1526
1527	2.227	.144	.087	314.3	9.9	32.9	-36.1		.063	2.445	184	1527
1528	2.415	.177	.143	188.3	142.9	37.7	-44.9		.156	2.173		1528
1530	2.249	.155	.094	20.4	281.8	33.4	-37.2		.060	2.401	180	1530
1531	2.628	.143	.234	77.8	280.4	38.9	-51.1		.411	2.014	140	1531
1532	3.005	.063	.166	131.4	325.9	63.4	-72.6		.977	1.827	2	1532
1533	3.013	.074	.177	170.5	162.9	62.3	-73.1		.949	1.789	2	1533
1534	2.730	.253	.171	110.1	46.5	51.3	-91.2	10				1534
1535	3.148	.191	.129	286.7	265.5	83.2	-110.8		.677	1.295		1535
1536	2.204	.151	.035	16.0	213.0	32.9	-35.8		.034	2.457	187	1536
1537	3.050	.259	.091	19.1	248.5	81.6	-119.8		.401	1.203		1537
1538	2.361	.173	.168	358.2	337.2	34.7	-41.9		.110	2.247	171	1538
1539	3.147	.151	.021	33.3	188.6	91.8	-105.6		.831	1.417	1A	1539
1540	2.850	.131	.200	41.4	280.7	51.3	-64.9	10	.376	1.550		1540
1541	2.769	.106	.089	188.7	348.7	56.7	-61.4		.645	1.947		1541
1542	3.095	.074	.060	11.7	232.1	82.6	-86.6		1.029	1.708		1542
1543	2.628	.296	.215	305.3	282.9	42.9	-69.5		-.002	1.610		1543
1544	2.373	.096	.045	77.3	51.3	38.8	-40.5		.334	2.404		1544
1545	2.771	.265	.046	145.3	34.2	60.5	-83.0		.204	1.518		1545
1546	3.172	.113	.293	123.7	197.0	45.9	-79.3		.942	1.497		1546
1547	2.645	.256	.226	34.3	294.0	42.0	-64.6		.115	1.693		1547
1548	2.788	.101	.273	187.4	118.3	38.9	-54.8		.647	1.964		1548
1549	2.231	.118	.086	114.2	87.4	33.0	-35.3		.128	2.502	182	1549
1550	2.548	.265	.148	19.3	65.6	43.4	-60.8		.028	1.791		1550
1551	2.395	.029	.052	326.1	113.8	39.5	-39.9		.517	2.540		1551
1552	3.010	.070	.173	62.6	3.5	62.7	-72.9		.958	1.801	2	1552
1553	2.907	.125	.037	146.1	118.3	68.5	-74.6		.715	1.754		1553
1554	2.620	.150	.228	349.7	223.0	39.3	-51.4		.389	2.004	140	1554
1555	2.690	.252	.131	41.4	344.8	51.3	-70.5	10	.103	1.594		1555
1556	3.420	.058	.257	349.0	92.6	61.9	-98.9		1.346	1.424		1556

NUMBER	A	E	SINI	WBAR	ANODE	DWBAR	DNODE	NRES	TOMARS	TOJUP	FAMILY	
1557	3.010	.070	.183	349.3	349.7	61.2	-72.3		.959	1.804	2	1557
1558	3.217	.039	.163	106.6	113.5	78.0	-89.8		1.244	1.691		1558
1559	2.390	.173	.069	178.6	313.8	39.2	-44.5		.162	2.206		1559
1560	2.684	.184	.136	28.2	287.9	49.9	-61.1		.328	1.811		1560
1561	3.191	.147	.093	254.5	240.8	90.7	-108.1		.865	1.381	104	1561
1562	2.226	.116	.082	187.8	136.3	32.9	-35.2		.126	2.509	182	1562
1563	2.191	.148	.096	160.2	53.1	31.6	-34.9		-.002	2.453	189	1563
1564	3.161	.166	.200	49.9	182.6	69.9	-98.9		.764	1.352		1564
1565	2.393	.242	.435	18.6	267.0	17.2	-42.5		.013	1.933		1565
1566	1.078					4.3	-21.9	5				1566
1567	3.219	.084	.288	173.4	49.4	47.0	-79.6		1.073	1.536		1567
1568	2.352	.197	.430	.6	143.1	15.8	-37.4		.114	2.098	168	1568
1569	3.152	.080	.195	336.7	100.1	67.4	-83.9		1.047	1.631		1569
1570	2.844	.043	.036	94.9	223.7	63.4	-64.3		.898	2.050	3	1570
1571	3.141	.112	.274	87.4	293.5	49.8	-78.9		.887	1.516	108	1571
1572	3.111	.168	.229	359.2	1.9	61.1	-90.9		.723	1.393		1572
1573	2.370	.212	.422	14.7	204.9	17.2	-39.3		.085	2.032		1573
1574	3.537	.059	.272	169.8	249.7	58.3	-108.0		1.434	1.306		1574
1575	2.375	.196	.430	201.0	209.9	16.0	-38.1		.135	2.074	168	1575
1576	3.135	.152	.022	58.7	234.4	90.4	-104.1		.817	1.429	1A	1576
1577	2.230	.141	.071	46.1	131.6	33.3	-36.2		.080	2.455	185	1577
1579	3.424	.132	.163	117.7	196.3	99.7	-132.9		1.089	1.175		1579
1580	2.196					4.3	167.2	5				1580
1581	3.164	.153	.025	188.9	111.9	93.6	-108.3		.837	1.392	1A	1581
1582	3.162	.141	.189	215.3	96.7	71.3	-94.9		.854	1.429		1582
1584	2.376	.195	.458	133.8	302.8	13.5	-37.5		.118	2.079		1584
1585	2.932	.156	.487	47.3	149.8	7.9	-56.4		.569	1.501		1585
1586	2.429	.144	.059	157.9	135.0	40.9	-44.7		.265	2.229	159	1586
1587	2.547	.156	.146	108.9	351.8	42.7	-50.3		.319	2.064	149	1587
1588	3.030	.049	.175	289.1	99.2	63.4	-73.1		1.045	1.847		1588
1589	2.417	.062	.077	35.9	91.4	39.9	-41.2		.457	2.436		1589
1590	2.230	.134	.088	258.0	228.7	32.9	-35.8		.089	2.465	182	1590
1591	2.393	.186	.413	335.5	173.0	17.3	-38.4	3	.215	2.062		1591
1592	2.768	.289	.212	276.7	109.2	49.6	-81.9		.080	1.431		1592
1593	2.225	.208	.157	296.8	122.9	30.9	-38.2		-.151	2.245		1593
1594	2.269	.155	.143	275.1	67.0	32.6	-37.5		.055	2.366		1594
1595	2.642	.083	.055	296.0	118.6	50.9	-52.9		.513	2.045	141	1595
1596	2.891	.112	.247	64.0	252.3	46.2	-62.9		.715	1.805	125	1596
1597	2.845	.095	.198	190.6	163.3	50.7	-61.6		.655	1.824		1597
1598	2.332	.097	.143	210.2	294.9	34.5	-37.8		.269	2.452		1598
1599	3.135	.105	.095	48.4	33.7	83.5	-92.9		.958	1.573		1599
1600	1.849	.027	.339	151.2	54.6	14.9	-20.7		.086	3.008		1600
1601	2.234	.118	.074	246.3	75.2	33.3	-35.6		.136	2.503	188	1601
1602	2.245	.154	.061	150.3	74.0	33.9	-37.2		.071	2.418		1602
1603	2.755	.048	.134	38.8	135.2	52.5	-56.4		.766	2.099		1603
1604	3.024	.066	.180	329.2	305.9	62.4	-73.3		.982	1.799	2	1604
1605	3.014	.076	.182	116.8	181.1	61.7	-73.1		.943	1.782	2	1605
1606	2.690	.283	.163	333.2	205.1	51.3	-75.6	10	.028	1.514		1606
1607	2.546	.265	.146	356.0	122.9	43.4	-60.7		.028	1.794		1607
1608	2.214	.119	.066	302.6	348.0	32.8	-34.9		.109	2.515	188	1608
1609	2.585					27.8	-43.9	6				1609
1610	2.202	.160	.038	26.2	339.3	32.8	-36.1		.010	2.438	187	1610
1611	3.187	.128	.095	291.9	246.0	89.3	-103.0		.925	1.447	104	1611
1612	3.102	.104	.313	186.6	316.3	39.1	-70.7		.893	1.622		1612
1613	2.736	.249	.164	41.4	266.4	51.3	-72.8	10	.143	1.539		1613
1614	2.996	.121	.240	169.9	167.8	51.7	-71.9		.685	1.576	120	1614
1615	3.113	.158	.025	49.7	197.8	88.1	-102.5		.779	1.434	1A	1615
1616	2.911	.033	.135	158.4	42.5	61.2	-65.1		.981	2.014	124	1616
1617	3.204	.156	.223	179.7	157.8	66.6	-99.0		.837	1.325	102	1617
1618	2.869	.046	.036	223.9	107.0	65.1	-66.1		.915	2.018	3	1618
1619	2.241	.151	.095	45.3	61.0	33.1	-36.8		-.061	2.417	180	1619
1620	1.244	.323	.282	252.6	336.3	9.6	-15.3		-.309	3.243		1620
1621	2.230	.123	.062	83.1	192.1	33.4	-35.7		.122	2.496	188	1621
1622	2.234	.153	.113	238.6	357.1	32.5	-36.5		.038	2.411	183	1622
1623	3.133	.155	.025	86.3	133.5	90.2	-104.5		.806	1.422	1A	1623
1624	3.180	.143	.021	159.2	170.8	95.2	-108.6		.879	1.407	1	1624
1625	3.183	.184	.327	236.4	322.2	41.6	-90.7		.698	1.260		1625
1626	2.364	.233	.446	77.9	283.7	15.7	-40.4		.016	1.998		1626
1627	1.864	.424	.149	297.3	141.4	22.4	-38.9		-.674	2.213		1627
1628	3.014	.103	.339	140.1	186.5	32.0	-63.1		.722	1.638		1628
1629	2.238	.162	.169	208.6	136.3	30.8	-36.4		-.039	2.328		1629
1630	3.030	.197	.073	155.3	44.5	78.3	-99.1		.588	1.406		1630
1631	2.235	.152	.125	331.1	13.2	32.2	-36.4		.033	2.406	183	1631
1632	2.656	.106	.109	319.3	210.4	49.5	-54.1		.534	2.048		1632
1633	3.169	.185	.029	178.2	124.9	95.5	-118.3		.734	1.288	107	1633
1634	2.246	.136	.120	261.7	92.3	32.6	-36.2		.089	2.440	183	1634
1635	2.855	.045	.037	277.8	217.8	64.1	-65.1		.903	2.033	3	1635
1636	2.235	.120	.082	70.5	176.5	33.2	-35.6		.129	2.495	182	1636
1637	3.070	.075	.241	219.3	16.8	53.5	-72.5		.974	1.712		1637
1638	2.749	.186	.022	273.2	261.5	58.4	-68.2		.410	1.756		1638

NUMBER	A	E	SINI	WBAR	ANODE	DWBAR	DNODE	NRES	TOMARS	TOJUP	FAMILY	
1639	2.574	.137	.163	84.4	320.8	42.5	-50.0		.397	2.077	149	1639
1640	2.289	.291	.123	350.8	347.7	34.3	-47.7		-.234	2.048		1640
1641	3.019	.073	.174	315.6	326.7	63.1	-73.8		.959	1.785	2	1641
1642	2.751	.095	.198	143.3	335.0	46.6	-56.0		.651	1.999		1642
1643	2.490	.163	.084	20.2	285.3	42.9	-48.7		.268	2.106	154	1643
1644	2.546	.178	.144	118.8	270.3	42.9	-51.9		.261	2.011	150	1644
1645	3.059	.077	.039	8.4	273.4	80.5	-83.8		.991	1.738	119	1645
1646	2.361	.099	.137	60.9	124.2	35.8	-39.1		.296	2.427	169	1646
1648	2.236	.194	.076	248.8	140.9	33.4	-38.7		-.036	2.328		1648
1649	3.021	.081	.176	168.3	150.8	63.1	-74.5		.933	1.757	2	1649
1650	2.437	.168	.057	241.6	215.1	41.3	-46.6		.213	2.161	24	1650
1651	2.180	.135	.098	156.1	193.5	31.1	-34.0		.008	2.478	189	1651
1652	2.251	.144	.069	244.5	252.1	34.0	-37.1		.098	2.432	185	1652
1653	2.610	.292	.108	36.3	299.9	48.7	-70.6		-.002	1.631		1653
1654	3.017	.052	.176	354.6	19.8	62.5	-72.2		1.023	1.851	2	1654
1655	2.783	.218	.151	83.7	117.3	54.5	-73.0		.322	1.619		1655
1656	1.878	.087	.410	123.5	181.1	11.4	-20.8		-.015	2.871	191	1656
1657	2.349	.189	.413	155.9	102.4	17.1	-37.0		.138	2.114		1657
1658	2.560	.173	.139	273.2	95.5	43.8	-52.4		.286	2.004	150	1658
1659	2.783	.200	.306	12.8	331.9	36.6	-62.9		.342	1.691		1659
1660	2.395	.212	.411	135.1	216.7	18.3	-40.5		.116	2.004		1660
1661	2.184	.120	.064	197.4	258.1	31.9	-33.9		.065	2.529	188	1661
1662	2.743	.135	.091	39.4	320.0	55.3	-61.8		.538	1.895	40	1662
1663	2.240	.131	.084	5.8	84.9	33.3	-36.1		.109	2.466	182	1663
1664	2.339	.263	.106	152.7	39.3	36.5	-48.0		-.110	2.062		1664
1665	2.414	.226	.174	105.8	90.9	36.4	-48.0		.027	2.059		1665
1666	2.185	.122	.059	352.9	258.5	32.0	-34.0		.063	2.524	188	1666
1667	2.190	.130	.070	252.7	83.5	32.0	-34.4		.049	2.502	188	1667
1668	2.806	.180	.076	345.4	175.5	60.4	-72.0		.468	1.706		1668
1669	3.140	.150	.026	194.6	319.3	90.8	-104.3		.828	1.429	1A	1669
1670	2.902	.093	.169	97.0	51.0	57.4	-67.0		.786	1.860		1670
1671	2.588	.249	.075	73.4	191.1	48.5	-63.2		.110	1.768	146	1671
1672	3.178	.257	.028	75.6	235.7	101.3	-149.2		.505	1.060		1672
1673	3.101	.157	.073	61.9	224.3	84.2	-99.3		.770	1.449		1673
1674	3.187	.139	.025	106.3	93.9	95.7	-108.5		.898	1.412	1	1674
1675	2.233	.149	.113	100.6	24.6	32.4	-36.3		.045	2.421	183	1675
1676	2.236	.181	.094	241.2	49.4	33.0	-37.9		-.014	2.355		1676
1677	2.532	.063	.263	259.6	335.6	32.4	-41.7		.493	2.296	153	1677
1678	3.165	.113	.186	133.6	347.8	71.3	-90.7		.951	1.513		1678
1679	3.125	.111	.319	240.6	180.6	38.5	-72.8		.882	1.573		1679
1680	2.724	.209	.056	230.2	81.0	56.3	-69.2		.323	1.718		1680
1681	2.698	.204	.107	106.9	94.3	52.6	-65.4		.284	1.741		1681
1682	2.239	.140	.075	333.7	315.8	33.5	-36.5		.091	2.449	185	1682
1683	2.735	.151	.228	305.1	324.0	43.4	-57.9		.482	1.863	140	1683
1684	3.092	.149	.044	240.0	112.0	84.9	-97.4		.790	1.482		1684
1685	1.368	.397	.244	237.2	185.5	-86.2	-18.7	13				1685
1686	3.167	.143	.024	286.5	306.7	93.6	-106.5		.873	1.420	1	1686
1687	3.158	.151	.025	60.9	91.6	92.8	-106.9		.839	1.406	1A	1687
1688	2.618	.220	.226	274.5	244.9	40.4	-58.0		.197	1.822		1688
1689	2.449	.176	.102	34.1	128.5	40.7	-47.3		.195	2.126		1689
1690	3.041	.050	.239	40.3	234.5	52.5	-69.0		1.009	1.801		1690
1691	3.165	.142	.025	48.7	233.4	93.3	-106.0		.874	1.426	1	1691
1692	2.788	.117	.052	294.3	222.2	59.6	-64.3		.636	1.900		1692
1693	2.804	.226	.207	299.2	63.6	50.4	-73.7		.265	1.534		1693
1694	2.396	.217	.186	13.5	8.8	35.1	-46.1		.026	2.103		1694
1695	2.784	.222	.312	4.2	227.1	36.5	-65.8		.269	1.625		1695
1696	2.262	.145	.100	175.6	15.2	33.7	-37.3		.094	2.416	180	1696
1697	2.374	.112	.108	83.4	324.7	37.4	-40.5		.284	2.370	165	1697
1698	3.154	.150	.029	158.6	342.0	92.2	-106.1		.840	1.414	1A	1698
1699	2.211	.112	.046	319.0	266.0	33.0	-34.7		.127	2.535	186	1699
1700	2.361	.188	.082	15.9	343.6	37.8	-43.9		.096	2.206		1700
1701	3.170	.124	.282	307.1	59.1	49.1	-82.2		.898	1.446		1701
1702	2.858	.102	.157	334.5	104.6	56.4	-65.0		.727	1.863		1702
1703	2.215	.117	.071	319.7	118.6	32.8	-34.4		.114	2.518	188	1703
1704	2.223	.136	.031	170.7	256.5	33.5	-35.9		.091	2.479	187	1704
1705	2.299	.193	.139	345.4	197.0	33.9	-40.8		.005	2.259	177	1705
1706	2.125	.110	.038	213.3	264.3	30.4	-31.8		-.008	2.564		1706
1707	2.219	.163	.070	66.6	354.2	32.9	-36.7		.016	2.413	184	1707
1708	2.916	.287	.131	85.4	200.2	67.0	-104.8		.212	1.280		1708
1709	2.378	.170	.149	344.2	294.9	36.1	-42.8		.137	2.241		1709
1710	2.322	.217	.147	330.5	352.4	34.5	-43.2		-.032	2.183		1710
1711	3.015	.069	.180	32.3	139.2	62.0	-72.8		.965	1.801	2	1711
1712	3.176	.151	.348	233.8	240.1	34.2	-80.4		.738	1.360		1712
1713	2.228	.136	.055	309.2	57.8	33.5	-36.1		.092	2.469	186	1713
1714	2.565	.160	.160	248.0	299.5	42.5	-51.1		.328	2.029	149	1714
1715	2.399	.245	.193	234.5	31.2	35.1	-48.4		-.045	2.029		1715
1716	2.733	.133	.120	196.1	250.3	53.1	-60.4		.519	1.897	133	1716
1717	2.195	.180	.113	113.7	335.7	31.3	-36.2		-.084	2.366		1717
1718	2.366	.224	.151	335.0	212.2	35.8	-45.7		-.010	2.125		1718
1719	2.657	.159	.276	27.6	318.9	35.7	-52.2		.362	1.937	137	1719

NUMBER	A	E	SINI	WBAR	ANODE	DWBAR	DNODE	NRES	TOMARS	TOJUP	FAMILY
1720	2.188	.130	.015	101.3	180.2	32.5	-34.5		.062	2.516		1720
1721	3.148	.060	.294	123.4	315.9	43.1	-71.6		1.093	1.695		1721
1722	2.514	.065	.094	129.4	178.9	43.4	-45.3		.536	2.324		1722
1723	3.013	.077	.179	168.4	155.4	62.1	-73.3		.939	1.778	2	1723
1724	2.711	.074	.208	129.3	169.5	44.0	-52.3		.677	2.102		1724
1725	2.903	.057	.037	334.2	131.4	67.6	-69.1		.913	1.953	3	1725
1726	2.787	.046	.076	246.8	241.5	58.2	-60.0		.837	2.094	38	1726
1727	1.854	.079	.349	101.0	135.5	14.5	-21.2		-.010	2.905		1727
1728	2.562	.071	.141	279.6	244.0	43.3	-46.9		.561	2.262		1728
1729	2.230	.092	.038	240.6	353.5	33.7	-34.9		.197	2.570		1729
1730	2.784	.178	.167	40.2	176.6	52.6	-66.9		.417	1.710		1730
1731	3.174	.081	.092	342.1	164.0	87.3	-94.5		1.070	1.607		1731
1732	3.012	.076	.177	.7	161.5	62.3	-73.3		.943	1.784	2	1732
1733	2.193	.136	.081	124.4	168.7	31.9	-34.6		.034	2.482	189	1733
1734	2.777	.191	.144	6.9	190.6	54.3	-69.0		.395	1.697	128	1734
1735	3.145	.086	.278	263.3	6.1	47.6	-75.4		.992	1.617	108	1735
1736	2.229	.161	.082	66.3	168.6	33.0	-36.9		.027	2.406	184	1736
1737	3.013	.081	.179	186.0	322.4	62.2	-73.6		.928	1.768	2	1737
1738	2.183	.135	.077	326.8	41.9	31.6	-34.3		.022	2.488	189	1738
1739	2.261	.102	.069	261.4	211.0	34.3	-36.1		.201	2.518	179	1739
1740	2.467	.151	.054	19.0	287.9	42.7	-47.1		.282	2.164	24	1740
1741	2.885	.039	.038	59.0	35.4	66.2	-67.1		.949	2.020	3	1741
1742	2.889	.059	.036	.8	180.2	66.6	-68.1		.897	1.960	3	1742
1743	2.470	.180	.117	182.9	199.2	41.0	-48.4		.199	2.094		1743
1744	2.229	.170	.070	173.7	19.6	33.3	-37.4		.012	2.389	184	1744
1745	2.846	.042	.038	96.5	70.1	63.5	-64.3		.903	2.051	3	1745
1747	1.709	.121	.420	249.3	270.0	9.9	-18.3		-.222	2.988		1747
1750	1.926	.116	.394	23.3	278.0	13.2	-22.7		-.033	2.756		1750
1751	2.790	.123	.161	11.0	246.0	52.6	-62.0		.584	1.856	127	1751
1752	2.238	.146	.077	334.0	240.4	33.4	-36.6		.075	2.435	184	1752
1753	3.015	.079	.183	259.4	54.3	61.6	-73.3		.934	1.772	2	1753
1755	3.093	.068	.177	153.1	163.8	67.2	-79.2		1.038	1.724	115	1755
1756	2.548	.197	.116	38.2	289.6	44.5	-54.2		.216	1.955	43	1756
1757	2.351	.166	.061	182.4	29.1	37.8	-42.3		.143	2.272		1757
1758	3.007	.063	.170	206.5	116.3	62.9	-72.4		.977	1.826	2	1758
1759	2.648	.253	.071	358.5	171.3	51.7	-68.7		.069	1.617		1759
1760	3.157	.149	.167	200.8	241.5	75.5	-98.1		.825	1.405	109	1760
1761	3.168	.258	.029	135.2	54.2	100.1	-147.5		.495	1.067		1761
1762	2.876	.046	.036	54.5	191.4	65.6	-66.7		.922	2.010	3	1762
1763	2.189	.138	.080	333.1	296.3	31.8	-34.6		.022	2.478	189	1763
1764	3.089	.155	.032	221.6	184.6	85.2	-98.3		.772	1.468	1	1764
1765	3.169	.068	.337	334.7	69.3	32.0	-68.9		.959	1.583		1765
1766	2.749	.047	.094	352.4	200.8	55.0	-57.2		.789	2.125	130	1766
1767	3.020	.066	.176	309.8	199.8	62.8	-73.2		.980	1.804	2	1767
1768	2.450	.154	.057	39.0	353.4	41.9	-46.4		.260	2.179	24	1768
1769	2.179	.084	.030	349.9	297.6	32.1	-33.0		.151	2.620		1769
1770	2.457	.058	.088	98.9	12.0	41.2	-42.7		.503	2.401		1770
1771	3.125	.140	.177	54.9	87.7	71.0	-91.7		.827	1.471	109	1771
1772	2.530	.133	.087	176.5	88.1	44.5	-48.9		.377	2.136		1772
1773	2.435	.148	.079	226.9	72.1	40.7	-45.1		.258	2.215		1773
1774	2.877	.058	.035	98.7	210.4	65.8	-67.3		.886	1.972	3	1774
1775	2.603	.155	.234	263.2	199.4	38.4	-50.8		.354	2.010	147	1775
1776	3.104	.044	.162	162.5	184.7	70.2	-79.7		1.127	1.789		1776
1777	2.626	.042	.066	144.1	320.7	49.7	-50.7		.684	2.247	143	1777
1778	3.146	.158	.023	224.1	116.9	91.8	-107.0		.807	1.397	1A	1778
1779	2.176	.147	.027	242.6	248.9	32.0	-34.6		.010	2.484		1779
1780	3.016	.067	.178	235.3	290.6	62.3	-72.9		.972	1.805	2	1780
1781	2.395	.081	.110	43.7	39.8	38.1	-40.3		.383	2.419		1781
1782	3.118	.164	.025	248.9	205.7	88.9	-104.5		.766	1.411	1A	1782
1783	2.662	.160	.209	151.9	197.2	42.5	-55.3		.390	1.931	138	1783
1784	2.405	.125	.010	262.6	108.7	40.4	-43.1		.297	2.297		1784
1785	2.236	.118	.078	162.7	277.7	33.3	-35.6		.137	2.500	182	1785
1786	3.021	.070	.178	2.0	10.7	62.6	-73.5		.967	1.790	2	1786
1787	3.002	.080	.176	203.3	304.2	62.0	-72.9		.920	1.781	2	1787
1788	3.111	.147	.020	289.6	244.3	87.8	-100.0		.813	1.469	1A	1788
1789	2.213	.140	.025	308.3	116.9	33.3	-35.7		.072	2.478	187	1789
1790	2.238	.154	.088	150.8	354.3	33.2	-36.9		.052	2.414	184	1790
1791	2.746	.150	.103	254.7	208.9	55.0	-63.2		.494	1.850		1791
1792	2.777	.243	.141	41.2	74.0	55.5	-77.0		.248	1.558		1792
1793	2.224	.144	.039	176.9	235.8	33.5	-36.2		.073	2.459	187	1793
1794	3.124	.198	.275	200.2	229.0	52.6	-93.8		.575	1.267		1794
1795	2.784	.195	.146	256.0	199.2	54.6	-70.0		.389	1.681	128	1795
1796	3.359	.071	.387	195.0	190.8	14.3	-74.7		1.280	1.378		1796

IV. SPECTRAL REFLECTANCES
OF THE ASTEROIDS

CLARK R. CHAPMAN
Planetary Science Institute

and

MICHAEL J. GAFFEY
University of Hawaii

Two TRIAD spectrophotometry files are here combined into a single table. For each asteroid the first line of data gives spectral reflectance parameters and quality codes as defined below. Succeeding lines for each object list the actual values of spectral reflectance from which the parameters were derived.

Visible and near-infrared relative reflectances were measured with a narrowband filter photometer by Chapman, McCord, and their associates, as described in the chapter by Chapman and Gaffey in Part V of this book. That chapter also lists all previous publications in which portions of these data have been published. All such published data are now recalibrated to new α Lyr/Sun ratios, many of the asteroids have been remeasured, while an additional 179 asteroids have been measured for the first time. The spectrophotometry file now contains data on 277 asteroids; all 277 spectra are plotted in the appendix of that chapter. For each asteroid, the following values are given for each filter in which the asteroid has been observed: wavelength (μm), reflectance (relative to unity at 0.56 μm), and error. For explanation of the errors and caveats about the data, refer to Chapman and Gaffey's chapter.

The spectral parameters provided for each asteroid have been derived from the spectra. The parameter definitions are those originally adopted by McCord and Chapman (*Astrophys. J.* 195: 553-562, 1975), and they are repeated here as follows: The terminology R_x means reflectance at wavelength x. R/B = $R_{0.7}/R_{0.4}$, a measure of "redness" through the visible. BEND = $(R_{0.56} - R_{0.4}) - (R_{0.73} - R_{0.56})$, a measure of curvature through the visible. IR = $(R_{1.05} - R_{0.73})$, a measure of infrared reflectance. DEPTH = the reflectance at the bottom of an infrared absorption feature divided by the highest reflectance on the short-wavelength side of the band; low values indicate a deep absorption, while DEPTH = 1.0 indicates that no band is present. The Center wavelength of a band, if present, is expressed in μm. The Bandwidth (in μm) is indicated, being an approximate measure of the full width at half maximum. The UV parameter is a qualitative measure of the steepness (1 = shallow, 3 = steep) of that portion of a spectrum shortwards of a prominent elbow (break in slope), if such an elbow is present; the wavelength of the Elbow is then given in μm. The final parameter is a qualitative measure of the depth of any small absorption feature near 0.65 μm, if present (from 0 = no band to 3 = prominent band).

Quality codes associated with the parameters are G (good) and Q (questionable). The latter suggest relatively noisy data in the pertinent wavelength range. A blank quality code means that a parameter is absent, because data in the pertinent wavelength range are either very noisy or altogether absent.

Acknowledgments. For institutional support and for the help of innumerable individuals in acquiring the data see the acknowledgments in the chapter by Chapman and Gaffey. In addition A. Hostetler deserves much credit for helping with the data reduction and preparation of these files for TRIAD.

```
NUMBER   R/B     BEND       IR   DEPTH  CENTER BANDW  ELBOW   UV   FEAT        NUMBER
-----------------------------------------------------------------------------------
  1    1.05 G 0.08 G -0.03 G 1.00 G                 0.40 G 3 G   0 G            1
     MICRON REFL   RERR MICRON REFL  RERR MICRON REFL  RERR MICRON REFL  RERR
      .330  .58   .04   .340  .70   .03   .355  .72  .03   .400  .91   .03
      .430  .96   .03   .470  .96   .03   .500  .95  .03   .540  .99   .03
      .570 1.00   .03   .600  .96   .03   .630  .96  .03   .670  .98   .03
      .700  .97   .03   .730  .96   .03   .765  .97  .03   .800  .95   .03
      .830 1.03   .03   .870  .91   .03   .900  .91  .03   .930  .98   .03
      .950  .92   .03   .970  .98   .03  1.000  .90  .03  1.030  .90   .03

  2    1.02 G 0.06 G -0.05 G 1.00 G                 0.40 G 2 G   0 G            2
      .330  .82   .03   .340  .87   .03   .355  .86  .03   .400  .98   .03
      .430  .98   .03   .470  .99   .03   .500  .98  .03   .540  .99   .03
      .570 1.00   .03   .600 1.00   .03   .630 1.00  .03   .670  .99   .03
      .700  .97   .03   .730  .98   .03   .765  .96  .03   .800 1.01   .03
      .830  .89   .03   .870  .91   .03   .900  .93  .03   .930  .94   .03
      .950  .95   .03   .970  .88   .03  1.000  .91  .03  1.030  .88   .03

  3    1.53 G 0.06 G -0.02 G 0.88 G 0.94 G 0.21 G               0 G            3
      .330  .58   .09   .340  .63   .08   .355  .71  .05   .400  .75   .03
      .430  .77   .04   .470  .89   .03   .500  .91  .03   .540  .99   .03
      .570 1.00   .03   .600 1.02   .03   .630 1.11  .03   .670 1.15   .03
      .700 1.20   .03   .730 1.14   .04   .765 1.20  .04   .800 1.19   .03
      .870 1.08   .03   .900 1.05   .03   .950 1.07  .03  1.000 1.08   .03
     1.060 1.17   .05  1.100 1.08   .08

  4    1.28 G 0.06 G -0.12 G 0.71 G 0.93 G 0.21 G               0 G            4
      .400  .82   .03   .430  .86   .03   .470  .93  .03   .500  .93   .03
      .540  .99   .03   .570 1.00   .03   .600 1.03  .03   .630 1.07   .03
      .670 1.05   .03   .700 1.10   .03   .730 1.10  .03   .765 1.09   .03
      .800 1.02   .03   .870  .87   .03   .900  .78  .03   .950  .79   .03
     1.000  .86   .03  1.060  .97   .03  1.100 1.16  .15

  5    1.64 G 0.07 G  0.04 G 0.87 G 0.93 G 0.15 G               0 G            5
      .340  .52   .10   .355  .80   .10   .400  .76  .03   .430  .75   .03
      .470  .84   .03   .500  .83   .03   .540  .90  .03   .570 1.02   .03
      .600 1.07   .03   .630 1.12   .03   .670 1.18  .03   .700 1.15   .03
      .730 1.23   .03   .765 1.21   .03   .800 1.31  .03   .870 1.22   .04
      .900 1.12   .03   .950 1.12   .04  1.000 1.25  .04  1.060 1.25   .05

  6    1.50 G 0.11 G  0.05 G 0.88 G 0.92 G 0.17 G               0 G            6
      .330  .55   .03   .340  .60   .05   .355  .70  .05   .400  .75   .03
      .430  .73   .03   .470  .84   .03   .500  .83  .03   .540  .94   .03
      .570  .99   .03   .600 1.01   .03   .630 1.08  .03   .670 1.04   .05
      .700 1.08   .03   .730 1.15   .04   .765 1.16  .05   .800 1.15   .04
      .870 1.06   .06   .900 1.02   .03   .950 1.04  .03  1.000 1.07   .03
     1.060 1.19   .03  1.100 1.27   .03

  7    1.67 G 0.10 G -0.09 G 0.88 G 1.02 G 0.36 G               0 G            7
      .330  .50   .05   .340  .63   .07   .355  .62  .03   .400  .73   .03
      .430  .73   .03   .470  .87   .03   .500  .89  .03   .540  .96   .03
      .570 1.00   .03   .600 1.08   .03   .630 1.12  .03   .670 1.16   .05
      .700 1.15   .03   .730 1.21   .09   .765 1.29  .09   .800 1.25   .03
      .870 1.06   .03   .900 1.12   .03   .950 1.15  .03  1.000 1.12   .03
     1.060 1.13   .04

  8    1.72 G 0.12 G -0.04 G 0.84 G 0.95 G 0.18 G               0 G            8
      .330  .48   .09   .340  .54   .04   .355  .64  .05   .400  .66   .03
      .430  .72   .03   .470  .81   .03   .500  .86  .03   .540  .94   .03
      .570 1.00   .03   .600 1.06   .03   .630 1.09  .03   .670 1.14   .03
      .700 1.20   .03   .730 1.18   .03   .765 1.19  .03   .800 1.20   .04
      .830 1.14   .04   .870 1.11   .03   .900 1.03  .04   .930 1.09   .03
      .950 1.00   .05   .970  .98   .03  1.000 1.07  .05  1.030 1.23   .06

  9    1.56 G 0.10 G  0.11 Q 0.98 Q 0.93 Q 0.12 Q               0 G            9
      .330  .52   .06   .340  .50   .03   .355  .63  .03   .400  .73   .03
      .430  .76   .03   .470  .85   .03   .500  .90  .03   .540  .96   .03
      .570 1.00   .03   .500 1.03   .03   .630 1.10  .03   .670 1.11   .03
      .700 1.14   .03   .730 1.15   .03   .765 1.19  .03   .800 1.22   .07
      .900 1.17   .03   .950 1.13   .04  1.000 1.25  .09  1.030 1.04   .18
     1.060 1.31   .15

 10    1.14 G 0.06 G -0.10 G 1.00 G                 0.43 G 2 G   0 G           10
      .330  .72   .05   .340  .78   .07   .355  .80  .05   .400  .92   .03
      .430  .90   .03   .470  .96   .03   .500  .97  .03   .540 1.01   .03
      .570 1.00   .03   .600  .98   .03   .630 1.03  .03   .670 1.03   .03
      .700 1.05   .03   .730 1.02   .03   .765 1.02  .03   .800 1.00   .03
      .830  .99   .03   .870  .96   .03   .900  .95  .04   .930  .96   .03
      .970  .91   .03  1.000  .91   .04  1.030  .91  .03  1.060  .95   .03

 11    1.54 G 0.12 G  0.02 G 0.93 G 0.93 G 0.19 G               0 G           11
      .330  .57   .08   .340  .61   .09   .355  .64  .04   .400  .75   .03
      .430  .77   .03   .470  .86   .03   .500  .91  .03   .540  .97   .03
      .570 1.00   .03   .600 1.06   .03   .630 1.11  .03   .670 1.11   .03
      .700 1.14   .03   .730 1.15   .03   .765 1.18  .03   .800 1.17   .03
      .870 1.07   .06   .900 1.09   .04   .950 1.14  .08  1.000 1.12   .05
     1.060 1.19   .08
```

```
NUMBER   R/B    BEND       IR   DEPTH  CENTER BANDW  ELBOW  UV   FEAT          NUMBER
------------------------------------------------------------------------------------
 12    1.67 G 0.14 G  0.07 Q 1.00 Q                           0 G             12
       MICRON REFL   RERR MICRON REFL   RERR MICRON REFL   RERR MICRON REFL   RERR
        .330  .40    .03   .340  .51    .08   .355  .56    .04   .400  .72    .03
        .430  .80    .03   .470  .85    .03   .500  .91    .03   .540  .96    .03
        .570 1.00    .03   .600 1.02    .03   .630 1.07    .04   .670 1.13    .03
        .700 1.15    .03   .730 1.17    .03   .755 1.20    .05   .800 1.25    .03
        .830 1.17    .03   .870 1.21    .10   .900 1.24    .03   .930 1.10    .06
        .950 1.32    .08   .970 1.22    .05  1.000 1.33    .05  1.060 1.15    .15

 13    1.17 G 0.21 G -0.02 G 1.00 G               0.47 Q 3 Q    0 G             13
        .330  .40    .17   .340  .58    .13   .355  .50    .08   .400  .81    .04
        .430  .92    .09   .470 1.09    .04   .500  .93    .05   .540 1.01    .03
        .570 1.00    .03   .600  .95    .05   .630  .96    .03   .670  .98    .03
        .700  .91    .05   .730  .95    .04   .765  .97    .03   .800  .92    .05
        .870  .99    .06   .900  .85    .05   .950  .93    .04  1.000  .87    .03
       1.060 1.01    .05  1.100  .80    .09

 14    1.54 G 0.09 G  0.07 G 0.86 G 0.94 G 0.14 G               0 G             14
        .330  .58    .03   .340  .63    .08   .355  .64    .05   .400  .75    .03
        .430  .78    .03   .470  .86    .03   .500  .90    .03   .540  .96    .03
        .570 1.00    .03   .600 1.04    .03   .630 1.08    .03   .670 1.11    .03
        .700 1.17    .03   .730 1.14    .03   .765 1.18    .03   .800 1.17    .03
        .830 1.18    .03   .870 1.07    .03   .900 1.04    .05   .930 1.02    .03
        .970 1.05    .03  1.000 1.14    .04  1.030 1.21    .05  1.060 1.23    .05

 15    1.53 G 0.09 G -0.07 G 0.92 G 0.98 G 0.25 G               1 G             15
        .330  .53    .08   .340  .63    .03   .355  .68    .03   .400  .73    .03
        .430  .78    .03   .470  .88    .03   .500  .88    .03   .540  .97    .03
        .570 1.00    .03   .600 1.05    .03   .630 1.09    .03   .670 1.06    .03
        .700 1.17    .03   .730 1.15    .03   .765 1.18    .03   .800 1.14    .03
        .870 1.07    .03   .900 1.11    .03   .950 1.11    .03  1.000 1.05    .04
       1.060 1.13    .06  1.100 1.13    .08

 16    1.23 G-0.02 G  0.12 G 0.97 Q 0.91 Q 0.06 Q               0 G             16
        .330  .87    .06   .340  .86    .06   .355  .86    .06   .400  .90    .04
        .430  .74    .04   .470  .95    .03   .500  .99    .04   .540  .96    .04
        .570 1.02    .03   .600 1.01    .04   .630 1.08    .05   .670 1.09    .04
        .700 1.12    .04   .730 1.10    .03   .765 1.14    .05   .800 1.13    .06
        .830 1.15    .05   .870 1.12    .04   .900 1.10    .04   .950 1.16    .06
       1.000 1.23    .04  1.060 1.25    .07  1.100 1.22    .11

 17    1.74 G 0.15 G  0.02 G 0.89 G 0.93 G 0.18 G               0 G             17
        .330  .53    .20   .340  .55    .18   .355  .58    .06   .400  .68    .03
        .430  .71    .03   .470  .80    .03   .500  .85    .03   .540  .95    .03
        .570 1.00    .03   .600 1.06    .03   .630 1.14    .03   .670 1.13    .03
        .700 1.17    .03   .730 1.17    .03   .765 1.21    .03   .800 1.20    .08
        .870 1.07    .08   .900 1.07    .08   .950 1.13    .10  1.000 1.11    .10
       1.060 1.22    .12  1.100 1.25    .20

 18    1.60 G 0.11 G -0.01 Q 0.88 Q 0.92 Q 0.19 Q               0 G             18
        .330  .51    .06   .340  .62    .07   .355  .75    .05   .400  .70    .03
        .430  .76    .03   .470  .87    .03   .500  .91    .03   .540  .95    .03
        .570 1.00    .03   .600 1.03    .03   .630 1.08    .03   .670 1.12    .03
        .700 1.15    .03   .730 1.15    .03   .765 1.17    .03   .800 1.08    .06
        .870  .99    .08   .900 1.05    .05   .950 1.04    .05  1.000 1.15    .06
       1.030  .98    .10  1.060 1.30    .12

 19    1.11 G 0.17 G -0.02 G 1.00 G               0.52 G 1 Q    0 G             19
        .330  .90    .16   .340  .69    .13   .355  .83    .09   .400  .86    .03
        .430  .89    .03   .470  .95    .03   .500  .96    .03   .540 1.01    .03
        .570 1.00    .03   .600  .99    .03   .630  .99    .03   .670  .98    .03
        .700  .96    .03   .730  .95    .03   .765  .97    .03   .800  .93    .03
        .900  .95    .03   .950  .95    .05  1.000  .93    .04  1.030  .94    .08
       1.060 1.06    .10

 20    1.50 G 0.09 G -0.05 G 0.88 G 0.90 G 0.16 G               0 G             20
        .330  .62    .03   .355  .68    .03   .400  .76    .03   .430  .80    .03
        .470  .86    .03   .500  .89    .03   .540  .96    .03   .570 1.00    .03
        .600 1.01    .03   .630 1.04    .03   .670 1.09    .03   .700 1.13    .03
        .730 1.15    .03   .765 1.15    .03   .800 1.13    .03   .830 1.07    .03
        .870 1.01    .03   .900 1.01    .04   .930 1.05    .06   .970 1.10    .05
       1.000 1.07    .03  1.030 1.10    .03  1.060 1.08    .03

 21    1.16 G 0.05 G  0.06 G 1.00 G                             0 G             21
        .330  .83    .18   .340  .85    .10   .355  .86    .08   .400  .92    .03
        .430  .90    .03   .470  .94    .03   .500  .95    .03   .540  .97    .03
        .570 1.00    .03   .600 1.02    .03   .630 1.03    .03   .670 1.02    .03
        .700 1.06    .03   .730 1.04    .03   .765 1.04    .03   .800 1.04    .04
        .870 1.00    .04   .900 1.04    .04   .950 1.11    .06  1.000 1.08    .08
       1.030 1.09    .10  1.060 1.15    .20
```

```
NUMBER   R/B   BEND     IR   DEPTH  CENTER BANDW  ELBOW UV  FEAT         NUMBER
--------------------------------------------------------------------------------
  22   1.19 G 0.00 G   0.20 Q                              0 G            22
       MICRON REFL  RERR MICRON REFL  RERR MICRON REFL  RERR MICRON REFL  RERR
       .330  .89   .03  .340  .82   .07  .355  .86   .03  .400  .93   .03
       .430  .93   .03  .470  .94   .03  .500  .94   .03  .540  .98   .03
       .570 1.00   .03  .600 1.03   .03  .630 1.04   .03  .670 1.05   .03
       .700 1.09   .03  .730 1.08   .03  .765 1.13   .03  .800 1.17   .03
       .830 1.16   .03  .900 1.06   .04  .950 1.18   .05 1.000 1.19   .06
      1.030 1.52   .12 1.060 1.38   .15

  23   1.54 G 0.07 G   0.11 Q 0.94 Q 0.92 Q 0.15 Q          0 G           23
       .330  .68   .06  .340  .59   .03  .355  .60   .03  .400  .76   .03
       .430  .78   .03  .470  .88   .03  .500  .90   .03  .540  .96   .03
       .570 1.00   .03  .600 1.05   .03  .630 1.10   .03  .670 1.09   .03
       .700 1.18   .03  .730 1.16   .03  .765 1.23   .03  .800 1.09   .08
       .900 1.13   .06  .950 1.12   .05 1.000 1.27   .10 1.030 1.26   .13
      1.060 1.41   .25

  24   1.07 G 0.14 G   0.02 Q 1.00 G         0.40 G 3 Q   1 G            24
       .330  .60   .10  .340  .67   .12  .355  .83   .09  .400  .91   .03
       .430  .86   .03  .470  .92   .04  .500  .91   .04  .540  .97   .04
       .570 1.00   .03  .600 1.00   .04  .630 1.01   .04  .670  .92   .04
       .700  .91   .04  .730  .97   .04  .765 1.04   .08  .800  .99   .07
       .830  .95   .05  .870  .93   .04  .900 1.05   .04  .930 1.08   .07
       .970 1.01   .07 1.000 1.01   .07 1.030  .85   .08

  25   1.89 G 0.13 G   0.11 G 0.83 G 0.89 G 0.15 G          0 G           25
       .340  .51   .24  .355  .73   .30  .400  .65   .03  .430  .71   .03
       .470  .83   .03  .500  .89   .03  .540  .97   .03  .570 1.01   .03
       .600 1.03   .03  .630 1.15   .03  .670 1.16   .03  .700 1.21   .03
       .730 1.23   .03  .765 1.27   .03  .800 1.23   .07  .870 1.04   .08
       .900 1.07   .08  .950 1.10   .08 1.000 1.24   .11 1.060 1.37   .21

  26   1.64 G 0.11 G                                        0 G           26
       .330  .50   .07  .355  .60   .03  .400  .67   .03  .430  .77   .03
       .470  .83   .03  .500  .85   .03  .540  .96   .03  .570 1.00   .03
       .600 1.02   .03  .630 1.05   .03  .670 1.15   .03  .700 1.17   .03
       .730 1.14   .03  .765 1.23   .03  .800 1.20   .03  .830 1.17   .03

  27   1.72 G 0.07 G  -0.02 G 0.86 G 0.94 G 0.21 G          2 G           27
       .340  .56   .06  .355  .61   .11  .400  .63   .03  .430  .68   .03
       .470  .84   .03  .500  .86   .03  .540  .96   .03  .570 1.00   .03
       .600 1.04   .03  .630 1.09   .03  .670 1.09   .03  .700 1.21   .03
       .730 1.23   .03  .765 1.26   .03  .800 1.19   .05  .900 1.05   .03
       .950 1.12   .03 1.000 1.08   .03 1.030 1.15   .03 1.060 1.37   .03

  28   1.63 G 0.15 G  -0.06 G 0.82 G 0.95 G 0.17 G          2 G           28
       .340  .49   .10  .355  .65   .09  .400  .64   .03  .430  .74   .03
       .470  .83   .03  .500  .86   .04  .540  .96   .04  .570 1.02   .04
       .600 1.02   .03  .630 1.06   .03  .670 1.03   .03  .700 1.15   .03
       .730 1.15   .03  .765 1.20   .03  .800 1.17   .03  .830 1.13   .03
       .870 1.01   .03  .900 1.05   .03  .950  .92   .08 1.000 1.07   .03
      1.060 1.13   .06 1.100 1.17   .10

  29   1.44 G 0.06 G   0.08 G 0.95 G 0.94 G 0.13 G          1 G           29
       .330  .65   .09  .340  .70   .08  .355  .77   .03  .400  .76   .03
       .430  .84   .04  .470  .91   .04  .500  .92   .03  .540  .99   .03
       .570 1.00   .03  .600 1.08   .03  .630 1.10   .04  .670 1.09   .04
       .700 1.16   .04  .730 1.18   .04  .765 1.18   .03  .800 1.19   .05
       .870 1.14   .03  .900 1.13   .03  .950 1.13   .04 1.000 1.18   .05
      1.060 1.27   .05 1.100 1.19   .06

  30   1.45 G 0.07 G   0.08 Q 1.00 Q                        0 G           30
       .330  .29   .20  .340  .90   .05  .355  .70   .05  .400  .74   .03
       .430  .74   .03  .470  .88   .03  .500  .85   .03  .540  .94   .03
       .570 1.02   .03  .600 1.01   .03  .630 1.07   .03  .670 1.12   .03
       .700 1.12   .03  .730 1.16   .03  .765 1.15   .03  .800 1.17   .10
       .900 1.19   .05  .950 1.23   .05 1.000 1.22   .05 1.030 1.08   .15
      1.060 1.28   .15

  31   1.14 G 0.11 G          1.00 G          0.52 G 1 G   0 G           31
       .330  .78   .03  .355  .83   .03  .400  .94   .03  .430  .91   .03
       .470  .97   .03  .500 1.00   .03  .540 1.03   .03  .570 1.00   .03
       .600 1.02   .03  .630 1.05   .03  .670 1.03   .03  .730 1.02   .03
       .765 1.04   .03  .800 1.06   .03  .830 1.00   .03  .870 1.02   .03
       .900 1.03   .03  .930 1.02   .03  .970 1.03   .03

  32   1.43 G 0.06 G  -0.03 G 0.90 G 0.94 Q 0.21 Q          0 G           32
       .330  .72   .14  .340  .63   .07  .355  .65   .03  .400  .77   .03
       .430  .78   .03  .470  .86   .03  .500  .90   .03  .540  .94   .03
       .570 1.00   .03  .600 1.03   .03  .630 1.07   .03  .670 1.08   .03
       .700 1.12   .03  .730 1.13   .03  .765 1.18   .03  .800 1.10   .04
       .900  .98   .03  .950 1.10   .03 1.000 1.06   .04 1.030 1.02   .10
      1.060 1.18   .10
```

```
NUMBER   R/B   BEND      IR   DEPTH  CENTER BANDW  ELBOW  UV   FEAT         NUMBER
----------------------------------------------------------------------------------
 34    1.14 G 0.10 G           1.00 G            0.45 G 2 G    0 G            34
      MICRON REFL  RERR MICRON REFL  RERR MICRON REFL  RERR MICRON REFL  RERR
       .330 .68  .03  .355 .75  .03  .400 .91  .03  .430 .89  .03
       .470 .99  .03  .500 .99  .03  .540 1.00 .03  .570 1.00 .03
       .600 .99  .03  .630 1.02 .03  .670 1.01 .03  .700 1.00 .03
       .730 1.00 .03  .765 1.06 .03  .800 1.04 .03  .830 1.01 .03
       .870 1.09 .03  .900 1.07 .03  .930 1.00 .03  .970 1.07 .03

 36    1.19 G 0.18 G                      0.43 G 3 G    0 G            36
       .330 .79  .06  .355 .61  .12  .400 .82  .06  .430 1.15 .06
       .470 1.04 .07  .500 .99  .06  .540 .98  .06  .570 .98  .23
       .600 1.08 .07  .630 .97  .06  .670 1.05 .06  .700 .98  .06
       .730 .99  .09  .765 1.02 .06  .800 .95  .06  .830 .96  .13
       .870 1.00 .06  .900 1.05 .57  .930 1.04 .11

 37    1.52 G 0.11 G  0.02 G 0.89 G 0.93 G 0.14 G                0 G          37
       .330 .65  .03  .355 .72  .03  .400 .72  .03  .430 .78  .03
       .470 .84  .03  .500 .91  .03  .540 .96  .03  .570 1.00 .03
       .600 1.02 .03  .630 1.05 .03  .670 1.10 .03  .700 1.16 .03
       .730 1.12 .03  .765 1.12 .03  .800 1.13 .03  .830 1.13 .03
       .870 1.09 .03  .900 1.00 .03  .930 1.05 .05  .970 1.07 .03
      1.000 1.11 .05 1.030 1.05 .03 1.060 1.24 .03

 39    1.71 G 0.13 G  0.01 G 0.95 G 0.96 G 0.17 G                0 G          39
       .330 .49  .03  .340 .54  .04  .355 .56  .03  .400 .68  .03
       .430 .75  .03  .470 .83  .03  .500 .89  .03  .540 .96  .03
       .570 1.00 .03  .600 1.04 .03  .630 1.10 .03  .670 1.11 .03
       .700 1.17 .03  .730 1.19 .03  .765 1.22 .03  .800 1.21 .03
       .830 1.20 .03  .870 1.18 .03  .900 1.16 .03  .930 1.21 .03
       .970 1.14 .03 1.000 1.16 .03 1.030 1.18 .04 1.060 1.25 .05

 40    1.66 G 0.11 G  0.12 G 0.93 G 0.93 G 0.12 G                0 G          40
       .330 .58  .15  .340 .60  .05  .355 .62  .05  .400 .71  .03
       .430 .73  .03  .470 .81  .03  .500 .87  .03  .540 .94  .03
       .570 1.00 .03  .600 1.04 .03  .630 1.11 .03  .670 1.13 .03
       .700 1.17 .03  .730 1.17 .03  .765 1.20 .03  .800 1.20 .03
       .830 1.19 .03  .870 1.12 .03  .900 1.10 .03  .930 1.16 .05
       .970 1.14 .05 1.000 1.20 .05 1.030 1.24 .05 1.060 1.38 .05

 41    1.11 G 0.17 G  0.16 G 1.00 G            0.52 G 1 G    0 G          41
       .330 .69  .03  .355 .75  .03  .400 .88  .03  .430 .91  .03
       .470 .89  .03  .500 .93  .05  .540 1.00 .04  .570 1.00 .03
       .600 .98  .03  .630 1.02 .03  .670 .95  .03  .700 1.00 .03
       .730 .91  .03  .765 .97  .03  .800 .99  .03  .830 .97  .03
       .870 1.08 .03  .900 1.05 .06  .930 1.05 .03  .970 1.05 .03
      1.030 1.17 .03 1.060 1.15 .04

 42    1.62 G 0.09 G                                        0 G          42
       .330 .47  .05  .355 .51  .03  .400 .71  .03  .430 .79  .03
       .470 .83  .03  .500 .91  .03  .540 .97  .03  .570 1.00 .03
       .600 1.04 .03  .630 1.08 .03  .670 1.14 .03  .700 1.20 .03
       .730 1.18 .03  .765 1.18 .03  .800 1.20 .03  .830 1.23 .03
       .870 1.27 .07  .900 1.28 .12

 43    1.63 G 0.13 G -0.16 G 0.80 G 0.96 G 0.25 G            0 G          43
       .330 .52  .03  .355 .59  .03  .400 .73  .03  .430 .80  .03
       .470 .88  .03  .500 .92  .03  .540 .97  .03  .570 1.00 .03
       .600 1.06 .03  .630 1.10 .03  .670 1.14 .03  .700 1.19 .03
       .730 1.16 .03  .765 1.18 .03  .800 1.15 .03  .830 1.07 .03
       .870 1.03 .03  .900 .99  .03  .930 .94  .04  .970 1.00 .03
      1.000 .97  .07 1.030 .99  .08 1.060 1.03 .06

 44    1.07 G -0.05 G  0.07 G 1.00 G                    0 G          44
       .330 .93  .04  .355 .93  .03  .400 .99  .03  .430 1.00 .03
       .470 .98  .03  .500 1.01 .03  .540 .99  .03  .570 1.00 .03
       .600 1.01 .03  .630 1.05 .03  .670 1.05 .03  .700 1.08 .03
       .730 1.04 .03  .765 1.06 .03  .800 1.08 .03  .830 1.03 .03
       .870 1.09 .03  .900 1.03 .03  .930 1.05 .07  .970 1.11 .04
      1.000 1.20 .05 1.030 1.09 .03 1.060 1.17 .07

 45    1.06 G -0.07 G                      0.40 G 2 G    0 G          45
       .330 .92  .06  .355 .90  .03  .400 1.01 .03  .430 1.02 .03
       .470 1.02 .03  .500 1.00 .03  .540 1.01 .03  .570 .98  .03
       .600 .99  .03  .630 1.04 .03  .670 1.06 .03  .700 1.09 .03
       .730 1.03 .03  .765 1.05 .06  .800 1.07 .03  .830 .99  .07
       .870 1.08 .13

 46    1.10 G 0.00 G           0.95 G 0.90 G 0.14 G          0 G          46
       .330 .90  .11  .355 .92  .11  .400 .94  .03  .430 .97  .03
       .470 .98  .03  .500 .95  .03  .540 .99  .03  .570 1.00 .03
       .600 1.00 .03  .630 1.04 .03  .670 1.02 .03  .700 1.05 .03
       .730 1.04 .03  .765 1.06 .03  .800 1.03 .03  .830 .99  .03
       .870 1.01 .07  .900 .98  .03  .930 .99  .05  .970 1.03 .03
      1.000 1.13 .04 1.030 .96  .04 1.060 .79  .18
```

```
NUMBER   R/B    BEND      IR   DEPTH  CENTER BANDW  ELBOW  UV   FEAT        NUMBER
--------------------------------------------------------------------------------
  48    1.22 G 0.25 G -0.02 G 1.00 G            0.48 G 2 Q   0 G             48
       MICRON REFL  RERR MICRON REFL  RERR MICRON REFL  RERR MICRON REFL  RERR
        .330  .68   .09   .340  .78   .15   .355  .67   .10   .400  .82   .03
        .430  .82   .03   .470  .96   .03   .500  .97   .03   .540 1.02   .03
        .570 1.00   .03   .600 1.02   .03   .630  .99   .03   .670  .97   .03
        .700  .96   .03   .730  .99   .03   .765  .92   .03   .800  .94   .03
        .830  .95   .03   .870 1.01   .03   .900  .98   .04   .930 1.02   .05
        .970  .94   .05  1.000  .93   .05  1.030  .99   .07  1.060  .91   .10

  51    1.31 G 0.28 G  0.17 G 1.00 G            0.50 G 3 G   1 G             51
        .330  .55   .33   .340  .54   .13   .355  .62   .06   .400  .78   .03
        .430  .79   .05   .470  .92   .03   .500  .97   .04   .540 1.00   .03
        .570 1.00   .03   .600  .99   .03   .630 1.03   .03   .670  .93   .04
        .700  .97   .03   .730  .98   .03   .765  .95   .06   .800 1.03   .03
        .870 1.04   .07   .900 1.07   .04   .950 1.10   .05  1.000 1.10   .03
       1.060 1.22   .15

  52    1.09 G 0.10 G  0.00 G 1.00 G            0.44 G 2 Q   0 G             52
        .330  .74   .09   .340  .73   .11   .355  .84   .07   .400  .94   .03
        .430  .93   .03   .470  .97   .03   .500  .99   .03   .540  .99   .03
        .570 1.00   .03   .600  .98   .03   .630 1.01   .03   .670 1.02   .03
        .700  .99   .03   .730  .96   .03   .765 1.03   .03   .800 1.03   .04
        .830  .99   .07   .870 1.04   .16   .900 1.00   .03   .930  .99   .08
        .950 1.06   .04   .970 1.03   .10  1.000 1.01   .04  1.030  .95   .10

  53    1.26 G 0.14 G                          0.51 G 2 G   0 Q             53
        .400  .85   .03   .430  .83   .04   .470  .96   .04   .500  .96   .06
        .540 1.05   .04   .570  .98   .04   .600 1.06   .04   .630  .97   .10
        .670 1.03   .04   .700 1.08   .07   .730 1.02   .04   .765 1.08   .05
        .800 1.04   .06

  54    1.25 G 0.12 G                          0.50 G 2 G   3 G             54
        .355  .74   .13   .400  .83   .05   .430  .90   .03   .470  .91   .03
        .500 1.00   .03   .518 1.01   .06   .570 1.00   .06   .800 1.01   .03
        .630  .94   .06   .670  .97   .03   .700 1.09   .03   .730 1.06   .03
        .765  .98   .05   .800 1.17   .08   .830 1.01   .08

  58    1.11 G 0.09 G  0.05 Q 1.00 Q            0.47 G 1 Q   0 G             58
        .330  .83   .12   .340 1.08   .15   .355  .78   .15   .400  .91   .04
        .430  .85   .04   .470  .98   .04   .500  .95   .03   .540  .97   .04
        .570 1.02   .04   .600  .97   .04   .630 1.00   .03   .670  .98   .03
        .700 1.02   .06   .730  .98   .04   .765  .97   .05   .800  .97   .05
        .830  .99   .05   .870 1.06   .05   .900  .97   .07   .930 1.20   .10
        .970 1.00   .08  1.000  .98   .15  1.030 1.12   .20

  60    1.56 G 0.17 G                                       1 G             60
        .340  .70   .30   .400  .68   .03   .430  .72   .03   .470  .85   .03
        .500  .89   .03   .540  .94   .03   .570 1.00   .03   .600 1.05   .03
        .630 1.07   .03   .670 1.02   .06   .700 1.12   .07   .730 1.10   .04
        .765 1.16   .06   .800 1.33   .10   .900 1.14   .05   .950 1.29   .07
       1.000 1.01   .10  1.060 1.05   .30  1.100 1.36   .30

  62    0.99 G 0.02 G                          0.40 G 3 G   2 G             62
        .330  .77   .03   .355  .87   .04   .400  .97   .03   .430 1.02   .03
        .470  .98   .03   .500  .98   .03   .540  .99   .03   .570 1.00   .03
        .600  .97   .03   .630  .95   .03   .670  .92   .03   .700 1.01   .03
        .730  .95   .03   .765  .95   .03   .800  .94   .03   .830  .93   .03

  63    2.01 G 0.13 G  0.01 G 0.90 G 0.95 G 0.15 G           0 G             63
        .330  .45   .04   .340  .47   .08   .355  .50   .06   .400  .62   .03
        .430  .57   .03   .470  .81   .03   .500  .90   .03   .540  .97   .03
        .570 1.00   .03   .600 1.05   .03   .630 1.16   .03   .670 1.19   .03
        .700 1.26   .03   .730 1.26   .04   .765 1.28   .04   .800 1.30   .06
        .830 1.27   .05   .870 1.21   .05   .900 1.20   .05   .930 1.17   .05
        .970 1.18   .04  1.000 1.23   .04  1.030 1.24   .05  1.060 1.32   .09

  64    1.15 G 0.01 G  0.04 G 1.00 G                         0 G             64
        .330  .93   .12   .355  .86   .05   .400  .93   .05   .430  .98   .03
        .470  .90   .03   .500  .89   .03   .540  .99   .05   .570  .96   .03
        .600 1.05   .03   .630 1.04   .03   .670 1.03   .03   .700 1.06   .06
        .730 1.05   .03   .765 1.08   .05   .800 1.11   .09   .830 1.10   .03
        .870 1.10   .03   .900 1.12   .03   .930 1.07   .05   .970 1.08   .03
       1.000 1.10   .07  1.030 1.10   .03  1.060 1.06   .03

  65    1.11 G 0.04 G  0.04 Q 1.00 Q            0.40 G 2 G   1 G             65
        .330  .69   .07   .355  .83   .03   .400  .93   .03   .430  .93   .05
        .470  .94   .03   .500  .89   .04   .540 1.03   .03   .570  .96   .04
        .600 1.03   .06   .630 1.06   .03   .670  .97   .03   .700 1.05   .08
        .730 1.03   .03   .765  .99   .06   .800 1.06   .04   .830 1.10   .08
        .870 1.06   .04   .900 1.05   .04   .930 1.21   .14   .970  .97   .12
       1.030 1.02   .15
```

```
NUMBER   R/B    BEND      IR   DEPTH  CENTER BANDW  ELBOW  UV   FEAT        NUMBER
------------------------------------------------------------------------------------
66    1.10 G 0.14 G                        0.40 G 3 G   1 G                   66
   MICRON REFL  RERR MICRON REFL  RERR MICRON REFL  RERR MICRON REFL  RERR
    .330  .59   .12   .355  .71   .06   .400  .91   .03   .430  .97   .03
    .470  .97   .05   .500  .95   .03   .540  .99   .07   .570 1.00   .03
    .600 1.04   .05   .630  .95   .03   .670  .99   .03   .700 1.05   .03
    .730  .97   .03   .765  .87   .05   .800  .98   .03   .830  .88   .09

68    1.58 G 0.08 G -0.09 G 0.91 G 0.99 G 0.26 G          0 G                 68
    .330  .19   .03   .340  .56   .15   .355  .77   .14   .400  .75   .03
    .430  .74   .05   .470  .84   .03   .500  .88   .03   .540  .94   .03
    .570  .98   .03   .600 1.09   .03   .630 1.08   .03   .670 1.11   .03
    .700 1.18   .03   .730 1.15   .04   .765 1.24   .03   .800 1.14   .03
    .870 1.15   .03   .900 1.08   .03   .950 1.11   .04  1.000 1.06   .03
   1.060 1.13   .03  1.100 1.09   .03

69    1.27 G 0.05 G          1.00 Q                       0 G                 69
    .330  .85   .03   .355  .91   .08   .400  .83   .04   .430  .87   .03
    .470  .94   .03   .500  .95   .03   .540 1.03   .03   .570  .98   .03
    .600 1.00   .03   .630 1.05   .03   .670 1.07   .03   .700 1.10   .04
    .730 1.07   .03   .765 1.12   .05   .800 1.08   .03   .830  .99   .03
    .870 1.12   .03   .900 1.07   .03   .930 1.11   .03   .970 1.11   .03
   1.000  .98   .09

71    1.43 G 0.15 G                                       0 G                 71
    .330  .66   .03   .355  .65   .03   .400  .73   .03   .430  .81   .03
    .470  .84   .03   .500  .87   .03   .540  .94   .03   .570 1.00   .03
    .600 1.01   .03   .630 1.02   .03   .670 1.05   .03   .700 1.08   .03
    .730 1.07   .03   .765 1.07   .03   .800 1.07   .03   .830 1.06   .03

78    1.13 G 0.11 G -0.01 G 1.00 G          0.45 G 3 G   0 G                  78
    .330  .65   .03   .355  .84   .08   .400  .91   .03   .430  .94   .03
    .470  .96   .03   .500  .97   .03   .540 1.03   .03   .570 1.00   .03
    .600  .98   .03   .630  .99   .03   .670 1.00   .03   .700 1.03   .03
    .730  .97   .03   .765 1.04   .03   .800 1.04   .03   .830  .98   .03
    .870  .98   .05   .900 1.04   .03   .930  .98   .03   .970  .98   .04
   1.000 1.04   .04  1.030  .95   .07  1.060  .97   .07

79    1.64 G 0.16 G 0.09 G 0.90 G 0.92 G 0.14 G          0 G                  79
    .330  .63   .16   .340  .47   .04   .355  .44   .03   .400  .70   .05
    .430  .74   .04   .470  .85   .06   .500  .87   .04   .540  .91   .03
    .570 1.03   .03   .600 1.04   .08   .630 1.12   .05   .670 1.08   .08
    .700 1.11   .03   .730 1.17   .04   .765 1.18   .04   .800 1.18   .05
    .870 1.09   .06   .900 1.06   .05   .950 1.09   .03  1.000 1.15   .05
   1.060 1.31   .07  1.100 1.30   .09

80    1.97 G 0.22 G -0.10 G 1.00 G                         1 G                80
    .340  .48   .12   .355  .43   .05   .400  .57   .03   .430  .64   .03
    .470  .81   .03   .500  .81   .03   .540  .93   .03   .570 1.00   .03
    .600 1.06   .03   .630 1.09   .03   .670 1.09   .03   .700 1.18   .03
    .730 1.19   .03   .765 1.17   .04   .800 1.07   .04   .830 1.13   .05
    .870 1.08   .08   .900 1.08   .06   .950 1.16   .06  1.000 1.10   .07
   1.060 1.07   .11  1.100 1.11   .10

82    1.56 G 0.07 G 0.03 Q 0.88 Q 0.92 Q 0.17 Q          0 G                  82
    .330  .53   .05   .340  .68   .10   .355  .79   .05   .400  .72   .04
    .430  .69   .05   .470  .90   .05   .500  .89   .05   .540  .95   .05
    .570 1.00   .05   .600 1.04   .03   .630 1.11   .03   .670 1.13   .03
    .700 1.15   .04   .730 1.15   .03   .765 1.26   .08   .800 1.16   .07
    .870 1.08   .06   .900 1.04   .06   .950 1.07   .07  1.000 1.18   .12
   1.060 1.21   .07

83    1.15 G 0.08 G                                       0 G                 83
    .330  .91   .03   .355  .79   .05   .400  .86   .03   .430  .94   .03
    .470  .90   .03   .500  .93   .03   .540  .95   .03   .570 1.00   .04
    .600 1.01   .03   .630 1.02   .03   .670  .99   .03   .700  .99   .05
    .730 1.01   .03   .765 1.07   .03   .800 1.00   .03   .830 1.02   .03

84    1.15 Q 0.17 Q                                       0 Q                 84
    .400  .85   .04   .430  .82   .03   .470  .92   .03   .500  .87   .03
    .540  .98   .03   .570 1.00   .04   .600  .98   .04   .630 1.04   .03
    .670  .93   .06   .700 1.07   .03   .730  .90   .08   .765  .86   .11
    .800 1.15   .16   .900  .76   .08   .950  .93   .13  1.000  .86   .15

85    1.07 G 0.07 G 0.01 G 0.95 Q 0.98 Q 0.08 Q 0.44 G 2 G  0 G             85
    .330  .85   .05   .340  .81   .05   .355  .89   .05   .400  .94   .03
    .430  .95   .03   .470 1.00   .03   .500  .98   .03   .540  .98   .03
    .570 1.01   .03   .600  .99   .03   .630 1.01   .03   .670  .99   .03
    .700 1.01   .03   .730  .97   .03   .765  .99   .03   .800  .98   .03
    .830  .95   .03   .870  .95   .03   .900 1.01   .03   .930 1.05   .05
    .970  .90   .03  1.000  .95   .03  1.030  .97   .03  1.060 1.08   .04
```

NUMBER	R/B	BEND	IR	DEPTH	CENTER	BANDW	ELBOW	UV	FEAT		NUMBER

87 1.08 G 0.02 G 0.10 Q 1.00 G — 0 G — **87**

MICRON	REFL	RERR	MICRON	REFL	RERR	MICRON	REFL	RERR	MICRON	REFL	RERR
.330	1.02	.08	.355	.89	.05	.400	.91	.04	.430	.94	.03
.470	.95	.03	.500	.95	.03	.540	.96	.03	.570	1.00	.03
.600	.99	.03	.630	.99	.03	.670	.99	.03	.700	1.02	.03
.730	1.01	.03	.765	1.03	.03	.800	1.01	.03	.830	.98	.03
.870	1.05	.04	.900	1.09	.05	.930	.98	.07	.970	1.07	.08
1.000	1.14	.06	1.030	1.16	.09	1.060	.85	.23			

88 1.16 G 0.09 G 0.00 G 1.00 G 0.43 G 2 Q 0 G **88**

MICRON	REFL	RERR	MICRON	REFL	RERR	MICRON	REFL	RERR	MICRON	REFL	RERR
.330	.66	.09	.340	.73	.12	.355	.78	.10	.400	.90	.03
.430	.89	.04	.470	.95	.03	.500	.97	.03	.540	.98	.03
.570	1.00	.03	.600	.98	.03	.630	1.00	.03	.670	1.00	.03
.700	1.02	.03	.730	1.00	.04	.765	1.00	.04	.800	1.00	.04
.830	.99	.04	.870	.94	.04	.930	.96	.05	.930	1.04	.08
.970	.98	.08	1.000	.99	.08	1.030	.99	.08	1.060	1.05	.10

89 1.56 G 0.09 G 0.16 G 0.91 G 0.95 G 0.09 G 0 G **89**

MICRON	REFL	RERR	MICRON	REFL	RERR	MICRON	REFL	RERR	MICRON	REFL	RERR
.330	.50	.08	.340	.55	.03	.355	.64	.03	.400	.76	.03
.430	.75	.03	.470	.91	.03	.500	.94	.03	.540	.95	.03
.570	1.00	.03	.600	1.04	.03	.630	1.12	.03	.670	1.12	.03
.700	1.16	.03	.730	1.15	.03	.765	1.19	.03	.800	1.20	.03
.870	1.21	.03	.900	1.17	.03	.950	1.08	.03	1.000	1.24	.03
1.060	1.40	.03	1.100	1.15	.05						

90 1.08 G 0.07 G 0.41 G 3 Q 0 G **90**

MICRON	REFL	RERR	MICRON	REFL	RERR	MICRON	REFL	RERR	MICRON	REFL	RERR
.330	.65	.08	.355	.76	.08	.400	.94	.05	.430	.97	.06
.470	1.02	.03	.500	.93	.03	.540	1.02	.03	.570	1.00	.03
.600	.98	.03	.630	1.00	.03	.670	1.00	.03	.700	1.06	.03
.730	.95	.03	.765	.99	.03	.800	.96	.03	.830	1.01	.04

92 1.15 G 0.01 G 0 G **92**

MICRON	REFL	RERR	MICRON	REFL	RERR	MICRON	REFL	RERR	MICRON	REFL	RERR
.330	.85	.03	.355	.91	.03	.400	.92	.03	.430	.93	.03
.470	.92	.03	.500	.93	.03	.540	.97	.03	.570	1.00	.03
.600	1.00	.03	.630	.99	.03	.670	1.04	.03	.700	1.07	.03
.730	1.04	.03	.765	1.07	.03	.800	1.05	.03	.830	1.03	.03

93 1.16 G 0.07 G 0 G **93**

MICRON	REFL	RERR	MICRON	REFL	RERR	MICRON	REFL	RERR	MICRON	REFL	RERR
.340	1.00	.16	.355	.72	.11	.400	.91	.05	.430	.83	.05
.500	.97	.06	.540	.97	.06	.570	1.00	.03	.600	.98	.06
.630	1.00	.06	.670	1.01	.11	.700	1.02	.13	.730	1.04	.05
.765	.98	.14	.800	1.06	.16	.870	.97	.23	.900	.93	.12
.950	1.10	.20	1.000	1.33	.29						

94 1.17 G 0.06 G 0.53 G 1 G 1 G **94**

MICRON	REFL	RERR	MICRON	REFL	RERR	MICRON	REFL	RERR	MICRON	REFL	RERR
.330	.74	.12	.355	.93	.06	.400	.86	.03	.430	.95	.03
.470	.96	.03	.500	.97	.03	.540	1.01	.03	.570	1.00	.03
.600	.99	.03	.630	.95	.03	.670	1.02	.03	.700	1.06	.03
.730	1.04	.03	.765	1.01	.03	.800	1.12	.05	.830	1.10	.07

97 1.12 G 0.01 G 0.09 G 1.00 G 0 G **97**

MICRON	REFL	RERR	MICRON	REFL	RERR	MICRON	REFL	RERR	MICRON	REFL	RERR
.330	.66	.30	.340	1.01	.12	.355	.90	.11	.400	.96	.03
.430	.95	.03	.470	.96	.03	.500	.94	.03	.540	.99	.05
.570	1.00	.03	.600	1.04	.03	.630	1.04	.03	.670	1.07	.05
.700	1.01	.03	.730	1.02	.04	.755	1.12	.04	.800	1.06	.05
.830	1.11	.03	.870	1.10	.03	.900	1.13	.03	.930	1.13	.03
.950	1.15	.10	.970	1.12	.04	1.000	1.10	.08	1.060	1.09	.10

105 1.09 G 0.08 G 0.13 G 1.00 G 0.48 G 1 G 0 G **105**

MICRON	REFL	RERR	MICRON	REFL	RERR	MICRON	REFL	RERR	MICRON	REFL	RERR
.330	.86	.03	.355	.80	.04	.400	.92	.03	.430	.95	.03
.470	.97	.03	.500	.99	.03	.540	.99	.03	.570	1.00	.03
.600	.98	.03	.630	1.01	.03	.670	1.00	.03	.700	1.00	.03
.730	.98	.03	.765	.98	.03	.800	.99	.03	.830	.99	.03
.870	1.02	.03	.900	1.06	.03	.930	1.02	.03	.970	1.04	.06
1.000	1.12	.03	1.030	1.06	.11	1.060	1.19	.18			

106 1.19 G 0.16 G -0.01 Q 0.95 Q 0.94 G 0.15 Q 0.50 G 2 G 0 G **106**

MICRON	REFL	RERR	MICRON	REFL	RERR	MICRON	REFL	RERR	MICRON	REFL	RERR
.330	.57	.03	.355	.71	.04	.400	.85	.03	.430	.89	.03
.470	.91	.03	.500	.98	.03	.540	1.02	.03	.570	1.00	.03
.600	.96	.03	.630	.95	.03	.670	.99	.03	.700	1.00	.03
.730	.97	.03	.765	1.03	.04	.800	1.00	.03	.830	.95	.03
.870	.94	.03	.900	.95	.06	.930	.89	.09	.970	.92	.05
1.000	1.08	.10	1.030	.92	.14	1.060	1.06	.07			

108 1.65 G 0.14 G 0.31 Q 0.98 Q 0.92 Q 0.09 Q 1 G **108**

MICRON	REFL	RERR	MICRON	REFL	RERR	MICRON	REFL	RERR	MICRON	REFL	RERR
.340	.49	.25	.355	.87	.20	.400	.71	.03	.430	.72	.03
.470	.81	.03	.500	.86	.03	.540	.94	.03	.570	1.02	.03
.600	1.01	.03	.630	1.10	.03	.670	1.03	.03	.700	1.18	.03
.730	1.15	.07	.765	1.15	.04	.800	1.17	.20	.900	1.15	.10
.950	1.12	.08	1.000	1.48	.15	1.030	1.48	.30	1.060	1.68	.30

```
NUMBER   R/B    BEND      IR   DEPTH  CENTER BANDW  ELBOW  UV   FEAT        NUMBER
--------------------------------------------------------------------------------
110    1.20 G 0.02 G -0.02 G 0.93 G 0.89 Q 0.16 Q                    0 G     110
       MICRON REFL  RERR MICRON REFL  RERR MICRON REFL  RERR MICRON REFL  RERR
        .330  .89   .03   .355  .84   .06   .400  .93   .03   .430  .92   .03
        .470  .96   .03   .500  .95   .03   .540  .98   .04   .570  .99   .05
        .600 1.02   .04   .630  .99   .03   .570 1.06   .04   .700 1.11   .05
        .730 1.08   .04   .765 1.07   .03   .800 1.08   .03   .830 1.06   .05
        .870  .93   .05   .900 1.00   .06   .930 1.06   .05   .970 1.04   .04
       1.000 1.02   .07  1.030 1.09   .06  1.060 1.06   .07

113    1.61 G 0.06 G                                              0 G        113
        .330  .53   .05   .355  .51   .09   .400  .76   .03   .430  .83   .05
        .470  .84   .03   .500  .91   .03   .540 1.00   .03   .570 1.00   .06
        .600 1.00   .03   .630 1.05   .03   .670 1.18   .03   .700 1.17   .03
        .730 1.22   .03   .765 1.19   .03   .800 1.13   .03   .830 1.08   .03
        .870 1.08   .05

115    1.55 G 0.07 G 0.24 Q 0.91 Q 0.91 Q 0.14 Q                  0 G        115
        .330  .66   .30   .340  .68   .20   .355  .70   .20   .400  .71   .03
        .430  .78   .03   .470  .87   .03   .500  .89   .03   .540  .95   .03
        .570  .99   .03   .600 1.04   .03   .630 1.12   .03   .670 1.15   .03
        .700 1.16   .03   .730 1.15   .03   .765 1.18   .03   .800 1.08   .04
        .870 1.15   .05   .900 1.01   .06   .950 1.08   .05  1.000 1.17   .06
       1.030 1.41   .15  1.060 1.53   .15

116    1.48 G 0.09 G -0.19 Q 0.83 Q 1.02 Q 0.30 Q                 0 G        116
        .330  .54   .08   .355  .64   .14   .400  .79   .03   .430  .89   .03
        .470  .95   .03   .500  .92   .03   .540 1.00   .03   .570  .97   .04
        .600 1.02   .03   .630 1.17   .03   .670 1.02   .03   .700 1.25   .06
        .730 1.12   .06   .765 1.14   .05   .800 1.12   .08   .830 1.13   .06
        .870 1.04   .05   .900 1.07   .04   .930  .86   .15   .970  .95   .10
       1.000 1.05   .10  1.030  .94   .10  1.060 1.00   .10

119    1.77 G 0.10 G -0.01 G 1.00 G                               0 G        119
        .330  .43   .03   .340  .55   .03   .355  .62   .03   .400  .64   .03
        .430  .70   .03   .470  .92   .03   .500  .87   .03   .540  .93   .03
        .570 1.00   .03   .600 1.06   .03   .630 1.12   .04   .670 1.12   .04
        .700 1.19   .04   .730 1.23   .05   .765 1.23   .05   .800 1.26   .03
        .830 1.18   .03   .870 1.24   .03   .900 1.27   .03   .930 1.23   .04
        .970 1.26   .04  1.000 1.24   .04  1.030 1.14   .05  1.060 1.21   .08

121    1.09 G 0.02 G                             0.40 G 3 G    0 G          121
        .330  .71   .03   .355  .71   .03   .400  .95   .03   .430  .97   .04
        .470  .97   .03   .500 1.03   .03   .540  .96   .03   .570 1.00   .03
        .600 1.00   .03   .630 1.00   .03   .570 1.03   .03   .700 1.04   .03
        .730  .99   .03   .765  .98   .03   .800 1.03   .03   .830 1.02   .03

122    1.63 G 0.17 G                                              0 G        122
        .330  .47   .08   .355  .51   .04   .400  .68   .03   .430  .81   .04
        .470  .85   .04   .500  .87   .04   .540  .96   .05   .570 1.00   .03
        .600 1.07   .05   .630 1.15   .05   .670 1.13   .07   .700 1.11   .04
        .730 1.14   .05   .765 1.10   .05   .800 1.22   .04   .830 1.16   .03

124    1.47 G 0.09 G                                              0 Q        124
        .330  .50   .09   .355  .64   .03   .400  .83   .03   .430  .88   .03
        .470  .86   .03   .500  .95   .05   .540 1.07   .03   .570 1.00   .03
        .600  .97   .06   .630 1.10   .05   .670 1.11   .04   .700 1.22   .07
        .730 1.14   .05   .765 1.08   .07   .800 1.15   .12   .830 1.25   .04

128    1.17 G 0.10 G        1.00 G         0.48 G 1 G    0 G                 128
        .330  .75   .03   .355  .82   .04   .400  .89   .03   .430  .91   .03
        .470  .97   .03   .500  .95   .03   .540 1.00   .03   .570 1.00   .03
        .600  .99   .03   .630 1.04   .03   .670 1.01   .03   .730 1.03   .03
        .765 1.05   .03   .800 1.02   .03   .830  .92   .03   .870  .98   .03
        .900 1.02   .03   .930 1.05   .04   .970 1.03   .03

129    1.14 G 0.07 G 0.02 G 1.00 G                           1 G            129
        .330  .87   .03   .355  .88   .03   .400  .89   .03   .430  .91   .03
        .470  .92   .03   .500  .97   .03   .540 1.02   .03   .570 1.00   .04
        .600 1.01   .03   .630 1.01   .04   .670 1.01   .03   .700 1.07   .05
        .730 1.06   .03   .765 1.02   .03   .800 1.05   .03   .830 1.06   .03
        .870 1.11   .03   .900 1.06   .03   .930  .99   .03   .970 1.09   .03
       1.030 1.15   .03  1.060 1.17   .04

130    1.15 G 0.13 G 0.17 Q 1.00 G                           0 G            130
        .340  .57   .30   .355  .86   .20   .400  .82   .04   .430  .88   .03
        .470  .91   .03   .500  .92   .03   .540  .96   .03   .570 1.00   .03
        .600  .98   .03   .630  .99   .03   .670  .98   .03   .700  .97   .03
        .730  .91   .05   .765 1.06   .04   .800 1.04   .04   .830 1.12   .03
        .870 1.06   .05   .900 1.06   .03   .930 1.07   .03   .950 1.04   .08
        .970 1.26   .05  1.000 1.09   .07  1.030 1.28   .08  1.060 1.11   .09
```

```
NUMBER   R/B   BEND       IR   DEPTH  CENTER BANDW  ELBOW  UV    FEAT          NUMBER
-----------------------------------------------------------------------------------
136    1.27 G 0.00 G         0.92 Q                              0 G            136
       MICRON REFL   RERR MICRON REFL   RERR MICRON REFL   RERR MICRON REFL   RERR
       .330  .80   .03  .355  .88   .04  .400  .86   .03  .430  .85   .03
       .470  .90   .03  .500  .87   .03  .540  .95   .03  .570 1.03   .03
       .600  .95   .03  .630 1.07   .03  .670 1.12   .05  .700 1.05   .03
       .730 1.12   .03  .765 1.17   .03  .800 1.19   .03  .830 1.10   .03
       .870 1.15   .05  .900 1.11   .08  .930 1.03   .07  .970 1.13   .03

139    1.15 G 0.11 G -0.02 G 1.00 G              0.47 G 1 Q    0 G            139
       .340  .59   .18  .355  .88   .04  .400  .89   .03  .430  .89   .03
       .470 1.00   .03  .500  .97   .03  .540 1.00   .03  .570 1.00   .03
       .600 1.04   .03  .630 1.03   .03  .670  .99   .03  .700 1.03   .03
       .730 1.02   .04  .765 1.00   .04  .800 1.01   .03  .830  .99   .03
       .870 1.01   .05  .900 1.03   .05  .950  .97   .07 1.000  .98   .07
      1.060 1.02   .10 1.100  .85   .11

140    1.31 G 0.03 G  0.09 G 1.00 G                           0 G            140
       .330  .76   .04  .340  .77   .05  .355  .85   .05  .400  .80   .03
       .430  .85   .03  .470  .93   .03  .500  .93   .03  .540  .98   .03
       .570 1.00   .03  .600 1.00   .03  .630 1.08   .03  .670 1.03   .03
       .700 1.10   .03  .730 1.15   .03  .765 1.11   .03  .800 1.17   .04
       .830 1.13   .03  .870 1.12   .03  .900 1.14   .03  .930 1.24   .04
       .970 1.30   .05 1.000 1.17   .05 1.030 1.15   .07

141    1.18 G 0.12 G  0.01 G 1.00 G              0.50 G 1 G    0 G            141
       .330  .73   .03  .340  .77   .03  .355  .81   .03  .400  .86   .03
       .430  .89   .03  .470  .96   .03  .500  .98   .03  .540 1.00   .03
       .570 1.00   .03  .600 1.00   .03  .630 1.01   .03  .670  .98   .03
       .700 1.01   .03  .730 1.02   .03  .765 1.04   .03  .800 1.04   .03
       .830 1.03   .03  .870  .99   .03  .900 1.04   .03  .930 1.09   .03
       .970 1.03   .03 1.000  .99   .03 1.030 1.06   .04 1.060 1.01   .06

144    1.16 G 0.15 G  0.00 G 1.00 G              0.54 G 1 G    0 G            144
       .330  .76   .03  .355  .82   .03  .400  .96   .03  .430  .98   .03
       .470  .93   .03  .500  .96   .03  .540  .99   .03  .570 1.00   .03
       .600  .98   .03  .630  .96   .03  .670  .99   .03  .700 1.01   .03
       .730  .95   .03  .765  .97   .03  .800  .96   .03  .830  .99   .03
       .870  .96   .03  .900  .95   .03  .930  .99   .04  .970 1.03   .03
      1.000  .94   .03 1.030  .99   .04 1.060  .97   .03

145    1.22 G 0.16 G -0.03 G 1.00 G              0.50 G 2 G    0 G            145
       .330  .70   .05  .340  .65   .08  .355  .76   .05  .400  .85   .03
       .430  .88   .03  .470  .96   .03  .500 1.01   .03  .540 1.00   .03
       .570 1.02   .03  .600 1.02   .03  .630 1.04   .03  .670  .99   .03
       .700 1.02   .03  .730 1.03   .03  .765 1.09   .03  .800 1.01   .03
       .830 1.02   .04  .870 1.02   .05  .900  .92   .05  .930 1.08   .05
       .970  .91   .05 1.000  .97   .05 1.030  .98   .07 1.060 1.04   .10

149    1.65 G 0.16 G                                          0 G            149
       .330  .62   .16  .355  .60   .10  .400  .68   .03  .430  .72   .03
       .470  .80   .03  .500  .90   .03  .540 1.00   .03  .570 1.00   .03
       .600 1.05   .03  .630 1.04   .03  .670 1.14   .03  .700 1.19   .04
       .730 1.12   .04  .765 1.19   .04  .800 1.14   .03  .830 1.12   .04

150    1.02 G 0.04 G  0.06 Q 1.00 G              0.39 G 3 G    0 G            150
       .330  .76   .07  .355  .90   .10  .400  .96   .03  .430 1.00   .03
       .470  .99   .04  .500  .97   .03  .540  .99   .03  .570 1.01   .03
       .600  .99   .03  .630  .99   .03  .670  .97   .03  .700  .98   .03
       .730 1.00   .03  .765 1.02   .03  .800  .98   .03  .830 1.05   .03
       .870 1.07   .04  .900 1.10   .06  .930 1.08   .04  .970 1.09   .05
      1.000 1.07   .08 1.030  .95   .22

156    1.12 G 0.14 G  0.13 G 1.00 G              0.52 G 1 G    0 G            156
       .330  .76   .06  .355  .75   .06  .400  .91   .03  .430  .93   .03
       .470  .96   .03  .500  .97   .03  .540 1.01   .03  .570 1.00   .03
       .600  .99   .03  .630  .97   .03  .670  .97   .03  .700  .98   .03
       .730  .97   .03  .765  .98   .03  .800 1.00   .03  .830  .96   .03
       .870 1.04   .03  .900 1.01   .04  .930 1.01   .03  .970 1.10   .03
      1.000 1.18   .03 1.030 1.07   .05 1.060 1.08   .10

158    1.44 G 0.10 G                                          0 G            158
       .330  .62   .05  .355  .70   .03  .400  .79   .03  .430  .86   .03
       .470  .86   .03  .500  .86   .03  .540  .97   .03  .570 1.00   .03
       .600 1.06   .03  .630 1.08   .03  .670 1.08   .03  .700 1.13   .05
       .730 1.14   .03  .765 1.14   .03  .800 1.09   .03  .830 1.04   .03

163    1.20 G 0.18 G                              0.50 G 2 Q    0 G            163
       .330  .69   .16  .340  .59   .10  .355  .61   .09  .400  .86   .05
       .430  .84   .03  .470  .93   .03  .500  .94   .03  .540  .98   .03
       .570 1.00   .03  .600  .98   .03  .630 1.00   .03  .670  .97   .04
       .700  .97   .03  .730 1.00   .04  .755  .97   .03  .800  .96   .05
       .830  .93   .05  .870  .95   .08  .900 1.00   .08  .950  .82   .14
      1.000 1.10   .12 1.060  .86   .17
```

NUMBER	R/B	BEND	IR	DEPTH	CENTER	BANDW	ELBOW	UV	FEAT	NUMBER

```
NUMBER  R/B    BEND      IR   DEPTH   CENTER BANDW  ELBOW  UV    FEAT        NUMBER
------------------------------------------------------------------------------------
 164    1.17 G 0.09 G                                  0.43 G 3 Q    1 G        164
        MICRON REFL   RERR MICRON REFL   RERR MICRON REFL   RERR MICRON REFL   RERR
         .330  .74    .10   .355  .69    .17   .400  .88    .06   .430 1.02     .03
         .470 1.00    .04   .500 1.00    .03   .540 1.00    .03   .570 1.00     .04
         .600 1.05    .03   .630  .95    .03   .670 1.05    .04   .700 1.06     .04
         .730 1.03    .03   .765 1.04    .03   .800 1.03    .03   .830 1.04     .06

 166    1.46 G 0.22 G  0.02 Q 0.91 G 0.96 Q 0.15 Q                  0 G         166
         .330  .74    .20   .340  .61    .15   .355  .49    .11   .400  .73     .03
         .430  .79    .04   .470  .89    .03   .500  .93    .03   .540  .97     .03
         .570 1.03    .03   .600  .98    .03   .630 1.04    .05   .670 1.07     .03
         .700 1.05    .03   .730 1.07    .03   .765 1.00    .05   .800 1.10     .07
         .830 1.13    .07   .870 1.06    .05   .900 1.05    .06   .930 1.00     .09
         .970  .98    .05  1.000 1.06    .04  1.030  .97    .20  1.060 1.16     .11

 167    1.48 G 0.09 G               0.81 Q                          0 G         167
         .355  .74    .10   .400  .72    .03   .430  .86    .04   .470  .87     .05
         .500  .92    .03   .540  .95    .03   .570 1.02    .06   .600  .92     .03
         .630 1.07    .05   .670 1.10    .04   .700 1.23    .05   .730 1.02     .03
         .765 1.24    .07   .800 1.20    .07   .830 1.05    .11   .870  .95     .08
         .900  .92    .06   .930 1.07    .13

 169    1.71 G 0.08 G                                               0 G         169
         .330  .45    .06   .355  .57    .06   .400  .68    .03   .430  .77     .03
         .470  .84    .03   .500  .90    .03   .540  .98    .03   .570 1.00     .03
         .600 1.06    .03   .630 1.07    .03   .670 1.17    .03   .700 1.19     .03
         .730 1.24    .03   .765 1.27    .03   .800 1.18    .03   .830 1.18     .05

 170    1.72 Q-0.05 Q               1.00 Q                          0 Q         170
         .400  .84    .05   .430  .72    .05   .470  .87    .03   .500  .91     .03
         .540  .96    .04   .570 1.00    .07   .600 1.09    .05   .630 1.19     .10
         .670 1.17    .12   .700 1.32    .06   .730 1.34    .06   .765 1.30     .06
         .800 1.37    .06   .830 1.35    .06   .870 1.40    .06   .900 1.31     .06
         .930 1.47    .10   .970 1.47    .08  1.000 1.48    .08

 175    1.10 G 0.06 G  0.92 Q 0.90 Q 0.09 Q 0.40 G 3 G    0 G       175
         .330  .66    .03   .355  .76    .05   .400  .95    .03   .430  .99     .03
         .470  .99    .03   .500 1.01    .03   .540  .98    .03   .570 1.00     .03
         .600 1.05    .03   .630 1.00    .03   .670 1.01    .03   .700 1.05     .03
         .730  .99    .03   .765 1.04    .03   .800 1.00    .03   .830 1.00     .04
         .870  .94    .04   .900  .89    .09   .930 1.01    .09   .970 1.06     .12
        1.000 1.14    .12  1.030 1.05    .24  1.060  .81    .10

 176    1.24 G 0.24 G  0.06 G 1.00 Q                0.48 G 3 G    2 G           176
         .330  .61    .08   .340  .57    .05   .355  .62    .06   .400  .80     .03
         .430  .85    .03   .470  .95    .03   .500  .96    .03   .540  .99     .03
         .570 1.02    .03   .600  .98    .03   .630  .96    .03   .670  .94     .03
         .700  .98    .03   .730  .98    .03   .765 1.00    .03   .800  .98     .04
         .830 1.02    .04   .870  .97    .03   .900 1.04    .03   .930 1.07     .03
         .970  .97    .04  1.000 1.08    .04  1.030  .92    .05  1.060 1.16     .08

 181    1.47 G 0.04 G               0.85 Q                          0 G         181
         .330  .54    .15   .340  .66    .14   .355  .91    .16   .400  .76     .03
         .430  .82    .03   .470  .92    .03   .500  .91    .03   .540  .98     .03
         .570 1.00    .03   .600 1.05    .03   .630 1.15    .03   .670 1.08     .03
         .700 1.19    .03   .730 1.15    .03   .765 1.24    .03   .800 1.16     .05
         .870 1.22    .09   .900 1.07    .05   .950 1.06    .10  1.000  .97     .10
        1.060 1.35    .35

 185    1.07 G 0.07 G                                               0 G         185
         .330  .85    .03   .355  .90    .03   .400  .91    .03   .430  .93     .03
         .470  .96    .03   .500  .95    .03   .540 1.00    .03   .570 1.02     .03
         .600  .99    .03   .630  .98    .03   .670 1.00    .03   .700 1.01     .03
         .730  .99    .03   .765 1.00    .03   .800  .98    .03   .830  .97     .03

 192    1.68 G 0.11 G  0.01 G 0.90 G 0.93 G 0.21 G                  0 G         192
         .330  .48    .15   .340  .48    .12   .355  .51    .07   .400  .75     .03
         .430  .73    .06   .470  .97    .03   .500  .83    .03   .540  .95     .04
         .570 1.00    .03   .600 1.03    .05   .630 1.13    .03   .670 1.11     .05
         .700 1.21    .03   .730 1.14    .04   .765 1.21    .03   .800 1.16     .03
         .870 1.07    .04   .900 1.08    .05   .950 1.08    .03  1.000 1.09     .03
        1.060 1.24    .04  1.100 1.23    .07

 194    1.27 G 0.14 G  0.09 G 1.00 Q                0.50 G 2 Q    0 G           194
         .330  .61    .08   .340  .57    .10   .355  .78    .15   .400  .82     .03
         .430  .84    .03   .470  .98    .03   .500  .95    .03   .540 1.01     .03
         .570 1.00    .03   .600 1.00    .03   .630 1.08    .03   .670 1.06     .03
         .700 1.02    .03   .730 1.06    .03   .765 1.06    .04   .800 1.08     .03
         .830 1.10    .03   .870 1.10    .03   .900 1.06    .03   .930 1.12     .05
         .970 1.09    .03  1.000 1.15    .03  1.030 1.01    .05  1.060 1.18     .06
```

```
NUMBER   R/B   BEND       IR   DEPTH  CENTER BANDW  ELBOW  UV   FEAT          NUMBER
-----------------------------------------------------------------------------------
196   1.58 G 0.15 G  0.00 G 1.00 Q                          0 G               196
      MICRON REFL  RERR MICRON REFL  RERR MICRON REFL  RERR MICRON REFL  RERR
       .330 .76   .12  .340 .42   .05  .355 .73   .03  .400 .71   .03
       .430 .73   .03  .470 .79   .03  .500 .86   .03  .540 .93   .03
       .570 1.01  .03  .600 1.02  .03  .630 1.03  .03  .670 1.09  .04
       .700 1.09  .04  .730 1.14  .04  .765 1.17  .04  .800 1.14  .06
       .870 1.12  .09  .900 1.14  .10  .950 1.11  .08 1.000 1.17  .10
      1.060 .91   .30

197   1.87 G 0.09 G        0.88 Q                         0 G                 197
       .330 .46   .14  .355 .46   .16  .400 .66   .03  .430 .67   .03
       .470 .82   .03  .500 .83   .03  .540 1.01  .03  .570 .96   .03
       .600 .98   .03  .630 1.12  .03  .670 1.07  .03  .700 1.28  .03
       .730 1.20  .07  .765 1.45  .07  .800 1.33  .07  .830 1.23  .05
       .870 1.40  .31  .900 1.15  .05  .930 1.14  .06  .970 1.38  .11

198   1.55 G 0.08 G                                        0 G                198
       .330 .57   .03  .355 .54   .03  .400 .74   .03  .430 .80   .03
       .470 .84   .03  .500 .90   .03  .540 .97   .03  .570 1.00  .03
       .600 1.05  .03  .630 1.08  .03  .670 1.09  .03  .700 1.21  .03
       .730 1.15  .03  .765 1.17  .03  .800 1.18  .04  .830 1.15  .03

200   1.12 G 0.07 G                                        0 G                200
       .400 .97   .05  .430 .93   .04  .470 .98   .03  .500 1.02  .03
       .540 1.03  .03  .570 1.00  .03  .600 1.03  .03  .630 1.00  .03
       .670 1.06  .04  .700 1.02  .08  .730 .99   .06  .765 1.14  .06
       .800 1.02  .12  .900 1.10  .09  .950 1.18  .15 1.000 1.02  .15
      1.030 .99   .35 1.060 1.26  .35

208   1.38 Q 0.10 Q                                        0 Q                208
       .400 .77   .06  .430 .90   .37  .470 .91   .06  .500 .87   .06
       .540 1.11  .06  .570 .93   .14  .600 1.06  .12  .630 1.04  .06
       .670 1.08  .06  .700 1.17  .32  .730 1.05  .09  .765 .90   .24
       .800 1.34  .12  .830 .74   .06

210   1.11 G 0.14 G                                        3 G                210
       .340 .93   .14  .355 .90   .10  .400 .83   .03  .430 .87   .03
       .470 .93   .03  .500 .93   .03  .540 .95   .04  .570 1.01  .04
       .600 1.02  .03  .630 .98   .04  .670 .92   .04  .700 .97   .04
       .730 .99   .03  .765 .94   .06  .800 .99   .05  .830 .97   .05
       .870 .87   .05  .900 .95   .05  .930 .87   .07  .970 .85   .06
      1.000 1.08  .06 1.030 1.08  .19

213   1.06 G 0.02 G -0.02 Q 1.00 Q                         0 G                213
       .330 .91   .09  .340 1.23  .10  .355 .97   .10  .400 1.01  .04
       .430 .94   .04  .470 .93   .03  .500 .94   .05  .540 .95   .04
       .570 1.05  .03  .600 .99   .05  .630 1.02  .03  .670 1.02  .07
       .700 1.05  .05  .730 1.02  .03  .765 .99   .05  .800 1.07  .04
       .830 .93   .05  .870 1.01  .07  .900 .95   .05  .930 .84   .10
       .970 1.09  .08 1.000 1.02  .07 1.030 .97   .09 1.060 .97   .12

216   1.19 G 0.01 G        1.00 G                          0 G                216
       .330 .90   .03  .355 .90   .03  .400 .89   .03  .430 .86   .03
       .470 .91   .03  .500 .94   .03  .540 1.01  .03  .570 1.00  .03
       .600 .97   .03  .630 1.03  .03  .670 1.04  .03  .700 1.08  .03
       .730 1.05  .03  .765 1.11  .03  .800 1.10  .03  .830 1.06  .03
       .870 1.12  .03  .900 1.15  .03  .930 1.07  .03  .970 1.12  .03

217   1.14 G 0.02 G  0.15 Q 1.00 G                         0 G                217
       .330 .95   .10  .355 .83   .05  .400 .91   .03  .430 .91   .03
       .470 .95   .03  .500 .95   .03  .540 .95   .03  .570 1.00  .03
       .600 1.01  .03  .630 1.02  .03  .670 1.33  .03  .700 1.06  .03
       .730 1.02  .03  .765 1.09  .03  .800 1.02  .03  .830 1.03  .03
       .870 1.06  .07  .900 1.13  .08  .930 1.11  .05  .970 1.21  .07
      1.000 1.22  .14 1.030 1.06  .05 1.060 1.28  .20

220   1.19 G 0.09 G                                        0 G                220
       .330 .81   .03  .355 .83   .03  .400 .85   .03  .430 .89   .03
       .470 .92   .03  .500 .94   .03  .540 .97   .03  .570 1.00  .03
       .600 1.00  .03  .630 .95   .03  .670 1.04  .03  .700 1.03  .03
       .730 1.01  .03  .765 1.03  .03  .800 1.03  .03  .830 1.14  .05

221   1.52 G 0.21 G -0.03 G 1.00 G                         0 G                221
       .330 .50   .06  .340 .51   .10  .355 .55   .07  .400 .73   .03
       .430 .75   .03  .470 .89   .03  .500 .90   .03  .540 1.01  .03
       .570 1.00  .03  .600 1.02  .03  .630 1.08  .03  .670 1.10  .03
       .700 1.09  .05  .730 1.09  .03  .765 1.05  .03  .800 1.12  .05
       .830 1.05  .05  .870 1.05  .04  .900 1.11  .04  .930 1.05  .05
       .970 1.02  .05 1.000 1.00  .05 1.030 1.10  .07
```

```
NUMBER   R/B    BEND      IR   DEPTH  CENTER BANDW  ELBOW  UV   FEAT        NUMBER
------------------------------------------------------------------------------------
230   1.55 G 0.05 G  0.10 G 0.94 G 0.90 G 0.11 G                 0 G          230
      MICRON REFL  RERR MICRON REFL  RERR MICRON REFL  RERR MICRON REFL  RERR
       .330  .62   .20  .340  .67   .06  .355  .69   .03  .400  .78   .03
       .430  .78   .03  .470  .86   .03  .500  .90   .03  .540  .97   .03
       .570 1.00   .03  .600 1.09   .03  .630 1.12   .03  .670 1.13   .03
       .700 1.18   .03  .730 1.20   .03  .765 1.25   .03  .800 1.25   .03
       .900 1.17   .03  .950 1.22   .03 1.000 1.26   .03 1.030 1.33   .04
      1.060 1.27   .05

236   1.56 G 0.08 G                                        0 G          236
       .330  .55   .09  .355  .73   .04  .400  .74   .03  .430  .78   .03
       .470  .87   .03  .500  .89   .03  .540  .99   .04  .570  .98   .03
       .600 1.03   .03  .630 1.06   .03  .670 1.12   .03  .700 1.10   .03
       .730 1.18   .03  .765 1.26   .03  .800 1.24   .05  .830 1.20   .03
       .870 1.30   .08  .900 1.28   .06  .930 1.15   .06  .970 1.34   .08

243   1.39 Q 0.08 Q                                  0 G          243
       .330  .87   .06  .355  .72   .03  .400  .77   .08  .430  .77   .05
       .470  .87   .06  .500  .88   .06  .540  .93   .09  .570 1.05   .05
       .600  .99   .05  .630 1.00   .07  .670 1.11   .08  .700 1.12   .06
       .730 1.08   .06  .765 1.20   .06  .800 1.30   .13

246   2.15 G 0.11 G        0.90 G                    0 G          246
       .330  .40   .03  .355  .45   .03  .400  .60   .03  .430  .67   .03
       .470  .79   .03  .500  .86   .03  .540  .94   .03  .570  .97   .03
       .600 1.08   .03  .630 1.13   .03  .670 1.26   .03  .700 1.28   .03
       .730 1.26   .03  .765 1.33   .03  .800 1.23   .03  .830 1.21   .03
       .870 1.23   .06  .900 1.20   .04  .930 1.19   .03  .970 1.17   .03

258   1.56 G 0.07 G                                  0 G          258
       .330  .51   .03  .355  .65   .03  .400  .75   .03  .430  .81   .03
       .470  .85   .03  .500  .88   .03  .540  .95   .03  .570 1.02   .03
       .600 1.03   .03  .630 1.05   .03  .670 1.12   .03  .700 1.19   .03
       .730 1.17   .03  .765 1.22   .03  .800 1.21   .03  .830 1.17   .03

262   1.85 Q 0.17 Q                                  0 Q          262
       .330  .07   .31  .355  .59   .22  .400  .65   .07  .430  .68   .04
       .470  .79   .06  .500  .77   .04  .540  .97   .04  .570 1.05   .14
       .600  .91   .03  .630 1.15   .07  .670 1.17   .08  .700 1.11   .08
       .730 1.06   .15  .765 1.23   .07  .800 1.32   .11  .830 1.23   .11
       .870 1.09   .12  .900 1.21   .25  .930  .79   .28  .970 1.07   .28

264   1.53 G 0.05 G                                  0 G          264
       .330  .69   .05  .355  .47   .33  .400  .74   .10  .430  .87   .03
       .470  .83   .03  .500  .89   .03  .540  .95   .08  .570 1.00   .03
       .600 1.08   .03  .630 1.09   .03  .670 1.15   .03  .700 1.20   .04
       .730 1.16   .05  .765 1.28   .04  .800 1.21   .03  .830 1.36   .04

268   1.10 G-0.01 G                     0.40 G 2 G    0 G          268
       .330  .84   .05  .355  .79   .07  .400  .93   .03  .430  .98   .03
       .470  .95   .03  .500  .93   .03  .540  .99   .03  .570 1.00   .03
       .600  .96   .03  .630  .96   .03  .670 1.02   .03  .700 1.05   .03
       .730  .99   .04  .765 1.15   .04  .800 1.05   .03  .830 1.13   .06

279   1.20 G 0.01 G                                  0 G          279
       .330  .97   .06  .355  .87   .07  .400  .89   .03  .430  .89   .04
       .470  .89   .03  .500  .93   .03  .540  .98   .04  .570 1.00   .03
       .600 1.00   .03  .630 1.01   .03  .670 1.05   .03  .700 1.10   .03
       .730 1.08   .06  .765 1.06   .06  .800 1.06   .03  .830 1.20   .12

281   1.52 G 0.13 G                                  0 G          281
       .330  .75   .06  .355  .58   .05  .400  .74   .03  .430  .79   .06
       .470  .82   .06  .500 1.00   .07  .540  .98   .03  .570 1.00   .05
       .600 1.06   .03  .630 1.12   .03  .670 1.06   .03  .700 1.14   .05
       .730 1.13   .03  .765 1.20   .05  .800 1.30   .09

293   1.28 G 0.28 G                     0.44 G 3 G    0 G          293
       .400  .72   .06  .430 1.00   .09  .470 1.03   .05  .500 1.00   .03
       .540  .99   .04  .570 1.02   .03  .600  .97   .03  .630 1.06   .03
       .670  .98   .03  .700  .92   .04  .730  .89   .08  .765 1.02   .06
       .800 1.06   .11  .830  .99   .10

308   1.50 Q 0.06 Q                                  0 G          308
       .330  .50   .08  .355  .77   .15  .400  .74   .04  .430  .89   .03
       .470  .84   .03  .500  .82   .03  .540  .98   .05  .570 1.00   .07
       .600 1.03   .04  .630 1.08   .06  .670 1.09   .03  .700 1.18   .07
       .730 1.18   .11  .765 1.04   .08  .800 1.31   .03  .830 1.17   .11

313   1.21 G 0.11 G -0.01 G 1.00 Q        0.48 G 2 G    0 G          313
       .330  .57   .06  .355  .81   .03  .400  .88   .03  .430  .89   .03
       .470  .96   .03  .500  .95   .04  .540  .97   .03  .570  .99   .03
       .600 1.03   .03  .630 1.00   .03  .670 1.01   .03  .700 1.06   .03
       .730 1.01   .03  .765 1.03   .03  .800 1.05   .03  .830 1.02   .03
       .870 1.01   .06  .900 1.13   .04  .930 1.02   .06  .970 1.04   .03
      1.000 1.02   .17 1.030 1.03   .13 1.060  .96   .10
```

```
NUMBER   R/B   BEND       IR   DEPTH  CENTER BANDW   ELBOW   UV    FEAT          NUMBER
---------------------------------------------------------------------------------------
 323    1.63 G 0.10 G                                            0 G              323
        MICRON REFL   RERR MICRON REFL   RERR MICRON REFL   RERR MICRON REFL  RERR
        .330  .73  .03  .355  .68  .04  .400  .72  .03  .430  .76  .03
        .470  .84  .03  .500  .92  .03  .540 1.00  .03  .570 1.00  .03
        .600 1.03  .04  .630 1.08  .03  .670 1.22  .03  .700 1.20  .03
        .730 1.13  .03  .765 1.20  .04  .800 1.26  .03

 324    1.09 G 0.10 G  0.04 Q                0.47 Q 1 Q      0 G              324
        .330  .68  .08  .340  .90  .20  .355 1.00  .18  .400  .90  .03
        .430  .87  .03  .470 1.02  .03  .500  .93  .03  .540  .99  .03
        .570  .99  .03  .600  .97  .03  .630 1.01  .03  .670  .94  .03
        .700  .99  .03  .730  .99  .03  .765  .94  .03  .800  .97  .05
        .830 1.00  .08  .870  .89  .13  .900  .76  .17  .930  .91  .08
        .950  .85  .10  .970 1.00  .08 1.000  .92  .08 1.030 1.07  .08

 325    1.22 G-0.01 G              1.00 Q                        0 G              325
        .330  .81  .05  .355  .95  .05  .400  .91  .03  .430  .90  .03
        .470  .99  .03  .500  .98  .03  .540 1.01  .03  .570  .97  .04
        .600 1.03  .03  .630 1.07  .03  .670 1.12  .03  .700 1.09  .04
        .730 1.08  .03  .765 1.14  .03  .800 1.18  .03  .830 1.05  .05
        .870 1.25  .07  .900 1.05  .04  .930 1.22  .08  .970 1.23  .09

 326    1.09 G 0.14 G                         0.48 G 1 Q      0 G              326
        .340 1.13  .30  .355  .89  .30  .400  .87  .03  .430  .87  .03
        .470  .97  .03  .500  .96  .03  .540 1.01  .04  .570 1.00  .04
        .600 1.02  .05  .630 1.02  .04  .670  .97  .07  .700  .96  .05
        .730  .98  .12  .765 1.19  .19

 335    1.01 G 0.03 G -0.07 Q 1.00 Q          0.36 G 3 Q      0 G              335
        .330  .76  .15  .340  .86  .06  .355  .99  .04  .400 1.01  .03
        .430  .93  .03  .470 1.00  .03  .500  .97  .03  .540 1.00  .03
        .570 1.00  .03  .600  .99  .03  .630 1.02  .03  .670  .97  .03
        .700 1.05  .03  .730  .95  .03  .765  .94  .03  .800  .94  .10
        .900  .89  .06  .950  .93  .05 1.000  .90  .10 1.030  .84  .25
       1.060  .92  .25

 337    1.32 Q 0.12 Q -0.04 G 1.00 Q                          0 Q              337
        .330  .87  .15  .340  .58  .33  .355 1.12  .13  .400  .90  .12
        .430  .77  .03  .470  .86  .03  .500  .79  .12  .540  .96  .04
        .570 1.00  .03  .600 1.13  .11  .630 1.04  .07  .670 1.10  .09
        .700 1.07  .03  .730 1.13  .03  .765 1.04  .03  .800 1.07  .07
        .870  .96  .03  .900 1.05  .03  .950 1.10  .07 1.000 1.06  .09
       1.060 1.04  .03 1.100 1.02  .03

 338    1.11 G 0.02 G                                            1 G              338
        .330  .88  .04  .355  .98  .06  .400  .95  .03  .430  .92  .03
        .470  .94  .03  .500  .95  .03  .540  .98  .03  .570 1.00  .03
        .600  .98  .03  .630  .97  .03  .670  .97  .03  .700 1.04  .03
        .730 1.01  .03  .765 1.09  .03  .800 1.09  .03  .830 1.08  .04

 339    1.40 G 0.12 G                                            0 G              339
        .330  .73  .04  .355  .58  .05  .400  .77  .03  .430  .88  .04
        .470  .90  .03  .500  .89  .03  .540  .98  .03  .570  .95  .03
        .600 1.06  .03  .630  .97  .03  .670 1.09  .03  .700 1.12  .03
        .730 1.04  .06  .765 1.09  .03  .800 1.03  .03  .830 1.07  .07

 340    1.58 G 0.11 G                                            0 Q              340
        .355  .56  .19  .400  .72  .04  .430  .76  .03  .470  .82  .04
        .500  .83  .04  .540 1.00  .04  .570 1.02  .04  .600  .94  .03
        .630 1.07  .04  .670 1.11  .05  .700 1.05  .04  .730 1.24  .09
        .765 1.16  .05  .800 1.04  .04  .830 1.18  .14

 341    1.75 G 0.18 G      0.93 G 0.90 G 0.13 G                   0 G              341
        .330  .54  .03  .355  .60  .03  .400  .65  .03  .430  .69  .03
        .470  .82  .03  .500  .89  .03  .540  .93  .03  .570 1.02  .03
        .600 1.06  .03  .630 1.09  .03  .670 1.14  .03  .700 1.08  .03
        .765 1.20  .03  .800 1.10  .03  .830 1.09  .03  .870 1.13  .03
        .900 1.09  .04  .930 1.07  .03  .970 1.15  .03 1.000 1.34  .03

 344    1.10 G 0.14 G  0.10 Q 1.00 Q           0.42 G 2 G      0 G              344
        .330  .75  .07  .355  .74  .07  .400  .92  .03  .430  .93  .03
        .470  .92  .03  .500  .97  .03  .540  .99  .03  .570 1.04  .03
        .600  .97  .03  .630  .95  .03  .670  .98  .03  .700  .99  .03
        .730  .96  .03  .765 1.00  .03  .800 1.00  .03  .830 1.02  .03
        .870  .93  .06  .900 1.06  .05  .930 1.11  .08  .970 1.08  .13
       1.000  .90  .09 1.030 1.14  .19 1.060 1.09  .13

 345    1.25 G 0.12 G                          0.45 G 3 G      0 G              345
        .330  .42  .07  .355  .60  .08  .400  .84  .05  .430  .89  .04
        .470  .97  .03  .500  .97  .04  .540  .95  .03  .570 1.00  .03
        .600 1.00  .03  .630  .95  .03  .670 1.03  .05  .700 1.04  .03
        .730 1.02  .03  .765 1.05  .03  .800 1.04  .05  .830  .94  .03
        .870  .97  .04  .900  .96  .03  .930  .94  .10
```

```
NUMBER  R/B    BEND        IR    DEPTH  CENTER BANDW  ELBOW  UV   FEAT      NUMBER
-------------------------------------------------------------------------------
347    1.26 G 0.07 G                                         O G           347
       MICRON REFL  RERR MICRON REFL  RERR MICRON REFL  RERR MICRON REFL  RERR
       .330  .88   .06  .355  .83   .05  .400  .82   .03  .430  .88   .03
       .470  .91   .03  .500 1.00   .03  .540  .98   .03  .570  .98   .03
       .600 1.07   .06  .630 1.05   .03  .670 1.09   .03  .700  .98   .03
       .730 1.12   .03  .765 1.11   .04  .800 1.02   .03  .830 1.09   .07
       .870 1.22   .04  .900 1.03   .04  .930  .99   .06  .970 1.10   .03

349    1.73 G 0.24 G -0.18 G 0.73 G 0.95 G 0.24 G              O G        349
       .340  .38   .09  .355  .53   .04  .400  .54   .03  .430  .69   .03
       .470  .81   .03  .500  .83   .03  .540  .96   .03  .570 1.00   .03
       .600 1.06   .03  .630 1.09   .03  .670 1.07   .03  .700 1.12   .03
       .730 1.12   .03  .765 1.15   .03  .800 1.04   .03  .830  .99   .04
       .870  .84   .04  .900  .83   .04  .950  .94   .04 1.000  .98   .05
      1.060  .93   .05 1.100 1.20   .12

354    1.91 G 0.14 G -0.19 G 1.00 G                           O G        354
       .330  .53   .17  .340  .44   .06  .355  .51   .03  .400  .63   .03
       .430  .68   .03  .470  .79   .03  .500  .86   .03  .540  .95   .03
       .570 1.00   .03  .600 1.04   .03  .630 1.11   .03  .670 1.16   .03
       .700 1.21   .03  .730 1.22   .03  .765 1.23   .03  .800 1.20   .04
       .900 1.11   .03  .950 1.08   .03 1.000 1.05   .03 1.030 1.07   .04
      1.060  .97   .05

356    1.31 G 0.14 G                                          O G        356
       .330  .40   .10  .340  .74   .10  .355  .79   .10  .400  .78   .03
       .430  .78   .04  .470  .93   .05  .500  .96   .03  .540  .98   .03
       .570 1.00   .10  .600 1.08   .03  .630 1.09   .03  .670 1.07   .03
       .700 1.07   .04  .730 1.09   .03  .765 1.04   .05

361    1.21 G-0.06 G                                          O G        361
       .330  .85   .10  .355  .96   .03  .400  .92   .04  .430  .95   .03
       .470  .93   .03  .500  .92   .03  .540 1.03   .03  .570 1.05   .04
       .600  .98   .04  .630 1.01   .04  .670 1.11   .05  .700 1.13   .05
       .730 1.12   .04  .765 1.18   .03  .800 1.16   .04  .830 1.20   .03

363    1.17 G 0.05 G         1.00 Q          0.40 Q 2 Q    1 G           363
       .355  .76   .04  .400  .90   .03  .430  .91   .03  .470  .94   .03
       .500  .98   .03  .540 1.00   .03  .570  .98   .03  .600 1.05   .03
       .630 1.00   .03  .670 1.03   .04  .700 1.03   .03  .730 1.08   .04
       .765 1.07   .03  .800 1.09   .04  .830 1.11   .04  .870  .96   .03
       .900 1.15   .03  .930 1.10   .07  .970 1.13   .12

365    1.18 G 0.01 G                      0.43 G 2 Q    0 Q             365
       .355  .87   .06  .400  .87   .17  .430 1.00   .07  .470  .99   .03
       .500  .99   .03  .540  .99   .10  .570  .98   .08  .600 1.03   .03
       .630 1.00   .06  .670 1.12   .04  .730 1.10   .04  .765 1.07   .03
       .800 1.14   .05  .830 1.34   .07

372    1.01 G 0.01 G                      0.40 Q 2 Q    0 G             372
       .330  .88   .17  .355  .89   .07  .400 1.00   .05  .430 1.06   .03
       .470  .95   .03  .500  .99   .03  .540  .95   .03  .570 1.00   .03
       .600 1.04   .03  .630 1.01   .03  .670 1.01   .03  .700 1.03   .03
       .730  .94   .03  .765 1.01   .03  .800  .92   .03  .830  .96   .03

374    1.49 G 0.05 G                                          O G        374
       .330  .49   .11  .355  .60   .08  .400  .80   .03  .430  .88   .03
       .470  .87   .03  .500  .97   .03  .540 1.01   .05  .570  .98   .06
       .600 1.05   .03  .630 1.11   .04  .670 1.17   .05  .700 1.17   .03
       .730 1.15   .03  .765 1.24   .03  .800 1.13   .12  .830 1.17   .06

375    1.09 G 0.03 G                      0.42 G 2 G    0 G             375
       .330  .79   .05  .355  .83   .03  .400  .92   .03  .430 1.01   .03
       .470  .95   .03  .500  .96   .03  .540  .96   .03  .570 1.00   .03
       .600 1.00   .03  .630 1.02   .03  .670 1.02   .03  .700 1.03   .03
       .730 1.00   .03  .765 1.01   .03  .800  .99   .03  .830  .98   .06

386    1.14 G 0.12 G -0.06 G 1.00 G                           O G        386
       .330  .76   .03  .355  .85   .03  .400  .87   .03  .430  .88   .03
       .470  .93   .03  .500  .94   .03  .540  .99   .03  .570 1.02   .03
       .600  .98   .03  .630  .99   .03  .670 1.00   .03  .700 1.01   .03
       .730  .97   .03  .765 1.04   .03  .800 1.00   .03  .830 1.01   .03
       .870 1.00   .03  .900  .94   .03  .930  .91   .03  .970  .99   .03
      1.000  .95   .03 1.030  .94   .05 1.060  .87   .03

389    1.43 G 0.07 G                                          1 G        389
       .330  .69   .03  .355  .71   .03  .400  .76   .03  .430  .84   .03
       .470  .90   .03  .500  .90   .03  .540  .96   .03  .570 1.02   .03
       .600 1.01   .03  .630 1.02   .03  .670 1.07   .03  .700 1.12   .03
       .730 1.16   .03  .765 1.18   .03  .800 1.13   .03  .830 1.12   .03
```

```
NUMBER   R/B   BEND        IR    DEPTH  CENTER BANDW  ELBOW  UV   FEAT          NUMBER
-------------------------------------------------------------------------------------
391    1.59 G 0.04 G                                           0 G              391
       MICRON REFL   RERR MICRON REFL  RERR MICRON REFL   RERR MICRON REFL  RERR
        .330 .70     .03   .355  .53   .03   .400 .75    .03   .430  .80    .03
        .470 .81     .03   .500  .92   .03   .540 .90    .03   .570 1.02    .03
        .600 1.07    .03   .630 1.13   .03   .670 1.18   .03   .700 1.17    .03
        .730 1.22    .03   .765 1.22   .05   .800 1.17   .03

402    1.38 G 0.15 G                                           2 G              402
        .340 .39     .23   .355  .74   .11   .400 .78    .03   .430  .78    .03
        .470 .86     .03   .500  .93   .03   .540 .97    .03   .570 1.00    .03
        .600 1.05    .03   .630 1.03   .03   .670 1.00   .05   .700 1.08    .04
        .730 1.08    .05   .765 1.11   .05   .800 1.05   .15   .950 1.19    .07
       1.000 1.01    .08  1.030  .79   .20  1.060  .97   .20

403    1.44 G 0.17 G                                           0 G              403
        .330 .54     .13   .355  .76   .10   .400 .72    .04   .430  .81    .04
        .470 .81     .03   .500  .93   .03   .540 .91    .07   .570 1.00    .03
        .600 1.03    .04   .630 1.03   .03   .670 1.08   .05   .700 1.08    .03
        .730 1.07    .03   .765  .99   .05   .800  .99   .06   .830 1.00    .03

409    1.26 G 0.04 G -0.08 G 1.00 G                            0 G              409
        .330 .78     .08   .340  .80   .11   .355 .76    .08   .400  .84    .07
        .430 .88     .03   .470  .94   .03   .500 .95    .03   .540  .96    .03
        .570 1.02    .03   .600  .99   .03   .630 1.12   .03   .670 1.07    .03
        .700 1.07    .03   .730 1.10   .03   .765 1.14   .03   .800 1.05    .03
        .830 1.11    .03   .870 1.12   .04   .900 1.05   .05   .930 1.10    .04
        .970 1.14    .04  1.000 1.02   .04  1.030  .98   .05  1.060  .96    .08

413    1.15 G-0.04 G                                           0 G              413
        .330 .84     .10   .355  .90   .11   .400 .99    .03   .430  .97    .03
        .470 .92     .03   .500  .89   .03   .540 1.01   .04   .570 1.00    .03
        .600 1.00    .04   .630 1.01   .03   .670 1.11   .04   .700 1.15    .03
        .730 1.01    .03   .765 1.12   .03   .800 1.10   .03   .830 1.18    .04

415    1.15 Q 0.00 Q  0.01 Q                                   0 Q              415
        .355 1.18    .09   .400  .87   .06   .430 .91    .08   .470  .89    .07
        .500 1.21    .10   .540  .97   .06   .570 .80    .10   .600  .94    .20
        .630 1.09    .06   .670 1.12   .05   .700 1.36   .09   .730 1.02    .14
        .765 1.14    .08   .800 1.05   .26   .830 1.27   .06   .870 1.05    .06
        .900 1.11    .06   .930 1.17   .08   .970 1.02   .08  1.000 1.09    .08
       1.030 1.14    .08  1.060  .97   .08

416    1.55 G 0.07 G                                           0 G              416
        .330 .55     .04   .355  .65   .03   .400 .74    .03   .430  .85    .03
        .470 .83     .03   .500  .90   .03   .540 .95    .03   .570 1.00    .03
        .600 1.04    .03   .630 1.10   .03   .670 1.17   .03   .700 1.18    .03
        .730 1.15    .03   .765 1.17   .03   .800 1.18   .03   .830 1.18    .03

419    1.08 G 0.00 G -0.07 G 1.00 G                            0 G              419
        .330 .94     .06   .355  .96   .05   .400 1.00   .04   .430  .96    .04
        .470 .96     .03   .500  .92   .03   .540 1.04   .05   .570 1.00    .06
        .600 1.03    .05   .630 1.04   .05   .670 1.06   .04   .700 1.09    .06
        .730 1.02    .04   .765 1.00   .03   .800  .97   .03   .830 1.02    .03
        .870 .94     .03   .900 1.00   .03   .900 .94    .04   .970  .97    .03
       1.000 1.00    .06  1.030  .83   .10  1.060  .89   .09

423    1.04 G 0.00 G                           0.40 G 2 G     0 G              423
        .330 .89     .04   .355  .84   .04   .400 .99    .03   .430 1.04    .03
        .470 1.01    .03   .500  .98   .03   .540 1.01   .03   .570 1.00    .03
        .600 1.01    .03   .630 1.00   .03   .570 1.01   .03   .700 1.08    .03
        .730 .97     .03   .765 1.02   .04   .800 .97    .03   .830  .97    .03

426    1.08 G 0.07 G                                           0 G              426
        .330 .94     .03   .355  .90   .03   .400 .91    .03   .430  .94    .03
        .470 .97     .03   .500  .95   .03   .540 1.01   .03   .570 1.02    .03
        .600 .98     .03   .630  .97   .03   .670 .99    .03   .700 1.02    .03
        .730 1.00    .03   .765 1.00   .03   .800 .98    .03   .830  .97    .06

433    1.73 G 0.19 G -0.05 G 0.90 G 0.94 G 0.20 G              0 G              433
        .330 .40     .08   .355  .48   .08   .430 .65    .04   .430  .77    .03
        .470 .86     .03   .500  .91   .03   .540 .99    .03   .570 1.00    .03
        .600 1.05    .03   .630 1.13   .03   .570 1.14   .03   .700 1.19    .03
        .730 1.19    .03   .765 1.21   .03   .800 1.18   .03   .830 1.14    .03
        .870 1.03    .03   .900 1.02   .03   .930 1.09   .03   .970 1.05    .03
       1.000 1.06    .03  1.030 1.09   .03

434    1.34 G 0.08 G               0.86 Q                       0 G             434
        .355 .64     .05   .400  .85   .03   .430 .95    .03   .470  .89    .03
        .500 .89     .05   .540  .95   .03   .570 1.01   .03   .600 1.02    .03
        .630 1.03    .03   .670 1.08   .03   .700 1.12   .08   .730 1.07    .07
        .765 1.15    .05   .800 1.15   .06   .830 1.03   .07   .870 1.04    .05
        .900 .95     .05   .930  .77   .12
```

```
NUMBER   R/B    BEND       IR    DEPTH  CENTER BANDW  ELBOW  UV   FEAT        NUMBER
------------------------------------------------------------------------------------
435    1.22 Q-0.04 Q                                                            435
       MICRON REFL  RERR MICRON REFL  RERR MICRON REFL  RERR MICRON REFL  RERR
       .330  .91   .22  .355  .88   .17  .400  .96   .05  .430  .77   .06
       .470  .98   .03  .500  .97   .05  .540  .85   .09  .570 1.03   .09
       .600 1.05   .03  .630  .76   .06  .670  .93   .10  .730 1.19   .12
       .765 1.27   .09  .800 1.14   .08  .830 1.40   .10

439    1.12 G 0.11 G                                          0 G              439
       .330  .82   .06  .355  .89   .09  .400  .86   .03  .430  .90   .03
       .470  .93   .03  .500  .95   .03  .540  .99   .03  .570 1.02   .03
       .600 1.00   .03  .630  .98   .03  .670 1.00   .03  .700 1.01   .03
       .730  .99   .03  .765 1.03   .03  .800  .97   .03  .830  .97   .11

441    1.34 G-0.03 G                                          0 G              441
       .330  .79   .18  .355  .99   .23  .400  .84   .13  .430  .84   .05
       .470 1.03   .05  .500  .95   .03  .540  .92   .03  .570 1.02   .03
       .600 1.04   .06  .630 1.08   .03  .670 1.08   .07  .700 1.20   .04
       .730 1.14   .03  .765 1.16   .03  .800 1.14   .03  .830 1.19   .03

446    2.76 G 0.24 G -0.07 Q                                  0 G              446
       .400  .47   .03  .430  .55   .03  .470  .73   .03  .500  .76   .03
       .540  .88   .03  .570 1.03   .03  .600 1.05   .03  .630 1.19   .03
       .670 1.23   .03  .700 1.29   .03  .730 1.29   .03  .765 1.28   .03
       .800 1.26   .03  .830 1.29   .03  .870 1.25   .03  .900 1.24   .04
       .930 1.32   .07  .970  .93   .09 1.000 1.12   .09 1.030 1.26   .11

453    1.73 G 0.13 G                                          0 G              453
       .330  .54   .03  .355  .53   .03  .400  .55   .03  .430  .73   .03
       .470  .82   .03  .500  .87   .03  .540  .94   .03  .570 1.00   .03
       .600 1.05   .03  .630 1.10   .03  .670 1.15   .03  .700 1.17   .03
       .730 1.18   .03  .765 1.22   .03  .800 1.14   .03  .830 1.07   .04

462    1.58 G 0.08 G -0.02 Q 1.00 Q                           0 G              462
       .330  .53   .20  .340  .75   .16  .355  .66   .12  .400  .70   .04
       .430  .72   .04  .470  .90   .04  .500  .88   .04  .540  .96   .05
       .570 1.02   .04  .600  .97   .04  .630 1.13   .03  .670 1.12   .05
       .700 1.09   .04  .730 1.19   .05  .765 1.35   .11  .800 1.22   .08
       .830 1.08   .07  .870 1.11   .06  .900 1.16   .04  .930 1.14   .10
       .970 1.20   .07 1.000 1.16   .07 1.030 1.18   .15 1.060 1.14   .12

468    1.06 G 0.07 G        1.00 G          0.40 Q 2 Q   0 G                   468
       .330  .82   .03  .355  .91   .03  .400  .96   .04  .430  .90   .03
       .470  .95   .03  .500 1.07   .05  .540 1.02   .03  .570 1.00   .03
       .600 1.05   .03  .630 1.03   .03  .670 1.00   .03  .730 1.02   .03
       .765 1.01   .03  .800 1.00   .03  .830  .90   .03  .870 1.00   .03
       .900 1.01   .04  .930  .98   .05  .970 1.01   .06 1.000 1.16   .09

471    1.48 G 0.11 G                                          0 G              471
       .330  .53   .03  .355  .53   .03  .400  .76   .03  .430  .81   .03
       .470  .85   .03  .500  .93   .03  .540  .96   .03  .570 1.00   .03
       .600 1.00   .03  .630 1.01   .03  .670 1.09   .03  .700 1.12   .03
       .730 1.09   .03  .765 1.17   .03  .800 1.15   .03  .830 1.09   .03

472    1.71 Q 0.01 Q                                          0 Q              472
       .330  .27   .14  .355  .60   .30  .400  .70   .04  .430  .81   .07
       .470  .97   .05  .500  .92   .04  .540  .97   .08  .570  .95   .11
       .600 1.13   .03  .630 1.07   .04  .670 1.23   .03  .700 1.19   .12
       .730 1.25   .09  .765 1.43   .07  .800 1.30   .03  .830 1.18   .14

481    1.25 G 0.19 G                                          0 G              481
       .330  .69   .10  .340  .50   .12  .355  .75   .09  .400  .82   .03
       .430  .84   .03  .470  .91   .03  .500  .91   .03  .540  .92   .03
       .570 1.03   .03  .600  .99   .05  .630  .99   .04  .670 1.00   .04
       .700 1.02   .04  .730  .99   .04  .765  .94   .05  .800  .97   .04
       .830  .97   .05  .870 1.02   .05  .900  .96   .08  .930 1.30   .12
       .970  .86   .08 1.000 1.00   .08 1.030 1.27   .10 1.060 1.05   .15

488    1.20 G 0.14 G        1.00 Q          0.51 G 1 G   0 G                   488
       .330  .72   .03  .355  .83   .03  .400  .87   .03  .430  .80   .03
       .470  .92   .03  .500  .98   .03  .540 1.00   .03  .570 1.00   .03
       .600 1.00   .03  .630  .97   .05  .670 1.05   .03  .730 1.01   .05
       .765 1.01   .03  .800  .99   .05  .830  .95   .04  .870 1.00   .04
       .900  .97   .05  .930 1.08   .05  .970  .99   .05 1.000  .90   .17

490    1.15 G 0.04 G                        0.45 G 2 G   0 G                   490
       .330  .72   .04  .355  .85   .07  .400  .90   .03  .430  .91   .03
       .470  .97   .03  .500  .95   .03  .540  .99   .03  .570 1.00   .03
       .600  .96   .03  .630 1.02   .03  .670 1.08   .03  .700 1.04   .03
       .730  .98   .03  .765 1.06   .04  .800 1.15   .03  .830 1.05   .03
```

```
NUMBER   R/B    BEND        IR    DEPTH  CENTER BANDW   ELBOW  UV    FEAT        NUMBER
------------------------------------------------------------------------------------------
496    1.75 G 0.26 G         0.89 G 0.90 G 0.17 G               0 G              496
       MICRON REFL  RERR MICRON REFL  RERR MICRON REFL  RERR MICRON REFL  RERR
        .330  .25   .10  .355  .46   .05  .400  .62   .03  .430  .72   .03
        .470  .85   .03  .500  .89   .03  .540  .96   .04  .570 1.00   .03
        .600 1.00   .03  .630 1.06   .03  .670 1.11   .03  .730 1.10   .03
        .765 1.12   .03  .800 1.05   .05  .830  .99   .05  .870 1.00   .04
        .900 1.02   .05  .930  .99   .05  .970 1.03   .06

505    1.12 G 0.03 G -0.04 G 0.88 G 0.95 G 0.20 G               0 G              505
        .330  .95   .10  .340  .78   .13  .400  .97   .04  .430  .91   .03
        .470  .99   .04  .500  .97   .04  .540 1.01   .04  .570 1.00   .04
        .600  .99   .04  .630 1.01   .04  .570 1.01   .04  .700 1.02   .04
        .730 1.04   .04  .765 1.10   .04  .800 1.10   .04  .830 1.05   .04
        .870 1.01   .04  .900  .97   .04  .930  .92   .05  .970  .94   .05
       1.000 1.07   .05 1.030  .96   .08 1.060 1.00   .10

510    1.36 G 0.04 G                                            0 G              510
        .330  .80   .03  .355  .82   .03  .400  .78   .03  .430  .84   .03
        .470  .88   .03  .500  .90   .03  .540  .99   .03  .570 1.00   .05
        .600 1.03   .03  .630 1.03   .03  .670 1.10   .03  .700 1.13   .03
        .730 1.13   .03  .765 1.16   .03  .800 1.14   .03  .830 1.12   .05

511    1.22 G 0.12 G -0.07 G 1.00 G         0.47 G 2 Q          0 G              511
        .330  .75   .20  .340  .65   .15  .355  .69   .10  .400  .86   .04
        .430  .89   .03  .470 1.03   .03  .500  .97   .03  .540  .99   .03
        .570  .99   .03  .600 1.00   .03  .630 1.02   .03  .670 1.04   .03
        .700 1.04   .03  .730 1.00   .03  .765 1.04   .03  .800  .97   .03
        .830 1.02   .05  .870 1.00   .03  .900  .97   .04  .930  .92   .05
        .950  .95   .04  .970  .99   .05 1.000  .90   .05 1.030  .98   .05

513    1.48 Q 0.07 Q                                            0 Q              513
        .355  .62   .03  .400  .75   .11  .430  .88   .03  .470  .81   .08
        .500  .96   .05  .540  .96   .12  .570 1.00   .10  .600 1.05   .06
        .630 1.05   .09  .670 1.12   .03  .700 1.21   .07  .730 1.14   .03
        .765 1.04   .05  .800 1.22   .10  .830 1.13   .18

526    1.13 G-0.03 G                       0.42 Q 3 Q          0 G              526
        .355  .50   .36  .400  .96   .15  .430 1.07   .06  .470  .98   .03
        .500 1.02   .04  .540 1.00   .07  .570  .97   .06  .600 1.02   .05
        .630 1.02   .03  .670 1.04   .04  .700 1.05   .11  .730 1.12   .03
        .765 1.04   .08  .800  .92   .04  .830  .84   .14

532    1.56 G 0.14 G -0.01 G 0.91 G 0.92 G 0.15 G               0 G              532
        .330  .51   .04  .340  .57   .03  .395  .63   .03  .400  .73   .03
        .430  .75   .03  .470  .86   .03  .500  .88   .03  .540  .96   .03
        .570 1.00   .03  .600 1.03   .03  .630 1.09   .03  .670 1.08   .03
        .700 1.13   .03  .730 1.13   .03  .765 1.14   .03  .800 1.14   .03
        .900 1.03   .03  .950 1.06   .03 1.000 1.08   .03 1.030 1.12   .03
       1.060 1.09   .05

554    1.15 G 0.13 G -0.01 G 1.00 G         0.50 G 1 G          0 G              554
        .330  .77   .05  .340  .79   .08  .355  .80   .04  .400  .90   .03
        .430  .88   .03  .470  .95   .03  .500  .98   .03  .540 1.00   .03
        .570 1.00   .03  .600 1.00   .03  .630 1.01   .03  .670  .98   .03
        .700 1.00   .03  .730  .99   .03  .765 1.03   .03  .800 1.03   .03
        .830 1.03   .03  .870 1.00   .03  .900 1.02   .03  .930 1.07   .03
        .970  .97   .03 1.000  .98   .10 1.030  .98   .10 1.060 1.02   .10

558    1.17 Q 0.29 Q                                            3 Q              558
        .355  .80   .35  .400  .76   .11  .430  .87   .10  .470  .84   .09
        .500  .80   .03  .540  .97   .06  .570 1.03   .04  .600  .96   .04
        .630  .95   .03  .670  .88   .04  .700  .90   .05  .730  .98   .03
        .765  .78   .08  .800  .75   .05  .830  .78   .06  .870  .68   .08
        .900  .77   .12  .930  .54   .08

560    1.08 Q 0.01 Q                                            3 Q              560
        .355  .74   .04  .400 1.04   .03  .430  .90   .05  .470  .91   .05
        .500  .93   .09  .540 1.10   .12  .570  .95   .12  .600 1.02   .05
        .630  .95   .12  .670  .95   .12  .700  .99   .13  .730 1.16   .09
        .765  .85   .18

562    1.37 Q 0.03 Q                                            0 Q              562
        .330  .10   .21  .355  .77   .17  .400  .90   .06  .430  .93   .03
        .470  .87   .05  .500  .97   .03  .540 1.03   .07  .570 1.03   .06
        .600  .91   .04  .630 1.01   .04  .570 1.12   .05  .700  .96   .07
        .730 1.30   .10  .765 1.15   .09  .800 1.26   .11  .830 1.07   .16
        .870  .93   .09  .900 1.53   .12  .930  .88   .08  .970  .88   .13

563    1.58 Q 0.04 Q 0.12 Q 0.88 Q 0.91 Q 0.11 Q               0 Q              563
        .330  .71   .08  .340  .58   .08  .355  .69   .08  .400  .62   .07
        .430 1.01   .08  .470  .94   .08  .500 1.02   .08  .540  .90   .08
        .570 1.00   .08  .600 1.11   .08  .630 1.10   .08  .670 1.29   .08
        .700 1.20   .08  .730 1.18   .08  .765 1.24   .08  .800 1.26   .08
        .870 1.07   .08  .900 1.04   .08  .950 1.18   .08 1.000 1.32   .08
       1.060 1.27   .08
```

```
NUMBER   R/B    BEND        IR   DEPTH  CENTER BANDW  ELBOW  UV   FEAT           NUMBER
-------------------------------------------------------------------------------------
574     1.68 G 0.03 G                                               0 G            574
        MICRON REFL  RERR MICRON REFL  RERR MICRON REFL  RERR MICRON REFL  RERR
        .330  .50   .11  .355  .57   .06  .400  .73   .03  .430  .79   .03
        .470  .84   .03  .500  .86   .03  .540  .95   .03  .570 1.00   .04
        .600 1.00   .03  .630 1.05   .03  .670 1.13   .03  .700 1.17   .04
        .730 1.23   .06  .765 1.34   .08  .800 1.29   .06  .830 1.18   .03

579     1.40 G 0.15 G                                               0 G            579
        .330  .60   .05  .355  .67   .04  .400  .77   .03  .430  .80   .03
        .470  .90   .03  .500  .91   .03  .540  .95   .03  .570 1.00   .03
        .600 1.01   .03  .630 1.00   .03  .670 1.09   .03  .700 1.02   .03
        .730 1.10   .03  .765 1.03   .03  .800 1.12   .03  .830 1.10   .03

582     1.64 G 0.20 G                                               0 G            582
        .330  .71   .03  .355  .62   .09  .400  .66   .03  .430  .72   .03
        .470  .76   .03  .500  .80   .03  .540  .98   .03  .570 1.00   .03
        .600 1.05   .03  .630 1.15   .05  .670 1.09   .03  .700 1.15   .05
        .730 1.10   .03  .765 1.13   .06  .800 1.18   .03

584     1.73 G 0.14 G       0.90 Q                                  0 G            584
        .340  .83   .20  .355  .69   .12  .400  .60   .03  .430  .70   .03
        .470  .80   .03  .500  .86   .03  .540  .90   .03  .570 1.00   .03
        .600 1.05   .03  .630 1.09   .03  .670 1.11   .03  .700 1.16   .03
        .730 1.18   .03  .765 1.14   .05  .800 1.34   .20  .900 1.11   .15
        .950 1.03   .15 1.000 1.12   .20

588     1.39 G 0.01 G       1.00 Q                                  0 G            588
        .330  .79   .09  .355  .81   .14  .400  .84   .03  .430  .83   .03
        .470  .92   .05  .500  .98   .04  .540  .97   .06  .570 1.00   .05
        .600 1.04   .05  .630 1.11   .06  .670 1.07   .04  .700 1.19   .04
        .730 1.17   .03  .765 1.07   .04  .800 1.24   .05  .830 1.30   .03
        .870 1.52   .10  .900 1.50   .06  .930 1.40   .13  .970 1.64   .13

599     1.45 G 0.06 G                                               0 G            599
        .330  .57   .03  .355  .67   .05  .400  .80   .03  .430  .84   .03
        .470  .88   .03  .500  .97   .04  .540  .96   .05  .570 1.00   .05
        .600 1.00   .03  .630 1.08   .03  .670 1.11   .05  .700 1.12   .04
        .730 1.15   .03  .765 1.17   .03  .800 1.20   .03  .830 1.31   .03

613     1.20 G 0.02 G                             0.41 G 3 Q        0 G            613
        .330  .81   .13  .355  .60   .08  .400  .88   .03  .430  .95   .03
        .470  .93   .03  .500  .94   .03  .540 1.00   .03  .570 1.00   .03
        .600  .97   .03  .630 1.12   .04  .670 1.03   .04  .700 1.11   .04
        .730 1.02   .03  .765 1.13   .06  .800 1.12   .08  .830 1.19   .06

617     1.22 G-0.04 G                                               0 G            617
        .330  .96   .05  .355  .92   .03  .400  .92   .03  .430  .92   .03
        .470  .99   .03  .500  .98   .03  .540 1.02   .03  .570  .97   .03
        .600 1.04   .04  .630 1.06   .03  .670 1.11   .04  .730 1.12   .04
        .765 1.14   .03  .800 1.14   .03  .830  .99   .05  .870 1.21   .06
        .900 1.03   .05  .930  .99   .05  .970 1.11   .09

624     1.51 G 0.00 G  0.32 Q 1.00 Q                                0 G            624
        .330  .47   .25  .340  .50   .20  .355  .66   .12  .400  .83   .05
        .430  .80   .04  .470  .91   .03  .500  .96   .03  .540  .95   .03
        .570 1.03   .04  .600  .99   .03  .630 1.12   .04  .670 1.17   .03
        .700 1.21   .06  .730 1.22   .04  .765 1.19   .05  .800 1.27   .06
        .830 1.34   .07  .870 1.56   .05  .900 1.20   .11  .930 1.44   .09
        .970 1.37   .17 1.000 1.63   .18 1.030 1.40   .25

628     1.64 G 0.04 G       1.00 G                                  0 G            628
        .330  .52   .03  .355  .67   .04  .400  .71   .03  .430  .79   .03
        .470  .87   .03  .500  .96   .03  .540 1.00   .03  .570  .98   .03
        .600 1.02   .03  .630 1.10   .03  .670 1.12   .03  .700 1.24   .03
        .730 1.19   .04  .765 1.23   .05  .800 1.34   .03  .830 1.24   .03
        .870 1.41   .06  .900 1.35   .06  .930 1.30   .04  .970 1.39   .03

639     1.46 G 0.06 G                                               0 G            639
        .330  .70   .03  .355  .78   .03  .400  .75   .03  .430  .81   .03
        .470  .84   .03  .500  .90   .03  .540  .95   .03  .570 1.03   .03
        .600  .99   .03  .630 1.03   .03  .670 1.10   .03  .700 1.14   .03
        .730 1.17   .07  .765 1.13   .03  .800 1.16   .05

648     1.18 Q 0.05 Q                                               0 Q            648
        .330  .58   .14  .355 1.10   .14  .400  .85   .05  .430  .98   .03
        .470  .96   .05  .500  .81   .03  .540  .96   .03  .570 1.00   .03
        .600 1.00   .05  .630 1.01   .03  .670 1.00   .03  .700 1.12   .05
        .730  .95   .06  .765 1.07   .07  .800  .92   .09  .830  .99   .03

654     1.24 G 0.22 G -0.01 G 1.00 G             0.51 G 2 G        0 G            654
        .340  .45   .14  .355  .77   .05  .400  .79   .03  .430  .81   .03
        .470  .94   .03  .500  .91   .03  .540  .99   .03  .570 1.00   .03
        .600  .99   .03  .630 1.01   .03  .670  .95   .03  .700  .96   .03
        .730  .98   .04  .765  .99   .05  .800  .97   .03  .830  .94   .03
        .870  .95   .05  .900  .95   .05  .950 1.05   .05 1.000  .93   .06
       1.060  .94   .10 1.100 1.10   .22
```

```
NUMBER   R/B   BEND       IR   DEPTH  CENTER BANDW  ELBOW  UV    FEAT        NUMBER
-----------------------------------------------------------------------------------
 660    1.57 Q 0.10 Q                                              0 Q
        MICRON REFL  RERR MICRON REFL  RERR MICRON REFL  RERR MICRON REFL  RERR
        .355  .78   .08  .400  .72   .05  .430  .75   .04  .470  .72   .10
        .500  .98   .06  .540  .71   .03  .570 1.13   .13  .600 1.03   .17
        .630 1.00   .06  .670 1.19   .08  .700 1.22   .09  .730 1.09   .14
        .765 1.32   .11  .800 1.17   .11

 674    1.61 G 0.13 G   0.07 G 0.95 G 0.92 G 0.15 G                0 G        674
        .330  .57   .03  .340  .61   .04  .355  .63   .03  .400  .69   .03
        .430  .74   .03  .470  .86   .03  .500  .87   .03  .540  .97   .03
        .570 1.00   .03  .600 1.04   .03  .630 1.09   .03  .670 1.09   .03
        .700 1.15   .03  .730 1.17   .03  .765 1.17   .04  .800 1.20   .03
        .830 1.14   .05  .870 1.12   .03  .900 1.13   .03  .930 1.13   .03
        .970 1.19   .03 1.000 1.11   .03 1.030 1.24   .04 1.060 1.27   .05

 676    1.24 G 0.13 G                                              0 G        676
        .330  .74   .03  .355  .91   .03  .400  .80   .03  .430  .84   .03
        .470  .90   .03  .500  .97   .06  .540  .98   .03  .570 1.02   .03
        .600  .98   .03  .630  .98   .03  .670 1.04   .03  .700 1.04   .03
        .730  .99   .03  .765 1.10   .03  .800 1.04   .03  .830 1.20   .12

 695    1.55 G 0.09 G        0.82 Q 0.97 Q 0.15 Q               0 G          695
        .330  .49   .08  .355  .54   .05  .400  .72   .03  .430  .82   .03
        .470  .86   .03  .500  .93   .03  .540  .93   .03  .570 1.00   .03
        .600 1.06   .03  .630 1.06   .03  .670 1.15   .03  .700 1.16   .03
        .730 1.16   .03  .765 1.17   .03  .800 1.21   .08  .830 1.12   .03
        .870 1.22   .19  .900 1.33   .28  .930  .91   .09  .970  .94   .06
       1.000  .86   .08 1.030 1.19   .14 1.060 1.02   .29

 696    1.19 G 0.04 G       1.00 G            0.47 G 1 G     0 G             696
        .330  .79   .03  .355  .88   .03  .400  .88   .03  .430  .91   .03
        .470 1.01   .03  .500  .98   .04  .540 1.01   .03  .570 1.00   .03
        .600 1.00   .03  .630 1.04   .03  .670 1.08   .03  .730 1.06   .03
        .765 1.12   .04  .800 1.02   .03  .830 1.01   .03  .870 1.05   .03
        .900 1.08   .05  .930 1.11   .03  .970  .95   .03 1.000  .88   .14

 704    1.08 G 0.05 G  -0.07 G 0.89 G 0.95 Q 0.22 Q               2 G        704
        .330  .88   .07  .340  .88   .06  .355  .95   .06  .400  .94   .05
        .430  .92   .05  .470  .94   .03  .500  .93   .03  .540 1.02   .03
        .570 1.00   .03  .600  .98   .03  .630  .96   .03  .670  .96   .05
        .700 1.03   .03  .730 1.02   .03  .765  .97   .04  .800  .97   .08
        .870  .92   .04  .900  .86   .04  .950  .91   .04 1.000  .92   .03
       1.060  .94   .03

 712    1.14 G 0.08 G  -0.01 G 1.00 G            0.42 G 2 G     0 G          712
        .330  .80   .03  .355  .78   .03  .400  .93   .03  .430  .91   .03
        .470  .97   .03  .500  .95   .03  .540 1.03   .04  .570 1.00   .03
        .600 1.01   .03  .630  .97   .03  .670 1.04   .03  .700 1.05   .03
        .730 1.01   .03  .765 1.03   .03  .800 1.02   .03  .830 1.05   .03
        .870 1.05   .03  .900 1.06   .05  .930 1.03   .05  .970 1.04   .04
       1.000 1.06   .03 1.030  .99   .06 1.060  .97   .07

 714    1.62 G 0.13 G  -0.02 Q 0.95 Q 0.94 Q 0.18 Q               0 G        714
        .330  .55   .10  .340  .64   .20  .355  .72   .10  .400  .70   .03
        .430  .65   .04  .470  .88   .03  .500  .91   .03  .540  .91   .03
        .570 1.02   .03  .600 1.05   .03  .630 1.12   .03  .670 1.11   .03
        .700 1.14   .03  .730 1.20   .04  .765 1.08   .04  .800 1.12   .04
        .830 1.12   .05  .870 1.18   .05  .900 1.03   .05  .930 1.16   .07
        .970 1.01   .06 1.000 1.24   .06 1.030 1.06   .07 1.060 1.18   .11

 739    1.10 G 0.02 G       1.00 Q                                 0 G        739
        .340  .88   .12  .355  .93   .07  .400  .96   .03  .430  .87   .03
        .470 1.02   .03  .500 1.01   .03  .540 1.00   .03  .570  .98   .03
        .600 1.04   .03  .630 1.02   .03  .570 1.04   .04  .700 1.00   .03
        .730 1.07   .03  .765 1.03   .07  .800 1.12   .09  .870  .94   .07
        .900 1.11   .12  .950 1.11   .11 1.000 1.15   .23

 741    1.15 G 0.02 G                                              0 G        741
        .355  .93   .25  .400  .92   .05  .430  .88   .03  .470  .93   .03
        .500  .96   .03  .540 1.02   .03  .570  .98   .04  .600 1.01   .03
        .630 1.02   .07  .670 1.03   .03  .700 1.05   .03  .730 1.03   .05
        .765 1.15   .06  .800 1.14   .07  .830 1.04   .07

 747    1.04 Q 0.10 Q   0.08 Q                   0.40 Q 3 Q     0 Q          747
        .355  .72   .07  .400  .95   .07  .430 1.03   .05  .470 1.01   .07
        .500  .94   .07  .540  .85   .07  .570 1.08   .07  .600  .99   .06
        .630 1.06   .07  .670  .93   .09  .700  .92   .07  .730  .87   .07
        .765 1.14   .18  .800  .95   .17  .830  .99   .07  .870  .89   .07
        .900  .96   .08  .930  .94   .11  .970  .99   .07 1.000 1.24   .09
       1.030 1.20   .07 1.060 1.10   .08
```

```
NUMBER   R/B   BEND      IR  DEPTH  CENTER BANDW  ELBOW   UV   FEAT        NUMBER
-----------------------------------------------------------------------------------
 750    0.92 Q-0.02 Q                                        0 Q           750
        MICRON REFL  RERR MICRON REFL  RERR MICRON REFL  RERR MICRON REFL  RERR
        .355 1.09  .18  .400 1.10  .10  .430 1.10  .09  .470  .98  .05
        .500 1.00  .03  .540  .87  .05  .570 1.08  .08  .600  .96  .03
        .630 1.08  .05  .670  .97  .03  .700  .92  .05  .730  .89  .06
        .765 1.01  .06  .800 1.03  .10  .830  .97  .12

 758    1.26 G 0.02 G          0.92 G 0.95 Q                 0 G           758
        .330  .78  .03  .355  .83  .03  .400  .86  .03  .430  .92  .03
        .470  .95  .03  .500  .97  .03  .540  .99  .03  .570 1.00  .03
        .600 1.02  .03  .630 1.04  .03  .670 1.08  .03  .730 1.12  .03
        .765 1.18  .03  .800 1.17  .04  .830 1.08  .03  .870 1.15  .03
        .900 1.08  .03  .930 1.05  .03  .970 1.09  .03 1.000 1.06  .08

 760    1.75 G 0.08 G          0.90 G                        0 G           760
        .330  .48  .03  .355  .58  .03  .400  .72  .03  .430  .73  .03
        .470  .86  .03  .500  .91  .03  .540  .99  .03  .570  .98  .03
        .600 1.05  .03  .630 1.07  .03  .670 1.19  .03  .700 1.17  .03
        .730 1.21  .03  .765 1.27  .03  .800 1.20  .03  .830 1.14  .03
        .870 1.23  .03  .900 1.11  .03  .930 1.12  .03  .970 1.13  .04

 770    1.89 G 0.22 G          0.82 Q                        0 Q           770
        .330  .22  .15  .355  .37  .23  .400  .60  .05  .430  .64  .06
        .470  .80  .06  .500  .88  .03  .540  .97  .06  .570 1.05  .05
        .600  .89  .04  .630 1.08  .03  .670 1.11  .05  .700 1.00  .06
        .730 1.27  .06  .765 1.18  .03  .800 1.11  .09  .830 1.12  .14
        .870 1.30  .12  .900  .98  .10  .930  .77  .13  .970 1.00  .09

 772    1.26 Q 0.15 Q                                        0 Q           772
        .380  .57  .07  .400  .87  .16  .430  .88  .06  .470  .87  .11
        .500 1.17  .06  .540 1.01  .16  .570 1.00  .05  .600  .94  .08
        .630 1.19  .09  .670 1.09  .04  .700 1.07  .06  .730  .85  .09
        .765 1.05  .13  .800  .74  .05  .830  .57  .11

 773    1.33 G-0.08 G                                        1 G           773
        .330  .81  .03  .355  .81  .03  .400  .88  .03  .430  .88  .03
        .470  .93  .03  .500  .99  .03  .540  .96  .03  .570 1.00  .03
        .600 1.01  .04  .630 1.04  .05  .670 1.06  .03  .700 1.18  .03
        .730 1.18  .03  .765 1.26  .03  .800 1.22  .03  .830 1.39  .09

 781    1.26 Q 0.08 Q                         0.42 Q 3 Q     0 Q           781
        .330  .55  .24  .355  .38  .10  .400  .85  .05  .430 1.01  .03
        .470 1.00  .07  .500  .98  .04  .540  .96  .09  .570 1.00  .08
        .600 1.08  .06  .630 1.03  .08  .670 1.16  .04  .700 1.06  .06
        .730 1.04  .06  .765 1.19  .12  .800 1.28  .15  .830 1.16  .14
        .900  .87  .27  .930  .81  .30  .970 1.08  .35

 782    1.85 Q 0.21 Q                                        0 Q           782
        .330  .59  .46  .355  .32  .17  .400  .69  .05  .430  .71  .06
        .470  .79  .05  .500  .84  .07  .540  .94  .04  .570 1.00  .08
        .600 1.05  .04  .630 1.03  .05  .670 1.25  .05  .700 1.17  .08
        .730 1.10  .10  .765 1.04  .17  .800 1.32  .17  .830 1.00  .13

 783    1.27 Q 0.16 Q                                        0 Q           783
        .330  .55  .41  .355  .49  .24  .400  .88  .07  .430  .85  .08
        .470  .89  .09  .500  .99  .04  .540  .99  .10  .570  .87  .04
        .600 1.07  .07  .630 1.07  .03  .670 1.12  .05  .700  .88  .06
        .730 1.02  .06  .765 1.14  .17  .800  .85  .17  .830 1.01  .16
        .870 1.11  .26  .900  .85  .25  .930 1.04  .28  .970  .99  .34

 785    1.02 G-0.02 G          1.00 G                        0 G           785
        .330 1.00  .07  .355 1.00  .06  .400  .98  .03  .430 1.02  .03
        .470  .97  .03  .500  .99  .03  .540  .99  .03  .570 1.00  .03
        .600 1.01  .03  .630 1.01  .03  .670 1.01  .03  .700 1.02  .03
        .730 1.01  .04  .765 1.03  .04  .800  .98  .04  .830  .98  .03
        .870 1.00  .03  .900  .99  .03  .930 1.06  .05

 790    1.15 G 0.07 G                                        0 G           790
        .330  .83  .04  .355  .84  .06  .400  .91  .03  .430  .92  .03
        .470  .91  .03  .500  .95  .03  .540  .98  .03  .570 1.03  .03
        .600  .97  .03  .630 1.02  .03  .670 1.03  .03  .700 1.02  .03
        .730 1.04  .03  .765 1.08  .03  .800 1.02  .03  .830 1.02  .03

 801    1.10 Q 0.02 Q                                        0 Q           801
        .330 1.03  .14  .355  .88  .10  .400  .92  .03  .430  .98  .08
        .470  .93  .07  .500  .93  .06  .540 1.06  .06  .570  .87  .08
        .600 1.05  .08  .630 1.02  .03  .670 1.15  .07  .700 1.01  .07
        .730  .92  .07  .765  .96  .18  .800 1.12  .15

 811    1.51 Q 0.23 Q                         0.47 Q 3 Q     1 Q           811
        .330  .25  .17  .355  .57  .25  .400  .69  .03  .430  .82  .03
        .470 1.01  .03  .500  .91  .03  .540 1.05  .04  .570  .97  .07
        .600 1.00  .05  .630  .89  .07  .670 1.07  .04  .700 1.04  .11
        .730 1.09  .14  .765 1.49  .07  .800 1.04  .13  .830  .99  .12
        .870  .93  .25  .900 1.29  .29  .930 1.29  .33  .970 1.24  .29
```

```
NUMBER   R/B   BEND       IR   DEPTH  CENTER  BANDW   ELBOW   UV   FEAT        NUMBER
-------------------------------------------------------------------------------------
839      1.66 G 0.09 G                                              0 G          839
    MICRON REFL  RERR MICRON REFL  RERR MICRON REFL  RERR MICRON REFL  RERR
        .330 .51  .07   .355  .55  .03   .400  .72  .03   .430  .76   .03
        .470 .85  .03   .500  .89  .03   .540  .93  .03   .570 1.00   .05
        .600 1.02 .03   .630 1.05  .03   .670 1.17  .03   .700 1.18   .05
        .730 1.15 .06   .765 1.18  .04   .800 1.16  .07   .830 1.12   .12

846      1.14 G 0.07 G            1.00 Q           0.52 Q 1 Q  0 G          846
        .330  .79 .03   .355  .98  .03   .400  .88  .03   .430  .99   .03
        .470  .97 .03   .500 1.00  .03   .540 1.03  .03   .570 1.00   .03
        .600  .99 .03   .630 1.00  .03   .670 1.07  .03   .730 1.04   .05
        .765 1.04 .05   .800 1.04  .03   .830  .90  .03   .870  .95   .03
        .900 1.04 .05   .930  .99  .05   .970 1.07  .08  1.000 1.19   .16

858      1.73 G 0.10 G                                              0 Q          858
        .330  .27 .15   .355  .65  .19   .400  .63  .03   .430  .78   .03
        .470  .82 .03   .500  .87  .03   .540 1.04  .05   .570 1.00   .06
        .600 1.01 .04   .630 1.08  .03   .670 1.10  .04   .700 1.29   .06
        .730 1.15 .08   .765 1.31  .06   .800 1.20  .09   .830 1.26   .12

884      1.18 G-0.04 G                                              3 Q          884
        .355  .89 .10   .400  .91  .03   .430  .94  .04   .470  .93   .03
        .500  .93 .03   .540  .95  .03   .570 1.00  .03   .600 1.06   .03
        .630  .99 .04   .670 1.03  .03   .700 1.09  .08   .730 1.24   .04
        .765 1.04 .03   .800 1.19  .06   .830 1.16  .26

887      1.69 G 0.17 G -0.05 Q 0.88 Q 0.93 Q 0.19 Q              0 G          887
        .330  .37 .09   .340  .43  .10   .355  .61  .10   .400  .70   .03
        .430  .75 .04   .470  .87  .03   .500  .87  .03   .540  .95   .03
        .570 1.00 .03   .600 1.04  .03   .630 1.11  .03   .670 1.13   .03
        .700 1.10 .04   .730 1.14  .03   .755 1.22  .03   .800 1.05   .05
        .830 1.10 .04   .870  .99  .06   .900 1.05  .07   .930  .99   .11
        .950 1.02 .07   .970 1.23  .14  1.000 1.09  .06  1.060 1.01   .06

895      1.09 G 0.00 G                                              0 G          895
        .355  .84 .12   .400 1.00  .05   .430  .96  .04   .470  .95   .05
        .500 1.03 .04   .540 1.02  .05   .570 1.00  .03   .600 1.02   .04
        .630 1.01 .04   .670 1.11  .05   .700 1.04  .03   .730 1.03   .04
        .765 1.13 .04   .800 1.00  .07   .830 1.13  .08   .870 1.02   .09
        .900 1.02 .07   .930  .84  .10

909      1.20 G 0.12 G            1.00 Q           0.54 G 1 G  0 G          909
        .330  .80 .05   .355  .98  .10   .400  .82  .03   .430  .87   .04
        .470  .87 .03   .500  .94  .05   .540  .99  .04   .570 1.00   .04
        .600  .99 .03   .630 1.00  .07   .670 1.02  .03   .730  .96   .04
        .765 1.05 .03   .800 1.03  .05   .830 1.01  .03   .870 1.03   .04
        .900 1.14 .03   .930 1.11  .05   .970 1.09  .09  1.000 1.21   .16

911      1.47 Q 0.06 Q                              0.49 Q 2 Q  0 G          911
        .355  .49 .30   .400  .80  .06   .430  .80  .06   .470  .95   .06
        .500 1.05 .07   .540  .97  .06   .570  .95  .06   .600 1.10   .05
        .630 1.10 .06   .670 1.14  .05   .700 1.13  .06   .730 1.14   .06
        .800 1.23 .07   .830 1.33  .10   .870 1.45  .10   .900 1.50   .15
        .970 1.20 .07

925      1.66 G 0.08 G            0.90 G                            0 G          925
        .330  .55 .04   .355  .60  .03   .400  .71  .03   .430  .79   .03
        .470  .89 .03   .500  .93  .05   .540  .99  .03   .570  .97   .03
        .600 1.07 .03   .630 1.11  .03   .670 1.17  .03   .700 1.18   .03
        .730 1.20 .03   .765 1.25  .04   .800 1.25  .03   .830 1.11   .03
        .870 1.13 .03   .900 1.23  .03   .930 1.10  .03   .970 1.09   .06

944      1.30 G-0.08 G 0.30 Q 1.00 Q                               0 G          944
        .330  .85 .13   .355  .84  .13   .400  .87  .05   .430  .90   .04
        .470  .93 .03   .500  .91  .03   .540  .98  .04   .570  .99   .03
        .600 1.02 .04   .630 1.06  .04   .670 1.12  .05   .700 1.16   .05
        .730 1.15 .05   .765 1.25  .05   .800 1.23  .07   .830 1.25   .08
        .870 1.16 .17   .900 1.41  .20   .930 1.21  .30   .970 1.63   .33
       1.000 1.60 .20  1.030 1.25  .27

959      1.31 Q 0.14 Q                                                           969
        .355  .68 .17   .400  .80  .10   .430  .71  .22   .470  .97   .08
        .500  .94 .08   .540  .86  .08   .570 1.28  .24   .600 1.07   .11
        .630 1.00 .08   .670  .96  .15   .730  .83  .08   .730 1.29   .09
        .765  .69 .14   .800 1.04  .19   .830 1.30  .27

976      1.19 G 0.15 G                              0.55 Q 1 Q  0 G          976
        .330  .70 .29   .355  .89  .09   .400  .88  .03   .430  .83   .05
        .470  .96 .03   .500  .93  .04   .540 1.04  .05   .570 1.02   .04
        .600  .92 .06   .630  .95  .04   .670  .99  .03   .700 1.00   .03
        .730 1.02 .07   .765  .93  .09   .800 1.10  .09   .830 1.14   .07
```

NUMBER	R/B	BEND	IR	DEPTH	CENTER BANDW	ELBOW	UV	FEAT	NUMBER

1015 1.11 G 0.07 G   0.39 G 3 G   0 G   1015

MICRON	REFL	RERR	MICRON	REFL	RERR	MICRON	REFL	RERR	MICRON	REFL	RERR
.330	.52	.14	.355	.77	.09	.400	.94	.04	.430	.97	.03
.470	.96	.03	.500	.98	.03	.540	.98	.05	.570	1.02	.03
.600	.98	.03	.630	1.01	.05	.670	1.05	.04	.700	1.04	.03
.730	1.00	.05	.765	.96	.03	.800	.96	.03	.830	1.06	.04

1019 1.77 Q 0.09 Q   0 Q   1019

.330	.46	.09	.355	.68	.07	.400	.63	.03	.430	.71	.04
.470	.80	.04	.500	.77	.05	.540	.92	.05	.570	1.00	.08
.600	1.03	.06	.630	1.07	.07	.670	1.21	.11	.700	1.14	.07
.730	1.25	.08	.765	1.23	.11	.800	1.17	.12	.830	1.08	.26

1025 1.27 G 0.16 G   0 G   1025

.330	.52	.10	.355	.73	.17	.400	.96	.06	.430	.86	.05
.470	.84	.03	.500	.93	.03	.540	.95	.05	.570	1.10	.05
.600	.92	.04	.630	1.01	.06	.670	1.10	.03	.700	1.02	.07
.730	1.05	.03	.755	1.05	.04	.800	1.09	.11	.830	.89	.07
.870	1.11	.08	.900	.89	.11	.930	.65	.12	.970	1.06	.14

1036 1.63 G 0.16 G   0.83 Q   0 G   1036

.330	.29	.45	.355	.55	.09	.400	.74	.03	.430	.77	.03
.470	.85	.03	.500	.90	.03	.540	.98	.03	.570	1.00	.03
.600	1.05	.03	.630	1.11	.03	.670	1.09	.03	.700	1.16	.03
.730	1.15	.04	.765	1.16	.04	.800	1.15	.05	.830	1.06	.06
.870	1.05	.08	.900	1.05	.07	.930	.94	.12			

1055 1.74 G 0.15 G   0.86 G   0 G   1055

.330	.47	.03	.355	.65	.03	.400	.65	.03	.430	.74	.03
.470	.79	.03	.500	.84	.03	.540	.99	.03	.570	1.04	.03
.600	.99	.03	.630	1.10	.04	.670	1.11	.04	.730	1.17	.04
.765	1.20	.04	.800	1.13	.04	.830	1.06	.03	.870	1.04	.03
.900	1.05	.08	.930	1.06	.05	.970	.96	.03	1.000	.88	.13

1058 1.57 G 0.07 G   1 G   1058

.330	.56	.22	.355	.67	.03	.400	.71	.05	.430	.77	.03
.470	.88	.03	.500	.88	.03	.540	.95	.03	.570	1.00	.05
.600	1.02	.03	.630	1.10	.03	.670	1.07	.03	.700	1.13	.03
.730	1.23	.03	.765	1.18	.04	.800	1.11	.03	.830	1.12	.07

1075 1.38 Q 0.08 Q   0 Q   1075

.355	.48	.34	.400	.79	.07	.430	.89	.03	.470	.97	.03
.500	.98	.03	.540	.92	.04	.570	1.00	.08	.600	1.03	.04
.630	1.12	.06	.670	1.07	.06	.700	1.03	.08	.730	1.17	.11
.765	1.07	.08	.800	1.14	.17	.830	1.02	.13			

1088 1.81 Q 0.24 Q   0 Q   1088

.330	.54	.16	.355	.41	.11	.400	.63	.03	.430	.67	.04
.470	.87	.05	.500	.81	.03	.540	.87	.04	.570	1.10	.08
.600	.99	.05	.630	.96	.06	.670	1.11	.04	.700	1.03	.05
.730	1.13	.06	.765	1.11	.10	.800	1.09	.08	.830	.78	.11
.870	.94	.15	.900	.57	.08	.930	1.03	.16	.970	.91	.20

1103 1.26 G 0.03 G   1.00 Q   0 G   1103

.330	.54	.43	.355	1.05	.27	.400	.85	.06	.430	.83	.05
.470	.96	.07	.500	.90	.05	.540	.97	.03	.570	1.00	.06
.600	.99	.03	.630	.95	.03	.670	1.09	.06	.700	1.08	.05
.730	1.08	.10	.765	1.09	.09	.800	1.03	.05	.830	1.00	.11
.870	.98	.13	.900	1.14	.11	.930	1.36	.10	.970	1.32	.16

1162 1.27 G 0.10 G   1.00 Q   0.56 G 3 Q   0 G   1162

.330	.64	.10	.355	.62	.12	.400	.90	.04	.430	.80	.05
.470	.95	.03	.500	.95	.03	.540	.99	.03	.570	1.05	.03
.600	.94	.03	.630	.98	.04	.670	1.07	.04	.700	1.00	.06
.730	1.10	.05	.765	1.13	.10	.800	1.10	.08	.830	1.00	.09
.870	1.06	.07	.900	1.02	.10	.930	1.16	.20	.970	1.15	.09

1172 1.34 G-0.05 G   1 G   1172

.330	1.02	.08	.355	.89	.06	.400	.81	.03	.430	.85	.03
.470	.86	.04	.500	.97	.03	.540	.99	.04	.570	.98	.03
.600	1.05	.06	.630	1.10	.04	.670	1.06	.03	.700	1.15	.06
.730	1.20	.03	.765	1.19	.05	.800	1.06	.10	.830	1.07	.39

1173 1.14 G-0.03 G   1 G   1173

.330	.99	.10	.355	1.07	.12	.400	.88	.05	.430	.85	.03
.470	.96	.03	.500	.95	.03	.540	1.02	.03	.570	1.00	.05
.600	.93	.03	.630	1.01	.03	.670	.98	.05	.700	1.08	.07
.730	1.11	.06	.765	.95	.05	.800	1.00	.08			

1199 1.35 G 0.00 G   0 G   1199

.330	.86	.45	.355	.93	.20	.400	.72	.04	.430	.92	.04
.470	.92	.03	.500	.93	.03	.540	.96	.05	.570	1.00	.04
.600	1.01	.03	.630	1.10	.05	.670	1.11	.03	.700	1.12	.06
.730	1.13	.05	.765	1.29	.04	.800	1.19	.12	.830	.98	.06

NUMBER	R/B	BEND	IR	DEPTH	CENTER	BANDW	ELBOW	UV	FEAT	NUMBER
1208	1.17 G	0.18 G								1208
1212	0.98 Q	0.16 Q								1212
1263	1.25 Q	0.12 Q							0 Q	1263
1284	1.45 G	0.04 G							0 G	1284
1317	1.11 Q	0.20 Q					0.40 Q	3 Q	0 Q	1317
1330	1.11 G	-0.02 G					0.39 Q	3 Q	3 Q	1330
1364	1.41 Q	0.12 Q							0 Q	1364
1449	1.76 G	0.13 G			1.00 Q				0 Q	1449
1493	0.96 G	0.07 G					0.39 Q	3 Q	2 G	1493
1512	1.32 G	-0.15 G			1.00 Q				0 Q	1512
1529	1.47 Q	-0.06 Q							0 Q	1529
1566	1.31 Q	0.14 Q							0 G	1566
1580	1.08 G	0.08 G			0.94 G		0.42 G	1 G	0 G	1580

1208   1.17 G   0.18 G

MICRON	REFL	RERR	MICRON	REFL	RERR	MICRON	REFL	RERR	MICRON	REFL	RERR
.330	.82	.10	.355	.96	.06	.400	.82	.03	.430	.87	.03
.470	.93	.06	.500	.91	.04	.540	1.03	.05	.570	1.05	.04
.600	1.00	.03	.630	1.06	.04	.670	.90	.05	.730	.96	.05
.765	1.04	.05	.800	.99	.10	.830	.87	.09	.870	.96	.09
.900	1.05	.17	.930	1.45	.22	.970	.89	.16			

1212   0.98 Q   0.16 Q

MICRON	REFL	RERR	MICRON	REFL	RERR	MICRON	REFL	RERR	MICRON	REFL	RERR
.330	1.20	.19	.355	.89	.18	.400	.93	.12	.430	.96	.09
.470	.90	.10	.500	.91	.06	.540	1.04	.09	.570	1.00	.10
.600	.95	.14	.630	.97	.08	.670	.82	.08	.700	1.04	.15
.730	.75	.17	.765	1.05	.32	.800	.87	.35			

1263   1.25 Q   0.12 Q   0 Q

MICRON	REFL	RERR	MICRON	REFL	RERR	MICRON	REFL	RERR	MICRON	REFL	RERR
.330	1.00	.46	.355	.71	.24	.400	.94	.07	.430	.88	.03
.470	.93	.05	.500	.97	.03	.540	.91	.04	.570	1.09	.05
.600	.94	.03	.630	1.10	.05	.670	1.00	.07	.700	1.03	.06
.730	1.09	.09	.765	1.04	.07	.800	.91	.07	.830	.89	.13

1284   1.45 G   0.04 G   0 G

MICRON	REFL	RERR	MICRON	REFL	RERR	MICRON	REFL	RERR	MICRON	REFL	RERR
.330	.67	.15	.355	.66	.09	.400	.84	.03	.430	.92	.03
.470	.87	.03	.500	.93	.04	.540	.99	.03	.570	1.00	.03
.600	1.05	.03	.630	1.12	.03	.670	1.10	.03	.700	1.14	.04
.730	1.21	.05	.765	1.17	.06	.800	1.21	.07	.830	1.24	.07

1317   1.11 Q   0.20 Q   0.40 Q   3 Q   0 Q

MICRON	REFL	RERR	MICRON	REFL	RERR	MICRON	REFL	RERR	MICRON	REFL	RERR
.355	.46	.13	.400	.87	.14	.430	.93	.11	.470	.94	.03
.500	.97	.08	.540	.98	.05	.570	.91	.16	.600	1.19	.13
.630	1.03	.12	.670	1.01	.05	.700	.95	.11	.730	.85	.04
.765	.87	.11	.800	.94	.17	.830	.85	.11			

1330   1.11 G   -0.02 G   0.39 Q   3 Q   3 Q

MICRON	REFL	RERR	MICRON	REFL	RERR	MICRON	REFL	RERR	MICRON	REFL	RERR
.330	.70	.05	.355	.90	.32	.400	1.05	.07	.430	1.02	.05
.470	.92	.03	.500	.95	.03	.540	.97	.06	.570	1.00	.04
.600	.99	.06	.630	.96	.03	.670	.99	.04	.700	1.04	.05
.730	1.15	.09	.765	1.05	.08	.800	1.04	.06	.830	1.24	.09

1364   1.41 Q   0.12 Q   0 Q

MICRON	REFL	RERR	MICRON	REFL	RERR	MICRON	REFL	RERR	MICRON	REFL	RERR
.330	.73	.21	.355	.66	.18	.400	.81	.08	.430	.93	.05
.470	.84	.05	.500	.84	.03	.540	.88	.08	.570	1.13	.09
.600	1.03	.04	.630	1.00	.12	.670	1.04	.06	.730	.86	.14
.765	1.22	.11	.800	1.16	.09	.830	.93	.10	.870	1.20	.21
.900	.97	.15	.930	1.06	.25						

1449   1.76 G   0.13 G   1.00 Q   0 Q

MICRON	REFL	RERR	MICRON	REFL	RERR	MICRON	REFL	RERR	MICRON	REFL	RERR
.330	.48	.06	.355	.65	.05	.400	.67	.03	.430	.72	.04
.470	.87	.03	.500	.86	.03	.540	.99	.05	.570	1.10	.06
.600	.99	.05	.630	1.06	.03	.670	1.11	.04	.700	1.28	.06
.730	1.14	.05	.765	1.22	.05	.800	1.25	.05	.830	1.12	.07
.870	1.36	.03	.900	1.27	.12	.930	1.27	.14	.970	1.39	.09

1493   0.96 G   0.07 G   0.39 Q   3 Q   2 G

MICRON	REFL	RERR	MICRON	REFL	RERR	MICRON	REFL	RERR	MICRON	REFL	RERR
.330	.65	.13	.355	.72	.24	.400	1.02	.03	.430	1.08	.06
.470	.96	.03	.500	1.01	.05	.540	1.06	.05	.570	1.00	.05
.600	1.04	.03	.630	.96	.05	.670	.94	.03	.700	.92	.03
.730	.97	.03	.755	1.01	.03	.800	.96	.05	.830	1.00	.05

1512   1.32 G   -0.15 G   1.00 Q   0 Q

MICRON	REFL	RERR	MICRON	REFL	RERR	MICRON	REFL	RERR	MICRON	REFL	RERR
.330	.79	.24	.355	.93	.13	.400	.98	.03	.430	.87	.03
.470	.99	.03	.500	.99	.03	.540	1.00	.03	.570	.97	.06
.600	1.00	.04	.630	1.14	.04	.670	1.09	.04	.700	1.27	.05
.730	1.23	.05	.765	1.27	.08	.800	1.26	.10	.830	1.22	.08
.870	1.34	.14	.900	1.33	.09	.930	1.34	.18	.970	1.52	.08

1529   1.47 Q   -0.06 Q   0 Q

MICRON	REFL	RERR	MICRON	REFL	RERR	MICRON	REFL	RERR	MICRON	REFL	RERR
.330	.56	.19	.400	1.00	.07	.430	.74	.04	.470	.88	.04
.500	.96	.03	.540	.85	.04	.570	.85	.05	.600	1.15	.06
.630	1.04	.06	.670	1.15	.04	.700	1.20	.04	.730	1.22	.07
.765	1.27	.04	.800	1.15	.05	.830	1.21	.12			

1566   1.31 Q   0.14 Q   0 G

MICRON	REFL	RERR	MICRON	REFL	RERR	MICRON	REFL	RERR	MICRON	REFL	RERR
.400	.82	.10	.430	.96	.06	.470	.88	.06	.500	.94	.05
.540	1.04	.06	.570	1.05	.05	.600	.98	.04	.630	1.06	.04
.670	1.05	.05	.700	1.06	.07	.730	1.10	.07	.765	1.06	.07
.800	.98	.09	.830	1.00	.08						

1580   1.08 G   0.08 G   0.94 G   0.42 G   1 G   0 G

MICRON	REFL	RERR	MICRON	REFL	RERR	MICRON	REFL	RERR	MICRON	REFL	RERR
.330	.83	.03	.355	.95	.03	.400	.95	.03	.430	.97	.03
.470	.98	.03	.500	.95	.03	.540	1.00	.03	.570	1.00	.03
.600	.99	.03	.630	1.00	.03	.670	1.00	.03	.700	1.02	.03
.730	.97	.03	.765	.99	.03	.800	.97	.03	.830	.96	.03
.870	.92	.03	.900	.93	.03	.930	.90	.03	.970	.91	.04

```
NUMBER   R/B   BEND      IR   DEPTH  CENTER BANDW  ELBOW  UV   FEAT        NUMBER
--------------------------------------------------------------------------------
1595    1.31 Q 0.22 Q                                       0 Q            1595
        MICRON REFL  RERR MICRON REFL  RERR MICRON REFL  RERR MICRON REFL  RERR
        .400  .78  .07  .430  .82  .05  .470  .88  .06  .500  .91  .03
        .540 1.08  .05  .570 1.00  .06  .600  .93  .03  .630 1.03  .07
        .670 1.02  .04  .700 1.03  .08  .730  .99  .05  .765 1.01  .12
        .800  .77  .10  .830 1.10  .20

1620    1.61 G 0.09 G                                       1 G            1620
        .330  .42  .16  .355  .64  .11  .400  .75  .06  .430  .80  .05
        .470  .82  .03  .500  .87  .03  .540  .91  .03  .570 1.02  .04
        .600 1.02  .05  .630 1.07  .03  .670 1.09  .05  .700 1.15  .05
        .730 1.22  .05  .765 1.13  .06  .800 1.15  .05  .830 1.10  .07

1636    1.74 Q 0.05 Q                                       0 Q            1636
        .330  .59  .08  .355  .52  .10  .400  .67  .08  .430  .76  .08
        .470  .83  .04  .500  .86  .04  .540  .95  .07  .570 1.05  .08
        .600 1.00  .03  .630 1.08  .04  .670 1.19  .06  .700 1.12  .08
        .730 1.31  .03  .765 1.45  .03  .800 1.23  .08

1645    1.26 Q 0.01 Q                                       0 Q            1645
        .355  .89  .27  .400  .82  .07  .430 1.05  .05  .470  .90  .03
        .500  .94  .03  .540 1.10  .03  .570 1.00  .08  .600  .94  .03
        .630  .99  .08  .670 1.03  .04  .700 1.18  .07  .730 1.06  .08
        .765 1.12  .09  .800 1.31  .15  .830 1.22  .15

1656    2.02 Q 0.27 Q                                       0 Q            1656
        .330  .22  .20  .355  .45  .11  .400  .54  .06  .430  .66  .03
        .470  .72  .03  .500  .79  .04  .540  .95  .08  .570 1.00  .03
        .600 1.02  .05  .630 1.18  .09  .670 1.13  .04  .700 1.12  .07
        .730 1.07  .07  .765 1.15  .03  .800 1.32  .12  .830 1.27  .06
        .870  .89  .13  .900 1.13  .24  .930  .97  .12  .970 1.15  .22

1685    1.75 G 0.15 G -0.11 G 0.82 G 0.97 G 0.18 G             0 G          1685
        .330  .72  .20  .340  .64  .10  .355  .58  .07  .400  .71  .03
        .430  .69  .04  .470  .85  .05  .500  .80  .05  .540  .95  .04
        .570 1.00  .05  .600 1.02  .03  .630 1.11  .04  .670 1.20  .04
        .700 1.17  .03  .730 1.13  .03  .765 1.16  .07  .800 1.13  .05
        .870 1.06  .06  .900 1.00  .05  .950  .97  .07 1.000  .95  .07
       1.060 1.18  .12

1717    1.59 Q-0.01 Q                                       0 Q            1717
        .330  .58  .16  .355  .62  .07  .400  .79  .05  .430  .92  .08
        .470  .85  .04  .500  .90  .06  .540  .96  .05  .570 1.00  .05
        .600 1.04  .03  .630 1.12  .06  .670 1.12  .07  .700 1.23  .05
        .730 1.23  .06  .765 1.37  .07  .800 1.21  .06  .830 1.49  .29

1727    1.54 Q 0.08 Q                                       1 Q            1727
        .330  .50  .17  .355  .56  .15  .400  .74  .07  .430  .85  .03
        .470  .87  .05  .500  .89  .07  .540 1.02  .07  .570  .97  .06
        .600 1.02  .06  .630 1.06  .04  .670 1.04  .03  .700 1.11  .08
        .730 1.15  .06  .765 1.34  .10  .800 1.28  .11  .830 1.20  .09
        .870 1.10  .14  .900 1.00  .14  .930 1.25  .21

1830    1.74 Q 0.09 Q                                                      1830
        .330  .48  .14  .355  .68  .19  .400  .75  .10  .430  .64  .07
        .470  .84  .03  .500  .84  .03  .540  .94  .04  .570 1.00  .06
        .600 1.18  .08  .630 1.14  .03  .670 1.09  .04  .700 1.29  .07
        .730 1.24  .12  .765 1.15  .14  .800  .95  .15  .830 1.24  .10
```

V. POLARIMETRY AND RADIOMETRY
OF THE ASTEROIDS

D. MORRISON
University of Hawaii

and

B. ZELLNER
University of Arizona

The first seventeen columns of this table give polarimetric parameters observed for the minor planets, and the remainder of the columns give thermal-radiometric results. The observational programs and the interpretations in terms of asteroid composition, surface texture, albedo, and diameter are discussed in the chapters by Dollfus and Zellner and by Morrison and Lebofsky in Part II.

The first three columns give the asteroid number, the polarimetric parameter P_{min}, and the quality code for P_{min}. The fourth through the eighth columns specify a filter passband, the inversion angle, its quality, the polarimetric slope, and its quality as measured in that passband; the ninth through the thirteenth columns give the same information where available for a second wavelength band. The filters are standard Blue, Green, or Visual, except for data measured with no filter as indicated by symbol X.

The fourteenth through seventeenth columns give the visual magnitude assumed for subsequent diameter computations, the polarimetric albedo according to the slope–albedo law, the implied diameter in km, and a reference to the source publication of the polarimetric data. Polarimetric albedos and diameters are derived only for albedos ≥ 0.06, and only from slope data in the visual or green filter passbands, with small color corrections to the visual wavelength in the latter case.

The eighteenth and succeeding columns list the radiometric diameter, the implied visual albedo, a quality code for the radiometric data, and references for the radiometric observations. The analyses are generally via the "standard" radiometric model described by Jones and Morrison (*Astron. J.* 79: 892-895, 1974) with thermal peaking parameter T_0 = 408 K, corresponding to $\beta \sim 0.9$ in the notation of Morrison and Lebofsky (see their chapter).

Throughout this table, quality code 1 means fragmentary data; e.g., a single observation or a result predating the high-quality surveys of recent years. Quality 2 applies to reliable, confirmed observations, and quality 3 is used for results that could hardly be improved.

There is no systematic discrepancy between the radiometric and polarimetric diameters listed here, but substantial discrepancies exist for some individual asteroids. The weighted average of albedos derived by the two methods, if of total weight 2 or higher, is used to derive adopted diameters in Table VII, below.

NUMBER	PMIN	Q	F1	ANG1	QA1	SLP1	Q1	F2	ANG2	QA2	SLP2	Q2	V	ALB	DIAM	REF	DIAM	ALB	Q	REFERENCES
1	1.72	3	B	18.2	3	0.253	3	G	18.1	3	0.257	3	3.63	0.060	1016.	P01	1014	.059	3	T10,11,12,13
2	1.38	2	B	18.6	3	0.194	3	G	18.1	3	0.228	2	4.36	0.066	622.	P01	589	.093	2	T10,11
3	0.76	3	B	19.6	3	0.107	3	G	20.2	3	0.098	3	5.65	0.152	232.	P01	247	.162	3	T 3, 4, 8,11
4	0.55	2	B	21.3	3	0.067	2	G	21.9	3	0.066	3	3.46	0.216	579.	P01	530	.255	2	T10,11,12,13
5	0.70	3	B	18.9	3	0.110	1	G	19.1	3	0.096	3	7.43	0.155	110.	P01	121	.125	2	T 3, 4,11
6	0.80	2	B	20.5	2	0.086	1	G	20.8	2	0.091	1	6.02	0.163	205.	P01	204	.162	2	T 3, 4,11
7	0.75	3	B	20.4	2	0.108	3	G	21.2	2	0.097	3	5.78	0.153	256.	P01	208	.196	3	T 3, 4, 8,11
8	0.68	2	B	19.3	2	0.103	2	G	20.0	3	0.104	2	6.73	0.146	156.	P01	162	.133	2	T 3, 4, 8,11
9	0.74	2	B	21.1	3	0.099	2	G	21.8	3	0.102	2	6.60	0.147	186.	P01	185	.118	3	T 8,10,11
10	1.50												5.64				430	.050	3	T10,11
11	0.73	2				0.123	2	G	18.9	2	0.124	2	6.95	0.123	154.	P01	157	.116	2	T 3, 4, 8,11
12	0.68	2	B	18.4	2	0.137	2						7.48				135	.098	2	T 8,10,11
13	2.10	2	B	22.0	2	0.258	2						6.81				244	.055	3	T10,11
14	0.82	2	B	20.2	2	0.116	2	G	21.7	1	0.257	1	6.71	0.142	160.	P01	259	.159	2	T 3, 4,11
15	0.72	1	B	23.1	2	0.120	1	G	20.5	2	0.105	2	5.50			P03	250	.163	3	T 3, 4,11
16	1.00	2	X	20.5	1	0.080	1						6.21				247	.094	3	T10,11
17	0.74	2	B	20.5	1	0.131	1	G	22.5	1	0.090	1	8.09				98	.108	2	T 3, 4, 8,11
18	0.80	3	B	21.4	1	0.098	3	G	21.6	3	0.101	3	6.61	0.147	165.	P01	162	.149	2	T 3, 4, 8,11
19	1.72	2	B	21.8	2	0.302	2	G	21.7	2	0.305	2	7.42	0.160	147.	P01	221	.037	3	T10,11
20	0.71	2	B	24.5	3	0.090	2	G	19.8	3	0.092	2	7.85	0.089	113.	P01	134	.189	2	T10,11
21	1.30	1	B		3	0.176	2	G	24.2	2	0.169	2	7.85				109	.106	2	T 3, 5,11
22	0.98	3											6.81				174	.108	2	T 3, 4,11
23	0.74	1	B	19.6	2	0.097	2						7.00				117	.200	2	T 3, 4,11
24	1.63																			
25	0.66	1	B	19.9	2	0.399	2						8.08				72	.194	2	T 3, 4,11
27	0.73	1											7.23				117	.162	2	T 3, 4,11
28	0.88	2	B	20.6	1	0.112	2	G	22.0	2	0.098	2	7.41	0.152	197.	P01	125	.120	1	T 8,11
29	0.78	2	B	19.3	1	0.109	2	G	19.8	1	0.104	2	6.18	0.146	95.	P01	200	.147	2	T 3, 4, 8,11
30	1.32	1											7.78				94	.152	2	T 3, 4, 8,11
31	0.63	2																		
36	0.73												8.78				111	.043	1	T 8,11
37	0.77												8.72				119	.039	1	T 5,11
38	0.78		B	20.5	3	0.098	2	G	21.0	3	0.090	2	7.42	0.167	155.	P99	95	.206	2	T 3, 4,11
39	1.88		B	20.7	2	0.106	2	G	20.8	2	0.100	2	6.55	0.150	117.	P01	157	.168	1	T 1, 8,10,11
40	0.64		B	17.3	2	0.133	3	G	18.4	2	0.129	2	7.33	0.119	110.	P01	116	.148	2	T 3, 4, 8,11
41													7.43				203	.045	2	T 3, 4, 8,11
42													7.71				94	.159	2	T10,11
43	0.31	3	B	18.5	2	0.052	1						8.15			P01	85	.130	1	T 3, 4,11
44													7.24				69	.480	2	T10,11
45													7.43				244	.030	3	T10,11
46	1.54												8.60			P01	131	.036	3	T10,11
47													8.23				158	.035	1	T 8,11

NUMBER	PMIN	Q	F1	ANG1	QA1	SLP1	Q1	F2	ANG2	QA2	SLP2	Q2	V	ALB	DIAM	REF	DIAM	ALB	Q	REFERENCES
51	1.96	2	B	20.3	1	0.292	1						7.66			P01	151	.062	3	T 2, 4, 8,11,12,13
52													6.61			P01	292	.047	2	T 8,10,11
54	1.95	3	B	22.0	1	0.300	1	G	22.2	1	0.357	1	7.95			P01	175	.037	2	T 8,10,11
56	1.47	1	B	20.0	2	0.316	3	G	19.7	2	0.318	3	8.59			P01	144	.031	2	T 8,10,11
57	0.71	1																		
58	1.70	1																		
60																				
63	0.70	2	B	19.4	3	0.124	2	G	19.8	3	0.102	2	8.81	0.150	96.	P01	51	.194	2	T 4,11
64	0.32	3	B	17.1	2	0.038	2	G	18.2	2	0.037	2	7.76	0.370	59.	P01	93	.159	3	T10,11
65	1.25	2											7.84			P01	56	.157	2	T 5,11
66													7.07			P01	309	.027	2	T 5,11
67													9.59				93	.029	1	T 5,11
68	0.68	1											8.52			P01	59	.194	1	T 8,11
69	1.10	1											7.15			P01	127	.147	3	T10,11
70	1.83	1											7.28			P01	108	.183	2	T 5,11,12,13
71	0.61	1											8.29			P01	151	.036	2	T 8,11
72													9.21				95	.039	1	T10,11
76													8.12				117	.071	1	T12,13
77													8.83				66	.115	1	T10,11
79													8.20				79	.145	3	T10,11
80	0.75	1											8.19			P01	83	.131	1	T 5,11
82													8.43				66	.169	2	T10,11
83													9.65			P01	87	.032	2	T 8,10,11
84	1.47	1	B	20.2	1	0.299	1	G	20.3	1	0.306	1	7.91			P01	147	.054	2	T10,11
85	1.49	2											7.34			P01	210	.045	1	T 5,11
88	1.36	1	X	22.1	1	0.119	1						6.84			P03	167	.114	2	T 3, 4,11
89	1.50	1											9.14				104	.035	2	T10,11
91	0.90	1											7.85			P99	168	.044	2	T 5,11
92	0.77	2											7.82			P01	190	.036	2	T 8,10,11
93	1.22	2											8.06				229	.019	2	T 8,11
94													7.72			P01	108	.121	2	T10,11
95	1.78	1											7.89				96	.131	1	T 8,11
97													8.64				126	.038	2	T 8,11
103													7.75				140	.070	1	T 8,11
105													7.18				213	.051	1	T10,11
106													9.24				75	.062	1	T10,11
107													8.07				47	.182	1	T 5,11
109													8.73					.244	1	T 5,11
110																				T12,13
113																				T10,11
114	1.24	1														P01				
115	0.71	2											7.71			P01	93	.162	2	T 8,10,11
116	0.70	1											7.98			P99	79	.175	2	T 5,11

NUMBER	PMIN	Q	F1	ANG1	QA1	SLP1	Q1	F2	ANG2	QA2	SLP2	Q2	V	ALB	DIAM	REF	DIAM	ALB	Q	REFERENCES
120													7.97				173	.037	2	T10,11
121	1.72	1											9.15			P99	47	.168	1	T10,11
123													8.35				74	.152	2	T 8,10,11
124	0.90	2											7.30			P99	113	.164	1	T 5,10,11
129													7.32				174	.067	2	T10,11
130													10.12				34	.128	2	T 5,11
131																	79	.124	1	T 8,10,11
135	1.06	1											8.38			P01	150	.038	1	T 8,11
137													8.26				172	.034	1	T 8,11
139	1.31	2	B	20.6	1	0.262	1						8.99			P01	103	.057	3	T10,11
140	1.78	3	B	21.3	3	0.306	3	G	20.6	1	0.330	1	8.63							
141													10.50			P01	52	.039	1	T12,13
142													8.17				131	.054	2	T10,11
144	1.86	2																		
145																				
146	1.05	1											8.26			P01	141	.043	1	T 8,11
153																				
158													9.88			P99	35	.159	2	T13
159													8.47				140	.036	1	T10,11
171													8.58				80	.099	2	T12,13
172													9.06				65	.097	1	T 8,10,11
182	0.64	2											9.29			P01	38	.223	2	T10,11,12,13
186													9.20				50	.145	3	T10,11
189													9.61		100.		42	.141	2	T10,11
192	0.75	2	B	19.6	3	0.102	3	G	19.8	3	0.084	3	7.48	0.180		P01	99	.185	3	T10,11
194													7.86				192	.033	2	T 8,10,11
196													6.82				161	.126	1	T 8,11
204													9.17				50	.146	3	T 5,11,13
208													8.99				48	.190	3	T 8,10,11
211													8.15				166	.034	1	T 8,11
219													9.78				39	.137	1	T13
221													7.97				95	.123	1	T10,11
224													8.84		106.		72	.097	3	T10,11
230	0.94	3	B	19.8	2	0.140	2	G	20.6	2	0.122	1	7.75	0.125		P01	125	.088	3	T10,11
238													8.26				154	.036	1	T10,11
241													7.83				200	.032	1	T 8,11
247													8.33				143	.039	3	T10,11
264													8.63				64	.149	2	T10,11
270	0.65	1											9.02			P99	51	.163	2	T10,11
306	0.66	1														P01				
308													8.33				138	.042	2	T 8,10,11
311													10.47				27	.158	1	T13
313													9.15				122	.033	2	T 8,11,12,13

NUMBER	PMIN	Q	F1	ANG1	QA1	SLP1	Q1	F2	ANG2	QA2	SLP2	Q2	V	ALB	DIAM	REF	DIAM	ALB	Q	REFERENCES	NUMBER
324	1.46	2	B	20.1	3	0.308	3	G	20.0	3	0.278	3	7.07			P01	252	.040	3	T10,11	324
326													9.41				80	.046	1	T 8,11	326
334	1.32	2														P99					334
338	0.98	1														P01					338
339													9.75			P01	45	.109	1	T13	339
345	1.55	2														P01					345
349	0.39	2		18.2	1								6.18			P01	145	.278	2	T 5,11	349
354	0.52	2											6.61			P99	154	.165	2	T10,11	354
356	1.50	1											8.51			P01	155	.028	2	T10,11	356
367	0.71	1														P99					367
374	0.85	1														P01					374
389	0.80	1														P01					389
397													9.59				50	.099	2	T10,11	397
404	1.94	2	B	19.9	1	0.313	1						9.22			P01	101	.034	1	T 8,11	404
410	1.28	1											8.43			P01	135	.040	1	T 8,11	410
415	0.77	1											9.48			P01	93	.032	1	T 5,11	415
416	1.40	1														P99					416
423																P02					423
433	0.70	3	B	19.4	3	0.100	3	V	20.6	3	0.082	3	11.0	0.170	20.	P99	22	.180	3	T 6	433
434	0.32	1											11.24				11	.428	2	T10,11	434
441													8.66				65	.140	1	T 8,11	441
444													8.07				166	.037	3	T 8,10,11	444
451	1.62	1											6.89			P01	279	.039	1	T 8,11	451
455													9.17				105	.033	1	T13	455
462													9.61				30	.269	1	T 3, 4,11	462
471	0.63	2											6.77			P01	144	.163	2	T 8,11	471
476													8.85				113	.039	1	T10,11	476
497													10.25				30	.152	1	T10,11	497
498	1.69	3	B	19.6	3	0.287	3	G	19.4	3	0.277	3	9.13			P01	71	.076	2	T10,11	498
511													6.48				323	.040	3	T13	511
513													10.04				40	.106	1	T10,11	513
516													8.45				64	.173	3	T13	516
529													10.52				29	.126	1	T 3, 4, 8,11,12,13	529
532	0.78	3	B	20.2	2	0.122	2						5.90			P01	220	.160	3	T13	532
534													10.04				30	.183	1	T13	534
537													9.02				135	.023	1	T 8,11	537
550													9.65				52	.086	1	T 3, 4,11	550
554													9.08				101	.039	1	T 5,11	554
558													9.13				64	.094	2	T 8,11	558
563	0.70	1	B	18.8	1	0.108	2						9.11			P01	55	.131	1	T10,11	563
584	0.64	1	B	17.4	2	0.114	3	G	19.1	3	0.108	2	8.79	0.142	61.	P01	51	.202	1	T 8,11	584
596													9.15				133	.021	2	T10,11	596
602	1.76	1											8.63			P01	137	.032	2	T 8,10,11	602

NUMBER	PMIN	Q	F1	ANG1	QA1	SLP1	Q1	F2	ANG2	QA2	SLP2	Q2	V	ALB	DIAM	REF	DIAM	ALB	Q	REFERENCES
617													8.49				158	.028	2	T 9,12,13
624	1.30	1											7.77			P99	233	.025	3	T 9,14
633													10.15				43	.082	2	T13
639													8.67				58	.175	1	T13
654	1.46	1	B	21.1	1	0.328	1	G	20.5	1	0.280	1	9.32			P01	51	.121	1	T 8,11
660													10.16				43	.081	1	T13
661													7.60				120	.107	1	T 1, 4
674	0.81	1											9.33			P01	71	.060	2	T 8,10,11
679													9.39				101	.029	1	T10,11
694													6.37				338	.043	2	T 8,10,11
704	1.45	1	G	15.8	3	0.338	3	G	15.7	3	0.305	2	10.14			P01	34	.135	1	T13
720													9.10				46	.193	2	T10,11
737	0.84	2											9.93				46	.086	2	T13
742													7.95				204	.027	2	T 3, 4,11
747	1.30	2											10.66			P01	40	.059	1	T13
766													11.73				15	.153	1	T 5,11
782													9.65				49	.099	2	T10,11
785													8.32				176	.026	2	T 5,11
790													9.88				54	.066	2	T13
798													7.97				141	.056	1	T 8,11
804													9.44				50	.112	1	T10,11
830																				
849	0.95	1											9.26			P99	34	.290	1	T 5,11
863	0.76	3	B	19.4	2	0.101	2						14.09			P01	4	.180	2	T 3, 4,11,12,13
887													11.13				32	.061	1	T13
890													8.15				155	.083	2	T12,13
911													10.90				32	.143	1	T12,13
958													12.89				7	.222	2	T10,11
1011	0.67	1														P99				
1052	0.69	1														P01				
1058																				
1075													10.55				30	.102	2	T13
1087													10.10				56	.051	2	T13
1129													10.69				32	.091	2	T13
1148													9.63				34	.083	2	T13
1162													8.54				37	.172	1	T12,13
1172													9.08				129	.039	1	T 9
1173													12.19				92	.047	1	T 9
1178													10.27				20	.057	1	T 5,11
1186													10.75				40	.083	1	T13
1199													10.46				34	.056	1	T13
1210													11.10				26	.087	1	T13
1223																		.071	1	T13

NUMBER	PMIN	Q	F1	ANG1	QA1	SLP1	Q1	F2	ANG2	QA2	SLP2	Q2	V	ALB	DIAM	REF	DIAM	ALB	Q	REFERENCES
1289													11.01				22	.146	2	T13
1362													11.29				31	.055	1	T12,13
1388													11.50				22	.090	1	T13
1416													10.91				33	.071	2	T13
1437													8.58				173	.021	1	T12,13
1512													9.87				84	.027	2	T12,13
1566			B			0.107	1	G			0.082	1	16.75	0.178	1.4	P01				T13
1567													10.04							
1580	1.58	2	B	19.6	1	0.292	1						15.91	0.187	5.4	P99	72	.033	2	T 5,11
1620			B			0.108	1	G	19.5	1	0.281	1			2.0	P01				
1685			B			0.130	1	V			0.074	1				P01				
2062			B	20.9	1	0.094	1						17.56			P99	.9	.200	1	T 7

REFERENCES

P01 ZELLNER, B. AND GRADIE, J. (1976). MINOR PLANETS AND RELATED OBJECTS. XX. POLARIMETRIC INDICATIONS OF ALBEDO AND COMPOSITION FOR 94 ASTEROIDS. ASTRON. J. 81, 262-280.

P02 ZELLNER, B. AND GRADIE, J. (1976). POLARIZATION OF THE REFLECTED LIGHT OF ASTEROID 433 EROS. ICARUS 28, 117-123.

P03 VEVERKA, J. (1973). OBSERVATIONS OF 9 METIS, 15 EUNOMIA, 89 JULIA, AND OTHER ASTEROIDS. ICARUS 19, 114-117. DATA REREDUCED BY ZELLNER.

P99 UNPUBLISHED DATA, UNIVERSITY OF ARIZONA.

T01 MATSON, D.L. (1972). INFRARED EMISSION FROM ASTEROIDS AT WAVELENGTHS OF 8.4, 10.5, AND 11.6 MICRONS. PH.D. DISSERTATION, CALTECH.

T02 CRUIKSHANK, D.P. AND MORRISON, D. (1973). RADII AND ALBEDOS OF ASTEROIDS 1, 2, 3, 4, 6, 15, 51, 433, AND 511. ICARUS 20, 477-481.

T03 MORRISON, D. (1974). RADIOMETRIC DIAMETERS AND ALBEDOS OF 40 ASTEROIDS. ASTROPHYS. J. 194, 203-212.

T04 CHAPMAN, C.R., MORRISON, D. AND ZELLNER, B. (1975). SURFACE PROPERTIES OF ASTEROIDS. A SYNTHESIS OF ASTEROID SPECTROPHOTOMETRY, RADIOMETRY, AND POLARIMETRY. ICARUS 25, 104-130.

T05 MORRISON, D. AND CHAPMAN, C.R. (1976) RADIOMETRIC DIAMETERS FOR AN ADDITIONAL 22 ASTEROIDS. ASTROPHYS. J. 204, 934.

T06 MORRISON, D. (1976). THE DIAMETER AND THERMAL INERTIA OF 433 EROS. ICARUS 28, 125-132.

T07 MORRISON, D., GRADIE, J.C. AND RIEKE, G.H. (1976). RADIOMETRIC DIAMETER AND ALBEDO OF THE REMARKABLE ASTEROID 1976AA. NATURE 260, 691.

T08 HANSEN, O.L. (1976). RADII AND ALBEDOS OF 84 ASTEROIDS FROM VISUAL AND INFRARED PHOTOMETRY. ASTRON. J. 81, 74-84.

T09 CRUIKSHANK, D.P. (1977). RADII AND ALBEDOS OF FOUR TROJAN ASTEROIDS AND JOVIAN SATELLITES 6 AND 7. ICARUS 30, 224-230.

T10 MORRISON, D. (1977). RADIOMETRIC DIAMETERS OF 84 ASTEROIDS FROM OBSERVATIONS IN 1974-1976. ASTROPHYS. J. 214, 667-677.

T11 MORRISON, D. (1977). ASTEROID SIZES AND ALBEDOS. ICARUS 31, 185-220.

T12 DEGEWIJ, J. (1978). PHOTOMETRY OF FAINT ASTEROIDS AND SATELLITES. PH. D. DISSERTATION, LEIDEN UNIVERSITY.

T13 GRADIE, J. C. (1978). AN ASTROPHYSICAL STUDY OF THE MINOR PLANETS IN THE EOS AND KORONIS ASTEROID FAMILIES. PH. D. DISSERTATION, UNIVERSITY OF ARIZONA.

T14 HARTMANN, W. K. AND CRUIKSHANK, D. P. (1978) THE NATURE OF TROJAN ASTEROID 624 HEKTOR. ICARUS 36, 353-366.

VI. LIGHTCURVE PARAMETERS OF ASTEROIDS

E. F. TEDESCO
University of Arizona

Lightcurve observations may be used to determine a synodic period, lightcurve amplitude, and phase coefficient (see the chapter by Burns and Tedesco in Part IV of this book). If high-quality observations were made at several different ecliptic longitudes, the sidereal period, pole orientation, shape, and sense of rotation can also be determined (see the chapter by Taylor). I have endeavored to list every asteroid known to have had one or more of the above quantities determined.

The first five columns list the asteroid number or provisional designation, the rotation period in hours, the least and the greatest photometric amplitudes observed, and a quality code. Symbols > and ? are used if the period is a lower limit or is uncertain, and an entry of 99.9999 indicates that the period is unknown but believed to be very long. The quality codes are subjective in nature and may be described as follows: Quality 0 indicates that a published observation exists, but no period or amplitude can be estimated; such observations do rule out the possibility that the asteroid is rapidly rotating and that it showed even a moderate amplitude (i.e., > 0.2 mag) at the time of observation. Quality 1 refers to results that are either a lower limit or an estimate based on observations covering less than a complete rotation; additional observations are needed. Quality 2 refers to secure results; i.e., the period is known to within approximately 0.1 hr. Quality 3 is used for periods known to about 0.01 hr. The listed period is *synodic* except for data of quality 4, for which the *sidereal* period is listed; this implies that all significant physical properties obtainable from lightcurve observations have been obtained.

Succeeding columns give the number of oppositions at which lightcurve observations were made, the ecliptic longitude and latitude of the asteroid's pole, the linear phase coefficient derived from lightcurve observations, reference numbers corresponding to publications cited at the end of the table, and an alphabetic key to remarks preceding the references.

The phase coefficients refer to observations made at the V wavelength of the UBV system, unless noted otherwise in the remarks. They are derived from photoelectric observations at phases greater than 7 deg, with explicit allowance for rotational lightcurve effects. The phase coefficients are reliable to within 0.002 mag/deg except when followed by a colon (:) or when noted otherwise in the remarks.

NUMBER	PERIOD	AMPMIN	AMPMAX	Q	OPP	POLE		BETA	REFERENCES	R
1	9.078		0.04	3	3			0.036	1 11 31 40 60 97 102	
2	7.88106	0.12 -	0.15	4	9	228	43	0.037	9 14 40 41 55 87 110 117	
3	7.213		0.15	2	3			0.025	12 31 40	
4	5.34213	0.10 -	0.14	3	9	139	47	0.025	12 13 16 18 23 30 36 39 45 90 94	
5	16.81184	0.21 -	0.27	4	4	148	9	0.014:	12 31 59 96	B
6	7.27445	0.06 -	0.19	4	4	5	50	0.027	1 5 31 34 110 115 118	
7	7.135	0.04 -	0.29	3	7	11	41	0.029	14 31 39 95 109	
8	13.6	0.01 -	0.04	2	2	157	10	0.028	1 31 59 64 109 113	
9	5.064	0.06 -	0.31	2	4	156	15	0.04 :	12 31 35 39 40 118	B
10	18.	0.09 -	0.21	2	3			0.03 :	7 40 122	B
11	10.67	0.07 -	0.12	2	3				109 110 117	
12	8.654	0.25 -	0.33	3	2			0.032:	106 122	
13	7.045		0.12	2	1				14	
14	11. ?	0.04 -	0.10	1	2				40 64 115	
15	6.0806	0.42 -	0.53	3	6			0.038	13 39 67 109	
16	4.303	0.12 -	0.32	3	2			0.028	97 103 109	
17	12.275	0.13 -	0.36	2	2				39 109	
18	11.573	0.15 -	0.35	1	2				31 42 116	
19	7.46		0.25	2	1				111 118	
20	8.0980	0.17 -	0.24	3	5			0.031	12 29 31 122	
21	6.133		0.15	2	1				14	R
22	4.147	0.14 -	0.30	2	5	215	45	0.031	1 31 77 110	
23	6.15		0.19	2	2				111 118	
24	8.369	0.11 -	0.14	2	2				100 111	
25	9.945		0.18	2	2				39 109	
26	13.13		0.12	1	1				74	
27	8.500		0.15	2	1				12	
28	16.523		0.21	1	1				111	
29	5.390	0.08 -	0.13	2	6			0.032	14 20 66 91 111	R
30	13.677		0.14	2	3				2 31 42 64	
31	5.54		0.08	2	1				119	
32	9.4431		0.20	2	1				79	
36	9.93		0.15	1	1				42 82	
37	4.5		0.10	1	1				73	
39	5.1382	0.18 -	0.54	3	13	121	37	0.026	31 39 40 65 109 110 115 118	
40	9.1358	0.22 -	0.28	2	2				31 40 43 50	
41	5.9878		0.38	3	1			0.053	70	
42	13.59		0.32	2	2				43 122	
43	5.7506	0.15 -	0.66	3	2			0.048	10 54 111	
44	6.44	0.22 -	0.50	2	5	105	30		4 12 31 40 59 64 88 118	
45	5.70	0.30 -	0.33	2	2				43 122	
46	20.5		0.12	2	1				119	
48	11.89		0.35	1	1				42 82	
49	10.42		0.18	2	1				76 100	
50		< 0.2 ?		0	1				50	L
51	7.785		0.14	2	2				14 110 115	R
52	11.2582		0.09	2	1				72 119	
54	7.04		0.12	2	1				111	
55	4.8043		0.24	2	1				83	
56	16.		0.06	2	1				43	R
59	13.690		0.10	2	1				21	
60	>30.	0.07 -	0.12	1	2				31 122	
61	11.45		0.30	2	1				117	R
63	9.297		0.47	3	1			0.035	64 71	
64		< 0.2 ?		0	1				50	L
65	6.56		0.06	1	1				82	
66		< 0.4		0	1				108	H
67	15.89		0.23	1	1				42	
68	14.85		0.15	1	1				42 82	
69	5.65		0.22	2	1				8	
71	11.213		0.10	2	1				54	R
77		< 0.4		0	1				108	H
78	8. ?		0.14	1	1				97	
79	5.979		0.05	3	1			0.032	68 80	A
80	>20.	> 0.07		1	1				73	
85	7. ?		0.15	1	1				118	
86		< 0.4		0	1				108	H
87	5.186		0.42	2	1				42 82	
88	6.0422		0.19	2	1				76	

NUMBER	PERIOD	AMPMIN	AMPMAX	Q	OPP	POLE	BETA	REFERENCES	R
89	11.3872		0.25	3	3		0.035	85 112 122	
90			< 0.4	1	1			108	H
91	6.025		0.15	2	1			42	
92	15.94		0.17	2	1			76	
95			< 0.4	0	1			108	H
97	16. ?		0.04	1	2			7 39 73	
100	>20. ?		0.05	1	1			82 100	
103	99.9999		< 0.02	1	2			43	
104	9.		0.28	1	1			100	
105	>20. ?		> 0.03	1	1			100	
107	4.56		0.53	2	1			7	
108	8.		0.2	1	1			2	V
110	10.9267	0.11 -	0.20	3	2		0.032	98	
112	15.783	0.14 -	0.50	2	1			46	R
113	>10.		> 0.20	1	1			122	
115	7.244		0.20	2	1			119	
116	13.7	0.3 -	0.6	1	2			2 42	
118	7.78		0.34	2	1			82	
121	9. ?		0.03	1	1			20	
125	4.0		0.35	1	1			100	
128	39.		0.10	2	2			19 75	
129	4.9572		0.32	3	2		0.024	70 122	A
131			< 0.2 ?	0	1			50	L
135	16.805		0.1	1	1			42	
139	20.9		0.19	2	1			37	
140			< 0.2 ?	0	1			50	L
142			< 0.2 ?	0	1			50	L
148	20.663		0.30	1	1			91	
150			< 0.2 ?	0	1			50	L
151			< 0.2 ?	0	1			50	I
162	14. ?		0.30	1	1			122	
164	27.3		0.07	2	1			86	
167	16.		0.24	1	1			100	
171	13.44		0.16	2	1			100	
173	5.93		0.04	1	1			83	
178			< 0.2 ?	0	1			50	L
182	80.		0.7	2	1			42	
184	6.7		0.25	1	1			100	
185	10.83		0.12	2	1			20	
186	19.58		0.5	1	2			2 50 52	R
190			< 0.2 ?	0	1			50	L
192	13.622	0.22 -	0.42	3	2		0.040	68 118	
194	15.67		0.27	1	1			74	
196	8.333		0.35	2	1			118	
198	16.			1	1			119	
200	19.		0.10	1	1			81	
209	8.		0.20	1	1			100	
210			< 0.2 ?	0	1			50	L
211			< 0.4	0	1			108	H
215			< 0.2 ?	0	1			50	L
216	5.394		0.40	2	1			73	
221	14.6		0.04	1	1			42	
222	7.0		0.33	1	1			100	
224	18.933		0.10	2	1			42	
230	7.996		0.10	2	1			118	
233	5. ?		> 0.2	1	1			42 63	R
234			0.2 ?	0	1			50	L
247	11.		0.12	1	1			42	
249			< 0.2 ?	0	1			50	L
254	6.0		0.56	1	1			50	P
267	5.90		0.21	2	1			50	P
268	6.1		0.15	1	1			100	
270	15.06		0.32	1	2			42 62 73	
273	>20.		0.65	1	1			100	
277			0.2 ?	0	1			50	L
281	4.3478	0.36 -	0.38	3	3		0.028	97	
283	6.908		0.31	2	1			89	
287	7.		> 0.2	1	1			74	
291	4.315		0.25	2	1			47	P
299			< 0.2 ?	0	1			50	L
302	>10.		> 0.3	1	1			50	P
304	18.3		0.20	2	1			42	
313	13.		0.23	1	1			73	
321	2.87		0.38	2	1			109	
323	10.		0.36	1	1			100	

NUMBER	PERIOD	AMPMIN	AMPMAX	Q	OPP	POLE	BETA	REFERENCES	R
324	8. ?		0.07	1	1			31	
328	>12.		> 0.15	1	1			50	P
332	7.0		0.32	1	1			50	P
333			< 0.4	0	1			108	H
334	6.1		0.16	2	1			46	
337	4.610		0.19	2	1			82	
340	7.7		0.17	1	1			50	P
345	16. ?		0.5	1	1			2	V
349	4.7012	0.31 -	0.40	3	3		0.022	14 100 120 122	A
352	7.0		0.25	1	1			50	P
353			< 0.2 ?	0	1			50	L
354	4.2772	0.07 -	0.30	3	4		0.020:	14 40 64 120 122	A
356	31.8		0.25	2	1			42	
357	>20.		> 0.08	1	1			100	
359	7.3		0.3	1	1			50	P
363	>10.		> 0.3	1	1			50	P
364	9.155		0.50	2	1			118	
366			< 0.4	0	1			108	H
372	5. ?		0.4	1	1			44	R
377	15.		0.16	1	1			100	
379	6.6		0.06	1	1			100	
383	6.4		0.17	1	1			100	
387	24.		> 0.1	1	1			82	
393	38.7		0.14	2	1			75	
395			< 0.2 ?	0	1			50	L
396	>12.		> 0.24	1	1			50	P
404	6. ?		> 0.2	1	1			122	
405	10.08		0.15	1	1			42	
409			< 0.4	1	1			108	H
433	5.2703	0.00 -	1.50	4	3	16	12 0.024	2 3 17 25 44 56	
								58 60 61 69 99	
441	10.35		0.13	2	1			42	
447			< 0.4	0	1			108	H
451	20. ?		0.1	1	2			97	
454	7.66		0.37	2	2			50	P
459	6.38		0.25	2	1			50	P
468	8.3		0.10	1	1			100	
470			< 0.2	0	1			108	H
471	7.113		0.12	2	1			53 73 93	R
499			< 0.2 ?	0	1			50	L
502	10.5		0.35	1	1			100	
505	6.5	0.1 -	0.17	1	1			42 50	
511	5.17	0.06 -	0.25	2	5	122	10	14 31 39 122	
513	5.23		0.45	1	1			100	
514	>20. ?		> 0.3	1	1			50	P
516	7.		0.15	1	1			42	
525			< 0.2	0	1			51	P
531			0.2 ?	0	1			50	L
532	9.406	0.08 -	0.18	2	3			14 40 43 64	
534	9.		0.35	1	1			100	
554	13.63		0.22	2	1			74	
558	10.		0.25	1	1			43	
563	6.26		0.2	1	1			82	
570			< 0.2 ?	0	1			50	L
579	13. ?		> 0.05	1	1			100	
592			< 0.2 ?	0	1			50	L
599	9.566		0.15	2	1			19 91	
621	>10.		> 0.4	1	1			50	P
624	6.9225	0.10 -	1.09	3	4	324	10	26	
632	4.6		0.4	1	1			50	P
633	>10.		> 0.3	1	1			50	P
641			> 0.20	1	1			50	L
654	31.90		0.3	2	1			78	
660	7.92		0.33	1	1			42	
666			< 0.2	0	1			50	L
675	7.7169		0.28	2	1			84	
677	>10.		< 0.1	1	1			50	P
679	8.4		0.07	1	1			82	
700	6.		0.4	1	1			51	P
704	8.723	0.05 -	0.11	3	2		0.044	55 105 118	
709	52.4		0.18	2	1			42	
716	>17.		> 0.3	1	1			51	P
720			< 0.2 ?	0	1			50	L
726			< 0.2 ?	0	1			50	L
736	6.7		0.32	1	1			100	

NUMBER	PERIOD	AMPMIN	AMPMAX	Q	OPP	POLE	BETA	REFERENCES	R
737	14.13		0.15	1	1			42	
747	8.		0.13	1	2			42 122	
750			0.2 ?	0	1			50	L
753	9.84		0.8	2	1			108	H
755			< 0.2	0	1			51	P
761			< 0.2 ?	0	1			50	L
776	23.0		> 0.15	1	1			82	
796	7.75		0.30	1	1			82	
818			< 0.4	0	1			108	H
828			< 0.2 ?	0	1			50	L
846	>24.		> 0.02	1	1			100	
852	4.56		0.31	2	1			100	
873	10.6		> 0.3	1	1			50	P
882			< 0.2 ?	0	1			50	L
887	73.97		0.37	1	2		0.042	28	R
905	10.		0.22	1	1			100	
911	7.	?	0.3	1	2			26	N
913			< 0.2 ?	0	1			50	L
914	>14.	?	< 0.02	1	1			100	
927			< 0.2 ?	0	1			50	L
939	>20.		> 0.20	1	1			100	
944	10.0644	0.35 -	0.60	2	1		0.047	101	A
952	7.51		0.13	2	1			82	
984	5.76		0.4	2	1			108	P
987	10.	?	> 0.3	1	1			51	R
1001			< 0.2 ?	0	1			50	L
1029	14.4		> 0.4	1	1			50	P
1051			< 0.2 ?	0	1			50	L
1058	>12.		> 0.3	1	1			122	
1059			0.2 ?	0	1			50	L
1077			< 0.2 ?	0	1			50	L
1094			0.2 ?	0	1			50	L
1100			< 0.2 ?	0	1			50	L
1144			< 0.2 ?	0	1			50	L
1177			0.2 ?	0	1			50	L
1186			< 0.2 ?	0	1			50	L
1207	8.1	0.5 -	0.71	1	2			50 51	P
1212	>16.		0.1	1	1			97	
1223	8.6		0.45	1	1			100	
1245	4.855		0.63	2	2			50 100	
1251			< 0.2 ?	0	1			50	L
1259	>12.		> 0.3	1	1			50	P
1267	5.50		0.5	1	1			51	P
1268			< 0.2 ?	0	1			50	L
1331	>10.		> 0.3	1	1			50	P
1336			< 0.2 ?	0	1			50	L
1340			0.2 ?	0	1			50	L
1350	6.0		> 0.3	1	1			50	P
1362	13.	?	> 0.20	1	1			97	
1416	4.3		0.4	1	1			50	P
1418			< 0.2 ?	0	1			50	P
1424			< 0.2 ?	0	1			50	L
1437	18.	? 0.35 -	0.42	1	3			26	N
1441			< 0.2 ?	0	1			50	L
1442			< 0.2 ?	0	1			50	L
1454			0.2 ?	0	1			50	L
1511			< 0.2 ?	0	1			50	L
1517			< 0.2 ?	0	1			50	L
1523	5.33		0.5	1	1			51	P
1536			< 0.2 ?	0	1			50	L
1562	8.2		0.4	1	1			51	P
1566	2.2730	0.05 -	0.22	3	1	49 00	0.032	33 57 114	
1576	6.7		0.2	1	1			50	P
1577			< 0.2 ?	0	1			50	L
1580	6.130	0.21 -	0.6	3	1	140 20	0.036	49 104	R
1584	10.		0.30	1	1			100	
1590	6.7		0.4	1	1			50	P
1601			< 0.2 ?	0	1			50	L
1604	8.	?	< 0.2	1	1			50	P
1615	18.		0.2	1	1			100	
1620	5.2233	1.20 -	2.03	4	2	200 60	0.030	5 24	
1634			< 0.2 ?	0	1			50	L
1641			0.2 ?	0	1			50	L
1672	>20.	?	> 0.20	1	1			50	P
1674	8.1		0.19	1	1			100	

NUMBER	PERIOD	AMPMIN	AMPMAX	Q	OPP	POLE	BETA	REFERENCES		R
1685	10.1956		0.80	4	1	200 55	0.037	27		
1687	6.3		0.75	1	1			100		
1702		< 0.2	?	0	1			50		L
1707	>10.	> 0.2		1	1			50		P
1715	>11.	> 0.6		1	1			50		P
1717		< 0.2		1	1			51		P
1753	8.81		0.22	1	1			50		P
1754		< 0.2	?	0	1			50		L
1771		< 0.2	?	0	1			50		L
1789	5.8		0.68	1	1			50		P
1793	7.0		0.39	1	1			50		P
1864	8.57		0.80	2	1			32		
1946	20.4	?	0.6	1	1			107		P
2062	99.9999		0.2	1	1		0.027	38		
2100	10.		0.15	1	1			7		
U1	7.0		0.31	1	1			50		P
U2		< 0.2	?	0	1			50		L
1975EA		< 0.2	?	0	1			50		L
1975GB		< 0.2	?	0	1			50		L
1975RB	9.02		0.3	2	1			48		P
1975RC		< 0.2	?	0	1			50		L
1976EA		< 0.2	?	0	1			50		L
1976EB	7.7		0.31	1	1			50		P
1976HA		< 0.2	?	0	1			50		L
1976UA	99.9999		0.01	1	1			121		
1977RA	5.9		0.84	1	1			6		
1978CA	3.761		0.80	2	1			22	92	
1978DA	8.		0.35	2	1			22	92	

REMARKS

A   AMPLITUDE INCREASES WITH PHASE ANGLE WITHIN A SINGLE OPPOSITION.
     (SEE BELOW UNDER ASTEROID NUMBER FOR ADDITIONAL REMARKS.)
B   BETA UNCERTAIN. (SEE BELOW UNDER ASTEROID NUMBER.)
H   AUTHOR (REF. 108) STATES: "...VARIABILITY COULD BE DEMONSTRATED IF
     THE RANGE WAS AT LEAST 0.4 MAG." EACH ASTEROID WAS OBSERVED BETWEEN
     5 TO 10 HOURS ON FROM 1 TO 3 NIGHTS WITHIN ONE WEEK. SEE BELOW
     UNDER ASTEROID NUMBERS FOR THE NUMBER OF NIGHTS EACH WAS OBSERVED.
L   PHOTOGRAPHIC OBSERVATIONS OF THIS ASTEROID WERE MADE BY LAGERKVIST
     WHO NOTED NO SIGNIFICANT VARIATION IN BRIGHTNESS DURING AN AVERAGE
     OBSERVING PERIOD OF 5.6 HOURS. 25 PERCENT OF ASTEROIDS IN THIS CLASS
     WERE OBSERVED ON MORE THAN ONE NIGHT. SEE ORIGINAL PUBLICATION
     (REF. 50) FOR FURTHER DETAILS.
N   LIGHTCURVE NOT PUBLISHED.
P   ONLY AVAILABLE LIGHTCURVES ARE FROM PHOTOGRAPHIC PHOTOMETRY.
V   ONLY AVAILABLE LIGHTCURVES WERE OBTAINED USING A VISUAL PHOTOMETER.
  5 BETA UNCERTAIN. VALUE DEPENDS UPON COMBINING OBSERVATIONS FROM TWO
     OPPOSITIONS EACH OF WHICH HAD POOR PHASE COVERAGE. (REF. 96).
  9 BETA UNCERTAIN. VALUES IN LITERATURE RANGE FROM 0.037 TO 0.049.
 10 BETA UNCERTAIN. VALUES BETWEEN 0.037 AND 0.049 SEEM TO SATISFY THE
     OBSERVATIONS EQUALLY WELL. (SEE REF. 11)
 21 JUDGING FROM THE LIGHTCURVE (REF. 14) THERE IS A SLIGHT CHANCE THE
     PERIOD COULD BE TWICE THE QUOTED VALUE.
 29 COMPLEX LIGHTCURVE. BETA DETERMINED BY E. F. TEDESCO USING DATA FROM
     7 DIFFERENT OPPOSITIONS. PHASE CURVE WILL BE PUBLISHED WITH LPL
     PAPER CURRENTLY IN PREPARATION (REF. 66).
 51 COMPLEX LIGHTCURVE.
 56 THE LISTED PERIOD IS THE MEAN OF 13.7 AND 19.0 HRS, THE TWO POSSIBLE
     PERIODS REPORTED BY THE AUTHORS (REF. 43).
 61 PERIOD MAY BE HALF THE QUOTED VALUE IF LIGHTCURVE IS SINGLY PERIODIC.
 66 2 NIGHTS. SEE REMARK H ABOVE.
 71 DISPLAYS COLOR VARIATION WITH ROTATION: 0.03 IN U-B AND 0.05 IN U-V.
     PERIOD ASSUMES ONE MAXIMUM AND MINIMUM AND HENCE MAY BE TWICE THE
     QUOTED VALUE. (REF. 54).
 77 2 NIGHTS. SEE REMARK H ABOVE.
 86 2 NIGHTS. SEE REMARK H ABOVE.
 90 1 NIGHT. SEE REMARK H ABOVE.
 95 2 NIGHTS. SEE REMARK H ABOVE.
112 AMPLITUDE VARIED FROM 0.14 TO 0.50 IN ONLY 45 DAYS.
186 VISUAL, PHOTOGRAPHIC AND TWO SHORT PHOTOELECTRIC LIGHTCURVES EXIST.
211 1 NIGHT. SEE REMARK H ABOVE.
233 PERIOD IS FROM VISUAL OBSERVATIONS MADE BY AMATEUR ASTRONOMERS (REF.
     63) AND CONFIRMED BY HARRIS' PHOTOELECTRIC OBSERVATIONS (REF. 42).
333 2 NIGHTS. SEE REMARK H ABOVE.
349 AMPLITUDE VARIES WITH PHASE ANGLE. SEE REF. 120 FOR DETAILS.

REMARKS

354 BETA UNCERTAIN. VALUE BASED ON THREE POINTS. AMPLITUDE VARIES WITH
    PHASE ANGLE. SEE REF. 120 FOR DETAILS.
366 1 NIGHT. SEE REMARK H ABOVE.
372 REPORTED PERIOD OF "ABOUT 0.1 DAY" HAS BEEN DOUBLED SINCE HARWOOD,
    AND OTHER EARLY ASTEROID LIGHTCURVE OBSERVERS ASSUMED THAT
    LIGHTCURVES WERE SINGLE PERIODIC UNLESS OBSERVED TO BE OTHERWISE.
409 2 NIGHTS. SEE REMARK H ABOVE.
447 1 NIGHT. SEE REMARK H ABOVE.
470 3 NIGHTS. SEE REMARK H ABOVE.
471 PRIMARY MAXIMUM IS DOUBLE PEAKED.
716 PERIOD ONLY PUBLISHED IN REF. 51. NO LIGHTCURVE PUBLISHED.
753 3 NIGHTS. SEE REMARK H ABOVE.
818 2 NIGHTS. SEE REMARK H ABOVE.
887 AUTHORS STATE: "THE COMPOSITE LIGHTCURVE INDICATES THAT THE
    AMPLITUDE OF ALINDA IS AT LEAST 0.32 MAG AND PERHAPS AS GREAT AS
    0.42 MAG." (REF. 28).
944 THE LARGE AMPLITUDE RANGE IS PROBABLY DUE MORE TO CHANGES IN SOLAR
    PHASE ANGLE THAN TO CHANGES IN THE PROJECTED AREA OF THE MINOR PLANET
    SINCE THE MAXIMUM CHANGE IN ASPECT WAS SMALL.
984 3 NIGHTS. SEE REMARK H ABOVE.
1207 PERIOD, AMPLITUDE GIVEN AS 7.7, 0.71 IN REF. 46 AND 8.4, 0.5 IN REF.
    51 (AT THE NEXT OPPOSITION). THE ADOPTED PERIOD IS THE MEAN OF THESE
    TWO VALUES.
1580 LARGE AMPLITUDE RANGE MAY BE DUE TO INCREASING SOLAR PHASE ANGLE
    RATHER THAN TO DIFFERENCES IN THE ASTEROID'S PROJECTED AREA. THE
    PHASE COEFFICIENT IS THE MEAN OF THAT OBTAINED FOR THE MAXIMA (0.033)
    AND THE MINIMA (ABOUT 0.040). THE POLE LOCATION IS AN EDUCATED GUESS
U1,U2 LOST ASTEROIDS FROM REF. 50 WHICH DID NOT RECEIVE PROVISIONAL
    DESIGNATIONS.

REFERENCES

1 AHMAD, I.I. (1954) PHOTOMETRIC STUDIES OF ASTEROIDS. IV. THE LIGHTCURVES
   OF CERES, HEBE, FLORA, AND KALLIOPE. ASTROPHYS. J. 120, 551-559.
2 BAILEY,S.I.(1913). OBSERVATIONS OF EROS AND OTHER ASTEROIDS. ANN. HARVARD
   COLL. OBS. 72, 165-189.
3 BEYER, M. (1953) DER LICHTWECHSEL UND DIE LAGE DER ROTATIONSACHSE DER
   PLANETEN 433 EROS WAHREND DER OPPOSITION 1951-52. ASTRON. NACHR. 281,
   121-130.
4 BIANCHI,E. AND PADOVA,E. (1920). LE VARIAZIONI DI LUCE DEL PIANETA (44)
   NYSA. MEM. SOC. ASTRON. ITAL. 1, 39-65.
5 BIANCHI,E. AND PADOVA,E. (1921). OSSERVAZIONI FOTOMETRICHE DI PIANETI.
   MEM. SOC. ASTRON. ITAL. 2, 45-54.
6 BOWELL, E. (1977). 1977RA. IAU CIRCULAR 3111.
7 BOWELL,E. PERSONAL COMMUNICATION.
8 BOWELL,E., MARTIN,L.J., THOMPSON,E.T., AND LUMME,K.(1979). THE PHASE
   FUNCTION OF 69 HESPERIA, TO BE SUBMITTED TO ASTRON. J.
9 BURCHI,R.(1972). SOME PHOTOMETRIC PARAMETERS OF THE MINOR PLANET 2 PALLAS.
   OSSERVATORIO ASTRONOMICO DI COLLURANIA TERAMO, NOTE E COMUNICAZIONI N. 49,
   1-8.
10 BURCHI,R. AND MILANO,L.(1974). PHOTOELECTRIC LIGHTCURVES OF THE MINOR
   PLANET 43 ARIADNE DURING THE 1972 OPPOSITION. ASTRON ASTROPHYS. SUPPL. 15,
   173-180.
11 CALDER,W.A. (1935). PHOTOELECTRIC PHOTOMETRY OF ASTEROIDS. HARVARD
   BULL. 904.
12 CHANG,Y. AND CHANG,C.(1962). PHOTOMETRIC INVESTIGATIONS OF SEVEN VARIABLE
   ASTEROIDS. ACTA ASTRON. SINICA 10, 101-110.
13 CHANG, Y. C. ET AL. (1959). THE LIGHT-CURVES OF MINOR PLANETS (4) VESTA AND
   (15) EUNOMIA. ACTA ASTRON. SINICA 7,207-
14 CHANG, Y. C. AND CHANG, C. S. (1963). PHOTOMETRIC INVESTIGATIONS OF
   VARIABLE ASTEROIDS. II. ACTA ASTRON. SINICA 11, 139-148.
15 CHEN,D., YANG,X., AND WU,Z.(1975). A PHOTOMETRIC STUDY OF EROS 433. ACTA
   ASTRONOMICA SINICA 16, 131-137.
16 CRISTESCU,C.(1972). THE LIGHT-CURVE FOR THE MINOR PLANET (4) VESTA, ST.
   CERC. ASTRON. 17, 177-181.
17 CRISTESCU,C. (1976). PHOTOELECTRIC LIGHTCURVES OF ASTEROID 433 EROS.
   ICARUS 28, 39-42.
18 CUFFEY, J. (1953). PALLAS, VESTA, CERES, AND VICTORIA: PHOTOELECTRIC
   PHOTOMETRY. ASTRON. J. 58, 212.
19 DEBEHOGNE,H., SURDEJ,A., AND SURDEJ,J. (1977). PHOTOELECTRIC LIGHTCURVES
   OF MINOR PLANETS 599 LUISA AND 128 NEMISIS DURING THE 1976 OPPOSITION.
   ASTRON. ASTROPHYS. SUPPL. 30, 375-379.
20 DEBEHOGNE,H., SURDEJ,A., AND SURDEJ,J. (1978) PHOTOELECTRIC LIGHTCURVES
   OF THE MINOR PLANETS 29 AMPHITRITE, 121 HERMIONE, AND 185 EUNIKE. ASTRON.
   ASTROPHYS. SUPPL. 32, 127-133.

REFERENCES

21 DEBEHOGNE, H., SURDEJ, A. AND SURDEJ, J. (1978). PHOTOELECTRIC
LIGHTCURVES AND ROTATION PERIOD OF THE MINOR PLANET 59 ELPIS.
ASTRON. ASTROPHYS. SUPPL. 33, 1-5.
22 DEGEWIJ, J. (1978). THE SIZES OF 1978CA AND 1978DA. THE MESSENGER NO. 13,5.
23 DEGEWIJ,J., AND ZELLNER,B. (1978). ASTEROID SURFACE VARIEGATION. LUNAR
AND PLANETARY SCIENCES IX, 235-237.
24 DUNLAP,J.L.(1974). MINOR PLANETS AND RELATED OBJECTS. XV. ASTEROID (1620)
GEOGRAPHOS. ASTRON. J. 79, 324-332.
25 DUNLAP,J.L. (1976). LIGHTCURVES AND THE AXIS OF ROTATION OF 433 EROS.
ICARUS 28, 69-78.
26 DUNLAP,J.L. AND GEHRELS,T.(1969). LIGHTCURVES OF A TROJAN ASTEROID. ASTRON.
JOURNL. 74, 796-803.
27 DUNLAP,J.L., GEHRELS,T. AND HOWES,M.L.(1973). MINOR PLANETS AND RELATED
OBJECTS. IX. PHOTOMETRY AND POLARIMETRY OF (1685) TORO. ASTRON. JOURNL. 78,
491-501.
28 DUNLAP, J.L. AND TAYLOR, R.(1979). MINOR PLANETS AND RELATED OBJECTS.
XXVII. LIGHTCURVES OF 887 ALINDA. ASTRON. J. 84, 269-273.
29 GEHRELS,T. (1956) PHOTOMETRIC STUDIES OF ASTEROIDS. V. THE LIGHTCURVE
AND PHASE FUNCTION OF 20 MASSALIA. ASTROPHYS. J. 123, 331-338.
30 GEHRELS, T. (1967) MINOR PLANETS. I. THE ROTATION OF VESTA. ASTRON.
JOURNAL 72, 929-938.
31 GEHRELS,T. AND OWINGS,D.(1962). PHOTOMETRIC STUDIES OF ASTEROIDS. IX.
ADDITIONAL LIGHTCURVES. ASTROPHYS. JOURNL. 135, 906-924.
32 GEHRELS,T., ROEMER,E. AND MARSDEN,B.G.(1971). MINOR PLANETS AND RELATED
OBJECTS. VII. ASTEROID 1971FA., ASTRON. JOURNL. 76, 607-608.
33 GEHRELS,T., ROEMER,E., TAYLOR,R.C. AND ZELLNER,B.H.(1970). ASTEROID (1566)
ICARUS. ASTRON. JOURNL. 75, 186-195.
34 GEHRELS,T. AND TAYLOR,R.C.(1977). MINOR PLANETS AND RELATED OBJECTS. XXII.
PHASE FUNCTIONS FOR (6) HEBE. ASTRON. JOURNL. 82, 229-237.
35 GICLAS, H.L. (1950) DIRECT PHOTOELECTRIC PHOTOMETRY. THE PROJECT FOR THE
STUDY OF PLANETARY ATMOSPHERES. LOWELL OBSERV. REPT. 5, PP. 47-63.
36 GICLAS, H.L. (1951) DIRECT PHOTOELECTRIC PHOTOMETRY. THE PROJECT FOR THE
STUDY OF PLANETARY ATMOSPHERES. LOWELL OBSERV. REPT. 9, PP. 33-72.
37 GOGUEN,J., VEVERKA,J.,ELLIOT,J.L. AND CHURCH,C.(1976). THE LIGHTCURVE AND
ROTATION PERIOD OF ASTEROID 139 JUEWA. ICARUS 29, 137-142.
38 GRADIE,J.C.(1976). PHYSICAL OBSERVATIONS OF OBJECT 1976 AA. BULL. AMERICAN
ASTRON. SOC. 8, 458-459.
39 GROENEVELD, I., AND KUIPER, G.P. (1954A) PHOTOMETRIC STUDIES OF ASTEROIDS.
I. ASTROPHYS. J. 120, 200-220.
40 GROENEVELD, I., AND KUIPER, G.P. (1954B) PHOTOMETRIC STUDIES OF ASTEROIDS.
II. ASTROPHYS. J. 120, 529-546.
41 GUNTHER,O. (1953). PHOTOMETRISCHE BEOBACHTUNGE DER PALLAS IN DER OPPOSITION
1951. ASTRON. NACHR. 281, 131-132.
42 HARRIS, A. PERSONAL COMMUNICATION.
43 HARRIS, A. AND YOUNG, J. (1979). PHOTOELECTRIC LIGHTCURVES OF ASTEROIDS 42,
45, 56, 103, 532 AND 558. ICARUS, 38, 100-105.
44 HARWOOD,M.(1921). EROS AND PALMA. HARVRD. BULL. 766.
45 HAUPT, H. (1958) PHOTOELEKTRISCH- PHOTOMETRISCHE STUDIE AN VESTA. MITT.
SONNOBS. KANZELHOHE 14, 172-173.
46 IMHOFF, K. PERSONAL COMMUNICATION.
47 LAGERKVIST,C.(1976). PHOTOGRAPHIC PHOTOMETRY OF THE ASTEROID 291 ALICE.
ICARUS 27, 157-160.
48 LAGERKVIST,C.(1976). PHOTOGRAPHIC PHOTOMETRY OF THE ASTEROID 1975RB. ICARUS
29, 143-146.
49 LAGERKVIST,C.-I. (1977). A PHOTOGRAPHIC LIGHTCURVE OF THE AMOR ASTEROID
1580 BETULIA. ICARUS 32, 233-234.
50 LAGERKVIST,C.(1978). PHOTOGRAPHIC PHOTOMETRY OF 110 MAIN-BELT ASTEROIDS.
ASTRON. ASTROPHYS. SUPPL. 31, 361-381. (SEE ALSO 1977). UPPSALA
ASTRONOMICAL OBSERVATORY REPORT NO. 9, 1-32.)
51 LAGERKVIST, C.-I. (1979). A LIGHTCURVE SURVEY OF ASTEROIDS WITH SCHMIDT
TELESCOPES: OBSERVATIONS OF NINE ASTEROIDS DURING OPPOSITIONS IN 1977,
ICARUS 38, 106-114.
52 LAGERKVIST, C.-I. AND PETTERSSON, B. (1978). 186 CELUTA: A SLOWLY SPINNING
ASTEROID. ASTRON. ASTROPHYS. SUPPL. 32, 339-342.
53 LUSTIG, G. (1977). DIE ROTATIONSPERIODE UND DIE PHOTOELEKTRISCHE LICHTKURVE
DES PLANETOIDEN 471 PAPAGENA. ASTRON. ASTROPHYS. SUPPL. 30, 117-119.
54 LUSTIG,G. AND DVORAK,R.(1975). PHOTOMETRISCHE UNTERSUCHUNGEN DER
PLANETOIDEN (43) ARIADNE UND (71) NIOBE. ACTA PHYSICA AUSTRIACA 43, 89-97.
55 LUSTIG,G. UND HAHN,G.(1976). PHOTOELEKTRISCHE LICHTKURVEN DER PLANETOIDEN
(2) PALLAS UND (704) INTERAMNIA. ACTA PHYSICA AUSTRIACA 44, 199-205.
56 MILLIS,R.L., BOWELL,E., AND THOMPSON,D.T. (1976). UBV PHOTOMETRY OF
ASTEROID 433 EROS. ICARUS 28, 53-68.
57 MINER, E., AND YOUNG, J. (1969) PHOTOMETRIC DETERMINATION OF THE ROTATION
PERIOD OF 1566 ICARUS. ICARUS 10, 436-440.

                          REFERENCES
58 MINER,E. AND YOUNG,J. (1976). FIVE-COLOR PHOTOELECTRIC PHOTOMETRY OF
   ASTEROID 433 EROS. ICARUS 28, 43-52.
59 PADOVA,E. (1921). OSSERVAZIONI FOTOMETRICHE DEI PIANETI (5) ASTREA, (44)
   NYSA E (8) FLORA. MEM. SOC. ASTRON. ITAL. 2, 82-92.
60 POP,V. AND CHIS,D. (1976). PHOTOELECTRIC OBSERVATIONS OF THE ASTEROIDS 433
   EROS AND 1 CERES. CONTRIB. ASTRON. OBS. UNIV. BABES-BOLYAI, CLUJ-NAPOCA,
   ROMANIA, PP. 33-40.
61 POP,V. AND CHIS,D. (1976). PHOTOELECTRIC LIGHTCURVES OF ASTEROID 433 EROS.
   ICARUS 28, 39-42.
62 PORTER, A. AND WALLINTINE, D. (1976). MINOR PLANET ROTATIONS REPORTED IN
   1975. MIN. PLAN. BULL. 3, 48-50.
63 PORTER,A., WALLENTINE,D., TEDESCO,E.F. AND PILCHER,F.(1976). LIGHT CURVES
   AND OTHER PHYSICAL OBSERVATIONS OF 233 ASTEROPE IN 1975. MINOR PLANET BULL.
   3, 47-48.
64 RIGOLLET,R. (1950). SUR LES CHANGEMENTS D'ECLAT A COURTE PERIODE DES
   PETITES PLANETES ET SUR LA VARIABILITE DE (63) AUSONIA. C.R. 230, 2077-2078
65 SATHER,R.E.(1976). MINOR PLANETS AND RELATED OBJECTS. XIX. SHAPE AND POLE
   ORIENTATION OF (39) LAETITIA, ASTRON. JOURNL. 81, 67-73.
66 SATHER,R.E., TAYLOR,R.C., AND TEDESCO,E.F.(1979). PERIOD, POLE AND PHASE
   FUNCTION OF (29) AMPHITRITE. TO BE SUBMITTED TO ASTRON. J.
67 SCALTRITI,F. AND ZAPPALA,V.(1975). A PHOTOMETRIC STUDY OF THE MINOR PLANET
   15 EUNOMIA. ASTRON. ASTROPHYS. SUPPL. 19, 249-255.
68 SCALTRITI,F. AND ZAPPALA,V.(1976). A PHOTOMETRIC STUDY OF THE MINOR PLANETS
   192 NAUSIKAA AND 79 EURYNOME. ASTRON. ASTROPHYS. SUPPL. 23, 167-179.
69 SCALTRITI,F. AND ZAPPALA,V. (1976). PHOTOMETRIC LIGHTCURVES AND POLE
   DETERMINATION OF 433 EROS. ICARUS 28, 29-36.
70 SCALTRITI,F. AND ZAPPALA,V.(1977). PHOTOELECTRIC PHOTOMETRY OF THE MINOR
   PLANETS 41 DAPHNE AND 129 ANTIGONE. ASTRON ASTROPHYS. 56, 7-11.
71 SCALTRITI,F. AND ZAPPALA,V.(1977). A PHOTOMETRIC STUDY OF THE MINOR PLANET
   (63) AUSONIA. ICARUS 31, 498-502.
72 SCALTRITI,F., AND ZAPPALA,V. (1977). ROTATION PERIOD OF THE ASTEROID 52
   EUROPA. ASTRON. ASTROPHIS. SUPPL. 30, 169-174.
73 SCALTRITI, F. AND ZAPPALA, V. (1978). PHOTOELECTRIC PHOTOMETRY OF ASTEROIDS
   37, 80, 97, 216, 270, 313 AND 471. ICARUS 34, 428-435.
74 SCALTRITI, F. AND ZAPPALA, V. (1979). PHOTOELECTRIC PHOTOMETRY AND ROTATION
   PERIODS OF THE ASTEROIDS 26 PROSERPINA, 194 PROKNE, 287 NEPTHYS, AND 554
   PERAGA. SUBMITTED TO ICARUS, NOVEMBER 1978.
75 SCALTRITI, F. ZAPPALA, V. AND SCHOBER, H. J. (1979). LONG ROTATION PERIODS
   OF THE ASTEROIDS 128 NEMESIS AND 393 LAMPETIA. ICARUS, 37, 133-141.
76 SCALTRITI, R. ZAPPALA, V. AND SCHOBER, H.J. (1979). PHOTOELECTRIC
   PHOTOMETRY AND ROTATION PERIODS OF THREE LARGE AND DARK ASTEROIDS
   49 PALES, 88 THISBE AND 92 UNDINA. ASTRON. ASTROPHYS. SUPPL., 36, 1-8.
77 SCALTRITI, F., ZAPPALA, V. AND STANZEL, R. (1978). LIGHTCURVES, PHASE
   FUNCTION AND POLE OF THE ASTEROID 22 KALLIOPE. ICARUS 34, 93-98.
78 SCHOBER,H.J.(1975). THE MINOR PLANET 654 ZELINDA: ROTATION PERIOD AND LIGHT
   CURVE. ASTRON. ASTROPHYS. 44, 85-89.
79 SCHOBER,H.J.(1976). PHOTOELECTRIC PHOTOMETRY OF THE MINOR PLANET 32 POMONA:
   COMPOSITE LIGHT CURVE AND THE SYNODIC PERIOD OF ROTATION. ASTRON.
   ASTROPHYS. 53, 115-119.
80 SCHOBER,H.J.(1976). THE PERIOD OF ROTATION AND THE PHOTOELECTRIC LIGHT
   CURVE OF THE MINOR PLANET 79 EURYNOME. ICARUS 28, 415-420.
81 SCHOBER,H.J.(1978). PHOTOELECTRIC LIGHTCURVE AND THE PERIOD OF ROTATION OF
   THE ASTEROID 200 DYNAMENE: A FURTHER OBJECT WITH LOW SPIN RATE. ASTRON.
   ASTROPHYS. SUPPL. 31, 175-178.
82 SCHOBER, H. J. PERSONAL COMMUNICATION.
83 SCHOBER, H. J. (1978). PHOTOMETRIC VARIATIONS OF THE MINOR PLANETS 55
   PANDORA AND 173 INO DURING THE OPPOSITION IN 1977: LIGHTCURVES AND ROTATION
   PERIODS. ASTRON. ASTROPHYS. SUPPL. 34, 377-381.
84 SCHOBER,H.J. AND DVORAK,R.(1975). ROTATION PERIOD AND PHOTOELECTRIC LIGHT
   CURVE OF THE MINOR PLANET 675 LUDMILLA. ASTRON. ASTROPHYS. 44, 81-84.
85 SCHOBER,H.J. AND LUSTIG,G.(1975). A PHOTOMETRIC INVESTIGATION OF THE
   ASTEROID (89) JULIA. ICARUS 25, 339-343.
86 SCHOBER,H.J., SCALTRITI,F. AND ZAPPALA,V.(1977). A POSSIBLE ROTATION PERIOD
   FOR THE MINOR PLANET 164 EVA. ICARUS 31, 175-179.
87 SCHROLL,A., HAUPT,H.F. AND MAITZEN,H.M.(1976). ROTATION AND PHOTOMETRIC
   CHARACTERISTICS OF PALLAS. ICARUS 27, 147-156.
88 SHATZEL, A.V. (1954) PHOTOMETRIC STUDIES OF ASTEROIDS. III. THE
   LIGHTCURVE OF 44 NYSA, ASTROPHYS. J. 120, 547-550.
89 STANZEL, R. (1979). LIGHTCURVES AND ROTATION PERIOD OF MINOR PLANET 283
   EMMA. ASTRON. ASTROPHYS. SUPPL. 34, 373-376.
90 STEPHENSON, C.B. (1951) THE LIGHTCURVE AND THE COLOR OF VESTA. ASTROPHYS.
   JOURNAL 114, 500-504.
91 DEBEHOGNE, H. PERSONAL COMMUNICATION.
92 SURDEJ,J. AND SURDEJ,A. (1978). PHOTOMETRIC OBSERVATIONS OF 1978CA AND
   1978DA. THE MESSENGER. NO. 13, 3-4.

REFERENCES

93 SURDEJ, J. AND SURDEJ, H. (1977). ROTATION PERIOD AND PHOTOELECTRIC LIGHT-
CURVES OF ASTEROID 471 PAPAGENA. ASTRON. ASTROPHYS. SUPPL. 30, 121-124.
94 TAYLOR,R.C.(1973). MINOR PLANETS AND RELATED OBJECTS. XIV. ASTEROID (4)
VESTA. ASTRON. JOURNL. 78, 1131-1139.
95 TAYLOR,R.C.(1977). MINOR PLANETS AND RELATED OBJECTS. XXIII. PHOTOMETRY OF
ASTEROID (7) IRIS. ASTRON. JOURNL. 82, 441-444.
96 TAYLOR, R. C. (1978). MINOR PLANETS AND RELATED OBJECTS. XXIV. PHOTOMETRIC
OBSERVATIONS FOR (5) ASTRAEA. ASTRON. JOURNL. 83, 201-204.
97 TAYLOR,R.C.,GEHRELS,T. AND CAPEN,R.C.(1976). MINOR PLANETS AND RELATED
OBJECTS. XXI. PHOTOMETRY OF EIGHT ASTEROIDS. ASTRON. JOURNL. 81, 778-786.
98 TAYLOR, R.C., GEHRELS, T., AND SILVESTER, A.B. (1971) MINOR PLANETS AND
RELATED OBJECTS. VI. ASTEROID (110) LYDIA. ASTRON. J. 76, 141-146.
99 TEDESCO,E.F. (1976). UBV LIGHTCURVES OF ASTEROID 433 EROS. ICARUS 28,
21-28.
100 TEDESCO,E.F.(1979). A PHOTOMETRIC INVESTIGATION OF THE COLORS, SHAPES, AND
SPIN RATES OF HIRAYAMA FAMILY ASTEROIDS. PH.D. DISSERTATION. NEW MEXICO
STATE UNIVERSITY. LAS CRUCES, NEW MEXICO.
101 TEDESCO,E.F. AND BOWELL,E. (1979). LIGHTCURVES AND UBV PHOTOMETRY OF 944
HIDALGO. TO BE SUBMITTED TO ASTRON. J.
102 TEDESCO,E., DRUMMOND,J., CANDY,M., BIRCH,P., NIKOLOFF,I., SCHOBER,H.,
ZAPPALA,V., AND SATHER,R.(1979). WORLDWIDE OBSERVATIONS OF (1) CERES
DURING THE 1975-1976 APPARITION. TO BE SUBMITTED TO ICARUS.
103 TEDESCO,E., DRUMMOND,J., CANDY,M., BIRCH,P., NIKOLOFF,I., ZAPPALA,V., AND
SATHER,R.(1979). WORLDWIDE OBSERVATIONS OF (16) PSYCHE DURING THE 1975-1976
APPARITION. TO BE SUBMITTED TO ICARUS.
104 TEDESCO,E., DRUMMOND,J., CANDY,M., BIRCH,P., NIKOLOFF,I. AND ZELLNER,B.
(1978). 1580 BETULIA: AN UNUSUAL ASTEROID WITH AN EXTRAORDINARY LIGHTCURVE,
ICARUS 35, 340-355.
105 TEMPESTI,P.(1975). PHOTOELECTRIC OBSERVATIONS OF THE MINOR PLANET 704
INTERAMNIA DURING ITS 1969 OPPOSITION. MEM. SOC. ASTRON. ITAL. 46, 397-405.
106 TEMPESTI,P. AND BURCHI,R.(1969). A PHOTOMETRIC RESEARCH ON THE MINOR PLANET
12 VICTORIA. NOTE E COMUNICAZIONI N. 48, OSSERVATORIO ASTRONOMICO DI
COLLURANIA TERAMO, 1-20.
107 VAN GENT,H.(1933). PERIOD, LIGHT CURVE, AND EPHEMERIS OF THE NEW ASTEROID
WITH VARIABLE BRIGHTNESS 1931 PH. BULL. ASTRON. INST. NETH. 7, 65-66.
108 VAN HOUTEN, C. J. (1962). AN INVESTIGATION OF ASTEROID LIGHT-CURVES ON
FRANKLIN-ADAMS PLATES. BULL. ASTRON. SOC. NETH. 16, 160-162.
109 VAN HOUTEN- GROENEVELD, I., AND VAN HOUTEN, C.J. (1958) PHOTOMETRIC
STUDIES OF ASTEROIDS. VII. ASTROPHYS. J. 127, 253-273.
110 VAN HOUTEN- GROENEVELD, I. AND VAN HOUTEN, C.J. (1972) IN PREPARATION.
111 VAN HOUTEN-GROENEVELD, I., VAN HOUTEN, C.J., AND ZAPPALA,V. (1979).
PHOTOELECTRIC PHOTOMETRY OF SEVEN ASTEROIDS. ASTRON. ASTROPHYS. SUPPL.,
35, 223-232.
112 VEVERKA,J.F.(1970). PHOTOMETRIC AND POLARIMETRIC STUDIES OF MINOR PLANETS
AND SATELLITES. PH.D. THESIS, 157-160. HARVARD UNIV.
113 VEVERKA, J. (1971) PHOTOPOLARIMETRIC OBSERVATIONS OF THE MINOR PLANET
FLORA. ICARUS 15, 454-460.
114 VEVERKA, J., AND LILLER, W. (1969) OBSERVATIONS OF ICARUS: 1968. ICARUS
10, 441-444.
115 WAMSTEKER, W. AND SATHER, R.E. (1974) MINOR PLANETS AND RELATED OBJECTS.
XVII. FIVE-COLOR PHOTOMETRY OF FOUR ASTEROIDS. ASTRON. J. 79, 1465-1470.
116 WELCH,D., BINZEL,R., AND PATTERSON,J. (1974). THE ROTATION PERIOD OF 18
MELPOMENE. MINOR PLANET BULL. 2, 20-21.
117 WOOD, H.J., AND KUIPER, G.P. (1963) PHOTOMETRIC STUDIES OF ASTEROIDS. X.
ASTROPHYS. J. 137, 1279-1285.
118 YANG,X.Y., ZHANG,Y.Y. AND LI,X.Q.(1965). PHOTOMETRIC OBSERVATIONS OF
VARIABLE ASTEROIDS. III. ACTA ASTRON. SINICA 13, 66
119 ZAPPALA,V. PERSONAL COMMUNICATION.
120 ZAPPALA,V., VAN HOUTEN-GROENEVELD,I., AND VAN HOUTEN,C.J.(1979). ROTATION
PERIOD AND PHASE CURVE OF THE ASTEROIDS 349 DEMBOWSKA AND 354 ELEONORA.
ASTRON. ASTROPHYS. SUPPL., 35, 213-221.
121 ZELLNER,B.(1977). PERSONAL COMMUNICATION.
122 UNPUBLISHED LPL OBSERVATIONS.

VII. MAGNITUDES, COLORS, TYPES
AND ADOPTED DIAMETERS
OF THE ASTEROIDS

E. BOWELL
Lowell Observatory

and

T. GEHRELS, B. ZELLNER
University of Arizona

This table presents the TRIAD files for absolute magnitude, UBV colors, and adopted types and diameters. As such it contains more blanks than any of the other tables. The blanks represent terrae incognitae and should stand as challenges to the observers.

The first six columns give the asteroid number, or preliminary designation if not yet numbered; the absolute magnitude B(1,0); a quality code for the magnitude; the mean phase of the magnitude observations; the linear phase coefficient used; and a reference for the magnitude data. The magnitude B(1,0) is on the true UBV system, and is computed according to

$$B(1,0) = B - 5 \log r\rho - (\text{PHASE}) \times (\text{COEFF})$$

when $\text{PHASE} \geq 7°$, or

$$B(1,0) = B - 5 \log r\rho + 0.538 - 0.134 \times (\text{PHASE})^{0.714}$$
$$- 7.0 \times (\text{COEFF})$$

when $\text{PHASE} < 7°$. (See chapter by Bowell and Lumme and the introduction by Gehrels in this book.) Distances are expressed in astronomical units, namely r measured between the asteroid and the sun and ρ between the asteroid and the earth at the time of observation. COEFF is assumed to be 0.039 mag/deg unless another value is indicated, such other value being obtained from extensive observations of the individual asteroid.

Photometric quality code E indicates an excellent source, from either photoelectric observations or from repeated observation in the Yerkes-McDonald Survey or the Palomar-Leiden Survey (see Gehrels' chapter in Part I of this book). S is used for survey quality, with observations made only photographically or from a single photoelectric observation. When an integer follows the letters E or S, it is the number of apparitions in which the magnitude observations have been made. If there is no letter, but for instance 1.8 for asteroid 31, it means that the source of the magnitude is not well known, but the QUAL given indicates some weight; for instance 0.6 means that the source is totally unknown, or it may be from some observatory that does not specialize in photometry but gives magnitudes only incidental to their astrometry. A colon indicates discordant data of good quality, but possibly having large lightcurve or aspect variations.

Succeeding columns list the U–B color index, its quality code, the B–V color and its quality, and references for the color data. Most of the color indices result from the weighted average of two or more observations on separate nights. Weight 0 applies for data considered uncertain, unreliable, or fragmentary (errors may exceed 0.05 mag), and weight 1 for data from a single night or from several nights disagreeing by 0.05 mag or more (errors unlikely to exceed 0.05 mag). Weight 2 indicates secure color indices, from two or more nights' concordant data (errors unlikely to exceed 0.03 mag), and weight 3 indicates results that can hardly be improved. No corrections for phase reddening have been applied; since the average phase of observation is about 10 deg, the listed values are likely to average 0.01 to 0.02 magnitude redder than the color indices at zero phase.

The final six columns give the taxonomic type, orbital element zone, an adopted diameter in km, the weight or quality of the diameter estimate, an apparent magnitude bin, and a data code as described below. The taxonomic system and the classification and diameter algorithms are discussed in Zellner's chapter in Part V, and the orbit zones are defined in Table III of that chapter.

The diameters in particular must not be used uncritically; values of weight 3 are based on actual polarimetric or radiometric observations of good quality, while those of weight 2 are generated with an albedo assumed on the basis of a fairly secure type classification, and those of weight 1 are based only on a computerized best guess at the albedo.

The magnitude bin is computed in half-magnitude steps according to the FORTRAN code

$$\text{BIN} = \text{IFIX}(2.0 * (\text{AMAG} - 5.0) + 1.0)$$

where AMAG is the apparent blue magnitude at mean opposition. The observational bias factor for any classified asteroid may be determined from the magnitude bin and Table IV of Zellner's chapter.

The final column is a four-digit code, of which the first digit gives the observational weight of the radiometric albedo (see TRIAD Table V); the second digit gives the total weight of the polarimetric albedo and P_{min} (also in Table V); the third digit is the sum of the weights of the B-V and U-B color indices (this table); and the fourth is the sum of weights of the spectrophotometric parameters R/B, BEND, and DEPTH (TRIAD Table IV). Thus the code gives at a glance the kinds and qualities of the observational data that went into the type and diameter analyses.

NUMBER	BMAG QUAL	PHASE	COEFF	REF	U-B	QU	B-V	QB	COLOR REFS	TYPE	ZONE	DIAM	WT	BIN	DATA	NUMBER
1	4.35 E7	9.66	0.038	M1	0.42	3	0.72	3	01 02 03 04 12	C	II	1025.	3	6	3369	1
2	5.01 E2	8.80	0.038	M1	0.29	3	0.55	3	01 04 09	U	I	583.	3	7	2269	2
3	6.47 E4	13.74	0.025	M1	0.42	3	0.82	3	01 02 03 04 99	S	II	249.	3	10	2669	3
4	4.24 E3	19.01	0.026	M1	0.49	3	0.78	3	01 03 04 99	U	I	555.	3	4	3569	4
5	8.26 E3	11.56	0.015	M1	0.42	2	0.83	3	01 02 04 21	S	II	106.	3	13	2559	5
6	6.85 E3	16.44	0.028	M1	0.39	?	0.83	3	01 02 04 11 19 99	S	I	206.	3	10	2359	6
7	6.61 E4	8.82	0.038	M1	0.47	3	0.84	3	01 02 04 20	S	I	222.	3	9	3669	7
8	7.61 E2	13.28	0.028	M1	0.46	2	0.88	2	01 02 03 04 07	S	FL	160.	3	10	2549	8
9	7.49 E2I	5.60	0.034	M1	0.50	3	0.87	3	01 03 04 07	S	I	168.	3	11	3467	9
10	6.32 E4	7.96		M1	0.31	2	0.69	3	01 04	C	III	443.	3	11	3259	10
11	7.79 E1	11.10	0.030	M1	0.41	3	0.83	3	01 02 03 04	S	I	155.	3	12	2569	11
12	8.37 E2I	10.17	0.030	M1	0.51	3	0.89	3	03 04 05	S	I	135.	3	12	2267	12
13	7.56 E3	17.00	0.045	M1	0.45	3	0.75	3	02 04 05	C	II	245.	3	12	3269	13
14	7.54 E4	4.30	0.023	M1	0.35	3	0.83	3	01 02 04 11	S	II	116.	3	13	2369	14
15	6.32 E3I	12.56	0.038	M1	0.44	3	0.82	3	01 02 04 37	S	II	261.	3	10	2169	15
16	6.91 E3I	10.47	0.028	M1	0.25	3	0.71	3	01 04 12	M	III	249.	3	12	3267	16
17	8.93 E3	5.22		M1	0.41	2	0.84	2	01 02 03 04	S	I	97.1	3	14	2249	17
18	7.45 S3	17.80		M1	0.37	2	0.84	2	01 02 03 04	S	I	164.	3	10	2647	18
19	8.16 E3	5.85		M1	0.38	3	0.74	3	02 03 04	C	I	226.	3	12	3269	19
20	7.59 E2	25.64	0.031	M1	0.43	3	0.82	2	01 04	S	I	140.	3	11	2459	20
21	8.56 E2	14.12	0.026	M1	0.19	3	0.71	3	01 02 04	M	I	114.	3	13	2369	21
22	7.52 E2	7.88	0.031	M1	0.27	3	0.71	3	01 02 03 04	M	III	179.	3	13	2366	22
23	7.87 E3	23.62		M1	0.45	3	0.87	3	01 02 03 04	S	II	118.	3	13	2167	23
24	7.90 E5	4.47	0.050	M1	0.34	3	0.69	3	02 04	C	TH	249.	2	15	169	24
25	9.01 E2	20.75		M1	0.51	3	0.93	3	01 04 99	S	PH	72.8	3	14	2069	25
26	8.78 E2	20.65	0.025	M1	0.55	2	0.90	2	04 35	S	II	87.9	2	14	46	26
27	8.11 E2	19.15		M1	0.51	3	0.89	3	03 04 05	S	I	118.	3	12	2169	27
28	8.27 E3	7.70		M1	0.48	2	0.86	2	04	S	II	109.	2	14	1149	28
29	7.02 E7	9.58	0.031	M1	0.42	3	0.84	3	01 02 03 04 13	S	II	199.	3	11	2369	29
30	8.66 E3	4.17		M1	0.45	2	0.88	2	03 04	S	I	95.4	3	13	2447	30
31	7.73 1.8			M1	0.32	1	0.67	1	04	C	III	270.	2	14	109	31
32	8.63 E2	8.81	0.038	M1	0.42	3	0.86	3	04	S	II	92.5	2	14	269	32
33	9.51 E1	8.01		M1	0.40	1	0.85	1	04	S	III	61.7	2	17	20	33
34	9.46 E2	13.04		M1	0.37	3	0.69	3	03 04	C	II	121.	2	16	1069	34
35	9.68 5.5			M1	0.32	1	0.71	1	04							35
36	9.46 E2	17.25		M1	0.35	2	0.73	2	02 04	C	II	124.	2	16	1042	36
37	8.26 E1	22.37		M1	0.44	2	0.84	3	01 02 04 05 14	S	II	95.7	3	13	2159	37
38	9.57 E1	13.05		M1	0.45	1	0.74	1	04	CU	II	118.	2	16	20	38
39	7.45 E5I	13.71	0.027	M1	0.49	3	0.89	3	01 02 03 04 06 11 99	S	II	158.	3	12	1369	39
40	8.17 E4	10.47		M1	0.43	3	0.85	3	01 02 04 14	S	I	118.	3	11	2469	40
41	8.16 E1	16.20	0.053	M1	0.35	3	0.73	3	04	C	II	204.	3	14	2169	41
42	8.58 E2	12.63		M1	0.46	3	0.87	3	04 05	S	I	104.	3	13	2466	42
43	9.02 E2I	5.04	0.047	M1	0.48	2	0.87	3	04 18	S	FL	77.6	3	13	1059	43
44	7.95 E4I	22.49	0.019	M1	0.24	3	0.70	3	01 04 05	E	NY	67.8	3	12	2369	44
45	8.11 E3	15.67		M1	0.28	3	0.68	3	04	U	II	250.	3	13	2066	45
46	9.30 E2	9.37	0.042	M1	0.23	3	0.70	3	02 03 04	C	III	156.	2	16	1020	46
47	8.88 E1	12.57		M1	0.29	1	0.65	1	03 04	U	III	156.	2	16	47	47
48	7.97 E2	11.73		M1	0.45	2	0.73	2	02 04	U	III	149.	1	15	49	48
49	8.72 E1	15.58		M1	0.38	2	0.74	2	04	C	III	175.	2	16	40	49
50	10.16 E1	2.19		M1	0.35	2	0.70	2	04	C	II	88.3	2	17	40	50
51	8.45 E2	15.97	0.042	M1	0.46	2	0.78	2	01 02 04 11	U	I	156.	3	13	3249	51
52	7.29 E2I	15.88		M1	0.35	3	0.68	3	01 02 03 04	C	III	291.	3	13	2069	52
53	9.70 E2	8.75	0.050	M1	0.32	3	0.71	2	04	C	II	110.	2	16	56	53
54	8.68 E3	10.19		M1	0.35	2	0.73	2	03 04	C	II	177.	3	15	2346	54
55	8.59 E1	8.77		M1	0.24	2	0.70	3	04 23	CMEU	II	185.	1	14	50	55
56	9.29 E1	9.28		M1	0.31	2	0.70	2	03 04	S	II	142.	3	15	2240	56
57	8.15 E1	15.29		M1	0.48	2	0.85	2	04	S	II	115.	2	15	140	57
58	9.81 E1	6.80		M1	0.38	3	0.68	2	04	C	II	165.	1	15	40	58
59	8.76 E1	21.47		M1	0.28	2	0.65	2	04	CEU	II	116.	2	14	40	59
60	9.68 E3	11.69		M1	0.46	3	0.85	3	01 04 05	S	I	51.5	3	15	2066	60
61	8.73 E3	4.62		M1	0.39	3	0.85	3	01 04 05	S	III	87.9	2	16	80	61
62	9.67 E1	17.86		M1	0.38	2	0.74	2	01 02 04	U	TH	68.6	1	18	46	62
63	8.67 E3I	11.41	0.035	M1	0.47	3	0.91	3	01 02 04	S	I	94.3	3	13	3469	63
64	8.61 E2	7.86		M1	0.27	3	0.77	2	02 04	E	II	59.9	3	14	2559	64
65	7.75 E1	10.07	0.027	M1	0.28	2	0.68	3	04	C	IV	311.	3	15	2257	65
66	10.31 E2	8.48		M1	0.35	2	0.71	2	04 05	C	II	82.8	2	18	1046	66
67	9.38 E2	13.10		M1	0.44	3	0.86	3	03 04 05	S	I	65.5	2	15	1060	67
68	7.99 E3	13.53		M1	0.49	3	0.84	2	02 04	S	II	128.	3	13	3159	68
69	7.97 E2	6.21	0.035	M1	0.23	2	0.69	3	01 04 14 15	M	III	108.	3	14	2157	69
70	9.03 E1	16.78		M1	0.41	2	0.74	2	04	S	II	153.	3	15	2240	70
71	8.30 E1	11.52		M1	0.40	1	0.93	2	04 99	S	II	136.	2	14	136	71
72	9.99 E3	6.18		M1	0.38	2	0.78	3	04	U	FL	60.3	1	15	1050	72
73	10.19 3.5			M1	0.43	1	0.82	1	04							73
74	9.62 S2	4.63		M1	0.32	1	0.69	2	04	C	II	113.	2	17	30	74
75	9.91 S1	8.03		M1	0.26	2	0.70	2	04	CMEU	II	99.1	1	17	1060	75
76	8.81 S2	14.43		M1	0.28	3	0.70	3	04	M	IV	79.1	2	17	1060	76
77	9.57 E2	20.75	0.032	M1	0.24	3	0.74	3	04	C	II	56.7	2	16	1060	77
78	9.11 E2	7.77	0.037	M1	0.37	3	0.71	3	02 04 12	C	II	144.	2	15	69	78
79	9.07 E2	19.80	0.032	M1	0.43	3	0.87	3	02 03 04	S	I	79.7	3	14	3069	79
80	9.11 E2	4.55		M1	0.50	2	0.92	2	04	U	I	84.2	3	13	2149	80
81	9.47 E2	8.52		M1	0.35	2	0.71	3	04	C	III	122.	2	17	50	81
82	9.24 E2	17.77		M1	0.37	3	0.81	3	04	S	II	66.4	3	16	2067	82
83	9.51 E1	18.33		M1	0.28	1	0.68	1	04	S	I	118.	2	15	126	83
84	10.39 E1	18.00		M1	0.44	1	0.73	3	02 03 04	S	II	86.6	3	16	2242	84
85	8.60 E1	17.24		M1	0.26	2	0.59	2	04	S	II	149.	3	14	2147	85
86	9.60 E2	10.88	0.031	M1	0.34	2	0.68	2	04	C	III	113.	2	18	40	86
87	7.88 E1	6.93		M1	0.24	2	0.59	2	04	CMEU	IV	251.	1	16	49	87
88	8.01 E2	5.79	0.036	M1	0.28	2	0.68	3	04	S	II	214.	3	13	2159	88
89	7.70 E2	25.20	0.038	M1	0.47	2	0.85	2	04	S	II	168.	3	12	2749	89
90	9.18 E1	9.05		M1	0.31	2	0.69	2	04	C	TH	138.	2	17	46	90
91	9.86 E1	4.58		M1	0.32	3	0.73	3	04 05	C	II	106.	3	16	2060	91
92	7.51 E1	16.76		M1	0.28	2	0.72	3	04	U	III	184.	1	14	256	92
93	8.55 E1	16.96		M1	0.32	1	0.70	1	04	C	III	170.	3	14	2226	93
94	8.47 E2	2.62		M1	0.31	2	0.65	2	04	S	III	191.	3	16	2046	94
95	8.80 E3	8.31		M1	0.38	2	0.73	2	02 03 04	C	III	168.	2	16	1140	95
96	8.65 E1	13.27		M1	0.34	2	0.77	2	04	U	III	111.	1	16	40	96
97	8.44 E2	22.33		M1	0.22	3	0.72	2	04	M	II	109.	3	14	2059	97
98	9.80 E1	21.20		M1	0.42	1	0.74	1	04	C	II	106.	2	17	20	98
99	10.40 E1	4.32		M1	0.31	2	0.70	2	04	C	II	79.1	2	18	40	99
100	8.79 E2	5.96		M1	0.34	1	0.63	2	04	SU	II	84.7	2	16	30	100
101	9.17 E1	21.35		M1	0.47	2	0.88	3	04	S	II	72.8	2	15	50	101
102	10.16 E2	8.62		M1	0.39	1	0.74	2	04	S	II	89.9	2	17	30	102
103	8.75 E1	8.75	0.032	M1	0.45	3	0.85	3	03 04 05	S	II	87.5	2	15	1060	103
104	9.23 E2	6.59		M1	0.34	3	0.68	3	04	C	II	134.	2	17	60	104
105	9.33 E2	18.28		M1	0.31	2	0.69	2	04	S	PH	134.	2	14	1049	105
106	8.50 E1	14.19		M1	0.46	2	0.75	3	04	U	III	118.	1	16	1057	106
107	7.88 E1	5.48		M1	0.29	2	0.70	2	04	S	IV	252.	2	16	1040	107
108	9.24 E1	5.57		M1	0.48	2	0.85	2	04	S	III	69.5	2	18	47	108
109	9.93 E1	9.49		M1	0.41	?	0.69	2	04	S	II	75.5	3	17	2040	109
110	8.78 E1	10.31	0.032	M1	0.30	3	0.71	3	01 03 04	U	II	102.	1	15	1069	110
111	8.82 E2	21.64		M1	0.39	3	0.70	3	04	C	II	156.	2	15	63	111
112	10.76 E1I	18.47		M1	0.29	2	0.70	2	04	C	I	67.0	2	17	40	112
113	9.64 E3	16.73		M1	0.51	2	0.93	2	04 05	U	I	48.6	3	15	2046	113
114	9.36 4.3			M1	0.37	?	0.77	2	04 05	C	II	131.	2	16	120	114
115	8.56 E2	16.34		M1	0.46	2	0.85	2	02 03 04	S	I	94.5	3	13	2247	115
116	8.80 E1I	4.20		M1	0.45	0	0.82	1	04	S	II	80.3	3	15	2117	116

NUMBER	BMAG	QUAL	PHASE	COEFF	REF	U-B	3U	B-V	3B	COLOR REFS	TYPE	ZONE	DIAM	WT	BIN	DATA	NUMBER
117	8.91	E1	11.64		M1	0.30	3	0.69	3	04	C	III	156.	2	16	60	117
118	9.87	3.9			M1	0.43	0	0.95	1	04							118
119	9.59	E1	6.61		M1	0.43	3	0.96	2	04	S	II	59.4	2	16	59	119
120	8.68	E2	15.14		M1	0.37	3	0.71	3	04	C	III	175.	3	16	2060	120
121	8.34	E1*	16.93		M1	0.44	1	0.75	2	04	C	IV	209.	2	16	136	121
122	8.84	E2	12.17		M1	0.41	1	0.78	0	01 04	S	III	83.6	2	17	16	122
123	10.00	E1	9.01		M1	0.39	2	0.85	2	04 05	S	II	49.0	2	17	1040	123
124	9.21	E2	9.70		M1	0.42	3	0.85	3	03 04	S	II	72.3	3	15	2066	124
125	9.79	E1	15.28		M1	0.24	2	0.67	2	04	CMEU	II	103.	1	17	40	125
126	10.39	E2	10.50		M1	0.49	2	0.87	3	04	S	I	41.3	2	17	50	126
127	9.46	S2	5.93		M1	0.35	2	0.70	2	04							127
128	8.48	E2	10.06		M1	0.42	0	0.70	1	04	CEU	II	191.	1	14	19	128
129	8.02	E3	13.85	0.030	M1	0.25	3	0.72	3	02 04 05	U	III	113.	3	14	2269	129
130	8.09	E3	18.15		M1	0.49	3	0.76	3	04 05	U	III	143.	1	15	1069	130
131	10.94	2.6			M1	0.39	2	0.74	2	04	SM	I	37.1	3	18	2000	131
132	10.21	1.8			M1												132
133	9.23	E1	6.24		M1	0.53	2	0.92	2	04	S	III	72.1	2	17	40	133
134	7.56	E1	16.89		M1	0.35	2	0.70	3	04	C	II	116.	2	16	50	134
135	9.07	E2	8.85		M1	0.27	2	0.69	2	02 04	M	I	79.3	3	14	2140	135
136	10.78	3.9			M1	0.23	1	0.74	1	04	MEJ	FL	41.1	1	17	27	136
137	8.98	E2	8.46		M1	0.31	2	0.72	2	04	C	III	153.	2	17	1040	137
138	9.89	E2	7.21		M1	0.49	2	0.88	2	04	S	I	52.2	2	16	40	138
139	8.80	E1	10.89		M1	0.29	1	0.70	1	03 04	C	II	165.	2	15	1229	139
140	9.33	E2	9.53		M1	0.30	2	0.72	2	04	C	II	105.	3	16	3049	140
141	9.55	2.0			M1	0.29	1	0.66	2	04	C	II	117.	2	16	309	141
142	11.11	E1	7.35		M1	0.25	3	0.61	3	04 14	U	NY	33.3	1	18	1060	142
143	10.14	E1	3.79		M1	0.34	2	0.74	2	04	C	II	90.8	2	18	40	143
144	8.91	E2	23.39		M1	0.40	2	0.73	3	04	C	II	132.	3	15	2059	144
145	9.21	E1	18.90		M1	0.37	1	0.70	1	04	C	II	137.	2	15	229	145
146	8.96	E1	18.85		M1	0.41	3	0.69	3	04	C	II	153.	2	15	1060	146
147	9.62	E1	3.43		M1	0.25	1	0.67	1	04	CMEU	III	112.	1	18	20	147
148	8.64	E1	11.28		M1	0.42	2	0.86	2	04	S	II	92.0	2	15	40	148
149	11.85	E1	1.97		M1	0.50	2	0.80	2	04	U	I	25.8	1	18	46	149
150	9.21	E2	6.69		M1	0.27	2	0.71	2	04	CEU	III	137.	1	17	40	150
151	10.39	E2	6.70		M1	0.50	2	0.87	3	04	S	II	41.3	2	17	50	151
152	9.49	E2	6.95		M1	0.45	1	0.88	2	04	S	III	62.8	2	18	30	152
153	8.40	E1	9.95		M1	0.29	2	0.66	2	04 14	U	HI	119.	1	18	140	153
154	8.37	2.3			M1												154
155	12.29	E1	12.52		M1	0.23	1	0.68	2	04	CMEU	II	32.8	1	22	30	155
156	9.69	E2	5.72		M1	0.30	3	0.69	2	04	C	II	109.	2	17	59	156
157	12.34	1.2			M1												157
158	10.57	E3*	9.46		M1	0.38	2	0.84	2	04	S	KO	37.6	3	19	2046	158
159	9.17	E1	9.18		M1	0.38	2	0.70	2	04	C	III	141.	3	17	2060	159
160	10.01	E2	4.70		M1	0.36	1	0.73	2	04	L	II	95.5	2	17	30	160
161	11.09	E1	18.46		M1	0.23	2	0.71	2	04	CMEU	I	100.	1	15	40	161
162	9.77	E2	7.77		M1	0.39	2	0.76	2	04	C	III	109.	2	18	40	162
163	10.63	E1	9.55		M1	0.33	2	0.70	2	04	C	I	71.1	2	17	46	163
164	9.64	E1	13.34		M1	0.32	3	0.68	3	04	C	III	111.	2	16	66	164
165	8.14	E1	15.68		M1	0.31	1	0.74	2	04	C	III	228.	2	15	30	165
166	10.86	E1	11.88		M1	0.42	1	0.72	1	14	U	II	39.3	1	19	29	166
167	10.24	E1	11.43		M1	0.41	1	0.81	1	14	S	KO	43.0	2	18	27	167
168	8.96	E1	4.71		M1	0.40	2	0.74	2	04	C	IV	156.	2	17	40	168
169	10.55	4.1			M1						S	I	38.0	2	17	6	169
170	10.62	1.2*			M1						U	II	45.5	1	18	3	170
171	9.26	E1	8.89		M1	0.34	2	0.68	2	04	U	TH	80.5	1	17	1040	171
172	9.96	E1	8.35		M1	0.49	2	0.90	3	02 04 05	S	I	65.6	3	16	2050	172
173	8.76	E1	9.60		M1	0.33	2	0.71	3	04 23	C	II	169.	2	15	50	173
174	9.50	E3	8.77		M1	0.48	2	0.85	2	04	S	III	61.9	2	17	40	174
175	9.29	E1	5.97		M1	0.33	2	0.70	2	04	U	III	80.2	1	18	47	175
176	9.42	1.3			M1						U	III	79.1	1	18	9	176
177	10.49	E1	6.26		M1	0.35	1	0.73	1	02	C	II	76.9	2	18	20	177
178	10.75	E1*	6.47	0.019	M1	0.49	2	0.90	3	04	S	I	35.5	2	18	50	178
179	9.29	E1	4.95		M1	0.41	3	0.84	3	04	S	III	67.6	2	17	60	179
180	11.41	E1	6.06		M1	0.45	2	0.83	2	04	S	II	25.3	2	20	40	180
181	8.86	E1	16.65		M1	0.36	1	0.81	2	04	S	III	81.3	2	16	37	181
182	10.18	E1	19.26		M1	0.43	2	0.89	2	01 02 03 04	S	I	45.9	2	16	1240	182
183	10.86	E1	3.66		M1	0.36	2	0.84	2	04	U	III	32.8	2	19	40	183
184	9.29	E2	5.52		M1	0.23	2	0.70	2	04 14	CMEU	TH	132.	1	17	40	184
185	8.49	E1	20.36		M1	0.33	2	0.67	2	04	C	II	188.	2	14	40	185
186	10.02	E2*	13.81		M1	0.32	3	0.77	2	04 14 16	U	I	49.1	3	16	2050	186
187	9.13	E1	15.62		M1	0.35	3	0.72	3	04	C	III	143.	2	16	60	187
188	9.94	S1	4.60		M1												188
189	10.49	E2	14.45		M1	0.50	2	0.91	2	04	S	I	42.8	3	17	2040	189
190	8.64	S1	3.20		M1												190
191	10.01	E1	9.77		M1	0.26	0	0.68	1	04							191
192	9.59	E2*	19.59	0.039	M1	0.49	3	0.92	3	02 04	S	I	98.9	3	13	3569	192
193	10.77	E1	4.80		M1												193
194	8.60	E1	17.44		M1	0.34	2	0.74	2	04	C	III	195.	3	14	2049	194
195	9.97	E2	4.02		M1	0.38	2	0.71	2	04	C	II	96.8	2	18	40	195
196	7.65	E2	2.07		M1	0.43	3	0.84	3	04 05	S	III	162.	3	14	2067	196
197	10.36	E1	4.91		M1	0.47	1	0.89	1	04	S	III	42.3	2	18	27	197
198	9.39	E1	19.33		M1	0.41	1	0.88	2	04	S	I	65.8	2	15	36	198
199	9.86	2.6			M1												199
200	9.23	E2	3.75	0.040	M1	0.38	2	0.73	3	04	C	II	137.	2	16	56	200
201	9.09	E1	4.34		M1	0.26	1	0.69	1	04	CMEU	II	144.	1	15	20	201
202	8.57	E2	8.43		M1	0.46	2	0.86	2	04	S	III	95.1	2	16	40	202
203	9.76	E2	5.27		M1	0.29	2	0.70	1	04	C	II	106.	2	17	30	203
204	9.96	E2	9.49		M1	0.42	2	0.81	2	03 04 05	S	II	49.0	2	17	1040	204
205	10.19	E1	9.90		M1	0.32	1	0.69	1	04	C	II	86.7	2	18	20	205
206	9.65	E3	5.76		M1	0.34	2	0.69	3	04	C	II	111.	2	17	50	206
207	10.95	E2	7.15		M1	0.36	2	0.73	2	04	U	I	62.2	2	17	40	207
208	9.82	E2*	2.08		M1	0.42	2	0.73	2	04 14	U	KO	46.2	3	18	3042	208
209	9.20	E2	5.86		M1	0.30	2	0.69	3	04	C	III	137.	2	17	50	209
210	10.16	E1	21.12		M1	0.24	2	0.66	2	04	C	II	86.7	2	18	46	210
211	8.87	E2	2.90		M1	0.36	3	0.72	3	04 05	C	III	168.	3	16	2060	211
212	9.27	E1	2.98		M1	0.29	1	0.70	1	04	C	II	133.	2	17	20	212
213	9.75	E1	13.19		M1	0.24	2	0.67	2	04	CMEU	II	105.	1	17	47	213
214	10.31	E2	4.06		M1	0.23	2	0.69	3	04	CMEU	II	82.0	1	17	50	214
215	10.70	E1	9.91		M1	0.45	2	0.84	2	04	S	III	35.3	2	19	40	215
216	8.04	E1*	19.97		M1	0.24	3	0.71	3	04	CMEU	III	236.	1	14	69	216
217	10.94	S2	3.35		M1						CMEU	III	61.6	1	20	9	217
218	9.61	E1	12.00		M1	0.42	1	0.85	1	04	S	II	58.9	2	16	20	218
219	10.57	E1	4.55		M1			0.79	0	04	SM	I	37.7	2	17	1000	219
220	12.18	S2	6.80		M1						CEU	I	34.8	1	20	6	220
221	8.74	E2	7.03		M1	0.38	2	0.81	2	04 05 14	U	EOS	98.0	3	16	2049	221
222	10.25	E1	4.95		M1	0.40	2	0.70	2	04	C	TH	84.7	2	19	40	222
223	11.06	S2	5.10		M1												223
224	9.59	E2	6.51		M1	0.20	3	0.75	3	04 05	M	II	56.5	2	16	1060	224
225	9.52	E1	20.61		M1			0.73	1	04							225
226	10.91	E1	2.50		M1												226
227	10.02	S2	4.00		M1												227
228	13.85	E1	6.60		M1												228
229	10.44	6.0			M1												229
230	8.62	E2	8.77	0.024	M1	0.43	2	0.97	3	02 03 04	S	I	116.	3	13	3559	230
231	10.49	5.1			M1												231
232	11.22	E1	7.12		M1	0.38	2	0.71	3	04	C	II	54.4	2	19	50	232

NUMBER	BMAG	QUAL	PHASE	COEFF	REF	U-B	OU	B-V	OB	COLOR REFS	TYPE	ZONE	DIAM	WT	BIN	DATA	NUMBER
233	9.27	E1	16.92		M1	0.34	1	0.79	1	04	SU	II	66.7	2	15	20	233
234	10.23	E2	9.65		M1	0.50	2	0.91	2	04	S	I	45.3	2	16	40	234
235	9.96	E1	4.50		M1	0.54	2	0.90	2	04	S	III	51.0	2	18	40	235
236	9.31	E2	8.72		M1	0.44	3	0.85	3	04	S	II	67.3	2	16	67	236
237	10.30	E1	2.86		M1	0.41	0	0.79	1	04							237
238	8.99	E2	12.79		M1	0.40	?	0.72	3	04	C	III	155.	3	16	2050	238
239	11.72	S1	3.20		M1												239
240	9.94	E1	18.75	0.039	M1	0.35	2	0.70	2	04	C	II	97.7	2	17	40	240
241	8.52	E2	8.88		M1	0.30	3	0.69	3	03 04	C	III	187.	2	15	1060	241
242	10.27	S1	5.70		M1												242
243	11.05	E2	3.34		M1	0.36	1	0.81	2	04	S	KO	29.6	2	20	32	243
244	13.40	S1	3.00		M1												244
245	8.82	E1	11.16		M1	0.48	2	0.83	2	04	S	III	84.3	2	16	40	245
246	9.84	E1	7.32		M1	0.57	1	0.96	1	04	U	II	70.1	1	17	29	246
247	9.04	E3	15.25		M1	0.26	2	0.69	3	02 04	C	II	143.	3	15	3050	247
248	11.18	E1	5.00		M1												248
249	12.16	S1	6.10		M1												249
250	8.29	E2	13.11		M1	0.25	3	0.72	3	04	CMEU	III	211.	1	15	60	250
251	11.20	S1	4.95		M1												251
252	10.62	S1	3.60		M1												252
253	11.39	S2	4.45		M1												253
254	13.16	E2*	4.92		M1	0.50	1	0.85	2	04	S	FL	11.4	2	21	30	254
255	11.32	E2	2.54		M1	0.25	2	0.68	1	04	CMEU	II	51.3	1	20	30	255
256	10.97	E2	4.25		M1												256
257	10.49	E1	6.98		M1	0.38	1	0.76	1	04	C	III	78.0	2	20	20	257
258	9.32	E1	8.76		M1	0.47	2	0.89	2	04	S	II	68.2	2	15	46	258
259	8.72	E1	12.41		M1	0.28	1	0.67	1	04	CMEU	III	169.	1	16	20	259
260	9.96	E1	4.72		M1	0.31	1	0.71	1	04	C	IV	97.3	2	20	20	260
261	10.28	E1	11.27		M1	0.35	2	0.71	2	04	C	I	83.9	2	16	40	261
262	12.78	E1	6.73		M1	0.53	1	0.84	1	04	U	II	17.1	1	22	22	262
263	11.58	S2*	4.30		M1												263
264	9.47	E2	7.64		M1	0.42	3	0.94	3	04	S	II	64.5	3	16	2066	264
265	12.69	1.3			M1												265
266	9.37	S2*	5.70		M1												266
267	11.65	S1	7.80		M1			0.80	1	04							267
268	9.16	E2*	8.40		M1	0.27	2	0.68	2	01 04	U	TH	84.7	1	17	26	268
269	11.22	2.1			M1												269
270	9.88	E3*	9.05		M1	0.53	3	0.87	3	04 05	S	FL	51.8	3	14	2160	270
271	10.78	E3	3.96		M1	0.33	2	0.71	1	04	C	III	66.7	2	20	30	271
272	11.7				M1												272
273	11.35	E3*	9.08		M1	0.38	2	0.76	2	04	C	PH	92.5	2	18	40	273
274	11.11	S1	6.10		M1												274
275	9.76	E2	5.87		M1	0.35	2	0.72	2	04	C	II	107.	2	17	40	275
276	9.44	E1	9.61		M1	0.28	2	0.69	3	04	CMEU	III	122.	1	18	50	276
277	11.09	E2	3.90		M1	0.47	1	0.75	1	04	U	KO	35.8	1	20	20	277
278	10.39	E1	3.50		M1												278
279	9.58	E3	7.17		M1	0.22	1	0.77	2	04	MEU	Z	72.4	1	21	36	279
280	11.86	2.9			M1												280
281	13.35	E3*	9.77		M1	0.49	2	0.95	2	02 12	U	FL	13.9	1	21	46	281
282	11.86	E1	8.91		M1	0.26	1	0.63	1	04	EU	I	23.8	1	19	20	282
283	9.67	E1	18.01		M1	0.30	1	0.71	1	04	C	III	111.	2	18	20	283
284	10.97	E1	8.19		M1	0.38	2	0.70	3	04	C	I	60.8	2	18	50	284
285	11.85	S1	1.90		M1												285
286	10.02	E2	7.94		M1	0.30	1	0.67	1	04	C	III	92.9	2	19	20	286
287	9.28	E1	21.12		M1	0.52	1	0.86	1	04	SU	I	68.5	2	14	20	287
288	11.04	E1	8.74		M1	0.40	1	0.84	1	04	S	II	30.2	2	19	20	288
289	10.72	2.0			M1												289
290	13.15	1.2			M1												290
291	12.70	2.0			M1												291
292	10.94	1.6			M1												292
293	10.92	E1	10.28		M1	0.35	2	0.72	3	04 05	U	III	38.2	1	20	56	293
294	11.18	E2	3.30		M1												294
295	11.33	E1	1.88		M1	0.50	2	0.85	2	04	S	II	26.5	2	20	40	295
296	13.7				M1												296
297	10.35	1.6			M1												297
298	12.28	E1	5.50		M1												298
299	12.75	S1	6.05		M1												299
300	10.95	S1	1.50		M1												300
301	11.10	S1	3.00		M1												301
302	11.76	E1	7.79		M1	0.28	2	0.67	2	04	CMEU	I	41.7	1	19	40	302
303	9.90	3.9			M1												303
304	10.75	E2	4.00		M1	0.25	2	0.71	2	04	CMEU	I	67.6	1	17	40	304
305	9.96	E1	9.69		M1	0.49	2	0.89	2	04	S	III	50.8	2	19	40	305
306	9.87	E1	8.37		M1	0.45	2	0.88	2	04	S	I	52.7	2	15	140	306
307	11.04	E1	8.58		M1	0.30	1	0.67	1	04	C	III	58.1	2	20	20	307
308	9.13	E2	14.94		M1	0.38	3	0.79	3	04	U	II	139.	3	16	2062	308
309	11.39	2.6			M1												309
310	11.55	S1	2.20		M1												310
311	11.08	E1	11.16		M1	0.43	1	0.83	1	14	S	KO	29.5	2	20	1020	311
312	10.00	E2	9.64		M1	0.41	2	0.84	1	04	S	II	48.7	2	17	30	312
313	9.87	E1	23.14		M1	0.33	?	0.72	2	04 05 14	C	I	108.	3	15	2047	313
314	10.85	S1	4.00		M1												314
315	13.77	3.6			M1												315
316	10.87	6.2			M1												316
317	10.42	E1	22.71		M1	0.21	1	0.69	1	04	MU	FL	47.4	1	16	20	317
318	10.24	E2	6.00		M1	0.29	1	0.68	1	04	C	III	84.3	2	19	20	318
319	11.3				M1												319
320	11.69	E1	3.40		M1												320
321	11.15	E3*	4.86		M1	0.41	2	0.79	2	01 04 14	U	KO	35.5	1	20	40	321
322	10.10	S1	10.80		M1												322
323	10.92	E1	13.20		M1	0.43	1	0.90	1	14	S	PH	32.8	2	18	26	323
324	7.76	E2	13.29	3.044	M1	0.30	3	0.70	3	02 04	C	II	256.	3	13	3266	324
325	9.76	E1	6.45		M1	0.24	1	0.70	1	04	MEU	III	64.6	1	18	27	325
326	10.10	E1	13.37		M1	0.30	2	0.69	2	04	C	PH	90.3	2	16	1046	326
327	11.30	E1	4.60		M1												327
328	9.91	1.7			M1	0.42	1	0.89	1	04							328
329	10.62	E1	10.81		M1	0.32	1	0.70	1	04	C	I	71.4	2	17	20	329
330	13.40	0.6			M1												330
331	10.60	E1	4.77		M1	0.30	2	0.70	2	04	C	III	72.1	2	20	40	331
332	10.42	S2*	4.60		M1												332
333	10.49	E1	7.07		M1	0.37	1	0.75	1	04	C	III	77.6	2	20	20	333
334	8.38	E2	4.72		M1	0.38	2	0.68	1	04 05	C	HI	199.	2	18	230	334
335	9.85	E2	3.15		M1	0.24	2	0.62	3	04	EU	I	59.7	1	16	57	335
336	10.74	E1	7.70		M1	0.29	2	0.75	2	04	C	FL	69.2	2	16	40	336
337	9.77	E1	7.55		M1	0.29	1	0.72	1	04	C	I	107.	2	15	23	337
338	9.71	E1	6.35		M1	0.22	1	0.71	1	04	M	III	52.5	2	17	126	338
339	10.04	E1	5.91		M1	0.41	2	0.79	2	04 14	U	EOS	91.9	1	19	1046	339
340	11.04	1.7			M1						S	III	30.3	2	19	6	340
341	12.50	S1	3.70		M1			0.92	1	01	S	FL	16.0	2	20	19	341
342	11.16	E2	7.00		M1	0.36	2	0.71	2	04	C	II	56.0	2	19	40	342
343	12.50	E1	7.55		M1	0.46	1	0.77	2	04	CU	I	31.0	2	21	30	343
344	9.08	E2	5.97		M1	0.40	2	0.72	2	04	C	II	142.	2	15	47	344
345	9.74	E1	7.71		M1	0.42	2	0.74	3	04	C	I	109.	2	15	256	345
346	8.39	E1	8.40		M1	0.48	2	0.83	2	04	S	III	102.	2	14	40	346
347	9.97	S1	4.70		M1						CEU	II	96.4	1	17	6	347
348	10.61	S1	3.45		M1												348

NUMBER	BMAG	QUAL	PHASE	COEFF	REF	U-B	QU	B-V	QB	COLOR REFS	TYPE	ZONE	DIAM	WT	BIN	DATA	NUMBER
349	7.15	E3†	7.28	0.022	M1	0.54	2	0.95	2	01 04 99	R	III	145.	3	12	2249	349
350	9.31	E1	8.12		M1	0.35	2	0.70	2	04	C	III	131.	2	17	40	350
351	10.07	E1	11.90		M1	0.40	2	0.84	2	04	S	II	47.2	2	18	40	351
352	11.07	E1	21.41		M1	0.55	2	0.92	2	04	S	FL	30.9	2	17	40	352
353	12.19	S2	2.87		M1												353
354	7.55	E4	13.30	0.020	M1	0.57	3	0.95	3	01 04 05	U	II	156.	3	13	2269	354
355	11.56	S1	2.30		M1												355
356	9.24	E3	18.02		M1	0.35	3	0.73	3	04	C	II	157.	3	16	2166	356
357	9.71	E2	8.05		M1	0.35	2	0.72	3	04	C	II	110.	2	18	50	357
358	10.29	S4	3.80		M1												358
359	10.05	S3†	8.38		M1	0.30	1	0.70	1	04	C	II	92.9	2	17	20	359
360	9.31	E1	8.25		M1	0.24	1	0.72	1	04	C	III	138.	3	17	2020	360
361	8.89	E1	5.86		M1	0.27	2	0.74	2	04 05	MEU	III	98.2	1	19	46	361
362	9.96	E1	15.13		M1	0.35	2	0.71	2	04	C	II	97.3	2	17	40	362
363	10.01	E2	4.23		M1	0.37	2	0.75	3	04	C	II	46.8	2	17	57	363
364	10.99	S1†	6.33		M1			0.92	0	03	SMRU	FL	39.3	1	17	1000	364
365	10.17	E2	10.60		M1	0.34	3	0.71	3	04 05	C	II	107.	3	18	2066	365
366	9.54	S1	6.00		M1												366
367	12.01	S2	3.20		M1						S	FL	19.4	2	19	100	367
368	11.03	E1	3.93		M1												368
369	9.51	E2	7.43		M1	0.25	2	0.72	2	04	CMEU	II	120.	1	16	40	369
370	11.61	E3	6.94		M1	0.29	1	0.71	1	04	C	I	45.5	2	19	20	370
371	9.84	E2	7.13		M1	0.51	1	0.82	1	04	U	II	65.8	1	17	20	371
372	8.43	0.5†			M1						CEU	III	196.	1	16	6	372
373	10.08	E1	7.06		M1	0.39	1	0.67	1	04	CU	III	90.3	2	19	20	373
374	10.07	S2	6.24		M1	0.48	1	0.86	1	05	S	II	47.6	2	18	126	374
375	8.36	E1	12.48†		M1	0.34	2	0.68	2	04	C	III	200.	2	15	46	375
376	10.62	E1	16.77		M1	0.45	2	0.89	2	04	S	FL	37.5	2	16	40	376
377	9.88	E2	13.39		M1	0.28	2	0.75	2	04	CMEU	II	103.	1	17	40	377
378	10.96	E1	3.82		M1	0.39	1	0.84	1	01	S	II	31.3	2	19	20	378
379	9.78	E1	3.76		M1	0.26	1	0.63	1	24	EU	TH	61.9	1	18	20	379
380	10.44	E1	2.90		M1			0.72	1	01							380
381	9.31	E1	9.00		M1	0.33	1	0.70	1	04	C	III	150.	3	18	2020	381
382	9.75	E1†	12.11		M1	0.24	1	0.69	2	04	CMEU	III	106.	1	18	30	382
383	10.90	E1	5.45		M1	0.32	1	0.71	1	04	C	TH	63.1	2	21	20	383
384	10.65	E1	8.26		M1	0.41	2	0.85	2	04 05	S	II	36.3	2	18	40	384
385	8.58	E1	19.03		M1	0.44	1	0.93	2	04	S	III	96.4	2	15	30	385
386	8.31	E1	19.25		M1	0.42	2	0.74	2	04	C	III	203.	3	15	2069	386
387	8.46	E1	19.92		M1	0.45	3	0.88	3	04	S	II	113.	3	14	2060	387
388	9.51	E1	7.10		M1	0.26	2	0.72	2	04	CMEU	III	120.	1	17	40	388
389	9.26	E1	7.15		M1	0.46	2	0.88	2	04	S	II	69.8	2	15	146	389
390	11.35	E2	4.89		M1	0.29	1	0.79	1	04	U	II	32.4	1	20	20	390
391	12.24	1.7			M1						S	A	17.3	?	20	6	391
392	10.85	S1	1.70		M1												392
393	8.11	E1	10.11		M1	0.33	3	0.74	3	04	C	II	117.	3	16	2050	393
394	10.97	E1	3.90		M1												394
395	11.43	E2	3.08		M1	0.38	2	0.73	3	04	C	II	49.9	2	20	50	395
396	10.83	S1	1.40		M1												396
397	10.40	E1	6.77		M1	0.37	2	0.81	3	04 05	S	II	50.8	3	18	2050	397
398	11.86	0.8			M1												398
399	10.22	S1	5.20		M1												399
400	11.31	3.7			M1												400
401	10.43	S1	3.77		M1												401
402	10.08	E2	9.34		M1	0.41	2	0.81	1	04	S	II	46.3	2	17	36	402
403	10.49	E1	5.20		M1			0.90	1	04	S	II	39.1	2	19	6	403
404	9.89	E1	3.58		M1	0.35	2	0.67	2	03 04	C	II	98.6	2	16	1040	404
405	9.39	E1	23.90		M1	0.39	2	0.70	3	04	C	II	126.	2	15	50	405
406	11.40	S1	3.43		M1												406
407	9.82	E1	14.96		M1	0.39	2	0.72	2	04	C	II	104.	2	16	40	407
408	10.69	S1	6.40		M1												408
409	8.47	E2	7.54		M1	0.34	2	0.72	3	02 04	C	II	194.	2	14	59	409
410	9.18	E2	8.15		M1	0.40	2	0.75	2	03 04 05	C	II	142.	2	16	1240	410
411	10.14	S2	4.95		M1												411
412	10.26	S2	7.80		M1						MEU	II	51.3	1	18	6	412
413	11.12	E1	11.29		M1	0.22	1	0.68	1	14	MEU	II	34.2	1	19	20	413
414	10.59	3.1			M1												414
415	10.20	E2	16.75		M1	0.23	2	0.71	3	02 04	C	II	87.1	2	18	1152	415
416	8.94	E2	12.50		M1	0.45	2	0.90	3	02 04	S	II	81.7	2	15	156	416
417	10.24	S1	7.70		M1												417
418	10.82	E1	7.62		M1	0.26	2	0.69	2	04	CMEU	II	64.9	1	18	40	418
419	9.31	E1	4.31		M1	0.24	1	0.62	1	04	EU	II	76.6	1	15	29	419
420	9.45	E1	4.23		M1	0.27	1	0.68	1	04	CMEU	IV	121.	1	19	20	420
421	12.93	1.0			M1												421
422	11.74	E2	3.07		M1	0.28	2	0.70	2	04	CMEU	I	42.6	1	18	40	422
423	8.25	E3†	7.16		M1	0.31	2	0.66	3	04	C	III	209.	2	15	156	423
424	10.71	S1	4.80		M1												424
425	9.36	4.9			M1												425
426	9.40	E1	10.49		M1	0.34	1	0.71	1	04	C	III	126.	2	17	26	426
427	10.48	S2	4.40		M1												427
428	13.03	S1	2.90		M1												428
429	10.72	S1	6.30		M1												429
430	11.47	S1	3.30		M1												430
431	9.98	5.8			M1												431
432	10.10	E1	14.64		M1	0.45	2	0.87	2	04	S	I	47.2	2	16	40	432
433	11.88	E1†		0.024	M1	0.50	3	0.88	3	02	S	A	20.0	3	13	3669	433
434	11.91	E1	23.58		M1	0.24	2	0.70	3	04 05	U	HU	11.6	3	17	2157	434
435	11.24	E2	7.66		M1	0.29	3	0.73	3	04	U	I	32.7	1	18	62	435
436	11.02	E1	5.63		M1												436
437	11.61	S4	3.80		M1												437
438	10.67	E1	12.61		M1	0.20	1	0.61	1	04	U	II	40.7	1	18	20	438
439	10.70	E1	5.88		M1	0.24	1	0.72	1	04	C	III	69.5	2	20	26	439
440	12.81	E1	13.57		M1												440
441	9.35	E3†	2.60		M1	0.27	2	0.69	3	04	M	II	65.6	3	16	2055	441
442	10.87	E1	4.23		M1												442
443	11.31	S3	6.60		M1												443
444	8.76	E1	6.76		M1	0.29	3	0.68	3	03 04	C	II	167.	2	15	1060	444
445	10.20	E1	19.14		M1	0.37	2	0.68	2	04	S	II	85.9	2	19	40	445
446	10.05	E2	6.52		M1	0.61	2	1.03	2	04	B	II	55.8	1	18	46	446
447	10.23	E2†	8.24		M1	0.36	1	0.78	1	04	U	III	54.3	1	19	20	447
448	11.19	E1	13.14		M1	0.30	3	0.66	3	14	CU	III	53.9	2	21	60	448
449	10.43	E1	6.23		M1	0.38	1	0.69	1	04	C	II	77.6	2	17	20	449
450	11.34	E1	9.85		M1	0.48	1	0.78	1	04	U	E0S	32.4	1	21	20	450
451	7.56	E1	6.47	0.035	M1	0.31	3	0.67	3	02 03 04 12	C	III	281.	3	14	3160	451
452	13.35	0.6			M1												452
453	11.88	5.7			M1						S	FL	20.6	2	18	6	453
454	10.13	E2	8.51		M1	0.35	2	0.65	2	04	CU	II	87.9	2	17	40	454
455	9.86	E2	5.60		M1	0.31	2	0.70	2	03 04	C	II	101.	2	17	1040	455
456	10.97	S1	5.15		M1												456
457	12.94	0.6			M1												457
458	10.78	E1	19.21		M1	0.47	1	0.88	1	04	S	III	34.7	2	20	20	458
459	11.55	S1	5.30		M1												459
460	11.85	E2	4.23		M1												460
461	11.38	E2	2.04		M1	0.31	2	0.61	2	14	U	TH	29.4	1	21	40	461
462	10.37	E2	7.86		M1	0.42	3	0.79	2	04 14	U	K0	50.8	1	19	1057	462
463	12.80	E2	9.97		M1	0.29	1	0.71	1	04	C	I	26.3	2	21	20	463
464	10.60	E1	4.10		M1			0.64	1	04							464

NUMBER	BMAG	QUAL	PHASE COEFF	REF	U-B	QU	B-V	QB	COLOR REFS	TYPE	ZONE	DIAM	WT	BIN	DATA	NUMBER
465	10.80	S1	2.25		M1											465
466	9.11	E1	6.38		M1	0.34	1	0.64	1	04	CU	IV	139.	2	18	20	466
467	11.94	S2	5.13		M1												467
468	10.61	E3	3.32		M1	0.31	2	0.67	3	04 14	C	TH	70.8	2	20	59	468
469	9.90	S2	3.47		M1												469
470	11.30	S1	6.30		M1												470
471	7.61	E2	10.34		M1	0.46	3	0.84	3	04	S	III	145.	3	13	2266	471
472	10.05	E2	13.43		M1	0.49	2	0.89	2	04	S	II	48.7	2	17	42	472
473	11.10	0.6			M1												473
474	11.82	2.5			M1												474
475	12.32	1.2			M1												475
476	9.60	E2	6.68		M1	0.40	1	0.72	1	03 04	C	II	115.	2	16	1020	476
477	11.27	E1	22.86		M1	0.43	0	0.87	2	04							477
478	9.12	E2	9.06		M1	0.44	3	0.85	3	04	S	III	73.8	2	17	60	478
479	10.77	S2*	4.27		M1												479
480	9.45	E1	14.31		M1	0.44	2	0.86	2	04	S	II	63.4	2	16	40	480
481	9.72	E1	8.05		M1	0.32	1	0.70	1	04	C	II	108.	2	17	6	481
482	9.96	E1	14.12		M1	0.46	1	0.87	1	04	S	III	50.3	2	18	20	482
483	9.57	1.7			M1												483
484	11.37	2.3			M1												484
485	9.66	2.9			M1												485
486	12.06	S2	7.87		M1												486
487	9.28	E7	7.68		M1	0.43	2	0.85	2	04	S	II	68.2	2	16	40	487
488	8.76	E1	8.81		M1	0.39	1	0.69	1	04	C	III	168.	2	16	27	488
489	9.31	E1	7.19		M1	0.36	1	0.69	2	04	C	III	130.	2	17	30	489
490	9.42	E1	8.66		M1	0.37	1	0.75	2	04	C	III	127.	2	18	36	490
491	9.91	2.5			M1												491
492	10.98	S3*	5.30		M1												492
493	11.75	2.8			M1												493
494	9.98	E3	5.82		M1	0.37	2	0.73	3	04	C	III	97.3	2	18	50	494
495	11.45	1.8			M1												495
496	12.92	3.8			M1							R	I	16.9	1	21	9	496
497	10.94	E1	1.14		M1	0.27	1	0.69	1	02 03	M	III	29.5	2	20	1020	497
498	9.89	E2	5.79		M1	0.42	2	0.76	2	01 04	U	II	71.7	3	17	2040	498
499	10.43	S2	2.67		M1												499
500	10.47	S1	5.90		M1												500
501	10.10	S1	4.00		M1												501
502	11.87	E1	17.98		M1	0.48	2	0.87	2	04	S	PH	20.9	2	19	40	502
503	10.10	E1	17.75		M1	0.32	1	0.72	1	04	C	II	91.6	2	17	20	503
504	11.16	S1	4.60		M1												504
505	10.04	2.8			M1	0.22	1	0.65	1	04	U	II	59.4	1	17	9	505
506	9.81	E1	16.91		M1	0.34	2	0.72	2	04	C	III	105.	2	18	40	506
507	10.56	E1	7.05		M1												507
508	9.21	E2	6.55		M1	0.33	3	0.73	3	04	C	III	139.	2	17	60	508
509	9.51	E2	9.48		M1	0.41	3	0.83	3	04	S	III	60.8	2	18	60	509
510	10.74	E1	10.10		M1	0.25	2	0.73	2	02	C	II	68.5	2	18	46	510
511	7.19	E5	10.71		M1	0.35	2	0.71	3	01 02 04	C	III	335.	3	13	3359	511
512	12.16	E1	29.56		M1	0.56	1	0.94	1	04	RU	A	23.9	1	19	20	512
513	10.79	E1*	6.81		M1	0.44	1	0.81	1	14	S	EOS	41.1	3	20	2022	513
514	9.93	E1	7.81		M1	0.24	1	0.72	1	04	CMEU	III	99.1	1	18	20	514
515	11.80	5.9			M1												515
516	9.19	E1	4.47		M1	0.27	3	0.74	3	04 05	M	II	65.0	3	15	3060	516
517	10.33	E2	2.93		M1	0.29	2	0.71	3	04	C	III	82.0	2	19	50	517
518	12.52	S1	9.50		M1												518
519	10.18	2.1			M1												519
520	11.54	E1	5.13		M1	0.57	1	0.74	1	04	U	EOS	29.0	1	21	20	520
521	9.23	E1	19.98		M1	0.36	2	0.71	3	04	C	II	136.	2	16	50	521
522	10.01	E1	6.82		M1	0.24	1	0.66	1	14	CEU	IV	92.9	1	20	20	522
523	10.69	S1	1.20		M1												523
524	10.87	E2	4.94		M1	0.32	2	0.72	2	04	C	II	64.3	2	19	40	524
525	13.73	S1	14.76		M1	0.56	1	0.95	1	14	RU	FL	11.6	1	22	20	525
526	10.83	3.3			M1							MEU	TH	39.4	1	20	6	526
527	11.38	S2	6.10		M1												527
528	10.16	S1	1.10		M1												528
529	11.08	E1	15.51		M1							SM	EOS	29.8	2	21	1000	529
530	10.21	E1	13.99		M1	0.28	1	0.69	2	04	CMEU	III	85.9	1	19	30	530
531	12.22	0.6			M1												531
532	6.76	E2	18.44		M1	0.42	2	0.86	3	01 03 04	S	II	219.	3	11	3359	532
533	10.79	E2	10.27		M1	0.42	1	0.87	2	04 14	S	III	34.4	2	20	33	533
534	10.84	E2*	3.63		M1	0.39	2	0.83	2	04 14	S	KO	33.0	2	20	1040	534
535	10.48	E1*	0.66		M1	0.39	1	0.74	2	04	C	II	77.6	2	18	30	535
536	8.99	E1	7.16		M1	0.31	1	0.71	1	04	C	IV	152.	2	18	20	536
537	10.00	3.4			M1			0.82	0	03	C	III	95.0	2	19	1000	537
538	10.49	4.1			M1												538
539	11.02	2.6			M1												539
540	12.05	E1	6.40		M1	0.48	1	0.90	1	01	S	FL	19.5	2	19	20	540
541	11.30	E1	3.35		M1												541
542	10.42	E1	13.09		M1	0.38	1	0.80	1	14	SU	III	39.4	2	19	20	542
543	10.65	S1	3.50		M1												543
544	11.26	S2	5.65		M1												544
545	9.77	E1	14.07		M1	0.30	1	0.69	2	04	C	III	105.	2	18	30	545
546	10.71	E1	17.17		M1	0.38	2	0.77	3	04 14	CU	II	70.8	2	18	50	546
547	10.49	E1	20.82		M1	0.25	1	0.76	1	04	MEU	II	47.4	1	18	20	547
548	12.47	2.3			M1												548
549	11.73	1.9			M1												549
550	10.50	E1	4.16		M1	0.39	1	0.85	1	03 05	S	II	38.9	2	18	1020	550
551	10.48	E2	3.67		M1	0.28	1	0.68	1	04	CMEU	III	75.5	1	19	20	551
552	10.84	S1	3.60		M1												552
553	13.45	S1	4.90		M1												553
554	9.81	E3	5.70	0.043	M1	0.36	2	0.71	3	02 04	C	I	104.	2	15	1059	554
555	11.60	S1	1.90		M1												555
556	10.47	3.1			M1	0.40	1	0.79	1	04							556
557	13.27	S1	2.20		M1												557
558	9.95	E1	5.00		M1							U	III	63.9	3	18	2002	558
559	10.39	E2	7.01		M1	0.39	2	0.73	2	04	C	III	80.5	2	18	60	559
560	11.67	S1	5.40		M1							CMEU	II	44.0	1	21	2	560
561	12.21	E1	7.31		M1			0.75	1	04							561
562	10.98	E3	4.98		M1	0.40	2	0.80	2	14	S	EOS	30.5	2	20	42	562
563	9.62	E2	15.34		M1	0.49	2	0.89	3	03 04	S	II	59.4	2	16	1153	563
564	11.43	E1	8.37		M1	0.31	1	0.73	1	04	C	II	49.9	2	20	20	564
565	11.96	E1	13.09		M1	0.42	2	0.81	3	04	SU	I	19.5	2	20	50	565
566	9.07	E3	4.79		M1	0.29	3	0.71	3	04 14	C	IV	147.	2	18	60	566
567	10.23	E2*	3.55		M1	0.31	1	0.65	1	04	CU	III	83.5	2	19	20	567
568	10.48	E1	7.27		M1												568
569	11.11	E2	2.91		M1	0.38	2	0.74	2	04	C	II	58.1	2	19	40	569
570	9.74	E2	0.47		M1	0.35	1	0.76	2	04	C	IV	110.	2	19	30	570
571	12.81	3.1			M1												571
572	11.97	E1*	11.39		M1	0.29	2	0.69	2	04	C	I	38.2	2	20	40	572
573	10.47	S1	4.50		M1												573
574	13.74	3.2			M1							S	I	8.7	2	22	6	574
575	12.27	S2	5.35		M1												575
576	11.00	S1	2.50		M1												576
577	10.91	S2	2.10		M1												577
578	10.43	1.7			M1												578
579	8.90	E3	6.59		M1	0.41	2	0.81	3	04	S	EOS	79.8	2	16	56	579
580	10.83	3.6			M1												580

NUMBER	BMAG	QUAL	PHASE COEFF	REF	U-B	2U	B-V	QB	COLOR REFS	TYPE	ZONE	DIAM	WT	BIN	DATA	NUMBER
581	10.79	E1	4.40	M1												581
582	10.28	E1	14.51	M1	0.56	1	0.89	1	04	S	II	43.8	2	17	26	582
583	10.19	S1	3.67	M1												583
584	9.70	E1	12.88	M1	0.53	2	0.91	2	02	S	I	56.8	3	15	2447	584
585	11.33	E1	19.34	M1	0.32	1	0.70	1	04	C	I	51.5	2	19	20	585
586	10.14	E1	4.68	M1	0.37	1	0.66	1	04	CU	III	87.5	2	19	20	586
587	13.48	0.8		M1												587
588	9.64	E2	3.19	M1	0.23	3	0.75	2	05	U	T	69.8	1	23	57	588
589	10.14	E1	5.38	M1	0.36	2	0.72	2	04	C	III	89.9	2	19	40	589
590	10.98	E1	12.10	M1												590
591	11.63	E1	3.77	M1	0.20	2	0.69	2	04	MU	II	27.2	1	20	40	591
592	10.70	S1	1.80	M1												592
593	10.29	E1	12.77	M1	0.31	2	0.66	1	04	CU	II	81.6	2	18	30	593
594	13.71	2.4		M1												594
595	9.16	2.0		M1												595
596	9.89	E2	6.93	M1	0.18	2	0.72	2	04	U	III	134.	3	18	2040	596
597	10.39	S1	0.30	M1												597
598	10.48	E1	20.79	M1	0.38	1	0.74	1	04	C	II	77.6	2	18	20	598
599	9.38	S1	3.00	M1							S	II	65.2	2	16	6	599
600	11.35	3.6		M1												600
601	10.59	E1	3.92	M1	0.24	1	0.66	1	04	CEU	III	71.1	1	20	20	601
602	9.32	E2	6.39	M1	0.32	3	0.69	3	04 05	C	III	139.	3	17	2160	602
603	13.46	1.3		M1												603
604	10.22	2.4		M1												604
605	10.52	2.3		M1												605
606	11.49	S1	4.50	M1												606
607	10.78	E1	4.53	M1												607
608	11.74	S2	3.45	M1												608
609	11.18	S3	4.18	M1												609
610	13.20	1.3		M1												610
611	10.28	E1	3.38	M1	0.41	1	0.82	2	04	S	III	42.5	2	19	30	611
612	12.29	0.6		M1												612
613	10.64	E1	7.71	M1	0.34	1	0.61	1	04	U	III	41.3	1	19	26	613
614	12.00	S2	5.60	M1												614
615	11.33	E1	5.43	M1	0.31	3	0.71	2	04	C	II	51.8	2	19	50	615
616	11.79	E1ı	14.23	M1	0.43	1	0.87	2	04	S	II	21.7	2	30	30	616
617	9.17	E1	6.18	M1	0.22	1	0.68	3	14	U	T	159.	3	22	2046	617
618	9.24	E2	7.06	M1	0.32	3	0.70	3	04	C	III	135.	2	1	60	618
619	10.95	E1	15.31	M1	0.46	1	0.86	1	04	S	II	31.8	2	18	20	619
620	12.26	S1	8.00	M1	0.24	1	0.68	1	04	CMEU	I	33.3	1	20	20	620
621	11.39	E1	13.58	M1	0.28	1	0.65	1	14	CEU	TH	49.0	1	21	20	621
622	11.67	2.3		M1												622
623	12.03	E1	10.66	M1	0.33	2	0.71	1	04	C	I	37.5	2	20	30	623
624	8.53	E2	4.44	M1	0.26	2	0.76	2	01	05	U	T	111.	3	21	3147	624
625	11.67	S1	7.16	M1												625
626	9.96	E1	19.04	M1	0.33	1	0.69	1	04	C	II	96.4	2	17	20	626
627	11.02	E2	2.49	M1	0.26	1	0.68	1	04	CMEU	III	58.9	1	20	20	627
628	10.27	E1	2.88	M1	0.30	1	0.82	1	04	U	II	54.0	1	17	29	628
629	10.80	2.6		M1												629
630	12.38	2.0		M1												630
631	9.81	E1	10.29	M1	0.45	3	0.86	3	04	S	II	53.7	2	17	60	631
632	13.37	S1	2.40	M1												632
633	10.77	E1	8.13	M1	0.42	1	0.79	1	14	U	EOS	46.7	3	20	2020	633
634	11.01	1.8		M1												634
635	9.98	E1	5.60	M1	0.35	1	0.64	1	04	CU	III	93.3	2	19	20	635
636	10.74	S2	3.85	M1												636
637	11.86	2.5		M1												637
638	10.85	S4ı	5.33	M1												638
639	9.28	E4	5.34	M1	0.48	3	0.85	3	04 14	S	EOS	68.2	2	17	1066	639
640	10.30	1.7		M1												640
641	13.58	2.3ı		M1												641
642	11.19	E1	4.43	M1	0.40	2	0.88	1	04	S	III	28.7	2	21	30	642
643	10.72	E1	9.37	M1	0.33	1	0.73	2	04 14	C	IV	69.2	2	21	30	643
644	11.71	3.1		M1	0.41	2	0.81	2	04	SU	II	21.9	2	20	40	644
645	11.18	E1	6.54	M1	0.41	1	0.86	2	04	S	III	28.6	2	21	30	645
646	14.20	1.0		M1												646
647	12.45	S2ı	4.92	M1	0.27	2	0.73	2	04	CMEU	I	31.2	1	21	40	647
648	10.43	E1	6.44	M1	0.28	1	0.68	1	04	CMEU	III	77.3	1	20	22	648
649	14.10	1.0		M1												649
650	13.47	3.7		M1												650
651	11.10	E2	6.18	M1	0.48	2	0.84	2	14	S	EOS	29.4	2	21	40	651
652	12.53	S1	2.30	M1												652
653	10.27	E1	12.78	M1	0.40	1	0.83	1	14	S	EOS	42.9	2	19	20	653
654	9.47	S1	10.06	M1	0.35	1	0.69	2	04	U	PH	73.5	1	14	139	654
655	10.70	S3	3.10	M1												655
656	10.85	S1	3.40	M1												656
657	11.72	S1	8.30	M1												657
658	11.61	E1	5.20	M1	0.36	1	0.87	1	01	SU	KO	23.5	2	21	20	658
659	9.69	S2ı	2.95	M1												659
660	10.20	E1	11.13	M1	0.48	2	0.86	2	03 04	S	II	44.9	2	17	1042	660
661	10.71	E2	8.25	M1	0.39	2	0.81	3	04 14	S	EOS	34.7	2	20	1050	661
662	11.54	E1	5.30	M1												662
663	10.10	E1	8.66	M1	0.34	2	0.69	2	04	C	III	90.3	2	19	40	663
664	10.96	E1	6.94	M1	0.31	1	0.69	1	14	C	III	60.8	2	21	20	664
665	9.62	S1	2.60	M1												665
666	11.87	S2	4.45	M1												666
667	10.19	S1	8.00	M1												667
668	13.22	S1	4.90	M1												668
669	11.33	E1	9.58	M1	0.47	1	0.82	1	14	SU	EOS	26.2	2	21	20	669
670	11.1			M1												670
671	11.43	E1	3.35	M1												671
672	12.48	S1	5.80	M1												672
673	11.25	E1	2.05	M1	0.43	1	0.78	1	04	C	II	55.5	2	20	20	673
674	8.57	E1	6.71	M1	0.43	2	0.88	1	04	S	III	95.9	2	15	1139	674
675	9.2			M1												675
676	10.52	S2	4.27	M1							C	III	74.8	2	20	6	676
677	10.81	S1	1.80	M1												677
678	10.6			M1												678
679	10.19	S1	13.22	M1	0.43	1	0.86	1	02 03	U	II	73.6	3	17	2020	679
680	10.27	E1	6.79	M1	0.27	2	0.69	2	04	CMEU	III	83.5	1	19	40	680
681	11.78	S1	4.10	M1												681
682	13.47	0.6		M1												682
683	9.53	S1	6.40	M1												683
684	11.79	S1	6.00	M1												684
685	12.83	S1	1.70	M1												685
686	10.78	E1	6.03	M1	0.46	1	0.84	1	04	S	II	34.0	2	18	20	686
687	12.95	1.6		M1												687
688	11.57	S2	3.20	M1							C	I	22.7	2	22	20	688
689	13.12	E1	14.96	M1	0.31	1	0.71	1	04	CEU	III	175.	1	16	40	689
690	8.63	E1	12.96	M1	0.28	2	0.66	2	04	C	III	85.1	2	19	20	690
691	10.28	E1	17.72	M1	0.31	1	0.74	1	04	S	IV	46.5	2	20	50	691
692	10.13	E1	13.98	M1	0.44	2	0.87	3	04 14	CU	III	81.3	2	19	20	692
693	10.42	E1	11.20	M1	0.41	1	0.78	1	04	C	II	94.6	2	17	1020	693
694	10.03	E2	6.94	M1	0.35	1	0.72	1	03 04	S	II	54.2	2	16	7	694
695	9.78	S1	7.70	M1							CMEU	III	85.1	1	19	9	695
696	10.24	5.4		M1												696

NUMBER	BMAG	QUAL	PHASE	COEFF	REF	U-B	QU	B-V	QB	COLOR REFS	TYPE	ZONE	DIAM	WT	BIN	DATA	NUMBER
697	10.62	E1	13.11		M1	0.39	1	0.73	1	04	C	III	72.4	2	19	20	697
698	11.80	1.5			M1												698
699	11.33	2.1			M1												699
700	12.36	2.2			M1												700
701	10.24	E2	5.49		M1	0.30	1	0.66	2	04	CU	III	83.5	2	19	30	701
702	8.17	E1	7.36		M1	0.31	3	0.66	2	04	CU	III	217.	2	15	50	702
703	13.59	1.9			M1												703
704	7.02	E2:	8.86	0.044	M1	0.25	3	0.64	3	02 03 04 05	U	III	338.	3	13	2169	704
705	9.36	E1	7.91		M1	0.28	2	0.71	2	04	CMEU	III	128.	1	17	40	705
706	12.00	1.3			M1												706
707	13.36	1.7			M1												707
708	11.76	E1	6.80		M1	0.38	2	0.88	2	04	SU	II	22.1	2	21	40	708
709	10.01	E1	12.32		M1	0.33	3	0.72	3	04	C	III	95.5	2	18	60	709
710	12.15	3.2			M1												710
711	12.54	1.7			M1												711
712	9.39	E1	11.34		M1	0.36	3	0.73	3	04	C	II	128.	2	15	69	712
713	9.81	E1	3.30		M1												713
714	10.21	E1	6.68		M1	0.43	2	0.89	2	04	S	II	45.3	2	17	47	714
715	11.13	S3:	5.97		M1												715
716	11.93	E2	5.52		M1	0.33	2	0.86	1	04	U	II	25.6	1	21	30	716
717	12.01	E2	3.80		M1	0.24	2	0.70	1	04	CMEU	III	37.7	1	23	30	717
718	10.73	2.6			M1												718
719	16.77	0.6			M1												719
720	10.89	E3	6.79		M1	0.44	1	0.81	3	04 14	S	KO	31.9	2	20	1040	720
721	10.31	E1	2.87		M1	0.25	1	0.77	1	04	MEU	IV	51.8	1	21	20	721
722	13.00	S1	5.20		M1												722
723	11.06	S1	2.80		M1												723
724	14.74	0.6			M1												724
725	12.27	1.5			M1												725
726	12.07	E1	3.10		M1												726
727	10.67	E1	14.37		M1	0.30	1	0.78	1	04	U	II	44.1	1	18	20	727
728	13.78	0.6			M1												728
729	10.35	E1	11.06		M1	0.38	1	0.78	1	14	U	II	51.1	1	18	20	729
730	14.66	1.4			M1												730
731	10.49	E2	7.10		M1	0.30	2	0.69	2	04	C	III	75.5	2	19	40	731
732	11.78	S1	7.70		M1												732
733	10.05	E1	8.65		M1	0.24	1	0.69	1	04	CMEU	IV	92.5	1	20	20	733
734	11.06	E1	4.60		M1												734
735	10.52	E1	6.89		M1	0.32	2	0.70	2	04	C	II	74.8	2	18	40	735
736	12.82	E1	4.62		M1	0.51	1	0.90	1	24	S	FL	13.7	2	20	20	736
737	9.94	E1	6.18		M1	0.39	0	0.83	1	03 04	S	II	45.4	3	17	2210	737
738	11.16	E1	1.57		M1	0.50	1	0.76	1	04	U	III	34.8	1	21	20	738
739	9.60	E1	9.49		M1	0.31	2	0.70	3	04	C	II	114.	2	16	57	739
740	9.93	E1	9.27		M1	0.33	1	0.68	2	04	C	III	97.3	2	18	30	740
741	11.47	S1	4.30		M1							CMEU	II	48.3	1	20	6	741
742	10.68	E1	5.05		M1	0.45	1	0.84	3	04 14	S	EOS	48.6	3	20	2040	742
743	11.28	E1	3.65		M1												743
744	11.14	E1	4.28		M1	0.16	1	0.66	1	04	U	III	33.6	1	21	20	744
745	10.99	2.1			M1												745
746	10.71	S2	3.70		M1												746
747	8.65	E1	9.37		M1	0.32	2	0.71	2	02	C	III	208.	3	16	2242	747
748	9.86	S2	2.73		M1												748
749	12.95	E1	5.48		M1	0.50	1	0.86	1	04	S	FL	12.6	2	21	20	749
750	12.98	E1	13.35		M1	0.22	1	0.61	1	14	EU	NY	14.1	1	22	22	750
751	9.62	E1	9.10		M1	0.39	2	0.69	3	04	C	II	113.	2	16	50	751
752	11.24	S1	7.00		M1												752
753	11.56	E2:	7.94		M1	0.49	1	0.94	1	04	SU	I	24.9	2	19	20	753
754	10.16	E1	12.57		M1	0.34	1	0.70	2	14	C	III	88.3	2	19	30	754
755	10.84	E1	2.49		M1	0.22	2	0.69	2	04	MEU	III	39.1	1	21	40	755
756	11.16	0.5			M1												756
757	11.32	S1	5.17		M1	0.20	1	0.75	1	04							757
758	9.17	E1	8.23		M1	0.42	1	0.74	1	34	U	III	86.3	1	17	29	758
759	11.62	S1	6.10		M1												759
760	9.37	E2	5.96		M1	0.52	2	0.95	1	04	S	III	68.5	2	18	39	760
761	11.83	S1	1.35		M1												761
762	9.27	E1	13.68		M1	0.31	1	0.66	2	04	CU	III	131.	2	17	30	762
763	13.42	S1	6.90		M1												763
764	10.48	E1	1.84		M1	0.40	2	0.72	2	04	C	III	76.9	2	20	40	764
765	14.01	0.6			M1												765
766	11.03	E3:	5.09		M1			0.81	1	04	C	EOS	62.2	2	20	1010	766
767	11.01	S3			M1												767
768	11.26	1.0			M1												768
769	10.14	3.8			M1												769
770	12.09	E1	8.78		M1	0.55	2	0.88	2	04	U	FL	24.0	1	19	47	770
771	11.68	S1			M1												771
772	9.25	E1	15.89		M1	0.36	2	0.67	2	04	C	III	132.	2	17	42	772
773	10.48	S1	7.55		M1							MEU	III	46.3	1	19	6	773
774	9.90	3.4			M1												774
775	11.54	E1	3.64		M1			0.84	1	04							775
776	8.59	E1	4.55		M1	0.39	2	0.71	3	04	C	III	183.	2	15	50	776
777	11.10	E1	4.00		M1												777
778	10.46	E1	6.62		M1	0.26	2	0.62	3	04 05	EU	III	45.1	1	20	50	778
779	9.65	0.5			M1												779
780	10.07	S2	9.95		M1												780
781	10.51	S1	6.30		M1							CEU	III	75.1	1	20	2	781
782	12.55	S1	6.80		M1							U	I	18.7	1	20	1002	782
783	12.03	S1	7.15		M1							C	I	37.3	2	20	2	783
784	10.22	S2	4.93		M1												784
785	10.29	E1	25.26		M1	0.17	2	0.64	3	04	U	II	49.5	3	17	2059	785
786	9.82	S2	5.45		M1												786
787	11.32	1.3			M1												787
788	9.46	2.1:			M1												788
789	12.18	S2	7.57		M1												789
790	9.04	E2	8.02		M1	0.34	2	0.72	2	04	C	IV	178.	3	18	2046	790
791	10.34	S2	5.85		M1												791
792	11.07	1.3			M1												792
793	11.35	E1	5.18		M1			0.82	1	04							793
794	12.28	E1	5.90		M1												794
795	10.93	S1	3.30		M1												795
796	10.16	E1	7.48		M1	0.27	1	0.70	2	02 04	CMEU	II	88.3	1	17	30	796
797	11.55	E2	6.95		M1	0.51	2	0.89	2	04	S	II	24.4	2	20	40	797
798	10.48	E1	12.32		M1							SM	EOS	61.1	3	19	2000	798
799	11.41	S1	4.83		M1												799
800	12.81	E1	3.58		M1	0.52	1	0.92	1	04	S	I	13.9	2	20	20	800
801	12.13	S1	6.00		M1							CMEU	II	35.6	1	21	2	801
802	13.51	2.0			M1												802
803	10.75	S1	1.60		M1												803
804	8.69	E1	10.29		M1	0.37	3	0.72	3	03 04	C	III	175.	2	15	1060	804
805	10.60	S1	6.00		M1												805
806	11.09	1.0			M1												806
807	11.65	E2	7.70		M1	0.46	2	0.85	2	14	S	EOS	22.7	2	22	40	807
808	10.68	S1	1.50		M1												808
809	13.12	S2	4.00		M1												809
810	14.11	S1			M1												810
811	11.89	E1	9.33		M1	0.52	1	0.88	1	14	U	KO	26.3	1	22	22	811
812	12.41	0.6			M1												812

NUMBER	BMAG	QUAL	PHASE	COEFF	REF	U-B	QU	B-V	QB	COLOR REFS	TYPE	ZONE	DIAM	WT	BIN	DATA	NUMBER
813	13.30	S1	6.30		M1												813
814	9.90	E2	6.37		M1	0.34	3	0.68	3	04 14	C	III	98.6	2	19	60	814
815	11.93	3.0			M1												815
816	11.32	3.6			M1												816
817	11.88	S1	4.80		M1												817
818	10.42	S1	5.20		M1												818
819	13.13	E1	5.67		M1												819
820	11.46	S1	3.10		M1												820
821	12.47	S1			M1												821
822	13.22	E1	6.14		M1			0.77	1	04							822
823	12.51	S2	6.15		M1												823
824	11.54	E1*	4.50		M1	0.41	1	0.85	1	04 14	S	II	24.1	2	21	20	824
825	12.86	E2*	7.93		M1	0.54	2	0.91	3	04	S	I	13.5	2	21	50	825
826	12.72	S1	4.90		M1												826
827	13.64	S1	2.10		M1												827
828	11.28	E2	5.31		M1			0.67	1	04							828
829	11.91	S3	5.45		M1												829
830	10.35	E2	7.32		M1	0.50	1	0.90	2	04 05	S	III	42.7	2	20	1030	830
831	13.48	0.6			M1												831
832	12.23	S2	3.80		M1												832
833	12.23	1.2			M1												833
834	10.41	E2	2.04		M1	0.47	1	0.75	2	04	U	III	49.0	1	20	30	834
835	12.20	S1	2.30		M1												835
836	14.27	0.8			M1												836
837	12.95	1.9			M1												837
838	11.24	E1	4.90		M1												838
839	11.79	S1	7.20		M1						S	II	21.5	2	20	6	839
840	10.50	0.6			M1												840
841	13.81	S1	4.50		M1												841
842	11.71	1.0			M1												842
843	14.22	0.6			M1												843
844	10.77	E1	4.83		M1												844
845	11.11	1.8			M1												845
846	11.31	E3	3.65		M1	0.39	1	0.61	1	04	CEU	TH	52.0	1	21	7	846
847	11.42	E1	3.59		M1	0.46	1	0.90	1	04	S	II	26.1	2	20	20	847
848	11.84	2.0			M1												848
849	9.00	E1	16.68		M1	0.26	2	0.70	2	04	M	III	72.4	2	17	140	849
850	10.61	E1	4.07		M1												850
851	12.90	S2	3.53		M1												851
852	11.12	S1	8.10		M1												852
853	12.76	E1	14.29		M1	0.29	1	0.73	2	04	C	I	27.0	2	21	30	853
854	13.45	E1	5.30		M1												854
855	13.02	S2	5.70		M1												855
856	11.91	6.6			M1												856
857	12.22	S1	11.46		M1	0.14	1	0.63	1	14	U	FL	20.1	1	19	20	857
858	11.22	S1	4.50		M1						S	II	27.9	2	20	6	858
859	10.98	S2	7.57		M1												859
860	10.75	2.3			M1						SM	II	34.7	2	19	1000	860
861	10.76	4.9			M1												861
862	11.23	1.4			M1												862
863	10.32	E1	9.78		M1	0.56	1	1.06	2	04	R	III	58.9	1	20	1030	863
864	14.2				M1												864
865	13.14	3.6			M1												865
866	10.35	3.2			M1												866
867	12.09	2.1			M1												867
868	11.21	E1	2.94		M1	0.36	1	0.71	1	04	C	II	54.7	2	20	20	868
869	13.22	2.8			M1												869
870	12.91	1.1			M1												870
871	13.69	2.6			M1												871
872	10.90	E1	5.58		M1	0.28	1	0.73	1	04	CMEU	II	63.7	1	19	20	872
873	12.05	S1	3.90		M1												873
874	10.85	S1	5.50		M1												874
875	12.80	S1	8.60		M1												875
876	11.96	E1	12.74		M1	0.43	1	0.81	2	14	SU	EOS	19.5	2	22	30	876
877	11.66	E2	8.47		M1	0.25	1	0.68	1	04	CMEU	NY	43.8	1	20	20	877
878	16.50	0.6			M1												878
879	12.67	0.6			M1												879
880	12.96	0.8			M1												880
881	13.50	1.1			M1												881
882	11.65	S1	2.35		M1												882
883	13.57	1.5			M1						S	FL	9.5	2	22	20	883
884	9.74	E1	3.81		M1	0.17	1	0.68	1	14	M	T	51.0	2	23	26	884
885	11.91	E1	4.05		M1												885
886	9.60	S1	4.50		M1												886
887	14.96	E2	22.22	0.042	M1	0.40	1	0.87	1	02	S	A	4.7	3	26	2327	887
888	10.62	E1	11.09		M1	0.50	3	0.88	3	04	S	II	37.3	2	18	60	888
889	12.60	S1	7.70		M1												889
890	11.77	E1	6.88		M1	0.33	2	0.77	1	14	U	EOS	26.4	1	22	1030	890
891	11.22	S3	6.00		M1												891
892	10.52	S1	5.90		M1												892
893	10.40	E1	7.36		M1	0.23	1	0.67	1	04	CMEU	III	78.0	1	19	20	893
894	10.88	S1	5.70		M1												894
895	9.74	2.5			M1						CMEU	III	107.	1	18	6	895
896	12.81	S1	7.00		M1												896
897	12.14	1.5			M1												897
898	13.37	1.3			M1												898
899	11.09	E2	5.72		M1	0.26	2	0.68	3	04 14	CMEU	III	57.0	1	20	50	899
900	12.87	3.0			M1												900
901	12.75	3.5*			M1												901
902	13.54	1.1			M1												902
903	10.75	3.6			M1												903
904	11.34	1.3			M1												904
905	12.71	E1	8.53		M1												905
906	10.63	3.2			M1												906
907	10.54	E1	9.10		M1												907
908	11.98	E1	8.65		M1												908
909	9.64	S1	3.80		M1						C	IV	112.	2	19	7	909
910	11.17	6.6			M1												910
911	8.92	E2	5.16		M1	0.22	2	0.77	3	02 05 14	U	T	98.2	1	22	1052	911
912	10.23	S2	6.40		M1												912
913	13.69	4.0			M1												913
914	10.52	E1	17.19		M1	0.33	1	0.71	1	24	C	PH	75.1	2	17	20	914
915	13.02	S1	4.00		M1												915
916	12.60	2.6			M1												916
917	12.55	E1	5.80		M1												917
918	11.90	S1	4.30		M1												918
919	12.40	E1	5.43		M1												919
920	12.25	E1	6.78		M1	0.30	1	0.80	1	14	U	II	21.5	1	21	20	920
921	11.10	S1	5.20		M1												921
922	13.01	E1	1.30		M1												922
923	12.65	1.0			M1												923
924	10.37	E2	6.23		M1	0.34	1	0.72	2	04 05	C	III	80.9	2	19	30	924
925	9.58	E2	15.40		M1	0.42	2	0.84	3	04	S	II	59.2	2	16	59	925
926	11.61	1.5			M1												926
927	10.52	E1	6.54		M1	0.34	1	0.67	1	04	C	III	73.8	2	20	20	927
928	10.72	3.1			M1												928

NUMBER	BMAG	QUAL	PHASE COEFF	REF	U-B 2U	B-V QB	COLOR REFS	TYPE	ZONE	DIAM	WT	BIN	DATA	NUMBER
929	13.45	S1	6.30	M1										929
930	12.49	1.1		M1										930
931	9.99	S1	4.70	M1										931
932	10.94	E1	9.28	M1	0.33 2	0.67 2 04		C	I	60.8	2	18	40	932
933	13.20	S1	7.00	M1										933
934	11.43	1.5t		M1										934
935	14.33	S1	3.80	M1										935
936	11.02	S2	2.97	M1										936
937	12.83	S2	2.90	M1										937
938	12.34	S1		M1										938
939	13.33	E1	7.65	M1	0.54 1	0.93 1 24		S	FL	11.0	2	22	20	939
940	10.13	E1	4.81	M1	0.36 1	0.58 1 04		U	IV	51.5	1	20	20	940
941	12.52	E1	6.10	M1	0.31 1	0.70 1 04		C	II	29.8	2	23	20	941
942	11.48	2.4		M1										942
943	10.81	E1	7.48	M1	0.39 1	0.78 1 04		U	III	41.3	1	20	20	943
944	11.58	E1t	12.63	0.050 M1	0.22 2	0.75 2 04 14 99		MEU	Z	28.6	1	28	47	944
945	11.10	E1	1.85	M1										945
946	11.44	E1	0.87	M1	0.32 3	0.67 3 04		C	TH	48.3	2	22	60	946
947	11.26	S1	4.90	M1										947
948	12.49	S1	6.07	M1										948
949	10.66	S2	3.00	M1										949
950	12.41	1.0		M1										950
951	12.90	S1	5.40	M1										951
952	10.20	S1	3.40	M1										952
953	11.47	S2	6.30	M1										953
954	10.84	E1	5.25	M1	0.31 2	0.62 2 14		U	TH	37.8	1	20	40	954
955	12.63	1.0		M1										955
956	13.40	S1	4.60	M1										956
957	10.92	S1	6.80	M1										957
958	11.53	E1	12.33	M2				ERU	HI	36.0	1	23	1000	958
959	11.80	0.8		M1										959
960	14.19	S2	7.65	M1										960
961	12.46	S2	6.90	M1										961
962	12.68	E1	4.53	M1										962
963	13.66	E1	10.51	M1	0.53 1	0.90 2 05		S	FL	9.3	2	22	30	963
964	12.04	S1	1.20	M1										964
965	11.29	S1	1.30	M1										965
966	11.06	E1	12.91	M1	0.45 2	0.87 3 04 05		S	II	30.3	2	19	50	966
967	13.61	S1	4.80	M1										967
968	11.15	E1	5.70	M1	0.37 1	0.87 1 04		SU	III	29.1	2	20	20	968
969	13.45	E1	6.27	M1	0.24 2	0.62 2 05		U	NY	11.4	1	23	42	969
970	13.43	1.6		M1										970
971	10.99	E1	5.52	M1										971
972	10.63	E1	3.40	M1										972
973	10.90	S3	4.95	M1										973
974	11.48	S1	2.30	M1										974
975	11.52	E1	6.97	M1	0.39 1	0.85 1 14		S	KO	24.3	2	21	20	975
976	10.32	S2	4.85	M1	0.25 1	0.74 1 01		C	III	83.5	2	20	26	976
977	10.73	E1	6.85	M1	0.39 2	0.71 2 05		C	III	68.2	2	20	40	977
978	10.65	E1	1.01	M1	0.24 1	0.67 2 04		CMEU	III	69.5	1	20	30	978
979	10.91	S2	2.70	M1										979
980	9.00	E1	9.26	M1	0.55 2	0.90 2 04		S	II	79.4	2	15	40	980
981	11.63	S2	7.58	M1	0.33 1	0.62 1 14		U	TH	26.3	1	22	20	981
982	11.35	S1	5.70	M1										982
983	10.60	E1	8.90	M1	0.28 2	0.74 1 04		CMEU	III	73.4	1	20	30	983
984	10.55	E1	3.47	M1										984
985	14.11	S1	5.20	M1										985
986	10.54	S2t	5.35	M1										986
987	10.52	E1	5.23	M1										987
988	12.39	0.8		M1										988
989	13.32	0.6		M1										989
990	12.69	S1	3.40	M1										990
991	12.05	E1	11.26	M1	0.37 1	0.66 1 14		CU	TH	36.3	2	23	20	991
992	11.97	S1	3.90	M1										992
993	13.66	1.1		M1										993
994	11.41	2.7		M1										994
995	11.17	E1	3.80	M1										995
996	11.73	1.8		M1		0.69 1 04								996
997	12.99	2.1		M1										997
998	12.11	0.6		M1										998
999	11.87	S1	3.10	M1										999
1000	11.29	0.6		M1										1000
1001	10.60	E1	3.83	M1	0.22 2	0.71 2 04		MEU	III	44.1	1	20	40	1001
1002	12.04	1.2		M1										1002
1003	11.23	2.9		M1										1003
1004	10.98	E1	1.01	M1	0.12 1	0.72 1 14		U	IV	37.2	1	22	20	1004
1005	10.81	S1	4.50	M1										1005
1006	12.72	S1	8.30	M1										1006
1007	12.60	S1	4.90	M1										1007
1008	11.63	S2	4.40	M1										1008
1009	16.82	0.6		M1										1009
1010	11.74	S1	1.90	M1										1010
1011	13.79	S1	27.73	M1	0.52 2	0.90 2 05		S	A	7.4	3	23	2040	1011
1012	13.25	E1	10.47	M1	0.25 1	0.64 1 14		CEU	I	20.7	1	23	20	1012
1013	10.98	E1	16.35	M1	0.36 2	0.74 1 04		C	II	61.6	2	19	30	1013
1014	12.85	S2	2.40	M1										1014
1015	9.99	E1	6.82	M1	0.32 2	0.69 3 14		C	III	95.0	2	19	56	1015
1016	13.28	S1	2.20	M1										1016
1017	12.17	1.1		M1										1017
1018	12.33	S1	5.85	M1										1018
1019	13.90	E1	16.26	M1	0.50 2	0.95 2 14		U	HU	10.8	1	21	42	1019
1020	12.11	1.1		M1										1020
1021	9.97	S1	8.90	M1	0.23 0	0.67 1 04								1021
1022	11.24	0.8		M1										1022
1023	10.77	E1	7.39	M1	0.49 2	0.74 2 04		U	III	41.3	1	20	40	1023
1024	11.66	S1	2.20	M1										1024
1025	14.04	0.8		M1					C	HU	14.8	2	21	6	1025
1026	14.52	0.8		M1										1026
1027	11.84	S2		M1										1027
1028	10.36	E1	2.57	M1										1028
1029	11.90	E2	6.90	M1	0.40 2	0.79 3 14		U	KO	25.1	1	22	50	1029
1030	11.52	S1	1.40	M1										1030
1031	10.51	E1	7.86	M1	0.32 1	0.68 1 14		C	III	74.5	2	19	20	1031
1032	10.98	S1	5.50	M1										1032
1033	12.14	S1	5.30	M1										1033
1034	13.66	2.3		M1										1034
1035	11.69	2.0		M1										1035
1036	10.43	E1	11.54	M1	0.41 3	0.83 3 04		S	A	39.8	2	18	67	1036
1037	15.05	0.8		M1										1037
1038	11.59	0.8		M1										1038
1039	12.30	S2	3.15	M1										1039
1040	11.09	S1	4.70	M1										1040
1041	11.10	S1	4.90	M1										1041
1042	11.23	S1	8.27	M1										1042
1043	11.00	E2	5.38	M1	0.45 1	0.70 1 01		S	III	31.6	2	21	20	1043
1044	11.95	E1	3.37	M1										1044

NUMBER	BMAG QUAL	PHASE COEFF	REF	U-B OU	B-V OB	COLOR REFS	TYPE	ZONE	DIAM	WT	BIN	DATA	NUMBER
1045	14.02 1.5		M1										1045
1046	11.49 E2	3.98	M1										1046
1047	13.19 S1	5.80	M1										1047
1048	10.64 E2	8.70	M1	0.32 3	0.71 3	04	C	II	71.1	2	19	60	1048
1049	11.75 0.8		M1										1049
1050	13.83 0.9		M1										1050
1051	10.94 S1	6.10	M1										1051
1052	13.14 E1	8.24	M1	0.54 1	0.90 1	05	S	FL	11.8	2	21	120	1052
1053	13.63 S1	7.30	M1										1053
1054	11.57 E1	5.70	M1										1054
1055	12.65 S1	6.20	M1				S	I	14.5	2	20	9	1055
1056	12.66 S1	5.70	M1										1056
1057	12.19 S2	2.75	M1										1057
1058	12.82 E1	6.00	M1				S	FL	13.4	2	20	106	1058
1059	12.27 2.41		M1										1059
1060	14.27 2.1		M1										1060
1061	11.98 2.9		M1										1061
1062	11.20 S3	3.95	M1										1062
1063	12.50 S2	3.70	M1										1063
1064	12.25 2.0		M1										1064
1065	13.78 2.11		M1										1065
1066	14.10 1.2		M1										1066
1067	11.85 S2	5.48	M1										1067
1068	12.33 S1		M1										1068
1069	10.72 2.5		M1										1069
1070	11.98 S1	2.50	M1										1070
1071	11.18 S2	3.47	M1										1071
1072	11.68 1.3		M1										1072
1073	12.52 S1	1.10	M1										1073
1074	11.25 S1	2.80	M1										1074
1075	11.22 E2	4.22	M1	0.37 2	0.76 2	14	U	EOS	33.6	3	21	2042	1075
1076	12.78 S31	5.85	M1	0.28 2	0.63 2	14	EU	NY	15.6	1	22	40	1076
1077	13.91 S1		M1										1077
1078	12.60 E1	4.17	M1										1078
1079	12.32 E1	11.78	M1	0.40 2	0.80 1	14	SU	KO	16.4	2	22	30	1079
1080	13.48 S1	6.40	M1										1080
1081	12.11 S2		M1										1081
1082	11.39 E1	8.33	M1	0.32 2	0.70 2	14	C	TH	50.1	2	22	40	1082
1083	13.91 1.0		M1										1083
1084	11.74 S21	5.55	M1										1084
1085	10.77 S2	5.17	M1										1085
1086	10.68 S1	5.55	M1										1086
1087	10.81 E2	4.93	M1	0.37 2	0.74 1	04 14	C	EOS	56.8	3	20	2030	1087
1088	12.61 E2	7.43	M1	0.57 1	0.94 2	04	D	FL	133.	1	20	32	1088
1089	11.88 E1	3110	M1										1089
1090	13.97 1.4		M1										1090
1091	11.85 0.9		M1										1091
1092	11.45 S1	3.60	M1										1092
1093	9.77 E2	9.88	M1	0.36 2	0.68 2	04 14	C	III	105.	2	18	40	1093
1094	13.28 E1	8.45	M1										1094
1095	11.28 2.0		M1										1095
1096	11.28 S1	5.80	M1										1096
1097	12.78 S1	2.00	M1										1097
1098	11.68 S1	3.50	M1										1098
1099	11.67 1.3		M1										1099
1100	12.42 E1	2.80	M1										1100
1101	11.98 1.3		M1										1101
1102	10.73 E2	4.95	M1	0.39 1	0.72 1	04	C	III	68.5	2	20	20	1102
1103	13.50 1.2		M1				CMEU	HU	19.0	1	20	7	1103
1104	13.50 1.5		M1										1104
1105	10.78 S1	6.60	M1										1105
1106	12.87 1.0		M1										1106
1107	10.32 S3	7.10	M1										1107
1108	12.34 1.61		M1										1108
1109	10.92 E2	5.17	M1	0.29 0	0.60 2	04							1109
1110	13.19 S1	5.80	M1										1110
1111	11.70 S31	5.32	M1		0.63 1	04							1111
1112	11.14 E2	6.49	M1	0.44 1	0.78 2	04	C	EOS	58.3	2	21	30	1112
1113	10.60 E1	5.20	M1										1113
1114	10.79 S2	6.30	M1										1114
1115	10.39 S2	5.40	M1										1115
1116	10.73 S1	5.10	M1										1116
1117	10.73 S1	7.20	M1										1117
1118	10.87 E1	5.20	M1										1118
1119	12.59 S1	3.40	M1										1119
1120	13.33 1.2		M1										1120
1121	12.52 0.8		M1										1121
1122	12.74 1.5		M1										1122
1123	12.65 S1	9.40	M1										1123
1124	11.88 S1	6.30	M1										1124
1125	14.20 0.8		M1										1125
1126	13.70 1.5		M1										1126
1127	11.89 E1	11.06	M1	0.30 2	0.70 2	14	C	II	39.8	2	20	40	1127
1128	11.89 E1	2.63	M1										1128
1129	11.11 E2	5.06	M1	0.41 1	0.78 1	14	U	EOS	37.7	3	21	2020	1129
1130	13.15 S1		M1										1130
1131	15.32 1.0		M1										1131
1132	11.82 S1		M1										1132
1133	13.22 2.5	1.70	M1										1133
1134	15.29 S1		M1										1134
1135	11.45 E1	5.50	M1										1135
1136	12.08 3.6	4.70	M1										1136
1137	11.98 S1		M1										1137
1138	12.20 1.7		M1										1138
1139	14.25 1.1		M1										1139
1140	11.47 E1	6.45	M1	0.48 2	0.91 2	04	S	II	25.6	2	20	40	1140
1141	14.51 0.8		M1										1141
1142	11.45 S1	3.80	M1										1142
1143	9.45 0.8	2.46	M1	0.24 2	0.78 1	05 14	EU	T	77.3	1	23	30	1143
1144	11.20 S1	2.50	M1										1144
1145	12.15 S2	4.50	M1										1145
1146	10.65 E1	9.73	M1	0.28 1	0.71 1	04	CMEU	III	70.9	1	20	20	1146
1147	13.09 S1	4.50	M1										1147
1148	11.13 S1	13.59	M1	0.46 1	0.88 1	14	S	EOS	41.0	3	21	2020	1148
1149	11.41 E2	4.85	M1										1149
1150	14.42 S1		M1										1150
1151	14.81 0.6		M1										1151
1152	12.23 3.6		M1										1152
1153	13.31 S2	2.50	M1										1153
1154	11.44 S1	2.20	M1										1154
1155	12.80 E1	7.15	M1										1155
1156	13.89 1.1		M1										1156
1157	11.19 S2	5.80	M1										1157
1158	12.05 S1	7.30	M1										1158
1159	12.62 S1	6.40	M1										1159
1160	12.77 1.5		M1										1160

NUMBER	BMAG QUAL	PHASE COEFF REF	U-B QU	B-V QB	COLOR REFS	TYPE	ZONE	DIAM	WT	BIN	DATA	NUMBER
1161	12.70 0.6	M1										1161
1162	10.43 E1	8.13 M1	0.40 1	0.80 1	14	U	HI	49.7	1	22	1027	1162
1163	11.72 S2	4.36 M1										1163
1164	14.03 1.3	M1										1164
1165	11.71 S1	1.70 M1										1165
1166	12.58 1.5	M1										1166
1167	10.95 E2	3.50 M1										1167
1168	13.01 1.9	M1										1168
1169	14.29 1.9	M1										1169
1170	13.11 2.0	M1										1170
1171	10.80 E2	2.08 M1	0.26 2	0.70 2	04	CMEU	TH	65.8	1	20	40	1171
1172	9.26 E1	3.47 M1	0.27 3	0.72 2	14	U	T	82.0	1	22	1056	1172
1173	9.82 E1	4.67 M1	0.24 2	0.74 1	14	U	T	64.0	1	23	1036	1173
1174	12.84 0.6	M1										1174
1175	11.53 S1	2.30 M1										1175
1176	12.18 1.5	M1										1176
1177	10.11 S1	2.40 M1										1177
1178	12.89 0.7	M1				C	II	25.1	2	23	1000	1178
1179	15.00 0.6	M1										1179
1180	10.13 4.7	M1										1180
1181	12.63 0.6	M1										1181
1182	12.50 S1	1.80 M1										1182
1183	12.98 S2	2.57 M1										1183
1184	12.32 S1	4.97 M1										1184
1185	13.20 S2	7.73 M1										1185
1186	10.25 E1	11.11 M1	0.43 1	0.79 1	14	U	EOS	53.7	1	19	1020	1186
1187	12.56 S2	1.80 M1										1187
1188	13.11 S1	3.83 M1										1188
1189	11.10 S2	5.07 M1										1189
1190	13.17 1.1	M1										1190
1191	11.64 1.9	M1										1191
1192	13.58 1.0	M1										1192
1193	13.22 1.9	M1										1193
1194	11.70 S1	4.40 M1										1194
1195	14.51 0.6	M1										1195
1196	11.48 E1	5.80 M1										1196
1197	11.26 S1	5.80 M1										1197
1198	16.69 0.6	M1										1198
1199	11.40 E1	5.06 M1	0.33 1	0.76 1	14	C	EOS	51.3	2	21	1026	1199
1200	11.75 E1	1.90 M1										1200
1201	12.56 S2	3.30 M1										1201
1202	11.31 1.8	M1										1202
1203	13.09 S2	9.30 M1										1203
1204	13.32 S1	1.50 M1										1204
1205	15.19 0.6	M1										1205
1206	11.38 1.64	M1										1206
1207	12.04 S2!	M1										1207
1208	9.79 S1	3.29 M1	0.38 1	0.67 1	14	C	T	103.	2	23	26	1208
1209	11.48 2.1	M1										1209
1210	11.09 E2	10.37 M1		0.83 1	14	SM	EOS	29.4	2	21	1010	1210
1211	12.01 S1	8.15 M1										1211
1212	8.52 1.9	2.24 M1	0.23 2	0.69 2	02	U	HI	114.	1	18	42	1212
1213	12.14 1.7	M1										1213
1214	12.04 2.8!	M1										1214
1215	11.81 1.9	M1		0.95 1	04							1215
1216	13.09 S1	2.80 M1	0.53 0	0.90 1	04							1216
1217	14.48 0.6	M1										1217
1218	14.14 S1	1.60 M1										1218
1219	13.15 S3	6.10 M1										1219
1220	12.24 2.1	M1										1220
1221	19.06 0.6	M1										1221
1222	13.16 0.6	M1										1222
1223	11.61 E2!	9.22 M1	0.40 1	0.84 2	14	S	KO	23.2	2	21	1030	1223
1224	12.53 E1	17.08 M1	0.41 1	0.90 1	04	SU	I	15.6	2	20	20	1224
1225	13.60 S2	M1										1225
1226	13.21 1.1	M1										1226
1227	11.45 S1	4.60 M1										1227
1228	12.71 1.0	M1										1228
1229	12.91 0.8	M1										1229
1230	14.64 0.6	M1										1230
1231	12.68 2.2	M1										1231
1232	11.28 E2	3.75 M1										1232
1233	12.32 1.2	M1										1233
1234	11.72 S1	M1										1234
1235	15.30 0.9	M1	0.33 1	0.75 1	14	C	HU	8.5	2	24	20	1235
1236	12.81 E1	6.07 M1										1236
1237	12.00 S1	4.30 M1										1237
1238	13.01 1.0	M1										1238
1239	13.67 S2	M1										1239
1240	10.90 S4	4.65 M1										1240
1241	10.47 E1	2.11 M1	0.29 1	0.75 1	14	C	III	78.3	2	20	20	1241
1242	11.32 S3	5.20 M1										1242
1243	10.94 S5!	6.79 M1										1243
1244	12.39 S2	7.20 M1										1244
1245	11.04 E2!	4.19 M1	0.53 1	0.84 2	04 14	U	KO	38.2	1	20	30	1245
1246	12.72 1.1!	M1										1246
1247	11.47 E1	3.29 M1	0.29 1	0.68 2	14	C	III	47.9	2	22	30	1247
1248	10.90 S2	8.70 M1										1248
1249	12.96 E1	5.03 M1										1249
1250	14.15 1.3	M1										1250
1251	11.58 E2	7.83 M1	0.17 1	0.65 1	04	U	II	27.3	1	20	20	1251
1252	12.04 E1	13.97 M1	0.42 1	0.89 1	14	S	Z	19.5	2	21	20	1252
1253	13.22 0.8	M1										1253
1254	11.54 3.3	M1										1254
1255	11.59 4.4	M1										1255
1256	10.81 S1	4.90 M1										1256
1257	12.82 S1	2.20 M1										1257
1258	11.61 E1	3.73 M1										1258
1259	11.89 S1	3.90 M1										1259
1260	12.87 2.3	M1										1260
1261	11.83 1.0	M1										1261
1262	11.29 1.5	M1										1262
1263	11.46 S1	20.02 M1	0.29 1	0.73 1	04	C	II	49.2	2	20	22	1263
1264	10.84 1.3	M1										1264
1265	11.03 0.6	M1										1265
1266	10.41 E3	3.24 M1	0.37 1	0.74 2	04 05	C	IV	80.2	2	20	30	1266
1267	13.53 S1	7.70 M1										1267
1268	10.01 E1	3.82 M1	0.23 1	0.67 1	05	CMEU	HI	93.3	1	21	20	1268
1269	9.72 3.6	2.30 M1										1269
1270	13.79 S1	3.35 M1										1270
1271	11.60 S1	3.70 M1										1271
1272	13.51 0.6	M1										1272
1273	14.05 S1	6.15 M1										1273
1274	12.99 E1	3.79 M1	0.53 1	0.90 2	04	S	FL	12.6	2	21	30	1274
1275	11.73 E2	7.23 M1	0.31 1	0.66 1	04	CU	II	42.1	2	21	20	1275
1276	11.83 1.0	M1										1276

NUMBER	BMAG	QUAL	PHASE COEFF	REF	U-B	QU	B-V	QB	COLOR REFS	TYPE	ZONE	DIAM	WT	BIN	DATA	NUMBER
1277	12.21	S2	1.40	M1												1277
1278	12.07	S1	7.00	M1												1278
1279	13.65	S1	6.20	M1												1279
1280	11.02	4.3		M1												1280
1281	12.59	S1	4.70	M1												1281
1282	11.15	E1	6.45	M1												1282
1283	11.92	0.6		M1												1283
1284	11.23	S2	7.70	M1						S	II	27.8	2	19	6	1284
1285	11.34	3.0		M1												1285
1286	11.80	S3†	6.82	M1	0.43	1	0.85	1	14	S	EOS	21.4	2	22	20	1286
1287	12.13	S3	5.03	M1												1287
1288	12.69	1.7†		M1												1288
1289	11.47	E1	3.77	M1	0.38	1	0.78	1	14	U	KO	25.2	3	21	2020	1289
1290	13.65	0.8		M1												1290
1291	11.36	S3	5.72	M1												1291
1292	12.54	S3	5.40	M1												1292
1293	15.13	S1		M1												1293
1294	11.59	4.6	7.50	M1												1294
1295	11.64	3.8		M1												1295
1296	12.62	1.6		M1												1296
1297	12.38	S1		M1												1297
1298	11.98	S1	2.40	M1												1298
1299	13.00	2.4	3.60	M1												1299
1300	12.29	3.6		M1												1300
1301	11.78	1.4		M1												1301
1302	11.88	2.9		M1												1302
1303	10.43	1.5		M1												1303
1304	10.33	S1	3.40	M1												1304
1305	11.51	S3	1.37	M1												1305
1306	10.75	E2	3.68	M1	0.40	1	0.85	1	04	S	III	34.7	2	20	20	1306
1307	13.17	1.1		M1												1307
1308	11.76	S2	1.10	M1												1308
1309	11.30	S1	1.40	M1												1309
1310	12.69	1.0		M1												1310
1311	13.77	1.6		M1												1311
1312	12.09	S1	3.30	M1												1312
1313	12.94	1.1		M1												1313
1314	13.82	S1	8.75	M1	0.46	1	0.87	1	05	S	I	8.5	2	23	20	1314
1315	11.08	S3	4.50	M1												1315
1316	14.79	0.6		M1												1316
1317	10.94	E1	5.94	M1	0.34	1	0.72	1	04 05	C	III	62.2	2	21	22	1317
1318	13.13	0.6		M1												1318
1319	11.71	1.1		M1												1319
1320	11.89	0.8		M1												1320
1321	11.37	S2	3.95	M1												1321
1322	11.11	1.0		M1												1322
1323	11.33	E1	4.53	M1												1323
1324	13.56	0.6		M1												1324
1325	13.21	1.2		M1												1325
1326	11.98	S1	5.73	M1	0.48	1	0.78	1	14	U	II	24.1	1	21	20	1326
1327	13.25	S1	2.40	M1												1327
1328	11.32	S1	8.12	M1	0.16	1	0.70	1	14	U	IV	31.5	1	23	20	1328
1329	11.37	2.2		M1	0.53	1	0.87	2	04	SU	II	26.3	2	20	30	1329
1330	11.12	S1	14.63	M1	0.17	1	0.67	1	14	M	III	26.9	2	21	26	1330
1331	11.16	S2	6.83	M1	0.35	1	0.64	1	14	CU	TH	54.2	2	21	20	1331
1332	11.32	S1		M1												1332
1333	12.74	S1	7.50	M1												1333
1334	11.09	E1	4.80	M1												1334
1335	14.89	5.0		M1												1335
1336	12.13	S1	1.80	M1												1336
1337	12.01	S1	5.90	M1												1337
1338	13.96	S1	4.70	M1												1338
1339	11.86	S1	8.61	M1	0.43	2	0.79	2	14	U	EOS	25.6	1	22	40	1339
1340	12.56	S1	2.73	M1												1340
1341	11.54	E1	12.25	M1	0.26	2	0.68	2	14	CMEU	II	46.3	1	20	40	1341
1342	13.35	0.8		M1												1342
1343	12.50	S2	4.00	M1												1343
1344	14.06	S1	1.90	M1												1344
1345	10.65	S1	7.76	M1	0.30	2	0.71	2	14	C	HI	70.8	2	23	40	1345
1346	12.37	0.9		M1												1346
1347	12.12	S2		M1												1347
1348	12.28	S2	3.77	M1												1348
1349	11.64	S2	4.30	M1												1349
1350	11.54	S1	5.90	M1												1350
1351	11.01	S1	0.80	M1												1351
1352	12.34	S2	2.60	M1												1352
1353	11.05	S1	4.30	M1												1353
1354	12.06	S1		M1												1354
1355	13.80	1.2		M1												1355
1356	11.33	S1	1.80	M1												1356
1357	12.03	S1	6.31	M1			0.73	1	04							1357
1358	12.84	2.0		M1												1358
1359	11.52	S1	13.46	M1	0.36	1	0.72	1	14	C	III	47.6	2	22	20	1359
1360	12.42	1.1		M1												1360
1361	11.93	0.8		M1												1361
1362	12.40	E2	11.46	M1	0.36	2	0.72	2	02	C	IV	31.8	2	24	1040	1362
1363	12.66	S1	2.30	M1												1363
1364	12.02	S1	4.40	M1						S	EOS	19.3	2	22	2	1364
1365	13.27	S1	6.00	M1												1365
1366	11.50	S1	5.20	M1												1366
1367	14.20	1.0		M1												1367
1368	11.92	2.5		M1												1368
1369	11.38	1.7		M1												1369
1370	14.86	0.6		M1												1370
1371	12.26	1.2		M1												1371
1372	12.70	4.0		M1												1372
1373	14.23	0.6		M1												1373
1374	14.70	0.8		M1												1374
1375	12.91	S2	4.80	M1												1375
1376	13.51	S1	6.60	M1												1376
1377	14.18	2.0		M1												1377
1378	13.30	S1	1.40	M1												1378
1379	12.01	S1	6.87	M1												1379
1380	13.08	0.6		M1												1380
1381	12.94	S2	5.08	M1												1381
1382	13.33	S1	4.50	M1												1382
1383	12.84	S1	3.00	M1												1383
1384	12.76	S2		M1												1384
1385	12.01	2.6	3.50	M1												1385
1386	14.68	0.6		M1												1386
1387	14.33	0.9		M1												1387
1388	12.01	S1	11.58	M1						SM	EOS	25.9	3	22	2000	1388
1389	12.67	S2	2.60	M1												1389
1390	10.15	E2†	4.36	M1	0.21	2	0.71	2	04 14	MU	IV	54.2	1	20	40	1390
1391	12.98	E1	8.53	M1	0.57	1	0.88	2	04 14	U	II	15.9	1	22	30	1391
1392	12.75	S1	6.15	M1	0.23	1	0.76	1	04	MEU	II	16.8	1	22	20	1392

NUMBER	BMAG	QUAL	PHASE COEFF	REF	U-B	QU	B-V	QB	COLOR REFS	TYPE	ZONE	DIAM	WT	BIN	DATA	NUMBER
1393	13.13	3.5!		M1												1393
1394	12.78	3.6		M1												1394
1395	12.67	0.8		M1												1395
1396	12.92	S2	4.65	M1												1396
1397	12.77	S2	2.40	M1												1397
1398	11.35	S1	5.60	M1												1398
1399	15.17	0.6		M1												1399
1400	12.88	0.6		M1												1400
1401	13.40	E1	6.21	M1	0.48	2	0.88	2	05	S	FL	10.4	2	22	40	1401
1402	14.52	0.6		M1												1402
1403	13.56	0.6		M1												1403
1404	10.17	0.9		M1												1404
1405	14.31	0.5		M1												1405
1406	12.42	S2		M1												1406
1407	12.29	S1	4.20	M1												1407
1408	12.00	2.2		M1												1408
1409	11.65	S1	2.80	M1												1409
1410	12.35	E1	6.23	M1												1410
1411	11.96	S2	3.45	M1												1411
1412	13.59	2.9		M1												1412
1413	12.43	E1	6.97	M1												1413
1414	13.74	0.6		M1												1414
1415	13.45	E1	7.82	M1	0.48	1	0.86	2	04	S	FL	10.0	2	22	30	1415
1416	11.60	S1	5.57	M1	0.41	1	0.79	1	14	U	EOS	32.4	3	2?	2020	1416
1417	12.25	S2	4.30	M1												1417
1418	13.14	S2!	4.25	M1												1418
1419	12.58	S2	4.50	M1												1419
1420	12.82	1.3		M1												1420
1421	11.43	S1	4.80	M1												1421
1422	13.7			M1												1422
1423	12.5			M1												1423
1424	10.56	S1	4.90	M1												1424
1425	12.76	1.8		M1												1425
1426	12.05	2.2		M1												1426
1427	11.81	S3	6.80	M1												1427
1428	11.43	S1	3.30	M1												1428
1429	13.16	S1		M1												1429
1430	13.16	1.2		M1												1430
1431	12.51	1.0		M1												1431
1432	13.28	S1	7.20	M1												1432
1433	12.76	4.2		M1												1433
1434	11.40	E2	3.22	M1	0.41	1	0.81	1	14	SU	EOS	25.2	2	21	20	1434
1435	14.70	0.6		M1												1435
1436	11.78	S1	5.50	M1												1436
1437	9.28	E1!	4.62	M1	0.24	2	0.70	2	01 05	C	T	132.	2	22	1040	1437
1438	12.70	1.0		M1												1438
1439	11.40	S1	9.71	M1	0.32	1	0.75	1	14	C	HI	51.0	2	24	20	1439
1440	12.80	2.1		M1												1440
1441	14.13	1.3		M1												1441
1442	12.53	2.8		M1												1442
1443	12.31	1.5		M1												1443
1444	12.14	0.6		M1												1444
1445	11.87	2.6		M1												1445
1446	13.87	2.8		M1												1446
1447	12.83	2.1		M1												1447
1448	14.29	0.5		M1												1448
1449	13.73	2.2		M1						S	FL	8.8	2	22	7	1449
1450	12.57	S1	2.60	M1												1450
1451	13.76	1.0		M1												1451
1452	13.00	E1		M1												1452
1453	13.84	1.3	16.05	M1	0.54	2	0.97	2	14	RU	HU	11.2	1	20	40	1453
1454	14.24	0.6		M1												1454
1455	14.41	E1		M1												1455
1456	11.87	0.8	15.37	M1	0.34	1	0.69	1	14	C	III	40.0	2	23	20	1456
1457	12.36	S1		M1												1457
1458	12.45	3.1	3.00	M1												1458
1459	11.86	2.0		M1												1459
1460	13.7			M1												1460
1461	11.00	E1	5.10	M1	0.22	1	0.72	1	14	MEU	III	36.8	1	21	20	1461
1462	12.09	0.5		M1												1462
1463	12.02	1.6		M1												1463
1464	12.22	S2	5.33	M1												1464
1465	12.14	0.6		M1												1465
1466	14.02	0.8		M1												1466
1467	9.60	E1	10.36	M1	0.36	2	0.72	2	04	C	IV	115.	2	19	40	1467
1468	14.51	S1	8.67	M1												1468
1469	10.84	S2	6.58	M1												1469
1470	12.2			M1												1470
1471	12.36	S4		M1												1471
1472	13.67	S1	5.20	M1												1472
1473	13.52	1.8		M1												1473
1474	13.50	S1	17.12	M1	0.20	1	0.63	3	04 14	U	A	11.?	1	24	40	1474
1475	14.09	0.8		M1												1475
1476	14.79	0.8		M1												1476
1477	12.59	E1	12.91	M1			0.72	1	04							1477
1478	13.49	1.0		M1												1478
1479	12.31	S1	6.20	M1			0.69	2	04							1479
1480	14.35	S1	5.10	M1												1480
1481	11.83	2.1		M1												1481
1482	11.95	S2	4.97	M1												1482
1483	12.54	S2	2.65	M1												1483
1484	12.22	0.8		M1												1484
1485	12.47	1.1		M1												1485
1486	14.51	S1	5.20	M1												1486
1487	11.97	S1	3.80	M1												1487
1488	12.0			M1												1488
1489	13.02	0.6		M1												1489
1490	12.66	1.7		M1												1490
1491	12.57	2.3		M1												1491
1492	14.33	2.7!		M1												1492
1493	12.68	E1	11.94	M1	0.28	1	0.62	1	04	EU	NY	16.2	1	21	26	1493
1494	13.88	2.7		M1												1494
1495	13.42	0.9		M1												1495
1496	13.52	S1	6.60	M1												1496
1497	12.90	S2		M1												1497
1498	13.00	0.7		M1												1498
1499	12.64	1.4		M1												1499
1500	14.24	E1	11.73	M1	0.52	1	0.92	1	05	S	FL	7.2	2	23	20	1500
1501	13.50	S1	5.95	M1												1501
1502	12.70	2.9!		M1												1502
1503	11.71	S2!	4.25	M1												1503
1504	12.88	E1	13.51	M1	0.42	1	0.88	1	04	S	I	13.2	2	2?	20	1504
1505	12.47	S1	6.40	M1												1505
1506	13.11	S1	8.00	M1												1506
1507	14.59	1.0		M1												1507
1508	13.15	1.5		M1												1508

NUMBER	BMAG	QUAL	PHASE	COEFF	REF	U-B	QU	B-V	QB	COLOR REFS		TYPE	ZONE	DIAM	WT	BIN	DATA	NUMBER
1509	14.00	1.8			M1													1509
1510	12.45	S2	3.27		M1													1510
1511	14.07	1.0			M1													1511
1512	10.57	E2	2.93		M1	0.16	1	0.70	1	14		U	HI	85.6	3	22	2027	1512
1513	14.36	1.3			M1													1513
1514	13.54	1.0			M1													1514
1515	13.85	0.6			M1													1515
1516	12.91	2.0			M1													1516
1517	12.12	3.7			M1													1517
1518	13.47	S1	4.50		M1													1518
1519	12.33	1.5'			M1													1519
1520	11.43	S1	4.40		M1													1520
1521	13.19	1.0			M1													1521
1522	13.64	1.4			41													1522
1523	13.31	S1	4.50		M1													1523
1524	11.82	S3	5.87		M1													1524
1525	13.57	2.2'			M1													1525
1526	14.73	3.3			M1													1526
1527	13.57	S1			M1													1527
1528	13.52	1.6			M1													1528
1529	11.08	E1	9.23		M1	0.32	1	0.76	2	14		U	HI	36.1	1	23	32	1529
1530	14.46	1.0			M1													1530
1531	13.02	0.8			M1													1531
1532	11.85	S3	2.85		M1	0.36	1	0.84	1	14		SU	EOS	20.8	2	22	20	1532
1533	12.06	E1	5.35		M1	0.48	1	0.80	1	14		U	EOS	23.4	1	72	20	1533
1534	12.96	S1	2.90		M1													1534
1535	12.80	0.9			M1													1535
1536	14.69	0.7			M1													1536
1537	13.06	2.4'			M1													1537
1538	15.48	0.6			M1													1538
1539	12.19	S1	5.20		M1													1539
1540	11.79	S2			M1													1540
1541	12.56	S1	6.20		M1													1541
1542	11.50	S1	3.57		M1													1542
1543	13.54	1.8'			M1													1543
1544	12.81	S2	2.60		M1													1544
1545	12.68	S2	3.10		M1													1545
1546	11.65	1.2			M1													1546
1547	11.80	E1	9.08		M1	0.34	2	0.78	2	04 05		U	II	26.2	1	20	40	1547
1548	12.84	S1			M1													1548
1549	13.60	1.3			M1													1549
1550	13.48	2.1			M1													1550
1551	13.58	E1	6.25		M1													1551
1552	12.63	1.1			M1													1552
1553	12.70	S2			M1													1553
1554	12.66	S1	3.30		M1													1554
1555	12.62	S1	6.35		M1													1555
1556	11.33	1.7			M1													1556
1557	12.34	S1	2.95		M1													1557
1558	11.51	S2	5.75		M1													1558
1559	13.06	2.2			M1													1559
1560	12.70	S1	6.20		M1													1560
1561	11.98	0.8			M1													1561
1562	13.37	1.8			M1													1562
1563	13.76	1.0			M1													1563
1564	12.09	2.1			M1													1564
1565	13.68	1.3			M1													1565
1566	17.55	2.5		0.032	M1	0.54	1	0.80	1	02		U	A	1.9	1	15	122	1566
1567	10.54	E1	4.59		M1			0.72	2	04		C	III	79.2	3	20	2020	1567
1568	13.12	1.7			M1													1568
1569	12.43	2.2'			M1													1569
1570	12.52	S1	2.70		M1													1570
1571	13.13	1.2			M1													1571
1572	11.14	S1	3.60		M1													1572
1573	13.90	1.1			M1													1573
1574	11.50	2.9			M1													1574
1575	13.90	0.8			M1													1575
1576	11.91	E1	0.85		M1			0.68	1	04								1576
1577	15.24	1.8			M1													1577
1578	11.50	S1	1.85		M1													1578
1579	11.02	1.1			M1													1579
1580	14.95	E1'	20.71	0.035	M1	0.27	2	0.66	2	05		U	A	5.8	1	25	249	1580
1581	11.29	S1			M1													1581
1582	12.99	1.1			M1													1582
1583	9.62	E1	5.70		M1	0.27	1	0.76	2	05		MEU	T	70.8	1	23	30	1583
1584	12.02	E1	8.85		M1	0.45	1	0.96	1	24		U	PH	25.7	1	20	20	1584
1585	11.69	1.9			M1													1585
1586	13.51	S1			M1													1586
1587	12.81	1.4			M1													1587
1588	12.06	S1			M1													1588
1589	13.18	S2	5.00		M1													1589
1590	12.92	S3	4.80		M1													1590
1591	13.18	3.9			M1													1591
1592	12.69	S1	8.00		M1													1592
1593	14.51	S1	10.10		M1													1593
1594	13.38	1.2			M1													1594
1595	12.87	S1'	12.66		M1	0.48	1	0.66	1	14		U	II	15.1	1	23	22	1595
1596	11.76	3.7			M1													1596
1597	13.27	1.6			M1													1597
1598	14.32	2.0			M1													1598
1599	12.09	S2	3.00		M1													1599
1600	14.17	3.5			M1													1600
1601	13.64	S1	5.80		M1													1601
1602	13.70	S1	10.73		M1	0.55	1	0.93	1	14		RU	FL	11.7	1	22	20	1602
1603	12.02	S2	3.43		M1													1603
1604	11.62	E2	4.08		M1	0.35	1	0.77	1	14		U	EOS	28.3	1	22	20	1604
1605	11.19	S2	4.13		M1													1605
1606	12.73	E1	8.05		M1													1606
1607	12.68	S2	6.60		M1													1607
1608	13.68	S1	2.40		M1													1608
1609	11.93	2.6			M1													1609
1610	14.67	S1			M1													1610
1611	11.84	S1			M1													1611
1612	12.10	1.3			M1													1612
1613	12.83	S1	5.60		M1													1613
1614	11.68	S4	6.95		M1													1614
1615	11.88	S2'	1.96		M1			0.71	2	04								1615
1616	12.29	S2			M1													1616
1617	12.03	1.2			M1													1617
1618	12.87	S1			M1													1618
1619	12.77	2.6'			M1													1619
1620	16.78	E1'	0.030	M1		0.50	2	0.87	2	22		S	A	2.2	2	19	146	1620
1621	12.80	E1	2.29		M1	0.47	1	0.90	1	04		S	FL	13.8	2	20	20	1621
1622	13.42	S1			M1													1622
1623	11.81	1.8			M1													1623
1624	11.9				M1													1624

NUMBER	BMAG	QUAL	PHASE COEFF	REF	U-B	QU	B-V	QB	COLOR REFS	TYPE	ZONE	DIAM	WT	BIN	DATA	NUMBER
1625	11.55	E1	5.09	M1	0.28	1	0.69	1	14	CMEU	III	46.3	1	22	20	1625
1626	13.6			M1												1626
1627	14.22	E1:	9.72	M1	0.45	2	0.86	2	05	S	A	7.0	2	21	40	1627
1628	11.56	1.6		M1												1628
1629	14.00	0.6		M1												1629
1630	12.52	E1		M1												1630
1631	13.57	1.9		M1												1631
1632	12.61	0.8		M1												1632
1633	11.54	0.7		M1												1633
1634	14.54	1.5		M1												1634
1635	12.72	0.5		M1												1635
1636	13.38	0.8		M1						S	FL	10.3	2	22	2	1636
1637	11.26	3.1:		M1												1637
1638	12.76	1.1		M1												1638
1639	11.92	E1	7.06	M1	0.37	2	0.68	2	04	C	II	38.9	2	20	40	1639
1640	14.6			M1												1640
1641	12.50	2.5		M1												1641
1642	12.50	3.8		M1												1642
1643	13.69	3.6		M1												1643
1644	12.11	0.5		M1												1644
1645	12.58	0.5		M1						CMEU	III	29.0	1	24	2	1645
1646	13.8			M1												1646
1647	11.5			M1												1647
1648	13.58	S1	15.91	M1	0.50	0	0.79	1	04							1648
1649	12.7			M1												1649
1650	12.55	E1	8.49	M1	0.22	1	0.66	2	14	EU	NY	17.5	1	21	30	1650
1651	13.37	3.0		M1												1651
1652	13.70	E1		M1												1652
1653	12.68	2.5		M1												1653
1654	12.04	2.7		M1												1654
1655	12.77	0.1		M1												1655
1656	14.2			M1						R	A	9.4	1	21	2	1656
1657	11.74	0.1		M1			0.86	1	04							1657
1658	12.73	S1	15.69	M1	0.61	1	0.96	2	14	RU	II	18.5	1	22	30	1658
1659	11.18	1.3		M1												1659
1660	14.13	0.9		M1												1660
1661	14.00	0.3		M1												1661
1662	12.97	0.3		M1												1662
1663	14.78	1.0		M1												1663
1664	13.7			M1												1664
1665	12.28	0.5		M1												1665
1666	13.4			M1												1666
1667	13.5			M1												1667
1668	13.46	0.1		M1												1668
1669	11.97	E1	10.12	M1	0.46	1	0.73	1	14	CU	TH	38.9	2	23	20	1669
1670	12.17	2.0		M1												1670
1671	12.94	2.0		M1												1671
1672	13.05	2.1		M1												1672
1673	12.1			M1												1673
1674	12.07	E1	1.07	M1												1674
1675	13.06	1.0		M1												1675
1676	14.13	0.5		M1												1676
1677	13.34	0.3		M1												1677
1678	11.98	0.8:		M1												1678
1679	11.54	2.3		M1												1679
1680	12.4			M1												1680
1681	12.71	E2	4.86	M1	0.45	2	0.88	1	04	S	II	14.3	2	23	30	1681
1682	14.5			M1												1682
1683	12.84	1.5		M1												1683
1684	12.7			M1												1684
1685	15.1	E1:	0.037	M1	0.47	2	0.88	2	02	S	A	4.7	2	18	49	1685
1686	11.92	1.5		M1												1686
1687	11.24	E1:	2.58	M1												1687
1688	13.3			M1												1688
1689	12.80	1.0		M1												1689
1690	11.8			M1												1690
1691	11.66	4.0		M1												1691
1692	12.38	2.0		M1												1692
1693	12.01	2.0	5.88	M1	0.38	1	0.77	1	04 14	CU	II	38.9	2	22	20	1693
1694	13.66	E2		M1	0.42	2	0.74	2	05	C	I	17.9	2	23	40	1694
1695	13.0			M1												1695
1696	14.30	0.5		M1												1696
1697	13.2			M1												1697
1698	12.4			M1												1698
1699	14.32	0.2		M1												1699
1700	13.62	1.0		M1												1700
1701	11.5			M1												1701
1702	12.06	2.0	8.03	M1	0.20	1	0.74	1	14	MU	III	22.8	1	22	20	1702
1703	14.4			M1												1703
1704	13.82	0.3		M1												1704
1705	14.2			M1												1705
1706	13.88	0.5		M1												1706
1707	13.66	E1	14.04	M1	0.53	1	0.87	1	05	SU	FL	9.2	2	22	20	1707
1708	12.9			M1												1708
1709	13.9			M1												1709
1710	14.54	0.5		M1												1710
1711	12.04	2.5		M1												1711
1712	11.0			M1												1712
1713	14.4			M1												1713
1714	12.74	0.5		M1												1714
1715	13.4			M1												1715
1716	13.0			M1												1716
1717	13.55	2.0		M1						U	FL	11.8	1	22	2	1717
1718	14.9			M1												1718
1719	12.5			M1												1719
1720	14.26	1.0		M1												1720
1721	11.9			M1												1721
1722	13.0			M1												1722
1723	11.06	E1	7.52	M1	0.44	1	0.76	1	14	C	EOS	60.0	2	20	20	1723
1724	12.0			M1												1724
1725	12.18	2.5		M1												1725
1726	13.0			M1												1726
1727	14.2			M1						S	HU	7.1	2	21	2	1727
1728	12.7			M1												1728
1729	13.49	1.0		M1												1729
1730	12.69	3.0		M1												1730
1731	11.0			M1												1731
1732	11.89	1.0		M1												1732
1733	14.13	2.5		M1												1733
1734	12.55	1.0		M1												1734
1735	10.70	2.0		M1												1735
1736	13.3			M1												1736
1737	12.1			M1												1737
1738	13.7			M1												1738
1739	13.7			M1												1739
1740	14.4			M1												1740

NUMBER	BMAG	QUAL	PHASE COEFF	REF	U-B	QU	B-V	QB	COLOR REFS	TYPE	ZONE	DIAM	WT	BIN	DATA	NUMBER
1741	12.6			M1												1741
1742	12.3			M1												1742
1743	13.48	E1		M1												1743
1744	14.86	E1		M1												1744
1745	13.17	E1		M1												1745
1746	10.9			M1												1746
1747	14.82	E1	3.02	M1	0.48	0	1.28	0	04							1747
1748	11.7			M1												1748
1749	11.2			M1												1749
1750	14.66	E2	8.58	M1	0.50	2	0.88	2	14	S	A	5.8	2	22	40	1750
1751	13.5			M1												1751
1752	14.6			M1												1752
1753	12.2			M1												1753
1754	10.70	E1	6.01	M1	0.26	1	0.62	1	14	EU	HI	40.4	1	23	20	1754
1755	12.01	E2	4.95	M1	0.36	1	0.92	2	14	U	III	25.4	1	23	30	1755
1756	13.9			M1												1756
1757	14.0			M1												1757
1758	12.0			M1												1758
1759	14.1			M1												1759
1760	12.6			M1												1760
1761	12.6			M1												1761
1762	12.8			M1												1762
1763	14.2			M1												1763
1764	12.5			M1												1764
1765	10.94	E1	7.12	M1	0.27	2	0.75	2	14	CMEU	III	63.1	1	21	40	1765
1766	13.5			M1												1766
1767	13.22	S1	10.86	M1	0.34	1	0.75	1	14	C	EOS	22.1	2	25	20	1767
1768	14.2			M1												1768
1769	14.0			M1												1769
1770	14.1			M1												1770
1771	11.2			M1												1771
1772	13.2			M1												1772
1773	13.3			M1												1773
1774	13.5			M1												1774
1775	13.3			M1												1775
1776	12.10	E1		M1												1776
1777	12.92	E1		M1												1777
1778	12.93	E1		M1												1778
1779	15.40	E1		M1												1779
1780	11.9			M1												1780
1781	14.0			M1												1781
1782	12.7			M1												1782
1783	12.6			M1												1783
1784	13.5			M1												1784
1785	13.9			M1												1785
1786	12.1			M1												1786
1787	12.4			M1												1787
1788	12.8			M1												1788
1789	14.3			M1												1789
1790	14.0			M1												1790
1791	13.1			M1												1791
1792	13.06	S1	2.25	M1	0.34	1	0.74	1	04	C	II	23.7	2	24	20	1792
1793	13.7			M1												1793
1794	11.9			M1												1794
1795	13.03	E1		M1												1795
1796	11.6			M1												1796
1797	13.9			M1												1797
1798	13.7			M1												1798
1799	12.4			M1												1799
1800	14.0			M1												1800
1801	12.3			M1												1801
1802	13.0			M1												1802
1803	13.3			M1												1803
1804	13.4			M1												1804
1805	12.5			M1												1805
1806	14.0			M1												1806
1807	14.0			M1												1807
1808	13.33	E1		M1												1808
1809	12.80	E1		M1												1809
1810	13.87	E1		M1												1810
1811	12.24	E1		M1												1811
1812	12.72	E1		M1												1812
1813	13.62	E1		M1												1813
1814	14.2			M1												1814
1815	12.4			M1												1815
1816	14.7			M1												1816
1817	13.3			M1												1817
1818	15.2			M1												1818
1819	12.0			M1												1819
1820	14.6			M1												1820
1821	14.8			M1												1821
1822	14.6			M1												1822
1823	14.1			M1												1823
1824	12.8			M1												1824
1825	12.9			M1												1825
1826	12.3			M1												1826
1827	13.49	E2I	5.88	M1			0.81	1	04							1827
1828	12.2			M1												1828
1829	13.7			M1												1829
1830	13.61	S1	14.92	M1	0.50	1	0.91	1	14	S	I	9.5	2	22	22	1830
1831	13.2			M1												1831
1832	11.6			M1												1832
1833	13.05	S1	8.89	M1												1833
1834	12.7			M1												1834
1835	12.7			M1												1835
1836	12.6			M1												1836
1837	14.9			M1												1837
1838	11.9			M1												1838
1839	12.7			M1												1839
1840	12.8			M1												1840
1841	11.7			M1												1841
1842	13.5			M1												1842
1843	12.6			M1												1843
1844	12.5			M1												1844
1845	12.9			M1												1845
1846	14.64	E1		M1												1846
1847	12.0			M1												1847
1848	11.8			M1												1848
1849	12.2			M1												1849
1850	14.2			M1												1850
1851	13.1			M1												1851
1852	11.8			M1												1852
1853	11.6			M1												1853
1854	13.6			M1												1854
1855	13.8			M1												1855
1856	13.4			M1												1856

NUMBER	BMAG	QUAL	PHASE COEFF	REF	U-B	QU	B-V	OB	COLOR REFS	TYPE	ZONE	DIAM	WT	BIN	DATA	NUMBER
1857	13.7			M1												1857
1858	12.8			M1												1858
1859	11.2			M1												1859
1860	12.6			M1												1860
1861	12.9			M1												1861
1862	17.0			M1												1862
1863	16.5			M1												1863
1864	15.96	E1:	15.00	M1	0.50	?	0.83	2	02	SU	A	3.1	2	21	40	1864
1865	17.5			M1												1865
1866	14.5			M1												1866
1867	9.42	E1	7.36	M1	0.26	?	0.74	2	14	CMEU	T	127.	1	23	40	1867
1868	10.75	E1		M1												1868
1869	12.32	E1		M1												1869
1870	11.9			M1												1870
1871	12.3			M1												1871
1872	11.5			M1												1872
1873	11.7			M1												1873
1874	12.1			M1												1874
1875	13.5			M1												1875
1876	16.0			M1												1876
1877	12.4			M1												1877
1878	12.4			M1												1878
1879	14.1			M1												1879
1880	12.7			M1												1880
1881	12.1			M1												1881
1882	12.1			M1												1882
1883	14.3			M1												1883
1884	14.3			M1												1884
1885	14.7			M1												1885
1886	12.7			M1												1886
1887	11.8			M1												1887
1888	13.1			M1												1888
1889	12.0			M1												1889
1890	12.5			M1												1890
1891	13.0			M1												1891
1892	13.2			M1												1892
1893	12.4			M1												1893
1894	13.4			M1												1894
1895	13.0			M1												1895
1896	14.8			M1												1896
1897	14.4			M1												1897
1898	13.3			M1												1898
1899	14.0			M1												1899
1900	13.5			M1												1900
1901	12.5			M1												1901
1902	10.6			M1												1902
1903	12.0			M1												1903
1904	12.8			M1												1904
1905	14.0			M1												1905
1906	13.8			M1												1906
1907	13.2			M1												1907
1908	12.5			M1												1908
1909	13.4			M1												1909
1910	11.7			M1												1910
1911	11.3			M1												1911
1912	13.14	E2		M1												1912
1913	12.3			M1												1913
1914	13.6			M1												1914
1915	19.3			M1												1915
1916	16.10	E1	12.70	M1	0.41	1	0.85	2	14	S	A	3.0	2	27	30	1916
1917	16.5			M1												1917
1918	12.3			M1												1918
1919	15.0			M1												1919
1920	15.6			M1												1920
1921	15.6			M1												1921
1922	12.9			M1												1922
1923	14.64	E1		M1												1923
1924	14.34	E1		M1												1924
1925	13.5			M1												1925
1926	13.2			M1												1926
1927	12.9			M1												1927
1928	13.9			M1												1928
1929	13.5			M1												1929
1930	12.3			M1												1930
1931	14.4			M1	0.32	1	0.69	1	10	C	II	12.5	2	25	20	1931
1932	14.6			M1												1932
1933	14.4			M1												1933
1934	14.0			M1												1934
1935	14.4			M1												1935
1936	12.5			M1												1936
1937	13.3			M1												1937
1938	14.0			M1												1938
1939	12.0			M1												1939
1940	12.5			M1												1940
1941	12.5			M1												1941
1942	14.2			M1												1942
1943	16.5			M1												1943
1944	14.8			M1												1944
1945	13.5			M1												1945
1946	14.0			M1												1946
1947	11.5			M1												1947
1948	13.5			M1												1948
1949	14.6			M1												1949
1950	14.0			M1												1950
1951	17.2			M1												1951
1952	11.34	E1	2.27	M1	0.31	1	0.74	1	04	C	III	52.2	2	21	20	1952
1953	12.9			M1												1953
1954	13.2			M1												1954
1955	12.6			M1												1955
1956	13.0			M1												1956
1957	12.0			M1												1957
1958	12.1			M1												1958
1959	14.0			M1												1959
1960	12.61	E1		M1												1960
1961	12.3			M1												1961
1962	13.3			M1												1962
1963	12.0			M1												1963
1964	14.46	E1		M1												1964
1965	13.42	E1		M1												1965
1966	15.14	E1		M1												1966
1967	14.0			M1												1967
1968	12.8			M1												1968
1969	12.6			M1												1969
1970	12.7			M1												1970
1971	13.4			M1												1971
1972	14.5			M1												1972

NUMBER	BMAG	QUAL	PHASE	COEFF	REF	U-B	QU	B-V	OB	COLOR REFS	TYPE	ZONE	DIAM	WT	BIN	DATA	NUMBER
1973	13.0				M1												1973
1974	13.1				M1												1974
1975	13.2				M1												1975
1976	13.8				M1												1976
1977	12.4				M1												1977
1978	14.2				M1												1978
1979	14.66	E1			M1												1979
1980	15.11	E1	26.67		M1	0.46	1	0.96	1	14	U	A	6.2	1	22	20	1980
1981	18.00	E2			M1												1981
1982	13.5				M1												1982
1983	14.0				M1												1983
1984	12.3				M1												1984
1985	12.3				M1												1985
1986	13.1				M1												1986
1987	12.9				M1												1987
1988	14.7				M1												1988
1989	13.3				M1												1989
1990	14.0				M1												1990
1991	14.6				M1												1991
1992	13.2				M1												1992
1993	13.5				M1												1993
1994	13.5				M1												1994
1995	13.7				M1												1995
1996	13.2				M1												1996
1997	14.4				M1												1997
1998	12.8				M1												1998
1999	12.0				M1												1999
2000	12.33	E1	24.12		M1	0.49	2	0.89	2	04	S	PH	17.1	2	20	40	2000
2001	14.0				M1												2001
2002	13.3				M1												2002
2003	12.91	E1			M1												2003
2004	13.9				M1												2004
2005	13.5				M1												2005
2006	14.1				M1												2006
2007	13.0				M1												2007
2008	11.2				M1												2008
2009	12.1				M1												2009
2010	12.2				M1												2010
2011	14.0				M1												2011
2012	14.3				M1												2012
2013	13.2				M1												2013
2014	12.9				M1												2014
2015	13.4				M1												2015
2016	12.5				M1												2016
2017	14.7				M1												2017
2018	15.6				M1												2018
2019	13.5				M1												2019
2020	11.7				M1												2020
2021	14.7				M1												2021
2022	12.3				M1												2022
2023	12.7				M1												2023
2024	14.4				M1												2024
2025	11.8				M1												2025
2026	14.5				M1												2026
2027	12.8				M1												2027
2028	15.2				M1												2028
2029	14.3				M1												2029
2030	14.7				M1												2030
2031	14.4				M1												2031
2032	12.7				M1												2032
2033	14.8				M1												2033
2034	14.0				M1												2034
2035	13.71	E1	10.26		M1			0.86	1	04							2035
2036	14.0				M1												2036
2037	14.8				M1												2037
2038	13.5				M1												2038
2039	14.0				M1												2039
2040	12.8				M1												2040
2041	13.64	E1			M1												2041
2042	14.03	E1			M1												2042
2043	12.1				M1												2043
2044	14.3				M1												2044
2045	13.4				M1												2045
2046	12.1				M1												2046
2047	15.0				M1												2047
2048	14.0				M1												2048
2049	16.2				M1												2049
2050	13.0				M1												2050
2051	13.0				M1												2051
2052	11.0				M1												2052
2053	12.7				M1												2053
2054	13.63	E1			M1												2054
2055	14.6				M1												2055
2056	13.5				M1												2056
2057	16.0				M1												2057
2058	12.0				M1												2058
2059	16.0				M1												2059
2060	6.0				M1												2060
2061	18.0				M1	0.35	1	0.76	1	08	C	A	2.5	2	30	20	2061
2062	18.36	E1	59.70	0.027	M1	0.46	1	0.93	1	05	S	A	1.1	2	12	1020	2062
2063	13.7				M1												2063
2064	15.0				M1												2064
2065	13.3				M1												2065
2066	14.1				M1												2066
2067	11.8				M1												2067
2068	12.8				M1												2068
2069	12.3				M1												2069
2070	14.7				M1												2070
2071	14.5				M1												2071
2072	13.2				M1												2072
2073	13.8				M1												2073
2074	14.9				M1												2074
2075	15.0				M1												2075
2076	15.5				M1												2076
2077	14.5				M1												2077
2078	14.0				M1												2078
2079	13.3				M1												2079
2080	14.7				M1												2080
2081	13.5				M1												2081
2082	13.78	E1			M1												2082
2083	14.13	S1	21.33		M1			0.64	1	04							2083
2084	13.6				M1												2084
2085	12.2				M1												2085
2086	13.1				M1												2086
2087	14.5				M1												2087
2088	14.3				M1												2088

NUMBER	BMAG	QUAL	PHASE COEFF	REF	U-B	QU	B-V	QB	COLOR REFS	TYPE	ZONE	DIAM	WT	BIN	DATA NUMBER
2089	12.22	S1	22.13	M1	0.40	1	0.82	2	04						2089
2090	11.5			M1											2090
2091	12.0			M1											2091
2092	12.7			M1											2092
2093	14.3			M1											2093
2094	13.3			M1											2094
2095	13.9			M1											2095
2096	14.3			M1											2096
2097	12.8			M1											2097
2098	13.3			M1											2098
2099	16.43	S1	16.40	M1	0.35	1	0.83	1	14						2099
2100	17.16	E1	3.29	M1	0.32	1	0.73	3	04						2100
2101	19.5			M1											2101
1975EA					0.50	1	0.77	1	24	U	X		0	30	20 1975EA
1975GB					0.50	1	0.85	1	24	S	X		0	30	30 1975GB
1975RB					0.41	1	0.80	1	24	SU	X		0	30	30 1975RB
1975U2					0.42	1	0.82	1	24	S	X		0	30	40 1975U2
1976EA					0.41	1	0.77	1	24	C	X		0	30	40 1976EA
1976UA					0.50	1	0.77	1	14	U	X		0	30	20 1976UA
1976YB							0.94	1	04						1976YB
1977CA							1.02	1	04						1977CA
1977RA					0.52	1	0.83	3	14	U	X		0	30	40 1977RA
1977VA					0.14	1	0.72	2	14	U	X		0	30	30 1977VA
1977VC					0.42	2	0.82	2	14	S	X		0	30	30 1977VC
1978CA					0.50	2	0.85	1	04 17	S	X		0	30	30 1978CA
1978DA					0.41	1	0.80	2	04 17	SU	X		0	30	30 1978DA
1978SB					0.41	2	0.77	2	04	C	X		0	30	40 1978SB

REFERENCES

M01 GEHRELS, T. AND TEDESCO, E. F. 1979. MINOR PLANETS AND RELATED OBJECTS.
XXVIII. ASTEROID MAGNITUDES AND PHASE RELATIONS. ASTRON. J. 84, IN PRESS.
M02 TABLE I OF THE CHAPTER BY DEGEWIJ AND VAN HOUTEN IN THIS BOOK.

C01 TAYLOR, R.C. 1971. PHOTOMETRIC OBSERVATIONS AND REDUCTIONS OF LIGHTCURVES
OF ASTEROIDS. IN "PHYSICAL STUDIES OF MINOR PLANETS", ED. T. GEHRELS,
NASA SP-267, PP. 117-131.
C02 ZELLNER, B., WISNIEWSKI, W.Z., ANDERSSON, L., AND BOWELL, E. 1975. MINOR
PLANETS AND RELATED OBJECTS. XVIII. UBV PHOTOMETRY AND SURFACE COMPOSITION.
ASTRON. J. 80:986-995.
C03 HANSEN, O.L. 1976. RADII AND ALBEDOS OF 84 ASTEROIDS FROM VISUAL AND
INFRARED PHOTOMETRY. ASTRON. J. 81:74-84.
C04 BOWELL, E. UNPUBLISHED UBV PHOTOMETRY.
C05 ZELLNER, B., ANDERSSON, L., AND GRADIE, J. 1977. UBV PHOTOMETRY
OF SMALL AND DISTANT ASTEROIDS. ICARUS 31:447-455.
C06 SATHER, R.E. 1976. MINOR PLANETS AND RELATED OBJECTS. XIX. SHAPE AND POLE
ORIENTATION OF 39 LAETITIA. ASTRON. J. 81:67-73.
C07 VEVERKA, J. 1970. PHOTOMETRIC AND POLARIMETRIC STUDIES OF MINOR PLANETS
AND SATELLITES. PH.D. DISSERTATION. HARVARD UNIV., CAMBRIDGE, MASS.
C08 RAKOS, K.D. 1960. LIGHT VARIATIONS OF THE FAST MOVING MINOR PLANET.
LOWELL OBS. BULL. 5:28-29.
C09 BURCHI, R. 1972. SOME PHOTOMETRIC PARAMETERS OF THE MINOR PLANET
2 PALLAS. MEM. SOC. ASTRON. ITAL. 43:27-32.
C10 DEGEWIJ, J., AND GRADIE, J. UNPUBLISHED OBSERVATIONS, JUNE 1977.
C11 WAMSTEKER, W., AND SATHER, R.E. 1974. MINOR PLANETS AND RELATED OBJECTS.
XVII. FIVE-COLOR PHOTOMETRY OF FOUR ASTEROIDS. ASTRON. J. 79:1465-1470.
C12 TAYLOR, R.C., GEHRELS, T., AND CAPEN, R.C. 1976. MINOR PLANETS AND
RELATED OBJECTS. XXI. PHOTOMETRY OF EIGHT ASTEROIDS. ASTRON. J. 81:778-786.
C13 DEBEHOGNE, H., SURDEJ, A., AND SURDEJ, J. 1978. PHOTOELECTRIC LIGHTCURVES
OF THE MINOR PLANETS 29 AMPHITRITE, 121 HERMIONE AND 185 EUNIKE. ASTRON.
ASTROPHYS. SUPPL. 33:1-5.
C14 DEGEWIJ, J., GRADIE, J., AND ZELLNER, B. 1978. MINOR PLANETS AND RELATED
OBJECTS. XXV. PHOTOMETRY OF 145 FAINT ASTEROIDS. ASTRON. J., 83:643-650.
C15 BOWELL, E., MARTIN, L.J., POUTANEN, M., AND THOMPSON, D.T. 1979.
PHOTOMETRY AND POLARIMETRY OF ATMOSPHERELESS BODIES. I. PHOTOELECTRIC
PHOTOMETRY OF 69 HESPERIA. TO BE SUBMITTED TO ASTRON. J.
C16 LAGERKVIST, C.-I., AND PETTERSON, B. 1978. 186 CELUTA, A SLOWLY SPINNING
ASTEROID. ASTRON. ASTROPHYS. SUPPL. 32:339-342.
C17 SURDEJ, J., AND SURDEJ, A. 1978. 1978CA AND 1978DA. IAUC 3185.
C18 BURCHI, R., AND MILANO, L. 1974. PHOTOELECTRIC LIGHTCURVES OF THE MINOR
PLANET 43 ARIADNE DURING THE 1972 OPPOSITION. ASTRON. ASTROPHYS. SUPPL.
15:173-180.
C19 GEHRELS, T., AND TAYLOR, R.C. 1977. MINOR PLANETS AND RELATED OBJECTS.
XXII. PHASE FUNCTIONS FOR 6 HEBE. ASTRON. J. 82:229-237.
C20 TAYLOR, R.C. 1977. MINOR PLANETS AND RELATED OBJECTS. XXIII. PHOTOMETRY
OF ASTEROID 7 IRIS. ASTRON. J. 82:441-444.
C21 TAYLOR, R.C. 1978. MINOR PLANETS AND RELATED OBJECTS. XXIV. PHOTOMETRIC
OBSERVATIONS FOR 5 ASTRAEA. ASTRON. J. 83:201-204.
C22 DUNLAP, J.L. 1974. MINOR PLANETS AND RELATED OBJECTS. XV. ASTEROID 1620
GEOGRAPHOS. ASTRON. J. 79:324-332.
C23 SCHOBER, H.J. 1978. PHOTOMETRIC VARIATIONS OF THE MINOR PLANETS 55
PANDORA AND 173 INO DURING THE OPPOSITION IN 1977. LIGHTCURVES AND ROTATION
PERIODS. ASTRON. ASTROPHYS. SUPPL. 34:377-381.
C24 TEDESCO, E.F. 1979. UNPUBLISHED OBSERVATIONS.
C99 EXHIBITS, OR IS THOUGHT TO EXHIBIT, ROTATIONAL VARIATIONS IN COLOR INDEX.
CF. DEGEWIJ, J., BOWELL, E., GAFFEY, M.J., GRADIE, J., TEDESCO, E.F., AND
ZELLNER, B. 1979. SPOTS ON ASTEROIDS. TO BE SUBMITTED TO ICARUS.

VIII. CIRCUMSTANCES OF MINOR PLANET DISCOVERY

FREDERICK PILCHER
Illinois College

The following column headings are used: No. = permanent number; Name = assigned name; pd = provisional designation; dd = discovery date; d = discoverer; dp = discovery place. The Notes column is used when two or more discoverers have names with combined length too great to fit in the discoverer column, to give a more complete description of programs involving several persons, and to reference cases in which two numbered planets were subsequently discovered to be identical and the number and name of one of these was reassigned to a newly-discovered planet. Notes have also been used to reference conflicting discovery claims and list important independent discoveries which are no longer regarded as official.

The discovery date is in local mean time prior to 1 January 1925, and in UT thereafter, and refers to the time of mid-exposure for planets discovered by photographic means. In many cases the permanent number was assigned only when several unnumbered planets observed in different years were found to be identical, often many years after the discovery photographs were taken. In this case the discovery date is the first of that series of photographic observations from which the preliminary orbit was computed, and the provisional designation is that associated with this particular set of observations. Often earlier observations exist, but they are considered prediscoveries.

The following literature has been examined comprehensively to determine the discovery data:

STRACKE, G., *Identifizierungsnachweis der Kleinen Planeten* (Berlin, 1938).

HERGET, P., *Names of Minor Planets* (University of Cincinnati Observatory, 1957, 1967).

Astronomische Nachrichten.
Astronomische Nachrichten Indices.
Monthly Notices of the Royal Astronomical Society.
Rechen-Institut Circulars.
Beobachtungs Zirkular.
Minor Planet Circulars.
Lick Research Surveys on Minor Planets.
Turku Informo.

Acknowledgments. The author wishes to thank the following people for valuable contributions to this work. B. Marsden has arduously searched the literature, resolved various errors and discrepancies, and has passed judgment on conflicting discovery claims. J. Meeus and M. Combes have prepared an earlier list of discovery data from which the present list was adapted and expanded, and have provided continuing advice and counsel in the preparation of the present list. Advice and counsel have also been provided graciously by J. LoGuirato.

No	Name	dd			d	dp	Notes
1	Ceres	1801	Jan	1	G. Piazzi	Palermo	
2	Pallas	1802	Mar	28	H. W. Olbers	Bremen	
3	Juno	1804	Sep	1	K. Harding	Lilienthal	
4	Vesta	1807	Mar	29	H. W. Olbers	Bremen	
5	Astraea	1845	Dec	8	K. L. Hencke	Driesen	
6	Hebe	1847	Jul	1	K. L. Hencke	Driesen	
7	Iris	1847	Aug	13	J. R. Hind	London	
8	Flora	1847	Oct	18	J. R. Hind	London	
9	Metis	1848	Apr	25	A. Graham	Markree	
10	Hygiea	1849	Apr	12	A. De Gasparis	Naples	
11	Parthenope	1850	May	11	A. De Gasparis	Naples	
12	Victoria	1850	Sep	13	J. R. Hind	London	
13	Egeria	1850	Nov	2	A. De Gasparis	Naples	
14	Irene	1851	May	19	J. R. Hind	London	1
15	Eunomia	1851	Jul	29	A. De Gasparis	Naples	
16	Psyche	1852	Mar	17	A. De Gasparis	Naples	
17	Thetis	1852	Apr	17	R. Luther	Düsseldorf	
18	Melpomene	1852	Jun	24	J. R. Hind	London	
19	Fortuna	1852	Aug	22	J. R. Hind	London	
20	Massalia	1852	Sep	19	A. De Gasparis	Naples	2
21	Lutetia	1852	Nov	15	H. Goldschmidt	Paris	
22	Kalliope	1852	Nov	16	J. R. Hind	London	
23	Thalia	1852	Dec	15	J. R. Hind	London	
24	Themis	1853	Apr	5	A. De Gasparis	Naples	
25	Phocaea	1853	Apr	6	J. Chacornac	Marseille	
26	Proserpina	1853	May	5	R. Luther	Düsseldorf	
27	Euterpe	1853	Nov	8	J. R. Hind	London	
28	Bellona	1854	Mar	1	R. Luther	Düsseldorf	
29	Amphitrite	1854	Mar	1	A. Marth	London	
30	Urania	1854	Jul	22	J. R. Hind	London	
31	Euphrosyne	1854	Sep	1	J. Ferguson	Washington	
32	Pomona	1854	Oct	26	H. Goldschmidt	Paris	
33	Polyhymnia	1854	Oct	28	J. Chacornac	Paris	
34	Circe	1855	Apr	6	J. Chacornac	Paris	
35	Leukothea	1855	Apr	19	R. Luther	Düsseldorf	
36	Atalante	1855	Oct	5	H. Goldschmidt	Paris	
37	Fides	1855	Oct	5	R. Luther	Düsseldorf	
38	Leda	1856	Jan	12	J. Chacornac	Paris	
39	Laetitia	1856	Feb	8	J. Chacornac	Paris	
40	Harmonia	1856	Mar	31	H. Goldschmidt	Paris	
41	Daphne	1856	May	22	H. Goldschmidt	Paris	
42	Isis	1856	May	23	N. R. Pogson	Oxford	
43	Ariadne	1857	Apr	15	N. R. Pogson	Oxford	
44	Nysa	1857	May	27	H. Goldschmidt	Paris	
45	Eugenia	1857	Jun	27	H. Goldschmidt	Paris	
46	Hestia	1857	Aug	16	N. R. Pogson	Oxford	
47	Aglaja	1857	Sep	15	R. Luther	Düsseldorf	
48	Doris	1857	Sep	19	H. Goldschmidt	Paris	
49	Pales	1857	Sep	19	H. Goldschmidt	Paris	
50	Virginia	1857	Oct	4	J. Ferguson	Washington	3

No	Name	dd			d	dp	Notes
51	Nemausa	1858	Jan	22	Laurent	Nîmes	
52	Europa	1858	Feb	4	H. Goldschmidt	Paris	
53	Kalypso	1858	Apr	4	R. Luther	Düsseldorf	
54	Alexandra	1858	Sep	10	H. Goldschmidt	Paris	
55	Pandora	1858	Sep	10	G. Searle	Albany	
56	Melete	1857	Sep	9	H. Goldschmidt	Paris	
57	Mnemosyne	1859	Sep	22	R. Luther	Düsseldorf	
58	Concordia	1860	Mar	24	R. Luther	Düsseldorf	
59	Elpis	1860	Sep	12	J. Chacornac	Paris	
60	Echo	1860	Sep	14	J. Ferguson	Washington	
61	Danaë	1860	Sep	9	H. Goldschmidt	Paris	4
62	Erato	1860	Sep	14	O. Lesser and W. Förster	Berlin	
63	Ausonia	1861	Feb	10	A. De Gasparis	Naples	
64	Angelina	1861	Mar	4	E. W. Tempel	Marseille	5
65	Cybele	1861	Mar	8	E. W. Tempel	Marseille	
66	Maja	1861	Apr	9	H. P. Tuttle	Cambridge	
67	Asia	1861	Apr	17	N. R. Pogson	Madras	6
68	Leto	1861	Apr	29	R. Luther	Düsseldorf	
69	Hesperia	1861	Apr	26	G. Schiaparelli	Milan	
70	Panopaea	1861	May	5	H. Goldschmidt	Paris	
71	Niobe	1861	Aug	13	R. Luther	Düsseldorf	
72	Feronia	1861	Apr	29	C. H. F. Peters	Clinton	
73	Klytia	1862	Apr	7	H. P. Tuttle	Cambridge	
74	Galatea	1862	Aug	29	E. W. Tempel	Marseille	
75	Eurydike	1862	Sep	22	C. H. F. Peters	Clinton	
76	Freia	1862	Oct	21	R. Luther	Copenhagen	7
77	Frigga	1862	Nov	12	C. H. F. Peters	Clinton	
78	Diana	1863	Mar	15	R. Luther	Düsseldorf	
79	Eurynome	1863	Sep	14	J. C. Watson	Ann Arbor	
80	Sappho	1864	May	2	N. R. Pogson	Madras	
81	Terpsichore	1864	Sep	30	E. W. Tempel	Marseille	
82	Alkmene	1864	Nov	27	R. Luther	Düsseldorf	
83	Beatrix	1865	Apr	26	A. De Gasparis	Naples	
84	Klio	1865	Aug	25	R. Luther	Düsseldorf	
85	Io	1865	Sep	19	C. H. F. Peters	Clinton	
86	Semele	1866	Jan	4	F. Tietjen	Berlin	
87	Sylvia	1866	May	16	N. R. Pogson	Madras	
88	Thisbe	1866	Jun	15	C. H. F. Peters	Clinton	
89	Julia	1866	Sep	6	E. Stéphan	Marseille	
90	Antiope	1866	Oct	1	R. Luther	Düsseldorf	
91	Aegina	1866	Nov	4	E. Stéphan	Marseille	
92	Undina	1867	Jul	7	C. H. F. Peters	Clinton	
93	Minerva	1867	Aug	24	J. C. Watson	Ann Arbor	
94	Aurora	1867	Sep	6	J. C. Watson	Ann Arbor	
95	Arethusa	1867	Nov	23	R. Luther	Düsseldorf	
96	Aegle	1868	Feb	17	J. Coggia	Marseille	
97	Klotho	1868	Feb	17	E. W. Tempel	Marseille	
98	Ianthe	1868	Apr	18	C. H. F. Peters	Clinton	
99	Dike	1868	May	28	A. Borrelly	Marseille	
100	Hekate	1868	Jul	11	J. C. Watson	Ann Arbor	8

No	Name	dd	d	dp	Notes
101	Helena	1868 Aug 15	J. C. Watson	Ann Arbor	
102	Miriam	1868 Aug 22	C. H. F. Peters	Clinton	
103	Hera	1868 Sep 7	J. C. Watson	Ann Arbor	
104	Klymene	1868 Sep 13	J. C. Watson	Ann Arbor	
105	Artemis	1868 Sep 16	J. C. Watson	Ann Arbor	
106	Dione	1868 Oct 10	J. C. Watson	Ann Arbor	
107	Camilla	1868 Nov 17	N. R. Pogson	Madras	
108	Hecuba	1869 Apr 2	R. Luther	Dusseldorf	
109	Felicitas	1869 Oct 9	C. H. F. Peters	Clinton	
110	Lydia	1870 Apr 19	A. Borrelly	Marseille	
111	Ate	1870 Aug 14	C. H. F. Peters	Clinton	
112	Iphigenia	1870 Sep 19	C. H. F. Peters	Clinton	
113	Amalthea	1871 Mar 12	R. Luther	Dusseldorf	
114	Kassandra	1871 Jul 23	C. H. F. Peters	Clinton	
115	Thyra	1871 Aug 6	J. C. Watson	Ann Arbor	
116	Sirona	1871 Sep 8	C. H. F. Peters	Clinton	
117	Lomia	1871 Sep 12	A. Borrelly	Marseille	
118	Peitho	1872 Mar 15	R. Luther	Dusseldorf	
119	Althaea	1872 Apr 3	J. C. Watson	Ann Arbor	
120	Lachesis	1872 Apr 10	A. Borrelly	Marseille	
121	Hermione	1872 May 12	J. C. Watson	Ann Arbor	
122	Gerda	1872 Jul 31	A. Borrelly	Clinton	
123	Brunhild	1872 Jul 31	C. H. F. Peters	Clinton	
124	Alkeste	1872 Aug 23	C. H. F. Peters	Clinton	
125	Liberatrix	1872 Sep 11	Prosper Henry	Paris	
126	Velleda	1872 Nov 5	Paul Henry	Paris	
127	Johanna	1872 Nov 5	Prosper Henry	Paris	9
128	Nemesis	1872 Nov 25	J. C. Watson	Ann Arbor	9
129	Antigone	1873 Feb 5	C. H. F. Peters	Clinton	
130	Elektra	1873 Feb 17	C. H. F. Peters	Clinton	
131	Vala	1873 May 24	C. H. F. Peters	Clinton	
132	Aethra	1873 Jun 13	J. C. Watson	Ann Arbor	
133	Cyrene	1873 Aug 16	J. C. Watson	Ann Arbor	
134	Sophrosyne	1873 Sep 27	R. Luther	Dusseldorf	
135	Hertha	1874 Feb 18	C. H. F. Peters	Clinton	
136	Austria	1874 Mar 18	J. Palisa	Pola	
137	Melibeoa	1874 Apr 21	J. Palisa	Pola	
138	Tolosa	1874 May 19	J. Perrotin	Toulouse	
139	Juewa	1874 Oct 10	J. C. Watson	Peking	9
140	Siwa	1874 Oct 13	J. Palisa	Pola	
141	Lumen	1875 Jan 13	Paul Henry	Paris	
142	Polana	1875 Jan 28	J. Palisa	Pola	
143	Adria	1875 Feb 23	J. Palisa	Pola	
144	Vibilia	1875 Jun 3	J. C. Watson	Clinton	9
145	Adeona	1875 Jun 3	J. C. Watson	Clinton	
146	Lucina	1875 Jun 8	A. Borrelly	Marseille	
147	Protogeneia	1875 Jul 10	L. Schulhof	Vienna	
148	Gallia	1875 Aug 7	Prosper Henry	Paris	
149	Medusa	1875 Sep 21	J. Perrotin	Toulouse	
150	Nuwa	1875 Oct 18	J. C. Watson	Ann Arbor	

No	Name	dd	d	dp	Notes
151	Abundantia	1875 Nov 1	J. Palisa	Pola	9
152	Atala	1875 Nov 2	Paul Henry	Paris	
153	Hilda	1875 Nov 7	J. Palisa	Pola	
154	Bertha	1875 Nov 4	Prosper Henry	Paris	9, 10
155	Scylla	1875 Nov 8	J. Palisa	Pola	
156	Xanthippe	1875 Nov 22	J. Palisa	Pola	
157	Dejanira	1875 Dec 1	A. Borrelly	Marseille	
158	Koronis	1876 Jan 4	V. Knorre	Berlin	
159	Aemilia	1876 Jan 26	Paul Henry	Paris	9
160	Una	1876 Feb 20	C. H. F. Peters	Clinton	
161	Athor	1876 Apr 19	J. C. Watson	Ann Arbor	
162	Laurentia	1876 Apr 21	Prosper Henry	Paris	9
163	Erigone	1876 Apr 26	J. Perrotin	Toulouse	
164	Eva	1876 Jul 12	Paul Henry	Paris	9
165	Loreley	1876 Aug 9	C. H. F. Peters	Clinton	
166	Rhodope	1876 Aug 15	C. H. F. Peters	Clinton	
167	Urda	1876 Aug 28	C. H. F. Peters	Clinton	
168	Sibylla	1876 Sep 28	J. C. Watson	Ann Arbor	
169	Zelia	1876 Sep 28	Prosper Henry	Paris	
170	Maria	1877 Jan 10	J. Perrotin	Toulouse	
171	Ophelia	1877 Jan 13	A. Borrelly	Marseille	
172	Baucis	1877 Feb 5	A. Borrelly	Marseille	
173	Ino	1877 Aug 1	A. Borrelly	Marseille	
174	Phaedra	1877 Sep 2	J. C. Watson	Ann Arbor	
175	Andromache	1877 Oct 1	J. C. Watson	Ann Arbor	
176	Iduna	1877 Oct 14	C. H. F. Peters	Clinton	11
177	Irma	1877 Nov 5	Paul Henry	Paris	
178	Belisana	1877 Nov 6	J. Palisa	Pola	
179	Klytaemnestra	1877 Nov 11	J. C. Watson	Ann Arbor	
180	Garumna	1878 Jan 29	J. Perrotin	Toulouse	
181	Eucharis	1878 Feb 2	Cottenot	Marseille	
182	Elsa	1878 Feb 7	J. Palisa	Pola	
183	Istria	1878 Feb 8	J. Palisa	Pola	
184	Dejopeja	1878 Feb 28	J. Palisa	Pola	
185	Eunike	1878 Mar 1	C. H. F. Peters	Clinton	9
186	Celuta	1878 Apr 6	Prosper Henry	Paris	
187	Lamberta	1878 Apr 11	J. Coggia	Marseille	
188	Montpensier	1878 Jun 18	J. Coggia	Clinton	
189	Phthia	1878 Sep 9	C. H. F. Peters	Clinton	
190	Ismene	1878 Sep 22	C. H. F. Peters	Clinton	
191	Kolga	1878 Sep 30	C. H. F. Peters	Clinton	
192	Nausikaa	1879 Feb 17	J. Palisa	Pola	
193	Ambrosia	1879 Feb 28	J. Coggia	Marseille	
194	Prokne	1879 Mar 21	C. H. F. Peters	Clinton	
195	Eurykleia	1879 Apr 19	J. Palisa	Pola	
196	Philomela	1879 May 14	C. H. F. Peters	Clinton	
197	Arete	1879 May 21	J. Palisa	Pola	
198	Ampella	1879 Jun 13	A. Borrelly	Marseille	
199	Byblis	1879 Jul 9	C. H. F. Peters	Clinton	
200	Dynamene	1879 Jul 27	C. H. F. Peters	Clinton	

No	Name	dd	d	dp	Notes
201	Penelope	1879 Aug 7	J. Palisa	Pola	
202	Chryseis	1879 Sep 11	C. H. F. Peters	Clinton	
203	Pompeja	1879 Sep 25	C. H. F. Peters	Clinton	
204	Kallisto	1879 Oct 8	J. Palisa	Pola	
205	Martha	1879 Oct 13	J. Palisa	Pola	
206	Hersilia	1879 Oct 13	C. H. F. Peters	Clinton	
207	Hedda	1879 Oct 17	J. Palisa	Pola	
208	Lacrimosa	1879 Oct 21	J. Palisa	Pola	
209	Dido	1879 Oct 22	C. H. F. Peters	Clinton	
210	Isabella	1879 Nov 12	J. Palisa	Pola	
211	Isolda	1879 Dec 10	J. Palisa	Pola	
212	Medea	1880 Feb 6	C. H. F. Peters	Clinton	
213	Lilaea	1880 Feb 16	C. H. F. Peters	Clinton	
214	Aschera	1880 Feb 29	J. Palisa	Pola	
215	Oenone	1880 Apr 7	V. Knorre	Berlin	
216	Kleopatra	1880 Apr 10	J. Palisa	Pola	
217	Eudora	1880 Aug 30	J. Coggia	Marseille	
218	Bianca	1880 Sep 4	J. Palisa	Pola	
219	Thusnelda	1880 Sep 30	J. Palisa	Pola	12
220	Stephania	1881 May 19	J. Palisa	Vienna	
221	Eos	1882 Jan 18	J. Palisa	Vienna	
222	Lucia	1882 Feb 9	J. Palisa	Vienna	
223	Rosa	1882 Mar 9	J. Palisa	Vienna	
224	Oceana	1882 Mar 30	J. Palisa	Vienna	
225	Henrietta	1882 Apr 19	J. Palisa	Vienna	
226	Weringia	1882 Jul 19	J. Palisa	Vienna	
227	Philosophia	1882 Aug 12	Paul Henry	Paris	9
228	Agathe	1882 Aug 19	J. Palisa	Vienna	
229	Adelinda	1882 Aug 22	J. Palisa	Vienna	
230	Athamantis	1882 Sep 3	K. De Ball	Bothkamp	
231	Vindobona	1882 Sep 10	J. Palisa	Vienna	
232	Russia	1883 Jan 31	J. Palisa	Vienna	
233	Asterope	1883 May 11	A. Borrelly	Marseille	
234	Barbara	1883 Aug 12	C. H. F. Peters	Clinton	
235	Carolina	1883 Nov 28	J. Palisa	Vienna	
236	Honoria	1884 Apr 26	J. Palisa	Vienna	
237	Coelestina	1884 Jun 27	J. Palisa	Vienna	
238	Hypatia	1884 Jul 1	V. Knorre	Berlin	
239	Adrastea	1884 Aug 18	J. Palisa	Vienna	
240	Vanadis	1884 Aug 27	A. Borrelly	Marseille	
241	Germania	1884 Sep 12	R. Luther	Dusseldorf	
242	Kriemhild	1884 Sep 22	J. Palisa	Vienna	
243	Ida	1884 Sep 29	J. Palisa	Vienna	
244	Sita	1884 Oct 14	J. Palisa	Vienna	
245	Vera	1885 Feb 6	N. R. Pogson	Madras	
246	Asporina	1885 Mar 6	A. Borrelly	Marseille	
247	Eukrate	1885 Mar 14	R. Luther	Dusseldorf	
248	Lameia	1885 Jun 5	J. Palisa	Vienna	
249	Ilse	1885 Aug 16	C. H. F. Peters	Clinton	
250	Bettina	1885 Sep 3	J. Palisa	Vienna	
251	Sophia	1885 Oct 4	J. Palisa	Vienna	
252	Clementina	1885 Oct 11	C. Perrotin	Nice	
253	Mathilde	1885 Nov 12	J. Palisa	Vienna	
254	Augusta	1886 Mar 31	J. Palisa	Vienna	
255	Oppavia	1886 Mar 31	J. Palisa	Vienna	
256	Walpurga	1886 Apr 3	J. Palisa	Vienna	
257	Silesia	1886 Apr 5	J. Palisa	Vienna	
258	Tyche	1886 May 4	R. Luther	Dusseldorf	
259	Aletheia	1886 Jun 28	C. H. F. Peters	Clinton	
260	Huberta	1886 Oct 3	J. Palisa	Vienna	
261	Prymno	1886 Oct 31	C. H. F. Peters	Clinton	
262	Valda	1886 Nov 3	J. Palisa	Vienna	
263	Dresda	1886 Nov 3	J. Palisa	Vienna	
264	Libussa	1886 Dec 22	C. H. F. Peters	Clinton	13
265	Anna	1887 Feb 25	J. Palisa	Vienna	
266	Aline	1887 May 17	J. Palisa	Vienna	
267	Tirza	1887 May 27	A. Charlois	Nice	14
268	Adorea	1887 Jun 8	A. Borrelly	Marseille	
269	Justitia	1887 Sep 21	J. Palisa	Vienna	
270	Anahita	1887 Oct 8	C. H. F. Peters	Clinton	
271	Penthesilea	1887 Oct 13	V. Knorre	Berlin	
272	Antonia	1888 Feb 4	A. Charlois	Nice	
273	Atropos	1888 Mar 8	J. Palisa	Vienna	
274	Philagoria	1888 Apr 3	J. Palisa	Vienna	
275	Sapientia	1888 Apr 15	J. Palisa	Vienna	
276	Adelheid	1888 Apr 17	J. Palisa	Vienna	
277	Elvira	1888 May 3	A. Charlois	Nice	
278	Paulina	1888 May 16	J. Palisa	Vienna	
279	Thule	1888 Oct 25	J. Palisa	Vienna	
280	Philia	1888 Oct 29	J. Palisa	Vienna	
281	Lucretia	1888 Oct 31	J. Palisa	Vienna	
282	Clorinde	1889 Jan 28	A. Charlois	Nice	
283	Emma	1889 Feb 8	A. Charlois	Nice	
284	Amalia	1889 Feb 29	A. Charlois	Nice	
285	Regina	1889 Aug 3	J. Palisa	Vienna	
286	Iclea	1889 Aug 25	J. Palisa	Vienna	
287	Nephthys	1889 Aug 25	C. H. F. Peters	Clinton	
288	Glauke	1890 Feb 20	R. Luther	Dusseldorf	
289	Nenetta	1890 Mar 10	A. Charlois	Nice	
290	Bruna	1890 Mar 20	J. Palisa	Vienna	
291	Alice	1890 Apr 25	J. Palisa	Vienna	
292	Ludovica	1890 Apr 25	J. Palisa	Vienna	
293	Brasilia	1890 May 20	A. Charlois	Nice	
294	Felicia	1890 Jul 15	A. Charlois	Nice	
295	Theresia	1890 Aug 17	J. Palisa	Vienna	
296	Phaetusa	1890 Aug 19	A. Charlois	Nice	
297	Caecilia	1890 Sep 9	A. Charlois	Nice	
298	Baptistina	1890 Sep 9	A. Charlois	Nice	
299	Thora	1890 Oct 6	J. Palisa	Vienna	
300	Geraldina	1890 Oct 3	A. Charlois	Nice	

No	Name	pd	dd	d	dp	Notes
301	Bavaria		1890 Nov 16	J. Palisa	Vienna	
302	Clarissa		1890 Nov 14	A. Charlois	Nice	
303	Josephina		1891 Feb 12	E. Millosevich	Rome	
304	Olga		1891 Feb 14	J. Palisa	Vienna	
305	Gordonia		1891 Feb 16	A. Charlois	Nice	
306	Unitas		1891 Mar 1	E. Millosevich	Rome	
307	Nike		1891 Mar 5	J. Palisa	Nice	
308	Polyxo		1891 Mar 31	A. Borrelly	Marseille	
309	Fraternitas		1891 Apr 6	J. Palisa	Vienna	
310	Margarita		1891 May 16	A. Charlois	Nice	
311	Claudia		1891 Jun 11	A. Charlois	Nice	
312	Pierretta		1891 Aug 28	A. Charlois	Nice	
313	Chaldaea		1891 Aug 30	A. Charlois	Nice	
314	Rosalia		1891 Sep 1	A. Charlois	Nice	
315	Constantia		1891 Sep 4	J. Palisa	Vienna	
316	Goberta		1891 Sep 8	A. Charlois	Nice	
317	Roxane		1891 Sep 11	A. Charlois	Nice	
318	Magdalena		1891 Sep 24	A. Charlois	Nice	
319	Leona		1891 Oct 8	A. Charlois	Nice	
320	Katharina		1891 Oct 11	J. Palisa	Vienna	
321	Florentina		1891 Oct 15	J. Palisa	Vienna	
322	Phaeo		1891 Nov 27	A. Borrelly	Marseille	15
323	Brucia		1891 Dec 22	M. Wolf	Heidelberg	
324	Bamberga		1892 Feb 25	J. Palisa	Vienna	
325	Heidelberga		1892 Mar 4	M. Wolf	Heidelberg	
326	Tamara		1892 Mar 19	J. Palisa	Vienna	
327	Columbia		1892 Mar 22	M. Wolf	Nice	
328	Gudrun		1892 Mar 18	M. Wolf	Heidelberg	
329	Svea	1892 X	1892 Mar 21	M. Wolf	Heidelberg	
330	Adalberta		1892 Mar 18	M. Wolf	Heidelberg	
331	Etheridgea		1892 Apr 1	A. Charlois	Nice	
332	Siri		1892 Mar 19	M. Wolf	Heidelberg	
333	Badenia	1892 A	1892 Aug 22	M. Wolf	Heidelberg	
334	Chicago	1892 L	1892 Aug 23	M. Wolf	Heidelberg	
335	Roberta	1892 C	1892 Sep 1	A. Staus	Heidelberg	
336	Lacadiera	1892 D	1892 Sep 19	A. Charlois	Nice	
337	Devosa	1892 E	1892 Sep 22	M. Wolf	Nice	
338	Budrosa	1892 F	1892 Sep 25	M. Wolf	Nice	
339	Dorothea	1892 G	1892 Sep 25	M. Wolf	Heidelberg	
340	Eduarda	1892 H	1892 Sep 25	M. Wolf	Heidelberg	
341	California	1892 J	1892 Sep 25	M. Wolf	Heidelberg	
342	Endymion	1892 K	1892 Oct 17	M. Wolf	Heidelberg	
343	Ostara	1892 M	1892 Nov 15	M. Wolf	Heidelberg	
344	Desiderata	1892 N	1892 Nov 15	A. Charlois	Nice	
345	Tercidina	1892 O	1892 Nov 23	A. Charlois	Nice	
346	Hermentaria	1892 P	1892 Nov 25	A. Charlois	Nice	
347	Pariana	1892 Q	1892 Nov 28	A. Charlois	Nice	
348	May	1892 R	1892 Nov 28	M. Wolf	Vienna	
349	Dembowska	1892 T	1892 Dec 9	A. Charlois	Heidelberg	
350	Ornamenta	1892 U	1892 Dec 14	A. Charlois	Nice	

No	Name	pd	dd	d	dp	Notes
351	Yrsa	1892 V	1892 Dec 16	M. Wolf	Heidelberg	
352	Gisela	1893 B	1893 Jan 12	M. Wolf	Heidelberg	
353	Ruperto-Carola	1893 F	1893 Jan 16	M. Wolf	Heidelberg	
354	Eleonora	1893 A	1893 Jan 17	A. Charlois	Nice	
355	Gabriella	1893 E	1893 Jan 20	A. Charlois	Nice	
356	Liguria	1893 G	1893 Jan 21	A. Charlois	Nice	
357	Ninina	1893 J	1893 Feb 11	A. Charlois	Nice	
358	Apollonia	1893 K	1893 Mar 8	A. Charlois	Nice	
359	Georgia	1893 M	1893 Mar 10	A. Charlois	Nice	
360	Carlova	1893 N	1893 Mar 11	A. Charlois	Nice	
361	Bononia	1893 P	1893 Mar 11	A. Charlois	Nice	
362	Havnia	1893 R	1893 Mar 12	A. Charlois	Nice	
363	Padua	1893 S	1893 Mar 17	A. Charlois	Nice	
364	Isara	1893 T	1893 Mar 21	A. Borrelly	Marseille	
365	Corduba	1893 V	1893 Mar 21	A. Charlois	Nice	
366	Vincentina	1893 W	1893 May 21	A. Charlois	Nice	
367	Amicitia	1893 AA	1893 May 19	A. Charlois	Nice	
368	Haidea	1893 AB	1893 Sep 15	A. Charlois	Nice	
369	Aeria	1893 AE	1893 Jul 4	A. Borrelly	Marseille	
370	Modestia	1893 AC	1893 Jul 16	A. Charlois	Nice	
371	Bohemia	1893 AD	1893 Jul 16	A. Charlois	Nice	
372	Palma	1893 AH	1893 Aug 19	A. Charlois	Nice	
373	Melusina	1893 AJ	1893 Sep 15	A. Charlois	Nice	
374	Burgundia	1893 AK	1893 Sep 18	A. Charlois	Nice	
375	Ursula	1893 AL	1893 Sep 18	A. Charlois	Nice	
376	Geometria	1893 AM	1893 Sep 18	A. Charlois	Nice	
377	Campania	1893 AN	1893 Sep 20	A. Charlois	Nice	
378	Holmia	1893 AP	1893 Dec 6	A. Charlois	Nice	
379	Huenna	1893 AQ	1894 Jan 8	A. Charlois	Nice	
380	Fiducia	1894 AR	1894 Jan 8	A. Charlois	Nice	
381	Myrrha	1894 AS	1894 Jan 10	A. Charlois	Nice	
382	Dodona	1894 AT	1894 Jan 29	A. Charlois	Nice	
383	Janina	1894 AU	1894 Jan 29	A. Charlois	Nice	
384	Burdigala	1894 AV	1894 Feb 11	F. Courty	Bordeaux	
385	Ilmatar	1894 AX	1894 Mar 1	M. Wolf	Heidelberg	
386	Siegena	1894 AY	1894 Mar 1	M. Wolf	Heidelberg	
387	Aquitania	1894 AZ	1894 Mar 5	F. Courty	Bordeaux	
388	Charybdis	1894 BA	1894 Mar 7	A. Charlois	Nice	
389	Industria	1894 BB	1894 Mar 8	A. Charlois	Nice	
390	Alma	1894 BC	1894 Mar 24	G. Bigourdan	Paris	
391	Ingeborg	1894 BE	1894 Nov 1	M. Wolf	Heidelberg	
392	Wilhelmina	1894 BF	1894 Nov 4	M. Wolf	Heidelberg	
393	Lampetia	1894 BG	1894 Nov 4	M. Wolf	Heidelberg	
394	Arduina	1894 BH	1894 Nov 19	A. Borrelly	Marseille	
395	Delia	1894 BK	1894 Nov 30	A. Charlois	Nice	
396	Aeolia	1894 BL	1894 Dec 1	A. Charlois	Nice	
397	Vienna	1894 BM	1894 Dec 19	A. Charlois	Nice	
398	Admete	1894 BN	1894 Dec 28	A. Charlois	Nice	
399	Persephone	1895 BP	1895 Feb 23	M. Wolf	Heidelberg	
400	Ducrosa	1895 BU	1895 Mar 15	A. Charlois	Nice	

No	Name	pd	dd	d	dp	Notes
401	Ottilia	BT	1895 Mar 16	M. Wolf	Heidelberg	
402	Chloe	BW	1895 Mar 21	A. Charlois	Nice	
403	Cyane	BX	1895 May 18	A. Charlois	Nice	
404	Arsinoë	BY	1895 Jun 20	A. Charlois	Nice	
405	Thia	BZ	1895 Jul 23	A. Charlois	Nice	
406	Erna	CB	1895 Aug 22	A. Charlois	Nice	
407	Arachne	CC	1895 Oct 13	M. Wolf	Heidelberg	
408	Fama	CD	1895 Oct 13	M. Wolf	Heidelberg	
409	Aspasia	CE	1895 Dec 9	A. Charlois	Nice	
410	Chloris	CH	1896 Jan 7	A. Charlois	Nice	
411	Xanthe	CJ	1896 Jan 7	A. Charlois	Nice	
412	Elisabetha	CK	1896 Jan 7	M. Wolf	Heidelberg	
413	Edburga	CL	1896 Jan 7	M. Wolf	Heidelberg	
414	Liriope	CN	1896 Jan 16	A. Charlois	Nice	
415	Palatia	CO	1896 Feb 7	M. Wolf	Heidelberg	
416	Vaticana	CS	1896 May 4	A. Charlois	Nice	
417	Suevia	CT	1896 May 6	M. Wolf	Heidelberg	
418	Alemannia	CV	1896 Sep 7	M. Wolf	Heidelberg	
419	Aurelia	CU	1896 Sep 7	A. Charlois	Nice	
420	Bertholda	CW	1896 Sep 7	M. Wolf	Heidelberg	
421	Zähringia	CZ	1896 Sep 7	M. Wolf	Heidelberg	
422	Berolina	DA	1896 Oct 8	G. Witt	Berlin	
423	Diotima	DB	1896 Sep 7	A. Charlois	Nice	
424	Gratia	DF	1896 Dec 31	M. Wolf	Nice	
425	Cornelia	DC	1896 Dec 28	A. Charlois	Nice	
426	Hippo	DH	1897 Aug 25	A. Charlois	Nice	
427	Galene	DJ	1897 Aug 27	A. Charlois	Nice	
428	Monachia	DK	1897 Nov 18	M. Villiger	Munich	
429	Lotis	DL	1897 Nov 23	M. Wolf	Nice	
430	Hybris	DM	1897 Dec 18	A. Charlois	Nice	
431	Nephele	DP	1897 Dec 18	A. Charlois	Nice	
432	Pythia	DQ	1897 Dec 18	G. Witt	Berlin	
433	Eros	DR	1898 Aug 13	G. Witt	Berlin	16
434	Hungaria	DR	1898 Sep 11	M. Wolf	Heidelberg	
435	Ella	DS	1898 Sep 11	M. Wolf and A. Schwassmann	Heidelberg	
436	Patricia	DT	1898 Sep 13	M. Wolf and A. Schwassmann	Heidelberg	
437	Rhodia	DW	1898 Jul 16	M. Wolf	Nice	
438	Zeuxo	DU	1898 Nov 8	A. Charlois	Nice	
439	Ohio	EB	1898 Oct 13	E. F. Coddington	Mount Hamilton	
440	Theodora	EC	1898 Oct 13	E. F. Coddington	Mount Hamilton	
441	Bathilde	ED	1898 Dec 8	A. Charlois	Nice	
442	Eichsfeldia	EE	1899 Feb 15	M. Wolf and A. Schwassmann	Heidelberg	
443	Photographica	EF	1899 Feb 17	M. Wolf and A. Schwassmann	Heidelberg	
444	Gyptis	EL	1899 Mar 31	J. Coggia	Marseille	
445	Edna	EX	1899 Oct 2	M. Wolf and A. Schwassmann	Mount Hamilton	
446	Aeternitas	ER	1899 Oct 27	M. Wolf and A. Schwassmann	Heidelberg	
447	Valentine	ES	1899 Oct 27	M. Wolf and A. Schwassmann	Heidelberg	
448	Natalie	ET	1899 Oct 27	M. Wolf and A. Schwassmann	Heidelberg	
449	Hamburga	EU	1899 Oct 31	M. Wolf and A. Schwassmann	Heidelberg	
450	Brigitta	EV	1899 Dec 10	M. Wolf and A. Schwassmann	Heidelberg	

No	Name	pd	dd	d	dp	Notes
451	Patientia	EY	1899 Dec 4	A. Charlois	Nice	
452	Hamiltonia	FD	1899 Dec 6	J. E. Keeler	Mount Hamilton	17
453	Tea	FA	1900 Feb 22	A. Charlois	Nice	
454	Mathesis	FC	1900 Mar 28	Schwassmann	Heidelberg	
455	Bruchsalia	FG	1900 May 22	M. Wolf and A. Schwassmann	Heidelberg	
456	Abnoba	FH	1900 Jun 4	M. Wolf	Heidelberg	
457	Alleghenia	FJ	1900 Sep 15	M. Wolf and A. Schwassmann	Heidelberg	18
458	Hercynia	FK	1900 Sep 21	M. Wolf and A. Schwassmann	Heidelberg	
459	Signe	FM	1900 Oct 22	M. Wolf	Heidelberg	
460	Scania	FN	1900 Oct 22	M. Wolf	Heidelberg	
461	Saskia	FP	1900 Oct 22	M. Wolf	Heidelberg	
462	Eriphyla	PQ	1900 Oct 22	M. Wolf	Heidelberg	
463	Lola	FS	1900 Oct 31	M. Wolf	Heidelberg	
464	Megaira	FV	1901 Jan 9	M. Wolf	Heidelberg	
465	Alekto	FW	1901 Jan 13	M. Wolf	Heidelberg	
466	Tisiphone	FX	1901 Jan 17	M. Wolf and L. Carnera	Heidelberg	19
467	Laura	FT	1901 Jan 18	M. Wolf	Heidelberg	
468	Lina	FZ	1901 Jan 18	M. Wolf	Heidelberg	
469	Argentina	GE	1901 Feb 20	L. Carnera	Heidelberg	
470	Kilia	GJ	1901 Apr 21	L. Carnera	Heidelberg	
471	Papagena	GN	1901 Jun 7	M. Wolf	Heidelberg	
472	Roma	GP	1901 Jul 11	L. Carnera	Heidelberg	
473	Nolli	GG	1901 Feb 13	M. Wolf	Heidelberg	
474	Prudentia	HU	1901 Feb 13	M. Wolf	Heidelberg	
475	Ocllo	HN	1901 Aug 14	D. Stewart	Arequipa	
476	Hedwig	OQ	1901 Aug 17	L. Carnera	Heidelberg	
477	Italia	QR	1901 Aug 23	L. Carnera	Heidelberg	
478	Tergeste	GU	1901 Sep 21	L. Carnera	Heidelberg	
479	Caprera	LU	1901 Nov 12	L. Carnera	Heidelberg	
480	Hansa	HZ	1901 May 21	M. Wolf and L. Carnera	Heidelberg	
481	Emita	HP	1902 Feb 12	L. Carnera	Heidelberg	
482	Petrina	HT	1902 Mar 3	M. Wolf	Heidelberg	
483	Seppina	HU	1902 Mar 4	M. Wolf	Heidelberg	
484	Pittsburghia	HX	1902 Apr 29	M. Wolf and L. Carnera	Heidelberg	
485	Genua	HZ	1902 May 7	L. Carnera	Heidelberg	
486	Cremona	JB	1902 May 11	L. Carnera	Heidelberg	
487	Venetia	JL	1902 Jul 9	M. Wolf	Heidelberg	
488	Kreusa	JG	1902 Jun 26	M. Wolf and L. Carnera	Heidelberg	
489	Comacina	JM	1902 Sep 3	L. Carnera	Heidelberg	
490	Veritas	JP	1902 Sep 3	M. Wolf	Heidelberg	
491	Carina	JQ	1902 Sep 3	M. Wolf	Heidelberg	
492	Gismonda	JR	1902 Sep 3	M. Wolf	Heidelberg	
493	Griseldis	JS	1902 Sep 7	M. Wolf	Heidelberg	
494	Virtus	JV	1902 Oct 7	M. Wolf	Heidelberg	
495	Eulalia	KG	1902 Oct 25	M. Wolf	Heidelberg	
496	Gryphia	KH	1902 Oct 25	M. Wolf	Heidelberg	
497	Iva	KS	1902 Nov 4	R. S. Dugan	Heidelberg	
498	Tokio	KU	1902 Dec 2	A. Charlois	Nice	
499	Venusia	EU	1902 Dec 24	M. Wolf	Heidelberg	
500	Selinur	LA	1903 Jan 16	M. Wolf	Heidelberg	

No	Name	pd		dd		d	dp	Notes
501	Urhixidur	1903	LB	1903	Jan 18	M. Wolf	Heidelberg	
502	Sigune	1903	LC	1903	Jan 19	M. Wolf	Heidelberg	
503	Evelyn	1903	LF	1903	Jan 19	R. S. Dugan	Heidelberg	
504	Cora	1902	LK	1902	Jun 30	R. I. Bailey	Arequipa	
505	Cava	1902	LL	1902	Aug 21	R. H. Frost	Arequipa	
506	Marion	1903	LN	1903	Feb 17	R. S. Dugan	Heidelberg	
507	Laodica	1903	LQ	1903	Feb 19	R. S. Dugan	Heidelberg	
508	Princetonia	1903	LD	1903	Apr 20	R. S. Dugan	Heidelberg	
509	Iolanda	1903	LR	1903	Apr 28	M. Wolf	Heidelberg	
510	Mabella	1903	LT	1903	May 20	R. S. Dugan	Heidelberg	
511	Davida	1903	LU	1903	May 30	R. S. Dugan	Heidelberg	
512	Taurinensis	1903	LV	1903	Jun 23	M. Wolf	Heidelberg	
513	Centesima	1903	LY	1903	Aug 24	M. Wolf	Heidelberg	
514	Armida	1903	MB	1903	Aug 24	M. Wolf	Heidelberg	
515	Athalia	1903	ME	1903	Sep 20	M. Wolf	Heidelberg	
516	Amherstia	1903	MG	1903	Sep 20	R. S. Dugan	Heidelberg	
517	Edith	1903	MH	1903	Sep 20	S. Dugan	Heidelberg	
518	Halawe	1903	HO	1903	Oct 20	S. Dugan	Heidelberg	
519	Sylvania	1903	HP	1903	Oct 20	S. Dugan	Heidelberg	
520	Franziska	1903	MV	1903	Oct 27	M. Wolf and P. Götz	Heidelberg	
521	Brixia	1904	NB	1904	Jan 10	S. Dugan	Heidelberg	
522	Helga	1904	NC	1904	Jan 10	M. Wolf	Heidelberg	
523	Ada	1904	ND	1904	Mar 20	S. Dugan	Heidelberg	
524	Fidelio	1904	NV	1904	Mar 14	M. Wolf	Heidelberg	
525	Adelaide	1908	EK^a	1908	Oct 21	J. H. Metcalf	Taunton	20
526	Jena	1904	NM	1904	Apr 14	M. Wolf	Heidelberg	
527	Euryanthe	1904	NR	1904	Apr 20	M. Wolf	Heidelberg	
528	Rezia	1904	NS	1904	Mar 20	M. Wolf	Heidelberg	
529	Preziosa	1904	NT	1904	Mar 20	M. Wolf	Heidelberg	
530	Turandot	1904	NV	1904	Apr 11	M. Wolf	Heidelberg	
531	Zerlina	1904	NW	1904	Apr 12	M. Wolf	Heidelberg	
532	Herculina	1904	NY	1904	Apr 20	M. Wolf	Heidelberg	21
533	Sara	1904	NZ	1904	Apr 19	S. Dugan	Heidelberg	
534	Nassovia	1904	OA	1904	Apr 14	S. Dugan	Heidelberg	
535	Montague	1904	OC	1904	May 7	S. Dugan	Heidelberg	
536	Merapi	1904	OF	1904	May 11	G. H. Peters	Washington	
537	Pauly	1904	OG	1904	Jul 1	A. Charlois	Nice	
538	Friederike	1904	OK	1904	Jul 18	P. Götz	Heidelberg	
539	Pamina	1904	OL	1904	Aug 2	M. Wolf	Heidelberg	
540	Rosamunde	1904	OC	1904	Aug 3	M. Wolf	Heidelberg	
541	Deborah	1904	OO	1904	Aug 4	M. Wolf	Heidelberg	
542	Susanna	1904	OQ	1904	Aug 15	P. Götz and A. Kopff	Heidelberg	
543	Charlotte	1904	OT	1904	Sep 11	P. Götz	Heidelberg	
544	Jetta	1904	OU	1904	Sep 11	P. Götz	Heidelberg	
545	Messalina	1904	OY	1904	Oct 3	P. Götz	Heidelberg	
546	Herodias	1904	PA	1904	Aug 4	P. Götz	Heidelberg	
547	Praxedis	1904	PB	1904	Oct 14	P. Götz	Heidelberg	
548	Kressida	1904	PC	1904	Oct 14	P. Götz	Heidelberg	
549	Jessonda	1904	PD	1904	Oct 15	M. Wolf	Heidelberg	
550	Senta	1904	PL	1904	Nov 16	M. Wolf	Heidelberg	

No	Name	pd		dd		d	dp
551	Ortrud	1904	PM	1904	Nov 16	M. Wolf	Heidelberg
552	Sigelinde	1904	PO	1904	Dec 14	M. Wolf	Heidelberg
553	Kundry	1904	PP	1904	Dec 27	M. Wolf	Heidelberg
554	Peraga	1905	PS	1905	Jan 8	P. Götz	Heidelberg
555	Norma	1905	PT	1905	Jan 14	M. Wolf	Heidelberg
556	Phyllis	1905	PW	1905	Jan 8	P. Götz	Heidelberg
557	Violetta	1905	PY	1905	Jan 26	M. Wolf	Heidelberg
558	Carmen	1905	QB	1905	Feb 9	M. Wolf	Heidelberg
559	Nanon	1905	QD	1905	Mar 8	M. Wolf	Heidelberg
560	Delila	1905	QF	1905	Mar 13	M. Wolf	Heidelberg
561	Ingwelde	1905	QG	1905	Mar 26	M. Wolf	Heidelberg
562	Salome	1905	QH	1905	Apr 3	M. Wolf	Heidelberg
563	Suleika	1905	QK	1905	Apr 6	P. Götz	Heidelberg
564	Dudu	1905	QM	1905	May 9	P. Götz	Heidelberg
565	Marbachia	1905	QN	1905	May 9	M. Wolf	Heidelberg
566	Stereoskopia	1905	QO	1905	May 28	P. Götz	Heidelberg
567	Eleutheria	1905	QP	1905	May 28	P. Götz	Heidelberg
568	Cheruskia	1905	QS	1905	Jul 26	P. Götz	Vienna
569	Misa	1905	QT	1905	Jul 27	J. Palisa	Vienna
570	Kythera	1905	QX	1905	Jul 30	M. Wolf	Heidelberg
571	Dulcinea	1905	QZ	1905	Sep 4	P. Götz	Heidelberg
572	Rebekka	1905	RB	1905	Sep 19	P. Götz	Heidelberg
573	Recha	1905	KC	1905	Sep 19	P. Götz	Heidelberg
574	Reginhild	1905	RD	1905	Sep 19	M. Wolf	Heidelberg
575	Renate	1905	RE	1905	Sep 19	M. Wolf	Heidelberg
576	Emanuela	1905	RF	1905	Sep 22	P. Götz	Heidelberg
577	Rhea	1905	RH	1905	Oct 20	M. Wolf	Heidelberg
578	Happelia	1905	RZ	1905	Nov 1	M. Wolf	Heidelberg
579	Sidonia	1905	SD	1905	Nov 19	A. Kopff	Heidelberg
580	Selene	1905	SE	1905	Dec 17	M. Wolf	Heidelberg
581	Tauntonia	1905	SH	1905	Dec 24	J. H. Metcalf	Taunton
582	Olympia	1906	SO	1906	Jan 23	A. Kopff	Heidelberg
583	Klotilde	1906	SP	1905	Dec 31	A. Kopff	Vienna
584	Semiramis	1906	SY	1906	Jan 15	A. Kopff	Heidelberg
585	Bilkis	1906	TA	1906	Feb 16	A. Kopff	Heidelberg
586	Thekla	1906	TC	1906	Feb 21	M. Wolf	Heidelberg
587	Hypsipyle	1906	TF	1906	Feb 22	M. Wolf	Heidelberg
588	Achilles	1906	TG	1906	Feb 22	M. Wolf	Heidelberg
589	Croatia	1906	TM	1906	Mar 3	M. Wolf	Heidelberg
590	Tomyris	1906	TO	1906	Mar 4	M. Wolf	Heidelberg
591	Irmgard	1906	TP	1906	Mar 14	A. Kopff	Heidelberg
592	Bathseba	1906	TS	1906	Mar 18	A. Kopff	Heidelberg
593	Titania	1906	TT	1906	Mar 20	A. Kopff	Heidelberg
594	Mireille	1906	TW	1906	Mar 27	A. Kopff	Heidelberg
595	Polyxena	1906	TZ	1906	Mar 27	A. Kopff	Heidelberg
596	Scheila	1906	UA	1906	Feb 21	A. Kopff	Heidelberg
597	Bandusia	1906	UB	1906	Apr 16	M. Wolf	Heidelberg
598	Octavia	1906	UC	1906	Apr 13	A. Kopff	Heidelberg
599	Luisa	1906	UJ	1906	Apr 25	J. H. Metcalf	Taunton
600	Musa	1906	UM	1906	Jun 14	J. H. Metcalf	Taunton

No	Name	pd	dd	d	dp	Notes
601	Northus	1906 UN	1906 Jun 21	M. Wolf	Heidelberg	
602	Marianna	1906 TE	1906 Feb 16	J. H. Metcalf	Taunton	
603	Timandra	1906 TJ	1906 Feb 16	J. H. Metcalf	Taunton	
604	Tekmessa	1906 TK	1906 Feb 16	J. H. Metcalf	Taunton	
605	Juvisia	1906 UU	1906 Aug 27	M. Wolf	Heidelberg	
606	Brangane	1906 VB	1906 Sep 18	A. Kopff	Heidelberg	
607	Jenny	1906 VC	1906 Sep 18	A. Kopff	Heidelberg	
608	Adolfine	1906 VD	1906 Sep 18	A. Kopff	Heidelberg	
609	Fulvia	1906 VF	1906 Sep 24	M. Wolf	Heidelberg	
610	Valeska	1906 VK	1906 Sep 26	M. Wolf	Heidelberg	
611	Valeria	1906 VL	1906 Sep 24	J. H. Metcalf	Taunton	
612	Veronika	1906 VN	1906 Oct 8	A. Kopff	Heidelberg	
613	Ginevra	1906 VP	1906 Oct 11	A. Kopff	Heidelberg	
614	Pia	1906 VQ	1906 Oct 11	A. Kopff	Heidelberg	
615	Roswitha	1906 VR	1906 Oct 11	A. Kopff	Heidelberg	
616	Elly	1906 VT	1906 Oct 17	A. Kopff	Heidelberg	
617	Patroclus	1906 VY	1906 Oct 17	A. Kopff	Heidelberg	
618	Elfriede	1906 VZ	1906 Oct 17	K. Lohnert	Heidelberg	
619	Triberga	1906 WC	1906 Oct 22	A. Kopff	Heidelberg	
620	Drakonia	1906 WE	1906 Oct 26	J. H. Metcalf	Taunton	
621	Werdandi	1906 WJ	1906 Nov 11	A. Kopff	Heidelberg	
622	Esther	1906 WP	1906 Nov 13	J. H. Metcalf	Taunton	
623	Chimaera	1907 XJ	1907 Jan 22	K. Lohnert	Heidelberg	
624	Hektor	1907 XM	1907 Feb 10	A. Kopff	Heidelberg	
625	Xenia	1907 XN	1907 Feb 11	A. Kopff	Heidelberg	
626	Notburga	1907 XO	1907 Feb 11	A. Kopff	Heidelberg	
627	Charis	1907 XS	1907 Mar 4	A. Kopff	Heidelberg	
628	Christine	1907 XT	1907 Mar 7	A. Kopff	Heidelberg	
629	Bernardina	1907 XU	1907 Mar 7	A. Kopff	Heidelberg	
630	Euphemia	1907 XW	1907 Mar 7	A. Kopff	Heidelberg	
631	Philippina	1907 YJ	1907 Mar 21	A. Kopff	Heidelberg	
632	Pyrrha	1907 YX	1907 Mar 27	A. Kopff	Heidelberg	
633	Zelima	1907 ZM	1907 May 12	A. Kopff	Heidelberg	
634	Ute	1907 ZN	1907 May 12	A. Kopff	Heidelberg	
635	Vundtia	1907 ZS	1907 Jun 9	K. Lohnert	Heidelberg	
636	Erika	1907 XP	1907 Feb 8	J. H. Metcalf	Taunton	
637	Chrysothemis	1907 YE	1907 May 11	J. H. Metcalf	Taunton	
638	Moira	1907 ZT	1907 May 25	J. H. Metcalf	Taunton	22
639	Latona	1907 ZT	1907 Jul 19	A. Kopff	Heidelberg	
640	Brambilla	1907 ZW	1907 Aug 29	J. H. Metcalf	Taunton	
641	Agnes	1907 ZX	1907 Sep 8	M. Wolf	Heidelberg	
642	Clara	1907 ZY	1907 Sep 8	M. Wolf	Heidelberg	
643	Scheherezade	1907 ZZ	1907 Sep 8	A. Kopff	Heidelberg	
644	Cosima	1907 AA	1907 Sep 7	A. Kopff	Heidelberg	
645	Agrippina	1907 AG	1907 Sep 13	J. H. Metcalf	Taunton	
646	Kastalia	1907 AC	1907 Sep 11	A. Kopff	Heidelberg	
647	Adelgunde	1907 AD	1907 Sep 11	A. Kopff	Heidelberg	
648	Pippa	1907 AE	1907 Sep 11	A. Kopff	Heidelberg	
649	Josefa	1907 AF	1907 Sep 11	A. Kopff	Heidelberg	
650	Amalasuntha	1907 AM	1907 Oct 4	A. Kopff	Heidelberg	

No	Name	pd	dd	d	dp	Notes
651	Antikleia	1907 AN	1907 Oct 4	A. Kopff	Heidelberg	
652	Jubilatrix	1907 AU	1907 Nov 27	J. Palisa	Vienna	
653	Berenike	1907 BK	1907 Nov 27	J. H. Metcalf	Taunton	
654	Zelinda	1908 BM	1908 Jan 4	A. Kopff	Heidelberg	
655	Briseis	1907 BF	1907 Nov 4	J. H. Metcalf	Taunton	
656	Beagle	1908 BU	1908 Jan 22	A. Kopff	Heidelberg	
657	Gunlöd	1908 BV	1908 Jan 23	A. Kopff	Heidelberg	
658	Asteria	1908 BW	1908 Jan 23	A. Kopff	Heidelberg	
659	Nestor	1908 CS	1908 Mar 23	M. Wolf	Heidelberg	
660	Crescentia	1908 CC	1908 Jan 8	J. H. Metcalf	Taunton	
661	Cloelia	1908 CL	1908 Feb 22	J. H. Metcalf	Taunton	
662	Newtonia	1908 CW	1908 Mar 30	J. H. Metcalf	Taunton	
663	Gerlinde	1908 DG	1908 Jun 24	A. Kopff	Heidelberg	
664	Judith	1908 DH	1908 Jun 24	A. Kopff	Heidelberg	
665	Sabine	1908 DK	1908 Jul 22	W. Lorenz	Heidelberg	
666	Desdemona	1908 DM	1908 Jul 23	A. Kopff	Heidelberg	
667	Denise	1908 DN	1908 Jul 27	A. Kopff	Heidelberg	
668	Dora	1908 DO	1908 Aug 4	A. Kopff	Heidelberg	
669	Kypria	1908 DR	1908 Aug 20	A. Kopff	Heidelberg	
670	Ottegebe	1908 DR	1908 Aug 20	A. Kopff	Taunton	
671	Carnegia	1908 DV	1908 Sep 21	J. Palisa	Heidelberg	
672	Astarte	1908 DY	1908 Sep 20	A. Kopff	Heidelberg	
673	Edda	1908 EA	1908 Sep 20	W. Metcalf	Heidelberg	23
674	Rachele	1908 EU	1908 Aug 30	W. Lorenz	Heidelberg	
675	Ludmilla	1908 DU	1908 Aug 20	J. H. Metcalf	Taunton	
676	Melitta	1909 FN	1909 Jan 16	P. Melotte	Greenwich	
677	Aaltje	1909 FR	1909 Jan 18	A. Kopff	Heidelberg	
678	Fredegundis	1909 FS	1909 Jan 22	W. Lorenz	Heidelberg	
679	Pax	1909 FY	1909 Jan 28	A. Kopff	Heidelberg	
680	Genoveva	1909 GW	1909 Apr 22	A. Kopff	Heidelberg	
681	Gorgo	1909 GZ	1909 May 13	A. Kopff	Heidelberg	
682	Hagar	1909 HA	1909 Jun 17	M. Wolf	Heidelberg	
683	Lanzia	1909 HC	1909 Jul 23	M. Wolf	Heidelberg	
684	Hildburg	1909 HD	1909 Aug 8	A. Kopff	Heidelberg	
685	Hermia	1909 HE	1909 Aug 12	W. Lorenz	Heidelberg	
686	Gersuind	1909 HF	1909 Aug 15	A. Kopff	Heidelberg	
687	Tinette	1909 HG	1909 Aug 16	J. Palisa	Vienna	
688	Melanie	1909 HH	1909 Aug 25	J. Palisa	Vienna	24
689	Zita	1909 HT	1909 Sep 12	J. Palisa	Vienna	
690	Wratislavia	1909 HZ	1909 Sep 19	J. H. Metcalf	Taunton	
691	Lehigh	1909 JG	1909 Dec 11	J. H. Metcalf	Heidelberg	
692	Hippodamia	1901 HD	1901 Nov 5	M. Wolf and A. Kopff	Heidelberg	
693	Zerbinetta	1909 HN	1909 Sep 21	A. Kopff	Heidelberg	
694	Ekard	1909 JA	1909 Nov 7	J. H. Metcalf	Taunton	
695	Bella	1909 JB	1909 Nov 7	J. H. Metcalf	Taunton	
696	Leonora	1910 JJ	1910 Jan 10	J. H. Metcalf	Taunton	
697	Galilea	1910 JO	1910 Feb 14	J. Helffrich	Heidelberg	
698	Ernestina	1910 JX	1910 Mar 5	J. Helffrich	Heidelberg	
699	Hela	910 KD	1910 Jun 5	J. Helffrich	Heidelberg	25
700	Auravictrix	910 KE	1910 Jun 5	J. Helffrich	Heidelberg	25

No	Name	pd	dd	d	dp	Notes
701	Oriola	1910 KN	1910 Jul 12	J. Helffrich	Heidelberg	
702	Alauda	1910 KQ	1910 Jul 16	J. Helffrich	Heidelberg	
703	Noëmi	1910 KT	1910 Oct 3	J. Palisa	Vienna	
704	Interamnia	1910 KU	1910 Oct 2	V. Cerulli	Teramo	
705	Erminia	1910 KV	1910 Oct 6	E. Ernst	Heidelberg	
706	Hirundo	1910 KX	1910 Oct 9	J. Helffrich	Heidelberg	
707	Steina	1910 LD	1910 Dec 22	M. Wolf	Heidelberg	
708	Raphaela	1911 LJ	1911 Feb 3	J. Helffrich	Heidelberg	
709	Fringilla	1911 LK	1911 Feb 3	J. Helffrich	Heidelberg	
710	Gertrud	1911 LM	1911 Feb 28	J. Palisa	Vienna	
711	Marmulla	1911 LN	1911 Mar 1	J. Palisa	Vienna	
712	Boliviana	1911 LO	1911 Mar 19	M. Wolf	Heidelberg	
713	Luscinia	1911 LS	1911 Apr 18	J. Helffrich	Heidelberg	
714	Ulula	1911 LW	1911 May 18	J. Helffrich	Heidelberg	
715	Transvaalia	1911 LX	1911 Apr 22	H. E. Wood	Johannesburg (UO)	26
716	Berkeley	1911 MD	1911 Jul 30	F. Kaiser	Heidelberg	
717	Wisibada	1911 MJ	1911 Aug 26	F. Kaiser	Heidelberg	
718	Erida	1911 MS	1911 Sep 29	J. Palisa	Vienna	
719	Albert	1911 MT	1911 Oct 3	J. Palisa	Vienna	
720	Bohlinia	1911 MW	1911 Oct 18	F. Kaiser	Heidelberg	
721	Tabora	1911 MZ	1911 Oct 18	F. Kaiser	Heidelberg	
722	Frieda	1911 NA	1911 Oct 18	J. Palisa	Vienna	
723	Walpurga	1911 NB	1911 Oct 21	J. Palisa	Vienna	
724	Hapag	1911 NC	1911 Oct 21	J. Palisa	Vienna	
725	Amanda	1911 ND	1911 Oct 21	J. Palisa	Vienna	
726	Joëlla	1911 NM	1911 Nov 22	J. H. Metcalf	Winchester	
727	Nipponia	1912 NT	1912 Feb 11	A. Massinger	Heidelberg	
728	Leontsis	1911 NU	1912 Feb 16	A. Massinger	Heidelberg	
729	Watsonia	1912 OD	1912 Feb 9	J. H. Metcalf	Winchester	
730	Athamantis	1912 OK	1912 Apr 10	J. Palisa	Vienna	
731	Sorga	1912 OQ	1912 Apr 15	A. Massinger	Heidelberg	
732	Tjilaki	1912 OR	1912 Apr 15	A. Massinger	Heidelberg	
733	Mocia	1912 OF	1912 Sep 16	M. Wolf	Heidelberg	
734	Benda	1912 PH	1912 Oct 11	J. Palisa	Vienna	
735	Marghanna	1912 PY	1912 Dec 9	H. Vogt	Heidelberg	
736	Harvard	1912 PZ	1912 Nov 16	J. H. Metcalf	Winchester	
737	Arequipa	1912 QB	1912 Dec 7	J. H. Metcalf	Winchester	
738	Alagasta	1913 QO	1913 Jan 7	F. Kaiser	Heidelberg	
739	Mandeville	1913 QR	1913 Feb 9	J. H. Metcalf	Winchester	
740	Cantabia	1913 QS	1913 Feb 10	J. H. Metcalf	Winchester	
741	Botolphia	1913 QT	1913 Feb 10	J. H. Metcalf	Winchester	
742	Edisona	1913 QU	1913 Feb 23	F. Kaiser	Heidelberg	
743	Eugenisis	1913 QV	1913 Feb 25	F. Kaiser	Heidelberg	
744	Aguntina	1913 QW	1913 Feb 26	J. Rheden	Vienna	
745	Mauritia	1913 QX	1913 Mar 1	F. Kaiser	Heidelberg	
746	Marlu	1913 QY	1913 Mar 1	F. Kaiser	Heidelberg	
747	Winchester	1913 RD	1913 Mar 7	J. H. Metcalf	Winchester	27
748	Simeisa	1913 RE	1913 Mar 14	G. Neujmin	Simeis	
749	Malzovia	1913 RF	1913 Apr 5	S. Belyavsky	Simeis	
750	Oskar	1913 RG	1913 Apr 28	J. Palisa	Vienna	
751	Faina	1913 RK	1913 Apr 28	G. Neujmin	Simeis	28
752	Sulamitis	1913 RL	1913 Apr 30	G. Neujmin	Simeis	28
753	Tiflis	1913 RM	1913 Apr 30	G. Neujmin	Simeis	
754	Malabar	1906 UT	1906 Aug 22	A. Kopff	Heidelberg	
755	Quintilla	1908 CZ	1908 Apr 6	J. H. Metcalf	Taunton	
756	Lilliana	1908 DC	1908 Apr 26	J. H. Metcalf	Taunton	
757	Portlandia	1908 EJ	1908 Sep 30	J. H. Metcalf	Taunton	
758	Mancunia	1912 PE	1912 May 18	H. E. Wood	Johannesburg (UO)	29
759	Vinifera	1913 SJ	1913 Aug 26	F. Kaiser	Heidelberg	
760	Massinga	1913 SL	1913 Aug 28	F. Kaiser	Heidelberg	
761	Brendelia	1913 SO	1913 Sep 8	F. Kaiser	Heidelberg	
762	Pulcova	1913 SQ	1913 Sep 3	G. Neujmin	Simeis	30
763	Cupido	1913 ST	1913 Sep 25	F. Kaiser	Heidelberg	
764	Gedania	1913 SU	1913 Sep 25	F. Kaiser	Heidelberg	
765	Mattiaca	1913 SV	1913 Sep 26	F. Kaiser	Heidelberg	
766	Moguntia	1913 SW	1913 Sep 29	F. Kaiser	Heidelberg	
767	Bondia	1913 SX	1913 Sep 23	J. H. Metcalf	Winchester	
768	Struveana	1913 SZ	1913 Oct 4	G. Neujmin	Simeis	
769	Tatjana	1913 TD	1913 Oct 2	G. Neujmin	Simeis	
770	Bali	1913 TX	1913 Oct 31	A. Massinger	Heidelberg	
771	Libera	1913 TO	1913 Nov 21	J. Rheden	Vienna	
772	Tanete	1913 TZ	1913 Dec 19	A. Massinger	Heidelberg	
773	Irmintraud	1913 TR	1913 Dec 22	F. Kaiser	Heidelberg	
774	Armor	1913 TW	1913 Dec 19	C. Le Morvan	Paris	
775	Lumière	1914 TX	1914 Jan 6	J. Lagrula	Nice	
776	Berbericia	1914 TP	1914 Jan 25	A. Massinger	Heidelberg	
777	Gutemberga	1914 TZ	1914 Jan 24	F. Kaiser	Heidelberg	
778	Theobalda	1914 UK	1914 Jan 25	F. Kaiser	Heidelberg	
779	Nina	1914 UB	1914 Jan 25	G. Neujmin	Simeis	
780	Armenia	1914 UC	1914 Jan 25	G. Neujmin	Simeis	
781	Kartvelia	1914 UF	1914 Jan 25	G. Neujmin	Simeis	
782	Montefiore	1914 UK	1914 Mar 18	J. Palisa	Vienna	
783	Nora	1914 UL	1914 Mar 18	J. Palisa	Vienna	
784	Pickeringia	1914 UM	1914 Mar 20	J. H. Metcalf	Winchester	
785	Zwetana	1914 UN	1914 Mar 20	A. Massinger	Heidelberg	
786	Bredichina	1914 UO	1914 Apr 20	F. Kaiser	Heidelberg	
787	Hoskwa	1914 UQ	1914 Apr 20	G. Neujmin	Simeis	
788	Hohensteina	1914 UR	1914 Apr 28	F. Kaiser	Heidelberg	
789	Lena	1914 UU	1914 Jun 24	G. Neujmin	Simeis	
790	Pretoria	1912 NW	1912 Jan 16	H. E. Wood	Johannesburg (UO)	29
791	Ani	1914 UV	1914 Jun 29	G. Neujmin	Simeis	
792	Metcalfia	1907 ZC	1907 Mar 20	J. H. Metcalf	Taunton	
793	Arizona	1907 ZD	1907 Apr 9	P. Lowell	Flagstaff	31
794	Irenaea	1914 VB	1914 Aug 27	J. Palisa	Vienna	
795	Fini	1914 VE	1914 Sep 26	J. Palisa	Vienna	
796	Sarita	1914 VH	1914 Oct 15	K. Reinmuth	Heidelberg	
797	Montana	1914 VR	1914 Nov 17	M. Thiele	Bergedorf	
798	Ruth	1914 VT	1914 Nov 21	M. Wolf	Heidelberg	
799	Gudula	1915 WO	1915 Mar 9	K. Reinmuth	Heidelberg	
800	Kressmannia	1915 WP	1915 Mar 20	M. Wolf	Heidelberg	

No	Name	pd	dd	d	dp	Notes
801	Helwerthia	1915 WQ	1915 Mar 20	M. Wolf	Heidelberg	
802	Epyaxa	1915 WR	1915 Mar 20	M. Wolf	Heidelberg	
803	Picka	1915 WS	1915 Mar 21	J. Palisa	Vienna	
804	Hispania	1915 WT	1915 Mar 20	J. Comas Solá	Barcelona	
805	Hormuthia	1915 WW	1915 Apr 17	M. Wolf	Heidelberg	
806	Gyldenia	1915 WX	1915 Apr 18	M. Wolf	Heidelberg	
807	Ceraskia	1915 WY	1915 Apr 18	M. Wolf	Heidelberg	
808	Merxia	1901 GY	1901 Oct 11	L. Carnera	Heidelberg	
809	Lundia	1915 XP	1915 Aug 11	M. Wolf	Heidelberg	
810	Atossa	1915 XQ	1915 Sep 8	M. Wolf	Heidelberg	
811	Nauheima	1915 XR	1915 Sep 8	M. Wolf	Heidelberg	
812	Adele	1915 XV	1915 Sep 8	S. Belyavsky	Simeis	
813	Baumeia	1915 XW	1915 Sep 28	M. Wolf	Heidelberg	32
814	Tauris	1916 YT	1916 Jan 2	G. Neujmin	Simeis	
815	Coppelia	1916 YU	1916 Feb 2	M. Wolf	Heidelberg	
816	Juliana	1916 YV	1916 Feb 8	M. Wolf	Heidelberg	
817	Annika	1916 YW	1916 Feb 6	M. Wolf	Heidelberg	
818	Kapteynia	1916 YZ	1916 Feb 21	M. Wolf	Heidelberg	
819	Barnardiana	1916 ZA	1916 Mar 3	M. Wolf	Heidelberg	
820	Adriana	1916 ZB	1916 Mar 30	M. Wolf	Heidelberg	
821	Fanny	1916 ZC	1916 Mar 31	M. Wolf	Heidelberg	
822	Lalage	1916 ZD	1916 Mar 31	M. Wolf	Vienna	
823	Sisigambis	1916 ZG	1916 Aug 31	J. Palisa	Vienna	
824	Anastasia	1916 ZH	1916 Mar 25	G. Neujmin	Simeis	33
825	Tanina	1916 ZL	1916 Mar 27	G. Neujmin	Simeis	34
826	Henrika	1916 ZO	1916 Apr 28	M. Wolf	Heidelberg	
827	Wolfiana	1916 ZX	1916 Aug 29	J. Palisa	Vienna	
828	Lindemannia	1916 ZY	1916 Aug 29	J. Palisa	Vienna	
829	Academia	1916 ZX	1916 Aug 25	G. Neujmin	Simeis	35
830	Petropolitana	1916 ZZ	1916 Aug 25	G. Neujmin	Simeis	36
831	Stateira	1916 AA	1916 Sep 20	M. Wolf	Heidelberg	
832	Karin	1916 AB	1916 Sep 20	M. Wolf	Heidelberg	
833	Monica	1916 AC	1916 Sep 20	M. Wolf	Heidelberg	
834	Burnhamia	1916 AD	1915 Sep 20	M. Wolf	Heidelberg	
835	Olivia	1916 AE	1916 Sep 23	M. Wolf	Heidelberg	
836	Jole	1916 AP	1916 Sep 23	M. Wolf	Heidelberg	
837	Schwarzschilda	1916 AG	1916 Sep 23	M. Wolf	Heidelberg	
838	Seraphina	1916 AH	1916 Sep 24	M. Wolf	Heidelberg	
839	Valborg	1916 AJ	1916 Sep 24	M. Wolf	Heidelberg	
840	Zenobia	1916 AK	1916 Sep 23	M. Wolf	Heidelberg	
841	Arabella	1916 AL	1916 Oct 1	M. Wolf	Heidelberg	
842	Kerstin	1916 AM	1916 Oct 1	M. Wolf	Heidelberg	
843	Nicolaia	1916 AN	1916 Sep 30	H. Thiele	Bergedorf	
844	Leontina	1916 AP	1916 Jun 4	J. Rheden	Vienna	
845	Naema	1916 AS	1916 Nov 16	M. Wolf	Heidelberg	
846	Lipperta	1916 AT	1916 Nov 26	K. Gyllenberg	Bergedorf	
847	Agnia	1915 XX	1915 Sep 2	G. Neujmin	Simeis	
848	Inna	1915 XS	1915 Sep 9	G. Neujmin	Simeis	37
849	Ara	1912 NY	1912 Feb 9	S. Belyavsky	Simeis	
850	Altona	1916 E24	1916 Mar 27	S. Belyavsky	Simeis	

No	Name	pd	dd	d	dp	Notes
851	Zeissia	1916 Σ26	1916 Apr 2	S. Belyavsky	Simeis	
852	Wladilena	1916 Σ27	1916 Apr 2	S. Belyavsky	Simeis	
853	Nansenia	1916 Σ28	1916 Apr 2	S. Belyavsky	Simeis	
854	Frostia	1916 Σ29	1916 Apr 3	S. Belyavsky	Simeis	
855	Newcombia	1916 ZP	1916 Apr 3	S. Belyavsky	Simeis	38
856	Backlunda	1916 Σ30	1916 Apr 3	S. Belyavsky	Simeis	
857	Glasenappia	1916 Σ33	1916 Apr 6	S. Belyavsky	Simeis	
858	El Djezaïr	1916 a	1916 May 26	F. Sy	Algiers	
859	Bouzaréah	1916 c	1916 Oct 2	F. Sy	Algiers	
860	Ursina	1917 BD	1917 Jan 22	M. Wolf	Heidelberg	39
861	Aïda	1917 BE	1917 Jan 28	M. Wolf	Heidelberg	
862	Franzia	1917 BF	1917 Jan 28	M. Wolf	Heidelberg	
863	Benkoela	1917 BH	1917 Feb 8	M. Wolf	Heidelberg	
864	Aase	1921 KE	1921 Sep 30	K. Reinmuth	Heidelberg	
865	Zubaida	1917 BO	1917 Feb 15	M. Wolf	Heidelberg	40
866	Fatme	1917 BQ	1917 Feb 25	M. Wolf	Heidelberg	
867	Kovacia	1917 BS	1917 Feb 25	M. Wolf	Vienna	
868	Lova	1917 BU	1917 Apr 26	M. Wolf	Heidelberg	
869	Mellena	1917 BV	1917 May 9	R. Schorr	Bergedorf	
870	Manto	1917 BX	1917 May 12	M. Wolf	Heidelberg	41
871	Amneris	1917 BY	1917 May 14	M. Wolf	Heidelberg	
872	Holda	1917 BS	1917 May 21	M. Wolf	Heidelberg	
873	Mechthild	1917 CA	1917 May 21	M. Wolf	Heidelberg	
874	Rotraut	1917 CC	1917 May 22	M. Wolf	Heidelberg	
875	Nymphe	1917 CF	1917 May 19	M. Wolf	Heidelberg	
876	Scott	1917 CH	1917 Jun 20	J. Palisa	Vienna	
877	Walkure	1915 Σ7	1915 Sep 13	G. Neujmin	Simeis	
878	Mildred	1916 f	1915 Sep 6	S. B. Nicholson	Mount Wilson	42
879	Ricarda	1917 CJ	1917 Sep 22	M. Wolf	Heidelberg	
880	Herba	1917 CK	1917 Jul 22	M. Wolf	Heidelberg	
881	Athene	1917 CL	1917 Jul 22	M. Wolf	Heidelberg	43
882	Swetlana	1917 CM	1917 Jul 22	G. Neujmin	Simeis	44
883	Matterania	1917 CP	1917 Sep 14	M. Wolf	Heidelberg	
884	Frakmus	1917 CQ	1917 Sep 22	F. Sy	Heidelberg	
885	Ulrike	1917 CX	1917 Sep 23	S. Belyavsky	Simeis	45
886	Washingtonia	1917 b	1917 Nov 16	G. H. Peters	Washington	
887	Alinda	1918 DB	1918 Jan 3	M. Wolf	Heidelberg	46
888	Parysatis	1918 DC	1918 Feb 2	M. Wolf	Heidelberg	47
889	Erynia	1918 DG	1918 Mar 5	M. Wolf	Heidelberg	
890	Waltraut	1918 DK	1918 Mar 11	M. Wolf	Heidelberg	
891	Gunhild	1918 DQ	1918 May 17	M. Wolf	Heidelberg	
892	Seeligeria	1918 DR	1918 May 31	M. Wolf	Heidelberg	
893	Leopoldina	1918 DS	1918 May 31	M. Wolf	Heidelberg	
894	Erda	1918 DT	1918 Jun 4	M. Wolf	Heidelberg	
895	Hello	1918 DU	1918 Jul 11	M. Wolf	Heidelberg	
896	Sphinx	1918 DV	1918 Aug 1	M. Wolf	Heidelberg	
897	Lysistrata	1918 DZ	1918 Aug 3	M. Wolf	Heidelberg	
898	Hildegard	1918 EA	1918 Aug 3	M. Wolf	Heidelberg	
899	Jokaste	1918 EB	1918 Aug 3	M. Wolf	Heidelberg	
900	Rosalinde	1918 EC	1918 Aug 10	M. Wolf	Heidelberg	

No	Name	pd	dd	d	dp	Notes
951	Gaspra	1916 ΥΑ5	1916 Jul 30	G. Neujmin	Simeis	
952	Cala	1916 ΣΣ1	1916 Oct 27	G. Neujmin	Simeis	
953	Painleva	1921 JT	1921 Apr 29	B. Jekhovsky	Algiers	
954	Li	1921 JU	1921 Aug 4	K. Reinmuth	Heidelberg	
955	Alstede	1921 JV	1921 Aug 5	K. Reinmuth	Heidelberg	
956	Elisa	1921 JW	1921 Aug 8	K. Reinmuth	Heidelberg	
957	Camelia	1921 JX	1921 Sep 7	K. Reinmuth	Heidelberg	
958	Asplinda	1921 KC	1921 Sep 28	K. Reinmuth	Heidelberg	
959	Arne	1921 KF	1921 Sep 30	K. Reinmuth	Heidelberg	
960	Birgit	1921 KH	1921 Oct 1	K. Reinmuth	Heidelberg	
961	Gunnie	1921 KH	1921 Oct 10	K. Reinmuth	Heidelberg	
962	Aslög	1921 KV	1921 Oct 25	K. Reinmuth	Heidelberg	
963	Iduberga	1921 KR	1921 Oct 26	K. Reinmuth	Heidelberg	
964	Subamara	1921 KS	1921 Oct 27	J. Palisa	Vienna	
965	Angelica	1921 KT	1921 Nov 4	J. Hartmann	La Plata	
966	Muschi	1921 KU	1921 Nov 9	W. Baade	Bergedorf	
967	Helionape	1921 KV	1921 Nov 9	W. Baade	Bergedorf	51
968	Petunia	1921 KY	1921 Nov 24	K. Reinmuth	Heidelberg	
969	Leocadia	1921 KZ	1921 Nov 18	S. Belyavsky	Simeis	
970	Primula	1921 LB	1921 Nov 29	K. Reinmuth	Heidelberg	
971	Alsatia	1921 LF	1921 Nov 23	A. Schaumasse	Nice	
972	Cohnia	1922 LK	1922 Jan 18	M. Wolf	Heidelberg	
973	Aralia	1922 LS	1922 Mar 18	K. Reinmuth	Heidelberg	
974	Lioba	1922 LS	1922 Mar 18	K. Reinmuth	Heidelberg	
975	Perseverantia	1922 LT	1922 Mar 18	J. Palisa	Vienna	
976	Benjamina	1922 LU	1922 Mar 27	B. Jekhovsky	Algiers	52
977	Philippa	1922 LV	1922 Apr 6	B. Jekhovsky	Algiers	
978	Aidamina	1922 LY	1922 May 18	S. Belyavsky	Simeis	
979	Ilsewa	1922 MC	1922 Jun 29	K. Reinmuth	Heidelberg	
980	Anacostia	1921 W 19	1921 Nov 21	G. H. Peters	Washington	
981	Martina	1917 ΣΣ2	1917 Sep 23	S. Belyavsky	Simeis	
982	Franklina	1922 MD	1922 May 21	H. E. Wood	Johannesburg (10)	
983	Gunila	1922 MH	1922 Jul 30	K. Reinmuth	Heidelberg	
984	Gretia	1922 MH	1922 Aug 27	K. Reinmuth	Heidelberg	
985	Rosina	1922 MO	1922 Oct 14	K. Reinmuth	Heidelberg	
986	Amelia	1922 MR	1922 Oct 19	J. Comas Solá	Barcelona	
987	Walli	1922 MR	1922 Oct 23	K. Reinmuth	Heidelberg	
988	Appella	1922 MT	1922 Nov 10	B. Jekhovsky	Algiers	
989	Schwassmannia	1922 MW	1922 Nov 18	A. Schwassmann	Bergedorf	
990	Yerkes	1922 MZ	1922 Nov 23	G. Van Biesbroeck	Williams Bay	
991	McDonalda	1922 NB	1922 Oct 24	O. Struve	Williams Bay	53
992	Swasey	1922 ND	1922 Oct 14	O. Struve	Williams Bay	
993	Moultona	1923 NJ	1923 Jan 12	G. Van Biesbroeck	Williams Bay	
994	Otthild	1923 NL	1923 Mar 18	K. Reinmuth	Heidelberg	
995	Sternberga	1923 NP	1923 Jun 8	S. Belyavsky	Simeis	
996	Hilaritas	1923 NM	1923 Mar 21	J. Palisa	Vienna	
997	Priska	1923 NR	1923 Jul 12	K. Reinmuth	Heidelberg	
998	Bodea	1923 NU	1923 Aug 6	K. Reinmuth	Heidelberg	
999	Zachia	1923 NW	1923 Aug 9	K. Reinmuth	Heidelberg	
1000	Piazzia	1923 NZ	1923 Aug 12	K. Reinmuth	Heidelberg	

No	Name	pd	dd	d	dp	Notes
901	Brunsia	1918 EE	1918 Aug 30	M. Wolf	Heidelberg	
902	Probitas	1918 EJ	1918 Sep 3	J. Palisa	Vienna	
903	Nealley	1918 EO	1918 Sep 13	J. Palisa	Vienna	
904	Rockefellia	1918 EO	1918 Oct 29	M. Wolf	Heidelberg	
905	Universitas	1918 ES	1918 Oct 30	A. Schwassmann	Bergedorf	
906	Repsolda	1918 ET	1918 Oct 30	A. Schwassmann	Bergedorf	
907	Rhoda	1918 EU	1918 Nov 12	M. Wolf	Heidelberg	
908	Buda	1918 EX	1918 Nov 30	M. Wolf	Heidelberg	
909	Ulla	1919 FA	1919 Feb 7	K. Reinmuth	Heidelberg	
910	Anneliese	1919 FB	1919 Mar 1	K. Reinmuth	Heidelberg	
911	Agamemnon	1919 FD	1919 Mar 19	K. Reinmuth	Heidelberg	
912	Maritima	1919 FJ	1919 Apr 27	A. Schwassmann	Bergedorf	
913	Otila	1919 FL	1919 May 19	K. Reinmuth	Heidelberg	
914	Palisana	1919 FN	1919 Jul 4	M. Wolf	Heidelberg	
915	Cosette	1918 b	1918 Dec 14	F. Gonnessiat	Algiers	
916	America	1915 ΣΣ1	1915 Aug 7	G. Neujmin	Simeis	
917	Lyka	1915 ΣΑ	1915 Sep 5	G. Neujmin	Simeis	
918	Itha	1919 FR	1919 Aug 22	K. Reinmuth	Heidelberg	
919	Ilsebill	1918 EQ	1918 Oct 30	M. Wolf	Heidelberg	
920	Rogeria	1919 FT	1919 Sep 1	K. Reinmuth	Heidelberg	
921	Jovita	1919 FV	1919 Sep 4	K. Reinmuth	Heidelberg	
922	Schlutia	1919 FW	1919 Sep 18	K. Reinmuth	Heidelberg	
923	Herluga	1919 GB	1919 Sep 30	K. Reinmuth	Heidelberg	
924	Toni	1919 GC	1919 Sep 20	K. Reinmuth	Heidelberg	
925	Alphonsina	1920 GM	1920 Jan 13	J. Comas Solá	Barcelona	
926	Imhilde	1920 GN	1920 Feb 15	K. Reinmuth	Heidelberg	
927	Ratisbona	1920 GO	1920 Feb 16	M. Wolf	Heidelberg	
928	Hildrun	1920 GP	1920 Feb 23	K. Reinmuth	Heidelberg	
929	Algunde	1920 GR	1920 Mar 10	K. Reinmuth	Heidelberg	
930	Westphalia	1920 GS	1920 Mar 10	W. Baade	Bergedorf	
931	Whittemora	1920 GU	1920 Mar 19	F. Gonnessiat	Algiers	48
932	Hooveria	1920 GV	1920 Mar 23	J. Palisa	Vienna	
933	Susi	1927 CH	1927 Feb 10	K. Reinmuth	Heidelberg	26
934	Thuringia	1920 HK	1920 Aug 15	W. Baade	Bergedorf	
935	Clivia	1920 HM	1920 Sep 7	K. Reinmuth	Heidelberg	
936	Kunigunde	1920 HN	1920 Sep 8	K. Reinmuth	Heidelberg	
937	Bethgea	1920 HO	1920 Sep 12	K. Reinmuth	Heidelberg	
938	Chlosinde	1920 HP	1920 Sep 9	K. Reinmuth	Heidelberg	
939	Isberga	1920 HR	1920 Oct 4	K. Reinmuth	Heidelberg	
940	Kordula	1920 HT	1920 Oct 10	K. Reinmuth	Heidelberg	
941	Murray	1920 HV	1920 Oct 10	J. Palisa	Vienna	49
942	Romilda	1920 HW	1920 Oct 11	K. Reinmuth	Heidelberg	50
943	Begonia	1920 HX	1920 Oct 20	K. Reinmuth	Heidelberg	
944	Hidalgo	1920 HZ	1920 Oct 31	W. Baade	Bergedorf	
945	Barcelona	1921 JB	1921 Feb 3	J. Comas Solá	Barcelona	
946	Poësia	1921 JC	1921 Feb 11	M. Wolf	Heidelberg	
947	Monterosa	1921 JD	1921 Feb 8	A. Schwassmann	Bergedorf	
948	Jucunda	1921 JE	1921 Mar 3	K. Reinmuth	Heidelberg	
949	Hel	1921 JK	1921 Mar 11	M. Wolf	Heidelberg	
950	Ahrensa	1921 JP	1921 Apr 1	K. Reinmuth	Heidelberg	

No Name	pd	dd	d	dp	Notes
1001 Gaussia	1923 OA	1923 Aug 8	S. Belyavsky	Simeis	
1002 Olbersia	1923 OB	1923 Aug 15	V. Albitzky	Simeis	
1003 Lilofee	1923 OK	1923 Sep 13	K. Reinmuth	Heidelberg	
1004 Belopolskya	1923 OS	1923 Sep 5	S. Belyavsky	Simeis	54
1005 Arago	1923 OT	1923 Sep 5	S. Belyavsky	Simeis	
1006 Lagrangea	1923 OU	1923 Sep 12	K. Reinmuth	Simeis	
1007 Pawlowia	1923 OX	1923 Oct 5	V. Albitzky	Simeis	
1008 La Paz	1923 PD	1923 Oct 31	M. Wolf	Heidelberg	
1009 Sirene	1923 PE	1923 Oct 31	K. Reinmuth	Heidelberg	
1010 Marlene	1923 PF	1923 Nov 12	K. Reinmuth	Heidelberg	
1011 Laodamia	1924 PK	1924 Jan 5	K. Reinmuth	Heidelberg	
1012 Sarema	1924 PM	1924 Jan 12	K. Reinmuth	Heidelberg	
1013 Tombecka	1924 PQ	1924 Jan 17	B. Jekhovsky	Algiers	
1014 Semphyra	1924 PN	1924 Jan 29	K. Reinmuth	Heidelberg	
1015 Christa	1924 QF	1924 Jan 31	K. Reinmuth	Heidelberg	
1016 Anitra	1924 QG	1924 Mar 31	K. Reinmuth	Heidelberg	
1017 Jacqueline	1924 QL	1924 Feb 4	B. Jekhovsky	Algiers	
1018 Arnolda	1924 QM	1924 Mar 3	K. Reinmuth	Heidelberg	
1019 Strackea	1924 QN	1924 Mar 7	K. Reinmuth	Heidelberg	
1020 Arcadia	1924 QV	1924 Mar 7	K. Reinmuth	Heidelberg	
1021 Flammario	1924 RG	1924 Mar 11	M. Wolf	Heidelberg	
1022 Olympiada	1924 RT	1924 Jun 23	V. Albitzky	Simeis	
1023 Thomana	1924 RU	1924 Jun 12	K. Reinmuth	Heidelberg	
1024 Hale	1923 YO13	1923 Dec 2	G. Van Biesbroeck	Williams Bay	
1025 Riema	1923 NK	1924 Aug 12	K. Reinmuth	Heidelberg	
1026 Ingrid	1923 NY	1923 Aug 13	K. Reinmuth	Heidelberg	
1027 Aesculapia	1923 YO11	1923 Nov 11	G. Van Biesbroeck	Williams Bay	
1028 Lydina	1923 PG	1923 Nov 6	V. Albitzky	Simeis	
1029 La Plata	1924 RK	1924 Apr 28	J. Hartmann	La Plata	
1030 Vitja	1024 RQ	1924 May 25	V. Albitzky	Simeis	
1031 Arctica	1924 RR	1924 Jun 6	S. Belyavsky	Simeis	
1032 Pafuri	1924 SA	1924 May 30	H. E. Wood	Johannesburg Bay (JO)	55
1033 Simona	1924 SM	1924 Sep 4	G. Van Biesbroeck	Williams Bay	
1034 Mozartia	1924 SS	1924 Sep 7	V. Albitzky	Simeis	
1035 Amata	1924 SW	1924 Sep 29	K. Reinmuth	Heidelberg	
1036 Ganymed	1924 TD	1924 Oct 23	W. Baade	Bergedorf	
1037 Davidweilla	1924 TF	1924 Oct 29	B. Jekhovsky	Algiers	
1038 Tuckia	1924 TK	1924 Nov 24	M. Wolf	Heidelberg	
1039 Sonneberga	1924 TL	1924 Nov 24	M. Wolf	Heidelberg	
1040 Klumpkea	1925 BD	1925 Jan 20	B. Jekhovsky	Algiers	
1041 Asta	1925 FA	1925 Mar 22	K. Reinmuth	Heidelberg	
1042 Amazone	1925 HA	1925 Apr 22	K. Reinmuth	Heidelberg	
1043 Beate	1925 HB	1925 Apr 22	K. Reinmuth	Heidelberg	
1044 Teutonia	1924 RO	1924 May 10	K. Reinmuth	Heidelberg	
1045 Michela	1924 TR	1924 Nov 19	G. Van Biesbroeck	Williams Bay	
1046 Edwin	1924 UA	1924 Dec 1	G. Van Biesbroeck	Williams Bay	
1047 Geisha	1924 TE	1924 Nov 29	K. Reinmuth	Heidelberg	
1048 Feodosia	1924 TP	1924 Nov 29	K. Reinmuth	Heidelberg	
1049 Gotho	1925 RB	1925 Sep 14	K. Reinmuth	Heidelberg	
1050 Meta	1925 KC	1925 Sep 14	K. Reinmuth	Heidelberg	

No Name	pd	dd	d	dp	Notes
1051 Merope	1925 SA	1925 Sep 16	K. Reinmuth	Heidelberg	56
1052 Belgica	1925 VD	1925 Nov 15	E. Delporte	Uccle	
1053 Vigdis	1925 WA	1925 Nov 16	M. Wolf	Heidelberg	
1054 Forsytia	1925 WD	1925 Nov 20	K. Reinmuth	Heidelberg	
1055 Tynka	1925 WG	1925 Nov 17	E. Buchar	Algiers	57
1056 Azalea	1924 QD	1924 Jan 31	K. Reinmuth	Heidelberg	
1057 Wanda	1925 QB	1925 Aug 16	G. Shajn	Simeis	
1058 Grubba	1925 MA	1925 Jun 22	G. Shajn	Simeis	
1059 Mussorgskia	1925 PA	1925 Jul 19	V. Albitzky	Simeis	
1060 Magnolia	1925 PA	1925 Aug 13	K. Reinmuth	Heidelberg	
1061 Paeonia	1925 TB	1925 Oct 10	K. Reinmuth	Heidelberg	58
1062 Ljuba	1925 TD	1925 Oct 11	S. Belyavsky	Simeis	
1063 Aquilegia	1925 XA	1925 Dec 6	K. Reinmuth	Heidelberg	
1064 Aethusa	1926 PA	1926 Aug 2	K. Reinmuth	Heidelberg	
1065 Amundsenia	1926 PD	1926 Aug 4	S. Belyavsky	Simeis	
1066 Lobelia	1926 RA	1926 Sep 3	K. Reinmuth	Heidelberg	
1067 Lunaria	1926 RG	1926 Sep 9	K. Reinmuth	Heidelberg	
1068 Nofretete	1926 RK	1926 Sep 13	E. Delporte	Uccle	
1069 Planckia	1927 BC	1927 Jan 28	M. Wolf	Heidelberg	
1070 Tunica	1926 RB	1926 Sep 1	K. Reinmuth	Heidelberg	
1071 Brita	1924 RE	1924 Mar 3	V. Albitzky	Simeis	
1072 Malva	1926 TA	1926 Oct 14	K. Reinmuth	Heidelberg	
1073 Gellivara	1923 OW	1923 Sep 14	J. Palisa	Vienna	
1074 Beljawskya	1926 KG	1926 Aug 29	S. Belyavsky	Simeis	
1075 Helina	1926 SC	1926 Sep 29	G. Neujmin	Simeis	
1076 Viola	1926 TE	1926 Oct 5	K. Reinmuth	Heidelberg	
1077 Campanula	1926 TK	1926 Oct 8	K. Reinmuth	Heidelberg	
1078 Mentha	1926 XB	1926 Dec 7	K. Reinmuth	Heidelberg	40
1079 Mimosa	1927 AD	1927 Jan 14	G. Van Biesbroeck	Williams Bay	
1080 Orchis	1927 QB	1927 Aug 30	K. Reinmuth	Heidelberg	
1081 Reseda	1927 QF	1927 Aug 31	K. Reinmuth	Heidelberg	
1082 Pirola	1927 UC	1927 Oct 28	K. Reinmuth	Heidelberg	
1083 Salvia	1928 BC	1928 Jan 26	K. Reinmuth	Heidelberg	
1084 Tamariwa	1926 CC	1926 Feb 21	S. Belyavsky	Simeis	
1085 Amaryllis	1927 QH	1927 Aug 31	K. Reinmuth	Heidelberg	
1086 Nata	1927 QL	1927 Aug 25	S. Belyavsky and N. Ivanov	Simeis	59
1087 Arabia	1927 RD	1927 Sep 2	K. Reinmuth	Heidelberg	
1088 Mitaka	1927 WA	1927 Nov 17	O. Oikawa	Tokyo	
1089 Tama	1927 WB	1927 Nov 17	O. Oikawa	Tokyo	
1090 Sumida	1928 DG	1928 Feb 20	O. Oikawa	Tokyo	
1091 Spiraea	1928 DT	1928 Feb 26	K. Reinmuth	Heidelberg	60
1092 Lilium	1924 PN	1924 Jan 12	K. Reinmuth	Heidelberg	
1093 Freda	1925 LA	1925 Jun 15	B. Jekhovsky	Algiers	
1094 Siberia	1926 CB	1926 Feb 12	S. Belyavsky	Simeis	
1095 Tulipa	1926 GS	1926 Apr 14	K. Reinmuth	Heidelberg	
1096 Reunerta	1928 OB	1928 Jul 21	H. E. Wood	Johannesburg (JO)	
1097 Vicia	1928 PC	1928 Aug 11	K. Reinmuth	Heidelberg	61
1098 Hakone	1928 RJ	1928 Sep 13	O. Oikawa	Tokyo	62
1099 Figneria	1928 RQ	1928 Sep 14	G. Neujmin	Simeis	
1100 Arnica	1928 SD	1928 Sep 22	K. Reinmuth	Heidelberg	

No	Name	pd	dd	d	dp	Notes
1101	Clematis	1928 SJ	1928 Sep 22	K. Reinmuth	Heidelberg	
1102	Pepita	1928 VA	1928 Nov 5	J. Comas Solá	Barcelona	
1103	Sequoia	1928 UR	1928 Sep 9	W. Baade	Bergedorf	
1104	Syringa	1928 XA	1928 Dec 9	K. Reinmuth	Heidelberg	
1105	Fragaria	1929 AB	1929 Jan 1	K. Reinmuth	Heidelberg	
1106	Cydonia	1929 CW	1929 Feb 5	K. Reinmuth	Heidelberg	
1107	Lictoria	1929 FB	1929 Mar 30	L. Volta	Pino Torinese	
1108	Demeter	1929 TO	1929 May 31	K. Reinmuth	Heidelberg	
1109	Tata	1929 CU	1929 Feb 5	K. Reinmuth	Heidelberg	
1110	Jaroslawa	1928 PD	1928 Aug 10	G. Neujmin	Simeis	
1111	Reinmuthia	1927 CO	1927 Feb 11	K. Reinmuth	Heidelberg	
1112	Polonia	1928 PE	1928 Aug 15	P. Shajn	Simeis	
1113	Katja	1928 QC	1928 Aug 15	P. Shajn	Simeis	
1114	Lorraine	1928 WA	1928 Nov 17	A. Schaumasse	Nice	63
1115	Sabauda	1928 XC	1928 Dec 13	L. Volta	Pino Torinese	
1116	Catriona	1929 GD	1929 Apr 5	C. Jackson	Johannesburg (UO)	
1117	Reginita	1927 KA	1927 May 24	J. Comas Solá	Barcelona	64
1118	Hanskya	1927 QD	1927 Aug 29	S. Belyavsky and N. Ivanov	Simeis	
1119	Euboea	1927 UB	1927 Oct 27	K. Reinmuth	Heidelberg	
1120	Cannonia	1928 RV	1928 Sep 11	P. Shajn	Simeis	
1121	Natascha	1928 RZ	1928 Sep 11	P. Shajn	Simeis	
1122	Neith	1928 SB	1928 Sep 17	E. Delporte	Uccle	
1123	Shapleya	1928 ST	1928 Sep 21	G. Neujmin	Simeis	
1124	Stereoskopia	1928 TB	1928 Oct 6	E. Delporte	Uccle	
1125	China	1957 UR1	1957 Oct 30	Y. C. Chang	Nanking	65
1126	Otero	1929 AC	1929 Jan 11	K. Reinmuth	Heidelberg	
1127	Mimi	1929 AJ	1929 Jan 13	S. Arend	Uccle	
1128	Astrid	1929 EB	1929 Mar 10	E. Delporte	Uccle	
1129	Neujmina	1929 PH	1929 Aug 8	P. Parchomenko	Simeis	
1130	Skuld	1929 RC	1929 Sep 2	K. Reinmuth	Heidelberg	
1131	Porzia	1929 RO	1929 Sep 10	K. Reinmuth	Heidelberg	
1132	Hollandia	1929 RB1	1929 Sep 13	H. van Gent	Johannesburg (LS)	
1133	Lugduna	1929 RC1	1929 Sep 13	H. van Gent	Johannesburg (LS)	
1134	Kepler	1929 SA	1929 Sep 25	M. Wolf	Heidelberg	
1135	Colchis	1929 TA	1929 Oct 3	G. Neujmin	Simeis	
1136	Mercedes	1929 UA	1929 Oct 30	J. Comas Solá	Barcelona	66
1137	Raissa	1929 WB	1929 Oct 27	G. Neujmin	Simeis	
1138	Attica	1929 WF	1929 Nov 22	K. Reinmuth	Heidelberg	
1139	Atami	1929 XE	1929 Dec 1	O. Oikawa and K. Kubokawa	Tokyo	
1140	Crimea	1929 YC	1929 Dec 30	G. Neujmin	Simeis	
1141	Bohmia	1930 AA	1930 Jan 4	M. Wolf	Heidelberg	
1142	Aetolia	1930 BC	1930 Jan 24	K. Reinmuth	Heidelberg	
1143	Odysseus	1930 BH	1930 Jan 28	K. Reinmuth	Heidelberg	
1144	Oda	1930 BJ	1930 Jan 28	K. Reinmuth	Heidelberg	
1145	Robelmonte	1929 CC	1929 Feb 3	E. Delporte	Uccle	
1146	Biarmia	1929 JF	1929 May 7	G. Neujmin	Simeis	
1147	Stavropolis	1929 LF	1929 Jun 11	G. Neujmin	Simeis	
1148	Rarahu	1929 NA	1929 Jul 5	A. Deutsch	Simeis	
1149	Volga	1929 PF	1929 Aug 1	E. Skvortsov	Simeis	
1150	Achaia	1929 RB	1929 Sep 2	K. Reinmuth	Heidelberg	67

No	Name	pd	dd	d	dp	Notes
1151	Ithaka	1929 RK	1929 Sep 8	K. Reinmuth	Heidelberg	
1152	Pawona	1930 AD	1930 Jan 8	K. Reinmuth	Heidelberg	
1153	Wallenbergia	1924 SL	1924 Sep 5	S. Belyavsky	Simeis	
1154	Astronomia	1927 CB	1927 Feb 8	K. Reinmuth	Heidelberg	
1155	Aenna	1928 BD	1928 Jan 26	K. Reinmuth	Heidelberg	
1156	Kira	1928 DA	1928 Feb 22	K. Reinmuth	Heidelberg	
1157	Arabia	1929 QC	1929 Aug 31	K. Reinmuth	Heidelberg	
1158	Luda	1929 QF	1929 Aug 31	G. Neujmin	Simeis	
1159	Granada	1929 RD	1929 Sep 2	K. Reinmuth	Heidelberg	
1160	Illyria	1929 RL	1929 Sep 9	K. Reinmuth	Heidelberg	
1161	Thessalia	1929 SF	1929 Sep 29	K. Reinmuth	Heidelberg	
1162	Larissa	1930 AC	1930 Jan 5	K. Reinmuth	Heidelberg	
1163	Saga	1930 BA	1930 Jan 20	K. Reinmuth	Heidelberg	
1164	Kobolda	1930 FB	1930 Mar 19	M. Wolf and M. Ferrero	Heidelberg	
1165	Imprinetta	1930 HM	1930 Apr 24	H. Van Gent	Johannesburg (LS)	68
1166	Sakuntala	1930 MA	1930 Jun 27	P. Parchomenko	Simeis	69
1167	Dubiago	1930 PB	1930 Aug 3	E. Skvortsov	Simeis	
1168	Brandia	1930 QA	1930 Aug 25	E. Delporte	Uccle	
1169	Alwine	1930 QH	1930 Aug 30	K. Reinmuth	Heidelberg	
1170	Siva	1930 SQ	1930 Sep 29	K. Reinmuth	Heidelberg	20
1171	Rusthawelia	1930 TA	1930 Oct 3	S. Arend	Uccle	
1172	Aneas	1930 UA	1930 Oct 17	K. Reinmuth	Heidelberg	
1173	Anchises	1930 UB	1930 Oct 17	K. Reinmuth	Heidelberg	
1174	Marmara	1930 UC	1930 Oct 17	K. Reinmuth	Heidelberg	
1175	Margo	1930 UD	1930 Oct 17	E. Delporte	Heidelberg	
1176	Lucidor	1930 VE	1930 Nov 15	E. Delporte	Uccle	
1177	Gonnessia	1930 WA	1930 Nov 24	L. Boyer	Algiers	
1178	Irmela	1931 EC	1931 Mar 13	M. Wolf	Heidelberg	
1179	Mally	1931 FD	1931 Mar 19	M. Wolf	Heidelberg	
1180	Rita	1931 GE	1931 Apr 9	K. Reinmuth	Heidelberg	
1181	Lilith	1927 CQ	1927 Feb 11	B. Jekhovsky	Algiers	
1182	Ilona	1927 EA	1927 Mar 3	K. Reinmuth	Heidelberg	
1183	Jutta	1930 DC	1930 Feb 22	K. Reinmuth	Heidelberg	
1184	Gaea	1926 RE	1926 Sep 4	M. Wolf	Heidelberg	
1185	Nikko	1927 WC	1927 Nov 17	O. Oikawa	Tokyo	
1186	Turnera	1929 PL	1929 Aug 1	C. Jackson	Johannesburg (100)	
1187	Afra	1929 XC	1929 Dec 6	K. Reinmuth	Heidelberg	
1188	Gothlandia	1930 SB	1930 Sep 30	J. Comas Solá	Barcelona	
1189	Terentia	1930 SG	1930 Sep 17	G. Neujmin	Simeis	
1190	Pelagia	1930 SL	1930 Sep 17	G. Neujmin	Simeis	
1191	Alfaterna	1931 CA	1931 Feb 11	L. Volta	Pino Torinese	
1192	Prisma	1931 FE	1931 Mar 17	A. Schwassmann	Bergedorf	70
1193	Africa	1931 HB	1931 Apr 24	C. Jackson	Johannesburg (100)	
1194	Aletta	1931 JG	1931 May 13	C. Jackson	Johannesburg (100)	
1195	Orangia	1931 KD	1931 May 24	C. Jackson	Johannesburg (100)	
1196	Sheba	1931 KE	1931 May 21	C. Jackson	Johannesburg (100)	
1197	Rhodesia	1931 LD	1931 Jun 9	C. Jackson	Johannesburg (100)	
1198	Atlantis	1931 RA	1931 Sep 7	K. Reinmuth	Heidelberg	
1199	Geldonia	1931 RF	1931 Sep 14	E. Delporte	Uccle	
1200	Imperatrix	1931 RH	1931 Sep 14	K. Reinmuth	Heidelberg	

No Name	pd	dd	d	dp	Notes
1201 Strenua	1931 RK	1931 Sep 14	K. Reinmuth	Heidelberg	
1202 Marina	1931 RL	1931 Sep 13	G. Neujmin	Simeis	71
1203 Nanna	1931 TA	1931 Oct 6	M. Wolf	Heidelberg	
1204 Renzia	1931 TE	1931 Oct 6	K. Reinmuth	Heidelberg	
1205 Ebella	1931 TB1	1931 Oct 6	K. Reinmuth	Heidelberg	
1206 Numerowia	1931 UH	1931 Oct 18	K. Reinmuth	Heidelberg	
1207 Ostenia	1931 VT	1931 Nov 15	K. Reinmuth	Heidelberg	
1208 Troilus	1931 YA	1931 Dec 31	K. Reinmuth	Heidelberg	
1209 Puuma	1927 HA	1927 Apr 22	K. Reinmuth	Heidelberg	
1210 Morosovia	1931 LB	1931 Jun 6	G. Neujmin	Simeis	
1211 Bressole	1931 XA	1931 Dec 2	L. Boyer	Algiers	
1212 Francette	1931 XC	1931 Dec 3	L. Boyer	Algiers	
1213 Algeria	1931 XD	1931 Dec 5	G. Reiss	Algiers	
1214 Richilde	1932 AA	1932 Jan 4	M. Wolf	Heidelberg	
1215 Boyer	1932 BA	1932 Jan 19	A. Schmitt	Algiers	
1216 Askania	1932 BL	1932 Jan 29	K. Reinmuth	Heidelberg	
1217 Maximiliana	1932 EC	1932 Mar 13	E. Delporte	Uccle	
1218 Aster	1932 BJ	1932 Jan 24	K. Reinmuth	Heidelberg	
1219 Britta	1932 CJ	1932 Feb 6	M. Wolf	Heidelberg	
1220 Crocus	1932 CU	1932 Feb 11	K. Reinmuth	Heidelberg	
1221 Amor	1932 EA1	1932 Mar 12	E. Delporte	Uccle	
1222 Tina	1932 LA	1932 Jun 11	E. Delporte	Uccle	
1223 Neckar	1931 TG	1931 Oct 6	K. Reinmuth	Heidelberg	
1224 Fantasia	1927 SD	1927 Aug 29	S. Belyavsky and N. Ivanov	Simeis	72
1225 Ariane	1930 HK	1930 Apr 23	W. van Gent	Johannesburg (LS)	73
1226 Golia	1930 HL	1930 Apr 22	H. van Gent	Johannesburg (LS)	74
1227 Geranium	1931 TD	1931 Oct 5	K. Reinmuth	Heidelberg	
1228 Scabiosa	1931 TU	1931 Oct 6	K. Reinmuth	Heidelberg	
1229 Tilia	1931 TP1	1931 Oct 9	K. Reinmuth	Heidelberg	
1230 Riceia	1931 TX1	1931 Oct 9	K. Reinmuth	Heidelberg	
1231 Auricula	1931 TE2	1931 Oct 10	E. Delporte	Heidelberg	
1232 Cortusa	1931 TF2	1931 Oct 10	E. Delporte	Heidelberg	
1233 Kobresia	1931 TG2	1931 Oct 16	K. Reinmuth	Heidelberg	
1234 Elyna	1931 UF	1931 Oct 18	K. Reinmuth	Heidelberg	
1235 Schorria	1931 UJ	1931 Oct 18	K. Reinmuth	Heidelberg	
1236 Tnala	1931 VX	1931 Nov 6	G. Neujmin	Simeis	
1237 Geneviève	1931 XB	1931 Dec 2	G. Reiss	Algiers	
1238 Predappia	1932 CA	1932 Feb 4	L. Volta	Pino Torinese	
1239 Queteleta	1932 CB	1932 Feb 7	E. Delporte	Uccle	
1240 Centenaria	1932 CD	1932 Feb 5	R. Schorr	Bergedorf	
1241 Dysona	1932 EB1	1932 Mar 4	H. E. Wood	Johannesburg (UO)	75
1242 Zambesia	1932 HL	1932 Apr 28	C. Jackson	Johannesburg (UO)	
1243 Pamela	1932 JE	1932 May 7	C. Jackson	Johannesburg (UO)	
1244 Deira	1932 KE	1932 May 25	C. Jackson	Johannesburg (UO)	
1245 Calvinia	1932 KF	1932 May 26	C. Jackson	Johannesburg (UO)	
1246 Chaka	1932 OA	1932 Jul 23	C. Jackson	Johannesburg (UO)	76
1247 Memoria	1932 QA	1932 Aug 30	M. Laugier	Uccle	
1248 Jugurtha	1932 KO	1932 Sep 1	C. Jackson	Johannesburg (UO)	
1249 Rutherfordia	1932 VB	1932 Nov 4	K. Reinmuth	Meudon	
1250 Galanthus	1933 BD	1933 Jan 25	K. Reinmuth	Heidelberg	

No Name	pd	dd	d	dp	Notes
1251 Hedera	1933 BE	1933 Jan 25	K. Reinmuth	Heidelberg	
1252 Celestia	1933 DG	1933 Feb 19	F. L. Whipple	Harvard	
1253 Frisia	1931 TV1	1931 Oct 9	K. Reinmuth	Heidelberg	
1254 Erfordia	1932 JA	1932 May 10	J. Hartmann	La Plata	
1255 Schilowa	1932 NC	1932 Jul 8	G. Neujmin	Simeis	
1256 Normannia	1932 PD	1932 Aug 8	K. Reinmuth	Heidelberg	
1257 Móra	1932 PE	1932 Aug 8	K. Reinmuth	Heidelberg	
1258 Sicilia	1932 PG	1932 Aug 8	K. Reinmuth	Heidelberg	
1259 Ógyalla	1933 BT	1933 Jan 29	K. Reinmuth	Heidelberg	
1260 Walhalla	1933 BW	1933 Jan 29	K. Reinmuth	Heidelberg	
1261 Legia	1933 FB	1933 Mar 23	E. Delporte	Uccle	
1262 Sniadeckia	1933 FE	1933 Mar 22	S. Arend	Uccle	
1263 Varsavia	1933 FF	1933 Mar 23	S. Arend	Uccle	
1264 Letaba	1933 HG	1933 Apr 21	C. Jackson	Johannesburg (UO)	
1265 Schweikarda	1911 MV	1911 Oct 18	P. Kaiser	Heidelberg	
1266 Tone	1927 BD	1927 Jan 23	Y. Oikawa	Tokyo	77
1267 Geertruida	1930 HD	1930 Apr 29	W. van Gent	Johannesburg (LS)	78
1268 Libya	1930 HJ	1930 Apr 29	C. Jackson	Johannesburg (UO)	
1269 Rollandia	1930 SH	1930 Sep 20	G. Neujmin	Simeis	
1270 Datura	1930 YE	1930 Dec 17	G. Van Biesbroeck	Williams Bay	79
1271 Isergina	1931 TN	1931 Oct 10	G. Neujmin	Simeis	
1272 Gefion	1932 TF1	1931 Oct 10	K. Reinmuth	Heidelberg	
1273 Dolores	1932 PF	1932 Aug 8	K. Reinmuth	Heidelberg	
1274 Delportia	1932 WC	1932 Nov 8	E. Delporte	Uccle	
1275 Cimbria	1932 MG	1932 Nov 30	K. Reinmuth	Heidelberg	
1276 Ucclia	1933 BA	1933 Jan 24	E. Delporte	Uccle	
1277 Dolores	1933 HA1	1933 Apr 18	G. Neujmin	Simeis	
1278 Kenya	1933 LA	1933 Jun 15	C. Jackson	Johannesburg (UO)	
1279 Uganda	1933 QB	1933 Aug 18	C. Jackson	Johannesburg (UO)	
1280 Baillauda	1933 QB	1933 Aug 18	E. Delporte	Uccle	
1281 Jeanne	1933 QJ	1933 Aug 24	S. Arend	Uccle	
1282 Utopia	1933 QM1	1933 Aug 17	C. Jackson	Johannesburg (UO)	
1283 Komsomolia	1925 SC	1925 Sep 25	V. Albitzky	Simeis	
1284 Latvia	1933 QP	1933 Jul 27	K. Reinmuth	Heidelberg	
1285 Julietta	1933 QF	1933 Aug 21	E. Delporte	Uccle	
1286 Banachiewicza	1933 QH	1933 Aug 25	S. Arend	Uccle	
1287 Lorcia	1933 QL	1933 Aug 25	S. Arend	Uccle	
1288 Santa	1933 QM	1933 Aug 26	E. Delporte	Uccle	
1289 Kutaïssi	1933 SO	1933 Sep 26	G. Neujmin	Simeis	
1290 Albertine	1933 QL1	1933 Aug 21	E. Delporte	Uccle	
1291 Phryne	1933 RA	1933 Sep 15	E. Delporte	Uccle	
1292 Luce	1933 SH	1933 Sep 17	P. Rigaux	Uccle	
1293 Sonja	1933 SO	1933 Sep 26	E. Delporte	Uccle	
1294 Antwerpia	1933 UB1	1933 Oct 24	E. Delporte	Uccle	
1295 Deflotte	1933 WD	1933 Nov 25	L. Boyer	Algiers	
1296 Audrée	1933 WE	1933 Nov 25	L. Boyer	Algiers	
1297 Quadea	1934 AD	1934 Jan 7	K. Reinmuth	Heidelberg	
1298 Nocturna	1934 AE	1934 Jan 7	K. Reinmuth	Heidelberg	
1299 Nocturna	1934 BA	1934 Jan 18	G. Reiss	Algiers	
1300 Marcelle	1934 CL	1934 Feb 10	G. Reiss	Algiers	

No	Name	pd	dd	d	dp	Notes
1301	Yvonne	1934 EA	1934 Mar 7	L. Boyer	Algiers	
1302	Werra	1924 SV	1924 Sep 28	K. Reinmuth	Heidelberg	
1303	Luthera	1928 FP	1928 Mar 16	A. Schwassmann	Bergedorf	
1304	Arosa	1928 KC	1928 May 21	K. Reinmuth	Heidelberg	
1305	Pongola	1928 OC	1928 Jul 19	H. E. Wood	Johannesburg (UO)	
1306	Scythia	1930 OB	1930 Jul 22	G. Neujmin	Simeis	
1307	Cimmeria	1930 UF	1930 Oct 17	G. Neujmin	Simeis	
1308	Halleria	1931 EB	1931 Mar 12	K. Reinmuth	Heidelberg	
1309	Hyperborea	1931 TO	1931 Oct 11	G. Neujmin	Simeis	
1310	Villigera	1932 DB	1932 Feb 28	A. Schwassmann	Bergedorf	80
1311	Knopfia	1933 FF_1	1933 Mar 24	K. Reinmuth	Heidelberg	
1312	Vassar	1933 OT	1933 Jul 27	G. Van Biesbroeck	Williams Bay	
1313	Berna	1933 QG	1933 Aug 24	S. Arend	Uccle	
1314	Paula	1933 SC	1933 Sep 16	S. Arend	Uccle	
1315	Bronislawa	1933 SF_1	1933 Sep 16	S. Arend	Uccle	
1316	Kasan	1933 WC	1933 Nov 17	G. Neujmin	Simeis	
1317	Silvretta	1935 RC	1935 Sep 1	K. Reinmuth	Heidelberg	81
1318	Nerina	1934 FG	1934 Mar 24	C. Jackson	Johannesburg (UO)	
1319	Disa	1934 FO	1934 Mar 19	C. Jackson	Johannesburg (UO)	
1320	Impala	1934 JG	1934 May 13	C. Jackson	Johannesburg (UO)	
1321	Majuba	1934 JH	1934 May 7	C. Jackson	Johannesburg (UO)	
1322	Coppernicus	1934 LA	1934 Jun 15	K. Reinmuth	Heidelberg	
1323	Tugela	1934 LD	1934 May 19	C. Jackson	Johannesburg (UO)	
1324	Knysna	1934 LL	1934 Jun 15	C. Jackson	Johannesburg (UO)	
1325	Inanda	1934 NR	1934 Jul 14	C. Jackson	Johannesburg (UO)	
1326	Losaka	1934 NS	1934 Jul 14	C. Jackson	Johannesburg (UO)	
1327	Namaqua	1934 RT	1934 Sep 7	C. Jackson	Johannesburg (UO)	
1328	Devota	1925 UA	1925 Oct 21	B. Jekhovsky	Algiers	
1329	Eliane	1933 FL	1933 Mar 23	E. Delporte	Uccle	
1330	Spiridonia	1925 DB	1925 Mar 17	V. Albitzky	Simeis	
1331	Solvejg	1933 QS	1933 Aug 25	G. Neujmin	Simeis	
1332	Marconia	1934 AA	1934 Jan 9	L. Volta	Pino Torinese	
1333	Cevenola	1934 DA	1934 Feb 20	O. Bancilhon	Algiers	
1334	Lundmarka	1934 OB	1934 Jul 16	K. Reinmuth	Heidelberg	
1335	Demoulina	1934 RE	1934 Sep 7	K. Reinmuth	Heidelberg	82
1336	Zeelandia	1934 RW	1934 Sep 9	H. Van Gent	Johannesburg (LS)	
1337	Gerarda	1934 RA_1	1934 Sep 9	H. Van Gent	Johannesburg (LS)	
1338	Duponta	1934 XA	1934 Dec 4	L. Boyer	Algiers	
1339	Désagneauxa	1934 XB	1934 Dec 4	L. Boyer	Algiers	
1340	Yvette	1934 YA	1934 Dec 27	L. Boyer	Algiers	
1341	Edmée	1935 BA	1935 Jan 27	E. Delporte	Uccle	
1342	Brabantia	1934 CV	1935 Feb 13	H. Van Gent	Johannesburg (LS)	
1343	Nicole	1935 FC	1935 Mar 29	L. Boyer	Algiers	
1344	Caubeta	1935 GA	1935 Apr 1	L. Boyer	Algiers	
1345	Potomac	1908 CG	1908 Feb 4	J. H. Metcalf	Taunton	
1346	Gotha	1929 CY	1929 Feb 5	K. Reinmuth	Heidelberg	
1347	Patria	1931 VW	1931 Nov 6	G. Neujmin	Simeis	
1348	Michel	1933 FD	1933 Mar 23	S. Arend	Uccle	
1349	Bechuana	1934 LJ	1934 Jun 13	C. Jackson	Johannesburg (UO)	
1350	Rosselia	1934 TA	1934 Oct 3	E. Delporte	Uccle	
1351	Uzbekistania	1934 TF	1934 Oct 5	G. Neujmin	Simeis	
1352	Wawel	1935 CE	1935 Feb 3	S. Arend	Uccle	
1353	Maartje	1935 CU	1935 Feb 13	H. Van Gent	Johannesburg (LS)	
1354	Botha	1935 GK	1935 Apr 3	C. Jackson	Johannesburg (UO)	
1355	Magoeba	1935 HE	1935 Apr 30	C. Jackson	Johannesburg (UO)	
1356	Nyanza	1935 JH	1935 May 3	C. Jackson	Johannesburg (UO)	
1357	Khama	1935 ND	1935 Jul 2	C. Jackson	Johannesburg (UO)	
1358	Gaika	1935 OB	1935 Jul 21	C. Jackson	Johannesburg (UO)	
1359	Prieska	1935 OC	1935 Jul 22	C. Jackson	Johannesburg (UO)	
1360	Tarka	1935 OD	1935 Jul 22	C. Jackson	Johannesburg (UO)	
1361	Leuschneria	1935 QA	1935 Aug 30	E. Delporte	Uccle	
1362	Griqua	1935 QG_1	1935 Jul 31	C. Jackson	Johannesburg (UO)	
1363	Herberta	1935 RA	1935 Aug 30	E. Delporte	Uccle	
1364	Safara	1935 VB	1935 Nov 18	L. Boyer	Algiers	
1365	Henyey	1928 RK	1928 Sep 9	M. Wolf	Heidelberg	
1366	Piccolo	1932 WA	1932 Nov 29	E. Delporte	Uccle	
1367		1934 NA	1934 Jul 3	C. Jackson	Johannesburg (UO)	83
1368	Numidia	1935 HD	1935 Apr 30	C. Jackson	Johannesburg (UO)	
1369	Ostanina	1935 QB	1935 Aug 27	P. Shajn	Simeis	
1370	Hella	1935 QG	1935 Aug 31	K. Reinmuth	Heidelberg	
1371	Resi	1935 QJ	1935 Aug 31	K. Reinmuth	Heidelberg	
1372	Haremari	1935 QK	1935 Aug 31	K. Reinmuth	Heidelberg	
1373	Cincinnati	1935 QN	1935 Aug 30	E. Hubble	Mount Wilson	
1374	Isora	1935 UA	1935 Oct 21	E. Delporte	Uccle	
1375	Alfreda	1935 UB	1935 Oct 22	E. Delporte	Uccle	
1376	Michelle	1935 UH	1935 Oct 29	G. Reiss	Algiers	
1377	Roberbauxa	1936 CD	1936 Feb 14	L. Boyer	Algiers	
1378	Leonce	1936 DB	1936 Feb 21	F. Rigaux	Uccle	
1379	Lomonosowa	1936 FC	1936 Mar 19	G. Neujmin	Simeis	84
1380	Volodia	1936 FM	1936 Mar 16	L. Boyer	Algiers	
1381	Danubia	1930 QJ	1930 Aug 20	E. Skvortsov	Simeis	
1382	Gerti	1925 BB	1925 Jan 21	K. Reinmuth	Heidelberg	
1383	Limburgia	1934 RV	1934 Sep 9	H. Van Gent	Johannesburg (LS)	
1384	Kniertje	1934 RX	1934 Sep 9	H. Van Gent	Johannesburg (LS)	
1385	Gelria	1935 MJ	1935 May 24	H. Van Gent	Johannesburg (LS)	
1386	Storeria	1935 PA	1935 Jul 28	G. Neujmin	Simeis	85
1387	Kama	1935 QD	1935 Aug 27	P. Shajn	Simeis	
1388	Aphrodite	1935 SS	1935 Sep 24	E. Delporte	Uccle	
1389	Onnie	1935 SS_1	1935 Sep 28	H. Van Gent	Johannesburg (LS)	
1390	Abastumani	1935 TA	1935 Oct 3	P. Shajn	Simeis	86
1391	Carelia	1936 DA	1936 Feb 16	Y. Väisälä	Turku	
1392	Pierre	1936 FO	1936 Mar 16	L. Boyer	Algiers	
1393	Sofala	1936 KD	1936 May 25	C. Jackson	Johannesburg (UO)	
1394	Algoa	1936 MD	1936 Jun 19	C. Jackson	Johannesburg (UO)	
1395	Aribeda	1936 OB	1936 Jul 16	K. Reinmuth	Heidelberg	
1396	Outeniqua	1936 PF	1936 Aug 9	C. Jackson	Johannesburg (UO)	
1397	Umtata	1936 PG	1936 Aug 9	C. Jackson	Johannesburg (UO)	
1398	Donnera	1936 QL	1936 Aug 26	Y. Väisälä	Turku	
1399	Teneriffa	1936 QY	1936 Aug 23	K. Reinmuth	Heidelberg	
1400	Tirela	1936 WA	1936 Nov 17	L. Boyer	Algiers	

No	Name	pd	dd	d	dp	Notes
1401	Luvonne	1935 UD	1935 Oct 22	E. Delporte	Uccle	
1402	Eri	1936 OC	1936 Jul 16	K. Reinmuth	Heidelberg	
1403	Idelsonia	1936 QA	1936 Aug 13	G. Neujmin	Simeis	
1404	Ajax	1936 QW	1936 Aug 17	Y. Väisälä	Turku	
1405	Sibelius	1936 RE	1936 Sep 12	Y. Väisälä	Turku	
1406	Komppa	1936 RF	1936 Sep 13	Y. Väisälä	Turku	
1407	Lindelöf	1936 WC	1936 Nov 21	Y. Väisälä	Turku	
1408	Trusanda	1936 WF	1936 Nov 23	K. Reinmuth	Heidelberg	
1409	Isko	1937 AK	1937 Jan 8	K. Reinmuth	Heidelberg	
1410	Margret	1937 AL	1937 Jan 8	K. Reinmuth	Heidelberg	
1411	Brauna	1937 BA	1937 Jan 8	K. Reinmuth	Heidelberg	
1412	Lagrula	1937 BA	1937 Jan 30	L. Boyer	Algiers	
1413	Roucarie	1937 CD	1937 Feb 12	L. Boyer	Algiers	
1414	Jérôme	1937 CE	1937 Feb 12	L. Boyer	Algiers	
1415	Malautra	1937 EA	1937 Mar 4	L. Boyer	Algiers	
1416	Renauxa	1937 EC	1937 Mar 4	L. Boyer	Algiers	
1417	Walinskia	1937 GH	1937 Apr 1	K. Reinmuth	Heidelberg	
1418	Fayeta	1903 RG	1903 Sep 22	P. Götz	Heidelberg	
1419	Danzig	1929 RF	1929 Sep 5	K. Reinmuth	Heidelberg	
1420	Radcliffe	1931 RJ	1931 Sep 14	K. Reinmuth	Heidelberg	
1421	Esperanto	1936 FQ	1936 Mar 18	Y. Väisälä	Turku	
1422	Strömgrenia	1936 QF	1936 Aug 23	J. Hunaerts	Uccle	
1423	Jose	1936 QM	1936 Aug 28	G. Neujmin	Simeis	
1424	Sundmania	1937 AJ	1937 Jan 9	Y. Väisälä	Turku	
1425	Tuorla	1937 GB	1937 Apr 3	K. Inkeri	Turku	
1426	Ruvuma	1937 GF	1937 Apr 1	M. Laugier	Nice	
1427	Mombasa	1937 KB	1937 May 16	C. Jackson	Johannesburg (10)	
1428	Monbasa	1937 ND	1937 Jul 1	C. Jackson	Johannesburg (10)	
1429	Pemba	1937 NH	1937 Jul 2	C. Jackson	Johannesburg (10)	
1430		1937 NK	1937 Jul 3	C. Jackson	Johannesburg (10)	
1431	Ethiopia	1937 QB	1937 Jul 29	C. Jackson	Johannesburg (10)	
1432	Germania	1937 PG	1937 Aug 1	C. Jackson	Johannesburg (10)	
1433	Germania	1937 UC	1937 Oct 30	E. Delporte	Uccle	
1434	Margot	1936 FD$_1$	1936 Mar 19	G. Neujmin	Simeis	
1435	Garlena	1936 WE	1936 Nov 23	K. Reinmuth	Heidelberg	
1436	Diomedes	1937 YA	1936 Dec 11	G. Kulin	Budapest	
1437		1937 FB	1937 Mar 3	K. Reinmuth	Heidelberg	
1438	Wendeline	1937 TC	1937 Oct 11	K. Reinmuth	Heidelberg	
1439	Vogtia	1937 TE	1937 Oct 11	K. Reinmuth	Heidelberg	
1440	Rostia	1937 TF	1937 Oct 11	K. Reinmuth	Heidelberg	
1441	Bolyai	1937 WA	1937 Nov 26	G. Kulin	Budapest	
1442	Corvina	1937 YF	1937 Dec 29	G. Kulin	Budapest	
1443	Ruppina	1937 YG	1937 Dec 29	K. Reinmuth	Heidelberg	
1444	Konkolya	1938 AE	1938 Jan 6	G. Kulin	Budapest	
1445		1938 AF	1938 Jan 6	G. Kulin	Budapest	
1446	Sillanpää	1938 BA	1938 Jan 26	Y. Väisälä	Turku	
1447	Utra	1938 BB	1938 Jan 26	Y. Väisälä	Turku	60
1448	Lindbladia	1938 DF	1938 Feb 16	Y. Väisälä	Turku	
1449	Virtanen	1938 DO	1938 Feb 16	Y. Väisälä	Turku	
1450	Raimonda	1938 DP	1938 Feb 20	Y. Väisälä	Turku	

No	Name	pd	dd	d	dp	Notes
1451	Granö	1938 DT	1938 Feb 22	Y. Väisälä	Turku	
1452		1938 DZ$_1$	1938 Feb 26	G. Kulin	Budapest	87
1453	Fennia	1938 ED$_1$	1938 Mar 8	Y. Väisälä	Turku	
1454	Kalevala	1936 DO	1936 Feb 16	Y. Väisälä	Turku	
1455	Mitchella	1937 LF	1937 Jun 5	A. Bohrmann	Heidelberg	
1456		1937 NG	1937 Jul 2	C. Jackson	Johannesburg (10)	
1457	Ankara	1937 PA	1937 Aug 1	K. Reinmuth	Heidelberg	88
1458	Mineura	1937 RC	1937 Sep 1	P. Rigaux	Uccle	
1459	Maguya	1937 VA	1937 Nov 4	G. Neujmin	Simeis	
1460	Baltia	1937 WC	1937 Nov 24	Y. Väisälä	Turku	
1461	Jean-Jacques	1937 YL	1937 Dec 30	M. Laugier	Nice	
1462	Zamenhof	1938 CA	1938 Feb 6	Y. Väisälä	Turku	
1463	Nordenmarkia	1938 CB	1938 Feb 6	Y. Väisälä	Turku	
1464	Armisticia	1939 VO	1939 Nov 11	G. Van Biesbroeck	Williams Bay	89
1465	Autonoma	1938 FA	1938 Mar 20	A. Wachmann	Bergedorf	
1466	Mündleria	1938 KA	1938 May 31	K. Reinmuth	Heidelberg	
1467	Mashona	1938 OE	1938 Jul 30	C. Jackson	Johannesburg (10)	
1468	Zomba	1938 PA	1938 Jul 23	C. Jackson	Johannesburg (10)	
1469	Linzia	1938 QD	1938 Aug 19	K. Reinmuth	Heidelberg	
1470	Carla	1938 SD	1938 Sep 17	A. Bohrmann	Heidelberg	
1471	Tornio	1938 SL$_1$	1938 Sep 16	Y. Väisälä	Turku	
1472	Muonio	1938 IQ	1938 Oct 18	Y. Väisälä	Turku	
1473	Ounas	1938 UT	1938 Oct 22	Y. Väisälä	Turku	
1474	Beira	1935 QY	1935 Aug 20	C. Jackson	Johannesburg (10)	
1475	Yalta	1935 SM	1935 Sep 21	P. Shajn	Simeis	
1476	Cox	1936 RA	1936 Sep 10	E. Delporte	Uccle	
1477	Bonsdorffia	1938 CC	1938 Feb 6	Y. Väisälä	Turku	90
1478	Ukhtri	1938 CF	1938 Feb 6	Y. Väisälä	Turku	91
1479	Inkeri	1938 DE	1938 Feb 16	Y. Väisälä	Turku	
1480	Aunus	1938 DK	1938 Feb 18	Y. Väisälä	Turku	
1481	Tübingia	1938 DR	1938 Feb 7	K. Reinmuth	Heidelberg	
1482	Sebastiana	1938 DA$_1$	1938 Feb 20	K. Reinmuth	Heidelberg	
1483	Hakoila	1938 DJ$_1$	1938 Feb 24	Y. Väisälä	Turku	
1484	Postrema	1938 HC	1938 Apr 29	G. Kulin	Budapest	
1485	Isa	1938 OB	1938 Jul 28	K. Reinmuth	Heidelberg	
1486	Marilyn	1938 QA	1938 Aug 23	E. Delporte	Uccle	
1487	Boda	1938 WC	1938 Nov 17	K. Reinmuth	Heidelberg	
1488	Aura	1938 XE	1938 Dec 15	Y. Väisälä	Turku	
1489		1939 GC	1939 Apr 12	G. Kulin	Budapest	
1490	Limpopo	1936 LB	1936 Jun 14	C. Jackson	Johannesburg (10)	
1491	Balduinus	1938 EJ	1938 Feb 23	E. Delporte	Uccle	
1492	Oppolzer	1938 FL	1938 Mar 23	Y. Väisälä	Turku	
1493	Sigrid	1938 QB	1938 Aug 26	E. Delporte	Uccle	
1494	Savo	1938 SJ	1938 Sep 16	Y. Väisälä	Turku	
1495	Helsinki	1938 SW	1938 Sep 21	Y. Väisälä	Turku	
1496	Turku	1938 SA$_1$	1938 Sep 22	Y. Väisälä	Turku	
1497	Tampere	1938 SB$_1$	1938 Sep 22	Y. Väisälä	Turku	
1498	Lahti	1938 SK$_1$	1938 Sep 16	Y. Väisälä	Turku	
1499	Pori	1938 UF	1938 Oct 16	Y. Väisälä	Turku	
1500	Jyväskylä	1938 UH	1938 Oct 16	Y. Väisälä	Turku	

No	Name	pd	dd	d	dp	Notes
1501	Baade	1938 UJ	1938 Oct 20	A. Wachmann	Bergedorf	
1502	Arenda	1938 WB	1938 Nov 17	K. Reinmuth	Heidelberg	
1503	Kuopio	1938 XD	1938 Dec 15	Y. Väisälä	Turku	
1504	Lappeenranta	1939 FM	1939 Mar 23	L. Oterma	Turku	
1505	Koranna	1939 HH	1939 Apr 21	C. Jackson	Johannesburg (100)	
1506	Xosa	1939 JC	1939 May 15	C. Jackson	Johannesburg (100)	
1507	Vaasa	1939 RD	1939 Sep 12	L. Oterma	Turku	
1508	Kent	1938 UP	1938 Oct 21	H. Alikoski	Turku	
1509	Esclangona	1939 YG	1938 Dec 21	A. Patry	Nice	92
1510	Charlois	1939 DC	1939 Feb 22	A. Patry	Nice	
1511	Daléra	1939 FB	1939 Mar 22	L. Boyer	Algiers	
1512	Oulu	1939 FE	1939 Mar 18	H. Alikoski	Turku	
1513		1940 EB	1940 Mar 10	G. Kulin	Budapest	
1514	Ricouxa	1906 UR	1906 Aug 22	M. Wolf	Heidelberg	
1515	Perrotin	1936 VG	1936 Nov 15	A. Patry	Nice	
1516	Henry	1938 BG	1938 Jan 28	A. Patry	Nice	
1517	Beograd	1938 FD	1938 Mar 20	M. B. Protitch	Belgrade	93
1518	Rovaniemi	1938 UA	1938 Oct 15	Y. Väisälä	Turku	
1519	Kajaani	1938 UB	1938 Oct 15	Y. Väisälä	Turku	
1520	Imatra	1938 UY	1938 Oct 22	Y. Väisälä	Turku	
1521	Seinäjoki	1938 UB$_1$	1938 Oct 22	Y. Väisälä	Turku	
1522	Kokkola	1938 WO	1938 Nov 18	L. Oterma	Turku	
1523	Pieksämäki	1939 BC	1939 Jan 18	Y. Väisälä	Turku	
1524	Joensuu	1939 SB	1939 Sep 18	Y. Väisälä	Turku	
1525	Savonlinna	1939 SC	1939 Sep 18	Y. Väisälä	Turku	
1526	Mikkeli	1939 TP	1939 Oct 7	Y. Väisälä	Turku	94
1527	Malmquista	1939 UG	1939 Oct 18	Y. Väisälä	Turku	
1528	Conrada	1940 CA	1940 Feb 12	K. Reinmuth	Heidelberg	
1529	Oterma	1938 BC	1938 Jan 26	Y. Väisälä	Turku	
1530	Rantaseppä	1938 SG	1938 Sep 16	Y. Väisälä	Turku	95
1531	Hartmut	1938 SH	1938 Sep 17	A. Bohrmann	Heidelberg	
1532	Inari	1938 SM	1938 Sep 16	Y. Väisälä	Turku	
1533	Saimaa	1939 BD	1939 Jan 30	Y. Väisälä	Turku	
1534	Näsi	1939 BK	1939 Jan 20	Y. Väisälä	Turku	
1535	Päijänne	1939 RG	1939 Sep 9	Y. Väisälä	Turku	
1536	Pielinen	1939 SE	1939 Sep 18	Y. Väisälä	Turku	
1537		1940 QA	1940 Aug 27	G. Stroomer	Budapest	
1538		1940 RF	1940 Sep 8	G. Kulin	Budapest	
1539	Borrelly	1940 UB	1940 Oct 29	A. Patry	Nice	
1540	Kevola	1938 WK	1938 Nov 18	L. Oterma	Turku	96
1541	Estonia	1939 CK	1939 Feb 12	Y. Väisälä	Turku	
1542	Schalén	1941 QE	1941 Aug 26	Y. Väisälä	Turku	
1543	Bourgeois	1941 SJ	1941 Sep 21	E. Delporte	Uccle	
1544	Vinterhansenia	1941 UK	1941 Oct 15	L. Oterma	Turku	
1545	Thernöe	1941 UW	1941 Oct 15	L. Oterma	Turku	
1546		1941 SG$_1$	1941 Feb 28	G. Kulin	Budapest	
1547		1929 CZ	1929 Feb 2	P. Bourgeois	Uccle	
1548	Palomaa	1935 FK	1935 Mar 26	Y. Väisälä	Turku	
1549	Mikko	1937 GA	1937 Apr 2	Y. Väisälä	Turku	
1550	Tito	1937 WD	1937 Nov 29	M. B. Protitch	Belgrade	
1551	Argelander	1938 DC$_1$	1938 Feb 24	Y. Väisälä	Turku	
1552	Bessel	1938 DE$_1$	1938 Feb 24	Y. Väisälä	Turku	
1553	Bauersfelda	1940 AD	1940 Jan 13	K. Reinmuth	Heidelberg	
1554	Yugoslavia	1940 RE	1940 Sep 6	M. B. Protitch	Belgrade	
1555	Dejan	1941 SA	1941 Sep 15	F. Rigaux	Uccle	
1556	Wingolfia	1942 AA	1942 Jan 14	K. Reinmuth	Heidelberg	
1557	Roehla	1942 AD	1942 Jan 14	K. Reinmuth	Heidelberg	
1558	Järnefelt	1942 BD	1942 Jan 20	L. Oterma	Turku	
1559	Kustaanheimo	1942 BF	1942 Jan 20	L. Oterma	Turku	
1560	Strattonia	1942 XB	1942 Dec 3	E. Delporte	Uccle	
1561	Fricke	1941 CG	1941 Feb 15	K. Reinmuth	Heidelberg	
1562	Gondolatsch	1943 EE	1943 Mar 7	K. Reinmuth	Heidelberg	
1563	Noel	1943 EG	1943 Mar 7	S. Arend	Uccle	
1564	Srbija	1936 TB	1936 Oct 15	M. B. Protitch	Belgrade	
1565	Lemaître	1948 WA	1948 Nov 25	S. Arend	Uccle	
1566	Icarus	1949 MA	1949 Jun 27	W. Baade	Palomar	
1567	Alikoski	1941 HN	1941 Apr 22	Y. Väisälä	Turku	
1568	Aisleen	1946 QB	1946 Aug 21	E. L. Johnson	Johannesburg (100)	
1569	Evita	1948 PA	1948 Aug 3	M. Itzigsohn	La Plata	
1570	Brunonia	1948 TX	1948 Oct 9	S. Arend	Uccle	
1571		1950 FJ	1950 Mar 20	M. Itzigsohn and A. Kwiek	Posen	
1572	Posnania	1949 SC	1949 Sep 27	J. Dobrzycki and A. Kwiek	Posen	
1573	Väisälä	1949 UA	1949 Oct 15	S. Arend	Uccle	
1574	Meyer	1949 FD	1949 Mar 22	L. Boyer	Algiers	
1575	Winifred	1950 HH	1950 Apr 20	R. C. Cameron	Indiana	
1576	Fabiola	1948 SA	1948 Sep 30	S. Arend	Uccle	
1577	Reiss	1949 BA	1949 Jan 19	L. Boyer	Algiers	
1578	Kirkwood	1951 AT	1951 Jan 10	Indiana University	Indiana	
1579	Herrick	1948 SB	1948 Sep 30	S. Arend	Uccle	
1580	Betulia	1950 KA	1950 May 22	E. L. Johnson	Johannesburg (100)	97
1581	Abanderada	1950 LA$_1$	1950 Jun 15	M. Itzigsohn	La Plata	
1582	Martir	1950 LY	1950 Jun 15	M. Itzigsohn	La Plata	
1583	Antilochus	1950 SA	1950 Sep 19	S. Arend	Uccle	
1584	Fuji	1927 CR	1927 Feb 7	O. Oikawa	Tokyo	
1585	Union	1947 RG	1947 Sep 7	E. L. Johnson	Johannesburg (100)	
1586	Thiele	1939 CJ	1939 Feb 13	A. Wachmann	Bergedorf	
1587	Kahrstedt	1933 FS$_1$	1933 Mar 25	K. Reinmuth	Heidelberg	
1588	Descamisada	1951 MH	1951 Jun 27	M. Itzigsohn	La Plata	
1589	Fanatica	1950 RK	1950 Sep 13	M. Itzigsohn	La Plata	
1590	Tsiolkovskaja	1933 NA	1933 Jul 1	G. Neujmin	Simeis	
1591	Baize	1951 KA	1951 May 31	S. Arend	Uccle	
1592	Mathieu	1951 LB	1951 Jun 1	S. Arend	Uccle	
1593	Fagnes	1951 LB$_1$	1951 Jun 1	S. Arend	Uccle	
1594	Danjon	1949 WA	1949 Nov 23	L. Boyer	Algiers	
1595		1930 ME	1930 Jun 19	C. Jackson and H. E. Wood	Johannesburg (100)	
1596		1951 EV	1951 Mar 8	M. Itzigsohn	La Plata	
1597	Laugier	1949 EB	1949 Mar 1	L. Boyer	Algiers	
1598	Paloque	1950 GA	1950 Feb 23	L. Boyer	Algiers	
1599	Giomus	1950 UA	1950 Nov 17	L. Boyer	Algiers	
1600	Vysotsky	1947 UC	1947 Oct 22	C. A. Wirtanen	Mount Hamilton	

No	Name	pd	dd	d	dp	Notes
1601	Patry	1942 KA	1942 May 18	L. Boyer	Algiers	
1602	Indiana	1950 GF	1950 Mar 14	Indiana University	Indiana	97
1603	Neva	1926 VH	1926 Nov 4	G. Neujmin	Simeis	
1604		1931 PH	1931 Mar 24	C. O. Lampland	Flagstaff (LO)	
1605		1936 CA	1936 Apr 13	P. Djurkovic	Uccle	
1606	Jekhovsky	1950 RH	1950 Sep 14	L. Boyer	Algiers	
1607	Mavis	1950 RA	1950 Sep 3	E. L. Johnson	Johannesburg (00)	98
1608		1951 NZ	1951 Sep 1	M. Itzigsohn	La Plata	
1609	Brenda	1951 NL	1951 Jul 10	E. L. Johnson	Johannesburg (00)	
1610	Mirnaya	1928 RT	1928 Sep 11	P. Shajn	Simeis	
1611	Beyer	1950 DJ	1950 Feb 17	K. Reinmuth	Heidelberg	
1612	Hirose	1950 BJ	1950 Jan 23	K. Reinmuth	Heidelberg	
1613	Smiley	1950 SD	1950 Jan 17	S. Arend	Uccle	
1614	Goldschmidt	1952 HA	1952 Apr 18	A. Schmitt	Uccle	
1615	Bardwell	1950 BW	1950 Jan 28	Indiana University	Indiana	97
1616	Filipoff	1950 EA	1950 Mar 15	L. Boyer	Algiers	
1617	Alschmitt	1952 FB	1952 Mar 20	L. Boyer	Algiers	
1618	Dawn	1948 NF	1948 Jul 5	E. L. Johnson	Johannesburg (00)	
1619	Ueta	1953 TA	1953 Oct 11	T. Mitani	Kwasan	
1620	Geographos	1951 RA	1951 Sep 14	A. Wilson and R. Minkowski	Palomar	
1621	Druzhba	1926 TH	1926 Oct 1	S. Belyavsky	Simeis	
1622	Chacornac	1952 EA	1952 Mar 15	A. Schmitt	Uccle	
1623	Vivian	1948 PL	1948 Aug 9	E. L. Johnson	Johannesburg (00)	
1624	Rabe	1931 TT1	1931 Oct 6	K. Reinmuth	Heidelberg	
1625	The NORC	1953 RB	1953 Sep 11	S. Arend	Uccle	
1626	Sadeya	1927 AA	1927 Jan 10	J. Comas Solá	Barcelona	
1627	Ivar	1929 SH	1929 Sep 25	K. Hertzsprung	Johannesburg (LS)	
1628	Strobel	1952 OG	1923 Jul 28	K. Reinmuth	Heidelberg	
1629	Pecker	1952 DA	1952 Feb 28	L. Boyer	Algiers	
1630	Milet	1952 BA	1952 Feb 28	L. Boyer	Algiers	
1631	Kopff	1936 UC	1936 Oct 11	Y. Väisälä	Turku	
1632	Siebohme	1941 DF	1941 Feb 26	S. B. Nicholson	Heidelberg	99
1633	Chimay	1929 RC	1929 Mar 3	S. Arend	Uccle	
1634		1935 QP	1935 Aug 7	C. Jackson	Johannesburg (00)	
1635	Bohrmann	1924 QM	1924 Mar 7	K. Reinmuth	Heidelberg	
1636	Porter	1950 RH	1950 Jan 24	K. Reinmuth	Heidelberg	
1637	Swings	1936 QO	1936 Aug 28	J. Hunaerts	Uccle	
1638		1935 JF	1935 May 2	C. Jackson	Johannesburg (00)	
1639	Bower	1951 KB	1951 Sep 12	S. Arend	Uccle	
1640		1951 QA	1951 Aug 31	S. Arend	Uccle	
1641	Hill	1935 OJ	1935 Jul 25	C. Jackson	Johannesburg (00)	
1642	Brown	1951 RU	1951 Sep 4	K. Reinmuth	Heidelberg	
1643	Raffia	1951 RQ	1951 Sep 4	K. Reinmuth	Heidelberg	
1644		1935 YA	1935 Dec 16	R. Carrasco	Madrid	
1645	Waterfield	1933 QJ	1933 Jul 24	K. Reinmuth	Heidelberg	
1646	Rosseland	1939 BG	1939 Jan 19	Y. Väisälä	Turku	
1647	Menelaus	1957 MK	1957 Jun 23	S. B. Nicholson	Mount Wilson	
1648	Shajna	1935 KF	1935 Sep 5	L. Boyer	Simeis	
1649	Fabre	1951 DE	1951 Feb 27	L. Boyer	Uccle	
1650	Heckmann	1937 TG	1937 Oct 11	K. Reinmuth	Heidelberg	

No	Name	pd	dd	d	dp	Notes
1651		1936 HD	1936 Apr 23	M. Laugier	Nice	
1652		1953 PA	1953 Aug 9	S. Arend	Berne	
1653	Yakhontovia	1937 RA	1937 Aug 30	G. Neujmin	Simeis	
1654	Bojeva	1931 TL	1931 Oct 8	P. Shajn	Turku	
1655		1929 WG	1929 Nov 28	J. Comas Solá	Johannesburg (00)	
1656	Suomi	1942 EC	1942 Mar 11	Y. Väisälä	Turku	
1657	Roemera	1961 EA	1961 Mar 6	P. Wild	Johannesburg (00)	
1658	Innes	1953 NA	1953 Jul 13	J. Bruwer	Heidelberg	
1659	Punkaharju	1940 YL	1940 Jan 1	Y. Väisälä	Heidelberg	
1660	Wood	1953 GA	1953 Apr 7	J. Bruwer	Johannesburg (00)	
1661	Granule	1916 ZE	1916 Mar 31	M. Wolf	Uccle	
1662	Hoffmann	1923 OD	1923 Sep 11	K. Reinmuth	Heidelberg	
1663	Van den Bos	1926 PE	1926 Aug 4	K. Reinmuth	Johannesburg (00)	
1664		1929 CD	1929 Feb 4	E. Wood	Johannesburg	
1665	Gaby	1930 DQ	1930 Feb 27	K. Reinmuth	Heidelberg	
1666	Van Gent	1930 OC	1930 Jul 22	H. Van Gent	Heidelberg	
1667	Pels	1930 SY	1930 Jul 16	K. Reinmuth	Heidelberg	
1668	Hanna	1933 OX	1933 Jul 24	K. Reinmuth	Belgrade	
1669	Dagmar	1934 RS	1934 Sep 7	H. Van Gent	Simeis	
1670	Minnaert	1934 RZ	1934 Sep 9	G. Neujmin	Uccle	
1671	Chaika	1934 TD	1934 Oct 3	E. Delporte	Heidelberg	
1672		1935 BD	1935 Jan 29	K. Reinmuth	Heidelberg	
1673	Van Houten	1937 TH	1937 Oct 11	K. Reinmuth	Belgrade	
1674	Groeneveld	1938 DS	1938 Feb 7	K. Reinmuth	Johannesburg (00)	
1675	Simonida	1938 FB	1938 Mar 20	M. B. Protitch	Belgrade	
1676		1939 LC	1939 May 15	C. Jackson	Johannesburg	
1677	Tycho Brahe	1940 RO	1940 Sep 2	Y. Väisälä	Turku	
1678	Hveen	1940 YH	1940 Dec 30	L. Oterma	Turku	
1679	Nevanlinna	1941 FR	1941 Mar 20	L. Oterma	Turku	
1680		1942 CH	1942 Feb 10	M. Laugier	Nice	
1681		1948 WE	1948 Nov 23	K. Reinmuth	Heidelberg	
1682	Karel	1949 PH	1949 Aug 16	S. Arend	Uccle	
1683		1950 SL	1950 Sep 19	M. Itzigsohn	La Plata	
1684		1951 QE	1951 Aug 23	C. A. Wirtanen	Mount Hamilton	
1685	Toro	1948 OA	1948 Jul 17	C. A. Wirtanen	Johannesburg (LS)	
1686	De Sitter	1935 SR1	1935 Sep 28	H. Van Gent	Johannesburg (LS)	100
1687	Glarona	1965 SC	1965 Sep 19	P. Wild	Berne	
1688		1951 EQ1	1951 Mar 3	M. Laugier	Heidelberg	
1689	Floria-Jan	1930 SO	1930 Sep 19	K. Reinmuth	Simeis	
1690		1948 VB	1948 Nov 8	G. Neujmin	Johannesburg (LS)	
1691	Oort	1956 RB	1956 Sep 9	H. Van Gent	Turku	
1692	Subbotina	1936 QD	1936 Aug 16	G. Neujmin	Turku	
1693	Hertzsprung	1935 LA	1935 May 5	L. Oterma	Uccle	
1694	Kaiser	1934 SB	1934 Sep 29	Y. Väisälä	Turku	
1695		1941 UO	1941 Oct 15	H. Alikoski	Belgrade	
1696		1939 FF	1939 Mar 18	E. Delporte		
1697		1940 RM	1940 Sep 11	Y. Väisälä		
1698		1934 CS	1934 Feb 10	P. Djurkovic		
1699		1941 QD	1941 Aug 26			
1700		1940 QC	1940 Aug 27			

No	Name	pd	dd	d	dp	Notes
1701		1953 NJ	1953 Jul 6	J. Churms	Johannesburg (UO)	
1702	Barry	1924 SH	1924 Jul 1	E. Hertzsprung	Johannesburg (LS)	
1703	Wachmann	1930 RB	1930 Sep 7	M. Wolf	Heidelberg	
1704	Wachmann	1924 QT	1924 Aug 16	K. Reinmuth	Heidelberg	
1705		1941 SL1	1941 Sep 29	L. Oterma	Turku	
1706	Dieckvoss	1931 TS	1931 Oct 5	K. Reinmuth	Heidelberg	
1707		1932 RL	1932 Sep 8	E. Delporte	Uccle	
1708		1929 XA	1929 Nov 30	J. Comas Solá	Barcelona	
1709	Ukraina	1925 QA	1925 Aug 16	G. Shajn	Simeis	
1710		1941 UF	1941 Oct 20	G. Kulin	Budapest	
1711		1935 BB	1935 Jan 29	E. Delporte	Uccle	
1712		1935 KC	1935 May 28	C. Jackson	Johannesburg (UO)	
1713	Bancilhon	1951 SC	1951 Sep 27	L. Boyer	Algiers	
1714	Sy	1951 OA	1951 Jul 25	L. Boyer	Algiers	
1715		1938 GK	1938 Apr 9	Y. Väisälä	Turku	
1716	Peter	1934 GF	1934 Apr 4	K. Reinmuth	Heidelberg	
1717		1954 AC	1954 Jan 8	S. Arend	Uccle	
1718		1942 XX	1942 Sep 14	K. Reinmuth	Turku	
1719	Jens	1950 DP	1950 Feb 17	K. Reinmuth	Heidelberg	
1720	Nields	1935 CQ	1935 Feb 7	K. Reinmuth	Heidelberg	
1721	Wells	1953 TD3	1953 Oct 3	Indiana University	Indiana	97
1722		1938 EG	1938 Mar 18	E. Delporte	Uccle	
1723		1936 FX	1936 Mar 18	Y. Väisälä	Turku	
1724		1932 DE	1932 Feb 28	E. Delporte	Uccle	
1725	CrAO	1930 SK	1930 Sep 20	G. Neujmin	Simeis	
1726	Hoffmeister	1933 GD	1933 Jul 24	K. Reinmuth	Heidelberg	
1727		1965 BA	1965 Jan 25	A. D. Andrews	Boyden	
1728	Goethe Link	1964 TO	1964 Oct 12	Indiana University	Indiana	97
1729	Beryl	1963 SL	1963 Sep 19	Indiana University	Indiana	
1730		1936 UA	1936 Oct 16	M. Laugier	Nice	
1731	Smuts	1948 PH	1948 Aug 9	E. L. Johnson	Johannesburg (UO)	
1732	Heike	1943 EY	1943 Mar 10	K. Reinmuth	Heidelberg	
1733	Silke	1938 DL1	1938 Feb 19	A. Bohrmann	Heidelberg	
1734	Zhongolovich	1928 TJ	1928 Oct 12	G. Neujmin	Simeis	
1735		1948 RJ1	1948 Sep 10	P. Shajn	Simeis	
1736	Floirac	1967 RA	1967 Sep 6	G. Soulié	Bordeaux	
1737	Severny	1966 TJ	1966 Oct 13	L. Chernykh	CRAO	
1738		1930 SP	1930 Sep 16	H. Van Gent	Johannesburg (LS)	
1739	Meyermann	1939 PF	1939 Aug 15	K. Reinmuth	Heidelberg	
1740		1939 GX	1939 Oct 16	Y. Väisälä	Turku	
1741	Giclas	1960 BC	1960 Jan 26	Indiana University	Indiana	97
1742	Schaifers	1934 RO	1934 Sep 9	K. Reinmuth	Heidelberg	
1743	Schmidt	4109 P-L	1960 Sep 24	PLS	Palomar-Leiden	101
1744	Harriet	6557 P-L	1960 Sep 24	PLS	Palomar-Leiden	101
1745	Ferguson	1941 SY1	1941 Sep 17	J. Willis	Washington	
1746	Brouwer	1963 RF	1963 Sep 14	Indiana University	Indiana	
1747	Wright	1947 NH	1947 Jul 4	C. A. Wirtanen	Mount Hamilton	
1748	Mauderli	1966 RA	1966 Sep 15	P. Wild	Berne	97
1749	Telamon	1949 SB	1949 Sep 23	K. Reinmuth	Heidelberg	97
1750	Eckert	1950 NA1	1950 Jul 15	K. Reinmuth	Heidelberg	
1751	Herget	1955 OC	1955 Jul 27	Indiana University	Indiana	97
1752		1930 OK	1930 Jul 22	H. Van Gent	Johannesburg (LS)	
1753		1934 JH	1934 May 10	H. Van Gent	Johannesburg (LS)	
1754	Cunningham	1935 FE	1935 Mar 29	E. Delporte	Uccle	
1755	Lorbach	1936 VD	1936 Nov 8	M. Laugier	Nice	
1756	Giacobini	1937 YA	1937 Dec 24	A. Patry	Nice	
1757		1939 PC	1939 Mar 17	Y. Väisälä	Turku	
1758		1942 DK	1942 Feb 22	L. Oterma	Turku	
1759	Kienle	1942 RF	1942 Sep 11	K. Reinmuth	Heidelberg	
1760	Sandra	1950 GB	1950 Apr 10	E. L. Johnson	Johannesburg (UO)	
1761	Edmondson	1952 FN	1952 Mar 30	Indiana University	Indiana	
1762	Russell	1953 TZ	1953 Oct 8	Indiana University	Indiana	
1763	Williams	1953 TN2	1953 Oct 13	Indiana University	Indiana	
1764	Cogshall	1953 VM1	1953 Nov 7	Indiana University	Indiana	
1765	Wrubel	1957 XB	1957 Dec 15	Indiana University	Indiana	
1766	Slipher	1962 RF	1962 Sep 24	Indiana University	Indiana	97
1767	Lampland	1962 RJ	1962 Sep 24	Indiana University	Indiana	97
1768	Appenzella	1965 SA	1965 Sep 23	P. Wild	Berne	
1769		1966 QP	1966 Aug 25	Z. Pereyra	Cordoba	
1770	Schlesinger	1967 JR	1967 May 10	C. Cesco and A. Klemola	El Leoncito	
1771	Makover	1968 BD	1968 Jan 24	L. Chernykh	CRAO	
1772	Gagarin	1968 CB	1968 Feb 6	L. Chernykh	CRAO	
1773	Rumpelstilz	1968 HE	1968 Apr 17	P. Wild	Berne	
1774	Kulikov	1968 UG1	1968 Oct 22	T. Smirnova	Berne	
1775	Zimmerwald	1969 JA	1969 May 13	P. Wild	Berne	
1776	Kuiper	2520 P-L	1960 Sep 24	PLS	Palomar-Leiden	101
1777	Gehrels	4007 P-L	1960 Sep 24	PLS	Palomar-Leiden	101
1778	Alfvén	4506 P-L	1960 Sep 24	PLS	Palomar-Leiden	101
1779		1950 LZ	1950 Jun 15	M. Itzigsohn	La Plata	
1780	Kippes	1906 UX	1906 Sep 12	A. Kopff	Heidelberg	
1781	Van Biesbroeck	1906 VV	1906 Oct 17	A. Kopff	Heidelberg	
1782	Schmeller	1931 TL1	1931 Oct 6	K. Reinmuth	Heidelberg	
1783		1935 FJ	1935 Mar 24	G. Neujmin	Simeis	
1784		1935 MG	1935 Jun 30	C. Jackson	Johannesburg (UO)	
1785	Wurm	1941 CD	1941 Feb 15	K. Reinmuth	Heidelberg	
1786		1948 TL	1948 Oct 9	H. Alikoski	Turku	
1787	Kiess	1948 SK	1948 Sep 19	S. Arend	Uccle	
1788		1952 QZ	1952 Aug 24	Indiana University	Indiana	97
1789	Dobrovolsky	1966 QC	1966 Aug 19	L. Chernykh	CRAO	
1790	Volkov	1967 ER	1967 Mar 9	L. Chernykh	CRAO	
1791	Patsayev	1967 RE	1967 Sep 4	T. Smirnova	CRAO	
1792	Reni	1968 HG	1968 Jan 28	L. Chernykh	CRAO	
1793	Zoya	1968 FA	1968 Feb 24	T. Smirnova	CRAO	
1794	Finsen	1970 GA	1970 Apr 4	J. Bruwer	Hartbeespoort	
1795	Woltjer	4010 P-L	1960 Sep 24	PLS	Palomar-Leiden	101
1796	Riga	1966 KB	1966 May 16	N. Chernykh	CRAO	
1797	Schaumasse	1936 VH	1936 Nov 15	A. Patry	Nice	
1798	Watts	1949 GC	1949 Apr 4	Indiana University	Indiana	97
1799	Koussevitzky	1950 OE	1950 Jul 25	Indiana University	Indiana	97
1800	Aguilar	1950 RJ	1950 Sep 12	M. Itzigsohn	La Plata	

No	Name	pd	dd	d	dp	Notes
1801	Zhang Heng	1952 SP$_1$	1952 Sep 23	M. Itzigsohn	La Plata	
1802		1964 TW$_1$	1964 Oct 6	Purple Mountain Obs.	Nanking	
1803	Zwicky	1967 CA	1967 Feb 6	P. Wild	Berne	
1804	Chebotarev	1967 GG	1967 Apr 6	T. Smirnova	CRAO	
1805	Dirikis	1970 GD	1970 Apr 1	L. Chernykh	CRAO	
1806		1971 UC	1971 Jun 13		Perth	
1807	Slovakia	1971 QA	1971 Aug 20	M. Antal	Skalnate Pleso	101
1808	Bellerophon	2517 P-L	1960 Sep 24	PLS	Palomar-Leiden	101
1809	Prometheus	2522 P-L	1960 Sep 24	PLS	Palomar-Leiden	101
1810	Epimetheus	4196 P-L	1960 Sep 24	PLS	Palomar-Leiden	101
1811	Bruwer	4576 P-L	1960 Sep 24	PLS	Palomar-Leiden	101
1812	Gilgamesh	4645 P-L	1960 Sep 24	PLS	Palomar-Leiden	101
1813	Imhotep	7589 P-L	1960 Oct 17	PLS	Palomar-Leiden	101
1814	Bach	1931 TS$_1$	1931 Oct 9	K. Reinmuth	Heidelberg	
1815	Beethoven	1932 CE$_1$	1932 Jan 27	K. Reinmuth	Heidelberg	
1816		1936 BD	1936 Jan 29	C. Jackson	Johannesburg (UO)	
1817		1939 MB	1939 Jun 20	C. Jackson	Johannesburg (UO)	
1818	Brahms	1939 PE	1939 Aug 15	K. Reinmuth	Heidelberg	
1819	Laputa	1948 AG	1948 Aug 2	E. L. Johnson	Johannesburg (UO)	
1820	Lohmann	1949 PO	1949 Aug 2	K. Reinmuth	Heidelberg	
1821	Waterman	1950 MB	1950 Jun 24	M. Itzigsohn	La Plata	
1822		1950 OO	1950 Jul 25	Indiana University	Indiana	97
1823	Gliese	1951 RD	1951 Sep 4	Indiana University	Indiana	
1824	Haworth	1952 FM	1952 Mar 30	Indiana University	Indiana	97
1825	Klare	1954 QH	1954 Aug 31	K. Reinmuth	Heidelberg	
1826	Miller	1955 RC$_4$	1955 Sep 14	Indiana University	Indiana	97
1827	Atkinson	1962 RK	1962 Sep 7	Indiana University	Indiana	97
1828	Kashirina	1966 PH	1966 Aug 14	L. Chernykh	CRAO	
1829	Dawson	1967 JJ	1967 May 6	C. Cesco and A. Klemola	El Leoncito	
1830	Pogson	1968 HA	1968 Apr 17	P. Wild	Berne	
1831	Nicholson	1968 HC	1968 Aug 17	P. Wild	Berne	
1832	Mrkos	1969 PC	1969 Aug 11	L. Chernykh	CRAO	
1833	Shmakova	1969 PN	1969 Aug 11	L. Chernykh	CRAO	
1834		1969 QP	1969 Aug 22	L. Kohoutek	Bergedorf	
1835	Gajdariya	1970 OE	1970 Jul 30	T. Smirnova	CRAO	
1836	Komarov	1971 OT	1971 Jul 26	N. Chernykh	CRAO	
1837	Osita	1971 QZ	1971 Aug 11	J. Gibson	El Leoncito	
1838	Ursa	1971 UC	1971 Oct 20	P. Wild	Berne	
1839	Ragazza	1971 UF	1971 Oct 26	P. Wild	Berne	
1840	Hus	1971 UY	1971 Oct 26	L. Kohoutek	Bergedorf	
1841	Masaryk	1971 UO$_1$	1971 Oct 26	L. Kohoutek	Bergedorf	
1842	Hynek	1972 AA	1972 Jan 14	L. Kohoutek	Bergedorf	
1843	Jarmila	1972 AB	1972 Jan 14	L. Kohoutek	Bergedorf	
1844	Susilva	1972 UB	1972 Oct 30	P. Wild	Berne	
1845	Helewalda	1972 UC	1972 Oct 30	P. Wild	Berne	
1846	Bengt	6553 P-L	1960 Sep 24	PLS	Palomar-Leiden	101
1847	Stobbe	1916 YX	1916 Feb 1	H. Thiele	Bergedorf	
1848		1933 QD	1933 Aug 18	E. Delporte	Uccle	
1849	Kresák	1942 AB	1942 Jan 14	K. Reinmuth	Heidelberg	
1850	Kohoutek	1942 EN	1942 Mar 23	K. Reinmuth	Heidelberg	

No	Name	pd	dd	d	dp	Notes
1851	Lacroute	1950 VA	1950 Nov 9	L. Boyer	Algiers	97
1852		1955 CA	1955 Apr 1	Indiana University	Indiana	97
1853		1957 XE	1957 Dec 15	Indiana University	Indiana	
1854	Skvortsov	1968 UE$_1$	1968 Oct 22	T. Smirnova	CRAO	
1855	Korolev	1969 TU$_1$	1969 Oct 8	L. Chernykh	CRAO	
1856	Riẓena	1969 TH$_1$	1969 Oct 8	L. Chernykh	CRAO	
1857	Parchomenko	1971 QS$_1$	1971 Aug 30	T. Smirnova	CRAO	
1858	Lobachevskij	1972 QL	1972 Aug 18	L. Zhuravleva	CRAO	
1859	Kovalevskaya	1972 RS$_2$	1972 Sep 4	L. Zhuravleva	CRAO	
1860	Barbarossa	1973 SK^2	1973 Sep 28	P. Wild	Berne	
1861	Komenský	1970 WB	1970 Nov 11	L. Kohoutek	Bergedorf	102
1862	Apollo	1932 HA	1932 Apr 24	K. Reinmuth	Heidelberg	
1863	Antinous	1948 EA	1948 Mar 7	C. A. Wirtanen	Mount Hamilton	
1864	Daedalus	1971 FA	1971 Mar 24	T. Gehrels	Palomar	
1865	Cerberus	1971 UA	1971 Oct 26	L. Kohoutek	Bergedorf	
1866	Sisyphus	1972 XA	1972 Dec 5	P. Wild	Berne	
1867	Deiphobus	1971 EA	1971 Mar 3	C. Cesco	El Leoncito	
1868	Thersites	2008 P-L	1960 Sep 24	PLS	Palomar-Leiden	101
1869	Philoctetes	4596 P-L	1960 Sep 24	PLS	Palomar-Leiden	101
1870	Glaukos	1971 FF	1971 Mar 24	C. J. Van Houten	Palomar-Leiden	103
1871	Astyanax	1971 FF	1971 Mar 24	C. J. Van Houten	Palomar-Leiden	103
1872	Helenos	1971 FG	1971 Mar 24	C. J. Van Houten	Palomar-Leiden	103
1873	Agenor	1971 FH	1971 Mar 24	L. Kohoutek	Simeis	
1874	Kacivelia	1924 SO	1924 Sep 5	L. Belyavsky	Simeis	
1875		1969 QO	1969 Aug 22	L. Kohoutek	Bergedorf	
1876	Napolitania	1970 BA	1970 Jan 31	C. Kowal	Palomar	103
1877	Marsden	1971 FC	1971 Mar 24	C. J. Van Houten	Palomar-Leiden	
1878		1933 QC	1933 Aug 18	E. Delporte	Johannesburg (LS)	
1879		1935 UN	1935 Oct 16	Van Gent	Johannesburg (LS)	
1880	McCrosky	1940 AN	1940 Jan 13	K. Reinmuth	Heidelberg	
1881	Shao	1940 PC	1940 Aug 3	K. Reinmuth	Heidelberg	
1882		1941 UJ	1941 Oct 15	J. Oterma	Turku	
1883		1942 XA	1942 Dec 4	Y. Väisälä	Turku	
1884	Skip	1943 EB$_1$	1943 Mar 2	M. Laugier	Nice	
1885	Herero	1948 AJ	1948 Feb 6	E. L. Johnson	Johannesburg (UO)	
1886	Lowell	1949 MP	1949 Jun 21	H. L. Giclas	Flagstaff (LO)	104
1887		1950 TD	1950 Oct 9	S. Arend	Uccle	
1888	zu Chong-Zhi	1964 VQ$_1$	1964 Nov 9	Purple Mountain Obs.	Nanking	
1889	Pakhmutova	1968 BE	1968 Jan 24	L. Chernykh	CRAO	
1890	Konoshenkova	1968 CD	1968 Feb 6	L. Chernykh	CRAO	
1891	Gondola	1969 RA	1969 Sep 11	P. Wild	Berne	
1892	Lucienne	1971 SD	1971 Sep 16	P. Wild	Berne	
1893	Jakoba	1971 UD	1971 Oct 20	P. Wild	Berne	
1894	Haffner	1971 UH	1971 Oct 16	L. Kohoutek	Bergedorf	
1895	Larink	1971 UZ	1971 Oct 16	L. Kohoutek	Bergedorf	
1896	Beer	1971 UC$_1$	1971 Oct 26	L. Kohoutek	Bergedorf	
1897	Hind	1971 UE$_1$	1971 Oct 26	L. Kohoutek	Bergedorf	
1898	Cowell	1971 UO	1971 Oct 26	L. Kohoutek	Bergedorf	
1899	Crommelin	1971 UR$_1$	1971 Oct 26	L. Kohoutek	Bergedorf	
1900	Katyusha	1971 YB	1971 Dec 16	T. Smirnova	CRAO	

No	Name	pd	dd	d	dp	Notes
1901	Moravia	1972 AD	1972 Jan 14	L. Kohoutek	Bergedorf	
1902	Shaposhnikov	1972 HU	1972 Apr 18	T. Smirnova	CRAO	
1903	Adzhimushkaj	1972 JH	1972 May 9	T. Smirnova	CRAO	
1904	Massevitch	1972 JM	1972 May 14	T. Smirnova	CRAO	
1905	Ambartsumian	1972 JZ	1972 May 14	T. Smirnova	CRAO	
1906	Naef	1972 RC	1972 Sep 5	P. Wild	Berne	
1907	Rudneva	1972 RC2	1972 Sep 11	N. Chernykh	CRAO	
1908	Pobeda	1972 RL2	1972 Sep 11	N. Chernykh	CRAO	
1909	Alekhin	1972 RK2	1972 Sep 4	L. Zhuravleva	CRAO	
1910	Mikhailov	1972 TZ1	1972 Oct 8	L. Zhuravleva	CRAO	
1911	Schubert	1973 UD	1973 Oct 25	P. Wild	Berne	
1912	Anubis	6534 P-L	1960 Sep 24	PLS	Palomar-Leiden	101
1913	Sekanina	1928 SF	1928 Sep 22	K. Reinmuth	Heidelberg	
1914	Loboda	1930 SB1	1930 Sep 28	H. Van Gent	Johannesburg (LS)	
1915	Quetzalcoatl	1953 EA	1953 Mar 9	A. G. Wilson	Palomar	
1916	Boreas	1953 RA	1953 Sep 1	S. Arend	Uccle	
1917	Cuyo	1968 AA	1968 Jan 1	C. Cesco and A. Samuel	El Leoncito	
1918		1968 UA	1968 Oct 19	G. Soulié	Bordeaux	
1919	Clemence	1971 SA	1971 Sep 16	J. Gibson and C. Cesco	El Leoncito	
1920	Sarmiento	1971 VO	1971 Nov 11	J. Gibson and C. Cesco	El Leoncito	
1921	Pala	1973 SE	1973 Sep 20	T. Gehrels	Palomar	
1922	Zulu	1949 HC	1949 Apr 25	E. L. Johnson	Johannesburg (UO)	
1923		4011 P-L	1960 Sep 24	PLS	Palomar-Leiden	101
1924		4023 P-L	1960 Sep 24	PLS	Palomar-Leiden	101
1925		1934 RY	1934 Sep 2	H. Van Gent	Johannesburg (LS)	
1926		1935 JA	1935 May 2	E. Delporte	Uccle	
1927		1936 FP	1936 Mar 18	R. Suvanto	Turku	
1928		1938 SO	1938 Sep 21	Y. Väisälä	Turku	
1929		1939 BS	1939 Jan 20	Y. Väisälä	Turku	
1930	Lucifer	1964 UA	1964 Oct 29	E. Roemer	Flagstaff (USNO)	
1931		1969 QB	1969 Aug 22	L. Kohoutek	Bergedorf	
1932		1971 UB1	1971 Oct 26	L. Kohoutek	Bergedorf	
1933	Tinchen	1972 AC	1972 Jan 14	L. Kohoutek	Bergedorf	
1934	Jeffers	1972 EA	1972 Dec 2	A. Klemola	El Leoncito	
1935	Lucerna	1973 RB	1973 Sep 2	P. Wild	Mount Hamilton	
1936	Lugano	1973 WD	1973 Nov 24	P. Wild	Berne	
1937	Locarno	1973 YA	1973 Dec 19	P. Wild	Berne	
1938	Lausanna	1974 WC	1974 Apr 19	P. Wild	Berne	
1939	Loretta	1974 CA	1974 Feb 17	C. Kowal	Palomar	
1940	Whipple	1975 CA	1975 Feb 2	Harvard College	Agassiz Station	105
1941	Wild	1931 TN1	1931 Oct 6	K. Reinmuth	Heidelberg	
1942	Jablunka	1972 SA	1972 Sep 30	L. Kohoutek	Bergedorf	
1943	Anteros	1973 EC	1973 Mar 10	J. Gibson	El Leoncito	
1944	Gunter	1925 RA	1925 Sep 14	K. Reinmuth	Heidelberg	
1945		1930 OL	1930 Jul 22	H. Van Gent	Johannesburg (LS)	
1946		1931 FH	1931 Aug 8	K. Reinmuth	Johannesburg (LS)	
1947		1935 EA	1935 Mar 3	Y. Väisälä	Turku	
1948		1935 GL	1935 Apr 1	Y. Väisälä	Johannesburg (UO)	
1949		1936 JH	1936 Jul 8	C. Jackson	Johannesburg (UO)	
1950	Wempe	1942 SO	1942 Mar 23	K. Reinmuth	Heidelberg	

No	Name	pd	dd	d	dp	Notes
1951	Lick	1949 OA	1949 Jul 26	C. A. Wirtanen	Mount Hamilton	
1952		1951 JC	1951 May 3	Indiana University	Indiana	97
1953		1951 UK	1951 Oct 29	Indiana University	Indiana	97
1954		1952 PH	1952 Aug 15	P. Shajn	Simeiz	
1955		1963 SR	1963 Sep 22	Indiana University	Indiana	97
1956	Artek	1969 TX1	1969 Oct 8	L. Chernykh	CRAO	
1957	Angara	1970 GF	1970 Apr 1	L. Chernykh	CRAO	
1958		1970 SB	1970 Sep 24	C. Cesco	El Leoncito	
1959	Karbysheva	1972 NB	1972 Jul 14	L. Zhuravleva	CRAO	
1960	Guisan	1973 UA	1973 Oct 25	P. Wild	Berne	
1961	Dufour	1973 WA	1973 Nov 19	P. Wild	Berne	
1962	Dunant	1973 WE	1973 Nov 24	P. Wild	Berne	
1963	Bezovec	1975 CB	1975 Feb 9	L. Kohoutek	Bergedorf	
1964		2007 P-L	1960 Sep 24	PLS	Palomar-Leiden	101
1965		2521 P-L	1960 Sep 24	PLS	Palomar-Leiden	101
1966		2552 P-L	1960 Sep 24	PLS	Palomar-Leiden	101
1967	Menzel	1905 RY	1905 Nov 1	M. Wolf	Heidelberg	
1968	Mehlreter	1932 BK	1932 Jan 29	K. Reinmuth	Heidelberg	
1969		1935 CG	1935 Feb 3	S. Arend	Uccle	
1970		1954 ER	1954 Mar 12	M. Itzigsohn	La Plata	
1971	Hagihara	1955 RD1	1955 Sep 14	Indiana University	Indiana	97
1972	Yi Xing	1964 VQ1	1964 Nov 9	Purple Mountain Obs.	Nanking	
1973		1968 OA	1968 Jul 18	C. Torres and S. Cofre	Cerro el Roble	
1974		1968 OE	1968 Jul 18	C. Torres and S. Cofre	Cerro el Roble	
1975	Pikelner	1969 PH	1969 Aug 11	L. Chernykh	CRAO	
1976	Kaverin	1970 GC	1970 Apr 1	L. Chernykh	CRAO	
1977	Shura	1970 QY	1970 Aug 30	T. Smirnova	CRAO	
1978		1971 LD	1971 Jun 13	—	Perth	
1979		2006 P-L	1960 Sep 24	PLS	Palomar-Leiden	101
1980	Tezcatlipoca	1950 LA	1950 Jun 19	A. G. Wilson	Palomar	106
1981	Midas	1973 EA	1973 Mar 6	C. Kowal	Palomar	
1982	Cline	1975 VA	1975 Nov 4	E. F. Helin	Palomar	
1983	Bok	1975 LB	1975 Jun 9	E. Roemer	Tucson	
1984		1926 TH	1926 Oct 10	S. Belyavsky	Simeiz	
1985	Hopmann	1929 AE	1929 Jan 13	K. Reinmuth	Heidelberg	
1986		1935 SV1	1935 Sep 28	H. Van Gent	Johannesburg (LS)	
1987		1952 SV	1952 Sep 11	P. Shajn	Simeiz	
1988	Delores	1952 SW	1952 Sep 28	Indiana University	Indiana	97
1989		1955 FG	1955 Mar 20	A. Paroubek	Skalnate Pleso	107
1990		1956 EE	1956 Mar 9	Indiana University	Indiana	97
1991		1967 JL	1967 May 6	C. Cesco and A. Klemola	El Leoncito	
1992	Jablunka	1968 OD	1968 Jul 18	C. Torres and S. Cofre	Cerro el Roble	
1993		1968 OH1	1968 Jul 25	G. Plougin and Y. Belyaev	Cerro el Roble	
1994		1961 TE	1961 Oct 4	Indiana University	Indiana	97
1995	Hajek	1971 UP1	1971 Oct 26	L. Kohoutek	Bergedorf	
1996	Adams	1961 UA	1961 Oct 16	Indiana University	Indiana	97
1997	Leverrier	1963 DK1	1963 Oct 16	Indiana University	Indiana	97
1998	Titius	1938 DR	1938 Feb 25	A. Bohrmann	Heidelberg	
1999	Hirayama	1973 DR	1973 Feb 27	L. Kohoutek	Bergedorf	
2000	Herschel	1960 OA	1960 Jul 29	J. Schubart	Sonneberg	

No	Name	pd	dd	d	dp	Notes
2001	Einstein	1973 EB	1973 Mar 5	P. Wild	Berne	
2002	Euler	1973 QQ$_1$	1973 Aug 29	T. Smirnova	CRAO	
2003	Harding	6559 P-L	1960 Sep 24	PLS	Palomar-Leiden	101
2004	Lexell	1973 SV$_2$	1973 Sep 2	N. Chernykh	CRAO	
2005	Hencke	1973 RA	1973 Sep 2	P. Wild	Berne	
2006	Polonskaya	1973 SB$_3$	1973 Sep 22	N. Chernykh	CRAO	
2007		1963 SQ	1963 Sep 22	Indiana University	Indiana	97
2008	Konstitutsiya	1973 SV$_4$	1973 Sep 27	L. Chernykh	CRAO	
2009	Voloshina	1968 UL	1968 Oct 22	T. Smirnova	CRAO	
2010	Chebyshev	1969 TL$_4$	1969 Oct 13	B. Burnasheva	CRAO	
2011	Veteraniya	1970 QB$_1$	1970 Aug 30	Purple Mountain Obs.	Nanking	
2012	Guo Shou-Jing	1964 TE$_2$	1964 Oct 9	—	Nanking	
2013		1971 UH$_4$	1971 Oct 22	—	Nanking	
2014	Vasilevskis	1973 JA	1973 May 2	A. Klemola	Mount Hamilton	
2015	Kachuevskaya	1972 RA$_3$	1972 Sep 4	L. Zhuravleva	CRAO	
2016	Heinemann	1938 SE	1938 Sep 18	A. Bohrmann	Heidelberg	
2017	Wesson	1903 NC	1903 Jul 2	M. Wolf	Heidelberg	
2018	Schuster	1931 UC	1931 Oct 21	K. Reinmuth	Heidelberg	
2019		1935 SX$_1$	1935 Sep 28	H. Van Gent	Johannesburg (LS)	97
2020		1936 FR	1936 Mar 18	Y. Väisälä	Turku	97
2021	Poincaré	1936 MA	1936 Jun 26	L. Boyer	Algiers	
2022	West	1938 CK	1938 Feb 7	K. Reinmuth	Heidelberg	
2023	Asaph	1952 SA	1952 Sep 16	Indiana University	Indiana	97
2024		1952 UR	1952 Oct 23	Indiana University	Indiana	97
2025		1953 LG	1953 Jun 6	J. Churms	Johannesburg (UO)	
2026	Cottrell	1955 FF	1955 Mar 30	Indiana University	Indiana	97
2027		1964 VR1	1964 Nov 9		Nanking	
2028		1968 OB$_1$	1968 Jul 18	C. Torres and S. Cofre	Cerro el Roble	
2029		1969 RB	1969 Sep 11	P. Wild	Berne	
2030	Belyaev	1969 RB	1969 Oct 8	L. Chernykh	CRAO	
2031	Bam	1969 TC$_2$	1969 Oct 8	L. Chernykh	CRAO	
2032	Ethel	1970 OH	1970 Jul 30	T. Smirnova	CRAO	
2033		1973 CA	1973 Feb 5	P. Wild	Berne	
2034		1973 EE	1973 Mar 5	P. Wild	Berne	
2035	Stearns	1973 SC	1973 Sep 21	J. Gibson	El Leoncito	
2036	Sheragul	1973 SY$_2$	1973 Sep 22	N. Chernykh	CRAO	
2037		1973 UB	1973 Oct 25	P. Wild	Berne	
2038		1973 WF	1973 Nov 24	P. Wild	Berne	
2039	Payne-Gaposchkin	1974 CA	1974 Feb 14	Harvard College	Agassiz Station	105
2040		1964 RA	1964 Oct 8	P. Wild	Berne	
2041		2523 P-L	1960 Sep 24	PLS	Palomar-Leiden	101
2042		4633 P-L	1960 Sep 24	PLS	Palomar-Leiden	101
2043		1936 TH	1936 Nov 12	G. Kulin	Budapest	
2044		1950 VR	1950 Nov 8	C. A. Wirtanen	Mount Hamilton	
2045		1964 BG	1964 Oct 8	P. Wild	Nanking	105
2046	Deornik	1968 UD$_1$	1968 Oct 22	T. Smirnova	CRAO	
2047		1971 UA$_1$	1971 Oct 26	L. Kohoutek	Bergedorf	
2048	Deornik	1973 QA	1973 Aug 27	E. F. Helin	Palomar	
2049		1973 SH	1973 Sep 25	T. Gehrels	Palomar	
2050	Francis	1974 KA	1974 May 28	E. F. Helin	Palomar	

No	Name	pd	dd	d	dp	Notes
2051	Chang	1976 UC	1976 Oct 23	Harvard College	Agassiz Station	105
2052		1976 UN	1976 Oct 24	R. M. West	ESO-La Silla	
2053		1976 UO	1976 Oct 24	R. M. West	ESO-La Silla	
2054		4097 P-L	1960 Sep 24	PLS	Palomar-Leiden	101
2055		1974 DB	1974 Feb 19	L. Kohoutek	Bergedorf	
2056	Nancy	1909 HP	1909 Oct 15	J. Helffrich	Heidelberg	
2057	Rosemary	1934 KQ	1934 Sep 7	K. Reinmuth	Heidelberg	
2058		1938 BH	1938 Jan 22	G. Kulin	Budapest	
2059		1963 UA	1963 Oct 16	Indiana University	Indiana	97
2060	Chiron	1977 UB	1977 Oct 18	C. Kowal	Palomar	
2061	Anza	1960 UA	1960 Oct 22	H. L. Giclas	Flagstaff (LO)	
2062	Aten	1976 AA	1976 Jan 7	E. F. Helin	Palomar	101
2063	Bacchus	1977 HB	1977 Apr 24	C. Kowal	Palomar	
2064	Thomsen	1942 RQ	1942 Sep 11	L. Oterma	Turku	97
2065		1959 RN	1959 Sep 9	Indiana University	Indiana	97
2066		1934 LB	1934 Jun 4	C. Jackson	Johannesburg (UO)	
2067	Aksnes	1936 DD	1936 Feb 23	Y. Väisälä	Turku	
2068	Dangreen	1948 AD	1948 Jan 8	M. Laugier	Nice	
2069		1955 FT	1955 Mar 29	Indiana University	Indiana	97
2070		1964 TQ	1964 Oct 17	Indiana University	Indiana	97
2071		1971 QS	1971 Aug 18	T. Smirnova	CRAO	
2072		1973 QE$_2$	1973 Aug 31	L. Kohoutek	Bergedorf	
2073		1974 AD	1974 Feb 21	E. F. Helin	Palomar	105
2074	Shoemaker	1974 UA	1974 Oct 29	E. F. Helin	Palomar	
2075		1974 VA	1974 Nov 9	—	El Leoncito	
2076	Levin	1974 VA	1974 Nov 16	Harvard College	Agassiz Station	105
2077		1974 VR1	1974 Dec 18	Purple Mountain Obs.	Nanking	
2078		1974 AD	1974 Jan 12	Purple Mountain Obs.	Nanking	
2079	Jacchia	1976 DG	1976 Feb 23	Harvard College	Agassiz Station	105
2080		1976 DG	1976 Feb 27	P. Wild	Berne	
2081		1976 DH	1976 Feb 27	P. Wild	Berne	
2082		7588 P-L	1960 Oct 17	ELS	Palomar-Leiden	101
2083	Smither	1973 WB	1973 Nov 29	E. F. Helin	Palomar	
2084		1973 UR	1973 Oct 24	S. Arend	Uccle	
2085		1965 YA	1965 Dec 20	Purple Mountain Obs.	Nanking	
2086		1966 BC	1966 Jan 20	Indiana University	Indiana	97
2087		1975 YC	1975 Dec 28	P. Wild	Berne	
2088		1976 DJ	1976 Feb 27	P. Wild	Berne	
2089		1977 VF	1977 Nov 9	M. G. Thomas	Flagstaff (LO)	
2090		1978 EA	1978 Mar 12	T. Urata	Yakiimo Station	
2091		1941 HO	1941 Apr 26	Y. Väisälä	Turku	
2092		1969 UP	1969 Oct 16	L. Chernykh	CRAO	
2093		1971 HX	1971 Apr 28	T. Smirnova	CRAO	
2094		1971 TC$_2$	1971 Oct 12	V. Piksaev	CRAO	
2095		6036 P-L	1960 Sep 24	PLS	Palomar-Leiden	101
2096		1939 UC	1939 Oct 18	Y. Väisälä	Turku	
2097		1953 PV	1953 Aug 11	K. Reinmuth	Heidelberg	
2098		1972 QE	1972 Aug 18	L. Zhuravleva	CRAO	
2099	Öpik	1977 VB	1977 Nov 8	E. F. Helin	Palomar	
2100	Ra-Shalom	1978 RA	1978 Sep 10	E. F. Helin	Palomar	108

No	Name	pd	dd	d	dp	Notes
2101	Adonis	1936 CA	1936 Feb 12	E. Delporte	Uccle	
2102	Tantalus	1975 YA	1975 Dec 27	C. Kowal	Palomar	
2103		1960 FL	1960 Mar 20	——	La Plata	
2104	Toronto	1963 PD	1963 Aug 15	K. W. Kamper	Tautenburg	109
2105	Gudy	1976 DA	1976 Feb 29	H.-E. Schuster	ESO-La Silla	
2106		1936 UF	1936 Oct 21	M. Laugier	Nice	
2107		1941 VA	1941 Nov 12	L. Oterma	Turku	
2108		1948 TR_1	1948 Oct 4	P. Shajn	Simeïs	
2109		1950 TH_2	1950 Oct 13	S. Arend	Uccle	
2110		1962 RD	1962 Sep 7	Indiana University	Indiana	97
2111		1969 LG	1969 Jun 13	T. Smirnova	CRAO	
2112		1972 MP	1972 Jul 13	T. Smirnova	CRAO	
2113		1972 RJ_2	1972 Sep 11	N. Chernykh	CRAO	
2114	Wallenquist	1976 HA	1976 Apr 19	C.-I. Lagerkvist	Mount Stromlo	110
2115		1976 UD	1976 Oct 24	R. M. West	ESO-La Silla	
2116		1976 UM	1976 Oct 24	R. M. West	ESO-La Silla	
2117		1978 AC	1978 Jan 9	R. M. West	ESO-La Silla	
2118		1978 PB	1978 Aug 5	H. L. Giclas	Anderson Mesa	
2119		1930 QG	1930 Aug 30	M. Wolf and M. Ferrero	Heidelberg	
2120		1967 RM	1967 Sep 9	T. Smirnova	CRAO	
2121		1971 ME	1971 Jun 27	T. Smirnova	CRAO	
2122		1971 XB	1971 Dec 14	T. Smirnova	CRAO	
2123		1973 SL_2	1973 Sep 22	N. Chernykh	CRAO	
2124		1974 MK	1974 Jun 20	——	El Leoncito	
2125		2005 P-L	1960 Sep 24	PLS	Palomar-Leiden	101

NOTES

1. Planet 9. Apr 26 is date of first measurement.
2. Planet 20. Independent discovery the next night by J. Chacornac, Marseille, was announced first.
3. Planet 50. Independent discovery Oct 19 by R. Luther, Düsseldorf, was announced first.
4. Planet 66. Apr 10, given by some authorities, is the civil date.
5. Planet 69. Apr 29 is date of first measurement.
6. Planet 71. Aug 14 is date of first measurement.
7. Planet 80. May 3 is date of first measurement.
8. Planet 100. Independent discovery July 18 by C. Wolf, Paris, was announced first.
9. Planets discovered by Prosper and Paul Henry. In their cooperative program it will probably never be known which of the Henry brothers discovered which planet. With strict impartiality, these discoveries were announced as having been made alternatively by Prosper and by Paul.
10. Planet 154. Initial announcement gave Nov 6.
11. Planet 179. Initial announcement gave Nov 12.
12. Planet 220. Initial announcement gave May 18.
13. Planet 264. Discovery possibly as early as Dec. 17.
14. Planet 268. Jun 9 is date of first measurement.
15. Planet 323. Dec 20 observation seems to be a prediscovery.
16. Planet 433. Independent discovery earlier the same night by A. Charlois, Nice.
17. Planet 457. Discovery telegram gave only Sep 16 observation.
18. Planet 468. Already photographed on Jan 13; it is not clear whether this should be regarded as a prediscovery image.
19. Planet 471. May 18 prediscovery observation by L. Carnera is in error.
20. Planets 525, 1171. The original planet 525 Adelaide, discovered 1904 Mar 14 by M. Wolf, Heidelberg, was observed only during the discovery opposition. In 1958 A. Patry, Nice, found that this object is identical with planet 1171 Rusthawelia, discovered 1930 Oct 3 by S. Arend, Uccle. This planet retains the latter designation 1171 Rusthawelia. The older designation 525 Adelaide has been given to a minor planet discovered 1908 Oct 21 by J. H. Metcalf and newly catalogued.
21. Planet 535. May 9, given by some authorities, is erroneous.
22. Planet 602. Independent discovery Feb 22 by A. Kopff, Heidelberg, was announced first.
23. Planet 673. Independent discovery Sep 21 by A. Kopff, Heidelberg, was announced first.
24. Planet 694. Independent discovery Nov 9 by J. Helffrich, Heidelberg, was announced first.
25. Planets 699, 700. Jun 3, given by some authorities, is erroneous.
26. Planets 715, 933. The original planet 715 Transvaalia, discovered 1911 Apr 22 by H. E. Wood, was recovered on 1920 Apr 23 as 1920 GZ and given the designation 933 Susi. The identity was discovered in 1928 and the older designation 715 Transvaalia was retained. In the same year the designation 933 Susi was assigned to a newly-discovered planet.
27. Planet 749. and G. Neujmin
28. Planets 752, 753. and M. Belyavsky.
29. Planets 758, 790. and K. Van der Spuy.
30. Planet 760. Independent discovery earlier the same night by G. Neujmin, Simeïs.
31. Planet 793. Independent discovery Apr 17 by J. H. Metcalf, Taunton, was announced first.
32. Planet 812. Independent discovery Sep 11 by M. Wolf, Heidelberg, was announced first.
33. Planet 824. Independent discovery Apr 1 by M. Wolf, Heidelberg, was announced first.
34. Planet 825. Independent discovery Apr 3 by M. Wolf, Heidelberg, was announced first.
35. Planet 829. Independent discovery Aug 31 by M. Wolf, Heidelberg, was announced first.
36. Planet 830. Independent discovery Sep 3 by M. Wolf, Heidelberg, was announced first.
37. Planet 848. Independent discovery Sep 9 by M. Wolf, Heidelberg, was announced first.
38. Planet 855. Independent discovery Apr 28 by M. Wolf, Heidelberg, was announced first.
39. Planet 859. Prediscovery observation Sep 20 by F. Gonnessiat, Algiers.
40. Planets 864, 1078. In 1958 A. Patry, Nice, found that planet 864 Aase, discovered 1917 Feb 13 by M. Wolf, Heidelberg, is identical with 1078 Mentha, discovered 1926 Dec 7 by K. Reinmuth, Heidelberg. This object retains the latter designation 1078 Mentha. In 1974 the designation 864 Aase was assigned to planet 1921 KE, discovered 1921 Sep 30 by K. Reinmuth and newly catalogued.
41. Planet 869. Independent discovery May 10 by M. Wolf, Heidelberg, was announced first.
42. Planet 878. and H. Shapley.
43. Planet 882. Independent discovery Aug 18 by M. Wolf, Heidelberg, was announced first.
44. Planet 883. Independent discovery earlier the same night by R. Schorr, Bergedorf. Both Wolf and Schorr initially thought the object was Encke's Comet, and Wolf was the first to realize that it was not.

45. Planet 885. Independent discovery Sep 26 by M. Wolf, Heidelberg, was announced first.
46. Planet 886. Independent discovery Nov 12 by M. Harwood, Nantucket.
47. Planet 887. The Feb 3 observation given by some authorities is a rediscovery.
48. Planet 931. and R. Jekhovsky. Independent discovery Mar 21 by K. Reinmuth, Heidelberg, was announced first.
49. Planet 942. Independent discovery later the same night by A. Schwassmann, Bergedorf.
50. Planet 943. Oct 12 observation by A. Schwassmann, Bergedorf, was apparently a prediscovery.
51. Planet 969. Independent discovery Nov 25 by K. Reinmuth, Heidelberg, was announced first.
52. Planet 978. Independent discovery May 30 by M. Wolf, Heidelberg, was announced first.
53. Planet 990. Independent discovery Dec 14 by M. Wolf, Heidelberg, was announced first.
54. Planet 1004. Independent discovery Sep 13 by K. Reinmuth, Heidelberg, was announced first.
55. Planet 1033. Independent discovery Sep 5 by S. Belyavsky, Simeïs, was announced first.
56. Planet 1051. Sep 15, given by some authorities, refers to local mean time.
57. Planet 1057. Independent discovery Aug 19 by K. Reinmuth, Heidelberg, was announced first.
58. Planet 1067. Sep 10, given by some authorities, refers to the local meridian.
59. Planet 1090. Independent discovery Feb 24 by K. Reinmuth, Heidelberg, was announced first.
60. Planets 1095, 1449. In 1966 C. Bardwell, Cincinnati, discovered that minor planet 1928 RC, discovered 1928 Feb 24 by K. Reinmuth and originally designated 1095 Tulipa, is identical with planet 1449 Virtanen, discovered 1938 Feb 20. This planet retains the latter designation 1449 Virtanen. The designation 1095 Tulipa has been reassigned to planet 1926 GS, discovered 1926 Apr 14 by K. Reinmuth and newly catalogued.
61. Planet 1098. Independent discovery Sep 9 by M. Wolf, Heidelberg, was announced first.
62. Planet 1099. Independent discovery the next night by M. Wolf, Heidelberg, was announced first.
63. Planet 1113. Independent discovery Aug 25 by K. Reinmuth, Heidelberg, was announced first.
64. Planet 1118. Independent discovery Aug 30 by K. Reinmuth, Heidelberg, was announced first.
65. Planet 1125. The original planet 1125 China, discovered 1928 Oct 25 as 1928 UF by Y. C. Chang at Williams Bay, has not been recovered since its discovery opposition. It was believed to have been recovered as 1957 UN$_1$ on 1957 Oct 30 by Chang at Nanking, but this identity is now known to be incorrect. In agreement with the discoverer, the designation 1125 will be assigned henceforth to the planet 1957 UN$_1$.
66. Planet 1137. Independent discovery Nov 21 by K. Reinmuth, Heidelberg, was announced first.
67. Planet 1148. The alternative spelling "Rarahu" has been given in the annual ephemeris volume from 1968 onward.
68. Planet 1164. Independent discovery Mar 20 by W. Baade, Bergedorf, was announced first.
69. Planet 1166. Independent discovery Jun 29 by K. Reinmuth, Heidelberg, was announced first.
70. Planet 1192. and A. Wachmann.
71. Planet 1202. Independent discovery Sep 15 by K. Reinmuth, Heidelberg, was announced first.
72. Planet 1223. Independent discovery Oct 11 by F. Rigaux, Uccle, was announced first.
73. Planet 1224. Independent discovery Sep 23 by E. Delporte, Uccle, was announced first.
74. Planet 1226. Some authorities give Apr 23.
75. Planet 1243. Subsequently reported as discovered on Apr 4 by E. L. Johnson.
76. Planet 1246. Subsequently reported as discovered on Jul 4.
77. Planet 1266. Independent discovery Jan 24 by G. Neujmin, Simeïs, was announced first.
78. Planet 1267. Independent discovery Apr 28 by K. Reinmuth, Heidelberg, was announced first.

79. Planet 1270. Independent discovery Dec 20 by M. Wolf, Heidelberg, was announced first.
80. and A. Wachmann.
81. Planets 787 and 1317. In 1938 G. Neujmin found that planet 1317 = 1934 FD, discovered 1934 Mar 19 by C. Jackson, Johannesburg (UO), is identical with 787 Moskva, discovered 1914 Apr 20 by Neujmin. The number 1317 was later given to planet 1935 KC.
82. Planet 1335. Independent discovery Sep 13 by E. Delporte, Uccle, was announced first.
83. Planet 1367. Independent discovery Jul 6 by K. Reinmuth, Heidelberg, was announced first.
84. Planet 1379. Independent discovery later the same night by P. Djurkovic, Uccle, was announced first. The alternative spelling "Lomonossowa" has been given in the annual ephemeris volume.
85. Planet 1386. Independent discovery Aug 2 by E. Delporte, Uccle, was announced first.
86. Planet 1390. Practically simultaneous independent discovery by C. Jackson, Johannesburg (UO).
87. Planet 1453. Confirmation Mar 23, shortly after an independent discovery by G. Neujmin, Simeïs.
88. Planet 1459. Independent discovery Nov 6 by A. Patry, Nice, was announced first.
89. Planet 1468. Independent discovery Aug 2 by L. Boyer, Algiers, was announced first.
90. Planet 1481. Independent discovery Feb 22 by Y. Väisälä, Turku, was announced first.
91. Planet 1482. Independent discovery Feb 26 by Y. Väisälä, Turku, was announced first.
92. Planet 1492. Independent discovery Oct 30 designated 1938 UO, by G. Kulin, Budapest.
93. Planet 1508. Independent discovery Oct 21 by E. Delporte, Uccle, was announced first.
94. Planet 1526. Confirmation Oct 20 by L. Oterma, Turku.
95. Planet 1530. Independent discovery Sep 18 by E. Delporte, Uccle, was announced first.
96. Planet 1540. Nov 16 observation seems to be a prediscovery.
97. Planets discovered by the Indiana Asteroid Program, Goethe Link Observatory, University of Indiana. This program was conceived and directed by F. K. Edmondson. The plate were blinked and measured astrometrically by B. Potter and, following her retirement, by D. Owings, and the photometry was performed under the direction of T. Gehrels. During the years 1947-1967 in which the plates were exposed a large number of people participated in various aspects of the program.
98. Planet 1640. Observation Aug 25 by C. Rogati, La Plata, should be considered a prediscovery.
99. Planet 1671. Observation 1926 Oct 5 as 1926 TH by K. Reinmuth, Heidelberg, should be considered a prediscovery.
100. Planet 1654. and R. Groeneveld.
101. These planets have all been discovered as a result of the Palomar-Leiden survey of faint minor planets and subsequently identified with planets observed at other oppositions. In Sep and Oct 1960 T. Gehrels exposed 130 plates with the 122 cm Schmidt camera at Palomar. In the following years C. J. Van Houten and I. Van Houten-Groeneveld measured these plates astrometrically and photometrically at Leiden. P. Herget, Cincinnati, computed the orbits of the planets found on the NORC computer, Dahlgren, Virginia, USA.
102. Planet 1867. First measurement Mar 26.
103. Planets 1870-1873, 1877. Planets discovered by C. J. Van Houten and I. Van Houten-Groeneveld Leiden, on plates exposed by T. Gehrels in a survey of faint Trojans with the 122 cm Schmidt camera at Palomar.
104. Planet 1866. and R. D. Schaldach.
105. Planet 1896. discovered by Harvard College at Agassiz station. The principal observers are R. E. McCrosky, C.-Y. Shao, G. Schwartz, and J. H. Bulger, with some assistance from other.
106. Planet 1985. and A. A. E. Wallenquist.
107. Planet 1986. and R. Podstanicka.
108. Planet 2095. and E. M. Shoemaker.
109. Planet 2104. on plates taken by S. Van den Bergh.
110. Planet 2114. at the Uppsala Southern Station.

Glossary,

Acknowledgments,

and Index

GLOSSARY[a]

a   semimajor axis of an orbit.

Å   Ångstrom = 10^{-8} cm.

α phase angle, which is the angle at the asteroid between the radius vectors to Earth and to the sun.

absolute   absolute magnitude, $B(1,0)$ at unit distances and zero
magnitude   phase, is defined by

$$B = B(1,0) + 5 \log r\Delta + F(\alpha)$$

where r is the distance from the sun and Δ from the earth, and $F(\alpha)$ is the phase function.

achondrite   a stony meteorite that lacks chondrules; most achondrites appear to be the products of igneous differentiation.

aeon (AE) 10^9 years.

Amor asteroids   asteroids having perihelion distance 1.017 AU $< q <$ 1.3 AU.

Apollo   asteroids having semimajor axis $a > 1.0$ AU, and perihelion
asteroids   distance $q < 1.017$ AU.

arcsec   second of arc.

aspect   angle between the rotation axis of the body and the radius vector to the earth.

[a]We have used various definitions from *Glossary of Astronomy of Astrophysics* by J. Hopkins (by permission of the University of Chicago Press, © 1976 by The University of Chicago) and from *Astrophysical Quantities* by C. W. Allen (London: Athlone Press, 1973).

asteroid — a moving object of stellar appearance, without any trace of cometary activity (but note that the same definition may also apply to a satellite or to a cometary core).

asteroid belt — a region of space lying between Mars and Jupiter, where the great majority of the asteroids is found.

Aten asteroids — asteroids having semimajor axis $a < 1.0$ AU, and aphelion distance $Q > 0.983$ AU.

AU — astronomical unit $= 1.496 \times 10^{13}$ cm.

$B(a,0)$ — the mean opposition magnitude as defined by

$$B(a,0) = B(1,0) + 5 \log a(a-1).$$

$B(1,0)$ — absolute magnitude (q.v.).

BEND — is a measure of the curvature on the visible part of the spectrum (see Chapman and Gaffey in Part VII of this book).

blackbody — an idealized body which absorbs radiation of all wavelengths incident on it. (Because it is a perfect absorber, it is also a perfect emitter.) The radiation emitted by a blackbody is a function of temperature only.

blackbody radiation — sometimes called thermal radiation. Radiation whose spectral intensity distribution is that of a blackbody in accordance with Planck's law.

blocking temperature — the temperature below which permanent magnetization is locked into ferromagnetic materials for long times.

Bond albedo — fraction of the total incident light reflected by a spherical body. It is equal to the phase integral, multiplied by the ratio of its brightness at zero phase angle to the brightness it would have if it were a perfectly diffusing disk.

breccia — rock composed of broken rock fragments cemented together by finer-grained material.

c — speed of light in a vacuum $= 3 \times 10^{10}$ cm sec^{-1}.

C_3 — injection energy into Earth-escape hyperbola; km^2 sec^{-2}.

CAI — calcium-aluminum rich inclusion.

chondrite — a stony meteorite usually characterized by the presence of chondrules, which are small spherical grains, usually composed of iron, aluminum, or magnesium silicates. Carbonaceous chondrites are characterized by the presence of carbon compounds, while their Type I or C1 contains no chondrules.

chondrules   small spherical grains varying from microscopic size to the size of a pea, usually composed of iron, aluminum, or magnesium silicates. They occur in abundance in primitive stony meteorites.

clast   a rock fragment produced by mechanical weathering of a larger rock and included in another rock.

cm   centimeter.

color index   the difference in magnitudes between any two spectral regions. Color index is always defined as the short-wavelength magnitude minus the long-wavelength magnitude. In the UBV system, the color of index for an AO star is defined as $B–V = U–B = 0$; it is negative for hotter stars and positive for cooler ones.

coma   the spherical region of diffuse gas, about 150,000 km in diameter, which surrounds the nucleus of a comet. Together, the coma and the nucleus form the comet's head.

comet   a diffuse body of gas and solid particles (such as CN, C_2, NH_3, and OH), which orbits the sun. The orbit is usually highly elliptical or even parabolic (average perihelion distance less than 1 AU; average aphelion distance, roughly 10^4 AU). Comets are unstable bodies with masses on the order of 10^{18} g whose average lifetime is about 100 perihelion passages. Periodic comets comprise only $\sim 4\%$ of all known comets.

comets, family of   an aggregation of comets with similar aphelion distances (e.g., Jupiter's family).

comets, group of   an aggregation of comets with identical orbits.

comets, nomenclature   when a newly discovered comet is confirmed, the IAU assigns an interim designation consisting of the year of discovery followed by a lowercase letter in order of discovery for that year. Frequently the discoverer's name precedes the designation – e.g., comet Bennett 1969i. If a reliable orbit is later established, the comet is given a permanent designation consisting of the year of perihelion passage followed by a Roman numeral in order of perihelion passage – e.g., comet Bennett 1970 II. If the comet is periodic, the letter P followed by the discoverer's (or computer's) name is used — e.g., comet 1910 II P/Halley.

commensurate orbits	a term applied to two bodies orbiting around a common barycenter when the period of one is an integral multiple of that of the other.
comminution	the reduction of a rock to progressively smaller particles by weathering, impacts, erosion, etc.
Coriolis effect	the acceleration which a body in motion experiences when observed in a rotating frame. This force acts at right angles to the direction of the angular velocity.
counterglow	also called Gegenschein. A very faint glow, $\sim 10°$ across, which can occasionally be seen, but only under the best of conditions, in a part of the sky opposite the sun.
CSM	classification of asteroids as described in the chapter by Zellner.
DEPTH	a measure of the Fe^{2+} absorption near 0.9 μm (see Chapman and Gaffey in Part VII of this book).
differentiation	the process of developing more than one rock type from a common reservoir of molten rock.
e'	proper eccentricity. Eccentricity (q.v.) corrected for effects of planetary perturbation.
eccentricity, e	of an elliptical orbit. The amount by which the orbit deviates from circularity: $e = c/a$, where c is the distance from the center to a focus and a is the semimajor axis. arc sin $e = \phi$ is also used.
elongation	the angle planet-Earth-Sun. Eastern elongations appear east of the sun in the evening; western elongations, west of the sun in the morning. An elongation of $0°$ is called conjunction; one of $180°$ is called opposition; and one of $90°$ is called quadrature.
ephemeris	(pl., ephemerides) a list of computed positions occupied by a celestial body over successive intervals of time.
Ephemeris	or EMP: *Ephemerides of Minor Planets,* published yearly by the Russian Academy of Sciences, Institute of Theoretical Astronomy, Leningrad, U.S.S.R.
eV	electron volt = 1.60×10^{-12} ergs.
$F(\alpha)$	phase function (see "absolute magnitude").

fall
a meteorite recovered immediately after falling to the earth; in some cases also applied to meteorites recovered months or years after observations of fall phenomena.

feldspars
common aluminous silicate minerals in meteorites and other rocks. *Plagioclase* feldspars are members of a solid solution series which varies continuously from sodium-rich to calcium-rich compositions. Names have been given to six composition ranges:

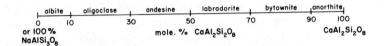

find
a meteorite which cannot be associated with an observed fall; find should also be applied to weathered meteorites recovered months or years after an observed fall.

fireball
see "meteor."

Fo$_{65}$
Forsterite concentration 65%. See olivine.

Fourier analysis
the analysis of a periodic function into its simple harmonic components.

Fourier theorem
any finite periodic motion may be analyzed into components, each of which is a simple harmonic motion of definite and determinable amplitude and phase.

FTS
Fourier transform spectroscopy (see the chapter by Larson and Veeder).

g
gram.

G
gauss = 1 oersted = 79.58 amp-turn/m.

Galilean satellites
the four largest satellites of Jupiter: Io (J I), Europa (J II), Ganymede (J III), and Callisto (J IV).

genomict breccia
breccia (q.v.) of which the components originated in distinct but genetically closely related rocks.

geometric albedo
ratio of the flux received from a planet to that expected from a perfectly reflecting Lambert disk of the same size at the same distance at zero phase angle.

GeV
giga electron volt = 10^9 eV.

gravitational constant, G
the constant of proportionality in the attraction between two unit masses a unit distance apart. $G = 6.668 \times 10^{-8}$ dyn cm^2 g^{-2}.

Greenschist facies metamorphism	a term referring to metamorphic rock containing abundant green minerals aligned in preferred orientations. On Earth these rocks are produced by subjecting surface rocks such as basalts to low-to-intermediate temperatures ($300-500°C$) and low-to-moderate hydrostatic pressures (3–8 kbars).
Gyr	giga-year $= 10^9$ yr.
Halley's comet	Its orbit was computed by Edmund Halley in 1704, at which time he predicted that the bright comet of 1682 would return in 1758 (Halley died in 1742, before he could see his prediction verified). Records of Halley's comet ($a = 17.8$ AU, $e = 0.967$, $i = 162°.3$, $P = 86.2$ yr, perihelion distance 0.587 AU) have been traced back to 240 B.C. Appearances: 1910, 1986, etc.
Hirayama family	asteroids with similar a, e', and i' (q.v.).
i	inclination of the orbit. The angle between an asteroid's orbit and the plane of the ecliptic (or between a satellite's orbit and the planet's equatorial plane).
i'	proper inclination. Inclination (q.v.) corrected for effects of planetary perturbation.
IAU	International Astronomical Union.
IAUC	International Astronomical Union Circulars (see Marsden's chapter).
IR	infrared.
IRAS	Infrared Astronomical Satellite (see the chapter by Morrison and Niehoff).
isochron	the term denotes a straight line containing all the points representing the same time or age. In geochronology it usually refers to a line constructed either by plotting the ratio $^{87}Sr/^{86}Sr$ as a function of the ratio $^{87}Rb/^{86}Sr$ or by plotting the ratio $^{207}Pb/^{204}Pb$ or $^{208}Pb/^{204}Pb$ as a function of the ratio $^{206}Pb/^{204}Pb$ for minerals or rocks of the same age.
JD	Julian Day number.
JSP	Jupiter scattered planetesimals.
Jy	Jansky; 1 Jy $= 10^{-26}$ W m^{-2} Hz^{-1}.

k Boltzmann constant = 1.38×10^{-16} erg deg^{-1}; alternately, = 8.62×10^{-5} eV deg^{-1}.

K   degrees Kelvin.

kcal   kilo-calorie = 4.185×10^{10} erg.

Kepler's laws 1. Each planetary orbit is an ellipse with the sun at one focus. 2. (law of areas) Equal areas are swept out in equal times. 3. (harmonic law) The square of the period is proportional to the cube of the distance. Newton's generalized formula for the third law is $P^2 = 4\pi^2 a^3 / [G(m_1 + m_2)]$.

keV   kilo-electron volt = 10^3 eV.

Kirkwood gaps voids in the asteroid belt where the orbital period of the asteroids is at certain fractions of the period of Jupiter.

km   kilometer = 10^5 cm.

kpc   kilo-parsec = 10^3 pc.

KREEP   from K for potassium, REE for rare-earth elements, P for phosphorus. Lunar basaltic material rich in radioactive elements.

kW   kilowatt = 10^{10} ergs sec^{-1}.

kyr 10^3 yr.

L_4, etc.   one of the Lagrangian points (see "Trojans").

$L_\odot$ solar luminosity = 3.826×10^{33} erg sec^{-1}.

Lagrangian   see "Trojans."
point

Lambert's law also called cosine law. The intensity of the light emanating in a given direction from a perfectly diffusing surface is proportional to the cosine of the angle of emission measured between the normal to the surface and the emitted ray.

lightcurve   magnitude values plotted as a function of time. Note that this plot does not necessarily have to show variability. Lightcurve amplitude: peak-to-peak value in magnitudes of a lightcurve showing variability.

Lommel-   a surface with large-scale roughness where shadowing
Seeliger   effects are important.
surface

μm   1 μm = 1 micrometer = 1 micron.

$M_\odot$ solar mass = 1.989×10^{33} g.

$M_\oplus$ mass of Earth = 5.976×10^{27} g.

mag   astronomical magnitude proportional to $-2.5 \log_{10} I$, where I is the intensity.

matrix   the smaller or finer-grained, continuous filling in the interstices between chondrules and larger mineral grains in chondritic meteorites.

MDS   McDonald Survey of Asteroids (see Gehrels' chapter).

megaregolith   regolith (q.v.) structure throughout the asteroid.

metamorphism   solid-state recrystallization and replacement of less stable by more stable phases as a result of the application of heat. Metamorphism in chondritic meteorites is much less severe than that observed in most terrestrial metamorphic rocks.

meteor   a "shooting star" the streak of light in the sky produced by the transit of a meteoroid through the earth's atmosphere; also the glowing meteoroid itself. The term "fireball" is sometimes used for a meteor approaching the brightness of Venus; the term "bolide" for one approaching the brightness of the full moon.

meteorite   extraterrestrial material which survives passage through the earth's atmosphere and reaches the earth's surface as a recoverable object (or objects).

meteorite classification   see Table I of the chapter by Wasson and Wetherill for the various classifications.

meteoroid   a small particle orbiting the sun in the vicinity of Earth.

MeV   million electron volts = 10^6 eV.

mG   milligauss = 10^{-3} gauss.

Mie theory   a theory of light scattering by small spherical particles.

minor satellite   a satellite of an asteroid.

monomict breccia   breccia (q.v.) of which all components originated in the same rock.

MPC   *Minor Planet Circulars* (see Marsden's chapter).

msec   millisecond = 10^{-3} sec.

Myr 10^6 yr.

nm   nanometer = 10^{-7} cm.

Ω	longitude of ascending node.
ω	argument of perihelion. Angular distance (measured in the plane of the asteroid's orbit and in the direction of motion) from the ascending node to the perihelion point.
$\tilde{\omega}$	longitude of perihelion = $\Omega + \omega$.
obliquity	the angle between a planet's axis of rotation and the pole of its orbit.
Ockham's razor	*Entia non sunt multiplicanda* ("entities are not to be multiplied"). A doctrine formulated by William of Ockham in the fourteenth century. Any hypothesis should be shorn of all unnecessary assumptions; if two hypotheses fit equally well, the one that makes the fewest assumptions should be chosen.
olivines	common rock-forming silicate minerals in meteorites and other rocks. The ratio of metal oxides (MgO and FeO) to SiO_2 is 2:1 in olivine. Olivines vary continuously in composition from Mg_2SiO_4, *forsterite,* to Fe_2SiO_4, *fayalite.* Iron and magnesium ions freely substitute for one another. A given composition is usually expressed in terms of mole % forsterite, e.g. Fo_{65} means 65% Mg_2SiO_4.
Oort cloud	a region extending to more than 100,000 AU from the sun, barely gravitationally bound, postulated as the birthplace of comets.
opposition	see elongation.
osculating orbit	the path that an asteroid would follow if it were subject only to the inverse-square attraction of the sun or other central body. In practice, secondary bodies such as Jupiter produce perturbations.
parent body	planet- or comet-like solar-system bodies in which meteorites were formed or stored.
pc	parsec; 1 parsec = the distance where 1 AU subtends 1 arcsec = 206,265 AU = 3.26 light-year = 3.086×10^{18} cm.
perihelion	the point in the asteroid orbit closest to the sun.
phase angle	α, the solar phase angle: the angle subtended at the (center of the) planet by the direction to Sun and observer.
Planck's blackbody formula	a formula that determines the distribution of intensity of radiation that prevails under conditions of thermal equilibrium at a temperature T: $B_\nu = (2h\nu^3/c^2)[\exp(h\nu/kT) -$

$1]^{-1}$ where h is Planck's constant and ν is the frequency.

planetesimal

bodies of intermediate (perhaps meter to 100 m) size, most of which finally accreted to larger bodies.

PLS

Palomar-Leiden Survey of Faint Asteroids (see Gehrels' chapter).

polymict breccia

breccia (q.v.) of which the components originated in two or more unrelated rocks.

Poynting-Robertson effect

a light pressure effect on small grains. The time it takes for particles to move into the sun $t = 7.0 \times 10^6\ r\,p\,a\,q$ yr (radius r in cm, density p in g cm^{-3}, a and q in AU) where a and q are the semimajor axis and perihelion distance of the initial particle orbit.

p-process

the name of the hypothetical nucleosynthetic process thought to be responsible for the synthesis of the rare heavy proton-rich nuclei which are bypassed by the r- and s-processes. It is manifestly less efficient (and therefore rarer) than the s- or r-process since the protons must overcome the Coulomb barrier, and may in fact work as a secondary process on the r- and s-process nuclei. It seems to involve primarily (p, γ) reactions above cerium (where neutron separation energies are low). The p-process is assumed to occur in supernova envelopes at a temperature $\gtrsim 10^9$ K and at densities $\lesssim 10^4$ g cm^{-3}.

precession

a slow, periodic conical motion of the rotation axis of a spinning body.

p_V

geometric albedo (q.v.); p_V, the geometric albedo with the V filter of the UBV system.

pyroxines

a group of common rock-forming silicates which have ratios of metal oxides (MgO, FeO or CaO) to SiO$_2$ of 1:1. These are called the metasilicates. Pure members of this group are MgSiO$_3$, *enstatite*, and FeSiO$_3$, *ferrosilite*. Pure CaSiO$_3$ does not crystallize with the pyroxene structure. Ca does

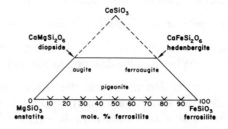

substitute for up to 50% of the Mg and Fe in the pyroxene structure. A truncated triangular composition diagram is used to illustrate these relationships.

q perihelion distance (q.v.).

Q aphelion distance. In the asteroid orbit, the most distant point from the sun.

$R_\odot$ solar radius = 6.96×10^{10} cm.

REE   rare earth elements = lanthanide series in the Periodic Table.

refractory   a material of high vaporization point, or the property of resisting heat.

regolith   the layer of fragmental incoherent rocky debris that nearly everywhere forms the surface terrain; it is produced by meteoritic impact on the surfaces of the planets, satellites or asteroids.

resonance   the selective response of any oscillating system to an external stimulus of the same frequency as the natural frequency of the system.

Reynolds number   a dimensionless number ($R = Lv/\nu$, where L is a typical dimension of the system, v is a measure of the velocities that prevail, and ν is the kinematic viscosity) that governs the conditions for the occurrence of turbulence in fluids.

rms   root mean square: the square root of the mean square value of a set of numbers.

Roche limit   the minimum distance at which a satellite under the influence of its own gravitation and that of a central mass, about which it is describing a Keplerian orbit, can be in equilibrium. For a satellite of negligible mass, zero tensile strength, and the same mean density as its primary, in a circular orbit around its primary, this critical distance, at which the satellite will break up, is 2.44 times the radius of the primary.

r-process   the capture of neutrons on a very rapid time scale (i.e., one in which a nucleus can absorb neutrons in rapid succession, so that regions of great nuclear instability are bridged), a theory advanced to account for the existence of all elements heavier than bismuth as well as the neutron-rich isotopes heavier than iron. The essential feature of the r-process is the release of great numbers of neutrons in a very short time (less than 100 sec). The presumed source for such a large flux of neutrons is a supernova, at the boundary between the neutron star and the ejected material.

Schmidt telescope a type of reflecting telescope (more accurately, a large camera) in which the coma produced by a spherical concave mirror is compensated for by a thin correcting lens placed at the opening of the telescope tube. The Palomar 122-cm Schmidt has a usable field of $6°$.

sidereal period the time it takes for a planet or satellite to make one complete circuit of its orbit ($360°$) relative to the stars.

SIRTF Space Lab Infrared Telescope Facility (see the chapter by Morrison and Niehoff).

solar wind a radial outflow of energetic charged particles from the solar corona, carrying mass and angular momentum away from the sun. Mean number density of solar wind (1971), 5 per cm^3; mean velocity at Earth 400 km sec^{-1}; mean magnetic field 5×10^{-5} gauss; mean electron temperature 20,000 K; mean ion temperature 10,000 K. The sun ejects $\sim 10^{-13}$ $M_\odot$ per year via the solar wind.

s-process (slow neutron capture) a process in which heavy, stable, neutron-rich nuclei are synthesized from iron-peak elements by successive captures of free neutrons in a weak neutron flux, so there is time for β-decay before another neutron is captured (cf., r-process). This is a slow but sure process of nucleosynthesis which is assumed to take place in the intershell regions during the red-giant phase of evolution, at densities up to 10^5 g cm^{-3} and temperatures of about 3×10^8 K (neutron densities assumed are 10^{10} cm^{-3}).

ST Space Telescope (see the chapter by Morrison and Niehoff).

Stokes parameters four parameters to describe fully a beam of polarized light. They involve the maximum intensity, the ellipticity, and the direction of polarization.

synodic period the period of revolution of one body about another with respect to the earth. (synodic period)$^{-1}$ $= \pm$ (sidereal period)$^{-1}$ $\mp$ (Earth's period)$^{-1}$.

TRIAD Tucson Revised Index of Asteroid Data (see Part VII of this book).

Trojans Trojan asteroids occur in two of the *Lagrangian Points*, (namely the ones preceding and following Jupiter in its orbit, equidistant from Sun and Jupiter) which may be defined as points in the orbital plane of two massive bodies in circular orbits around a common center of gravity where a third body of negligible mass can remain in equilibrium. (see the chapters by Gehrels and by Degewij and van Houten).

UBV system — a system of stellar magnitudes devised by Johnson and Morgan at the Yerkes Observatory which consists of measuring an object's apparent magnitude through three color filters: the ultraviolet (U) at 3600 Å; the blue (B) at 4200 Å; and the "visual" (V) in the green-yellow spectral region at 5400 Å. It is defined so that, for AO stars, $B-V = U-B = 0$; it is negative for hotter stars and positive for cooler stars. Filters at other wavelengths are also used and indicated with letters $R, I, H, J, K, L, M,$ etc.

Universal Time (UT) — the local mean time of the prime meridian. It is the same as Greenwich mean time, counted from 0 hr beginning at Greenwich mean midnight.

UV — ultraviolet.

van der Waals forces — the relatively weak attractive forces operative between neutral atoms and molecules.

Widmanstätten pattern — a geometric pattern found in some iron meteorites, consisting of groups of parallel lamellae crossing each other at various angles.

xenolith — an inclusion or fragment not genetically related to the rock in which it is found.

zodiacal light — a faint glow that extends away from the sun in the ecliptic plane; it is caused by scattering of sunlight by interplanetary particles.

ACKNOWLEDGMENTS

The following people helped to make this book possible, in organizing, writing, refereeing or otherwise. The members of the Organizing Committee and Chairpersons are indicated with an asterisk (*).

C. C. Allen, Lunar and Planetary Laboratory, University of Arizona, Tucson, Ariz.
*E. Anders, Enrico Fermi Institute for Nuclear Studies, University of Chicago, Chicago, Ill.
L. Andersson, Lunar and Planetary Laboratory, University of Arizona, Tucson, Ariz.
*J. R. Arnold, Chemistry Department, University of California at San Diego, La Jolla, Calif.
J. C. Aubele, Lunar and Planetary Laboratory, University of Arizona, Tucson, Ariz.
R. D. Austin, Mount John University Observatory, Lake Tekapo, New Zealand.
D. Bender, Jet Propulsion Laboratory, Pasadena, Calif.
G. F. Benedict, Department of Astronomy, University of Texas, Austin, Texas.
C. Y. Benkheiri, Laboratoire de Minéralogie, Muséum National d'Histoire Naturelle, Paris, France.
R. P. Binzel, Astronomy Department, Macalester College, St. Paul, Minn.
P. Birch, Perth Observatory, Bickley, Australia.
C. Blanco, Osservatorio Astrofisico di Catania, Catania, Italy.
M. Blander, Argonne National Laboratory, Argonne, Ill.
D. D. Bogard, NASA Johnson Space Flight Center, Houston, Texas.
*E. Bowell, Lowell Observatory, Flagstaff, Ariz.
*W. Boynton, Lunar and Planetary Laboratory, University of Arizona, Tucson, Ariz.
A. Brahic, Observatoire de Paris, Meudon, France.
A. Brecher, Physics Department, Wellesley College, Wellesley, Mass.
J. Breton, Centre National d'Etudes Spatiales, Toulouse, France.
D. E. Brownlee, Department of Astronomy, University of Washington, Seattle, Wash.
*W. Brunk, NASA Headquarters, Washington, D.C.
T. E. Bunch, Space Science Division, NASA Ames Research Center, Moffett Field, Calif.
*J. A. Burns, Cornell University, Ithaca, New York.
*A. G. W. Cameron, Harvard-Smithsonian Center for Astrophysics, Cambridge, Mass.
D. B. Campbell, National Astronomy and Ionosphere Center, Arecibo, Puerto Rico.
H. Campins, Lunar and Planetary Laboratory, University of Arizona, Tucson, Ariz.
N. P. Carleton, Smithsonian Astrophysical Observatory, Tucson, Ariz.
A. Carusi, Laboratorio Astrofisica Spaziale, Frascati, Italy.

S. Catalano, Osservatorio Astrofisico di Catania, Catania, Italy.
J. Caubel, Centre National d'Etudes Spatiales, Toulouse, France.
A. Cazenave, Groupe de Recherches de Géodésie Spatiale, Toulouse, France.
T. Chandrasekhar, Physical Research Laboratory, Ahmedabad, India.
*C. R. Chapman, Planetary Science Institute, Tucson, Ariz.
M. J. Cintala, Department of Geological Science, Brown University, Providence, Rhode Island.
W. J. Cocke, Steward Observatory, Tucson, Ariz.
G. J. Consolmagno, Harvard-Smithsonian Center for Astrophysics, Cambridge, Mass.
A. F. Cook, Harvard-Smithsonian Center for Astrophysics, Cambridge, Mass.
A. Coradini, Laboratorio di Astrofisica Spaziale, Rome, Italy.
G. Crozaz, Department of Earth and Planetary Science, Washington University, St. Louis, Missouri.
D. Cruikshank, Institute for Astronomy, Honolulu, Hawaii.
P. Cruvellier, Laboratoire d'Astronomie Spatiale, Marseille, France.
D. R. Davis, Planetary Science Institute, Tucson, Ariz.
H. Debehogne, Observatoire Royal de Belgique, Brussels, Belgium.
K. Denomy, Lunar and Planetary Laboratory, University of Arizona, Tucson, Ariz.
J. Degewij, Lunar and Planetary Laboratory, University of Arizona, Tucson, Ariz.
J. R. Dickel, Astronomy Department, University of Illinois, Urbana, Ill.
J. Dingell, Planetary Science Institute, Tucson, Ariz.
*A. Dollfus, Observatoire de Paris, Meudon, France.
*M. J. Drake, Lunar and Planetary Laboratory, University of Arizona, Tucson, Ariz.
M. B. Duke, NASA Johnson Space Flight Center, Houston, Texas.
S. Dunbar, Princeton University Observatory, Princeton, New Jersey.
D. W. Dunham, Systems Sciences Division, Computer Science Corp., Silver Spring, Maryland.
Y. Dupuis, Centre National d'Etudes Spatiales, Toulouse, France.
F. K. Edmondson, Astronomy Department, Indiana University, Bloomington, Ind.
J. L. Elliot, Department of Earth and Planetary Science, Massachusetts Institute of Technology, Cambridge, Mass.
E. Everhart, Physics Department and Chamberlin Observatory, University of Denver, Denver, Colo.
S. Fair, Alltype, Tucson, Ariz.
F. P. Fanale, Jet Propulsion Laboratory, Pasadena, Calif.
P. Farinella, Osservatorio Astronomico di Brera, Merate, Italy.
M. Feierberg, Lunar and Planetary Laboratory, University of Arizona, Tucson, Ariz.
H. Ferguson, Harvard-Smithsonian Center for Astrophysics, Cambridge, Mass.
U. Fink, Lunar and Planetary Laboratory, University of Arizona, Tucson, Ariz.
F. A. Franklin, Harvard-Smithsonian Center for Astrophysics, Cambridge, Mass.
O. G. Franz, Lowell Observatory, Flagstaff, Ariz.
K. Fredriksson, Department of Mineral Science, Smithsonian Institution, Washington D.C.
A. L. Friedlander, Science Applications, Inc., Schaumburg, Ill.
C. Froeschlé, Observatoire de Nice, Nice, France.
M. J. Gaffey, Institute for Astronomy, University of Hawaii, Honolulu, Hawaii.
D. E. Gault, NASA Ames Research Center, Moffett Field, Calif.
*T. Gehrels, Lunar and Planetary Laboratory, University of Arizona, Tucson, Ariz.
H. Giclas, Lowell Observatory, Flagstaff, Ariz.
P. Goldreich, California Institute of Technology, Pasadena, Calif.
A. Goswami, Physical Research Laboratory, Ahmedabad, India.
J. Gradie, Laboratory for Planetary Studies, Cornell University, Ithaca, New York.
*R. Greenberg, Planetary Science Institute, Tucson, Ariz.
J. Gunter, 1411 N. Mangum St., Durham, North Carolina.
J. S. Hall, RR 3 Red Butte Dr., Sedona, Ariz.
B. W. Hapke, Department of Earth and Planetary Science, University of Pittsburgh, Pittsburgh, Penn.
R. S. Harrington, U. S. Naval Observatory, Washington, D.C.

A. W. Harris, Jet Propulsion Laboratory, Pasadena, Calif.
*W. K. Hartmann, Planetary Science Institute, Tucson, Ariz.
*H. Haupt, Universitäts-Sternwarte, Graz, Austria.
J. W. Head, Department of Geological Science, Brown University, Providence, Rhode Island.
E. F. Helin, California Institute of Technology, Pasadena, Calif.
F. Herbert, Lunar and Planetary Laboratory, University of Arizona, Tucson, Ariz.
J. M. Herndon, Department of Chemistry, University of California at San Diego, La Jolla, Calif.
B. T. Herrick, 13500 Mulholland Drive, Beverly Hills, Calif.
R. Hodgson, 316 S. Main Ave., Sioux Center, Iowa.
K. R. Housen, Lunar and Planetary Laboratory, University of Arizona, Tucson, Ariz.
W. B. Hubbard, Lunar and Planetary Laboratory, University of Arizona, Tucson, Ariz.
W. H. Ip, Max-Planck Institut für Aeronomie, Katlenburg-Lindau, W. Germany.
R. J. Jurgens, Jet Propulsion Laboratory, Pasadena, Calif.
R. A. Kassander, Vice President of Research, University of Arizona, Tucson, Ariz.
C. E. KenKnight, Lunar and Planetary Laboratory, University of Arizona, Tucson, Ariz.
J. J. Kerridge, Institute of Geophysics, University of California, Los Angeles, Calif.
E. A. King, Department of Geology, University of Houston, Houston, Texas.
T. V. King, Department of Geology, University of Houston, Houston, Texas.
W. J. Klepczynski, U. S. Naval Observatory, Washington, D.C.
B. K. Kothari, Department of Chemistry, University of California at San Diego, La Jolla, Calif.
*C. T. Kowal, California Institute of Technology, Pasadena, Calif.
Y. Kozai, Tokyo Astronomical Observatory, Tokyo, Japan.
*Ľ. Kresák, Astronomicky Ústav Slovenskej Akadémie, Bratislava, Czechoslovakia.
V. G. Kruchinenko, Astronomical Observatory, Kiev State University, Kiev, USSR.
C. I. Lagerkvist, Astronomiska Observatoriet, Uppsala, Sweden.
B. Lago, Groupe de Recherches de Géodésie Spatiale, Toulouse, France.
*D. Lal, Physical Research Laboratory, Ahmedabad, India.
J. V. Lambert, New Mexico State University, Las Cruces, New Mex.
H. H. Lane, National Science Foundation, Washington, D.C.
J. W. Larimer, Department of Geology and Center for Meteorite Studies, Arizona State University, Tempe, Ariz.
H. P. Larson, Lunar and Planetary Laboratory, University of Arizona, Tucson, Ariz.
M. Leake, Lunar and Planetary Laboratory, University of Arizona, Tucson, Ariz.
L. A. Lebofsky, Lunar and Planetary Laboratory, University of Arizona, Tucson, Ariz.
E. H. Levy, Lunar and Planetary Laboratory, University of Arizona, Tucson, Ariz.
W. Liller, Harvard-Smithsonian Center for Astrophysics, Cambridge, Mass.
J. C. Lorin, Laboratoire de Minéralogie-Cristallographie, Paris, France.
K. Lumme, Observatory and Astrophysics Laboratory, University of Helsinki, Tähtitorninmäki, Finland.
J. MacConnell, Fundacion Centro de Investigacion de Astronomia "Francisco J. Duarte", Mérida, Venezuela.
P. Maley, NASA Johnson Space Flight Center, Houston, Texas.
S. Marinus, Lunar and Planetary Laboratory, University of Arizona, Tucson, Ariz.
*B. G. Marsden, Harvard-Smithsonian Center for Astrophysics, Cambridge, Mass.
D. L. Matson, Jet Propulsion Laboratory, Pasadena, Calif.
M. S. Matthews, Lunar and Planetary Laboratory, University of Arizona, Tucson, Ariz.
M. Maurette, Laboratoire René Bernas, Orsay, France.
*T. B. McCord, Institute for Astronomy, University of Hawaii, Honolulu, Hawaii.
L. McFadden, Institute for Astronomy, University of Hawaii, Honolulu, Hawaii.
G. McLaughlin, Lunar and Planetary Laboratory, University of Arizona, Tucson, Ariz.
S. C. McMillan, Arizona Foundation, University of Arizona, Tucson, Ariz.
H. Y. McSween, Department of Geological Science, University of Tennessee, Knoxville, Tenn.
D. A. Mendis, Department of Applied Physics and Information Science, University of California at San Diego, La Jolla, Calif.

R. L. Millis, Lowell Observatory, Flagstaff, Ariz.

J. W. Minear, NASA Johnson Space Flight Center, Houston, Texas.

E. Miner, Earth and Space Science Division, Jet Propulsion Laboratory, Pasadena, Calif.

J. F. Minster, Institut de Physique du Globe, Laboratoire de Géochimie, Paris, France.

D. Mittlefehldt, Lunar and Planetary Laboratory, University of Arizona, Tucson, Ariz.

S. Monroe, Alltype, Tucson, Ariz.

*D. Morrison, Institute for Astronomy, University of Hawaii, Honolulu, Hawaii.

J. P. Murphy, NASA Ames Research Center, Moffett Field, Calif.

B. Nagy, Department of Geosciences, University of Arizona, Tucson, Ariz.

H. Newsom, Lunar and Planetary Laboratory, University of Arizona, Tucson, Ariz.

J. Niehoff, Science Applications, Inc., Schaumberg, Ill.

*I. Nikoloff, Perth Observatory, Bickley, Australia.

S. Nozette, Lunar and Planetary Laboratory, University of Arizona, Tucson, Ariz.

*B. O'Leary, Department of Physics, Princeton University, Princeton, New Jersey.

S. J. Ostro, Department of Earth and Planetary Science, Massachusetts Institute of Technology, Cambridge, Mass.

M. W. Ovenden, Department of Geophysics and Astronomy, University of British Columbia, Canada.

*T. Owen, Earth and Planetary Science Department, State University of New York at Stony Brook, Stony Brook, New York.

D. Owings, Astronomy Department, Indiana University, Bloomington, Ind.

S. Peale, Department of Physics, University of California at Santa Barbara, Santa Barbara, Calif,

P. Pellas, Laboratoire de Minéralogie-Cristallographie, Paris, France.

A. Perret, Centre National d'Etudes Spatiales, Toulouse, France.

G. Pettengill, Department of Earth and Planetary Science, Massachusetts Institute of Technology, Cambridge, Mass.

F. Pilcher, Illinois College, Jacksonville, Ill.

M. Price, Science Applications, Inc., Tucson, Ariz.

*J. Rahe, Astronomisches Institut, Universität Erlangen-Nürnberg, Bamberg, W. Germany.

R. S. Rajan, Department of Terrestrial Magnetism, Carnegie Institution of Washington, Washington, D.C.

H. J. Reitsema, Lunar and Planetary Laboratory, University of Arizona, Tucson, Ariz.

R. T. Reynolds, NASA Ames Research Center, Moffett Field, Calif.

*N. Richter, Post Ulla bei Weizel, Weimar, E. Germany.

H. Rickman, Astronomiska Observatoriet, Uppsala, Sweden.

S. T. Ridgway, Kitt Peak National Observatory, Tucson, Ariz.

E. Roemer, Lunar and Planetary Laboratory, University of Arizona, Tucson, Ariz.

S. K. Runcorn, School of Physics, The University, Newcastle upon Tyne, England.

V. S. Safronov, Schmidt Institute of Physics of the Earth, Moscow, USSR.

R. Sather, Lunar and Planetary Laboratory, University of Arizona, Tucson, Ariz.

F. Scaltriti, Osservatorio Astronomica di Torino, Pino Torinese, Italy.

H. J. Schober, Universitäts-Sternwarte, Graz, Austria.

H. Scholl, Astronomisches Rechen-Institute, Heidelberg, W. Germany.

*J. Schubart, Astronomisches Rechen-Institut, Heidelberg, W. Germany.

E. R. D. Scott, Department of Terrestrial Magnetism, Carnegie Institution of Washington, Washington, D.C.

A. Scribot, Centre National d'Etudes Spatiales, Toulouse, France.

D. W. Sears, Department of Metallurgy, University of Manchester, Manchester, United Kingdom.

I. I. Shapiro, Department of Earth and Planetary Science, Massachusetts Institute of Technology, Cambridge, Mass.

P. J. Shelus, Department of Astronomy, University of Texas, Austin, Texas.

L. M. Sherbaum, Astronomical Observatory, Kiev State University, Kiev, USSR.

E. M. Shoemaker, U.S. Geological Survey, Flagstaff, Ariz.

M. J. Simmons, Lunar and Planetary Laboratory, University of Arizona, Tucson, Ariz.

A. N. Simonenko, USSR Academy of Science Committee on Meteorites, Moscow, USSR.

N. Sjolander, Astronomiska Observatoriet, Uppsala, Sweden.
B. A. Smith, Lunar and Planetary Laboratory, University of Arizona, Tucson, Ariz.
H. Smith, Lunar and Planetary Laboratory, University of Arizona, Tucson, Ariz.
T. S. Smith, Lunar and Planetary Laboratory, University of Arizona, Tucson, Ariz.
C. P. Sonett, Lunar and Planetary Laboratory, University of Arizona, Tucson, Ariz.
G. Sprock, Alltype, Tucson, Ariz.
N. Stevens, Alltype, Tucson, Ariz.
C. Stoll, Lunar and Planetary Laboratory, University of Arizona, Tucson, Ariz.
E. Stolper, Department of Geological Science, Harvard University, Cambridge, Mass.
D. Storzer, Laboratoire de Minérologie, Muséum National d'Histoire Naturelle, Paris, France.
A. Stout, Lunar and Planetary Laboratory, University of Arizona, Tucson, Ariz.
S. Sutton, Alltype, Tucson, Ariz.
H. Takeda, Mineralogical Institute, University of Tokyo, Tokyo, Japan.
G. E. Taylor, Royal Greenwich Observatory, Herstmonceux Castle, Hailsham, England.
R. C. Taylor, Lunar and Planetary Laboratory, University of Arizona, Tucson, Ariz.
*E. F. Tedesco, Lunar and Planetary Laboratory, University of Arizona, Tucson, Ariz.
P. Thomas, Laboratory for Planetary Studies, Cornell University, Ithaca, New York.
C. Tombaugh, Astronomy Department, Mesilla Park, New Mex.
M. Townsend, University of Arizona Press, Tucson, Ariz.
A. B. Valsecchi, Laboratorio Astrofisica Spaziale, Frascati, Italy.
T. C. Van Flandern, U.S. Naval Observatory, Washington, D.C.
*C. J. van Houten, Sterrewacht, Leiden, the Netherlands.
I. van Houten-Groeneveld, Sterrewacht, Leiden, the Netherlands.
G. J. Veeder, Jet Propulsion Laboratory, Pasadena, Calif.
C. Vesely, Lunar and Planetary Laboratory, University of Arizona, Tucson, Ariz.
*J. Veverka, Laboratory for Planetary Studies, Cornell University, Ithaca, New York.
F. Vilas, Lunar and Planetary Laboratory, University of Arizona, Tucson, Ariz.
J. Wacker, Lunar and Planetary Laboratory, University of Arizona, Tucson, Ariz.
W. Wamsteker, European Southern Observatory, Santiago, Chile.
L. H. Wasserman, Lowell Observatory, Flagstaff, Ariz.
*J. Wasson, Institute of Geophysics and Planetary Physics, University of California, Los Angeles, Calif.
A. B. Weaver, Executive Vice President, University of Arizona, Tucson, Ariz.
S. J. Weidenschilling, Planetary Science Institute, Tucson, Ariz.
*G. W. Wetherill, Department of Terrestrial Magnetism, Carnegie Institution of Washington, Washington, D.C.
F. L. Whipple, Harvard-Smithsonian Center for Astrophysics, Cambridge, Mass.
E. A. Whitaker, Lunar and Planetary Laboratory, University of Arizona, Tucson, Ariz.
W. Wiesel, Department of Aeronautics and Astronautics, Air Force Institute of Technology, Wright-Patterson Air Force Base, Ohio.
M. Wijesinghe, Lunar and Planetary Laboratory, University of Arizona, Tucson, Ariz.
*L. L. Wilkening, Lunar and Planetary Laboratory, University of Arizona, Tucson, Ariz.
B. A. Wilking, Lunar and Planetary Laboratory, University of Arizona, Tucson, Ariz.
J. G. Williams, Jet Propulsion Laboratory, Pasadena, Calif.
L. Wilson, Department of Environmental Science, University of Lancaster, Lancaster, United Kingdom.
R. J. Wolfe, U.S. Geological Survey, Flagstaff, Ariz.
H. J. Wood, Astronomy Department, Indiana University, Bloomington, Ind.
J. A. Wood, Harvard-Smithsonian Center for Astrophysics, Cambridge, Mass.
S. P. Worden, Air Force Cambridge Research Laboratory, Sacramento Peak Observatory, Sunspot, New Mex.
C. F. Yoder, Jet Propulsion Laboratory, Pasadena, Calif.
J. Young, University of Arizona Press, Tucson, Ariz.
V. Zappalà, Osservatorio Astronomica di Torino, Pino Torinese, Italy.
*B. H. Zellner, Lunar and Planetary Laboratory, University of Arizona, Tucson, Ariz.

INDEX

Asteroid satellites. *See* satellites of asteroids
Aten objects. *See* Apollo *and under* asteroid
BEND parameter, 371, 785
Bias corrections, 790
Binary asteroids. *See* satellites of asteroids
Breccias, 68, 568, 593, 611, 746, 920, 936

Carbonaceous meteorite, 746 *ff.*
Chaotic orbit, 438
Chiron. *See under* asteroid
Chondritic meteorite, 809, 928
Chondrites (carbonaceous), 707, 742, 746 *ff.*, 896 *ff.*
Chondrule, 815, 824, 838, 909, 1000
Chronology of collision, 558 *ff.*
Classifications, 134, 690, 738, 783 *ff.*
Coagulation, 986